Financial Mathematics

A Comprehensive Treatment

CHAPMAN & HALL/CRC
Financial Mathematics Series

Aims and scope:
The field of financial mathematics forms an ever-expanding slice of the financial sector. This series aims to capture new developments and summarize what is known over the whole spectrum of this field. It will include a broad range of textbooks, reference works and handbooks that are meant to appeal to both academics and practitioners. The inclusion of numerical code and concrete real-world examples is highly encouraged.

Published Titles

American-Style Derivatives; Valuation and Computation, *Jerome Detemple*
Analysis, Geometry, and Modeling in Finance: Advanced Methods in Option Pricing, *Pierre Henry-Labordère*
Computational Methods in Finance, *Ali Hirsa*
Credit Risk: Models, Derivatives, and Management, *Niklas Wagner*
Engineering BGM, *Alan Brace*
Financial Mathematics: A Comprehensive Treatment, *Giuseppe Campolieti and Roman N. Makarov*
Financial Modelling with Jump Processes, *Rama Cont and Peter Tankov*
Interest Rate Modeling: Theory and Practice, *Lixin Wu*
Introduction to Credit Risk Modeling, Second Edition, *Christian Bluhm, Ludger Overbeck, and Christoph Wagner*
An Introduction to Exotic Option Pricing, *Peter Buchen*
Introduction to Risk Parity and Budgeting, *Thierry Roncalli*
Introduction to Stochastic Calculus Applied to Finance, Second Edition, *Damien Lamberton and Bernard Lapeyre*
Monte Carlo Methods and Models in Finance and Insurance, *Ralf Korn, Elke Korn, and Gerald Kroisandt*
Monte Carlo Simulation with Applications to Finance, *Hui Wang*
Nonlinear Option Pricing, *Julien Guyon and Pierre Henry-Labordère*
Numerical Methods for Finance, *John A. D. Appleby, David C. Edelman, and John J. H. Miller*
Option Valuation: A First Course in Financial Mathematics, *Hugo D. Junghenn*
Portfolio Optimization and Performance Analysis, *Jean-Luc Prigent*
Quantitative Finance: An Object-Oriented Approach in C++, *Erik Schlögl*

Quantitative Fund Management, *M. A. H. Dempster, Georg Pflug, and Gautam Mitra*

Risk Analysis in Finance and Insurance, Second Edition, *Alexander Melnikov*

Robust Libor Modelling and Pricing of Derivative Products, *John Schoenmakers*

Stochastic Finance: An Introduction with Market Examples, *Nicolas Privault*

Stochastic Finance: A Numeraire Approach, *Jan Vecer*

Stochastic Financial Models, *Douglas Kennedy*

Stochastic Processes with Applications to Finance, Second Edition, *Masaaki Kijima*

Structured Credit Portfolio Analysis, Baskets & CDOs, *Christian Bluhm and Ludger Overbeck*

Understanding Risk: The Theory and Practice of Financial Risk Management, *David Murphy*

Unravelling the Credit Crunch, *David Murphy*

Proposals for the series should be submitted to one of the series editors above or directly to:
CRC Press, Taylor & Francis Group
3 Park Square, Milton Park
Abingdon, Oxfordshire OX14 4RN
UK

Chapman & Hall/CRC FINANCIAL MATHEMATICS SERIES

Financial Mathematics

A Comprehensive Treatment

Giuseppe Campolieti
Roman N. Makarov

CRC Press
Taylor & Francis Group
Boca Raton London New York

CRC Press is an imprint of the
Taylor & Francis Group, an **informa** business
A CHAPMAN & HALL BOOK

CRC Press
Taylor & Francis Group
6000 Broken Sound Parkway NW, Suite 300
Boca Raton, FL 33487-2742

Printed on acid-free paper
Version Date: 20150708

International Standard Book Number-13: 978-1-4398-9242-8 (Hardback)

Visit the Taylor & Francis Web site at
http://www.taylorandfrancis.com

and the CRC Press Web site at
http://www.crcpress.com

To our families

Contents

List of Figures and Tables

List of Algorithms

Preface

Objectives and Audience

This book has evolved from several mathematics courses that the authors have taught mainly within the bachelor's and master's programs in financial mathematics at Wilfrid Laurier University. The contents of this book are a culmination of course material that spans over a decade of the authors' teaching experiences, as well as course and curriculum development, in financial mathematics programs at both undergraduate and master's graduate levels. The material has been tested and refined through years of classroom teaching experience. As the title suggests, this book is a comprehensive, self-contained, and unified treatment of the main theory and application of mathematical methods behind modern day financial mathematics. In writing this book, the authors have really strived to create a single volume that can be used as a complete standard university textbook for several interrelated courses in financial mathematics at the undergraduate as well as graduate levels. As such, the authors have aimed to introduce both the financial theory and the relevant mathematical methods in a mathematically rigorous, yet student-friendly and engaging style, that includes an abundance of examples, problem exercises, and fully worked out solutions. In contrast to most published single volumes on the subject of financial mathematics, this book presents multiple problem solving approaches and hence bridges together related comprehensive techniques for pricing different types of financial derivatives. The book contains a rather complete and in-depth comprehensive coverage of both discrete-time and continuous-time financial models that form the cornerstones of financial derivative pricing theory. This book also provides a self-contained introduction to stochastic calculus and martingale theory, which are important cornerstones in quantitative finance. The material in many of the chapters is presented at a level that is mainly accessible to undergraduate students of mathematics, finance, actuarial science, economics, and other related quantitative fields. The textbook covers a breadth of material, from beginner to more advanced levels, that is required, i.e., absolutely essential, in the core curriculum courses on financial mathematics currently taught at the second, third, and senior year undergraduate levels at many universities across the globe. As well, a significant portion of the more advanced material in the textbook is meant to be used in courses at the master's graduate level. These courses include formal derivative pricing theory, stochastic calculus, and courses in simulation (Monte Carlo) and other numerical methods. The combination of analytical and numerical methods for solving various derivative pricing problems can also be a useful reference for researchers and practitioners in quantitative finance.

The book has the following key features:

- comprehensive treatment covering a complete undergraduate program in financial mathematics as well as some master's level courses in financial mathematics;
- student-friendly presentation with numerous fully worked out examples and exercise problems in every chapter;

- in-depth coverage of both discrete-time and continuous-time theory and methodology;
- mathematically rigorous and consistent, yet simple, style that bridges various basic and more advanced concepts and techniques;
- judicious balance of financial theory, mathematical, and computational methods.

Guide to Material

This book is divided into four main parts with each part consisting of several chapters. There are a total of eighteen chapters, and every chapter (with the exception of Chapter 9 on general probability theory and Chapter 18 on numerical applications) ends with a comprehensive and exhaustive set of exercises of varying difficulty. Part I is an introduction to pricing and management of financial securities. This part has four chapters. Chapter 1 introduces the reader to time value of money, compounding interest, and the basic concepts of fixed income markets. Chapter 2 introduces basic derivative securities and the concept of arbitrage. Chapter 3 covers standard theoretical topics of portfolio management and only requires some very basic linear algebra and optimization. Chapter 4 presents more formal definitions and gives a thorough discussion on basic options theory, including payoff replication, hedging, put-call parity relations, forwards and futures contracts, swaps, American options, and other contracts.

Part II is devoted to discrete-time financial modelling. Chapters 5–8 of Part II can be considered as a complete course on discrete-time asset pricing. Part II introduces the main financial concepts in risk-neutral pricing theory, which are further developed in Part III on continuous-time theory. Chapter 5 covers the financial and formal mathematical underpinnings of the single-period (Arrow-Debreu) economic model. Chapter 6 lays down the foundation for stochastic processes in discrete time, which is then essential for Chapter 7. The latter chapter covers the multi-period binomial market model and is considered the centerpiece of discrete-time financial derivative pricing. All the important concepts for pricing and hedging standard European, as well as path-dependent derivatives within this model, are presented. A general multi-asset, multi-period, discrete-time model is covered in Chapter 8. This chapter also presents the two fundamental theorems of asset pricing and equivalent martingale measures for discrete-time derivative pricing.

Part III is essentially the second half of the textbook and is a major part that is devoted to continuous-time modelling. In fact, on its own, Part III (i.e., Chapters 9 to 16) can be considered as a complete text in continuous-time financial mathematics for senior undergraduates and master's level students. Chapter 9 is a stand-alone chapter that summarizes the main theoretical concepts in formal probability theory as it relates to measure theory. This also provides some mathematical foundation for later chapters that deal with continuous time modelling and stochastic calculus. Chapter 10 lays down the foundation for standard Brownian motion. Chapter 11 is a comprehensive coverage of stochastic (Itô) calculus that is required for a large portion of the material in the rest of the book chapters. Chapters 12 and 13 are the main chapters on continuous-time derivative pricing theory, which also include the Black-Scholes-Merton theory of European option pricing. The central concepts of dynamic hedging and replication are presented. Chapter 12 deals with derivative pricing and hedging in the Black-Scholes-Merton model with a single risky asset. Chapter 12 also covers path-dependent derivative pricing within the Black-Scholes-Merton framework. Chapter 13 extends the theory and methodology to derivative pricing and

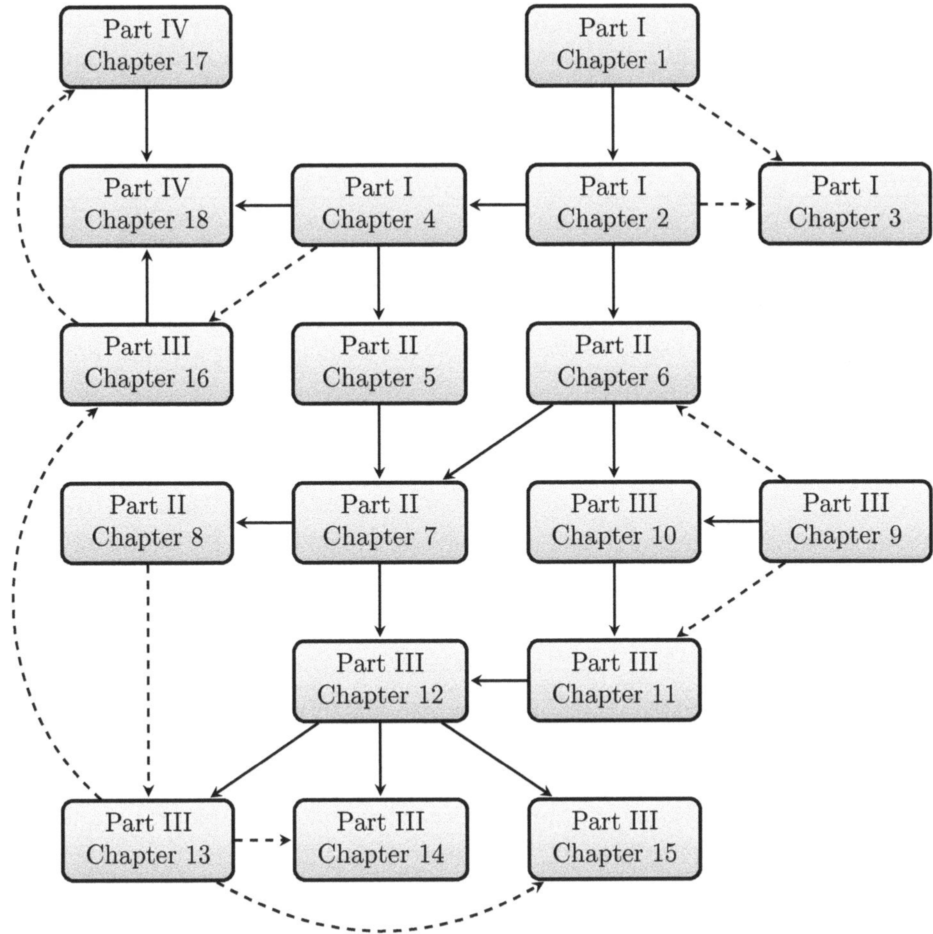

FIGURE: Guide to material.

hedging with multiple underlying assets as well as the valuation of cross-currency options. The chapter combines different techniques for pricing multi-asset financial derivatives. Various option pricing formulae are then derived. Chapter 13 also presents risk-neutral asset pricing theory within a mathematically rigorous framework that incorporates equivalent martingale measures and change of numéraire methods for pricing. Chapter 14 is devoted to pricing American options on a single asset. Chapter 15 covers interest-rate modelling and derivative pricing for fixed-income products. Chapter 16 introduces some alternative asset price models, including the local volatility model and solvable state-dependent volatility (e.g., the CEV diffusion) models; stochastic volatility models; jump-diffusion and pure jump processes and variance gamma models.

Part IV concludes the book with Chapters 17 and 18. Chapter 17 is a self-contained exposition of various Monte Carlo and simulation methods that are relevant for simulating financial assets and pricing financial derivatives by simulation. Chapter 18 presents some specific algorithms for the numerical pricing of financial derivatives under various models.

The inter-relationship among the different chapters is summarized in the figure, which represents a flow chart of the material in the textbook. Each solid arrow indicates a strong connection between the material in the respective chapters, i.e., when a chapter is viewed as a prerequisite for the other. A dashed arrow indicates that a chapter is relevant but not necessarily a prerequisite for the other. Finally, the table below is a reference guide for instructors. It displays five different courses for which this book can be adopted as a required textbook at both the undergraduate and graduate levels. The relevant chapters for each course and the basic prerequisites are indicated in the table.

Course	Chapters	Prerequisites
Introduction to Financial Mathematics	1–4	Calculus, Linear Algebra, Elementary Probability Theory
Discrete-Time Derivative Pricing	2, 4–8	Calculus, Linear Algebra, Probability Theory
Stochastic Calculus	6, 9–11	Analysis, Probability Theory
Continuous-Time Derivative Pricing	10–16	Analysis, Probability Theory, Differential Equations
Introduction to Computational Finance	2, 4, 16–18	Calculus, Linear Algebra, Probability Theory, Numerical Methods

Acknowledgements

We would like to thank all our past and current undergraduate and graduate students for their valuable comments, feedback, and advice. We are also grateful for the feedback we have received from our colleagues in the Mathematics Department at Wilfrid Laurier University.

Giuseppe Campolieti and Roman N. Makarov

Waterloo, Ontario
January 2014

Part I

Introduction to Pricing and Management of Financial Securities

Chapter 1

Mathematics of Compounding

1.1 Interest and Return

1.1.1 Amount Function and Return

Any rational investor prefers a dollar in the pocket today to a dollar in her pocket one year from now. If an investor lends money to a borrower, the investor expects to be compensated for the use of the money. *Interest* is a compensation that a borrower of capital pays to a lender of capital for its use. If an initial amount P grows to an amount V over time, then the difference $I = V - P$ is *interest.* This situation is illustrated in Figure 1.1. For investments, it is often called interest earned, but it goes by other names such as return on investment or coupon payment. We assume that a nonnegative amount, and usually a positive amount, of interest is paid. The other crucial assumption is that there is no risk involved in this operation. So in this chapter we only deal with risk-free financial instruments. For example, the capital is deposited in a risk-free bank account or invested in government bonds (here, we neglect the possibility of government default on financial obligations).

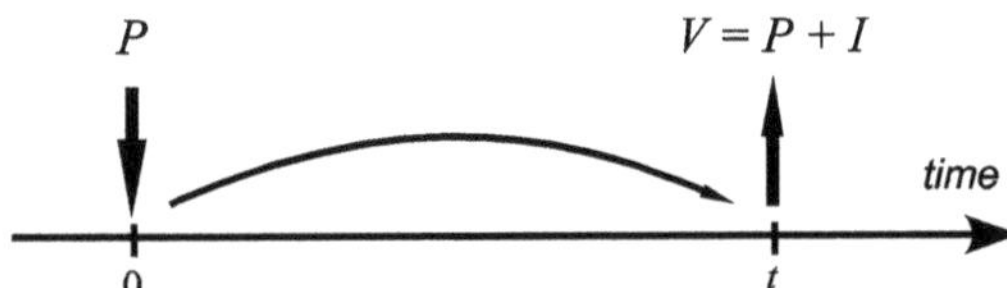

FIGURE 1.1: A time diagram for a simple cash flow. The interest is earned at time t. The accumulated value V is a sum of the principal P and interest I.

There are various explanations for the existence of interest, including the following.

The time value of money. Generally, people prefer to have money now rather than the same amount of money at some later day.

Inflationary expectations. The actual cost of the same amount changes in time.

Alternative investments. A lender of capital no longer has the option of immediately using the money invested. Interest compensates a lender for this loss of choices.

The following notation will be used.

- P denotes the initial capital borrowed or invested. It is called the *principal* or the *present value* of capital.
- $V(t)$ denotes the accumulated value at time $t \geqslant 0$, called the *value function.* Its initial value is equal to the principal, $V(0) = P$.

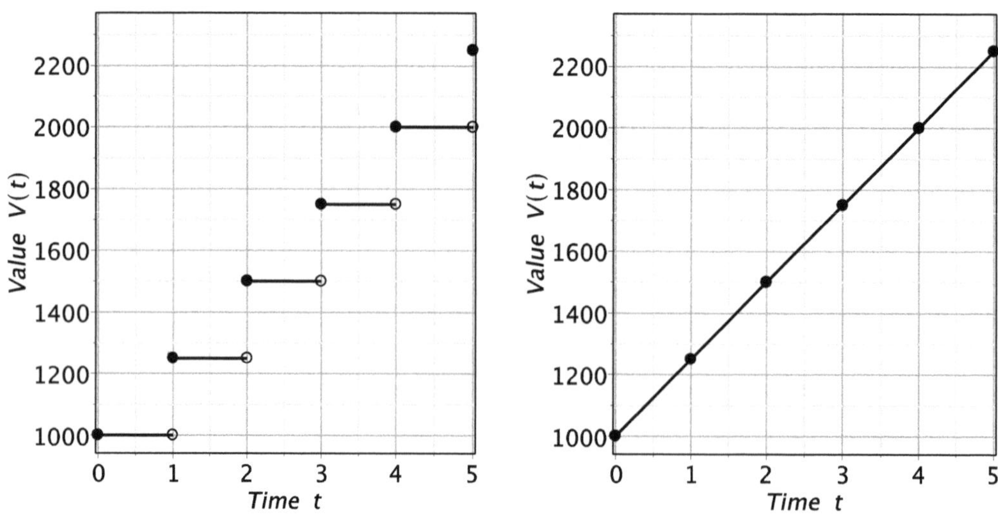

(a) Interest is paid at the end of each year.

(b) Interest is paid continuously such that the amount of interest earned over a period of time is proportional to the length of the period.

FIGURE 1.2: An investment of \$1,000 grows by a constant amount of \$250 each year for five years.

t is the time measured in years or, more generally, in time periods; typically, one period is one year, although it can be one month, one week, one day, etc. In practice, the duration of one period relates to the frequency at which interest is earned.

Example 1.1. An investment of \$1,000 grows by a constant amount of \$250 each year for five years. What does the graph of the value function $V(t)$ look like if

(a) interest is only paid at the end of each year;

(b) interest is paid continuously so that V is a linear function.

Solution. In case (a), the value function is a piecewise step function with jumps at the end of every year, i.e., $V(t) = V(t-1) + 250$ for all $t \geqslant 1$. In case (b), the value function is a linear function of time t, i.e., $V(t) = V(0) + ct$ for some parameter c. Since $V(1) = V(0) + 250$, the parameter c is equal to 250, and, therefore, $V(t) = 1000 + 250t$. The plots of the value functions are given in Figure 1.2. □

Typically, the amount your investment is worth at time t is proportional to the principal P you deposited at time 0. Let $A(t)$ denote the accumulated value for principal \$1. This function is called the *accumulation function*. So, one dollar invested at time 0 grows to $A(t)$ at time t. Then the accumulated value for principal P is given by

$$V(t) = P\,A(t). \tag{1.1}$$

Consider an investment with the value function $V(t)$, $t \geqslant 0$. The *total return* on the investment is the ratio of the amount received at the end of a period to the amount invested at the beginning of the period,

$$\text{total return} = \frac{\text{amount received}}{\text{amount invested}}\,.$$

Thus, the total return for the time interval $[s,t]$ with $0 \leqslant s < t$, denoted $R_{[s,t]}$, on an investment commencing at time s and terminating at time t is

$$R_{[s,t]} = \frac{V(t)}{V(s)}\,. \tag{1.2}$$

Note that the total return is a function of a period of time rather than a function of an instantaneous time moment.

The *rate of return* or *rate of interest* is the ratio of the amount of interest earned during the period to the investment value at the beginning of the period,

$$\text{rate of return} = \frac{\text{interest earned}}{\text{amount invested}}\,.$$

The rate of return for the time interval $[s,t]$, denoted $r_{[s,t]}$, is given by

$$r_{[s,t]} = \frac{V(t)-V(s)}{V(s)} = \frac{V(t)}{V(s)} - 1\,. \tag{1.3}$$

It can be expressed in terms of the accumulation function. For example, for the time interval of t periods from the date of the investment we have

$$r_{[0,t]} = \frac{V(t)-V(0)}{V(0)} = \frac{P\,A(t)-P}{P} = A(t)-1\,.$$

It is clear that the two notions are related by

$$R = 1 + r\,.$$

Equations (1.2) and (1.3) can be rewritten as

$$V(t) = R_{[s,t]}\,V(s) = (1 + r_{[s,t]})\,V(s)\,. \tag{1.4}$$

Note that upper- and lowercase letters, such as R and r, are used for total returns and rates of return, respectively. Often, for simplicity, the term *return* is used for both notions. Let n be a positive integer. The interval $[n-1,n]$ is the nth year (or the nth period, in general). Denote by R_n and r_n the total return and rate of return during the nth year from the date of investment, respectively. That is, we have

$$R_n := R_{[n-1,n]} = \frac{V(n)}{V(n-1)} \quad \text{and} \quad r_n := r_{[n-1,n]} = \frac{V(n)-V(n-1)}{V(n-1)} \tag{1.5}$$

for $n = 1,2,3,\ldots$. The accumulated value $V(n)$ can be now written as follows:

$$\begin{aligned} \frac{V(n)-V(n-1)}{V(n-1)} &= r \\ V(n)-V(n-1) &= rV(n-1) \\ V(n) &= (1+r)V(n-1), \quad \text{with } r = r_n\,. \end{aligned} \tag{1.6}$$

For two nonnegative integers n and m with $n < m$, the amount $V(m)$ accumulated by the end of period m can be written in two different ways:

$$V(m) = (1 + r_{[n,m]})\,V(n)$$

and

$$V(m) = (1 + r_m)\, V(m-1) = (1 + r_{m-1})\,(1 + r_m)\, V(m-2) = \dots$$
$$= (1 + r_{n+1}) \cdots (1 + r_{m-1})\,(1 + r_m) V(n)\,. \tag{1.7}$$

Therefore, one-period returns and aggregate returns are related as follows:

$$1 + r_{[n,m]} = (1 + r_{n+1})\,(1 + r_{n+2}) \cdots (1 + r_m)\,, \tag{1.8}$$
$$R_{[n,m]} = R_{n+1}\, R_{n+2} \cdots R_m\,. \tag{1.9}$$

Example 1.2 (The rate of return). Bill invested \$200 for 3 years in an account with the accumulation function $A(t) = 0.1t^2 + 1$. Find annual returns r_1, r_2, and r_3.

Solution. The accumulated values and returns at the end of years 1, 2, and 3 are, respectively,

$$V(1) = 200 \cdot (0.1 \cdot 1^2 + 1) = 220 \implies r_1 = \frac{V(1) - V(0)}{V(0)} = \frac{220 - 200}{200} = \frac{1}{10} = 10\%,$$
$$V(2) = 200 \cdot (0.1 \cdot 2^2 + 1) = 280 \implies r_2 = \frac{V(2) - V(1)}{V(1)} = \frac{280 - 220}{220} = \frac{3}{11} \cong 27.27\%,$$
$$V(3) = 200 \cdot (0.1 \cdot 3^2 + 1) = 380 \implies r_3 = \frac{V(3) - V(2)}{V(2)} = \frac{380 - 280}{280} = \frac{5}{14} \cong 35.71\%.$$

□

Example 1.3 (The accumulated value). Let $V(3) = 1000$ and $R_n = \frac{3n+2}{2n+2}$, $n = 1, 2, 3, \dots$. Find $V(6)$.

Solution. By using (1.4) and (1.9), we obtain

$$V(6) = V(3)\, R_4\, R_5\, R_6 = 1000 \cdot \frac{7}{5} \cdot \frac{17}{12} \cdot \frac{10}{7} = \$2{,}833.33\,.$$ □

1.1.2 Simple Interest

For simple interest, the rate of return $r_{[0,t]}$ is proportional to time t measured in years. That is, there exists a positive constant r, called a *simple interest rate* (per year), so that $r_{[0,t]} = r\,t$ for all $t \geqslant 0$. In particular, the rate of return for year 1 is $r_1 = r_{[0,1]} = r$. The interest $I(t)$ earned on the original principal P during the time period of length t is then calculated by the simple formula

$$I(t) = r_{[0,t]}\, P = r\, t\, P\,.$$

The accumulated value after t years will be

$$V(t) = P + I(t) = P + rtP = (1 + rt)P\,, \quad t \geqslant 0\,. \tag{1.10}$$

Clearly, the accumulated value $V(t)$ is a linear function of time t (see Figure 1.2b).

To find the initial capital (the principal) whose accumulated value at time t is given, we invert the formula (1.10) to obtain

$$P = V(0) = \frac{V(t)}{1 + rt} = (1 + rt)^{-1} V(t)\,. \tag{1.11}$$

This number is called the *present* or *discounted value* of the amount $V(t)$, and $(1+rt)^{-1}$ is a *discount factor at a simple interest rate* r. Formula (1.11) allows us to find the original principal P invested at time 0.

Example 1.4. How long will it take \$3,000 to earn \$60 interest at 6%?

Solution. We have $P = 3000$, $I = 60$, $r = 0.06$, and then

$$t = \frac{I}{Pr} = \frac{60}{3000 \cdot 0.06} = \frac{1}{3} \text{ years} = 4 \text{ months}. \qquad \square$$

Example 1.5. Treasury bills (T-bills) are popular short-term securities issued by the Federal Government of Canada with maturities of 1, 3, 6, or 12 months. T-bills are issued in different denominations, or face values. The face value of a T-bill is the amount the government guarantees it will pay on the maturity date. There is no interest stated on a T-bill. Instead, to determine its purchase price, you need to discount the face value to the date of sale at an interest rate that is determined by market conditions.

Suppose that a 6-month T-bill with a face value of \$25,000 is purchased by an investor who wishes to yield 3.80%. What price is paid?

Solution. We have $V = 25000$, $t = \frac{6}{12}$, and $r = 0.038$. Therefore, the investor will pay

$$P = V \cdot (1 + rt)^{-1} = 25000 \cdot \left(1 + 0.038 \cdot \frac{1}{2}\right)^{-1} = \$24{,}533.86$$

for the T-bill and receive \$25,000 in six months. $\square$

Let us study the return on an account earning interest at rate r. The distinction between the interest rate and the rate of return is that the interest rate refers to a period of one year (i.e., interest per annum) and is independent of the actual duration of an investment, whereas the return reflects both the interest rate and the length of time the investment is held. For all $0 \leqslant t < s$, we have

$$r_{[t,s]} = \frac{V(s) - V(t)}{V(t)} = \frac{(1 + rs)P - (1 + rt)P}{(1 + rt)P} = \frac{(s - t)r}{1 + rt}.$$

In particular, for the first year, we have $r_1 = r_{[0,1]} = r$. The rate of return during the nth year from the date of investment is

$$r_n = r_{[n-1,n]} = \frac{r}{1 + (n - 1)\, r}.$$

Clearly, these rates form a decreasing sequence that converges to zero as n approaches ∞. In other words, if the investment earns simple interest at the same rate every year, the effective annual rate of return is decreasing. To guarantee constant annual returns, the interest should be reinvested, as is demonstrated in the next section.

1.1.3 Periodic Compound Interest

Assume that the interest earned at a constant rate $i > 0$ is automatically *reinvested*, i.e., it will be added to the investment periodically (e.g., annually, semi-annually, quarterly, weekly, daily). In this situation, we are dealing with *periodic compounding*. The interest is said to be *compounded* or to be *converted*.

Example 1.6. Determine the compound interest earned on \$1,000 for 1 year at an annual rate of 8% compounded quarterly and compare it with the simple interest earned on the same amount for 1 year at 8% per annum.

Solution. Since the compounding period is one quarter, the interest rate per period is equal to $\frac{3}{12} \cdot 8\% = 2\%$. In the case of simple interest, the amount of interest earned during each quarter is fixed and equal to $\$1{,}000 \cdot 0.02 = \$20$. In the case of compounded interest, the interest earned during one quarter depends on the amount invested at the beginning of the period. The interest on the investment of V dollars is $I = V \cdot 0.08 \cdot \frac{1}{4} = V \cdot 0.02$. Results of our calculations are given in Table 1.3.

TABLE 1.3: Calculation of compound interest.

	Compound Interest		Simple Interest
At end of	Interest Earned	Accumulated Value	Accumulated Value
period 1	$1,000.00 · 0.02 = $20.00	$1,020.00	$1,020.00
period 2	$1,020.00 · 0.02 = $20.40	$1,040.40	$1,040.00
period 3	$1,040.40 · 0.02 = $20.81	$1,061.21	$1,060.00
period 4	$1,061.21 · 0.02 = $21.22	$1,082.43	$1,080.00

In summary, the compound interest earned on \$1,000 for 1 year at 8% compounded quarterly is \$82.43, whereas the simple interest on \$1,000 for 1 year at 8% is equal to $\$1{,}000 \cdot 0.08 \cdot 1 = \$80$. □

Let δt denote the time between two consecutive interest conversions (this time interval is called an *interest conversion period*). Assume that there are m interest conversion periods per year, so $\delta t = \frac{1}{m}$. The interest earned during one period is

$$I = V\, i\, \delta t = V\, \frac{i}{m},$$

where V is the amount invested at the beginning of the period, and i is the annual interest rate. Thus, the interest rate per period of δt is equal to $\frac{i}{m}$. Denote this one-period rate by j. Let us find the accumulated value $V(n\,\delta t)$ at the end of period $n = 1, 2, \ldots$. At the beginning, we have $V(0) = P$. By compounding the interest at the end of each period, we obtain the following accumulated values:

$$\begin{array}{ll}
\text{at the end of the 1st period} & V(\delta t) = V(0) + jV(0) = (1+j)V(0) = (1+j)P\,, \\
\text{at the end of the 2nd period} & V(2\,\delta t) = V(\delta t) + jV(\delta t) = (1+j)V(\delta t) = (1+j)^2 P\,, \\
\text{at the end of the 3rd period} & V(3\,\delta t) = (1+j)V(2\,\delta t) = (1+j)^3 P\,, \\
 & \vdots \\
\text{at the end of the } n\text{th period} & V(n\,\delta t) = (1+j)V((n-1)\,\delta t) = (1+j)^n P\,.
\end{array}$$

Therefore, the accumulated value at time $t = n\,\delta t$ is

$$V(t) = \left(1 + \frac{i}{m}\right)^n P\,. \tag{1.12}$$

The quantity m is called the *frequency of compounding*. Commonly used frequencies of compounding are $m = 1$ for annual compounding, $m = 2$ for semi-annual compounding, $m = 4$ for quarterly compounding, $m = 12$ for monthly compounding, $m = 52$ for weekly compounding, and $m = 365$ for daily compounding. The length of time between two consecutive interest calculations is called the *interest conversion period* or just *interest period*. Let $i^{(m)}$ denote a *nominal* interest rate that is compounded (convertible) m times per year. Note that $i^{(m)}$ is stated as an annual rate of interest. The interest rate compounded annually is simply denoted by i, i.e., $i^{(1)} \equiv i$. The interest rate per period of $\delta t = 1/m$

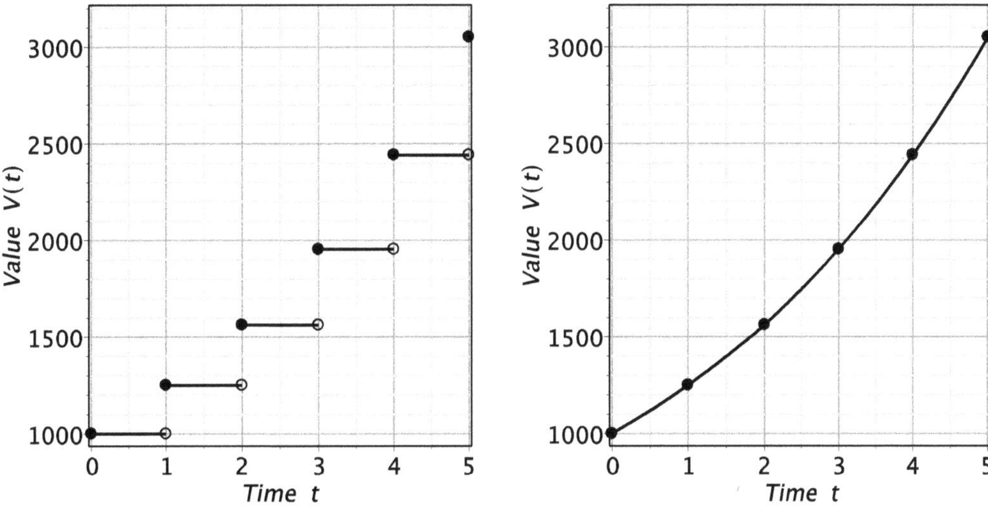

(a) Interest is paid and then compounded at the end of each year.

(b) Interest is paid and then compounded continuously such that the amount function grows exponentially.

FIGURE 1.4: An investment of \$1,000 grows at a rate of 25% per year for five years.

years is equal to $i^{(m)}/m$. The accumulated value of \$1 at the end of m interest conversion periods (i.e., at the end of one year) is

$$\left(1+\frac{i^{(m)}}{m}\right)^m .$$

Thus, after t years the principal P grows to

$$V(t) = \left(1+\frac{i^{(m)}}{m}\right)^{mt} P . \tag{1.13}$$

Here, mt gives the cumulative number of periods. The factor $\left(1+\frac{i^{(m)}}{m}\right)^m$ is called the *accumulation factor* over a one-year period, and it is in fact the accumulated value of \$1 at the end of one year with m as the compounding frequency. The annual nominal rate is meaningless until we specify the frequency m. At the same nominal rate the accumulated value depends on this frequency; it increases with increasing m. In particular, under interest compounded at the end of every year, the accumulation function is $V(t) = (1+i)^t P$ (see Figure 1.4).

To find the return on a deposit attracting interest periodically compounded, we use the general formula $r_{[s,t]} = \frac{V(t)-V(s)}{V(s)}$ and readily arrive at:

$$r_{[s,t]} = \frac{V(t)-V(s)}{V(s)} = \frac{V(t)}{V(s)} - 1 = \left(1+\frac{i^{(m)}}{m}\right)^{(t-s)m} - 1 \text{ for } 0 \leqslant s < t .$$

For the nth year the rate of return is

$$r_n = r_{[n-1,n]} = \left(1+\frac{i^{(m)}}{m}\right)^m - 1 .$$

As is seen, the annual rate of return r_n does not depend on n and is a constant quantity for any given m. It is clear that the return on a deposit subject to periodic compounding is not additive, i.e., $R_{[t_1,t_2]} + R_{[t_2,t_3]} \neq R_{[t_1,t_3]}$ and $r_{[t_1,t_2]} + r_{[t_2,t_3]} \neq r_{[t_1,t_3]}$ for $t_1 < t_2 < t_3$ in general. For example, take $m = 1$, then $r_{[0,1]} = r_{[1,2]} = r$ and

$$r_{[0,2]} = (1+r)^2 - 1 = 2r + r^2 \neq 2r = r_{[0,1]} + r_{[1,2]} .$$

Example 1.7. Find the total return and rate of return over two years under quarterly compounding with 8% per annum.

Solution. We have

$$R_{[0,2]} = \left(1 + \frac{0.08}{4}\right)^{2\cdot 4} = 1.02^8 = 1.17166 .$$

Therefore, $r_{[0,2]} = R_{[0,2]} - 1 = 17.166\%$. □

In business transactions it is frequently necessary to determine what principal P now will accumulate at a given interest rate to a specified amount $V(t)$ at a specified future date t. From the fundamental formula (1.13), we obtain

$$P = \frac{V(t)}{(1 + i^{(m)}/m)^{mt}} = \left(1 + \frac{i^{(m)}}{m}\right)^{-mt} V(t) . \tag{1.14}$$

P is called the *discounted value or present value* of $V(t)$. The process of determining P from $V(t)$ is called *discounting*. Similarly, we can find the value $V(s)$ of an investment at any intermediate date $0 < s < t$ given the value $V(t)$ at some fixed future time t:

$$V(s) = \frac{V(t)}{(1 + i^{(m)}/m)^{(t-s)m}} .$$

1.1.4 Continuous Compound Interest

Consider a fixed nominal rate of interest r compounded at different frequencies m. Let us find the limiting value of the accumulated value of \$1 over a one-year period as $m \to \infty$:

$$\lim_{m\to\infty} \left(1 + \frac{r}{m}\right)^m = \left(\lim_{m\to\infty} \left(1 + \frac{r}{m}\right)^{\frac{m}{r}}\right)^r$$

(denote r/m by x, which converges to 0 as $m \to \infty$)

$$= \left(\lim_{x\searrow 0} (1+x)^{\frac{1}{x}}\right)^r = \mathrm{e}^r .$$

[Note: The number $\mathrm{e} = 2.71828\ldots$ is the base of the natural logarithm; it is occasionally called Euler's number.] Therefore, the accumulated value of principal P compounded continuously at rate r over a period of t years (note that t may be fractional) is given by

$$V(t) = \lim_{m\to\infty} \left(1 + \frac{r}{m}\right)^{mt} P = \left[\lim_{m\to\infty} \left(1 + \frac{r}{m}\right)^m\right]^t P = \mathrm{e}^{rt} P. \tag{1.15}$$

We will denote the continuously compounded interest rate by r rather than by $i^{(\infty)}$. The discounted value P for given $V(t)$, r, and t is

$$P = \mathrm{e}^{-rt} V(t). \tag{1.16}$$

Being given the annual rate of return $i = \frac{V(1)-V(0)}{V(0)}$, one can find the continuously compounded rate r as follows: $r = \ln(1+i)$.

1.1.5 Equivalent Rates

Two nominal compound interest rates are said to be *equivalent* if they yield the same accumulated value at the end of one year, and hence at the end of any number of years. For example, the rate i compounded annually and rate $i^{(m)}$ compounded m per year are equivalent iff

$$1+i=\left(1+\frac{i^{(m)}}{m}\right)^m.$$

Example 1.8. Determine what rate $i^{(4)}$ is equivalent to (a) $i^{(12)}=6\%$, (b) $i^{(2)}=6\%$.

Solution.

(a) Equate the two accumulation factors, $(1+0.06/12)^{12}=(1+i^{(4)}/4)^4$, and solve for the rate $i^{(4)}$:

$$i^{(4)}=4\cdot((1+0.06/12)^{12/4}-1)=4\cdot((1.005)^3-1)\cong 6.030\%.$$

(b) We have $(1+0.06/2)^2=(1+i^{(4)}/4)^4$, hence

$$i^{(4)}=4\cdot((1+0.06/2)^{2/4}-1)=4\cdot(\sqrt{1.03}-1)\cong 5.956\%.$$ □

For a given nominal rate $i^{(m)}$ compounded m times per year or for a rate r compounded continuously, we define the corresponding annual *effective* rate, i, to be that rate which, if compounded annually, will produce the same interest. In other words, i is the annual rate of interest that is equivalent to $i^{(m)}$ or r. Note that equivalent compound rates have the same annual effective rate. To determine i, we compare the accumulated values of \$1 at the end of one year; thus

$$i=\begin{cases}\left(1+\frac{i^{(m)}}{m}\right)^m-1 & \text{for a rate compounded } m \text{ times per year,}\\ e^r-1 & \text{for a rate compounded continuously.}\end{cases} \tag{1.17}$$

Example 1.9. You wish to invest a sum of money for a number of years and have narrowed your choices to the following three investments:

A at the interest rate $i^{(2)}=10.35\%$;

B at the interest rate $i^{(12)}=10.15\%$;

C at the interest rate $i^{(4)}=10.25\%$.

Which investment should you choose?

Solution. To decide which investment is best, you need to calculate the equivalent rate compounded with the same frequency (e.g., annually) for each investment. So, let us calculate the annual effective rate of interest for each investment:

A: $i=\left(1+\frac{0.1035}{2}\right)^2-1\cong 10.6178\%$;

B: $i=\left(1+\frac{0.1015}{12}\right)^{12}-1\cong 10.6358\%$;

C: $i=\left(1+\frac{0.1025}{4}\right)^4-1\cong 10.6508\%$.

It turns out that investment **C** has the highest annual effective rate, so you should choose this investment. □

Example 1.10. Determine rates $i^{(2)}$ and $i^{(12)}$ that are equivalent to $r = 6\%$.

Solution. We equate the annual accumulation factors at each interest rate to obtain

$$(1+i^{(2)}/2)^2 = e^{0.06} \qquad (1+i^{(12)}/12)^{12} = e^{0.06}$$
$$i^{(2)}/2 = e^{0.06/2} - 1 \qquad i^{(12)}/12 = e^{0.06/12} - 1$$
$$i^{(2)} \cong 6.091\% \qquad i^{(12)} \cong 6.015\%$$

□

We can generalize the result presented above and state that for rates of interest that are all equivalent we have

$$i^{(1)} > i^{(2)} > i^{(4)} > i^{(12)} > i^{(365)} > i^{(\infty)}, \tag{1.18}$$

where $i^{(1)} \equiv i$ and $i^{(\infty)} \equiv r$.

Proposition 1.1. *For a fixed annual effective rate i, the compound interest rates $i^{(m)}$, where $m = 1, 2, 3, \ldots$, all equivalent to i, form a decreasing sequence that converges to the rate $r = \ln(1+i)$ as $m \to \infty$.*

Proof. The interest rate $i^{(m)}$ is equivalent to the annual effective rate i iff $(1+i^{(m)}/m)^m = 1+i$ holds. This defines the interest rate $i(m) \equiv i^{(m)}$ as a function of $m > 0$:

$$i(m) = \left((1+i)^{1/m} - 1\right) m = \left(e^{r/m} - 1\right) m.$$

Hence, its derivative w.r.t. m is

$$i'(m) = i(m)/m - (r/m)e^{r/m} = i(m)/m - (1+i(m)/m)\ln(1+i(m)/m).$$

Let us prove that the derivative $i'(m)$ is strictly negative for all $m > 0$.

- To show that $(1+i(m)/m)\ln(1+i(m)/m) > i(m)/m$ holds for all $m > 0$, it suffices to prove that the inequality $\ln(1+x) > x/(1+x)$ holds for all $x > 0$.
- Notice that if two functions $f(x)$ and $g(x)$ are such that $f(0) \geqslant g(0)$ and $f'(x) > g'(x)$ holds for all $x > 0$, then $f(x) > g(x)$ is valid for all $x > 0$.
- Let $f(x) := \ln(1+x)$ and $g(x) := x/(1+x) = 1 - (1+x)^{-1}$. Their derivatives are $f'(x) = (1+x)^{-1}$ and $g'(x) = (1+x)^{-2}$, respectively.
- Clearly, $(1+x)^{-1} > (1+x)^{-2}$ holds for all $x > 0$. Since $f(0) = g(0) = 0$, we obtain that $\ln(1+x) > x/(1+x)$ is valid for all $x > 0$.

Using l'Hôpital's rule (LHR) gives

$$\lim_{m\to\infty} i^{(m)} = \lim_{m\to\infty} \frac{e^{r/m} - 1}{1/m} = \lim_{t\to 0} \frac{e^{rt} - 1}{t} \overset{\text{(LHR)}}{=} \lim_{t\to 0} re^{rt} = r \lim_{t\to 0} e^{rt} = r.$$

□

Proposition 1.2. *The accumulation function $A(m) = \left(1 + \frac{r}{m}\right)^m$ for a fixed nominal interest rate r compounded with frequency m over a one-year period is a strictly increasing function of m that converges to e^r as $m \to \infty$.*

Proof. Differentiate the function $A(m)$ w.r.t. m to obtain

$$A'(m) = (1+r/m)^m \left(\ln(1+r/m) - \frac{r/m}{1+r/m}\right).$$

Using the inequality $\ln(1+x) > x/(1+x)$ with $x = r/m > 0$ gives that $A'(m) > 0$. Therefore, $\{A(m)\}_{m \geqslant 1}$ is a strictly increasing sequence that converges to e^r as $m \to \infty$. □

As follows from the above proposition, continuous compounding produces higher accumulated value than periodic compounding with any frequency m, that is, $\left(1+\frac{r}{m}\right)^m < \mathrm{e}^r$ for all $m \geqslant 1$.

Example 1.11. Consider an investment at: (a) simple interest rate 6%, (b) annual compound interest rate 6%, and (c) continuous compound interest rate 6%. Find the time that is necessary to double the original principal.

Solution. For each investment, we find time t so that the respective accumulation function equals 2.

(a) For simple interest: $1 + 0.06t = 2$, hence $t = 1/0.06 \cong 16.67$ years.

(b) For interest compounded annually: $(1 + 0.06)^t = 2$, hence $t = \frac{\ln 2}{\ln 1.06} \cong 11.9$ years.

(c) For interest compounded continuously: $\mathrm{e}^{0.06t} = 2$, hence $t = \frac{\ln 2}{0.06} \cong 11.55$ years. □

Let us summarize the above three methods of calculating interest. The *accumulation function* $A(t) = \frac{V(t)}{V(0)} = R_{[0,t]}$ has the following form:

$$A(t) = \begin{cases} 1 + rt & \text{for a rate of simple interest,} \\ \left(1 + \frac{i^{(m)}}{m}\right)^{mt} & \text{for a rate of interest compounded periodically,} \\ \mathrm{e}^{rt} & \text{for a rate of interest compounded continuously.} \end{cases} \tag{1.19}$$

In what follows, we will also use the *discounting function* $D(t) = \frac{V(0)}{V(t)} = \frac{1}{A(t)}$ taking the form:

$$D(t) = \begin{cases} (1 + rt)^{-1} & \text{for a rate of simple interest,} \\ \left(1 + \frac{i^{(m)}}{m}\right)^{-mt} & \text{for a rate of interest compounded periodically,} \\ \mathrm{e}^{-rt} & \text{for a rate of interest compounded continuously.} \end{cases} \tag{1.20}$$

1.1.6 Continuously Varying Interest Rates

Consider the case of continuously compounded interest but with a rate that is changing in time. Let T be the time horizon of investment; let $r(t)$ denote the *instantaneous* interest rate at time $t \in [0, T]$. Consider a time interval $[t, t + \delta t]$ with small $\delta t > 0$. Suppose that the rate $r(t)$ is approximately constant on the interval $[t, t + \delta t]$. Then the amount $V(t)$ invested at time t will grow to $V(t + \delta t) = V(t)\mathrm{e}^{r(t)\,\delta t}$ at time $t + \delta t$.

Let the principal P be invested at time 0. To determine $V(t)$ in terms of the principal and the instantaneous interest rate function $r(t)$ for $0 \leqslant t \leqslant T$, we split the time interval $[0, T]$ in N equal subintervals of length $\delta t = T/N$ and apply the above approach. For times $t_k = k\,\delta t$ with $k = 0, 1, \ldots, N$, we obtain

$$\begin{aligned} V(t_1) &= V(t_0)\mathrm{e}^{r(t_0)\,\delta t} = P\mathrm{e}^{r(t_0)\,\delta t} \\ V(t_2) &= V(t_1)\mathrm{e}^{r(t_1)\,\delta t} = P\mathrm{e}^{(r(t_0)+r(t_1))\,\delta t} \\ V(t_3) &= V(t_2)\mathrm{e}^{r(t_2)\,\delta t} = P\mathrm{e}^{(r(t_0)+r(t_1)+r(t_2))\,\delta t} \\ &\vdots \\ V(t_N) &= P\mathrm{e}^{(r(t_0)+r(t_1)+\cdots+r(t_{N-1}))\,\delta t} = P\exp\left(\sum_{k=0}^{N-1} r(t_k)\,\delta t\right). \end{aligned}$$

Recognizing the Riemann sum in the last equation with the time step $\delta t = T/N$ and taking the limit as $N \to \infty$, we have

$$\lim_{N\to\infty} \sum_{k=0}^{N} r(t_k)\,\delta t = \int_0^T r(t)\,\mathrm{d}t\,.$$

Therefore, the value at the maturity time T is given by

$$V(T) = P\exp\left(\int_0^T r(t)\,\mathrm{d}t\right). \tag{1.21}$$

If $r(t)$ is constant and equal to r for all $t \in [0, T]$, then (1.21) reduces to the usual formula for the accumulated value under continuous compounding:

$$V(T) = P\exp\left(\int_0^T r\,\mathrm{d}t\right) = P\exp\left(r\int_0^T \mathrm{d}t\right) = P\exp(rT)\,.$$

Let $y(T)$ denote the average of the instantaneous interest rates from time 0 to time T:

$$y(T) := \frac{1}{T}\int_0^T r(t)\,\mathrm{d}t\,.$$

Then, (1.21) takes a familiar form:

$$V(T) = P\,\mathrm{e}^{y(T)T}.$$

The interest rate $y(T)$ is called the *yield rate* or *spot rate* for maturity T.

It is possible to derive (1.21) in a different way. Consider an investment with principal $V(0) = P$ and accumulated value $V(t)$ at time $t > 0$. The nominal rate of interest compounded m times per year and evaluated at time t is given by

$$i^{(m)}(t) = \frac{V(t+\frac{1}{m}) - V(t)}{\frac{1}{m}V(t)}\,.$$

Let $\delta t = \frac{1}{m}$, so that, as $m \to \infty$ and $\delta t \to 0$, we have

$$r(t) := \lim_{m\to\infty} i^{(m)}(t) = \lim_{\delta t\to 0} \frac{V(t+\delta t) - V(t)}{\delta t\, V(t)} = \frac{V'(t)}{V(t)} = \frac{\mathrm{d}\ln V(t)}{\mathrm{d}t}\,. \tag{1.22}$$

Thus, $r(t)$ is a measure of the relative instantaneous rate of growth at time t. It is sometimes called the *force of interest.* Integrating $r(t) = \frac{\mathrm{d}\ln V(t)}{\mathrm{d}t}$ from 0 to T gives

$$\int_0^T r(t)\,\mathrm{d}t = \int_0^T \frac{\mathrm{d}\ln V(t)}{\mathrm{d}t}\,\mathrm{d}t = \ln V(T) - \ln V(0) = \ln\left(\frac{V(T)}{V(0)}\right),$$

and then (1.21) follows by exponentiation.

Example 1.12. Calculate the accumulated value of \$1,000 at the end of 4 years if $r(t) = 0.05 + 0.1t$.

Solution. The accumulation function is

$$A(T) = \exp\left(\int_0^T r(t)\,\mathrm{d}t\right) = \exp\left(\int_0^T (0.05 + 0.1t)\,\mathrm{d}t\right) = \mathrm{e}^{0.05T + 0.05T^2}$$

for every $T \geqslant 0$. Thus, after 4 years we have

$$A(4) = \mathrm{e}^{0.05\cdot 4 + 0.05\cdot 4^2} = \mathrm{e} \cong 2.71828182$$

and $V(4) = P\,A(4) = 1000\,\mathrm{e} \cong \$2{,}718.28$. □

1.2 Time Value of Money and Cash Flows

1.2.1 Equations of Value

All financial decisions must take into account the basic idea that money has its time value. In order to compare or add different amounts of money we must place the amounts at the same point in time, called the *focal date.* The mathematics of finance deals with *dated values.* In general, we compare dated values by the following definition of equivalence.

Definition 1.1. Assume the periodic compounding of interest. Amount $\$X$ due on a given date is *equivalent* to $\$Y$ due n interest conversion periods later, if $Y = X(1+j)^n$ or $X = Y(1+j)^{-n}$, where $j = i^{(m)}/m$ and $i^{(m)}$ is a given compound interest rate. The following time diagram illustrates dated values equivalent to a given dated value X.

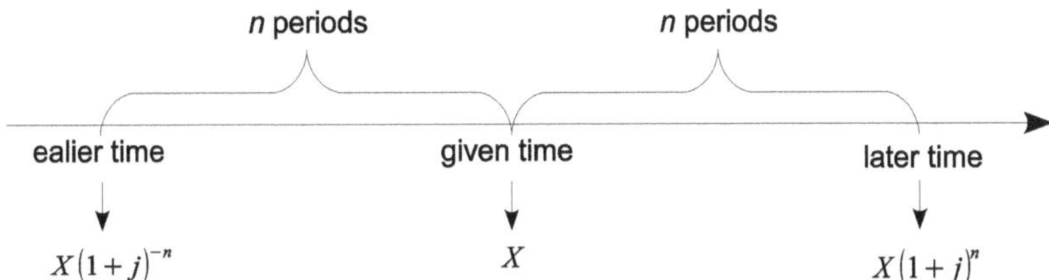

FIGURE 1.5: Equivalence at a given compound interest rate.

We say that two sets of payments are equivalent at a given interest rate if the dated values of the sets, on any common date, are equal. An equation stating that the dated values of two sets of payments are equal is called an *equation of value.*

Example 1.13. A debt of $5,000 is due at the end of 3 years. Determine an equivalent debt due at the end of (a) 3 months, (b) 3 years and 9 months. Assume quarterly compounding with $i^{(4)} = 12\%$.

Solution. We arrange the data on a time diagram below where one period is 3 months.

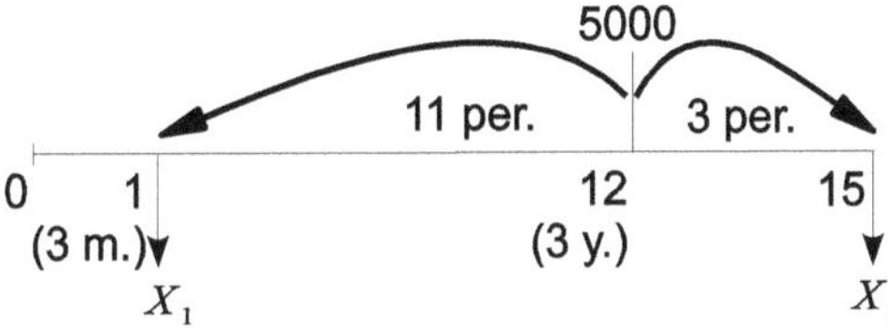

By the definition of equivalence:

$$\text{(a) } X_1 = \$5{,}000 \cdot (1+0.12/4)^{-11} = \$5{,}000 \cdot (1.03)^{-11} \cong \$3{,}612.11;$$
$$\text{(b) } X_2 = \$5{,}000 \cdot (1+0.12/4)^{3} = \$5{,}000 \cdot (1.03)^{3} \cong \$5{,}463.64.$$

□

Example 1.14. A debt of $5,000 is due at the end of 5 years. It is proposed that $\$X$ be paid now and with another $\$X$ paid in 10 years' time to liquidate the debt. Calculate the value of X if the effective interest rate is 12% for the first 6 years and 8% for the next 4 years.

Solution. The value of the first payment at time $t = 5$ is $X \cdot (1.12)^5$. The value of the second payment at time $t = 5$ is $X \cdot (1.08)^{-4} \cdot (1.12)^{-1}$. We equate the sum of these two values and the value of the debt of \$5,000 at $t = 5$ to obtain

$$5000 = X \cdot (1.12)^5 + X \cdot (1.08)^{-4} \cdot (1.12)^{-1} = X \cdot (1.76234 + 0.656277) = 2.41862\, X\,.$$

Hence, $X = \$5{,}000/2.41862 \cong \$2{,}067.30$. □

Example 1.15. Suppose you can buy a house for \$84,000 cash (option 1) or for payments of \$50,000 now, \$20,000 in 1 year, and \$20,000 in 2 years (option 2). If money is worth $i^{(12)} = 9\%$, which option is more profitable?

Solution. The discounted value of option 1 is \$84,000. Using monthly compounding, the discounted value of option 2 is

$$\begin{aligned}&\$50{,}000 + \$20{,}000 \cdot (1 + 0.09/12)^{-12} + \$20{,}000 \cdot (1 + 0.09/12)^{-24}\\ &\cong \$50{,}000 + \$18{,}284.76 + \$16{,}716.63 = \$85{,}001.39.\end{aligned}$$

Clearly, option 1 is more preferable than option 2. □

Example 1.16. If you file your taxes late, the Canada Revenue Agency charges you an immediate 5% late penalty, as well as 5% compounded daily on your outstanding balance. Express my total penalty as a single rate (compounded daily) assuming I pay 26 weeks late.

Solution. If I pay n days late, every dollar I originally owed swells to

$$1.05 \cdot (1 + 0.05/365)^n\,.$$

Therefore, on a per dollar basis my total debt after $n = 26 \cdot 7 = 182$ days is

$$1.05 \cdot (1 + 0.05/365)^{182}\,.$$

Had I been charged the single rate $i^{(365)}$, my total debt would have been $(1+i^{(365)}/365)^{182}$. Equating and solving for $i^{(365)}$ we get

$$1.05 \cdot (1 + 0.05/365)^{182} = (1 + i^{(365)}/365)^{182}\,,$$

which can be rearranged to find

$$i^{(365)} = 365 \cdot \left(1.05^{1/182} \cdot (1 + 0.05/365) - 1\right) \cong 0.14787\,.$$

Therefore my effective penalty rate is approximately 14.79%. □

Remark. If I pay n days late, then my penalty rate is

$$i_n^{(365)} = 365 \cdot \left(1.05^{1/n}(1 + 0.05/365) - 1\right)\,.$$

Observe that $\lim_{n\to\infty} i_n^{(365)} = 0.05$. Thus, if I pay extremely late, the initial 5% penalty is dwarfed by the compound interest. Of course, this does not mean that it is a good idea to pay extremely late.

Remark. Suppose you are offered a Guaranteed Investment Certificate (GIC) that promises 4% compounded annually with an upfront service fee of 1% of the amount invested. If invested for n years, the effective rate that you earn on the GIC can be determined by solving $0.99 \cdot (1 + r)^n = \left(1 + i^{(n)}\right)^n$, which yields $i^{(n)} = 0.99^{1/n} \cdot (1 + r) - 1$. It is easy to prove (do this as an exercise) that $i^{(n)}$ is less than r and converges to r as $n \to \infty$.

Example 1.17. You have access to an account that pays $100r\%$ per annum, compounded annually. You are given two options to repay a debt. Option A is an immediate payment of \$30,000; option B is the payment of \$16,000 per year for each of the next two years. For what values of r would you choose option B?

Solution. The present value of B, in thousands of dollars, is

$$P_B(r) = \frac{16}{1+r} + \frac{16}{(1+r)^2}.$$

The present value of A, in thousands of dollars, is $P_A(r) = 30$. It is only sensible to choose option B provided $P_B(r) < P_A(r)$, which occurs if and only if

$$30 > \frac{16}{1+r} + \frac{16}{(1+r)^2} \iff 30(1+r)^2 > 16(1+r) + 16,$$

which clearly occurs if and only if $30r^2 + 44r - 2 > 0$. The roots of this quadratic are $r \cong -1.510$ and $r \cong 0.0441$, and since the leading coefficient of the quadratic is positive we should therefore choose option B if r exceeds the larger root. In summary, we choose A if the interest rate is less than (approximately) 4.41% and choose B otherwise. □

1.2.2 Deterministic Cash Flows

From a broad point of view, an investment can be defined as a sequence of expenditures and receipts spanning a period of time. Let all expenses and incomes be denominated in cash. The series of cash payments over a period of time is called a *cash flow stream.* Suppose that the total time interval $[0, T]$ is divided into n subintervals. Let the payments of $C_0, C_1, \ldots, C_n$ dollars be respectively received on the dates $t_0, t_1, \ldots, t_n$ so that $0 = t_0 < t_1 < t_2 < \ldots < t_n = T$. Table 1.6 represents such a scenario.

TABLE 1.6: Cash flow stream.

Dates	t_0	t_1	$\ldots$	t_{n-1}	t_n
Cash Flows	C_0	C_1	$\ldots$	C_{n-1}	C_n

A cash flow stream can also be represented as a pair $(\mathbf{C}, \mathbf{T})$ of two vectors:

$$\mathbf{C} = [C_0, C_1, \ldots, C_n] \text{ and } \mathbf{T} = [t_0, t_1, \ldots, t_n].$$

Every coin has two sides. If one party holds the cash flow stream $\mathbf{C}$, then the opposite party holds the stream $-\mathbf{C} = [-C_0, -C_1, \ldots, -C_n]$. Here, we assume that if $C_k > 0$, then the payment of C_k dollars is received at time t_k (an in-flow of cash); if $C_k < 0$, then the payment of $|C_k|$ dollars is made to the counter-party at time t_k (an out-flow of cash).

To visualize a cash flow stream $(\mathbf{C}, \mathbf{T})$, one can also use a time diagram which is constructed as follows. First, draw a horizontal line, which represents time increasing from the present (denoted by 0) as we are moving from left to right. After that, draw short vertical lines that start on the horizontal line. Those that go up represent cash coming in (positive cash flows, or receipts), while those that go down represent cash going out (negative cash flows, or disbursements). In general, without knowing the sign of C_k we cannot, on the time diagram, correctly indicate whether C_k should be above or below the time line. The cash flow stream for compounding is given in in Figure 1.7.
Here, the time is measured in periods; $j = i^{(m)}/m$ is the interest rate for one period. This rate is assumed to be constant. The cash flow stream for discounting is represented in Figure 1.8.

Dates	0	n
Cash Flows	$-P$	$P(1+j)^n$

FIGURE 1.7: The cash flow stream for compounding.

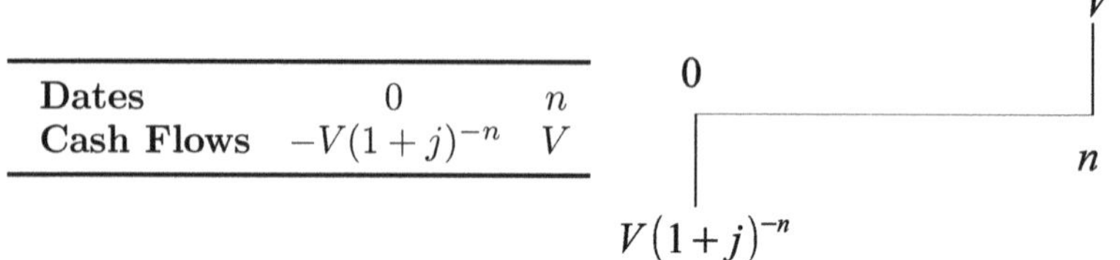

Dates	0	n
Cash Flows	$-V(1+j)^{-n}$	V

FIGURE 1.8: The cash flow stream for compounding.

Definition 1.2. The *net present value*, NPV, of an investment is the difference between the present value of the cash inflows ($C_k > 0$) and the present value of the cash outflows ($C_k < 0$), that is,

$$\mathrm{NPV}(\mathbf{C}) = C_0 + D(t_1)C_1 + D(t_2)C_2 + \cdots + D(t_n)C_n, \tag{1.23}$$

where d is the discounting function given by (1.20).

For example, let us assume that the lengths of the periods (t_{k-1}, t_k), $k = 1, 2, \ldots, n$, are equal, and interest is periodically compounded at interest rate j. Then

$$\mathrm{NPV}(\mathbf{C}) = C_0 + C_1(1+j)^{-1} + C_2(1+j)^{-2} + \cdots + C_n(1+j)^{-n}.$$

The interest rate used is known as the cost of capital and can be considered as the cost of borrowing money by the business or the rate of return that an investor may obtain if the money is invested with security. The NPV can be used to make a business decision. If NPV $\geqslant 0$, we can conclude that the rate of return from the cash in-flows is greater than or equal to the cost of the cash out-flows and the project has economic merit and should proceed. If NPV < 0, the project should not proceed. When attempting to choose between two investments with the same levels of risk, the investor generally chooses the one with the higher net present value. If both investments have the same net present value and the same time interval, then the investor is said to be indifferent between the investments.

Example 1.18 (Calculating the NPV). An investor is considering two investments with the following annual cash flows:

Years	0	1	2	3
Cash Flows (Investment 1)	−$13,000	$5,000	$6,000	$7,000
Cash Flows (Investment 2)	−$13,000	$7,000	$4,800	$6,000

Which is the better investment if the prevailing annual compound interest rate is (a) $i = 4.5\%$; (b) $i = 9\%$?

Solution.
(a) At 4.5%, we have

$$\begin{aligned}\text{NPV of Investment 1} &= -13000 + \frac{5000}{1+0.045} + \frac{6000}{(1+0.045)^2} + \frac{7000}{(1+0.045)^3} \\ &= \$3{,}413.14, \\ \text{NPV of Investment 2} &= -13000 + \frac{7000}{1+0.045} + \frac{4800}{(1+0.045)^2} + \frac{6000}{(1+0.045)^3} \\ &= \$3{,}351.85.\end{aligned}$$

Thus, at $i = 4.5\%$, Investment 1 is a better choice.
(b) At 9%, we have

$$\begin{aligned}\text{NPV of Investment 1} &= -13000 + \frac{5000}{1+0.09} + \frac{6000}{(1+0.09)^2} + \frac{7000}{(1+0.09)^3} \\ &= \$2{,}042.52, \\ \text{NPV of Investment 2} &= -13000 + \frac{7000}{1+0.09} + \frac{4800}{(1+0.09)^2} + \frac{6000}{(1+0.09)^3} \\ &= \$2{,}095.18.\end{aligned}$$

Thus, at $i = 9\%$, Investment 2 is a better choice. □

1.3 Annuities

One of the main types of a cash flow stream is an *annuity*, which is defined as a sequence of periodic payments, usually equal, made at equal intervals of time. There are many examples of annuities in the financial world: mortgage payments on a home, car loan payments, payments on rent, dividends, payments on instalment purchases, etc. The following standard terminology is used.

- The time between successive payments of an annuity is called the *payment interval.*
- The time from the beginning of the first payment interval to the end of the last payment interval is called the *term* of an annuity.
- The payments of an *ordinary annuity* (also known as an *annuity immediate*) are made at the end of the payment intervals. When the payments are made at the beginning of the payment intervals, the annuity is called an *annuity due.*
- When the payment interval and interest conversion period coincide, the annuity is called a *simple annuity,* otherwise it is a *general annuity.*

We define the *accumulated value* of an annuity as the equivalent dated value of the set of payments due at the end of the term. Similarly, the *discounted value* of an annuity is defined as the equivalent dated value of the set of payments due at the beginning of the term. Throughout this section, we shall use the following notations.

C denotes the periodic payment of the annuity.

n is the number of interest compounding periods during the term of an annuity. In the case of a simple annuity, n equals the total number of payments.

j denotes interest rate per conversion period (assume $j > 0$).

V_A is the future (accumulated) value of an annuity at the end of the term.

P_A is the present (discounted) value of an annuity.

1.3.1 Simple Annuities

1.3.1.1 Ordinary Annuities

The present value P_A of an ordinary simple annuity is defined as the equivalent dated value of the set of payments due at the beginning of the term, i.e., *one period before the first payment.* The accumulated value V_A of an ordinary simple annuity is defined as the equivalent dated value of the set of payments due at the end of the term, i.e., *on the date of the last payment.* In Figure 1.9 we display an ordinary simple annuity on a time diagram. The cash flow streams for the repayment of a loan amounting to P_A and for the accumulation of a fund amounting to V_A are given in Tables 1.10 and 1.11, respectively.

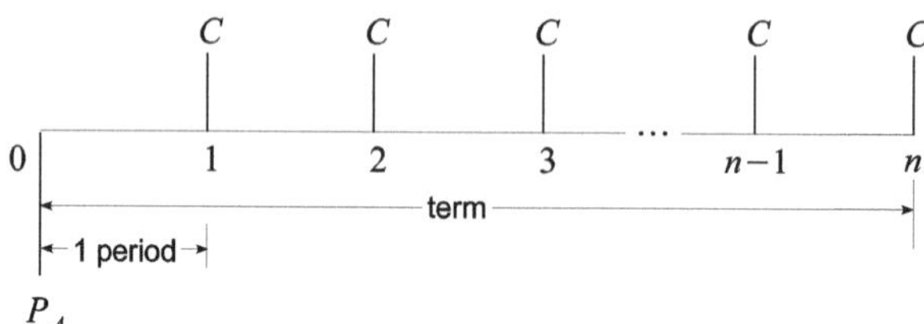

FIGURE 1.9: The time diagram for an ordinary simple annuity.

TABLE 1.10: Cash flow streams for the repayment of a loan amounting to P_A.

Dates	0	1	...	$n-1$	n
Cash Flows	P_A	$-C$	...	$-C$	$-C$

TABLE 1.11: Cash flow streams for the accumulation of a fund amounting to V_A.

Dates	0	1	...	$n-1$	n
Cash Flows	0	$-C$	...	$-C$	$V_A - C$

Now we develop a formula for the accumulated value V_A of an ordinary simple annuity of n payments of $\$C$ using the sum of a geometric progression. Let $V(k)$ denote the accumulated value at the end of period k. At the end of 3 periods, we have

Period	Investment	Interest	Balance at the end of the period
1	C	0	$C = V(1)$
2	C	$jV(1)$	$C + (1+j)V(1) = C + (1+j)C = V(2)$
3	C	$jV(2)$	$C + (1+j)V(2) = C + (1+j)C + (1+j)^2C = V(3)$
			$\vdots$

At the end of n periods, we obtain

$$V_A = V(n) = C + (1+j)C + (1+j)^2C + (1+j)^3C + \cdots + (1+j)^{n-1}C\,. \tag{1.24}$$

Recall the formula for the sum of a geometric progression. For real and positive a and $q \neq 1$, and natural m, we have

$$a + aq + aq^2 + \cdots + aq^m = a\sum_{k=0}^{m} q^k = a\,\frac{q^{m+1}-1}{q-1}\,.$$

Applying this formula to a geometric progression in (1.24) with $m = n-1$ terms, whose first term is $a = C$ and common ratio is $q = 1 + j$, we obtain the accumulated value V_A of an ordinary simple annuity of n payments of $\$C$ each:

$$V_A = C\,\frac{(1+j)^n - 1}{j}\,. \tag{1.25}$$

To calculate the periodic payment we solve equation (1.25) for C to obtain

$$C = V_A \frac{j}{(1+j)^n - 1}\,.$$

It is possible to derive the discounted value P_A in several different ways. First, let us consider the amortization of a loan by regular payments of $\$C$. At the end of 2 periods, we have

Period	Payment	Interest	Balance at the end of the period
0			$P_A = V(0)$
1	C	$jV(0)$	$(1+j)V(0) - C = (1+j)P_A - C = V(1)$
2	C	$jV(1)$	$(1+j)V(1) - C = (1+j)^2P_A - (1+j)C - C = V(2)$
			$\vdots$

At the end of n periods, we obtain

$$V(n) = (1+j)^n P_A \underbrace{- C - (1+j)^1C - (1+j)^2C - \ldots - (1+j)^{n-1}C}_{=-V_A} = (1+j)^n P_A - V_A = 0\,.$$

Therefore,

$$V_A = (1+j)^n P_A. \tag{1.26}$$

Note that this relation also follows directly from the fact that P_A and V_A are both dated values of the same set of payments and thus they are equivalent. Therefore, the basic formula for the discounted value P_A of an ordinary simple annuity of n payments of $\$C$ each is

$$\begin{aligned} P_A &= (1+j)^{-n}V_A = C\,(1+j)^{-n}\frac{(1+j)^n - 1}{j} = C\,\frac{1-(1+j)^{-n}}{j} \\ &= C\,\frac{1-(1+j)^{-n}}{j}\,. \end{aligned} \tag{1.27}$$

It is convenient to use the following standard notation:

$$a_{\overline{n}|j} = \frac{1-(1+j)^{-n}}{j}\,.$$

This number is called an *annuity symbol* (read "a angle n at j"). It is the present value of an ordinary simple annuity of n payments of \$1 each. The expressions for P_A and V_A take the following concise forms:

$$P_A = C\, a_{\overline{n}|j}, \quad V_A = C\, a_{\overline{n}|j}\,(1+j)^n\,. \tag{1.28}$$

The formulae (1.25) and (1.27) can also be derived by using the fact that the NPV of the respective cash flow stream is zero:

$$\mathrm{NPV}(P_A, \underbrace{-C, \ldots, -C}_{n \text{ payments}}) = P_A - CD(1) - CD(2) - \ldots - CD(n) = 0$$

$$\Longrightarrow P_A = C \sum_{k=1}^{n} (1+j)^{-k} = C\,\frac{1-(1+j)^{-n}}{j}$$

$$\mathrm{NPV}(0, \underbrace{-C, \ldots, -C}_{n \text{ payments}} + V_A) = -CD(1) - CD(2) - \ldots - CD(n) + V_A D(n) = 0$$

$$\Longrightarrow V_A = \frac{C}{D(n)} \sum_{k=1}^{n} (1+j)^{-k} = C\,(1+j)^n \frac{1-(1+j)^{-n}}{j} = C\,\frac{(1+j)^n - 1}{j}\,.$$

Here, $D(k) = (1+j)^{-k}$ is the discounting function.

Example 1.19. You make deposits of \$1,500 every six months starting today into a fund that pays interest at $i^{(2)} = 7\%$. How much do you have in your fund immediately after the 30th deposit?

Solution. We have $C = 1500$, $n = 30$, and the interest rate $0.07/2 = 0.035$ per half year. Since we have been asked to determine the accumulated value on the date of the 30th deposit, we deal with an ordinary simple annuity. Therefore,

$$V_A = 1500 \cdot \frac{(1.035)^{30} - 1}{0.035} \cong \$77{,}434.02\,.$$

□

Example 1.20. It is estimated that a machine will need replacing 10 years from now at a cost of \$80,000. How much must be put aside each year to provide that amount of money if the company's savings earn interest at an 8% annual effective rate?

Solution. Assume that an amount of $\$C$ is deposited at the end of each of 10 years. So we deal with an ordinary simple annuity. The accumulated value is

$$V_A = C\,\frac{(1.08)^{10} - 1}{0.08} = \$80{,}000.$$

Hence, $C \cong \$5{,}522.36$. □

Example 1.21. A used car is purchased for \$2,000 down and \$200 a month for 6 years. Interest is at $i^{(12)} = 10\%$.

(a) Determine the price of the car.

(b) Assuming no payments are missed, what single payment at the end of 2 years will completely pay off the debt?

Solution.
(a) We have that the price P of the car is a sum of the down payment and the present value P_A of a simple annuity paying \$200 a month:

$$P = 2000 + P_A = 2000 + 200\,\frac{1-(1.008333)^{-72}}{0.008333} \cong \$12{,}795.73\,.$$

(b) The value of the debt at the end of 2 years is given by the sum of one regular monthly payment and the present value of $48 = 4 \cdot 12$ payments left:

$$200 + 200\,\frac{1-(1.008333)^{-48}}{0.008333} \cong \$8{,}085.63\,.$$

□

1.3.1.2 Annuities Due

An *annuity due* is an annuity whose periodic payments are due at the beginning of each payment interval. The term of an annuity due starts at the time of the first payment and ends one payment period after the date of the last payment. The diagram below shows the simple case of an annuity due for n payments when the payment intervals and interest periods coincide.

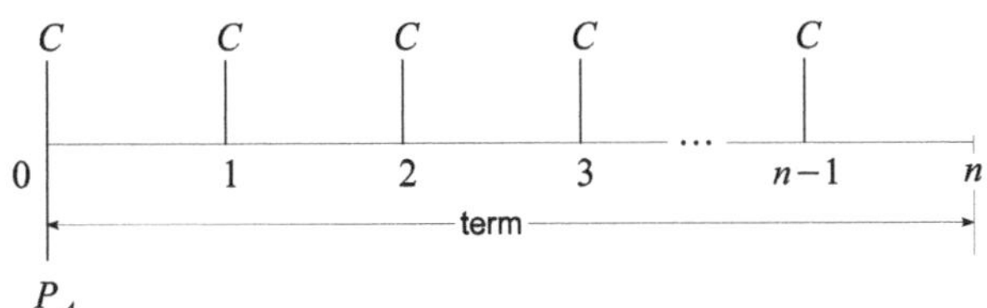

FIGURE 1.12: The time diagram for a simple annuity due.

Annuities due (and deferred annuities presented below) may be handled using the concept of an equation of values. The accumulated value of the payments on the date of the last nth payment (which is time $n-1$) is $C\,a_{\overline{n}|j}(1+j)^n$. We then accumulate this amount for one interest period to obtain the accumulated value V_A of an annuity due:

$$V_A = C\,a_{\overline{n}|j}\,(1+j)^{n+1}\,. \tag{1.29}$$

The discounted value of the payments one interest period before the first payment is $Ca_{\overline{n}|j}$. We then accumulate this amount for one interest period to determine the present value P_A of an annuity due:

$$P_A = C\,a_{\overline{n}|j}\,(1+j)\,. \tag{1.30}$$

Example 1.22. Determine the discounted value and the accumulated value of \$400 payable semi-annually at the beginning of each half-year over 10 years if interest is 8% per year payable semi-annually.

Solution. We deal with an annuity due. There are $n = 20$ payment periods. The interest rate over one period of 6 months is $0.08/2 = 0.04$. Therefore,

$$V_A = 400\,\frac{1.04^{20}-1}{0.04}\,(1+0.04) = 400 \cdot 30.9692 \cong \$12{,}387.68\,,$$
$$P_A = 400\,\frac{1-1.04^{-20}}{0.04}\,(1+0.04) = 400 \cdot 14.1339 \cong \$5{,}653.58\,.$$

□

1.3.1.3 Deferred Annuities

A *deferred annuity* is an annuity whose first payment is due some time later than the end of the first interest period. Thus, an ordinary deferred annuity is an ordinary simple annuity whose term is deferred for, say, k periods. The time diagram below depicts this case.

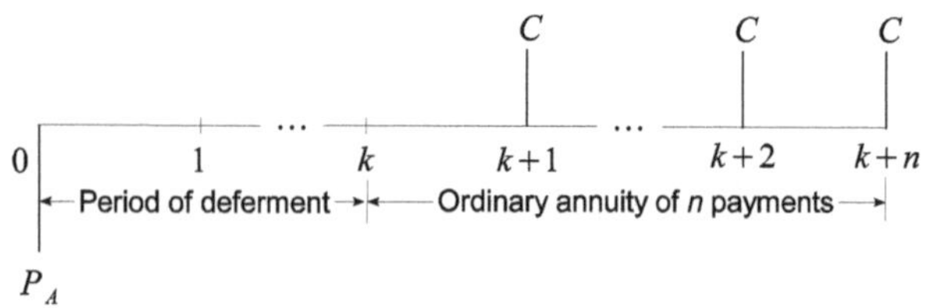

FIGURE 1.13: The time diagram for a deferred annuity.

Deferred annuities may be handled using the concept of an equation of values. The period of deferment is k periods and the first payment of the ordinary annuity is at time $k+1$. This is because the term of an ordinary annuity starts one period before its first payment. Hence, the discounted value one period before the first payment (time k) is $C\,a_{\overline{n}|j}$; the discount factor for k periods is $(1+j)^{-k}$. Therefore, the present value P_A of an ordinary deferred annuity is given by

$$P_A = C\,a_{\overline{n}|j}\,(1+j)^{-k}\,. \tag{1.31}$$

Example 1.23. A used car sells for \$9,550. Brent wishes to pay for it in 18 monthly instalments, the first due in three months from the day of purchase. If 12% compounded monthly is charged, determine the size of the monthly payment.

Solution. The discounted value of the deferred annuity is

$$9550 = C\,\frac{1-1.01^{-18}}{0.01}\,1.01^{-2} = C\cdot 11.033308\,.$$

Hence, the monthly payment is $C = \frac{9550}{11.033308} \cong \$865.56$. □

1.3.2 Determining the Term of an Annuity

In some problems, the accumulated value V_A or the discounted value P_A, the periodic payment C, and the rate j are specified. This leaves the number of payments n to be determined. Note that n can be found by using logarithms:

$$V_A = C\,\frac{(1+j)^n-1}{j} \implies (1+j)^n = \frac{jV_A}{C}+1 \implies n = \frac{\ln(jV_A/C+1)}{\ln(1+j)}$$

$$P_A = C\,\frac{1-(1+j)^{-n}}{j} \implies (1+j)^{-n} = 1-\frac{jP_A}{C} \implies n = -\frac{\ln(1-jP_A/C)}{\ln(1+j)}.$$

Usually, being given V_A or P_A, C, and j, we cannot find an integer number of periods n for the annuity. It is necessary to make the concluding payment differ from C in order to have equivalence. There are two ways, as follows.

Procedure 1. The last payment is increased by a sum that will make the payments equivalent to the accumulated value V_A or the discounted value P_A. This increase is sometimes referred to as a *balloon* payment.

Procedure 2. A smaller concluding payment is made one period after the last full payment. The smaller concluding payment is sometimes referred to as a *drop* payment.

Example 1.24. A debt of $4,000 bears interest at $i^{(2)} = 8\%$. It is to be repaid by semi-annual payments of $400. Determine the number of full payments needed. Use both Procedure 1 and Procedure 2.

Solution. We have $P_A = 4000$, $C = 400$, $i^{(2)}/2 = 0.08/2 = 0.04$, and we want to calculate n. Substituting in equation (1.27), we obtain:

$$\frac{1-(1+0.04)^{-n}}{0.04} = 10 \implies -n\ln(1.04) = \ln(0.6) \implies n = 13.024.$$

Procedure 1. There will be 12 regular payments and a final payment. Let X be the balloon payment that will be added to the last regular payment to make the payments equivalent to the discounted value $P_A = 4000$. Using time 0 as the focal date, we obtain the following equation of value:

$$400\,\frac{1-(1+0.04)^{-13}}{0.04} + X\,(1+0.04)^{-13} = 4000$$
$$\implies X = 1.04^{13}\cdot(4000 - 400\,\frac{1-1.04^{-13}}{0.04})$$
$$\implies X = 5.7409\cdot 1.04^{13} \cong \$9.56.$$

So the last, 13th payment, will be $400 + 9.56 = \$409.56$.

Procedure 2. There will be 13 full payments and a final smaller payment. Let Y be the size of a smaller concluding payment (the drop payment). Using time 0 as the focal date, we obtain the following equation of value:

$$400\,\frac{1-1.04^{-13}}{0.04} + Y\,1.04^{-14} = 4000 \implies Y = 5.7409\cdot 1.04^{14} \cong \$9.94.$$ □

1.3.3 General Annuities

Let us consider annuities for which payments are made more or less frequently than interest is compounded. Such a series of payments is called a *general annuity.* One way to solve general annuity problems is to replace the given interest rate by an equivalent rate for which the interest compounding period is the same as the payment period. Another approach used in solving a general annuity problem is to replace the given payment by equivalent payments made on the stated interest conversion dates. As a result, for both these approaches, a general annuity problem is transformed into a simple annuity problem.

Example 1.25 (Interest is compounded more frequently than payments are made)**.** Joe deposits $200 at the beginning of each year into a bank account that earns interest at $i^{(4)} = 6\%$. How much money will be in his bank account at the end of 5 years?

Solution. First, determine the rate j per year equivalent to the rate of $0.06/4$ per quarter:

$$1+j = (1+0.06/4)^4 \implies j = 1.015^4 - 1 \implies j = 6.136355\%$$

Second, calculate the accumulated value V_A of an ordinary annuity due with $C = \$200$, $n = 5$, and $j = 6.136355\%$:

$$V_A = 200\left(\frac{(1+j)^6 - 1}{j} - 1\right) = 200\cdot 5.99931 \cong \$1{,}199.86.$$ □

Example 1.26 (Payments are made more frequently than interest is compounded).

A car is purchased by paying \$2,000 down and then \$300 each quarter for 3 years. If the interest on the loan was $i^{(2)} = 9.2\%$, what did the car sell for?

Solution. First, determine the rate j per quarter equivalent to 0.092/2 per half-year:

$$(1+j)^4 = (1+0.092/2)^2 \implies 1+j = \sqrt{1.046} \implies j = 2.27414\%$$

Second, calculate the discounted value P of an ordinary simple annuity with $C = \$300$, $n = 3 \cdot 4 = 12$, and $j = 2.27414\%$:

$$P_A = 300\,\frac{1-(1+j)^{-12}}{j} = 300 \cdot 10.39946 \cong \$3{,}119.84.$$

The sale price of the car is $P = \$2{,}000 + P_A = \$5{,}119.84$. □

Example 1.27. Suppose the following.

- You have access to an account paying $100r\%$ compounded annually for the indefinite future.
- Your salary will grow by $100g\%$ per year for as long as you work, and your salary is paid in a lump sum at the end of each year and you just got paid.
- You plan to retire T years from now.
- You expect to live for M years after retirement.
- Upon retirement, you would like an annual income equal to $100p\%$ of your final salary, adjusted for inflation.
- Inflation is expected to remain stable at $100i\%$ for the indefinite future.
- You will save a fixed percentage $100q\%$ of your salary each year until retirement.

Here, r, g, p, i, and q are positive real parameters; T and M are positive integer parameters. What is the minimum value of q that will allow you to achieve your retirement goals?

Solution. Suppose I currently earn S (just paid), so that next year I will earn $S(1+g)$, the year after $S(1+g)^2$, etc. I am depositing the fraction q of this each year for the next T years and will earn a return of r each year. Thus, the moment I retire (i.e., the moment I make my final deposit), I will have

$$qS(1+g)(1+r)^{T-1} + qS(1+g)^2(1+r)^{T-2} + \ldots + qS(1+g)^{T-1}(1+r) + qS(1+g)^T\,.$$

This is the same as

$$qS(1+g)^T\left[\frac{(1+r)^{T-1}}{(1+g)^{T-1}} + \frac{(1+r)^{T-2}}{(1+g)^{T-2}} + \cdots + \frac{(1+r)}{(1+g)} + 1\right]$$

and if we let $1+h = \frac{1+r}{1+g}$ (so that $h = \frac{1+r}{1+g} - 1$) we can recognize the term in square brackets as the accumulated amount

$$s_{\overline{T}|h} := \frac{(1+h)^T - 1}{h} = (1+h)^T\, a_{\overline{T}|h}\,.$$

Thus, upon retirement, I will have

$$qS(1+g)^T s_{\overline{T}|h} \,.$$

Exactly one year after my final deposit, I will withdraw $pX(1+i)$, where $X = S(1+g)^T$ is my final salary (remember that I am withdrawing fraction p of my final salary, and that I am adjusting that final salary for inflation). The year after I will withdraw $pX(1+i)^2$, etc. As I plan to make M withdrawals, I will need at least

$$\frac{pX(1+i)}{1+r} + \frac{pX(1+i)^2}{(1+r)^2} + \cdots + \frac{pX(1+i)^M}{(1+r)^M} = pS(1+g)^T a_{\overline{j}|M}$$

in the account at date T, where $\frac{1}{1+j} = \frac{1+i}{1+r}$ (so that $j = \frac{1+r}{1+i} - 1$). Thus we must ensure that $qS(1+g)^T s_{\overline{h}|T} \geqslant pS(1+g)^T a_{\overline{j}|M}$, equivalently

$$q \geqslant p \cdot \frac{a_{\overline{j}|M}}{s_{\overline{h}|T}} \,. \qquad \square$$

1.3.4 Perpetuities

A *perpetuity* is an annuity whose payments begin on a fixed date and continue forever. Examples of perpetuities are the series of interest payments from a sum of money invested permanently at a certain interest rate, or a scholarship paid from an endowment on a perpetual basis. We shall discuss an *ordinary simple perpetuity*, that is, when a lump sum is invested and a series of level periodic payments is made, with the first payment made at the end of the first interest period and payments continuing forever. It is meaningless to speak about the accumulated value of a perpetuity. The discounted value, however, is well defined as the equivalent dated value of the set of payments at the beginning of the term of the perpetuity.

Let j denote the interest rate per period. The discounted value P of an ordinary simple perpetuity must be equivalent to the set of payments C, as shown on the diagram below.

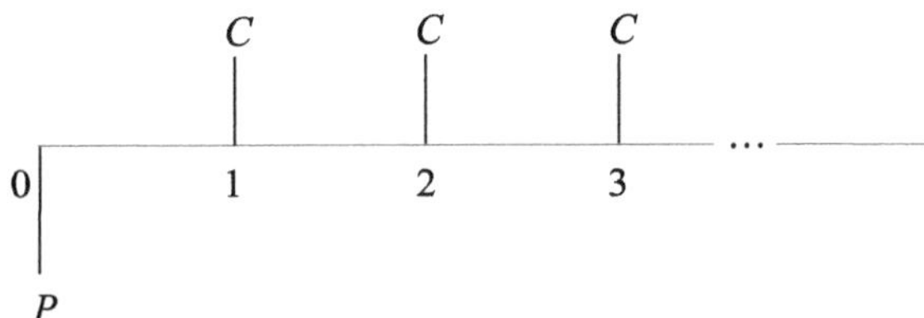

FIGURE 1.14: The time diagram for a perpetuity.

From the equation of value we get

$$P = C(1+j)^{-1} + C(1+j)^{-2} + C(1+j)^{-3} + \dots \,. \tag{1.32}$$

We know that the sum of an infinite geometric progression can be expressed as

$$a + aq + aq^2 + aq^3 + \ldots = \frac{a}{1-q} \quad \text{if } -1 < q < 1 \,.$$

The expression in (1.32) is a geometric progression with $a = C(1+j)^{-1}$ and $q = (1+j)^{-1}$. Clearly, $0 < q < 1$ for $j > 0$, hence

$$P = \frac{C(1+j)^{-1}}{1-(1+j)^{-1}} = \frac{C}{(1+j)-1} = \frac{C}{j} \,.$$

Alternatively, it is evident that P will perpetually provide $C = Pj$ as interest payments on the invested capital P at the end of each interest period as long as it remains invested at rate j per period.

Example 1.28. How much money is needed to establish a scholarship fund paying $1,500 annually if the fund will earn interest at $i = 6\%$ and the first payment will be made (a) at the end of the first year, (b) immediately, (c) 5 years from now?

Solution.

(a) We have an ordinary simple perpetuity with $C = 1500$ and $j = 0.06$; therefore, we calculate $P = \frac{\$1{,}500}{0.06} = \$25{,}000$.

(b) We have a simple perpetuity due and $P = \frac{C}{j} + R$. So, we calculate $P = \$25{,}000 + \$1{,}500 = \$26{,}500$.

(c) We have a simple perpetuity deferred $k = 4$ periods for which $P = \frac{C}{j}(1+j)^{-k}$ (we set the focal date 4 years from now). Therefore, $P = \$25{,}000 \cdot 1.06^{-4} = \$19{,}802.34$. □

1.3.5 Continuous Annuities

Consider a general annuity in which the payments are made more frequently than interest is compounded. Let there be m payments made throughout each of n interest periods. Interest is periodically compounded at interest rate j. Each payment is equal to $\$\frac{1}{m}$ for a total of $1 per period. The present value of such an annuity is

$$\sum_{k=1}^{nm} \frac{1}{m}(1+j)^{-\frac{k}{m}} = \frac{1}{m} a_{\overline{nm}|j_m},$$

where $j_m = (1+j)^{\frac{1}{m}} - 1$. By letting the value of m approach infinity, we obtain a continuous annuity in which payments are made continuously for a total of $1 per period. The present value of such an annuity, denoted $\bar{a}_{\overline{n}|j}$, is

$$\bar{a}_{\overline{n}|j} = \lim_{m\to\infty} \frac{1}{m} a_{\overline{nm}|j_m} = \lim_{m\to\infty} \sum_{k=1}^{nm} \frac{1}{m}(1+j)^{-\frac{k}{m}} = \lim_{h\to 0} \sum_{k=1}^{nm} h(1+j)^{-hk} \quad (\text{where } h := \frac{1}{m})$$

(note that we deal with a limit of Riemann sums)

$$= \int_0^n (1+j)^{-t}\,\mathrm{d}t = \frac{-(1+j)^{-t}}{\ln(1+j)}\Big|_0^n = \frac{1-(1+j)^{-n}}{\ln(1+j)} = \frac{1-(1+j)^{-n}}{j_\infty}, \tag{1.33}$$

where $j_\infty = \ln(1+j)$ is an equivalent interest rate compounded continuously. Here, one unit of time is one period. Hence, j_∞ is a continuous compound rate per period. Equation (1.33) takes the following concise form:

$$\bar{a}_{\overline{n}|j} = \frac{j}{j_\infty} a_{\overline{n}|j}.$$

If the payment is $\$C$ per period payable continuously, the present value P_A and accumulated value V_A are, respectively,

$$P_A = C\bar{a}_{\overline{n}|j} \quad \text{and} \quad V_A = C\bar{a}_{\overline{n}|j}(1+j)^n.$$

In a similar fashion, we can derive the present value of an annuity in which payments are made continuously and interest is compounded continuously at rate j_∞:

$$\begin{aligned}\bar{a}_{\overline{n}|j_\infty} &= \lim_{m\to\infty}\sum_{k=1}^{nm}\frac{1}{m}\mathrm{e}^{-j_\infty\frac{k}{m}} \\ &= \int_0^n \mathrm{e}^{-j_\infty t}\,\mathrm{d}t = -\frac{\mathrm{e}^{-j_\infty t}}{j_\infty}\Big|_0^n = \frac{1-\mathrm{e}^{-j_\infty n}}{j_\infty}\,. \end{aligned} \tag{1.34}$$

Alternatively, using the equivalency $1+j = \mathrm{e}^{-j_\infty}$, we can derive (1.34) directly from (1.33).

Finally, let us consider a general case with payments made continuously at a varying rate. Let $c(t)$ be the intensity of payments at time t. So the payment made throughout an infinitesimally small time interval $[t, t+\mathrm{d}t]$ is $c(t)\,\mathrm{d}t$. The present value of this payment is $(1+j)^{-t}c(t)\,\mathrm{d}t$ in the case of periodic compounding. To find the present value P_A of such an annuity, we only need to sum up the present values of all payments. Since there are infinitely many of such infinitesimally small payments, the summation becomes an integral from 0 to n:

$$P_A = \int_0^n c(t)(1+j)^{-t}\,\mathrm{d}t\,.$$

If interest is compounded continuously at a constant rate j_∞, then

$$P_A = \int_0^n c(t)\mathrm{e}^{-j_\infty t}\,\mathrm{d}t\,.$$

Finally, if interest is compounded continuously at a varying rate $j_\infty(t)$, we have

$$P_A = \int_0^n c(t)\mathrm{e}^{-\int_0^t j_\infty(u)\,\mathrm{d}u}\,\mathrm{d}t\,.$$

1.4 Bonds

1.4.1 Introduction and Terminology

Bonds are a method used to borrow money from a large number of investors to raise funds for financing long-term debts. From the investor's point of view, bonds provide steady periodic interest payments along with returning the amount borrowed at some point in the future. A bond is a written contract between the issuer (borrower) and the investor (lender) that specifies the following.

- The *face value*, or the *denomination*, denoted F, of the bond (usually is a multiple of 100).
- The *redemption date*, or *maturity date*, denoted T, is the date on which the loan will be repaid. In most cases, the *redemption value*, denoted W, of a bond is the same as the face value (i.e., $W = F$), and in such cases we say the bond is *redeemed at par*.
- The *bond rate*, or *coupon rate*, denoted c, is the rate at which the bond pays interest on its face value at equal time intervals until the maturity date. For example, an 8% bond with semiannual coupons has $c = 0.04$ or, equivalently, $c^{(2)} = 0.08$.

The following notations will be used in calculating bond prices:

P denotes the purchase price of a bond (its present value).

C is the amount of the coupon; it is a fraction of the face value of the bond, i.e., $C = Fc$, where c is the periodic coupon rate. If the coupons are paid with the frequency m, then we say that the bond pays interest at the nominal rate $c^{(m)}$; the periodic coupon rate is $c = c^{(m)}/m$.

n is the number of coupon/interest payment periods; i.e., we assume here that the bond rate and interest rate have the same conversion period.

j denotes the yield rate per interest period, often called the yield to maturity, i.e., the interest rate actually earned by the investor, assuming that the bond is held until it matures. If the interest is compounded with the frequency m at the annual nominal rate $i^{(m)}$, then $j = i^{(m)}/m$.

When a bond is sold, its present value is reported as a percent of its face value. For example, the price of a \$1,000 bond may be reported as 98, which means that the market value of the bond is (98/100)\$1,000 = \$980. A price of 100 means that the value of the bond is equal to its face value. There are two main examples of bonds: (1) *saving bonds* (such as Canada Savings Bonds) can be cashed in at any time before the redemption date, and you will receive the full face value plus accrued interest; (2) *marketable bonds* (such as corporate or Government of Canada bonds) do not allow the bond owner to cash the bond in before maturity. Marketable bonds can be sold on the bond market, where the price you receive will be influenced by current interest rates.

1.4.2 Zero-Coupon Bonds

The simplest case of a bond is a *zero-coupon bond*, which pays no coupons but instead just returns the investor an amount equal to the face value of the bond on the date of maturity. In this case the coupon rate c is zero. Why would anyone buy such a bond? The point is that the investor pays a smaller amount than the face value of the bond. Given the interest rate, the present value of such a bond can be easily computed. Suppose that a bond with face value F dollars is maturing in T years, and the annual effective rate is i; then the present value (purchase price) of the bond is

$$P = (1+i)^{-T} F .$$

In reality, the opposite happens: bonds are freely traded and their prices are determined by the markets, whereas the interest rate is implied by the bond prices. The implied annual effective rate is then

$$i = \left(\frac{F}{P}\right)^{1/T} - 1 .$$

For simplicity, let us consider a unit bond whose face value is equal to one unit of the home (domestic) currency, i.e., $F = 1$. Denote the purchase price of a unit bond at time $t \leqslant T$ by $Z(t,T)$, where times t and T are measured in years. In particular, $Z(0,T) = P$ is the present value of the bond at time $t = 0$, and $Z(T,T) = 1$ is equal to the face value. Let us summarize the pricing formulae for a zero-coupon bond.

- If interest is compounded annually at rate i, then

$$Z(t,T) = (1+i)^{-(T-t)}.$$

- Using periodic compounding with frequency m, we have

$$Z(t,T) = \left(1 + \frac{i^{(m)}}{m}\right)^{-m(T-t)}.$$

- In the case of continuous compounding, we obtain

$$Z(t,T) = \mathrm{e}^{-(T-t)r}.$$

- If the instantaneous interest rate is a function of time, i.e., the interest rate at time $s \in [t,T]$ is $r = r(s)$, then

$$Z(t,T) = \mathrm{e}^{-\int_t^T r(s)\,\mathrm{d}s}.$$

  Note that the instantaneous interest rate is readily obtained by differentiating the log-price of a bond:

$$r(t) = \frac{\partial \ln Z(t,T)}{\partial t} = \frac{1}{Z(t,T)} \frac{\partial Z(t,T)}{\partial t}.$$

  The rate $r(t)$ is sometimes referred to as the *short rate*.

For comparing different investment opportunities, we can use the *yield rate* or *spot rate*, denoted $y(t,T)$, which is defined implicitly as the yield to maturity of a zero-coupon bond:

$$Z(t,T) = \mathrm{e}^{-y(t,T)\,(T-t)}.$$

The spot rate can be viewed as an average interest rate for the time interval $[t,T]$. For time varying interest rates, we have $y(t,T) = \frac{1}{T-t}\int_t^T r(s)\,\mathrm{d}s$. Typically, the yield rate is considered as a function of the time to maturity $\tau = T - t$, so we can also use the notation $y(\tau) := y(t,T)$.

1.4.3 Coupon Bonds

For a coupon bond, the buyer will receive two types of payments, namely, a coupon payment Fc at the end of each interest period (with the length of $\delta t = \frac{1}{m}$ years) and the redemption value W on the redemption date. Thus, a coupon bond can be viewed as a cash flow stream of the form

$$[-P,\ \underbrace{Fc, Fc, \ldots, Fc}_{n \text{ coupon payments}} + W].$$

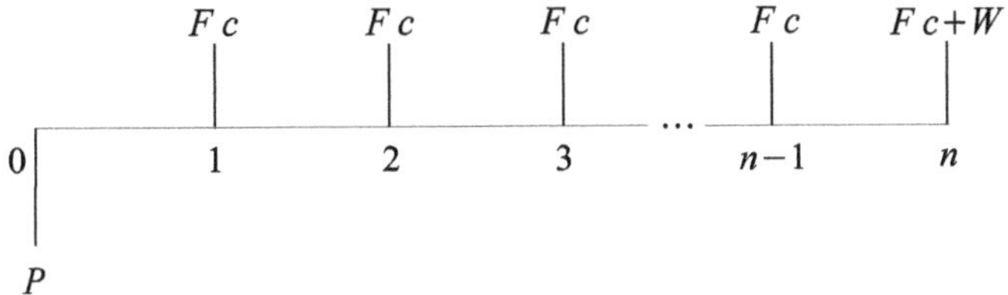

FIGURE 1.15: The time diagram for a coupon bond.

First, let us assume that the interest is compounded periodically at rate j, and the coupons are paid at the interest conversion dates. To determine the price paid to buy the bond, we discount each cash flow to the date of sale at rate j. The purchase price P is the sum of the discounted value of all coupons and the discounted value of the redemption value:

$$P = Fc\left(\frac{1}{1+j} + \frac{1}{(1+j)^2} + \cdots + \frac{1}{(1+j)^n}\right) + \frac{W}{(1+j)^n} \tag{1.35}$$

$$= Fca_{\overline{n}|j} + W(1+j)^{-n}. \tag{1.36}$$

The formula (1.36) can be simplified by eliminating the factor $(1+j)^{-n}$ as follows:

$$P = Fca_{\overline{n}|j} + W(1 - ja_{\overline{n}|j}) = W + (Fc - Wj)a_{\overline{n}|j}\,. \tag{1.37}$$

Hence, if $P > W$ (the purchase price of a bond exceeds its redemption value), the bond is said to have been purchased *at a premium.* The size of the premium is given by

$$\text{Premium} = P - W = (Fc - Wj)a_{\overline{n}|j}\,.$$

That is, a premium occurs when $Fc > Wj$. For par value bonds (i.e., $W = F$) the bond is purchased at a premium when $c > j$. Similarly, if $P < W$ (the purchase price is less than the redemption value) the bond is said to have been purchased *at a discount.* The size of the discount is

$$\text{Discount} = W - P = (Wj - Fc)a_{\overline{n}|j}\,.$$

That is, a discount occurs when $Fc < Wj$. In other words, each coupon is less than the interest desired by the investor. For par value bonds (i.e., $W = F$) the bond is purchased at a discount when $j > c$. Formula (1.37) is often more efficient than (1.36) since it requires only the calculation of $a_{\overline{n}|j}$. It also tells us whether the bond is purchased at a premium (the purchase price of a bond exceeds its redemption value) or at a discount.

Suppose a bond is redeemable at par (that is, $W = F$). Then its price is equal to the face value (i.e., $P = F$) iff the coupon rate and yield rate coincide (i.e., rate c = rate j). In this case, the bond is said to be *purchased at par.* The proof is straightforward. Assume $W = F$ in (1.36), then

$$P = W \iff ca_{\overline{n}|j} + (1+j)^{-n} = 1 \iff c = \frac{1-(1+j)^{-n}}{a_{\overline{n}|j}} = j\,.$$

Example 1.29. A \$1,000 bond that pays interest at $c^{(2)} = 8\%$ is redeemable at par at the end of 5 years. Determine the purchase price to yield an investor 10% compounded semi-annually.

Solution. We know that

the redemption value is $W = F = \$1{,}000$,

the coupon rate is $c = 0.08/2 = 0.04$ and the coupon is hence $C = Fc = \$40$,

the number of coupon payments is $n = 5 \cdot 2 = 10$,

the yield rate is $j = 0.1/2 = 0.05$.

The purchase price P to yield $i^{(2)} = 10\%$ is the sum of the discounted values of coupons and of the redemption value:

$$P = 40\,\frac{1-(1+0.05)^{-10}}{0.05} + F\,(1+0.05)^{-10} = 308.87 + 613.91 = \$922.78\,.$$

Since the buyer is buying the bond for less than the redemption value, that is, $P < F$, we say that the bond is purchased at a discount. □

Example 1.30. A corporation issues 15-year bonds redeemable at par. Under the contract, interest payments will be made at the rate $c^{(2)} = 10\%$. The bonds are priced to yield 8% per annum compounded monthly. What is the issue price of a \$1,000 bond?

Solution. We have $F = W = \$1{,}000$, $c = 0.1/2 = 0.05$, $n = 15 \cdot 2 = 30$. Calculate rate j per half-year equivalent to $i^{(12)} = 8\%$:

$$(1+j)^2 = \left(1 + \frac{0.08}{12}\right)^{12} \implies j = \left(1 + \frac{0.08}{12}\right)^{6} - 1 \implies j = 0.040672622\,.$$

Now the purchase price is

$$\begin{aligned} P &= W + (Fc - Wj)a_{\overline{30}|j} = 1000 + (50 - 40.672622)a_{\overline{30}|j} \\ &= 1000 + 9.327378 \cdot 17.15168 = \$1{,}159.98\,. \end{aligned}$$

□

One of the most important properties of a coupon bond is the relation between its price and the yield rate. Bonds are freely traded and their prices are determined by the markets. Therefore, the yield rate of a bond is implied by the purchase price. Clearly, the higher the yield rate, the lower the bond price. Let us formally prove this fact.

Theorem 1.3 (The Price-Yield Theorem). *The price and yield of a coupon bond move in opposite directions.*

Proof. Let us think of P as a function of the rate j. If we differentiate (1.35) with respect to j, we find that

$$\frac{\mathrm{d}P}{\mathrm{d}j} = Fc\left(-\frac{1}{(1+j)^2} - \frac{2}{(1+j)^3} - \cdots - \frac{n}{(1+j)^{n+1}}\right) - \frac{nW}{(1+j)^{n-1}},$$

which is negative, so P is a decreasing function of j. □

So far, we discussed the pricing of a coupon bond in the situation where the desired yield rate is given and is constant during the lifetime of the bond. Now we consider the general case where the yield rate varies with time and is implicitly defined by values of zero-coupon bonds. First, note that a coupon payment of \$$C$ due at time t gives the same cash flow as that of C zero-coupon bonds maturing at time t. Therefore, a coupon bond can be viewed as a portfolio of zero-coupon bonds with different maturity times. The present value of a coupon paid at time t is $C \cdot Z(0,t)$. The present value of the redemption value paid at time T is $W \cdot Z(0,T)$. Therefore, the purchase price P of a coupon bond (i.e., its value at time 0) can be represented as a weighted sum of zero-coupon bonds' values:

$$\begin{aligned} P &= C\,Z(0,\delta t) + C\,Z(0, 2\,\delta t) + \cdots + C\,Z(0,(n-1)\,\delta t) + (C+W)\,Z(0,T) \\ &= \sum_{k=1}^{n} C\,Z(0,k\,\delta t) + W\,Z(0,T)\,, \end{aligned} \tag{1.38}$$

where δt is the length of the coupon/interest payment period and $T = n\,\delta t$ is the maturity time.

For each zero-coupon bond with maturity time t we can find the annual yield rate $y(t)$ implied by the bond price $Z(0,t)$ as follows:

$$Z(0,t) = \mathrm{e}^{-y(t)\,t} \implies y(t) = -\frac{\ln Z(0,t)}{t}\,.$$

The formula (1.38) takes the form

$$P = \sum_{k=1}^{n} \mathrm{e}^{-y_k\, k\, \delta t}\, C + \mathrm{e}^{-y_n\, T}\, W\,, \tag{1.39}$$

where $y_k = y(k\, \delta t)$ is the yield rate (at time 0) for the zero-coupon bond maturing at time $t = k\, \delta t$. Commonly the rates are tied to interest rates on short-term bonds or money market accounts. Typically, interest rates are expected to go up, i.e., $y(s) \leqslant y(t)$ for $s < t$.

There are bonds that have variable coupon rates. If it is the case, then (1.39) takes the form:

$$P = \sum_{k=1}^{n} \mathrm{e}^{-y_k\, k\, \delta t}\, C_k + \mathrm{e}^{-y_n\, T}\, W\,, \tag{1.40}$$

where C_k is the coupon paid at time $t = k\, \delta t$.

1.4.4 Serial Bonds, Strip Bonds, and Callable Bonds

A set of bonds issued at the same time but having different maturity dates is called *serial bonds*. Serial bonds can be thought of simply as several bonds covered under one bond contract. The present value of the entire issue of the bonds is just the sum of the present values of the individual bonds.

Example 1.31. To finance an expansion of services, the City of Waterloo issued \$30,000,000 of serial bonds on March 15, 2013. Bond interest at 4% per annum is payable half yearly on March 15 and September 15, and the contract provides for redemption as follows: \$10,000,000 of the issue to be redeemed March 15, 2018; \$10,000,000—March 15, 2023; \$10,000,000—March 15, 2028. Calculate the purchase price of the issue to the public to yield $i = 4\%$ on those bonds redeemable in 5 years and $i = 5\%$ on the remaining bonds.

Solution. This issue can be viewed as a sequence of three coupon bonds redeemable at par. The face value of each bond is $F = \$10{,}000{,}000$. The coupon rate is $i = 0.02$; each coupon is $C = 200{,}000$. Let us find the present values.

1. The bond is redeemed March 15, 2018. There are $n = 5 \cdot 2 = 10$ coupon payments. The yield rate $j = 0.04/2 = 0.02$ is equal to the coupon rate. Therefore, the bond is purchased at par. The purchase price is $P_1 = \$10{,}000{,}000$.

2. The bond is redeemed March 15, 2023. There are $n = 10 \cdot 2 = 20$ coupons. The yield rate is $j = 0.05/2 = 0.025 > i$; the bond is purchased at a discount. The purchase price is

$$P_2 = 10{,}000{,}000 + (200{,}000 - 10{,}000{,}000 \cdot 0.25) a_{\overline{20}|0.025} = \$9{,}220{,}541.87\,.$$

3. The bond is redeemed March 15, 2028. There are $n = 15 \cdot 2 = 30$ coupons. The yield rate is $j = 0.025$. The purchase price is

$$P_3 = 10{,}000{,}000 + (200{,}000 - 10{,}000{,}000 \cdot 0.25) a_{\overline{30}|0.025} = \$8{,}953{,}485.37\,.$$

The purchase price of the issue is $P = P_1 + P_2 + P_3 = \$28{,}174{,}027.24$. □

Some investors may separate coupons from the principal of coupon bonds, so that different investors may receive the principal (i.e., the redemption value) and each of the coupon payments. The coupons and remainder are sold separately. This creates a supply of new zero-coupon bonds. This method of creating zero-coupon bonds is known as stripping and the contracts are known as *strip bonds.* **STRIPS** stands for **S**eparate **T**rading of **R**egistered **I**nterest and **P**rincipal **S**ecurities. Recall that the discounted value of the redemption value is $W(1+j)^{-n}$. Each coupon has the discounted value $Fc(1+j)^{-k}$, where k is the number of interest conversion periods from now to the time the coupon is paid. Thus, the present value of coupons is $Fca_{\overline{n}|j}$.

Example 1.32. Investor **A** buys a \$2,500 10-year bond paying interest at $c^{(2)} = 6\%$, redeemable at 105, to yield $i^{(2)} = 7\%$. She sells the coupons to investor **B**, who wishes to yield $i^{(2)} = 6.25\%$, and she sells the strip bond to investor **C**, who wishes to yield $i^{(4)} = 6.5\%$. What profit does investor **A** make?

Solution. The face value is $F = \$2{,}500$. The redemption value is $W = 2500 \cdot 1.05 = \$2{,}625$. Investor **A** (with the yield rate $j = 7\%/2 = 3.5\%$) pays

$$P_{\mathbf{A}} = 2625 + (2500 \cdot 0.03 - 2635 \cdot 0.035) \cdot a_{\overline{20}|0.035} = \$2{,}385.17.$$

Investor **B** (with the yield rate $j = 6.25\%/2 = 3.125\%$) pays for the coupons

$$P_{\mathbf{B}} = 2500 \cdot 0.03 \cdot a_{\overline{20}|0.03125} = \$1{,}103.02.$$

Investor **C** pays for the strip bond

$$P_{\mathbf{C}} = 2625 \cdot \left(1 + \frac{0.065}{4}\right)^{-40} = \$1{,}377.55\,.$$

The profit of investor **A** is

$$P_{\mathbf{B}} + P_{\mathbf{C}} - P_{\mathbf{A}} = \$1{,}103.02 + \$1{,}377.55 - \$2{,}385.17 = \$95.40\,. \qquad \square$$

A *callable bond* is a bond that allows the issuer to pay off the loan (or a fraction of the loan) at any of a set of designated *call dates.* If interest rates decline, a bond issuer would call a bond early, pay off the old issue, and replace it with a new series of bonds with a lower coupon rate. A bond is said to have a *European option* if it has a single call date prior to maturity. A bond has an *American option* if it is callable at *any* date following the lockout period (the period before the first call date). When an investor calculates the price of a callable bond, the investor must determine a price that will guarantee the desired yield regardless of the call date. *Putable* (or *extendible*, or *retractable*) bonds are bonds that allow the bond owner (not the issuer) to redeem the bond at a time other than the stated redemption date.

Example 1.33. A \$5,000 callable bond matures on September 1, 2023, at par. It is callable on September 1, 2018 at \$5,250. Interest on the bond is $c^{(2)} = 6\%$. Calculate the price on September 1, 2013, to yield an investor $i^{(2)} = 5\%$.

Solution.

Scenario 1: The bond is called prior to maturity. The present value is

$$P_1 = 5250 + (5000 \cdot 0.03 - 5250 \cdot 0.025) \cdot a_{\overline{10}|0.025} = \$5{,}414.10\,.$$

Scenario 2: The bond is not called prior to maturity. The present value is

$$P_2 = 5000 + 5000 \cdot (0.03 - 0.025) \cdot a_{\overline{20}|0.025} = \$5{,}389.73\,.$$

To guarantee that the buyer will yield $i^{(2)} = 5\%$, the purchase price must be the smallest of the two prices:

$$P = \min\{P_1, P_2\} = \min\{5414.10, 5389.73\} = \$5{,}389.73\,. \qquad \square$$

1.5 Yield Rates

1.5.1 Internal Rate of Return and Evaluation Criteria

The interest rate that produces a zero NPV of some cash flow stream $[C_0, C_1, \ldots, C_n]$ is called the *internal rate of return* or simply the *rate of return*. Its calculation is another financial tool that can be used to determine whether or not to proceed with a project. For the case with equal time periods and when the interest is periodically compounded, the internal rate of return j_{int} solves the equation $\text{NPV}(j_{\text{int}}) = 0$, where

$$\text{NPV}(j) = C_0 + C_1(1+j)^{-1} + C_2(1+j)^{-2} + \cdots + C_n(1+j)^{-n} = \sum_{k=0}^{n} C_k d^k$$

with $d := (1+j)^{-1}$. Notice that multiple solutions are possible.

Let j_{cc} be the cost of capital (the risk-free interest rate), and let j_{int} be the internal rate of return.

- If $j_{\text{cc}} < j_{\text{int}}$, then $\text{NPV}(j_{\text{cc}}) > 0$ and the project will return a profit.
- If $j_{\text{cc}} = j_{\text{int}}$, then $\text{NPV}(j_{\text{cc}}) = 0$.
- If $j_{\text{cc}} > j_{\text{int}}$, then $\text{NPV}(j_{\text{cc}}) < 0$ and the project will not return the required rate of return j_{cc}.

Example 1.34 (Multiple Solutions). Find the rate of return for

(a) an investment with the principal of 100 that yields returns of 70 at the end of each of two periods; i.e., find the rate of return for the cash flow stream $\mathbf{C} = [-100, 70, 70]$;

(b) the cash flow stream $\mathbf{C} = [a\,b, -a-b, 1]$.

Solution.

(a) The rate of the return j solves the equation

$$100 = (1+j)^{-1}70 + (1+j)^{-2}70\,.$$

Denoting $d = (1+j)^{-1}$ and solving the quadratic equation $70d^2 + 70d - 100 = 0$ for d, we obtain that $d = \frac{-7\pm\sqrt{329}}{14}$. Since $70+70 > 100$ we are looking for a positive rate j solving the equation. Since $j > 0$ implies that $0 < d < 1$, we obtain the solution $d = 0.795597$. Thus, the internal rate of return is $j = \frac{1}{d} - 1 \cong 25.69\%$.

(b) The net present value is given by

$$\text{NPV} = ab - (a+b)d + d^2 = (d-a)(d-b),$$

where $d = (1+j)^{-1}$. The NPV is zero if $d = a$ or $d = b$, and it is not uniquely determined when $a \neq b$. Furthermore, the NPV of the cash flow stream $-\mathbf{C}$ has the same zeros. It is therefore not clear how to use the internal rate to determine which of the two cash flows is preferable. □

The equation $\text{NPV}(j_{\text{int}}) = 0$ can be solved analytically (for short cash flow streams) or numerically. As is shown in the previous example, this equation may have multiple solutions. However, in some special cases, the internal rate j_{int} is uniquely determined by the cash flows.

Proposition 1.4. *If $C_0 > 0$ and $C_k < 0$ for $k = 1, 2, \ldots, n$ (or $C_0 < 0$ and $C_k > 0$ for $k = 1, 2, \ldots, n$), then the internal rate j_{int} is uniquely determined. In this case, the rate is positive, i.e., the discount factor $d = (1 + j_{int})^{-1} < 1$, iff*

$$|C_0| < \sum_{k=1}^{n} |C_k|.$$

Proof. Define $P(d) := C_0 + C_1 d + \cdots + C_n d^n$. Then, in the first case,

$$P(d) = |C_0| - (|C_1|d + \cdots + |C_n|d^n).$$

Clearly, $P(d)$ is a decreasing function for $d \geqslant 0$ where $P(0) > 0$. If $|C_0| < \sum_{k=1}^{n} |C_k|$ holds, then $P(1) < 0$. Hence, $P(d)$ has a unique zero $d_0 \in (0, 1)$. Solving $d_0 = (1 + j_{\text{int}})^{-1}$ gives us the rate. □

1.5.2 Determining Yield Rates for Bonds

One of the fundamental problems related to bonds is to determine what rate of return a bond will give to the buyer when bought for a given price P. In practice, the market price of a bond is often given without stating the yield rate. The yield of the bond reflects expectations of future rates. Bonds of different maturities can have different yields implied by the market. The investor is interested in determining the true rate of return on investment. Based on the yield rate, the investor can decide whether a particular bond is an attractive investment or can determine which of several bonds available is the best investment.

1.5.2.1 Zero-Coupon Bonds

The purchase price of a zero-coupon bonds defines the implied yield rate. In the case of continuous compounding, the annual yield rate y is given by

$$Z(t, T) = \mathrm{e}^{-y\,(T-t)} \implies y = -\frac{\ln Z(t, T)}{T - t}.$$

If interest is compounded annually, then

$$Z(t, T) = (1 + y)^{-(T-t)} \implies y = Z(t, T)^{-(T-t)^{-1}}.$$

The yield rate obtained is a measure of the return for a single payment made at maturity to the holder of a zero-coupon bond. The yield rate is a function of time to maturity. Recall that we use the notation $y(t, T)$ to denote the yield rate (or the spot rate) at time t for a zero-coupon bond maturing at time T (time is measured in years). If we plot the yield rates of zero-coupon bonds with different maturities, we obtain the *yield curve*, which is discussed in Subsection 1.5.4.

1.5.2.2 Coupon Bonds

A direct method for determining the yield rate for a coupon bond involves the solution of the nonlinear algebraic equation $P = Fca_{\overline{n}|j} + W(1 + j)^{-n}$ for the periodic rate j. A typical question is whether this equation has a solution j for a given a purchase price P. The following theorem settles this issue.

Theorem 1.5 (The Yield Existence Theorem). *If the purchase price P of a coupon bond satisfies the inequality*

$$0 < P < nFc + W,$$

then there exists a yield rate j such that $P = Fca_{\overline{n}|j} + W(1 + j)^{-n}$ holds.

Proof. We know that P is a continuous decreasing function of $j \in \mathbb{R}_+$. Furthermore,

$$\lim_{j\to\infty} P(j) = 0 \quad \text{and} \quad \lim_{j\searrow 0} P(j) = nFc + W\,.$$

Hence, there is a positive real j such that P satisfies the inequality above. □

Note that the annual yield rate y can be expressed in terms of the periodic yield rate j as follows:

$$1 + y = (1+j)^m \implies y = (1+j)^m - 1\,.$$

Here, m is the number of coupon (or interest) periods per year.

1.5.3 Approximation Methods

1.5.3.1 The Method of Averages

The method of averages calculates an approximate value of the yield rate j as a ratio of the average interest payment to the average amount invested. If n is the number of interest periods from the date of sale until the redemption date, then the average interest payment is

$$\frac{nFc + W - P}{n},$$

and the average amount invested is $\frac{W+P}{2}$. The approximate value of j is then given by the ratio

$$j \cong \frac{\text{average interest payment}}{\text{average amount invested}} = \frac{\frac{nFc+W-P}{n}}{\frac{W+P}{2}}\,.$$

Example 1.35. A \$500 bond, paying interest at $c^{(2)} = 9.5\%$, redeemable at par on August 15, 2018, is quoted at 109.50 on August 15, 2006. Compute an approximate value of the yield rate compounded semi-annually.

Solution. We have $F = W = 500$. The purchase price is $P = 500 \cdot 1.095 = \$547.50$ since the bond is sold on a bond interest date. If the buyer holds the bond until maturity, she will receive 24 coupon payments of $\$500 \cdot 0.095/2 = \$23.75$ each plus the redemption payment of \$500. In total the buyer will receive $24 \cdot \$23.75 + \$500 = \$1{,}070$. The net gain $\$1{,}070 - \$547.50 = \$522.50$ is realized over 24 interest periods, so that the average interest per period is $\$522.50/24 = \$21.77$. The average amount invested is $(\$547.50 + \$500)/2 = \$523.75$. The approximate value of the yield rate is

$$j = \frac{21.77}{523.75} = 0.0416 = 4.16\% \text{ or } i^{(2)} = 8.32\%\,.$$

□

1.5.3.2 The Method of Interpolation

The method of interpolation is based on the following approach. First, calculate two adjacent nominal rates so that the market price of the bond lies between the prices determined by those two rates. To find two rates bracketing the yield rate, one can use the method of averages. Second, determine the actual yield rate using linear interpolation between the two adjacent rates.

Recall the idea of linear interpolation. Given two points $(a, f(a))$ and $(b, f(b))$ and a number c such that $a < c < b$, how can one estimate $f(c)$? Without any other information, we simply join the two given points with a straight line, $y = g(x)$, and then approximate

$f(c)$ with the value $g(c)$. The line $y = g(x)$ that joins the points $(a, f(a))$ and $(b, f(b))$ passes through $(a, f(a))$ and has slope $\frac{f(b)-f(a)}{b-a}$; therefore its equation is given by

$$g(x) - f(a) = \frac{f(b) - f(a)}{b - a}(x - a).$$

With $x = c$, where $a < c < b$, we have

$$f(c) \approx g(c) = \frac{f(b) - f(a)}{b - a}(c - a) + f(a),$$

or in the form:

$$f(c) \approx g(c) = \frac{b - c}{b - a}f(a) + \frac{c - a}{b - a}f(b).$$

Other (exact or more accurate numerical) methods of determining yield rates are based on the use of financial calculators or software such as Maple, Excel, etc.

Example 1.36. Compute the yield rate as in the previous example by the method of interpolation.

Solution. By the method of averages we determined that the yield rate is approximately $j = 4.16\%$. Now we compute the purchase prices to yield $j = 4\%$ and $j = 4.75\%$ compounded semi-annually. To yield $j = 4\%$ the purchase price is

$$P = 500 + (23.75 - 20)a_{\overline{24}|0.04} = \$557.18.$$

If the bond interest rate is $c = 9.5\%/2 = 4.75\%$, and the yield rate $j = 4.75\%$, the purchase price is equal to the redemption value: $P = 500$. Assuming that the purchase price P is a linear function of the yield rate j, we obtain

$$\frac{j - 4.75}{4 - 4.75} = \frac{547.50 - 500}{557.18 - 500}$$

with solution $j = \frac{547.50-500}{557.18-500}(4 - 4.75) + 4.75 = 4.75 - 0.75\frac{47.5}{57.18} = 4.127\%$. So the yield rate is $i^{(2)} = 8.254\%$.
The bond price for a yield $j = 4.127\%$ is $P = 500 + (23.75 - 20.635)a_{\overline{24}|4.127} = \$547.24$. □

Example 1.37. An investor purchases a \$1,000 bond, paying interest at $c^{(2)} = 8\%$ and redeemable at par in 10 years. The bond will yield $i = 7\%$ if held to maturity. After holding the bond for 3 years, the bond is sold to yield the new holder $i^{(2)} = 6\%$. Calculate the investor's yield rate $i^{(2)}$ over the 3-year investment period.

Solution. Using $Fc = 1000 \cdot 0.08/2 = 40$, $W = 1000$, $n = 10 \cdot 2 = 20$, and $j = 0.035$, we find the purchase price:

$$\text{Purchase Price} = 40a_{\overline{20}|0.035} + 1000(1.035)^{-20} = \$1{,}071.06.$$

After 3 years the selling price can be calculated using $Fc = 40$, $W = 1000$, $n = 7 \cdot 2 = 14$, and $j = 0.03$:

$$\text{Selling Price} = 40a_{\overline{14}|0.03} + 1000(1.03)^{-14} = \$1{,}112.96.$$

Now consider the 3-year investment. The investor pays \$1,071.06 for the bond, receives a \$400 coupon payment every 6 months for 3 years, and then obtains \$1,112.96 when the bond is sold. The equation of value at rate j per half-year is

$$1071.06 = 40a_{\overline{6}|j} + 1112.96(1+j)^{-6}.$$

Using the method of averages, an approximate value of the yield rate j is

$$j \cong \frac{6 \cdot 40 + 1112.96 - 1071.06}{(1112.96 + 1071.06)/2} = 0.043 \text{ or } i^{(2)} = 8.6\%\,.$$

By linear interpolation we obtain a more accurate answer. The bond prices for the two yields of 8% and 9% are

$$P(i^{(2)} = 8\%) = 40a_{\overline{6}|0.04} + 1112.96(1.04)^{-6} = \$1{,}089.27\,,$$
$$P(i^{(2)} = 9\%) = 40a_{\overline{6}|0.045} + 1112.96(1.045)^{-6} = \$1{,}060.95\,.$$

Hence for the yield rate $i^{(2)}$ we use interpolation,

$$\frac{i^{(2)} - 8}{9 - 8} = \frac{1071.06 - 1089.27}{1060.95 - 1089.27} = \frac{18.21}{28.32} = 0.643$$

giving the yield rate $i^{(2)} = 8.643\%$. □

1.5.3.3 Numerical Methods

Although the method of interpolation provides a better approximation of the yield rate in comparison with the method of averages, the result is still inaccurate. The approximation error can be significantly reduced by employing a numerical algorithm such as Newton's method for solving nonlinear equations. The Newton iterative rule for finding a zero of a continuously differentiable function $f = f(x)$ takes the following form:

$$x_{k+1} = x_k - \frac{f(x_k)}{f'(x_k)}, \quad k = 0, 1, 2, \ldots. \tag{1.41}$$

If initial approximation x_0 is sufficiently close to a zero of the function f, then the sequence $\{x_k\}_{k \geqslant 0}$ produced by (1.41) converges to a solution of $f(x) = 0$.

To determine the yield rate for a given bond price P, we need to solve the equation

$$W + (Fc - Wj)a_{\overline{n}|j} = P$$

for the rate j. Define the function $B(j) := W + (Fc - Wj)a_{\overline{n}|j}$ and rewrite the above equation as $B(j) - P = 0$. According to Theorem 1.5, we have $B(0) = nFc + W$, $B(\infty) = 0$, and $B'(y) < 0$. Thus, for any given price $P \in (0, nFc + W)$, there exists a unique yield rate $j \in (0, \infty)$ such that $P = B(j)$.

Newton's method can be applied to find a zero of the function $f(x) := B(x) - P$. An initial approximation of the rate j can be calculated with the method of averages. Such an approximation can be used as a starting value j_0 within Newton's method. To carry out Newton's method it is worth noting that

$$\frac{\mathrm{d}a_{\overline{n}|j}}{\mathrm{d}j} = \frac{n\,(1+j)^{-n-1}}{j} - \frac{a_{\overline{n}|j}}{j}\,.$$

Hence, the derivative of the bond value function B is

$$B'(j) = (Fc - Wj)\,\frac{n\,(1+j)^{-n-1}}{j} - Fc\,\frac{a_{\overline{n}|j}}{j}\,.$$

Let us describe the method of solving for the corresponding yield to maturity. Newton's method begins with an initial guess j_0 and recursively computes approximate rates j_n by

$$j_{n+1} = j_n - \frac{f(j_n)}{f'(j_n)} = j_n - \frac{B(j_n) - P}{B'(j_n)}, \quad n = 0, 1, \ldots.$$

The rate of convergence of Newton's method is quadratic since $B'(j) \neq 0$ for all $j > 0$. So, after several iterations, the current approximation j_n is very close to the desired solution.

1.5.4 The Yield Curve

For many reasons (credit risk, liquidity, etc.) cash flows occurring at different points in time need to be discounted using different rates. Yield curves are often used to determine what rates should be used. Such a curve is a graph of the yield-to-maturity of a bond plotted against tenor (the number of years to maturity). The value of the spot (yield) rate is based on the assumption that the investor holds the bond until redemption and that the yield-to-maturity is the same yield rate used in pricing the bond. The market price reflects today's expectation of interest rates. The effective interest implied by the market prices determines the yield of the bond. Bonds of different maturities can have different yields implied by the market.

The yield curve is an instrument used for analysis of the bond market. The shape of the yield curve reflects the market's expectation of where interest rates are heading. As the date of maturity of a bond approaches, there is less and less uncertainty left. The principal of the bond is known and will be paid on the redemption date, and the interest rate is unlikely to move much in a short period of time. A bond of longer maturity will be exposed to more uncertainty (i.e., it is riskier) than one of short maturity. We can therefore expect long-dated bonds to have higher yields to compensate investors for this additional risk. So the yield curve is upward sloping. This is a so-called *normal yield curve* with upward sloping. The market expects that interest rates are likely to rise in the future. However, sometimes shorter maturities have higher yields so that the yield curve is known as downward sloping. The market predicts that the interest rates will fall. In such situations, the yield curve is said to be *inverted*.

There are well-established markets in the major government bonds so an investor need not hold the bond until maturity. Instead, the investor can sell it on the market. A normal yield curve highlights the fact that if you purchase a bond, hold on to the investment for a period of time, and then sell it, it is likely that your selling yield will be lower than your purchase yield. This provides investors with the opportunity to obtain a yield in excess of the yield-to-maturity.

The yield curve is used for pricing coupon bonds according to the following three basic steps. Consider a coupon bond issued at time $t_0 = 0$. Let $t_1, t_2, \ldots, t_n$ be the coupon dates so that $t_0 < t_1 < t_2 < \ldots < t_n$, with respective coupons $C_1, C_2, \ldots, C_n$, and let the redemption value W be paid at maturity $T = t_n$.

Step 1. Determine the annual yield rates $y(t_1), y(t_2), \ldots, y(t_n)$ that correspond to the coupon dates $t_1, t_2, \ldots, t_n$.

Step 2. Evaluate the prices of zero-coupon bonds using $Z(t_0, t_k) = \mathrm{e}^{-y(t_k)(t_k - t_0)}$ (for continuously compounded yield rates) or $Z(t_0, t_k) = (1 + y(t_k))^{-(t_k - t_0)}$ (for annually compounded yield rates) for each $k = 1, 2, \ldots, n$.

Step 3. Evaluate the present (time t_0) value of the coupon bond by simply discounting all the future cash flows back to current time t_0:

$$P = C_1 \cdot Z(t_0, t_1) + C_2 \cdot Z(t_0, t_2) + \cdots + C_{n-1} \cdot Z(t_0, t_{n-1}) + (C_n + W) \cdot Z(t_0, t_n). \quad (1.42)$$

Example 1.38. Assume we have the interest rate structure as given in Table 1.16. Assume annual compounding. Calculate the price of a 5-year par value $1,000 bond if it is

(a) a zero-coupon bond;

(b) a coupon bond with rate $c = 4\%$.

TABLE 1.16: Term structure of interest rates

Time to Maturity	Spot Rate
1 year	1.50%
2 year	1.50%
3 year	1.75%
4 year	2.00%
5 year	3.00%

Solution.
(a) The price of the zero-coupon bond is equal to the discounted value of the redemption value paid at maturity. To determine the price, we discount the value $W = 1000$ using the 5-year spot rate of 4%:

$$P = 1000 \cdot 1.03^{-5} \cong \$862.61.$$

(b) To determine the price of the coupon bond, we use (1.42). The amount of each coupon payment is $C = 1000 \cdot 0.04 = \$40$. The cash flows received from the coupon bond are no different from the cash flows received from a portfolio of four zero-coupon bonds with face value $40 maturing in one, two, three, and four years, respectively, and one zero-coupon bond with face value $1040 maturing in five years. Using the spot rates from Table 1.16 gives

$$\begin{aligned} P &= 40 \cdot 1.015^{-1} + 40 \cdot 1.015^{-2} + 40 \cdot 1.0175^{-3} + 40 \cdot 1.02^{-4} + 1040 \cdot 1.03^{-5} \\ &\cong \$1{,}050.27. \end{aligned}$$

□

Example 1.39. Suppose that the continuously compounded yield curve on a zero-coupon bond with t years to maturity is

$$y(t) = 0.05 - 0.03\left(1 - \mathrm{e}^{-t}\right)/t\,, \quad t \geqslant 0\,.$$

How much would you pay for a 5-year bond that pays an annual coupon of 3% on a face value of $100?

Solution. The cash flows that I receive from the coupon bond (assuming zero credit risk) are no different from the cash flows that I would receive from a portfolio of (i) one zero-coupon bond with face value $3 maturing in one year, (ii) one zero-coupon bond with face value $3 maturing in two years, ..., (v) one zero-coupon bond with face value $103 maturing in five years. Therefore I should pay the same amount for the coupon bond as I would for the portfolio. The cost of the portfolio is

$$\begin{aligned} C &= 3\mathrm{e}^{-y(1)\cdot 1} + 3\mathrm{e}^{-y(2)\cdot 2} + 3\mathrm{e}^{-y(3)\cdot 3} + 3\mathrm{e}^{-y(4)\cdot 4} + 103\mathrm{e}^{-y(5)\cdot 5} \\ &\cong 3\mathrm{e}^{-0.031} + 3\mathrm{e}^{-2\cdot 0.037} + 3\mathrm{e}^{-3\cdot 0.04} + 3\mathrm{e}^{-4\cdot 0.043} + 103\mathrm{e}^{-5\cdot 0.044} \\ &\cong \$93.52\,. \end{aligned}$$

Therefore we should be willing to pay \$93.52 for the bond. Using the method of averages we get an approximate yield of

$$\frac{(3 \cdot 5 + 100 - 93.53)/5}{(93.52 + 100)/2} \cong 0.04439 \,. \qquad \square$$

If we only know the actual yields for a very small number of tenors, we can fit a curve to the data to determine what the yield on an arbitrary tenor should be. For example, Figure 1.17 illustrates actual yield curves using data from Government of Canada bonds as of January 28, 2013. Using the quadratic fit we would estimate that the yield on a 15-year bond issued by the Government of Canada should be $y(15) = 2.43938$, or approximately 2.44% compounded semi-annually.

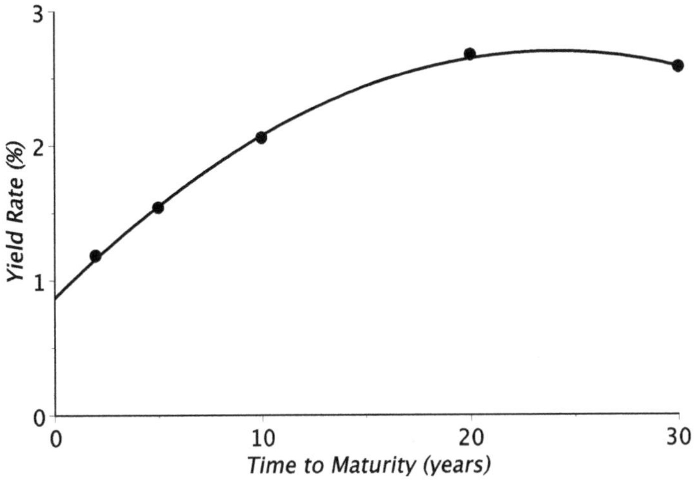

FIGURE 1.17: Yields of several Government of Canada zero-coupon bonds (with maturities 2, 5, 10, 20, and 30 years) and a quadratic fit $y(t) = 0.87008 + 0.15202t - 0.00316t^2$ to the data.

A popular method for fitting yield curves used in practice is a so-called "Nelson–Siegel" curve

$$y(t) = \alpha_0 + \alpha_1 \left(\frac{1 - \mathrm{e}^{-t/\beta}}{t/\beta} \right) + \alpha_2 \left(\frac{1 - \mathrm{e}^{-t/\beta}}{t/\beta} - \mathrm{e}^{-t/\beta} \right) ,$$

where $\alpha_0, \alpha_1, \alpha_2, \beta$ are parameters to be calibrated to the data (for example, using the method of least squares or another curve fitting technique).

Example 1.40. Evaluate $\lim_{t\to\infty} y(t)$ and $\lim_{t\to 0} y(t)$ for a Nelson–Siegel curve. Use your answers to interpret the parameters α_0 and α_1.

Solution. We use l'Hôpital's rule to obtain that

$$\lim_{t\to 0} \frac{1 - \mathrm{e}^{-t/\beta}}{t/\beta} = \lim_{t\to 0} \mathrm{e}^{-t/\beta} = 1 \,,$$

and therefore

$$\lim_{t\to 0} \left(\frac{1 - \mathrm{e}^{-t/\beta}}{t/\beta} - \mathrm{e}^{-t/\beta} \right) = 1 - 1 = 0 \,.$$

Thus, $y(0) = \alpha_0 + \alpha_1$ and we can interpret the sum of these parameters as the so-called "short rate" at time 0. It is obvious that

$$\lim_{t\to\infty}\left(\frac{1-e^{-t/\beta}}{t/\beta}\right) = 0 = \lim_{t\to\infty}\left(\frac{1-e^{-t/\beta}}{t/\beta} - e^{-t/\beta}\right),$$

and therefore $\lim_{t\to\infty} y(t) = \alpha_0$. So we can interpret α_0 as the long rate, and $-\alpha_1$ therefore represents how steep the yield curve is, in the sense that it measures the differential cost of long-term borrowing relative to short-term borrowing. □

1.6 Exercises

Exercise 1.1. Suppose that the return for the first year is r_1 and the return for the second year is r_2. Show that the return for the entire period of two years equals $r_1 + r_2 + r_1 r_2$.

Exercise 1.2. Show that the return on a deposit subject to periodic compounding is not additive, i.e., in general, $R_{[t_1,t_2]} + R_{[t_2,t_3]} \neq R_{[t_1,t_3]}$ for $t_1 < t_2 < t_3$.

Exercise 1.3. Prove that $V(t) = P(1+i)^t$ with $P, i > 0$ is an increasing, convex function of time $t \geqslant 0$.

Exercise 1.4. Find the effective annual rate when the nominal rate of 10% is

(a) compounded semi-annually;

(b) compounded monthly;

(c) compounded continuously.

Exercise 1.5. Let interest be paid at a nominal rate of 15% per year. How long will it take for a principal to double if the interest is

(a) compounded annually;

(b) compounded quarterly;

(c) compounded continuously?

Exercise 1.6. Assume a fixed yearly compounding rate $i > 0$. Find a formula that gives the number of years required to increase initial funds by n times for $n = 2, 3, 4, \ldots$.

Exercise 1.7. Construct a time diagram that represents the following cash flow stream and find the rate of return.

0	1	2
\$1,000	−\$500	−\$600

Exercise 1.8. Find the closed-form expression for $S_n = \sum_{k=1}^{n} x^{k-1} = 1 + x + x^2 + \cdots + x^{n-1}$ by first showing that $xS_n - S_n = x^n - 1$ and then solving for S_n.

Exercise 1.9. You are considering three different investments:

(a) one paying 7% compounded quarterly;

(b) one paying 7.1% compounded annually;

(c) one paying 6.9% compounded continuously.

Which investment has the highest effective annual rate of return?

Exercise 1.10. A firm can buy a piece of equipment for \$200,000 paid immediately, or in three instalments: \$70,000 paid now, \$70,000 in one year, and \$70,000 in two years time. Which option is better if money can be invested at a nominal rate of 6% compounded monthly?

Exercise 1.11. Which of the following projects should a firm choose if each proposal costs \$50,000 and the cost of capital is $i = 8\%$? The projects provide the following end-of-year net cash flows:

Years	1	2	3	4	5
Project A	\$20,000	\$10,000	\$5,000	\$10,000	\$20,000
Project B	\$5,000	\$20,000	\$20,000	\$20,000	\$5,000

Exercise 1.12. A used car sells for \$9,550. You wish to pay for it in 18 monthly instalments, the first due on the day of purchase. If 12% compounded monthly is charged, determine the size of the monthly payment.

Exercise 1.13. A couple is thinking of buying a new house by amortizing a loan. They can afford to pay \$1,000 a month for the first 10 years and \$1,200 a month for the next 10 years. If the current interest rate is $i^{(12)} = 6\%$, then how much can they borrow?

Exercise 1.14. A company is able to borrow money at $i^{(4)} = 9\frac{1}{2}\%$. It is considering the purchase of a machine costing \$110,000, which will save the company \$5,000 at the end of every quarter for 7 years and may be sold for \$14,000 at the end of the term. Should it borrow the money to buy the machine?

Exercise 1.15. Tom and Jack purchased bonds on the same day. Both bonds are redeemable at par and have a yield of $i = 6\%$ and a face value of \$10,000.

(a) If Tom's bond pays annual coupons at the rate of 8% and it has 15 years to maturity, then how much does he pay for it?

(b) If Jack's bond has 10 years to maturity and he pays \$11,487.75 for it, then what is the annual coupon rate?

Exercise 1.16. Jack invests \$2,000 in Bond A and \$2,000 in Bond B. On the same day Tom invests \$2,000 in Bond C and \$8,000 in Bond D. At the end of six months Bond A is worth \$2,050, Bond B is worth \$2,100, Bond C is worth \$2,040, and Bond D is worth \$8,360. Confirm that Jack has a higher (internal) rate of return on each of his bonds than Tom, but Tom has a higher (internal) rate of return on his total portfolio.

Exercise 1.17. A \$1,000 callable bond pays interest at $c^{(2)} = 8\%$ and matures at 105 in 20 years. It may be called at 110 at the end of year 5, or at 107.5 at the end of year 15.

(a) Determine the price of the bond to yield at least $i^{(2)} = 7\%$.

(b) Estimate, using first the method of averages and then linear interpolation, the yield rate earned by the buyer if the bond is held to maturity.

Exercise 1.18. On September 15, 2013, you were given a set of yield rates for various terms to maturity, generated using pricing data for Government of Canada bonds and treasury bills. Use the yield rates from the table below to price a coupon bond (issued on September 15, 2013) with the face value $F = \$1{,}000$ and the coupon rate $c = 2.5\%$. The bond is redeemable at par at the end of 5 years and it pays five annual coupons.

Time t	1	2	3	4	5
Rate $y(t)$	2.395%	2.487%	2.623%	2.758%	2.882%

Exercise 1.19. Show that

(a) the rate of return corresponding to a nominal rate $i^{(m)}$ compounded m times a year does not depend on the number of years it is invested;

(b) the rate of return corresponding to a simple interest investment rate of r depends on the number of years it is invested.

Exercise 1.20. Prove that for a fixed nominal interest rate i, the accumulation function $A(m) = (1 + \frac{i}{m})^m$ for the interest rate compounded m times a year is a strictly increasing function of $m \geqslant 1$. [*Hint*: Use the inequality $\ln(1 + x) > \frac{x}{1+x}$ for $x > 0$.]

Exercise 1.21. A \$1,000 bond paying interest at $c^{(2)} = 6\%$ matures at par on August 1, 2013. If this bond is quoted at 92 on August 1, 2011, determine the yield rate compounded semi-annually by

(a) using the method of averages;

(b) using the method of interpolation.

Exercise 1.22. Find the yield curve $y(t)$ and the present value of a zero-coupon bond $Z(0, t) = \mathrm{e}^{-y(t)\,t}$ if the instantaneous interest rate is

$$r(t) = \frac{1}{1+t}\,a + \frac{t}{1+t}\,b$$

with $a, b > 0$.

Chapter 2

Primer on Pricing Risky Securities

2.1 Stocks and Stock Price Models

2.1.1 Underlying Assets and Derivative Securities

Let us start with a basic classification of financial instruments. First, we differentiate between underlying assets (or securities) and derivative financial contracts. A *financial security* is a legal claim to some future benefit. Bonds, bank deposits, common stocks, and the like are all examples of financial securities. In contrast, a general *financial contract* links nominally two (or more) parties. Such a contract specifies conditions under which payments or payoffs are to be made between the parties. The main example is *derivative contracts* whose payoff depends on the value of another financial variable such as price of a stock, price of a bond, market index, or exchange rate called *underlying assets.* Examples include forward contracts, futures, swaps, and options.

In Chapter 1, we discussed various types of financial assets and investments, such as bank accounts, bonds, annuities, mortgages, and other types of loans. They all have one thing in common. It is the assumption that all future payments are guaranteed. However, in any investment portfolio, where a certain amount is invested today and future cash flows are returned over time, there is always the possibility that not all future payments will actually be received or the amount of future payments is not certain. For most investments, future cash flows are contingent upon the financial position of the company issuing the investment. On the other hand, there are many financial assets (or securities), such as stocks, foreign currencies, or commodities, whose future market values are unpredictable, because they depend on the choices and decisions made by a great number of agents acting under conditions of uncertainty. Such assets can be viewed as "risky" assets in comparison with "risk-free" assets with certain future cash flows. There are several types of risk involved in the purchase and sale of financial assets. These include economic risk, interest rate sensitivity, the possibility of company failure, company management problems, competition with other companies, and governmental rulings that may negatively affect the company. It is reasonable to assume that future prices of risky assets depend on random factors. That is, the asset values can be treated as random variables.

A corporation that needs funds for its development or expansion may issue *stock* to investors. Stock represents ownership of a corporation. Owners of *common stock* have voting rights and are entitled to the earnings of the company after all obligations are paid. If an investor purchases shares of stock in the company, then such an investor assumes a large amount of risk in return for the possibility of growth of the company and a corresponding increase in the value of the shares. It is important to realize that the corporation receives its money when the stock shares are issued. Any trading after that point takes place between the shareholders and the people wishing to purchase the stock, and does not directly represent a profit or loss to the company. Investors who purchase stock may receive dividends periodically (usually quarterly). Thus, the investor may profit by either an increase in the value of the stock or by receipt of dividends. The price that an investor is willing to pay for a share of common stock is based upon the investor's expectations regarding dividends and the future price of the stock.

In order to buy or sell shares of stocks, the investor uses an investment firm registered with the appropriate governmental agencies. The fee or commission charged for the transaction is an important consideration. The person actually making the transactions for the firm is called a broker. If an investor believes that a stock is going to decrease in value, then the investor may borrow shares from the broker (if such a transaction is possible), and then sell the stock. Such a financial operation is called *short selling.* If the stock decreases in value, then the investor may purchase an equivalent number of shares at the lower price and use those shares to repay the loan.

Stock market *indices* are used to compute an "average" price for groups of stocks. A stock market index attempts to mirror the performance of the group of stocks it represents through the use of one number, the index. Indices may represent the performance of all stocks in an exchange or a smaller group of stocks, such as an industrial or technological sector of the market. In addition, there are foreign and international indices. Examples are Dow Jones, Standard and Poor's, and NASDAQ indices.

The most well-known example of a derivatives is an option contract, which is defined just below.

Definition 2.1. An *option* is a contract that gives its buyer the right, but not the obligation, to buy (for a *call option*) or to sell (for a *put option*) a specified asset at a specified price (called the *exercise price* or *strike price*) on or before a specified date (called the *expiry date* or *maturity date*).

An option is an example of a derivative financial contract, whose value is derived from the values of other underlying assets. Options can be written for numerous products, such as gold, wheat, tulip bulbs, foreign exchange, movie scripts, stocks, etc. The purchase price of an option is called the *premium.* An option is said to be *in the money* if exercising the option yields a profit, excluding the premium. It is *out of the money* if exercising the option is unprofitable. If the purchase price of the underlying asset is equal to the exercise price, then the option is said to be *at the money.*

In this chapter we deal with stock options (also called *equity options*), which give the holder the right to buy (or to sell) a stock for a specified price during a specified time period. Thus, the buyer of an option has the right, but not the obligation, to buy or sell the stock. A *call* option gives the right to buy one unit (share) of the underlying stock at the strike price. A *put* option gives the right to sell one unit of the underlying stock at the strike price. The seller (writer) of an option must sell or buy the asset at the strike price once the option is exercised. An *American option* gives its holder the right to buy or sell an asset for the strike price at any time before or at the expiration date. A *European option* gives the same right, except the option may only be exercised on the expiration date. The terms American and European refer to the type of option, not the geographical region where the options are bought or sold.

2.1.2 Basic Assumptions for Asset Price Models

In practice, financial mathematicians generally make the following assumptions when dealing with asset price models.

Not moving the market. Our actions do not affect the market prices. In other words, we can buy or sell any amount of assets without affecting their prices. Clearly, this is not true for a free market, since an increasing demand moves market prices up.

Liquidity. At any time we can buy or sell as much as we wish at the market price without being forced to wait until a counter-party can be found. A liquid asset can be sold rapidly, any time within market hours. Cash is the most liquid asset. A market is liquid when there are ready and willing buyers and sellers at all times.

Shorting. We can have a negative amount of an asset by selling assets we do not hold. In this case we say that a *short position* is taken or that the asset is *shorted.* A short position in bonds means that the investor borrows cash and the interest rate is determined by the bond prices. A short position in stock means that the investor borrows the stock, sells it, and uses the proceeds to make other investments. The opposite of going "short," i.e., holding an asset, is sometimes called being "long" in the asset.

Fractional quantities. We are able to purchase fractional quantities of any asset. It is a reasonable assumption when the size of a typical financial transaction is sufficiently larger than the smallest unit one can hold.

No transactions costs. We can buy and sell assets without paying any additional fees. In the market, one of the typical ways to collect transaction costs is that buy and sell prices differ slightly. Such a difference in prices is called a *bid–ask spread.* The size of the bid–ask spread is closely related to liquidity. For a very liquid asset, the bid–ask spread will be quite small.

Stochastic prices. The future prices of financial assets are uncertain. Thus, we deal with stochastic asset price models. All factors that can affect the outcome of an economy are commonly called risk factors. We assume that all prices are equilibrium prices. We are not concerned with modelling the mechanism by which prices equilibrate.

Model assumptions. At the time we introduce asset price models, some assumptions on the models will be stated.

Basic Components of a Financial Model:

- **The time horizon** $T > 0$ is a date at which trading activity stops.
- **Trading dates** are calendar dates between the initial time $t = 0$ and time horizon $t = T$ at which trading can be allowed to take place.
- **A state of the world** ω characterizes the "real-world" state, which is relevant to the economic environment we wish to model. Such outcomes ω are also called *market scenarios* or *states of the world.* The set of all feasible scenarios or states of the world, denoted by Ω, is called the *state space.*
- **Tradable base assets** are underlying assets available for trading. The price of any derivative instrument depends on the current price or history of prices of the corresponding underlying asset.
- **Trading rules** such as allowance of short selling, presence of taxes and transaction costs, should be specified.

Financial models can be categorized as either of two general types: continuous-time or discrete-time. For a continuous-time model, the allowable time moments and trading dates form a continuous interval $[0, T]$ with specified initial and final calendar times 0 and T, respectively. For discrete-time models, assets are observed at only a finite set of calendar dates $\{0, 1, \ldots, T\}$. The discrete-time models can be further subdivided into *single period* models with only two relevant trading dates consisting of the current date $t = 0$ and the maturity date $t = T$ and multi-period models, where trading takes place at several dates. Financial models can also be categorized with respect to the state space Ω, which can be finite or infinite. In this section, Ω is assumed to be finite.

Consider some asset (or financial security) denoted by A or S. The price of asset A at time t will be denoted by A_t (or $A(t)$ in some cases). The price is assumed to be positive for all times t. The collection of asset prices indexed by time, $\{A_t\}_{t\in\mathbf{T}}$, is called the *price process* of asset A. Here, we have $\mathbf{T} = \{0, T\}$ for a single-period model, $\mathbf{T} = \{0, 1, \ldots, T\}$ for a multi-period discrete-time model, and $\mathbf{T} = [0, T]$ for a continuous-time model. At the present time $t = 0$, the current price A_0 is known to all investors. In general, the future values of any asset are uncertain. From the mathematical point of view, the prices A_t, $0 \leqslant t \leqslant T$, are positive random variables on the state space Ω. Therefore, the price process $\{A_t\}_{t\in\mathbf{T}}$ is nothing but a sequence of random variables indexed by time. Such a sequence of random variables is called a *random* or *stochastic process.* The notation $A_t(\omega)$ is used to denote the price at time t given that the market follows scenario $\omega \in \Omega$. Fixing ω gives a particular realization of the asset price process for the given scenario. Such a realization $A_t(\omega)$ considered as a function of time $t \in \mathbf{T}$ is called a *sample asset price path* or simply a *sample path.*

2.2 Basic Price Models

2.2.1 A Single-Period Binomial Model

Consider an investor who is faced with an economy whose future state is not known for certain. Let us construct the simplest possible model that describes such an economy. First we suppose that there are only two dates: a current date labeled 0 and a future (maturity) date labeled T. Since the economy is uncertain, there should be at least two future states, each corresponding to one of two possible scenarios or outcomes. Let these two states in our model be denoted by ω^- and ω^+. For example, these states can respectively represent "bad" news and "good" news. So our first financial model is a *single-period* model with the state space $\Omega = \{\omega^-, \omega^+\}$.

Now we come back to the investor. The wealth V of the investment in underlying assets is a function of time: $V = V_t$, $t \in \{0, T\}$. It is reasonable to assume that the *initial wealth* V_0 of the investment is known at time $t = 0$. However, the *terminal wealth* V_T at the maturity date is uncertain and is a function of state $\omega \in \Omega$. There are two states; hence V_T can take on two possible values, $V_T(\omega^-)$ and $V_T(\omega^+)$.

To further develop our model, we first need to quantify the chances of finding our economy in either of the possible states of Ω. Second, we need to specify what assets are available to form an investment portfolio. Assume that state ω^+ occurs with probability $p \in (0, 1)$. Thus, state ω^- occurs with probability $1 - p$. So the terminal wealth V_T can be viewed as a random variable on the finite probability space $(\Omega, \mathcal{F}, \mathbb{P})$ with Ω as a scenario set and $\mathbb{P}$ as a probability distribution function defined on the collection of events $\mathcal{F} = \{\emptyset, \{\omega^-\}, \{\omega^+\}, \Omega\}$ as follows:

$$\mathbb{P}(E) = \begin{cases} 0 & \text{if } E = \emptyset, \\ p & \text{if } E = \{\omega^+\}, \\ 1 - p & \text{if } E = \{\omega^-\}, \\ 1 & \text{if } E = \Omega = \{\omega^-, \omega^+\}. \end{cases}$$

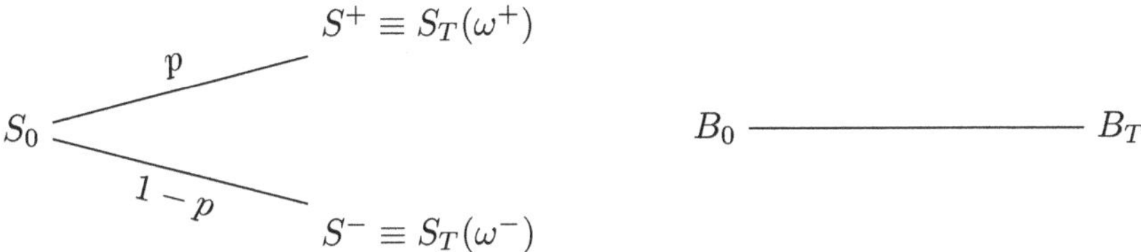

FIGURE 2.1: A schematic representation of a single-period model with two scenarios. B is a risk-free asset. S is a risky asset with two time-T prices, $S^{\pm} = S_T(\omega^{\pm})$.

We note that $\mathbb{P}(E)$ represents the *real-world (physical) probability* of event E. We will see shortly that the fair price of a derivative contract does *not* depend on the real-world probabilities $\{p, 1-p\}$.

We consider the case with only two tradable base assets, namely, a risk-free bond B and a risky asset such as a stock S. Let S_t and B_t, $t \in \{0, T\}$, be the respective time-t prices of the stock and the bond. The initial prices S_0 and B_0 are positive constants. The terminal prices S_T and B_T are positive random variables on the probability space $(\Omega, \mathcal{F}, \mathbb{P})$. The bond is a risk-free asset iff the variable B_T is certain, i.e., $B_T(\omega^+) = B_T(\omega^-) \equiv B_T$. The stock is a risky asset iff the variable S_T is uncertain, i.e., $S_T(\omega^+) \neq S_T(\omega^-)$. This is depicted in Figure 2.1, where for convenience we take $S^+ \equiv S_T(\omega^+) > S_T(\omega^-) \equiv S^-$.

A *static portfolio* in the bond and stock is a pair of real numbers $(x, y) \in \mathbb{R}^2$ that represents the positions (fixed in time) of the investment in the stock and bond:

$$x = \#\text{ of shares or units in the stock} \quad \text{and} \quad y = \#\text{ of shares or units in the bond.}$$

If $x > 0$ (or $y > 0$) then the position in the stock (or bond) is said to be *long*. If $x < 0$ (or $y < 0$) then the position in the stock (or bond) is said to be *short*. Hence, the value of a portfolio (x, y), denoted by $\Pi^{(x,y)}$, is given by

$$\Pi_t^{(x,y)} = xS_t + yB_t, \quad t \in \{0, T\}.$$

Note that in this economic model the investor takes on positions (x, y) in the two base assets at initial time $t = 0$ and holds these positions until the end of the period at time $t = T$. The initial value $\Pi_0 = \Pi_0^{(x,y)}$ is constant, while the terminal value Π_T is a generally nonconstant random variable on $(\Omega, \mathcal{F}, \mathbb{P})$. Note that the position x represents the number of units held in the risky asset. Hence, Π_T is constant iff $x = 0$, in which case the portfolio contains only positions in the bond.

Before proceeding further with the discussion of this two-state single period economy, we note the basic model assumptions as follows.

Divisibility. Positions in the base assets, x and y, may have noninteger value.

Liquidity. There are no bounds on x and y. That is, any asset can be bought or sold on demand at the market price in arbitrary quantities.

Short Selling. The positions x and y may be negative. In this case we say that a *short position* in the asset is taken or that the asset is *shorted* (otherwise, we say that an investor has a *long position* in the asset). A short position in the bond means that the investor borrows cash and the interest rate is determined by the bond prices. A short position in the stock means that the investor borrows shares of the stock, sells them, and uses the proceeds to make other investments.

Solvency. The portfolio value must be nonnegative at all times, $\Pi_t \geqslant 0$ for $t \in \{0, T\}$. A portfolio satisfying this condition is said to be *admissible.*

Clearly, there are many possible ways to form an investment portfolio. While the initial wealth of such portfolios is the same, the terminal values may have different distributions. Different investments can be compared by calculating their returns. Return on investment, is the ratio of the terminal value of the investment to the initial value of the investment. The total return on the portfolio (x, y) from time 0 to time T, denoted by R_Π, can be expressed as a weighted sum of returns on the base assets:

$$\begin{aligned} R_\Pi &= \frac{\Pi_T^{(x,y)}}{\Pi_0^{(x,y)}} = \frac{xS_T + yB_T}{xS_0 + yB_0} = \underbrace{\frac{xS_0}{xS_0 + yB_0}}_{\equiv w_1} \underbrace{\frac{S_T}{S_0}}_{\equiv R_S} + \underbrace{\frac{yB_0}{xS_0 + yB_0}}_{\equiv w_2} \underbrace{\frac{B_T}{B_0}}_{\equiv R_B} \\ &= w_1 R_S + w_2 R_B \,, \end{aligned} \tag{2.1}$$

where $R_S := \frac{S_T}{S_0}$ and $R_B := \frac{B_T}{B_0}$ are the total returns on the stock and bond, respectively. Note that the return on the bond is nonrandom, i.e., $R_B = R_B(\omega^\pm)$ is a constant. The weights w_1 and w_2, with $w_1 + w_2 = 1$, correspond to the respective fractions of initial wealth invested in the stock and bond. Since the sum of the weights is one, the rate of return on the portfolio (x, y) is a weighted sum of the respective rates of return on the stock and bond:

$$\begin{aligned} r_\Pi &= R_\Pi - 1 = w_1 R_S + w_2 R_B - 1 = w_1(r_S + 1) + w_2(r_B + 1) - 1 \\ &= w_1 r_S + w_2 r_B \,. \end{aligned} \tag{2.2}$$

As noted above, it is convenient in what follows to denote the stock price at maturity in the two possible states as $S^\pm \equiv S_T(\omega^\pm)$ and let $S^+ > S^-$. The return on the stock is the random variable whose value on the two states we can simply denote by $r_S^\pm \equiv r_S(\omega^\pm)$, where $r_S^\pm = S^\pm / S_0 - 1$ and $S^\pm = S_0(1 + r_S^\pm)$. From (2.2), we see that the rate of return on the portfolio, r_Π, is a random variable with two possible values: $r_\Pi(\omega^+) = w_1 r_S^+ + w_2 r_B$ for the higher return outcome and $r_\Pi(\omega^-) = w_1 r_S^- + w_2 r_B$ for the lower return outcome. Hence, the return on any portfolio without short selling (with positive x, y) always falls in between the largest possible return and smallest possible return on the stock and bond:

$$\min\{r_B, r_S^-, r_S^+\} \leqslant r_\Pi \leqslant \max\{r_B, r_S^-, r_S^+\} \,.$$

The terminal value of a portfolio in base assets is a function of the form $\Pi_T \colon \Omega \to \mathbb{R}$. Any such function is called a *payoff* function (or payoff for short). A nonnegative payoff is called a *claim.* A payoff is in fact a random variable defined on the probability space $(\Omega, \mathcal{F}, \mathbb{P})$. Every contract C in our economic model can be specified by the terminal payoff C_T and by the initial market price C_0 at which the contract is traded. Clearly, a sum (or linear combination) of several payoffs is again a payoff function. Therefore, we can consider a portfolio of contracts, which, in fact, is another contract. A typical financial problem is the evaluation of the fair initial price of a contract with a given terminal payoff. This can be done by constructing a portfolio whose terminal value is equivalent to the payoff of the contract. Such a portfolio is said to *replicate* the payoff. Being given a replicating portfolio, the fair initial price of the contract is simply equal to the initial cost of setting up the portfolio. An important question that arises, and which we now answer, is as follows.

> How do we *replicate* the payoff of a specified target financial contract with a portfolio in the two base assets B and S? In other words, how do we form an investment portfolio (x, y) whose terminal value coincides with the payoff of the contract in every state?

The claims that we wish to replicate are generally contingent (derivative) claims with uncertain payoff dependent on the outcome. In the two-state economy, any payoff C_T has two possible values, $C^+ \equiv C_T(\omega^+)$ and $C^- \equiv C_T(\omega^-)$. For a contingent payoff, $C^- \neq C^+$ holds; otherwise, when $C^- = C^+$, the payoff is said to be *deterministic.* To replicate a claim C_T with a portfolio in B and S, we must form a so-called *replicating portfolio* (x, y) such that $\Pi_T^{(x,y)}(\omega) = C_T(\omega)$ for both outcomes $\omega = \omega^+$ and $\omega = \omega^-$. The replication is equivalent to solving a system of two linear equations:

$$\begin{aligned} xS^+ + yB_T &= C^+, \\ xS^- + yB_T &= C^-. \end{aligned}$$

Since $S^+ \neq S^-$ in the model, this system has a unique solution:

$$x = \frac{C^+ - C^-}{S^+ - S^-}, \quad y = \frac{C^- S^+ - C^+ S^-}{B_T(S^+ - S^-)}. \tag{2.3}$$

This solution states that, given an arbitrary claim with payoffs $C^\pm$ in the two possible outcomes $\omega^\pm$, we can form a unique replicating portfolio (x, y) with x, y given by (2.3) where $\Pi_T^{(x,y)}(\omega^\pm) = C^\pm$. We can rewrite (2.3) in terms of the initial prices S_0 and B_0, the return on the bond, where $B_T = (1 + r_B)B_0$, and the return on the stock in the two states, where $S^\pm = (1 + r_S^\pm)S_0$, as follows:

$$x = \frac{C^+ - C^-}{S_0(r_S^+ - r_S^-)}, \quad y = \frac{C^-(1 + r_S^+) - C^+(1 + r_S^-)}{B_0(1 + r_B)(r_S^+ - r_S^-)}. \tag{2.4}$$

As we now see, and as discussed later in Section 2.3 and more formally and mathematically in-depth in later chapters, replication is the key to fair pricing and valuation of derivative contracts. By replicating the exact payoff structure of a target contract, by means of a portfolio in tradable assets, we are arriving at the fair price of the contract, which is given by the initial cost of setting up the replicating portfolio: $C_0 = \Pi_0^{(x,y)}$. In particular, by substituting the above values for x, y, we can represent the initial value of the replicating portfolio, and hence the fair price C_0 of the derivative contract, in the following equivalent ways:

$$\begin{aligned} C_0 = \Pi_0^{(x,y)} &= xS_0 + yB_0 = \frac{C^+ - C^-}{S^+ - S^-} S_0 + \frac{C^- S^+ - C^+ S^-}{B_T(S^+ - S^-)} B_0 \\ &= \frac{B_0}{B_T}\left[\left(\frac{(B_T/B_0)S_0 - S^-}{S^+ - S^-}\right)C^+ + \left(\frac{S^+ - (B_T/B_0)S_0}{S^+ - S^-}\right)C^-\right] \\ &= \frac{1}{1 + r_B}\left[\left(\frac{(1 + r_B)S_0 - S^-}{S^+ - S^-}\right)C^+ + \left(\frac{S^+ - (1 + r_B)S_0}{S^+ - S^-}\right)C^-\right] \\ &= \frac{1}{1 + r_B}\left[\left(\frac{(1 + r_B)S_0 - (1 + r_S^-)S_0}{(r_S^+ - r_S^-)S_0}\right)C^+ + \left(\frac{(1 + r_S^+)S_0 - (1 + r_B)S_0}{(r_S^+ - r_S^-)S_0}\right)C^-\right] \\ &= \frac{1}{1 + r_B}\left[\left(\frac{r_B - r_S^-}{r_S^+ - r_S^-}\right)C^+ + \left(\frac{r_S^+ - r_B}{r_S^+ - r_S^-}\right)C^-\right]. \end{aligned} \tag{2.5}$$

It is clear from (2.5) that the value of the replicating portfolio, and hence the initial fair price of the derivative contract, does not depend on the real-world probabilities of the outcomes $\omega^\pm$. Later we shall formally introduce the notions of arbitrage and risk-neutral probabilities, which will bring a more complete meaning to the result encapsulated in (2.5).

Example 2.1. Suppose $B_0 = \$100$, $B_T = \$110$, $S_0 = \$100$, $S^+ = \$120$, $S^- = \$90$.

(a) Determine a portfolio (x, y) whose value at time T is given by $\Pi_T(\omega^+) = \$930$ and $\Pi_T(\omega^-) = \$780$.

(b) Find the initial value Π_0 of the portfolio constructed in (a) and the rate of return, r, on the portfolio.

(c) Assuming that $\mathbb{P}(\omega^+) = \frac{2}{5}$, find the expected rate of return $\mathrm{E}[r]$ and the expected terminal value $\mathrm{E}[\Pi_T]$.

Solution.

(a) Apply (2.3) with $C^+ = \Pi_T(\omega^+) = \$930$, $C^- = \Pi_T(\omega^-) = \$780$ to obtain:

$$x = \frac{930 - 780}{120 - 90} = 5, \;\; y = \frac{780 \cdot 120 - 930 \cdot 90}{110 \cdot (120 - 90)} = 3.$$

(b) From (2.5) we have $\Pi_0 = xS_0 + yB_0 = 5 \cdot 100 + 3 \cdot 100 = \$800$. Equation (2.2) gives $r(\omega^\pm) = \frac{\Pi_T(\omega^\pm) - \Pi_0}{\Pi_0}$, hence

$$r(\omega^+) = (930 - 800)/800 = 0.1625 = 16.25\%,$$
$$r(\omega^-) = (780 - 800)/800 = -0.025 = -2.5\%.$$

(c) The real-world expected return on the portfolio given that $\mathbb{P}(\omega^+) = p = \frac{2}{5}$ and $\mathbb{P}(\omega^-) = 1 - p = \frac{3}{5}$ is

$$\mathrm{E}[r] = pr(\omega^+) + (1-p)r(\omega^-) = \frac{2}{5} \cdot 0.1625 + \frac{3}{5} \cdot (-0.025) = 0.05 = 5\%\,.$$

The expected terminal value is

$$\mathrm{E}[\Pi_T] = p\Pi_T(\omega^+) + (1-p)\Pi_T(\omega^-) = \frac{2}{5} \cdot 930 + \frac{3}{5} \cdot 780 = \$840\,.$$

This number is consistent with the fact that the expected return is

$$\mathrm{E}[r] = \mathrm{E}[\Pi_T/\Pi_0 - 1] = \mathrm{E}[\Pi_T]/\Pi_0 - 1, \text{ giving } \mathrm{E}[\Pi_T] = \Pi_0(1 + \mathrm{E}[r])\,.$$

As a check: $840 = 800 \cdot (1 + 0.05)$. □

In a discrete-time financial model one is generally interested in basic characteristics of the model, such as whether or not the following statements hold true:

(1) There exists a replicating portfolio for every arbitrary derivative contract or claim.

(2) Every replicating portfolio is unique for a given claim.

(3) If there are different replicating portfolios for a given claim, then they all have the same initial cost.

(4) The initial cost of a portfolio replicating a *positive* claim C_T (i.e., $C_T \geqslant 0$ and $C_T(\omega) > 0$ for at least one $\omega \in \Omega$) is necessarily positive.

In the above simplest two-state model with two base assets we have already seen that statements 1 and 2 hold; hence statement 3 is irrelevant. Statement 4 can be guaranteed once we impose the extra condition that $r_S^- < r_B < r_S^+$, which is equivalent to requiring that there is no *arbitrage* in the model. This is discussed in detail in Section 2.3.

2.2.2 A Discrete-Time Model with a Finite Number of States

Let us make our model more realistic by increasing the number of observation periods and adding more states. We assume that trading can take place at discretely monitored times $t = 0, 1, 2, \ldots, T$. That is, the time horizon T is an integer, and time is measured in periods. One observation period may correspond to one year, one week, one day, or even one second. The state space Ω is finite and contains M states of the world $\omega^1, \omega^2, \ldots, \omega^M$. Consider an asset such as a stock ($A = S$) or a bond ($A = B$) with the price A_t monitored at times $t = 0, 1, 2, \ldots, T$. Assume that $A_t > 0$ for all t. At the present time $t = 0$, the current price A_0 is assumed to be known. The future prices A_t for $t > 0$ remain uncertain until information about the market state is revealed. So, mathematically, for every fixed t, price A_t is a positive random variable on a finite probability space.

2.2.2.1 Asset Returns

Consider an asset A (or a portfolio of assets) with the price process $\{A_t\}_{t=0,1,\ldots,T}$. The total return and the rate of return on the asset A over a time interval $[s, t]$ with $0 \leqslant s < t$, respectively denoted by $R^A_{[s,t]}$ and $r^A_{[s,t]}$, are random variables defined by

$$R^A_{[s,t]} := \frac{A_t}{A_s} \quad \text{and} \quad r^A_{[s,t]} := \frac{A_t - A_s}{A_s},$$

respectively. They are related by $R_A = r_A + 1$. Often, for simplicity, the term *return* is used for both notions. Returns on risky assets (or portfolios of risky assets) are uncertain and are therefore functions of $\omega \in \Omega$, e.g., $R^A_{[s,t]}(\omega) = \frac{A_t(\omega)}{A_s(\omega)}$. For returns on asset A over a single period $[t-1, t]$ where $t = 1, 2, \ldots, T$, we use notations $R^A_t := R^A_{[t-1,t]}$ and $r^A_t := r^A_{[t-1,t]}$. For simplicity and when the context is clear, we will omit the superscript A and will simply denote the total return by R and the rate of return by r.

Now the asset prices can be written in terms of single period returns on the stock. For every $t = 1, 2, \ldots, T$, we have that

$$r_t = R_t - 1 = \frac{A_t - A_{t-1}}{A_{t-1}} = \frac{A_t}{A_{t-1}} - 1 \implies A_t = R_t\, A_{t-1} \quad \text{and} \quad A_t = (1 + r_t)\, A_{t-1}.$$

By applying successively this rule to $A_{t-1}, A_{t-2}, \ldots, A_1$, we obtain

$$\begin{aligned} A_t = (1 + r_t)\, A_{t-1} &= (1 + r_{t-1})\,(1 + r_t)\, A_{t-2} = \ldots \\ &= (1 + r_1)\,(1 + r_2) \cdots (1 + r_t)\, A_0. \end{aligned}$$

Equivalently, we have $A_t = R_1\, R_2 \cdots R_t\, A_0$ for $t = 0, 1, \ldots, T$. Therefore, the dynamics of asset prices can also be described by asset returns and initial price A_0. Note that the aggregate returns $r_{[s,t]} = \frac{A_t - A_s}{A_s}$ and $R_{[s,t]} = \frac{A_t}{A_s}$ on asset A from time s to time t with $0 \leqslant s < t \leqslant T$ respectively satisfy

$$1 + r_{[s,t]} = (1 + r_{s+1})\,(1 + r_{s+2}) \cdots (1 + r_t) \quad \text{and} \quad R_{[s,t]} = R_{s+1}\, R_{s+2} \cdots R_t. \tag{2.6}$$

Example 2.2. Consider a market that assumes three possible scenarios: $\Omega = \{\omega^1, \omega^2, \omega^3\}$. Suppose that stock S takes on the following values over a two-period interval:

Scenario, ω	S_0	S_1	S_2
ω^1(boom)	100	120	150
ω^2(stability)	100	105	100
ω^3(recession)	100	80	60

Find the returns r_1, r_2 and compare them with $r_{[0,2]}$.

Solution. The one-period returns $r_1 = \frac{S_1 - S_0}{S_0}$ and $r_2 = \frac{S_2 - S_1}{S_1}$ and the two-period return $r_{[0,2]} = \frac{S_2 - S_0}{S_0}$ take on the following values:

Scenario	r_1	r_2	$r_{[0,2]}$
ω^1	20%	25%	50%
ω^2	5%	-4.76%	0%
ω^3	-20%	-25%	-40%

As is seen from the table above, the returns satisfy (2.6), which takes the following form for a two-period model: $1 + r_{[0,2]} = (1 + r_1)(1 + r_2)$. □

The returns on a risky asset are random variables. If the scenario probabilities are given, then expected values of returns can be calculated. Suppose that the probabilities $p_k := \mathbb{P}(\omega^k)$ are known for every $\omega^k \in \Omega$. Suppose that the probabilities p_k are strictly positive and sum up to one:

$$p_1, p_2, \ldots, p_M > 0, \quad p_1 + p_2 + \cdots + p_M = 1.$$

For any given scenario ω^k, $k = 1, 2, \ldots, M$, the return $r_{[s,t]}(\omega^k)$ is known. The expected return for the period $[s,t]$ with $0 \leqslant s < t \leqslant T$ can be calculated as follows:

$$\mathrm{E}[r_{[s,t]}] = r_{[s,t]}(\omega^1) \cdot p_1 + r_{[s,t]}(\omega^2) \cdot p_2 + \cdots + r_{[s,t]}(\omega^M) \cdot p_M . \tag{2.7}$$

If the one-period returns $r_{s+1}, r_{s+2}, \ldots, r_t$ are *independent* random variables, then the expected aggregate return can be expressed in terms of expected one-period returns as follows:

$$1 + \mathrm{E}[r_{[s,t]}] = \left(1 + \mathrm{E}[r_{s+1}]\right) \left(1 + \mathrm{E}[r_{s+2}]\right) \cdots \left(1 + \mathrm{E}[r_t]\right).$$

Suppose that the one-period returns r_t, $t \geqslant 1$, are independent and identically distributed (i.i.d.) random variables with the common expected value of a one-period return, $r_A := \mathrm{E}[r_1]$. Then, we obtain

$$1 + \mathrm{E}\left[r_{[s,t]}\right] = \left(1 + \mathrm{E}[r_1]\right)^{t-s} = (1 + r_A)^{t-s}.$$

Since $A_t = (1 + r_{[0,t]})\, A_0$ and A_0 is certain, the expected asset price at time t is

$$\mathrm{E}[A_t] = (1 + \mathrm{E}[r_{[0,t]}])\, A_0 = (1 + r_A)^t\, A_0. \tag{2.8}$$

Note that this expression is very similar to the formula of the accumulated value under periodic compounding of interest. In practice it may be difficult to estimate probability distributions of returns. What can be easily computed from historical data is an average return over a certain period. As a result, one can estimate expected future cash flows.

The *log-return* on asset A over a time interval $[s,t]$, denoted $L^A_{[s,t]}$, is given by

$$L^A_{[s,t]} := \ln\left(\frac{A_t}{A_s}\right) = \ln R^A_{[s,t]} = \ln(1 + r^A_{[s,t]}).$$

A single-period log-return, denoted L^A_t, is

$$L^A_t := L^A_{[t-1,t]}, \quad t = 1, 2, \ldots, T.$$

Since $R^A_{[s,t]} = R^A_{s+1}\, R^A_{s+2} \cdots R^A_t$, we obtain that the log-returns are additive:

$$L^A_{[s,t]} = L^A_{s+1} + L^A_{s+2} + \cdots + L^A_t.$$

The bond prices B_t, $t \geqslant 0$, are nonrandom (deterministic). In other words, they do not depend on the world state: $B_t(\omega) = B_t$ for all $\omega \in \Omega$ and $t \geqslant 0$. The returns on the bond $r_t = \frac{B_t - B_{t-1}}{B_{t-1}}$, $t \geqslant 1$, are deterministic as well. We can express the bond prices B_t, $t \geqslant 1$, in terms of the initial price B_0 and one-period returns as

$$B_t = B_0(1 + r_1)(1 + r_2) \cdots (1 + r_t).$$

Assuming that the one-period returns r_t all have the same constant value r_B, we arrive at a formula that is analogous to (2.8):

$$B_t = B_0(1 + r_B)^t. \tag{2.9}$$

Equations (2.8) and (2.9) enlighten us on how to compare performances of a risky asset and risk-free asset—we can simply compare the expected one-period returns on assets of interest.

In summary, to construct a discrete-time financial model we need to know the initial price and the probability distribution of one-period returns for each base asset. For a model with a finite number of states, all returns are discrete random variables defined on a common probability space. Usually we can distinguish one underlying whose returns are certain, e.g., a risk-free bond. A more detailed analysis of single-period models and multi-period models will be respectively done in later chapters. In the next section, we present the binomial tree model, which is the simplest example of a multi-period model.

2.2.3 Introducing the Binomial Tree Model

In this subsection we introduce a discrete-time model with one risky stock S and one risk-free asset B such as a bond (or a bank account). Assume that there are T periods. The one-period return on the risk-free asset is denoted by $r \equiv r_B > 0$. In fact, r is a risk-free interest rate compounded periodically. The bond prices are $B_t = B_0(1+r)^t$, $t = 1, 2, \ldots, T$. Assume that the one-period returns $R_t \equiv R_t^S$, $t = 1, 2, \ldots, T$, on the stock are i.i.d. random variables such that

$$R_t = \begin{cases} u & \text{with probability } p, \\ d & \text{with probability } 1 - p, \end{cases} \tag{2.10}$$

where the factors d and u are such that $0 < d < u$, and $p \in (0, 1)$ is the probability that the stock price moves up. Notice that the average one-period return on the stock is given by

$$\mathrm{E}[R_t] = p\,u + (1 - p)\,d\,.$$

Since $R_t = S_t / S_{t-1}$, the stock price S_t at time t is expressed in terms of the price S_{t-1} at the previous time moment as follows:

$$S_t = R_t\,S_{t-1} = \begin{cases} S_{t-1}u & \text{with probability } p, \\ S_{t-1}d & \text{with probability } 1 - p. \end{cases} \tag{2.11}$$

In other words, at each time t the stock price S_t can move up by a factor u or down by a factor d (relative to S_{t-1}). The inequality $d > 0$ guarantees the positiveness of stock prices, i.e., $S_t > 0$ for all $t \geqslant 1$, provided that $S_0 > 0$. Equivalently, we may work with rates of return, $r_t = R_t - 1$. Equations (2.10)–(2.11) take the form

$$S_t = (1 + r_t^S)\,S_{t-1}, \quad \text{where } r_t^S = \begin{cases} u - 1 & \text{with probability } p, \\ d - 1 & \text{with probability } 1 - p. \end{cases}$$

A useful way to represent (2.11) is to write

$$S_t = u^{X_t} d^{1-X_t} S_{t-1}, \quad \text{where} \quad X_t = \begin{cases} 1 & \text{with probability } p, \\ 0 & \text{with probability } 1-p. \end{cases} \tag{2.12}$$

Note that X_t is a Bernoulli random variable. For each t, the variable X_t is a function of the return R_t. Since the returns on the stock are independent, the variables X_t are independent as well. In other words, $\{X_t\}_{t\geqslant 1}$ is a collection of i.i.d. Bernoulli random variables with probability p of success. Iterating Equation (2.12) gives

$$\begin{aligned} S_1 &= S_0 u^{X_1} d^{1-X_1}, \\ S_2 &= S_1 u^{X_2} d^{1-X_2} = S_0 u^{X_1+X_2} d^{2-(X_1+X_2)}, \\ &\vdots \\ S_n &= S_0 u^{X_1+X_2+\cdots+X_n} d^{n-(X_1+X_2+\cdots+X_n)}. \end{aligned}$$

The last expression gives the stock price at time $t = n$ in terms of its initial price and the sum $X_1 + X_2 + \cdots + X_n$. The sum of n i.i.d. Bernoulli random variables, denoted Y_n, can be interpreted as the number of successes in n independent trials with probability p of success on each trial. It has the binomial distribution; the probability mass function of $Y_n \sim Bin(n,p)$ is

$$b(k;n,p) = \mathbb{P}(Y_n = k) = \binom{n}{k} p^k (1-p)^{n-k}, \quad k = 0,1,2,\ldots,n.$$

Thus, at time $t = n$ we have

$$S_n = S_0 u^{Y_n} d^{n-Y_n}, \quad \text{where} \quad Y_n \sim Bin(n,p).$$

Note that the stock price S_n can only take on a value in the set

$$\{S_{n,k} := S_0 u^k d^{n-k} \ : \ k = 0,1,\ldots,n\}.$$

By the equivalence of events $\{S_n = S_{n,k}\}$ and $\{Y_n = k\}$, the probability distribution of S_n is then

$$\mathbb{P}(S_n = S_{n,k}) = \mathbb{P}(Y_n = k) = \binom{n}{k} p^k (1-p)^{n-k}, \quad k = 0,1,\ldots,n. \tag{2.13}$$

Equation (2.13) tells us that the stock price S_n at time $t = n$ can admit $n+1$ different values, i.e., the values $S_{n,k}$, $k = 0,1,\ldots,n$. There are $\binom{n}{k}$ different scenarios (n-step price paths) with exactly k upward and $n-k$ downward price moves that produce the same (terminal) stock price $S_{n,k}$ at time n. The set Ω of all the possible scenarios can be compactly represented by the n-step *recombining binomial tree* or *binomial lattice* (see Figure 2.2). Each n-step path or scenario leading to $S_{n,k}$ has equal probability $p^k(1-p)^{n-k}$ of occurring. Every such path starts at S_0, moves up a total of k times and down a total of $n-k$ times. Since there are $\binom{n}{k}$ paths leading to the same value $S_{n,k}$, then summing this over values $k = 0,1,\ldots,n$ must give the total number of all possible paths, which is 2^n. This, of course, corresponds to the binomial formula:

$$2^n = (1+1)^n = \sum_{k=0}^{n} \binom{n}{k} = \sum_{k=0}^{n} \left\{ \begin{array}{c} \#\text{ of scenarios with } k \text{ upward and} \\ n-k \text{ downward price moves} \end{array} \right\}.$$

For the sake of comparison, let us consider three two-step binomial tree models under three different assumptions on the probability distributions of returns.

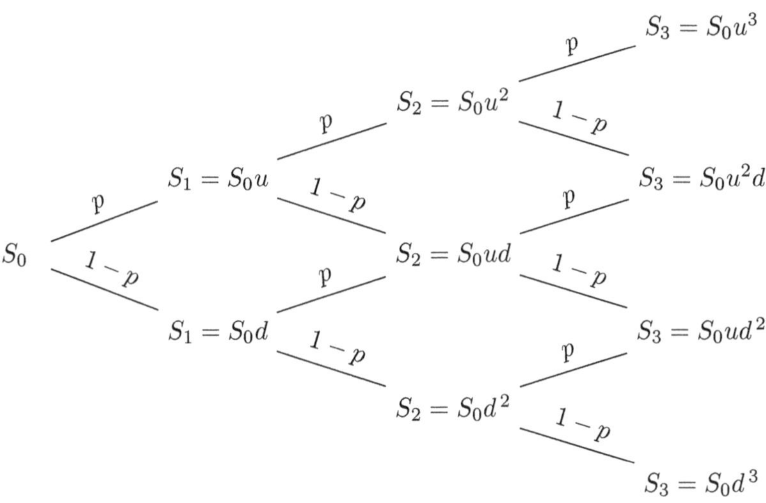

FIGURE 2.2: A schematic representation of the binomial lattice with three periods.

Example 2.3. Let $S_0 = 100$. Find the distributions of prices S_1 and S_2 assuming one of the following.

(a) The returns R_1 and R_2 are i.i.d. and $R_t = \begin{cases} 1.1 & \text{with prob. } 1/2, \\ 0.9 & \text{with prob. } 1/2. \end{cases}$

(b) The returns R_1 and R_2 are independent, but not identically distributed:

$$R_1 = \begin{cases} 1.1 & \text{with prob. } 1/2, \\ 0.9 & \text{with prob. } 1/2; \end{cases} \qquad R_2 = \begin{cases} 1.15 & \text{with prob. } 1/3, \\ 0.9 & \text{with prob. } 2/3. \end{cases}$$

(c) The returns R_1 and R_2 are not independent and not identically distributed:

$$R_1 = \begin{cases} 1.1 & \text{with prob. } 1/2, \\ 0.9 & \text{with prob. } 1/2; \end{cases}$$

$$R_2 | \{R_1 = 1.1\} = \begin{cases} 1.15 & \text{with prob. } 1/3, \\ 0.9 & \text{with prob. } 2/3; \end{cases}$$

$$R_2 | \{R_1 = 0.9\} = \begin{cases} 1.1 & \text{with prob. } 1/4, \\ 0.85 & \text{with prob. } 3/4. \end{cases}$$

Solution. Since the probability distribution of R_1 is the same for all three cases, the distribution of S_1 is the same as well:

$$S_1 = S_0 \, R_1 = \begin{cases} 100 \cdot 1.1 & \text{with prob. } 1/2 \\ 100 \cdot 0.9 & \text{with prob. } 1/2 \end{cases} = \begin{cases} 110 & \text{with prob. } 1/2 \\ 90 & \text{with prob. } 1/2. \end{cases}$$

The stock price at the end of period 2 is $S_2 = S_1 \, R_2$. So we find the probability distribution of S_2 for each of the three cases based on the values of R_2 and S_1.

(a) In this case, we have a two-period binomial tree model discussed just above with parameters $u = 1.1$, $d = 0.9$, and $p = 1/2$. Applying (2.13) gives

$$S_2 = S_{2,2} = 100 \cdot 1.1^2 = 121 \qquad \text{with probability } \binom{2}{0} 0.5^2 = 0.25,$$

$$S_2 = S_{2,1} = 100 \cdot 1.1 \cdot 0.9 = 99 \qquad \text{with probability } \binom{2}{1} 0.5^2 = 0.5,$$

$$S_2 = S_{2,0} = 100 \cdot 0.9^2 = 81 \qquad \text{with probability } \binom{2}{2} 0.5^2 = 0.25.$$

(b) Given that $S_1 = 110$, the price $S_2 = 110\,R_2$ takes one of two values: $110 \cdot 1.15 = 126.5$ or $110 \cdot 0.9 = 99$. If $S_1 = 90$, then $S_2 = 90\,R_2$ is either $90 \cdot 1.15 = 103.5$ or $90 \cdot 0.9 = 81$. So there are four distinct values for S_2. We can compute the probabilities by using the independence of the returns R_1 and R_2, i.e., using that $P(R_1 = x, R_2 = y) = \mathbb{P}(R_1 = x) \cdot \mathbb{P}(R_2 = y)$ for all x and y (hence, S_1 and R_2 are independent random variables):

$$\begin{aligned}
\mathbb{P}(S_2 = 126.5) &= \mathbb{P}(R_1 = 1.1, R_2 = 1.15) = \mathbb{P}(R_1 = 1.1) \cdot \mathbb{P}(R_2 = 1.15) = \frac{1}{2} \cdot \frac{1}{3} = \frac{1}{6}, \\
\mathbb{P}(S_2 = 103.5) &= \mathbb{P}(R_1 = 0.9, R_2 = 1.15) = \mathbb{P}(R_1 = 0.9) \cdot \mathbb{P}(R_2 = 1.15) = \frac{1}{2} \cdot \frac{1}{3} = \frac{1}{6}, \\
\mathbb{P}(S_2 = 99) &= \mathbb{P}(R_1 = 1.1, R_2 = 0.9) = \mathbb{P}(R_1 = 1.1) \cdot \mathbb{P}(R_2 = 0.9) = \frac{1}{2} \cdot \frac{2}{3} = \frac{1}{3}, \\
\mathbb{P}(S_2 = 81) &= \mathbb{P}(R_1 = 0.9, R_2 = 0.9) = \mathbb{P}(R_1 = 0.9) \cdot \mathbb{P}(R_2 = 0.9) = \frac{1}{2} \cdot \frac{2}{3} = \frac{1}{3}.
\end{aligned}$$

(c) If $R_1 = 1.1$ (and hence $S_1 = 110$), then $S_2 = 110 \cdot R_2$ takes on the value $110 \cdot 1.15 = 126.5$ or $110 \cdot 0.9 = 99$. If $R_1 = 1.1$ (and hence $S_1 = 90$), then $S_2 = 90 \cdot R_2$ is either $90 \cdot 1.1 = 99$ or $90 \cdot 0.85 = 76.5$. The distribution of S_2 is given by conditioning on R_1:

$$\begin{aligned}
\mathbb{P}(S_2 = 126.5) &= \mathbb{P}(R_1 = 1.1, R_2 = 1.15) \\
&= \mathbb{P}(R_2 = 1.15 \mid R_1 = 1.1) \cdot \mathbb{P}(R_1 = 1.1) = \frac{1}{3} \cdot \frac{1}{2} = \frac{1}{6}, \\
\mathbb{P}(S_2 = 99) &= \mathbb{P}(R_1 = 0.9, R_2 = 1.1) + \mathbb{P}(R_1 = 1.1, R_2 = 0.9) \\
&= \mathbb{P}(R_2 = 1.1 \mid R_1 = 0.9) \cdot \mathbb{P}(R_1 = 0.9) \\
&\quad + \mathbb{P}(R_2 = 0.9 \mid R_1 = 1.1) \cdot \mathbb{P}(R_1 = 1.1) = \frac{1}{4} \cdot \frac{1}{2} + \frac{2}{3} \cdot \frac{1}{2} = \frac{11}{24}, \\
\mathbb{P}(S_2 = 76.5) &= \mathbb{P}(R_1 = 0.9, R_2 = 0.85) \\
&= \mathbb{P}(R_2 = 0.85 \mid R_1 = 0.9) \cdot \mathbb{P}(R_1 = 0.9) = \frac{3}{4} \cdot \frac{1}{2} = \frac{3}{8}.
\end{aligned}$$

□

Example 2.4. Suppose that the stock price follows a binomial tree with current price $S_0 = \$100$. Find the factors d and u if (a) $S_1 \in \{\$120, \$90\}$; (b) $\min_\omega S_2(\omega) = \$64$ and $\max_\omega S_2(\omega) = \$121$. These are simple examples of a two-period binomial model calibration.

Solution.

(a) In one period, the stock prices are $S^+ = S_0 u = 100u = 120$ and $S^- = S_0 d = 100d = 90$. Solve these equations for d and u to obtain $d = 0.9$ and $u = 1.2$.

(b) In two periods:

$$\min_\omega S(2, \omega) = S_0 \min_{0 \leqslant k \leqslant 2} u^k d^{2-k} = S_0 \min\{d^2, ud, u^2\} = S_0 d^2 = 100 d^2$$

and

$$\max_{\omega} S(2,\omega) = S_0 \max_{0 \leqslant k \leqslant 2} u^k d^{2-k} = S_0 \max\{d^2, ud, u^2\} = S_0 u^2 = 100u^2.$$

Hence, $u = \sqrt{1.21} = 1.1$ and $d = \sqrt{0.64} = 0.8$. □

2.2.4 Self-Financing Investment Strategies in the Binomial Model

Here we give a brief introduction to the concept of a self-financing trading strategy within the binomial model. Recall that a static portfolio (x, y) in the two base assets is an investment consisting of fixed x positions in the stock (the risky asset) and fixed y positions in the bond or money market account (the risk-free asset). For such a portfolio there is no trading, or re-balancing, of the positions over all time periods. In contrast, in the binomial model trading is allowed to take place at the beginning of each period so that the positions held in the two base assets are generally not static over time. That is, within each period the positions are allowed to be *re-balanced* at the beginning of the period and are subsequently held fixed until the end of the period. The model allows for trading at the discrete times $t = 0, 1, \ldots, T$. Imagine that an investor begins by setting up a portfolio (x_0, y_0) at time $t = 0$ and holds it until the end of the first period. At time $t = 1$, trading is allowed and the investor re-balances the portfolio positions to (x_1, y_1) and this new portfolio is held to the end of the second period at time $t = 2$. At time $t = 2$, trading takes place and the investor again re-balances the positions to (x_2, y_2) and holds this portfolio until time $t = 3$, and so on until maturity time $t = T$. The sequence of portfolios $(x_0, y_0), (x_1, y_1), \ldots, (x_{T-1}, y_{T-1})$ forms what is called a *trading* or *investment strategy* (with maturity time T) in the binomial model. This is also referred to as a *portfolio strategy* in the two base assets.

At each time t, *after* re-balancing, the portfolio value is

$$\Pi_t = x_t S_t + y_t B_t.$$

During the period from time t to $t + 1$ the portfolio positions are held static and the new portfolio value, *before* re-balancing, at time $t + 1$ is

$$\Pi_{t+1} = x_t S_{t+1} + y_t B_{t+1}.$$

Note that the change in value for the one period is due solely to the change in prices of the base assets:

$$\Pi_{t+1} - \Pi_t = x_t(S_{t+1} - S_t) + y_t(B_{t+1} - B_t). \tag{2.14}$$

The first term on the right of this equation gives the change in portfolio value due to the change in the share price of the stock. Assuming the number of shares x_t is positive (i.e., long the stock), then $x_t(S_{t+1} - S_t)$ corresponds to a gain (or loss) if the share price increases (or decreases) whereas the opposite is true if x_t is negative (short the stock). The second term gives the change in value due to the bond component. If $y_t > 0$, then $y_t(B_{t+1} - B_t) > 0$ is a gain or earning based on the interest paid by the bond or a bank account during the period. If $y_t < 0$, the position in the bond is negative and this gives a loss corresponding to borrowing money from a bank account. At time $t + 1$, the portfolio value *after re-balancing* must be

$$\Pi_{t+1} = x_{t+1} S_{t+1} + y_{t+1} B_{t+1}.$$

At this point we bring in the important notion of self-financing within the above strategy. So far, we allowed for the positions to be re-balanced at each time step without any restriction. Of course, any decision made by the investor about when to change the asset positions and what assets to sell or to buy is only based on the historical information about the market currently available. Generally, the investor may alter the positions at any time by selling some assets and investing the proceeds in others. However, we now enforce the condition that there cannot be any consumption or injection of funds within the strategy at any time past initial time $t_0 = 0$. In other words, after initially setting up the portfolio, we only allow for self-financed strategies. Hence, at each time t, the portfolio value before re-balancing the positions must be the same as its value after re-balancing. The above two expressions for Π_{t+1} must therefore be equal and this gives rise to the so-called *self-financing condition (s.f.c.)*:

$$S_{t+1}(x_{t+1} - x_t) + B_{t+1}(y_{t+1} - y_t) = 0 \tag{2.15}$$

for all $0 \leqslant t \leqslant T-1$. The first term corresponds to the cost (at time $t+1$) of re-balancing in the stock with share price S_{t+1} and position change $\delta x_t := x_{t+1} - x_t$. The second term gives the re-balancing cost (at time $t+1$) for the bond holdings where the position change is $\delta y_t := y_{t+1} - y_t$. Note that (i) $\delta x_t < 0$ iff $\delta y_t > 0$ and (ii) $\delta x_t > 0$ iff $\delta y_t < 0$. In case (i) re-balancing involves selling $|\delta x_t|$ stock shares at price S_{t+1} and investing the proceeds in buying δy_t bonds at price B_{t+1}. In case (ii) re-balancing involves selling $|\delta y_t|$ bonds at price B_{t+1} and investing this amount in buying δx_t stock shares at price S_{t+1}.

Let us denote the one-period changes in the portfolio value and asset prices by

$$\delta\Pi_t := \Pi_{t+1} - \Pi_t\,, \quad \delta S_t := S_{t+1} - S_t\,, \quad \delta B_t := B_{t+1} - B_t\,,$$

respectively. Since $S_{t+1} = S_t + \delta S_t$ and $B_{t+1} = B_t + \delta B_t$, the above s.f.c. takes the form

$$(S_t + \delta S_t)\,\delta x_t + (B_t + \delta B_t)\,\delta y_t = 0\,. \tag{2.16}$$

In later chapters we shall see that this form is similar to the s.f.c. for continuous-time modelling of positions and prices. If we now compute the change in the re-balanced portfolio values from time t to $t+1$, we see that the s.f.c. is equivalent to the statement that this change is due only to the changes in the asset prices:

$$\begin{aligned}
\delta\Pi_t &= x_{t+1}S_{t+1} + y_{t+1}B_{t+1} - (x_tS_t + y_tB_t)\\
&= (x_t + \delta x_t)(S_t + \delta S_t) + (y_t + \delta y_t)(B_t + \delta B_t) - (x_tS_t + y_tB_t)\\
&= x_t\,\delta S_t + y_t\delta B_t + (S_t + \delta S_t)\,\delta x_t + (B_t + \delta B_t)\,\delta y_t\\
&= x_t\,\delta S_t + y_t\delta B_t\,.
\end{aligned}$$

This recovers (2.14) above and follows by enforcing the s.f.c. in (2.16) in the last expression.

We can write the s.f.c. in terms of the one-period ($t \to t+1$) returns, where $S_{t+1} = (1 + r_{t+1}^S)\,S_t$ and $B_{t+1} = (1 + r_B)\,B_t$. Given the positions x_t, y_t at time t, then by (2.14)

$$\Pi_{t+1} = x_t\,(1 + r_{t+1}^S)\,S_t + y_t\,(1 + r_B)\,B_t$$

and the s.f.c. takes the form

$$S_t(1 + r_{t+1}^S)(x_{t+1} - x_t) + B_t(1 + r_B)(y_{t+1} - y_t) = 0\,. \tag{2.17}$$

In particular, assume the rate of return r_{t+1}^S is one of two known values r_S^+ or r_S^-, i.e., two scenarios are possible for one period. Then, given knowledge of S_t, B_t, r_B, x_t, and y_t, the above s.f.c. gives a linear relation between the position in the stock and that in the bond at time $t+1$ for either ($+$ or $-$) stock return scenarios:

$$S_t\,(1 + r_S^{\pm})\,(x_{t+1}^{\pm} - x_t) + B_t\,(1 + r_B)\,(y_{t+1}^{\pm} - y_t) = 0\,, \tag{2.18}$$

where the time $t+1$ portfolio (x_{t+1}, y_{t+1}) can be explicitly denoted by (x_{t+1}^+, y_{t+1}^+) in case

$r_{t+1}^S = r_S^+$ and by (x_{t+1}^-, y_{t+1}^-) for $r_{t+1}^S = r_S^-$. Note that if we are given another independent linear relation between the positions at time $t+1$, then the portfolio (x_{t+1}, y_{t+1}) is uniquely given. The example below describes such a situation where we impose an added independent condition on the positions besides the s.f.c.

Example 2.5. Consider a one-period binomial model with $r_B = 10\%$, $r_S^- = -10\%$, $r_S^+ = 20\%$, $S_0 = \$100$, and $B_0 = \$10$. Construct a self-financing strategy with an initial value of \$1000 such that 50% of the wealth is always invested in risk-free bonds.

Solution. We solve this problem by applying the above formulae for a single period $t = 0$ to $t+1 = 1$. Initially, $\Pi_0 = 1000 = x_0 S_0 + y_0 B_0$, such that $y_0 B_0 = 0.5 \cdot 1000 = 500$. Hence, $y_0 = \frac{500}{B_0} = 50$ units of the bond and $x_0 = \frac{500}{S_0} = 5$ shares of stock, i.e., $(x_0, y_0) = (5, 50)$. At time 1, $S_1 = S_0(1 + r_S)$ and $\Pi_1 = x_1 S_1 + y_1 B_1 = x_1 S_0(1 + r_S) + y_1 B_0(1 + r_B)$. Since our strategy is constrained to have 50% equally invested in the stock and in the bond, then $x_1 S_0(1 + r_S) = y_1 B_0(1 + r_B)$. Combining this relation with the above one-period s.f.c. in (2.18) (with $t = 0$) gives us two linear equations in the two unknowns x_1, y_1:

$$S_0(1 + r_S)x_1 + B_0(1 + r_B)y_1 = \Pi_1,$$
$$S_0(1 + r_S)x_1 - B_0(1 + r_B)y_1 = 0,$$

where $\Pi_1 = x_0 S_0(1 + r_S) + y_0 B_0(1 + r_B)$. The solution is:

$$x_1 = \frac{\Pi_1}{2S_0(1 + r_S)}, \quad y_1 = \frac{\Pi_1}{2B_0(1 + r_B)}.$$

(a) For $r_S = r_S^+ = 0.2$, $S_0(1+r_S) = 120$, $B_0(1+r_B) = 11$, $\Pi_1 = 5 \cdot 120 + 50 \cdot 110 = 1150$, giving the portfolio $(x_1, y_1) = (\frac{115}{24}, \frac{575}{11}) \equiv (x_1^+, y_1^+)$. In this case, $\delta x \equiv x_1 - x_0 = \frac{115}{24} - 5 = -\frac{5}{24}$ and $\delta y \equiv y_1 - y_0 = \frac{575}{11} - 50 = \frac{25}{11}$. Hence, at time $t = 1$, re-balancing involves selling $\frac{5}{24}$ stock shares at price $S_1 = \$120$ and investing the proceeds in buying $\frac{25}{11}$ bonds at price $B_1 = \$110$.

(b) For $r_S = r_S^- = -0.1$, $S_0(1+r_S) = 90$, $\Pi_1 = 5 \cdot 90 + 50 \cdot 110 = 1000$, giving the portfolio $(x_1, y_1) = (\frac{50}{9}, \frac{500}{11}) \equiv (x_1^-, y_1^-)$. Since $\delta x = \frac{50}{9} - 5 = \frac{5}{9}$ and $\delta y = \frac{500}{11} - 50 = -\frac{50}{11}$, we re-balance by buying $\frac{5}{9}$ stock shares at price $S_1 = \$90$ and finance this by selling off $\frac{50}{11}$ bonds at price $B_1 = \$110$.

Finally, note that in both cases we have the s.f.c. $\delta x\, S_1 + \delta y\, B_1 = 0$. □

Let us end this subsection with the definition of an admissible strategy.

Definition 2.2. An *admissible strategy* is a self-financing strategy with nonnegative values for all dates from time zero until maturity.

At any date the holder of an admissible strategy will have no potential liabilities which he or she would not able to honour. In particular, since the terminal value will be nonnegative in any state of the world, the liquidation of the terminal portfolio will not result in a loss.

2.2.5 Log-Normal Pricing Model

The binomial tree model has apparent disadvantages as a discrete-time and discrete-price model. We shall remove these restrictions by passing to the continuous-time limit from the binomial tree model. As a result, we will obtain a continuous model of the stock price $S(T)$. Although $S(T)$ is uncertain at time 0, we assume that the mathematical expectation μ and variance σ^2 of the log-return on the stock over the time interval $[0, T]$ given by

$$\mu := \frac{1}{T} \mathrm{E}\left[\ln \frac{S(T)}{S(0)}\right] \text{ and } \sigma^2 := \frac{1}{T} \mathrm{Var}\left(\ln \frac{S(T)}{S(0)}\right)$$

can be estimated from historical observations.

For each $N \geqslant 1$, consider an N-period binomial tree model with trading dates

$$0, \delta_N, 2\delta_N, \ldots, N\delta_N = T$$

spaced uniformly in $[0, T]$ with the time step $\delta_N := \frac{T}{N}$. Let $\{S_t^{(N)}\,;\, t = 0, 1, \ldots, N\}$ denote the stock price process in the N-period recombining binomial tree model; let the initial price be $S_0 \equiv S(0)$. Recall that the stock prices are given by a product of the initial stock price and single-period returns on the stock S,

$$S_t^{(N)} = S_0 \prod_{k=1}^{t} R_k^{(N)} .$$

The returns $R_k^{(N)}$, $k = 1, 2, \ldots, N$, are i.i.d. random variables having the common probability distribution

$$R_k^{(N)} = \begin{cases} u_N & \text{with probability } p_N, \\ d_N & \text{with probability } 1 - p_N. \end{cases}$$

The next step is to parametrize the binomial tree model. The upward and downward factors, u_N and d_N, can be obtained by matching the first two moments of the stock price returns. In the binomial model, the aggregate log-returns on the stock, $L_{[0,N]}^{(N)} := \ln \frac{S_N^{(N)}}{S_0}$, have the following first two moments:

$$\mathrm{E}\left[L_{[0,N]}^{(N)}\right] = N\,\mathrm{E}\left[L_1^{(N)}\right] = N\,\left(\ln(u_N)p_N + \ln(d_N)(1 - p_N)\right), \tag{2.19}$$

$$\mathrm{Var}\left[L_{[0,N]}^{(N)}\right] = N\,\mathrm{Var}\left(L_1^{(N)}\right) = N\,\left(\ln(u_N)^2 p_N + \ln(d_N)^2(1 - p_N)\right) . \tag{2.20}$$

Equating the respective moments of the log-returns $\ln \frac{S(T)}{S(0)}$ and $L_{[0,N]}^{(N)}$ gives us two equations:

$$\ln(u_N)p_N + \ln(d_N)(1 - p_N) = \mu\,\delta_N \tag{2.21}$$

$$\ln(u_N)^2 p_N + \ln(d_N)^2(1 - p_N) = \sigma^2\,\delta_N. \tag{2.22}$$

Note that there are three unknowns, i.e., u_N and d_N and the probability p_N. Hence, we need a third equation. A convenient choice is the symmetry condition

$$u_N \cdot d_N = 1. \tag{2.23}$$

The following solution is the most commonly used in binomial models:

$$u_N = e^{\sigma\sqrt{\delta_N}}, \tag{2.24}$$

$$d_N = e^{-\sigma\sqrt{\delta_N}}, \tag{2.25}$$

$$p_N = \frac{1}{2} + \frac{1}{2}\frac{\mu}{\sigma}\sqrt{\delta_N}. \tag{2.26}$$

Under this parametrization, the log-returns on the stock have the following properties:

$$\mathrm{E}\left[L_{[0,N]}^{(N)}\right] = N\,\mu\,\delta_N = \mu T, \tag{2.27}$$

$$\mathrm{Var}\left[L_{[0,N]}^{(N)}\right] = N\,\sigma^2\,\delta_N + N\mu^2\,\delta_N^2 = \sigma^2 T + (\mu T)^2/N\,. \tag{2.28}$$

That is, the solution (2.24)–(2.26) satisfies (2.19)–(2.20) up to $\mathcal{O}\left(\delta t\right)$. In other words, the binomial tree models we constructed here conserve the expected log-return on the stock and asymptotically (as $N \to \infty$) conserve the variance of the log-return.

The binomial price $S_N^{(N)}$ at the end of the Nth period is a discrete random variable taking on a value in the set $\{S_{N,k} = S_0 u_N^k d_N^{N-k}\ ;\ k = 0, 1, \ldots, N\}$. As the number of periods N increases to ∞, the length δ_N of one period approaches zero. Moreover, the density of the points $S_{N,k}$ increases and their range expands. Let us find the limiting distribution of the binomial prices. At the maturity date T, we define the limiting asset price:

$$S(T) := \lim_{N\to\infty} S_N^{(N)} = \lim_{N\to\infty} S(0)\exp\left(L_{[0,N]}^{(N)}\right).$$

To understand the distribution of the limiting price $S(T)$, we take a closer look at the distribution of the log-return $L_{[0,N]}^{(N)} = L_1^{(N)} + L_2^{(N)} + \cdots + L_N^{(N)}$. The one-step log-returns $L_k^{(N)}$, $k = 1, 2, \ldots, N$, are i.i.d. random variables. For each value of N, we introduce a sequence of i.i.d. Bernoulli random variables $X_k^{(N)}$, $k = 1, 2, \ldots, N$, having the following two-point probability distribution:

$$X_k^{(N)} = \begin{cases} 1 & \text{with probability } p_N, \\ 0 & \text{with probability } 1 - p_N. \end{cases}$$

We have that $X_k^{(N)} = 1$ and $1 - X_k^{(N)} = 0$ with probability p_N, and $X_k^{(N)} = 0$ and $1 - X_k^{(N)} = 1$ with probability $1 - p_N$. Therefore, we can express the log-return $L_k^{(N)}$ in terms of $X_k^{(N)}$ as follows:

$$L_k^{(N)} = \ln(u_N)\, X_k^{(N)} + \ln(d_N)\,(1 - X_k^{(N)}) = \ln\left(\frac{u_N}{d_N}\right) X_k^{(N)} + \ln(d_N)\,.$$

As a result, the aggregate log-return is given by

$$L_{[0,N]}^{(N)} = \sum_{k=1}^{N} L_k^{(N)} = \ln\left(\frac{u_N}{d_N}\right) \sum_{k=1}^{N} X_k^{(N)} + N\ln(d_N)\,.$$

Denote $Y_N := \sum_{k=1}^N X_k^{(N)}$. A sum Y_N of N i.i.d. Bernoulli variables has the binomial probability distribution: $Y_N \sim \mathit{Bin}(N, p_N)$. By the de Moivre–Laplace Theorem 2.1 (provided below), for large N the distribution of Y_N is approximately normal with mean Np_N and variance $Np_N(1 - p_N)$. Therefore, for large N, the distribution of the log-return $L_{[0,N]}^{(N)} = \ln(u_N/d_N)Y_N + N\ln(d_N)$ is also approximately normal with mean μT and variance $\sigma^2 T$ (see formulae (2.27) and (2.28)). In the limiting case, we obtain that the probability distribution of $\lim_{N\to\infty} L_{[0,N]}^{(N)}$ is $\mathit{Norm}(\mu T, \sigma^2 T)$. Thus, the limiting stock price $S(T) = \lim_{N\to\infty} S(0)\exp(L_{[0,N]}^{(N)})$ has the *log-normal probability distribution* and admits the following representation:

$$S(T) = S(0)\, \mathrm{e}^{\mu T + \sigma\sqrt{T} Z}, \quad \text{where } Z \sim \mathit{Norm}(0, 1). \tag{2.29}$$

The parameter μ is called the *drift parameter*; σ is the *volatility parameter*.

Here and below, $\mathit{Norm}(a, b^2)$ denotes the normal probability distribution with mean a and variance b^2. The cumulative distribution function (CDF) of the standard normal distribution $\mathit{Norm}(0, 1)$, denoted $\mathcal{N}$ (or Φ in some other texts), is

$$\mathcal{N}(z) := \frac{1}{\sqrt{2\pi}} \int_{-\infty}^{z} \mathrm{e}^{-x^2/2}\, \mathrm{d}x. \tag{2.30}$$

If Z is a standard normal variate, then for any real a and b, the random variable $a + bZ$ has the *Norm*(a, b^2) probability distribution. Hence, the CDF F of *Norm*(a, b^2) is

$$F(z) = \mathbb{P}(a + bZ \leqslant z) = \mathbb{P}\left(Z \leqslant \frac{z-a}{b}\right) = \mathcal{N}\left(\frac{z-a}{b}\right) = \frac{1}{\sqrt{2\pi}b}\int_{-\infty}^{z} \mathrm{e}^{-\frac{(x-a)^2}{2b^2}}\,\mathrm{d}x\,.$$

A rigorous justification of (2.29) is based on the following theorem, which we are giving without a proof.

Theorem 2.1 (Moivre–Laplace). *Consider a sequence $\{p_n\}_{n\geqslant 1}$ in $(0,1)$ that converges to $p \in (0,1)$ as $n \to \infty$. Let $\{Y_n\}_{n\geqslant 1}$ be a sequence of independent binomial random variables with $Y_n \sim Bin(n, p_n)$. Then the sequence of rescaled (normalized) random variables*

$$Y_n^* := \frac{Y_n - \mathrm{E}[Y_n]}{\sqrt{\mathrm{Var}(Y_n)}} = \frac{Y_n - np_n}{\sqrt{np_n(1-p_n)}}$$

converges weakly (in distribution) to a standard normal variable.

Recall that a sequence of random variables $\{Y_n\}_{n\geqslant 1}$ converges in distribution to another random variable Y, denoted $Y_n \xrightarrow{d} Y$, as $n \to \infty$, if the CDF's of Y_n converge to the CDF of Y, as $n \to \infty$, i.e., for almost all $x \in \mathbb{R}$, we have $F_{X_n}(x) \to F_X(x)$, as $n \to \infty$.

Corollary 2.2. *Suppose that a sequence $\{Y_n\}_{n\geqslant 1}$ converges weakly to a standard normal random variable: $Y_n \xrightarrow{d} \mathrm{Norm}(0,1)$, as $n \to \infty$. Consider two converging sequences of real numbers: $a_n \to a$ and $b_n \to b \neq 0$, as $n \to \infty$. Then*

$$a_n + b_n Y_n \xrightarrow{d} \mathrm{Norm}(a, b^2), \quad \textit{as } n \to \infty.$$

We have a sequence of binomial random variables $Y_N \sim Bin(N, p_N)$, where the probability $p_N = \frac{1}{2} + \frac{1}{2}\frac{\mu\sqrt{T}}{\sigma\sqrt{N}} \to \frac{1}{2}$, as $N \to \infty$. Therefore, by Theorem 2.1,

$$\frac{Y_N - \mathrm{E}[Y_N]}{\sqrt{\mathrm{Var}(Y_N)}} = \frac{Y_N - Np_N}{\sqrt{Np_N(1-p_N)}} \xrightarrow{d} \mathrm{Norm}(0,1), \text{ as } N \to \infty.$$

On the other hand, for the log-returns $L_N \equiv L_{[0,N]}^{(N)}$, we have the identity

$$\frac{L_N - \mathrm{E}[L_N]}{\sqrt{\mathrm{Var}(L_N)}} = \frac{Y_N - \mathrm{E}[Y_N]}{\sqrt{\mathrm{Var}(Y_N)}} \tag{2.31}$$

the proof of which is left as an exercise for the reader. Thus, $L_N^* := \frac{L_N - \mathrm{E}[L_N]}{\sqrt{\mathrm{Var}(L_N)}} \to Norm(0,1)$, as $N \to \infty$. Since we can express L_N in terms of L_N^*,

$$L_N = \mathrm{E}[L_N] + L_N^*\sqrt{\mathrm{Var}(L_N)} = \mu T + L_N^*\sqrt{\sigma^2 T + (\mu T)^2/N},$$

and $\sqrt{\sigma^2 T + (\mu T)^2/N} \to \sigma\sqrt{T}$, as $N \to \infty$, then, by the corollary, we have

$$L_N \xrightarrow{d} Norm(\mu T, \sigma^2 T), \text{ as } N \to \infty\,.$$

To summarize, in the binomial tree model, the stock price $S_N^{(N)}$ is a **discrete** random variable, which is a function of a binomial random variable. In the log-normal model, the stock price $S(T)$ is a **continuous** random variable having the log-normal probability distribution. As seen in Figure 2.3, the shape of the probability distribution of binomial prices is close to that of log-normal prices.

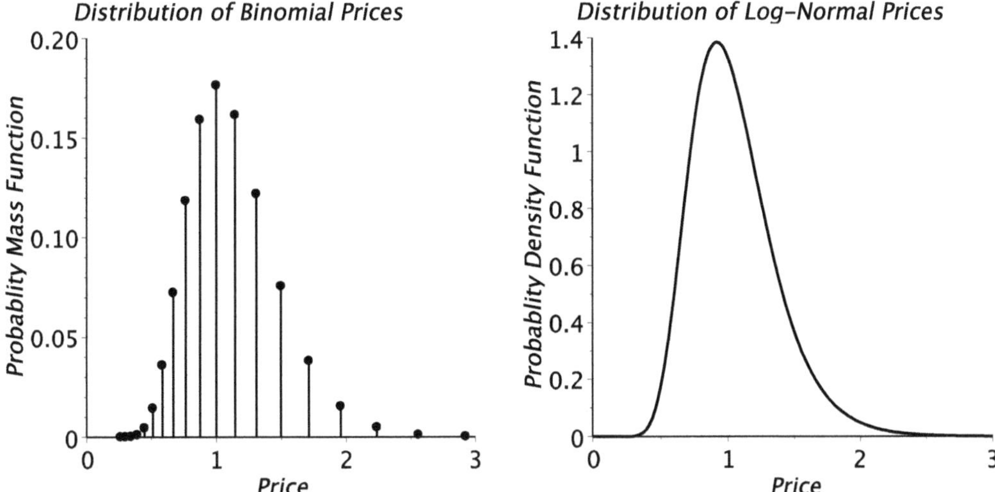

FIGURE 2.3: The probability distributions of asset prices in the binomial tree model (a) and log-normal model (b). The initial price is $S_0 = 1$; the time to maturity is $T = 1$; the binomial tree model has $N = 20$ periods; the model parameters are $\mu = 1\%$ and $\sigma = 30\%$.

Example 2.6 (The log-normal distribution)**.** Find the cumulative distribution function (CDF) and the probability density function (PDF) of the log-normal price

$$S(T) = S_0 \mathrm{e}^{\mu T + \sigma\sqrt{T}Z} \text{ with } T > 0,$$

where $Z \sim Norm(0, 1)$.

Solution. First, obtain the CDF F of $S(T)$:

$$\begin{aligned} F(s) &= \mathbb{P}(S(T) \leqslant s) = \mathbb{P}\left(S_0 \mathrm{e}^{\mu T + \sigma\sqrt{T}Z} \leqslant s\right) \\ &= \mathbb{P}\left(\mu T + \sigma\sqrt{T}Z \leqslant \ln(s/S_0)\right) = \mathbb{P}\left(Z \leqslant \frac{\ln(s/S_0) - \mu T}{\sigma\sqrt{T}}\right) \\ &= \mathcal{N}\left(\frac{\ln(s/S_0) - \mu T}{\sigma\sqrt{T}}\right) \quad \text{for} \quad s > 0, \end{aligned}$$

where $\mathcal{N}$ is the *standard normal* CDF defined by (2.30): $\mathcal{N}(z) = \frac{1}{\sqrt{2\pi}} \int_{-\infty}^{z} \mathrm{e}^{-x^2/2}\,\mathrm{d}x$. The standard normal PDF is given by $\mathcal{N}'(z) = \frac{1}{\sqrt{2\pi}}\mathrm{e}^{-z^2/2}$. Differentiating the CDF $F(s)$ w.r.t. s gives us the PDF f of the log-normal price:

$$\begin{aligned} f(s) = F'(s) &= \mathcal{N}'\left(\frac{\ln(s/S_0) - \mu T}{\sigma\sqrt{T}}\right) \cdot \frac{\mathrm{d}}{\mathrm{d}s}\left(\frac{\ln(s/S_0) - \mu T}{\sigma\sqrt{T}}\right) \\ &= \frac{1}{s\sigma\sqrt{2\pi T}} \mathrm{e}^{-(\ln(s/S_0) - \mu T)^2/(2\sigma^2 T)}, \quad s > 0. \end{aligned}$$

□

2.3 Arbitrage and Risk-Neutral Pricing

An *arbitrage opportunity* arises when someone can buy an asset at a low price to immediately sell it for a higher price. For example, such a combination of matching deals can be done by taking advantage of an asset price difference between two or more markets. Both buying in one market and selling on the other must occur simultaneously to avoid exposure to any type of market risk. In practice, such simultaneous transactions are only possible with assets and financial products which can be traded electronically. The prices should not change before both transactions are complete; the cost of transport, storage, transaction, or insurance should not eliminate the arbitrage opportunity. In other words, arbitrage is a risk-free opportunity of gaining money.

A trader who engages in arbitrage is called an *arbitrageur.* Arbitrage opportunities are often hard to come by, due to transaction costs, the costs involved with finding an arbitrage opportunity, and the number of people who are also looking for such opportunities. Arbitrage profits are generally short-lived, as the buying and selling of assets will change the price of those assets in such a way as to eliminate that arbitrage opportunity. This is particularly the case in an efficient market.

Arbitrageurs can often be found in currency markets. Such financial markets have the advantage of being quite liquid, so we do not take the risk of acquiring an asset that may take some time to sell. The transaction costs are minimal for large currency transactions. Since foreign currency markets are an ideal environment for arbitrageurs, arbitrage opportunities tend to be very limited, as any discrepancies in exchange rates tend to be corrected quite quickly by investors trying to exploit those differences.

While arbitrage opportunities may exist in financial markets, in what follows we assume that the financial models we deal with do not allow for arbitrage. All asset prices are equilibrium prices and all arbitrage opportunities are eliminated. We are going to develop a non-arbitrage pricing theory. In a market model that admits arbitrage, wealth can be created from nothing. Thus, it is reasonable to assume that the financial model of consideration does not admit arbitrage opportunities. Let us start with a basic definition of arbitrage without reference to any model.

Definition 2.3. An *arbitrage opportunity* is a trading strategy that costs nothing to begin with (i.e., zero initial capital is used to set up) and has no chance of incurring any loss, but has a nonzero chance of making a gain.

This is reminiscent of a free lottery ticket. One of the fundamental properties of a good mathematical model for a financial market is that it does not allow for arbitrage.

2.3.1 The Law of One Price

As is mentioned in Section 2.2, replication is a key to pricing derivatives. Indeed, the following theorem states that if the future values of any two assets (or two portfolios of base assets) at some time are equal to each other in all possible market scenarios, then the present prices of these assets must be the same as well. Therefore, being given a derivative security which can be replicated by a portfolio or trading strategy in base assets (it means that the future values of the derivative and the portfolio are equal in all market states), the initial price of the derivative has to be equal to the initial value of the replicating portfolio in the absence of arbitrage.

Theorem 2.3 (Law of One Price). *Assume that the market is arbitrage free. Let there be two assets X and Y whose respective initial prices are X_0 and Y_0. Suppose at some time $T > 0$ the prices of X and Y are equal in all states of the world: $X_T(\omega) = Y_T(\omega)$, for all states $\omega \in \Omega$. Then $X_0 = Y_0$.*

Proof. We shall show that if $X_0 \neq Y_0$, then there exists an arbitrage. Without loss of generality, we suppose that $X_0 > Y_0$ (if $X_0 < Y_0$ then we may just relabel X and Y). Let us construct an arbitrage portfolio in these two assets. Starting with \$0, we first borrow and sell one unit of X and realize $\$X_0$. We then buy one unit of Y, costing us $\$Y_0$. Both transactions give us a positive amount $\$(X_0 - Y_0)$, which we can keep in cash or invest in a risk-free asset. So at time zero we have a portfolio of one unit of Y, negative one unit of X, and the cash amount of $\$(X_0 - Y_0)$. The initial value of this portfolio is zero:

$$\Pi_0 = -X_0 + Y_0 + (X_0 - Y_0) = 0\,.$$

Note that this portfolio requires no initial investment.

At time T, we sell the unit of Y to obtain $\$Y_T$. We buy and return the unit of X. This costs $\$X_T$. Since $X_T = Y_T$, the net cost of these two trades is zero. However, we still have the positive cash amount $X_0 - Y_0$ (and possible interest earned), and hence we have exhibited an arbitrage opportunity:

$$\Pi_T = -X_T + Y_T + X_0 - Y_0 = X_0 - Y_0 > 0\,.$$

Therefore, to eliminate the arbitrage, we must have $X_0 = Y_0$. □

In this proof we have assumed that there are no transaction costs in carrying out the trades required, short sells are allowed, and the assets involved can be bought and sold at any time at will.

2.3.2 A First Look at Arbitrage in the Single-Period Binomial Model

In the single-period binomial model of a financial market considered in the previous section, an arbitrage strategy simply reduces to an arbitrage portfolio. Consider a portfolio (x, y) in stock S and bond B with initial value $\Pi_0 = \Pi_0^{(x,y)} = xS_0 + yB_0$ and terminal values $\Pi_T(\omega^\pm) = \Pi_T^{(x,y)}(\omega^\pm) = xS_T(\omega^\pm) + yB_T \equiv xS^\pm + yB_T$. The above definition of arbitrage hence implies that (x, y) will be an arbitrage portfolio when the following conditions are met:

(a) $\Pi_0 = 0$,

(b) $\Pi_T(\omega^\pm) \geqslant 0$, and $\Pi_T(\omega^+) > 0$ or $\Pi_T(\omega^-) > 0$.

Note that condition (b) can be stated using probabilities:

(b) $\mathbb{P}(\Pi_T \geqslant 0) = 1$ and $\mathbb{P}(\Pi_T > 0) > 0$.

Hence, the single-period binomial model admits an arbitrage iff there exists a portfolio (x, y) satisfying conditions (a) and (b). As Theorem 2.4 shows, there is no such arbitrage portfolio (x, y) when the return on the bond falls strictly in between the higher and lower returns on the stock.

Theorem 2.4 (Arbitrage: Single-period binomial model). *The single-period binomial model admits no arbitrage iff* $r_S^- < r_B < r_S^+$.

Proof. First, note that condition $r_S^- < r_B < r_S^+$ is equivalent to $S^- < (1 + r_B)S_0 < S^+$. So we can formulate our proof using either returns or prices.

- Consider any zero-cost, nontrivial portfolio (x, y), i.e., $\Pi_0^{(x,y)} = xS_0 + yB_0 = 0$. This implies that the portfolio (x, y) has either form $(x, -xS_0/B_0)$ with $x > 0$ or $(-yB_0/S_0, y)$ with $y > 0$. The first portfolio type corresponds to buying the stock and borrowing money while the second is a portfolio short in stock and positively invested in a bond.

- For a portfolio of the form $(x, -xS_0/B_0)$, we have: $\Pi_T^{(x,y)}(\omega^\pm) = x[S^\pm - (1+r_B)S_0]$. In the worst case scenario $\Pi_T^{(x,y)}(\omega^-) = x[S^- - (1+r_B)S_0]$. Hence $S^- \geqslant (1+r_B)S_0$ implies $\Pi_T^{(x,y)}(\omega^\pm) \geqslant 0$ and $\Pi_T(\omega^+) = x[S^+ - (1+r_B)S_0] > 0$, since $S^- < S^+$. Therefore, such a portfolio is an arbitrage unless $S^- < (1+r_B)S_0$.

- Similarly, for the second type we have $\Pi_T^{(x,y)}(\omega^\pm) = y(B_0/S_0)[(1+r_B)S_0 - S^\pm]$. In the best case scenario $\Pi_T^{(x,y)}(\omega^+) = y(B_0/S_0)[(1+r_B)S_0 - S^+]$. Hence $S^+ \leqslant (1+r_B)S_0$ implies $\Pi_T^{(x,y)}(\omega^\pm) \geqslant 0$ and $\Pi_T^{(x,y)}(\omega^-) = y(B_0/S_0)[(1+r_B)S_0 - S^-] > 0$, since $S^- < S^+$. The portfolio is an arbitrage unless $S^+ > (1+r_B)S_0$.

Hence, by combining the two cases, we have shown that there is no arbitrage iff

$$S^- < (1+r_B)S_0 < S^+. \qquad \square$$

Note that the result of Theorem 2.4 can be formulated in terms of asset values: there is no arbitrage iff $\frac{S^-}{S_0} < \frac{B_T}{B_0} < \frac{S^+}{S_0}$. If market prices do not allow for a profitable arbitrage, then the prices are said to constitute an arbitrage-free market. Later we shall see that the assumption that there is no arbitrage is used in quantitative finance to calculate unique (no-arbitrage) prices for derivatives that can be replicated.

In conclusion, let us demonstrate that if the initial price C_0 of a claim is equal to the initial cost Π_0 of the replicating portfolio in (2.5), then there is no arbitrage. In other words, if the actual initial price is less than or greater than the price in (2.5), then there is an arbitrage portfolio in the base assets B and S and the claim C. One way to argue this statement is to apply the Law of One Price. Indeed, since the payoff C_T is identical to that of the replicating portfolio, Π_T, the present values C_0 and Π_0 have to be the same, or else an arbitrage exists. On the other hand, we can always form an arbitrage portfolio when $C_0 \neq \Pi_0$, as demonstrated in the following example.

Example 2.7. Consider a single-period binomial model with $r_B = 10\%$, $r_S^- = -10\%$, $r_S^+ = 20\%$, $S_0 = \$100$, and $B_0 = \$10$. Contract C has the following payoff: $C^- = 0$ and $C^+ = \$50$.

(a) Find portfolio (x, y) that replicates the payoff (C^-, C^+) and then calculate its initial cost $\Pi_0 = \Pi_0^{(x,y)}$.

(b) Suppose that $C_0 > \Pi_0^{(x,y)}$, where (x, y) is the portfolio replicating the contract. Construct an arbitrage portfolio.

Solution. Applying (2.4) gives us the replicating portfolio:

$$x = \frac{50 - 0}{100 \cdot (0.2 - (-0.1))} = \frac{5}{3} \text{ and } y = \frac{0 \cdot (1+0.2) - 50 \cdot (1 - 0.1)}{10 \cdot (1+0.1) \cdot (0.2 - (-0.1))} = -\frac{150}{11}.$$

The initial value of the portfolio is

$$\Pi_0^{(x,y)} = xS_0 + yB_0 = \frac{5}{3} \cdot 100 - \frac{150}{11} \cdot 10 = \frac{1000}{33} \cong \$30.303\,.$$

Suppose that the actual price of the contract C_0 is larger than Π_0. Write and sell the contract for C_0. Use the proceeds to buy $x = \frac{5}{3}$ shares of stock. If $C_0 < xS_0 = \frac{500}{3}$, then borrow $\frac{50}{3} - \frac{C_0}{10}$ bonds; otherwise, we invest the balance in bonds. So the portfolio (x, y, z) contains $x = \frac{5}{3}$ shares of stock S, $y = \frac{C_0}{10} - \frac{50}{3}$ bonds B, and $z = -1$ contracts C. The initial cost is zero. At the end of the period, we sell stock and pay C_T to the holder of the contract. The balance is positive for every market scenario whenever $C_0 > \frac{1000}{33}$:

$$\Pi_T^{(x,y,z)} = \underbrace{\frac{5}{3}S_T - \frac{150}{11}B_T - C_T}_{=0} + \left(\frac{150}{11} - \frac{50}{3} + \frac{C_0}{10}\right) B_T = \frac{11}{10}\left(C_0 - \frac{1000}{33}\right). \qquad \square$$

2.3.3 Arbitrage in the Binomial Tree Model

In a multi-period model, there is more flexibility for an investment portfolio. The investor may alter the positions in the portfolio at any time by selling some assets and investing the proceeds in others. Therefore, the definition of an arbitrage investment strategy is a bit different in the comparison with the definition of an arbitrage portfolio for the single-period case.

Definition 2.4. An admissible strategy such that $\Pi_0 = 0$ and $\mathbb{P}(\Pi_t > 0) > 0$ for some $t = 1, 2, \ldots$ is called an *arbitrage* (strategy).

The cost of setting up an arbitrage strategy is zero. The self-financing condition means that there are no injections of funds at intermediate dates. The admissibility guarantees that the holder will not face a potential loss. At the maturity date, there is no loss since the terminal value is nonnegative. Moreover, there exists at least one scenario where liquidating the portfolio will result in a positive gain. In summary, an arbitrage strategy is a possibility of having a potential gain at no cost and without potential losses. Since the wealth of an admissible strategy is always nonnegative, the definitions of an arbitrage portfolio and an arbitrage strategy are the same for the single-period binomial model.

Theorem 2.5. *The binomial tree model admits no arbitrage iff $d < 1 + r < u$.*

Proof. First, consider the case of a one-period binomial tree (i.e., $T = 1$). As was justified in Theorem 2.4, the rate r of interest on a risk-free investment has to satisfy $d < 1 + r < u$, or else an arbitrage possibility would arise. Indeed, for the one-step returns on the stock we have $1 + r_S^- = d$ and $1 + r_S^+ = u$; the one-step return on the bond is r. There is no arbitrage iff $r_S^- < r_B < r_S^+$, which is equivalent to $d < 1 + r < u$.

Now let us consider a multi-period binomial model. Suppose that $1 + r \leqslant d$ or $u \leqslant 1 + r$ holds. To construct an arbitrage portfolio, we proceed as follows. At time 0, construct a portfolio which is long in stock if $1 + r \leqslant d$ or short in stock if $u \leqslant 1 + r$ with zero initial value. At time 1, close the position in stock and invest the proceeds in risk-free bonds. As a result of these manipulations, we obtain a positive amount of cash invested in bonds.

Let $d < 1 + r < u$ and suppose that there is an arbitrage strategy, i.e., there is a self-financing strategy with zero initial value such that $\Pi_t \geqslant 0$ for all $t \geqslant 0$ with probability 1 and $\Pi_T > 0$ with nonzero probability at maturity time T. Find the smallest time $t > 0$ for which $\Pi_t(\omega) > 0$ at some state ω. Since each state in the model is a path in the binomial tree, we can find a one-step subtree with two branches, so that $\Pi_{t-1} = 0$ at its root, $\Pi_t \geqslant 0$ at each node growing out of this root with $\Pi_t > 0$ in at least one of these nodes. Note that the path ω is passing through the root and the node where $\Pi_t > 0$. In the one-step case this is impossible if $d < 1 + r < u$, leading to a contradiction. $\square$

2.3.4 Risk-Neutral Probabilities

Although the future prices of stock are unknown with certainty, it is natural to compare the expected return on stock and the risk-free rate of return. The expected stock prices under the real-world probability function $\mathbb{P}$ are given by

$$\mathrm{E}[S_t] = S_0(1 + \mathrm{E}[r_S])^t\,, \quad t = 0, 1, 2, \ldots \tag{2.32}$$

where r_S denotes a one-period rate of return on the stock. Since $r_S = u-1$ with probability p and $r_S = d-1$ with probability $1-p$, we obtain

$$\mathrm{E}[r_S] = p(u-1) + (1-p)(d-1) = pu + (1-p)d - 1\,.$$

Suppose that the amount S_0 is invested in a risk-free bank account. It will grow to $S_0(1+r)^t$ after t steps, where r is the compound risk-free interest rate. Clearly, to compare the expected return on the stock, $\mathrm{E}[S_t/S_0]$, and the risk-free return, $(1+r)^t$, we only need to compare the average one-step rate $\mathrm{E}[r_S]$ on the stock and the risk-free rate r. There exist three main types of investors.

- A typical *risk-averse* investor requires that $\mathrm{E}[r_S] > r$, arguing that she or he should be rewarded with a higher expected return as a compensation for risk.
- A *risk-seeker* investor may be attracted by the reverse situation when $\mathrm{E}[r_S] < r$, if a risky return is very high with small nonzero probability and low with large probability.
- A border situation of a market in which $\mathrm{E}[r_S] = r$ is referred to as *risk-neutral.*

We now introduce a new probability function $\widetilde{\mathbb{P}}$ with the probabilities of one-period upward and downward moves of the stock price $\widetilde{\mathbb{P}}(\text{up}) = \tilde{p}$ and $\widetilde{\mathbb{P}}(\text{down}) = 1-\tilde{p}$, respectively, such that the risk-neutrality condition

$$\widetilde{\mathrm{E}}[r_S] = \tilde{p}u + (1-\tilde{p})d - 1 = r \tag{2.33}$$

is satisfied. This implies that

$$\tilde{p} = \frac{1+r-d}{u-d} \text{ and } 1-\tilde{p} = \frac{u-r-1}{u-d}\,. \tag{2.34}$$

We shall call $\tilde{p}$ and $1-\tilde{p}$ the *risk-neutral probabilities* of the stock price upward and downward moves, respectively. The corresponding probability function is called the risk-neutral probability function (or measure) and is denoted by $\widetilde{\mathbb{P}}$. Here $\widetilde{\mathrm{E}}$ denotes the mathematical expectation with respect to the probability function $\widetilde{\mathbb{P}}$; it is called the *risk-neutral expectation.*

Theorem 2.6. *The binomial tree model admits no arbitrage iff there exists the risk-neutral probability* $\tilde{p} \in (0,1)$.

Proof. It is clear from (2.34) that $0 < \tilde{p} < 1$ iff $d < 1+r < u$. The latter is a necessary and sufficient condition of the absence of arbitrage. □

To explain why $\tilde{p}$ is called a risk-neutral probability, we are going to compare the real-world expected return $\mathrm{E}[r_S]$ and the risk-neutral expected return $\widetilde{\mathrm{E}}[r_S] = r$. Let us define the risk of the investment in the stock to be the standard deviation of the one-step return r_S:

$$\sigma_S = \sqrt{\mathrm{Var}(r_S)} = \sqrt{\mathrm{E}[r_S^2] - (\mathrm{E}[r_S])^2}\,.$$

This parameter is often called the *volatility* of stock price return. It follows that

$$\sigma_S^2 = \mathrm{Var}(r_S) = (u-d)^2 p(1-p). \tag{2.35}$$

Let us compare the expected returns $\mathrm{E}[r_S]$ and $\widetilde{\mathrm{E}}[r_S]$:

$$\mathrm{E}[r_S] - \widetilde{\mathrm{E}}[r_S] = up + d(1-p) - u\tilde{p} - d(1-\tilde{p}) = (p-\tilde{p})(u-d). \tag{2.36}$$

Let us assume that $\mathrm{E}[r_S] \geqslant r$, that is, the expected return is not less than the risk-free return. Combining (2.35) with (2.36) gives

$$\mathrm{E}[r_S] - r = \frac{p-\tilde{p}}{\sqrt{p(1-p)}}\sigma_S\,.$$

We say that one asset is riskier than another when it has a higher volatility of return. If the volatility is zero (i.e., we deal with a risk-free asset), then the expected return is just r; if the volatility is nonzero, then we have a higher expected return. This result fits well with reality—if you want a higher expected return you must take on more risk. However, when $p = \tilde{p}$, i.e., we deal with a risk-neutral market, the expected return is always r no matter what value the volatility σ has.

In reality, the risk-neutral probability $\tilde{p}$ has no relation to the real-world probability p. However, the risk-neutral probability function is of great practical importance to us with respect to computing no-arbitrage prices of derivative contracts. Let us consider a single-period binomial model. The fair price C_0 of a derivative contract with payoffs $C^{\pm} = C_T(\omega^{\pm})$ in the two possible outcomes $\omega^{\pm}$ is given by (2.5). By using the notation $r_B = r$, $r_S^+ = u-1$, and $r_S^- = d-1$, we can rewrite (2.5) as follows:

$$C_0 = \frac{1}{1+r}\left[\frac{1+r-d}{u-d}C^+ + \frac{u-r-1}{u-d}C^-\right].$$

Substituting (2.34) in the above equation gives us a simple valuation formula:

$$C_0 = \frac{1}{1+r}\left[\tilde{p}\,C^+ + (1-\tilde{p})\,C^-\right] = \frac{1}{1+r}\widetilde{\mathrm{E}}\left[C_T\right] = \frac{B_0}{B_T}\widetilde{\mathrm{E}}\left[C_T\right]. \tag{2.37}$$

In other words, the no-arbitrage price of claim C is given by a risk-neutral expectation of the discounted future payoff function. The discounting factor is $\frac{B_0}{B_T} = \frac{1}{1+r_B}$. The interesting fact is that this formula works for more sophisticated models and general payoff functions.

2.3.5 Martingale Property

Equation (2.37) can be rewritten as follows:

$$\widetilde{\mathrm{E}}\left[\frac{C_T}{B_T}\right] = \frac{C_0}{B_0}. \tag{2.38}$$

In this case, the process $\left\{\frac{C_t}{B_t}\right\}_{t\in\{0,T\}}$ is said to be a *martingale* under the risk-neutral probability function $\widetilde{\mathbb{P}}$. Now we proceed to the multi-period case.

It follows from (2.32) that the expectation of S_t with respect to the risk-neutral probability function $\widetilde{\mathbb{P}}$ is

$$\widetilde{\mathrm{E}}[S_t] = S_0(1+r)^t, \tag{2.39}$$

since $\widetilde{\mathrm{E}}[r_S] = r$. In other words, the expected return on the stock under $\widetilde{\mathbb{P}}$ is equal to the risk-free return over the same time interval from 0 to t.

Equation (2.39) can be extended to any time step in the binomial tree model. Suppose that t time steps have passed and the stock price has changed from S_0 to S_t. Let us find the risk-neutral expectation of the price S_{t+1} given the price S_t. Formally, we need to find the *conditional expectation* of S_{t+1} given S_t. We can write $S_{t+1} = S_t\, R^S_{t+1}$. Since in the binomial tree model all single-period returns are i.i.d., the distribution of the return R^S_{t+1} does not depend on the time t. The risk-neutral expectation of R^S_{t+1} is equal to $1+r$. The expectation of S_{t+1} given S_t is

$$\widetilde{\mathrm{E}}[S_{t+1} \mid S_t] = \widetilde{\mathrm{E}}[S_t\, R^S_{t+1} \mid S_t] = S_t\widetilde{\mathrm{E}}[R^S_{t+1} \mid S_t] = S_t\widetilde{\mathrm{E}}[R^S_{t+1}] = S_t\,(1+r). \tag{2.40}$$

The above derivation is based on the following two facts. First, S_t is given and hence can be taken out of the expectation. Second, since stock returns are mutually independent, R^S_{t+1} is independent of S_t and hence the last expectation becomes an unconditional one. Now we introduce *discounted stock prices*:

$$\overline{S}_t = \frac{S_t}{B_t} = \frac{S_t}{B_0(1+r)^t}, \quad t \geqslant 0 .$$

Dividing both sides of (2.40) by the bond price B_{t+1} and using $B_{t+1} = B_t\,(1+r)$ give

$$\widetilde{\mathrm{E}}\left[\frac{S_{t+1}}{B_{t+1}} \mid S_t\right] = \frac{S_t(1+r)}{B_{t+1}} = \frac{S_t(1+r)}{B_t(1+r)} = \frac{S_t}{B_t} .$$

Thus, (2.40) takes the form

$$\widetilde{\mathrm{E}}\left[\overline{S}_{t+1} \,|\, \overline{S}_t\right] = \overline{S}_t, \quad t \geqslant 0 \tag{2.41}$$

We say that the discounted stock price process $\{\overline{S}_t\}_{t\geqslant 0}$ is a *martingale* under the risk-neutral probability measure $\widetilde{\mathbb{P}}$.

2.3.6 Risk-Neutral Log-Normal Model

In Subsection 2.2.5, we derived the log-normal price model as the limiting case of a sequence of binomial tree models as the number of periods N approaches infinity. Let r be the risk-free interest rate under continuous compounding. The equivalent one-period interest rate r_N in the binomial tree model with N periods is $r_N = \mathrm{e}^{r\delta_N} - 1$, where $\delta_N = \frac{T}{N}$. As was demonstrated in Subsection 2.2.5, the log-return $L^{(N)}_{[0,N]} = \ln(S^{(N)}_N / S_0)$ is approximately normal, as $N \to \infty$. Let us find the parameters of the limiting normal distribution under the risk-neutral probability measure. In the N-period model, the risk-neutral probability of the upward movement of the stock price is

$$\tilde{p}_N = \frac{r_N + 1 - d_N}{u_N - d_N} = \frac{\mathrm{e}^{\delta_N r} - \mathrm{e}^{-\sqrt{\delta_N}\sigma}}{\mathrm{e}^{\sigma\sqrt{\delta_N}} - \mathrm{e}^{-\sigma\sqrt{\delta_N}}} .$$

Introduce the risk-neutral probability function $\widetilde{\mathbb{P}}_N$ with the probability $\tilde{p}_N$ of an upward movement. Under this probability function, the normalized log-return, $L^*_N := \frac{L_N - \mathrm{E}[L_N]}{\sqrt{\mathrm{Var}(L_N)}}$, where $L_N \equiv L^{(N)}_{[0,N]}$, is expressed in terms of a binomial random variable with probability $\tilde{p}_N$ of success, $Y_N \sim Bin(N, \tilde{p}_N)$, as given in (2.31). It is not difficult to compute the expectation and variance of the log-returns under the risk-neutral probability $\widetilde{\mathbb{P}}_N$:

$$\mathrm{E}_{\widetilde{\mathbb{P}}_N}[L_N] = (2\tilde{p}_N - 1)\sigma\sqrt{NT}, \quad \mathrm{Var}_{\widetilde{\mathbb{P}}_N}(L_N) = \tilde{p}_N(1-\tilde{p}_N)4\sigma^2 T .$$

To find the distribution of $\lim_{N\to\infty} L_N$, we need to know the limiting values of the expectation and variance.

Proposition 2.7. *As $N \to \infty$, we have the following limiting behaviour:*

$$\tilde{p}_N \to \frac{1}{2}, \quad \mathrm{E}_{\widetilde{\mathbb{P}}_N}[L_N] \to rT - \frac{1}{2}\sigma^2 T, \quad \mathrm{Var}_{\widetilde{\mathbb{P}}_N}(L_N) \to \sigma^2 T.$$

Proof. The proof is left as an exercise for the reader. □

By the de Moivre–Laplace theorem, the probability distribution of log-returns converges weakly to the normal distribution as $N \to \infty$. Under the risk-neutral probability, the asymptotic distribution of L_N, as $N \to \infty$, is $Norm((r - \frac{1}{2}\sigma^2)T, \sigma^2 T)$. Therefore, in the limiting case, the risk-neutral probability distribution of the stock price $S(T) = \lim_{N\to\infty} S_N^{(N)}$ is the log-normal distribution:

$$S(T) = S(0)\,\mathrm{e}^{(r-\sigma^2/2)T+\sigma\sqrt{T}Z}, \tag{2.42}$$

where $Z \sim Norm(0,1)$. The interesting fact is that the limiting distribution does not depend on the real-world expected return μ on the stock. In the risk-neutral binomial tree model, the expected return on the stock is the same as that of the risk-free bond. It is not difficult to check that in the limiting case we observe the same behaviour for the risk-neutral log-normal price model:

$$\widetilde{\mathrm{E}}[S(T)] = S(0)\,\mathrm{e}^{rT}.$$

2.4 Value at Risk

In the beginning of this chapter, we defined a risky asset as that with uncertain future cash flows. Examples of such assets include stocks, derivative contracts, defaultable bonds, and similar contingent claims subject to default risk. To distinguish risky and risk-free assets, we need to take a look at the distribution of their returns. The return of a risky asset is uncertain. Hence, from the mathematical point of view, it may be viewed as a random variable with nonzero variance. The return on a risk-free asset is certain, so its variance is zero. Therefore, the risk associated with an asset (with return R) can be measured by computing the standard deviation of the return on the asset: $\sigma = \sqrt{\mathrm{Var}(R)}$. Clearly, a risk-averse investor prefers an asset with lower σ. However, the value of σ may not tell us how large the loss may be. The variance and expectation of the return on a risky asset alone define the shape of the profit and loss distribution only when the asset return has a normal distribution (or Student's t-distribution).

One can use other market *risk metrics* to measure the uncertainty in the portfolio return. For example, one may be interested in the probability of loss L on a specific financial asset (or a portfolio of assets) over some period of time being less than a given amount. That is, one may wish to evaluate the probability $\mathbb{P}(L < A)$ for a given loss L and amount A. Let us reverse the question and find an amount A so that the probability of a loss not exceeding this amount is equal to a given probability, say 95% (although we may consider another confidence level such as 90% or 99%). That is, find A such that $\mathbb{P}(L \leqslant A) = 95\%$. This value is referred to as *Value at Risk* and denoted by VaR.

The VaR is a measure of the risk of loss on a specific portfolio of financial assets. For a given portfolio, probability level, and time horizon, VaR is defined as a *threshold level* such that the probability that the loss on the portfolio over the given time horizon exceeds this level is equal to the given probability.

VaR has two basic parameters: the *significance level* denoted $\alpha \in (0,1)$ (or *confidence level* denoted $1-\alpha$) and the *risk horizon* denoted h, which is the period of time over which we measure the potential loss. Traditionally, h is measured in trading days. Common parameters for VaR are 1% and 5% significance levels and 1-day and 10-day risk horizons, although other combinations are also in use. When VaR is computed, it is assumed that the current position in the portfolio of interest will remain unaltered over the chosen time period. Let Π_t denote the value of the portfolio at the time t. The value of the same portfolio at the future time $t+h$, discounted to time t, is $Z(t,t+h)\Pi_{t+h}$, where $Z(t,t+h)$ is the price of a unit zero-coupon bond that matures at the time $t+h$. For example, for the case with continuously compounded interest, we have that $Z(t,t+h) = \mathrm{e}^{-jh}$ (where j is the daily interest rate). So we use a zero-coupon bond as a discounting factor. The discounted profit-and-loss (P&L) over a risk horizon of h days is

$$G_h := Z(t,t+h)\Pi_{t+h} - \Pi_t.$$

In other words, G_h is the present value of the gain from an investment; hence, $-G_h$ is the present value of the loss. Since the future value Π_{t+h} is uncertain, the discounted P&L is a random variable. To calculate the VaR of the portfolio, we need to know the distribution of this random variable.

Given the significance level α and risk horizon h, the $100\alpha\%$ h-day VaR is defined as the present value of the largest possible loss amount that would be exceeded with only a probability α over an h-day time period:

$$\mathrm{VaR}_{\alpha,h} = -\inf\{x \in \mathbb{R} : \mathbb{P}(G_h > x) \leqslant 1-\alpha\} = -\inf\{x \in \mathbb{R} : F_{G_h}(x) \geqslant \alpha\}, \tag{2.43}$$

where $F_{G_h}(x) = \mathbb{P}(G_h \leqslant x)$ is the CDF of G_h. For a continuous price model with a continuous and strictly monotonic distribution function F_{G_h}, it is possible to solve the equation $F_{G_h}(x) = \alpha$ for x and then the VaR is defined as the α-quantile of the discounted h-day P&L distribution:

$$\mathrm{VaR}_{\alpha,h} = -g_{\alpha,h}, \text{ where } \mathbb{P}(G_h \leqslant g_{\alpha,h}) = \alpha.$$

Note that for a portfolio loss in h days, the present value change $g_{\alpha,h}$ is negative and hence the VaR value is positive, i.e., VaR measures losses and is reported as a positive amount that corresponds to the loss.

When VaR is estimated from a P&L distribution, it is given in value terms. However, one may prefer to analyze the return distribution rather than the P&L distribution. In this case, the VaR is expressed as a percentage of the current value of the portfolio. The discounted h-day return on the portfolio is

$$\frac{Z(t,t+h)\Pi_{t+h} - \Pi_t}{\Pi_t}.$$

To calculate the VaR, we first find $r_{\alpha,h}$, the α-quantile of the return distribution:

$$\mathbb{P}\left(\frac{Z(t,t+h)\Pi_{t+h} - \Pi_t}{\Pi_t} < r_{\alpha,h}\right) = \alpha,$$

and then the VaR is given by

$$\mathrm{VaR}_{\alpha,h} = \begin{cases} -r_{\alpha,h} & \text{as a percentage of the portfolio value } \Pi_t, \\ -r_{\alpha,h}\Pi_t & \text{as a quantity in value terms.} \end{cases}$$

Example 2.8. Assume that the discounted P&L is a normal random variable with mean μ and variance σ^2. Calculate 1% VaR.

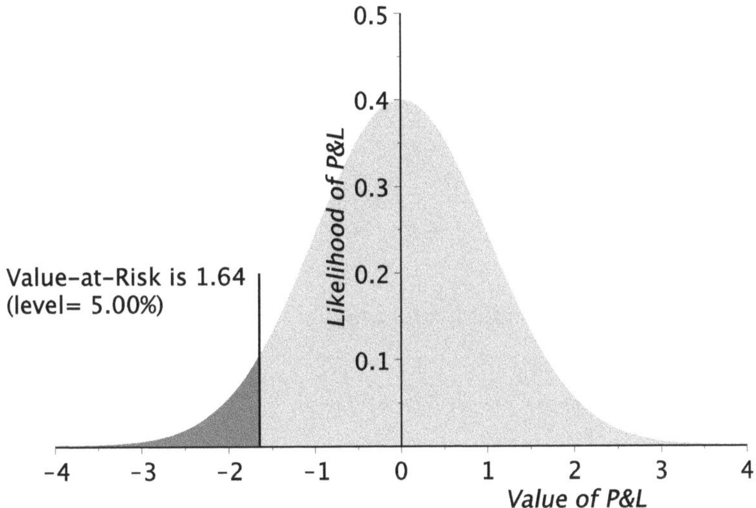

FIGURE 2.4: The Value-at-Risk diagram for a standard normal Profit-and-Loss PDF. The light-grey area to the right of the line represents 95% of the total area under the curve. The dark-grey area to the left of the line represents 5% of the total area under the curve.

Solution. We have that the discounted P&L is $G = \mu + \sigma Z$ for some $Z \sim Norm(0,1)$. We need to find g so that

$$\alpha = 0.01 = \mathbb{P}(G < g) = \mathbb{P}\left(\frac{G-\mu}{\sigma} < \frac{g-\mu}{\sigma}\right) = \mathbb{P}\left(Z < \frac{g-\mu}{\sigma}\right) = \mathcal{N}\left(\frac{g-\mu}{\sigma}\right).$$

Let us find the 0.01-quantile x for the standard normal distribution. We use the table of the standard normal CDF to find x such that $\mathcal{N}(x) = 0.01$. By symmetry, $\mathcal{N}(-x) = 1 - 0.01 = 0.99$. From the table we have $\mathcal{N}(z = 2.33) = 0.9901$ as the closest value. Hence, $x = -z = -2.33$, and the VaR value is given by $-g = -(\mu + \sigma x) = -\mu + 2.33\sigma$. Therefore, among investments whose gains are normally distributed, the VaR criterion would select the one having the largest value of $-\mu + 2.33\sigma$. □

The VaR with significance level α gives us a value that has only a $100\%\alpha$ chance of being exceeded by the loss from an investment. However, this value does not tell us what the actual loss may be. It has been suggested that the conditional expected loss given that it exceeds the VaR is a better metric of the risk. This conditional expected loss is called the *conditional value at risk* or CVaR. The CVaR criterion is to choose the investment having the smallest CVaR.

Example 2.9. Assume that the discounted P&L is a normal random variable with mean μ and variance σ^2. Calculate 1% CVaR.

Solution. We have that the discounted P&L is $G = \mu + \sigma Z$ with $Z \sim Norm(0,1)$. The

CVaR is given by

$$\begin{aligned}\text{CVaR} &= \mathrm{E}[-G \mid -G > \text{VaR}] = \mathrm{E}[-G \mid -G > -\mu + 2.33\sigma] \\ &= \mathrm{E}\left[\sigma\left(\frac{-G+\mu}{\sigma}\right) - \mu \,\middle|\, \frac{-G+\mu}{\sigma} > 2.33\right] \\ &= \sigma\mathrm{E}\left[\frac{-G+\mu}{\sigma} \,\middle|\, \frac{-G+\mu}{\sigma} > 2.33\right] - \mu = \sigma\mathrm{E}[Z \mid Z > 2.33] - \mu\,.\end{aligned}$$

For a standard normal random variable Z we have

$$\begin{aligned}\mathrm{E}[Z \mid Z > a] &= \frac{\mathrm{E}[Z\,\mathbb{I}_{\{Z>a\}}]}{\mathbb{P}(Z>a)} = \frac{1}{\mathbb{P}(Z>a)}\int_a^\infty z\,n(z)\,\mathrm{d}z \\ &= \frac{1}{\mathbb{P}(Z>a)}\int_a^\infty \frac{1}{\sqrt{2\pi}}\mathrm{e}^{-z^2/2}\,\mathrm{d}(z^2/2) = \frac{1}{\sqrt{2\pi}\mathbb{P}(Z>a)}\mathrm{e}^{-a^2/2}\end{aligned}$$

for any real a. Hence, we obtain that

$$\text{CVaR} = \sigma\frac{1}{\sqrt{2\pi}\,0.01}\mathrm{e}^{-2.33^2/2} - \mu \cong -\mu + 2.64\sigma,$$

since $\mathbb{P}(Z > 2.33) \cong 0.01$. □

2.5 Dividend Paying Stock

Consider a stock (with the price process $\{S(t)\}_{t\geqslant 0}$) that pays dividends. Every moment a dividend payment is made, the price of one share instantaneously drops down by the amount of the dividend payment or otherwise an arbitrage opportunity would arise. Indeed, one can buy a share of stock right before a dividend payment is made, receive the payment, and then immediately sell the share. There is an arbitrage profit if the stock price is not adjusted when the dividend payment is made. Here, we assume that there is no delay between the ex-dividend date and the date when shareholders receive dividend payments. Note that in the U.S., the Internal Revenue Service (IRS) defines the ex-dividend date as "the first date following the declaration of a dividend on which the buyer of a stock is not entitled to receive the next dividend payment."

Suppose that a dividend payment $\mathrm{div}(t_*)$ is made at time t_*. This payment can be given as a monetary amount or as a percentage of the spot price, i.e., $\mathrm{div}(t_*) = \mathrm{d}_* S(t_*)$ with the dividend percentage $0 \leqslant \mathrm{d}_* \leqslant 1$. The price of one share immediately after the dividend payment must be $S(t_*) - \mathrm{div}(t_*) = S(t_*) - \mathrm{d}_* S(t_*) = (1-\mathrm{d}_*)S(t_*)$ or otherwise an arbitrage opportunity exists. We illustrate this idea with a single-period model. Let S be the stock price at the beginning of period. At the end of period, the (pre-dividend) price is Su with probability p or Sd with probability $1-p$. After the dividend is paid, the price goes down by the dividend amount to become $Su(1-\mathrm{d}_*)$ or $Sd(1-\mathrm{d}_*)$. This situation is illustrated in Figure 2.5.

Suppose that the dividend on a single share is used to purchase

$$\frac{\mathrm{d}_* S(t_*)}{(1-\mathrm{d}_*)S(t_*)} = \frac{\mathrm{d}_*}{1-\mathrm{d}_*}$$

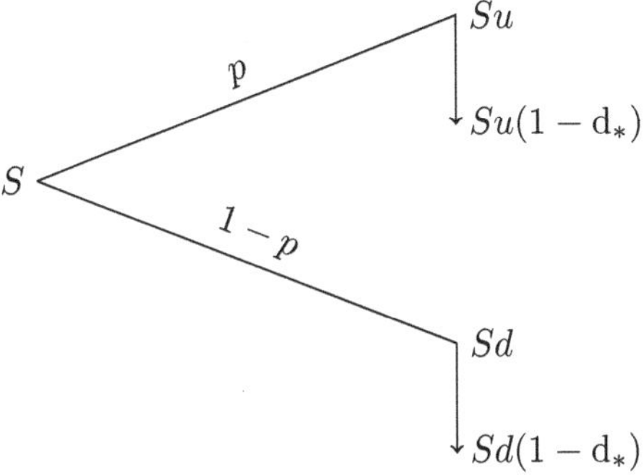

FIGURE 2.5: A single-period binomial model for a stock with dividends.

additional shares at time t_*. So, each share in our portfolio will grow to $1 + \frac{d_*}{1-d_*} = \frac{1}{1-d_*}$ shares right after time t_*. The market value of one share is

$$\Pi_t = \begin{cases} S(t) & \text{if } t < t_*, \\ \frac{1}{1-d_*} S(t) & \text{if } t \geqslant t_*. \end{cases}$$

So, we may use the same asset price model without making dividend adjustments, if the dividends are assumed to be reinvested in the stock.

Let the dividends be paid at times $t_k = k\Delta t$, $k = 1, 2, \ldots, m$, distributed evenly with the step size $\Delta t = \frac{T}{m}$. Let d_m be the dividend percentage. Starting with one share at time 0, the market value of our portfolio at time T after the mth dividend payment is $\Pi_T = \frac{1}{(1-d_m)^m} S(T)$. Let $m \to \infty$, so that $d_m/\Delta t \to q$ with $q > 0$, then

$$\frac{1}{(1 - d_m)^m} \to e^{qT}.$$

The stock is said to pay dividends continuously at a rate of q. If the dividends are reinvested in the stock, then an investment in one share held at time 0 will increase to become e^{qT} shares at time T. Therefore, we need to start with e^{-qT} shares at time 0 to obtain 1 share at time T.

2.6 Exercises

Exercise 2.1. At time 0, the value of a risk-free bond is $B_0 = 100$, and the stock price is $S_0 = 100$. Suppose that the annual risk-free interest rate is $r = 5\%$, and the one-year return on the stock is

$$r_S = \begin{cases} 10\% & \text{with probability } 60\% \\ -5\% & \text{with probability } 40\% \end{cases}$$

(a) Find positions x and y so that the wealth $\Pi_t = xS_t + yB_t$ of the portfolio (x, y) at time $t = T = 1$ is

$$\Pi_T = \begin{cases} \$1000 & \text{if the stock price goes up} \\ \$1500 & \text{if the stock price goes down} \end{cases}$$

(b) What is the expected return of the portfolio over the first year?

Exercise 2.2. A variant of the one-price theorem. Assume that there are no arbitrage portfolios. Suppose there are two assets X and Y with initial prices X_0 and Y_0. At some time $T > 0$, let $X_T(\omega) \geqslant Y_T(\omega)$ hold for all states of the world and $X_T(\omega') > Y_T(\omega')$ hold for at least one state ω'. Prove that $X_0 > Y_0$. [Hint: Suppose the converse is true and then construct an arbitrage portfolio.]

Exercise 2.3. Consider two investments with respective wealth functions V_t and W_t, with $t \in [0, T]$. Suppose that $V_0 > W_0$ and $\mathbb{P}(V_T \leqslant W_T) = 1$. Find an arbitrage opportunity.

Exercise 2.4. In the setting of Exercise 2.1

(a) find the risk-neutral probabilities $\{\tilde{p}, 1 - \tilde{p}\}$ for this binomial model;

(b) verify that under the risk-neutral probabilities we have

$$\widetilde{\mathrm{E}}\left[\frac{S_1}{B_1}\right] = \frac{S_0}{B_0}.$$

In the latter case, the discounted stock price process S_t/B_t is said to be a martingale.

Exercise 2.5. Consider a single-period market model with a risk-free asset B, such that $B_0 = 10$ and $B_1 = 11$, and a risky asset, the price of which can follow three possible scenarios:

Scenario	S_0	S_1
ω^1	50	70
ω^2	50	55
ω^3	50	40

(a) Show that the model admits no arbitrage opportunities.

(b) Find risk-neutral probabilities $\tilde{p}_i = \widetilde{\mathbb{P}}(\omega^i)$ of the scenarios, such that

$$\widetilde{\mathrm{E}}[S_1/B_1] = S_0/B_0$$

holds. Find the general solution that will depend on a variable parameter. Find the range for that parameter so that $\tilde{p}_1$, $\tilde{p}_2$, $\tilde{p}_3$ define a probability function.

Exercise 2.6. Consider a single-period market model with three states of the world and two base assets: a risky stock S and risk-free money market account A. Suppose that the possible stock prices at time $t = T$ are as follows:

$$S_T = \begin{cases} S^u & \text{with probability } p_1 > 0, \\ S^m & \text{with probability } p_2 > 0, \\ S^d & \text{with probability } p_3 = 1 - p_1 - p_2 > 0, \end{cases}$$

where $0 < S^d < S^m < S^u$. Let the initial investment in a risk-free bond be equal to the current stock price, i.e., $B_0 = S_0$, Prove that at time $t = T$ we have that $S^d < B_T < S^u$ or else an arbitrage possibility would arise. In the latter case, construct an arbitrage portfolio.

Exercise 2.7. Consider a market with a risk-free bond B, for which $B_0 = 50$, $B_1 = 55$, and $B_2 = 60$, and a risky stock with the spot price $S_0 = 50$. Suppose that the stock price at times $t = 1$ and $t = 2$ can follow four possible scenarios:

Scenario	S_1	S_2
ω^1	60	70
ω^2	60	55
ω^3	45	45
ω^4	45	40

(a) Find an arbitrage investment strategy if there are no restrictions on short selling.

(b) Is there an arbitrage opportunity if no short selling of the risky asset is allowed?

Exercise 2.8. Given the bond and stock prices in Exercise 2.7, is there an arbitrage strategy if short selling of stock is allowed, but transaction costs of 5% of the transaction volume apply whenever stock is traded (purchased or sold)?

Exercise 2.9. Given the bond and stock prices in Exercise 2.7, except that now assume $S_2(\omega^2) = S_2(\omega^3) = 50$, and the probabilities $\mathbb{P}(\omega^1) = \frac{17}{33}$, $\mathbb{P}(\omega^2) = \frac{5}{33}$, $\mathbb{P}(\omega^3) = \frac{10}{33}$, and $\mathbb{P}(\omega^4) = \frac{1}{33}$, show that the discounted stock price process S_t/B_t, $t = 0, 1, 2$, is a martingale under this measure $\mathbb{P}$, i.e., show that $\mathrm{E}\left[\frac{S_{t+1}}{B_{t+1}} \mid S_t\right] = \frac{S_t}{B_t}$ for $t = 0, 1$.

Exercise 2.10. Consider a binomial model with $r_B = 10\%$, $r_S^- = -10\%$, $r_S^+ = 20\%$, $S_0 = 100$, and $B_0 = 10$. On the (x, y)-plane draw a domain representing the set of all portfolios (x, y) that lead to a self-financing one-step strategy with $x_1 = x$ stock shares and $y_1 = y$ bonds.

Exercise 2.11. Consider a one-period binomial tree model with $r_B = 5\%$, $r_S^- = -10\%$, $r_S^+ = 15\%$, $S_0 = \$100$, and $B_0 = \$100$. Construct a self-financing strategy with an initial value of \$10,000 such that, at all times, the stock investment is twice that invested in risk-free bonds.

Exercise 2.12. Prove Proposition 2.7.

Exercise 2.13. Let $Z \sim \mathit{Norm}(0, 1)$. Find the mathematical expectation of $X = e^{aZ+b}$ with $a, b \in \mathbb{R}$. Use the result obtained to find the variance $\mathrm{Var}(X) = \mathrm{E}[X^2] - \mathrm{E}[X]^2$.

Exercise 2.14. Consider the log-normal price model $S(T) = S(0)\mathrm{e}^{\mu T + \sigma\sqrt{T}Z}$, where $Z \sim \mathit{Norm}(0, 1)$, with drift parameter $\mu = 0.02$ and volatility parameter $\sigma = 0.2$. If $S_0 = 100$, find (a) $\mathrm{E}[S(5)]$, (b) $\mathbb{P}(S(5) > 100)$, (c) $\mathbb{P}(S(5) < 110)$.

Exercise 2.15. Consider the log-normal price model with the risk-neutral dynamic

$$S(T) = S(0)\mathrm{e}^{(r-\sigma^2/2)T+\sigma\sqrt{T}Z},$$

where $Z \sim \mathit{Norm}(0, 1)$. Show that $\widetilde{\mathrm{E}}[\mathrm{e}^{-rT}S(T)] = S(0)$.

Exercise 2.16. To calibrate an N-period binomial tree model, one needs to solve the simultaneous equations (2.21)–(2.23).

(a) Find the exact solution to (2.21)–(2.23).

(b) Let the condition $u_N d_N = 1$ in (2.23) be replaced by $p_N = \frac{1}{2}$. Find u_N and d_N satisfying (2.21)–(2.22) and this new condition.

Chapter 3

Portfolio Management

3.1 Expected Utility Functions

3.1.1 Utility Functions

Suppose we have different investment opportunities to choose from. These investments may affect our future wealth. For example, the task is to allocate an initial capital among several risky assets to form an investment portfolio. Once we decide on such a risky investment, the future wealth becomes uncertain so it follows some probability distribution. The investment selection procedure can be reduced to the optimization of the probability distribution of the uncertain future wealth. If the outcomes from all alternatives were certain, then we would select the investment that produces the largest return. For the case with uncertain outcomes, we may be interested in minimizing the variance of the respective probability distribution but other criteria can be applied. So we need a systematic way to rank random wealth levels. In the case of alternatives with uncertain outcomes, we introduce some score function that is calculated as an expected value of a so-called *utility function*.

Suppose we are given a function $u\colon \mathbb{R} \to \mathbb{R}$ so that each possible investment can be assessed by computing the expected utility value $\mathrm{E}[u(V)]$ of the future wealth V. In other words, the value of an investment can be measured by the expected value of the utility of its consequences. To compare possible alternatives, we first compute $\mathrm{E}[u(V)]$ for each possible wealth function V and then choose an alternative with the greatest expected utility value. The specific utility function used depends on individual investment preferences, risk tolerance, and individual financial environment. The simplest example of a utility is the linear function $u(x) = x$. Whoever uses this utility function ranks uncertain wealths by their expected values.

Here are some of the most commonly used utility functions (see Figure 3.1):

- the logarithmic (log) utility function $u(x) = \ln x$;
- the exponential utility function $u(x) = -e^{-ax}$ with $a > 0$;
- the power utility function $u(x) = x^a$ with $0 < a < 1$;
- the quadratic utility function $u(x) = x - ax^2$ with $0 < a$ defined for $x < \frac{1}{2a}$.

As you can see, some utility functions can take negative values. This negativity does not matter since an investor ranks investments using relative values. Moreover, the addition of a constant to a utility function and the multiplication of a utility function by a positive constant do not affect the rankings. If for some utility function u and two investments V_1 and V_2 we have that $\mathrm{E}[u(V_1)] \leqslant \mathrm{E}[u(V_2)]$, then for any $a \in \mathbb{R}$ and $b > 0$ we obtain

$$\mathrm{E}[a + bu(V_1)] = a + b\mathrm{E}[u(V_1)] \leqslant a + b\mathrm{E}[V_2] = \mathrm{E}[a + bu(V_2)]\,.$$

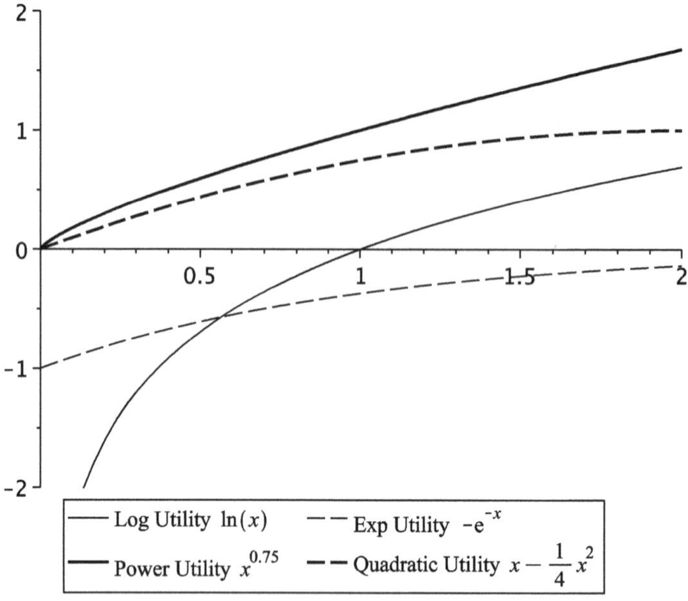

FIGURE 3.1: Sample plots of four commonly used utility functions.

In general, given a utility function u, we can define another utility function $\tilde{u}$ of the form

$$\tilde{u}(x) = a + bu(x)$$

with $b > 0$. This new utility function $\tilde{u}$ is said to be *equivalent* to u. Equivalent utility functions give identical rankings of investment opportunities.

Another example of a utility function is the linear utility $u(x) = a + bx$, $b > 0$. This function reflects expectations of a *risk-indifferent* investor. For any random wealth V we have

$$\mathrm{E}[u(V)] = \mathrm{E}[a + bV] = a + b\mathrm{E}[V] = u(\mathrm{E}[V]),$$

hence the linear utility function has no preference for a deterministic wealth or for a random wealth provided that expected wealths are the same.

Example 3.1. An investor with total capital W can invest any amount between 0 and W. If an amount is invested, then the same amount is either won or lost with respective probabilities p and $1-p$. In other words, with probability p, the investor doubles the initial investment; with probability $1-p$, the investor loses all the invested money. What amount should be invested if the log utility function $u(V) = \ln V$ is utilized for ranking alternatives?

Solution. Let the amount of xW for some $0 \leqslant x \leqslant 1$ be invested. The investor's final fortune $V(x)$ is either $W + xW$ or $W - xW$ with respective probabilities p and $1-p$. Hence the expected utility of the final wealth is

$$\begin{aligned}\mathrm{E}[u(V(x))] &= \ln(W + xW)\,p + \ln(W - xW)\,(1-p) \\ &= \ln((1+x)W)\,p + \ln((1-x)W)\,(1-p) \\ &= \ln(1+x)\,p + \ln(1-x)\,(1-p) + \ln W\,.\end{aligned}$$

To find the optimal value of x, let us differentiate $\mathrm{E}[u(V(x))]$ with respect to x and then find zeros of the derivative obtained:

$$\frac{\mathrm{d}}{\mathrm{d}x}\mathrm{E}[u(V(x))] = \frac{\mathrm{d}}{\mathrm{d}x}\left(p\ \ln(1+x) + (1-p)\ \ln(1-x)\right)$$
$$= \frac{p}{1+x} - \frac{1-p}{1-x} = \frac{2p-(1+x)}{1-x^2}\,.$$

If $p \in (0, \frac{1}{2}]$, then the derivative is strictly negative for all $x \in (0, 1)$ and the expected utility attains its maximum value at $x = 0$. In this case, the risk to lose the invested amount is too high, and it is reasonable to invest nothing. If $p \in (\frac{1}{2}, 1)$, then the derivative is zero at $x^* = 2p-1 \in (0, 1)$. The second derivative of the expected utility function is negative at x^*, hence $x^* = 2p - 1 \in (0, 1)$ is the point of maximum of $V(x)$. Therefore, $100(2p-1)\%$ of the initial capital is to be invested. For example, for $p = 70\%$, the investor shall invest 40% of the fortune. □

Example 3.2. The Saint Petersburg Paradox, originally proposed by Nicolaus Bernoulli, is a classical example of how utility functions are used in the decision making process. Consider a game of chance where a fixed fee is paid to enter, and then a fair coin will be tossed repeatedly until a head first appear ending the game. The payoff starts at \$1 and then is doubled every time a tail appears. As a result, the player wins $\$2^{k-1}$ if a head first appears on the kth toss ($k = 1, 2, 3, \ldots$). How much should the player be willing to pay to enter such a game?

Solution. First, let us find the expected value of the payoff. We deal with a sequence of independent trials where the probability of success (i.e., a head occurs) is $\frac{1}{2}$. With probability $p_1 = \frac{1}{2}$, a head first appears on the first toss and the player wins \$1; with probability $p_2 = \frac{1}{4}$, a head first appears on the second toss and the player wins \$2; with probability $p_3 = \frac{1}{8}$, a head first appears on the third toss and the player wins \$4, etc. The probability that a head first appears on the kth toss is $p_k = 2^{-k}$; the payoff is then $\$2^{k-1}$. Therefore, the expected payoff is then

$$E = \frac{1}{2}\cdot 1 + \frac{1}{4}\cdot 2 + \frac{1}{8}\cdot 4 + \cdots + \frac{1}{2^k}2^{k-1} + \cdots = \frac{1}{2} + \frac{1}{2} + \frac{1}{2} + \cdots = \infty\,.$$

The expected win for the player of this game is an infinite amount of money. So no matter how large is the fee paid to enter this game, the player will eventually make a profit in the long run repeatedly playing this game. The classical solution to this "paradox" is to

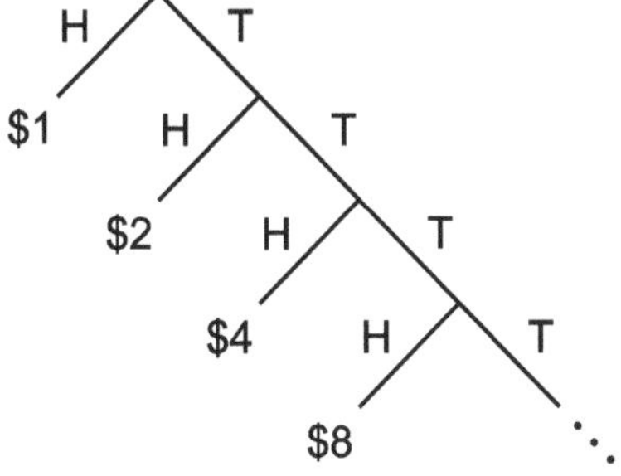

FIGURE 3.2: An outcome tree for the St. Petersburg game. The game consists of a series of coin tosses offering a 50% chance of winning \$1, a 25% chance of \$2, a 12.5% chance of \$4, and so on. The gamble may continue indefinitely.

assume that one's valuation of money is different from its face value and depends on his or her wealth. Let us apply the logarithmic utility model to find a reasonable price c charged to enter the game. Let the initial wealth of the player be denoted V_0. The expected log utility function of the total wealth $V = V(c)$ after playing the game is

$$\mathrm{E}[\ln V] = \sum_{k=1}^{\infty} \ln(V_0 + 2^{k-1} - c)\,\frac{1}{2^k} < \infty\,.$$

A rational player is willing to play the game only if the game will not decrease the expected utility of the wealth:

$$\mathrm{E}[\ln V] \geqslant \mathrm{E}[\ln V_0]\,.$$

After plotting the expected change in utility,

$$\mathrm{E}[\ln V] - \mathrm{E}[\ln V_0] = \mathrm{E}[\ln V - \ln V_0] = \mathrm{E}[\ln(V/V_0)]\,,$$

as a function of the cost c, we observe that $\mathrm{E}[\ln(V(c)/V_0)]$ is a strictly decreasing function of c and there exists a maximum cost c^* so that any price $c < c^*$ gives a positive expected change in utility. Such a cost c^* depends on the initial capital V_0 and can be found by solving $\mathrm{E}[\ln(V(c)/V_0)] = 0$ for c. For example, a person with \$2 in his pocket is willing to pay up to \$2, a person with \$1000 is willing to pay up to \$5.97, and a millionaire is willing to pay up to \$10.94. □

3.1.1.1 Risk Aversion

Utility functions are constructed based on the following principles.

Principle 1. Investors prefer more to less. If there are two certain wealths V_1 and V_2, then an investor prefers the larger one, i.e., $V_1 < V_2$ implies $u(V_1) < u(V_2)$. Hence, u is an increasing function.

Principle 2. Investors are averse to risk. Positive deviations ΔV from average wealth V cannot compensate for equally large and equally probable negative deviations $-\Delta V$ from average wealth, i.e., $u(V) - u(V - \Delta V) > u(V + \Delta V) - u(V)$. Therefore,

$$u(V) > \frac{u(V + \Delta V) + u(V - \Delta V)}{2}\,. \tag{3.1}$$

The inequality in (3.1) holds true for all V if u is a concave function. The left-hand side, $u(x) - u(V - \Delta V)$, is "the pain of losing ΔV dollars," and the right-hand side, $u(V + \Delta V) - u(V)$, is "the joy of winning ΔV dollars." The inequality says that the pain of losing outweighs the joy of winning, alternatively that we react more severely to a loss then we do to a gain of the same magnitude.

Suppose that there are two alternatives for future wealth: the first provides either x or y each with a probability of $\frac{1}{2}$, the second gives $\frac{1}{2}x + \frac{1}{2}y$ with certainty. Although both alternatives have the same expected value, a risk-averse investor prefers the certain wealth of $\frac{1}{2}x + \frac{1}{2}y$ to a 50-50 chance of x and y: $u(\frac{1}{2}x + \frac{1}{2}y) \geqslant \frac{1}{2}u(x) + \frac{1}{2}u(y)$.

Recall that a function u defined on an interval $[a, b]$ is said to be *concave* if for any α with $0 \leqslant \alpha \leqslant 1$ and any $x, y \in \mathbb{R}$ there holds

$$u(\alpha x + (1 - \alpha)y) \geqslant \alpha u(x) + (1 - \alpha)u(y)\,.$$

A function u is said to be *convex* on $[a, b]$ if the function $-u$ is concave on $[a, b]$. That is, if for any α with $0 \leqslant \alpha \leqslant 1$ and any $x, y \in \mathbb{R}$ there holds

$$u(\alpha x + (1 - \alpha)y) \leqslant \alpha u(x) + (1 - \alpha)u(y)\,.$$

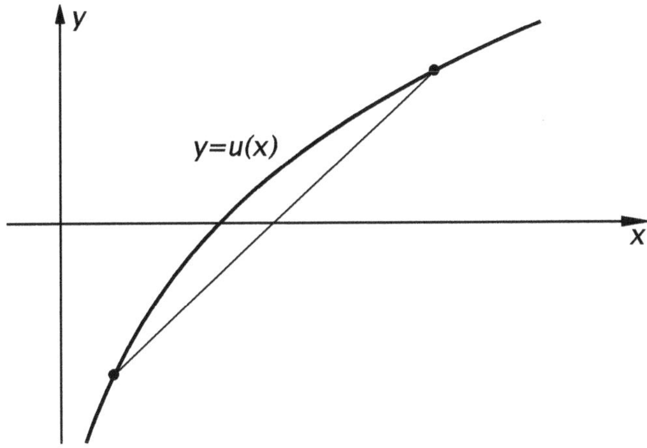

FIGURE 3.3: The concave (or convex-upward) plot of a typical risk-averse utility function. As is seen, every curve segment of a concave plot lies above a chord connecting the endpoints of the segment.

A twice differentiable function u is concave (convex) on an interval $[a, b]$ if its second derivative u'' is nonpositive (nonnegative) on $[a, b]$. A utility function is said to be *risk-averse* (on an interval $[a, b]$) if it is concave (on the interval $[a, b]$). A concave function is depicted in Figure 3.3. For a twice-differentiable utility function, the risk-averse condition means that the second derivative of the utility function is nonpositive.

Recall Jensen's inequality: let u be a concave function, then for any random variable V,

$$\mathrm{E}[u(V)] \leqslant u(\mathrm{E}[V]) . \tag{3.2}$$

This means that a risk-averse investor prefers a certain wealth of W to an uncertain wealth V with the same expected value $\mathrm{E}[V] = W$. This observation relates to the notion of the *certainty equivalent.*

The certainty equivalent of an uncertain wealth V is defined as the amount of a constant wealth C that has the utility level equal to the expected utility of V:

$$u(C) = \mathrm{E}[u(V)] . \tag{3.3}$$

Clearly, the certainty equivalent is the same for all equivalent functions. Combining (3.2) and (3.3) gives that the certainty equivalent is always less than the expected value of the wealth for a risk-averse investor with a concave utility function:

$$u(C) \leqslant u(\mathrm{E}[V]) \implies C \leqslant \mathrm{E}[V] .$$

Let us represent the uncertain return V in the form $V = W + \epsilon$ where W is the initial capital and ϵ is a zero-mean risk. A natural way to measure risk aversion is to ask how much an investor is ready to pay to get rid of the zero-mean risk ϵ. This price, called a *risk premium* and denoted π, is defined implicitly by

$$\mathrm{E}[u(W + \epsilon)] = u(W - \pi) . \tag{3.4}$$

Let us consider a small risk ϵ. Expanding the left- and right-hand sides of (3.4) in Taylor's approximations gives

$$\begin{aligned}\mathrm{E}[u(W+\epsilon)] &\approx \mathrm{E}\left[u(W)+\epsilon u'(W)+\frac{\epsilon^2}{2}u''(W)\right] \\ &= u(W)+\mathrm{E}[\epsilon]\,u'(W)+\frac{\mathrm{E}[\epsilon^2]}{2}\,u''(W) = u(W)+\frac{\sigma_\epsilon^2}{2}\,u''(W)\end{aligned}$$

and

$$u(W-\pi) \approx u(W)-\pi u'(W)\,,$$

respectively, where $\sigma_\epsilon^2 := \mathrm{E}[\epsilon^2]$ is the variance of ϵ. Substituting these back into (3.4), we obtain

$$\pi \approx \frac{\sigma_\epsilon^2}{2}A_u(W)\,,$$

where

$$A_u(W) := -\frac{u''(W)}{u'(W)}$$

is the *Arrow–Pratt absolute risk aversion coefficient.* We say that investor 1 (with utility function u_1) is more risk averse than investor 2 (with utility function u_2) if for the same initial wealth W and zero-mean risk ϵ, the risk premium π_1 paid by investor 1 is larger than the risk premium π_2 of investor 2, or, equivalently, $A_{u_1}(W) > A_{u_2}(W)$.

The degree of risk aversion can be viewed as a measure of the magnitude of concavity of the utility function: the stronger the bend in the function, the larger the risk aversion coefficient A. For example, the risk-aversion coefficient for a linear utility function, $u(V) = a + bV$, is zero. The coefficient A is normalized by the derivative u' that appears in the denominator. This makes A independent of linear transformations of the utility function u. Indeed, for any reals a and $b \neq 0$ we have

$$-\frac{(a+bu(x))''}{(a+bu(x))'} = -\frac{bu''(x)}{bu'(x)} = -\frac{u''(x)}{u'(x)}\,.$$

The coefficient function $A(W)$ shows how risk aversion changes with the wealth level. It is usually argued that absolute risk aversion should be a decreasing function of wealth. That is, many investors are willing to take more risk when they are financially secure. For example, a lottery to gain or lose \$100 is potentially life-threatening for an investor with initial wealth $W = 101$, whereas it is negligible for an investor with wealth $W = 100{,}000$. The former individual should be willing to pay more than the latter for the elimination of such a risk. Thus, we may require that the risk premium associated with any risk is decreasing in wealth. It can be shown that this holds if and only if the Arrow–Pratt absolute risk aversion coefficient is decreasing in wealth. This requirement means that

$$A'(W) = -\frac{u'''(W)u'(W)-u''(W)^2}{u'(W)^2} < 0\,.$$

A necessary condition for this to hold is $u'''(W) > 0$.

As a specific example, consider the exponential utility function $u(x) = -\mathrm{e}^{-ax}$. Differentiate it to obtain

$$u'(x) = a\mathrm{e}^{-ax} \quad\text{and}\quad u''(x) = -a^2\mathrm{e}^{-ax}\,.$$

Therefore, we have $A(x) = -u''(x)/u'(x) = a$. In this case, the risk aversion remains constant as wealth increases. As another example, consider the power utility function $u(x) = x^a$ with $0 < a < 1$. We have $u'(x) = ax^{a-1}$ and $u''(x) = a(a-1)x^{a-2}$. Thus, $A(x) = (1-a)/x$. So risk aversion decreases as wealth increases. Similarly, for the logarithmic utility function $u(x) = \ln x$, we have $u'(x) = 1/x$, $u''(x) = -1/x^2$, and $A(x) = 1/x$.

3.1.2 Mean-Variance Criterion

Suppose that the optimal investment opportunity is chosen by maximizing the expected utility of the wealth. Let us show how the utility maximization method reduces to the mean-variance criterion when an optimal investment is selected by maximizing the expected wealth and minimizing the variance of the wealth. Suppose that the final wealth follows a normal probability distribution and the investor uses an exponential utility function $u(x) = -e^{-ax}$ with $a > 0$. Recall that the mathematical expectation of an exponential function of a normal random variable Z can be expressed in terms of the expected value and variance of Z as follows:

$$\mathrm{E}[e^Z] = e^{\mathrm{E}[Z]+\mathrm{Var}(Z)/2}.$$

If the wealth V is normal, then $-aV$ is normal as well with mean $\mathrm{E}[-aV] = -a\mathrm{E}[V]$ and variance $\mathrm{Var}(-aV) = a^2\,\mathrm{Var}(V)$. Therefore, the expected utility of the wealth V is

$$\mathrm{E}[u(V)] = -\exp\left(-a\mathrm{E}[V] + a^2\,\mathrm{Var}(V)/2\right) = -\exp\left(-a(\mathrm{E}[V] - a\,\mathrm{Var}(V)/2)\right).$$

The exponential function is increasing. Thus, the expected utility is maximized by choosing an investment that maximizes $\mathrm{E}[V] - a\,\mathrm{Var}(V)/2$. This means that alternative investments can be ranked by comparing their means and variances. If there are two investments so that $\mathrm{E}[V_1] \geqslant \mathrm{E}[V_2]$ and $\mathrm{Var}(V_1) \leqslant \mathrm{Var}(V_2)$, then the first investment results in a larger expected utility than does the second: $\mathrm{E}[u(V_1)] \geqslant \mathrm{E}[u(V_2)]$.

One can arrive at the same conclusion for the case of a quadratic utility function $u(x) = x - ax^2$ with $a > 0$. Assuming that the wealth V satisfies $V < \frac{1}{2a}$, the expected utility $\mathrm{E}[u(V)]$ is maximized by selecting an investment with a larger expected wealth and smaller variance $\mathrm{Var}(V)$.

To deal with a general utility function u, let us consider the Taylor expansion of u about the point $\mathrm{E}[V]$:

$$u(V) \approx u(\mathrm{E}[V]) + u'(\mathrm{E}[V])(V - \mathrm{E}[V]) + \frac{1}{2}u''(\mathrm{E}[V])(V - \mathrm{E}[V])^2.$$

Taking expectations gives

$$\begin{aligned}\mathrm{E}[u(V)] &\approx u(\mathrm{E}[V]) + u'(\mathrm{E}[V])\mathrm{E}[V - \mathrm{E}[V]] + \frac{1}{2}u''(\mathrm{E}[V])\mathrm{E}\left[(V - \mathrm{E}[V])^2\right] \\ &= u(\mathrm{E}[V]) + u''(\mathrm{E}[V])\,\mathrm{Var}[V]/2.\end{aligned} \tag{3.5}$$

Here we use that $\mathrm{E}[V - \mathrm{E}[V]] = \mathrm{E}[V] - \mathrm{E}[V] = 0$ and $\mathrm{E}\left[(V - \mathrm{E}[V])^2\right] = \mathrm{Var}(V)$. Therefore, a reasonable approximation to the optimal investment is given by an investment that maximizes

$$u(\mathrm{E}[V]) + u''(\mathrm{E}[V])\,\mathrm{Var}[V]/2.$$

Suppose that $u''(x)$ is a nondecreasing function in x. Then, since $u''(x) \leqslant 0$, an optimal investment V can be again selected by both maximizing the expected value $\mathrm{E}[V]$ and minimizing the variance $\mathrm{Var}(V)$. Recall that the standard deviation $\sigma_V = \sqrt{\mathrm{Var}(V)}$ characterizes the risk associated with the investment V. Therefore, the mean-variance criterion tells us that the optimal investment is attained by maximizing the expected value of the wealth and minimizing the risk.

3.2 Portfolio Optimization for Two Assets

3.2.1 Portfolio of Two Assets

In a one-period setting, let us consider a model with two risky assets A_t^1 and A_t^2, where $t \in \{0, T\}$. Each asset, labelled by $i = 1, 2$, is characterized by its initial value A_0^i and the respective single-period return $r_i = \frac{A_T^i - A_0^i}{A_0^i}$. At that we have $A_T^i = A_0^i(1 + r_i)$. The risky returns r_1 and r_2 (as well as the terminal asset prices A_T^1 and A_T^2) are random variables defined on a common probability space with state space Ω and probability function $\mathbb{P}$.

Let us form a portfolio $[x_1, x_2]^\top$ by purchasing x_1 shares of asset 1 and x_2 shares of asset 2. The initial wealth of such a portfolio is $V_0 = x_1 A_0^1 + x_2 A_0^2$. The (one-period) rate of return r_V is then given by

$$\begin{aligned} r_V &= \frac{V_T - V_0}{V_0} = \frac{x_1(A_T^1 - A_0^1) + x_2(A_T^2 - A_0^2)}{V_0} \\ &= \frac{x_1(A_T^1 - A_0^1)}{A_0^1}\frac{A_0^1}{V_0} + \frac{x_2(A_T^2 - A_0^2)}{A_0^2}\frac{A_0^2}{V_0} \\ &= \frac{x_1 A_0^1}{V_0} r_1 + \frac{x_2 A_0^2}{V_0} r_2. \end{aligned}$$

Introduce the following weights:

$$w_1 = \frac{x_1 A_0^1}{V_0} \quad \text{and} \quad w_2 = \frac{x_2 A_0^2}{V_0}$$

which are called the *allocation weights* of funds between the two underlying assets. In other words, $100 w_i\%$ of the initial wealth is invested in asset $i = 1, 2$. By the definition of a wealth function, the weights add up to one: $w_1 + w_2 = 1$. If short selling is allowed, then one of the weights may be negative and, hence, the other weight is greater than one. For a portfolio without short selling, both weights are between zero and one. Being given the values of returns r_i and weights w_i, the total wealth at the end of the period is

$$V_T = (1 + r_V)V_0 = (1 + w_1 r_1 + w_2 r_2)V_0 = (w_1(1 + r_1) + w_2(1 + r_2))V_0 \,.$$

A portfolio with weights $[w_1, w_2]^\top$ can be characterized by the expected return and the variance of the return. Since $r_V = w_1 r_1 + w_2 r_2$, we have that

$$\begin{aligned} \mathrm{E}[r_V] &= \mathrm{E}[w_1 r_1] + \mathrm{E}[w_2 r_2] \\ &= w_1 \mathrm{E}[r_1] + w_2 \mathrm{E}[r_2] \qquad (3.6) \\ \mathrm{Var}(r_V) &= \mathrm{Var}(w_1 r_1) + \mathrm{Var}(w_2 r_2) + 2\,\mathrm{Cov}(w_1 r_1, w_2 r_2) \\ &= w_1^2\,\mathrm{Var}(r_1) + w_2^2\,\mathrm{Var}(r_2) + 2 w_1 w_2\,\mathrm{Cov}(r_1, r_2) \\ &= w_1^2\,\mathrm{Var}(r_1) + w_2^2\,\mathrm{Var}(r_2) + 2 w_1 w_2\,\mathrm{Corr}(r_1, r_2)\sqrt{\mathrm{Var}(r_1)}\sqrt{\mathrm{Var}(r_2)}\,. \qquad (3.7) \end{aligned}$$

Here, we define the *coefficient of correlation* between two random variables as follows:

$$\mathrm{Corr}(r_1, r_2) = \frac{\mathrm{Cov}(r_1, r_2)}{\sqrt{\mathrm{Var}(r_1)\,\mathrm{Var}(r_2)}} \in [-1, 1]\,.$$

Note that if the variance of one of the random variables is zero then the correlation coefficient is undefined.

Proposition 3.1. *The variance of the return on a portfolio without short selling (i.e., both w_1 and w_2 are nonnegative) cannot exceed the greater of the variances of the underlying asset returns:*

$$0 \leqslant \operatorname{Var}(r_V) \leqslant \max\{\operatorname{Var}(r_1), \operatorname{Var}(r_2)\}.$$

Proof. Since the value of the correlation coefficient is always between −1 and 1, from (3.7) we obtain that

$$\begin{aligned}
\operatorname{Var}(r_V) &\leqslant w_1^2 \operatorname{Var}(r_1) + w_2^2 \operatorname{Var}(r_2) + 2w_1 w_2 \sqrt{\operatorname{Var}(r_1)}\sqrt{\operatorname{Var}(r_2)} \\
&\leqslant \left(w_1\sqrt{\operatorname{Var}(r_1)} + w_2\sqrt{\operatorname{Var}(r_2)}\right)^2 \\
&\leqslant (w_1 + w_2)^2 \max\{\operatorname{Var}(r_1), \operatorname{Var}(r_2)\} = \max\{\operatorname{Var}(r_1), \operatorname{Var}(r_2)\}.
\end{aligned}$$

On the other hand, the variance is always a nonnegative quantity. □

Introduce the following notation for the expected returns, the variances of returns, and the correlation coefficient:

$$\mu_i = \mathrm{E}[r_i],\ \sigma_i^2 = \operatorname{Var}(r_i);\ \ (i = 1, 2);\ \ \rho_{12} = \operatorname{Corr}(r_1, r_2).$$

Moreover, denote $\mu_V = \mathrm{E}[r_V]$ and $\sigma_V^2 = \operatorname{Var}(r_V)$. In this notation, Equations (3.6) and (3.7) take the respective forms:

$$\mu_V = w_1\mu_1 + w_2\mu_2 \ \text{ and } \ \sigma_V^2 = w_1^2\sigma_1^2 + w_2^2\sigma_2^2 + 2\rho_{12}w_1w_2\sigma_1\sigma_2\,. \tag{3.8}$$

Example 3.3. Consider two risky assets with the following probability distributions of their returns:

Scenario ω	Probability $\mathbb{P}(\omega)$	Return r_1	Return r_2
ω^1	0.1	−20%	30%
ω^2	0.6	5%	10%
ω^3	0.3	10%	−20%

Calculate the expected returns, μ_i, standard deviations, σ_i, and correlation coefficient of returns, ρ_{12}.

Solution. To compute the mathematical expectation of a random variable X on a finite sample space Ω, we use the formula

$$\mathrm{E}[X] = \sum_{\omega\in\Omega} X(\omega)\mathbb{P}(\omega)\,.$$

The expected returns are

$$\mu_1 = \mathrm{E}[r_1] = \sum_{i=1}^{3} r_1(\omega^i)\mathbb{P}(\omega^i) = (-0.2)\cdot 0.1 + 0.05\cdot 0.6 + 0.1\cdot 0.3 = 0.04 = 4\%\,,$$

$$\mu_2 = \mathrm{E}[r_2] = \sum_{i=1}^{3} r_2(\omega^i)\mathbb{P}(\omega^i) = 0.3\cdot 0.1 + 0.1\cdot 0.6 + (-0.2)\cdot 0.3 = 0.03 = 3\%\,.$$

Using the fact that $\mathrm{Var}(X) = \mathrm{E}[(X - \mathrm{E}[X])^2]$ and $\mathrm{Cov}(X,Y) = \mathrm{E}[(X - \mathrm{E}[X])(Y - \mathrm{E}[Y])]$, we similarly obtain:

$$\begin{aligned}\mathrm{Var}(r_1) &= (-0.2 - 0.04)^2 \cdot 0.1 + (0.05 - 0.04)^2 \cdot 0.6 + (0.1 - 0.04)^2 \cdot 0.3 = 0.0069\,,\\ \mathrm{Var}(r_2) &= (0.3 - 0.03)^2 \cdot 0.1 + (0.1 - 0.03)^2 \cdot 0.6 + (-0.2 - 0.03)^2 \cdot 0.3 = 0.02610\,,\\ \mathrm{Cov}(r_1, r_2) &= (-0.2 - 0.04) \cdot (0.3 - 0.03) \cdot 0.1 + (0.05 - 0.04) \cdot (0.1 - 0.03) \cdot 0.6\\ &\quad + (0.1 - 0.04) \cdot (-0.2 - 0.03) \cdot 0.3 = -0.0102\,.\end{aligned}$$

The standard deviations are

$$\sigma_1 = \sqrt{\mathrm{Var}(r_1)} = \sqrt{0.0069} \cong 0.08307\,,\ \sigma_2 = \sqrt{\mathrm{Var}(r_2)} = \sqrt{0.02610} \cong 0.16156\,.$$

The correlation coefficient ρ_{12} is

$$\rho_{12} = \frac{\mathrm{Cov}(r_1, r_2)}{\sqrt{\mathrm{Var}(r_1)\,\mathrm{Var}(r_2)}} \cong \frac{-0.0102}{0.08307 \cdot 0.16156} \cong -0.76007 \cong -76\%\,. \qquad \square$$

Example 3.4. Find an optimal allocation of the initial wealth $V_0 = 1000$ between two risky assets from Example 3.3 when attempting to maximize the expected value, $\mathrm{E}[u(V_T)]$, of an exponential utility function $u(x) = 1 - \mathrm{e}^{-0.01x}$ of the wealth V_T.

Solution. Let the weights of a portfolio V in the two assets be $w_1 = x$ and $w_2 = 1 - x$, respectively. The return on such a portfolio is $r_V(x) = xr_1 + (1 - x)r_2$. At the end of the period, the portfolio value is $V_T = V_0(1 + r_V)$. Now we can find the optimal allocation by solving the following maximization problem:

$$\mathrm{E}[u(V_T)] = \mathrm{E}\left[1 - \mathrm{e}^{-0.01V_T}\right] = \mathrm{E}\left[1 - \mathrm{e}^{-0.01V_0(1+r_V)}\right] = 1 - \mathrm{E}\left[\mathrm{e}^{-10(1+xr_1+(1-x)r_2)}\right] \to \max_x .$$

It is equivalent to minimizing $\mathrm{E}[\mathrm{e}^{-10(1+xr_1+(1-x)r_2)}]$ w.r.t. x. Evaluate the mathematical expectation:

$$\begin{aligned}\mathrm{E}[\mathrm{e}^{-10(1+xr_1+(1-x)r_2)}] &= \sum_{i=1}^{3} p_i\, \mathrm{e}^{-10(1+xr_1(\omega^i)+(1-x)r_2(\omega^i))}\\ &= 0.1\mathrm{e}^{-13+5x} + 0.6\mathrm{e}^{-11+0.5x} + 0.3\mathrm{e}^{-8-3x}\,.\end{aligned}$$

Differentiate the expected value w.r.t. x and equate the obtained derivative to zero:

$$0.5\mathrm{e}^{-13+5x} + 0.3\mathrm{e}^{-11+0.5x} - 0.9\mathrm{e}^{-8-3x} = 0\,.$$

The resulting equation can be solved numerically to yield the optimal value $x^* \cong 0.67431$, where the expected utility function attains its maximum value. Therefore, the optimal allocation weights are $w_1 \cong 67.431\%$ and $w_2 \cong 32.569\%$. $\square$

3.2.2 Portfolio Lines

Consider two risky assets with respective returns r_1 and r_2. It is a typical situation when the joint probability distribution of the returns is unknown. However, it may be possible to estimate the moments of the returns from historical data. Suppose we only know the expected returns and variances of the returns, μ_i, σ_i^2, $i = 1, 2$, and the correlation coefficient ρ_{12}. Every portfolio in these assets can be characterized by its expected return and variance of its return.

On the (σ, μ)-plane, a portfolio V with allocation weights $[w_1,\, w_2]^\top$ is represented by a point whose coordinates (σ_V, μ_V) are calculated by (3.8). Let us find a set of points on the (σ, μ)-plane that describes all possible portfolios in the two underlying assets. Since $w_1 + w_2 = 1$, all portfolios can be parameterized by a single variable $x \in \mathbb{R}$: $w_1 = x$ and $w_2 = 1 - x$. Therefore, the set of all possible portfolios can be represented by a *portfolio line* (which can shrink to a single point in some extreme cases). Equations (3.8) can be rewritten as follows:

$$\mu_V(x) = x\mu_1 + (1-x)\mu_2, \quad \sigma_V^2(x) = x^2\sigma_1^2 + (1-x)^2\sigma_2^2 + 2x(1-x)\sigma_1\sigma_2\rho_{12} \tag{3.9}$$

with $x \in (-\infty, \infty)$. For portfolios without short selling (i.e., $w_1 \geqslant 0$ and $w_2 \geqslant 0$), we have that $0 \leqslant x \leqslant 1$.

3.2.2.1 Case with $|\rho_{12}| = 1$

First, let $\rho_{12} = 1$. Then from (3.9) we obtain that the variance of return on portfolio V is given by $\sigma_V^2(x) = (x\sigma_1 + (1-x)\sigma_2)^2$, and hence $\sigma_V(x) = |x\sigma_1 + (1-x)\sigma_2|$. The portfolio line is described by

$$\sigma_V(x) = |x(\sigma_1 - \sigma_2) + \sigma_2| \text{ and } \mu_V(x) = x\mu_1 + (1-x)\mu_2, \quad x \in \mathbb{R}.$$

Let us assume that $\mu_1 \neq \mu_2$ and $\sigma_1 \neq \sigma_2$ (we leave the other cases as exercises for the reader). We can solve the second equation for x to obtain $x = \frac{\mu_V - \mu_2}{\mu_1 - \mu_2}$. Substituting this expression in the formula for σ_V gives us the following relationship:

$$\sigma_V = \left|\sigma_2 + (\sigma_1 - \sigma_2)\frac{\mu_V - \mu_2}{\mu_1 - \mu_2}\right| \implies \sigma_V = \left|\frac{\sigma_1 - \sigma_2}{\mu_1 - \mu_2}\mu_V + \frac{\sigma_2\mu_1 - \sigma_1\mu_2}{\mu_1 - \mu_2}\right|.$$

As we can see, the standard deviation σ_V is a piecewise-linear function of μ_V:

$$\sigma_V = \begin{cases} a\mu_V + b & \text{if } \mu_V \geqslant -\frac{b}{a}, \\ -(a\mu_V + b) & \text{if } \mu_V < -\frac{b}{a}, \end{cases} \quad \text{where } a := \frac{\sigma_1 - \sigma_2}{\mu_1 - \mu_2} \text{ and } b := \frac{\sigma_2\mu_1 - \sigma_1\mu_2}{\mu_1 - \mu_2}.$$

The plot of σ_V as a function of μ_V is a broken line with two half-lines. It is interesting to observe that there is a portfolio with zero variance (i.e., a risk-free portfolio). Indeed, $\sigma_V = 0$ iff $\mu_V = \frac{\sigma_2\mu_1 - \sigma_1\mu_2}{\sigma_1 - \sigma_2}$. The weights $w_1 = x$ and $w_2 = 1 - x$ can be obtained by solving the equation $x\sigma_1 + (1-x)\sigma_2 = 0$ for x. Hence, the weights of a risk-free portfolio are

$$\widehat{w}_1 = \frac{\sigma_2}{\sigma_2 - \sigma_1} \text{ and } \widehat{w}_2 = \frac{\sigma_1}{\sigma_1 - \sigma_2}. \tag{3.10}$$

One of the weights is negative, hence short selling is necessary to construct a risk-free portfolio.

Now let us find what part of the portfolio line corresponds to portfolios without short selling. The portfolios with weights $(0, 1)$ and $(1, 0)$ are the endpoints of such a set. By changing x from 0 to 1, we continuously move the point along the line of portfolios without short selling from one endpoint to the other. Since the portfolio with $\sigma_V = 0$ has a negative weight, the no-short-selling line is a segment lying on one of two rays. The final result of our analysis is presented in Figure 3.4a.

Similarly, we can construct the portfolio line for the case with $\rho_{12} = -1$. The portfolio line is described by

$$\sigma_V(x) = |x(\sigma_1 + \sigma_2) - \sigma_2| \text{ and } \mu_V(x) = x\mu_1 + (1-x)\mu_2 \text{ with } x \in \mathbb{R}.$$

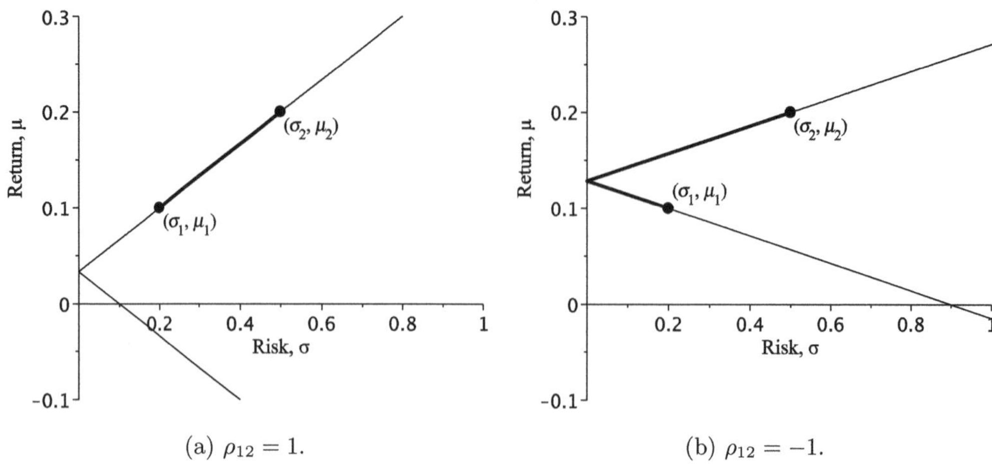

FIGURE 3.4: A typical portfolio line for the case with $|\rho_{12}| = 1$. The bold part indicates portfolios without short selling.

By excluding x from the above equations, we obtain

$$\sigma_V = \left| \frac{\sigma_1 + \sigma_2}{\mu_1 - \mu_2} \mu_V - \frac{\sigma_2 \mu_1 + \sigma_1 \mu_2}{\mu_1 - \mu_2} \right| .$$

Again, the plot of σ_V as a function of μ_V is a broken line. Now $\sigma_V = 0$ iff $\mu_V = \frac{\sigma_2\mu_1 + \sigma_1\mu_2}{\sigma_1+\sigma_2}$. The weights of the risk-free portfolio are

$$\widehat{w}_1 = \frac{\sigma_2}{\sigma_1 + \sigma_2} \quad \text{and} \quad \widehat{w}_2 = \frac{\sigma_1}{\sigma_1 + \sigma_2} . \tag{3.11}$$

Both weights are positive, so no short selling is required to construct a portfolio with a zero variance of return. The no-short-selling line is a broken line segment lying on both half-lines. The result of our analysis is given in Figure 3.4b.

3.2.2.2 Case with $|\rho_{12}| < 1$

Excluding x from (3.9) and expressing σ^2 as a function of μ gives

$$\sigma^2 = \frac{(\mu - \mu_2)^2}{(\mu_1 - \mu_2)^2} \sigma_1^2 + \frac{(\mu - \mu_1)^2}{(\mu_1 - \mu_2)^2} \sigma_2^2 - 2 \frac{(\mu - \mu_1)(\mu - \mu_2)}{(\mu_1 - \mu_2)^2} \rho_{12} \sigma_1 \sigma_2 .$$

After doing some algebra, we can bring this equation to the form:

$$\begin{aligned}
\sigma^2 &= A\mu^2 - 2B\mu + C, \text{ where} \\
A &= \frac{\sigma_1^2 - 2\rho_{12}\sigma_1\sigma_2 + \sigma_2^2}{(\mu_1 - \mu_2)^2}, \\
B &= \frac{\mu_1\sigma_2^2 + \mu_2\sigma_1^2 - 2\rho_{12}\sigma_1\sigma_2(\mu_1 + \mu_2)}{(\mu_1 - \mu_2)^2}, \\
C &= \frac{(\mu_1\sigma_2)^2 + (\mu_2\sigma_1)^2 - 2\rho_{12}\sigma_1\sigma_2\mu_1\mu_2}{(\mu_1 - \mu_2)^2} .
\end{aligned}$$

The curve defined by the above equation is a hyperbola. Indeed, let us rewrite the equation as follows: $\sigma^2 = A(\mu - \frac{B}{A})^2 + D$, where $D = C - \frac{B^2}{A}$. By changing variables from (σ, μ) to

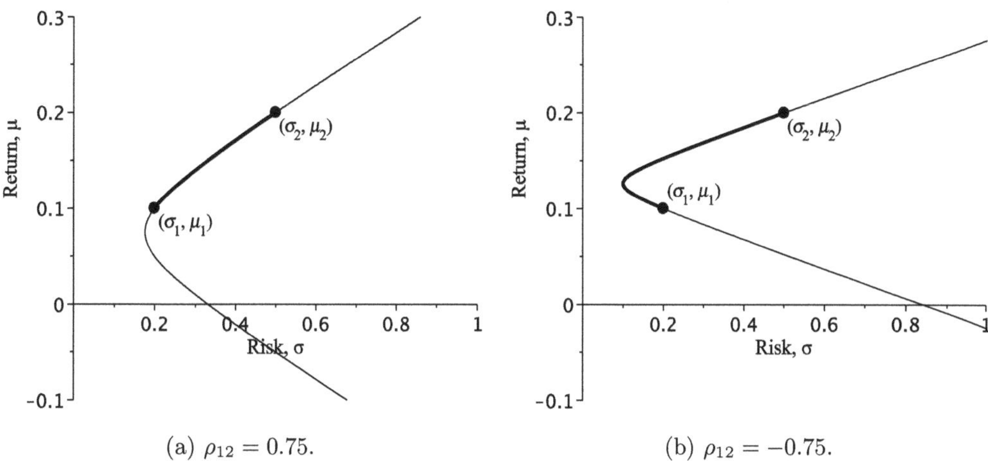

(a) $\rho_{12} = 0.75$. (b) $\rho_{12} = -0.75$.

FIGURE 3.5: A typical portfolio line for the case with $-1 < \rho_{12} < 1$. The bold part indicates portfolios without short selling.

$(x = \frac{\sigma}{\sqrt{D}}, y = \frac{\sqrt{A}}{\sqrt{D}}\mu - \frac{B}{\sqrt{AD}})$, one can easily obtain the canonical equation of a hyperbola: $x^2 - y^2 = 1$. A typical portfolio line is given in Figure 3.5.

As is seen from Figures 3.4 and 3.5, the plot of a portfolio line is a hyperbola for the case with $\rho_{12} \in (-1, 1)$ and a broken line for the extreme case with $|\rho_{12}| = 1$. The evolution of a portfolio line when $\mu_{1,2}$ and $\sigma_{1,2}$ are fixed and ρ_{12} is changing from -1 to 1 is represented in Figure 3.6.

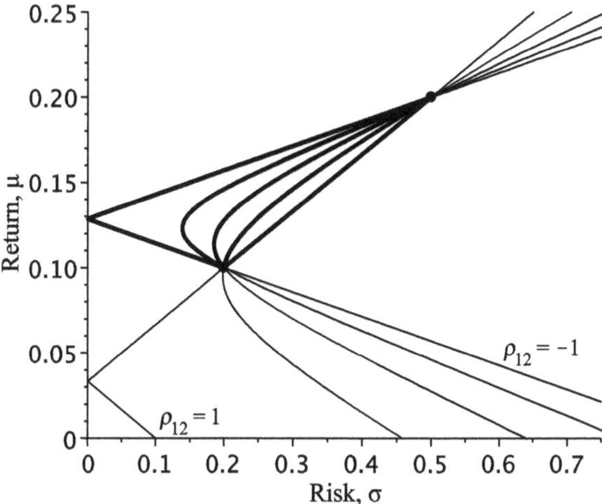

FIGURE 3.6: Portfolio lines for varying ρ_{12} and fixed μ's and σ's. The parameter ρ_{12} is changing from -1 to 1 with the step size of 0.5. The bold parts indicate portfolios without short selling.

3.2.2.3 Case with a Risk-Free Asset

Suppose that one of two assets (say, asset 2) in our portfolio is risk-free, that is, the variance of its return is zero: $\sigma_2^2 = \mathrm{Var}(r_2) = 0$. Hence, the return r_2 is constant: $r_2 \equiv r$.

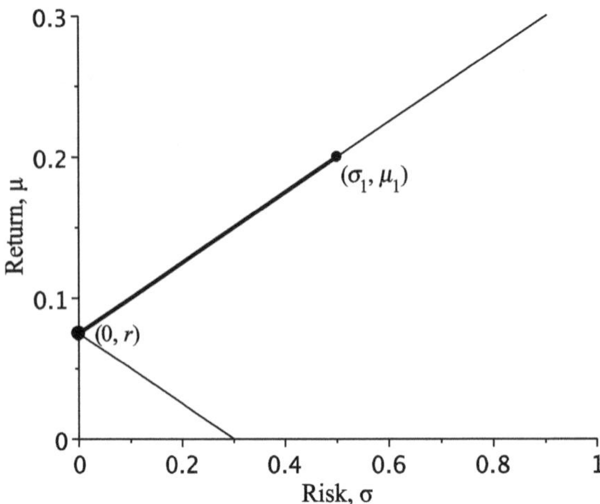

FIGURE 3.7: Portfolio line for one risky and one risk-free asset. The risk-free rate of return is r. The bold part indicates portfolios without short selling.

The formula of the variance in (3.9) reduces to $\sigma_V^2(x) = x^2\sigma_1^2$ or just $\sigma_V(x) = |x|\sigma_1$. So the standard deviation σ_V of such a portfolio depends on the weight w_1 of the risky asset as follows: $\sigma_V = |w_1|\sigma_1$. The portfolio line is described by a piecewise linear function: $\sigma_V = \sigma_1 \left|\frac{\mu_V - r}{\mu_1 - r}\right|$. Thus, the portfolio plot is a broken line with its vertex at the point that corresponds to the risk-free asset (see Figure 3.7).

3.2.3 The Minimum Variance Portfolio

As is seen in Figures 3.4 and 3.6, there is always a portfolio with the smallest possible variance σ_V^2. We already found risk-free portfolios with zero variance for the case with $|\rho_{12}| = 1$. Let us now find the general solution to this problem.

Theorem 3.2. *Suppose that $|\rho_{12}| < 1$ or $\sigma_1 \neq \sigma_2$ holds. The portfolio with the minimum variance is attained at*

$$\widehat{w}_1 = \frac{\sigma_2^2 - \rho_{12}\sigma_1\sigma_2}{\sigma_1^2 + \sigma_2^2 - 2\rho_{12}\sigma_1\sigma_2} \quad \textit{and} \quad \widehat{w}_2 = \frac{\sigma_1^2 - \rho_{12}\sigma_1\sigma_2}{\sigma_1^2 + \sigma_2^2 - 2\rho_{12}\sigma_1\sigma_2}. \tag{3.12}$$

The variance of the portfolio is

$$\sigma_{\mathrm{mv}}^2 = \frac{(1 - \rho_{12}^2)\sigma_1^2\sigma_2^2}{\sigma_1^2 + \sigma_2^2 - 2\rho_{12}\sigma_1\sigma_2}. \tag{3.13}$$

Proof. By differentiating the variance σ_V^2 given by (3.9) w.r.t. x and equating the derivative to zero, we obtain the following linear equation for x:

$$\frac{\mathrm{d}\sigma_V^2}{\mathrm{d}x} = 2(\sigma_1^2 + \sigma_2^2 - 2\rho_{12}\sigma_1\sigma_2)x - 2(\sigma_2^2 - \rho_{12}\sigma_1\sigma_2) = 0\,.$$

The solution is

$$x_0 = \frac{\sigma_2^2 - \rho_{12}\sigma_1\sigma_2}{\sigma_1^2 + \sigma_2^2 - 2\rho_{12}\sigma_1\sigma_2}. \tag{3.14}$$

Thus, for the weights $\widehat{w}_1 = x_0$ and $\widehat{w}_2 = 1 - x_0$, we immediately obtain (3.12). Since the second derivative of σ_V^2 w.r.t. x is positive everywhere:

$$\frac{\mathrm{d}^2\sigma_V^2}{\mathrm{d}x^2} = 2(\sigma_1^2 + \sigma_2^2 - 2\rho_{12}\sigma_1\sigma_2) > 0\,,$$

the variance σ_V^2 attains its smallest value at x_0. Substituting x_0 in (3.8) gives us Equation (3.13) for the minimum variance.

Clearly, the formulae in (3.12) and (3.13) work for both cases when $|\rho_{12}| = 1$ or $|\rho_{12}| < 1$. If $|\rho_{12}| = 1$, then $\sigma_{\mathrm{mv}}^2 = 0$ in (3.13) and the formulae in (3.12) reduce to (3.10) or (3.11) depending on the sign of ρ_{12}. □

3.2.3.1 Case without Short Selling

While proving Theorem 3.2, we did not take into account whether short sells are allowed. Let us find the minimum variance portfolio without short sells, i.e., with nonnegative weights. The function σ_V^2 in (3.9) attains its minimum value on $[0,1]$ either at one of the boundary points $x \in \{0,1\}$ or at the point x_0 given by (3.14) provided $0 < x_0 < 1$. Both weights $\widehat{w}_1$ and $\widehat{w}_2$ in (3.12) are positive iff $\rho_{12}\sigma_2 < \sigma_1$ and $\rho_{12}\sigma_1 < \sigma_2$, or, equivalently, if $\rho_{12} < \min\{\frac{\sigma_1}{\sigma_2}, \frac{\sigma_2}{\sigma_1}\}$. If that is the case, then it is possible to construct a portfolio without short selling with risk lower than that of any of the individual assets.

Otherwise, when $\rho_{12} \geqslant \min\{\frac{\sigma_1}{\sigma_2}, \frac{\sigma_2}{\sigma_1}\}$, the minimum variance portfolio without short selling is composed of shares of only one of the assets. If $\sigma_1 < \sigma_2$ (hence $x_0 > 1$), then the portfolio has only shares of asset 1 and its variance is σ_1^2. If $\sigma_2 < \sigma_1$ (hence $x_0 < 0$), then the portfolio has only shares of asset 2 and its variance is σ_2^2. For the special case with $\sigma_1 = \sigma_2$ and $\rho_{12} = 1$, the variance is the same for any portfolio: $\sigma_V^2 = \sigma_1^2 = \sigma_2^2$.

3.2.4 Selection of Optimal Portfolios

A typical problem of a risk manager is the selection of an optimal portfolio. Let us consider a portfolio with two risky assets. Suppose that the returns r_1 and r_2 follow a bivariate normal distribution, which is characterized by five parameters, namely, the expected returns, μ_1 and μ_2, the variances of returns, σ_1^2 and σ_2^2, and the correlation coefficient $\rho_{12} = \mathrm{Corr}(r_1, r_2)$.

There are many criteria that can be used to select an optimal portfolio. We consider three examples: minimization of the risk, maximization of an expected utility function of the return, and minimization of the probability of loss. All examples will be illustrated with the following data:

$$\mu_1 = 10\%,\ \sigma_1 = 20\%,\ \mu_2 = 15\%,\ \sigma_2 = 40\%, \quad \text{and} \quad \rho_{12} = -20\%. \tag{3.15}$$

3.2.4.1 Minimum Variance Portfolio

The variance of the terminal portfolio value is

$$\mathrm{Var}(V_T) = \mathrm{Var}\left(V_0\left(1 + r_V\right)\right) = V_0\left(1 + \mathrm{Var}(r_V)\right).$$

So, minimization of $\mathrm{Var}(V_T)$ is equivalent to minimization of $\sigma_V^2 = \mathrm{Var}(r_V)$. Let us find the weights $\widehat{w}_1 = x_0$ and $\widehat{w}_2 = 1 - x_0$ that minimize the variance of the portfolio return:

$$x_0 = \frac{\sigma_2^2 - \rho_{12}\sigma_1\sigma_2}{\sigma_1^2 + \sigma_2^2 - 2\rho_{12}\sigma_1\sigma_2} = \frac{0.4^2 - (-0.2)\cdot 0.2 \cdot 0.4}{0.2^2 + 0.4^2 - 2\cdot(-0.2)\cdot 0.2\cdot 0.4} \cong 0.7586\,.$$

Thus, $\widehat{w}_1 \cong 75.86\%$ and $\widehat{w}_2 \cong 24.14\%$. The expected return and variance of return are, respectively,

$$\mu_V = 0.1 \cdot 0.7586 + 0.15 \cdot 0.2414 \cong 0.1121 = 11.21\%,$$
$$\sigma_V^2 = 0.2^2 \cdot 0.7586^2 + 0.4^2 \cdot 0.2414^2 + 2 \cdot (-0.2) \cdot 0.2 \cdot 0.4 \cdot 0.7586 \cdot 0.2414 \cong 0.02648.$$

Thus, $\sigma_V \cong 0.1627 = 16.27\%$. Notice that $\mu_1 < \mu_V < \mu_2$ but $\sigma_V < \min\{\sigma_1, \sigma_2\} = \min\{0.2, 0.4\} = 0.2$. We managed to decrease the risk by diversifying the portfolio.

3.2.4.2 Maximum Expected Utility Portfolio

Let us find a portfolio that maximizes the mathematical expectation of the exponential utility function, $u(V) = 1 - \mathrm{e}^{-\alpha V}$ with $\alpha > 0$, of the wealth $V_T = (1 + r_V)V_0$ with some initial capital $V_0 > 0$:

$$\mathrm{E}[u(V_T)] = \mathrm{E}\big[1 - \mathrm{e}^{-\alpha V_0 (1+r_V)}\big] \to \max. \tag{3.16}$$

Choosing the fraction that maximizes utility is slightly more complicated. If we invest fraction x in the high-risk asset, then the terminal wealth will be $V_T(x) = (1 + r_V(x))V_0$, where we recall that the rate of return $r_V(x)$ is normally distributed with mean $\mu_V(x)$ and variance $\sigma_V^2(x)$. Using (3.9) gives $\mu_V(x) = \mu_2 + x(\mu_1 - \mu_2)$ and $\sigma_V^2(x) = Ax^2 + 2Bx + C$, where

$$A = \sigma_1^2 + \sigma_2^2 - 2\rho_{12}\sigma_1\sigma_2, \quad B = (\rho_{12}\sigma_1\sigma_2 - \sigma_2^2), \quad C = \sigma_2^2.$$

The expected utility is

$$\mathrm{E}\left[1 - \mathrm{e}^{-\alpha V_T(x)}\right] = 1 - \mathrm{E}\left[\mathrm{e}^{-\alpha V_T(x)}\right] = 1 - \mathrm{e}^{-\alpha V_0}\mathrm{E}\left[\mathrm{e}^{-\alpha V_0 r_V(x)}\right]$$
$$= 1 - \mathrm{e}^{-\alpha V_0}\mathrm{e}^{-\alpha V_0 \mu_V(x) + \alpha^2 V_0^2 \sigma_V^2(x)/2}.$$

It therefore suffices to maximize the function

$$\mu_V(x) - \frac{\alpha V_0 \sigma_V^2(x)}{2}.$$

Substituting the above expressions for μ_V and σ_V gives us the following target function to be maximized:

$$f(x) := \mu_2 + x(\mu_1 - \mu_2) - \frac{\alpha V_0}{2}(Ax^2 + 2Bx + C).$$

Differentiate f with respect to x and equate the derivative to zero to obtain

$$f'(x) = (\mu_1 - \mu_2) - \alpha V_0\,(Ax + B) = 0.$$

The solution is

$$x^* = -\frac{B}{A} + \frac{1}{A}\frac{\mu_1 - \mu_2}{\alpha V_0} = \frac{(\mu_1 - \mu_2)/(\alpha V_0) + \sigma_2^2 - \rho_{12}\sigma_1\sigma_2}{\sigma_1^2 + \sigma_2^2 - 2\rho_{12}\sigma_1\sigma_2}.$$

Since $f''(x) = -\alpha V_0 A < 0$, the function f is a concave function. Therefore, f attains its maximum at x^*. Suppose that $\alpha V_0 = 1$. Now we can do computations for our problem with the data in (3.15). The optimal weights are

$$\widehat{w}_1 = x^* \cong 0.5431 \quad \text{and} \quad \widehat{w}_2 = 1 - x^* \cong 0.4569.$$

The expected return is $\mu_V(x_0) \cong 12.28\%$; the volatility of return is $\sigma_V(x_0) \cong 19.30\%$.

One can consider other utility functions. However, in most cases we need to use a computational method to find the maximum of an expected utility function. The Taylor series approximation (3.5) can also be applied, as is demonstrated in the next example. Let us find an optimal portfolio when attempting to maximize the expected value of the square-root utility function of $V_T = (1 + r_V)V_0$:

$$\mathrm{E}[u(V_T)] = V_0\,\mathrm{E}\left[\sqrt{1+r_V}\right] \to \max\,. \tag{3.17}$$

Expand $\sqrt{1+r}$ in a Taylor series about the point $r = \mu_V(x)$ to obtain

$$\mathrm{E}\left[\sqrt{1+r_V(x)}\right] \approx \sqrt{1+\mu_V(x)} - \frac{1}{8}(1+\mu_V(x))^{-3/2}\sigma_V^2(x)\,,$$

where $r_V(x) = xr_1 + (1-x)r_2$, and $\mu_V(x)$ and $\sigma_V^2(x)$ are given by (3.9). Now the maximization problem (3.17) reduces to

$$f(x) := \sqrt{1 + x\mu_1 + (1-x)\mu_2} - \frac{x^2\sigma_1^2 + (1-x)^2\sigma_2^2 + 2x(1-x)\sigma_1\sigma_2\rho_{12}}{8(1 + x\mu_1 + (1-x)\mu_2)^{3/2}} \to \max_x\,.$$

Equating the derivative of the function f to zero and solving numerically the equation obtained give the optimal weights: $\widehat{w}_1 \cong 25.64\%$ and $\widehat{w}_2 \cong 74.36\%$. The resulting portfolio has the following expected return and volatility of return: $\mu_V \cong 13.72\%$ and $\sigma_V \cong 29.16\%$.

Remark. Let us assume without loss of generality that $\sigma_1 < \sigma_2$. We would then expect that $\mu_1 < \mu_2$, otherwise no risk-averse investor would ever want to purchase the high-risk asset. To minimize the volatility σ_V we should invest $x_0 = -\frac{B}{A}$ in the low-risk asset. To maximize the expected exponential utility, we invest $x^* = -\frac{B}{A} + \frac{1}{A}\frac{\mu_1-\mu_2}{\alpha V_0}$. Clearly, we have

$$x^* = x_0 + \frac{1}{A}\frac{\mu_1 - \mu_2}{\alpha V_0} < x_0\,,$$

since $\mu_1 < \mu_2$ holds. There are several important insights here. First, we see that maximizing expected utility is *not* the same as minimizing volatility, even for a risk-averse investor. Risk-averse investors are willing to take on a certain amount of risk *provided they are adequately compensated.* To see this, note that the difference $x_0 - x^*$ is increasing in $\mu_2 - \mu_1$; the greater the compensation being offered the more the risk-averse investor will allocate to the riskier asset. Risk aversion is therefore not the same as complete risk avoidance. Finally, if V_0 is large, then $x_0 - x^*$ will be small; the increase in wealth is simply not worth the possibility of losing large sums when marginal returns to wealth are small (as they are for wealthy individuals). Finally, we can observe that if σ_2 is large, then A is large, and the difference $x_0 - x^*$ is small.

3.2.4.3 Minimum Loss-Probability Portfolio

Suppose we wish to find the allocation weights when attempting to minimize the probability that the return on the portfolio is less than a certain threshold r_0:

$$\mathbb{P}(r_V \leqslant r_0) \to \min\,.$$

Given that r_1 and r_2 follow a bivariate normal distribution, the probability distribution of r_V is $Norm\big(\mu_V(x), \sigma_V^2(x)\big)$. Therefore,

$$\begin{aligned}\mathbb{P}(r_V(x) \leqslant r_0) &= \mathbb{P}\left(\frac{r_V(x) - \mu_V(x)}{\sigma_V(x)} \leqslant \frac{r_0 - \mu_V(x)}{\sigma_V(x)}\right)\\ &= \mathbb{P}\left(Z \leqslant \frac{r_0 - \mu_V(x)}{\sigma_V(x)}\right) = \mathcal{N}\left(\frac{r_0 - \mu_V(x)}{\sigma_V(x)}\right),\end{aligned}$$

where Z denotes a standard normal random variable and $\mathcal{N}$ is a standard normal CDF. Since a normal CDF is a strictly increasing function of its argument, it is sufficient to solve

$$\frac{\mu_V(x) - r_0}{\sigma_V(x)} \to \max_x . \tag{3.18}$$

In fact, Equation (3.18) relates to the so-called *Sharpe ratio*. The Sharpe ratio is a measure of the excess return (or risk premium) per unit of risk in an investment portfolio. It is named after William Forsyth Sharpe. The Sharpe ratio is defined as

$$\frac{\mathrm{E}[r_V - r_0]}{\sqrt{\mathrm{Var}(r_V - r_0)}} , \tag{3.19}$$

where r_0 is the return on a benchmark asset, such as the risk-free rate of return, $\mathrm{E}[r_V - r_0]$ is the expected value of the excess of the portfolio return r_V over the benchmark return, and $\mathrm{Var}(r_V - r_0)$ is the variance of the excess return. Since r_0 is constant, we have

$$\mathrm{E}[r_V - r_0] = \mathrm{E}[r_V] - r_0 = \mu_V - r_0 \quad \text{and} \quad \mathrm{Var}(r_V - r_0) = \mathrm{Var}(r_V) = \sigma_V^2 .$$

The Sharpe ratio is used to characterize how well the return of a portfolio compensates the investor for the risk taken. When comparing two portfolios against the same benchmark asset, the portfolio with the higher Sharpe ratio gives more return for the same level of risk. Investors are often advised to pick investments with high Sharpe ratios.

The solution to (3.18) can be obtained by using standard methods of calculus: differentiate the left-hand side of (3.18) w.r.t. x, equate the derivative to zero, and then solve the obtained equation for x. As a result, we obtain the following allocation weights:

$$\begin{aligned} \widehat{w}_1 &= \frac{(\mu_1 - r_0)\sigma_2^2 - (\mu_2 - r_0)\rho_{12}\sigma_1\sigma_2}{(\mu_1 - r_0)\sigma_2^2 + (\mu_2 - r_0)\sigma_1^2 - (\mu_1 + \mu_2 - 2r_0)\rho_{12}\sigma_1\sigma_2} , \\ \widehat{w}_2 &= \frac{(\mu_2 - r_0)\sigma_1^2 - (\mu_1 - r_0)\sigma_1\sigma_2\rho_{12}}{(\mu_1 - r_0)\sigma_2^2 + (\mu_2 - r_0)\sigma_1^2 - (\mu_1 + \mu_2 - 2r_0)\sigma_1\sigma_2\rho_{12}} . \end{aligned} \tag{3.20}$$

Let the risk-free rate be $r_0 = 5\%$. Substituting (3.15) into (3.20) gives us the following solution: $\widehat{w}_1 = \frac{2}{3}$ and $\widehat{w}_2 = \frac{1}{3}$. The expected return and volatility of the portfolio return are $\mu_V \cong 11.67\%$ and $\sigma_V \cong 16.87\%$, respectively.

3.3 Portfolio Optimization for N Assets

3.3.1 Portfolios of Several Assets

Consider a market model with N different assets $A_t^1, A_t^2, \ldots, A_t^N$, where $t \in \{0, T\}$. The return on the ith asset is $r_i = \frac{A_T^i - A_0^i}{A_0^i}$. Suppose a portfolio is constructed from these base assets. Let x_i be the number of shares of asset i with $i = 1, 2, \ldots, N$. The time-t portfolio value is $V_t = \sum_{i=1}^N x_i A_t^i$ for $t \in \{0, T\}$. The return on the portfolio is a linear combination of the returns on the assets:

$$\begin{aligned} r_V = \frac{V_T - V_0}{V_0} &= \sum_{i=1}^N \frac{x_i(A_T^i - A_0^i)}{V_0} \\ &= \sum_{i=1}^N \frac{x_i A_0^i}{V_0} \frac{A_T^i - A_0^i}{A_0^i} = \sum_{i=1}^N \frac{x_i A_0^i}{V_0} r_i . \end{aligned}$$

Define the allocation weights $w_i = \frac{x_i A_0^i}{V_0}$ with $i = 1, 2, \ldots, N$ of funds between the N base assets. The formula for the return r_V takes the following compact form:

$$r_V = w_1 r_1 + w_2 r_2 + \cdots + w_N r_N = \sum_{i=1}^{N} w_i r_i \,.$$

Let us denote

$$\mathbf{w} := \begin{bmatrix} w_1 & w_2 & \cdots & w_N \end{bmatrix}^\top \in \mathbb{R}^N \,.$$

Clearly, the sum of the weights is one. This fact can be written in vector form:

$$\mathbf{u}^\top \mathbf{w} = 1, \text{ where } \mathbf{u} := \begin{bmatrix} 1 & 1 & \cdots & 1 \end{bmatrix}^\top \in \mathbb{R}^N \,. \tag{3.21}$$

Here $\mathbf{x}^\top$ denotes the transpose of a vector $\mathbf{x}$. Here, we operate with column vectors.

We denote $\mu_i = \mathrm{E}[r_i]$—the expected return on asset i, $\sigma_i^2 = \mathrm{Var}(r_i)$—the variance of the return on asset i, and $c_{ij} = \mathrm{Cov}(r_i, r_j)$—the covariance between returns r_i and r_j for $i, j = 1, 2, \ldots, N$. The expected returns and covariances between returns can be respectively arranged into an $N \times 1$ column vector and an $N \times N$ matrix:

$$\mathbf{m} := \begin{bmatrix} \mu_1 \\ \mu_2 \\ \vdots \\ \mu_N \end{bmatrix} \quad \text{and} \quad \mathbf{C} := \begin{bmatrix} c_{11} & c_{12} & \cdots & c_{1N} \\ c_{21} & c_{22} & \cdots & c_{2N} \\ \vdots & \vdots & \ddots & \vdots \\ c_{N1} & c_{N2} & \cdots & c_{nn} \end{bmatrix} .$$

The matrix $\mathbf{C}$ is called a *covariance matrix*. The covariance $\sigma_{XY} \equiv \mathrm{Cov}(X, Y)$ of two random variables X and Y can be factorized into a product of the standard deviations, σ_X and σ_Y, and the coefficient of correlation between X and Y denoted by $\mathrm{Corr}(X, Y) \equiv \rho_{XY}$ as follows:

$$\sigma_{XY} = \sigma_X \, \rho_{XY} \, \sigma_Y \,.$$

Therefore, the covariance matrix $\mathbf{C}$ can be represented as a product of a diagonal matrix filled with standard deviations of returns, $\sigma_i := \sqrt{\mathrm{Var}(r_i)}$, and a correlation matrix whose entries are coefficients of correlation between returns, $\rho_{ij} \equiv \mathrm{Corr}(r_i, r_j)$, $i, j = 1, 2. \ldots, N$:

$$\mathbf{C} = \begin{bmatrix} \sigma_1 & 0 & \cdots & 0 \\ 0 & \sigma_2 & \cdots & 0 \\ \vdots & \vdots & \ddots & \vdots \\ 0 & 0 & \cdots & \sigma_N \end{bmatrix} \begin{bmatrix} 1 & \rho_{12} & \cdots & \rho_{1N} \\ \rho_{21} & 1 & \cdots & \rho_{2N} \\ \vdots & \vdots & \ddots & \vdots \\ \rho_{N1} & \rho_{N2} & \cdots & 1 \end{bmatrix} \begin{bmatrix} \sigma_1 & 0 & \cdots & 0 \\ 0 & \sigma_2 & \cdots & 0 \\ \vdots & \vdots & \ddots & \vdots \\ 0 & 0 & \cdots & \sigma_N \end{bmatrix} .$$

Here, we use the fact that $\mathrm{Corr}(X, X) = 1$ for every random variable X, hence $\rho_{ii} = 1$ for all $i = 1, 2, \ldots, N$.

The covariance matrix is symmetric (i.e., $\mathbf{C} = \mathbf{C}^\top$) and positive definite, i.e., $\mathbf{w}^\top \mathbf{C} \mathbf{w} > 0$ for every nonzero vector $\mathbf{w} \in \mathbb{R}^N$. Since $\mathbf{C}$ is positive definite, it is a nonsingular matrix and hence its inverse matrix $\mathbf{C}^{-1}$ exists. There exist several necessary and sufficient criteria to determine if a symmetric real matrix $\mathbf{C}$ is positive definite, including the following.

- All eigenvalues of $\mathbf{C}$ are positive.
- All the leading principal minors are positive. The kth leading principal minor of $\mathbf{C}$ is the determinant of the upper-left k-by-k corner of $\mathbf{C}$, where $k = 1, 2, \ldots, N$. This criterion is known as *Sylvester's criterion*.

- There exists a unique lower triangular matrix $\mathbf{L}$, with strictly positive diagonal elements, that allows the factorization of $\mathbf{C}$ into $\mathbf{C} = \mathbf{L}\mathbf{L}^\top$. Such a factorization is called the *Cholesky factorization.*

Note that in general C can be a semi-positive definite matrix, meaning that $\mathbf{w}^\top \mathbf{C}\mathbf{w} \geqslant 0$ for all $\mathbf{w} \in \mathbb{R}^N$.

Let us find the mathematical expectation and variance of r_V by applying well-known equations for the mathematical expectation and variance of a sum of (dependent) random variables. The expected return on the portfolio V with weights $\mathbf{w}$ is

$$\mu_V = \mathrm{E}[r_V] = \mathrm{E}\left[\sum_{i=1}^N w_i r_i\right] = \sum_{i=1}^N \mathrm{E}[w_i r_i] = \sum_{i=1}^N w_i \mu_i\,; \tag{3.22}$$

the variance of r_V is

$$\begin{aligned}\sigma_V^2 &= \mathrm{Var}(r_V) = \mathrm{Var}\left(\sum_{i=1}^N w_i r_i\right)\\ &= \mathrm{Cov}\left(\sum_{i=1}^N w_i r_i, \sum_{j=1}^N w_j r_j\right) = \sum_{i=1}^N\sum_{j=1}^N \mathrm{Cov}(w_i r_i, w_j r_j)\\ &= \sum_{i=1}^N\sum_{j=1}^N w_i w_j c_{ij}\,.\end{aligned} \tag{3.23}$$

The above equations can be written in matrix-vector form:

$$\mu_V = \mathbf{m}^\top \mathbf{w}; \tag{3.24}$$

$$\sigma_V^2 = \mathbf{w}^\top \mathbf{C}\mathbf{w}\,. \tag{3.25}$$

Note that we do not assume any probability distribution for the vector of returns. Our analysis of portfolios is entirely based on the knowledge of the vector of expected returns $\mathbf{m}$ and covariance matrix $\mathbf{C}$.

In the next sections, we shall solve the following two problems.

1. Find a portfolio with the minimum variance. It will be called the *minimum variance portfolio.*

2. Find a portfolio with the minimum variance among all portfolios whose expected return is fixed and equal to a given number. We may obtain different solutions for portfolios with or without short sells. The set of such portfolios parameterized by the expected return is called the *minimum variance (portfolio) line.*

3.3.2 The Minimum Variance Portfolio

To find the minimum variance portfolio, we need to solve

$$f(\mathbf{w}) := \mathbf{w}^\top \mathbf{C}\mathbf{w} \to \min_{\mathbf{w}} \tag{3.26}$$

subject to the constraint

$$\mathbf{u}^\top \mathbf{w} = 1\,. \tag{3.27}$$

Let us use the method of Lagrange multipliers. First, we find the critical points of the function

$$F(\mathbf{w},\lambda) := \mathbf{w}^\top \mathbf{C}\mathbf{w} - \lambda(\mathbf{u}^\top \mathbf{w} - 1).$$

The partial derivatives of F with respect to w_i for $i = 1, 2, \ldots, N$ are

$$\begin{aligned}\frac{\partial F}{\partial w_i}(\mathbf{w},\lambda) &= \frac{\partial}{\partial w_i}\left(\sum_{i=1}^{N}\sum_{j=1}^{N} w_i w_j c_{ij} - \lambda \sum_{i=1}^{N} w_i + \lambda\right) \\ &= 2\sum_{j=1}^{N} w_j c_{ij} - \lambda.\end{aligned}$$

Equating them to zero gives us the following linear equations:

$$2\sum_{j=1}^{N} w_j c_{ij} - \lambda = 0 \quad \text{for all} \quad i = 1, 2, \ldots, N.$$

Let $\mathbf{c}_j^\top$ denote the jth row of matrix $\mathbf{C}$. Then the above equations can be rewritten in vector form:

$$2\mathbf{c}_j^\top \mathbf{w} - \lambda = 0 \quad \text{for} \quad i = 1, 2, \ldots, N.$$

Finally, we have $2\mathbf{C}\mathbf{w} - \lambda\mathbf{u} = \mathbf{0}$. Multiplying both parts by the inverse matrix $\mathbf{C}^{-1}$ from the left gives

$$2\mathbf{C}^{-1}\mathbf{C}\mathbf{w} - \lambda\mathbf{C}^{-1}\mathbf{u} = 2\mathbf{w} - \lambda\mathbf{C}^{-1}\mathbf{u} \text{ and } \mathbf{C}^{-1}\mathbf{0} = \mathbf{0} \implies 2\mathbf{w} - \lambda\mathbf{C}^{-1}\mathbf{u} = \mathbf{0}.$$

Solve this equation for $\mathbf{w}$ to obtain that $\mathbf{w} = \frac{\lambda}{2}\mathbf{C}^{-1}\mathbf{u}$. The only missing variable is λ. Substitute the expression for $\mathbf{w}$ in the constraint (3.27) to obtain

$$1 = \frac{\lambda}{2}\mathbf{u}^\top \mathbf{C}^{-1}\mathbf{u} \implies \lambda = \frac{2}{\mathbf{u}^\top \mathbf{C}^{-1}\mathbf{u}}.$$

Finally, we obtain the weight vector for the minimum variance portfolio:

$$\mathbf{w}_{\text{mv}} = \frac{\mathbf{C}^{-1}\mathbf{u}}{\mathbf{u}^\top \mathbf{C}^{-1}\mathbf{u}}. \tag{3.28}$$

Since the matrix of second derivatives of the function $f(\mathbf{w})$ is $2\mathbf{C}$ (which is positive definite), the function $F(\mathbf{w},\lambda)$ is a concave function of $\mathbf{w}$ for every value of λ. Therefore, the function $f(\mathbf{w})$ has a minimum at $\mathbf{w}_{\text{mv}}$. The minimum variance can be computed by putting the weights $\mathbf{w}_{\text{mv}}$ in (3.23):

$$\sigma_{\text{mv}}^2 = \mathbf{w}_{\text{mv}}^\top \mathbf{C}\mathbf{w}_{\text{mv}} = \frac{1}{\mathbf{u}^\top \mathbf{C}^{-1}\mathbf{u}}.$$

As an example, let us consider the case of two assets. The covariance matrix $\mathbf{C}$ and its inverse can be written using $\sigma_i = \sqrt{\operatorname{Var}(r_i)}$, $i = 1, 2$, and $\rho_{12} = \operatorname{Corr}(r_1, r_2)$:

$$\mathbf{C} = \begin{bmatrix} \sigma_1^2 & \rho_{12}\sigma_1\sigma_2 \\ \rho_{12}\sigma_1\sigma_2 & \sigma_2^2 \end{bmatrix} \text{ and } \mathbf{C}^{-1} = \frac{1}{1-\rho_{12}^2}\begin{bmatrix} \frac{1}{\sigma_1^2} & -\frac{\rho_{12}}{\sigma_1\sigma_2} \\ -\frac{\rho_{12}}{\sigma_1\sigma_2} & \frac{1}{\sigma_2^2} \end{bmatrix}.$$

The weight vector is then

$$\begin{aligned}\mathbf{w}_{\text{mv}}^\top = [w_1,\ w_2] &= \frac{1}{\frac{1}{\sigma_1^2} - \frac{2\rho_{12}}{\sigma_1\sigma_2} + \frac{1}{\sigma_2^2}}\left[\frac{1}{\sigma_1^2} - \frac{\rho_{12}}{\sigma_1\sigma_2},\ \frac{1}{\sigma_2^2} - \frac{\rho_{12}}{\sigma_1\sigma_2}\right] \\ &= \left[\frac{\sigma_2^2 - \rho_{12}\sigma_1\sigma_2}{\sigma_1^2 - 2\rho_{12}\sigma_1\sigma_2 + \sigma_2^2},\ \frac{\sigma_1^2 - \rho_{12}\sigma_1\sigma_2}{\sigma_1^2 - 2\rho_{12}\sigma_1\sigma_2 + \sigma_2^2}\right].\end{aligned}$$

The resulting expression is identical to that of (3.12).

3.3.3 The Minimum Variance Portfolio Line

Now, we consider a set of portfolios with fixed expected return μ, i.e., $\mathbf{m}^\top\mathbf{w} = \mu$. To find the minimum variance portfolio in such a set, we need to minimize $f(\mathbf{w}) := \mathbf{w}^\top\mathbf{C}\mathbf{w}$ subject to the constraints

$$\mathbf{u}^\top\mathbf{w} = 1 \quad\text{and}\quad \mathbf{m}^\top\mathbf{w} = \mu\,. \tag{3.29}$$

As a result, we obtain a family of minimum variance portfolios $\mathbf{w} = \widehat{\mathbf{w}}(\mu)$ parameterized by μ. On the risk-return plot, such a family is represented by a continuous line called the *minimum variance line.*

Again, to find the equation of the minimum variance line we apply the method of Lagrange multipliers. Introduce the function

$$G(\mathbf{w}, \lambda_1, \lambda_2) := \mathbf{w}^\top\mathbf{C}\mathbf{w} - \lambda_1(\mathbf{u}^\top\mathbf{w} - 1) - \lambda_2(\mathbf{m}^\top\mathbf{w} - \mu)\,,$$

where λ_1 and λ_2 are Lagrange multipliers. Differentiate G w.r.t. weight w_i and equate the derivative to zero:

$$\frac{\partial G}{\partial w_i} = 2\sum_{j=1}^{N} w_j c_{ij} - \lambda_1 - \lambda_2\mu_i = 0 \quad\text{for}\quad i = 1, 2, \ldots, N\,.$$

The above simultaneous linear equations can be expressed in matrix-vector form:

$$2\mathbf{C}\mathbf{w} - \lambda_1\mathbf{u} - \lambda_2\mathbf{m} = 0.$$

By solving for the weights $\mathbf{w}$, we obtain:

$$\mathbf{w} = \mathbf{C}^{-1}\left(\frac{\lambda_1}{2}\mathbf{u} + \frac{\lambda_2}{2}\mathbf{m}\right) = \frac{\lambda_1}{2}\mathbf{C}^{-1}\mathbf{u} + \frac{\lambda_2}{2}\mathbf{C}^{-1}\mathbf{m}. \tag{3.30}$$

The constraints (3.29) are revealed from equations $\frac{\partial G}{\partial \lambda_i} = 0$ for $i = 1, 2$. Now substitute this expression for $\mathbf{w}$ into the constraints (3.29) to obtain the following system of equations:

$$\begin{cases} \frac{1}{2}\mathbf{u}^\top\mathbf{C}^{-1}\mathbf{u}\lambda_1 + \frac{1}{2}\mathbf{u}^\top\mathbf{C}^{-1}\mathbf{m}\lambda_2 = 1, \\ \frac{1}{2}\mathbf{m}^\top\mathbf{C}^{-1}\mathbf{u}\lambda_1 + \frac{1}{2}\mathbf{m}^\top\mathbf{C}^{-1}\mathbf{m}\lambda_2 = \mu. \end{cases} \tag{3.31}$$

Recall that a 2-by-2 system of linear equations

$$\begin{cases} a_{11}x_1 + a_{12}x_2 = b_1, \\ a_{21}x_1 + a_{22}x_2 = b_2 \end{cases}$$

admits a unique solution $x_1 = \frac{1}{D}\begin{vmatrix} b_1 & a_{12} \\ b_2 & a_{22} \end{vmatrix}$ and $x_2 = \frac{1}{D}\begin{vmatrix} a_{11} & b_1 \\ a_{21} & b_2 \end{vmatrix}$ with $D := \begin{vmatrix} a_{11} & a_{12} \\ a_{21} & a_{22} \end{vmatrix}$ provided $D \neq 0$. Here, $\begin{vmatrix} a & b \\ c & d \end{vmatrix} = ad - bc$ denotes the determinant of a 2-by-2 matrix. Solve the system (3.31) for λ_1 and λ_2 and then plug the solution into (3.30) to obtain the final formula for the portfolio weights:

$$\widehat{\mathbf{w}} = \frac{1}{D}\begin{vmatrix} 1 & \mathbf{u}^\top\mathbf{C}^{-1}\mathbf{m} \\ \mu & \mathbf{m}^\top\mathbf{C}^{-1}\mathbf{m} \end{vmatrix}\mathbf{C}^{-1}\mathbf{u} + \frac{1}{D}\begin{vmatrix} \mathbf{u}^\top\mathbf{C}^{-1}\mathbf{u} & 1 \\ \mathbf{m}^\top\mathbf{C}^{-1}\mathbf{u} & \mu \end{vmatrix}\mathbf{C}^{-1}\mathbf{m} \tag{3.32}$$

with $D := \begin{vmatrix} \mathbf{u}^\top\mathbf{C}^{-1}\mathbf{u} & \mathbf{u}^\top\mathbf{C}^{-1}\mathbf{m} \\ \mathbf{m}^\top\mathbf{C}^{-1}\mathbf{u} & \mathbf{m}^\top\mathbf{C}^{-1}\mathbf{m} \end{vmatrix} \neq 0$. The determinants in (3.32) are linear functions of

μ. Therefore, the weights of portfolios on the minimum variance line depend on μ linearly as well: $\widehat{\mathbf{w}} = \mu\mathbf{a} + \mathbf{b}$ with

$$\mathbf{a} := \frac{\mathbf{u}^\top \mathbf{C}^{-1}\mathbf{u}\mathbf{C}^{-1}\mathbf{m} - \mathbf{u}^\top \mathbf{C}^{-1}\mathbf{m}\mathbf{C}^{-1}\mathbf{u}}{D},$$

$$\mathbf{b} := \frac{\mathbf{m}^\top \mathbf{C}^{-1}\mathbf{m}\mathbf{C}^{-1}\mathbf{u} - \mathbf{m}^\top \mathbf{C}^{-1}\mathbf{u}\mathbf{C}^{-1}\mathbf{m}}{D}.$$

This observation allows us to describe the shape of the minimum variance line. Let us select two different portfolios with respective weights $\mathbf{w}'$ and $\mathbf{w}''$ on the line. Then the minimum variance line consists of portfolios with weights $\mathbf{w} = x\mathbf{w}' + (1-x)\mathbf{w}''$ for $x \in \mathbb{R}$. Indeed, the weights of the two chosen portfolios satisfy $\mathbf{w}' = \mu'\mathbf{a} + \mathbf{b}$ and $\mathbf{w}'' = \mu''\mathbf{a} + \mathbf{b}$ for some $\mu' \neq \mu''$. Every linear combination of the weights $\mathbf{w}'$ and $\mathbf{w}''$ satisfies the same equation:

$$\mathbf{w} = x\mathbf{w}' + (1-x)\mathbf{w}'' = (x\mu' + (1-x)\mu'')\mathbf{a} + (x + (1-x))\mathbf{b} = \mu_x\mathbf{a} + \mathbf{b}$$

with $\mu_x = x\mu' + (1-x)\mu''$. Conversely, for every $\mu \in \mathbb{R}$ there exist $x \in \mathbb{R}$ so that $\mu = x\mu' + (1-x)\mu''$. Therefore, the portfolios with weights $x\mathbf{w}' + (1-x)\mathbf{w}''$, $x \in \mathbb{R}$, exhaust the whole minimum variance line. This result means that the minimum variance line has the same shape as that describing a set of portfolios constructed from two assets. The shape of the line (which is a hyperbola) does not depend on the number of assets. The set of admissible portfolios is represented by a planar domain bounded by the minimum variance line. The shape of this domain is known as the *Markowitz bullet*. All elementary portfolios consisting of individual assets lie inside the bullet, as shown in Figure 3.8.

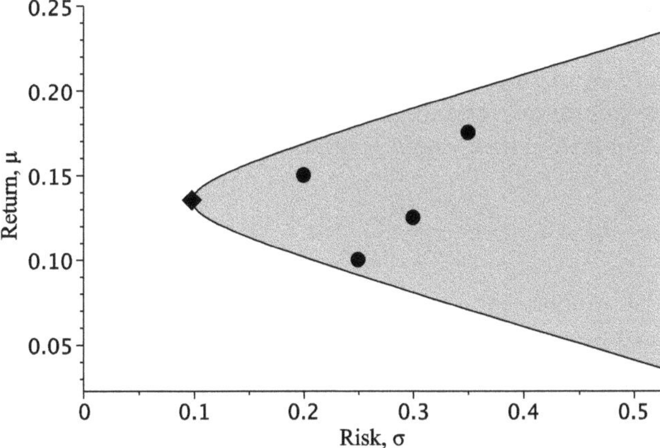

FIGURE 3.8: The set of admissible portfolios (the Markowitz bullet) in four underlying assets (which are marked by solid circles) bounded by the minimum variance line. The minimum variance portfolio is marked by a diamond.

Example 3.5. Let us consider a portfolio in three underlying assets whose expected returns, standard deviations of returns, and correlations between returns are as follows:

$$\begin{aligned} &\mu_1 = 0.1, && \sigma_1 = 0.2, && \rho_{12} = \rho_{21} = -0.2, \\ &\mu_2 = 0.15, && \sigma_2 = 0.3, && \rho_{23} = \rho_{32} = 0.2, \\ &\mu_3 = 0.3, && \sigma_3 = 0.4, && \rho_{31} = \rho_{13} = -0.4. \end{aligned}$$

(a) Find the minimum variance portfolio.

(b) Find the minimum variance portfolio line.

Solution. First, to apply the formulae in (3.28) and (3.32), we arrange the expected returns μ_i in a vector $\mathbf{m}$ and construct the covariance matrix $\mathbf{C}$ with entries $C_{ij} = \sigma_i \sigma_j \rho_{ij}$:

$$\mathbf{m} = \begin{bmatrix} 0.10 \\ 0.15 \\ 0.30 \end{bmatrix}, \quad \mathbf{C} = \begin{bmatrix} 0.040 & -0.012 & -0.032 \\ -0.012 & 0.090 & 0.024 \\ -0.032 & 0.024 & 0.160 \end{bmatrix}.$$

The matrix $\mathbf{C}$ is positive definite, hence the inverse matrix $\mathbf{C}^{-1}$ exists:

$$\mathbf{C}^{-1} \cong \begin{bmatrix} 30.3030 & 2.5252 & 5.6818 \\ 2.5252 & 11.7845 & -1.2626 \\ 5.6818 & -1.2626 & 7.5758 \end{bmatrix}.$$

The weights of the minimum variance portfolio are

$$\mathbf{w}_{\text{mv}} = \frac{\mathbf{u}^\top \mathbf{C}^{-1}}{\mathbf{u}^\top \mathbf{C}^{-1}\mathbf{u}} \cong \begin{bmatrix} 0.6060 & 0.2053 & 0.1887 \end{bmatrix}^\top.$$

The expected return μ_{mv} and standard deviation (the risk) of the return σ_{mv} of the minimum variance portfolio are

$$\mu_{\text{mv}} = \mathbf{m}^\top \mathbf{w}_{\text{mv}} \cong 0.1480 \quad \text{and} \quad \sigma_{\text{mv}} = \sqrt{\mathbf{w}_{\text{mv}}^\top \mathbf{C} \mathbf{w}_{\text{mv}}} \cong 0.1254.$$

To describe the minimum variance line, we need to find the weight vectors for two portfolios on the line. We found one of them—the minimum variance portfolio. Since $\mu_{\text{mv}} \neq 0$, the other portfolio on the minimum variance line to be selected can be a portfolio with zero expected return. To find its weights, apply E quation (3.32) where we put $\mu = 0$ to obtain

$$\mathbf{w}_0 = \frac{\mathbf{m}^\top \mathbf{C}^{-1}\mathbf{m}}{D} \mathbf{C}^{-1}\mathbf{u} - \frac{\mathbf{m}^\top \mathbf{C}^{-1}\mathbf{u}}{D} \mathbf{C}^{-1}\mathbf{m} \cong \begin{bmatrix} 1.1459 & 0.4721 & -0.6180 \end{bmatrix}^\top.$$

Now the portfolios with weights $x\mathbf{w}_{\text{mv}} + (1-x)\mathbf{w}_0$, $x \in \mathbb{R}$, exhaust the minimum variance line. □

Since $w_3 = 1 - w_1 - w_2$, all portfolios in three basis assets from the above example can be described by the weights w_1 and w_2. On the (w_1, w_2)-plane, every portfolio line is represented by a straight line. For example, the line given by the equation $w_1 = 0$ represents all portfolios in the basis assets 2 and 3 only; the line $w_1 + w_2 = 1$ represents all portfolios in the basis assets 1 and 2 only, etc. Figure 3.9 visualizes the set of admissible portfolios from Example 3.5 on the (w_1, w_2)-plane (the left plot) and on the (σ, μ)-plane (the right plot). The bold line represents the minimum variance line; the minimum variance portfolio is marked by a diamond.

3.3.4 Case without Short Selling

The case without short selling is very similar to that considered in the previous section. No short selling means that all positions in an investment portfolio have to be nonnegative. We can find the minimum variance portfolio line and the minimum variance portfolio by solving respective quadratic programming problems that have one additional condition: the

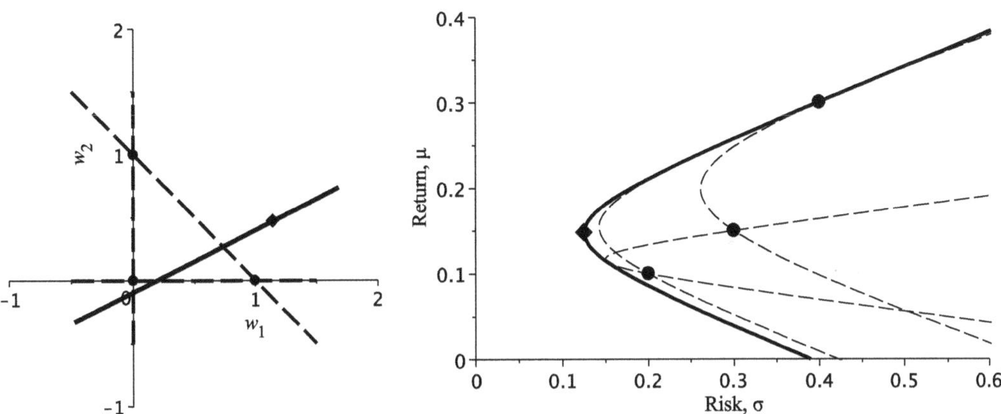

FIGURE 3.9: The minimum variance line from Example 3.5 is plotted as a bold line. The minimum variance portfolio is marked by a diamond. The basis assets are represented by solid circles. The dashed lines represent portfolio lines.

weights w_i are now nonnegative. To find the minimum variance portfolio, we need to solve (3.26):

$$f(\mathbf{w}) := \mathbf{w}^\top \mathbf{C}\mathbf{w} \to \min_{w}$$

subject to the constraints

$$\mathbf{u}^\top \mathbf{w} = 1, \ \mathbf{w} \geqslant \mathbf{0}. \tag{3.33}$$

Here, the meaning of $\mathbf{w} \geqslant \mathbf{0}$ is that all $w_i \geqslant 0$. To obtain a family of minimum variance portfolios parameterized by the expected return μ, we need to minimize $f(\mathbf{w})$ subject to the constraints

$$\mathbf{u}^\top \mathbf{w} = 1, \ \mathbf{m}^\top \mathbf{w} = \mu, \ \text{ and } \ \mathbf{w} \geqslant \mathbf{0}. \tag{3.34}$$

The constraints in (3.33) and (3.34) are almost the same as are in (3.27) and (3.29), respectively. The quadratic problems can be solved numerically. Computer systems such as MAPLE™, MATHEMATICA™, and MATLAB™ can be applied to solve the minimization problems (3.26)–(3.33) and (3.26)–(3.34).

Let us consider the case with three assets from Example 3.5. The weights can be parameterized by two real variables $w_1, w_2 \in [0,1]$ with $w_1 + w_2 \leqslant 1$ and hence $w_3 = 1 - w_1 - w_2 \geqslant 0$. On the (w_1, w_2)-plane, the set of admissible portfolios is represented by a triangle with vertices $(0,0)$, $(0,1)$, and $(1,0)$. Clearly, the expected return on a portfolio with nonnegative weights $\mathbf{w}$ is bounded above and below by $\max \mu_i$ and $\min \mu_i$, respectively. Therefore, the minimum variance line is a bounded curve on the (σ, μ)-plane. It connects two points corresponding to the assets with the lowest μ and highest μ, respectively. The set of admissible portfolios (with nonnegative weights) is represented by a planar domain bounded by the minimum variance line and portfolio lines (without short selling) corresponding to different pairs of the underlying assets (see Figure 3.10).

3.3.5 Efficient Frontier and Capital Market Line

Given the choice between two risky assets, a rational investor will choose an asset with higher expected return μ and lower risk σ.

Definition 3.1. An asset with (μ_1, σ_1) is said to *dominate* another asset with (μ_2, σ_2) whenever $\mu_1 \geqslant \mu_2$ and $\sigma_1 \leqslant \sigma_2$. A portfolio in risky assets is called *efficient* if there is no

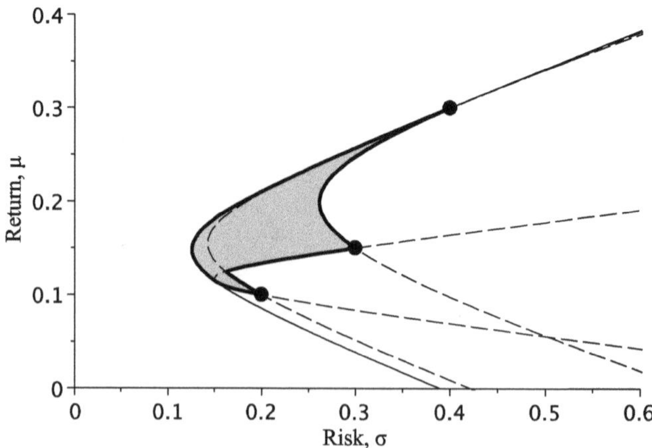

FIGURE 3.10: The set of admissible portfolios in three underlying assets without short selling.

other portfolio, except itself, that dominates it. The set of efficient portfolios among all attainable portfolios is called the *efficient frontier.*

In particular, an efficient portfolio has the highest expected return μ among all attainable portfolios with the same level of risk σ and has the lowest σ among all attainable portfolios with the same μ.

Let us consider the case with two risky assets. The set of admissible portfolios is represented by a portfolio line on the (σ, μ)-plane. The line is passing through the two base assets (σ_1, μ_1) and (σ_2, μ_2). As was proved in the previous section, there is a portfolio with the minimum possible variance σ_{mv}^2 given in (3.13). For every $\sigma > \sigma_{\text{mv}}$, there are two portfolios on the portfolio line, (σ, μ_1) and (σ, μ_2), with $\mu_1 < \mu_2$. A rational investor would choose the portfolio (σ, μ_2) with a higher expected return. Therefore, in the case of two risky assets, the efficient frontier is the upper half of the portfolio line with the minimum variance portfolio $(\sigma_{\text{mv}}, \mu_{\text{mv}})$ as an endpoint. If one of the two assets is risk-free, then the portfolio line is a broken line with its vertex at the risk-free asset. The efficient frontier is the upper half-line. Both cases are represented in Figure 3.11.

In the situation with multiple risky assets ($N > 2$), the set of admissible portfolios (the Markowitz bullet) is a planar domain on the (σ, μ)-plane bounded by the minimum variance line. Fix the value of $\sigma \geqslant \sigma_{\text{mv}}$ and consider all admissible portfolios V with the standard deviation $\sigma_V = \sigma$. On the (σ, μ)-plane, this set is a segment enclosed by the minimum variance line. By maximizing the expected return, we find that the efficient portfolios are all lying on the upper half of the minimum variance line (see Figure 3.12a).

Finally, let us assume that one risk-free asset labelled B with the rate of return r is available in addition to N risky assets. Let $100\alpha\%$ of the capital be allocated in the risk-free asset and $100(1-\alpha)\%$ is a risky portfolio:

$$r_V = \alpha r + (1-\alpha) \underbrace{\sum_{i=1}^{N} w_i r_i}_{=:r_{\mathcal{M}}} = \alpha r + (1-\alpha) r_{\mathcal{M}},$$

where $w_1, \ldots, w_N$ are the allocation weights for the risky assets (so that $w_1 + \cdots + w_N = 1$ holds).

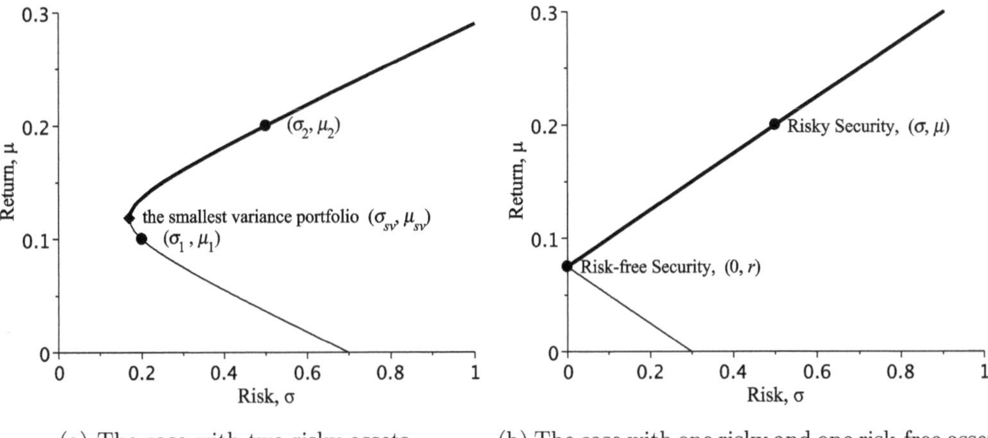

(a) The case with two risky assets. (b) The case with one risky and one risk-free asset

FIGURE 3.11: The efficient frontier (the bold line) for two assets.

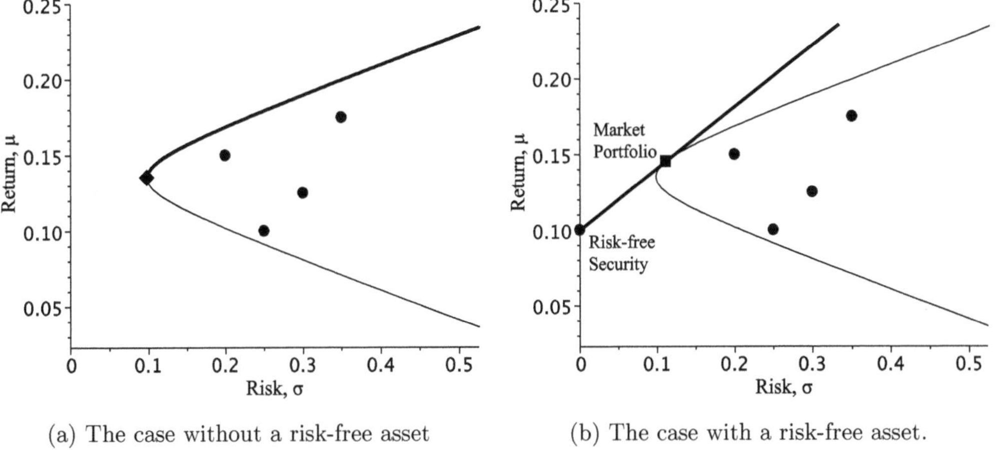

(a) The case without a risk-free asset (b) The case with a risk-free asset.

FIGURE 3.12: The efficient frontier (the bold line) constructed from multiple risky assets.

As is shown in Subsection 3.2.2, all portfolios with $r_V = \alpha r + (1-\alpha) r_{\mathcal{M}}$ consisting of one risk-free and one risky asset (the risky portfolio $V_{\mathcal{M}}$ of an investment portfolio can be considered as a new asset) form a broken line having upper and lower half-lines with the common vertex at the point with coordinates $(0, r)$. The efficient frontier constructed from such portfolios is the upper half-line like that in Figure 3.11. By taking a risky portfolio $V_{\mathcal{M}}$ anywhere in the Markowitz bullet, we can construct the set of admissible portfolios that is represented on the (σ, μ)-plane by a cone bounded by two half-lines, as is shown in Figure 3.13.

The efficient frontier of the portfolios containing a risk-free asset in addition to N risky ones is the upper half-line which is passing through the point representing the risk-free asset and tangent to the minimum variance line. Indeed, to minimize the risk, the portfolio $V_{\mathcal{M}}$ in risky assets has to be selected on the minimum variance line. To maximize the return, the portfolio $V_{\mathcal{M}}$ has to be selected so that the upper half-line has the largest possible slope. If the risk-free return r is not too high, the largest possible slope is achieved when the upper half-line is tangent to the Markowitz bullet. If r is too high, then the efficient

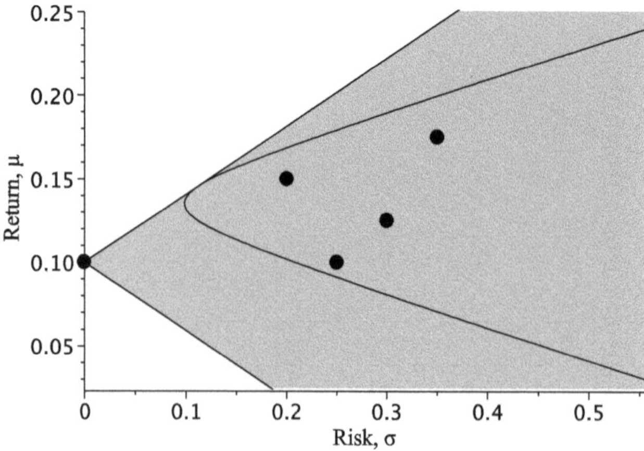

FIGURE 3.13: The set of admissible portfolios constructed from four risky assets and one risk-free asset.

frontier is obtained in the limiting case as the portfolio $V_{\mathcal{M}}$ selected on the upper half of the minimum variance line goes to infinity. The efficient frontier is no longer tangent to the Markowitz bullet, but is parallel to its asymptote (recall that the shape of the bullet is a hyperbola). The tangency point with coordinates $(\sigma_{\mathcal{M}}, \mu_{\mathcal{M}})$ is the so-called *market portfolio*. The efficient frontier is called the *capital market line*. Every rational investor forming her portfolio with a risk-free asset with return r and risky assets available on the market selects the portfolio on this line. Figure 3.12b shows the market line for portfolios with one risk-free and several risky assets.

The weights of the market portfolio are

$$\mathbf{w}_{\mathcal{M}} = \frac{(\mathbf{m}^\top - r\mathbf{u}^\top)\mathbf{C}^{-1}}{\mathbf{u}^\top \mathbf{C}^{-1}(\mathbf{m} - r\mathbf{u})}.$$

The expected return, $\mu_{\mathcal{M}}$, and variance of return, $\sigma_{\mathcal{M}}^2$, of the market portfolio can be found by using (3.24) and (3.25). The capital market line that starts at the risk-free asset (represented by the point $(0, r)$ on the (σ, μ)-plane) and passes through the market portfolio with expected return $\mu_{\mathcal{M}}$ and standard deviation of return $\sigma_{\mathcal{M}}$ satisfies the equation

$$\frac{\mu - r}{\mu_{\mathcal{M}} - r} = \frac{\sigma - 0}{\sigma_{\mathcal{M}} - 0} \iff \mu = r + \frac{\mu_{\mathcal{M}} - r}{\sigma_{\mathcal{M}}}\sigma.$$

3.4 The Capital Asset Pricing Model

The Capital Asset Pricing Model (CAPM) attempts to relate r_i, the return on asset i, to $r_{\mathcal{M}}$, the return of the entire market, which can be measured by some index such as Standard and Poor's index of 500 stocks (S&P500). In the Markowitz portfolio model, the market portfolio can be used as a good approximation to such a market index. Indeed, every rational investor will select a portfolio on the capital market line since it is the efficient frontier constructed from a risk-free asset and several risky assets. Therefore, every investor will be holding a portfolio with the same relative proportions of risky assets. This means that for each risky asset its weight in the market portfolio is equal to the relative share of the asset in the whole market.

The CAPM assumes that the dependence between r_i and $r_{\mathcal{M}}$ takes the following form:

$$r_i = r + \beta_i(r_{\mathcal{M}} - r) + \epsilon_i\,, \tag{3.35}$$

where β_i is a constant called the *beta factor* for asset i, r is a risk-free rate of return, and ϵ_i is a residual random variable having a normal distribution with mean zero. The residual ϵ_i is assumed to be independent of $r_{\mathcal{M}}$.

There are several ways to compute beta factors.

(1) Suppose that the joint probability distribution of r_i and $r_{\mathcal{M}}$ is given. Compute the covariance of r_i and $r_{\mathcal{M}}$ by employing (3.35) and using the independence of $r_{\mathcal{M}}$ and ϵ_i:

$$\operatorname{Cov}(r_i, r_{\mathcal{M}}) = \underbrace{\operatorname{Cov}(r, r_{\mathcal{M}})}_{=0} + \beta_i \underbrace{\operatorname{Cov}(r_{\mathcal{M}}, r_{\mathcal{M}})}_{=\operatorname{Var}(r_{\mathcal{M}})} - \beta_i \underbrace{\operatorname{Cov}(r, r_{\mathcal{M}})}_{=0} + \underbrace{\operatorname{Cov}(\epsilon_i, r_{\mathcal{M}})}_{=0} = \beta_i \operatorname{Var}(r_{\mathcal{M}}\,).$$

Therefore, the beta factor of asset i is given by

$$\beta_i = \frac{\operatorname{Cov}(r_i, r_{\mathcal{M}})}{\operatorname{Var}(r_{\mathcal{M}})}\,. \tag{3.36}$$

(2) Consider a market model with a set of market scenarios Ω. Suppose that for each market scenario $\omega \in \Omega$, the values of returns on asset i and the market portfolio $\mathcal{M}$ are given. We can plot the value of $r_i(\omega^j)$ against $r_{\mathcal{M}}(\omega)$ for each $\omega \in \Omega$ and then find the line of best fit, also known as the regression line. Employ the model $r_i = \alpha + \beta r_{\mathcal{M}} + \epsilon_i$. So the residual random variable $\epsilon_i \colon \Omega \to \mathbb{R}$ is the difference between the actual return r_i and the predicted return $\alpha + \beta r_{\mathcal{M}}$. The line of best fit is defined by

$$\mathrm{E}[\epsilon_i^2] \to \min_{\alpha,\beta}\,.$$

The expected value of ϵ_i^2 is given by

$$\mathrm{E}[\epsilon_i^2] = \mathrm{E}[r_i^2] - 2\beta\mathrm{E}[r_i r_{\mathcal{M}}] + \beta^2\mathrm{E}[r_{\mathcal{M}}^2] + \alpha^2 - 2\alpha\mathrm{E}[r_i] + 2\alpha\beta\mathrm{E}[r_{\mathcal{M}}]\,.$$

A necessary condition for a minimum of $\mathrm{E}[\epsilon_i^2]$ as a function of α and β is that the partial derivatives w.r.t. α and β should be zero at the point of minimum, (α_i, β_i):

$$\begin{aligned}
\frac{\partial}{\partial \alpha}\mathrm{E}[\epsilon_i^2] = 0 &\iff \alpha + \beta\mathrm{E}[r_{\mathcal{M}}] = \mathrm{E}[r_i], \\
\frac{\partial}{\partial \beta}\mathrm{E}[\epsilon_i^2] = 0 &\iff \alpha\mathrm{E}[r_{\mathcal{M}}] + \beta\mathrm{E}[r_{\mathcal{M}}^2] = \mathrm{E}[r_i r_{\mathcal{M}}]\,.
\end{aligned}$$

As a result, we obtain a system of linear equations that can be solved to find

$$\beta_i = \frac{\operatorname{Cov}(r_i, r_{\mathcal{M}})}{\operatorname{Var}(r_{\mathcal{M}})}, \quad \alpha_i = \mathrm{E}[r_i] - \beta_i\mathrm{E}[r_{\mathcal{M}}]\,.$$

Note that for the beta factor we obtained the same expression as that in (3.36).

(3) Suppose that historical data of returns on some portfolio V and the market portfolio M, $\{r_V^{(j)}, r_{\mathcal{M}}^{(j)}\}_{j=1,2,\dots,N}$, are available. Let us find the line of best fit by minimizing the sum of squared residuals:

$$\sum_{j=1}^{N} \left(r_V^{(j)} - (\alpha + \beta r_{\mathcal{M}}^{(j)})\right)^2 \to \min_{\alpha,\beta}\,.$$

The solution to this minimization problem is

$$\beta_i = \frac{N\sum_j r_V^{(j)} r_{\mathcal{M}}^{(j)} - \left(\sum_j r_V^{(j)}\right)\left(\sum_j r_{\mathcal{M}}^{(j)}\right)}{N\sum_j (r_{\mathcal{M}}^{(j)})^2 - \left(\sum_j r_{\mathcal{M}}^{(j)}\right)^2}, \quad \alpha_i = \frac{\sum_j r_V^{(j)} - \beta_i \sum_j r_{\mathcal{M}}^{(j)}}{N}.$$

The beta factors for individual assets can be computed by (3.36) or from historical data. The beta factor of a portfolio V in N assets with weights $w_1, \ldots, w_N$ is given by

$$\beta_V = w_1\beta_1 + \cdots + w_N\beta_N.$$

Indeed, the covariance function is bilinear; therefore

$$\begin{aligned} \beta_V &= \frac{\operatorname{Cov}(r_V, r_{\mathcal{M}})}{\operatorname{Var}(r_V)} = \frac{\operatorname{Cov}(w_1 r_1 + \cdots + w_n r_N, r_{\mathcal{M}})}{\operatorname{Var}(r_V)} \\ &= \frac{w_1 \operatorname{Cov}(r_1, r_{\mathcal{M}}) + \cdots + w_N \operatorname{Cov}(r_N, r_{\mathcal{M}})}{\operatorname{Var}(r_V)} = w_1\beta_1 + \cdots + w_N\beta_N\,. \end{aligned}$$

Clearly, the beta factor of the market portfolio is equal to one.

By taking the mathematical expectation of both parts of (3.35), we obtain

$$\mu_i = r + \beta_i(\mu_{\mathcal{M}} - r)\,,$$

where $\mu_i = \mathrm{E}[r_i]$ and $\mu_{\mathcal{M}} = \mathrm{E}[r_{\mathcal{M}}]$. The expected return plotted against the beta factor of any portfolio will form a straight line on the (β, μ)-plane, called the *asset market line.*

3.5 Exercises

Exercise 3.1. Show that the functions $u_1(x) = \ln x$ and $u_2(x) = 1 - e^{-ax}$ with $a > 0$ both satisfy the definition of a utility function, i.e., each of them is an increasing, concave function.

Exercise 3.2. Show that the functions $u_3(x) = x^a$ with $0 < a < 1$ and $u_4(x) = x - bx^2$ with $b > 0$ and $x < \frac{1}{2b}$ both satisfy the definition of a utility function, i.e., each of them is an increasing, convex-upward function.

Exercise 3.3. An investor with capital W can invest an amount $V = aW$ for some $0 \leqslant a \leqslant 1$. If V is invested, then after one year the invested amount is doubled with probability p or lost with probability $1-p$. Suppose that the remaining capital $W - aW$ can be put in a risk-free bank account to earn interest at an annual rate of interest r. How much should be invested by an investor using:

(a) a log utility function $u(V) = \ln V$,

(b) an exponential utility function $u(V) = 1 - \mathrm{e}^{-0.1V}$?

Exercise 3.4. Consider an investment of \$1000 in two risky assets whose returns follow a bivariate normal distribution with the following expected values and standard deviations:

$$\mu_1 = 0.1,\ \sigma_1 = 0.2,\ \mu_2 = 0.15,\ \sigma_1 = 0.3.$$

The correlation coefficient between the returns is $\rho = -0.5$.

(a) Suppose that the allocation weights of an investment portfolio for assets 1 and 2 are, respectively, $w_1 = x$ and $w_2 = 1 - x$ for some $x \in \mathbb{R}$. Show that the terminal value V_T of a portfolio is normal. Find the expected value and variance of V_T.

(b) Find the optimal portfolio when employing the utility function

$$u(V) = 1 - \mathrm{e}^{-0.01V}.$$

Exercise 3.5. Consider a market model with three scenarios $\{\omega^1, \omega^2, \omega^3\}$ and two risky assets with returns r_1 and r_2. Let the probabilities of the scenarios and values of the returns be as follows:

ω	$\mathbb{P}(\omega)$	$r_1(\omega)$	$r_2(\omega)$
ω^1	0.5	10%	5%
ω^2	0.3	5%	10%
ω^3	0.2	15%	−5%

Find the expected values and standard deviations of the returns. Find the coefficient of correlation between r_1 and r_2.

Exercise 3.6. Show that the optimal allocation weights $\{w_i\}$ of one's investment portfolio $V_t = \sum_{i=1}^{N} w_i A_t^i$, $t \in \{0, T\}$, that correspond to amounts invested in each asset do not depend on the initial capital V_0 when attempting to maximize the mathematical expectation of:

(a) a log utility function $u(V_T) = \ln V_T$,

(b) a power utility function $u(V_T) = (V_T)^a$ with $0 < a < 1$.

In other words, the maximization of $\mathrm{E}[u(V_T)]$ reduces to the maximization of $\mathrm{E}\left[u\left(\frac{V_T}{V_0}\right)\right]$.

Exercise 3.7. Plot portfolio lines with and without short selling for the case with two assets if

(a) $|\rho_{12}| = 1$, $\mu_1 = \mu_2$, and $\sigma_1 \neq \sigma_2$,

(b) $|\rho_{12}| = 1$, $\mu_1 \neq \mu_2$, and $\sigma_1 = \sigma_2$,

(c) $\mu_1 = \mu_2$, and $\sigma_1 = \sigma_2$.

Exercise 3.8. Consider three assets whose returns have the following standard deviation and correlation coefficients:

$$\sigma_1 = 0.2,\ \sigma_2 = 0.25,\ \sigma_3 = 0.15,\ \rho_{12} = -0.4,\ \rho_{13} = 0.3,\ \rho_{23} = 0.7.$$

Obtain the covariance matrix $\mathbf{C}$.

Exercise 3.9. Compute the weights in the minimum variance portfolio constructed using the assets in Exercise 3.7. Also compute the expected return and standard deviation of the minimum variance portfolio.

Exercise 3.10. Show that

$$\mathbf{C} = \begin{bmatrix} 1 & 0.75 & -0.3 \\ 0.75 & 1 & 0.5 \\ -0.3 & 0.5 & 1 \end{bmatrix}$$

cannot be a covariance matrix.

Exercise 3.11. Suppose that the risk-free return is $r = 3\%$. Find the weights in the market portfolio constructed from the three assets in Example 3.5. Compute the expected return and standard deviation of the return of the market portfolio.

Chapter 4

Primer on Derivative Securities

A *derivative* is a financial contract whose value depends on (or derives from) the values of other basis variables such as stock prices, bond values, interest rates, exchange rates, commodity prices, market indices, etc. Such basis variables are called *underlyings*. There are three broad categories of traders interested in trading derivatives contracts:

hedgers who use derivatives to reduce the risk that they face from potential future movements in market variables;

speculators who use derivatives to bet on the future directions of market variables;

arbitrageurs who take offsetting positions in two or more instruments to lock in a risk-free profit.

In this chapter we consider two main derivative contracts, namely, *forwards* and *options*.

4.1 Forward Contracts

A *forward contract* is an agreement to buy or to sell an asset for a fixed price on a fixed future date, all specified in advance. The fixed price paid for the asset is called the *forward price*; the fixed date is called the *delivery time*. A forward contract is a direct agreement between two parties. Both parties are *obliged* to fulfil the contract. The party to the contract who agrees to sell the asset is said to enter into a *short forward position*. The other party who has to buy the asset is said to take a *long forward position*. The exchange flows between the two parties are illustrated in Figure 4.1. The asset may be physically delivered to the buyer, or the contract may be settled by paying the difference between the forward price and the actual asset price in cash. The main motivation for entering into a forward contract is to become independent of the uncertain future price of a risky asset. There is no premium paid for entering into a forward contract.

Let $F(t,T)$ with $t < T$ denote the delivery price for a forward contract, which is agreed upon at time t, and whose delivery time is T. Let $S(t)$ denote the time-t price of the underlying asset on which the forward contract is written. The payoff of the long forward contract is $S(T) - F(t,T)$ and the payoff is $F(t,T) - S(T)$ for the short forward contract. At the delivery time T there are two possibilities:

1. If $F(t,T) < S(T)$, then the party taking a long forward position benefits from a positive payoff. The instant profit is $S(T) - F(t,T)$ since the asset bought (at time T) at the forward price $F(t,T)$ can be immediately sold for $S(T)$. The party with a short position has negative payoff $F(t,T) - S(T)$ by selling the asset below the market price $S(T)$; this represents a loss in the amount of $S(T) - F(t,T)$.

2. If $F(t,T) > S(T)$, the situation is exactly reversed. The party taking a short forward

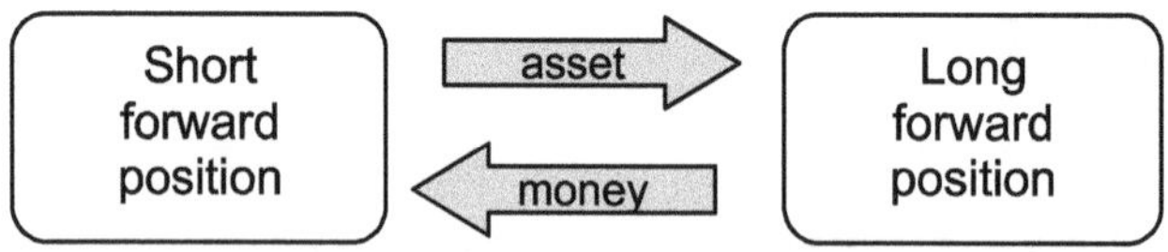

FIGURE 4.1: A forward contract diagram.

position benefits from a positive payoff since the asset is sold at price $F(t,T)$ which is above its market price $S(T)$. The payoff to the party taking a long position is negative, representing a loss in the amount of $F(t,T) - S(T)$.

The payoff diagrams for the two parties respectively taking short and long forward positions are illustrated in Figure 4.2.

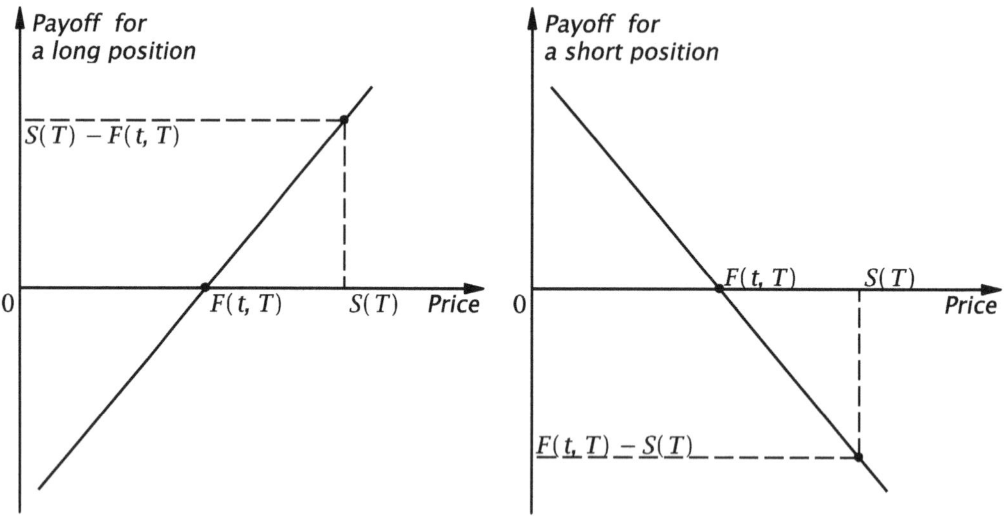

FIGURE 4.2: Payoff diagrams for the long and short positions in a forward contract.

4.1.1 No-Arbitrage Evaluation of Forward Contracts

Consider a market with a risk-free bank account B (with a constant interest rate r) and a risky stock S. Suppose that the discounting function $D(t) = \frac{B(0)}{B(t)}$ and the accumulation function $A(t) = \frac{1}{D(t)} = \frac{B(t)}{B(0)}$ for the bank account are given by (1.20) and (1.19), respectively. Thus, for any time period $[t,T]$ with $0 \leqslant t \leqslant T$ we have

- $B(T) = B(t)(1+r)^{T-t}$ and $D(T-t) = (1+r)^{-(T-t)}$ if the interest is compounded annually;
- $B(T) = B(t)(1+r/m)^{(T-t)m}$ and $D(T-t) = (1+r/m)^{-(T-t)m}$ if the interest is compounded m times per year;
- $B(T) = B(t)\mathrm{e}^{r(T-t)}$ and $D(T-t) = \mathrm{e}^{-r(T-t)}$ if the interest is compounded continuously.

In what follows, we assume that the bank account allows us to borrow or invest any amount of money at the same interest rate. We can also make or withdraw an investment at any moment without losing interest or paying an additional fee.

Suppose that a forward contract on the stock is written at time t with delivery time T with $0 \leqslant t < T$. Let us find the forward price of such a contract which guarantees the absence of arbitrage.

4.1.1.1 Forward Price for a Stock Paying No Dividends

Theorem 4.1. *For a stock paying no dividends, the forward price at time $t < T$ is*

$$F(t,T) = \frac{S(t)}{D(T-t)} \tag{4.1}$$

or an arbitrage opportunity would arise otherwise. In particular, if the interest is compounded continuously, then the no-arbitrage forward price is $F(t,T) = S(t)\mathrm{e}^{r(T-t)}$.

Proof. Suppose that $F(t,T) > \frac{S(t)}{D(T-t)}$. Let us form the following portfolio at time t: borrow $S(t)$ dollars from the risk-free bank account, buy one share of stock for $S(t)$, and enter into a short forward contract (with the agreement to sell one share of stock for $F(t,T)$). The initial value of such a portfolio is zero. At the delivery time T, we sell the stock for $F(t,T)$ and pay $\frac{S(t)}{D(T-t)}$ to return the amount borrowed with interest. As a result, we end up with a positive risk-free profit of $F(t,T) - \frac{S(t)}{D(T-t)}$. This contradicts the no-arbitrage principle.

Let $F(t,T) < \frac{S(t)}{D(T-t)}$. Then we take a short position in stock (borrow one share), sell the share for $S(t)$ and invest the proceeds in the bank account, and enter into a long forward contract (with the agreement to sell one share of stock for $F(t,T)$). Again the initial value of the portfolio is zero. At the delivery time T, we buy one stock share for $F(t,T)$ and close out the short position in the stock; we cash out the risk-free investment with interest to get $\frac{S(t)}{D(T-t)}$. As a result, we end up with a positive risk-free profit of $\frac{S(t)}{D(T-t)} - F(t,T)$. Again, this contradicts the no-arbitrage principle. Therefore, the only possibility is that the forward price $F(t,T)$ is equal to $\frac{S(t)}{D(T-t)}$. □

Since for any $t < T$ the discount factor $D(T-t)$ is less than 1, the forward price is always larger than the spot price: $F(t,T) = \frac{S(t)}{D(T-t)} > S(t)$. As $t \to T$, the difference $F(t,T) - S(t)$ converges to zero, i.e., $F(T,T) = S(T)$. A visualization of this is given in Figure 4.3, which depicts a sample path simulation of a stock price process and its corresponding forward price up to maturity T. Note that $S(t)$ is always below $F(t,T)$ until T, at which point they coincide.

Example 4.1. Consider a 10-month forward contract on stock with a current price of \$500. Assume that the risk-free rate of interest is $r_{12} = 6\%$.

(a) Show that the no-arbitrage forward price is $F(0, \frac{10}{12}) = \$525.57$.

(b) Find an arbitrage opportunity if $F(0, \frac{10}{12}) = \$500$.

Solution.

(a) Applying (4.1) for monthly compounding gives the no-arbitrage forward price:

$$F(0, \frac{10}{12}) = 500(1 + 0.06/12)^{10} \cong \$525.57.$$

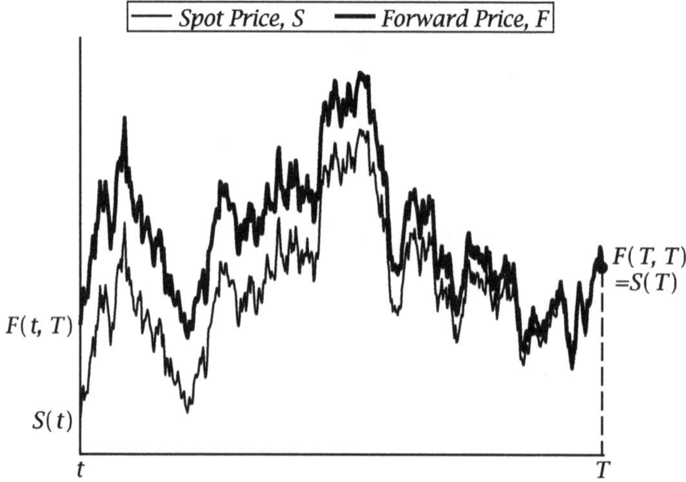

FIGURE 4.3: The spot and forward prices.

(b) Take a short position in stock by borrowing one share; sell the share for $S(0) = \$500$ and invest the proceeds in the bank account; enter into a long forward contract (with the agreement to sell one share of stock for $F(0, \frac{10}{12}) = \$500$). At the delivery time $T = \frac{10}{12}$, we cash out the risk-free investment with interest to obtain $500(1 + 0.06/12)^{10} \cong \$525.57$, buy one stock share for \$500, and close out the short position in the stock. As a result, we end up with a positive arbitrage profit of $525.57 - 500 = \$25.57$. □

4.1.1.2 Forward Price for a Stock Paying Dividends

Suppose that the stock pays dividends. Let the dividend payments

$$\text{div}(t_1), \text{div}(t_2), \ldots, \text{div}(t_m)$$

be made on each share of stock at times $t_1, t_2, \ldots, t_m$, respectively, where $t < t_1 < t_2 < \ldots < t_m < T$. The present (time-$t$) value of the dividends paid over $[t, T]$, denoted $\text{div}(t, T)$, is given by

$$\text{div}(t, T) = \text{div}(t_1)D(t_1 - t) + \text{div}(t_2)D(t_2 - t) + \cdots + \text{div}(t_m)D(t_m - t),$$

where $D(t)$ is the discounting function. The formula (4.1) for the forward price can be modified by subtracting the present value of the dividend payments from the spot price $S(t)$. Let us consider the following strategy: Buy one share of the asset and enter into a short forward contract to sell the asset for $F(t, T)$ at time T. The net present value of such a strategy (at time t) is $\text{NPV}(t) = -S(t) + \text{div}(t, T) + F(t, T)\, D(T - t)$. There is no arbitrage iff the NPV is zero. Therefore, $F(t, T) = (S(t) - \text{div}(t, T))/D(T - t)$. The no-arbitrage forward price can also be justified by constructing arbitrage strategies when $F(t, T) \neq (S(t) - \text{div}(t, T))/D(T - t)$, similar to those in the proof of Theorem 4.1.

Theorem 4.2. *For a stock paying dividends, the no-arbitrage forward price at time $t < T$ is*

$$F(t, T) = \frac{S(t) - \text{div}(t, T)}{D(T - t)}. \tag{4.2}$$

Proof. First, suppose that $F(t,T) > \frac{S(t)-\text{div}(t,T)}{D(T-t)}$. Then at time t we form the following portfolio. Borrow $S(t)$ dollars from the bank account, buy one share of stock for $S(t)$, and enter into a short forward contract. The initial value of such a portfolio is zero. Cash all dividend payments and invest them in the risk-free bank account. At the delivery time T, we sell the stock for $F(t,T)$, collect $\frac{\text{div}(t,T)}{D(T-t)}$, and pay $\frac{S(t)}{D(T-t)}$ to return the amount borrowed with interest. The risk-free profit is $F(t,T) + \frac{\text{div}(t,T)}{D(T-t)} - \frac{S(t)}{D(T-t)} > 0$. This contradicts the no-arbitrage principle.

Let $F(t,T) < \frac{S(t)-\text{div}(t,T)}{D(T-t)}$. Then we take a short position in stock (borrow one share), sell it for $S(t)$, and invest the proceeds in the bank account. Then we enter into a long forward contract. Again the initial value of the portfolio is zero. Every time a dividend payment is due, we borrow the necessary amount and pay a dividend to the stockholder. As a result, by time T we owe $\frac{\text{div}(t,T)}{D(T-t)}$. At the delivery time T, we buy one share of S for $F(t,T)$ and close out the short position in the stock; we cash out the risk-free investment with interest $\frac{S(t)}{D(T-t)}$ and clear the loan. As a result, we end up with a positive risk-free profit of $\frac{S(t)}{D(T-t)} - \frac{\text{div}(t,T)}{D(T-t)} - F(t,T) > 0$. □

Example 4.2. Consider a 10-month forward contract on stock with a price of \$500. Assume that the risk-free rate of interest is $r_{12} = 6\%$. Let a dividend payment of \$50 be paid 5 months after the initialization of the forward contract.

(a) Find the no-arbitrage forward price $F(0, \frac{10}{12})$.

(b) Find an arbitrage opportunity if $F(0, \frac{10}{12}) = \$510$.

Solution.

(a) The present value of the dividend is

$$V(0, 10/12) = 50 \cdot (1 + 0.06/12)^{-5} = 50 \cdot (1 + 0.005)^{-5} \cong \$48.77,$$

since the 1-month interest rate is $r_{12}/12 = 0.06/12 = 0.005$. Therefore, the no-arbitrage forward price is

$$F(0, 10/12) = (500 - 48.77) \cdot (1 + 0.005)^{10} \cong \$474.31.$$

(b) Borrow \$500 by making two loans: $50 \cdot 1.005^{-5} = \$48.77$ for 5 months and $\$500 - \$48.77 = \$451.23$ for 10 months. Buy one stock share for \$500. Enter into a forward contract to sell the stock in 10 months for \$510. In 5 months, collect the dividend payment of \$50 and use it to repay the first loan. In 10 months, sell the stock share for \$510. Repay the second loan: $451.23 \cdot 1.005^{10} \cong \$474.31$. The risk-free profit realized is $\$510 - \$474.31 = \$35.69$. □

4.1.2 Value of a Forward Contract

Consider a long forward contract with delivery time T and forward price $F(0,T)$ that is initiated at time 0. Initially, the value of such a forward contract is zero: $V(0) = 0$. At the delivery date, the value is equal to the payoff: $V(T) = S(T) - F(0,T)$. As time goes by, the spot price of the underlying asset will change; hence the value of the forward contract, $V(t)$, will vary as well.

Theorem 4.3. *The no-arbitrage value $V(t)$, $0 \leqslant t \leqslant T$, of a long forward contract that is initiated at time $t = 0$ and has delivery time T and delivery price $F(0,T)$ is given by*

$$V(t) = (F(t,T) - F(0,T))\, D(T-t). \tag{4.3}$$

Proof. At time t, consider two portfolios: one only has a long forward contract initiated at time 0 with value $V(t)$; the other consists of the long forward contract initiated at time t (whose value is zero at time t) and the risk-free investment of $(F(t,T) - F(0,T))D(T-t)$. At the delivery time T, both portfolios have the same value:

$$S(T) - F(0,T) = S(T) - F(t,T) + \frac{(F(t,T) - F(0,T))D(T-t)}{D(T-t)}.$$

By the Law of One Price, the time-t values of these portfolios have to be the same or else an arbitrage opportunity exists. □

The value $V(t)$ of the long forward contract can be expressed in terms of stock prices by combining Equation (4.3) with (4.1) or (4.2). Consider several important cases.

- For a stock paying no dividends, Equations (4.1) and (4.3) give

$$V(t) = S(t) - S(0)\frac{D(T-t)}{D(T)} = S(t) - \frac{S(0)}{D(t)}.$$

  In particular, if interest is compounded continuously, we have $V(t) = S(t) - S(0)\,\mathrm{e}^{rt}$.

- For a stock paying dividends, (4.3) and (4.2) give

$$V(t) = S(t) - \frac{S(0) - \operatorname{div}(0,t)}{D(t)},$$

  where $\operatorname{div}(0,t)$ is the time-0 value of dividend payments paid over $[0,t]$.

Example 4.3. Let $S(0) = \$100$, $T = 1$ year, $r = 5\%$ compounded continuously. The long forward contract is exchanged at time 0.

(a) Find the forward price $F(0,1)$.

(b) If $S(0.5) = \$110$, what is the value $V(0.5)$ of the long forward contract?

Solution.

(a) Applying (4.1), find $F(0,1) = 100\,\mathrm{e}^{0.05\cdot 1} \cong \$105.13$.

(b) The forward price at time $t = 0.5$ is $F(0.5,1) = 110\,\mathrm{e}^{0.05\cdot 0.5} \cong \$112.79$. Therefore, the value $V(0.5)$ of the long forward contract that has been initiated at time 0 is

$$V(0.5) = (F(0.5,1) - F(0,1))\,\mathrm{e}^{-0.05\cdot(1-0.5)} \cong \$7.47.$$ □

One can consider a forward contract initiated with a fixed delivery price K that may differ from the forward price $F(t,T)$ given by (4.1).

Theorem 4.4. *The no-arbitrage value $V_K(t)$, $0 \leqslant t \leqslant T$, of a long forward contract initiated at time $t = 0$ and having the delivery time T and delivery price K is given by*

$$V_K(t) = (F(t,T) - K)\,D(T-t).$$

The proof is very similar to that of Theorem 4.3 and is left as an exercise for the reader (see Exercise 4.3). As we can see, the initial time-0 price of such a forward contract is nonzero iff the delivery price $K \neq F(0,T)$.

4.2 Basic Options Theory

4.2.1 Concept of an Option Contract

The holder of a forward contract is **obliged** to trade at the maturity of the contract. Unless the position is closed before the delivery date (i.e., the contract is sold to another party), the holder of a long forward must take possession of the asset regardless of whether the asset has risen or fallen (or pay the difference in prices).

An *option* contract gives the holder the **right**, not the obligation, to trade in the future at a previously agreed price. A *European call option* is a contract giving the holder the right (not the obligation) to buy an underlying asset for a price K fixed in advance, called the *exercise price* or *strike price*, at a specified future time T, called the *exercise time*, or *expiry time*, or *maturity time*. A *European put option* is a contract giving the holder the right to sell an underlying asset for an agreed exercise price K at an expiry time T. Since the holder of a European call (put) option can only exercise his or her right and sell (buy) the asset at the time T when the option is expiring, there is no difference between the exercise time and expiry time for a European-type derivative. There is a difference between these two time moments for *American options*, which can be exercised at a **time** prior to the expiry (maturity) date.

At the delivery time there are two possible scenarios for the holder of a standard European option:

$S(T) \geqslant K$**:** The holder of a call option will exercise the option. The payoff at time T is $S(T) - K$. The holder of a put option will not exercise the option. The payoff is zero.

$S(T) \leqslant K$**:** The holder of a put option will exercise the option. The payoff at time T is $K - S(T)$. The holder of a call option will not exercise the option so the payoff is zero.

The payoff function Λ of the holder of a European option is a nonnegative and piecewise-linear function:

$$\Lambda_{Call}(S(T)) = \begin{cases} S(T) - K & \text{if } S(T) > K, \\ 0 & \text{if } S(T) \leqslant K, \end{cases} \quad \Lambda_{Put}(S(T)) = \begin{cases} K - S(T) & \text{if } S(T) < K, \\ 0 & \text{if } S(T) \geqslant K. \end{cases}$$

We can write the payoff functions in compact form:

$$\Lambda_{Call}(x) = (x - K)^+ \quad \text{and} \quad \Lambda_{Put}(x) = (K - x)^+,$$

where $(x)^+ := \max\{x, 0\}$. The call and put payoffs as functions of the stock price at maturity $S(T)$ are plotted in Figure 4.4. The diagram in Figure 4.4 illustrates the following terminology which is often used. An option is said to be *in the money* if its payoff function is positive, hence the option is worth exercising. Otherwise an option is said to be *out of the money* when the payoff function is zero; an option is said to be *at the money* when the strike price is equal to the price of the underlying security (asset). When an option is in the money, it does not mean that the option is overall profitable, since the positive payoff may still be less than the premium paid for the option contract.

It costs a trader nothing to enter into a forward contract (with the no-arbitrage delivery price), whereas the purchase of an option requires an up-front payment which is the option value at inception. Indeed, if no premium is paid, the option holder will lose no money and would make a positive profit (with some positive probability) whenever the option is in

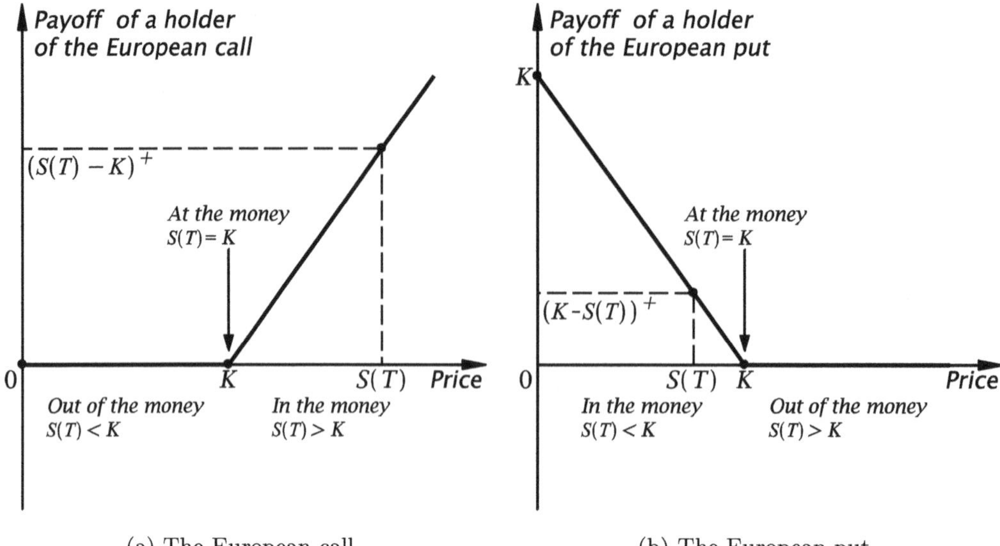

(a) The European call.

(b) The European put.

FIGURE 4.4: Payoff diagrams for European options.

the money. According to the no-arbitrage principle, the price paid for an option has to be nonnegative, although the option price is strictly positive in most cases. Let $C^E \geqslant 0$ and $P^E \geqslant 0$ denote the price paid by a buyer of a European call and put option, respectively. For convenience, suppose that the option is written at time 0. By taking into account the time value of the price, we have the following formulae for the overall gain (profit) of an option buyer at time T:

$$\text{Profit}_{Call} = (S(T) - K)^+ - C^E e^{rT} \text{ and } \text{Profit}_{Put} = (K - S(T))^+ - P^E e^{rT}.$$

The profit of a holder of the European call or put option is plotted in Figure 4.5. Here and later in this section, we assume that the interest is compounded continuously at a risk-free rate r, i.e., $D(t) = e^{-rt}$.

Similarly to a forward contract, there are two parties to every option contract: the buyer of an option, who has the right to exercise the option, and the writer, who is obliged to deliver the underlying asset if the option will be exercised at maturity. The buyer takes the long position; the writer takes the short position. This situation is illustrated in Figure 4.6 for both call and put options. The writer receives a cash premium up front but has potential liabilities later. The writer's profit or loss is the reverse of that for the purchaser (the holder) of the option.

Example 4.4. Find the expected value of the gain for a holder of a European put option with strike price $K = \$100$ and exercise date $T = 0.5$ years, if the risk-free continuously compounded rate is $r = 10\%$, the current price of the underlying security is $S(0) = \$95$, the option is bought for \$7, and the price $S(0.5)$ may take one of four values: \$80, \$90, \$100, \$110, with equal probability.

Solution. The expected value of the gain (or loss) to the option holder is the difference between the expected payoff and the appreciated value of the option price paid. The option is in the money only when $S(0.5) = \$80$ or $S(0.5) = \$90$. The respective payoffs are

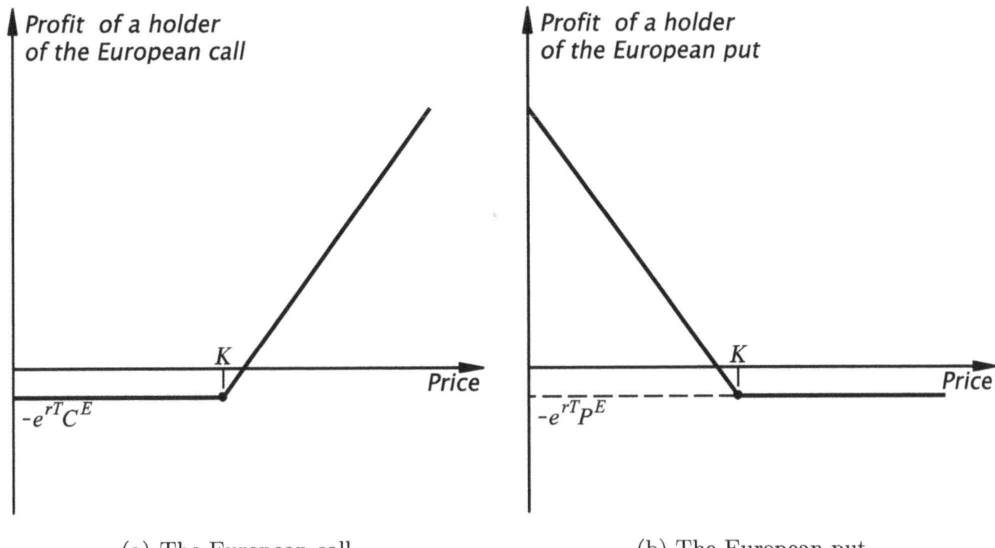

(a) The European call. (b) The European put.

FIGURE 4.5: Profit diagrams for European options.

$(100 - 80)^+ = \$20$ and $(100 - 90)^+ = \$10$. The expected gain is

$$\begin{aligned}\mathrm{E}[\text{Gain}] &= -P^E\,\mathrm{e}^{rT} + \mathrm{E}[\text{Payoff}] \\ &= -7\,\mathrm{e}^{0.1\cdot 0.5} + 20 \cdot 0.25 + 10 \cdot 0.25 \cong -7.36 + 5 + 2.50 = \$0.14. \qquad \square\end{aligned}$$

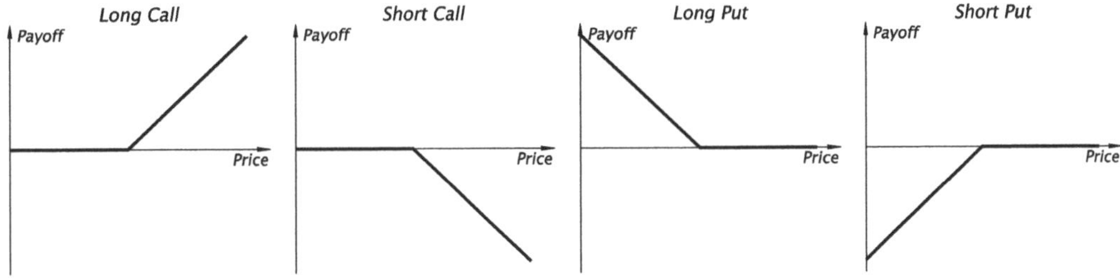

FIGURE 4.6: Payoff diagrams for long and short European call and put options.

4.2.2 Put-Call Parities

We now derive an important relation between the no-arbitrage prices of European put and call options called the *put-call parity*. A similar relation can be obtained for other pairs of put and call options. Such a parity can be used to find the price of one option (call or put) given the price of the other option. Also, the parity can be used to find out if an arbitrage opportunity exists. Consider the following two portfolios:

Portfolio 1 consists of one European call and the risk-free investment with the time-0 value of $K\mathrm{e}^{-rT}$;

Portfolio 2 consists of one European put and one share of stock.

Assume that both options have the same expiry date T and exercise price K. At the expiry time T, both portfolios are worth $\max(S(T), K)$. Indeed, if $S(T) \geqslant K$, then the two portfolios have the following values at maturity: $\Pi_1(T) = S(T) - K + K = S(T)$ and $\Pi_2(T) = 0 + S(T) = S(T)$. If $S(T) \leqslant K$, then $\Pi_1(T) = 0 + K = K$ and $\Pi_2(T) = K - S(T) + S(T) = K$. By the law of one price, if $\Pi_1(T,\omega) = \Pi_2(T,\omega)$ for any outcome $\omega \in \Omega$, then $\Pi_1(0) = \Pi_2(0)$. The portfolios must therefore have identical values today:

$$\underline{C^E + K\mathrm{e}^{-rT}} = \Pi_1(0) = \Pi_2(0) = \underline{P^E + S(0)}.$$

As a result, we obtain the *put-call parity*

$$C^E - P^E = S(0) - K\,\mathrm{e}^{-rT}. \tag{4.4}$$

Assuming that interest is compounded annually, the parity takes the following form:

$$C^E - P^E = S(0) - K\,(1+r)^{-T}.$$

The put-call parity can also be written in terms of a risk-free ZCB price:

$$C^E - P^E = S(0) - K\,Z(0,T).$$

Recall that the time value of a long forward contract with delivery price K and delivery time T is given by

$$V_K(t) = (F(t,T) - K)\mathrm{e}^{-r(T-t)},\ 0 \leqslant t \leqslant T.$$

In particular, if $K \neq F(0,T)$, then such a contact has a nonzero initial value

$$V_K(0) = (F(0,T) - K)\,\mathrm{e}^{-rT} = (\mathrm{e}^{rT}S(0) - K)\,\mathrm{e}^{-rT} = S(0) - K\mathrm{e}^{-rT}.$$

At the expiry date, the payoff of such a long forward contract can be represented as a sum of payoffs of a long call option and short put option (see Figure 4.7). At time 0, the portfolio with one long call and one short put should have the same value as a long forward contract. Therefore,

$$C^E - P^E = S(0) - K\mathrm{e}^{-rT}.$$

In the case of a stock paying dividends, we have $V_K(0) = S(0) - \mathrm{div}(0,T) - Ke^{-rT}$. Thus,

$$C^E - P^E = S(0) - \mathrm{div}(0,T) - K\mathrm{e}^{-rT}. \tag{4.5}$$

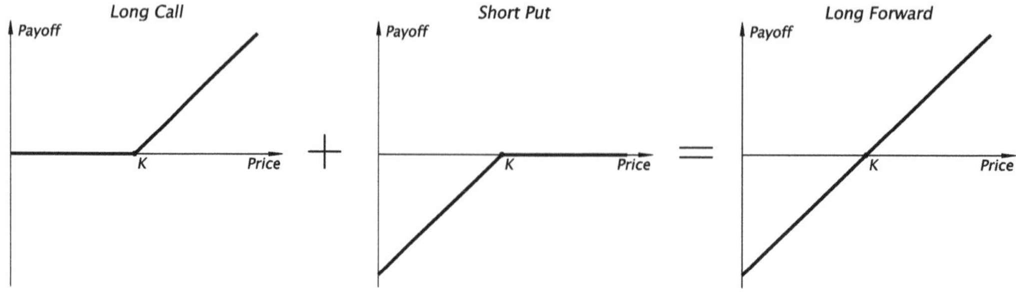

FIGURE 4.7: The representation of a long forward payoff as a sum of a long European call and a short European put. The strike price equals the delivery price.

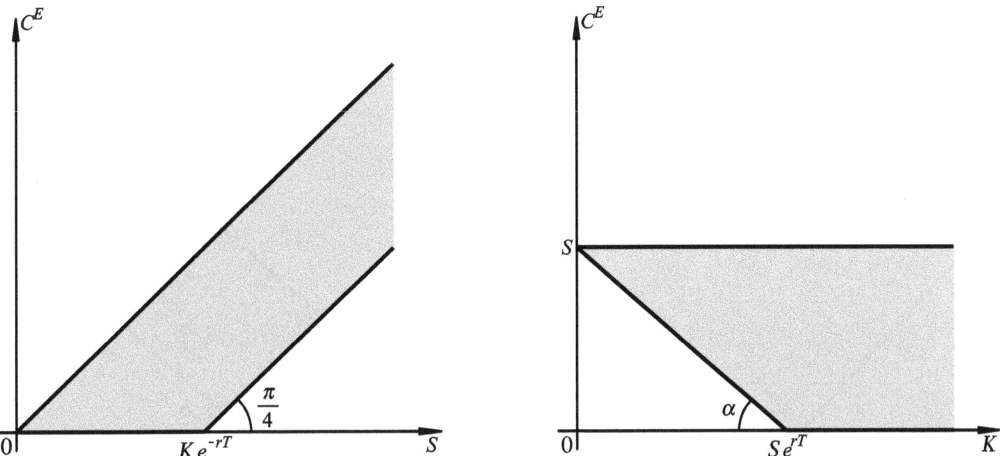

FIGURE 4.8: Lower and upper bounds on the European call price. Here, S denotes the spot price, and $\alpha = \arctan \mathrm{e}^{-rT}$.

4.2.3 Properties of European Options

The option price depends on a number of variables such as

variables describing the contract: the strike price K and expiry time T;

variables describing the underlying: the initial price $S(0)$ and dividend yield q;

market variables: the interest rate r.

One can derive various properties of the prices C^E and P^E solely based on the no-arbitrage principle. Such properties are general and independent of the pricing model. For example, one can obtain bounds on the option price (as a function of $S(0)$ and K), prove monotonicity of the option price (as a function of $S(0)$, K, and T), and prove convexity of the option price (as a function of $S(0)$ and K).

Theorem 4.5 (Upper and lower bounds on European options prices). *The no-arbitrage prices of the European call, C^E, and European put, P^E, satisfy*

$$(S(0) - K\mathrm{e}^{-rT})^+ \leqslant C^E < S(0), \tag{4.6}$$

$$(K\mathrm{e}^{-rT} - S(0))^+ \leqslant P^E < K\mathrm{e}^{-rT}. \tag{4.7}$$

Proof. Suppose that $C^E \geqslant S(0)$. Let us sell a call option and buy one stock share. The balance $C^E - S(0) \geqslant 0$ is invested in the risk-free bank account. At the expiry time T, we sell the stock for $\min\{S(T), K\}$. Our arbitrage profit is $\min\{S(T), K\} + (C^E - S(0))\,\mathrm{e}^{rT} > 0$. Thus, the no-arbitrage principle leads to the upper bound on C^E.

Using the put-call parity and nonnegativity of option prices, we obtain the lower bounds on the option prices:

$$\begin{cases} C^E \geqslant 0, \\ C^E = S(0) - K\mathrm{e}^{-rT} + P^E \geqslant S(0) - K\mathrm{e}^{-rT} \end{cases} \implies C^E \geqslant \max\{0, S(0) - K\mathrm{e}^{-rT}\},$$

$$\begin{cases} P^E \geqslant 0, \\ P^E = K\mathrm{e}^{-rT} - S(0) + C^E \geqslant K\mathrm{e}^{-rT} - S(0) \end{cases} \implies P^E \geqslant \max\{0, K\mathrm{e}^{-rT} - S(0)\}.$$

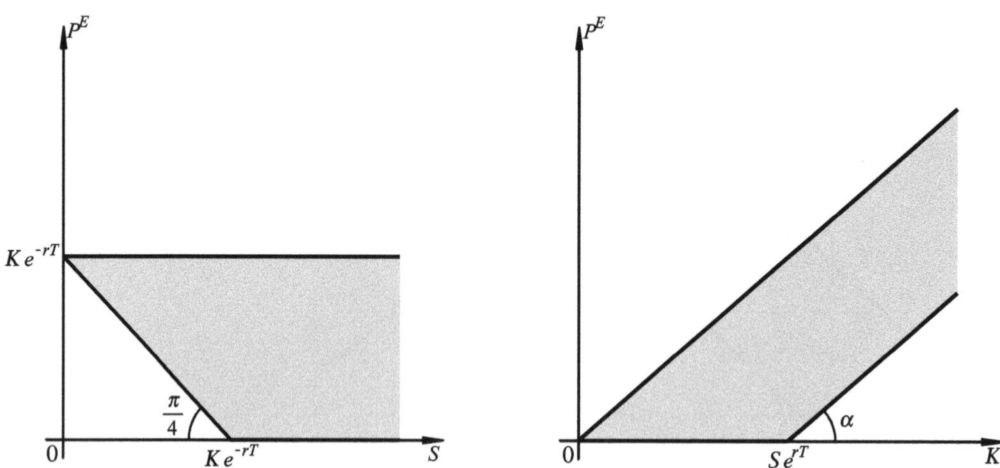

FIGURE 4.9: Lower and upper bounds on the European put price. Here, S denotes the spot price, and $\alpha = \arctan \mathrm{e}^{-rT}$.

The upper bound on C^E and the put-call parity give us the upper bound on P^E. □

The results of Theorem 4.5 are illustrated in Figures 4.8 and 4.9.

Theorem 4.6 (Monotonicity of European option prices). *The no-arbitrage price C^E (P^E) of the European call (put) is a nondecreasing (nonincreasing) function of the spot price S and a nonincreasing (nondecreasing) function of the strike price K:*

$$K' < K'' \implies C^E(K') \geqslant C^E(K'') \;\&\; P^E(K') \leqslant P^E(K''),$$
$$S' < S'' \implies C^E(S') \leqslant C^E(S'') \;\&\; P^E(S') \geqslant P^E(S'').$$

Proof. Suppose that $C^E(K') < C^E(K'')$ for some $K' < K''$. Write and sell a call with strike K''; buy a call with strike K'; invest the difference $C^E(K'') - C^E(K')$ without risk. The payoff of the combination of two options is nonnegative: $(S - K'')^+ \leqslant (S - K')^+$ since $K' < K''$. At the expiry time T, the total arbitrage profit is

$$(C^E(K'') - C^E(K'))\,\mathrm{e}^{rT} + (S(T) - K')^+ - (S(T) - K'')^+ > 0.$$

Similarly, we prove that $P^E(K') \leqslant P^E(K'')$: suppose the converse and then form an arbitrage portfolio with a long put struck at K'', a short put struck at K', and a risk-free investment of $P^E(K') - P^E(K'')$.

Suppose that $C^E(S') > C^E(S'')$ for some $S' < S''$. Let $S(0)$ be the initial stock price. Consider two portfolios with $x' = \frac{S'}{S(0)}$ and $x'' = \frac{S''}{S(0)}$ shares of stock, respectively. The respective costs of the portfolios are $S' = x'S(0)$ and $S'' = x''S(0)$. Proceed as follows: write and sell a call on a portfolio with x' shares of stock; buy a call on a portfolio with x'' shares; invest the balance $C^E(S') - C^E(S'')$ without risk. Both options have the same strike price and expiry time. Again, the payoff of the combination of two options is nonnegative. Indeed, if $S(T) < \frac{K}{x''}$, then both options will not be exercised; if $\frac{K}{x''} \leqslant S(T) < \frac{K}{x'}$, then exercise the long call, the short call will not be exercised; if $\frac{K}{x'} \leqslant S(T)$, then both options are exercised, we cover the liability of the short call by the proceeds of the long call. At the maturity time T, the cumulative arbitrage profit is

$$(C^E(S') - C^E(S''))\,\mathrm{e}^{rT} + (x''S(T) - K)^+ - (x'S(T) - K)^+ > 0.$$

Similarly, we prove that $P^E(S') \geqslant P^E(S'')$ if $S' < S''$. □

4.2.4 Early Exercise and American Options

An *American call option* is a contract giving the holder the right (not the obligation) to buy an underlying asset for a strike price K (fixed in advance) at *any time* between now and the expiry time T. An *American put option* is a contract giving the holder the right (not the obligation) to sell an underlying asset for an agreed strike price K at *any time* between now and the expiry time T. The main difference between a European-style derivative and an American-style derivative is that the latter can be exercised at any time up to and including the expiry time, whereas the former can only be exercised at the expiry time. American options are similar to callable bonds that can be redeemed at some point before the date of maturity.

4.2.4.1 Relation to European Option Prices

Theorem 4.7. *Consider European and American options with the same strike price K and expiry time T. Since the American option gives at least the same rights as the corresponding European counterpart, we have*

$$C^E \leqslant C^A \quad \text{and} \quad P^E \leqslant P^A.$$

Proof. Suppose that $C^E > C^A$. Let us form the following portfolio: buy one American call (worth C^A), and write and sell one European call (worth C^E). The difference $C^E - C^A > 0$ is invested in a risk-free money market account. Now we wait until the expiry date. If the stock price at the expiry date is larger than the strike price, then exercise the American option and sell the stock share to the holder of the European call for the same price K. If $S(T) \leqslant K$, then both options are not exercised. At time T withdraw the investment. The final balance $(C^E - C^A)\,\mathrm{e}^{rT} > 0$ is a risk-free profit. Similarly, we prove that $P^E \leqslant P^A$. □

Consider European and American call options with the same strike price K and expiry time T on a *nondividend* paying stock. Theorem 4.7 tells us that $C^A \geqslant C^E$. In fact, an American call is worth exactly the same as a European call with the same term. Indeed, suppose that $C^A > C^E$. Form the following portfolio: write and sell one American call (for C^A) and buy one European call (for C^E), investing the balance $C^A - C^E$ without risk.

Scenario 1: The American call is exercised at time $t \leqslant T$.

- Borrow one stock share and sell it for K to settle your obligation as writer of the call.
- Invest K at rate r.
- At the expiry date, use the European call to buy the share for K and close your short position in stock.
- Your arbitrage profit is $(C^A - C^E)\,\mathrm{e}^{rT} + K\,(\mathrm{e}^{r(T-t)} - 1) > 0$.

Scenario 2: The American call is not exercised at all. You will end up with the European call and your arbitrage profit is $(C^A - C^E)\mathrm{e}^{rT} + (S(T) - K)^+ > 0$.

Therefore, the American call price should be equal to that of the European call with the same term: $C^E = C^A$. So it is not wise to exercise an American call early. Note that the situation is different for a dividend-paying stock.

4.2.4.2 Put-Call Parity Estimate

Consider American put and call options with the same strike K and expiry T written on a stock paying no dividends (hence, $C^E = C^A$). Let us first obtain the upper bound:

$$C^A - P^A = C^E - P^A \leqslant C^E - P^E = S(0) - Ke^{-rT}.$$

Now let us show that $C^A - P^A \geqslant S(0) - K$. Suppose that it fails. That is,

$$C^A - P^A < S(0) - K \iff P^A - C^A + S(0) > K.$$

Form the following portfolio:

- write and sell a put (for P^A);
- buy a call (for C^A);
- sell short one share (for $S(0)$);
- invest the balance $P^A - C^A + S(0) > K$ without risk.

There are two scenarios.

Scenario 1: The put is exercised at time $t \leqslant T$.

- Withdraw K from the risk-free investment to buy a share of stock (from the holder of the put).
- Use the share to close out the short position in stock.
- We still have the call and a positive cash balance:

$$(P^A - C^A + S(0))e^{rt} - K > Ke^{rt} - K \geqslant 0.$$

Scenario 2: The put has not been exercised at all.

- At time T we exercise the call option, buy one share of stock for K, and close out the short position in stock.
- The balance at time T is

$$(P^A - C^A + S(0))e^{rT} - K > Ke^{rT} - K = K(e^{rT} - 1) > 0.$$

As a result, we proved the put-call parity estimate for American call and put options:

$$S(0) - K \leqslant C^A - P^A \leqslant S(0) - Ke^{-rT}. \tag{4.8}$$

With the help of the put-call parity estimate, we can obtain bounds on the American option prices. Let us consider the case of a stock paying no dividends. As we proved before, the European and American calls with the same strike and expiry have the same price, i.e., $C^A = C^E$. Therefore, the bounds on the European call price are valid for the American call price C^A as well. Consider the American put. Since one can purchase and then immediately exercise the American option, the price cannot be less than the immediate payoff (otherwise the risk-free profit is $(K - S(0))^+ - P^A$). Thus, $P^A \geqslant (K - S(0))^+$. By using (4.8), we obtain that

$$P^A \leqslant C^A + K - S(0) = C^E + K - S(0) \leqslant S(0) + K - S(0) = K.$$

Therefore, $P^A \leqslant K$. By combining both of the bounds, we obtain

$$(K - S(0))^+ \leqslant P^A \leqslant K.$$

4.2.4.3 Monotonicity Properties of American Option Prices

Similar to the European case, one can derive various properties of C^A and P^A based on the no-arbitrage principle. These properties are general and independent of the asset price model used. For example,

- the option price is a monotonic function of $S = S(0)$, K, or T:

$$\left.\begin{array}{lll} C^A(S) \nearrow & C^A(K) \searrow & C^A(T) \nearrow \\ P^A(S) \searrow & P^A(K) \nearrow & P^A(T) \nearrow \end{array}\right\} \quad \text{as } S \nearrow,\ K \nearrow,\ T \nearrow;$$

- the option price is a convex function of $S = S(0)$ (or K).

4.2.5 Nonstandard European Options

Standard European options can be used as building blocks to create more sophisticated financial instruments. An investor with specific views on the future behavior of stock prices may be interested in derivatives with payoff profiles that are different from those of the standard European call and put options. In principle, any continuous piecewise payoff function can be manufactured from European call and put payoffs with different strikes. So being given a specific payoff, we can design a portfolio of securities with the same payoff function.

A *spread strategy* involves taking a position in two or more options of the same type. An investor who expects the stock price to rise may form the following portfolio: Buy a call option with strike price K_1 and then, to reduce the premium paid for the call option, write and sell another call option with the same exercise date but with the strike price $K_2 > K_1$. The resulting portfolio is called a *bull spread* (see Figure 4.10).

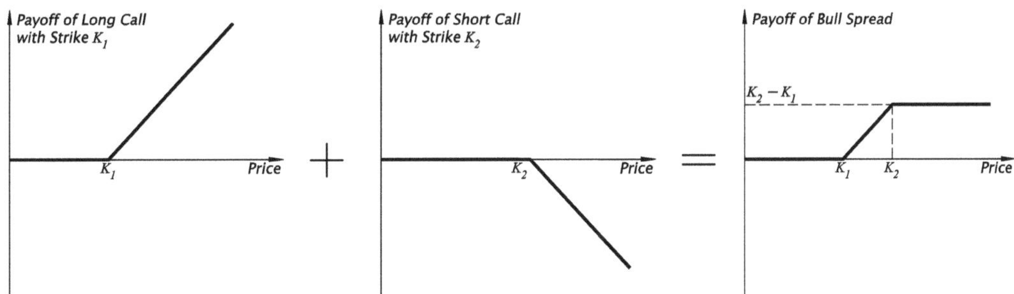

FIGURE 4.10: The bull spread.

The payoffs are described in the table below.

Stock Price S:	Long Call Payoff:	Short Call Payoff:	Total Payoff:
$S \leqslant K_1$	0	0	0
$K_1 < S < K_2$	$S - K_1$	0	$S - K_1$
$K_2 \leqslant S$	$S - K_1$	$K_2 - S$	$K_2 - K_1$

The payoff is positive for high future stock prices. The no-arbitrage price of the bull spread is equal to $C^E(K_1) - C^E(K_2)$.

A *bear spread* would satisfy an investor who believes that the stock price will decline. The bear spread is equivalent to a portfolio that has one long put option with strike price

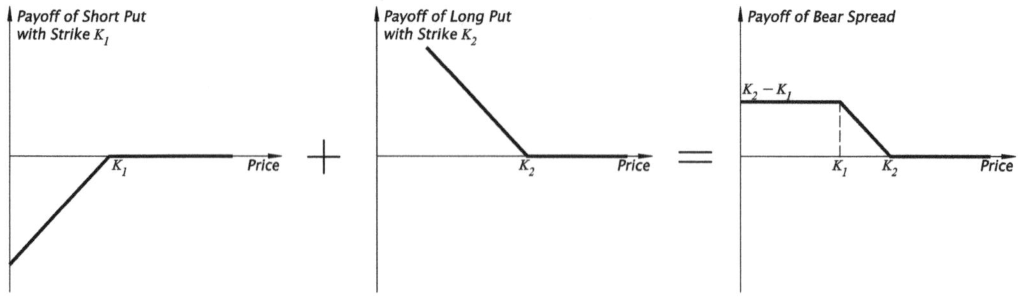

FIGURE 4.11: The bear spread.

K_2 and one short put option with strike $K_1 < K_2$ (see Figure 4.11). Both options have the same exercise date. The payoffs are described in the table below.

Stock Price S:	Short Put Payoff:	Long Put Payoff:	Total Payoff:
$S \leqslant K_1$	$S - K_1$	$K_2 - S$	$K_2 - K_1$
$K_1 < S < K_2$	$K_2 - S$	0	$K_2 - S$
$K_2 \leqslant S$	0	0	0

The payoff is positive for low future stock prices. The no-arbitrage price of the bear spread is equal to $P^E(K_2) - P^E(K_1)$.

An investor who expects that the future stock price will change insignificantly and stay in an interval $[K_1, K_3]$ may choose a *butterfly spread* that is constructed from three options of the same kind expiring on the same date with strike prices K_i, $i = 1, 2, 3$, so that $K_1 < K_2 < K_3$. One way to construct a butterfly spread is to combine one long call with strike K_1, two short calls with strike K_2, and one long call with strike K_3 (see Figure 4.12). Assuming that $K_2 = \frac{K_1+K_3}{2}$, the payoff function of the butterfly spread is

$$\Lambda_{BS}(S) = \begin{cases} 0 & \text{if } S \leqslant K_1, \\ S - K_1 & \text{if } K_1 < S \leqslant K_2, \\ K_3 - S & \text{if } K_2 < S \leqslant K_3, \\ 0 & \text{if } S > K_3. \end{cases}$$

The no-arbitrage price is $C^E(K_1) + C^E(K_3) - 2C^E(K_2)$.

A *combination* is an option portfolio that involves taking a position in both calls and puts on the same underlying security. Some examples are as follows.

Straddle combines one put and one call with the same strike and expiry date.

Strip combines one long call and two long puts with the same strike and expiry date.

Strap combines two long calls and one long put with the same strike and expiry date.

Strangle is a combination of a put and a call with the same expiry date but different strike prices.

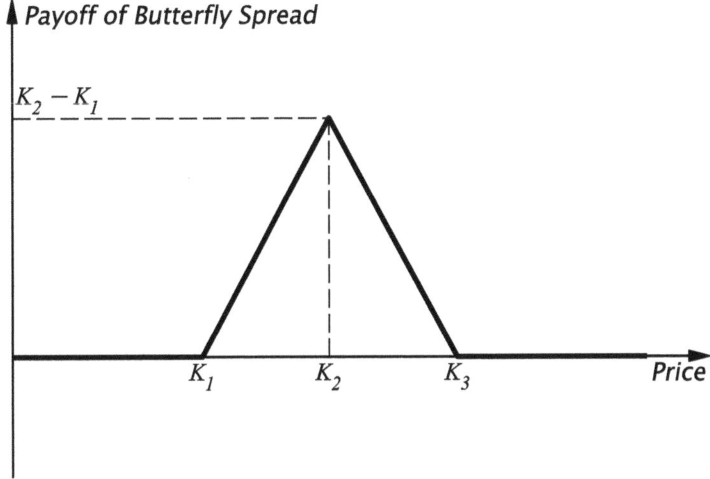

FIGURE 4.12: The butterfly spread.

4.3 Basics of Option Pricing

Consider a European-style derivative D on an asset S, whose payoff Λ is a function of the asset price only at maturity time T. Examples include:

- European call option with $\Lambda(S) = (S - K)^+$;
- European put option with $\Lambda(S) = (K - S)^+$;
- long forward contract with $\Lambda(S) = S - K$;
- spreads and combinations such as a straddle, strip, strap, and strangle;
- binary call and put options (cash-or-nothing options) with respective payoffs

$$\Lambda(S) = \mathbb{I}_{\{S \geqslant K\}} = \begin{cases} 1 & \text{if } S \geqslant K, \\ 0 & \text{otherwise,} \end{cases} \quad \text{and} \quad \Lambda(S) = \mathbb{I}_{\{S \leqslant K\}} = \begin{cases} 1 & \text{if } S \leqslant K, \\ 0 & \text{otherwise.} \end{cases}$$

In this section we will consider both discrete-time and continuous-time asset price models, respectively represented by the binomial tree model and the log-normal model. Let S_t and $S(t)$ denote the time-t asset price in discrete time and continuous time, respectively. The time-t derivative price will be denoted by V_t and $V(t)$, respectively. One can prove that at every time $t \in [0, T]$, the value of such derivatives is a function of the current asset price $S(t)$ (or S_t) and calendar time t: $V(t) = V(t, S(t))$ (or $V_t = V_t(S_t)$). Since the future asset price is a random variable, the derivative price is also random (i.e., uncertain) at any time $t > 0$. We are interested in finding the initial price of the derivative, $V(0) = V_0(S(0))$ (or $V_0 = V_0(S_0)$).

4.3.1 Pricing of European-Style Derivatives in the Binomial Tree Model

4.3.1.1 Replication and Pricing of Options in the Single-Period Binomial Model

Consider a single-period binomial model with two states of the world: $\Omega = \{\omega^+, \omega^-\}$. At time $t = 0$ the stock price is S_0; at time $t = 1$ the stock price equals $S^+ := S_1(\omega^+) = S_0 u$ with the probability p and $S^- := S_1(\omega^-) = S_0 d$ with the probability $1 - p$. Recall that p and $1 - p$ are called the physical or real-world probabilities. The time values of a risk-free bond are B_0 and $B_1 = B_0(1 + r)$. Here, r is a risk-free one-period interest rate; $d - 1$ and $u - 1$ are, respectively, downward and upward one-period returns on the stock S.

Suppose we wish to find a no-arbitrage (fair) price of a derivative with the payoff $\Lambda(S)$ at maturity time $T = 1$. In Section 2.2, we constructed a portfolio (x, y) in the stock and bond that replicates the payoff function as $xS_1(\omega^\pm) + yB_1 = \Lambda(S_1(\omega^\pm))$ by solving the system of two linear equations

$$\begin{cases} xS^+ + yB_1 = \Lambda(S^+) \\ xS^- + yB_1 = \Lambda(S^-). \end{cases}$$

The solution gives the replicating portfolio:

$$\begin{cases} x = \dfrac{\Lambda(S^+) - \Lambda(S^-)}{S^+ - S^-} = \dfrac{\Lambda(S^+) - \Lambda(S^-)}{S_0(u - d)} \\[2ex] y = \dfrac{\Lambda(S^-)S^+ - \Lambda(S^+)S^-}{B_1(S^+ - S^-)} = \dfrac{\Lambda(S^-)u - \Lambda(S^+)d}{B_0(1 + r)(u - d)}. \end{cases} \tag{4.9}$$

The portfolio (x, y) and the derivative have the same value at time 1. According to the law of one price they should have the same value at time 0 or else an arbitrage opportunity will arise, i.e.,

$$\forall \omega \in \Omega \ \Pi_1^{(x,y)}(\omega) = V_1(S_1(\omega)) = \Lambda(S_1(\omega)) \implies V_0(S_0) = \Pi_0^{(x,y)}.$$

Thus, the initial value of the derivative is

$$\begin{aligned} V_0(S_0) = \Pi_0^{(x,y)} &= xS_0 + yB_0 = \frac{\Lambda(S^+) - \Lambda(S^-)}{S_0(u - d)} S_0 + \frac{\Lambda(S^-)u - \Lambda(S^+)d}{B_0(1 + r)(u - d)} B_0 \\ &= \frac{\Lambda(S^+) - \Lambda(S^-)}{u - d} + \frac{\Lambda(S^-)u - \Lambda(S^+)d}{(1 + r)(u - d)} \\ &= \frac{1}{1 + r} \left(\Lambda(S^+) \frac{(1 + r) - d}{u - d} + \Lambda(S^-) \frac{u - (1 + r)}{u - d} \right) \\ &= \frac{1}{1 + r} \left(\Lambda(S_0 u) \frac{1 + r - d}{u - d} + \Lambda(S_0 d) \frac{u - r - 1}{u - d} \right). \end{aligned}$$

Notice the above expression can be written as a mathematical expectation:

$$\begin{aligned} V_0(S_0) &= \frac{1}{1 + r} \left(\Lambda(S_0 u) \tilde{p} + \Lambda(S_0 d) (1 - \tilde{p}) \right) \\ &= \frac{1}{1 + r} \left(V_1(S_0 u) \tilde{p} + V_1(S_0 d) (1 - \tilde{p}) \right) \qquad (4.10) \\ &= \frac{1}{1 + r} \widetilde{\mathrm{E}}[V_1(S(1))] \qquad (4.11) \end{aligned}$$

where $\tilde{p} := \frac{1+r-d}{u-d}$ and $1 - \tilde{p} = \frac{u-r-1}{u-d}$. This formula is called the *risk-neutral pricing formula.* The numbers $\tilde{p}$ and $1 - \tilde{p}$ are called the *risk-neutral probabilities.* They exist iff $d < 1 + r < u$.

Example 4.5 (Replication and pricing in one period). Consider a single-period binomial model with the following parameters: $S_0 = \$100$, $B_0 = \$1$, $d = 0.9$, $u = 1.15$, $r = 0.05$. Replicate a call option with strike price $K = \$95$ and expiry time $T = 1$. Find the option price.

Solution. Application of (4.9) gives positions x and y of the replicating portfolio for the European call option:

$$x = \frac{(100 \cdot 1.15 - 95)^+ - (100 \cdot 0.9 - 95)^+}{100 \cdot (1.15 - 0.9)} = \frac{20 - 0}{100 \cdot 0.25} = \frac{4}{5},$$
$$y = \frac{0 \cdot 1.15 - 20 \cdot 0.9}{1 \cdot 1.05 \cdot (1.15 - 0.9)} = \frac{-18}{1.05 \cdot 0.25} \cong -68.5714.$$

So to replicate the European call, we need $\frac{4}{5}$ shares of stock and a loan in the amount of \$68.57. The current price of the call option is equal to the portfolio value:

$$C_0^E = xS_0 + yB_0 \cong \frac{4}{5} \cdot 100 - 68.5714 \cdot 1 \cong \$11.43.$$

On the other hand, $\tilde{p} = (1.05 - 0.9)/(1.15 - 0.9) = 0.6$, $1 - \tilde{p} = 0.4$ so the risk-neutral pricing formula in (4.10) gives

$$C_0^E = \frac{1}{1.05} \cdot (20 \cdot 0.6 + 0 \cdot 0.4) = \frac{20 \cdot 0.6}{1.05} \cong \$11.43.$$ □

It is easy to see that a mispriced derivative security leads to arbitrage. Consider a derivative contract with payoff Λ in a single-period binomial model. If the trading price of the derivative, say $\widehat{V}_0$, is not equal to the risk-neutral price V_0 then one can construct an arbitrage portfolio (x, y, z) with x stock shares, y dollars in the interest-bearing risk-free account, and z derivative contracts as follows. Let $\widehat{V}_0 > V_0$. Form a portfolio $\Pi^{(x,y,z)}$ with $x = \Delta$, $y = \beta + \widehat{V}_0 - V_0$, and $z = -1$, where

$$\Delta = \frac{\Lambda(S^+) - \Lambda(S^-)}{S_0(u - d)} \quad \text{and} \quad \beta = \frac{\Lambda(S^-)u - \Lambda(S^+)d}{(1 + r)(u - d)}.$$

The initial wealth is zero. At expiry, $\Pi_1 = (\widehat{V}_0 - V_0)(1 + r) > 0$. Let $\widehat{V}_0 < V_0$. Form a portfolio with x, y, z that are negative for the case with $\widehat{V}_0 > V_0$, i.e., the following portfolio: $(-\Delta, -\beta + (V_0 - \widehat{V}_0), 1)$. Again, the initial wealth is zero. At expiry, $\Pi_1 = (V_0 - \widehat{V}_0)(1+r) > 0$.

4.3.1.2 Pricing in the Binomial Tree Model

Now we turn our attention to a multi-period binomial tree model. Recall that at time $t \in \{0, 1, 2, \dots\}$, the stock price has the probability distribution given in (2.13):

$$S_t = S_0 u^n d^{t-n} \text{ with probability } \binom{t}{n} p^t (1 - p)^{t-n} \text{ for } n = 0, 1, 2, \dots, t,$$

where S_0 is the initial stock price. The time-t bond value is $B_t = B_0(1 + r)^t$.

Our main objective is the pricing of a derivative contract with maturity time $T \in \{1, 2, \dots\}$. Let V_t denote the no-arbitrage derivative price at time $t \in \mathbf{T}$ where $\mathbf{T} = \{0, 1, \dots, T\}$. These prices form a derivative price process $\{V_t\}_{t \in \mathbf{T}}$, which is a stochastic process meaning that for any time t, the value V_t is a random variable defined on the state space Ω. So we write $V_t = V_t(\omega)$ with $\omega \in \Omega$.

To find the initial no-arbitrage price of the derivative, V_0, one can again use replication.

However, a static replicating portfolio is *not sufficient* for a multi-period model. To replicate the whole derivative price process, we need to apply a self-financing dynamic portfolio strategy in the stock and the bond. An admissible self-financing strategy $\{(x_t, y_t)\}_{t\in\{0,1,\ldots,T-1\}}$ is said to *replicate the derivative* if the terminal portfolio value, Π_T, and the terminal derivative value, V_T, coincide for every market scenario $\omega \in \Omega$:

$$x_{T-1}(\omega)S_T(\omega) + y_{T-1}(\omega)B_T = V_T(\omega).$$

By the law of one price, the initial price of the derivative has to be equal to the initial value of the replicating strategy:

$$V_0 = x_0 S_0 + y_0 B_0.$$

Additionally, for every time $t \in \mathbf{T}$, the replicating strategy and the derivative must have the same value for all market scenarios: $V_t = x_t S_t + y_t B_t$. This problem will be investigated in full detail in later chapters. At this point our goal is simply to obtain a closed-form formula for the initial derivative price. So we will use a more straightforward approach.

Consider a European-style derivative contract with payoff function Λ, which is contingent on the stock price process. At the maturity time T, $V_T = \Lambda(S_T)$ holds. As will be demonstrated in later chapters, the time-t price V_t of any European-style derivative is a function of the time-t asset price: $V_t = V_t(S_t)$.

In the binomial tree model, the evolution of a stock price process is described by a recombining binomial tree. Such a tree can be viewed as a combination of single-period subtrees with two branches, where $S_{t+1} = S_t u$ for upward and $S_{t+1} = S_t d$ for downward branches. For each subtree, we can use the one-period pricing formula (4.10). Therefore, in the binomial tree model the derivative prices can be computed using backward-in-time recurrence:

$$V_t(S_t) = \frac{1}{1+r}\left(\tilde{p}V_{t+1}(S_t u) + (1-\tilde{p})V_{t+1}(S_t d)\right) \tag{4.12}$$

for $t = T-1, T-2, \ldots, 0$. A T-period binomial tree has $n+1$ nodes at any time $n \in \{0,1,\ldots,T\}$ and hence has $\sum_{n=0}^{T}(n+1) = \frac{(T+1)(T+2)}{2}$ nodes in total. The goal is to calculate the derivative value for each node. Before beginning to apply the above recurrence, the derivative prices $V_T(S_T)$ at maturity for each of $T+1$ nodes $S_{T,m} \equiv S_0 u^m d^{T-m}$ with $m = 0, 1, \ldots, T$ are simply given by the payoff function:

$$V_T(S_{T,m}) = \Lambda(S_{T,m}).$$

For the first backward time step, we set $t = T-1$ in (4.12) and obtain T derivative prices $V_{T-1}(S_{T-1})$ for $S_{T-1} = S_{T-1,m} \equiv S_0 u^m d^{T-1-m}$ with $m = 0, 1, \ldots, T-1$. In similar fashion, we apply (4.12) to obtain the derivative prices at times $t = T-2, \ldots, 1, 0$. For example, by applying (4.12) $T-n$ times (for any $0 \leqslant n \leqslant T$) we arrive at $n+1$ derivative prices $V_n(S_{n,m})$ at time $t = n$ on the nodes $S_{n,m} := S_0 u^n d^{n-m}$ with $m = 0, 1, \ldots, n$. After applying the recurrence relation T times we arrive at a single price $V_0 = V_0(S_0)$. The process is illustrated in Figure 4.13.

To better clarify how one applies the recurrence in (4.12), we now consider a two-period binomial tree model. To compute the derivative prices $V_t \equiv V_t(S_t)$, $t = 0, 1, 2$, we proceed backward in time starting at the maturity $T = 2$. For every price S_2, derivative prices are equal to payoff values: $V_2(S_2) = \Lambda(S_2)$. At time $t = 1$, the stock price takes one of two possible values: $S_0 u$ or $S_0 d$. The derivative prices can be computed using the one-period

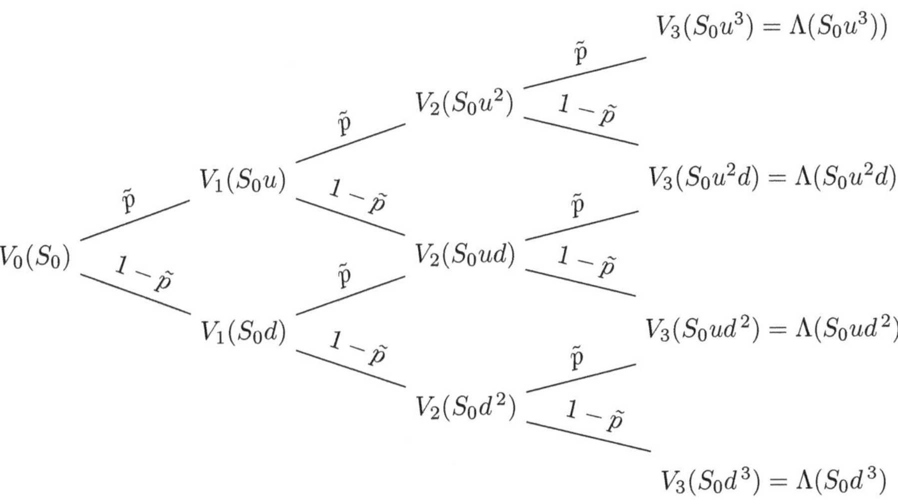

FIGURE 4.13: Valuation of a European derivative on a binomial tree model with three periods.

pricing formula as follows:

$$\begin{aligned} V_1(S_0u) &= \frac{1}{1+r}\left(\tilde{p}V_2(S_0uu) + (1-\tilde{p})V_2(S_0ud)\right) \\ &= \frac{1}{1+r}\left(\tilde{p}\Lambda(S_0u^2) + (1-\tilde{p})\Lambda(S_0ud)\right), \\ V_1(S_0d) &= \frac{1}{1+r}\left(\tilde{p}V_2(S_0du) + (1-\tilde{p})V_2(S_0dd)\right) \\ &= \frac{1}{1+r}\left(\tilde{p}\Lambda(S_0du) + (1-\tilde{p})\Lambda(S_0d^2)\right). \end{aligned}$$

Here, we used $V_2(S) = \Lambda(S)$. Finally, at time $t = 0$, we again apply the one-period pricing formula to obtain the final expression for the no-arbitrage pricing formula:

$$\begin{aligned} V_0(S_0) &= \frac{1}{1+r}\left(\tilde{p}V_1(S_0u) + (1-\tilde{p})V_1(S_0d)\right) \\ &= \frac{1}{(1+r)^2}\left(\tilde{p}^2\Lambda(S_0u^2) + 2\tilde{p}(1-\tilde{p})\Lambda(S_0ud) + (1-\tilde{p})^2\Lambda(S_0d^2)\right). \end{aligned}$$

Example 4.6 (Pricing in two periods)**.** Consider the two-period binomial model from Example 4.5. Find the no-arbitrage prices C_0^E and C_1^E of the call price with the strike price $K = \$95$ and expiry time $T = 2$.

Solution. First, compute the risk-neutral probabilities:

$$\tilde{p} = \frac{1.05 - 0.9}{1.15 - 0.9} = 0.6, \quad 1 - \tilde{p} = 0.4.$$

Second, we calculate the prices $C_1^E(S_1)$ when $S_1 = S_0 u = \$115$ and $S_1 = S_0 d = \$90$:

$$\begin{aligned}
C_1^E(90) &= \frac{1}{1+r}(\Lambda(90\,u)\,\tilde{p} + \Lambda(90\,d)\,(1-\tilde{p})) \\
&= \frac{1}{1.05}\cdot((90\cdot 1.15 - 95)^+ \cdot 0.6 + (90\cdot 0.9 - 95)^+\cdot 0.4) \\
&= \frac{8.5\cdot 0.6 + 0\cdot 0.4}{1.05} \cong \$4.85714, \\
C_1^E(115) &= \frac{1}{1.05}\cdot((115\cdot 1.15 - 95)^+ \cdot 0.6 + (115\cdot 0.9 - 95)^+\cdot 0.4) \\
&= \frac{37.25\cdot 0.6 + 8.5\cdot 0.4}{1.05} \cong \$24.52381.
\end{aligned}$$

Third, calculate $C_0^E(100)$:

$$\begin{aligned}
C_0^E(100) &= \frac{1}{1+r}(C_1^E(100\,u)\,\tilde{p} + C_1^E(100\,d)\,(1-\tilde{p})) \\
&\cong \frac{1}{1.05}\cdot(24.52381\cdot 0.6 + 4.85714\cdot 0.4) \cong \$15.86394.
\end{aligned}$$

□

Theorem 4.8 (The Binomial Derivative Price Formula). *In the binomial tree model, the no-arbitrage initial price of a non-path-dependent derivative with payoff $\Lambda(S_T)$, at (discrete) time $T \geqslant 1$, is given by*

$$V_0(S_0) = \frac{1}{(1+r)^T}\sum_{n=0}^{T}\binom{T}{n}\tilde{p}^n\,(1-\tilde{p})^{T-n}\,\Lambda\left(S_0 u^n d^{T-n}\right). \tag{4.13}$$

Proof. We prove the assertion by induction. The formula (4.13) is valid for $T = 1$. Suppose it is valid for some $T \geqslant 1$. Let us prove that it holds for $T+1$. For $T = 1$, we have

$$V_0(S_0) = \frac{1}{1+r}\left[V_1(S_0 u)\,\tilde{p} + V_1(S_0 d)\,(1-\tilde{p})\right]. \tag{4.14}$$

That is, the price V_0 is a weighted sum of $V_1(S_0 u)$ and $V_1(S_0 d)$, which are the prices of the derivative issued at time $t = 1$ and expiring at time $t = T+1$ provided that the current stock price is, respectively, $S_0 u$ and $S_0 d$. Since there are T periods from the issue time $t = 1$ to expiry time $t = T+1$, we can use (4.13) to evaluate the derivatives:

$$\begin{aligned}
V_1(S_0 d) &= \frac{1}{(1+r)^T}\sum_{n=0}^{T}\binom{T}{n}\tilde{p}^n\,(1-\tilde{p})^{T-n}\,\Lambda\left(S_0 u^n d^{T-n+1}\right), \\
V_1(S_0 u) &= \frac{1}{(1+r)^T}\sum_{n=0}^{T}\binom{T}{n}\tilde{p}^n\,(1-\tilde{p})^{T-n}\,\Lambda\left(S_0 u^{n+1} d^{T-n}\right).
\end{aligned} \tag{4.15}$$

By combining (4.14) and (4.15), we obtain

$$\begin{aligned}
V_0(S_0) = \frac{1}{(1+r)^{T+1}}\Bigg[&\tilde{p}^0\,(1-\tilde{p})^{T+1}\binom{T}{0}\Lambda\left(S_0 u^0 d^{T+1}\right) \\
&+ \tilde{p}^{T+1}\,(1-\tilde{p})^0\binom{T}{T}\Lambda\left(S_0 u^{T+1} d^0\right) \\
&+ \sum_{n=1}^{T}\left(\binom{T}{n-1} + \binom{T}{n}\right)\tilde{p}^n\,(1-\tilde{p})^{T+1-n}\,\Lambda\left(S_0 u^n d^{T+1-n}\right)\Bigg].
\end{aligned}$$

Using the identities $\binom{T}{0} = \binom{T}{T} = 1$ and $\binom{T}{n-1} + \binom{T}{n} = \binom{T+1}{n}$ gives the derivative price formula for $T+1$ time periods:

$$V_0(S_0) = \frac{1}{(1+r)^{T+1}} \sum_{n=0}^{T+1} \binom{T+1}{n} \tilde{p}^n (1-\tilde{p})^{(T+1)-n} \Lambda\left(S_0 u^n d^{(T+1)-n}\right). \qquad \square$$

As follows from (2.13), the asset price S_T can be expressed in terms of a binomial random variable $S_T = S_0 u^{Y_T} d^{T-Y_T}$, where $Y_T \sim Bin(T, \tilde{p})$, so that

$$Y_T = n \text{ with } \tilde{\mathbb{P}}\text{-probability } \binom{T}{n} \tilde{p}^n (1-\tilde{p})^{T-n} \text{ for all } n = 0, 1, 2, \ldots, T.$$

Therefore, the option price formula in (4.13) admits a very compact form:

$$V_0(S_0) = \frac{1}{(1+r)^T} \widetilde{\mathrm{E}}\left[\Lambda\left(S_0 u^{Y_T} d^{T-Y_T}\right)\right] = \frac{1}{(1+r)^T} \widetilde{\mathrm{E}}\left[\Lambda(S_T)\right]. \tag{4.16}$$

Example 4.7 (Pricing two nonstandard options). Consider a binomial tree model with $S_0 = \$100$, $u = 1.2$, $d = 0.9$, and $r = 0.1$. Find the no-arbitrage present value of the following European options:

(a) the butterfly option with the payoff function

$$\Lambda_B(S) = \begin{cases} 0, & \text{if } S \notin [100, 140], \\ 140 - S, & \text{if } S \in [120, 140], \\ S - 100, & \text{if } S \in [100, 120); \end{cases}$$

(b) the binary (cash-or-nothing) call option with the payoff function

$$\Lambda_D(S) = \begin{cases} 0, & \text{if } S < 120, \\ 20, & \text{if } S \geqslant 120. \end{cases}$$

Both options are exercised at time $T = 3$.

Solution. The risk-neutral probabilities are

$$\tilde{p} = \frac{1 + 0.1 - 0.9}{1.2 - 0.9} = \frac{0.2}{0.3} = \frac{2}{3}, \quad 1 - \tilde{p} = \frac{1}{3}.$$

Application of Equations (4.13) and (4.16) gives us the price of the butterfly option:

$$\begin{aligned}
\frac{1}{(1+r)^3} \widetilde{\mathrm{E}}[\Lambda_B(S_3)] &= \frac{1}{1.1^3} \sum_{n=0}^{3} \binom{3}{n} \left(\frac{2}{3}\right)^n \left(\frac{1}{3}\right)^{3-n} \Lambda_B\left(100 \cdot 1.2^n \cdot 0.9^{3-n}\right) \\
&= \frac{1}{1.1^3} \cdot \left(\frac{1}{27} \cdot \Lambda_B(172.8) + 3 \cdot \frac{2}{27} \cdot \Lambda_B(129.6) + 3 \cdot \frac{4}{27} \cdot \Lambda_B(97.2) + \frac{8}{27} \cdot \Lambda_B(72.9)\right) \\
&= \frac{1}{1.331} \cdot \left(\frac{1}{27} \cdot 0 + 3 \cdot \frac{2}{27} \cdot 10.4 + 3 \cdot \frac{4}{27} \cdot 0 + \frac{8}{27} \cdot 0\right) \\
&= \frac{1}{1.331} \cdot \frac{2}{9} \cdot 10.4 \cong \$1.7364.
\end{aligned}$$

The pricing formula for the cash-or-nothing call option is simplified as follows:

$$\begin{aligned}\frac{1}{(1+r)^3}\widetilde{\mathrm{E}}[\Lambda_D(S_3)] &= \frac{1}{1.1^3}\widetilde{\mathrm{E}}[20\cdot\mathbb{I}_{\{S_3\geqslant 120\}}] \\ &= \frac{1}{1.1^3}\cdot 20\cdot\widetilde{\mathbb{P}}(S_3\geqslant 120) = \frac{20}{1.1^3}\cdot\left(\widetilde{\mathbb{P}}(S_3=172.8)+\widetilde{\mathbb{P}}(S_3=129.6)\right) \\ &= \frac{20}{1.1^3}\cdot\left(\frac{1}{27}+\frac{2}{9}\right) = \frac{20\cdot 7}{1.1^3\cdot 27}\cong \$3.89571. \qquad \square\end{aligned}$$

For standard European call and put options, the option price formula in (4.13) can be written in a simpler form. Introduce the cumulative distribution function (CDF for short) $\mathcal{B}$ of the binomial probability distribution $Bin(n,p)$:

$$\mathcal{B}(m;n,p) = \sum_{k=0}^{m} \mathscr{b}(k;n,p) = \sum_{k=0}^{m}\binom{n}{k}p^k(1-p)^{n-k},\ m=0,1,\ldots,n.$$

Here $\mathscr{b}$ denotes the probability mass function (PMF) of the binomial distribution; it is given by $\mathscr{b}(k;n,p) = \binom{n}{k}p^k(1-p)^{n-k}$ for $k=0,1,\ldots,n$. Recall that the CDF F of a random variable X is defined by $F(x)=\mathbb{P}(X\leqslant x)$; the PMF of a discrete random variable X is $p(x)=\mathbb{P}(X=x)$.

Theorem 4.9 (The Cox–Ross–Rubinstein (CRR) Option Price Formula). *In the binomial tree model, the no-arbitrage initial price of standard European call and put options with strike price K and expiry time T are, respectively, given by*

$$C_0^E(S_0) = S_0(1-\mathcal{B}(m_T;T,\hat{p})) - \frac{K}{(1+r)^T}(1-\mathcal{B}(m_T;T,\tilde{p}))\,, \tag{4.17}$$

$$P_0^E(S_0) = \frac{K}{(1+r)^T}\mathcal{B}(m_T;T,\tilde{p}) - S_0\mathcal{B}(m_T;T,\hat{p})\,, \tag{4.18}$$

where $\tilde{p}=\frac{1+r-d}{u-d}$, $\hat{p}=\frac{d}{1+r}\tilde{p}$, and

$$m_T = \max\left\{m\ :\ 0\leqslant m\leqslant T;\ S_0\,u^m\,d^{T-m}\leqslant K\right\} = \left\lfloor\frac{\ln(K/S_0)-T\ln d}{\ln(u/d)}\right\rfloor. \tag{4.19}$$

Proof. The proof is left as an exercise for the reader. A more general result is proved in later chapters. $\square$

4.3.2 Pricing of American Options in the Binomial Tree Model

Consider a T-period binomial tree model. Let $\Lambda(S)$ be the payoff function of an American derivative security, which is expiring at time T. The option can be exercised at any time $t\in\{0,1,\ldots,T\}$ with the immediate payoff $\Lambda(S_t)$. Denote by $V_t^A(S_t)$ the value of the American derivative at time t given that the option has not been exercised previously.

At the expiry time $t=T$ the option is worth $V_T^A(S_T)=\Lambda(S_T)$ given that it has not been exercised until maturity. At time $t=T-1$, the option holder has the choice to exercise the option immediately with the **intrinsic value** $\Lambda(S_{T-1})$ or to postpone until time T, when the value of the option will become $\Lambda(S_T)$. The **continuation value** can be computed by considering a one-step European contingent claim to be priced at time $T-1$. Its value at time $T-1$ at stock price $S_{T-1}=S$ is given by

$$\frac{1}{1+r}\left(\tilde{p}\Lambda(Su)+(1-\tilde{p})\Lambda(Sd)\right) = \frac{1}{1+r}\widetilde{\mathrm{E}}\left[V_T^A(S_T)\mid S_{T-1}=S\right].$$

At time $T-1$ the American option should be worth the higher of the intrinsic value and continuation value:

$$V_{T-1}^{A}(S) = \max\left\{\Lambda(S), \frac{1}{1+r}\widetilde{\mathrm{E}}\left[V_T^A(S_T) \mid S_{T-1} = S\right]\right\}.$$

Therefore, at every time $t = T-1, T-2, \ldots, 0$, the option holder has the choice between immediately exercising with the intrinsic value

$$V_t^{\mathrm{intr}}(S_t) = \Lambda(S_t)$$

or postponing with the continuation value

$$V_t^{\mathrm{cont}}(S_t) = \frac{1}{1+r}\widetilde{\mathrm{E}}\left[V_{t+1}^A(S_{t+1}) \mid S_t\right].$$

Thus, we obtain the following recursive pricing formulae:

$$\begin{aligned} V_T^A(S_T) &= \Lambda(S_T) \\ V_t^A(S_t) &= \max\{V_t^{\mathrm{intr}}(S_t), V_t^{\mathrm{cont}}(S_t)\} = \max\left\{\Lambda(S_t), \frac{1}{1+r}\widetilde{\mathrm{E}}\left[V_{t+1}^A(S_{t+1}) \mid S_t\right]\right\} \\ &= \max\left\{\Lambda(S_t), \frac{1}{1+r}\left[\tilde{p}V_{t+1}^A(S_n u) + (1-\tilde{p})V_{t+1}^A(S_t d)\right]\right\}, \\ &\quad \text{for } t = T-1, T-2, \ldots, 0. \end{aligned}$$

At time 0 we obtain the initial no-arbitrage price $V_0^A(S_0)$.

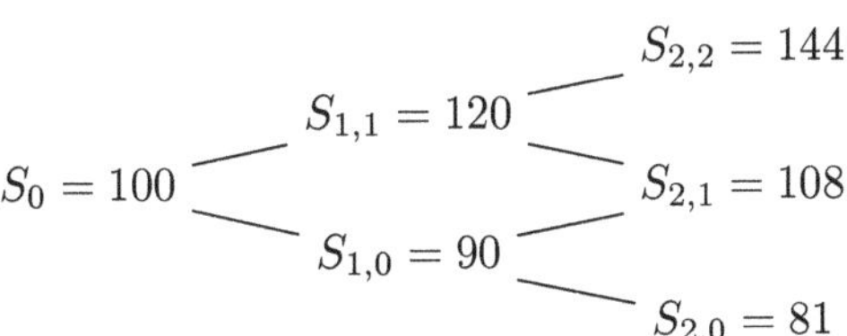

FIGURE 4.14: Stock prices.

Example 4.8 (The case without dividends). Find no-arbitrage prices of the American put and European put options with the common strike price $K = \$100$ both expiring at time $T = 2$ on a stock with the initial price $S_0 = \$100$ in a two-period binomial tree model with $u = 1.2$, $d = 0.9$, and $r = 0.1$.

Solution. First, construct the binomial tree with two periods (see Figure 4.14) and find the risk-neutral probabilities: $\tilde{p} = \frac{2}{3}$ and $1 - \tilde{p} = \frac{1}{3}$ (see Figure 4.14). Second, calculate prices of the put options starting from the expiry date $T = 2$ and going backward in time:

$$
\begin{aligned}
n = 2: \quad & P_2^E(81) = P_2^A(81) = (100 - 81)^+ = \$19 \\
& P_2^E(108) = P_2^A(108) = (100 - 108)^+ = \$0 \\
& P_2^E(144) = P_2^A(144) = (100 - 144)^+ = \$0, \\
n = 1: \quad & P_1^E(90) = \frac{1}{1.1}\left(\frac{2}{3}P_2^E(108) + \frac{1}{3}P_2^E(81)\right) \cong \$5.75758 \\
& P_1^E(120) = \frac{1}{1.1}\left(\frac{2}{3}P_2^E(144) + \frac{1}{3}P_2^E(108)\right) = \$0 \\
& P_1^A(90) = \max\left\{(100 - 90)^+, \frac{1}{1.1}\left(\frac{2}{3}P_2^A(108) + \frac{1}{3}P_2^A(81)\right)\right\} \\
& \qquad = \max\{10, 5.75758\} = \$10 \\
& P_1^A(120) = \max\left\{(100 - 120)^+, \frac{1}{1.1}\left(\frac{2}{3}P_2^A(144) + \frac{1}{3}P_2^A(108)\right)\right\} \\
& \qquad = \max\{0, 0\} = \$0, \\
n = 0: \quad & P_0^E(100) = \frac{1}{1.1}\left(\frac{2}{3}P_1^E(120) + \frac{1}{3}P_1^E(90)\right) = \$1.74472 \\
& P_0^A(100) = \max\left\{(100 - 100)^+, \frac{1}{1.1}\left(\frac{2}{3}P_1^A(120) + \frac{1}{3}P_1^A(90)\right)\right\} \\
& \qquad = \max\left\{0, \frac{10}{3 \cdot 1.1}\right\} \cong \$3.03030.
\end{aligned}
$$

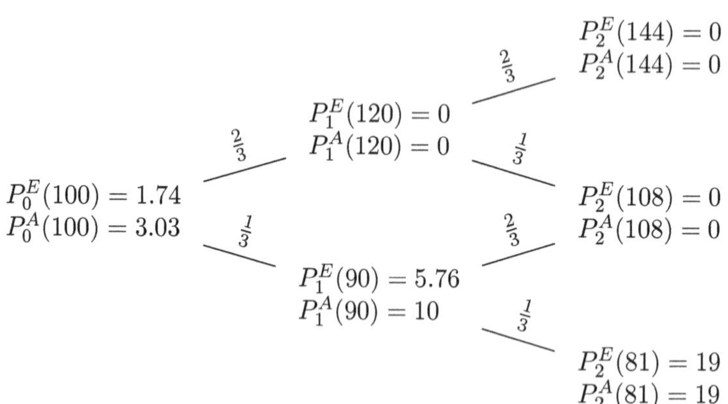

FIGURE 4.15: Pricing in a two-period binomial tree model.

The solution can be represented in the form of a binomial tree (see Figure 4.15). The American put option should be exercised whenever the continuation value $V_{\text{cont}}(t, S)$ falls below the intrinsic value $(K - S)^+$. In this case, the American option should be exercised early at time $t = 1$ if $S_1 = 90$. □

4.3.3 Option Pricing in the Log-Normal Model: The Black–Scholes–Merton Formula

In this section we present the European option pricing formula under the log-normal model. The culminating point will be the derivation of the famous Black–Scholes–Merton formulae for the no-arbitrage prices of standard European call and put options.

The log-normal pricing model can be obtained as the limiting case of a sequence of suitably scaled binomial tree models. Consider a sequence of binomial prices $S_N^{(N)}$ first presented in Section 2.2.5. In the risk-neutral setting, the prices $S_N^{(N)}$ converge in distribution (as $N \to \infty$) to the log-normal price $S(T)$ defined by

$$S(T) = S_0 \mathrm{e}^{(r-\sigma^2/2)T+\sigma\sqrt{T}Z} \text{ with } Z \sim Norm(0,1).$$

Recall that S_0 is the initial asset price, r is the risk-free interest rate compounded continuously, T is the maturity time (in years), and σ is the (annual) volatility of the asset price. Let us investigate if it is possible to obtain the no-arbitrage option price formula for the log-normal model by proceeding to the continuous-time limit in (4.16), (4.17), and (4.18).

The first approach to be considered is based on a property of the convergence in distribution and is stated below without a proof.

Theorem 4.10. *Suppose $X_n \xrightarrow{d} X$, as $n \to \infty$. Let $f : \mathbb{R} \to \mathbb{R}$ be a bounded and continuous function. Then*

$$\mathrm{E}[f(X_n)] \to \mathrm{E}[f(X)], \quad \text{as } n \to \infty.$$

For the binomial tree model, we obtained a general no-arbitrage price formula (4.16) of a European-type derivative with payoff Λ. By applying (4.16) to the sequence of scaled binomial tree models with terminal prices $S_N^{(N)}$, $N \geqslant 1$, we obtain a sequence of no-arbitrage derivative prices:

$$V_0^{(N)} = (1+r_N)^{-N} \widetilde{\mathrm{E}}\left[\Lambda(S_N^{(N)})\right].$$

Suppose that Λ is a continuous and bounded function such that the European put payoff $\Lambda_P(S) = (K-S)^+$. Since the sequence of binomial prices converges in distribution to the log-normal price, we immediately obtain that

$$\widetilde{\mathrm{E}}\left[\Lambda(S_N^{(N)})\right] \to \widetilde{\mathrm{E}}[\Lambda(S(T))], \text{ as } N \to \infty.$$

By the definition of $r_N = \mathrm{e}^{r\delta_N} - 1$, the discount factor is $(1+r_N)^{-N} = \mathrm{e}^{rT}$. Therefore, we have that

$$V_0^{(N)} \to V(0) := \mathrm{e}^{-rT}\widetilde{\mathrm{E}}[\Lambda(S(T))] = \mathrm{e}^{-rT}\widetilde{\mathrm{E}}\left[\Lambda(S_0\mathrm{e}^{(r-\sigma^2/2)T+\sigma\sqrt{T}Z})\right],$$

as N goes to ∞. Since for each N the price $V_0^{(N)}$ admits no-arbitrage, we may expect that in the limiting case the derivative price $V(0)$ admits no-arbitrage as well. A rigorous proof of this fact will be provided in Parts II and III. The general formula for the no-arbitrage price of a European-style derivative under the log-normal model becomes

$$V(0, S_0) = \mathrm{e}^{-rT}\widetilde{\mathrm{E}}[\Lambda(S(T))].$$

Let us apply the above formula to find the no-arbitrage price of a European put:

$$\begin{aligned} P^E(0, S_0) &= \widetilde{\mathrm{E}}\left[\mathrm{e}^{-rT}\left(K - S_0\mathrm{e}^{(r-\frac{1}{2}\sigma^2)T+\sigma\sqrt{T}Z}\right)^+\right] \\ &= \widetilde{\mathrm{E}}\left[\left(\mathrm{e}^{-rT}K - S_0\mathrm{e}^{-\frac{1}{2}\sigma^2 T+\sigma\sqrt{T}Z}\right)^+\right]. \end{aligned}$$

Now the trick is to get rid of the function $(x)^+$ inside the expectation in the above formula by using the representation $(x)^+ = x\mathbb{I}_{\{x\geqslant 0\}}$. The condition $\mathrm{e}^{-rT}K - S_0\mathrm{e}^{-\frac{1}{2}\sigma^2T+\sigma\sqrt{T}Z} \geqslant 0$ is simplified as follows:

$$S_0\mathrm{e}^{-\frac{1}{2}\sigma^2T+\sigma\sqrt{T}Z} \leqslant \mathrm{e}^{-rT}K \tag{4.20}$$

$$\iff \quad -\frac{1}{2}\sigma^2T + \sigma\sqrt{T}Z \leqslant -rT + \ln\left(\frac{K}{S_0}\right)$$

$$\iff \quad Z \leqslant -\frac{\ln\left(\frac{S_0}{K}\right) + rT - \frac{1}{2}\sigma^2T}{\sigma\sqrt{T}}. \tag{4.21}$$

Introduce the notation:

$$d_\pm = \frac{\ln\left(\frac{S_0}{K}\right) + (r \pm \frac{1}{2}\sigma^2)T}{\sigma\sqrt{T}}. \tag{4.22}$$

Note that $d_+ - d_- = \sigma\sqrt{T}$ holds. Now, the European put price formula takes the form

$$\begin{aligned} P^E(0,S_0) &= \widetilde{\mathrm{E}}\left[\left(\mathrm{e}^{-rT}K - S_0\mathrm{e}^{-\frac{1}{2}\sigma^2T+\sigma\sqrt{T}Z}\right)\mathbb{I}_{\{Z\leqslant -d_-\}}\right] \\ &= \widetilde{\mathrm{E}}\left[\mathrm{e}^{-rT}K\mathbb{I}_{\{Z\leqslant -d_-\}}\right] - \widetilde{\mathrm{E}}\left[S_0\mathrm{e}^{-\frac{1}{2}\sigma^2T+\sigma\sqrt{T}Z}\mathbb{I}_{\{Z\leqslant -d_-\}}\right] \\ &= \mathrm{e}^{-rT}K\widetilde{\mathrm{E}}\left[\mathbb{I}_{\{Z\leqslant -d_-\}}\right] - S_0\widetilde{\mathrm{E}}\left[\mathrm{e}^{-\frac{1}{2}\sigma^2T+\sigma\sqrt{T}Z}\mathbb{I}_{\{Z\leqslant -d_-\}}\right]. \end{aligned} \tag{4.23}$$

There are two expectations in (4.23) to take care of. The first one is easy to compute:

$$\widetilde{\mathrm{E}}\left[\mathbb{I}_{\{Z\leqslant -d_-\}}\right] = \widetilde{\mathbb{P}}(Z \leqslant -d_-) = \mathcal{N}(-d_-).$$

Recall that $\mathcal{N}(x)$ denotes the cumulative distribution function (CDF) of the standard normal distribution $Norm(0,1)$ given by $\mathcal{N}(z) = \frac{1}{\sqrt{2\pi}}\int_{-\infty}^{z}\mathrm{e}^{-\frac{x^2}{2}}\,\mathrm{d}x$. The other expectation in (4.23) can be calculated by using the brute force of calculus:

$$\widetilde{\mathrm{E}}\left[\mathrm{e}^{-\frac{1}{2}\sigma^2T+\sigma\sqrt{T}Z}\mathbb{I}_{\{Z\leqslant -d_-\}}\right] = \int_{-\infty}^{-d_-}\mathrm{e}^{-\frac{1}{2}\sigma^2T+\sigma\sqrt{T}z}\mathcal{N}'(z)\,\mathrm{d}z$$

$$= \int_{-\infty}^{-d_-}\frac{1}{\sqrt{2\pi}}\mathrm{e}^{-\frac{1}{2}\sigma^2T+\sigma\sqrt{T}z-\frac{z^2}{2}}\,\mathrm{d}z = \int_{-\infty}^{-d_-}\frac{1}{\sqrt{2\pi}}\mathrm{e}^{-\frac{1}{2}(z-\sigma\sqrt{T})^2}\,\mathrm{d}z$$

$$= \int_{-\infty}^{-(d_-+\sigma\sqrt{T})}\frac{1}{\sqrt{2\pi}}\mathrm{e}^{-\frac{z^2}{2}}\,\mathrm{d}z = \mathcal{N}(-(d_-+\sigma\sqrt{T})) = \mathcal{N}(-d_+).$$

The reader will note that this also follows directly by simply using the identity (A.2) in Appendix. As a result, we obtain the Black–Scholes–Merton formula for the no-arbitrage price of a European put option:

$$P^E(0,S_0) = \mathrm{e}^{-rT}K\mathcal{N}(-d_-) - S_0\mathcal{N}(-d_+). \tag{4.24}$$

The formula for a call option price can be obtained by using the put-call parity (4.4) as follows:

$$C^E(0,S_0) = P^E(0,S_0) + S_0 - \mathrm{e}^{-rT}K = S_0(1-\mathcal{N}(-d_+)) - \mathrm{e}^{-rT}K(1-\mathcal{N}(-d_-)).$$

By using the identity $\mathcal{N}(x) = 1 - \mathcal{N}(-x)$, we finally obtain the pricing formula for a European call option:

$$C^E(0,S_0) = S_0\mathcal{N}(d_+) - \mathrm{e}^{-rT}K\mathcal{N}(d_-). \tag{4.25}$$

In the case of standard European call and put options under the binomial tree model, the general option price formula in (4.16) reduces to the CRR option price formula (4.17) or (4.18). Thus, another way to derive the Black–Scholes–Merton formula is to directly proceed to the limiting case in the CRR option price formula as the number of periods N approaches ∞. To illustrate this idea, let us consider the case of the European put option. For the sequence of scaled binomial tree models introduced in Section 2.2.5, the CRR option price formula in (4.18) takes the form

$$P_0^{(N)}(S_0) = \frac{K}{(1+r_N)^N}\mathcal{B}(m_N(K); N, \tilde{p}_N) - S_0\mathcal{B}(m_N(K); N, \hat{p}_N),$$

where

$$\tilde{p}_N = \frac{e^{\delta_N r} - e^{-\sqrt{\delta_N}\sigma}}{e^{\sigma\sqrt{\delta_N}} - e^{-\sigma\sqrt{\delta_N}}}, \quad \hat{p}_N = \frac{e^{-\sqrt{\delta_N}\sigma}}{e^{\delta_N r}}\tilde{p}_N,$$

and

$$m_N(K) = \left\lfloor \frac{\ln(K/S_0) - N\ln d_N}{\ln(u_N/d_N)} \right\rfloor = \left\lfloor \frac{\ln(K/S_0) + N\sigma\sqrt{\delta_N}}{2\sigma\sqrt{\delta_N}} \right\rfloor.$$

As $N \to \infty$, the probabilities converge to one half: $\tilde{p}_N \to \frac{1}{2}$ and $\hat{p}_N \to \frac{1}{2}$. According to the de Moivre–Laplace theorem 2.1, the sequence of standardized binomial random variables $Y_n^* := \frac{Y_n - \mathrm{E}[Y_n]}{\sqrt{\mathrm{Var}(Y_n)}}$ (where expectations are calculated under either $\widetilde{\mathbb{P}}$ or $\widehat{\mathbb{P}}$) converges in distribution to a standard normal random variable. That is, for each $x \in \mathbb{R}$, $F_{Y_n^*}(x) \to \mathcal{N}(x)$. When the limiting distribution is the normal one, we can use the following property given without a proof.

Theorem 4.11. *Suppose $X_n \xrightarrow{d} Z$, as $n \to \infty$, with $Z \sim \mathrm{Norm}(0,1)$. Let a real sequence $\{a_n\}_{n\geqslant 1}$ converge to $a \in \mathbb{R}$. Then $F_{X_n}(a_n) \to \mathcal{N}(a)$, as $n \to \infty$.*

To apply the above theorem, we need to find $\lim_{N\to\infty} \frac{m_N(K) - \mathrm{E}[Y_N]}{\sqrt{\mathrm{Var}(Y_N)}}$ under $\widetilde{\mathbb{P}}$ and $\widehat{\mathbb{P}}$.

Proposition 4.12.

$$\frac{m_N(K) - N\tilde{p}_N}{\sqrt{N\tilde{p}_N(1-\tilde{p}_N)}} \to -d_- \quad \textit{and} \quad \frac{m_N(K) - N\hat{p}_N}{\sqrt{N\hat{p}_N(1-\hat{p}_N)}} \to -d_+, \quad \textit{as } N \to \infty.$$

Proof. The proof is left as an exercise for the reader. □

By combining all the above results, we obtain:

$$\mathcal{B}(m_N(K); N, \tilde{p}_N) = \widetilde{\mathbb{P}}(Y_N \leqslant m_N(K)) = \widetilde{\mathbb{P}}\left(Y_N^* \leqslant \frac{m_N(K) - N\tilde{p}_N}{\sqrt{N\tilde{p}_N(1-\tilde{p}_N)}}\right) \to \mathcal{N}(-d_-),$$

$$\mathcal{B}(m_N(K); N, \hat{p}_N) = \widehat{\mathbb{P}}(Y_N \leqslant m_N(K)) = \widehat{\mathbb{P}}\left(Y_N^* \leqslant \frac{m_N(K) - N\hat{p}_N}{\sqrt{N\hat{p}_N(1-\hat{p}_N)}}\right) \to \mathcal{N}(-d_+).$$

This proves that the CRR put option prices converge to the Black–Scholes price of the put option:

$$P_0^{(N)} \to Ke^{-rT}\mathcal{N}(-d_-) - S_0\mathcal{N}(-d_+), \text{ as } N \to \infty.$$

Formulae for the time-t value of the European call and put options can be obtained from (4.24) and (4.25) by making changes: $T \to T - t$ and $S_0 \to S_t$. That is, the maturity time

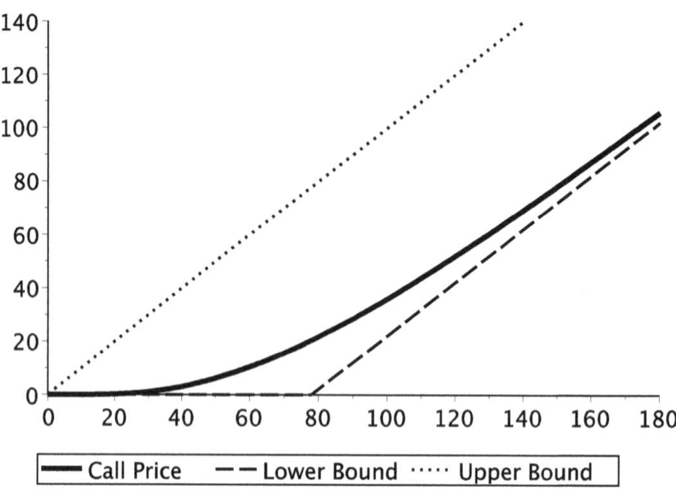

FIGURE 4.16: The Black–Scholes price of the European call option as a function of the initial stock price S. The parameters used are: $K = \$100$, $T = 5$, $\sigma = 30\%$, $r = 5\%$. The upper bound is S; the lower bound is $(S - K\mathrm{e}^{-rT})^+$.

is replaced by the time to maturity, and the initial stock value is replaced by the time-t stock price denoted S_t. The resulting formulae are

$$C^E(t, S_t) = S_t \mathcal{N}(d_+) - \mathrm{e}^{-r(T-t)} K \mathcal{N}(d_-), \tag{4.26}$$

$$P^E(t, S_t) = \mathrm{e}^{-r(T-t)} K \mathcal{N}(-d_-) - S_t \mathcal{N}(-d_+), \tag{4.27}$$

where

$$d_\pm = \frac{\ln(S_t/K) + (r \pm \sigma^2/2)(T-t)}{\sigma\sqrt{T-t}}. \tag{4.28}$$

Typical plots of Black–Scholes prices of European call and put options along with upper and lower bounds are provided in Figures 4.16 and 4.17.

4.3.4 Greeks and Hedging of Options

The writer of a European option receives a cash premium upfront but has potential liabilities later on the option exercise date. The writer's profit or loss is the reverse of that for the purchaser of the option and is given by $C^E \mathrm{e}^{rT} - (S(T) - K)^+$, where C^E is the premium received from the purchaser of the option. The writer can invest the premium in risk-free bonds. If the option will end up deep in the money when $S(T) > K$, the writer of the option will be exposed to the risk of a large loss. Theoretically, the loss of the writer may be unlimited. For a European put option, the writer's loss is limited: $P^E \mathrm{e}^{rT} - (K - S(T))^+ \geqslant P^E \mathrm{e}^{rT} - K$, but it still may be very large compared to the premium P^E. The writer of an option may reduce this risk over a small time interval by forming a suitable portfolio in the underlying security called a hedge or a hedging portfolio. For a binomial model, such a portfolio is constructed by replicating the option payoff. In reality, it is impossible to hedge in a perfect way by designing a single (static) portfolio to be held for the whole period. The hedge has to be rebalanced dynamically to adapt it to changes

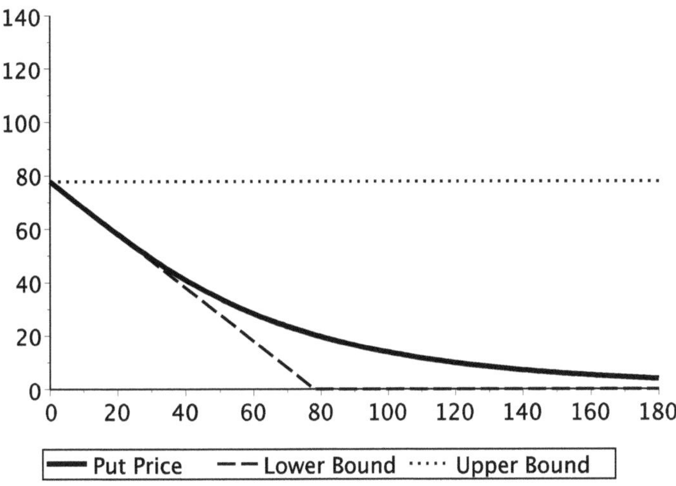

FIGURE 4.17: The Black–Scholes price of the European put option as a function of the initial stock price S. The parameters used are $K = \$100$, $T = 5$, $\sigma = 30\%$, $r = 5\%$. The upper bound is $K\mathrm{e}^{-rT}$; the lower bound is $(K\mathrm{e}^{-rT} - S)^+$.

in risk factors that affect the option value. This leads to a hedging strategy that will be studied in more detail in Parts II and III. In this section we discuss static hedging over a single short time interval.

4.3.4.1 Delta of a Derivative in the Binomial Model

Consider a binomial tree model with a risk-free bank account and a risky stock. The current price of the stock is denoted by S. The no-arbitrage time-0 price, $V_0(S)$, of a derivative contract is given by the following risk-neutral one-step pricing formula:

$$V_0(S) = \frac{1}{1+r}\widetilde{\mathrm{E}}[V_1(S_1)] = \frac{1}{1+r}\left(\tilde{p}V_1(Su) + (1-\tilde{p})V_1(Sd)\right),$$

where $\tilde{p} := \frac{r+1-d}{u-d}$ is the risk-neutral probability, u and d are binomial price factors, and $V_1(S_1)$ is the value of the derivative at the end of the first period. Alternatively, we can construct a replicating portfolio with Δ_0 shares of stock and β_0 dollars in a risk-free money account:

$$\begin{cases} Su\Delta_0 + (1+r)\beta_0 = V_1(Su) \\ Sd\,\Delta_0 + (1+r)\,\beta_0 = V_1(Sd) \end{cases} \implies \begin{cases} \Delta_0 = \dfrac{V_1(Su) - V_1(Sd)}{Su - Sd} \\ \beta_0 = \dfrac{V_1(Sd)u - V_1(Su)d}{(1+r)(u-d)} \end{cases}$$

The no-arbitrage price of the derivative is $V_0(S) = \Delta_0 S + \beta_0$.

A writer sells the derivative contract. To hedge her short position, she buys Δ_0 shares of stock and invests the remainder without risk. The initial wealth of the portfolio with Δ_0 shares of stock, β_0 dollars in a risk-free money account, and one short derivative is zero. At time 1, the wealth becomes

$$\Pi_1 = \begin{cases} Su\Delta_0 + (1+r)\beta_0 - V_1(Su) = 0 & \text{if } S_1 = Su, \\ Sd\Delta_0 + (1+r)\beta_0 - V_1(Sd) = 0 & \text{if } S_1 = Sd. \end{cases}$$

As is seen, in either case the time-1 value of the portfolio is zero. So the portfolio hedges all future risks over the first period. At the end of the first period, the portfolio has to be re-balanced to hedge over the period that follows and so on. As a result, the writer constructs a discrete-time delta hedging strategy, $\{\Delta_n\}_{n\geqslant 0}$. This topic will be address in more detail in later chapters.

4.3.4.2 Delta of a Derivative in the Log-Normal Model

Recall that the log-normal model can be obtained as a limiting case of the binomial tree model when the number of periods N goes to infinity. Consider the parametrization of a binomial tree model from Section 2.2.5:

$$\begin{aligned} u &= \mathrm{e}^{\sigma\sqrt{\delta}}, \\ d &= \mathrm{e}^{-\sigma\sqrt{\delta}}, \\ \tilde{p} &= \frac{\mathrm{e}^{\delta r} - \mathrm{e}^{-\sqrt{\delta}\sigma}}{\mathrm{e}^{\sigma\sqrt{\delta}} - \mathrm{e}^{-\sigma\sqrt{\delta}}}, \end{aligned}$$

where $\delta > 0$ is the length of one period of time. Suppose that the binomial model is applied on the single period $[t, t+\delta]$ with $t \geqslant 0$. The number of stock shares in the hedging portfolio at time t is

$$\Delta_t \approx \frac{V(t+\delta, S\mathrm{e}^{\sigma\sqrt{\delta}}) - V(t+\delta, S\mathrm{e}^{-\sigma\sqrt{\delta}})}{S\mathrm{e}^{\sigma\sqrt{\delta}} - S\mathrm{e}^{-\sigma\sqrt{\delta}}},$$

where $V(t,s)$ is the value of the derivative at time t when the time-t price of the stock is s. To determine, under the log-normal price model, the number of stock shares in the hedging strategy, we need to let δ go to zero. L'Hôpital's rule and the chain rule for differentiating a two-variable function give

$$\begin{aligned} &\lim_{\delta\to 0} \frac{V(t+\delta, S\mathrm{e}^{\sigma\sqrt{\delta}}) - V(t+\delta, S\mathrm{e}^{-\sigma\sqrt{\delta}})}{S\mathrm{e}^{\sigma\sqrt{\delta}} - S\mathrm{e}^{-\sigma\sqrt{\delta}}} \\ &= \lim_{h\to 0} \frac{V(t+h^2, S\mathrm{e}^{\sigma h}) - V(t+h^2, S\mathrm{e}^{-\sigma h})}{S\mathrm{e}^{\sigma h} - S\mathrm{e}^{-\sigma h}} \\ &= \lim_{h\to 0} \frac{S\sigma\mathrm{e}^{\sigma h}\frac{\partial}{\partial s}V(t+h^2, s)|_{s=S\mathrm{e}^{\sigma h}} + S\sigma\mathrm{e}^{-\sigma h}\frac{\partial}{\partial s}V(t+h^2, s)|_{s=S\mathrm{e}^{-\sigma h}}}{S\sigma\mathrm{e}^{\sigma h} + S\sigma\mathrm{e}^{-\sigma h}} \\ &= \frac{\lim_{h\to 0}\mathrm{e}^{\sigma h}\frac{\partial}{\partial s}V(t+h^2, S\mathrm{e}^{\sigma h}) + \lim_{h\to 0}\mathrm{e}^{-\sigma h}\frac{\partial}{\partial s}V(t+h^2, S\mathrm{e}^{\sigma h})}{\lim_{h\to 0}\mathrm{e}^{\sigma h} + \lim_{h\to 0}\mathrm{e}^{-\sigma h}} \\ &= \frac{\partial}{\partial s}V(t,s)|_{s=S} = \frac{\partial}{\partial S}V(t,S). \end{aligned}$$

Therefore, a portfolio with $\Delta_t = \frac{\partial}{\partial S}V(t,S)$ stock shares allows the writer to hedge the risk associated with a short position in the derivative. Such a hedge works only over a short time interval while the stock price does not deviate too much from its initial value S. The partial derivative of the price $V(t,S)$ w.r.t. the current stock price S is called the *delta* of the derivative security.

Consider a European call option in the log-normal model. The no-arbitrage price of a European call at time 0 is given by

$$C^E(S) \equiv C^E(0,S) = S\mathcal{N}(d_+) - K\mathrm{e}^{-rT}\mathcal{N}(d_-),$$

where $d_\pm = \dfrac{\ln(S/K) + (r \pm \sigma^2/2)T}{\sigma\sqrt{T}}$ and S is the current stock price. The delta of the call option is given by the derivative of the price $C^E(S)$ w.r.t. S:

$$\frac{\partial}{\partial S} C^E(S) = \mathcal{N}(d_+) + S\frac{\partial}{\partial S}\mathcal{N}(d_+) - K\mathrm{e}^{-rT}\frac{\partial}{\partial S}\mathcal{N}(d_-).$$

The partial derivatives of $\mathcal{N}(d_\pm)$ w.r.t. S are given by

$$\frac{\partial}{\partial S}\mathcal{N}(d_+) = \mathcal{N}'(d_+)\frac{\partial}{\partial S}d_+(S) = \mathcal{N}'(d_+)\frac{1}{S\sigma\sqrt{T}},$$
$$\frac{\partial}{\partial S}\mathcal{N}(d_-) = \mathcal{N}'(d_-)\frac{\partial}{\partial S}d_-(S) = \mathcal{N}'(d_-)\frac{1}{S\sigma\sqrt{T}},$$

where we use the identity $\frac{\partial d_\pm}{\partial S} = \frac{\partial}{\partial S}\left(\frac{\ln(S)}{\sigma\sqrt{T}}\right) = \frac{1}{S\sigma\sqrt{T}}$. As a result, we have

$$\frac{\partial}{\partial S} C^E(S) = \mathcal{N}(d_+) + \frac{1}{\sigma\sqrt{T}}\left(\mathcal{N}'(d_+) - \frac{K}{S}\mathrm{e}^{-rT}\mathcal{N}'(d_-)\right). \tag{4.29}$$

The derivative of the normal CDF gives the normal PDF: $\mathcal{N}'(x) = \frac{1}{\sqrt{2\pi}}\mathrm{e}^{-x^2/2}$. Applying this formula in (4.29) gives

$$\frac{K}{S}\mathrm{e}^{-rT}\mathcal{N}'(d_-) = \frac{1}{\sqrt{2\pi}}\mathrm{e}^{-\ln(S/K)-rT-d_-^2/2} = \frac{1}{\sqrt{2\pi}}\mathrm{e}^{-d_+^2/2} = \mathcal{N}'(d_+).$$

This simplifies the expression in (4.29) to finally give us a formula for the delta of a European call option in the log-normal model as:

$$\mathrm{delta}_{C^E} = \frac{\partial C^E}{\partial S} = \mathcal{N}(d_+).$$

4.3.4.3 Delta Hedging

Consider a portfolio whose value $\Pi = \Pi(S)$ is a function of the current stock price S. Suppose that the price changes from S to $S + \delta S$ over a short time interval. According to Taylor's Theorem, the value of the portfolio is approximately

$$\Pi(S + \delta S) \cong \Pi(S) + \frac{\mathrm{d}\Pi(S)}{\mathrm{d}S}\,\delta S.$$

Hence, the change in value of the portfolio is

$$\delta\Pi(S) := \Pi(S + \delta S) - \Pi(S) \cong \frac{\mathrm{d}\Pi(S)}{\mathrm{d}S}\,\delta S.$$

The risk associated with this change in value is

$$\sqrt{\mathrm{Var}(\delta\Pi(S))} \cong \left|\frac{\mathrm{d}\Pi(S)}{\mathrm{d}S}\right|\sqrt{\mathrm{Var}(\delta S)}.$$

The risk will be eliminated if $\frac{\mathrm{d}}{\mathrm{d}S}\Pi(S) = 0$. Note that the derivative $\frac{\mathrm{d}}{\mathrm{d}S}\Pi(S)$ is called the *delta of the portfolio.*

Delta hedging is the process of reducing the risk associated with price changes in the underlying security by keeping the delta of the portfolio as close to zero as possible. This can be achieved by offsetting long and short positions. For example, a short call position may be delta-hedged by a long position in the underlying security. Such a portfolio is called *delta-neutral.*

Consider a portfolio (x, y, z) composed of x shares of stock with current price S, a cash amount y dollars invested without risk, and a short derivative security (i.e., $z = -1$) to be hedged with the initial wealth

$$\Pi(S) \equiv \Pi_0^{(x,y,z)} = xS + y - V(S),$$

where $V(S)$ is the current price of the derivative. Then, the delta of the portfolio is $\frac{\mathrm{d}}{\mathrm{d}S}\Pi(S) = x - \frac{\mathrm{d}}{\mathrm{d}S}V(S)$. If $\frac{\mathrm{d}}{\mathrm{d}S}\Pi(S) = 0$, then $x = \frac{\mathrm{d}}{\mathrm{d}S}V(S)$, where $\frac{\mathrm{d}}{\mathrm{d}S}V(S)$ is the delta of the derivative security. For example, construct a portfolio that hedges the short position in a European call option:

$$(x, y, z) = (\mathcal{N}(d_+), y, -1), \text{ where } d_+ = d_+(S).$$

For any cash amount y, the delta of this portfolio is zero. Consequently, its value does not vary much under small changes about the initial value S of the stock share. It is convenient to choose y so that the initial wealth is zero: $S\mathcal{N}(d_+) + y - C^E(S) = 0$. Application of the Black–Scholes formula for $C^E(S)$ gives

$$y = C^E(S) - \mathcal{N}(d_+)S = -K\mathrm{e}^{-rT}\mathcal{N}(d_-).$$

Example 4.9 (Delta-neutral portfolio). An investor writes and sells 1000 one-month call options with strike price $K = \$50$ on stock with an initial price $S_0 = \$50$. Assuming a log-normal price model with $r = 5\%$ and $\sigma = 20\%$, construct a delta-neutral portfolio with initial wealth zero.

Solution. Application of (4.25) with $S_0 = \$50$, $K = \$50$, $T = \frac{1}{12}$, $r = 0.05$, and $\sigma = 0.2$ gives the price of one call option, C^E:

$$\begin{aligned}
d_+ &= \frac{\ln\left(\frac{50}{50}\right) + (0.05 + 0.2^2/2) \cdot \frac{1}{12}}{0.2\sqrt{1/12}} \cong 0.1010363, \\
d_- &= d_+ - 0.2\sqrt{1/12} \cong 0.0433013, \\
\mathcal{N}(d_+) &\cong 0.54024, \\
\mathcal{N}(d_-) &\cong 0.51727, \\
C^E &= 50\mathcal{N}(d_-) - \mathrm{e}^{-0.05/12} 50\mathcal{N}(d_+) \cong \$1.25603.
\end{aligned}$$

The price of 1000 call options is $1000\,C^E \cong 1000 \cdot 1.25603 = \$1{,}256.03$. The delta of one European call is $\Delta = \mathcal{N}(d_+) \cong 0.54024$. Therefore, to construct the hedge, the investor buys $1000\,\Delta = 540.24$ stock shares for $50 \cdot 540.24 = \$27{,}012$ dollars. To cover expenses, the investor borrows $\$27{,}012 - \$1{,}256.03 = \$25{,}755.97$. So the hedging portfolio consists of $x = 540.24$ stock shares, $y = -25{,}755.97$ dollars in the form of a risk-free loan, and $z = -1000$ short call options. The time-0 value of the portfolio is zero. □

4.3.4.4 Greeks

As was demonstrated in the previous section, the delta of an option is given by the derivative of the option price w.r.t. the current price S of the underlying. The delta of an option is the sensitivity of the option price to a change in the price of the underlying security. Similarly, we can introduce other so-called *Greek parameters* that measure the sensitivity of a derivative such as an option (or a portfolio of securities) w.r.t. a small change in a given underlying parameter. In the log-normal model, European put and call options are characterized by four parameters: the current underlying price S, current time t, interest rate r, and volatility σ (note that the strike price K and exercise time T are fixed once the option is written). Consider a portfolio consisting of the underlying security and some European-type derivatives based on this security. The most commonly used Greeks for such a portfolio with the value function $V = V(t, S; \sigma, r)$ are described just below.

delta$_V$ $= \frac{\partial V}{\partial S}$ is the sensitivity of the portfolio value to a change in the price of the underlying security.

gamma$_V$ $= \frac{\partial^2 V}{\partial S^2}$ is the sensitivity of the portfolio's delta to a change in the price of the underlying security.

theta$_V$ $= \frac{\partial V}{\partial t}$ is the sensitivity of the portfolio value to a *negative* change in time to maturity, $T - t$.

vega$_V$ $= \frac{\partial V}{\partial \sigma}$ is the sensitivity of the portfolio value to a change in volatility.

rho$_V$ $= \frac{\partial V}{\partial r}$ is the sensitivity of the portfolio value to the interest rate.

For small changes δS, δt, $\delta\sigma$, δr of the respective variables, the change in value of the portfolio value is approximated by using Taylor's Theorem:

$$\begin{aligned}
\delta V &= V(t + \delta t, S + \delta S; \sigma + \delta\sigma, r + \delta r) - V(t, S; \sigma, r) \\
&\cong \frac{\partial V}{\partial S}\delta S + \frac{1}{2}\frac{\partial^2 V}{\partial S^2}(\delta S)^2 + \frac{\partial V}{\partial t}\delta t + \frac{\partial V}{\partial \sigma}\delta\sigma + \frac{\partial V}{\partial r}\delta r \\
&= \text{delta}_V\,\delta S + \frac{1}{2}\,\text{gamma}_V\,(\delta S)^2 + \text{theta}_V\,\delta t + \text{vega}_V\,\delta\sigma + \text{rho}_V\,\delta r.
\end{aligned}$$

To immunize the portfolio against small changes of a selected variable, we let the corresponding Greek be equal to zero. The resulting portfolio is said to be *neutral* relative to the Greek selected. In particular, a *delta-neutral* portfolio is protected against small changes of the underlying security price; a *vega-neutral* portfolio is not sensitive to volatility movements.

The delta-gamma approximation

$$\delta V \cong \text{delta}_V\,\delta S + \frac{1}{2}\,\text{gamma}_V\,(\delta S)^2$$

is often used in historical Value-at-Risk (VaR) calculations for portfolios that include options.

The Black–Scholes formula (4.26) allows us to evaluate the Greeks explicitly for a single European call option in the same manner as it was done for the delta:

$$\begin{aligned}
\text{delta}_{CE} &= \mathcal{N}(d_+) \\
\text{gamma}_{CE} &= \frac{1}{S\sigma\sqrt{T-t}}\mathcal{N}'(d_+) \\
\text{theta}_{CE} &= -\frac{S\sigma}{2\sqrt{T-t}}\mathcal{N}'(d_+) - rK\mathrm{e}^{-r(T-t)}\mathcal{N}(d_-) \\
\text{vega}_{CE} &= S\sqrt{T-t}\mathcal{N}'(d_+) \\
\text{rho}_{CE} &= (T-t)K\mathrm{e}^{-r(T-t)}\mathcal{N}(d_-)
\end{aligned}$$

where $d_\pm$ is given in (4.28). From the above formulae, one can obtain the following relation:

$$\text{theta}_{C^E} + r\,S\,\text{delta}_{C^E} + \frac{\sigma^2 S^2}{2}\,\text{gamma}_{C^E} - r\,C^E = 0.$$

Therefore, the price $C \equiv C^E$ of a European call (as well as the price of any European-style derivative security) satisfies the following partial differential equation (PDE) known as the *Black–Scholes PDE*:

$$\frac{\partial C}{\partial t} + \frac{1}{2}\sigma^2 S^2 \frac{\partial^2 C}{\partial S^2} + rS\frac{\partial C}{\partial S} - rC = 0. \tag{4.30}$$

In Chapter 12, we will see that Equation (4.30) is derived for a continuous-time asset price model by using a self-financing portfolio replication strategy and a no-arbitrage argument.

The Greeks for put options can be calculated in the same manner by differentiating the expression in (4.27) or via put-call parity. Given a European call and put option for the same underlying, strike price, and time to maturity, and with no dividend yield, the sum of the absolute values of the delta of each option will be 1. This is due to the put-call parity: a long call plus a short put (a call minus a put) replicates a forward whose delta is equal to 1. Indeed, let us differentiate the put-call parity (4.4) w.r.t. S:

$$C^E(S) - P^E(S) = S - Ke^{-rT},$$
$$\frac{\partial}{\partial S}C^E(S) - \frac{\partial}{\partial S}P^E(S) = \frac{\partial}{\partial S}S - \frac{\partial}{\partial S}Ke^{-rT},$$
$$\text{delta}_{C^E} - \text{delta}_{P^E} = 1.$$

Therefore, the delta of a European put is

$$\text{delta}_{P^E} = \text{delta}_{C^E} - 1 = -(1 - \mathcal{N}(d_+)) = -\mathcal{N}(-d_+).$$

4.3.5 Black–Scholes Equation

The log-normal pricing model is obtained as a limiting case of binomial tree models when the number of periods approaches infinity. In previous sections, we derived the Black–Scholes pricing formulae for European call and put options as the limit of the respective binomial pricing formulae. Let us now demonstrate how the Black–Scholes PDE (4.30) can be deduced from a single period recurrence relationship for the option value when the length of a period converges to zero. Note that the Black–Scholes equation will also be derived based on the replication argument in Chapter 12. All required tools from stochastic calculus will be introduced in Chapter 11.

Consider a European option with expiration time T and payoff function $\Lambda(S)$. Let $V(t,S)$ denote the option value at time $t \in (0,T]$ for the current stock price S. Assume that the stock price process follows the log-normal model. Let σ be the annual volatility of the underlying asset, and r be the continuously compounded risk-free rate of interest. On a small time interval $[t-\delta, t]$ with $\delta = \frac{1}{n}$, where n is the number of periods per year, the continuous-time model can be approximated by a single-period binomial model with factors $u_n = e^{\sigma\sqrt{\delta}}$ and $d_n = e^{-\sigma\sqrt{\delta}}$, and rate $r_n = e^{r\delta} - 1$. Additionally, suppose the stock pays dividends continuously at a rate (yield) of q. The dividend payment accumulated during the time period of length δ is approximately equal to $100(1-e^{-q\delta})\%$ of the current stock price. Let S be the stock price just before the dividend payment. Then, according to the no-arbitrage principle, the asset price just after the dividend payment is equal to $Se^{-q\delta}$. The final risk-neutral dynamics of the stock price in a single-period binomial period is illustrated in Figure 4.18

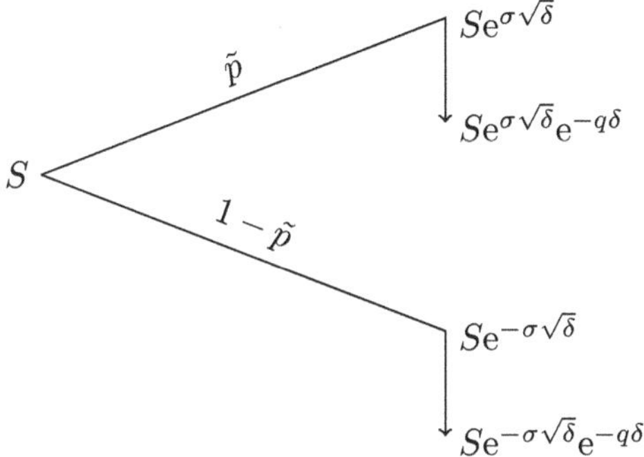

FIGURE 4.18: A single-period binomial approximation for a stock with continuous dividends.

The single-period binomial approximation of the option price (on the interval $[t-\delta, t]$) can be written as follows:

$$e^{-r\delta}V(t-\delta, S) = \tilde{p}_n\, V\big(t, Se^{\sigma\sqrt{\delta}-q\delta}\big) + (1-\tilde{p}_n)\, V\big(t, Se^{\sigma\sqrt{\delta}-q\delta}\big)\,, \tag{4.31}$$

where $\tilde{p}_n = \frac{1+r_n-d_n}{u_n-d_n}$. Let us find expansions of the risk-neutral probabilities $\tilde{p}_n$ and $1-\tilde{p}_n$ for small δ. Using the truncated Maclaurin expansion of the exponential function,

$$e^x = 1 + x + \frac{x^2}{2} + \frac{x^3}{6} + \frac{x^4}{24} + \mathcal{O}(x^5)\,,$$

we obtain

$$\begin{aligned} u_n &= e^{\sigma\sqrt{\delta}} = 1 + \sigma\delta^{1/2} + \frac{\sigma^2}{2}\delta + \frac{\sigma^3}{6}\delta^{3/2} + \frac{\sigma^4}{24}\delta^2 + \mathcal{O}(\delta^{5/2})\,, \\ d_n &= e^{-\sigma\sqrt{\delta}} = 1 - \sigma\delta^{1/2} + \frac{\sigma^2}{2}\delta - \frac{\sigma^3}{6}\delta^{3/2} + \frac{\sigma^4}{24}\delta^2 + \mathcal{O}(\delta^{5/2})\,, \end{aligned}$$

and hence

$$u_n - d_n = 2\sigma\delta^{1/2}\left(1 + \frac{\sigma^2}{6}\delta + \mathcal{O}(\delta^2)\right) \implies \frac{1}{u_n - d_n} = \frac{\delta^{-1/2}}{2\sigma}\left(1 - \frac{\sigma^2}{6}\delta + \mathcal{O}(\delta^2)\right),$$

where we also use the geometric series $\frac{1}{1+a\delta+\mathcal{O}(\delta^2)} = 1 - a\delta + \mathcal{O}(\delta^2)$ with $0 < a\delta < 1$. Therefore, the risk-neutral probabilities are given by

$$\begin{aligned} \tilde{p}_n &= \frac{\delta^{-1/2}}{2\sigma}\left((1+r\delta) - (1 - \sigma\delta^{1/2} + \frac{\sigma^2}{2}\delta + \mathcal{O}(\delta^{3/2}))\right)\left(1 - \frac{\sigma^2}{6}\delta + \mathcal{O}(\delta^2)\right) \\ &= \frac{1}{2\sigma}\left(\sigma + (r - \sigma^2/2)\delta^{1/2} + \frac{\sigma^3}{6}\delta + \mathcal{O}(\delta^{3/2})\right)\left(1 - \frac{\sigma^2}{6}\delta + \mathcal{O}(\delta^2)\right) \\ &= \frac{1}{2} + \left(r - \frac{\sigma^2}{2}\right)\frac{\delta^{1/2}}{2\sigma} + \mathcal{O}(\delta^{3/2})\,, \qquad (4.32) \\ 1 - \tilde{p}_n &= \frac{1}{2} - \left(r - \frac{\sigma^2}{2}\right)\frac{\delta^{1/2}}{2\sigma} + \mathcal{O}(\delta^{3/2})\,. \qquad (4.33) \end{aligned}$$

Now, apply Taylor's formula to the option value function to obtain

$$V(t-\delta,S) = V - V_t\delta + \mathcal{O}(\delta^2)\,, \tag{4.34}$$

$$\begin{aligned} V(t, Se^{\sigma\sqrt{\delta}-q\delta}) &= V\Big(t, S + \big(\sigma\sqrt{\delta} - q\delta + \sigma^2\delta/2 + \mathcal{O}(\delta^{3/2})\big)S\Big) \\ &= V + \Big(\sigma\sqrt{\delta} - q\delta + \sigma^2\delta/2\Big) SV_S + (\sigma^2\delta/2)S^2V_{SS} + \mathcal{O}(\delta^{3/2})\,, \end{aligned} \tag{4.35}$$

$$\begin{aligned} V(t, Se^{-\sigma\sqrt{\delta}-q\delta}) &= V\Big(t, S + \big(-\sigma\sqrt{\delta} - q\delta + \sigma^2\delta/2 + \mathcal{O}(\delta^{3/2})\big)S\Big) \\ &= V + \Big(-\sigma\sqrt{\delta} - q\delta + \sigma^2\delta/2\Big) SV_S + (\sigma^2\delta/2)S^2V_{SS} + \mathcal{O}(\delta^{3/2})\,, \end{aligned} \tag{4.36}$$

where $V \equiv V(t,S)$, $V_t \equiv \frac{\partial V(t,S)}{\partial t}$, $V_S \equiv \frac{\partial V(t,S)}{\partial S}$, and $V_{SS} \equiv \frac{\partial^2 V(t,S)}{\partial S^2}$. Substituting (4.32)–(4.33) and (4.34)–(4.36) in (4.31) gives

$$\begin{aligned} &\big(1 + r\delta + \mathcal{O}(\delta^2)\big)\ \big(V - \delta V_t + \mathcal{O}(\delta^2)\big) \\ &= \Bigg[\ \left(\frac{1}{2} + \left(r - \frac{\sigma^2}{2}\right)\frac{\delta^{1/2}}{2\sigma} + \mathcal{O}(\delta^{3/2})\right) \\ &\qquad \times \left(V + \left(\sigma\sqrt{\delta} - q\delta + \frac{\sigma^2\delta}{2}\right) SV_S + \frac{\sigma^2\delta}{2}S^2V_{SS} + \mathcal{O}(\delta^{3/2})\right) \\ &\quad + \left(\frac{1}{2} - \left(r - \frac{\sigma^2}{2}\right)\frac{\delta^{1/2}}{2\sigma} + \mathcal{O}(\delta^{3/2})\right) \\ &\qquad \times \left(V + \left(-\sigma\sqrt{\delta} - q\delta + \frac{\sigma^2\delta}{2}\right) SV_S + \frac{\sigma^2\delta}{2}S^2V_{SS} + \mathcal{O}(\delta^{3/2})\right)\ \Bigg]. \end{aligned}$$

Simplify the above expression and collect terms of the same magnitude (w.r.t. δ) to obtain

$$\begin{aligned} &V - V_t\delta + rV\delta + \mathcal{O}(\delta^2) \\ &= V + \underbrace{\left(\frac{\sigma}{2}SV_S - \frac{\sigma}{2}SV_S + \left(r - \frac{\sigma^2}{2}\right)\frac{V}{2\sigma} - \left(r - \frac{\sigma^2}{2}\right)\frac{V}{2\sigma}\right)}_{=0}\sqrt{\delta} \\ &\quad + \underbrace{\left(\left(r - \frac{\sigma^2}{2}\right) SV_S + \left(\frac{\sigma^2}{2} - q\right) SV_S + \frac{\sigma^2}{2}\delta S^2V_{SS}\right)}_{=(r-q)SV_S + \sigma^2S^2/2V_{SS}}\delta + \mathcal{O}(\delta^{3/2})\,. \end{aligned}$$

Cancel V on both sides and divide both sides by δ to obtain

$$V_t + \frac{\sigma^2}{2}S^2V_{SS} + (r-q)SV_S - rV + \mathcal{O}(\sqrt{\delta}) = 0\,.$$

Now, taking a limit as $\delta \to 0$ gives the Black–Scholes equation

$$\frac{\partial V}{\partial t} + \frac{\sigma^2}{2}S^2\frac{\partial^2 V}{\partial S^2} + (r-q)S\frac{\partial V}{\partial S} - rV = 0 \tag{4.37}$$

where $V = V(t,S)$ for $0 \leqslant t \leqslant T$ and $S > 0$. The equation is accompanied by the terminal condition at maturity:

$$V(T,S) = \Lambda(S) \text{ for } S > 0\,.$$

4.4 Exercises

Exercise 4.1. Consider a 12-month long forward contract on a stock currently priced at \$90. The risk-free interest rate is 7% per annum, compounded continuously. Suppose that dividend payments of \$8 per share are expected after 4 months and \$7 after 8 months.

(a) Determine the forward price at time 0.

(b) What is the value of this forward contract 6 months from now if the stock price at that time is \$95?

(c) Suppose that the value of this forward contract 6 months from now is \$5. Determine whether there is an arbitrage opportunity. If one exists in this situation, construct an arbitrage portfolio and find the profit realized.

Exercise 4.2. Suppose that the stock pays dividends continuously at a rate of $q > 0$ and the dividends are reinvested in the stock. An investment in one share held at time 0 will increase to become $\mathrm{e}^{q\,T}$ shares at time $T > 0$. Therefore, we need to start with $\mathrm{e}^{-q\,T}$ shares at time 0 to obtain 1 share at time T. Assuming that interest is compounded continuously at rate r, show that the no-arbitrage forward price at time 0 is

$$F(0,T) = S(0)\mathrm{e}^{(r-q)T}. \tag{4.38}$$

Exercise 4.3. Prove the formula from Theorem 4.4 for the no-arbitrage value of a long forward contract initiated at time $t = 0$ and having delivery time T and delivery price K:

$$V_K(t) = (F(t,T) - K)\,D(T-t).$$

Exercise 4.4. Suppose that the lending rate r_ℓ and the borrowing rate r_b are different for investors so that $r_\ell < r_b$. Show that the no-arbitrage forward price $F(0,T)$ satisfies

$$S(0)\mathrm{e}^{r_\ell T} \leqslant F(0,T) \leqslant S(0)\mathrm{e}^{r_b T}.$$

Exercise 4.5. Prove by an arbitrage argument that $P^E(K)$ is a nondecreasing function of strike K. That is, if $K' < K''$, then $P^E(K') \leqslant P^E(K'')$ holds.

Exercise 4.6. Prove by an arbitrage argument that if the stock pays dividends whose discounted value is $\mathrm{div}(0,T)$, then

$$\max\{S(0) - K\mathrm{e}^{-rT} - \mathrm{div}(0,T), S(0) - K\} \leqslant C^A,$$
$$\max\{K\mathrm{e}^{-rT} + \mathrm{div}(0,T) - S(0), K - S(0)\} \leqslant P^A.$$

Exercise 4.7. Suppose that the stock pays dividends between time 0 and the expiry time T, whose discounted value (at time 0) is $\mathrm{div}(0,T)$. Show that

$$S(0) - \mathrm{div}(0,T) - K \leqslant C^A - P^A \leqslant S(0) - K\mathrm{e}^{-rT}.$$

Exercise 4.8. Show that if the stock pays dividends continuously at a rate q, then

$$S(0)\mathrm{e}^{-qT} - K \leqslant C^A - P^A \leqslant S(0) - K\mathrm{e}^{-rT}.$$

Exercise 4.9. Prove by an arbitrage argument that $P^A(S)$ is a nonincreasing function of the current price $S = S(0)$. That is, if $S' < S''$, then $P^A(S') \geqslant P^A(S'')$ holds.

Exercise 4.10. Obtain payoff functions for a straddle, strip, strap, and strangle. Assume a long position is taken.

Exercise 4.11. Consider a single-period binomial model from Example 4.5. Replicate the put option with strike price $K = \$95$ and expiry time $T = 1$. Find the option price.

Exercise 4.12. Let the asset price process $\{S_t\}_{t=0,1,\ldots,T}$ follow the binomial tree model with the parameters $S_0 = \$50$, $u = 1.2$, $d = 0.9$. Suppose that the risk-free interest rate is $r = 0.1$.

(a) Find the values of the replicating portfolios for a put and a call with strike price $K = \$55$ and exercise time $T = 1$.

(b) Compute the values of the put and call options from (a) using with the replicating portfolios constructed in (a).

(c) Compute the value of a European derivative security with the payoff function

$$\Lambda(S) = \begin{cases} 0 & \text{if } S \leqslant 40 \text{ or } S \geqslant 60, \\ S - 40 & \text{if } 40 < S < 50, \\ 60 - S & \text{if } 50 \leqslant S < 60. \end{cases}$$

and exercise time $T = 2$ using the risk-neutral pricing formula.

Exercise 4.13. Compute the price of an American put with strike price $K = \$13$ expiring at time $T = 2$ on a stock with $S(0) = \$12$ in a binomial tree model with $u = 1.1$, $d = 0.95$, and $r = 0.03$.

Exercise 4.14. Derive the Black–Scholes formula (4.25) for the price of a European call option by calculating the risk-neutral expectation:

$$\mathrm{e}^{-rT}\widetilde{\mathrm{E}}[(S(T) - K)^+] = \mathrm{e}^{-rT}\widetilde{\mathrm{E}}\left[\left(S(0)\mathrm{e}^{(r-\frac{1}{2}\sigma^2)T+\sigma\sqrt{T}Z} - K\right)^+\right].$$

Exercise 4.15. Prove the formulae for the Greeks of a European call option.

Exercise 4.16. Derive the formulae for the Greeks of a European put option.

Exercise 4.17. The share-or-nothing call and put options with strike K at time T have the respective payoff functions $f(S) = S\,\mathbb{I}_{\{S>K\}}$ and $f(S) = S\,\mathbb{I}_{\{S<K\}}$.

(a) Assuming the Black–Scholes model, find the non-arbitrage prices of the share-or-nothing put and call options.

(b) Find the put-call parity for the share-or-nothing options.

Exercise 4.18. Consider the Black–Scholes price formula for the price $C^E = C^E(t, S; \sigma)$ of a vanilla European call on a stock paying dividends. Find the following limits:

$$\lim_{S\to 0+} C^E, \quad \lim_{S\to\infty} C^E, \quad \lim_{t\to T-} C^E, \quad \lim_{\sigma\to 0+} C^E, \quad \lim_{\sigma\to\infty} C^E.$$

Exercise 4.19. Consider the log-normal asset price model with $S_0 = \$80$, $\sigma = 30\%$, and $r = 6\%$. Construct (a) a delta-gamma neutral portfolio and (b) a delta-rho neutral portfolio to hedge a short position on 500 calls expiring after 90 days with strike price \$80, taking as an additional component a call option expiring after 120 days with strike price \$85. Find the changes in value of these portfolios after 5 days if one of the following three scenarios will happen: (i) the stock drops to \$75; (ii) the interest rate jumps to 9%; (iii) the stock drops to \$75 and the interest rate jumps to 9%.

Exercise 4.20. In the framework of the log-normal asset price model consider a strangle option with the payoff function

$$\Lambda(S) = \begin{cases} K_1 - S & \text{if } S \leqslant K_1, \\ 0 & \text{if } K_1 < S < K_2, \\ S - K_2 & \text{if } S \geqslant K_2. \end{cases}$$

(a) Plot the payoff function.

(b) Find a portfolio consisting of standard European calls and puts that replicates the payoff.

(c) Find the arbitrage-free price by using the Black–Scholes price formulae for vanilla options.

(d) Find the delta.

Exercise 4.21. If the interest rate r increases, then how will the present value of a forward contract $V_0(S_0, K)$ on an underlying stock with spot S_0 and strike K change? Show explicit details. Assume simple interest.

Exercise 4.22. Assume the interest rate is zero and also assume that both call and put options with respective values $C_t(S_t, K)$ and $P_t(S_t, K)$ are differentiable functions of the strike K.

(a) Derive a relation between $\partial C_t(S_t, K)/\partial K$ and $\partial P_t(S_t, K)/\partial K$ for $t = 0$ and $t = T$.

(b) Find the values of these partial derivatives at maturity $t = T$.

[Hint: Use the put-call parity.]

Exercise 4.23. Assume the spot is S_0 and the simple interest rate is r. Show that the call and put options both struck at $(1 + rT)S_0$ and both expiring at time $t = T$ are of equal value at present time $t = 0$.

Exercise 4.24. Assume a single-period economy with two states. The price of the forward contract struck at K can be shown to be given by $V_t = V_t(S_t, K) = S_t - (1 + r(T-t))^{-1}K$, where S_t is the stock price at time $t \in \{0, T\}$. Find a replicating portfolio in the stock and the zero-coupon bond, with nominal value assumed as $B_T = 1$ for the bond. Verify that the portfolio is indeed a hedge (or replicating) portfolio.

Exercise 4.25. A stock is worth \$200 today and will be worth either \$190 or \$220 tomorrow with equal probability. Assuming a zero interest rate, obtain the prices of call options struck at \$190, \$200, and \$220.

Exercise 4.26. Using a similar integration procedure as was shown for the call option price, provide a complete derivation of the Black–Scholes pricing formula of a European put under the log-normal model. Assume strike K, spot S, time to maturity T, interest rate r, and volatility σ.

Exercise 4.27. A European *binary* option is a so-called "all-or-nothing" claim on an underlying asset. For example, one share of a *cash-or-nothing binary call* has a payoff of exactly one dollar if the asset price ends up above the strike and zero otherwise, i.e., the payoff function is

$$\Lambda(S) = \mathbb{I}_{\{S \geqslant K\}} \equiv \begin{cases} 1 & \text{if } S \geqslant K, \\ 0 & \text{if } S < K. \end{cases}$$

Similarly, a *cash-or-nothing binary put* has payoff $\mathbb{I}_{\{S<K\}}$. Assume the asset price process $\{S(t)\}_{t \geqslant 0}$ is geometric Brownian motion.

(a) Derive the (Black–Scholes) exact pricing formulas for both the binary call $C(S,T)$ and the put $P(S,T)$ in terms of the spot S and time to maturity T.

(b) Give the relationship between the binary put and call price where both options have the same strike K and maturity T.

(c) Derive the exact formula for the Greek delta of the binary put and call: $\Delta_c = \partial C/\partial S$ and $\Delta_p = \partial P/\partial S$.

Exercise 4.28. Consider the European-style option with payoff

$$\Lambda(S) = \mathbb{I}_{\{K_1 \leqslant S \leqslant K_2\}} \equiv \begin{cases} 1 & \text{if } K_1 \leqslant S \leqslant K_2, \\ 0 & \text{otherwise,} \end{cases}$$

where $0 < K_1 < K_2$. Assume the geometric Brownian motion (log-normal) model for the stock price process $\{S(t)\}_{t \geqslant 0}$.

(a) Find the option value as a function of spot S, strikes K_1 and K_2, expiration time T, rate r and volatility σ. [Hint: You may decompose the payoff in terms of binary options with appropriate indicator functions.]

(b) Derive formulas for the following sensitivities of the option value in (a): $\Delta = \partial V/\partial S$, $\Gamma = \partial^2 V/\partial S^2$, and $\Theta = \partial V/\partial T$.

Part II

Discrete-Time Modelling

Chapter 5

Single-Period Arrow–Debreu Models

5.1 Specification of the Model

5.1.1 Finite-State Economy. Vector Space of Payoffs. Securities

We now turn to a multinomial generalization of the one-period binomial market model. Although single-period models cannot give a realistic representation of a complex, dynamically changing stock market, we use such models to illustrate many important economic principles. Let us begin with the main assumptions.

- Any economic activity such as trading and consumption takes place only at two times: the initial time $t = 0$ and the terminal time $t = T$.
- The economic environment at time $t = 0$ is completely known.
- At time $t = T$, the economy can be in one of M different states of the world, denoted by ω^j, $j = 1, 2 \ldots, M$, which constitute the state space

  $$\Omega = \{\omega^1, \omega^2, \ldots, \omega^M\}.$$

- For each state, or outcome, $\omega^j \in \Omega$, there exists an occurrence probability $p_j > 0$ (called a state probability) such that

  $$p_1 + p_2 + \cdots + p_M = 1$$

  holds. These probabilities constitute the *real-world or physical probability measure* $\mathbb{P}\colon 2^\Omega \to [0, 1]$, where $\mathbb{P}(\omega^j) \equiv \mathbb{P}(\{\omega^j\}) = p_j$, $j = 1, 2, \ldots, M$, and

  $$\mathbb{P}(E) = \sum_{\omega \in E} \mathbb{P}(\omega) \text{ for any event } E \subset \Omega.$$

  The probability function is known in advance, but the future state of the world is unknown at time $t = 0$ and is only revealed at time $t = T$.

The above model, specified by the set of market scenarios Ω and the probability distribution function $\mathbb{P}$, is called a *multinomial* one-period model. In fact, the pair $(\Omega, \mathbb{P})$ defines a *finite probability space*. At first glance, a one-period model seems to be unrealistic. However, one-period models are useful for modelling the case with an investor pursuing a *buy-and-hold* strategy. The investor sets up an investment portfolio at time $t = 0$ and holds it to liquidate at time $t = T$. More importantly for this text, such models offer an ideal setting for introducing some of the important concepts underlying asset pricing theory in mathematical finance.

Any financial contract in a single-period model is defined by its initial price and the payoff function at terminal time T. Let us begin with the definition of a payoff.

Definition 5.1. The *payoff function* X of a financial contract is a function $X : \Omega \to \mathbb{R}$. The value $X(\omega)$ is the payment due at time T when the state of the world $\omega \in \Omega$ is reached. Since the state ω is uncertain, X is a random variable defined on the finite probability space $(\Omega, \mathbb{P})$. The M-dimensional vector

$$\mathbf{X} \equiv X(\Omega) := \left[X(\omega^1), \quad X(\omega^2), \quad \ldots \quad , X(\omega^M)\right]$$

is called a *payoff vector* or a *cash-flow vector.*

The simplest example of financial contracts is Arrow–Debreu securities (named after Kenneth Arrow and Gérard Debreu). There are M such securities, each corresponding to a different market scenario. The Arrow–Debreu (AD for short) payoff $\mathcal{E}^j$, $j = 1, 2, \ldots, M$, pays one unit of currency (or one unit of another *numéraire*[1]) if the state of the world ω^j is reached and zero otherwise. As a random variable, the jth AD security corresponds to the indicator random variable $\mathcal{E}^j = \mathbb{I}_{\{\omega^j\}}$, i.e.,

$$\mathcal{E}^j(\omega) := \mathbb{I}_{\{\omega^j\}}(\omega) = \begin{cases} 0, & \text{if } \omega \neq \omega^j, \\ 1, & \text{if } \omega = \omega^j. \end{cases}$$

Recall that $\mathbb{I}_A$ is the indicator random variable for event A; that is, $\mathbb{I}_A(\omega) = 1$ if $\omega \in A$ and $\mathbb{I}_A(\omega) = 0$ if $\omega \notin A$.

Proposition 5.1. *The set of payoffs denoted by $\mathcal{L}(\Omega)$ is a vector space. The AD securities $\{\mathcal{E}^1, \mathcal{E}^2, \ldots, \mathcal{E}^M\}$ form a vector-space basis of $\mathcal{L}(\Omega)$.*

Proof. Clearly, any linear combination of two payoffs is again a payoff. Indeed, for any $a, b \in \mathbb{R}$ and $X, Y \in \mathcal{L}(\Omega)$, the function $aX + bY$ defined by $(aX + bY)(\omega) := aX(\omega) + bY(\omega)$ is a payoff from $\mathcal{L}(\Omega)$. The set $\mathcal{L}(\Omega)$ contains a zero payoff $X \equiv 0$. Therefore, $\mathcal{L}(\Omega)$ satisfies the definition of a vector space. Now, take any $X \in \mathcal{L}(\Omega)$. We have the following representation:

$$X(\omega) = \sum_{j=1}^{M} X(\omega^j)\mathbb{I}_{\{\omega^j\}}(\omega) \text{ for } \omega \in \Omega,$$

where the sum on the right-hand side may only contain one nonzero term. Therefore, the payoff X can be expressed as a linear combination of AD securities:

$$X = x_1\mathcal{E}^1 + x_2\mathcal{E}^2 + \cdots + x_M\mathcal{E}^M, \text{ where } x_j := X(\omega^j).$$

Finally, let us prove that the functions $\mathcal{E}^1, \mathcal{E}^2, \ldots, \mathcal{E}^M$ are linearly independent. Suppose that $a_1\mathcal{E}^1 + a_2\mathcal{E}^2 + \cdots + a_M\mathcal{E}^M \equiv 0$ holds for some real numbers $a_1, a_2, \ldots, a_M$. In this case, for every $j = 1, 2, \ldots, M$, we have $a_1\mathcal{E}^1(\omega^j) + \cdots + a_M\mathcal{E}^M(\omega^j) = a_j = 0$, since $\mathcal{E}^i(\omega^j) = 1$ iff $i = j$. Therefore, the linear combination $a_1\mathcal{E}^1 + a_2\mathcal{E}^2 + \cdots + a_M\mathcal{E}^M$ is zero iff all a_j are zero. □

In what follows, we shall use the following terminology. A vector $\mathbf{v} \in \mathbb{R}^n$ is said to be:

strictly positive (denoted $\mathbf{v} \gg 0$) if all its entries are positive ($v_i > 0$ for all $i = 1, 2, \ldots, n$);

positive (denoted $\mathbf{v} > 0$) if all its entries are nonnegative and at least one entry is strictly positive ($v_i \geqslant 0$ for all $i = 1, 2, \ldots, n$ and $v_j > 0$ for at least one j);

[1] A *numéraire* is a basic standard by which values are measured. Normally, we use dollars or another currency to compare values of various objects. However, other numéraires can be used, such as a bushel of grain, an ounce of gold, a barrel of oil, or a cowrie shell.

nonnegative (denoted $\mathbf{v} \geqslant 0$) if all its entries are nonnegative ($v_i \geqslant 0$ for all $i = 1, 2, \ldots, n$).

So, we have the following inclusions:

$$\{\text{strictly positive vectors}\} \subseteq \{\text{positive vectors}\} \subseteq \{\text{nonnegative vectors}\}.$$

The same terminology will be used for discrete random variables defined on a finite probability space. For example, a random variable $X \colon \Omega \to \mathbb{R}$ is said to be positive if the vector of all its values $X(\omega)$, $\omega \in \Omega$ is positive. That is, $X(\omega) \geqslant 0$ for all $\omega \in \Omega$ and $X(\omega^*) > 0$ for at least one ω^*.

Definition 5.2. A *contingent claim*, or a claim for short, is any nonnegative payoff X, i.e., $X(\omega) \geqslant 0$ holds for all $\omega \in \Omega$. We will denote the set of all claims by $\mathcal{L}^+(\Omega)$.

For example, in the binomial model with $M = 2$ states, every payoff X is represented by a vector $[x_1, x_2] \in \mathbb{R}^2$, where $x_1 = X(\omega^1)$ and $x_2 = X(\omega^2)$. Every claim is given by a positive vector in $\mathbb{R}^2$. That is, $\mathcal{L}(\Omega) = \mathbb{R}^2$ and $\mathcal{L}^+(\Omega) = \mathbb{R}^2_+$, where $\mathbb{R}_+ \equiv [0, \infty)$. Note that the set of claims $\mathcal{L}^+(\Omega)$ is convex since a linear combination of claims with nonnegative weights a, b is also a claim:

$$\forall a, b \in [0, \infty) \;\; X, Y \in \mathcal{L}^+(\Omega) \implies aX + bY \in \mathcal{L}^+(\Omega).$$

Definition 5.3. An *asset* or a (financial) security, denoted by S, is described by

- its initial price $S_0 > 0$ at time $t = 0$, which is constant for all $\omega \in \Omega$,
- its payoff $S_T \in \mathcal{L}^+(\Omega)$ at time $t = T$ (that is, the security pays $S_T(\omega)$ at time T if the market is in a given state ω).

We also say that asset S has the *price process* $\{S_t\}_{t \in \{0,T\}}$, which is simply a function

$$S \colon \{0, T\} \times \Omega \to [0, \infty).$$

An asset is said to be *risky* if there exists at least two states of the world, say ω^j and ω^k, such that $S_T(\omega^j) \neq S_T(\omega^k)$. So, the payoff of a risky asset, S_T, is a nonconstant random variable and hence its value $S_T(\omega)$ depends on the outcome ω realized at time T, i.e., its value is uncertain at time 0. Otherwise, if the asset has a given value $S_T(\omega)$ that is the same for all $\omega \in \Omega$, it is called a *risk-free* asset, i.e., it is constant on all outcomes and hence has a certain time-T value. Such an asset represents a bank account, a money market account, or a risk-free zero-coupon bond. We will denote a risk-free asset by B. Since the payoff of a risk-free asset is the same for all scenarios, its return is constant. Let $r := \frac{B_T - B_0}{B_0} = \frac{B_T}{B_0} - 1 > -1$ denote the one-period risk-free return. Then the time-T value of the risk-free asset is $B_T = B_0(1 + r)$, with $B_0 > 0$ being the initial (time-0) value of the asset.

The immediate result of Proposition 5.1 is that the payoff function of security S can be *replicated* by Arrow–Debreu payoffs. That is, for every financial contract there exists a portfolio of AD securities such that the payoff function of the portfolio is the same as that of the contract:

$$S_T(\omega) = s_1\mathcal{E}^1(\omega) + s_2\mathcal{E}^2(\omega) + \cdots + s_M\mathcal{E}^M(\omega) \text{ for all } \omega \in \Omega,$$

where the real numbers $s_1, s_2, \ldots, s_M$ are the portfolio positions in the respective AD securities.

5.1.2 Initial Price Vector and Payoff Matrix

Let there be N base assets $S^1, S^2, \ldots, S^N$ traded on a market. For many applications risky assets are stocks and risk-free ones are bonds or bank accounts. The initial prices $S_0^i > 0$, $i = 1, 2, \ldots, N$, are known; the terminal values of the respective N assets, $S_T^1, \ldots, S_T^N$, are random variables whose values are generally uncertain at time 0. Typically, one of these N assets can be risk-free, i.e., its terminal payoff is a constant random variable with future value at time T being independent of $\omega \in \Omega$. A one-period N-by-M model with N base assets and M states of the world is described by:

- the *(initial) N-by-1 price vector* $\mathbf{S}_0 = \left[S_0^1, S_0^2, \cdots, S_0^N\right]^\top$;
- the N-by-M *payoff matrix* $\mathbf{S}_T(\Omega)$ (also called the *cash-flow matrix*, or the *dividend matrix*) given by

$$\left[\mathbf{S}_T(\omega^1) \,|\, \mathbf{S}_T(\omega^2) \,|\, \cdots \,|\, \mathbf{S}_T(\omega^M)\right] = \begin{bmatrix} S_T^1(\omega^1) & S_T^1(\omega^2) & \cdots & S_T^1(\omega^M) \\ S_T^2(\omega^1) & S_T^2(\omega^2) & \cdots & S_T^2(\omega^M) \\ \vdots & \vdots & \ddots & \vdots \\ S_T^N(\omega^1) & S_T^N(\omega^2) & \cdots & S_T^N(\omega^M) \end{bmatrix}.$$

  That is, the (i,j)-entry of $\mathbf{S}_T(\Omega)$ is $S_T^i(\omega^j)$—the amount paid by the ith asset in state w^j at time T; the jth column is denoted by $\mathbf{S}_T(\omega^j)$, i.e., the vector whose components correspond to the time-T value of the assets in the given state ω^j. The ith row of the payoff matrix is the *payoff vector* $S^i(\Omega)$ of the ith asset. In what follows, we also denote the payoff matrix by $\mathbf{D}$, where $D_{i,j} = S_T^i(\omega^j)$.

Both $\mathbf{S}_0$ and $\mathbf{S}_T(\Omega)$ are known to all investors. However, the state of the world, ω, is not known in advance at time $t = 0$. Thus, $\mathbf{S}_T = [S_T^1, S_T^2, \cdots, S_T^N]^\top$ is a random N-by-1 vector defined on Ω. The model is depicted in Figure 5.1.

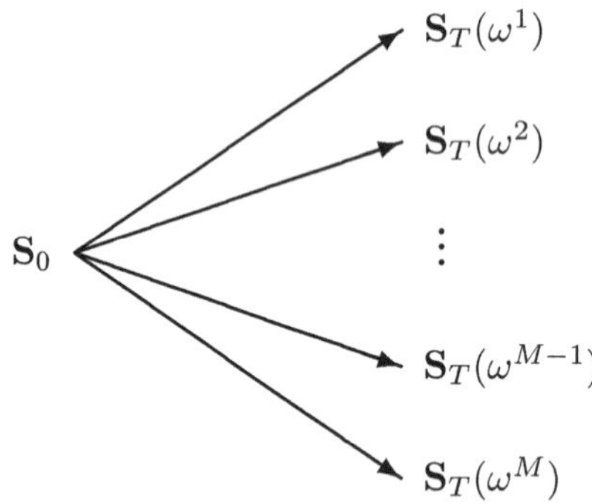

FIGURE 5.1: A scenario tree for the Arrow–Debreu model with M states of the world.

Example 5.1 (Single-period binomial model)**.** Consider a 2-by-2 economy with two states (one up and one down state), $\Omega = \{\omega^1, \omega^2\} \equiv \{\omega^+, \omega^-\}$, and two base assets. The first asset $S^1 \equiv B$ is a risk-free zero-coupon bond paying 1 unit of currency at time T; the second asset $S^2 \equiv S$ is a risky stock. Hence, the price vector $\mathbf{S}_T = [B_T, S_T]^\top$ has first component $S_T^1(\omega^+) = S_T^1(\omega^-) \equiv B_T = 1$. The second component is the time-T stock price, $S_T^2 \equiv S_T$, which is assumed to be given by $S_T(\omega^+) = S_0 u$ and $S_T(\omega^-) = S_0 d$ in the respective states, where $0 < d < u$ are, respectively, down- and up-movement stock price factors and $r > 0$ is the risk-free interest rate. The price dynamics of the two assets is depicted by the following diagram:

$$\mathbf{S}_0 = \begin{bmatrix} (1+r)^{-1} \\ S_0 \end{bmatrix} \quad \nearrow \mathbf{S}_T(\omega^+) = \begin{bmatrix} 1 \\ S_0 u \end{bmatrix}, \quad \searrow \mathbf{S}_T(\omega^-) = \begin{bmatrix} 1 \\ S_0 d \end{bmatrix}$$

Thus, the initial price vector and the payoff matrix are, respectively,

$$\mathbf{S}_0 = \begin{bmatrix} B_0 \\ S_0 \end{bmatrix} = \begin{bmatrix} (1+r)^{-1} \\ S_0 \end{bmatrix},$$

$$\mathbf{D} = \left[\ \mathbf{S}_T(\omega^+) \mid \mathbf{S}_T(\omega^-)\ \right] = \begin{bmatrix} B_T(\omega^+) & B_T(\omega^-) \\ S_T(\omega^+) & S_T(\omega^-) \end{bmatrix} = \begin{bmatrix} 1 & 1 \\ S_0 u & S_0 d \end{bmatrix}.$$

5.1.3 Portfolios of Base Securities

A market portfolio is simply a collection of assets traded on the market. To describe a portfolio in our model, the number of units held in the portfolio needs to be calculated for every base asset. Since there are N base assets, any portfolio can be represented by a vector with N entries.

Definition 5.4. A *portfolio* of N base assets is a 1-by-N vector $\boldsymbol{\varphi} = [\varphi_1, \varphi_2, \cdots, \varphi_N] \in \mathbb{R}^N$, where φ_i, called the position in the ith asset, is the number of units of the ith base asset held in the portfolio. If $\varphi_i > 0$, then the position is said to be *long*; if $\varphi_i < 0$, then the position is said to be *short*. The *portfolio value* denoted Π_t^{φ} or $\Pi_t[\boldsymbol{\varphi}]$ gives the total value of the portfolio $\boldsymbol{\varphi}$ at time $t \in \{0, T\}$. The initial (time-0) value is just the inner product[2] of the vectors $\boldsymbol{\varphi}$ and $\mathbf{S}_0$:

$$\Pi_0^{\varphi} = \boldsymbol{\varphi}\,\mathbf{S}_0 = \varphi_1 S_0^1 + \varphi_2 S_0^2 + \cdots + \varphi_N S_0^N . \tag{5.1}$$

The terminal (time-T) value is a function of $\omega \in \Omega$ given by

$$\Pi_T^{\varphi}(\omega) = \boldsymbol{\varphi}\,\mathbf{S}_T(\omega) = \varphi_1 S_T^1(\omega) + \varphi_2 S_T^2(\omega) + \cdots + \varphi_N S_T^N(\omega) . \tag{5.2}$$

The initial and terminal portfolio values constitute a portfolio value process $\{\Pi_t^{\varphi}\}_{t\in\{0,T\}}$. The terminal value Π_T^{φ} of a portfolio is a random variable defined on Ω. In the state ω^j, the terminal portfolio value is given by the inner product of the portfolio vector $\boldsymbol{\varphi}$ and the jth column of the cash-flow matrix: $\Pi_T(\omega^j) = \boldsymbol{\varphi}\,\mathbf{S}_T(\omega^j)$. Thus, the 1-by-$M$ payoff vector of portfolio terminal values is given by a product of the 1-by-N portfolio position vector and the N-by-M payoff matrix:

$$\Pi_T^{\varphi}(\Omega) := \left[\Pi_T^{\varphi}(\omega^1), \Pi_T^{\varphi}(\omega^2), \cdots, \Pi_T^{\varphi}(\omega^M)\right] = \boldsymbol{\varphi}\,\mathbf{D}.$$

As noted above, any real random variable (or payoff) $X : \Omega \to \mathbb{R}$ has the representation $X = \sum_{j=1}^{M} X(\omega^j)\mathcal{E}^j$, i.e., $X(\omega) = \sum_{j=1}^{M} X(\omega^j)\mathcal{E}^j(\omega)$, for any $\omega \in \Omega$. Hence, the random variable corresponding to the terminal portfolio value is also expressible in terms of Arrow–Debreu securities as follows:

$$\Pi_T^{\varphi}(\omega) = \sum_{j=1}^{M} \Pi_T^{\varphi}(\omega^j)\,\mathcal{E}^j(\omega) = \sum_{j=1}^{M} \left(\boldsymbol{\varphi}\,\mathbf{S}_T(\omega^j)\right)\mathcal{E}^j(\omega), \quad \omega \in \Omega.$$

[2]Note that the matrix product $\boldsymbol{a}\,\boldsymbol{b}$ is a scalar and is equivalent to the usual dot (inner) product of two vectors where $\boldsymbol{a}$ is 1-by-N and $\boldsymbol{b}$ is N-by-1.

Since the right-hand sides of (5.1) and (5.2) are linear functions of the portfolio vector, Π_t^{φ} is a linear function of φ, as is stated in the following proposition.

Proposition 5.2. *The portfolio value Π_t^{φ} with $t \in \{0, T\}$ is a linear function of $\varphi \in \mathbb{R}^N$.*

Proof. Consider two portfolios $\varphi, \psi \in \mathbb{R}^N$. Then, for any $a, b, \in \mathbb{R}$, the value of the combined portfolio $a\varphi + b\psi$ has time-T value

$$\begin{aligned}
\Pi_T^{a\varphi+b\psi}(\omega) &= \sum_{j=1}^{M} \left((a\varphi + b\psi)\mathbf{S}_T(\omega^j)\right) \mathcal{E}^j(\omega) \\
&= \sum_{j=1}^{M} \left(a\varphi\mathbf{S}_T(\omega^j) + b\psi\mathbf{S}_T(\omega^j)\right) \mathcal{E}^j(\omega) \\
&= a\sum_{j=1}^{M} \left(\varphi\mathbf{S}_T(\omega^j)\right) \mathcal{E}^j(\omega) + b\sum_{j=1}^{M} \left(\psi\mathbf{S}_T(\omega^j)\right) \mathcal{E}^j(\omega) \\
&= a\Pi_T^{\varphi}(\omega) + b\Pi_T^{\psi}(\omega),
\end{aligned}$$

for all $\omega \in \Omega$. The initial value of the combined portfolio is

$$\begin{aligned}
\Pi_0^{a\varphi+b\psi} &= (a\varphi + b\psi)\mathbf{S}_0 \\
&= a\varphi\mathbf{S}_0 + b\psi\mathbf{S}_0 = a\Pi_0^{\varphi} + b\Pi_0^{\psi}.
\end{aligned}$$

That is, $\Pi_t^{a\varphi+b\psi} = a\Pi_t^{\varphi} + b\Pi_t^{\psi}$, for each $t \in \{0, T\}$. □

5.2 Analysis of the Arrow–Debreu Model

5.2.1 Redundant Assets and Attainable Securities

Suppose that there exists a nonzero portfolio φ such that $\Pi_T^{\varphi}(\omega) = 0$ for all $\omega \in \Omega$. Let $\varphi_j \neq 0$ for some j. Then, the claim of the jth asset, S_T^j, can be expressed as a linear combination of the other asset claims:

$$\sum_{i=1}^{N} \varphi_i S_T^i = 0 \implies \varphi_j S_T^j = -\sum_{i \neq j} \varphi_i S_T^i \implies S_T^j = \sum_{i \neq j} \left(-\frac{\varphi_i}{\varphi_j}\right) S_T^i.$$

If this is the case, the asset S^j is said to be *redundant.* There exists redundancy iff the payoff vector of one base asset is a linear combination of payoffs of other base assets. That is, there exists a redundant asset iff one row of the dividend matrix $\mathbf{D}$ is a linear combination of other rows of $\mathbf{D}$. Therefore, there exists a redundant base asset iff rank$(\mathbf{D}) < N$. A redundant asset can be found by finding a nontrivial solution φ to the system of M linear equations

$$\varphi\mathbf{D} = \mathbf{0} \iff \begin{cases}
\varphi_1 S_T^1(\omega_1) + \varphi_2 S_T^2(\omega_1) + \cdots + \varphi_N S_T^N(\omega_1) = 0, \\
\varphi_1 S_T^1(\omega_2) + \varphi_2 S_T^2(\omega_2) + \cdots + \varphi_N S_T^N(\omega_2) = 0, \\
\vdots \\
\varphi_1 S_T^1(\omega_M) + \varphi_2 S_T^2(\omega_M) + \cdots + \varphi_N S_T^N(\omega_M) = 0.
\end{cases}$$

Let us consider the case where $N = M$. The payoff matrix $\mathbf{D}$ is then a square one. Recall that the rank of a square matrix is less than its size iff the matrix is singular. Therefore, it is sufficient to calculate the determinant of the payoff matrix and apply the following criterion:

$$\det(\mathbf{D}) = 0 \iff \text{there is a redundant asset.}$$

Example 5.2. Consider a 3-by-3 economy (with 3 assets and 3 states) with given payoff matrix

$$\mathbf{D} = \begin{bmatrix} 10 & 8 & 6 \\ 13 & 7 & 0 \\ 7 & 9 & 12 \end{bmatrix}. \tag{5.3}$$

Find a redundant base asset, if any.

Solution. Let us calculate the determinant of the payoff matrix in (5.3):

$$\det(\mathbf{D}) = 10 \cdot 7 \cdot 12 + 13 \cdot 6 \cdot 9 + 7 \cdot 8 \cdot 0 - 7 \cdot 7 \cdot 6 - 9 \cdot 10 \cdot 0 - 13 \cdot 8 \cdot 12 = 0.$$

Hence there is redundancy. Indeed, the sum of the second and third rows equals twice the first row. Thus, $S^1_T = \frac{1}{2}S^2_T + \frac{1}{2}S^3_T$, i.e., asset S^1 is redundant (or, equivalently, S^2 is redundant since $S^2_T = 2S^1_T - S^3_T$). □

The terminal value of a portfolio in base assets is a payoff function. In other words, each portfolio $\boldsymbol{\varphi} \in \mathbb{R}^N$ generates a payoff $X = \Pi^{\boldsymbol{\varphi}}_T$ from the vector space $\mathcal{L}(\Omega)$. Let us reverse this situation and find out if it is possible to represent a given payoff $X \in \mathcal{L}(\Omega)$ as the terminal value $\Pi^{\boldsymbol{\varphi}}_T$ of some portfolio $\boldsymbol{\varphi} \in \mathbb{R}^N$.

Definition 5.5. A payoff $X \in \mathcal{L}(\Omega)$ is said to be *attainable* if there exists a portfolio $\boldsymbol{\varphi} \in \mathbb{R}^N$ such that $\Pi^{\boldsymbol{\varphi}}_T(\omega) = X(\omega)$ for all $\omega \in \Omega$. Any such portfolio $\boldsymbol{\varphi}$ is called a *hedge* or a *replicating portfolio* for the payoff X. The set of attainable payoffs in a given N-by-M economy will be denoted by $\mathcal{A}(\Omega)$.

The set of attainable payoffs $\mathcal{A}(\Omega)$ is a subset of $\mathcal{L}(\Omega)$. Moreover, $\mathcal{A}(\Omega)$ is a vector subspace of $\mathcal{L}(\Omega)$. Indeed, thanks to Proposition 5.2, if X and Y are elements of $\mathcal{A}(\Omega)$, then any linear combination of X and Y is an element of $\mathcal{A}(\Omega)$. Indeed, let the portfolio $\boldsymbol{\varphi}_X$ and $\boldsymbol{\varphi}_Y$ replicate the payoffs X and Y, respectively. Then, the portfolio $a\boldsymbol{\varphi}_X + b\boldsymbol{\varphi}_Y$ replicates the payoff $aX + bY$:

$$\Pi_T[a\boldsymbol{\varphi}_X + b\boldsymbol{\varphi}_Y] = a\Pi_T[\boldsymbol{\varphi}_X] + b\Pi_T[\boldsymbol{\varphi}_Y] = aX + bY.$$

Finally, the zero payoff $X \equiv 0$ is attainable since the terminal value of a zero portfolio $\mathbf{0} = [0, 0, \ldots, 0] \in \mathbb{R}^N$ is zero:

$$\Pi^{\mathbf{0}}_T = \sum_{i=1}^{N} 0 \cdot S^i_T \equiv 0.$$

To find a portfolio vector $\boldsymbol{\varphi} = [\varphi_1, \varphi_2, \ldots, \varphi_N] \in \mathbb{R}^N$ that replicates the payoff vector $\mathbf{X} = [x_1, x_2, \ldots, x_M] \in \mathbb{R}^M$, we need to solve the following system of linear equations:

$$\begin{cases} \varphi_1 S^1_T(\omega^1) + \varphi_2 S^2_T(\omega^1) + \cdots + \varphi_N S^N_T(\omega^1) = X(\omega^1) \\ \varphi_1 S^1_T(\omega^2) + \varphi_2 S^2_T(\omega^2) + \cdots + \varphi_N S^N_T(\omega^2) = X(\omega^2) \\ \vdots \\ \varphi_1 S^1_T(\omega^M) + \varphi_2 S^2_T(\omega^M) + \cdots + \varphi_N S^N_T(\omega^M) = X(\omega^M) \end{cases} \tag{5.4}$$

This system can be written in a compact vector-matrix form:

$$\boldsymbol{\varphi}\mathbf{D} = \mathbf{X}.$$

If the solution to the system exists, then the payoff X is attainable; otherwise it is said to be *unattainable.* Note that the system (5.4) may have multiple solutions. If that is the case, there are infinitely many replicating portfolios for the payoff X.

Replication is a key component to pricing financial securities. The initial price of any asset with an attainable payoff $X \in \mathcal{A}(\Omega)$ can be evaluated by replicating X in the base assets as follows:

(a) Find a portfolio φ_X in the base assets that replicates X, i.e., $\Pi_T[\varphi_X] = X$.

(b) Set the initial price $\pi_0(X)$ of the security with payoff X equal to the initial cost $\Pi_0[\varphi_X]$ of setting up the portfolio φ_X:

$$\pi_0(X) = \Pi_0[\varphi_X].$$

To proceed with this approach, we need to answer the following questions.

1. Under what condition does there exist a unique portfolio that replicates a given payoff and hence gives the initial security price uniquely?
2. Can every payoff be replicated and hence every claim's initial price be evaluated based on replication?
3. Suppose that a given payoff is replicated nonuniquely. Under what condition do all portfolios replicating the same payoff have the same initial value?

The following lemma answers the first question. The other questions are to be investigated in the next sections.

Lemma 5.3. *Every attainable payoff $X \in \mathcal{A}(\Omega)$ has a unique replicating portfolio $\varphi \in \mathbb{R}^N$ iff there are no redundant base assets.*

Proof. The base assets are nonredundant iff their payoffs are linearly independent. That is,

$$\Pi_T^{\varphi} = \varphi_1 S_T^1 + \varphi_2 S_T^2 + \cdots + \varphi_N S_T^N \equiv 0 \text{ iff } \boldsymbol{\varphi} = \mathbf{0}.$$

If the only portfolio replicating the zero payoff vector is a zero portfolio, then there is no redundancy. Consider any two portfolios φ and ψ that are assumed to replicate the same payoff X. Then, $\Pi_T^{\varphi} = X = \Pi_T^{\psi}$. Moreover, by linearity, this is the case iff the difference portfolio $\varphi - \psi$ has a zero payoff, i.e.,

$$\Pi_T^{\varphi} = \Pi_T^{\psi} \iff \Pi_T^{\varphi-\psi} \equiv 0.$$

Thus, there are no redundant base assets iff $\varphi - \psi = \mathbf{0} \iff \varphi = \psi$, i.e., any attainable payoff X is replicated by a unique portfolio in the base assets iff there are no redundant base assets. □

Corollary 5.4. *If there exists a redundant base asset, then every attainable payoff has infinitely many replicating portfolios.*

Proof. Suppose that there is a redundant base asset. Then, there exists a nonzero portfolio ψ_0 such that $\Pi_T^{\psi_0} = 0$. Therefore, for any portfolio φ replicating a payoff X, the portfolio $\varphi_\lambda := \varphi + \lambda\psi_0$ (with arbitrary constant $\lambda \in \mathbb{R}$) replicates X as well:

$$\Pi_T^{\varphi_\lambda} = \Pi_T^{\varphi} + \lambda\Pi_T^{\psi_0} = X + \lambda \cdot 0 = X.$$

Hence, we have an infinite family of portfolios (parametrized by $\lambda \in \mathbb{R}$) replicating the same payoff X. □

5.2.2 Completeness of the Model

In this subsection, we investigate if any (arbitrary) payoff can be replicated by a portfolio in base assets. This is in essence the question of whether or not the market model is complete, as defined just below.

Definition 5.6. A market model is said to be *complete* if every payoff is attainable, i.e., if $\mathcal{L}(\Omega) = \mathcal{A}(\Omega)$ holds.

Since it is impossible to verify every payoff for attainability, we require some simple criterion for market completeness, as we now discuss.

Lemma 5.5. *The following statements are equivalent.*

(1) every payoff is attainable: $\mathcal{L}(\Omega) = \mathcal{A}(\Omega)$*;*

(2) every claim is attainable: $\mathcal{L}^+(\Omega) \subset \mathcal{A}(\Omega)$*;*

(3) every Arrow–Debreu security is attainable: $\mathcal{E}^j \in \mathcal{A}(\Omega)$*, for all* $j = 1, 2, \ldots, M$.

Proof. Let us prove that (1) implies (2), (2) implies (3), and (3) implies (1).

(1) $\Longrightarrow$ (2): Obviously, $\mathcal{L}^+(\Omega) \subset \mathcal{L}(\Omega) = \mathcal{A}(\Omega)$.

(2) $\Longrightarrow$ (3): This follows since, for every j, $\mathcal{E}^j \in \mathcal{L}^+(\Omega) \subset \mathcal{A}(\Omega)$.

(3) $\Longrightarrow$ (1): According to Proposition 5.1, every payoff can be represented as a linear combination of AD securities, i.e., $X = \sum_{j=1}^{M} x_j \mathcal{E}^j$. By assumption, there exists a portfolio φ_j that replicates $\mathcal{E}^j$, for each $j = 1, 2, \ldots, M$. Then, the portfolio defined by $\varphi = \sum_{j=1}^{M} x_j \varphi_j$ replicates X, since

$$\Pi_T^{\varphi} = \sum_{j=1}^{M} x_j \Pi_T^{\varphi_j} = \sum_{j=1}^{M} x_j \mathcal{E}^j = X. \qquad \square$$

In conclusion, the market is complete iff the base security payoffs $S_T^1, S_T^2, \ldots, S_T^N$ span the vector space $\mathcal{L}(\Omega)$. That is, the N payoff vectors

$$\begin{aligned} \mathbf{v}_1 &= \left[S_T^1(\omega^1), S_T^1(\omega^2), \ldots, S_T^1(\omega^M)\right] \\ \mathbf{v}_2 &= \left[S_T^2(\omega^1), S_T^2(\omega^2), \ldots, S_T^2(\omega^M)\right] \\ &\vdots \\ \mathbf{v}_N &= \left[S_T^N(\omega^1), S_T^N(\omega^2), \ldots, S_T^N(\omega^M)\right] \end{aligned}$$

span $\mathbb{R}^M$, the space of all 1-by-M real vectors. The dimension of $\mathcal{L}(\Omega)$ is M, hence for completeness to hold there has to be at least M base assets with linearly independent payoffs. Therefore, the market is complete iff $\operatorname{rank}(\mathbf{D}) \geqslant M$. As was proved in Lemma 5.3, every attainable payoff is replicated uniquely iff $\operatorname{rank}(\mathbf{D}) \geqslant N$. Since $\mathbf{D}$ is an N-by-M matrix, its rank does not exceed $\min\{N, M\}$. Therefore, the one-period model is complete and every payoff is replicated uniquely iff $\mathbf{D}$ is a square (i.e., $N = M$), nonsingular matrix. However, in a realistic asset price model, the number of market scenarios, M, is very large. On the other hand, the number of base assets is relatively small since it is otherwise impractical for an investor to operate with a large number of underlying securities. Thus, for a realistic one-period model we have that $M \gg N$, and hence such a model is incomplete.

Example 5.3. Consider the following 2-by-4 model with $N = 2$ base assets and $M = 4$ market scenarios:

$$\mathbf{S}_0 = \begin{bmatrix} 5 \\ 10 \end{bmatrix}, \quad \mathbf{D} = \begin{bmatrix} 6 & 6 & 4 & 2 \\ 12 & 8 & 8 & 12 \end{bmatrix}.$$

(a) Check if the model is complete and/or nonredundant.

(b) Is the payoff $\mathbf{X} = \begin{bmatrix} 0, 4, 0, -8 \end{bmatrix}$ attainable?

Solution. There are four states and only two base assets, $M = 4 > N = 2$, and hence the market is incomplete. Calculate the rank of the cash-flow matrix $\mathbf{D}$. Since $\det \begin{bmatrix} 6 & 6 \\ 12 & 8 \end{bmatrix} \neq 0$, we have that $\operatorname{rank}(\mathbf{D}) = 2$, i.e., the payoff vectors (obtained from the two rows of $\mathbf{D}$), $\mathbf{v}_1 = [6, 6, 4, 2]$ and $\mathbf{v}_2 = [12, 8, 8, 12]$, are linearly independent. Therefore, there are no redundant base assets. To find a portfolio $\varphi = \begin{bmatrix} \varphi_1, \varphi_2 \end{bmatrix}$ replicating X, we solve the following system of linear equations:

$$\varphi\mathbf{D} = \mathbf{X} \iff \begin{cases} 6\varphi_1 + 12\varphi_2 = 0 \\ 6\varphi_1 + 8\varphi_2 = 4 \\ 4\varphi_1 + 8\varphi_2 = 0 \\ 2\varphi_1 + 12\varphi_2 = -8 \end{cases} \iff \begin{cases} \varphi_1 = 2 \\ \varphi_2 = -1 \end{cases}$$

Thus, X is attainable and is replicated by the unique portfolio $\varphi = \begin{bmatrix} 2, -1 \end{bmatrix}$. The portfolio value process is $\Pi_t = 2S_t^1 - S_t^2$, $t \in \{0, T\}$. Let us verify that $\Pi_T(\omega) = X(\omega)$ for all $\omega \in \Omega$:

$$\begin{aligned} \Pi_T(\omega_1) &= 2 \cdot 6 - 12 = 0 = X(\omega_1) && \checkmark \\ \Pi_T(\omega_2) &= 2 \cdot 6 - 8 = 4 = X(\omega_2) && \checkmark \\ \Pi_T(\omega_3) &= 2 \cdot 4 - 8 = 0 = X(\omega_3) && \checkmark \\ \Pi_T(\omega_4) &= 2 \cdot 2 - 12 = -8 = X(\omega_4) && \checkmark \end{aligned}$$

□

Let us summarize all criteria for investigating if a single-period market model is complete and/or has redundant base assets.

- If $\operatorname{rank}(\mathbf{D}) < M$, then the market model is incomplete. Hence, not every payoff can be replicated by a portfolio in base assets.
- If $\operatorname{rank}(\mathbf{D}) < N$, then there are redundant base assets. Hence, a zero claim can be replicated by a nonzero portfolio. Every attainable payoff has infinitely many replicating portfolios.
- If $\operatorname{rank}(\mathbf{D}) = N = M$, then the market is complete and free of redundant base assets. Every payoff is attainable, and the portfolio replicating a payoff is unique. Note that a square payoff matrix $\mathbf{D}$ has a full rank iff $\det(\mathbf{D}) \neq 0$.

5.3 No-Arbitrage Asset Pricing

5.3.1 The Law of One Price

Let us come back to the problem of calculating the initial price of a given security. A security with an attainable payoff can be priced by constructing a portfolio in base assets that replicates the payoff function. If there are no redundant securities, then such a replicating portfolio is unique for every attainable payoff. However, a payoff may be replicated nonuniquely. If this is the case, we expect that all replicating portfolios yield the same initial price. We say that *the Law of One Price* holds if for every attainable payoff all replicating portfolios have the same initial value, i.e., for any two portfolios $\boldsymbol{\varphi}, \boldsymbol{\psi} \in \mathbb{R}^N$ we have

$$\Pi_T^{\boldsymbol{\varphi}} = \Pi_T^{\boldsymbol{\psi}} \implies \Pi_0^{\boldsymbol{\varphi}} = \Pi_0^{\boldsymbol{\psi}}. \tag{5.5}$$

Thus, if the Law of One Price holds, then we can define the initial fair price $S_0 = \pi_0(S)$ of a security S with an attainable payoff S_T by

$$\pi_0(S) := \Pi_0^{\boldsymbol{\psi}} \text{ for any } \boldsymbol{\psi} \in \mathbb{R}^N \text{ such that } \Pi_T^{\boldsymbol{\psi}} = S_T. \tag{5.6}$$

We will refer to this approach as *pricing via replication.*

Consider two securities with attainable payoffs, X and Y. Let their initial prices be $\pi_0(X)$ and $\pi_0(Y)$, respectively. We say that *linear pricing* holds if for every choice of constants a and b, the security with payoff $aX+bY$ has the initial price of $a\pi_0(X)+b\pi_0(Y)$. Clearly, the Law of One Price implies linear pricing. Indeed, let portfolios $\boldsymbol{\varphi}_X$ and $\boldsymbol{\varphi}_Y$ replicate X and Y, respectively. Then, the portfolio $a\boldsymbol{\varphi}_X + b\boldsymbol{\varphi}_Y$ replicates $aX + bY$. According to the Law of One Price, the initial price of the attainable payoff $aX + bY$ is

$$\pi_0(aX + bY) = \Pi_0[a\boldsymbol{\varphi}_X + b\boldsymbol{\varphi}_Y] = a\Pi_0[\boldsymbol{\varphi}_X] + b\Pi_0[\boldsymbol{\varphi}_Y] = a\pi_0(X) + b\pi_0(Y).$$

That is, the *pricing functional* $\pi_0\colon \mathcal{A}(\Omega) \to \mathbb{R}$ is linear.

It is impossible to check the condition (5.5) for all replicating portfolios. The following lemma provides a simple criterion for verifying if the Law of One Price holds.

Lemma 5.6. *The Law of One Price holds iff,* $\forall \boldsymbol{\varphi} \in \mathbb{R}^N$,

$$\Pi_T^{\boldsymbol{\varphi}} = 0 \implies \Pi_0^{\boldsymbol{\varphi}} = 0.$$

That is, the Law of One Price holds iff every portfolio $\boldsymbol{\varphi}$ *that replicates the zero claim has zero initial value.*

Proof.
The necessity part. Suppose that $\boldsymbol{\varphi}$ and $\boldsymbol{\psi}$ replicate the same payoff X:

$$\Pi_T^{\boldsymbol{\varphi}} = \Pi_T^{\boldsymbol{\psi}} = X.$$

Then, the portfolio $\boldsymbol{\theta} = \boldsymbol{\varphi} - \boldsymbol{\psi}$ replicates the zero claim. By assumption, if $\Pi_T^{\boldsymbol{\theta}} = 0$ then $\Pi_0^{\boldsymbol{\theta}} = 0$. Therefore,

$$0 = \Pi_0^{\boldsymbol{\theta}} = \Pi_0^{\boldsymbol{\varphi}-\boldsymbol{\psi}} = \Pi_0^{\boldsymbol{\varphi}} - \Pi_0^{\boldsymbol{\psi}} \implies \Pi_0^{\boldsymbol{\varphi}} = \Pi_0^{\boldsymbol{\psi}}.$$

The sufficiency part. Let the Law of One Price hold. Suppose that there exists a portfolio $\boldsymbol{\theta}$ with $\Pi_0^{\boldsymbol{\theta}} \neq 0$ and $\Pi_T^{\boldsymbol{\theta}} = 0$. Take any nonzero attainable payoff X and a portfolio $\boldsymbol{\varphi}$ replicating the payoff of X. For $\lambda \in \mathbb{R}$, define $\boldsymbol{\varphi}_\lambda = \boldsymbol{\varphi} + \lambda\boldsymbol{\theta}$. Then, $\forall\, \lambda \in \mathbb{R}$,

$$\Pi_T^{\boldsymbol{\varphi}_\lambda} = \Pi_T^{\boldsymbol{\varphi}} + \lambda\Pi_T^{\boldsymbol{\theta}} = X + \lambda \cdot 0 = X.$$

However, the initial prices vary with λ:

$$\Pi_0^{\boldsymbol{\varphi}_\lambda} = \Pi_0^{\boldsymbol{\varphi}} + \lambda\Pi_0^{\boldsymbol{\theta}} \neq \Pi_0^{\boldsymbol{\varphi}}, \quad \text{for } \lambda \neq 0.$$

Thus, we can construct infinitely many portfolios replicating the same payoff and having different initial values. The supposition contradicts the Law of One Price. □

Example 5.4. Verify if the Law of One Price holds for a 3-by-3 single-period model with

$$\mathbf{S}_0 = \begin{bmatrix} 10 \\ 10 \\ 10 \end{bmatrix} \quad \text{and} \quad \mathbf{D} = \begin{bmatrix} 5 & 10 & 15 \\ 5 & 10 & 15 \\ 15 & 10 & 5 \end{bmatrix}.$$

Solution. Find all portfolios replicating the zero claim. The general solution to $\boldsymbol{\varphi}\mathbf{D} = \mathbf{0}$ is

$$\boldsymbol{\varphi} = [-t, t, 0], \text{ where } t \in \mathbb{R}.$$

The intitial value $\Pi_0^{\boldsymbol{\varphi}}$ of the portfolio $\boldsymbol{\varphi}$ is

$$\boldsymbol{\varphi}\,\mathbf{S}_0 = -10t + 10t = 0.$$

Since $\Pi_0^{\boldsymbol{\varphi}} = 0$ for all $t \in \mathbb{R}$, the Law of One Price holds. □

If there exist many portfolios replicating the same payoff X but having different initial values (i.e., the Law of One Price is violated), then it seems to be reasonable to define the fair initial price of X as the lowest cost for which we can replicate the target payoff. On the other hand, as follows from the next proposition, if the Law of One Price is violated, then for any attainable payoff there exists a replicating portfolio with arbitrarily low (or large) initial value. Thus, the fair price $\pi_0(X)$ cannot be defined meaningfully.

Proposition 5.7. *If the Law of One Price does not hold, then for every attainable payoff X and any real c_0 there exists a portfolio replicating X with the initial cost c_0.*

Proof. If the Law of One Price does not hold, then there exists $\boldsymbol{\theta}$ so that $\Pi_T^{\boldsymbol{\theta}} = 0$ and $\Pi_0^{\boldsymbol{\theta}} \neq 0$. Let $\boldsymbol{\varphi}$ replicate payoff X. For any $c_0 \in \mathbb{R}$, set $\lambda = \frac{c_0 - \Pi_0^{\boldsymbol{\varphi}}}{\Pi_0^{\boldsymbol{\theta}}}$. Then $\boldsymbol{\varphi}_\lambda = \boldsymbol{\varphi} + \lambda\boldsymbol{\theta}$ replicates X and $\Pi_0^{\boldsymbol{\varphi}_\lambda} = c_0$. □

By comparing the results of Lemmas 5.3 and 5.6 we conclude that the nonredundancy condition implies the Law of One Price. However, the converse is generally not true. Indeed, let us consider a model with two states of the world and two base assets having the following cash-flow matrix and initial price vector:

$$\mathbf{D} = \begin{bmatrix} 1 & 2 \\ 1 & 2 \end{bmatrix} \quad \text{and} \quad \mathbf{S}_0 = \begin{bmatrix} 1 \\ 1 \end{bmatrix}.$$

Clearly, the two base assets S^1 and S^2 are redundant with the same payoff vector $[1, 2]$. Consider any portfolio $\boldsymbol{\varphi} = [\varphi_1, \varphi_2]$ in S^1 and S^2 with zero terminal value:

$$\Pi_T^{\boldsymbol{\varphi}} = \varphi_1 S_T^1 + \varphi_2 S_T^2 = 0 \iff \varphi_1 + \varphi_2 = 0.$$

Since $\Pi_0^{\boldsymbol{\varphi}} = \varphi_1 S_0^1 + \varphi_2 S_0^2 = \varphi_1 + \varphi_2$, we have that $\Pi_T^{\boldsymbol{\varphi}} = 0 \implies \Pi_0^{\boldsymbol{\varphi}} = 0$. According to Lemma 5.6, the Law of One Price holds. Hence, this is a model with redundant assets and for which the Law of One Price holds. To conclude, market models for which the Law of One Price holds include models with and without redundant base assets. The next section ties together the Law of One Price with the concept of arbitrage in a market model.

5.3.2 Arbitrage

There exist two types of arbitrage opportunities. One type of arbitrage is an investment that gives a positive reward at time 0 and has no future cost at time T. Another type of arbitrage is an investment with zero initial value and nonnegative future cost that has a positive probability of yielding a strictly positive payoff at time T. Since any investment in a single-period model is a portfolio in base assets, we have the following formal definition of arbitrage.

Definition 5.7. An *arbitrage portfolio* is a portfolio φ such that one of the following two alternatives holds:

$$\Pi_0^{\varphi} = 0 \text{ and } \Pi_T^{\varphi} > 0, \tag{5.7}$$

or

$$\Pi_0^{\varphi} < 0 \text{ and } \Pi_T^{\varphi} \geqslant 0. \tag{5.8}$$

In other words, an arbitrage portfolio has zero initial value (i.e., there is no cost to set it up) and offers a potential gain with no potential liabilities at time $t = T$. Alternatively, an arbitrage opportunity is provided by a portfolio with a negative initial value and nonnegative terminal value (i.e., there are no potential liabilities). Conditions (5.7) and (5.8) can be written as inequalities

$$\varphi\,\mathbf{S}_0 \leqslant 0, \quad \varphi\,\mathbf{S}_T(\omega^j) \geqslant 0, \quad \text{for all } j = 1, 2, \ldots, M, \tag{5.9}$$

where at least one inequality is strict. Note that the last inequality is equivalently written as $\varphi\,\mathbf{D} \geqslant 0$.

If a market model admits no arbitrage, then any two portfolios replicating the same payoff must have the same initial value. Indeed, suppose that there exist two portfolios replicating the same payoff but having different initial values. By reversing positions in one portfolio (with a larger initial price) and combining it with the other portfolio, we can construct an arbitrage portfolio with a negative initial value and terminal value zero. Let us formally prove this fact.

Lemma 5.8. *No arbitrage implies the Law of One Price.*

Proof. Assume the absence of arbitrage opportunities. Recall that the Law of One Price holds when $\Pi_T^{\varphi} = 0 \implies \Pi_0^{\varphi} = 0$, for all $\varphi \in \mathbb{R}^N$. Suppose that there exists a portfolio ψ_0 such that $\Pi_T[\psi_0] = 0$ and $\Pi_0[\psi_0] \neq 0$. Take any attainable claim $X > 0$ (i.e., $X(\omega) \geqslant 0$ for all ω, and $X(\omega^*) > 0$ for some ω^*), and let φ_X replicate X. The portfolio $\varphi_\lambda := \varphi_X + \lambda\psi_0$ replicates X for any $\lambda \in \mathbb{R}$. Choose $\lambda = \lambda_0$ so that $\Pi_0[\varphi_{\lambda_0}] = 0$:

$$\Pi_0[\varphi_X + \lambda_0\psi_0] = \Pi_0[\varphi_X] + \lambda_0\Pi_0[\psi_0] = 0 \iff \lambda_0 = -\frac{\Pi_0[\varphi_X]}{\Pi_0[\psi_0]}.$$

Thus, φ_{λ_0} is an arbitrage portfolio, since $\Pi_0[\varphi_{\lambda_0}] = 0$ and $\Pi_T[\varphi_{\lambda_0}] = \Pi_T[\varphi_X] = X > 0$. This proves that violation of the Law of One Price implies the existence of an arbitrage portfolio. Hence, the lemma is proven. □

As follows from the Law of One Price, the initial price S_0 of a security with an attainable payoff S_T has to be equal to the initial cost of a replicating portfolio or else there exists arbitrage, as is given in (5.6). Indeed, if the security is priced at $S_0 \neq \Pi_0[\varphi_S]$, where φ_S replicates S_T, then an arbitrage portfolio combining the security S and base assets can be created. If $S_0 < \Pi_0[\varphi_S]$, then we buy the security S and form the portfolio $-\varphi_S$. If $S_0 > \Pi_0[\varphi_S]$, then we sell the asset and form the portfolio φ_S. In both cases, our proceeds at time 0 are positive, and the terminal value is zero. So the assumption $S_0 \neq \Pi_0[\varphi_S]$ leads to arbitrage. Thus, the price $S_0 = \pi_0(S)$ given by (5.6) is called the *no-arbitrage price* of security S.

5.3.3 The First Fundamental Theorem of Asset Pricing

It can be difficult to find an arbitrage portfolio. Hence, it is reasonable to first investigate whether a model admits arbitrage. If the Law of One Price is violated, then, according to

Lemma 5.8, there exists an arbitrage opportunity. However, we cannot make any conclusion if the Law of One Price holds. The *first fundamental theorem of asset pricing* (FTAP) provides a necessary and sufficient condition for the absence of arbitrage. Although we are proving the FTAP for a finite single-period economy, a very similar result is true for discrete-time multiperiod models and continuous-time models.

Theorem 5.9 (The first FTAP). *There are no arbitrage portfolios iff there exists a strictly positive solution* $\mathbf{\Psi} \in \mathbb{R}^M$ *to the linear system of equations*

$$\mathbf{D}\,\mathbf{\Psi} = \mathbf{S}_0. \tag{5.10}$$

That is, there exists $\mathbf{\Psi} = \left[\Psi_1, \Psi_2, \cdots, \Psi_M\right]^\top \gg 0$ *such that*

$$\sum_{j=1}^{M} S_T^i(\omega^j)\,\Psi_j = S_0^i,\ \forall i = 1, 2, \ldots, N.$$

Before proving this theorem, let us illustrate it with the following example.

Example 5.5. Find an arbitrage portfolio (if any) for the model with three states and three assets specified as follows:

$$\mathbf{S}_0 = \begin{bmatrix} 1 \\ 1 \\ 1 \end{bmatrix}, \quad \mathbf{D} = \begin{bmatrix} 1 & 1 & 1 \\ 1 & 0 & 0 \\ 0 & 1 & 0 \end{bmatrix}.$$

Solution. Solving $\mathbf{D}\,\mathbf{\Psi} = \mathbf{S}_0$ gives $\mathbf{\Psi} = [1, 1, -1]^\top$. Since not all Ψ_j's are positive, according to Theorem 5.9, there exists an arbitrage. Let us try to replicate the Arrow–Debreu security $\mathcal{E}^3$ having payoff vector $\boldsymbol{\mathcal{E}}^3 = [\mathcal{E}^3(\omega^1),\, \mathcal{E}^3(\omega^2),\, \mathcal{E}^3(\omega^3)] = [0,\, 0,\, 1]$:

$$\boldsymbol{\varphi}\,\mathbf{D} = \boldsymbol{\mathcal{E}}^3 \iff \begin{cases} \varphi_1 + \varphi_2 = 0 \\ \varphi_1 + \varphi_3 = 0 \\ \varphi_1 = 1 \end{cases} \iff \begin{cases} \varphi_1 = 1 \\ \varphi_2 = -1 \\ \varphi_3 = -1 \end{cases}$$

The replicating portfolio is $\boldsymbol{\varphi} = [1, -1, -1]$. Its initial value is negative:

$$\Pi_0^{\varphi} = \boldsymbol{\varphi}\,\mathbf{S}_0 = 1 - 1 - 1 = -1 < 0.$$

Since $\Pi_T^{\varphi} = \mathcal{E}^3 > 0$, the solution $\boldsymbol{\varphi}$ is an arbitrage portfolio. □

5.3.3.1 The First FTAP: Sufficiency Part

Proof. Suppose that $\mathbf{\Psi}$ solves (5.10). Then, the initial value of a portfolio $\boldsymbol{\varphi} \in \mathbb{R}^N$ is

$$\Pi_0^{\varphi} = \boldsymbol{\varphi}\,\mathbf{S}_0 = \boldsymbol{\varphi}\,(\mathbf{D}\,\mathbf{\Psi}) = (\boldsymbol{\varphi}\,\mathbf{D})\ \mathbf{\Psi} = \Pi_T^{\varphi}(\Omega)\,\mathbf{\Psi} = \sum_{j=1}^{M} \Pi_T^{\varphi}(\omega^j)\,\Psi_j. \tag{5.11}$$

That is, the initial value of a portfolio is equal to a product of the terminal payoff vector and the state-price vector. Let $\boldsymbol{\varphi}^*$ be an arbitrage portfolio. Hence, $\Pi_0^{\varphi^*} \leqslant 0$, $\Pi_T^{\varphi^*} \geqslant 0$, and $\Pi_0^{\varphi^*} < 0$ or $\Pi_T^{\varphi^*}(\omega^k) > 0$ for some $k \in \{1, \ldots, M\}$ holds. Applying (5.11) to $\boldsymbol{\varphi}^*$ gives

$$\Pi_0^{\varphi^*} = \sum_{j \neq k}^{M} \Pi_T^{\varphi^*}(\omega^j)\,\Psi_j + \Pi_T^{\varphi^*}(\omega^k)\,\Psi_k. \tag{5.12}$$

In (5.12), either the left-hand side is negative and the right-hand side is nonnegative, or the left-hand side is zero and the right-hand side is positive. So, we have a contradiction. Hence, the existence of a strictly positive solution to (5.10) implies that there are no arbitrage portfolios. □

5.3.3.2 The First FTAP: Necessity Part

Proof. The set $\mathbb{R}_+^{M+1} = \{\mathbf{x} \in \mathbb{R}^{M+1} : \mathbf{x} \geqslant 0\}$ is a closed convex cone in the vector space $\mathbb{R}^{M+1}$. That is,

- $\mathbb{R}_+^{M+1}$ containing its boundary points (it is a closed set),
- for all $\mathbf{x}, \mathbf{y} \in \mathbb{R}_+^{M+1}$ the segment $[\mathbf{x}, \mathbf{y}]$ of a straight line connecting $\mathbf{x}$ and $\mathbf{y}$ is contained in $\mathbb{R}_+^{M+1}$ (it is a convex set),
- for all $\mathbf{x} \in \mathbb{R}_+^{M+1}$ the ray $\{\lambda \mathbf{x} : \lambda \geqslant 0\}$ is contained in $\mathbb{R}_+^{M+1}$ (it is a cone).

Introduce another subset L of vectors in $\mathbb{R}^{M+1}$ defined by

$$L = \left\{ \left[-\boldsymbol{\theta}\mathbf{S}_0, \boldsymbol{\theta}\mathbf{S}_T(\omega^1), \cdots, \boldsymbol{\theta}\mathbf{S}_T(\omega^M)\right] : \boldsymbol{\theta} \in \mathbb{R}^N \right\},$$

where $-\boldsymbol{\theta}\mathbf{S}_0 = -\sum_{i=1}^N \theta_i S_0^i$ and $\boldsymbol{\theta}\mathbf{S}_T(\omega^j) = \sum_{i=1}^N \theta_i S_T^i(\omega^j)$, $j = 1, 2, \ldots, M$. Clearly, L is a linear (vector) subspace of $\mathbb{R}^{M+1}$, where for any $a, b \in \mathbb{R}$,

$$\mathbf{x}, \mathbf{y} \in L \implies a\mathbf{x} + b\mathbf{y} \in L.$$

Suppose that $\boldsymbol{\theta} \in \mathbb{R}^N$ is an arbitrage portfolio, i.e., $\boldsymbol{\theta}\mathbf{S}_0 < 0$ and $\boldsymbol{\theta}\mathbf{D} \geqslant 0$ (or $\boldsymbol{\theta}\mathbf{S}_0 \leqslant 0$ and $\boldsymbol{\theta}\mathbf{D} > 0$). Then there is a point in $L \cap \mathbb{R}_+^{M+1}$ corresponding to the arbitrage portfolio $\boldsymbol{\theta}$ and vice versa—any point in $L \cap \mathbb{R}_+^{M+1}$ corresponds to an arbitrage opportunity. Therefore, nonexistence of an arbitrage portfolio means that the subspace L and the cone $\mathbb{R}_+^{M+1}$ intersect only at the origin $\mathbf{0} = [0, 0, \cdots, 0] \in \mathbb{R}^{M+1}$. The rest of the proof is based on the so-called separating hyperplane theorem, which, being applied to this situation, states the following. There exists a hyperplane $H \subset \mathbb{R}^{M+1}$ (i.e., a linear subspace of dimensions M) that separates $\mathbb{R}^{M+1}$ into two half-spaces H^+ and H^- such that

$$\mathbb{R}_+^{M+1} \subseteq H^+, \quad L \subseteq H^-, \quad H^+ \cap H^- = H.$$

In other words, the cone $\mathbb{R}_+^{M+1}$ lies on one side of H and the subspace L lies on the other side of H. The general equation for a hyperplane in $\mathbb{R}^{M+1}$ passing through the origin is

$$\boldsymbol{\lambda}\mathbf{x}^\top = 0 \iff \lambda_0 x_0 + \lambda_1 x_1 + \cdots + \lambda_M x_M = 0,$$

where $\boldsymbol{\lambda} = [\lambda_0, \lambda_1, \cdots, \lambda_M]$ is a normal vector and $\boldsymbol{x} = [x_0, x_1, \cdots, x_M] \in \mathbb{R}^{M+1}$. The concept of separation can be expressed as follows: either $\boldsymbol{\lambda}\mathbf{x}^\top > \boldsymbol{\lambda}\mathbf{y}^\top$ or $\boldsymbol{\lambda}\mathbf{x}^\top < \boldsymbol{\lambda}\mathbf{y}^\top$ holds for all $\mathbf{x} \in \mathbb{R}_+^{M+1} \setminus \{\mathbf{0}\}$ and all $\mathbf{y} \in L$. In particular, the set $\{\boldsymbol{\lambda}\mathbf{y}^\top : \mathbf{y} \in L\}$ is bounded from above or below. This is possible iff $\boldsymbol{\lambda}\mathbf{y}^\top = 0$, i.e., L is contained in H. To show this, suppose that there exists $\mathbf{y} \in L$ such that $\boldsymbol{\lambda}\mathbf{y}^\top > 0$. Since L is a vector space, $\mathbf{y} \in L \implies a\mathbf{y} \in L$ for every $a \in \mathbb{R}$. Then the set $\{a\boldsymbol{\lambda}\mathbf{y}^\top : a \in \mathbb{R}\} = \mathbb{R}$ is unbounded. We arrive at a contradiction. On the other hand, since $\boldsymbol{\lambda}\mathbf{x}^\top > 0$ for every $\mathbf{x} \in \mathbb{R}_+^{M+1}$ (if $\boldsymbol{\lambda}\mathbf{x}^\top < 0$ then just replace $\boldsymbol{\lambda}$ by $-\boldsymbol{\lambda}$), all λ_i's are positive. Indeed, for each $j = 0, 1, \ldots, M$, take $\mathbf{x} = \mathbf{e}_j := [0, \cdots, 0, \underbrace{1}_{j\text{th}}, 0, \cdots, 0] \in \mathbb{R}^{M+1}$ to obtain that $\boldsymbol{\lambda}\mathbf{e}_j^\top = \lambda_j > 0$. Since $L \subset H$,

$$-\lambda_0 \boldsymbol{\theta}\mathbf{S}_0 + \sum_{j=1}^M \lambda_j \boldsymbol{\theta}\mathbf{S}_T(\omega^j) = 0$$

holds for every portfolio $\boldsymbol{\theta} \in \mathbb{R}^N$. Setting $\boldsymbol{\theta} = \mathbf{e}_i \in \mathbb{R}^N$, for each $i = 1, 2, \ldots, N$, we obtain

$$-\lambda_0 S_0^i + \sum_{j=1}^M \lambda_j S_T^i(\omega^j) = 0.$$

This implies that

$$-\lambda_0 \mathbf{S}_0 + \sum_{j=1}^{M} \lambda_j \mathbf{S}_T(\omega^j) = \mathbf{0} \iff \mathbf{S}_0 = \sum_{j=1}^{M} \underbrace{\frac{\lambda_j}{\lambda_0}}_{\equiv \Psi_j} \mathbf{S}_T(\omega^j).$$

In matrix form, we have

$$\mathbf{D}\,\boldsymbol{\Psi} = \mathbf{S}_0, \text{ where } \boldsymbol{\Psi}^\top = \left[\Psi_1, \Psi_2, \cdots, \Psi_M\right] = \left[\tfrac{\lambda_1}{\lambda_0}, \tfrac{\lambda_2}{\lambda_0}, \cdots, \tfrac{\lambda_M}{\lambda_0}\right] \gg 0. \qquad \square$$

Example 5.6. Consider a 3-by-3 model with initial price vector $\mathbf{S}_0 = [1, 5, 10]^\top$ and payoff matrix

$$\text{(a) } \mathbf{D} = \begin{bmatrix} 1 & 1 & 1 \\ 1 & 6 & 15 \\ 12 & 8 & 6 \end{bmatrix} \quad \text{(b) } \mathbf{D} = \begin{bmatrix} 1 & 1 & 1 \\ 4 & 6 & 8 \\ 12 & 8 & 4 \end{bmatrix} \quad \text{(c) } \mathbf{D} = \begin{bmatrix} 1 & 1 & 1 \\ 3 & 5 & 5 \\ 10 & 10 & 15 \end{bmatrix}.$$

Find an arbitrage opportunity, if any.

Solution. First, solve the matrix equation $\mathbf{D}\,\boldsymbol{\Psi} = \mathbf{S}_0$ for each matrix $\mathbf{D}$:

(a) $\boldsymbol{\Psi} = \left[\frac{8}{13}, \frac{2}{13}, \frac{3}{13}\right]^\top$;

(b) $\boldsymbol{\Psi} = \left[\frac{1}{2} + t, \frac{1}{2} - 2t, t\right]^\top$, where $t \in \mathbb{R}$;

(c) $\boldsymbol{\Psi} = [0, 1, 0]^\top$.

In case (a), the solution is unique and strictly positive. In case (b), the solution is strictly positive iff $0 < t < \frac{1}{4}$. In case (c), the solution has nonpositive components. Therefore, the model is arbitrage free only in cases (a) and (b). Let us find an arbitrage opportunity in case (c). Replicate the AD security $\mathcal{E}^1$ by solving the replication equation

$$\boldsymbol{\varphi}\,\mathbf{D} = \mathcal{E}^1, \text{ where } \mathcal{E}^1 = [1, 0, 0].$$

The solution $\boldsymbol{\varphi} = \left[\frac{5}{2}, -\frac{1}{2}, 0\right]$ is an arbitrage portfolio since its initial value is zero,

$$\Pi_0^{\varphi} = \boldsymbol{\varphi}\,\mathbf{S}_0 = \frac{5}{2} \cdot 1 + \left(-\frac{1}{2}\right) \cdot 5 = 0,$$

the terminal value $\Pi_T^{\varphi}(\omega) = \mathcal{E}^1(\omega)$ is nonnegative for all $\omega \in \Omega$, and $\Pi_0^{\varphi}(\omega^1) = 1$ is strictly positive with nonzero probability. Alternatively, we can find another arbitrage portfolio $\boldsymbol{\psi} = \left[-2, 0, \frac{1}{5}\right]$ that replicates $\mathcal{E}^3$. We have $\Pi_0^{\psi} = 0$ and $\Pi_T^{\psi} = \mathcal{E}^3 > 0$. $\square$

5.3.3.3 A Geometric Interpretation of the First FTAP

We now give a geometric interpretation of Theorem 5.9. Define the set $C \subset \mathbb{R}^N$ by

$$C := \left\{ \sum_{j=1}^{M} x_j\, \mathbf{S}_T(\omega^j) \;:\; x_j \geqslant 0,\; 1 \leqslant j \leqslant M \right\}.$$

It is a convex closed cone, which is contained in the span of the M column vectors of the payoff matrix, $\mathbf{S}_T(\omega^j)$, $j = 1, 2, \ldots, M$. The absence of arbitrage means that the initial price vector $\mathbf{S}_0 \in \mathbb{R}^N$ lies in the interior of C. Indeed, suppose that $\mathbf{S}_0$ lies in the exterior or on the boundary of the cone C. The separating hyperplane theorem guarantees that there exists a hyperplane described by the equation

$$\boldsymbol{\theta}\mathbf{x}^\top = 0, \quad \mathbf{x} \in \mathbb{R}^N,$$

for some $\boldsymbol{\theta} \in \mathbb{R}^N$, which would separate the ray generated by $\mathbf{S}_0$ from the interior of the cone C. This fact implies the following inequalities:

$$\boldsymbol{\theta}\mathbf{S}_0 \leqslant 0 \text{ and } \boldsymbol{\theta}\mathbf{S}_T(\omega^j) \geqslant 0 \text{ for all } j = 1, 2, \ldots, M,$$

with at least one product positive for some j. That is, $\boldsymbol{\theta}$ is an arbitrage portfolio.

For example, consider the two-state (binomial) model with two securities: the risk-free bond $S^1 \equiv B$ and risky stock $S^2 \equiv S$. The initial price vector and payoff matrix are, respectively, given by

$$\mathbf{S}_0 = \begin{bmatrix} (1+r)^{-1} \\ S_0 \end{bmatrix} \quad \text{and} \quad \mathbf{D} = \left[\ \mathbf{S}_T(\omega^+) \mid \mathbf{S}_T(\omega^-)\ \right] = \begin{bmatrix} 1 & 1 \\ S_0 u & S_0 d \end{bmatrix}.$$

The cone $C \subset \mathbb{R}^2$ is generated by strictly positive linear combinations of the vectors $\mathbf{S}_T(\omega^+)$ and $\mathbf{S}_T(\omega^-)$. The no-arbitrage condition means that the vector $\mathbf{S}_0$ lies inside the cone C. As follows from Figure 5.2, this is possible iff

$$S_0 d < (1+r)S_0 < S_0 u \iff d < 1 + r < u.$$

This criterion can be generalized on any model with two base assets and arbitrarily many states of the world. The vector $\mathbf{S}_0$ lies inside the cone C iff

$$\min_{\omega \in \Omega} \frac{S_T^2(\omega)}{S_T^1(\omega)} < \frac{S_0^2}{S_0^1} < \max_{\omega \in \Omega} \frac{S_T^2(\omega)}{S_T^1(\omega)}.$$

5.3.4 Risk-Neutral Probabilities

An important component of Theorem 5.9 is the M-dimensional vector $\boldsymbol{\Psi} \gg 0$ which solves $\mathbf{D}\,\boldsymbol{\Psi} = \mathbf{S}_0$. As follows from (5.11), the initial value of a portfolio $\varphi \in \mathbb{R}^N$ is equal to the product of the terminal payoff vector and the state-price vector, $\Pi_0^\varphi = \Pi_T^\varphi(\Omega)\,\boldsymbol{\Psi}$. Therefore, the no-arbitrage initial value $\pi_0(X)$ of an attainable payoff X replicated by φ_X can be calculated as follows:

$$\pi_0(X) = \Pi_0[\varphi_X] = \sum_{j=1}^{M} \Pi_T[\varphi_X](\omega^j)\,\Psi_j = \sum_{j=1}^{M} X(\omega^j)\,\Psi_j = \mathbf{X}\,\boldsymbol{\Psi}. \tag{5.13}$$

As is seen from (5.13), a replicating portfolio is not required to calculate the no-arbitrage initial value (i.e., the fair price) of a claim with an attainable payoff. We only need to find the vector $\boldsymbol{\Psi}$ and then multiply it by the payoff vector to find $\pi_0(X)$. The solutions $\Psi_j \in (0, \infty)$, $j = 1, 2, \ldots, M$, are called *state prices*. To explain such a name, consider an Arrow–Debreu security $\mathcal{E}^j = \mathbb{I}_{\{\omega^j\}}$ having a nonzero payoff of one unit only for the state ω^j. According to (5.13), the initial price of $\mathcal{E}^j$ is

$$\pi_0(\mathcal{E}^j) = \sum_{k=1}^{M} \mathcal{E}^j(\omega^k)\,\Psi_k = \sum_{k=1}^{M} \mathbb{I}_{\{\omega^j\}}(\omega^k)\,\Psi_k = \Psi_j. \tag{5.14}$$

Thus, the financial meaning of the solutions $\Psi_j \in (0, \infty)$, $j = 1, 2, \ldots, M$, is that they are

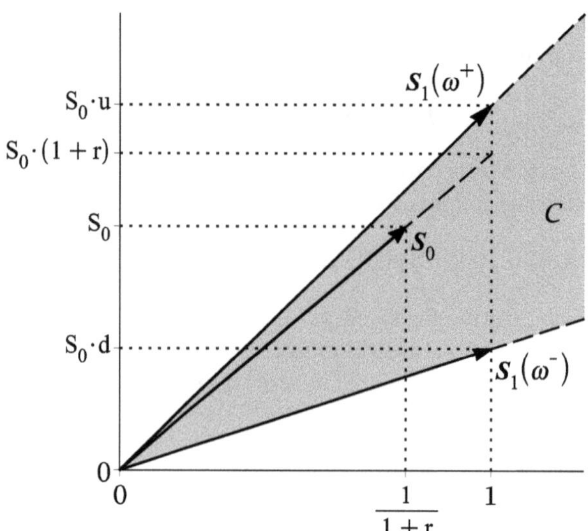

FIGURE 5.2: The no-arbitrage condition means that the initial price vector $\mathbf{S}_0$ has to lie inside the cone $C \subset \mathbb{R}^N_+$ generated by the M column vectors of the payoff matrix $\mathbf{D}$. This situation is illustrated in the case where $M = N = 2$.

no-arbitrage prices of the Arrow–Debreu securities. In particular, Ψ_j can also be thought of as the price of an insurance contract that pays one unit of currency in case scenario ω^j occurs.

The pricing formula (5.13) can also be derived using (5.14) and the principle of linear pricing. According to Proposition 5.1, every payoff X can be represented as a portfolio in AD securities:

$$X = x_1\mathcal{E}^1 + x_2\mathcal{E}^2 + \cdots + x_M\mathcal{E}^M \quad \text{for some} \quad x_1, x_2, \ldots, x_M \in \mathbb{R}.$$

Thus, the initial value $\pi_0(X)$ is the same as that of the portfolio in the AD securities:

$$\begin{aligned}\pi_0(X) &= x_1\pi_0(\mathcal{E}^1) + x_2\pi_0(\mathcal{E}^2) + \cdots + x_M\pi_0(\mathcal{E}^M) \\ &= x_1\Psi_1 + x_2\Psi_2 + \cdots + x_M\Psi_M = \mathbf{X}\,\mathbf{\Psi}.\end{aligned}$$

Consider a risk-less asset B that pays one unit of currency at the terminal time (i.e., B is a zero-coupon bond). According to (5.13), since $B_T(\omega^j) = 1$ for all $j = 1, \ldots, M$, the initial price of the payoff $B_T \equiv 1$ is

$$B_0 := \pi_0(B_T) = \sum_{j=1}^{M} \Psi_j.$$

Let us set $1+r = \left(\sum_{j=1}^M \Psi_j\right)^{-1}$. Hence $B_0 = \frac{1}{1+r}$. The quantity r is in fact the single-period return on B:

$$\frac{B_T - B_0}{B_0} = \frac{B_T}{B_0} - 1 = \frac{1}{(1+r)^{-1}} - 1 = r.$$

Now we are ready to rewrite the security pricing formula (5.13) in a more familiar form.

First, note that the positive state prices Ψ_j, $j = 1, 2, \ldots, M$, can be normalized to give us a new set of probabilities:

$$\tilde{p}_j := \frac{\Psi_j}{\sum_{i=1}^{M} \Psi_i}, \quad j = 1, 2, \ldots, M. \tag{5.15}$$

Clearly, all $\tilde{p}_j$'s are positive and sum to unity. These probabilities are called *risk-neutral probabilities* and they have no relation to the real-world (physical) probabilities $\{p_j = \mathbb{P}(\omega^j)\}$. The real-world probabilities can be estimated from historical data, whereas the risk-neutral probabilities are not observed. Let $\widetilde{\mathbb{P}}$ denote the *risk-neutral probability measure* defined by $\widetilde{\mathbb{P}}(\omega^j) = \tilde{p}_j$ for $j = 1, 2, \ldots, M$. The risk-neutral probability of any event $E \subseteq \Omega$ is then given by

$$\widetilde{\mathbb{P}}(E) = \sum_{\omega \in E} \widetilde{\mathbb{P}}(\omega) = \sum_{\omega^j \in E} \tilde{p}_j.$$

The mathematical expectation of a random variable $X \in \mathcal{L}(\Omega)$ with respect to $\widetilde{\mathbb{P}}$, denoted by $\widetilde{\mathrm{E}}[X]$, is defined by

$$\widetilde{\mathrm{E}}[X] = \sum_{j=1}^{M} X(\omega^j)\, \tilde{p}_j.$$

Using (5.15), the initial price in (5.13) can now be written as the mathematical expectation of the discounted payoff function with respect to the risk-neutral probability measure:

$$\begin{aligned}\pi_0(X) &= \sum_{j=1}^{M} X(\omega^j)\, \Psi_j = \left(\sum_{k=1}^{M} \Psi_k\right) \sum_{j=1}^{M} X(\omega^j)\, \frac{\Psi_j}{\sum_{k=1}^{M} \Psi_k} \\ &= \frac{1}{1+r} \sum_{j=1}^{M} X(\omega^j)\, \tilde{p}_j = \frac{1}{1+r} \widetilde{\mathrm{E}}[X].\end{aligned} \tag{5.16}$$

Here, $\frac{X}{1+r}$ is the discounted value of X. That is, it is the present (time-0) value of the payoff paid at time T. If we apply (5.16) to a portfolio that only contains one unit of the base asset S^i, then we have that

$$S_0^i = \frac{1}{1+r} \widetilde{\mathrm{E}}[S_T^i]. \tag{5.17}$$

Therefore, the risk-neutral expectation of the return on S^i is

$$\widetilde{\mathrm{E}}\left[\frac{S_T^i - S_0^i}{S_0^i}\right] = \frac{\widetilde{\mathrm{E}}[S_T^i]}{S_0^i} - 1 = (1+r) - 1 = r.$$

In other words, in the risk-neutral probability measure, the expected return of any base security is equal to the risk-free interest rate r on the bond B. Since $B_0/B_T = (1+r)^{-1}$, we can rewrite (5.17) as

$$\frac{S_0^i}{B_0} = \widetilde{\mathrm{E}}\left[\frac{S_T^i}{B_T}\right] \tag{5.18}$$

for every base asset S^i, $i = 1, \ldots, N$. In other words, the discounted asset price processes $\left\{\overline{S}_t^i := \frac{S_t^i}{B_t}\right\}_{t \in \{0,T\}}$, $i = 1, 2, \ldots, N$, are martingales under the risk-neutral probability measure $\widetilde{\mathbb{P}}$. The subject of martingales will be introduced and covered in depth in the next chapter. It suffices to note here that the discounted price processes $\left\{\overline{S}_t^i\right\}_{t \in \{0,T\}}$ are examples of single-period stochastic processes. A single-period stochastic process, say

$\{X_t\}_{t\in\{0,T\}}$, is a called a martingale with respect to the measure $\widetilde{\mathbb{P}}$ if $\widetilde{\mathrm{E}}[X_T] = X_0$. That is, our best prediction of the future time-T value (i.e., the expected future value with respect to the given measure) of the process is the same as the current time-0 value.

To summarize, if the single-period market model is arbitrage-free, then there exists a risk-neutral probability measure $\widetilde{\mathbb{P}}$. Conversely, we can construct a state-space vector from the risk-neutral probabilities as

$$\Psi_j = \frac{\tilde{p}_j}{1+r}, \quad j = 1, 2, \ldots, M.$$

Hence, the existence of $\widetilde{\mathbb{P}}$ implies no arbitrage. Therefore, Theorem 5.9 can now be formulated as follows.

Theorem 5.10 (The first FTAP—the 2nd version). *There are no arbitrage portfolios in a single-period N-by-M model iff there exist probabilities $\{\tilde{p}_j > 0 \ : \ j = 1, 2, \ldots, M\}$ such that the discounted asset price processes $\left\{\overline{S}^i_t\right\}_{t\in\{0,T\}}$, $i = 1, 2, \ldots, N$, are all martingales with respect to the probability measure $\widetilde{\mathbb{P}}$. Such a probability measure is called the risk-neutral probability measure.*

Let us come back to the binomial model from Example 5.1 with two states and two assets: $S^1_t = B_t$ and $S^2_t = S_t$. We readily solve for the state-price vector as follows:

$$\begin{bmatrix} 1 & 1 \\ S_0 u & S_0 d \end{bmatrix} \begin{bmatrix} \Psi_1 \\ \Psi_2 \end{bmatrix} = \begin{bmatrix} (1+r)^{-1} \\ S_0 \end{bmatrix} \iff \begin{cases} \Psi_1 + \Psi_2 = \frac{1}{1+r} \\ u\,\Psi_1 + d\,\Psi_2 = 1 \end{cases}$$

$$\iff \begin{cases} \Psi_1 = \frac{(1+r)-d}{(1+r)(u-d)} \\ \Psi_2 = \frac{u-(1+r)}{(1+r)(u-d)} \end{cases} \tag{5.19}$$

The solution $\boldsymbol{\Psi} = [\Psi_1, \ \Psi_2]^\top$ in (5.19) is strictly positive iff $d < 1+r < u$. If $1+r \leqslant d$, then an arbitrage portfolio can be obtained by shorting the bond (for example, $\varphi_1 = -S_0(1+r)$ and $\varphi_2 = 1$). If $1+r \geqslant u$, then an arbitrage portfolio can be obtained by shorting the stock (for example, $\varphi_1 = S_0(1+r)$ and $\varphi_2 = -1$). The reader may verify that Π^φ_0 is zero and Π^φ_T is positive in both cases. Finally, we can calculate the risk-neutral probabilities for the binomial model:

$$\tilde{p} \equiv \tilde{p}_1 = \frac{\Psi_1}{\Psi_1 + \Psi_2} = \frac{(1+r)-d}{u-d}, \quad 1 - \tilde{p} \equiv \tilde{p}_2 = \frac{\Psi_2}{\Psi_1 + \Psi_2} = \frac{u-(1+r)}{u-d}.$$

The probabilities are strictly positive iff the no-arbitrage condition $d < 1+r < u$ holds. Finally, note that the state prices are themselves given by the discounted (risk-neutral) expected value of the Arrow–Debreu payoffs:

$$\Psi_j = \frac{1}{1+r}\widetilde{\mathrm{E}}[\mathcal{E}^j]\,, \; j = 1, \ldots, M.$$

5.3.5 The Second Fundamental Theorem of Asset Pricing

We are now ready to present the second fundamental theorem of asset pricing which states a sufficient and necessary condition of completeness of a market model. We give two versions of the theorem: first in terms of state prices and then in terms of risk-neutral probabilities.

Theorem 5.11 (The second FTAP). *Assuming absence of arbitrage, there exists a unique solution to the state-price equation, $\boldsymbol{\Psi} \gg 0$, iff the market is complete.*

Theorem 5.12 (The second FTAP—the 2nd version). *Assuming absence of arbitrage, there exists a unique set of risk-neutral probabilities* $\{\tilde{p}_j > 0 \ : \ j = 1, 2, \ldots, M\}$ *iff the market is complete.*

Proof. The no-arbitrage assumption implies that there exists a strictly positive solution $\boldsymbol{\Psi}$ to the matrix equation $\mathbf{D}\,\boldsymbol{\Psi} = \mathbf{S}_0$. If the market is complete, then $\operatorname{rank}(\mathbf{D}) = M$. Hence the solution is unique. Conversely, let the solution $\boldsymbol{\Psi} \gg 0$ be unique. We will argue that the market is complete by contradiction. If the market is not complete, then $\operatorname{rank}(\mathbf{D}) < M$. Therefore, there exists a nontrivial solution $\boldsymbol{\lambda} \neq \mathbf{0}$ to the matrix equation $\mathbf{D}\,\boldsymbol{\lambda} = \mathbf{0}$. Since $\Psi_j > 0$ for all $1 \leqslant j \leqslant M$, there exists a sufficiently small constant $a \neq 0$ such that $\Psi_j + a\lambda_j > 0$ for all $1 \leqslant j \leqslant M$, i.e., $\boldsymbol{\Psi} + a\boldsymbol{\lambda} \gg 0$. Moreover, this vector solves

$$\mathbf{D}\,(\boldsymbol{\Psi} + a\boldsymbol{\lambda}) = \mathbf{D}\,\boldsymbol{\Psi} + a\mathbf{D}\,\boldsymbol{\lambda} = \mathbf{S}_0 + a\,\mathbf{0} = \mathbf{S}_0.$$

Thus, $\boldsymbol{\Psi} + a\boldsymbol{\lambda}$ is another state-price vector. This contradicts our hypothesis that the state-price vector is unique. □

Example 5.7. Verify if the single-period model from Example 5.6 is complete.

Solution. In cases (a) and (b), there exists a positive solution to the state-price equation $\mathbf{D}\,\boldsymbol{\Psi} = \mathbf{S}_0$. Hence, there are no arbitrage opportunities. In case (a), the solution $\boldsymbol{\Psi}$ is unique; therefore, the model is complete. In case (b), there are infinitely many positive solutions parametrized by $t \in (0, \frac{1}{4})$; therefore, the model is incomplete. Indeed, in case (b) the determinant of the payoff matrix is zero; hence, the model is incomplete. For example, the payoff of the at-the-money call option on asset S^2 is unattainable since the system $\boldsymbol{\varphi}\,\mathbf{D} = [0, 0, 2]$ is inconsistent. □

5.3.6 Investment Portfolio Optimization

A no-arbitrage price of a derivative security is calculated by taking the risk-neutral expectation of the payoff function. So, a model that does not admit arbitrage opportunities has two probability measures, the actual (real-world) measure and the risk-neutral measure. We only use the latter in no-arbitrage pricing. As is demonstrated in previous chapters, the real-world measure is used in asset management and risk management. However, as is shown below, the risk-neutral measure also plays an important role when finding an optimal allocation.

Consider the following investment problem for a nonarbitrage, complete model. Being given an initial capital W_0, we find a portfolio $\boldsymbol{\varphi} = \mathbb{R}^N$ with initial value $\Pi_0^{\boldsymbol{\varphi}} = W_0$ that maximizes the expected utility of the terminal value, $\mathrm{E}[u(\Pi_T^{\boldsymbol{\varphi}})]$, where u is a utility function, i.e., u is a nondecreasing and concave function. That is, we find the solution to the following constrained optimization problem:

$$\text{maximize } \mathrm{E}[u(\Pi_T^{\boldsymbol{\varphi}})] = \sum_{j=1}^{M} u(\Pi_T^{\boldsymbol{\varphi}}(\omega^j))\,p_j = \sum_{j=1}^{M} u\left(\sum_{i=1}^{N} \varphi_i S_T^i(\omega^j)\right) p_j \text{ w.r.t. } \boldsymbol{\varphi} \in \mathbb{R}^N \tag{5.20}$$

$$\text{subject to } \Pi_0^{\boldsymbol{\varphi}} = \sum_{i=1}^{N} \varphi_i S_0^i = W_0. \tag{5.21}$$

Since the market model is complete, the solution of the problem (5.20)–(5.21) can be split into two steps: first, we find the terminal value of the optimal portfolio $\Pi_T^{\boldsymbol{\varphi}}$, i.e., a (terminal) payoff X, that maximizes the expected utility $\mathrm{E}[u(X)]$; second, we find the optimal portfolio vector $\boldsymbol{\varphi}$ that replicates the payoff X. To obtain a constraint equation on X, we use the fact that the discounted value process is a martingale under the risk-neutral probability measure:

$$\frac{1}{1+r}\tilde{\mathrm{E}}\left[\Pi_T^{\varphi}\right] = \Pi_0^{\varphi} = W_0.$$

Denoting $X = \Pi_T^{\varphi}$ and using state prices gives

$$\frac{1}{1+r}\tilde{\mathrm{E}}[X] = \frac{1}{1+r}\sum_{j=1}^{M} x_j\,\tilde{p}_j = \sum_{j=1}^{M} x_j \Psi_j = W_0.$$

The optimization problem (5.20)–(5.21) now takes the form

$$\mathrm{E}[u(X)] = \sum_{j=1}^{M} u(x_j)\,p_j \to \max_{\mathbf{X}\in\mathbb{R}^M} \tag{5.22}$$

$$\text{subject to } \mathbf{X}\,\boldsymbol{\Psi} = \sum_{j=1}^{M} x_j \Psi_j = W_0. \tag{5.23}$$

Here, $\mathbf{X} = [x_1, x_2, \ldots, x_M]$ denotes the payoff vector for $X \in \mathcal{L}(\Omega)$, i.e., $X(\omega^j) = x_j$ for all $j = 1, 2, \ldots, M$. Let $\mathbf{X}^*$ be a solution to (5.22)–(5.23). Solving the matrix-vector equation $\Pi_T^{\varphi} = \mathbf{X}^*$ gives the optimal portfolio φ^*. Under the completeness assumption, the solution φ^* exists and is unique.

Example 5.8. Consider a single-period model with two scenarios $\{\omega^+, \omega^-\}$ and two base assets, namely, a risky stock with prices $S_T(\omega^+) = 12$, $S_T(\omega^-) = 8$, $S_0 = 10$, and an at-the-money call option on the stock with initial price $C_0 = 1$. Suppose that the real-world probabilities are $p = \mathbb{P}(\omega^+) = \frac{1}{4}$ and $1-p = \mathbb{P}(\omega^-) = \frac{3}{4}$. Find the optimal portfolio (φ_1, φ_2) with initial value $W_0 = 100$ that maximizes $\mathrm{E}\left[\sqrt{\Pi_T[(\varphi_1, \varphi_2)]}\right]$.

Solution. The strike price of the call option is $K = S_0$; the payoff is $C_T(\Omega) = (S_T(\Omega) - S_0)_+ = [2, 0]$. Hence, the initial price vector and payoff matrix are, respectively, given by

$$\mathbf{S}_0 = \begin{bmatrix} 10 \\ 1 \end{bmatrix}, \quad \mathbf{D} = \begin{bmatrix} 12 & 8 \\ 2 & 0 \end{bmatrix}.$$

Solving the matrix-vector equation, $\mathbf{D}\,\boldsymbol{\Psi} = \mathbf{S}_0$ for $\boldsymbol{\Psi} = [\Psi_1, \Psi_2]^\top$, gives the state prices $\Psi_1 = \Psi_2 = \frac{1}{2}$.

First, we find the payoff vector $\mathbf{X} = [x_1, x_2]$ that solves the optimization problem

$$\mathrm{E}[\sqrt{X}] = \frac{1}{4}\sqrt{x_1} + \frac{3}{4}\sqrt{x_2} \to \max_{x_1, x_2}$$

$$\text{subject to } x_1\Psi_1 + x_2\Psi_2 = \frac{x_1 + x_2}{2} = 100.$$

The optimal value of x_1 is a point of maximum of the function

$$f(x) = \frac{1}{4}\sqrt{x} + \frac{3}{4}\sqrt{200 - x}.$$

Equating the derivative $f'(x) = \frac{1}{8\sqrt{x}} - \frac{3}{8\sqrt{200-x}}$ to zero and solving the equation obtained gives the solution $x = 20$. Since the second derivative $f''(x)$ is strictly negative for all $x \in (0, 200)$, the function f attains its maximum at $x = 20$. Thus, the terminal payoff of the optimal portfolio is given by

$$x_1 = 20 \text{ and } x_2 = 180.$$

The Profit&Loss realized is $X - W_0$, which is equal to -80 in state ω^+ and 80 in state ω^-.

Second, find portfolio $[\varphi_1, \varphi_2]$ replicating the payoff $\mathbf{X} = [20, 180]$. Solve the system of replication equations:

$$\begin{cases} 12\varphi_1 + 2\varphi_2 = 20 \\ 8\varphi_1 + 0\varphi_2 = 180 \end{cases} \implies \begin{cases} \varphi_1 = \frac{45}{2} \\ \varphi_2 = -125 \end{cases}$$

So, the optimal allocation portfolio is $[\varphi_1, \varphi_2] = [\frac{45}{2}, -125]$. This means that we should sell 125 call contracts and purchase 22.5 units of stock. □

To find a general solution to the optimization problem (5.22)–(5.23), we apply the method of Lagrange multipliers. The Lagrangian is

$$L(\mathbf{X}, \lambda) = \sum_{j=1}^{M} u(x_j)\, p_j - \lambda \left(\sum_{j=1}^{M} x_j \Psi_j - W_0 \right).$$

Differentiating L w.r.t. $x_1, x_2, \ldots, x_M$, and λ, and equating the derivatives obtained to zero gives the following simultaneous equations:

$$\frac{\partial L}{\partial x_j} = u'(x_j)\, p_j - \lambda \Psi_j = 0, \quad j = 1, 2, \ldots, M, \tag{5.24}$$

$$\frac{\partial L}{\partial \lambda} = W_0 - \sum_{j=1}^{M} x_j \Psi_j = 0. \tag{5.25}$$

Suppose that the derivative of the utility function, u', is strictly monotone, i.e., it is a strictly increasing function everywhere it is finite. Additionally, we assume that the range of possible values of $u'(x)$ includes all positive reals. Examples of such utility functions include $\ln x$ and x^γ with $\gamma \in (0, 1)$. Under the above assumptions, the equation $u'(x) = y$ has a unique solution for every $y \in (0, \infty)$. Thus, we can define the inverse function v for u' with the property that $u'(v(y)) = y$ for all $y \in (0, \infty)$.

Now, solving the equations in (5.24) individually gives

$$x_j = v\left(\frac{\lambda\, \Psi_j}{p_j}\right), \quad j = 1, 2, \ldots, M. \tag{5.26}$$

So, we have a formula for the optimal payoff in terms of the multiplier λ. Substituting (5.26) into (5.25) gives

$$\sum_{j=1}^{M} v\left(\frac{\lambda\, \Psi_j}{p_j}\right) \Psi_j = W_0. \tag{5.27}$$

Solving (5.27) for λ and substituting the solution λ^* into (5.26) gives the optimal payoff vector

$$x_j^* = v\left(\frac{\lambda^*\, \Psi_j}{p_j}\right), \quad j = 1, 2, \ldots, M. \tag{5.28}$$

Let us show that the solution X^* with values in (5.28) maximizes $\mathrm{E}[u(X)]$. That is, let us show that for every payoff $X \in \mathcal{L}(\Omega)$ chosen such that $\pi_0(X) = \mathbf{X}\, \mathbf{\Psi} = W_0$ we have

$$\mathrm{E}[u(X)] \leqslant \mathrm{E}[u(X^*)].$$

Consider the function $f(x) := u(x) - yx$ with a positive parameter y. Clearly, $x = v(y)$

maximizes f. Indeed, solving $f'(x) = u'(x) - y = 0$ gives $x = v(y)$; since $f''(x) = u''(x) \leqslant 0$ for all x, $x = v(y)$ is a point of maximum. Therefore, we have

$$u(x) - yx \leqslant u(v(y)) - yv(y) \quad \text{for all} \quad x. \tag{5.29}$$

Replacing x and y in (5.29) by x_j and $\frac{\lambda^* \Psi_j}{p_j}$, respectively, multiplying both parts of the inequality by p_j, and then using (5.28) gives

$$u(x_j)p_j - \lambda^* \Psi_j x_j \leqslant u\left(x_j^*\right) p_j - \lambda^* \Psi_j x_j^* \text{ for all } j = 1, 2, \ldots, M.$$

Adding all the above M inequalities up gives

$$\sum_{j=1}^{M} u(x_j)p_j - \lambda^* \sum_{j=1}^{M} \Psi_j x_j \leqslant \sum_{j=1}^{M} u\left(x_j^*\right) p_j - \lambda^* \sum_{j=1}^{M} \Psi_j x_j^*.$$

This is equivalent to

$$\mathrm{E}[u(X)] - \lambda \pi_0(X) \leqslant \mathrm{E}[u(X^*)] - \lambda \pi_0(X^*).$$

Since the payoffs X and X^* have the same initial value equal to W_0, we obtain that $\mathrm{E}[u(X)] \leqslant \mathrm{E}[u(X^*)]$. That is, X^* maximizes $\mathrm{E}[u(X)]$ and hence it is the payoff of the optimal portfolio. Finally, note that the difference $X^*(\omega^j) - W_0$ is the actual gain (if positive) or loss (if negative) realized in state ω^j. Under the no-arbitrage condition, the gain $X - W_0$ (if it is not identically zero) has to change its sign: to be negative in at least one state and positive in another state.

Example 5.9. Consider a 3-by-3 model with

$$\mathbf{D} = \begin{bmatrix} 1 & 1 & 1 \\ 1 & 6 & 15 \\ 12 & 8 & 6 \end{bmatrix} \quad \text{and} \quad \mathbf{S}_0 = \begin{bmatrix} 1 \\ 5 \\ 10 \end{bmatrix}.$$

Suppose that the real-world-probabilities are $p_1 = p_3 = \frac{1}{4}$ and $p_2 = \frac{1}{2}$. Find the optimal portfolio φ that maximizes $\mathrm{E}[\ln(\Pi_T[\varphi])]$ and has initial value $W_0 = 1$. Find the gain/loss realized.

Solution. By solving $\mathbf{D}\,\mathbf{\Psi} = \mathbf{S}_0$, we obtain the state-price vector $\mathbf{\Psi} = \left[\frac{8}{13}, \frac{2}{13}, \frac{3}{13}\right]^\top$. It is strictly positive and unique, hence the model is arbitrage-free and complete. The utility function $u(x) = \ln x$ satisfies the criteria stated above since the derivative $u'(x) = \frac{1}{x}$ varies from 0 to ∞ as $x \nearrow \infty$ and $x \searrow 0$, respectively. Substituting the inverse of the derivative, $v(y) = \frac{1}{y}$, into (5.28) gives the optimal payoff $\mathbf{X}^*$ in terms of λ^*:

$$x_j = \frac{p_j}{\lambda^* \Psi_j},\ j = 1, 2, 3 \implies \mathbf{X}^* = \frac{1}{\lambda^*} \left[\frac{13}{32}, \frac{13}{4}, \frac{13}{12}\right].$$

Solving (5.27) with $W_0 = 1$ for λ gives

$$1 = \sum_{j=1}^{3} \frac{p_j}{\lambda \Psi_j} \Psi_j = \frac{1}{\lambda} \sum_{j=1}^{3} p_j = \frac{1}{\lambda} \implies \lambda^* = 1.$$

Therefore, the optimal payoff vector is $\mathbf{X}^* = \left[\frac{13}{32}, \frac{13}{4}, \frac{13}{12}\right]$. Solving the matrix-vector equation $\varphi \mathbf{D} = \mathbf{X}^*$ gives the optimal allocation portfolio:

$$\varphi^* = \left[\frac{853}{48}, -\frac{53}{96}, -\frac{269}{192}\right].$$

The gain of the investment is $\mathbf{X}^* - W_0 = \left[-\frac{19}{32}, \frac{9}{4}, \frac{1}{12}\right]$. So the return on the investment is only positive in states ω^2 and ω^3. □

5.4 Pricing in an Incomplete Market

5.4.1 A Trinomial Model of an Incomplete Market

The simplest single-period incomplete market model is a model with three states of the world, i.e., $\Omega = \{\omega^1, \omega^2, \omega^3\} \equiv \{\omega^+, \omega^0, \omega^-\}$, and two assets: B and S. The risk-free asset B provides the rate of return r and it pays \$1 at maturity:

$$B_0 = \frac{1}{1+r}, \quad B_T = 1.$$

The risky asset S admits three possible cash flows at time T:

$$S_T(\omega) = \begin{cases} S_0 u & \text{if } \omega = \omega^1, \\ S_0 m & \text{if } \omega = \omega^2, \\ S_0 d & \text{if } \omega = \omega^3, \end{cases}$$

where $S_0 > 0$ is the initial price and $0 < d < m < u$ are price factors. The initial price vector and the cash-flow matrix are, respectively, given by

$$\mathbf{S}_0 = \begin{bmatrix} (1+r)^{-1} \\ S_0 \end{bmatrix} \quad \text{and} \quad \mathbf{D} = \begin{bmatrix} B_T(\omega^1) & B_T(\omega^2) & B_T(\omega^3) \\ S_T(\omega^1) & S_T(\omega^2) & S_T(\omega^3) \end{bmatrix} = \begin{bmatrix} 1 & 1 & 1 \\ S_0 u & S_0 m & S_0 d \end{bmatrix}.$$

$\mathbf{D}$ has rank 2 since $\begin{vmatrix} 1 & 1 \\ S_0 u & S_0 m \end{vmatrix} = S_0(m-u) \neq 0$. The two payoff vectors [1, 1, 1] and $[S_0 u,\ S_0 m,\ S_0 d]$ are clearly independent. Therefore, we conclude that there are no redundant base assets but the market is incomplete as the two vectors do not span $\mathbb{R}^3$. We wish to find all strictly positive solutions $\boldsymbol{\Psi}$ to the equation $\mathbf{D}\,\boldsymbol{\Psi} = \mathbf{S}_0$. This is a linear system of two equations and three unknowns. Generally, the solution represents a line in $\mathbb{R}^3$ corresponding to the intersection of two planes:

$$\begin{cases} \Psi_1 + \Psi_2 + \Psi_3 = (1+r)^{-1} \\ u\Psi_1 + m\Psi_2 + d\Psi_3 = 1 \end{cases} \iff \begin{cases} \Psi_1 = \dfrac{(1+r)-d}{(1+r)(u-d)} - \dfrac{m-d}{u-d}c \\ \Psi_2 = c \\ \Psi_3 = \dfrac{u-(1+r)}{(1+r)(u-d)} - \dfrac{u-m}{u-d}c \end{cases} \tag{5.30}$$

We set $c > 0$, giving $\Psi_2 > 0$. The limiting value of $\boldsymbol{\Psi}$ as $c \searrow 0$ is

$$\Psi_2 \searrow 0, \quad \Psi_1 \nearrow \frac{(1+r)-d}{(1+r)(u-d)}, \quad \text{and} \quad \Psi_3 \nearrow \frac{u-(1+r)}{(1+r)(u-d)}.$$

The limiting values of Ψ_1 and Ψ_3, as $c \searrow 0$, are strictly positive iff $d < 1+r < u$ holds. Therefore, there exists a strictly positive solution $\boldsymbol{\Psi} = \left[\Psi_1,\ \Psi_2,\ \Psi_3\right]^\top$ iff $d < 1+r < u$. Indeed, if $1+r \leqslant d$, then $\Psi_1 < 0$ for any $\Psi_2 = c > 0$; if $u \leqslant 1+r$, then $\Psi_3 < 0$ for any $\Psi_2 = c > 0$. From (5.30), we obtain all values of c such that $\Psi_j > 0$, $j = 1, 2, 3$:

$$\Psi_1 > 0 \iff c < \frac{(1+r)-d}{(1+r)(m-d)},$$

$$\Psi_2 > 0 \iff c > 0,$$

$$\Psi_3 > 0 \iff c < \frac{u-(1+r)}{(1+r)(u-m)}.$$

Thus, the solution $\boldsymbol{\Psi} = \boldsymbol{\Psi}(c)$ from (5.30) is strictly positive iff

$$c \in (0, c_{\max}), \quad \text{where } c_{\max} := \frac{1}{1+r} \min\left\{\frac{(1+r)-d}{m-d}, \frac{u-(1+r)}{u-m}\right\}.$$

Such a set of positive solutions, denoted by $\{\boldsymbol{\Psi}\}$, is an open segment of a line in $\mathbb{R}^3$. Indeed, according to (5.30), $\boldsymbol{\Psi}(c)$ is a linear function of c. Hence all solutions lie on a straight line. Since $0 \leqslant \Psi_j \leqslant (1+r)^{-1}$ for $j = 1, 2, 3$, the set $\{\boldsymbol{\Psi}\}$ is bounded. The coordinates of the two endpoints of $\{\boldsymbol{\Psi}\}$, denoted by $\boldsymbol{\Psi}(0)$ and $\boldsymbol{\Psi}(c_{\max})$ are obtained by setting $c = 0$ and $c = c_{\max}$ in (5.30), respectively:

$$\boldsymbol{\Psi}(0) = \left[\frac{(1+r)-d}{(1+r)(u-d)}, 0, \frac{u-(1+r)}{(1+r)(u-d)}\right]^\top, \tag{5.31}$$

$$\boldsymbol{\Psi}(c_{\max}) = \begin{cases} \left[\frac{(1+r)-m}{(1+r)(u-m)}, \frac{u-(1+r)}{(1+r)(u-m)}, 0\right]^\top, & \text{if } m \leqslant 1+r, \\ \left[0, \frac{(1+r)-d}{(1+r)(m-d)}, \frac{m-(1+r)}{(1+r)(m-d)}\right]^\top, & \text{if } m \geqslant 1+r. \end{cases} \tag{5.32}$$

All the solutions of $\{\boldsymbol{\Psi}\}$ can be parametrized as $\boldsymbol{\Psi}(v\, c_{\max}) = (1-v)\boldsymbol{\Psi}(0) + v\boldsymbol{\Psi}(c_{\max})$, where $v = {}^{c}/_{c_{\max}} \in (0, 1)$.

The binomial model (with two states) is recovered as a limiting case of the trinomial model as $\Psi_2 \to 0$. By setting $c = 0$ in (5.30), we obtain

$$\Psi_1 = \frac{(1+r)-d}{(1+r)(u-d)}, \quad \Psi_2 = 0, \quad \Psi_3 = \frac{u-(1+r)}{(1+r)(u-d)}.$$

The formulae of the state prices Ψ_1 and Ψ_3 are exactly the same as those of the binomial state prices in (5.19). Two other extreme cases are obtained by setting either Ψ_1 or Ψ_3 to zero:

- Let $\Psi_1 \searrow 0$. Then $\Psi_3 = \frac{1}{1+r}\frac{m-(1+r)}{m-d}$ and $\Psi_2 = \frac{1}{1+r}\frac{(1+r)-d}{m-d}$. If $d < 1+r < m$, then both Ψ_2 and Ψ_3 are positive.
- Let $\Psi_3 \searrow 0$. Then $\Psi_1 = \frac{1}{1+r}\frac{(1+r)-m}{u-m}$ and $\Psi_2 = \frac{1}{1+r}\frac{u-(1+r)}{u-m}$. If $m < 1+r < u$, then both Ψ_1 and Ψ_2 are positive.

Alternatively, one can recover the binomial model in the limiting case as $m \searrow d$ or $m \nearrow u$.

For every positive state-price vector $\boldsymbol{\Psi} \in \{\boldsymbol{\Psi}\}$ we can identify a respective risk-neutral probability measure with the probabilities $\tilde{p}_j = (1+r)\Psi_j$, $j = 1, 2, 3$. Therefore, any incomplete arbitrage-free model has infinitely many risk-neutral probability measures. For the trinomial model described above, the collection of risk-neutral probability measures (i.e., the probability mass functions) can be denoted as vectors $\tilde{\mathbf{p}} = [\tilde{p}_1, \tilde{p}_2, \tilde{p}_3]$, forming an open linear segment in $\mathbb{R}^3$, whose endpoints are

$$\tilde{\mathbf{p}}(0) = (1+r)\boldsymbol{\Psi}(0) \quad \text{and} \quad \tilde{\mathbf{p}}(c_{\max}) = (1+r)\boldsymbol{\Psi}(c_{\max}). \tag{5.33}$$

Hence, we have a collection of risk-neutral measures and risk-neutral probability vectors that we denote by $\{\widetilde{\mathbb{P}}\}$ and $\{\tilde{\mathbf{p}}\}$, respectively.

Example 5.10. Consider a single-period trinomial model with two assets B and S where $S_0 = \$100$, $d = 0.8$, $m = 1.1$, $u = 1.4$, $r = 0.2$. Show that the model is arbitrage-free. Find all strictly positive state-price vectors and risk-neutral probability vectors.

Solution. The initial price vector and payoff matrix are, respectively,

$$\mathbf{S}_0 = \begin{bmatrix} \frac{5}{6} \\ 100 \end{bmatrix} \quad \text{and} \quad \mathbf{D} = \begin{bmatrix} 1 & 1 & 1 \\ 140 & 110 & 80 \end{bmatrix}.$$

Let us find the general solution to the state-price equation $\mathbf{D}\,\boldsymbol{\Psi} = \mathbf{S}_0$:

$$\left[\begin{array}{ccc|c} 1 & 1 & 1 & \frac{5}{6} \\ 140 & 110 & 80 & 100 \end{array}\right] \sim \left[\begin{array}{ccc|c} 1 & \frac{1}{2} & 0 & \frac{5}{9} \\ 0 & \frac{1}{2} & 1 & \frac{5}{18} \end{array}\right] \implies \begin{cases} \Psi_1 = \frac{5}{9} - \frac{c}{2}, \\ \Psi_2 = c, \\ \Psi_3 = \frac{5}{18} - \frac{c}{2}. \end{cases}$$

The state-price vector

$$\boldsymbol{\Psi}(c) = \left[\frac{5}{9} - \frac{c}{2},\, c,\, \frac{5}{18} - \frac{c}{2}\right]^\top \tag{5.34}$$

is strictly positive iff $0 < c < \frac{5}{9}$. Therefore, there is no arbitrage. Alternatively, we can notice the condition $d < 1 + r < u$ is satisfied and hence the model is arbitrage-free.

Now, we find the risk-neutral probabilities. Applying (5.31)–(5.32) and (5.33) gives the following endpoints of the collection of risk-neutral probability vectors $\{\tilde{\mathbf{p}}\}$:

$$\tilde{\mathbf{p}}(0) = \left[\frac{(1+r)-d}{u-d},\, 0,\, \frac{u-(1+r)}{u-d}\right]^\top = \left[\frac{1.2-0.8}{1.4-0.8},\, 0,\, \frac{1.4-1.2}{1.4-0.8}\right]^\top = \left[\frac{2}{3},\, 0,\, \frac{1}{3}\right]^\top,$$

$$\tilde{\mathbf{p}}\left(\tfrac{5}{9}\right) = \left[\frac{(1+r)-m}{u-m},\, \frac{u-(1+r)}{u-m},\, 0\right]^\top = \left[\frac{1.2-1.1}{1.4-1.1},\, \frac{1.4-1.2}{1.4-1.1},\, 0\right]^\top = \left[\frac{1}{3},\, \frac{2}{3},\, 0\right]^\top.$$

The collection $\{\tilde{\mathbf{p}}\}$ of risk-neutral probability vectors can be parametrized by a single variable as follows:

$$\tilde{\mathbf{p}}\left(\tfrac{5v}{9}\right) = (1-v)\tilde{\mathbf{p}}(0) + v\tilde{\mathbf{p}}\left(\tfrac{5}{9}\right) = \left[\frac{2-v}{3},\, \frac{2v}{3},\, \frac{1-v}{3}\right]^\top, \quad v \in (0,1). \tag{5.35}$$

Alternatively, the solution (5.35) can be obtained by normalizing the state-price vector in (5.34) and applying the change of variables $c = \frac{5}{9}v$:

$$\frac{\boldsymbol{\Psi}(c)}{\sum_{j=1}^{3} \Psi_j(c)} = \left[\frac{2}{3} - \frac{3c}{5},\, \frac{6c}{5},\, \frac{1}{3} - \frac{3c}{5}\right]^\top = \left[\frac{2-v}{3},\, \frac{2v}{3},\, \frac{1-v}{3}\right]^\top, \quad v \in (0,1). \qquad \square$$

5.4.2 Pricing Nonattainable Payoffs: The Bid-Ask Spread

Assume that the Law of One Price holds. Then the no-arbitrage initial price of any financial claim with an attainable payoff is unique and is given by the initial value of any portfolio in the base assets that replicates the payoff. We can use (5.13) to price an attainable payoff X in the trinomial model. Alternatively, the pricing formula in (5.16) is used with the risk-neutral probabilities $\tilde{p}_j = (1+r)\Psi_j$, $j = 1, 2, 3$. Any choice of state-price vector $\boldsymbol{\Psi}$, or the corresponding risk-neutral probability measure $\widetilde{\mathbb{P}}$, gives us the same value for the initial price $\pi_0(X)$.

Example 5.11. Consider the trinomial model from Example 5.10. Find the no-arbitrage initial price of a long forward contract F with the delivery (strike) price $K = \$100$.

Solution. The payoff of a long forward contract is attainable as it is replicated by a portfolio consisting of one long position in the stock and a loan in the amount of K:

$$F_T := S_T - K = S_T - K\,B_T.$$

The initial price F_0 of the forward contract is unique and equal to

$$F_0 = S_0 - K\,B_0 = S_0 - \frac{K}{1+r} = 100 - \frac{100}{1.2} = \frac{50}{3} \cong \$16.67.$$

On the other hand, we know that the risk-neutral pricing formula (5.16) must give us the same initial value of the contract regardless of the choice of the risk-neutral probability measure $\widetilde{\mathbb{P}}(c)$ with $c \in (0, \frac{5}{9})$. Using the probabilities in (5.35) gives

$$\begin{aligned}
F_0 = \frac{1}{1+r}\mathrm{E}^{\widetilde{\mathbb{P}}(c)}[F_T] &= \frac{1}{1+r}\sum_{j=1}^{3}\left(S_T(\omega^j) - K\right)\tilde{p}_j(c) \\
&= \frac{1}{1.2}\left((80-100)\cdot\frac{1-v}{3} + (110-100)\cdot\frac{2v}{3} + (140-100)\cdot\frac{2-v}{3}\right) \\
&= \frac{-20(1-v)+20v+40(2-v)}{3.6} = \frac{-20+20v+20v+80-40v}{3.6} \\
&= \frac{60}{3.6} = \frac{50}{3} \cong \$16.67,
\end{aligned}$$

where $v = \frac{9c}{5} \in (0,1)$. Alternatively, using the pricing formula (5.13) and solution (5.34) gives

$$\begin{aligned}
F_0 &= \sum_{j=1}^{3} F_T(\omega^j)\,\Psi_j(c) = (-20)\cdot\left(\frac{5}{18} - \frac{c}{2}\right) + 10\cdot c + 40\cdot\left(\frac{5}{9} - \frac{c}{2}\right) \\
&= (10+10-20)c + \frac{200-50}{9} = \frac{50}{3}.
\end{aligned}$$

□

To price a nonattainable payoff, we can formally apply (5.13), or equivalently the asset pricing formula in (5.16). However, the no-arbitrage initial price of $X \notin \mathcal{A}(\Omega)$ is now not unique since there are infinitely many no-arbitrage state-price vectors $\boldsymbol{\Psi}$, or equivalently infinitely many risk-neutral measures. There are several approaches to deal with incomplete market models. One approach is to complete the market by including extra tradeable securities into the market model for which the initial value is known. It should be evident that the additional security should not be a redundant one as this would not change the original space of attainable payoffs. For example, we can add in some other security such as an option on the underlying risky asset or stock with known market value. We now give an example of how this is accomplished.

Example 5.12. Consider the incomplete trinomial model from Example 5.10.

(a) Show that the payoff of the European call option with strike $K = \$100$ is nonattainable.

(b) Assume that the initial price of the call option from (a) is \$20. Add this option in the trinomial model as a third base asset. Show that the new model with three base assets is complete and arbitrage-free.

Solution.
(a) First, calculate the values of the European call payoff $X = (S_T - K)^+$:

$$\begin{aligned} X(\omega^1) &= (S_T(\omega^1) - K)^+ = (140 - 100)^+ = 40, \\ X(\omega^2) &= (S_T(\omega^2) - K)^+ = (110 - 100)^+ = 10, \\ X(\omega^3) &= (S_T(\omega^3) - K)^+ = (80 - 100)^+ = 0. \end{aligned}$$

Second, we try to replicate X by a portfolio in B and S. That is, find (if possible) a portfolio vector $\varphi \in \mathbb{R}^2$ such that

$$\varphi \mathbf{D} = \mathbf{X} \iff \begin{cases} \varphi_1 + 140\varphi_2 = 40 \\ \varphi_1 + 110\varphi_2 = 10 \\ \varphi_1 + 80\varphi_2 = 0 \end{cases} \tag{5.36}$$

The rank of the augmented coefficient matrix of the system in (5.36) is 3 and hence is not equal to rank$(\mathbf{D}) = 2$ since

$$\begin{vmatrix} 1 & 140 & 40 \\ 1 & 110 & 10 \\ 1 & 80 & 0 \end{vmatrix} = -600 \neq 0.$$

Therefore, the system of linear equations in (5.36) is inconsistent, i.e., it does not have a solution.
(b) Now assume that the European call is the third base asset. The initial price vector and payoff matrix of the new 3-by-3 model are, respectively,

$$\widehat{\mathbf{S}}_0 = \begin{bmatrix} \frac{5}{6} \\ 100 \\ 20 \end{bmatrix} \quad \text{and} \quad \widehat{\mathbf{D}} \equiv \widehat{\mathbf{S}}_T(\Omega) = \begin{bmatrix} 1 & 1 & 1 \\ 140 & 110 & 80 \\ 40 & 10 & 0 \end{bmatrix}.$$

The payoff matrix $\widehat{\mathbf{D}}$ has a full rank so the 3-by-3 model does not have redundant assets and is complete. To prove the absence of arbitrage, we need to show that the solution $\mathbf{\Psi}$ to the linear system $\widehat{\mathbf{D}}\,\mathbf{\Psi} = \widehat{\mathbf{S}}_0$ is strictly positive. Find $\mathbf{\Psi}$ as follows:

$$\begin{cases} \Psi_1 + \Psi_2 + \Psi_3 = \frac{5}{6} \\ 140\Psi_1 + 110\Psi_2 + 80\Psi_3 = 100 \\ 40\Psi_1 + 10\Psi_2 + 0\Psi_3 = 20 \end{cases} \iff \begin{cases} \Psi_1 = \frac{4}{9} \\ \Psi_2 = \frac{2}{9} \\ \Psi_3 = \frac{1}{6} \end{cases}$$

As we can see, the state-price vector $\mathbf{\Psi} = \left[\frac{4}{9}, \frac{2}{9}, \frac{1}{6}\right]^\top$ is strictly positive. Therefore, the 3-by-3 model is arbitrage-free. The risk-neutral probabilities are

$$\tilde{p}_1 = (1+r)\Psi_1 = \frac{8}{15}, \quad \tilde{p}_2 = (1+r)\Psi_2 = \frac{4}{15}, \quad \tilde{p}_3 = (1+r)\Psi_3 = \frac{1}{5}. \qquad \square$$

Another approach to pricing a security with an unattainable payoff X is to find two attainable payoffs X^d and X^u so that $X^d(\omega) \leqslant X(\omega) \leqslant X^u(\omega)$ for all $\omega \in \Omega$. Then, the no-arbitrage price $\pi_0(X)$ is bounded:

$$\pi_0(X^d) < \pi_0(X) < \pi_0(X^u).$$

Indeed, let the portfolios φ^d and φ^u replicate X^d and X^u, respectively. Suppose that $\pi_0(X^d) \geqslant \pi_0(X)$, then there exists the following arbitrage opportunity. At time $t = 0$

we buy the security and sell the portfolio φ^d. Our proceeds, which we keep in cash, are nonnegative:

$$\Pi_0[\varphi^d] - \pi_0(X) = \pi_0(X^d) - \pi_0(X) \geqslant 0.$$

At time $t = T$ the net position is nonnegative in all states and strictly positive in some states:

$$X - \Pi_T[\varphi^d] = X - X^d > 0.$$

Indeed, since $X^d \leqslant X$, $X^d \in \mathcal{A}(\Omega)$, and $X \notin \mathcal{A}(\Omega)$, we have that $X^d(\omega) < X(\omega)$ in at least one state ω. Hence, this is an arbitrage portfolio with a positive payoff equal to $(\pi_0(X^d) - \pi_0(X)) + X - X^d$. Similarly, there would be an arbitrage opportunity if the security with payoff X were found to be selling in the market at a price $\pi(X) \geqslant \pi_0(X^u)$.

A *super-replicating portfolio* φ with the terminal value

$$\Pi_T^{\varphi} \geqslant X \tag{5.37}$$

covers the liabilities of the writer of the security with payoff X. So the writer will choose φ with minimal initial cost Π_0^{φ} subject to the constraints (5.37). As a result, the writer obtains the following linear programming problem (the super-replication problem):

$$\Pi_0^{\varphi} \to \min_{\varphi \in \mathbb{R}^N} \text{ subject to } \Pi_T^{\varphi} \geqslant X. \tag{5.38}$$

Since the risk-free security is strictly positive, the linear programming problem is feasible. There always exists a risk-free portfolio that satisfies the constraints (5.37). The problem (5.38) is equivalent to the minimization of $\pi_0(Y)$ over the set of dominating attainable claims for X:

$$\pi_0(Y) \to \min_{Y \in \mathcal{D}_X}, \text{ where } \mathcal{D}_X := \{Y \in \mathcal{A}(\Omega) \,:\, X \leqslant Y\}. \tag{5.39}$$

The set $\mathcal{D}_X \neq \emptyset$ includes all payoffs that *dominate* the claim's payoff X.

The problem (5.38) or (5.39) can be solved graphically or numerically. Let us denote the optimal solution to (5.38) by φ^u. Then $\pi_0^u(X) := \Pi_0[\varphi^u]$ is an upper bound for a no-arbitrage initial price $\pi_0(X)$ for the derivative or claim with payoff X. If the writer were selling the security at a price higher than $\pi_0^u(X)$, then another agent could form an arbitrage portfolio.

Let us now look at the matter from the buyer's perspective. Let the portfolio $\varphi = \varphi^d$ solve the linear programming problem (the sub-replication problem):

$$\Pi_0^{\varphi} \to \max_{\varphi \in \mathbb{R}^N} \text{ subject to } \Pi_T^{\varphi} \leqslant X. \tag{5.40}$$

The problem (5.40) is equivalent to the maximization of $\pi_0(Y)$ over the set of attainable claims dominated by X:

$$\pi_0(Y) \to \max_{Y \in \mathcal{M}_X}, \text{ where } \mathcal{M}_X := \{Y \in \mathcal{A}(\Omega) \,:\, Y \leqslant X\}. \tag{5.41}$$

The set $\mathcal{M}_X \neq \emptyset$ includes all payoffs that are *dominated* by the claim's payoff X. Denote the initial value of the portfolio φ^d by $\pi_0^d(X)$. The buyer of the derivative security with payoff X will not agree to pay more than $\pi_0^d(X)$ since it will be more beneficial to purchase the portfolio in base assets rather than the derivative. As a result we obtain a lower bound for a no-arbitrage price of $\pi_0(X)$. In summary, the initial price $\pi_0(X)$ for the claim with payoff X must satisfy the inequality relation

$$\pi_0^d(X) < \pi_0(X) < \pi_0^u(X). \tag{5.42}$$

Otherwise, an arbitrage opportunity will arise. The interval $\left(\pi_0^d(X), \pi_0^u(X)\right)$ is called the *bid-ask spread* for the initial price of a derivative security with payoff X.

The ask price $\pi_0^u(X)$ and the bid price $\pi_0^d(X)$ can also be calculated by respectively taking the maximum and the minimum of the discounted expectation of the payoff function over the set of possible risk-neutral measures:

$$\pi_0^u(X) = \sup_{\widetilde{\mathbb{P}} \in \{\widetilde{\mathbb{P}}\}} \left\{ \frac{1}{1+r} \sum_{i=1}^{M} \tilde{p}_i X(\omega^i) \right\} = \sup_{\boldsymbol{\Psi} \in \{\boldsymbol{\Psi}\}} \mathbf{X}\,\boldsymbol{\Psi}, \tag{5.43}$$

$$\pi_0^d(X) = \inf_{\widetilde{\mathbb{P}} \in \{\widetilde{\mathbb{P}}\}} \left\{ \frac{1}{1+r} \sum_{i=1}^{M} \tilde{p}_i X(\omega^i) \right\} = \inf_{\boldsymbol{\Psi} \in \{\boldsymbol{\Psi}\}} \mathbf{X}\,\boldsymbol{\Psi}. \tag{5.44}$$

Here, $\{\boldsymbol{\Psi}\}$ and $\{\widetilde{\mathbb{P}}\}$ denote the collection of no-arbitrage state-price vectors and the corresponding collection of risk-neutral probabilities measures, respectively.

Consider the case of the trinomial model. State-price vectors form a finite segment in $\mathbb{R}^3$. Thus, the range of possible no-arbitrage initial prices of a nonattainable payoff is also a finite segment. Since the price $\pi_0 = \mathbf{X}\,\boldsymbol{\Psi}$ is a linear function of the vector $\boldsymbol{\Psi}$, these bounds are attained at the endpoints (5.31)–(5.32) of the state-price set $\{\boldsymbol{\Psi}\}$.

Example 5.13. Consider the trinomial (incomplete) model from Example 5.10. Find the bid-ask spread for the initial price of the call option with strike $K = \$100$.

Solution. The first approach is to solve the super- and sub-replication problems given by (5.38) and (5.40), respectively,

$$\begin{cases} \Pi_0[\boldsymbol{\varphi}^d] \to \max_{\boldsymbol{\varphi}^d \in \mathbb{R}^2} \\ \Pi_T[\boldsymbol{\varphi}^d] \leqslant X \end{cases} \iff \begin{cases} \frac{5}{6}\varphi_1^d + 100\varphi_2^d \to \max_{[\varphi_1^d, \varphi_2^d] \in \mathbb{R}^2} \\ \varphi_1^d + 80\varphi_2^d \leqslant 0, \\ \varphi_1^d + 110\varphi_2^d \leqslant 10, \\ \varphi_1^d + 140\varphi_2^d \leqslant 40 \end{cases} \tag{5.45}$$

$$\begin{cases} \Pi_0[\boldsymbol{\varphi}^u] \to \min_{\boldsymbol{\varphi}^u \in \mathbb{R}^2} \\ \Pi_T[\boldsymbol{\varphi}^u] \geqslant X \end{cases} \iff \begin{cases} \frac{5}{6}\varphi_1^u + 100\varphi_2^u \to \min_{[\varphi_1^u, \varphi_2^u] \in \mathbb{R}^2} \\ \varphi_1^u + 80\varphi_2^u \geqslant 0, \\ \varphi_1^u + 110\varphi_2^u \geqslant 10 \\ \varphi_1^u + 140\varphi_2^u \geqslant 40 \end{cases} \tag{5.46}$$

The solutions to the linear programming problems (5.45) and (5.46) are, respectively,

$$\boldsymbol{\varphi}^d = \left[-100,\, 1\right] \quad \text{and} \quad \boldsymbol{\varphi}^u = \left[-53\tfrac{1}{3},\, \tfrac{2}{3}\right].$$

The bid price π_0^d and ask price π_0^u of the call option are then given by

$$\pi_0^d = \Pi_0[\boldsymbol{\varphi}^d] = -100 \cdot \frac{5}{6} + 1 \cdot 100 = \frac{50}{3} \cong \$16.67,$$
$$\pi_0^u = \Pi_0[\boldsymbol{\varphi}^u] = -53\frac{1}{3} \cdot \frac{5}{6} + \frac{2}{3} \cdot 100 = \frac{200}{9} \cong \$22.22.$$

Thus, the bid-ask spread is $(16.67, 22.22)$.

The other approach is to find the bid-ask spread by computing the extreme values of the risk-neutral expectation of the discounted payoff. Since the mathematical expectation of a random variable from $\mathcal{L}(\Omega)$ is a linear function of the state probabilities, the maximum and minimum in (5.43) and (5.44) are attained at the endpoints of the domain $\{\widetilde{\mathbb{P}}\}$. Therefore, we just need to calculate the expected value

$$\pi_0(c) = \frac{1}{1+r} \mathrm{E}^{\widetilde{\mathbb{P}}(c)}[(S_T - K)^+] = \frac{1}{1+r} \sum_{j=1}^{3} \tilde{p}_j(c)(S_T(\omega^j) - K)^+$$

using risk-neutral probabilities given by (5.35) with extreme values $c = 0$ and $c = \frac{5}{9}$. The bid price is the smaller of the two prices calculated, and the ask price is the larger one. The above expectation is calculated explicitly in the measure $\widetilde{\mathbb{P}}(v)$:

$$\pi_0(\tfrac{5v}{9}) = \frac{5}{6} \cdot \left(0 \cdot \frac{1-v}{3} + 10 \cdot \frac{2v}{3} + 40 \cdot \frac{2-v}{3}\right) = \frac{50(4-v)}{9}, \quad 0 \leqslant v \leqslant 1.$$

The extreme values are $\pi_0(v = 0) = \frac{200}{9}$ and $\pi_0(v = 1) = \frac{50}{3} = \frac{150}{9}$. The bid price is $\min\{\frac{200}{9}, \frac{50}{3}\} = \frac{50}{3}$, and the ask price is $\max\{\frac{200}{9}, \frac{50}{3}\} = \frac{200}{9}$. As required, both prices agree with the values obtained using the first approach. □

In the conclusion of this section, let us consider an example of a single-period model with four states of the world and two tradable assets.

Example 5.14. Let the initial price vector and payoff matrix be given by

$$\mathbf{S}_0 = \begin{bmatrix} 10 \\ 10 \end{bmatrix} \quad \text{and} \quad \mathbf{D} = \begin{bmatrix} 11 & 11 & 11 & 11 \\ 6 & 8 & 12 & 14 \end{bmatrix}.$$

(a) Show that the market is arbitrage-free. Find the general solution for the state prices and illustrate it with a diagram.

(b) Find the bid-ask spread for the put option on the risky asset with strike price $K = 11$.

Solution. The general solution to the state-price equation $\mathbf{D}\,\mathbf{\Psi} = \mathbf{S}_0$ is a function of two free parameters:

$$\begin{cases} \Psi_1 = 2x + 3y - \frac{15}{11} \\ \Psi_2 = \frac{25}{11} - 3x - 4y \\ \Psi_3 = x \\ \Psi_4 = y \end{cases}$$

The solution is strictly positive if

$$0 < x < \frac{25}{33} \quad \text{and} \quad \max\left\{\frac{5}{11} - \frac{2}{3}x, 0\right\} < y < \frac{24}{44} - \frac{3}{4}x \tag{5.47}$$

holds. The domain of admissible solutions, $D = \{[x, y] \;:\; \mathbf{\Psi}(x, y) \gg 0\}$, can be illustrated by a diagram on the (x, y)-plane (Figure 5.3).

The no-arbitrage initial price π_0 of the put option with strike price $K = 11$ is given by a product of the payoff vector

$$\mathbf{P} = (K - S(\Omega))^+ = [5, 3, 0, 0]$$

and the state-price vector $\mathbf{\Psi}$:

$$\pi_0 = \mathbf{P}\,\mathbf{\Psi} = 5 \cdot \left(2x + 3y - \frac{15}{11}\right) + 3 \cdot \left(\frac{25}{11} - 3x - 4y\right) = x + 3y.$$

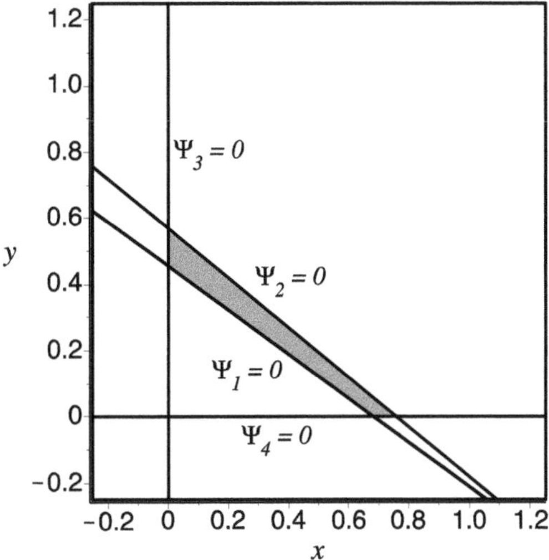

FIGURE 5.3: The domain of admissible solutions $\mathbf{\Psi}(x, y) \gg 0$.

The bid and ask prices are, respectively, the minimum and maximum values attained by the function $x + 3y$ in the domain D as is specified in (5.47). These values can be found by solving respective linear programming problems. Alternatively, we can use the fact that a linear function attains its extreme values at the boundary. The boundary of the domain D is piecewise linear. Hence, by applying the same principle one more time, we can conclude that the maximum and minimum values of $x + 3y$ in D are attained at the corner points

$$\mathbf{q}_1 = \left[0, \frac{15}{33}\right], \quad \mathbf{q}_2 = \left[0, \frac{25}{44}\right], \quad \mathbf{q}_3 = \left[\frac{15}{22}, 0\right], \quad \mathbf{q}_4 = \left[\frac{25}{33}, 0\right].$$

Calculate the value of π_0 at each of these points:

$$\pi_0(\mathbf{q}_1) = \frac{15}{11}, \quad \pi_0(\mathbf{q}_2) = \frac{75}{44}, \quad \pi_0(\mathbf{q}_3) = \frac{15}{22}, \quad \pi_0(\mathbf{q}_4) = \frac{25}{33}.$$

The smaller price is the bid price; the larger price is the ask price:

$$\min_{i=1,2,3,4} \pi_0(\mathbf{q}_i) = \frac{15}{22} \cong 0.68182, \qquad \max_{i=1,2,3,4} \pi_0(\mathbf{q}_i) = \frac{75}{55} \cong 1.70455.$$

Thus, the bid-ask spread for the put option is $(0.68182, 1.70455)$. □

5.5 Change of Numéraire

5.5.1 The Concept of a Numéraire Asset

Suppose that there exists a risk-free asset $\{B_t\}_{t \in \{0,T\}}$ with a fixed rate of return r. Note that for a risk-free bond we assume nonnegative interest rates so that $r \geqslant 0$. As was shown in the previous section, the market model is arbitrage-free iff there exists a probability function $\widetilde{\mathbb{P}}$ called a *risk-neutral probability measure* or *martingale probability measure* such that the discounted price processes $\left\{\frac{S_t^i}{B_t}\right\}_{t\in\{0,T\}}$ for $i = 1, 2, \ldots, N$ are martingales with respect to $\widetilde{\mathbb{P}}$.

As a process, the prices of base assets divided by the value of the risk-free account are martingales within a risk-neutral probability measure. There is nothing preventing us from also expressing prices of assets relative to another choice of available strictly positively valued security. The security may be any one of the base assets, or any strictly positively valued portfolio in the base assets, or even a derivative security.

Definition 5.8. A *numéraire*, or *numéraire asset*, or *accounting unit*, denoted $\{g_t\}_{t\in\{0,T\}}$, is any strictly positive asset price process, i.e., with payoff $g_T \gg 0$ and initial price $g_0 > 0$.

Hence, for any numéraire, we have $g_t(\omega) > 0$ for all time $t \geqslant 0$ and all outcomes $\omega \in \Omega$. The numéraire is used for measuring the relative worth of any asset or a portfolio of assets. The ratio process $\{\frac{S_t}{g_t}\}_{t\geqslant 0}$ is called the value process of asset S discounted by g or relative to g. The ratio $\frac{S_t}{g_t}$ gives the number of units of the numéraire g that can be exchanged for one unit of asset S at time t. For example, in the binomial model both the bond B and stock S can be chosen as a numéraire.

5.5.2 Change of Numéraire in a Binomial Model

Consider a single-period binomial model with two states of the world $\Omega = \{\omega^+, \omega^-\}$ and two base securities B and S such that

$$B_0 = (1+r)^{-1}, \quad B_T = 1, \quad S_T(\omega) = (d\,\mathbb{I}_{\{\omega^-\}}(\omega) + u\,\mathbb{I}_{\{\omega^+\}}(\omega))S_0,$$

where $0 < d < u$. The market is arbitrage-free and complete if $d < 1 + r < u$. Let $\widetilde{\mathbb{P}}$ be a risk-neutral probability measure with $\tilde{p} := \widetilde{\mathbb{P}}(\omega^+) = \frac{(1+r)-d}{u-d}$ and $1 - \tilde{p} = \widetilde{\mathbb{P}}(\omega^-) = \frac{u-(1+r)}{u-d}$. Under $\widetilde{\mathbb{P}}$, the base asset price processes discounted by the numéraire $g = B$ are martingales:

$$\widetilde{\mathrm{E}}\left[\frac{S_T}{B_T}\right] = \frac{S_0}{B_0}, \quad \widetilde{\mathrm{E}}\left[\frac{B_T}{B_T}\right] = 1 = \frac{B_0}{B_0}.$$

Moreover, any portfolio $[\Delta, \beta]$ with Δ shares in the stock S and β units of the bond B discounted by the value of asset B is a martingale:

$$\widetilde{\mathrm{E}}\left[\frac{\Delta S_T + \beta B_T}{B_T}\right] = \Delta\widetilde{\mathrm{E}}\left[\frac{S_T}{B_T}\right] + \beta\widetilde{\mathrm{E}}\left[\frac{B_T}{B_T}\right] = \Delta\frac{S_0}{B_0} + \beta = \frac{\Delta S_0 + \beta B_0}{B_0}.$$

In summary, there is no arbitrage in the binomial model iff there exists a martingale probability measure $\widetilde{\mathbb{P}}^{(B)}$ for the numéraire $g = B$. The no-arbitrage binomial model is complete, hence $\widetilde{\mathbb{P}}^{(B)}$ is unique. Is it possible to use the stock price as a numéraire? Is there a unique risk-neutral probability function $\widetilde{\mathbb{P}}^{(S)}$ for the numéraire $g = S$ under which the price processes for all base assets are martingales? Both questions are answered positively just below. There exists $\widetilde{\mathbb{P}}^{(B)}$ with bond B as numéraire iff there exists $\widetilde{\mathbb{P}}^{(S)}$ with stock S as numéraire. The value process of any attainable claim discounted by numéraire $g = B$ or $g = S$ is a martingale under $\widetilde{\mathbb{P}}^{(B)}$ or $\widetilde{\mathbb{P}}^{(S)}$, respectively. Thus, these probability measures are called *martingale measures*.

The martingale probability measures $\widetilde{\mathbb{P}}^{(B)}$ and $\widetilde{\mathbb{P}}^{(S)}$ are said to be *equivalent*. That is, any event E with zero probability under one measure has zero probability under the other equivalent measure, i.e., $\widetilde{\mathbb{P}}^{(B)}(E) = 0$ iff $\widetilde{\mathbb{P}}^{(S)}(E) = 0$ for all $E \subseteq \Omega$. So, two equivalent measures are consistent in their assignment of nonzero probability values to all events having nonzero probability, although the probability values assigned to the events will generally differ. Hence, both $\widetilde{\mathbb{P}}^{(B)}$ and $\widetilde{\mathbb{P}}^{(S)}$ are called *equivalent martingale measures.* The formal definition of equivalent martingale measures in the context of multiple base assets within the single-period setting is given in the next section. Note that the martingale measures $\widetilde{\mathbb{P}}^{(B)}$ and $\widetilde{\mathbb{P}}^{(S)}$ and the real-world measure $\mathbb{P}$ are equivalent to each other; however, $\mathbb{P}$ is not an equivalent martingale measure!

For the binomial case, we only need to show that there exists probability $\tilde{p} = \widetilde{\mathbb{P}}^{(B)}(\omega^+) \in (0,1)$ iff there exists probability $\tilde{q} = \widetilde{\mathbb{P}}^{(S)}(\omega^+) \in (0,1)$. Let us work out the martingale probability function $\widetilde{\mathbb{P}}^{(S)}$. In this case the numéraire asset price at time t is the stock price, $g_t \equiv S_t$. Hence, the stock price process discounted by g_t is a constant and hence is automatically a martingale, i.e., $\frac{S_t}{g_t} \equiv \frac{S_t}{S_t} \equiv 1$, $t \in \{0, T\}$. The bond price B_T discounted by S_T is given by

$$\frac{B_T(\omega)}{S_T(\omega)} = \frac{(1+r)B_0}{S_T(\omega)} = \begin{cases} \frac{1+r}{d}\frac{B_0}{S_0} & \text{if } \omega = \omega^-, \\ \frac{1+r}{u}\frac{B_0}{S_0} & \text{if } \omega = \omega^+. \end{cases}$$

So we have that $\widetilde{\mathrm{E}}^{(S)}\left[\frac{B_T}{S_T}\right] = \frac{B_0}{S_0}$ iff

$$\frac{1+r}{u}\frac{B_0}{S_0}\tilde{q} + \frac{1+r}{d}\frac{B_0}{S_0}(1-\tilde{q}) = \frac{B_0}{S_0} \iff \frac{1+r}{u}\tilde{q} + \frac{1+r}{d}(1-\tilde{q}) = 1.$$

Here $\widetilde{\mathrm{E}}^{(S)}[\,\cdot\,]$ denotes the mathematical expectation under the probability $\widetilde{\mathbb{P}}^{(S)}$. Solving the above equation for $\tilde{q}$:

$$\left(\frac{1+r}{u} - \frac{1+r}{d}\right)\tilde{q} = 1 - \frac{1+r}{d} \iff \frac{(1+r)(u-d)}{ud}\tilde{q} = \frac{(1+r)-d}{d}$$
$$\iff \tilde{q} = \frac{(1+r)-d}{u-d}\,\frac{u}{1+r}$$

The probability measure $\widetilde{\mathbb{P}}^{(S)}$ is defined by the probabilities $\tilde{q}$ and $1-\tilde{q}$, which can be expressed in terms of the probabilities for measure $\widetilde{\mathbb{P}}^{(B)}$, $\tilde{p} = \frac{(1+r)-d}{u-d}$ and $1-\tilde{p}$, as follows:

$$\tilde{q} = \frac{(1+r)-d}{u-d} \cdot \frac{u}{1+r} = \tilde{p} \cdot \frac{u}{1+r},$$
$$1-\tilde{q} = \frac{u-(1+r)}{u-d} \cdot \frac{d}{1+r} = (1-\tilde{p}) \cdot \frac{d}{1+r}.$$

Clearly, $\tilde{q} \in (0,1)$ iff $d < 1+r < u$. The latter condition is equivalent to $\tilde{p} \in (0,1)$. So assuming no-arbitrage, $\widetilde{\mathbb{P}}^{(B)}$ and $\widetilde{\mathbb{P}}^{(S)}$ are equivalent probability measures. Therefore, in the binomial model there exists an equivalent martingale measure (EMM) w.r.t. the stock, denoted by $\widetilde{\mathbb{P}}^{(S)}$, and an EMM w.r.t. the bond, denoted by $\widetilde{\mathbb{P}}^{(B)}$, iff the market admits no arbitrage. The probability measures are unique and hence the model is complete.

Every attainable asset price process that can be replicated by some portfolio value process Π_t^{φ}, $t \in \{0, T\}$, for some φ, discounted by the numéraire (the bond or stock) is then a martingale w.r.t. an equivalent martingale measure. Thus, the unique initial price of any (derivative) asset can be calculated by using any appropriate choice of equivalent martingale measure. The initial price $\pi_0(X)$ of an attainable payoff X is equivalently given by

$$\begin{aligned}
\pi_0(X) &= g_0\, \widetilde{\mathrm{E}}^{(g)}\left[\frac{X}{g_T}\right] \quad \text{(for any choice of numéraire } g\text{)} \\
&= B_0\, \widetilde{\mathrm{E}}^{(B)}\left[\frac{X}{B_T}\right] = \frac{1}{1+r}\left[X(\omega^+)\,\tilde{p} + X(\omega^-)\,(1-\tilde{p})\right] \quad (\text{for } g = B) \\
&= S_0\, \widetilde{\mathrm{E}}^{(S)}\left[\frac{X}{S_T}\right] = \frac{X(\omega^+)}{u}\,\tilde{q} + \frac{X(\omega^-)}{d}\,(1-\tilde{q}) \quad (\text{for } g = S).
\end{aligned}$$

5.5.3 Change of Numéraire in a Multinomial Model

Consider the general case of a single-period model with M states of the world and N base assets. Any attainable generic asset g with a strictly positive price process can be used as a numéraire. Since g is attainable, its value g_t at any time t corresponds to the value of a replicating portfolio for g:

$$g_t = \Pi_t[\boldsymbol{\theta}^{(g)}] = \sum_{i=1}^{N} \theta_i^{(g)}\, S_t^i, \quad t \in \{0, T\}.$$

Since a numéraire asset must have strictly positive value, the portfolio in the base assets, $\boldsymbol{\theta}^{(g)} \in \mathbb{R}^N$, is chosen such that $g_0 > 0$ and $g_T(\omega) > 0$ for all $\omega \in \Omega$. For the ith base asset to qualify as a potential numéraire asset, we need only require that it has a strictly positive payoff (i.e., the respective row in the payoff matrix is strictly positive) and a positive initial price. It is convenient to define base asset price processes discounted by the numéraire asset price g_t as

$$\overline{S}_t^i := \frac{S_t^i}{g_t}, \quad t \in \{0, T\}, \quad i = 1, 2, \ldots, N.$$

$\overline{S}_t^i$ is the number of units of the numéraire that can be purchased at time t for the same amount needed to buy one unit of the ith base asset. Note that the discounted price of the numéraire asset itself is equal to 1. For a given portfolio $\boldsymbol{\varphi} \in \mathbb{R}^N$ we can consider its *discounted value process relative to the numéraire* g:

$$\overline{\Pi}_t^{\varphi} := \frac{\Pi_t^{\varphi}}{g_t} = \sum_{i=1}^{N} \varphi_i\, \overline{S}_t^i.$$

Clearly, $\overline{\Pi}_t[\,\cdot\,]$ is a linear function, i.e., for any two portfolios $\boldsymbol{\varphi}, \boldsymbol{\psi} \in \mathbb{R}^N$ and constants $a, b \in \mathbb{R}$, we have

$$\overline{\Pi}_t[a\boldsymbol{\varphi} + b\boldsymbol{\psi}] = a\overline{\Pi}_t^{\varphi} + b\overline{\Pi}_t[\boldsymbol{\psi}].$$

Definition 5.9. An *equivalent martingale measure* (EMM) $\widetilde{\mathbb{P}} \equiv \widetilde{\mathbb{P}}^{(g)}$ for a given numéraire asset $\{g_t\}_{t\geqslant 0}$ is a probability measure defined on the state space Ω such that the following conditions are fulfilled.

- The probability measure $\widetilde{\mathbb{P}}$ is equivalent to the real-world probability measure $\mathbb{P}$, i.e., $\widetilde{\mathbb{P}}(E) = 0$ iff $\mathbb{P}(E) = 0$ for all $E \subseteq \Omega$. Note that for finite or countable Ω, two probability measures, $\mathbb{P}$ and $\widetilde{\mathbb{P}}$, are equivalent when $\widetilde{\mathbb{P}}(\omega) = 0$ iff $\mathbb{P}(\omega) = 0$ for all $\omega \in \Omega$. Since $\mathbb{P}(\omega) > 0$ for every outcome ω, the measures $\mathbb{P}$ and $\widetilde{\mathbb{P}}$ are equivalent if $\widetilde{\mathbb{P}}(\omega) > 0$ for all $\omega \in \Omega$.

- The discounted base asset price process $\left\{\overline{S}^i_t := \frac{S^i_t}{g_t}\right\}_{t\geqslant 0}$ is a martingale w.r.t. $\widetilde{\mathbb{P}}$ for every base asset S^i, $i = 1, 2, \ldots, N$, i.e., for a single period model we have that

$$\widetilde{\mathrm{E}}^{(g)}[\overline{S}^i_T] = \widetilde{\mathrm{E}}^{(g)}\left[\frac{S^i_T}{g_T}\right] = \frac{S^i_0}{g_0} = \overline{S}^i_0. \tag{5.48}$$

Here, $\widetilde{\mathrm{E}}^{(g)}[\,\cdot\,]$ denotes the expectation w.r.t. $\widetilde{\mathbb{P}}^{(g)}$.

Suppose that a risk-free asset is used as a numéraire asset. Let r denote the risk-free rate of return on a risk-free numéraire: $r = \frac{g_T - g_0}{g_0}$. Then, the martingale condition (5.48) takes the form:

$$\frac{1}{1+r}\widetilde{\mathrm{E}}[S^i_T] = S^i_0.$$

We should note here that the formal definition of a martingale process for a multi-period model is presented in Chapter 6, as it will later come into play when we discuss asset pricing in a multi-period setting in Chapter 7. Within a single-period trading model, the next result tells us that the existence of an EMM for a given choice of numéraire is equivalent to saying that the value of any portfolio in the base assets discounted by the numéraire asset value is a martingale w.r.t. the EMM.

Lemma 5.13. *Consider a single-period model with M states and N base assets. Let g be a numéraire asset and $\widetilde{\mathbb{P}}^{(g)}$ be a probability measure equivalent to $\mathbb{P}$. Then $\widetilde{\mathbb{P}}^{(g)}$ is an equivalent martingale measure iff for every portfolio $\varphi \in \mathbb{R}^N$ the discounted value process, $\{\overline{\Pi}_t \equiv \overline{\Pi}^{\varphi}_t\}_{t\in\{0,T\}}$, is a martingale w.r.t. $\widetilde{\mathbb{P}}^{(g)}$.*

Proof. Let $\widetilde{\mathbb{P}}^{(g)}$ be an equivalent martingale measure. Then,

$$\begin{aligned}\widetilde{\mathrm{E}}^{(g)}\left[\overline{\Pi}_T\right] &= \sum_{\omega\in\Omega}\overline{\Pi}_T(\omega)\widetilde{\mathbb{P}}^{(g)}(\omega) = \sum_{j=1}^{M}\sum_{i=1}^{N}\varphi_i\overline{S}^i_T(\omega^j)\widetilde{\mathbb{P}}^{(g)}(\omega^j)\\ &= \sum_{i=1}^{N}\varphi_i\sum_{j=1}^{M}\overline{S}^i_T(\omega^j)\widetilde{\mathbb{P}}^{(g)}(\omega^j) = \sum_{i=1}^{N}\varphi_i\widetilde{\mathrm{E}}^{(g)}[\overline{S}^i_T] = \sum_{i=1}^{N}\varphi_i\overline{S}^i_0 = \overline{\Pi}_0.\end{aligned}$$

The converse follows by assumption. That is, for every $i = 1, 2, \ldots, N$ consider a portfolio that only contains one unit of the ith base asset, i.e., $\overline{\Pi}_t = \overline{S}^i_t$. Then, the process $\left\{\overline{S}^i_t\right\}_{t\in\{0,T\}}$ is a martingale w.r.t. $\widetilde{\mathbb{P}}^{(g)}$. □

According to Lemma 5.13, the discounted value process of a financial claim with an attainable payoff X is a martingale w.r.t. $\widetilde{\mathbb{P}}^{(g)}$. Let $\varphi_X \in \mathbb{R}^N$ be a replicating portfolio such that $\Pi_T[\varphi_X] = X$. Then, the initial price of a claim with payoff X is calculated as follows:

$$\frac{\pi^{(g)}_0(X)}{g_0} = \frac{\Pi_0[\varphi_X]}{g_0} = \widetilde{\mathrm{E}}^{(g)}\left[\frac{\Pi_T[\varphi_X]}{g_T}\right] = \widetilde{\mathrm{E}}^{(g)}\left[\frac{X}{g_T}\right] \implies \pi^{(g)}_0(X) = g_0\widetilde{\mathrm{E}}^{(g)}\left[\frac{X}{g_T}\right]. \tag{5.49}$$

Since the initial value of a replicating portfolio is unique, the initial price $\pi^{(g)}_0(X)$ does not depend on the choice of g and $\widetilde{\mathbb{P}}^{(g)}$. If the payoff vector is positive, then the initial price is positive as well:

$$\mathbf{X} > 0 \implies \pi^{(g)}_0(X) > 0.$$

The functional $\pi^{(g)}_0$ is defined on the set of attainable payoffs $\mathcal{A}(\Omega)$ but it can be extended on $\mathcal{L}(\Omega)$ in the case of an incomplete market. For example, we may set the initial price of $X \in \mathcal{L}(\Omega) \setminus \mathcal{A}(\Omega)$ equal to the ask price:

$$\pi_0^{(g),u}(X) := \inf_{Y \in \mathcal{D}_X} \pi_0^{(g)}(Y), \text{ where } \mathcal{D}_X = \{Y \in \mathcal{A}(\Omega) : X \leqslant Y\}.$$

As was proved above, the no-arbitrage condition is equivalent to the existence of a (strictly positive) state-price vector $\mathbf{\Psi}$ that solves $\mathbf{D\Psi} = \mathbf{S}_0$. Consider a numéraire asset g and respective martingale measure $\widetilde{\mathbb{P}}^{(g)}$ with probabilities $\tilde{p}_j = \widetilde{\mathbb{P}}^{(g)}(\omega_j)$, $j = 1, 2, \ldots, M$. Let us find the relationship between $\widetilde{\mathbb{P}}^{(g)}$ and a state-price vector $\mathbf{\Psi}$. By the definition of an EMM, we have:

$$\sum_{j=1}^{M} \frac{S_T^i(\omega^j)}{g_T(\omega^j)} \tilde{p}_j = \frac{S_0^i}{g_0} \implies \sum_{j=1}^{M} \frac{g_0 \tilde{p}_j}{g_T(\omega^j)} S_T^i(\omega^j) = S_0^i, \; i = 1, 2, \ldots, N.$$

Let us denote $\Psi_j = \frac{g_0 \tilde{p}_j}{g_T(\omega^j)}$. Then, as before, we have

$$\sum_{j=1}^{M} S_T^i(\omega^j)\, \Psi_j = S_0^i, \; i = 1, 2, \ldots, N \iff \mathbf{D\,\Psi} = \mathbf{S}_0.$$

Therefore, the existence of an EMM implies the existence of a strictly positive state-price vector. Since a numéraire asset has strictly positive prices, we have that $\Psi_j > 0$ if $\tilde{p}_j > 0$. Conversely, let $\Psi_j > 0$, $j = 1, 2, \ldots, M$, be the state prices. Define $\tilde{p}_j = \frac{g_T(\omega^j)}{g_0} \Psi_j$, $j = 1, 2, \ldots, M$. All $\tilde{p}_j$ are positive and they sum up to one if g is an attainable asset, i.e., if $g_t = \Pi_t[\boldsymbol{\theta}^{(g)}]$ for some portfolio $\boldsymbol{\theta}^{(g)} \in \mathbb{R}^N$. Indeed,

$$\begin{aligned}
\sum_{j=1}^{M} \Psi_j g_T(\omega^j) &= \sum_{j=1}^{M} \Psi_j \Pi_T[\boldsymbol{\theta}^{(g)}](\omega^j) = \sum_{j=1}^{M} \Psi_j \sum_{i=1}^{N} \boldsymbol{\theta}_i^{(g)} S_T^i(\omega^j) \\
&= \sum_{i=1}^{N} \boldsymbol{\theta}_i^{(g)} \sum_{j=1}^{M} \Psi_j S_T^i(\omega^j) = \sum_{i=1}^{N} \boldsymbol{\theta}_i^{(g)} S_0^i \\
&= \Pi_0[\boldsymbol{\theta}^{(g)}] = g_0 \implies \sum_{j=1}^{M} \Psi_j \frac{g_T(\omega^j)}{g_0} = \sum_{j=1}^{M} \tilde{p}_j = 1.
\end{aligned}$$

Therefore, $\tilde{p}_j$, $j = 1, 2, \ldots, M$, are probabilities. As a result, we can restate the first and second fundamental theorems of asset pricing in terms of equivalent martingale measures as follows.

Theorem 5.14 (The first and second FTAPs—the 3rd version).

1. *There are no arbitrage opportunities iff there exists an equivalent martingale measure with respect to a given numéraire asset.*

2. *Assuming absence of arbitrage, there exists a unique equivalent martingale measure with respect to a given numéraire asset iff the market is complete.*

Example 5.15. Consider a 3-by-3 model from Example 5.12 with

$$\mathbf{S}_0 = \begin{bmatrix} \frac{5}{6} \\ 100 \\ 20 \end{bmatrix} \quad \text{and} \quad \mathbf{D} = \begin{bmatrix} 1 & 1 & 1 \\ 140 & 110 & 80 \\ 40 & 10 & 0 \end{bmatrix}.$$

Find the EMM $\widetilde{\mathbb{P}}^{(g)}$ using one of the base assets as numéraire.

Solution. To find the EMM $\widetilde{\mathbb{P}}^{(g)}$ we use the martingale condition (5.48) written for each base assets. As a result, we obtain a system of equations on the probabilities $\tilde{p}_j^{(i)} = \widetilde{\mathbb{P}}^{(S^i)}(\omega^j)$, $j = 1, 2, 3$, $i = 1, 2, 3$:

$$\begin{cases} \dfrac{S_T^1(\omega^1)}{g_T(\omega^1)}\tilde{p}_1 + \dfrac{S_T^1(\omega^2)}{g_T(\omega^2)}\tilde{p}_2 + \dfrac{S_T^1(\omega^3)}{g_T(\omega^3)}\tilde{p}_3 = \dfrac{S_0^1}{g_0} \\ \dfrac{S_T^2(\omega^1)}{g_T(\omega^1)}\tilde{p}_1 + \dfrac{S_T^2(\omega^2)}{g_T(\omega^2)}\tilde{p}_2 + \dfrac{S_T^2(\omega^3)}{g_T(\omega^3)}\tilde{p}_3 = \dfrac{S_0^2}{g_0} \\ \dfrac{S_T^3(\omega^1)}{g_T(\omega^1)}\tilde{p}_1 + \dfrac{S_T^3(\omega^2)}{g_T(\omega^2)}\tilde{p}_2 + \dfrac{S_T^3(\omega^3)}{g_T(\omega^3)}\tilde{p}_3 = \dfrac{S_0^3}{g_0} \end{cases}$$

First, consider the case with $g = S^1$:

$$\begin{cases} \tilde{p}_1^{(1)} + \tilde{p}_2^{(1)} + \tilde{p}_3^{(1)} = 1 \\ 140\tilde{p}_1^{(1)} + 110\tilde{p}_2^{(1)} + 80\tilde{p}_3^{(1)} = \frac{100}{5/6} \\ 40\tilde{p}_1^{(1)} + 10\tilde{p}_2^{(1)} + 0\tilde{p}_3^{(1)} = \frac{20}{5/6} \end{cases} \implies \begin{cases} \tilde{p}_1^{(1)} = \frac{8}{15} \\ \tilde{p}_2^{(1)} = \frac{4}{15} \\ \tilde{p}_3^{(1)} = \frac{1}{5} \end{cases} \tag{5.50}$$

Second, consider the case with $g = S^2$:

$$\begin{cases} \dfrac{1}{140}\tilde{p}_1^{(2)} + \dfrac{1}{110}\tilde{p}_2^{(2)} + \dfrac{1}{80}\tilde{p}_3^{(2)} = \dfrac{5/6}{100} \\ \tilde{p}_1^{(2)} + \tilde{p}_2^{(2)} + \tilde{p}_3^{(2)} = 1 \\ \dfrac{40}{140}\tilde{p}_1^{(2)} + \dfrac{10}{110}\tilde{p}_2^{(2)} + \dfrac{0}{80}\tilde{p}_3^{(2)} = \dfrac{20}{100} \end{cases} \implies \begin{cases} \tilde{p}_1^{(2)} = \frac{28}{45} \\ \tilde{p}_2^{(2)} = \frac{11}{45} \\ \tilde{p}_3^{(2)} = \frac{2}{15} \end{cases} \tag{5.51}$$

Note that the third asset cannot serve as numéraire since its payoff vector is not strictly positive. ☐

Suppose that there are two choices, g and f, for the numéraire asset. For the two numéraire assets, let $\widetilde{\mathbb{P}}^{(g)}$ and $\widetilde{\mathbb{P}}^{(f)}$ be the respective equivalent martingale measures (EMMs) and $\widetilde{\mathrm{E}}^{(g)}[\,\cdot\,]$ and $\widetilde{\mathrm{E}}^{(f)}[\,\cdot\,]$ denote the respective mathematical expectations under the two measures. We are interested in the relationship between the two sets of probabilities $\tilde{p}_j^{(g)} := \widetilde{\mathbb{P}}^{(g)}(\omega^j)$ and $\tilde{p}_j^{(f)} := \widetilde{\mathbb{P}}^{(f)}(\omega^j)$, $j = 1, 2, \ldots, M$. For this purpose, consider any attainable claim having value $\Pi_t = \Pi_t^{\varphi}$, $t \in \{0, T\}$, $\varphi \in \mathbb{R}^N$. Applying (5.49) with EMMs $\widetilde{\mathbb{P}}^{(g)}$ and $\widetilde{\mathbb{P}}^{(f)}$ gives the no-arbitrage initial value of the claim

$$\Pi_0 = g_0\widetilde{\mathrm{E}}^{(g)}\left[\frac{\Pi_T}{g_T}\right] = f_0\widetilde{\mathrm{E}}^{(f)}\left[\frac{\Pi_T}{f_T}\right] \implies \widetilde{\mathrm{E}}^{(g)}\left[\frac{\Pi_T}{g_T}\right] = \widetilde{\mathrm{E}}^{(f)}\left[\frac{g_T/g_0}{f_T/f_0}\frac{\Pi_T}{g_T}\right]. \tag{5.52}$$

Let us define a new random variable

$$\varrho \equiv \varrho^{(f\to g)} := \frac{g_T/g_0}{f_T/f_0}. \tag{5.53}$$

The variable ϱ is called the *Radon–Nikodym derivative.* As follows from (5.52), it acts as a re-weighting factor that enters into the expected value of a random payoff X when changing probability measures in the expectation from one EMM into another EMM:

$$\widetilde{\mathrm{E}}^{(g)}[X] = \widetilde{\mathrm{E}}^{(f)}[\varrho X]. \tag{5.54}$$

Equivalently, and more explicitly,

$$\sum_{j=1}^{M} X(\omega^j)\tilde{p}_j^{(g)} = \sum_{j=1}^{M} \varrho(\omega^j)X(\omega^j)\tilde{p}_j^{(f)}. \tag{5.55}$$

Note the following properties of a Radon–Nikodym derivative.

Proposition 5.15 (Properties of a Radon–Nikodym derivative).

(1) The variable $\varrho^{(f\to g)}$ is strictly positive;

(2) The expected value of the Radon–Nikodym derivative under $\widetilde{\mathbb{P}}^{(f)}$ is one,

$$\widetilde{\mathrm{E}}^{(f)}\left[\varrho^{(f\to g)}\right] = 1\,.$$

Proof. The first property is obvious since for every ω, the value $\varrho(\omega)$ is a ratio of positive values. The last property follows directly from (5.54) by setting $X \equiv 1$. □

As follows from (5.55), the probabilities $\{\tilde{p}_j^{(g)}\}$ and $\{\tilde{p}_j^{(f)}\}$ relate to each other by

$$\tilde{p}_j^{(g)} = \varrho^{(f\to g)}(\omega^j)\,\tilde{p}_j^{(f)} = \frac{g_T(\omega^j)/g_0}{f_T(\omega^j)/f_0}\,\tilde{p}_j^{(f)}, \quad j = 1, 2, \ldots, M, \tag{5.56}$$

or, equivalently,

$$\varrho^{(f\to g)}(\omega) = \frac{\widetilde{P}^{(g)}(\omega)}{\widetilde{P}^{(f)}(\omega)} \text{ for all } \omega \in \Omega\,. \tag{5.57}$$

Hence, being given the risk-neutral probabilities of one EMM and the Radon–Nikodym derivative, we can calculate the state probabilities of the other EMM using the relation (5.56).

Note that the variable $\frac{1}{\varrho^{(f\to g)}}$ is also a Radon–Nikodym derivative $\varrho^{(g\to f)}$ that allows for switching from $\widetilde{\mathbb{P}}^{(g)}$ to $\widetilde{\mathbb{P}}^{(f)}$. For any payoff $X \in \mathcal{L}(\Omega)$ we have the following property, which is symmetric to that in (5.54):

$$\widetilde{\mathrm{E}}^{(f)}[X] = \widetilde{\mathrm{E}}^{(g)}\left[\frac{1}{\varrho^{(f\to g)}}\,X\right] = \widetilde{\mathrm{E}}^{(g)}\left[\varrho^{(g\to f)}\,X\right].$$

If the market is incomplete and arbitrage-free, then there are many equivalent $\widetilde{\mathbb{P}}^{(g)}$-martingale measures for a given numéraire g. Hence, $\tilde{p}_j^{(f)}$ and $\tilde{p}_j^{(g)}$ are not necessarily uniquely related by the above expression (5.56). However, given a set of probabilities $\{\tilde{p}_j^{(f)}\}$ for a martingale measure $\widetilde{\mathbb{P}}^{(f)}$ there exists a set $\{\tilde{p}_j^{(g)}\}$ for a martingale measure $\widetilde{\mathbb{P}}^{(g)}$ using (5.56). The relationship and hence the martingale measure $\widetilde{\mathbb{P}}^{(g)}$ for any given numéraire g is unique iff the market is complete.

Example 5.16 (Example 5.15 continued). Find the Radon–Nikodym derivative $\varrho^{(S^1\to S^2)}$.

Solution. The Radon–Nikodym derivative $\varrho \equiv \varrho^{(S^1\to S^2)}$ given by

$$\varrho(\omega) = \frac{S_T^2(\omega)/S_0^2}{S_T^1(\omega)/S_0^1} = \begin{cases} \frac{7}{6} & \omega = \omega^1 \\ \frac{11}{12} & \omega = \omega^2 \\ \frac{2}{3} & \omega = \omega^3 \end{cases}$$

allows for expressing the probabilities $\tilde{p}_j^{(1)} \equiv \tilde{p}_j^{(S^1)}$ in terms of $\tilde{p}_j^{(2)} \equiv \tilde{p}_j^{(S^2)}$ and vice versa. For example, we have $\tilde{p}_j^{(2)} = \varrho(\omega^j)\,\tilde{p}_j^{(1)}$. Indeed, using the solutions (5.50) and (5.51) gives

$$\varrho(\omega^1)\,\tilde{p}_1^{(1)} = \frac{7}{6}\cdot\frac{8}{15} = \frac{28}{45} = \tilde{p}_1^{(2)} \quad \checkmark$$
$$\varrho(\omega^2)\,\tilde{p}_2^{(1)} = \frac{11}{12}\cdot\frac{4}{15} = \frac{1}{45} = \tilde{p}_2^{(2)} \quad \checkmark$$
$$\varrho(\omega^3)\,\tilde{p}_3^{(1)} = \frac{2}{3}\cdot\frac{1}{5} = \frac{2}{15} = \tilde{p}_3^{(2)} \quad \checkmark$$

□

The Radon–Nikodym derivative can be defined for any pair of equivalent probability measures that are not required to relate to particular numéraires. Let us consider a finite sample space Ω on which two equivalent probability measures $\mathbb{P}$ and $\widehat{\mathbb{P}}$ are defined. We assume that $\mathbb{P}$ and $\widehat{\mathbb{P}}$ both give positive probability to every element of the sample space, so we can calculate the quotient

$$\varrho(\omega) \equiv \varrho^{\mathbb{P}\to\widehat{\mathbb{P}}}(\omega) = \frac{\widehat{\mathbb{P}}(\omega)}{\mathbb{P}(\omega)}$$

for every $\omega \in \Omega$ (compare it with (5.57)). The random variable ϱ is called the *Radon–Nikodym derivative of* $\widehat{\mathbb{P}}$ *with respect to* $\mathbb{P}$. As proved above in Proposition 5.15 and in equation (5.54), the variable ϱ has the following properties:

(a) $\varrho > 0$ with probability 1;

(b) $\mathrm{E}[\varrho] = 1$;

(c) for any random variable Y, we have $\widehat{\mathrm{E}}[Y] = \mathrm{E}[\varrho Y]$.

Here, $\mathrm{E}[\,\cdot\,]$ and $\widehat{\mathrm{E}}[\,\cdot\,]$ denote the mathematical expectation operators under $\mathbb{P}$ and $\widehat{\mathbb{P}}$, respectively.

5.6 Exercises

Exercise 5.1. Consider a standard binomial model with the state space $\Omega = \{\omega^+, \omega^-\}$ and two base securities, risk-free bond B and risky stock S.

(a) Find the portfolios $[\beta^\pm, \Delta^\pm]$ in the base assets B and S that replicate the payoffs of Arrow–Debreu securities $\mathcal{E}^\pm = \mathbb{I}_{\{\omega^\pm\}}$.

(b) Find the no-arbitrage prices $\pi_0(\mathcal{E}^\pm)$.

(c) For an arbitrary payoff function X, we can write

$$X(\omega) = X(\omega^+)\mathcal{E}^+(\omega) + X(\omega^-)\mathcal{E}^-(\omega).$$

Since the replicating portfolio (β_X, Δ_X) is given by

$$[\beta_X, \Delta_X] = X(\omega^+)[\beta^+, \Delta^+] + X(\omega^-)[\beta^-, \Delta^-]$$

derive the formula for $[\beta_X, \Delta_X]$ using the result of (a).

(d) Using the results of (b) and (c), derive the formula for $\pi_0(X)$.

(e) Find the no-arbitrage price $\pi_0(X)$ for the two payoffs $X = \max(S_T, K)$ and $X = \max(S_T - K, 0)$ with $K > 0$. [Hint: Consider three cases: 1) $\frac{K}{S_0} \leqslant d$; 2) $d < \frac{K}{S_0} < u$; 3) $u \leqslant \frac{K}{S_0}$.]

Exercise 5.2. Consider a standard binomial model. To hedge a short position in a claim C with payoff $[C^+, C^-]$, the investor forms a portfolio with Δ shares of stock. Find the optimal value of Δ that minimizes the variance of the terminal value $\Pi_T = -C_T + \Delta\, S_T$. What is the variance of Π_T when Δ is optimal?

Exercise 5.3. Assume a 2-by-2 economy with the state space $\Omega = \{\omega^1, \omega^2\}$ and two contingent assets S_t^1 and S_t^2, $t \in \{0, T\}$. Assume that $S_0^1 > 0$ and $S_0^2 > 0$. Let $S_T^i(\omega^j) = S_0^i\, R_j^i$ for some positive constants R_j^i, $i, j = 1, 2$.

(a) Derive a simple condition (in terms of R_j^i) for the completeness of this market model.

(b) Find the risk-neutral probabilities $\tilde{p} = \widetilde{\mathbb{P}}(\omega^1)$ and $1 - \tilde{p} = \widetilde{\mathbb{P}}(\omega^2)$ such that the discounted price process $\frac{S_t^2}{S_t^1}$ is a martingale, i.e.

$$\widetilde{\mathrm{E}}\left[\frac{S_T^2}{S_T^1}\right] = \frac{S_0^2}{S_0^1}.$$

(c) Assume that the market is complete (i.e., the condition derived in (a) holds). Find the portfolio (φ_1, φ_2) that replicates an arbitrary payoff X. Find the formula for $\pi_0(X)$.

Exercise 5.4 (A variant of the Law of One Price)**.** Suppose that there are no arbitrage opportunities. Let X_t and Y_t, $t \in \{0, T\}$, be two securities such that $X_T(\omega) \geqslant Y_T(\omega)$ for all $\omega \in \Omega$ and $X_T(\omega^*) > Y_T(\omega^*)$ for at least one $\omega^* \in \Omega$. Prove that $X_0 > Y_0$.

Exercise 5.5. Consider a 3-by-3 market model with the following payoff matrix:

$$\mathbf{D} = \begin{bmatrix} 23 & 33 & 9 \\ 15 & 19 & 21 \\ 7 & 5 & 33 \end{bmatrix}.$$

(a) Show that the market is not complete.

(b) Find any redundant base asset and represent its payoff vector as a linear combination of the payoffs of the other two assets.

Exercise 5.6. Consider a one-period binomial model. Assume that $B_0 = \$0.9$, $B_T = \$1$, $S_0 = \$100$, and that the two possible values of S_T are \$90 and \$105.

(a) Is this model arbitrage-free? Use the geometric interpretation of the first FTAP to verify this. Provide a diagram.

(b) If the model is arbitrage-free, find a state-price vector and risk-neutral probabilities. If the model admits arbitrage, find an arbitrage portfolio.

Exercise 5.7. Consider a one-period, arbitrage-free binomial model. Find a replicating portfolio for a forward contract with strike price K. Find the initial value of the forward contract.

Exercise 5.8. Three assets A, B, and C have market prices and payoffs as given in the table below:

Asset	Price	Payoff in state 1	Payoff in state 2
A	\$70	\$50	\$100
B	\$60	\$30	\$120
C	\$80	\$38	\$112

(a) Construct a portfolio $[\varphi_A, \varphi_B]$ in assets A and B that replicates the payoff of asset C.

(b) Find the present time-0 value of the replicating portfolio. Is there a possible arbitrage in this market? Explain.

(c) Determine whether or not assets A and B form the basis for a complete arbitrage-free two-state market. If so, find the risk-neutral probabilities.

Exercise 5.9. Consider a single-period economy with three states ω^j, $j = 1, 2, 3$ and three base assets, a zero-coupon bond with interest rate $r = 0.1$, a stock S with spot price $S_0 = 45$, and a call option on the stock S with strike $K = 40$ and maturity T. Assume that the call option has current price $C_0 = C_0(S_0, K) = 10$ and that the stock can attain terminal values of 60, 50, and 30 in the respective states 1, 2, and 3.

(a) Define appropriate base asset price processes $\{S_t^i\}_{t\in\{0,T\}}$, $i = 1, 2, 3$, and provide the payoff matrix $\mathbf{D}$. Show that the market is complete.

(b) Using the three base assets, find the three replicating portfolios $\varphi^{(j)}$, $j = 1, 2, 3$ for the Arrow–Debreu securities that pay one unit of account in the respective jth state. From this determine the risk-neutral probabilities $\tilde{p}_j$, $j = 1, 2, 3$.

(c) Find a replicating portfolio in the three base assets for a put option struck at $K = 40$ and determine its initial price $P_0 = P_0(S_0, K)$.

(d) Re-price the put option from part (c) but this time simply employ the asset pricing formula (i.e., discounted expected value) using appropriate risk-neutral probabilities. Is this price the same as that obtained in part (c)? If so, why?

Exercise 5.10. A stock currently trades at \$100. In one month its price will either be \$125, \$100, or \$75. I sell you a call option on this stock, struck at \$95, for \$11. I hedge my exposure by purchasing Δ shares, borrowing $100\Delta - 11$ in order to fund the purchase. The simple rate of interest is 12%.

(a) What will my profit/loss be in one month?

(b) Is it possible for me to completely hedge my exposure? Explain.

Exercise 5.11. Consider a 3-by-2 one-period model with three assets: a bond B and two stocks S^i, $i = 1, 2$. Assume that the bond sells for \$1 at $t = 0$ and pays \$$R$ at $t = T$. For stock 1, we have $S_0^1 = 10$ and two possible prices of \$12 and \$8, respectively, at $t = T$. The initial price of stock 2 is \$8, and S_T^2 has two possible prices of \$$Z$ and \$4. Here, $Z \geqslant 0$ and $R \geqslant 1$ are parameters.

(a) Show that there are redundant base assets. Represent stock 2 as a portfolio of the bond and stock 1.

(b) Verify that this model is complete for all choices of $Z \geqslant 0$ and $R \geqslant 1$.

(c) For what values of Z and R is the model arbitrage-free?

(d) Suppose that a call on stock 1 with a strike price of \$9 has a price of \$2. Find Z and R. Find a unique state-price vector and risk-neutral probabilities.

Exercise 5.12. Consider the following 4-by-4 Arrow–Debreu model:

$$\mathbf{S}_0 = \begin{bmatrix} 9/10 \\ 1 \\ 1/4 \\ 1/4 \end{bmatrix} \quad \text{and} \quad \mathbf{D} = \begin{bmatrix} 1 & 1 & 1 & 1 \\ 0 & 1 & 0 & 3 \\ 1 & 0 & 0 & 0 \\ 0 & 1 & 0 & 0 \end{bmatrix}$$

(a) Show that the model is arbitrage-free and complete.

(b) Find a unique state-price vector and risk-neutral probabilities.

(c) Find the initial price of a put option on S_T^2 with strike price $K = 2$.

(d) Find the initial price of a call option on the maximum of S_T^2, S_T^3, and S_T^4 with strike price $K = 2$.

(e) Find the initial price of a put option on the portfolio $\varphi = [0,\ 1,\ 1,\ 1]$ with strike price $K = 3$ (the payoff at $t = T$ is $(K - (S_T^2 + S_T^3 + S_T^4))^+$).

Exercise 5.13. Consider a one-period model with three states and two assets, a risk-free bond and a stock. Assume that $B_0 = 100$, $B_T = 105$, $S_0 = 10$, and $S_T \in \{8, 9, 12\}$.

(a) Determine if the model is arbitrage-free and complete.

(b) Find the general solution for the state prices: a unique solution if the market is complete, or a range of solutions if the market is incomplete.

(c) Consider a call option on the stock with strike price $K = 10$. Find the bid-ask spread at $t = 0$.

(d) Suppose the market price of the call option is zero. Construct an arbitrage portfolio.

Exercise 5.14. Determine if the following Arrow–Debreu models with $M = 4$ and $N = 3$ are arbitrage-free. If yes, find a state-price vector (just one). If not, find an arbitrage portfolio.

(a) $\mathbf{S}_0 = \begin{bmatrix} 1 \\ 2 \\ 10 \end{bmatrix}, \quad \mathbf{D} = \begin{bmatrix} 1 & 1 & 1 & 1 \\ 12 & 3 & 0 & 0 \\ 0 & 0 & 0 & 10 \end{bmatrix};$

(b) $\mathbf{S}_0 = \begin{bmatrix} 1 \\ 2 \\ 10 \end{bmatrix}, \quad \mathbf{D} = \begin{bmatrix} 1 & 1 & 1 & 1 \\ 12 & 3 & 0 & 0 \\ 0 & 0 & 0 & 20 \end{bmatrix}.$

Exercise 5.15. Consider a trinomial single-period model with stock S and zero-coupon bond B. Let $B_0 = \frac{1}{1+r}$, $B_T = 1$, and

$$\frac{S_T(\omega)}{S_0} = \begin{cases} u, & \text{if } \omega = \omega^1, \\ m, & \text{if } \omega = \omega^2, \\ d, & \text{if } \omega = \omega^3. \end{cases}$$

The model is incomplete. Find the bid-ask spreads for the three Arrow–Debreu securities.

Exercise 5.16. Consider a single-period model with four states of the world and two tradable assets. Let the initial price vector and payoff matrix be given by

$$\mathbf{S}_0 = \begin{bmatrix} 10 \\ 10 \end{bmatrix} \quad \text{and} \quad \mathbf{D} = \begin{bmatrix} 11 & 11 & 11 & 11 \\ 8 & 10 & 11 & 14 \end{bmatrix}.$$

(a) Show that the market is arbitrage-free. Find the general solution for the state prices and illustrate it with a diagram.

(b) Find the bid-ask spread for the put on the risky asset with strike price $K = 11$.

Exercise 5.17. Assume a two-state economy with two contingent base assets such that

$$\mathbf{S}_0 = \begin{bmatrix} 10 \\ 15 \end{bmatrix} \quad \text{and} \quad \mathbf{D} = \begin{bmatrix} 5 & 15 \\ 20 & 10 \end{bmatrix}.$$

(a) Show that market is complete and there are no redundant base assets.

(b) Find the equivalent martingale measure for the numéraire $g = S^1$.

(c) Find the equivalent martingale measure for the numéraire $g = S^2$.

(d) Calculate and compare the risk-neutral prices of the call on the maximum $\max(S_1^1, S_1^2)$ with strike $K = 15$ using first the EMM $\widetilde{\mathbb{P}}^{(S^1)}$ and then $\widetilde{\mathbb{P}}^{(S^2)}$.

Exercise 5.18. Consider a single-period economy with three states and two base assets, a zero-coupon bond B with interest rate $r = 0.1$ and a stock S with spot price $S_0 = 45$ and terminal values of 60, 50, and 30 in the respective states ω^1, ω^2, and ω^3.

(a) Determine the payoff matrix. Show that the market is incomplete.

(b) Show that the market is arbitrage-free. Provide the general formula for the state prices.

(c) Consider a call option on the stock with strike $K = 40$ and maturity T. Find the bid-ask spread for its initial price C_0.

(d) Assume that the call option has initial price $C_0 = 10$. Consider this option as the third base asset (i.e., we make the market complete). Provide the upgraded payoff matrix. Show that the new market model is complete. Is it arbitrage-free?

(e) For the market with three base assets introduced in part (d), find the unique fair prices of the Arrow–Debreu securities that pay one unit of currency in the respective states of the world. From this determine the risk-neutral probabilities $\tilde{p}_j$, $j = 1, 2, 3$.

(f) For the market with three base assets introduced in part (d), consider a put option struck at $K = 40$. Determine its initial price P_0 by using the asset pricing formula with the risk-neutral probabilities from part (e).

Exercise 5.19. Consider the following 4-by-3 market model:

$$\mathbf{S}_0 = \begin{bmatrix} 1 \\ 1 \\ 1 \\ 2 \end{bmatrix} \quad \text{and} \quad \mathbf{D} = \begin{bmatrix} 1 & 0 & 1 \\ 2 & 1 & 0 \\ 0 & 3 & 1 \\ 2 & 0 & 2 \end{bmatrix}$$

(a) Verify that this model is arbitrage-free and complete by finding a unique state-price vector.

(b) What is the risk-free rate r of return in this model? Replicate the risk-free asset.

(c) Find the price at $t = 0$ for the call option on the maximum of S^1, S^2, S^3, and S^4, with strike price $K = 2$.

(d) Find the price at $t = 0$ for the put option on the arithmetic average of S^1, S^2, S^3, and S^4, with strike price $K = 2$.

Exercise 5.20. Assume a two-state economy. Consider a market with two risky assets S_t^1 and S_t^2, $t \in \{0, T\}$, whose initial prices are $S_0^1 = 10$ and $S_0^2 = 25$ and terminal values are $S_T^1(\omega^1) = 5$, $S_T^1(\omega^2) = 15$, $S_T^2(\omega^1) = c$, $S_T^2(\omega^2) = c + 10$, respectively. Here c is some positive parameter.

(a) For what values of c is the market complete?

(b) Find the equivalent martingale measure $\widetilde{\mathbb{P}}^{(g)}$ for the numéraire $g = S^1$. Provide the general solution (with arbitrary c).

(c) For what values of the parameter c is the market arbitrage-free?

Exercise 5.21. Show that if a market model is incomplete, then at least one Arrow–Debreu security is not attainable.

Exercise 5.22. Assume a two-state single period economy with two base assets, the zero-coupon bond with initial price B_0 and a stock with price S_t at time $t \in \{0, T\}$. Let $r > 0$ be the return on the bond. Denote $S_+ = S_T(\omega^1)$ and $S_- = S_T(\omega^2)$. Assume that $S_- < S_+$. Let $q_j = \widetilde{\mathbb{P}}^{(g)}(\omega^j)$, $j = 1, 2$, be the probabilities for the martingale measure with g as numéraire asset defined by the price process $g_t = \alpha S_t + \beta B_t$, $t \in \{0, T\}$. Assume the positions α, β are chosen such that $g_t(\omega) > 0$ for all t and all ω.

(a) Show that:

$$q_1 = \frac{\alpha S_+ + \beta(1 + r)B_0}{\alpha S_0 + \beta B_0} \left(\frac{S_0 - S_-(1 + r)^{-1}}{S_+ - S_-} \right)$$

and

$$q_2 = \frac{\alpha S_- + \beta(1 + r)B_0}{\alpha S_0 + \beta B_0} \left(\frac{S_+(1 + r)^{-1} - S_0}{S_+ - S_-} \right).$$

(b) Verify that $q_1 + q_2 = 1$.

(c) Show that $q_1, q_2 > 0$ from the no-arbitrage condition.

(d) Determine the current price of the two Arrow–Debreu securities by explicitly taking expectations with respect to the $\widetilde{\mathbb{P}}^{(g)}$ measure.

Exercise 5.23. Assume a three-state single period economy with three base assets, the zero-coupon bond with price $S_t^1 = B_t$, a stock with price $S_t^2 = S_t$, and a call on the stock with price $S_t^3 = C_t = C_t(S_t, K)$, $t \in \{0, T\}$. Let $S_T(\omega^1) = S_+$, $S_T(\omega^2) = S_0$, $S_T(\omega^3) = S_-$ and assume that the call is struck at $K = S_0$ where $S_+ > S_0 > S_-$.

(a) Set $g = S$ as the numéraire asset price process and write down the linear system of three equations in the probabilities $q_j = \widetilde{\mathbb{P}}^{(g)}(\omega^j)$, $j = 1, 2, 3$, corresponding to the martingale conditions for the relative base asset prices. Leave the equations expressed in terms of S_+, S_-, S_0, C_0, and the bond return r.

(b) Now assume $C_0 = 20/3$, $S_- = 20$, $S_0 = 40$, $S_+ = 60$, $r = 0$ and hence from part (a) determine the probabilities q_1, q_2, q_3.

(c) Assuming the same parameters as in part (b), find the initial price of a call with strike 30 and maturity T in this economy.

Exercise 5.24. Consider a single-period economy having three assets and two states with initial price vector and payoff matrix:

$$\mathbf{S}_0 = \begin{bmatrix} 1 \\ 10 \\ S_0^3 \end{bmatrix}, \quad \mathbf{D} = \begin{bmatrix} 1.05 & 1.05 \\ 20 & 5 \\ 10 & 0 \end{bmatrix}.$$

(a) Find the state-price vector $\boldsymbol{\Psi}$ and the corresponding risk-neutral probabilities. Determine the no-arbitrage price S_0^3.

(b) Find the portfolio $[\varphi_1, \varphi_2, 0]$, consisting of only nonzero positions in the first two base assets, that replicates any arbitrary claim with cash-flow vector $[c_1, c_2]$ in the respective states $\{\omega^1, \omega^2\}$. Based on this, determine the initial price of such a claim. Express your answers in terms of c_1 and c_2.

Exercise 5.25. Consider a three-state single-period model with three base assets, a stock with the price S_t, a European call struck at K_1 with the price C_t^1, and a European call struck at K_2 with the price C_t^2, where $t \in \{0, T\}$. Both options expire at time T. Let $S_T(\omega^1) = S_+$, $S_T(\omega^2) = S_0$, $S_T(\omega^3) = S_-$, where $0 < S_- < S_0 < S_+$. Suppose that $K_1 = S_-$ and $K_2 = S_0$.

(a) Provide the payoff matrix $\mathbf{D}$ in terms of S_+, S_0, S_- only.

(b) Determine if the market is complete and free of redundant securities.

(c) Determine the three replicating portfolios $\boldsymbol{\varphi}^{(j)} = [\varphi_1^{(j)}, \varphi_2^{(j)}, \varphi_3^{(j)}]$ for the respective Arrow–Debreu securities $\mathcal{E}^j = \mathbb{I}_{\{\omega^j\}}$, $j = 1, 2, 3$.

(d) Without computing the prices C_0^1 and C_0^2, determine which one is larger. Explain.

(e) Suppose that $S_+ = 3$, $S_0 = 2$, $S_- = 1$, $C_0^1 = 1$. Find the upper and lower bounds on the no-arbitrage price C_0^2.

Exercise 5.26. Consider a single-period economy with sample space $\Omega = \{\omega^1, \omega^2, \omega^3\}$. Assume an equivalent martingale measure $\widetilde{\mathbb{P}}$ with cash as numéraire is given by the state probabilities $\widetilde{\mathbb{P}}(\omega^j) = \tilde{p}_j = 1/3$, for all $j = 1, 2, 3$. Note that cash is an asset equivalent to a money market account (or bond) with no interest. Two assets X and Y in this economy have payoff vectors $X_T(\Omega) = [0,\ 1,\ 2]$ and $Y_T(\Omega) = [1,\ 2,\ 6]$.

(a) Let $[\varphi_X, \varphi_Y]$ be a portfolio where φ_X and φ_Y are fixed (and unknown) positions in assets X and Y, respectively. Find the value of the ratio φ_X/φ_Y that makes the initial value of this portfolio equal zero. Hint: Use the $\widetilde{\mathbb{P}}$-martingale measure conditions.

(b) Determine the probabilities $\hat{p}_j := \widetilde{\mathbb{P}}^{(g)}(\omega^j)$, $j = 1, 2, 3$, for the equivalent martingale measure with $g = Y$ as numéraire asset.

Exercise 5.27. Consider a trinomial model with the state space $\Omega = \{\omega^d, \omega^m, \omega^u\}$ and two base assets, risky asset S and risk-free asset B. To hedge the short position in a claim C with payoff $[C^d, C^m, C^u]$ where the values C^d, C^m, C^u are all different, the investor forms a portfolio with Δ shares of stock.

(a) Find the optimal value of Δ that minimizes the variance of the terminal value $\Pi_T = -C_T + \Delta\, S_T$.

(b) Show that the variance of Π_T is always strictly positive, so the Δ-hedging cannot entirely eliminate the risk associated with such a risky claim in this trinomial model.

(c) Show that the optimal value of Δ can be expressed as a linear combination of optimal Δ's calculated for three binomial submodels with respective state spaces $\Omega_1 = \{\omega^d, \omega^u\}$, $\Omega_2 = \{\omega^m, \omega^u\}$, and $\Omega_3 = \{\omega^d, \omega^m\}$ (see Exercise 5.2).

Exercise 5.28. Consider a general N-by-M model with dividend matrix $\mathbf{D}$. Form a portfolio in base securities, $\varphi \in \mathbb{R}^N$, that hedges the short position in a claim with terminal payoff C_T. Show that the optimal portfolio vector φ that minimizes the variance of $-C_T + \Pi_T^{\varphi}$ is given by

$$\varphi = \mathbf{B}\,\mathbf{\Sigma}^{-1},$$

where $\mathbf{B} = \left[\operatorname{Cov}(C_T, S_T^i)\right]_{i=1,\ldots,N}$ and $\mathbf{\Sigma} = \left[\operatorname{Cov}(S_T^i, S_T^j)\right]_{i,j=1,\ldots,N}$.

Chapter 6

Introduction to Discrete-Time Stochastic Calculus

The term *stochastic* means random. Because it is usually used together with the term *process*, it makes people think of something that changes in a random way over time. The term *calculus* refers to ways to calculate things or find objects that can be calculated (like derivatives in the differential calculus). *Stochastic Calculus* is the study of stochastic processes through a collection of special methods such as stochastic differential equations and stochastic integrals used to find probability distributions and to compute quantitative characteristics. Many fundamental concepts of stochastic calculus like filtration, conditioning, martingale, and stopping time are easy to introduce in a discrete-time setting. This chapter deals with discrete-time stochastic processes although many concepts will be defined in a general way.

6.1 A Multi-Period Binomial Probability Model

6.1.1 The Binomial Probability Space

6.1.1.1 A Sample Space

A *sample space* is a set of all possible outcomes of some experiment with an uncertain result. Examples of such *random experiments* include tossing coins and rolling dice. We shall denote a sample space by Ω. The elements $\omega \in \Omega$ are called *outcomes*. The goal of this section is the construction and characterization of a sample space that describes the multi-period binomial price model. Consider a risky asset such as a stock with initial price $S_0 > 0$. Suppose the stock price $\{S_k\}_{k=0,1,2,\dots,N}$ follows a binomial tree model with T periods that has been introduced in Chapter 2. Recall that the stock price is defined by a recurrence relation

$$S_n = \begin{cases} u\,S_{n-1} & \text{with probability } p, \\ d\,S_{n-1} & \text{with probability } 1-p, \end{cases} \tag{6.1}$$

where $p \in (0,1)$, $k = 1, 2, \dots, N$, and u and d are, respectively, the up-factor and down-factor satisfying $0 < d < u$. Let outcome ω_k describe the stock price dynamics in the kth period from time $k-1$ to time k as follows: $\omega_k = \mathsf{U}$ if the price moves up, and $\omega_k = \mathsf{D}$ if the price moves down. We shall call ω_k a *market move* in the kth period. During N periods, the stock price changes N times; thus its evolution is described by N market moves $\omega_1, \dots, \omega_N$. Each possible market scenario, denoted by ω, is an N-step path in the binomial tree, and it can be represented by a string of D's and U's of length N:

$$\omega = (\omega_1, \omega_2, \dots, \omega_N), \text{ where } \omega_k \in \{\mathsf{D}, \mathsf{U}\},\ 1 \leqslant k \leqslant N.$$

The set of all strings of D's and U's of length N is a Cartesian product

$$\Omega_N = \prod_{k=1}^{N} \{\mathsf{D}, \mathsf{U}\} = \{(\omega_1, \omega_2, \ldots, \omega_N) \;:\; \omega_k \in \{\mathsf{D}, \mathsf{U}\},\; 1 \leqslant k \leqslant N\}.$$

To simplify the notation, we shall write $\omega_1\omega_2\ldots\omega_N$ instead of $(\omega_1, \omega_2, \ldots, \omega_N)$. For instance, the sample spaces for N =1-, 2-, and 3-period models are, respectively,

$$\begin{aligned}
\Omega_1 &= \{\mathsf{D}, \mathsf{U}\};\\
\Omega_2 &= \{\mathsf{DD}, \mathsf{DU}, \mathsf{UD}, \mathsf{UU}\};\\
\Omega_3 &= \{\mathsf{DDD}, \mathsf{DDU}, \mathsf{DUD}, \mathsf{DUU}, \mathsf{UDD}, \mathsf{UDU}, \mathsf{UUD}, \mathsf{UUU}\}.
\end{aligned}$$

Note that Ω_N contains 2^N elements since there are N periods and two outcomes are possible in each period. The set Ω_N is called a *sample space* of the (N-period) binomial tree model whose elements ω^k, $1 \leqslant k \leqslant 2^N$, correspond to all possible market scenarios. In fact, each string from Ω_N can be equivalently viewed as a possible outcome of N tosses of a coin whose two sides are labelled U (a "head") and D (a "tail"), respectively. Thus, Ω_N is a sample space for the Bernoulli experiment with N independent tosses of a coin.

Any subset E of a sample space Ω is called an *event*. For instance, in the 3-period model

$$E = \{\mathsf{UUU}, \mathsf{UUD}, \mathsf{UDU}, \mathsf{UDD}\} \subset \Omega_3$$

represents the event that the market moves up in the first period. [Note: we shall use $\subset$ to mean subset where $A \subset B$ means every element in A is also in B.] We say that an event E *occurs* if the actual outcome ω belongs to E. We can consider a collection of all possible events (subsets including the empty set) of Ω. Such a collection is denoted 2^Ω and called the *power set* of Ω. If Ω is a finite set with $|\Omega|$ elements, then the power set 2^Ω contains $2^{|\Omega|}$ elements (events) since for every element of Ω there are two possibilities: to be a member of a particular event, or not to be. Therefore, the power set 2^{Ω_N} of the binomial sample space Ω_N contains 2^{2^N} elements.

6.1.1.2 Random Variables

Typically, the actual outcome $\omega \in \Omega \equiv \Omega_N$ of a random experiment is not observable, but some information about ω can be revealed via values of *random variables* that are (set) functions that map Ω to $\mathbb{R}$. For example, we introduce two special random variables on Ω:

$$\#\mathsf{D}(\omega) = \text{number of D's in } \omega, \tag{6.2}$$
$$\#\mathsf{U}(\omega) = \text{number of U's in } \omega, \tag{6.3}$$

$\forall \omega \in \Omega$. In some cases, we will need to calculate the number of D's and U's in the first k market moves. For this reason, we also define the random variables

$$\mathsf{D}_k(\omega) = \text{number of D's in } \omega_1\omega_2\ldots\omega_k = \#\mathsf{D}(\omega_1\omega_2\ldots\omega_k), \tag{6.4}$$
$$\mathsf{U}_k(\omega) = \text{number of U's in } \omega_1\omega_2\ldots\omega_k = \#\mathsf{U}(\omega_1\omega_2\ldots\omega_k). \tag{6.5}$$

for $1 \leqslant k \leqslant N$ and $\mathsf{D}_0 \equiv \mathsf{U}_0 \equiv 0$. There are two obvious properties:

1. $\mathsf{U}_k(\omega), \mathsf{D}_k(\omega) \in \{0, 1, \ldots, k\}$,

2. $\mathsf{U}_k(\omega) + \mathsf{D}_k(\omega) = k$,

for all $\omega \in \Omega_N$. That is, the random variables U_k and $\mathsf{D}_k = k - \mathsf{U}_k$ are linearly dependent.

In the recombining binomial tree model, the main objects are stock prices $\{S_k\}_{k=0,1,\ldots,N}$ defined via the recurrence relation in (6.1). These stock prices are functions of the market scenario $\omega = \omega_1\omega_2\ldots\omega_N$. For instance, there are two distinct one-step market moves with two distinct values for S_1; four distinct two-step market moves with a total of three possible values for S_2; eight distinct three-step market moves with a total of four possible values for S_3:

$$S_1(\omega) = \begin{cases} u\,S_0 & \text{if } \omega_1 = \mathsf{U} \\ d\,S_0 & \text{if } \omega_1 = \mathsf{D} \end{cases}$$

$$S_2(\omega) = \begin{cases} u\,S_1(\omega) & \text{if } \omega_2 = \mathsf{U} \\ d\,S_1(\omega) & \text{if } \omega_2 = \mathsf{D} \end{cases} = \begin{cases} u^2\,S_0 & \text{if } \omega_1\omega_2 = \mathsf{UU} \\ u\,d\,S_0 & \text{if } \omega_1\omega_2 \in \{\mathsf{DU}, \mathsf{UD}\} \\ d^2\,S_0 & \text{if } \omega_1\omega_2 = \mathsf{DD} \end{cases}$$

$$S_3(\omega) = \begin{cases} u\,S_2(\omega) & \text{if } \omega_3 = \mathsf{U} \\ d\,S_2(\omega) & \text{if } \omega_3 = \mathsf{D} \end{cases} = \begin{cases} u^3\,S_0 & \text{if } \omega_1\omega_2\omega_3 = \mathsf{UUU} \\ u^2\,d\,S_0 & \text{if } \omega_1\omega_2\omega_3 \in \{\mathsf{DUU}, \mathsf{UDU}, \mathsf{UUD}\} \\ u\,d^2\,S_0 & \text{if } \omega_1\omega_2\omega_3 \in \{\mathsf{DDU}, \mathsf{DUD}, \mathsf{UDD}\} \\ d^3\,S_0 & \text{if } \omega_1\omega_2\omega_3 = \mathsf{DDD} \end{cases}$$

Clearly, distinct multi-step market moves can result in the same value for the stock price; e.g., there are eight possible outcomes $\omega_1\omega_2\omega_3$ for a three-step move, but only four different stock prices S_3 at time 3.

Using the random variables $\#\mathsf{D}$ and $\#\mathsf{U}$ allows us to rewrite the recurrence relation (6.1) in a compact form:

$$S_k(\omega) = S_{k-1}(\omega)u^{\#\mathsf{U}(\omega_k)}d^{\#\mathsf{D}(\omega_k)}, \quad 1 \leqslant k \leqslant N. \tag{6.6}$$

By recursively applying this formula n times, we have the following expression for the stock price at time n:

$$\begin{aligned} S_n(\omega) &= S_{n-1}(\omega)u^{\#\mathsf{U}(\omega_n)}d^{\#\mathsf{D}(\omega_n)} \\ &= S_{n-2}(\omega)u^{\#\mathsf{U}(\omega_{n-1})+\#\mathsf{U}(\omega_n)}d^{\#\mathsf{D}(\omega_{n-1})+\#\mathsf{D}(\omega_n)} \\ &\;\vdots \\ &= S_0u^{\sum_{k=1}^n \#\mathsf{U}(\omega_k)}d^{\sum_{k=1}^n \#\mathsf{D}(\omega_k)} = S_0u^{\#\mathsf{U}(\omega_1\omega_2\ldots\omega_n)}d^{\#\mathsf{D}(\omega_1\omega_2\ldots\omega_n)} \\ &= S_0u^{\mathsf{U}_n(\omega)}d^{\mathsf{D}_n(\omega)} = S_0u^{\mathsf{U}_n(\omega)}d^{n-\mathsf{U}_n(\omega)} \end{aligned} \tag{6.7}$$

for any $\omega = \omega_1\omega_2\ldots\omega_N \in \Omega_N$ and $1 \leqslant n \leqslant N$. This formula shows us that the stock price S_n depends only on the first n market moves $\omega_1, \omega_2, \ldots, \omega_n$. Since it is clear that all random variables in the binomial model are functions of outcome ω, we will omit ω in some cases to simplify the notation. For instance, (6.7) gives the expression for S_n in terms of random variables $\mathsf{U}_n, \mathsf{D}_n$:

$$S_n = S_0u^{\mathsf{U}_n}d^{\mathsf{D}_n}, \quad 1 \leqslant n \leqslant N.$$

As is seen from (6.7), the binomial price S_n can only take on a value in the set (support)

$$\{S_{n,k} := S_0u^kd^{n-k};\ k = 0, 1, \ldots, n\}.$$

Equation (6.7) can be used to represent S_n in terms of S_m, the stock value at an earlier time m, $0 \leqslant m \leqslant n$. Later we shall use this for conditioning on a particular value of the random variable S_m which has support $\{S_{m,\ell} = S_0 u^\ell d^{m-\ell};\ \ell = 0, 1, \ldots, m\}$. Writing $S_n(\omega) = S_m(\omega)\frac{S_n(\omega)}{S_m(\omega)}$ and employing (6.7) gives

$$\begin{aligned} S_n(\omega) &= S_m(\omega)\frac{S_0 u^{\mathsf{U}_n(\omega)} d^{\mathsf{D}_n(\omega)}}{S_0 u^{\mathsf{U}_m(\omega)} d^{\mathsf{D}_m(\omega)}} = S_m(\omega)\, u^{\mathsf{U}_n(\omega)-\mathsf{U}_m(\omega)} d^{\mathsf{D}_n(\omega)-\mathsf{D}_m(\omega)} \\ &= S_m(\omega)\, u^{\#\mathsf{U}(\omega_{m+1}\omega_{m+2}\ldots\omega_n)} d^{\#\mathsf{D}(\omega_{m+1}\omega_{m+2}\ldots\omega_n)}. \end{aligned} \tag{6.8}$$

Since

$$\#\mathsf{U}(\omega_{m+1}\ldots\omega_n) \in \{0, 1, \ldots, n-m\} \text{ and } \#\mathsf{D}(\omega_{m+1}\ldots\omega_n) = (n-m) - \#\mathsf{U}(\omega_{m+1}\ldots\omega_n),$$

the random variable $S_n(\omega)$ conditional on $S_m(\omega) = S_m$ has support

$$\{S_m u^k d^{n-m-k}\ :\ k = 0, 1, \ldots, n-m\}.$$

These points give the possible nodes (n, S_n) attainable at time n, after $(n-m)$ market moves on the binomial tree, given that we start at a node (m, S_m) at time m.

Suppose that Ω is a finite sample space. Then every real-valued random variable X on Ω takes on a finite number of possible values, say $\{x_1, \ldots, x_K\}$. Such a finite set $X(\Omega) = \{x_1, x_2, \ldots, x_K\}$ is called the *range or support* of X. For each x_j in the range of X we can form the event

$$\{X = x_j\} \equiv \{\omega \in \Omega\ :\ X(\omega) = x_j\} = X^{-1}(\{x_j\}).$$

In what follows we will also denote the inverse image of a singleton set $X^{-1}(\{x\})$ by $X^{-1}(x)$ where $x \in \mathbb{R}$, e.g., $X^{-1}(\{x_j\}) \equiv X^{-1}(x_j)$. Since the range of a real-valued random variable is a subset of $\mathbb{R}$, we can also consider events such as

$$\{X \leqslant x\} \equiv \{\omega \in \Omega\ :\ X(\omega) \leqslant x\} = X^{-1}((-\infty, x]).$$

The following example introduces an important special kind of random variable that corresponds to the first (passage or hitting) time of the stock price for a given upper level. This kind of random variable has range $\mathbb{R} \cup \{\infty\}$ and is covered in more detail in Section 6.3.7.

Example 6.1. Suppose that $S_0 = 4$, $d = \frac{1}{2}$, and $u = 2$. Let

$$\tau(\omega) = \min\{n \in \{0, 1, 2, 3\}\ :\ S_n(\omega) \geqslant 6\}.$$

Find the event $\{\tau \leqslant 3\} := \{\omega\ :\ \tau(\omega) \leqslant 3\}$ as a subset of the 3-period sample space Ω_3.

Solution. First note that $S_0 < 6$, hence $\tau \geqslant 1$. Since Ω_3 includes only eight outcomes, we can find $\tau(\omega)$ for each $\omega \in \Omega_3$. For instance, if $\omega = \mathsf{DUU}$, then $S_1(\omega) = S_0 d = 2 < 6$, $S_2(\omega) = S_0 ud = 4 < 6$, $S_3(\omega) = S_0 u^2 d = 8 > 6$. Hence, $\tau(\mathsf{DUU}) = 3$. For $\omega = \mathsf{UUU}$, $S_1(\omega) = S_0 u = 8 > 6$ so $\tau(\mathsf{UUU}) = 1$. In fact, for any outcome of the form $\omega = \mathsf{U}\omega_2\omega_3$ (where the first move is up) we have $S_1(\omega) = S_0 u = 8$ and hence $\tau(\mathsf{U}\omega_2\omega_3) = 1$. In this way, we find

$$\begin{aligned} &\tau(\mathsf{UUU}) = \tau(\mathsf{UUD}) = \tau(\mathsf{UDU}) = \tau(\mathsf{UDD}) = 1, \\ &\tau(\mathsf{DUU}) = 3,\ \tau(\mathsf{DUD}) = \infty,\ \tau(\mathsf{DDU}) = \infty,\ \tau(\mathsf{DDD}) = \infty. \end{aligned}$$

By convention, if for some $\omega^* \in \Omega_3$ $S_n(\omega^*) < 6$, for all $n = 0, 1, 2, 3$, then we set $\tau(\omega^*) = \infty$; i.e., the process never reaches the value 6 for the given outcome. In summary,

$$\{\tau \leqslant 3\} = \{\mathsf{UUU}, \mathsf{UUD}, \mathsf{UDU}, \mathsf{UDD}, \mathsf{DUU}\}.$$

This is equivalent to $A_{\mathsf{U}} \cup \{\mathsf{DUU}\}$ where $A_{\mathsf{U}} = \{\omega\ :\ \omega_1 = \mathsf{U}\}$ is the event that the first move is up. Alternatively, since $\{\tau \leqslant 3\} = \cup_{n=1}^{3}\{\tau \leqslant n\} = \cup_{n=1}^{3}\{S_n \geqslant 6\}$, we can also use the representation:

$$\begin{aligned}\{\tau \leqslant 3\} &= \{S_1 \geqslant 6 \text{ or } S_2 \geqslant 6 \text{ or } S_3 \geqslant 6\} = \{S_1 \geqslant 6\} \cup \{S_2 \geqslant 6\} \cup \{S_3 \geqslant 6\} \\ &= A_{\mathsf{U}} \cup \{\mathsf{UUU}, \mathsf{UUD}\} \cup \{\mathsf{UUU}, \mathsf{UUD}, \mathsf{UDU}, \mathsf{DUU}\} \\ &= \{\mathsf{UUU}, \mathsf{UUD}, \mathsf{UDU}, \mathsf{UDD}, \mathsf{DUU}\}.\end{aligned}$$

□

6.1.1.3 Probability Measure

Definition 6.1. A real-valued function $\mathbb{P}$ defined on the set of all subsets $\mathcal{F} = 2^{\Omega}$ of a sample space Ω is called a *probability measure* on a sample space Ω if it satisfies the following properties:

1. $\mathbb{P}(E) \geqslant 0$ for every event $E \in \mathcal{F}$,
2. $\mathbb{P}(\Omega) = 1$,
3. $\mathbb{P}(\cup_{i\geqslant 1} E_i) = \sum_{i\geqslant 1} \mathbb{P}(E_i)$ for any countable collection $\{E_i\}_{i\geqslant 1}$, $E_i \in \mathcal{F}$, of pairwise disjoint events.

These properties are called the *Kolmogorov axioms.*

Recall that a family of sets is pairwise disjoint or mutually disjoint if every two sets in the family are disjoint, i.e., $i \neq j \implies E_i \cap E_j = \emptyset$. Suppose that Ω is a finite sample space. Then, Axiom 3, called the *countable additivity property,* can be simplified as follows:

3*. $\mathbb{P}(E_1 \cup E_2) = \mathbb{P}(E_1) + \mathbb{P}(E_2)$ for any two disjoint events $E_1, E_2 \in \mathcal{F}$.

Indeed, if Ω is finite, then 2^{Ω} is finite as well. So every collection of subsets of Ω consists of finitely many different events. If events $E_1, E_2, \ldots, E_n$ are mutually disjoint, then by applying Axiom 3* we have

$$\mathbb{P}(\cup_{i=1}^{n} E_i) = \mathbb{P}(\cup_{i=1}^{n-1} E_i) + \mathbb{P}(E_n) = \mathbb{P}(\cup_{i=1}^{n-2} E_i) + \mathbb{P}(E_{n-1}) + \mathbb{P}(E_n) = \cdots = \sum_{i=1}^{n} \mathbb{P}(E_i).$$

All outcomes (elements) of a finite sample space Ω can be enumerated:

$$\Omega = \{\omega^1, \omega^2, \ldots, \omega^M\}.$$

The probability (likelihood) of each outcome is specified by the numbers

$$p_i = \mathbb{P}(\{\omega^i\}) \equiv \mathbb{P}(\omega^j) \in [0,1], \quad j = 1, 2 \ldots, M.$$

We assume that all probabilities p_i are strictly positive (if $p_j = 0$, then the outcome ω^j can be removed from Ω as an impossible one). Since $\mathbb{P}(\Omega) = 1$ by Axioms 2 and 3, we have

$$1 = \mathbb{P}(\Omega) = p_1 + p_2 + \cdots + p_M.$$

For the case with a finite sample space Ω, every event $E \in \mathcal{F} = 2^{\Omega}$ can be described by specifying all its elements:

$$E = \{\omega^{j_1}, \omega^{j_2}, \ldots, \omega^{j_k}\} \text{ with } 1 \leqslant j_1 < j_2 < \cdots < j_k \leqslant M. \tag{6.9}$$

Therefore, the probability of E can be calculated by using the additivity property:

$$\mathbb{P}(E) = \mathbb{P}\left(\cup_{\ell=1}^{k}\{\omega^{j_\ell}\}\right) = \sum_{\ell=1}^{k} \mathbb{P}(\{\omega^{j_\ell}\}) = \sum_{\ell=1}^{k} p_{j_\ell}\,. \tag{6.10}$$

Thus, for a finite Ω with $|\Omega| = M$ outcomes, the probability measure $\mathbb{P}\colon 2^{\Omega} \to [0,1]$ can be defined by (6.9) and (6.10), where the probabilities $p_j > 0$, $j = 1, 2, \ldots, M$, sum to one:

$$p_1 + p_2 + \cdots + p_M = 1.$$

Definition 6.2. A *finite probability space* is a triplet $(\Omega, \mathcal{F}, \mathbb{P})$, which consists of a finite nonempty sample space Ω and a probability measure $\mathbb{P}$ defined on the set of subsets $\mathcal{F} = 2^{\Omega}$.

Let us return to the sample space $\Omega = \Omega_N$ of the binomial tree model where $|\Omega_N| = M = 2^N$. Our goal is to define the probability measure $\mathbb{P}$. Since Ω_N is finite, it is sufficient to define $\mathbb{P}(\omega)$ for any single market scenario $\omega \in \Omega_N$ and then apply the above approach to compute $\mathbb{P}(E)$ for any $E \subset \Omega_N$. It is useful to consider two special complementary events, for any given $k \in \{1, 2, \ldots, N\}$:

$$\{\omega_k = \mathsf{U}\} \equiv \{\omega \in \Omega_N \ : \ \omega_k = \mathsf{U}\} = \{\text{all outcomes with } k\text{th move as up}\}$$

and

$$\{\omega_k = \mathsf{D}\} \equiv \{\omega \in \Omega_N \ : \ \omega_k = \mathsf{D}\} = \{\text{all outcomes with } k\text{th move as down}\}.$$

By assumed independence of successive moves, it follows trivially that the respective probabilities of these events are $\mathbb{P}(\omega_k = \mathsf{U}) = p > 0$ and $\mathbb{P}(\omega_k = \mathsf{D}) = 1 - p > 0$. Hence, we can write

$$\mathbb{P}(\{\omega_k = \omega_*\}) = p^{\#\mathsf{U}(\omega_*)}(1-p)^{\#\mathsf{D}(\omega_*)}, \quad \omega_* \in \{\mathsf{D}, \mathsf{U}\}.$$

Moreover, for every $k, m \in \{1, 2, \ldots, N\}$, $k \neq m$, the events $\{\omega_k = \omega_*\}$ and $\{\omega_m = \omega_*\}$, $\omega_* \in \{\mathsf{D}, \mathsf{U}\}$, are independent. Then, for any particular outcome $\omega^* = \omega_1^* \omega_2^* \cdots \omega_N^* \in \Omega_N$, we have

$$\begin{aligned}
\mathbb{P}(\omega^*) &= \mathbb{P}(\{\omega_1 = \omega_1^*\} \cap \{\omega_2 = \omega_2^*\} \cap \cdots \cap \{\omega_N = \omega_N^*\}) \\
&= \prod_{k=1}^{N} \mathbb{P}(\{\omega_k = \omega_k^*\}) = \prod_{k=1}^{N} p^{\#\mathsf{U}(\omega_k^*)}(1-p)^{\#\mathsf{D}(\omega_k^*)} \\
&= p^{\#\mathsf{U}(\omega_1^*)+\cdots+\#\mathsf{U}(\omega_N^*)}(1-p)^{\#\mathsf{D}(\omega_1^*)+\cdots+\#\mathsf{D}(\omega_N^*)} = p^{\#\mathsf{U}(\omega^*)}(1-p)^{\#\mathsf{D}(\omega^*)}.
\end{aligned}$$

It is not difficult to show that, as is required by Axiom 2,

$$\mathbb{P}(\Omega_N) = \mathbb{P}(\{\omega \ : \ \omega \in \Omega_N\}) = \sum_{\omega \in \Omega_N} \mathbb{P}(\omega) = 1, \tag{6.11}$$

for any finite integer $N \geqslant 1$. One way is by induction. Indeed, if $N = 1$ then we simply have

$$\sum_{\omega_1 \in \{\mathsf{U}, \mathsf{D}\}} \mathbb{P}(\omega_1) = \mathbb{P}(\mathsf{D}) + \mathbb{P}(\mathsf{U}) = p + (1-p) = 1.$$

Now assuming (6.11) holds for N, it follows that the formula holds true for $N+1$ as well:

$$\begin{aligned}
\sum_{\omega \in \Omega_{N+1}} \mathbb{P}(\omega) &= \sum_{\omega \in \Omega_N} \mathbb{P}(\omega\mathsf{D}) + \sum_{\omega \in \Omega_N} \mathbb{P}(\omega\mathsf{U}) \\
&= \sum_{\omega \in \Omega_N} \mathbb{P}(\omega)\mathbb{P}(\{\omega_{N+1} = \mathsf{D}\}) + \sum_{\omega \in \Omega_N} \mathbb{P}(\omega)\mathbb{P}(\{\omega_{N+1} = \mathsf{U}\}) \\
&= \sum_{\omega \in \Omega_N} \mathbb{P}(\omega) \cdot (1-p) + \sum_{\omega \in \Omega_N} \mathbb{P}(\omega) \cdot p = (1 - p + p) \sum_{\omega \in \Omega_N} \mathbb{P}(\omega) = 1.
\end{aligned}$$

An alternative proof, as follows, is also instructive. Let

$$E_k = \{\mathsf{U}_N = k\} \equiv \{\omega \in \Omega_N \ : \ \mathsf{U}_N(\omega) = k\}$$
$$= \{\text{all outcomes with } k \text{ up moves in total}\}.$$

Then, the entire outcome space is simply a union of such exclusive events, $\Omega_N = \cup_{k=0}^{N} E_k$. As noted below, $\mathbb{P}(\mathsf{U}_N = k) = \binom{N}{k} p^k (1-p)^{N-k}$. Hence, by Axiom 3 and the binomial expansion:

$$\mathbb{P}(\Omega_N) = \sum_{k=0}^{N} \mathbb{P}(E_k) = \sum_{k=0}^{N} \mathbb{P}(\mathsf{U}_N = k) = \sum_{k=0}^{N} \binom{N}{k} p^k (1-p)^{N-k} = 1.$$

The probability of any event $E \subset \Omega_N$ is given by

$$\mathbb{P}(E) = \sum_{\omega \in E} \mathbb{P}(\omega) = \sum_{\omega \in E} p^{\mathsf{U}_N(\omega)} (1-p)^{\mathsf{D}_N(\omega)}.$$

Thus, $\mathbb{P}$ is a probability measure defined on $\mathcal{F}_N = 2^{\Omega_N}$. The triplet $(\Omega_N, \mathcal{F}_N, \mathbb{P})$ is a finite probability space called the *N-period binomial probability space.*

Example 6.2. Consider the random variable τ from Example 6.1. Suppose that $p = \frac{1}{4}$. Find $\mathbb{P}(\{\tau \leqslant 3\})$.

Solution. From the solution in Example 6.1, we have

$$\begin{aligned}
\mathbb{P}(\{\tau \leqslant 3\}) &= \mathbb{P}(\{\mathsf{UUU}, \mathsf{UUD}, \mathsf{UDU}, \mathsf{UDD}, \mathsf{DUU}\}) \\
&= p^3 + p^2(1-p) + p^2(1-p) + p(1-p)^2 + p^2(1-p) \\
&= p^3 + 3p^2(1-p) + p(1-p)^2 \\
&= \left(\frac{1}{4}\right)^3 + 3\left(\frac{1}{4}\right)^2\left(\frac{3}{4}\right) + \left(\frac{1}{4}\right)\left(\frac{3}{4}\right)^2 = \frac{19}{64}.
\end{aligned}$$ □

Example 6.3. Find the probability distributions of D_n and U_n for $n = 1, 2, \ldots, N$. For what values of p, do U_n and D_n have the same distribution, i.e., find p such that $\mathsf{U}_n \stackrel{d}{=} \mathsf{D}_n$?

Solution. The variables D_n and U_n, respectively, give the number of D's and the number of U's in the first n ω_j's. So they only depend on $\omega_1, \omega_2, \ldots, \omega_n$. For every $n = 1, 2, \ldots, N$ and $k = 0, 1, \ldots, n$, there exist $\binom{n}{k}$ scenarios in Ω_n that have exactly k U's. The probability of each particular scenario with k U's is $p^k(1-p)^{n-k}$. Therefore, U_n has the binomial distribution $Bin(n, p)$ where

$$\mathbb{P}(\mathsf{U}_n = k) = \binom{n}{k} p^k (1-p)^{n-k}.$$

Similarly, D_n has the binomial distribution $Bin(n, 1-p)$. Thus, $\mathsf{U}_n \stackrel{d}{=} \mathsf{D}_n$ iff $p = 1 - p$, i.e., iff $p = \frac{1}{2}$. □

6.1.2 Random Processes

Definition 6.3. A *random process* (or a *stochastic process*) is a collection of random variables that are all defined on a common sample space Ω and indexed by $t \in \mathbf{T}$:

$$\{X_t(\omega)\}_{t \in \mathbf{T}}, \text{ so that } \forall t \in \mathbf{T}\ X_t \colon \Omega \to \mathbb{R}.$$

$\mathbf{T}$ is an index set. Typically, $\mathbf{T}$ is a subset of the positive real half-line since t is usually referred to as time. The random process is said to be a *discrete-time* process if $\mathbf{T} = \{0, 1, 2, \ldots\}$; it is said to be a *continuous-time* process if $\mathbf{T}$ is an interval, e.g., $\mathbf{T} = [0, \infty)$. Note that $\mathbf{T}$ can be bounded or unbounded. For a given (fixed) outcome $\omega \in \Omega$, $X_t(\omega)$ considered as function of time $t \in \mathbf{T}$ is called a *sample path* or a *trajectory* of the stochastic process.

Let us consider two important examples of discrete-time stochastic processes defined on the binomial sample space Ω_N.

6.1.2.1 Binomial Price Process and Path Probabilities

The binomial prices defined in (6.7) constitute a discrete-time stochastic process:

$$\{S_n\}_{n=0,1,\ldots,N}, \quad \text{where } S_n(\omega) = S_0 u^{\mathsf{U}_n(\omega)} d^{\mathsf{D}_n(\omega)},\ 1 \leqslant n \leqslant N,\ \omega \in \Omega_N. \tag{6.12}$$

Moreover, random variables D_n and U_n that respectively count D's and U's in ω are also considered random processes:

$$\{\mathsf{U}_n\}_{n=0,1,\ldots,N} \text{ and } \{\mathsf{D}_n\}_{n=0,1,\ldots,N}.$$

As is shown in Example 6.3, U_n and D_n have the binomial probability distribution, $\mathsf{U}_n \sim Bin(n,p)$ and $\mathsf{D}_n \sim Bin(n, 1-p)$, for any $n = 1, 2, \ldots, N$, with probability mass functions

$$\mathbb{P}(\mathsf{U}_n = k) = \binom{n}{k} p^k (1-p)^{n-k}, \quad \mathbb{P}(\mathsf{D}_n = k) = \binom{n}{k} p^{n-k}(1-p)^k, \quad 0 \leqslant k \leqslant n.$$

Therefore, the probability mass function (PMF) of S_n is given by

$$\mathbb{P}(S_n = S_{n,k}) = \binom{n}{k} p^k (1-p)^{n-k}, \quad S_{n,k} = S_0 u^k d^{n-k}, \quad 0 \leqslant k \leqslant n \leqslant N.$$

A path of a binomial price process passes through the nodes of a binomial tree: $\{(n,k) : 0 \leqslant k \leqslant n \leqslant N\}$. At the node (n,k), the process S_n takes on the value $S_{n,k}$. Let us calculate the *transition probability* that a path of the binomial price process will pass through the node (n,k) given that it passes through (m,ℓ) with $0 \leqslant m < n$, $0 \leqslant \ell \leqslant m$, $0 \leqslant k \leqslant n$, and $\ell \leqslant k$. To get from (m,ℓ) to (n,k), the process makes $k-\ell$ up moves and $(n-m)-(k-\ell)$ down moves. Hence, $S_{n,k} = S_{m,\ell} u^{k-\ell} d^{(n-m)-(k-\ell)}$. There are $\binom{n-m}{k-\ell} = \frac{(n-m)!}{(k-\ell)!(n-m-k+\ell)!}$ distinct paths in the binomial tree that connect the nodes (m,ℓ) and (n,k). From independence of one-step moves, we obtain that the probability for any one particular path is $p^{k-\ell}(1-p)^{n-m-k+\ell}$. Since each particular path has the same probability of occurring, and each path represents a mutually exclusive event, we have

$$\mathbb{P}\left(S_n = S_{n,k} \mid S_m = S_{m,\ell}\right) = \binom{n-m}{k-\ell} p^{k-\ell}(1-p)^{n-m-k+\ell}.$$

This expression can also be obtained by operating with events in terms of the ω_j's and by using the relation (6.8) as well as the independence of random variables $\frac{S_n}{S_m}$ and S_m:

$$\begin{aligned}
\mathbb{P}\left(S_n = S_{n,k} \mid S_m = S_{m,\ell}\right) &= \mathbb{P}\left(S_m \frac{S_n}{S_m} = S_{n,k} \middle| S_m = S_{m,\ell}\right) \\
&= \mathbb{P}\left(\frac{S_n}{S_m} = \frac{S_{n,k}}{S_{m,\ell}}\right) \\
&= \mathbb{P}\left(u^{\#\mathsf{U}(\omega_{m+1}\ldots\omega_n)} d^{(n-m)-\#\mathsf{U}(\omega_{m+1}\ldots\omega_n)} = u^{k-\ell} d^{(n-m)-(k-\ell)}\right) \\
&= \mathbb{P}\left(\#\mathsf{U}(\omega_{m+1}\cdots\omega_n) = k-\ell\right) = \binom{n-m}{k-\ell} p^{k-\ell}(1-p)^{n-m-k+\ell}.
\end{aligned}$$

The last expression follows since $\#\mathsf{U}(\omega_{m+1}\dots\omega_n) \sim Bin(n-m,p)$.

6.1.2.2 Random Walk

A random walk is a mathematical formalization of a process that consists of taking successive random steps of size 1 upwards or downwards. For $\omega = \omega_1\omega_2\cdots\omega_N \in \Omega_N$ we define random variables

$$X_n(\omega) = \begin{cases} 1 & \text{if } \omega_n = \mathsf{U}, \\ -1 & \text{if } \omega_n = \mathsf{D}, \end{cases} \qquad n = 1, 2, \dots, N.$$

Consider the stochastic process $\{M_n\}_{n=0,1,2,\dots,N}$ defined as follows:

$$M_0 = 0, \quad M_n = \sum_{k=1}^{n} X_k, \; n = 1, 2, \dots, N.$$

Like the time-n binomial price S_n, the variable M_n is a function of the first n market moves $\omega_1, \omega_2, \dots, \omega_n$. $\{M_n\}_{n\geqslant 0}$ is called a *random walk*. It admits a recurrence representation:

$$M_{n+1} = \sum_{k=1}^{n+1} X_k = M_n + X_{n+1} = \begin{cases} M_n + 1 & \text{with probability } p, \\ M_n - 1 & \text{with probability } 1-p. \end{cases}$$

If $p = \frac{1}{2}$, then $\{M_n\}_{n\geqslant 0}$ is called a *symmetric random walk* since at every step it may go equally likely upwards or downwards.

Since $\mathsf{U}_n = \sum_{k=1}^{n}(X_k)^+$ and $\mathsf{D}_n = \sum_{k=1}^{n}(X_k)^-$ (using the notation $(x)^+ = \max\{x, 0\}$ and $(x)^- = \max\{-x, 0\}$), we can express M_n in terms of U_n and D_n as follows:

$$M_n = \sum_{k=1}^{n} X_k = \sum_{k=1}^{n} (X_k)^+ - \sum_{k=1}^{n} (X_k)^- = \mathsf{U}_n - \mathsf{D}_n.$$

This is clearly the difference between the number of upward and downward moves. Since $\mathsf{D}_n + \mathsf{U}_n = n$, we also write M_n in terms of only D_n or U_n:

$$M_n = n - 2\mathsf{D}_n = 2\mathsf{U}_n - n.$$

Thus, by equivalence of events $\{\mathsf{U}_n = k\} = \{M_n = 2k - n\}$, the PMF of M_n is given by

$$\mathbb{P}(M_n = 2k - n) = \binom{n}{k} p^k (1-p)^{n-k}, \quad k = 0, 1, \dots, n\,.$$

Example 6.4. Construct the sample path of a particular random walk and find the path probability for $\omega^* = \mathsf{DUUDDUDD} \in \Omega_8$.

Solution. A random walk starts at the origin, i.e., $M_0 = 0$. Find recursively the sample values $M_n = M_n(\omega^*)$ for $n = 1, 2, \dots, 8$:

$$\begin{aligned} M_1 &= M_0 + X_1 = 0 - 1 = -1, & M_2 &= M_1 + X_2 = -1 + 1 = 0, \\ M_3 &= M_2 + X_3 = 0 + 1 = 1, & M_4 &= M_3 + X_4 = 1 - 1 = 0, \\ M_5 &= M_4 + X_5 = 0 - 1 = -1, & M_6 &= M_5 + X_5 = -1 + 1 = 0, \\ M_7 &= M_6 + X_7 = 0 - 1 = -1, & M_8 &= M_7 + X_8 = -1 - 1 = -2. \end{aligned}$$

Figure 6.1 demonstrates the sample paths of the process $\{M_n\}$ and $\{\mathsf{U}_n\}$ for the outcome ω^*. The path probability is

$$\mathbb{P}(\omega^*) = p^{\#\mathsf{U}(\omega^*)}(1-p)^{\#\mathsf{D}(\omega^*)} = p^3(1-p)^5. \qquad \square$$

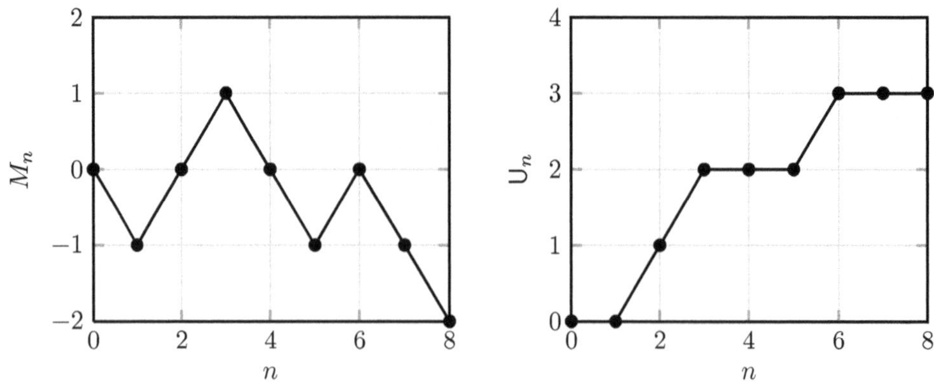

FIGURE 6.1: Sample paths of the random walk $\{M_n\}$ and the process $\{\mathsf{U}_n\}$ in an eight-period model for $\omega = \mathsf{DUUDDUDD}$.

6.2 Information Flow

The pricing of derivative securities is based on contingent claims. We need a way to mathematically model the arriving information on which our future decisions can be based. In the binomial tree model, that information is the knowledge of all market moves between the initial and future dates. We hence introduce, in what follows, the concepts of partition, σ-algebra, and filtration.

6.2.1 Partitions and Their Refinements

Suppose that some random experiment is performed, and its actual outcome ω is unknown. However, we might be given some information that is enough to narrow down the possible value of ω. We can then create a list of events that are sure to contain the actual outcome, and other sets that are sure not to contain it. These sets are *resolved* by the available information. This leads us to the following definition.

Definition 6.4. A *partition* $\mathcal{P}$ of a sample space $\Omega \neq \emptyset$ is a (set) collection of mutually disjoint nonempty subsets whose union is Ω. That is,

$$\mathcal{P} = \{A_i\}_{i\geqslant 1} = \{A_1, A_2, \ldots\}, \text{ so that } \cup_{i\geqslant 1} A_i = \Omega, \text{ and } i \neq j \implies A_i \cap A_j = \emptyset.$$

The subsets A_i are called *atoms* of the partition $\mathcal{P}$.

If the information about the actual outcome ω is available in the form of a partition, then we are able to say which event from the collection has occurred. There are three trivial examples of partitions:

(a) $\mathcal{P}_0 = \{\Omega\}$;

(b) $\mathcal{P}_E = \{E, E^{\mathsf{C}}\}$ where $E \cup E^{\mathsf{C}} = \Omega$;

(c) $\mathcal{P}_\Omega = \{\{\omega\} : \omega \in \Omega\}$.

The partition $\mathcal{P}_0$ represents the absence of any information regarding the actual outcome ω, and $\mathcal{P}_\Omega$ represents the full information about ω. The partition $\mathcal{P}_E$ means that we are only able to say whether the event E has occurred or not.

Example 6.5. Construct a partition representing the available information in each case below.

(a) Roll two dice. Suppose that the sum of values on the facing-up sides is known.

(b) Toss three coins. Suppose that the total number of heads is known.

Solution.
(a) The sample space consists of 36 pairs on integers: $\Omega = \{(i,j) \;:\; 1 \leqslant i,j \leqslant 6\}$. The sum of values on the facing-up sides is an integer between 2 and 12. So the partition $\mathcal{P}$ contains 11 elements: $\mathcal{P} = \{A_2, A_3, \ldots, A_{12}\}$, where $A_k = \{(i,j) \;:\; 1 \leqslant i,j \leqslant 6 \text{ and } i+j=k\}$:

$$\begin{aligned}
A_2 &= \{(1,1)\},\\
A_3 &= \{(1,2),(2,1)\},\\
A_4 &= \{(1,3),(2,2),(3,1)\},\\
A_5 &= \{(1,4),(2,3),(3,2),(4,1)\},\\
A_6 &= \{(1,5),(2,4),(3,3),(4,2),(5,1)\},\\
A_7 &= \{(1,6),(2,5),(3,4),(4,3),(5,2),(6,1)\},\\
A_8 &= \{(2,6),(3,5),(4,4),(5,3),(6,2)\},\\
A_9 &= \{(3,6),(4,5),(5,4),(6,3)\},\\
A_{10} &= \{(4,6),(5,5),(6,4)\},\\
A_{11} &= \{(5,6),(5,6)\},\\
A_{12} &= \{(6,6)\},
\end{aligned}$$

where $\cup_i A_i = \Omega$ and $A_j \cap A_k = \emptyset$ if $j \neq k$.

(b) The coin toss sample space consists of eight combinations of H's and T's:

$$\Omega_3 = \{TTT, TTH, THT, THH, HTT, HTH, HHT, HHH\}.$$

The number of heads can be any integer between 0 and 3. Therefore, the partition $\mathcal{P} = \{A_0, A_1, A_2, A_3\}$ consists of four atoms:

$$A_0 = \{TTT\},\ A_1 = \{TTH, THT, HTT\},\ A_2 = \{THH, HTH, HHT\},\ A_3 = \{HHH\}.$$

□

The next example illustrates a succession of three partitions where each new partition is obtained by subdividing (breaking down) the atoms in the previous one.

Example 6.6. Consider a three-period binomial model. Represent the available information in the form of a partition if (a) nothing is known; (b) the first market move ω_1 is known; (c) the first two market moves ω_1, ω_2 are known.

Solution.

(a) $\mathcal{P}_0 = \{\Omega_3\}$.

(b) $\mathcal{P}_1 = \{\{\mathsf{DDD}, \mathsf{DDU}, \mathsf{DUD}, \mathsf{DUU}\}, \{\mathsf{UDD}, \mathsf{UDU}, \mathsf{UUD}, \mathsf{UUU}\}\} := \{A_\mathsf{D}, A_\mathsf{U}\}$.

(c) $\mathcal{P}_2 = \{\{\mathsf{DDD}, \mathsf{DDU}\}, \{\mathsf{DUD}, \mathsf{DUU}\}, \{\mathsf{UDD}, \mathsf{UDU}\}, \{\mathsf{UUD}, \mathsf{UUU}\}\}$
$:= \{A_{\mathsf{DD}}, A_{\mathsf{DU}}, A_{\mathsf{UD}}, A_{\mathsf{UU}}\}$.

Note: $A_{\mathsf{D}} \cup A_{\mathsf{U}} = \Omega_3$, $A_{\mathsf{D}} = A_{\mathsf{DD}} \cup A_{\mathsf{DU}}$, $A_{\mathsf{U}} = A_{\mathsf{UD}} \cup A_{\mathsf{UU}}$. Hence, the one atom in $\mathcal{P}_0$ is given by the union of the atoms in $\mathcal{P}_1$, and each atom in $\mathcal{P}_1$ is in turn given by a union of two atoms in $\mathcal{P}_2$. □

6.2.1.1 Partition Generated by a Random Variable

Consider a random variable X defined on a sample space Ω. If the set Ω is finite, then any random variable on Ω is a discrete random variable with range $X(\Omega) \equiv \mathsf{S}_X = \{x_1, x_2, \ldots, x_K\}$. The partition *generated by* X is the collection

$$\mathcal{P}(X) = \{X^{-1}(x_1), X^{-1}(x_2), \ldots, X^{-1}(x_K)\},$$

i.e., with atoms $A_i = X^{-1}(x_i) \equiv \{\omega \in \Omega \ : \ X(\omega) = x_i\}$, $i = 1, \ldots, K$. Note that the random variable also has the representation in terms of indicator functions w.r.t. the atoms: $X = \sum_i x_i \mathbb{I}_{A_i}$. As an example, consider picking a card at random from a deck of playing cards with four suits $\clubsuit, \diamondsuit, \heartsuit, \spadesuit$. We can represent the suit of the card selected by an integer-valued random variable defined by $X(\{\heartsuit\}) = 1$, $X(\{\diamondsuit\}) = 2$, $X(\{\clubsuit\}) = 3$, $X(\{\spadesuit\}) = 4$. Then the partition generated by X consists of four sub-collections of cards where each sub-collection contains cards of the same suit, i.e., $\mathcal{P}(X) = \{X^{-1}(1), X^{-1}(2), X^{-1}(3), X^{-1}(4)\} = \{\{\heartsuit\}, \{\diamondsuit\}, \{\clubsuit\}, \{\spadesuit\}\}$ where the set of all cards is $\Omega = \{\heartsuit\} \cup \{\diamondsuit\} \cup \{\clubsuit\} \cup \{\spadesuit\}$.

In the next example we work out the respective partitions generated by two random variables on the three-period binomial model.

Example 6.7. Consider a three-period binomial model. Find the partitions $\mathcal{P}(S_2)$ and $\mathcal{P}(Y)$ generated by the stock price at time 2 and the Bernoulli random variable $Y := \mathbb{I}_{\{\omega_2 = \mathsf{U}\}}$ that only takes on a nonzero value of 1 when the second market move is up.

Solution. The price $X \equiv S_2$ admits three possible values: $X \in \{S_0 d^2, S_0 ud, S_0 u^2\}$. The partition $\mathcal{P}(X) \equiv \mathcal{P}(S_2)$ has three atoms:

$$\begin{aligned} X^{-1}(S_0 d^2) &= \{\omega_1 = \mathsf{D}, \omega_2 = \mathsf{D}\} = \{\mathsf{DDD}, \mathsf{DDU}\} = A_{\mathsf{DD}}, \\ X^{-1}(S_0 ud) &= \{\omega_1 = \mathsf{D}, \omega_2 = \mathsf{U}\} \cup \{\omega_1 = \mathsf{U}, \omega_2 = \mathsf{D}\} \\ &= \{\mathsf{DUD}, \mathsf{DUU}, \mathsf{UDD}, \mathsf{UDU}\} = A_{\mathsf{DU}} \cup A_{\mathsf{UD}}, \\ X^{-1}(S_0 u^2) &= \{\omega_1 = \mathsf{U}, \omega_2 = \mathsf{U}\} = \{\mathsf{UUD}, \mathsf{UUU}\} = A_{\mathsf{UU}}. \end{aligned}$$

Since $Y \in \{0, 1\}$, then $\mathcal{P}(Y)$ has two atoms:

$$\begin{aligned} Y^{-1}(0) &= \{\omega_2 = \mathsf{D}\} = \{\mathsf{DDD}, \mathsf{DDU}, \mathsf{UDD}, \mathsf{UDU}\}, \\ Y^{-1}(1) &= \{\omega_2 = \mathsf{U}\} = \{\mathsf{DUD}, \mathsf{DUU}, \mathsf{UUD}, \mathsf{UUU}\}. \end{aligned}$$

□

Consider a collection of discrete random variables, $X_k(\Omega) \to \mathbb{R}$, $k = 1, 2 \ldots, M$, defined on a common sample space Ω. Let each X_k have support $X_k(\Omega) = \mathsf{S}_k$ and hence the random vector $\mathbf{X} = (X_1, X_2, \ldots, X_M)$ has support $\mathsf{S}_{\mathbf{X}} \equiv \{(x_1, \ldots, x_M) \ : \ x_k \in \mathsf{S}_k, k = 1, \ldots, M\}$. We may then define a partition generated by the random vector $\mathbf{X}$ as the collection $\mathcal{P}(\mathbf{X}) \equiv \mathcal{P}(X_1, X_2, \ldots, X_M) = \{A_{\mathbf{x}}; \mathbf{x} \in \mathsf{S}_{\mathbf{X}}\}$ with each atom $A_{\mathbf{x}} \equiv A_{(x_1, \ldots, x_M)}$ given by

$$A_{(x_1, \ldots, x_M)} := \{X_1 = x_1, \ldots, X_M = x_M\} \equiv \{\omega \in \Omega \ : \ \mathbf{X}(\omega) = \mathbf{x}\}.$$

In the N-period binomial model, we allow the market to move N times (or toss a coin

N times) to obtain Ω_N—the set of 2^N possible sequences $\omega = \omega_1 \ldots \omega_N$ of N-tuples of U's and D's. Consider the vector $(X_1, \ldots, X_n)$, $1 \leqslant n \leqslant N$ of i.i.d. Bernoulli random variables $X_i = \mathbb{I}_{\{\omega_i = \mathsf{U}\}}$, $1 \leqslant i \leqslant n$. That is, X_i takes on value 1 if the ith market move ω_i is up and zero if it's down. Hence, any given sequence $\omega_1^*, \ldots, \omega_n^*$ that specifies (resolves) the first n moves can be equivalently described in terms of the sequence of 0's and 1's, i.e., the event $\{\omega_1 = \omega_1^*, \ldots, \omega_n = \omega_n^*\}$ is equivalent to $\{X_1 = x_1^*, \ldots, X_n = x_n^*\}$ where $x_1^* = \mathbb{I}_{\{\omega_1^* = \mathsf{U}\}}, \ldots, x_n^* = \mathbb{I}_{\{\omega_n^* = \mathsf{U}\}}$. Hence, the atoms and the partitions that are generated by $(X_1, \ldots, X_n)$ and by the subsets of ω's that resolve the first n moves are equivalent. In particular, we have the atoms $A_{\omega_1^* \ldots \omega_n^*} \equiv A_{(x_1^*, \ldots, x_n^*)}$ of the partition $\mathcal{P}_n \equiv \mathcal{P}(X_1, X_2, \ldots, X_n)$:

$$A_{\omega_1^* \ldots \omega_n^*} := \{\omega = \omega_1 \ldots \omega_N \in \Omega_N \ : \ \omega_1 = \omega_1^*, \ldots, \omega_n = \omega_n^*\}, \tag{6.13}$$

for every $\omega_j^* \in \{\mathsf{D}, \mathsf{U}\}, j = 1, \ldots, n$. That is, each such atom consists of fixing the first n letters in ω with all others $\omega_{n+1}, \ldots, \omega_N$ allowed to be U or D. Since $|\Omega_n| = 2^n$, the partition $\mathcal{P}_n$ contains 2^n atoms. For example, the first partition $\mathcal{P}_1 = \{A_{\mathsf{D}}, A_{\mathsf{U}}\} = \{A_{\omega_1} : \omega_1 \in \{\mathsf{U}, \mathsf{D}\}\}$ resolves the first market move, the second $\mathcal{P}_2 = \{A_{\mathsf{DD}}, A_{\mathsf{DU}}, A_{\mathsf{UD}}, A_{\mathsf{UU}}\} = \{A_{\omega_1\omega_2} : \omega_1, \omega_2 \in \{\mathsf{U}, \mathsf{D}\}\}$ resolves the first two market moves, the third partition is given by $\mathcal{P}_3 = \{A_{\mathsf{DDD}}, A_{\mathsf{DDU}}, A_{\mathsf{DUD}}, A_{\mathsf{DUU}}, A_{\mathsf{UDD}}, A_{\mathsf{UDU}}, A_{\mathsf{UUD}}, A_{\mathsf{UUU}}\}$ and this resolves the first three moves, and so on. Finally, for $n = N$ the partition $\mathcal{P}_N$ contains 2^N atoms where each atom has only one element:

$$\mathcal{P}_N = \{\{\omega^1\}, \{\omega^2\}, \ldots, \{\omega^{2^N}\}\}.$$

The kth member $\{\omega^k\}$ of this collection is the singleton set $\{\omega_1^{(k)} \ldots \omega_N^{(k)}\}$ representing the complete N-period path $\#k$ among 2^N possible distinct paths. This partition represents the full information about the actual market scenario ω where all N moves are resolved.

Note that the partitions $\mathcal{P}_n$ can be generated by the first n stock prices as well:

$$\mathcal{P}_n = \mathcal{P}(S_1, S_2, \ldots, S_n), \quad 1 \leqslant n \leqslant N.$$

This is seen, for example, by noting that the value of ω_j can be revealed from the ratio $\frac{S_j}{S_{j-1}}$. In particular, knowing S_1 reveals a value $\omega_1 = \omega_1^*$. Knowing S_1 and S_2 then reveals the string $\omega_1\omega_2 = \omega_1^*\omega_2^*$. Iterating gives us that knowing the values $S_1, S_2, \ldots, S_{n-1}$ is knowledge of $\omega_1^* \ldots \omega_{n-1}^*$ and combining this with knowing S_n gives $\omega_1^* \ldots \omega_n^*$. Hence, knowledge of the first n binomial prices is equivalent to knowing the first n market moves.

6.2.1.2 Refinements of Partitions

Let $\mathcal{P}$ and $\mathcal{Q}$ be two partitions of a sample space Ω. $\mathcal{Q}$ is said to be a *refinement* of $\mathcal{P}$, denoted $\mathcal{P} \preceq \mathcal{Q}$, if every atom of $\mathcal{P}$ is expressible as a union of atoms of $\mathcal{Q}$. So to speak, partition $\mathcal{Q}$ is obtained by breaking down the atoms of $\mathcal{P}$. By splitting any atom of a partition into two or more disjoint and exhaustive subsets, we can obtain a refinement of the original partition. For example, by slicing a pizza, one can obtain a sequence of refinements (see Figure 6.2).

Another example of a sequence of refinements is the following:

$$\text{body} \preceq \text{molecules} \preceq \text{atoms} \preceq \text{elementary particles}.$$

Given a sequence of random variables $\{X_k\}_{k \geqslant 1}$, we can generate a refinement as follows:

$$\mathcal{P}(X_1) \preceq \mathcal{P}(X_1, X_2) \preceq \mathcal{P}(X_1, X_2, X_3) \preceq \ldots$$

Clearly, every addition of a random variable can only increase the amount of available information.

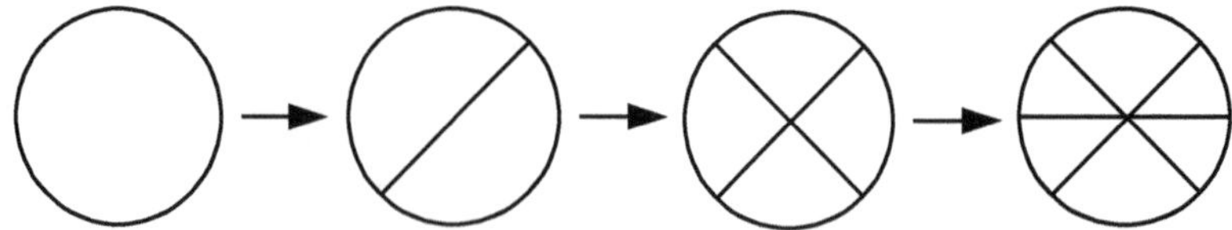

FIGURE 6.2: Slicing a pizza produces a new refinement with every new cut made.

A partition represents information about a stochastic process available to us at a particular moment. A refinement of a partition represents a transition from one information level to another level that is more detailed. As time passes, we learn more and more about the process and its history. The information is accumulated and catalogued using a sequence of partitions. A sequence of refinements,

$$\mathcal{P}_0 \preceq \mathcal{P}_1 \preceq \cdots \preceq \mathcal{P}_N,$$

is called an *information structure*. The collection of partitions $\{\mathcal{P}_n\}_{n=0,\dots,N}$ where each $\mathcal{P}_n$ is defined by the atoms in (6.13) is an information structure for the N-period binomial model. The information in this case is based on market moves. In the beginning, the actual state of the model is unknown. This absence of information is represented by the partition $\mathcal{P}_0 = \{\Omega_N\}$. At the end of each period, a new portion of information (i.e., another market move) is revealed. At the end of period n we know the first n market moves $\omega_1, \omega_2, \dots, \omega_n$. Clearly, for every $n = 1, 2, \dots, N-1$, the partition $\mathcal{P}_n$ is a refinement of $\mathcal{P}_{n+1}$. At time $t = 1$, we have our first refinement $\mathcal{P}_0 \preceq \mathcal{P}_1 = \{A_{\mathsf{D}}, A_{\mathsf{U}}\}$; at time $t = 2$, we have $\mathcal{P}_1 \preceq \mathcal{P}_2 = \{A_{\mathsf{DD}}, A_{\mathsf{DU}}, A_{\mathsf{UD}}, A_{\mathsf{UU}}\}$, and so on. Therefore, the partitions form an information structure, i.e., $\mathcal{P}_0 \preceq \mathcal{P}_1 \preceq \cdots \preceq \mathcal{P}_N$. Figure 6.3 illustrates the information structure for Ω_3.

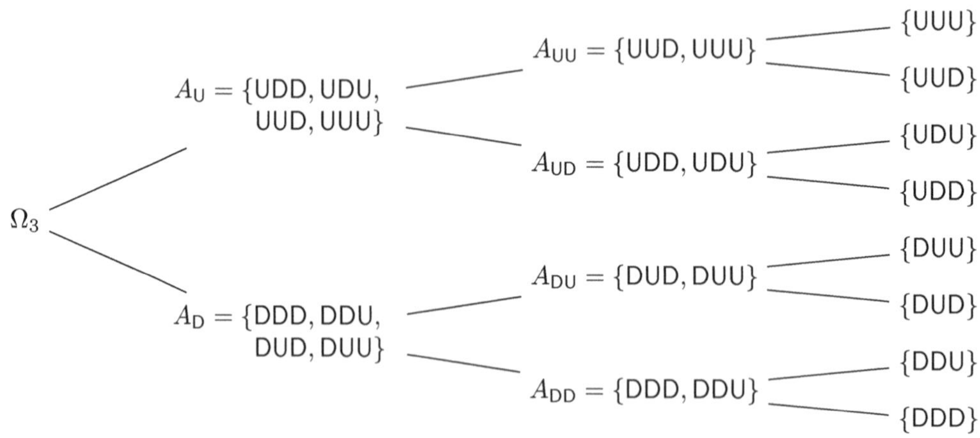

FIGURE 6.3: An information structure for Ω_3.

6.2.2 Sigma-Algebras

Suppose that we are given a partition $\mathcal{P}$ of a sample space Ω that represents the set of available information. We are able to say which event from collection $\mathcal{P}$ has occurred, but we might not know the actual outcome. Clearly, if we can say whether events A and/or B has occurred, then we can say whether events A^{C}, $A \cap B$, and $A \cup B$ occurred. Based on the available information, we can construct a collection of all "observable" events. That is, we are able to comment on the occurrence of every event from such a collection. The collection of all possible events of interest forms a collection of sets that is known as a σ-algebra.

Definition 6.5. A σ-algebra (or σ-field) $\mathcal{F}$ on a nonempty set Ω is a collection of subsets of Ω with the following properties:

1. $\emptyset \in \mathcal{F}$;
2. $E \in \mathcal{F} \implies E^{\mathsf{C}} \in \mathcal{F}$;
3. for every countable collection $\{E_i\}_{i\geqslant 1} \in \mathcal{F}$, we have $\cup_{i\geqslant 1} E_i \in \mathcal{F}$.

The pair $(\Omega, \mathcal{F})$ is called a *measurable space.*

In other words, a σ-algebra on a set Ω is a nonempty collection $\mathcal{F}$ of subsets of Ω (including Ω itself) that is closed under taking complementation and countable unions of its members. As follows from Properties 2 and 3 and De Morgan's Laws, a σ-algebra is closed under countable intersections of its elements as well:

$$\{E_i\}_{i\geqslant 1} \in \mathcal{F} \implies \{E_i^{\mathsf{C}}\}_{i\geqslant 1} \in \mathcal{F} \implies \cup_{i\geqslant 1} E_i^{\mathsf{C}} \in \mathcal{F} \implies (\cap_{i\geqslant 1} E_i)^{\mathsf{C}} \in \mathcal{F} \implies \cap_{i\geqslant 1} E_i \in \mathcal{F}.$$

Notice that if Ω is a finite sample space, then it is sufficient to say that

3*. for every pair $E_1, E_2 \in \mathcal{F}$ we have $E_1 \cup E_2 \in \mathcal{F}$,

since there are only finitely many different subsets of a finite set. Note that a collection of subsets of Ω is called an *algebra* if it contains Ω and is closed w.r.t. the formation of complements and *finite* unions. Thus, in the case of a finite set Ω, every algebra on Ω is a σ-algebra on Ω.

The following are examples of σ-algebras.

(a) The minimal σ-algebra consisting only of the empty set and the set Ω: $\mathcal{F}_0 = \{\emptyset, \Omega\}$. It is also called a *trivial σ-algebra.*

(b) Let E be a subset of Ω such that $E \neq \emptyset$ and $E^{\mathsf{C}} \neq \emptyset$. Then there exists a smallest σ-algebra containing E: $\mathcal{F}_E = \{\emptyset, E, E^{\mathsf{C}}, \Omega\}$.

(c) The power set of Ω: $2^{\Omega} = \{E : E \subset \Omega\} \cup \emptyset$.

An interesting fact is that an intersection of any two σ-algebras on the same set Ω is again a σ-algebra on Ω, as is proved in Theorem 6.1. In contrast, a union of two σ-algebras is not necessarily a σ-algebra (and may not even be an algebra). As a simple example, consider $\Omega = \{a, b, c\}$ with three elements. Let $A = \{a\}$ and $B = \{b\}$. Then the union of $\mathcal{F}_A = \{\emptyset, A, A^{\mathsf{C}}, \Omega\}$ and $\mathcal{F}_B = \{\emptyset, B, B^{\mathsf{C}}, \Omega\}$ is

$$\mathcal{F}_A \cup \mathcal{F}_B = \{\emptyset, A, B, A^{\mathsf{C}}, B^{\mathsf{C}}, \Omega\} = \{\emptyset, \{a\}, \{b\}, \{b, c\}, \{a, c\}, \{a, b, c\}\}.$$

This collection is not an algebra (hence not a σ-algebra) since it does not contain $A \cup B = \{a, b\}$ and $\{a, b\}^{\mathsf{C}} = \{c\}$. However, the intersection $\mathcal{F}_A \cap \mathcal{F}_B = \{\emptyset, \Omega\}$ is the trivial σ-algebra. The result just below tells us that any intersection over a collection of (countable or uncountable) σ-algebras is again a σ-algebra.

Theorem 6.1. *An intersection of a family of σ-algebras on Ω is a σ-algebra on Ω.*

Proof. Let $\mathcal{C}$ be a family of σ-algebras on Ω. Let $\mathcal{F}_{\mathcal{C}} = \bigcap\{\mathcal{F} : \mathcal{F} \in \mathcal{C}\}$ be the intersection of all members of $\mathcal{C}$. Using the fact that every $\mathcal{F} \in \mathcal{C}$ is a σ-algebra, then we have:

1. $\emptyset \in \mathcal{F}$ for all $\mathcal{F} \in \mathcal{C}$, so $\emptyset \in \mathcal{F}_{\mathcal{C}}$.
2. If $E \in \mathcal{F}_{\mathcal{C}}$, then $E \in \mathcal{F}$ and hence $E^{\complement} \in \mathcal{F}$ for all $\mathcal{F} \in \mathcal{C}$. Therefore, $E^{\complement} \in \mathcal{F}_{\mathcal{C}}$.
3. If $E_k \in \mathcal{F}_{\mathcal{C}}$, then $E_k \in \mathcal{F}$ for $k = 1, 2 \ldots$, and hence $\cup_{k\geqslant 1} E_k \in \mathcal{F}$ for all $\mathcal{F} \in \mathcal{C}$. Thus, $\cup_{k\geqslant 1} E_k \in \mathcal{F}_{\mathcal{C}}$. □

Note that any intersection of σ-algebras is a nonempty collection since it includes at least two sets: $\emptyset$ and Ω. Theorem 6.1 now also allows us to construct the *smallest σ-algebra that includes a given collection* $\mathcal{A}$ of subsets of Ω. Let $\mathcal{C}(\mathcal{A})$ denote the family of all σ-algebras on Ω that contain $\mathcal{A}$. That is, let $\mathcal{C}(\mathcal{A}) \equiv \{\mathcal{F} : \mathcal{A} \subset \mathcal{F}$ and $\mathcal{F}$ is a σ-algebra on $\Omega\}$. Note that the power set 2^{Ω} is the collection of all subsets of Ω (including $\emptyset$ and Ω) and hence $\mathcal{A} \subset 2^{\Omega}$. Since 2^{Ω} is itself a σ-algebra, the collection $\mathcal{C}(\mathcal{A})$ is nonempty, i.e., $\mathcal{F} = 2^{\Omega}$ is one such $\mathcal{F}$ in $\mathcal{C}(\mathcal{A})$. The *σ-algebra generated by a collection* $\mathcal{A} \subset 2^{\Omega}$, denoted as $\sigma(\mathcal{A})$, is defined by the intersection:

$$\sigma(\mathcal{A}) := \bigcap\{\mathcal{F} : \mathcal{F} \in \mathcal{C}(\mathcal{A})\}.$$

By Theorem 6.1, $\sigma(\mathcal{A})$ is a σ-algebra, and by construction it is the smallest σ-algebra that contains all sets (events) from $\mathcal{A}$.

Example 6.8. Let A and B be nonempty subsets of Ω such that $A \cap B = \emptyset$. Find $\sigma(\{A, B\})$.

Solution. We can construct the smallest σ-algebra $\mathcal{F} = \sigma(\mathcal{A}) \equiv \sigma(\{A, B\})$ by requiring that the sets $A, B \in \mathcal{F}$. As well, their complements and intersections (or unions) must be in $\mathcal{F}$. Listing all the elements gives:

$$\sigma(\{A, B\}) = \{\emptyset, A, B, A^{\complement}, B^{\complement}, A \cup B, A^{\complement} \cap B^{\complement}, \Omega\}.$$

It is simple to check that $\sigma(\{A, B\})$ satisfies the properties in Definition 6.5 and that it is the smallest $\mathcal{F}$ such that $\{A, B\} \subset \mathcal{F}$. □

A very important example of a σ-algebra that is central to measure theory and general probability theory is the *Borel σ-algebra* on $\mathbb{R}$, denoted $\mathcal{B}(\mathbb{R})$. Every member of the Borel σ-algebra, $B \in \mathcal{B}(\mathbb{R})$, is called a *Borel set.* The pair $(\mathbb{R}, \mathcal{B}(\mathbb{R}))$ is called a *Borel space.* The collection $\mathcal{B}(\mathbb{R})$ of all Borel sets in $\mathbb{R}$ is the smallest σ-algebra on $\mathbb{R}$ that contains all intervals (including zero length intervals or points) in $\mathbb{R}$. One way to form this collection is to start with all intervals and then add in all countable unions, countable intersections, and relative complements. $\mathcal{B}(\mathbb{R})$ is the σ-algebra generated by all the intervals in $\mathbb{R}$:

$$\mathcal{B}(\mathbb{R}) = \bigcap\{\mathcal{F} : \mathcal{F} \text{ is a } \sigma\text{-algebra containing all intervals in } \mathbb{R}\}.$$

To gain a better understanding of $\mathcal{B}(\mathbb{R})$ we now observe that it has equivalent representations as the σ-algebra generated by the collection of all open sets in $\mathbb{R}$, or all intervals of the form (a, b), or $[a, b]$, or (a, ∞), or $[a, \infty)$, or $(-\infty, b)$, or $(-\infty, b]$. That is, the Borel σ-algebra $\mathcal{B}(\mathbb{R})$ is equivalently given by each $\sigma(\mathcal{A}_i)$, $i = 1, \ldots, 6$, where

$$\mathcal{A}_1 = \{(a, b) : a, b \in \mathbb{R}\},\ \mathcal{A}_2 = \{[a, b] : a, b \in \mathbb{R}\},\ \mathcal{A}_3 = \{(a, \infty) : a \in \mathbb{R}\},$$
$$\mathcal{A}_4 = \{[a, \infty) : a \in \mathbb{R}\},\ \mathcal{A}_5 = \{(-\infty, b) : b \in \mathbb{R}\},\ \mathcal{A}_6 = \{(-\infty, b] : b \in \mathbb{R}\}.$$

In fact, any open set in $\mathbb{R}$ is a countable union of open intervals, and hence this gives $\mathcal{B}(\mathbb{R}) = \sigma(\mathcal{O})$, where $\mathcal{O}$ is the set of all open sets in $\mathbb{R}$. Other representations of $\mathcal{B}(\mathbb{R})$ are also possible. It is not difficult to prove that $\sigma(\mathcal{A}_1) = \ldots = \sigma(\mathcal{A}_6)$, i.e., are all equivalent representations of $\mathcal{B}(\mathbb{R})$. For instance, the equivalences $\sigma(\mathcal{A}_1) = \sigma(\mathcal{A}_2)$ and $\sigma(\mathcal{A}_1) = \sigma(\mathcal{A}_5)$ follow by the respective identities for countable intersections and unions over open intervals:

$$[a,b] = \bigcap_{n=1}^{\infty} (a - \frac{1}{n}, b + \frac{1}{n}) \quad \text{and} \quad (a,b) = \bigcup_{n=1}^{\infty} (-\infty, b) \setminus (-\infty, a + \frac{1}{n})$$

where $A \setminus B \equiv A \cap B^{\mathsf{C}}$. By closure of a σ-algebra with respect to taking countable unions and intersections, the sets on the right-hand side are all in the $\mathcal{B}(\mathbb{R})$. The other equivalences follow by employing similar identities.

In $\mathbb{R}^n$, $n \geqslant 2$, the Borel σ-algebra is denoted by $\mathcal{B}_n \equiv \mathcal{B}(\mathbb{R}^n)$. The Borel sets $B \in \mathcal{B}_n$ are formed as n-outer products of Borel sets in $\mathbb{R}$, i.e., $\mathcal{B}_n = \mathcal{B} \times \ldots \times \mathcal{B}$ where $\mathcal{B} \equiv \mathcal{B}(\mathbb{R})$. In particular, every Borel set $B \in \mathcal{B}_n$ has the form $B_1 \times \ldots \times B_n$, where $B_1, \ldots, B_n \in \mathcal{B}$. So these sets are n-outer products of all intervals (open, closed, semi-open, or points) $I_1, \ldots, I_n \subset \mathbb{R}$. For example, these sets include n-dimensional rectangles $B = [a_1, b_1] \times \ldots \times [a_n, b_n]$ where $a_1 < b_1, \ldots, a_n < b_n$, single points in $(a_1, \ldots, a_n) \in \mathbb{R}^n$, lines and hyperplanes, etc. For $n = 2$ dimensions, the Borel sets are outer products $I_1 \times I_2$ on the plane $\mathbb{R}^2$ where I_1, I_2 are any intervals.

The concept of Borel functions is important in the defining property of random variables and their expectation. A single variable real-valued function $g\colon \mathbb{R} \to \mathbb{R}$ is said to be *Borel measurable (or simply a Borel function)* on $\mathbb{R}$ if

$$g^{-1}(B) \equiv \{x \in \mathbb{R} : g(x) \in B\} \in \mathcal{B}(\mathbb{R}) \text{ for all } B \in \mathcal{B}(\mathbb{R}).$$

That is, its pre-image of any Borel set in $\mathbb{R}$ is a Borel set in $\mathbb{R}$. Borel functions include practically all (and for our purposes all) types of real-valued functions. For example, this includes every continuous or piecewise continuous function, indicator functions of Borel sets, simple functions over Borel sets, the limit of a sequence of Borel functions, and the list goes on! It is actually quite challenging to come up with a function that is not a Borel function. Similarly, a real-valued function of two variables, $g : \mathbb{R}^2 \to \mathbb{R}$, is said to be a Borel function if $g^{-1}(B) \equiv \{(x,y) \in \mathbb{R}^2 \colon g(x,y) \in B\} \in \mathcal{B}_2$ for every $B \in \mathcal{B}$. For any dimension $n \geqslant 1$, $g : \mathbb{R}^n \to \mathbb{R}$ is a Borel function if $g^{-1}(B) \equiv \{(x_1, \ldots, x_n) \in \mathbb{R}^n \colon g(x_1, \ldots, x_n) \in B\} \in \mathcal{B}_n$ for all $B \in \mathcal{B}(\mathbb{R})$, i.e., the pre-image of g of any Borel set B in $\mathbb{R}$ is a Borel set in $\mathbb{R}^n$.

6.2.2.1 Construction of a Sigma-Algebra from a Partition

In the previous section we introduced the concept of a σ-algebra generated by a collection $\mathcal{A}$ of subsets of some given set (sample space) Ω. We can hence consider the special case in which the collection corresponds to a partition $\mathcal{P}$ of Ω, i.e., $\mathcal{A} = \mathcal{P}$ is a collection of atoms. In this situation the σ-algebra generated by the collection is denoted by $\sigma(\mathcal{P})$ and it is the smallest σ-algebra containing the collection $\mathcal{P}$. Since a partition consists of mutually disjoint and exhaustive sets in Ω, then $\sigma(\mathcal{P})$ corresponds to the power set $2^{\mathcal{P}}$ of $\mathcal{P}$. The result just below states this formally.

Theorem 6.2. *Consider a partition $\mathcal{P} = \{A_i\}_{i \in \mathcal{I}}$ of a set Ω, where $\mathcal{I}$ is a finite or countably infinite index set: $\mathcal{I} = \{1, 2, \ldots, M\}$ or $\mathcal{I} = \mathbb{N}$. The smallest σ-algebra, $\sigma(\mathcal{P})$, generated by $\mathcal{P}$ consists of sets of the form*

$$E = \bigcup_{i \in I} A_i, \tag{6.14}$$

where $I \subset \mathcal{I}$ is a set of indices including $I = \emptyset$ and $I = \mathcal{I}$.

Proof. Let $\mathcal{F}$ be a collection of sets of the form (6.14). First, by taking in (6.14) a set

$I = \{i\}$ with only one index, we have that $\mathcal{F}$ includes the partition $\mathcal{P}$. Second, let us show that $\mathcal{F}$ is a σ-algebra. If we take $I = \emptyset$ and $I = \mathcal{I}$ then we obtain that $E = \emptyset$ and $\mathrm{E} = \Omega$ are both in $\mathcal{F}$. If $E \in \mathcal{F}$, then $E = \cup_{i \in I} A_i$ for some $I \subset \mathcal{I}$. Hence $E^{\mathsf{C}} = \cup_{i \in \mathcal{I} \setminus I} A_i \in \mathcal{F}$ since $\mathcal{I} \setminus I$ is again a set of indices. Finally, let $E_1, E_2, \ldots \in \mathcal{F}$. For each E_j, there exists a set of indices I_j. The union of all E_j's is again a set of the form (6.14):

$$\bigcup_{j \geqslant 1} E_j = \bigcup_{j \geqslant 1} \bigcup_{i \in I_j} A_i = \bigcup_{i \in \cup_{j \geqslant 1} I_j} A_i$$

where $\cup_{j \geqslant 1} I_j \subset \mathcal{I}$. Finally, removing any element from $\mathcal{F}$ leads to violating the property that a σ-algebra is closed under taking countable unions. Therefore, $\mathcal{F}$ is the smallest σ-algebra generated by the partition $\mathcal{P}$. □

For instance, the σ-algebra generated by the partition $\mathcal{P} = \{E, E^{\mathsf{C}}\}$ for some $E \subset \Omega$ is $\sigma(\mathcal{P}) = \{\emptyset, E, E^{\mathsf{C}}, \Omega\} = 2^{\mathcal{P}}$. The σ-algebra constructed in Example 6.8 is generated by the partition $\mathcal{P} = \{A, B, A^{\mathsf{C}} \cap B^{\mathsf{C}}\}$ (since $A \cap B = \emptyset$). As we can see from (6.14), every element of $\sigma(\mathcal{P})$ is formed from atoms of $\mathcal{P}$. For a finite partition $\mathcal{P}$, the total number of possible combinations is $2^{|\mathcal{P}|}$. Hence, $|\sigma(\mathcal{P})| = 2^{|\mathcal{P}|}$ as must be the case since $\sigma(\mathcal{P}) = 2^{\mathcal{P}}$.

As an explicit construction, let us consider the N-period binomial model. We have seen that the partitions $\mathcal{P}_n, 0 \leqslant n \leqslant N$ are generated by the first n market moves with 2^n atoms in $\mathcal{P}_n$ given by (6.13):

$$\mathcal{P}_n = \mathcal{P}(\{A_{\omega_1 \ldots \omega_n} : \omega_1, \ldots, \omega_n \in \{\mathsf{U}, \mathsf{D}\}\}). \tag{6.15}$$

We can now construct the σ-algebras $\mathcal{F}_n = \sigma(\mathcal{P}_n) = 2^{\mathcal{P}_n}, 0 \leqslant n \leqslant N$, as the power sets generated by such partitions. For time $n = 0$ (no market moves) we have the trivial case $\mathcal{F}_0 = \sigma(\mathcal{P}_0) = \sigma(\Omega_N) = \{\emptyset, \Omega_N\}$. For time $n = 1$ (first market move is resolved):

$$\mathcal{F}_1 = \sigma(\mathcal{P}_1) = \sigma(\{A_{\mathsf{U}}, A_{\mathsf{D}}\}) = \{\emptyset, A_{\mathsf{U}}, A_{\mathsf{D}}, \Omega_N\}.$$

For time $n = 2$ (first two market moves are resolved)

$$\begin{aligned} \mathcal{F}_2 &= \sigma(\mathcal{P}_2) = \sigma(\{A_{\mathsf{UU}}, A_{\mathsf{UD}}, A_{\mathsf{DU}}, A_{\mathsf{DD}}\}) \\ &= \{\emptyset, A_{\mathsf{U}}, A_{\mathsf{D}}, A_{\mathsf{UU}}, A_{\mathsf{UD}}, A_{\mathsf{DU}}, A_{\mathsf{DD}}, A_{\mathsf{UU}}^{\mathsf{C}}, A_{\mathsf{UD}}^{\mathsf{C}}, A_{\mathsf{DU}}^{\mathsf{C}}, A_{\mathsf{DD}}^{\mathsf{C}}, \\ &\quad A_{\mathsf{UU}} \cup A_{\mathsf{DU}}, A_{\mathsf{UU}} \cup A_{\mathsf{DD}}, A_{\mathsf{UD}} \cup A_{\mathsf{DU}}, A_{\mathsf{UD}} \cup A_{\mathsf{DD}}, \Omega_N\}. \end{aligned}$$

This collection contains all the possible events of interest in which we can resolve the first two market moves (without any information on the third and subsequent moves). Note that $\mathcal{F}_1 \subset \mathcal{F}_2$, i.e., $\mathcal{F}_2$ includes all the events in $\mathcal{F}_1$. For example, A_{U} is the event that the first move is up, A_{UU} is the event that the first two moves are up, $A_{\mathsf{UU}}^{\mathsf{C}}$ is the event that the first two moves are both not up, $A_{\mathsf{UU}} \cup A_{\mathsf{DD}}$ is the event that the first two moves are either both up or both down, $(A_{\mathsf{UU}} \cup A_{\mathsf{DD}})^{\mathsf{C}} = A_{\mathsf{UD}} \cup A_{\mathsf{DU}}$ is the event that the first two moves are different, and so on.

Note that $\mathcal{F}_0$ has $2^{2^0} = 2$ events, $\mathcal{F}_1$ has $2^{2^1} = 4$ events, $\mathcal{F}_2$ has $2^{2^2} = 2^4 = 16$ events. For $n = 3$, then $\mathcal{F}_3 = \sigma(\mathcal{P}_3)$ has $2^{2^3} = 2^8 = 256$ events. For any $0 \leqslant n \geqslant N$, $\mathcal{F}_n = 2^{\mathcal{P}_n}$ is the power set of $\mathcal{P}_n$ having $2^{|\mathcal{P}_n|} = 2^{2^n}$ subsets (events) $E \subset \Omega_N$. At the terminal time $n = N$, all possible moves are resolved and $\mathcal{F}_N = 2^{\mathcal{P}_N}$ contains all the possible 2^{2^N} events of interest.

6.2.2.2 Sigma-Algebra Generated by a Random Variable

We recall that by observing the value of a random variable X, we may comment on the actual value of $\omega \in \Omega$. Formally, a real-valued[1] random variable $X : \Omega \to \mathbb{R}$ on a probability

[1] Some random variables can also take on infinite values ∞ or $-\infty$. These are defined on the extended real axis. That is, $X^{-1}(\infty) \equiv \{\omega \in \Omega \colon X(\omega) = \infty\}$ and $X^{-1}(-\infty) \equiv \{\omega \in \Omega \colon X(\omega) = -\infty\}$ are

space $(\Omega, \mathcal{F}, \mathbb{P})$ is defined as a set function such that $X^{-1}((-\infty, b])$ is in $\mathcal{F}$, for all $b \in \mathbb{R}$:

$$\{X \in (-\infty, b]\} \equiv \{\omega \in \Omega \; : \; -\infty < X(\omega) \leqslant b\} \in \mathcal{F}.$$

Based on the properties of the Borel sets, we can re-state this definition using any type of interval or simply as $X^{-1}(B) \equiv \{X \in B\} \in \mathcal{F}$ for all $B \in \mathcal{B}(\mathbb{R})$. We therefore interpret a random variable to be an *$\mathcal{F}$-measurable* function. Later we shall precisely define the measurability of a random variable. We will see that X is said to be *$\mathcal{F}$-measurable* if $\sigma(X) \subset \mathcal{F}$ where $\sigma(X)$ denotes the σ-algebra generated by X. At this point we need to introduce the concept of σ-algebras generated by random variables.

All possible information extracted from X can be represented in the form of a σ-algebra, denoted $\sigma(X)$. We say that $\sigma(X)$ is generated by X. That is, $\sigma(X)$ should contain all events of interest to us with respect to X. Such events should include all those of the form $\{X \leqslant b\}$, $\{X \geqslant a\}$, $\{a < X \leqslant b\}$, and so on. More generally, all events $\{X \in B\}$, $B \in \mathcal{B}(\mathbb{R})$, should be contained in $\sigma(X)$. This leads us to the following definition.

Definition 6.6. The σ-algebra, denoted by $\sigma(X)$, generated by a random variable $X : \Omega \to \mathbb{R}$ is the σ-algebra generated by the collection of all subsets of Ω of the form $\{X \in B\}$, $B \in \mathcal{B}(\mathbb{R})$.

Note that this tells us that generally $\sigma(X)$ is the smallest σ-algebra containing all subsets of Ω of the form $X^{-1}(J) \equiv \{X \in J\}$ for every interval J in $\mathbb{R}$. For a discrete random variable X, this therefore corresponds to simply defining $\sigma(X) = \sigma(\mathcal{P}(X))$, i.e., as the σ-algebra generated by the partition $\mathcal{P}(X)$. Indeed, let $\mathcal{I} = \{1, 2, \ldots, M\}$ or $\mathcal{I} = \mathbb{N}$ where the support of X is the countable set of numbers $\{x_1, x_2, \ldots\}$. Given any interval J, the set $X^{-1}(J) = \bigcup_{i \in \mathcal{J}} X^{-1}(x_i) = \bigcup_{i \in \mathcal{J}} A_i$ for some subset of indicies $\mathcal{J} \subset \mathcal{I}$. That is, every set $X^{-1}(J)$ is a union of a sub-collection of atoms $\{A_i\}_{i \in \mathcal{J}}$ in $\mathcal{P}(X)$. Hence, the σ-algebra generated by all the sets $X^{-1}(J)$ is the same as the σ-algebra generated by the partition $\mathcal{P}(X)$. So in this case $\sigma(X) = \sigma(\mathcal{P}(X)) = 2^{\mathcal{P}(X)}$.

Example 6.9. Consider Example 6.7. Find $\sigma(S_2)$ and $\sigma(Y)$.

Solution. Using the respective partitions $\mathcal{P}(S_2)$ and $\mathcal{P}(Y)$ in Example 6.7 we have:

$$\begin{aligned}\sigma(S_2) = \sigma(\mathcal{P}(S_2)) &= \sigma(\{A_{\mathsf{DD}}, A_{\mathsf{DU}} \cup A_{\mathsf{UD}}, A_{\mathsf{UU}}\}) \\ &= \{\emptyset, A_{\mathsf{DD}}, A_{\mathsf{UU}}, A_{\mathsf{DU}} \cup A_{\mathsf{UD}}, A^{\mathsf{C}}_{\mathsf{DD}}, A^{\mathsf{C}}_{\mathsf{UU}}, A_{\mathsf{DD}} \cup A_{\mathsf{UU}}, \Omega_3\}\end{aligned}$$

and

$$\sigma(Y) = \sigma(\{\{\omega_2 = \mathsf{D}\}, \{\omega_2 = \mathsf{U}\}\}) = \{\emptyset, \{\omega_2 = \mathsf{D}\}, \{\omega_2 = \mathsf{U}\}, \Omega_3\} \square$$

Similarly, a sigma algebra can be generated from multiple random variables such as $X_1, \ldots, X_M$ or even by a countable or uncountable collection of random variables. In the case with M discrete random variables, $\sigma(X_1, \ldots, X_M) = \sigma(\mathcal{P}(X_1, \ldots, X_M))$. In the case where we have M random variables that are not necessarily discrete-valued, then $\sigma(X_1, \ldots, X_M)$ is the smallest σ-algebra containing all subsets of the form

$$\{\omega \in \Omega \; : \; X_i(\omega) \in B\} \text{ where } B \in \mathcal{B}(\mathbb{R}),\ 1 \leqslant i \leqslant M.$$

More generally, we consider a countable or uncountable index set Λ and define $\sigma(\{X_\lambda\}_{\lambda \in \Lambda})$ as the smallest σ-algebra generated by the collection of random variables $\{X_\lambda\}_{\lambda \in \Lambda}$. This σ-algebra consists of all events of the form $\{X_\lambda \in B\}$, $B \in \mathcal{B}(\mathbb{R})$, $\lambda \in \Lambda$.

---

mutually exclusive events in $\mathcal{F}$ for which we can assign respective probabilities of occurrence, $\mathbb{P}(X = \infty)$ and $\mathbb{P}(X = -\infty)$.

6.2.3 Filtration

As time passes, we acquire more information about the actual state of a stochastic model. At any instant, the current level of information available to us is represented by a partition or a σ-algebra. The flow of information can be modeled by a sequence of partitions that we named an information structure. Since it is more convenient to work with σ-algebras, we can similarly define a sequence of σ-algebras that accumulates more and more knowledge as time increases. We have seen this explicitly in the binomial model where more and more information about events is available with each new market move, where $\mathcal{F}_1 \subset \mathcal{F}_2 \subset \ldots \mathcal{F}_N$. The collection of such σ-algebras is an example of what is called a *filtration.* The following is a formal definition in the continuous-time case.

Definition 6.7. A *filtration* $\mathbb{F}$ is a sequence of σ-algebras $\{\mathcal{F}_t\}_{0\leqslant t\leqslant T}$ over a set Ω such that $\mathcal{F}_s \subset \mathcal{F}_t$ for all $0 \leqslant s \leqslant t \leqslant T$.

This definition implies that information can only increase (not decrease) with time. In the discrete-time case, a filtration is a finite sequence of σ-algebras $\{\mathcal{F}_n\}_{0\leqslant n\leqslant N} \equiv \{\mathcal{F}_0, \mathcal{F}_1, \ldots, \mathcal{F}_N\}$ with the property $\mathcal{F}_{n-1} \subset \mathcal{F}_n$, for all $1 \leqslant n \leqslant N$, i.e.,

$$\mathcal{F}_0 \subset \mathcal{F}_1 \subset \cdots \subset \mathcal{F}_N.$$

6.2.3.1 Construction of a Filtration from an Information Structure

Consider an information structure $\mathcal{P}_0 \preceq \mathcal{P}_1 \preceq \cdots \preceq \mathcal{P}_N$. Define $\mathcal{F}_n = \sigma(\mathcal{P}_n)$, $0 \leqslant n \leqslant N$. Then, the σ-algebras $\mathcal{F}_0, \mathcal{F}_1, \ldots, \mathcal{F}_N$ form a filtration. It is sufficient to prove the following assertion.

Proposition 6.3. *Suppose that $\mathcal{P}$ and $\mathcal{Q}$ are two partitions of Ω such that $\mathcal{P} \preceq \mathcal{Q}$. Then $\sigma(\mathcal{P}) \subset \sigma(\mathcal{Q})$.*

Proof. First, enumerate all atoms in the partitions $\mathcal{P}$ and $\mathcal{Q}$:

$$\mathcal{P} = \{A_i\}_{i\geqslant 1}, \quad \mathcal{Q} = \{B_i\}_{i\geqslant 1}.$$

Take any $E \in \sigma(\mathcal{P})$. Since $\sigma(\mathcal{P}) = \sigma(\{A_i\}_{i\geqslant 1})$, there exists an index set $I_E \subset \mathbb{N}$ such that $E = \bigcup_{i\in I_E} A_i$. Moreover, $\mathcal{P} \preceq \mathcal{Q}$ implies that for every $A_i \in \mathcal{P}$ there is a countable collection of atoms of $\mathcal{Q}$, $\{B_j\}_{j\geqslant I_i}$, for some index set $I_i \subset \mathbb{N}$, such that $A_i = \bigcup_{j\in I_i} B_j$. Hence, event E can be represented as a union of atoms of $\mathcal{Q}$:

$$E = \bigcup_{i\in I_E} A_i = \bigcup_{i\in I_E} \bigcup_{j\in I_i} B_j.$$

This means that E is an element of $\sigma(\mathcal{Q})$ by definition. □

By Proposition 6.3, for all $1 \leqslant n \leqslant N$ we have $\mathcal{F}_{n-1} \subset \mathcal{F}_n$ since $\mathcal{P}_{n-1} \preceq \mathcal{P}_n$. Therefore, $\{\mathcal{F}_n\}_{0\leqslant n\leqslant N}$ is a filtration. This type of filtration is a sequence of power sets on finer and finer partitions. Explicit examples are the σ-algebras $\mathcal{F}_n = \sigma(\mathcal{P}_n)$ with $\mathcal{P}_n$ in (6.15).

6.2.3.2 Construction of a Filtration from a Stochastic Process: Natural Filtration

A natural method to obtain a filtration is to generate σ-algebras from a sequence of random variables. Consider a stochastic process $\{X_t\}_{0\leqslant t\leqslant T}$ over Ω. Let

$$\mathcal{F}_t^X = \sigma(\{X_u : 0 \leqslant u \leqslant t\})$$

be the smallest σ-algebra generated by all paths up to time t: $\{X_u\}_{0 \leqslant u \leqslant t}$. So $\mathcal{F}_t^X$ contains all information available from the observation of the process up to time t. Note that, since time $t \in [0, T]$ is continuous, then $\mathcal{F}_t^X$ is the σ-algebra generated by the uncountable collection of random variables $\{X_u\}_{0 \leqslant u \leqslant t}$. In the discrete-time case, $\mathcal{F}_n^X = \sigma(\{X_k : k = 0, \ldots, n\}) \equiv \sigma(X_0, X_1, \ldots, X_n)$. Clearly, $\mathcal{F}_s^X \subset \mathcal{F}_t^X$ for all $0 \leqslant s < t \leqslant T$. The filtration $\{F_t^X\}_{0 \leqslant t \leqslant T}$ is called a *natural filtration* of the process $\{X_t\}_{0 \leqslant t \leqslant T}$. As is demonstrated in the following example, σ-algebras generated *only from the time-t value* X_t of a stochastic process (rather than from all path values from 0 to t) may not form a filtration.

Example 6.10. Show that σ-algebras generated by binomial prices, $\{\mathcal{F}_n\}_{0 \leqslant n \leqslant 2}$ where $\mathcal{F}_n = \sigma(S_n)$, do not constitute a filtration.

Solution. Consider the N-period model for any $N \geqslant 2$. Since the range of S_1 is $\{S_0 d, S_0 u\}$, the partition $\mathcal{P}(S_1)$ consists of two atoms:

$$\{S_1 = S_0 d\} = \{\omega \in \Omega_N : \omega_1 = \mathsf{D}\} \equiv A_{\mathsf{D}}$$
$$\{S_1 = S_0 u\} = \{\omega \in \Omega_N : \omega_1 = \mathsf{U}\} \equiv A_{\mathsf{U}}.$$

Similarly, the partition $\mathcal{P}(S_2)$ contains three atoms:

$$\{S_2 = S_0 d^2\} = \{\omega_1 = \omega_2 = \mathsf{D}\} \equiv A_{\mathsf{DD}},$$
$$\{S_2 = S_0 ud\} = \{\omega_1 = \mathsf{D}, \omega_2 = \mathsf{U}\} \cup \{\omega_1 = \mathsf{U}, \omega_2 = \mathsf{D}\} \equiv A_{\mathsf{DU}} \cup A_{\mathsf{UD}},$$
$$\{S_2 = S_0 u^2\} = \{\omega_1 = \omega_2 = \mathsf{U}\} \equiv A_{\mathsf{UU}}.$$

As we can see, not all (in fact none) of the atoms of $\mathcal{P}(S_1)$ can be obtained as a union of atoms of $\mathcal{P}(S_2)$. Therefore, $\mathcal{P}(S_2)$ is not a refinement of $\mathcal{P}(S_1)$, and $\sigma(S_1) \not\subset \sigma(S_2)$. □

This example points out that by only observing the stock price at one point in time after two or more moves, we are missing information about all the previous history (i.e., all the previous moves) of the stock price path up to that time.

As a main example of a natural filtration, re-consider the binomial model on the sample space Ω_N. The natural filtration $\{F_n\}_{0 \leqslant n \leqslant N}$ can be obtained from the binomial price process $\{S_n\}$, the random walk $\{M_n\}$, the Bernoulli random variables $X_n = \mathbb{I}_{\{\omega_n = \mathsf{U}\}}$, or directly from the market moves $\{\omega_n\}$, $0 \leqslant n \leqslant N$:

$$\mathcal{F}_0 = \{\emptyset, \Omega_N\}$$
$$\mathcal{F}_n = \sigma(S_1, \ldots, S_n) = \sigma(M_1, \ldots, M_n) = \sigma(\omega_1, \ldots, \omega_n), \quad 1 \leqslant n \leqslant N,$$

where we are denoting $\sigma(\omega_1, \ldots, \omega_n) \equiv \sigma(\mathcal{P}_n)$ with $\mathcal{P}_n$ in (6.15). The σ-algebra $\mathcal{F}_n = 2^{\mathcal{P}_n}$, $1 \leqslant n \leqslant N$, is generated from the partition $\mathcal{P}_n$ having 2^n atoms of the form (6.13) and $\mathcal{F}_n$ has 2^{2^n} events. As discussed above, $\mathcal{F}_1 = \sigma(\{A_{\mathsf{D}}, A_{\mathsf{U}}\})$, $\mathcal{F}_2 = \sigma(\{A_{\mathsf{DD}}, A_{\mathsf{DU}}, A_{\mathsf{UD}}, A_{\mathsf{UU}}\})$, until $\mathcal{F}_N = 2^{\Omega_N}$.

6.2.4 Filtered Probability Space

Now it is time to revisit the definition of a probability space. In Definition 6.2, it is assumed that the sample space Ω is finite and the probability can be calculated for every possible event in Ω. In other words, the probability measure $\mathbb{P}$ was defined as a function that maps $\mathcal{F} = 2^{\Omega}$ to $[0, 1]$. It is a typical situation when the set of measurable events (for which we can compute their probability of occurrence) is smaller than a power set. However, such a set must be a σ-algebra to match the Kolmogorov axioms. The power set 2^{Ω} is one example of a σ-algebra, but it corresponds to the ultimate case where every event (i.e., every subset of Ω) can be measured by $\mathbb{P}$. So we need to add a σ-algebra $\mathcal{F}$ to the pair $(\Omega, \mathbb{P})$. The pair $(\Omega, \mathcal{F})$ by itself is called a *measurable space.* To complete the picture, we need a probability function for "measuring" events (which are subsets of Ω and simply elements in $\mathcal{F}$). In other words, a probability space includes a measuring tool (i.e., a probability measure $\mathbb{P}$) together with $(\Omega, \mathcal{F})$, where the collection of all measurable events are contained in the σ-algebra $\mathcal{F}$ on Ω. This then leads us to the following definition.

Definition 6.8. The triple $(\Omega, \mathcal{F}, \mathbb{P})$ is called a *probability space* where

- Ω is a sample space,
- $\mathcal{F}$ is a σ-algebra on Ω,
- $\mathbb{P}\colon \mathcal{F} \to [0,1]$ is a probability measure that satisfies the Kolmogorov axioms in (6.1).

Note that a finite probability space is a particular case of a general probability space with $\mathcal{F} = 2^{\Omega}$. Our main working examples stem from the binomial probability spaces $(\Omega_N, \mathcal{F}_k, \mathbb{P}_k)$, where $\mathcal{F}_k = \sigma(\mathcal{P}_k)$ for some $0 \leqslant k \leqslant N$. That is, $\mathcal{F}_k$ contains all the information about the first k periods (or moves). For each $k = 0, \ldots, N$, we can define a probability measure $\mathbb{P}_k\colon \mathcal{F}_k \to [0,1]$ by

$$\mathbb{P}_k(E) = \sum_{\omega_1,\ldots,\omega_k \in \{\mathsf{U},\mathsf{D}\}\,:\,\omega \in E} p^{\mathsf{U}_k(\omega)}(1-p)^{\mathsf{D}_k(\omega)}, \tag{6.16}$$

for every nonempty $E \in \mathcal{F}_k$ and with $\mathbb{P}_k(\emptyset) = 0$. Note that the sum is over the first k ω's and is restricted to outcomes $\omega = \omega_1 \ldots \omega_k\, \omega_{k+1} \ldots \omega_N \in \Omega_N$ contained in E. It is readily seen that $\mathbb{P}_k$ is a probability measure on $(\Omega_N, \mathcal{F}_k)$ where $\mathbb{P}_k(\Omega_N) = 1$. Indeed, setting $E = \Omega_N$ removes the restriction $\omega \in E$ and the above sum evaluates to its maximum value of unity (note that this also corresponds to $\mathbb{P}_k(\Omega_k) = 1$). It is important to note that the measure $\mathbb{P}_k$ is the probability measure $\mathbb{P} \equiv \mathbb{P}_N$ restricted to events that are in $\mathcal{F}_k = \sigma(\mathcal{P}_k)$ which is the σ-algebra on Ω_N generated by the 2^k atoms, each denoted by $A_{\omega_1 \ldots \omega_k}$ (see (6.15)). To see this, observe that any $E \in \mathcal{F}_k$ is expressible as

$$E = E \cap \Omega_N = \bigcup_{\omega_1,\ldots,\omega_k \in \{\mathsf{U},\mathsf{D}\}} (E \cap A_{\omega_1 \ldots \omega_k}) = \bigcup_{\omega_1,\ldots,\omega_k \in \{\mathsf{U},\mathsf{D}\}\,:\,\omega \in E} A_{\omega_1 \ldots \omega_k}.$$

The last expression is a union over the restricted sub-collection of atoms in $\mathcal{P}_k$ that collectively define E. Taking the probability $\mathbb{P}(E)$ then recovers (6.16) by

$$\begin{aligned}
\mathbb{P}\left(A_{\omega_1 \ldots \omega_k}\right) &= \sum_{\omega_{k+1},\ldots,\omega_N \in \{\mathsf{U},\mathsf{D}\}} \mathbb{P}(\{\omega_1 \ldots \omega_k \omega_{k+1} \ldots \omega_N\}) \\
&= p^{\mathsf{U}_k(\omega)}(1-p)^{\mathsf{D}_k(\omega)} \left(\sum_{\omega_{k+1},\ldots,\omega_N \in \{\mathsf{U},\mathsf{D}\}} p^{\#\mathsf{U}(\omega_{k+1} \ldots \omega_N)}(1-p)^{\#\mathsf{D}(\omega_{k+1} \ldots \omega_N)} \right) \\
&= p^{\mathsf{U}_k(\omega)}(1-p)^{\mathsf{D}_k(\omega)}
\end{aligned}$$

where the quantity in brackets is $\mathbb{P}(\omega_{k+1} \in \{\mathsf{U},\mathsf{D}\}) \times \ldots \times \mathbb{P}(\omega_N \in \{\mathsf{U},\mathsf{D}\}) = 1 \times \ldots \times 1 = 1$.

In Chapter 9 we provide a discussion on general probability theory and probability distributions for more general random variables that include continuous random variables. A probability space is a natural framework for dealing with "static" random objects like random variables that do not change in time. To work with stochastic processes we need to introduce a "dynamic" version of a probability space.

Definition 6.9. A *filtered probability space* is a quadruple $(\Omega, \mathcal{F}, \mathbb{P}, \{\mathcal{F}_t\}_{t\geqslant 0})$ where the first three objects form a probability space $(\Omega, \mathcal{F}, \mathbb{P})$ and $\{\mathcal{F}_t\}_{t\geqslant 0}$ is a filtration such that $\mathcal{F}_t \subset \mathcal{F}$ for all $t \geqslant 0$.

Again, the main working model of this chapter is the binomial model with the filtered probability space $(\Omega_N, \mathcal{F}_N, \mathbb{P}_N, \{\mathcal{F}_n\}_{0\leqslant n\leqslant N})$, where all the required ingredients are constructed as discussed above.

6.3 Conditional Expectation and Martingales

6.3.1 Measurability of Random Variables and Processes

Measurability is an important concept that arises in the theory of measure and integration. Recall that a measurable space is a pair $(\Omega, \mathcal{F})$ consisting of a nonempty set Ω and a σ-algebra $\mathcal{F}$ on Ω; such subsets are said to be *measurable*. A function between measurable spaces is said to be *measurable* if its pre-image on each measurable set is measurable.

Definition 6.10. Let $X\colon \Omega \to \mathbb{R}$ be a random variable on Ω, and let $\mathcal{F}$ be a σ-algebra of subsets of Ω. X is said to be *$\mathcal{F}$-measurable* if $\sigma(X) \subset \mathcal{F}$.

Alternatively, X is $\mathcal{F}$-measurable iff $\{\omega \in \Omega \,:\, X(\omega) \in B\} \in \mathcal{F}$ for all $B \in \mathcal{B}(\mathbb{R})$. A discrete random variable X is $\mathcal{F}$-measurable if $\{\omega \in \Omega \,:\, X(\omega) = x\} \in \mathcal{F}$ for all $x \in X(\Omega)$. The image $X(\Omega) = \{X(\omega) \in \mathbb{R} \,:\, \omega \in \Omega\}$ is the support or range of X. If $\sigma(X) \subset \mathcal{F}$, then all information about X is contained in $\mathcal{F}$. Therefore, $\mathcal{F}$-measurability of X means that we can calculate the value of X if we know what events of $\mathcal{F}$ have occurred. Suppose that a σ-algebra $\mathcal{F}$ is generated by a partition $\mathcal{P}$. To verify whether X is $\mathcal{F}$-measurable, it is enough to check if X is constant on every atom of $\mathcal{P}$.

Here are some examples of measurable random variables.

(a) Any random variable X is measurable relative to the σ-algebra $\sigma(X)$ generated by X.

(b) A constant random variable is $\mathcal{F}_0$-measurable, where $\mathcal{F}_0 = \{\emptyset, \Omega\}$ is a trivial σ-algebra.

(c) In the N-period binomial model, S_n and M_n are $\mathcal{F}_n$-measurable for all $0 \leqslant n \leqslant N$.

In what follows, it will always be assumed that every random variable defined on a probability space $(\Omega, \mathcal{F}, \mathbb{P})$ is $\mathcal{F}$-measurable. It is a reasonable assumption, since otherwise it may be not possible to calculate all probabilities relating to a random variable that is not $\mathcal{F}$-measurable. For example, consider a probability space $(\Omega, \mathcal{F}, \mathbb{P})$ corresponding to a single roll of a fair die: $\Omega = \{1, 2, 3, 4, 5, 6\}$. Suppose that the σ-algebra of events, $\mathcal{F}$, is generated by observing whether the facing-up value of the die is odd or even: $\mathcal{F} = \{\emptyset, \Omega, \{1, 3, 5\}, \{2, 4, 6\}\}$. Define $\mathbb{P}$ on $\mathcal{F}$ as usual: $\mathbb{P}(\emptyset) = 0$, $\mathbb{P}(\Omega) = 1$, $\mathbb{P}(\{1, 3, 5\}) = \mathbb{P}(\{2, 4, 6\}) = \frac{1}{2}$. It is easy to verify that the random variable $X(\omega) = \omega$ that gives us the value of the die is not $\mathcal{F}$-measurable. Therefore, it is impossible to compute probabilities $\mathbb{P}(X = k)$, $1 \leqslant k \leqslant 6$, as we are missing information on the probabilities for obtaining any given number on the die.

It is a common situation when one random variable is measurable relative to the σ-algebra generated by another random variable. Let Y be $\sigma(X)$-measurable. Since $\sigma(Y) \subset \sigma(X)$, we may expect that being given the value of X we can calculate the value of Y. Indeed, the Doob–Dynkin theorem states that if Y is $\sigma(X)$-measurable, then there exists a Borel function f such that $Y = f(X)$. Recall from above that $f\colon \mathbb{R} \to \mathbb{R}$ is a Borel function on $\mathbb{R}$ if $f^{-1}(B) \in \mathcal{B}(\mathbb{R})$ for all $B \in \mathcal{B}(\mathbb{R})$.

Example 6.11. Consider a two-period binomial model Ω_2 with σ-algebras $\mathcal{F}_1 = \sigma(\omega_1)$ and $\mathcal{F}_2 = \sigma(\omega_1, \omega_2)$. Define two random variables X and Y on Ω_2 as follows:

$$X(\mathsf{UU}) = X(\mathsf{DU}) = 1, \; X(\mathsf{UD}) = X(\mathsf{DD}) = -1,$$
$$Y(\mathsf{UU}) = Y(\mathsf{UD}) = 1, \; Y(\mathsf{DD}) = Y(\mathsf{DU}) = -1.$$

Show that Y is $\mathcal{F}_1$-measurable and X is not.

Solution. Clearly, X and Y are both $\mathcal{F}_2$-measurable since $\mathcal{F}_2$ contains all information about the outcome $\omega_1\omega_2$. $\mathcal{F}_1 = \{\emptyset, A_{\mathsf{D}}, A_{\mathsf{U}}, \Omega_2\}$ is generated by ω_1. Since $Y^{-1}(1) = \{\mathsf{UD}, \mathsf{UU}\} = A_{\mathsf{U}} \in \mathcal{F}_1$ and $Y^{-1}(-1) = \{\mathsf{DD}, \mathsf{DU}\} = A_{\mathsf{D}} \in \mathcal{F}_1$, Y is $\mathcal{F}_1$-measurable. For X we observe that $X^{-1}(1) = \{\mathsf{DU}, \mathsf{UU}\} \notin \mathcal{F}_1$. Thus, X is not $\mathcal{F}_1$-measurable. □

Proposition 6.4. *Consider an N-period binomial model with the natural filtration*

$$\{\mathcal{F}_n = \sigma(\omega_1, \ldots, \omega_n)\}_{0 \leqslant n \leqslant N}.$$

The time-n binomial price S_n, $0 \leqslant n \leqslant N$, is $\mathcal{F}_k$-measurable iff $n \leqslant k$.

Proof. The random variable S_n is a function of $\omega_1, \ldots, \omega_n$. Therefore, S_n is $\mathcal{F}_n$-measurable. If $k > n$ then $\mathcal{F}_k \supset \mathcal{F}_n$. Thus S_n is $\mathcal{F}_k$-measurable for all $k \geqslant n$. Suppose that $0 \leqslant k < n$. The σ-algebra $\mathcal{F}_k$ is generated by a partition $\mathcal{P}_k$. Since S_n is not constant on atoms of $\mathcal{P}_k$, it is not $\mathcal{F}_k$-measurable. □

Similarly, one can show that the time-n value of the random walk process M_n, $0 \leqslant n \leqslant N$, is $\mathcal{F}_k$-measurable iff $n \leqslant k$ (see Exercise 6.13).

Definition 6.11. Let $\mathbb{F} = \{\mathcal{F}_t\}_{0 \leqslant t \leqslant T}$ be a filtration over Ω. A random process $\{X_t\}_{0 \leqslant t \leqslant T}$ on Ω is said to be *adapted* to filtration $\mathbb{F}$ if X_t is $\mathcal{F}_t$-measurable for every t.

Since, S_n and M_n are both $\mathcal{F}_n$-measurable, for all $0 \leqslant n \leqslant N$, then processes $\{S_n\}_{0 \leqslant n \leqslant N}$ and $\{M_n\}_{0 \leqslant n \leqslant N}$ are adapted to their respective natural filtrations $\{\sigma(S_0, S_1, \ldots, S_n)\}_{0 \leqslant n \leqslant N}$ and $\{\sigma(M_0, M_1, \ldots, M_n)\}_{0 \leqslant n \leqslant N}$. In general, every stochastic process $\{X_t\}_{0 \leqslant t \leqslant T}$ is adapted to its natural filtration $\{\mathcal{F}_t^X\}_{0 \leqslant t \leqslant T}$ since, by definition, $\mathcal{F}_t^X \equiv \sigma(\{X_u : 0 \leqslant u \leqslant t\})$ and this obviously implies $\sigma(X_t) \subset \mathcal{F}_t^X$, i.e., X_t is $\mathcal{F}_t^X$-measurable.

6.3.2 Conditional Expectations

In what follows we *assume a finite probability space*, although the theory is presented more generally in Chapter 9 to include random variables on uncountable probability spaces within a single consistent framework using Lebesgue integration. We remark that all the important results obtained below are valid if we replace the summations by appropriate integrals. To simplify the discussion, we hence consider a finite probability space $(\Omega, \mathcal{F}, \mathbb{P})$ with $\Omega = \{\omega^1, \ldots, \omega^M\}$ and an $\mathcal{F}$-measurable random variable X on Ω. The discussion below follows identically for a countably infinite space ($M = \infty$) where we have infinite summations. Since Ω is finite (or countably infinite), $X \colon \Omega \to \mathbb{R}$ is a discrete random variable that admits the following representation:

$$X = \sum_{k=1}^{M} X(\omega^k) \, \mathbb{I}_{\{\omega^k\}}$$

with indicator random variable $\mathbb{I}_{\{\omega^k\}} = \mathbb{I}_{\{\omega=\omega^k\}}$, i.e., $\mathbb{I}_{\{\omega^k\}}(\omega) = 1$, if $\omega = \omega^k$, and 0 if $\omega \neq \omega^k$. This is one way to express the random variable X as a simple function by using a sum over all outcomes. Note that X may be a nonzero constant (or zero) on more than one outcome. The (unconditional) expectation of X (w.r.t. the probability measure $\mathbb{P}$) is

$$\mathrm{E}[X] = \sum_{k=1}^{M} X(\omega^k)\,\mathrm{E}[\mathbb{I}_{\{\omega^k\}}] = \sum_{k=1}^{M} X(\omega^k)\,\mathbb{P}(\omega^k) = \sum_{\omega\in\Omega} X(\omega)\,\mathbb{P}(\omega) \tag{6.17}$$

where $\mathrm{E}[\mathbb{I}_{\{\omega^k\}}] = \mathbb{P}(\{\omega^k\}) \equiv \mathbb{P}(\omega^k)$. Note that this formula can also be written in a more traditional form by using the disjoint sets defined by $A_x := \{\omega \in \Omega \,:\, X(\omega) = x\}$ for all nonzero x values in the finite support S_X of X, i.e., $X = \sum_{x\in\mathsf{S}_X} x\,\mathbb{I}_{A_x}$, giving

$$\mathrm{E}[X] = \sum_{x\in\mathsf{S}_X} x\,\mathrm{E}[\mathbb{I}_{A_x}] = \sum_{x\in\mathsf{S}_X} x\,\mathbb{P}(A_x) = \sum_{x\in\mathsf{S}_X} x\,\mathbb{P}(X = x).$$

The mathematical expectation of a random variable X represents the best prediction of X given no additional information about X or the actual outcome $\omega \in \Omega$. Being given some additional information about X or ω, we may obtain best predictions of X conditional on the available information. The simplest case is the conditional expectation of X given an event $B \in \mathcal{F}$. Such an expected value is denoted by $\mathrm{E}[X \mid B]$. In a more general case, the available information can be represented by a σ-algebra $\mathcal{G}$ of subsets of Ω. The best prediction of X given the information in $\mathcal{G}$ is a conditional expectation of X given $\mathcal{G}$, denoted by $\mathrm{E}[X \mid \mathcal{G}]$. In fact, such an expected value is generally not a number but a $\mathcal{G}$-measurable random variable on Ω. Later, we shall define different versions of a conditional expectation.

6.3.2.1 Conditioning on an Event

Recall that the *conditional probability* of event $A \in \mathcal{F}$ given event $B \in \mathcal{F}$ is

$$\mathbb{P}(A \mid B) = \frac{\mathbb{P}(A \cap B)}{\mathbb{P}(B)},$$

provided $\mathbb{P}(B) \neq 0$. The *conditional expectation* of X conditioned to event $B \in \mathcal{F}$ is defined as

$$\mathrm{E}[X \mid B] := \sum_{\omega\in\Omega} X(\omega)\,\mathbb{P}(\omega \mid B) \tag{6.18}$$

where $\mathbb{P}(\omega \mid B) \equiv \mathbb{P}(\{\omega\} \mid B)$ is the conditional probability of a given outcome $\omega \in \Omega$ given B. This conditional probability is expressible as

$$\mathbb{P}(\omega \mid B) = \frac{\mathbb{P}(\{\omega\} \cap B)}{\mathbb{P}(B)} = \frac{\mathbb{P}(\omega)}{\mathbb{P}(B)}\,\mathbb{I}_B(\omega),$$

where we use the property that $\{\omega\} \cap B = \emptyset$ if $\omega \notin B$, and $\{\omega\} \cap B = \{\omega\}$ if $\omega \in B$. Therefore, the formula in (6.18) can be rewritten as follows:

$$\begin{aligned}\mathrm{E}[X \mid B] &= \sum_{\omega\in\Omega} X(\omega)\,\frac{\mathbb{P}(\omega)}{\mathbb{P}(B)}\,\mathbb{I}_B(\omega)\\ &= \frac{1}{\mathbb{P}(B)} \sum_{\omega\in\Omega} X(\omega)\,\mathbb{I}_B(\omega)\,\mathbb{P}(\omega)\\ &= \frac{1}{\mathbb{P}(B)} \sum_{\omega\in B} X(\omega)\,\mathbb{P}(\omega) = \frac{\mathrm{E}[X\,\mathbb{I}_B]}{\mathbb{P}(B)}.\end{aligned} \tag{6.19}$$

We know that the expectation of an indicator function of an event equals the probability of the event: $\mathrm{E}[\mathbb{I}_A] = \mathbb{P}(A)$, for $A \in \mathcal{F}$. It follows by (6.19) that the same result holds for the conditional expected value:

$$\mathrm{E}[\mathbb{I}_A \mid B] = \frac{\mathrm{E}[\mathbb{I}_A \, \mathbb{I}_B]}{\mathbb{P}(B)} = \frac{\mathrm{E}[\mathbb{I}_{A \cap B}]}{\mathbb{P}(B)} = \frac{\mathbb{P}(A \cap B)}{\mathbb{P}(B)} = \mathbb{P}(A \mid B).$$

Since the (unconditional) expectation is a linear functional, so is the conditional expectation.

Proposition 6.5. *Let $B \in \mathcal{F}$ be such that $\mathbb{P}(B) \neq 0$. For any reals c_i and random variables X_i, $i = 1, 2$, we have*

$$\mathrm{E}[c_1 X_1 + c_2 X_2 \mid B] = c_1 \mathrm{E}[X_1 \mid B] + c_2 \mathrm{E}[X_2 \mid B].$$

Proof. Using (6.19) and the linearity of the operator E:

$$\begin{aligned}\mathrm{E}[c_1 X_1 + c_2 X_2 \mid B] &= \frac{\mathrm{E}[(c_1 X_1 + c_2 X_2)\, \mathbb{I}_B]}{\mathbb{P}(B)} = \frac{c_1 \mathrm{E}[X_1 \mathbb{I}_B] + c_2 \mathrm{E}[X_2 \mathbb{I}_B]}{\mathbb{P}(B)} \\ &= c_1 \frac{\mathrm{E}[X_1 \, \mathbb{I}_B]}{\mathbb{P}(B)} + c_2 \frac{\mathrm{E}[X_2 \, \mathbb{I}_B]}{\mathbb{P}(B)} = c_1 \mathrm{E}[X_1 \mid B] + c_2 \mathrm{E}[X_2 \mid B].\end{aligned}$$ □

Example 6.12. Consider a four-period binomial model with $p = \frac{1}{2}$. Find $\mathrm{E}[\mathsf{U}_4 \mid B]$ for $B = \{\text{at least two } \mathsf{U}\text{'s}\}$.

Solution. The event $B \subset \Omega_4$ can be characterized by listing all elements $\omega = \omega_1\omega_2\omega_3\omega_4$ of its complement:

$$B^{\mathsf{C}} = \{\omega \mid \mathsf{U}_4(\omega) \leqslant 1\} = \{\mathsf{DDDD}, \mathsf{DDDU}, \mathsf{DDUD}, \mathsf{DUDD}, \mathsf{UDDD}\}.$$

Note that all outcomes in Ω_4 are all equally likely since $p = 1 - p = \frac{1}{2}$; the probability of each particular scenario is $\left(\frac{1}{2}\right)^4 = \frac{1}{16}$. First, calculate $\mathbb{P}(B)$:

$$\mathbb{P}(B) = 1 - \mathbb{P}(B^{\mathsf{C}}) = 1 - 5 \cdot \frac{1}{16} = \frac{11}{16}.$$

Second, apply (6.19) to calculate $\mathrm{E}[\mathsf{U}_4 \mid B]$:

$$\begin{aligned}\mathrm{E}[\mathsf{U}_4 \, \mathbb{I}_B] &= \sum_{\omega \in B} \mathsf{U}_4(\omega)\, \mathbb{P}(\omega) = \sum_{\omega : \mathsf{U}_4(\omega) \geqslant 2} \mathsf{U}_4(\omega)\, \mathbb{P}(\omega) = \sum_{k=2}^{4} k \cdot \mathbb{P}(\mathsf{U}_4 = k) \\ &= \frac{1}{16} \cdot \left(2 \cdot \binom{4}{2} + 3 \cdot \binom{4}{3} + 4 \cdot \binom{4}{4}\right) = \frac{1}{16} \cdot (2 \cdot 6 + 3 \cdot 4 + 4 \cdot 1) = \frac{28}{16} = \frac{7}{4}.\end{aligned}$$

Thus, $\mathrm{E}[\mathsf{U}_4 \mid B] = \mathrm{E}[\mathsf{U}_4 \, \mathbb{I}_B]/\mathbb{P}(B) = \frac{7/4}{11/16} = \frac{28}{11}$. □

The concept of the conditional expectation allows us to generalize the law of total probability. Let $\mathcal{P}$ be a partition of Ω, i.e., $\Omega = \cup_{A \in \mathcal{P}} A$, and let X be a random variable on Ω. Since $\mathbb{I}_\Omega = \sum_{A \in \mathcal{P}} \mathbb{I}_A$, then the expectation of X can be calculated as a sum of products of conditional expectations of X with respect to every atom $A \in \mathcal{P}$ and the probability of the event corresponding to every atom:

$$\mathrm{E}[X] = \mathrm{E}[X \, \mathbb{I}_\Omega] = \sum_{A \in \mathcal{P}} \mathrm{E}[X \, \mathbb{I}_A] = \sum_{A \in \mathcal{P}} \mathrm{E}[X \mid A]\, \mathbb{P}(A). \tag{6.20}$$

6.3.2.2 Conditioning on a Sigma-Algebra

Let X be a random variable of a probability space $(\Omega, \mathcal{F}, \mathbb{P})$ and let $\mathcal{G}$ be a sub-σ-algebra of $\mathcal{F}$, i.e., $\mathcal{G}$ is a σ-algebra on Ω and $\mathcal{G} \subset \mathcal{F}$. Hence, X is $\mathcal{F}$-measurable. If X were also $\mathcal{G}$-measurable then the information contained in $\mathcal{G}$ is sufficient to determine a value for X. In particular, if X were $\mathcal{G}$-measurable and $\mathcal{G}$ were generated by a given partition then X would be a constant on the atoms of the partition. However, more generally, when X is not $\mathcal{G}$-measurable the information in $\mathcal{G}$ can only be used to provide an estimate of X. Such an estimate is the expectation of X conditioned on information in $\mathcal{G}$ and forms the basis of the following definition, which generalizes the concept of a conditional expectation.

Definition 6.12. A random variable Y is called the *conditional expectation* of X given $\mathcal{G}$, denoted by $Y = \mathrm{E}[X \mid \mathcal{G}]$, if

(i) Y is a $\mathcal{G}$-measurable random variable, i.e., $\sigma(Y) \subset \mathcal{G}$;

(ii) $\mathrm{E}[X\, \mathbb{I}_B] = \mathrm{E}[Y\, \mathbb{I}_B]$ for every event $B \in \mathcal{G}$.

We remark that property (i) tells us that $\mathrm{E}[X \mid \mathcal{G}]$ is itself a random variable that is constant on every atom of a partition that generates $\mathcal{G}$. Property (ii) is sometimes referred to as the *partial averaging property*. That is, the expected value of X restricted to any given event in $\mathcal{G}$ is the same as the expected value of $\mathrm{E}[X \mid \mathcal{G}]$ restricted to the same given event in $\mathcal{G}$. Note that for $B = \Omega$, property (ii) implies the nested expectation identity $\mathrm{E}[X] = \mathrm{E}[\mathrm{E}[X \mid \mathcal{G}]]$, i.e., the expected value of X is the same as the expected value of X that has been conditioned on any information set $\mathcal{G} \subset \mathcal{F}$.

Three simple examples of conditional expectations with respect to a sub-σ-algebra are as follows.

(a) If $\mathcal{G} = \mathcal{F}_0 \equiv \{\emptyset, \Omega\}$, then $\mathrm{E}[X \mid \mathcal{G}] = \mathrm{E}[X]$.

(b) For a constant random variable $X \equiv C$, $\mathrm{E}[C \mid \mathcal{G}] = C$.

(c) If X is $\mathcal{G}$-measurable, then $\mathrm{E}[X \mid \mathcal{G}] = X$.

To show that a given Y is $\mathrm{E}[X \mid \mathcal{G}]$ we need to verify that properties (i) and (ii) hold for Y and given $\mathcal{G}$. For (a), we let $Y = \mathrm{E}[X]$ and $\mathcal{G} = \mathcal{F}_0$. Since Y is a constant, then (i) holds since $\sigma(Y) = \mathcal{F}_0$ implies that Y is $\mathcal{F}_0$-measurable. Property (ii) is now verified for every $B \in \mathcal{F}_0$, i.e. for $B = \emptyset$ and for $B = \Omega$. Since $\mathbb{I}_\emptyset = 0$, then $B = \emptyset$ gives $\mathrm{E}[X\, \mathbb{I}_\emptyset] = 0 = \mathrm{E}[Y\, \mathbb{I}_\emptyset]$. Since $\mathbb{I}_\Omega = 1$, then $B = \Omega$ gives $\mathrm{E}[X\, \mathbb{I}_\Omega] = \mathrm{E}[X] = Y = \mathrm{E}[Y\, \mathbb{I}_\Omega]$. For (b), we let $Y = C$. Hence, property (i) holds since $\sigma(Y) = \sigma(Y^{-1}(C)) = \sigma(\Omega) = \mathcal{F}_0 \subset \mathcal{G}$ for any $\mathcal{G}$. Property (ii) holds in the obvious manner since $Y = X = C$. For case (c), we let $Y = X$. So properties (i) and (ii) follow automatically since X is $\mathcal{G}$-measurable implies Y is $\mathcal{G}$-measurable and the expectations in (ii) are equivalent.

Example 6.13. Consider a random variable X on $(\Omega, \mathcal{F}, \mathbb{P})$. Suppose that $\mathcal{G} = \{\emptyset, A, A^{\mathsf{C}}, \Omega\}$ with nonempty $A \in \mathcal{F}$. Show that

$$\mathrm{E}[X \mid \mathcal{G}](\omega) = \mathrm{E}[X \mid A]\, \mathbb{I}_A(\omega) + \mathrm{E}[X \mid A^{\mathsf{C}}]\, \mathbb{I}_{A^{\mathsf{C}}}(\omega) = \begin{cases} \mathrm{E}[X \mid A] & \text{if } \omega \in A, \\ \mathrm{E}[X \mid A^{\mathsf{C}}] & \text{if } \omega \in A^{\mathsf{C}}. \end{cases}$$

solution. Let $Y = \mathrm{E}[X \mid A]\, \mathbb{I}_A + \mathrm{E}[X \mid A^{\mathsf{C}}]\, \mathbb{I}_{A^{\mathsf{C}}}$. Note that $\mathrm{E}[X \mid A]$ and $\mathrm{E}[X \mid A^{\mathsf{C}}]$ are two constants. Hence, if $\mathrm{E}[X \mid A] \neq \mathrm{E}[X \mid A^{\mathsf{C}}]$ then $\sigma(Y) = \sigma(A) = \mathcal{G}$; otherwise Y is constant and $\sigma(Y) = \mathcal{F}_0$. In either case, Y is $\mathcal{G}$-measurable. We now prove that $\forall B \in \mathcal{G}$, $\mathrm{E}[Y\, \mathbb{I}_B] = \mathrm{E}[X\, \mathbb{I}_B]$.

$B = \emptyset$: $\mathrm{E}[Y\,\mathbb{I}_\emptyset] = \mathrm{E}[X\,\mathbb{I}_\emptyset] = 0$.

$B = \Omega$: Using $\mathrm{E}[\mathbb{I}_A] = \mathbb{P}(A)$ and applying the total probability law in (6.20) gives

$$\mathrm{E}[Y\,\mathbb{I}_\Omega] = \mathrm{E}[Y] = \mathrm{E}[X \mid A] \cdot \mathbb{P}(A) + \mathrm{E}[X \mid A^{\complement}] \cdot \mathbb{P}(A^{\complement}) = \mathrm{E}[X] = \mathrm{E}[X\,\mathbb{I}_\Omega].$$

$B = A(\text{ or } A^{\complement})$: Since $Y\,\mathbb{I}_B = \mathrm{E}[X \mid B]\,\mathbb{I}_B$, we have

$$\mathrm{E}[Y\,\mathbb{I}_B] = \mathrm{E}[X \mid B] \cdot \mathbb{P}(B) = \mathrm{E}[X\,\mathbb{I}_B].$$

Hence, $\mathrm{E}[X \mid \mathcal{G}] = \mathrm{E}[X \mid A]\,\mathbb{I}_A + \mathrm{E}[X \mid A^{\complement}]\,\mathbb{I}_{A^{\complement}}$. □

Example 6.14. Consider a three-period binomial model. Find $\mathrm{E}[\mathsf{U}_3 \mid \sigma(\omega_1)]$.

Solution. The σ-algebra $\sigma(\omega_1)$ is of the form considered in Example 6.13:

$$\sigma(\omega_1) = \{\emptyset, A_{\mathsf{D}}, A_{\mathsf{U}}, \Omega\}.$$

Therefore, $\mathrm{E}[\mathsf{U}_3 \mid \sigma(\omega_1)] = \mathrm{E}[\mathsf{U}_3 \mid \{\omega_1 = \mathsf{U}\}]\,\mathbb{I}_{A_{\mathsf{U}}} + \mathrm{E}[\mathsf{U}_3 \mid \{\omega_1 = \mathsf{D}\}]\,\mathbb{I}_{A_{\mathsf{D}}}$. We calculate the two expectations conditional on the respective events of the up move and the down move:

$$\begin{aligned}
&\mathrm{E}[\mathsf{U}_3 \mid \{\omega_1 = \mathsf{U}\}] = \frac{\mathrm{E}[\mathsf{U}_3\,\mathbb{I}_{\{\omega_1=\mathsf{U}\}}]}{\mathbb{P}(\omega_1 = \mathsf{U})}\\
&= \frac{1}{p}\left\{\mathsf{U}_3(\mathsf{UDD})p(1-p)^2 + [\mathsf{U}_3(\mathsf{UDU}) + \mathsf{U}_3(\mathsf{UUD})]p^2(1-p) + \mathsf{U}_3(\mathsf{UUU})p^3\right\}\\
&= (1-p)^2 + 4p(1-p) + 3p^2 = 1 + 2p\\
&\mathrm{E}[\mathsf{U}_3 \mid \{\omega_1 = \mathsf{D}\}] = \frac{\mathrm{E}[\mathsf{U}_3\,\mathbb{I}_{\{\omega_1=\mathsf{D}\}}]}{\mathbb{P}(\omega_1 = \mathsf{D})}\\
&= \frac{1}{(1-p)}\left\{\mathsf{U}_3(\mathsf{DDD})(1-p)^3 + [\mathsf{U}_3(\mathsf{DDU}) + \mathsf{U}_3(\mathsf{DUD})]p(1-p)^2 + \mathsf{U}_3(\mathsf{DUU})(1-p)p^2\right\}\\
&= 2p(1-p) + 2p^2 = 2p.
\end{aligned}$$

Thus, $\mathrm{E}[\mathsf{U}_3 \mid \sigma(\omega_1)] = (1+2p)\,\mathbb{I}_{A_{\mathsf{U}}} + 2p\,\mathbb{I}_{A_{\mathsf{D}}}$. Note that we have the nested property:

$$\mathrm{E}[\mathrm{E}[\mathsf{U}_3 \mid \sigma(\omega_1)]] = (1+2p)\,\mathbb{P}(A_{\mathsf{U}}) + 2p\,\mathbb{P}(A_{\mathsf{D}}) = (1+2p)p + 2p(1-p) = 3p = \mathrm{E}[\mathsf{U}_3].$$ □

The above definition of $\mathrm{E}[X \mid \mathcal{G}]$ seems to be formal and nonconstructive. However, as is demonstrated in Example 6.13, it is possible to explicitly construct the random variable $\mathrm{E}[X \mid \mathcal{G}]$ for some special cases of $\mathcal{G}$. Below we provide such an explicit construction of the conditional expectation for any σ-algebra generated by a *countable partition* of the sample space.

Theorem 6.6. *Let X be a random variable on $(\Omega, \mathcal{F}, \mathbb{P})$ and $\mathcal{P} = \{A_i\}_{i\geqslant 1}$ be a countable partition of Ω with $\mathcal{G} = \sigma(\mathcal{P}) \subset \mathcal{F}$. Then,*

$$\mathrm{E}[X \mid \mathcal{G}] = \sum_{i\geqslant 1} \mathrm{E}[X \mid A_i]\,\mathbb{I}_{A_i}. \tag{6.21}$$

Proof. Let Y equal the right-hand side of (6.21). Note that Y has the form of a simple function, $Y = \sum_i y_i\,\mathbb{I}_{A_i}$, with constants $y_i = \mathrm{E}[X \mid A_i]$. Hence, if all y_i's are distinct, the σ-algebra generated by Y is $\sigma(Y) = \sigma(\{Y^{-1}(y_i)\}_{i\geqslant 1}) = \sigma(\{A_i\}_{i\geqslant 1}) = \sigma(\mathcal{P}) = \mathcal{G}$ and Y is $\mathcal{G}$-measurable. Otherwise, if not all y_i's are distinct then some of the subsets $Y^{-1}(y_i)$ will be unions of atoms in $\mathcal{P}$, i.e., the partition $\{Y^{-1}(y_i)\}_{i\geqslant 1} \preceq \mathcal{P}$. In this case, $\sigma(Y)$ will be a proper subset of $\sigma(\mathcal{P})$ and we still have that Y is $\mathcal{G}$-measurable. Take any $B \in \mathcal{G}$. By definition, any element of $\mathcal{G} = \sigma(\mathcal{P})$ is a countable union of atoms:

$$B = \bigcup_{j \geqslant 1} A_{i_j} \text{ for some } i_1 < i_2 < \cdots$$

Since all atoms are mutually disjoint, the indicator of B is a sum of indicators of A_{i_j}: $\mathbb{I}_B = \sum_{j \geqslant 1} \mathbb{I}_{A_{i_j}}$. It follows that $\mathrm{E}[Y\,\mathbb{I}_B] = \mathrm{E}[X\,\mathbb{I}_B]$:

$$\begin{aligned}
\mathrm{E}[Y\,\mathbb{I}_B] &= \sum_{j \geqslant 1} \mathrm{E}\left[Y\,\mathbb{I}_{A_{i_j}}\right] \\
&= \sum_{j \geqslant 1} \mathrm{E}\left[\mathrm{E}[X \mid A_{i_j}] \cdot \mathbb{I}_{A_{i_j}}\right] = \sum_{j \geqslant 1} \mathrm{E}[X \mid A_{i_j}] \cdot \mathrm{E}[\mathbb{I}_{A_{i_j}}] \\
&= \sum_{j \geqslant 1} \mathrm{E}[X \mid A_{i_j}] \cdot \mathbb{P}(A_{i_j}) \\
&= \sum_{j \geqslant 1} \mathrm{E}[X\,\mathbb{I}_{A_{i_j}}] = \mathrm{E}\left[X \cdot \sum_{j \geqslant 1} \mathbb{I}_{A_{i_j}}\right] = \mathrm{E}[X\,\mathbb{I}_B].
\end{aligned}$$

Here, we used the property that $Y\,\mathbb{I}_{A_k} = \mathrm{E}[X \mid A_k] \cdot \mathbb{I}_{A_k}$ for all $A_k \in \mathcal{P}$. □

6.3.2.3 Conditioning on a Random Variable

Let X and Y be two random variables on the same probability space $(\Omega, \mathcal{F}, \mathbb{P})$. Formally, the conditional expectation of X given Y is equal to the conditional expectation of X given the σ-algebra generated by Y:

$$\mathrm{E}[X \mid Y] := \mathrm{E}[X \mid \sigma(Y)].$$

Equivalently, for every $\omega \in \Omega$,

$$\mathrm{E}[X \mid Y](\omega) := \mathrm{E}[X \mid Y = Y(\omega)].$$

Note that in this case the sub-σ-algebra, $\mathcal{G} = \sigma(Y) \subset \mathcal{F}$, contains all the information about Y only. The random variable $\mathrm{E}[X \mid Y]$ is $\sigma(Y)$-measurable. If the joint distribution of the pair (X, Y) is known, then one can obtain the conditional distribution of X given the value of Y. Then $\mathrm{E}[X \mid Y = y]$ is the expected value of X relative to the conditional distribution of X given $Y = y$. In any standard text on probability theory, the reader can find formulas of such conditional expectations for the cases with discrete or continuous random variables. For a discrete random variable, $Y \in Y(\Omega) \equiv \{y_k; k \geqslant 1\}$, the atoms $A_k := \{Y = y_k\} \equiv \{\omega \in \Omega : Y(\omega) = y_k\}$, $k \geqslant 1$, generate $\sigma(Y) = \sigma(\{A_k\}_{k \geqslant 1})$. Applying (6.21) gives the representation

$$\mathrm{E}[X \mid Y] = \sum_{k \geqslant 1} \mathrm{E}[X \mid Y = y_k]\,\mathbb{I}_{\{Y = y_k\}}. \tag{6.22}$$

That is, for every outcome $\omega \in \{Y = y_k\}$ the random variable $\mathrm{E}[X \mid Y]$ takes on a constant value given by the conditional expectation $\mathrm{E}[X \mid Y = y_k]$. Hence, given that Y has range $\{y_k; k \geqslant 1\}$ then $\mathrm{E}[X \mid Y]$ has range given by the set of values $\{\mathrm{E}[X \mid Y = y_k]; k \geqslant 1\}$.

Example 6.15. Consider the four-period binomial model. Define two random variables

$$X(\omega) \equiv \#\text{ of U's before the first D in } \omega \in \Omega_4 \text{ and } Y \equiv \mathsf{U}_4.$$

1. Find $\mathrm{E}[X \mid Y]$.
2. Calculate $\mathbb{P}(\mathrm{E}[X \mid Y] \leqslant 2)$.

Solution. $\mathsf{U}_4 \in \{0,1,2,3,4\}$ and we apply (6.22) by finding the conditional expectations $\mathrm{E}[X \mid Y=k]$ for $0 \leqslant k \leqslant 4$:

$$\mathrm{E}[X \mid Y=k] = \frac{1}{\mathbb{P}(Y=k)} \sum_{\omega\,:\,Y(\omega)=k} X(\omega)\mathbb{P}(\omega).$$

$k=0$: Since X is zero on the set $\{Y=0\} = \{\mathsf{DDDD}\}$, we have $\mathrm{E}[X \mid Y=0] = 0$.

$k=1$: The probability $\mathbb{P}(Y=1) = \mathbb{P}(\{\mathsf{UDDD}, \mathsf{DUDD}, \mathsf{DDUD}, \mathsf{DDDU}\}) = 4p(1-p)^3$. Note that $X(\mathsf{UDDD}) = 1$ and $X(\omega)$ is zero for other $\omega \in \{Y=1\}$. Therefore,

$$\mathrm{E}[X \mid Y=1] = \frac{1}{4p(1-p)^3} X(\mathsf{UDDD})\,\mathbb{P}(\mathsf{UDDD}) = \frac{1}{4}.$$

$k=2$: By listing all $\omega \in \Omega_4$ for which $Y(\omega)=2$ and then calculating $X(\omega)$:

$$\mathrm{E}[X \mid Y=2] = \frac{1}{\mathbb{P}(Y=2)} \sum_{\omega\,:\,Y(\omega)=2} X(\omega)\,\mathbb{P}(\omega) = \frac{1}{6p^2(1-p)^2} 4p^2(1-p)^2 = \frac{2}{3}.$$

$k=3$: Similarly, we obtain $\mathrm{E}[X \mid Y=3] = \frac{3}{2}$.

$k=4$: Since $\{Y=4\} = \{\mathsf{UUUU}\}$ and $X(\mathsf{UUUU}) = 4$, we have $\mathrm{E}[X \mid Y=4] = 4$.

In summary,

$$E[X \mid Y](\omega) = \begin{cases} 0 & \text{if } Y(\omega)=0 \\ \frac{1}{4} & \text{if } Y(\omega)=1 \\ \frac{2}{3} & \text{if } Y(\omega)=2 \\ \frac{3}{2} & \text{if } Y(\omega)=3 \\ 4 & \text{if } Y(\omega)=4 \end{cases}$$

This gives the random variable $E[X \mid Y] = \frac{1}{4}\mathbb{I}_{\{Y=1\}} + \frac{2}{3}\mathbb{I}_{\{Y=2\}} + \frac{3}{2}\mathbb{I}_{\{Y=3\}} + 4\,\mathbb{I}_{\{Y=4\}}$. Finally, we can compute the probability:

$$\mathbb{P}(\mathrm{E}[X \mid Y] \leqslant 2) = 1 - \mathbb{P}(\mathrm{E}[X \mid Y] > 2) = 1 - \mathbb{P}(\mathrm{E}[X \mid Y] = 4) = 1 - \mathbb{P}(Y=4) = 1 - p^4. \quad \square$$

6.3.3 Properties of Conditional Expectations

Consider a probability space $(\Omega, \mathcal{F}, \mathbb{P})$. Suppose that X, X_1, and X_2 are $\mathcal{F}$-measurable random variables on Ω, and let $\mathcal{G}$ be a sub-σ-algebra of $\mathcal{F}$. We now derive some important identities that are satisfied by expectations conditional on such σ-algebras. Although the properties below are valid in more general cases, for simplicity of proofs we assume that $\mathcal{G}$ is generated by some countable partition. In particular, we will simply assume a finite number of atoms in the partition $\mathcal{P} = \{A_1, A_2, \ldots, A_K\}$ of the sample space Ω, i.e., $\mathcal{G} = \sigma(\mathcal{P})$.

6.3.3.1 Linearity

For any real c_1 and c_2,

$$\mathrm{E}[c_1 X_1 + c_2 X_2 \mid \mathcal{G}] = c_1 \mathrm{E}[X_1 \mid \mathcal{G}] + c_2 \mathrm{E}[X_2 \mid \mathcal{G}]. \tag{6.23}$$

Proof. By using the linearity of the expectation conditional on any event and by (6.21):

$$\begin{aligned}\mathrm{E}[c_1X_1 + c_2X_2 \mid \mathcal{G}] &= \sum_{k=1}^{K} \mathrm{E}[c_1X_1 + c_2X_2 \mid A_k]\, \mathbb{I}_{A_k} \\ &= c_1 \sum_{k=1}^{K} \mathrm{E}[X_1 \mid A_k]\, \mathbb{I}_{A_k} + c_2 \sum_{k=1}^{K} \mathrm{E}[X_2 \mid A_k]\, \mathbb{I}_{A_k} \\ &= c_1\mathrm{E}[X_1 \mid \mathcal{G}] + c_2\mathrm{E}[X_2 \mid \mathcal{G}].\end{aligned}$$

□

6.3.3.2 Independence

First, let us define what it means for a pair of random variables and a pair of σ-algebras to be independent w.r.t. a given probability measure $\mathbb{P}$.

Definition 6.13. (Pairwise independence)

1. Two random variables X_1 and X_2 are said to be independent w.r.t. $\mathbb{P}$ if

$$\mathbb{P}(X_1 \in B_1, X_2 \in B_2) = \mathbb{P}(X_1 \in B_1)\, \mathbb{P}(X_2 \in B_2)$$

   for all $B_1, B_2 \in \mathcal{B}(\mathbb{R})$.

2. Two σ-algebras $\mathcal{G}_1, \mathcal{G}_2 \subset \mathcal{F}$ are said to be independent w.r.t. $\mathbb{P}$ if $\mathbb{P}(A_1 \cap A_2) = \mathbb{P}(A_1)\mathbb{P}(A_2)$ for all $A_1 \in \mathcal{G}_1$ and $A_2 \in \mathcal{G}_2$.

3. A random variable X and σ-algebra $\mathcal{G}$ are said to be independent w.r.t. $\mathbb{P}$ if one of the two equivalent conditions holds:

   (i) for every $A \in \mathcal{G}$, $\mathbb{I}_A$ and X are independent random variables w.r.t. $\mathbb{P}$;

   (ii) $\sigma(X)$ and $\mathcal{G}$ are independent σ-algebras w.r.t. $\mathbb{P}$.

As is seen, the independence between X and $\mathcal{G}$ means that we do not gain any information about X if we know $\mathcal{G}$ and vice versa.

Suppose that a random variable X and a σ-algebra $\mathcal{G}$ are independent. Then, the conditional expectation of X given $\mathcal{G}$ is equal to the unconditional expectation of X:

$$\mathrm{E}[X \mid \mathcal{G}] = \mathrm{E}[X]. \tag{6.24}$$

Proof. For every $A \in \mathcal{G}$, X and $\mathbb{I}_A$ are independent. Therefore,

$$\mathrm{E}[X\, \mathbb{I}_A] = \mathrm{E}[X]\, \mathrm{E}[\mathbb{I}_A] = \mathrm{E}[\mathrm{E}[X]\, \mathbb{I}_A].$$

Clearly, $\mathrm{E}[X]$ is $\mathcal{G}$-measurable since it is a constant. By definition of $\mathrm{E}[X \mid \mathcal{G}]$, the assertion is proved. □

Since any random variable X is independent of the trivial σ-algebra $\mathcal{F}_0 = \{\emptyset, \Omega\}$, we have that $\mathrm{E}[X \mid \mathcal{F}_0] = \mathrm{E}[X]$. Let X_1 and X_2 be independent random variables. Then,

$$\mathrm{E}[X_1 \mid X_2] \equiv \mathrm{E}[X_1 \mid \sigma(X_2)] = \mathrm{E}[X_1] \text{ and } \mathrm{E}[X_2 \mid X_1] = \mathrm{E}[X_2].$$

For independent random variables, we then also have $\mathrm{E}[X_1X_2] = \mathrm{E}[X_1]\mathrm{E}[X_2]$.

The above definition of independence extends naturally to an arbitrary (countable or uncountable) collection of random variables and σ-algebras. We present the definition for a countable collection of random variables and σ-algebras, as follows. Note that we simply generalize parts 1 and 2 in the above definition.

Definition 6.14. (Independence for a collection) Let n be a positive integer and consider a collection of random variables $X_1, X_2, \ldots, X_n$ defined on the probability space $(\Omega, \mathcal{F}, \mathbb{P})$ and a collection of sub-σ-algebras $\mathcal{G}_1, \mathcal{G}_2, \ldots, \mathcal{G}_n \subset \mathcal{F}$.

1. $X_1, X_2, \ldots, X_n$ are said to be (mutually) independent w.r.t. $\mathbb{P}$ if, for every subcollection of indices $1 \leqslant i_1 < i_2 < \ldots < i_k \leqslant n$, $1 \leqslant k \leqslant n$,

$$\mathbb{P}(X_{i_1} \in B_1, X_{i_2} \in B_2, \ldots, X_{i_k} \in B_k) = \mathbb{P}(X_{i_1} \in B_1) \cdot \mathbb{P}(X_{i_2} \in B_2) \cdots \mathbb{P}(X_{i_k} \in B_k)$$

   for all $B_1, B_2, \ldots, B_k \in \mathcal{B}(\mathbb{R})$.

2. $\mathcal{G}_1, \mathcal{G}_2, \ldots, \mathcal{G}_n$ are said to be (mutually) independent w.r.t. $\mathbb{P}$ if, for every subcollection of indices $1 \leqslant i_1 < i_2 < \ldots < i_k \leqslant n$, $1 \leqslant k \leqslant n$,

$$\mathbb{P}(A_{i_1} \cap A_{i_2} \cap \ldots \cap A_{i_k}) = \mathbb{P}(A_{i_1}) \cdot \mathbb{P}(A_{i_2}) \cdots \mathbb{P}(A_{i_k})$$

   for all $A_{i_1} \in \mathcal{G}_{i_1}, A_{i_2} \in \mathcal{G}_{i_2}, \ldots, A_{i_k} \in \mathcal{G}_{i_k}$.

A countable sequence of random variables $X_1, X_2, \ldots$ and sub-σ-algebras $\mathcal{G}_1, \mathcal{G}_2, \ldots$, are (respectively) independent w.r.t. $\mathbb{P}$ if properties 1 and 2 hold for all $n \geqslant 1$.

We note that it suffices to take the Borel sets in property 1 to be of the form $B_i = (-\infty, x_i]$, $x_i \in \mathbb{R}$, since these semi-open infinite intervals also generate the Borel σ-algebra $\mathcal{B}(\mathbb{R})$.

6.3.3.3 Taking out What Is Known

Suppose that Y is $\mathcal{G}$-measurable, then in the conditional expectation $\mathrm{E}[XY \mid \mathcal{G}]$, the variable Y can be taken out of the expectation:

$$\mathrm{E}[X\,Y \mid \mathcal{G}] = Y\,\mathrm{E}[X \mid \mathcal{G}]. \tag{6.25}$$

In particular, letting $X \equiv 1$ in (6.25) we obtain

$$\mathrm{E}[Y \mid \mathcal{G}] = Y\,\mathrm{E}[1 \mid \mathcal{G}] = Y.$$

Proof. To prove (6.25), let $\mathcal{G} = \sigma(\mathcal{P})$. Since Y is $\mathcal{G}$-measurable it is constant on every atom $A \in \mathcal{P}$. Thus, by (6.21) we have

$$\begin{aligned}\mathrm{E}[X\,Y \mid \mathcal{G}] &= \sum_{A\in\mathcal{P}} \mathrm{E}[X\,Y \mid A]\,\mathbb{I}_A = \sum_{A\in\mathcal{P}} \frac{\mathrm{E}[X\,Y\,\mathbb{I}_A]}{\mathbb{P}(A)}\,\mathbb{I}_A \\ &= \sum_{A\in\mathcal{P}} Y\,\frac{\mathrm{E}[X\,\mathbb{I}_A]}{\mathbb{P}(A)}\,\mathbb{I}_A = Y \sum_{A\in\mathcal{P}} \mathrm{E}[X \mid A]\,\mathbb{I}_A = Y\,\mathrm{E}[X \mid \mathcal{G}].\end{aligned}$$ □

For an alternate proof of this result see Exercise (6.16). This result is readily extended as follows. Suppose that the σ-algebra $\mathcal{G}$ in (6.25) is generated by another random variable Z, and that Y is a function of Z: $\mathcal{G} = \sigma(Z)$ and $Y = f(Z)$. Clearly, Y is $\mathcal{G}$-measurable since $Y = f(Z)$ implies $\sigma(Y) \subset \sigma(Z)$. Therefore,

$$\mathrm{E}[X\,f(Z) \mid \sigma(Z)] \equiv \mathrm{E}[X\,f(Z) \mid Z] = f(Z)\,\mathrm{E}[X \mid Z].$$

Consider now the case of a function of two or more random variables in which some of the variables are $\mathcal{G}$-measurable and the rest are independent of $\mathcal{G}$. Then, we have a general version of "taking out what is known" whereby the expectation of the function, conditional on $\mathcal{G}$, is turned into an unconditional expectation. We will see later that such a property turns out to be quite useful. We now state and prove this result in the case of two random variables as follows. Note that here we give the proof for the case that the conditioning σ-algebra, $\mathcal{G}$, is generated by a countable partition of Ω, with discrete random variables. However, this result holds in the more general case of any type of random variable and σ-algebra $\mathcal{G}$. For a proof of this result in the case of continuous random variables we refer the reader to Chapter 9.

Proposition 6.7 (Independence Proposition for Two Random Variables). *Let X and Y be random variables on $(\Omega, \mathcal{F}, \mathbb{P})$. Suppose that X is $\mathcal{G}$-measurable and that Y is independent of the σ-algebra $\mathcal{G} \subset \mathcal{F}$. Let $h : \mathbb{R}^2 \to \mathbb{R}$ be a Borel function. Then*

$$\mathrm{E}[h(X, Y) \mid \mathcal{G}] = g(X),$$

with function $g : \mathbb{R} \to \mathbb{R}$ given by the unconditional expectation $g(x) := \mathrm{E}[h(x, Y)]$.

Proof. Let $\mathcal{G} = \sigma(\mathcal{P})$, $\mathcal{P} = \{A_i\}_{i \geqslant 1}$. Then, by (6.21) of Theorem 6.6:

$$\mathrm{E}[h(X, Y) \mid \mathcal{G}] = \sum_{i \geqslant 1} \mathrm{E}[h(X, Y) \mid A_i] \, \mathbb{I}_{A_i} = \sum_{i \geqslant 1} \frac{\mathrm{E}[h(X, Y) \, \mathbb{I}_{A_i}]}{\mathbb{P}(A_i)} \mathbb{I}_{A_i}.$$

Any given outcome $\omega \in \Omega$ must be in only one atom, say $\omega \in A_k$ for some value $k \geqslant 1$. Since X is $\mathcal{G}$-measurable, it is constant on a given atom of $\mathcal{G}$, i.e., say $X(\omega) = x_k$ and hence $h(X, Y) \, \mathbb{I}_{A_k} = h(x_k, Y) \, \mathbb{I}_{A_k}$. Moreover, the atoms are mutually exclusive so that $\mathbb{I}_{A_i}(\omega) = \delta_{ik}$. Note that Y and $\mathbb{I}_{A_k}$ are independent and hence $h(x_k, Y)$ and $\mathbb{I}_{A_k}$ are independent random variables. By combining these facts, the above random variable $\mathrm{E}[h(X, Y) \mid \mathcal{G}]$ evaluated on ω is

$$\mathrm{E}[h(X, Y) \mid \mathcal{G}](\omega) = \frac{\mathrm{E}[h(x_k, Y) \, \mathbb{I}_{A_k}]}{\mathbb{P}(A_k)} = \frac{\mathrm{E}[h(x_k, Y)] \, \mathrm{E}[\mathbb{I}_{A_k}]}{\mathbb{P}(A_k)} = \mathrm{E}[h(x_k, Y)].$$

By definition of g, the last expression is $g(x_k) = g(X(\omega)) = g(X)(\omega)$. Since the above argument holds for arbitrary ω, we have $\mathrm{E}[h(X, Y) \mid \mathcal{G}] = g(X)$. $\square$

We now remark on the way this result is used in practice and its essence. We begin by fixing the random variable X to some ordinary variable (parameter) x in $h(X, Y)$ and compute the function $g(x)$ as the unconditional expectation, $g(x) = \mathrm{E}[h(x, Y)]$. After having obtained $g(x)$, we put back the random variable X in place of x, giving the random variable $g(X)$, which is the same as $\mathrm{E}[h(X, Y) \mid \mathcal{G}]$. That is, conditioning on the information in $\mathcal{G}$ allows us to hold X as constant (as a given value x) and then the conditional expectation of $h(x, Y)$ given the information in $\mathcal{G}$ becomes an unconditional expectation since Y, and hence $h(x, Y)$, is independent of $\mathcal{G}$. Note that when $\mathcal{G} = \sigma(X)$ we obtain the well-known formula used for computing the expectation of a function of two random variables, say $h(X, Y)$, conditional on one of the random variables, say X, where Y is independent of X (i.e., Y is independent of $\sigma(X)$):

$$\mathrm{E}[h(X, Y) \mid \sigma(X)] \equiv \mathrm{E}[h(X, Y) \mid X] = g(X)$$

where $g(x) = \mathrm{E}[h(X, Y) \mid X](x) \equiv \mathrm{E}[h(X, Y) \mid X = x] = \mathrm{E}[h(x, Y)]$.

Proposition 6.7 extends to the multidimensional case in the obvious manner. Since this is an important result that is also used in further chapters, we state it in a proposition as follows. A similar proof as given above for Proposition 6.7 in the case of $\mathcal{G}$ being generated by a countable partition, with discrete random variables, is left as an exercise for the reader. A general proof of this result, particularly in the case of continuous random variables, follows in Chapter 9.

Proposition 6.8 (Independence Proposition for Several Random Variables). *Consider two random vectors* $\mathbf{X} = (X_1, \ldots, X_m)$ *and* $\mathbf{Y} = (Y_1, \ldots, Y_n)$*, where all components of* $\mathbf{X}$ *are assumed* $\mathcal{G}$*-measurable and all components of* $\mathbf{Y}$ *are assumed independent of* $\mathcal{G}$*. Let* $h : \mathbb{R}^{m+n} \to \mathbb{R}$ *be a Borel function. Then,* $\mathrm{E}[h(\mathbf{X}, \mathbf{Y}) \mid \mathcal{G}] = g(\mathbf{X})$ *where* $g : \mathbb{R}^m \to \mathbb{R}$ *is defined by the unconditional expectation* $g(\mathbf{x}) := \mathrm{E}[h(\mathbf{x}, \mathbf{Y})]$*, i.e.,*

$$\mathrm{E}[h(X_1, \ldots, X_m, Y_1, \ldots, Y_n) \mid \mathcal{G}] = g(X_1, \ldots, X_m)$$

where $g(x_1, \ldots, x_m) := \mathrm{E}[h(x_1, \ldots, x_m, Y_1, \ldots, Y_n)]$.

As in the case of two variables, this result is applied by assigning ordinary variables $X_1 = x_1, \ldots, X_m = x_m$ in $h(X_1, \ldots, X_m, Y_1, \ldots, Y_n)$ and computing the function g as the unconditional expectation of the random variable $h(x_1, \ldots, x_m, Y_1, \ldots, Y_n)$ which is a function of random variables $Y_1, \ldots, Y_n$ with $x_1, \ldots, x_m$ as parameters assigned to the first m arguments of h. Then, $\mathrm{E}[h(\mathbf{X}, \mathbf{Y}) \mid \mathcal{G}]$ is the random variable $g(X_1, \ldots, X_m)$. An important case is when $\mathcal{G} = \sigma(\mathbf{X}) \equiv \sigma(X_1, \ldots, X_m)$, i.e., conditioning on a random vector $\mathbf{X}$ which is independent of random vector $\mathbf{Y}$. We obtain the known formula for $\mathrm{E}[h(\mathbf{X}, \mathbf{Y}) \mid \sigma(\mathbf{X})] \equiv \mathrm{E}[h(\mathbf{X}, \mathbf{Y}) \mid \mathbf{X}]$:

$$\mathrm{E}[h(\mathbf{X}, \mathbf{Y}) \mid \mathbf{X}] = g(\mathbf{X}) \text{ where } g(\mathbf{x}) = \mathrm{E}[h(\mathbf{X}, \mathbf{Y}) \mid \mathbf{X} = \mathbf{x}] = \mathrm{E}[h(\mathbf{x}, \mathbf{Y})].$$

6.3.3.4 Tower Property (Iterated Conditioning)

Let $\mathcal{H}$ be a sub-σ-algebra of $\mathcal{G}$, i.e., assume $\mathcal{H} \subset \mathcal{G} \subset \mathcal{F}$. Then,

$$\mathrm{E}[\mathrm{E}[X \mid \mathcal{G}] \mid \mathcal{H}] = \mathrm{E}[X \mid \mathcal{H}]. \tag{6.26}$$

This property is proven in an elegant manner as follows, allowing the random variables to be discrete or continuous, and the σ-algebras $\mathcal{G}$ and $\mathcal{H}$ to be countable or uncountable.

Proof. Denote $Y = \mathrm{E}[X \mid \mathcal{G}]$ and $Z = \mathrm{E}[X \mid \mathcal{H}]$. By the definition of an expectation conditional on the respective σ-algebras $\mathcal{G}$ and $\mathcal{H}$,

$$\forall A \in \mathcal{G}: \quad \mathrm{E}[X\,\mathbb{I}_A] = \mathrm{E}[Y\,\mathbb{I}_A],$$
$$\forall A \in \mathcal{H}: \quad \mathrm{E}[X\,\mathbb{I}_A] = \mathrm{E}[Z\,\mathbb{I}_A].$$

Since $\mathcal{H} \subset \mathcal{G}$, $A \in \mathcal{H}$ implies $A \in \mathcal{G}$ and we hence have

$$\forall A \in \mathcal{H}: \quad \mathrm{E}[Y\,\mathbb{I}_A] = \mathrm{E}[X\,\mathbb{I}_A] = \mathrm{E}[Z\,\mathbb{I}_A].$$

Since Z is $\mathcal{H}$-measurable, then Z is $\mathrm{E}[Y \mid \mathcal{H}]$ and hence

$$\mathrm{E}[X \mid \mathcal{H}] = \mathrm{E}[\mathrm{E}[X \mid \mathcal{G}] \mid \mathcal{H}].$$

□

By taking $\mathcal{H} = \mathcal{F}_0 = \{\emptyset, \Omega\}$, we can recover the *law of double expectation*:

$$\mathrm{E}[\mathrm{E}[X \mid \mathcal{G}]] = \mathrm{E}[\mathrm{E}[X \mid \mathcal{G}] \mid \mathcal{F}_0] = \mathrm{E}[X \mid \mathcal{F}_0] = \mathrm{E}[X]. \tag{6.27}$$

For two random variables X and Y, the law takes the form:

$$\mathrm{E}[\mathrm{E}[X \mid Y]] = \mathrm{E}[X].$$

By definition, $\mathrm{E}[X \mid \mathcal{H}]$ is $\mathcal{H}$-measurable. Since $\mathcal{H} \subset \mathcal{G}$, then $\mathrm{E}[X \mid \mathcal{H}]$ is a $\mathcal{G}$-measurable random variable. Hence this corresponds to case (c) just after Definition 6.12 where $\mathrm{E}[X \mid \mathcal{H}]$ now plays the role of X. Thus, the tower property (6.26) also works if we swap $\mathcal{G}$ and $\mathcal{H}$:

$$\mathrm{E}[\mathrm{E}[X \mid \mathcal{H}] \mid \mathcal{G}] = \mathrm{E}[X \mid \mathcal{H}].$$

6.3.4 Conditioning in the Binomial Model

Now we can apply the above properties of conditional expectations to the *filtered* binomial probability space $(\Omega_N, \mathcal{F}, \mathbb{P}, \mathbb{F})$. Recall that

$$\Omega_N = \{\omega = \omega_1\omega_2\ldots\omega_N \ : \ \omega_i \in \{\mathsf{D}, \mathsf{U}\}, 1 \leqslant i \leqslant N\},$$
$$\mathcal{F} = 2^{\Omega_N},$$
$$\mathbb{F} = \{\mathcal{F}_n = \sigma(\omega_1, \omega_2, \ldots, \omega_n)\}_{n=0,1,2,\ldots,N}, \text{ and } \mathcal{F}_0 = \{\emptyset, \Omega_N\},\ \mathcal{F}_N \equiv \mathcal{F},$$
$$\mathbb{P}(\omega) = p^{\#\mathsf{U}(\omega)}(1-p)^{\#\mathsf{D}(\omega)}.$$

It proves very convenient in what follows to use the more compact notation

$$\mathrm{E}_n[X] \equiv \mathrm{E}[X \mid \mathcal{F}_n],$$

where X is an $\mathcal{F}$-measurable (i.e., $\mathcal{F}_N$-measurable) random variable on Ω_N. For $n = 0$, the conditional expectation reduces to an unconditional one: $\mathrm{E}_0[X] = \mathrm{E}[X \mid \mathcal{F}_0] = \mathrm{E}[X]$. For the case $n = N$, the property in (6.25) gives us $\mathrm{E}_N[X] = \mathrm{E}[X \mid \mathcal{F}_N] = X$. Note that $\mathrm{E}_n[X]$ is an $\mathcal{F}_n$-measurable random variable and $\mathcal{F}_{n-1} \subset \mathcal{F}_n$, for every $n = 1, \ldots, N$. The tower property (6.26) therefore takes the following compact form:

$$\mathrm{E}_n[\mathrm{E}_m[X]] = E_n[X], \text{ for } 0 \leqslant n \leqslant m \leqslant N. \tag{6.28}$$

Recall that each $\mathcal{F}_n = \sigma(\mathcal{P}_n)$ is generated by atoms $A_{\omega_1^*\ldots\omega_n^*} = \{\omega_1 = \omega_1^*, \ldots, \omega_n = \omega_n^*\}$, where each corresponds to fixing the first n market moves. In accordance with (6.21), $\mathrm{E}_n[X]$ is then a simple random variable:

$$\mathrm{E}_n[X] \equiv \mathrm{E}[X \mid \mathcal{F}_n] = \sum_{\omega_1^*,\ldots,\omega_n^* \in \{\mathsf{U},\mathsf{D}\}} \mathrm{E}[X \mid A_{\omega_1^*\ldots\omega_n^*}]\, \mathbb{I}_{A_{\omega_1^*\ldots\omega_n^*}}. \tag{6.29}$$

Hence, for every $\omega = \omega^* \equiv \omega_1^* \ldots \omega_N^* \in \Omega_N$, i.e., for every outcome, the number $\mathrm{E}_n[X](\omega^*)$ is a function of only the first n moves where:

$$\mathrm{E}_n[X](\omega^*) = \mathrm{E}[X \mid A_{\omega_1^*\ldots\omega_n^*}]\,. \tag{6.30}$$

This dependence on *only the first n ω's in every outcome* is represented as follows:

$$\mathrm{E}_n[X](\omega) = \mathrm{E}_n[X](\omega_1, \omega_2, \ldots, \omega_n)$$

where it is a function of $\omega_1, \ldots, \omega_n$. [Note: sometimes (as in Chapter 7) we shall simply write $Y(\omega_1, \omega_2, \ldots, \omega_n) \equiv Y(\omega_1\omega_2\ldots\omega_n)$ (without the use of commas) for any $\mathcal{F}_n$-measurable random variable Y, e.g. for $Y = \mathrm{E}_n[X]$, defined on the above probability space.] The unconditional expectation of X can be calculated using (6.17):

$$\mathrm{E}[X] = \sum_{\omega \in \Omega_N} X(\omega)\mathbb{P}(\omega).$$

From (6.30) and applying (6.19), we have a direct formula for calculating $\mathrm{E}_n[X](\omega^*)$ for any given outcome:

$$\mathrm{E}_n[X](\omega^*) = \mathrm{E}_n[X](\omega_1^*, \ldots, \omega_n^*) = \frac{\mathrm{E}[X\, \mathbb{I}_{A_{\omega_1^*\ldots\omega_n^*}}]}{\mathbb{P}(A_{\omega_1^*\ldots\omega_n^*})}. \tag{6.31}$$

By the definition of the unconditional expectation, the numerator involves a sum over all $\omega = \omega_1 \ldots \omega_n\, \omega_{n+1} \ldots \omega_N$, but with indicator function fixing the first n values, i.e.,

$\mathbb{I}_{A_{\omega_1^*\cdots\omega_n^*}}(\omega) \equiv \mathbb{I}_{\{\omega_1=\omega_1^*,\ldots,\omega_n=\omega_n^*\}}(\omega) = 1$, if $\omega_1 = \omega_1^*, \ldots, \omega_n = \omega_n^*$ and equals zero otherwise. Using this fact, the expectation sum over $\omega_1, \ldots, \omega_n, \omega_{n+1}, \ldots, \omega_N \in \{\mathsf{D}, \mathsf{U}\}$ collapses to a sum over only the last $N - n$ market moves $\omega_{n+1}, \ldots, \omega_N \in \{\mathsf{D}, \mathsf{U}\}$, with the first n fixed to $\omega_1^* \ldots \omega_n^*$:

$$\mathrm{E}[X\,\mathbb{I}_{A_{\omega_1^*\cdots\omega_n^*}}] = \sum_{\omega_{n+1},\ldots,\omega_N\in\{\mathsf{D},\mathsf{U}\}} X(\omega_1^* \ldots \omega_n^*\,\omega_{n+1} \ldots \omega_N)\,\mathbb{P}(\omega_1^* \ldots \omega_n^*\,\omega_{n+1} \ldots \omega_N)\,.$$

Since $\mathbb{P}(\omega_1^* \ldots \omega_n^*\,\omega_{n+1} \ldots \omega_N) = \mathbb{P}(A_{\omega_1^*\cdots\omega_n^*}) \cdot \mathbb{P}(\omega_{n+1} \ldots \omega_N)$, with the sequence of moves $\omega_{n+1} \ldots \omega_N$ having probability $\mathbb{P}(\omega_{n+1} \ldots \omega_N) \equiv p^{\#\mathsf{U}(\omega_{n+1}\ldots\omega_N)}(1-p)^{\#\mathsf{D}(\omega_{n+1}\ldots\omega_N)}$, then (6.31) gives

$$\mathrm{E}_n[X](\omega_1^*, \ldots, \omega_n^*) = \sum_{\omega_{n+1},\ldots,\omega_N\in\{\mathsf{D},\mathsf{U}\}} X(\omega_1^* \ldots \omega_n^*\,\omega_{n+1} \ldots \omega_N)\,\mathbb{P}(\omega_{n+1} \ldots \omega_N). \tag{6.32}$$

We can remove the $*$ on all ω's and thereby state that for any $\omega = \omega_1 \ldots \omega_N \in \Omega_N$:

$$\mathrm{E}_n[X](\omega_1, \ldots, \omega_n) = \sum_{\omega_{n+1},\ldots,\omega_N\in\{\mathsf{D},\mathsf{U}\}} X(\omega_1 \ldots \omega_n\,\omega_{n+1} \ldots \omega_N)\,\mathbb{P}(\omega_{n+1} \ldots \omega_N). \tag{6.33}$$

Let us apply this to binomial prices S_n, $0 \leqslant n \leqslant N$. Here we define i.i.d. Bernoulli *Bin*$(1, p)$ random variables $\mathsf{B}_m \stackrel{d}{=} \mathbb{I}_{\{\omega_m=\mathsf{U}\}} = \mathsf{U}_m - \mathsf{U}_{m-1}$, for $1 \leqslant m \leqslant N$, where $\mathsf{U}_0 \equiv 0$. Hence, B_n is a function of only the nth market move ω_n, i.e., it is $\mathcal{F}_n$-measurable and independent of $\mathcal{F}_{n-1} = \sigma(\omega_1, \omega_2, \ldots, \omega_{n-1}) = \sigma(\mathsf{B}_1, \ldots, \mathsf{B}_{n-1})$.

1. For $0 \leqslant n < N$, we have by (6.6):

$$\begin{aligned} \mathrm{E}_n[S_{n+1}] &= \mathrm{E}_n\left[S_n\, u^{\mathsf{B}_{n+1}}\, d^{1-\mathsf{B}_{n+1}}\right] \\ &\quad \big(S_n \text{ is } \mathcal{F}_n\text{-measurable} \implies (6.25) \text{ is applied}\big) \\ &= S_n\, \mathrm{E}_n\left[u^{\mathsf{B}_{n+1}}\, d^{1-\mathsf{B}_{n+1}}\right] \\ &\quad \big(\mathsf{B}_{n+1} \text{ is independent of } \mathcal{F}_n \implies (6.24) \text{ is applied}\big) \\ &= S_n\, \mathrm{E}\left[u^{\mathsf{B}_{n+1}}\, d^{1-\mathsf{B}_{n+1}}\right] \\ &= S_n\,(u\,p + d\,(1-p)). \end{aligned} \tag{6.34}$$

2. For $0 \leqslant n < k \leqslant N$ we similarly have

$$\begin{aligned} \mathrm{E}_n[S_k] &= \mathrm{E}_n\left[S_n \prod_{m=n+1}^{k} u^{\mathsf{B}_m}\, d^{1-\mathsf{B}_m}\right] = S_n \mathrm{E}_n\left[\prod_{m=n+1}^{k} u^{\mathsf{B}_m}\, d^{1-\mathsf{B}_m}\right] \\ &\quad \big(\mathsf{B}_{n+1}, \ldots, \mathsf{B}_k \text{ are all mutually independent and independent of } \mathcal{F}_n\big) \\ &= S_n \prod_{m=n+1}^{k} \mathrm{E}\left[u^{\mathsf{B}_m}\, d^{1-\mathsf{B}_m}\right] = S_n\,(u\,p + d\,(1-p))^{k-n}. \end{aligned} \tag{6.35}$$

On the other hand, (6.35) can be derived from (6.34) by applying the tower property (6.28):

$$\begin{aligned} \mathrm{E}_n[S_k] &= \mathrm{E}_n[\mathrm{E}_{n+1}[S_k]] = \cdots = \mathrm{E}_n[\mathrm{E}_{n+1}[\cdots \mathrm{E}_{k-2}[\mathrm{E}_{k-1}[S_k]]\cdots]] \\ &= \mathrm{E}_n[\mathrm{E}_{n+1}[\cdots \mathrm{E}_{k-2}[S_{k-1}\,(up + d(1-p))]\cdots]] = \cdots \\ &= S_n\,(up + d(1-p))^{k-n}. \end{aligned}$$

6.3.5 Sub-, Super-, and True Martingales

Recall that a stochastic process $X \equiv \{X_t\}_{t\in\mathbf{T}}$ is a collection of random variables (indexed by a set $\mathbf{T} \subset [0,\infty)$) on a filtered probability space $(\Omega, \mathcal{F}, \mathbb{P}, \mathbb{F} = \{\mathcal{F}_t\}_{t\in\mathbf{T}})$ so that $\forall t \in \mathbf{T}$ X_t is $\mathcal{F}_t$-measurable, i.e., X is adapted to the filtration. Recall that X is a discrete-time process if $\mathbf{T} = \{0, 1, 2, \ldots\}$; it is a continuous-time process if $\mathbf{T} = [0,\infty)$ or an interval in $[0,\infty)$. We now give a formal definition of important classes of processes that have many financial applications.

Definition 6.15. Consider a filtered probability space $(\Omega, \mathcal{F}, \mathbb{P}, \mathbb{F})$ and a stochastic process $\{M_t\}_{t\in\mathbf{T}}$ adapted to the filtration $\mathbb{F} = \{\mathcal{F}_t\}_{t\in\mathbf{T}}$ so that $\mathrm{E}[|M_t|] < \infty$ for every $t \in \mathbf{T}$ (in this case the process is said to be *integrable*). A stochastic process $\{M_t\}_{t\in\mathbf{T}}$ is called:

(a) a *martingale* if $\mathrm{E}[M_{t+s} \mid \mathcal{F}_t] = M_t$ (it has no tendency to fall or to rise),

(b) a *sub-martingale* if $\mathrm{E}[M_{t+s} \mid \mathcal{F}_t] \geqslant M_t$ (it has no tendency to fall),

(c) a *super-martingale* if $\mathrm{E}[M_{t+s} \mid \mathcal{F}_t] \leqslant M_t$ (it has no tendency to rise),

for all $t, s \in \mathbf{T}$. In cases where the inequalities are strict, we say that the process is a *strict sub-martingale* or *strict super-martingale*, respectively. The processes that are really of interest to us are either discrete-time processes or continuous-time processes where time values are all real numbers $s, t \geqslant 0$ or $0 \leqslant s, t \leqslant T$ for some fixed time $T > 0$.

We note that in the strict sense of the above definition, which includes processes having a continuous state space, relations (a)–(c) are meant to hold true with probability one, i.e., almost surely (a.s.). Moreover, a process is a martingale (or sub- or super-martingale) w.r.t. a given filtration and probability measure. A process may be a martingale w.r.t. a given filtration and measure, but it may not necessarily be a martingale if either the filtration or measure is changed. So, more precisely, we say that $\{M_t\}_{t\geqslant 0}$ is a $(\mathbb{P}, \mathbb{F})$-martingale, i.e., w.r.t. a measure $\mathbb{P}$ and filtration $\mathbb{F}$. In what follows in this chapter we simply fix the probability measure $\mathbb{P}$, with expectation E implied w.r.t. this measure $\mathbb{P}$, and then it suffices to say that the process is a martingale w.r.t. (or relative to) a given filtration $\mathbb{F}$.

When the filtration is fixed (e.g. to a specified natural filtration) and we change probability measures, then we shall refer to a process as a martingale w.r.t. a given measure, i.e., we say that the process $\{M_t\}_{t\geqslant 0}$ is a $\mathbb{P}$-martingale or a martingale under measure $\mathbb{P}$. In Chapter 5, we already saw an example of measure changes from the physical measure, say $\mathbb{P}$, into other risk-neutral measures, e.g., $\widetilde{\mathbb{P}}$, where we simply had a single-period stochastic process with discrete-time parameter having only an initial value $t = 0$ and a terminal value $T > 0$. In that case the conditional expectations w.r.t. information at time $t = 0$ are w.r.t. the trivial filtration $\mathcal{F}_0$, and hence the martingale condition is simply stated as an unconditional expectation. That is, $\{M_t\}_{t=0,T}$ is a $\mathbb{P}$-martingale if $\mathrm{E}[M_T] = M_0$, since $\mathrm{E}[M_T \mid \mathcal{F}_0] \equiv \mathrm{E}[M_T]$. For example, the risk-neutral measure $\widetilde{\mathbb{P}} \equiv \widetilde{\mathbb{P}}^{(B)}$ discussed in Chapter 5 was defined such that $\widetilde{\mathrm{E}}[S_T^i/B_T] = S_0^i/B_0$, for all discounted security price processes $\{S_t^i/B_t\}_{t=0,T}$, where the expectation is taken w.r.t. risk-neutral measure $\widetilde{\mathbb{P}}$. Hence, according to the above definition, each of these discounted price processes is a $\widetilde{\mathbb{P}}$-martingale.

In the discrete-time setting, the definition of a sub-, or super-martingale can be simplified thanks to the tower property. A discrete-time stochastic process $\{M_n\}_{n=0,1,2,\ldots}$ with a finite expectation of its absolute value, $\mathrm{E}[|M_n|] < \infty$ for all $n \geqslant 0$, is called a martingale relative to a filtration $\{\mathcal{F}_n\}_{n\geqslant 0}$ if $\mathrm{E}_n[M_{n+1}] = M_n$ for all $n \geqslant 0$. Indeed, by applying the tower property (6.28), we obtain

$$M_n = \mathrm{E}_n[M_{n+1}] = \mathrm{E}_n[\mathrm{E}_{n+1}[M_{n+2}]] = \mathrm{E}_n[M_{n+2}] = \ldots = \mathrm{E}_n[M_{n+m}]$$

for all $n, m = 0, 1, 2, \ldots$. This leads to the following result, which tells us that the expectation of a martingale is constant over time. In other words, the expected value of a martingale is conserved.

Proposition 6.9. *Let $\{M_n\}_{n\geqslant 0}$ be a martingale. Then*

$$\mathrm{E}[M_n] = \mathrm{E}[M_0] \text{ for all } n = 1, 2, \ldots.$$

Proof. By the above application of the tower property we have $M_k = \mathrm{E}_k[M_{k+n}]$, for all $n, k \geqslant 0$. Hence, $\mathrm{E}[M_k] = \mathrm{E}[\mathrm{E}_k[M_{k+n}]] = \mathrm{E}[M_{k+n}]$. In particular, for $k = 0$, $\mathrm{E}[M_0] = \mathrm{E}[M_n]$. □

We note that in most cases M_0 is a known constant so that $M_0 = \mathrm{E}[M_n], n \geqslant 0$. The notion of a sub-, or super-martingale can be described in terms of fairness of a game. Suppose that we start playing a game with an initial capital W_0. Let W_n be the total winning after n rounds of the game. The game is considered to be fair if $W := \{W_n\}_{n=0,1,2\ldots}$ is a martingale; it is a favorable game if W is a sub-martingale; it is an unfavorable game if W is a super-martingale.

6.3.5.1 Examples

(a) Let X be an integrable random variable, i.e., $\mathrm{E}[|X|] < \infty$. Consider a filtration $\mathbb{F} = \{\mathcal{F}_t\}_{t\geqslant 0}$. Define $X_t := \mathrm{E}[X \mid \mathcal{F}_t]$ for $t \geqslant 0$. Then $\{X_t\}_{t\geqslant 0}$ is a martingale relative to $\mathbb{F}$.

Solution. First, show that the process is integrable:

$$\mathrm{E}[|X_t|] = \mathrm{E}[|E[X \mid \mathcal{F}_t]|] \leqslant \mathrm{E}[\mathrm{E}[|X| \mid \mathcal{F}_t]] = \mathrm{E}[|X|] < \infty.$$

Applying the tower property gives

$$\mathrm{E}[X_{t+s} \mid \mathcal{F}_t] = \mathrm{E}[\mathrm{E}[X \mid \mathcal{F}_{t+s}]|\mathcal{F}_t] = \mathrm{E}[X \mid \mathcal{F}_t] = X_t.$$

Finally, note that $X_t := \mathrm{E}[X \mid \mathcal{F}_t]$ is $\mathcal{F}_t$-measurable. □

(b) Consider a sequence of integrable i.i.d. random variables, $\{Y_n\}_{n=1,2,\ldots}$, with $\mathrm{E}[Y_1] = \mu$ and $\mathrm{E}[|Y_1|] < \infty$. Define $\mathcal{F}_0 = \{\emptyset, \Omega\}$ and $\mathcal{F}_n = \sigma(Y_1, \ldots, Y_n)$ for $n = 1, 2, \ldots$ and hence form the natural filtration $\mathbb{F} = \{\mathcal{F}_n\}_{n\geqslant 0}$.

(i) Let $X_n := \sum_{k=1}^{n} Y_k$, $n = 1, 2, \ldots$ and $X_0 = 0$. The process $\{X_n\}_{n\geqslant 0}$ is a martingale w.r.t. $\mathbb{F}$ iff $\mu = 0$. It is a strict sub-martingale or a strict super-martingale iff $\mu > 0$ or $\mu < 0$, respectively.

Proof. Note that $\mathrm{E}[|X_n|] \leqslant \mathrm{E}[|Y_1|] + \cdots + \mathrm{E}[|Y_n|] < \infty$. Since X_n is a function of $Y_1, \ldots, Y_n$, X_n is $\mathcal{F}_n$-measurable for all $n \geqslant 0$. Noting that $X_{n+1} = X_n + Y_{n+1}$ and applying the linearity property of conditional expectations:

$$\mathrm{E}[X_{n+1} \mid \mathcal{F}_n] = \mathrm{E}[X_n \mid \mathcal{F}_n] + \mathrm{E}[Y_{n+1} \mid \mathcal{F}_n] = X_n + \mathrm{E}[Y_{n+1}] = X_n + \mu,$$

for $n \geqslant 0$. □

For example, the random walk process is given by

$$X_n = \sum_{k=1}^{n} Y_k,\ n \geqslant 1, X_0 = 0 \text{ with } Y_k = \begin{cases} 1 & \text{with probability } p \\ -1 & \text{with probability } 1-p \end{cases},\ k \geqslant 1.$$

In this case $\mu = \mathrm{E}[Y_1] = 2p - 1$. Therefore, the random walk is a martingale, strict sub-martingale, or strict super-martingale if $p = \frac{1}{2}$ (i.e., it is a symmetric random walk), $p > \frac{1}{2}$, or $p < \frac{1}{2}$, respectively. Similarly, the log-price process in the binomial model is defined by $X_0 = 0$ and

$$X_n := \ln \frac{S_n}{S_0} = \sum_{k=1}^{n} Y_k,\ n \geqslant 1, \quad \text{with } Y_k = \begin{cases} \ln u & \text{with probability } p, \\ \ln d & \text{with probability } 1-p, \end{cases} \quad k \geqslant 1.$$

Here $\mu = \mathrm{E}[Y_1] = p \ln u + (1-p) \ln d = 0$. If $0 < d < 1 < u$, then the log-price process is a martingale iff $p = \ln(\frac{1}{d}) / \ln(\frac{u}{d})$.

(ii) The process $X_n := \sum_{k=1}^{n} Y_k - \mu\, n$, $n \geqslant 1$, $X_0 = 0$, is a martingale for any μ.

(iii) Suppose that $\mathrm{E}[Y_1] = 0$ and $\mathrm{E}[Y_1^2] = \sigma^2 < \infty$. Then the process started at X_0 and given by $X_n := \left(\sum_{k=1}^{n} Y_k\right)^2 - \sigma^2\, n$, $n \geqslant 1$, is a martingale. That is, squaring a symmetric random walk and subtracting its variance gives a martingale.

Proof.

$$\begin{aligned} X_{n+1} &= \left(\sum_{k=1}^{n} Y_k + Y_{n+1}\right)^2 - \sigma^2 (n+1) \\ &= \underbrace{\left(\sum_{k=1}^{n} Y_k\right)^2 - \sigma^2\, n}_{=X_n} + 2 Y_{n+1} \sum_{k=1}^{n} Y_k + Y_{n+1}^2 - \sigma^2 \end{aligned}$$

$$\begin{aligned} \mathrm{E}[X_{n+1} \mid \mathcal{F}_n] &= \mathrm{E}[X_n \mid \mathcal{F}_n] + 2\mathrm{E}[Y_{n+1} \sum_{k=1}^{n} Y_k \mid \mathcal{F}_n] + \mathrm{E}[Y_{n+1}^2 - \sigma^2 \mid \mathcal{F}_n] \\ &= X_n + 2 \underbrace{\mathrm{E}[Y_{n+1}]}_{=0} \sum_{k=1}^{n} Y_k + \underbrace{\mathrm{E}[Y_{n+1}^2 - \sigma^2]}_{=0} = X_n. \end{aligned}$$

□

(c) The binomial price process $\{S_n\}_{n \geqslant 1}$ is given by

$$S_n = S_0 \prod_{k=1}^{n} Z_k,\ n \geqslant 1, \quad \text{with } Z_k = \begin{cases} u & \text{with probability } p, \\ d & \text{with probability } 1-p, \end{cases} \quad k \geqslant 1.$$

It is a martingale iff $up + d(1-p) = 1$. Indeed, since $S_{n+1} = S_n Z_{n+1}$,

$$\mathrm{E}_n[S_{n+1}] = S_n \mathrm{E}_n[Z_{n+1}] = S_n(up + d(1-p)) = S_n \text{ iff } up + d(1-p) = 1.$$

Suppose that $0 < d < 1 < u$. Then, $\{S_n\}_{n \geqslant 0}$ is a martingale iff $p = \frac{1-d}{u-d}$.

6.3.6 Classification of Stochastic Processes

The following classes of stochastic processes can be considered.

1. X is said to be a *process with independent increments* if $\forall n \in \mathbb{N}$ and $\forall t_1, t_2, \ldots, t_N \in \mathbf{T}$ so that $0 \leqslant t_1 < t_2 < \cdots < t_n$, the increments $X_{t_{i+1}} - X_{t_i}$, $1 \leqslant i \leqslant n-1$, are jointly independent random variables.

2. X is said to be a *process with stationary increments* if $\forall s, t \in \mathbf{T}$ with $s > t$ the probability distribution of the increment $X_s - X_t$ depends only on the length $s - t$ of the interval $[t, s]$ and not on s and t separately.

3. X is said to be a *martingale* with respect to the filtration $\mathbb{F}$ if $\mathrm{E}[|X_t|] < \infty$ and $\mathrm{E}[X_s \mid \mathcal{F}_t] = X_t$ for every $t, s \in \mathbf{T}$ with $t \leqslant s$.

4. X is said to be a *Markov process or Markovian* if, given the value of X_t, $t \in \mathbf{T}$, the probability distribution of X_s, $s > t$, does not depend on the past history of values $\{X_u : u \in \mathbf{T}, u < t\}$. In other words,

$$\mathbb{P}(X_s \in B \mid \mathcal{F}_t^X) = \mathbb{P}(X_s \in B \mid X_t), \quad \forall B \in \mathcal{B}(\mathbb{R}).$$

This property can also be stated as $\mathbb{P}(X_s \leqslant x \mid \mathcal{F}_t^X) = \mathbb{P}(X_s \leqslant x \mid X_t)$, for any $x \in \mathbb{R}$. Equivalently, X is a Markov process if for any Borel function f and time index $t \in \mathbf{T}$ there exists a (Borel) function g (that may generally be defined in terms of f and s, t) such that

$$\mathrm{E}[f(X_s) \mid \mathcal{F}_t^X] = g(X_t), \quad \forall s \in \mathbf{T},\ s \geqslant t.$$

It should be noted here that the Markov property is defined by conditioning on the natural filtration, i.e., on F_t^X. However, if the above conditional expectation properties hold for any filtration, i.e., by conditioning on $\mathcal{F}_t$ instead of F_t^X, with X still assumed adapted to the filtration, then the process is Markov. This follows by the tower property since $F_t^X \subset F_t$ and $g(X_t)$ is F_t^X-measurable. Indeed, the condition $\mathrm{E}[f(X_s) \mid \mathcal{F}_t] = g(X_t)$ implies that $\mathrm{E}[f(X_s) \mid \mathcal{F}_t^X] = \mathrm{E}[\mathrm{E}[f(X_s) \mid \mathcal{F}_t] \mid \mathcal{F}_t^X] = \mathrm{E}[g(X_t) \mid \mathcal{F}_t^X] = g(X_t)$, $s \geqslant t$.

For a Markov process, its distribution at any future time is completely determined from the information contained only in the sub-σ-algebra $\sigma(X_t)$ generated by the process at the current time t. For a *discrete-time Markov process* $X = \{X_n\}_{n=0,1,\ldots}$, it follows that $\mathrm{E}[f(X_m) \mid \mathcal{F}_n] \equiv \mathrm{E}_n[f(X_m)] = g(X_n)$, for all $0 \leqslant n \leqslant m$. Hence, by the tower property, a discrete-time process is Markov iff the (one-step ahead) condition $\mathrm{E}_n[f(X_{n+1})] = g(X_n)$ holds for all $n = 0, 1, \ldots$. This gives us a very practical way to verify the Markov property when we make use of Proposition 6.7.

The above definition extends to the multidimensional case in the following obvious manner. Let $\mathbf{X} = \{\mathbf{X}_t := (X_t^1, \ldots, X_t^m)\}_{t \geqslant 0}$ be an $\mathbb{R}^m$-valued joint process, for integer $m \geqslant 1$. Then, we say that this vector process is Markovian or a Markov process if, given any Borel function $f : \mathbb{R}^m \to \mathbb{R}$, we have

$$\mathrm{E}[f(\mathbf{X}_s) \mid \mathcal{F}_t^{\mathbf{X}}] = g(\mathbf{X}_t), \forall s > t,$$

for some $g : \mathbb{R}^m \to \mathbb{R}$. If time is discrete then the vector process $\{\mathbf{X}_n := (X_n^1, \ldots, X_n^m)\}_{n \geqslant 0}$ is said to be Markovian if $\mathrm{E}_n[f(\mathbf{X}_{n+1})] = g(\mathbf{X}_n)$, for all $n = 0, 1, \ldots$.

Example 6.16. Show that the binomial log-price process $\{\ln S_n\}_{n \geqslant 0}$ is a Markov process with stationary and independent increments.

Solution. Denote $X_n := \ln S_n$. Then $X_{n+1} = \ln S_{n+1} = \ln S_n + Y_{n+1} = X_n + Y_{n+1}$, with i.i.d. random variables $Y_k, k \geqslant 1$, taking on value $Y_k = \ln u$ with probability p and value $\ln d$ with probability $1 - p$. The natural filtration is given by $\mathcal{F}_n = \sigma(S_1, \ldots, S_n) = \sigma(Y_1, \ldots, Y_n)$. Hence, Y_{n+1} and $\mathcal{F}_n$ are independent and X_n is $\mathcal{F}_n$-measurable, so we may use Proposition 6.7 (i.e., take $\mathcal{G} = \mathcal{F}_n, X = X_n, Y = Y_{n+1}, f(x, y) = h(x + y)$). That is, for any Borel function h:

$$\mathrm{E}_n[h(X_{n+1})] = \mathrm{E}_n[h(X_n + Y_{n+1})] = g(X_n)$$

where $g : \mathbb{R} \to \mathbb{R}$ is defined by $g(x) := \mathrm{E}\big[h(x + Y_{n+1})\big]$. Hence, the log-price process $\{X_n\}_{n \geqslant 0}$ is Markov. Now, consider any two adjacent intervals $[n-1, n]$ and $[n, n+1]$. Then, $X_{n+1} - X_n = Y_{n+1}$ and $X_n - X_{n-1} = Y_n$ are independent and it clearly follows that process X has independent increments. Consider arbitrary intervals $[m + \ell, n + \ell]$, $\ell \geqslant 0$, of the same length $(n - m)$, with integers $n > m \geqslant 0$. The stationarity property now follows since $X_{n+\ell} - X_{m+\ell} = \sum_{k=m+\ell+1}^{n+\ell} Y_k$ is a sum of $(n-m)$ i.i.d. random variables having the same distribution as $Y_1 + \ldots + Y_{(n-m)}$. □

Example 6.17. Let $\{Z_k\}_{k\in\mathbb{N}}$ be a sequence of i.i.d. random variables. Construct a piecewise-constant continuous-time process $\{X_t\}_{t\geqslant 0}$ as follows:

$$X_t = \sum_{k=1}^{\lfloor t \rfloor} Z_k, \ t > 0, \quad X_0 \equiv 0.$$

Show that X is a Markov process with independent increments. What condition guarantees that X is a martingale w.r.t. its natural filtration? [Note that increments of X are non-stationary in general.]

Solution. Take any two indices $t, s \in \mathbf{T}$ with $t < s$. The increment of X over $[t, s]$ is

$$X_s - X_t = \sum_{k=\lfloor t \rfloor+1}^{\lfloor s \rfloor} Z_k.$$

So, $X_s - X_t$ is zero if $(t, s]$ does not contain integers, i.e., if $\lfloor t \rfloor = \lfloor s \rfloor$ then $X_s - X_t = \sum_{k=\lfloor t \rfloor+1}^{\lfloor t \rfloor} Z_k \equiv 0$; otherwise it is a sum of Z_k, $k \in (t, s] \cap \mathbb{N}$. Therefore, for every selection of times $0 \leqslant t_1 < t_2 < \cdots < t_n$, all nonzero increments are sums of different selections of Z_k's and hence are jointly independent random variables. Moreover, X_s can be written as a sum of two independent random variables: X_t and $Y \equiv \sum_{k=\lfloor t \rfloor+1}^{\lfloor s \rfloor} Z_k$. Note that X_t is clearly $\mathcal{F}_t^X$-measurable and Y is independent of $\sigma(Z_1, \ldots, Z_{\lfloor t \rfloor}) = \sigma(X_1, \ldots, X_{\lfloor t \rfloor}) = \mathcal{F}_{\lfloor t \rfloor}^X = \mathcal{F}_t^X$. Using Proposition 6.7 (set $\mathcal{G} = \mathcal{F}_t^X, X = X_t, f(x, y) = h(x+y)$) gives (for any Borel function h):

$$\mathrm{E}[h(X_s) \mid \mathcal{F}_t^X] = \mathrm{E}\big[h(X_t + Y) \mid \mathcal{F}_t^X\big] = g(X_t)$$

where $g : \mathbb{R} \to \mathbb{R}$ is defined by $g(x) := \mathrm{E}\big[h(x + Y)\big]$. Hence the process is Markov. Putting $h(x) = x$, $g(x) = \mathrm{E}[x + Y] = x + \mathrm{E}[Y]$,

$$\mathrm{E}[X_s \mid \mathcal{F}_t^X] = X_t + \mathrm{E}[Y] = X_t + \sum_{k=\lfloor t \rfloor+1}^{\lfloor s \rfloor} \mathrm{E}[Z_k]\,.$$

Hence, the process is a martingale w.r.t. its natural filtration iff $\mathrm{E}[Z_k] = 0$. □

6.3.7 Stopping Times

Let us fix a filtered probability space $(\Omega, \mathcal{F}, \mathbb{P}, \mathbb{F})$. There is a class of random variables that play an important role in financial modelling and derivative pricing. These random variables are known as "stopping times." The formal definition of a stopping time random variable (given just below) is quite general. That is, given any nonnegative number (time) $t \geqslant 0$, then a random variable $\mathcal{T}$ is a stopping time with respect to a filtration $\mathbb{F} = \{\mathcal{F}_t\}_{t\geqslant 0}$ if the event $\{\mathcal{T} \leqslant t\}$ is $\mathcal{F}_t$-measurable. For example, if we let $\mathcal{T}$ represent the first time that the share price of a stock has passed a certain level, then the information contained in $\mathcal{F}_t$ is enough to know whether or not $\mathcal{T} \leqslant t$, for any given value of time $t \geqslant 0$, i.e., the event that it took less than or equal to a given time t (or greater than time t) for the stock to pass a certain level is contained in $\mathcal{F}_t$. The event $\{\mathcal{T} \leqslant t\}$ and its complement $\{\mathcal{T} > t\}$ are in the σ-algebra $\mathcal{F}_t$. The following is a general definition.

Definition 6.16. A random variable $\mathcal{T}\colon \Omega \to [0,\infty)\cup\{\infty\}$ is called a stopping time w.r.t. a filtration $\{\mathcal{F}_t\}_{t\geqslant 0}$ if the event $\{\mathcal{T}\leqslant t\}\equiv\{\omega\in\Omega\,:\,\mathcal{T}(\omega)\leqslant t\}\in\mathcal{F}_t$ for every $t\geqslant 0$.

In the discrete-time setting, stopping times take on integer values and as a consequence there is an equivalent definition contained in the following proposition.

Proposition 6.10. *In the discrete-time case, $\mathcal{T}\colon\Omega\to\mathbb{N}_0\cup\{\infty\}$, where $\mathbb{N}_0=\{0,1,2,\ldots\}$, is a stopping time iff $\{\mathcal{T}=n\}\in\mathcal{F}_n$ for every $n\in\mathbb{N}_0$.*

Proof. Suppose that, $\forall n\in\mathbb{N}_0$, $\{\mathcal{T}\leqslant n\}\in\mathcal{F}_n$. If $n=0$, then $\{\mathcal{T}\leqslant 0\}=\{\mathcal{T}=0\}\in\mathcal{F}_0$. Now let $n\geqslant 1$. Then, $\{\mathcal{T}\leqslant n\}\in\mathcal{F}_n$ and $\{\mathcal{T}\leqslant n-1\}\in\mathcal{F}_{n-1}\subset\mathcal{F}_n$. Since $\mathcal{F}_n$ is a σ-algebra, $\{\mathcal{T}=n\}=\{\mathcal{T}\leqslant n\}\setminus\{\mathcal{T}\leqslant n-1\}\in\mathcal{F}_n$. Conversely, suppose that $\forall n\in\mathbb{N}_0$ $\{\mathcal{T}=n\}\in\mathcal{F}_n$. Therefore, $\{\mathcal{T}=k\}\in\mathcal{F}_k\subset\mathcal{F}_n$ for every $k=0,1,\ldots,n$. Again, by using the fact that $\mathcal{F}_n$ is a σ-algebra, we have

$$\{\mathcal{T}\leqslant n\}=\bigcup_{k=0}^{n}\{\mathcal{T}=k\}\in\mathcal{F}_n. \qquad \square$$

Denoting $x\wedge y:=\min\{x,y\}$ and $x\vee y:=\max\{x,y\}$, we have the following basic properties of any stopping times.

Proposition 6.11. *Suppose that $\mathcal{T}$, $\mathcal{T}_1$, and $\mathcal{T}_2$ are stopping times. Then, $\mathcal{T}\wedge m$, $m\in\mathbb{N}_0$, $\mathcal{T}_1\wedge\mathcal{T}_2$, and $\mathcal{T}_1\vee\mathcal{T}_2$ are stopping times as well.*

The proof of Proposition 6.11 is left as an exercise for the reader (see Exercises 6.27 and 6.28).

Some important examples of stopping times are so-called first passage times or hitting times of a process. For a discrete-time stochastic process $\{X_n\}_{n\geqslant 0}$ the *first hitting time to a level ℓ* is defined as the smallest nonnegative integer value of time such that the process attains the value ℓ:

$$\mathcal{T}_\ell:=\min\{n\geqslant 0\,:\,X_n=\ell\}. \tag{6.36}$$

We put $\mathcal{T}_\ell=\infty$ if the set $\{\mathcal{T}_\ell<\infty\}=\emptyset$. Hence, $\mathcal{T}_\ell$ is allowed to be infinite, i.e., $\mathcal{T}_\ell:\Omega\to\mathbb{N}_0\cup\{\infty\}$. If $\mathbb{P}(\mathcal{T}_\ell<\infty)=1$ then we say that $\mathcal{T}_\ell$ is (almost surely) finite.

The *first passage time up to level ℓ*, $\mathcal{T}_\ell^+$, and the *first passage time down to level ℓ*, $\mathcal{T}_\ell^-$, are both in $\mathbb{N}_0\cup\{\infty\}$ and are defined by

$$\mathcal{T}_\ell^+:=\min\{n\geqslant 0\,:\,X_n\geqslant\ell\}\ \text{ and }\ \mathcal{T}_\ell^-:=\min\{n\geqslant 0\,:\,X_n\leqslant\ell\}.$$

We put $\mathcal{T}_\ell^+=\infty$ if $\{\mathcal{T}_\ell^+<\infty\}=\emptyset$ and, similarly, we put $\mathcal{T}_\ell^-=\infty$ if $\{\mathcal{T}_\ell^-<\infty\}=\emptyset$. For the respective $\pm$ cases, if $\mathbb{P}(\mathcal{T}_\ell^\pm<\infty)=1$ then we say that $\mathcal{T}_\ell^\pm$ is (almost surely) finite. We note also that if the process is defined only for finite integer times $0\leqslant n\leqslant N$, for some fixed integer $N>0$, then we set $\{\mathcal{T}_\ell^+>N\}\equiv\{\mathcal{T}_\ell^+=\infty\}$, $\{\mathcal{T}_\ell^->N\}\equiv\{\mathcal{T}_\ell^-=\infty\}$ and $\{\mathcal{T}_\ell>N\}\equiv\{\mathcal{T}_\ell=\infty\}$. The meaning here is that if the process has not attained level ℓ by the maximum allowed finite time N then it will never attain (or takes an "infinite time" to attain) level ℓ.

Example 6.18. Consider the three-period binomial model with $S_0=1$, $u=2$, and $d=\frac{1}{2}$. Show that $\mathcal{T}_2^+$ and $\mathcal{T}_{\frac{1}{4}}^-$ are stopping times w.r.t. the natural filtration $\{\mathcal{F}_n=\sigma(S_0,S_1,\ldots,S_n)\}_{n\geqslant 0}$ generated by the stock price process $S\equiv\{S_n\}_{n\geqslant 0}$.

Solution. It is sufficient to show that $\{\mathcal{T} = k\} \in \mathcal{F}_k$ for $k = 0, 1, 2, 3$.

$$\{\mathcal{T}_2^+ = 0\} = \emptyset \in \mathcal{F}_0, \qquad \{\mathcal{T}_{\frac{1}{4}}^- = 0\} = \emptyset \in \mathcal{F}_0,$$
$$\{\mathcal{T}_2^+ = 1\} = A_{\mathsf{U}} \in \mathcal{F}_1, \qquad \{\mathcal{T}_{\frac{1}{4}}^- = 1\} = \emptyset \in \mathcal{F}_1,$$
$$\{\mathcal{T}_2^+ = 2\} = \emptyset \in \mathcal{F}_2, \qquad \{\mathcal{T}_{\frac{1}{4}}^- = 2\} = A_{\mathsf{DD}} \in \mathcal{F}_2,$$
$$\{\mathcal{T}_2^+ = 3\} = \{\mathsf{DUU}\} \in \mathcal{F}_3, \qquad \{\mathcal{T}_{\frac{1}{4}}^- = 3\} = \emptyset \in \mathcal{F}_3.$$

□

Figure 6.4 illustrates the values of the random variables $\mathcal{T}_b^+$ and $\mathcal{T}_a^-$, where $a < b$, for two (paths) outcomes on the binomial stock price model with six periods.

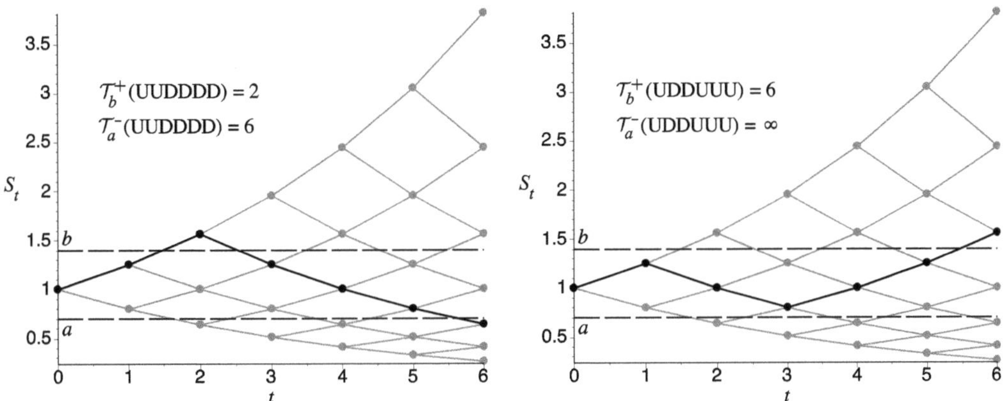

FIGURE 6.4: First passage time values $\mathcal{T}_b^+(\omega)$ and $\mathcal{T}_a^-(\omega)$ in the case of the six-period binomial model ($N = 6$) with parameters $u = 1.25$ and $d = 1/u = 0.8$, for upper level $b = 1.4$ and lower level $a = 0.7$, for two particular sample paths: (left) $\omega = \mathsf{UUDDDD}$ and (right) $\omega = \mathsf{UDDUUU}$.

First-passage times are examples of stopping times. The precise claim and its proof, for the case of a discrete-time process, is given just below. Before proving the result, we define two other important and related random variables. One is the *running (or sampled) maximum*

$$M_n := \max\{X_k : k = 0, 1, \ldots, n\}$$

and the other is the *running (or sampled) minimum*

$$m_n := \min\{X_k : k = 0, 1, \ldots, n\}$$

of the process observed from time zero to time $n \geqslant 0$. These are now related to the first passage times by noting that $\{\mathcal{T}_\ell^+ \leqslant n\}$ is the event that the process takes at most n time steps to reach or go above level ℓ and $\{\mathcal{T}_\ell^- \leqslant n\}$ is the event that it takes at most n time steps to reach or go below level ℓ. Hence, we have the equivalence of events:

$$\{\mathcal{T}_\ell^+ \leqslant n\} = \{X_0 \geqslant \ell\} \cup \{X_1 \geqslant \ell\} \cup \ldots \cup \{X_n \geqslant \ell\} = \{\max(X_0, \ldots, X_n) \geqslant \ell\} \equiv \{M_n \geqslant \ell\}$$

and

$$\{\mathcal{T}_\ell^- \leqslant n\} = \{X_0 \leqslant \ell\} \cup \{X_1 \leqslant \ell\} \cup \ldots \cup \{X_n \leqslant \ell\} = \{\min(X_0, \ldots, X_n) \leqslant \ell\} \equiv \{m_n \leqslant \ell\}.$$

Proposition 6.12. *The first passage times $\mathcal{T}_\ell^\pm$ for a real-valued discrete-time process $\{X_n\}_{n\geqslant 0}$ are stopping times w.r.t. its natural filtration $\{\mathcal{F}_n := \sigma(X_0, X_1, \ldots, X_n)\}_{n\geqslant 0}$.*

Proof. Note that $\{X_k \geqslant \ell\} \in \mathcal{F}_n$ and $\{X_k \leqslant \ell\} \in \mathcal{F}_n$, since X_k is $\mathcal{F}_n$-measurable for every $0 \leqslant k \leqslant n$. Now, fix any integer $n \geqslant 0$. By the above equivalence of events we immediately have $\{\mathcal{T}_\ell^+ \leqslant n\} = \cup_{k=0}^n \{X_k \geqslant \ell\} \in \mathcal{F}_n$ and $\{\mathcal{T}_\ell^- \leqslant n\} = \cup_{k=0}^n \{X_k \leqslant \ell\} \in \mathcal{F}_n$ by closure under countable unions of sets in $\mathcal{F}_n$. □

The above definitions extend to continuous-time real-valued processes $X \equiv \{X(t)\}_{t \geqslant 0}$. The *first hitting time to a level* $\ell \in \mathbb{R}$ is defined as the smallest nonnegative real value of time such that the process attains the value ℓ :

$$\mathcal{T}_\ell := \inf\{t \geqslant 0 \,:\, X(t) = \ell\}\,. \tag{6.37}$$

We put $\mathcal{T}_\ell = \infty$ if the set $\{\mathcal{T}_\ell < \infty\} = \emptyset$, that is, if there are no paths ω for which $\mathcal{T}_\ell < \infty$. More importantly, if $\mathbb{P}(\mathcal{T}_\ell < \infty) = 0$, then we say that the first hitting time to ℓ is (almost surely or with probability one) infinite. The first hitting time $\mathcal{T}_\ell$ is hence a random variable with mapping $\mathcal{T}_\ell : \Omega \to [0, \infty) \cup \{\infty\}$.

The *first passage time up*, $\mathcal{T}_\ell^+$, *and the first passage time down*, $\mathcal{T}_\ell^-$, to a level $\ell \in \mathbb{R}$, are defined as follows:

$$\mathcal{T}_\ell^+ := \inf\{t \geqslant 0 \,:\, X(t) \geqslant \ell\} \;\text{ and }\; \mathcal{T}_\ell^- := \inf\{t \geqslant 0 \,:\, X(t) \leqslant \ell\} \tag{6.38}$$

where we set $\mathcal{T}_\ell^+ = \infty$ if $\{t \geqslant 0 \,:\, X(t) \geqslant \ell\} = \emptyset$ and, similarly, we set $\mathcal{T}_\ell^- = \infty$ if $\{t \geqslant 0 \,:\, X(t) \leqslant \ell\} = \emptyset$. Let $X(0) = x_0$ be the initial value of the process and assume the process has continuous paths (such as Brownian motion). Then, we clearly have $\mathcal{T}_\ell^+ = \mathcal{T}_\ell$ if $x_0 \leqslant \ell$ (hitting or passing above level ℓ) and $\mathcal{T}_\ell^- = \mathcal{T}_\ell$ if $x_0 \geqslant \ell$ (hitting or passing below level ℓ). That is, for processes with continuous paths the first hitting time corresponds to the appropriate first passage time.

Some continuous-time processes are defined only up to a finite time T, i.e., $\{X(t)\}_{0 \leqslant t \leqslant T}$. Then, for each of the above respective first passage times we mean $\{\mathcal{T}_\ell = \infty\} \equiv \{\mathcal{T}_\ell > T\}$ or $\{\mathcal{T}_\ell^+ = \infty\} \equiv \{\mathcal{T}_\ell^+ > T\}$ or $\{\mathcal{T}_\ell^- = \infty\} \equiv \{\mathcal{T}_\ell^- > T\}$. We therefore observe that the first passage times are the smallest nonnegative (and possibly infinite) values of time such that the process hits or goes, respectively, above or below the given level ℓ. In the trivial respective cases where the process starts at $X(0) = x_0 \geqslant \ell$ (or, respectively, $x_0 \leqslant \ell$) then $\mathcal{T}_\ell^+ = 0$ (or, respectively, $\mathcal{T}_\ell^- = 0$). By combining the two types of first passage times, we can also consider the *first exit time* $\mathcal{T}_{(a,b)}$ from a region (a, b), assuming the process begins at an interior point $x_0 \in (a, b)$ and $\mathcal{T}_{(a,b)} = 0$ in the case $x_0 \notin (a, b)$. We have $\mathcal{T}_{(a,b)} = \mathcal{T}_a^- \wedge \mathcal{T}_b^+ = \inf\{t \geqslant 0 \,:\, X(t) \leqslant a \text{ or } X(t) \geqslant b\}$.

The sampled maximum of the process X, up to a time $t \geqslant 0$, is defined by

$$M(t) := \sup\{X(u) : 0 \leqslant u \leqslant t\} \tag{6.39}$$

and its sampled minimum is defined by

$$m(t) := \inf\{X(u) : 0 \leqslant u \leqslant t\}\,. \tag{6.40}$$

Note that the process is sampled (observed) continuously in time from time zero to time $t \geqslant 0$. As in the above discrete-time case, we have a simple relationship between the first passage times and these two extreme values of the process. That is, the event that the process takes more than a given time t to first pass above level ℓ is the same as the event that its observed maximum value up to time t is less than ℓ. Similarly, the event that the process takes more than a time t to first pass below level ℓ is the event that its observed minimum value up to time t is greater than ℓ. Precisely in terms of events, we have the equivalences:

$$\{\mathcal{T}_\ell^+ > t\} = \{X(u) < \ell, \text{ for all } u \leqslant t\} \equiv \{M(t) < \ell\} \tag{6.41}$$

and

$$\{\mathcal{T}_\ell^- > t\} = \{X(u) > \ell, \text{ for all } u \leqslant t\} \equiv \{m(t) > \ell\}. \tag{6.42}$$

Based on these relations we can readily prove that, in the case of a continuous-time process, the first passage or hitting times $\mathcal{T}_\ell^\pm$, and hence $\mathcal{T}_\ell$, are stopping times w.r.t. the natural filtration of the process.

Proposition 6.13. *The first passage times $\mathcal{T}_\ell^\pm$ defined for a real-valued $\{X(t)\}_{t\geqslant 0}$ are stopping times w.r.t. the natural filtration $\mathbb{F}^X = \{\mathcal{F}_t^X\}_{t\geqslant 0}$.*

Proof. We first realize that the event $\{M(t) < \ell\} \equiv \cap_{0\leqslant u\leqslant t}\{X(u) < \ell\} \in \mathcal{F}_t^X$ by the very definition of the natural filtration $\mathbb{F}^X$, i.e., $\mathcal{F}_t^X$ contains all events $\{X(u) \in B\}$, for any Borel set B and real times $0 \leqslant u \leqslant t$. [Note: We can equivalently write $\cap_{0\leqslant u\leqslant t}\{X(u) < \ell\}$ as a countable intersection using only rational time values $\cap_{0\leqslant u\leqslant t:u\in\mathbb{Q}}\{X(u) < \ell\}$, which is a set in $\mathcal{F}_t^X$ by closure under countable unions of sets $\{X(u) < \ell\} \in \mathcal{F}_t^X$.]. Then, $\{\mathcal{T}_\ell^+ > t\} = \{M(t) < \ell\}$ gives $\{\mathcal{T}_\ell^+ > t\} \in \mathcal{F}_t^X$ and hence $\{\mathcal{T}_\ell^+ > t\}^{\complement} = \{\mathcal{T}_\ell^+ \leqslant t\} \in \mathcal{F}_t^X$. The proof that $\{\mathcal{T}_\ell^- \leqslant t\} \in \mathcal{F}_t^X$ follows in the same manner and we leave it to the reader. □

According to its name, a stopping time is used to stop a stochastic process. Let $\{X(t)\}_{t\geqslant 0}$ be adapted to a filtration $\mathbb{F} = \{\mathcal{F}_t\}_{t\geqslant 0}$, and $\mathcal{T}$ be a stopping time w.r.t. $\mathbb{F}$. The process $X(t \wedge \mathcal{T})$ is called a *stopped process* and is defined by

$$X(t \wedge \mathcal{T})(\omega) \equiv X(t \wedge \mathcal{T}(\omega), \omega) := \begin{cases} X(t, \omega) & \text{if } t < \mathcal{T}(\omega) \\ X(\mathcal{T}(\omega), \omega) & \text{if } t \geqslant \mathcal{T}(\omega) \end{cases}$$

for every $\omega \in \Omega$. We can represent this compactly in terms of indicator functions on events: $X(t \wedge \mathcal{T}) = X(t)\mathbb{I}_{\{t<\mathcal{T}\}} + X(\mathcal{T})\mathbb{I}_{\{t\geqslant\mathcal{T}\}}$. Note that stopping the process does not mean that time itself stops. What stopping really means is that, for all times $t \geqslant \mathcal{T}$, the process is kept constant (i.e., it is fixed, pinned down, or frozen) to the value $X(\mathcal{T}, \omega)$ that it took on at the stopping time $\mathcal{T} = \mathcal{T}(\omega)$ for a realization (path) $\omega \in \Omega$.

As an example, consider the 6-period binomial model with parameters $S_0 = 1, u = 1.25, d = 1/u = 0.8$, as shown in Figure 6.4. For the choice of stopping time $\mathcal{T} = \mathcal{T}_b^+$ we have $\mathcal{T}(\omega) \equiv \mathcal{T}_b^+(\omega) = 2$ for all $\omega \in A_{\mathsf{UU}}$, i.e., the first passage time above level $b = 1.4$ is $t = 2$ time steps for all 16 stock price trajectories (scenarios) with the first two upward moves. In particular, $\mathcal{T}(\mathsf{UUDDDD}) = \mathcal{T}(\mathsf{UUUDDD}) = \mathcal{T}(\mathsf{UUDUDD}) = \ldots = \mathcal{T}(\mathsf{UUUUUU}) = 2$. For each $\omega \in A_{\mathsf{UU}}$, the stopped stock price process, denoted by $\hat{S}_t := S_{t\wedge\mathcal{T}}$, has the path $\hat{S}_0(\omega) = S_0 = 1, \hat{S}_1(\omega) = S_1 = 1.5$ and $\hat{S}_t(\omega) = S_2 = 1.5625$ for $t = 2, 3, 4, 5, 6$.

Clearly, the stopped process is adapted to $\mathbb{F}$. Moreover, if the original process is a martingale w.r.t. $\mathbb{F}$, then the stopped process is also a martingale. This observation follows from a general statement about stopping times called the Optional Sampling Theorem. Below we state the theorem and give the proof for a discrete-time process $\{X_t\}_{t\geqslant 0}$. However, the theorem also holds for continuous-time processes $\{X(t)\}_{t\geqslant 0}$.

Theorem 6.14 (Doob's Optional Sampling Theorem). *Consider a stopping time $\mathcal{T}$ and a stochastic process $\{X_t\}_{t\geqslant 0}$, both adapted to a filtration $\mathbb{F} = \{\mathcal{F}_t\}_{t\geqslant 0}$. If the process $\{X_t\}_{t\geqslant 0}$ is a martingale (or supermartingale, or submartingale) w.r.t. $\mathbb{F}$, then the stopped process $\{X_{t\wedge\mathcal{T}}\}_{t\geqslant 0}$ is also a martingale (or supermartingale, or submartingale, respectively).*

Proof. Let $\{X_t\}_{t\geqslant 0}$ be a martingale w.r.t. filtration $\mathbb{F}$, then

$$\mathrm{E}[X_t \mid \mathcal{F}_{t-1}] = X_{t-1}, \quad t = 1, 2, \ldots$$

For the stopped process we have $X_{t\wedge\mathcal{T}} = \mathbb{I}_{\{\mathcal{T}<t\}}X_{\mathcal{T}} + \mathbb{I}_{\{\mathcal{T}\geqslant t\}}X_t$. Since the event $\{\mathcal{T} < t\} =$

$\cup_{n=1}^{t-1}\{\mathcal{T} = n\}$, i.e., $\mathbb{I}_{\{\mathcal{T}<t\}} = \sum_{n=1}^{t-1} \mathbb{I}_{\{\mathcal{T}=n\}}$, then

$$\begin{aligned}
\mathrm{E}[X_{t\wedge\mathcal{T}} \mid \mathcal{F}_{t-1}] &= \sum_{n=1}^{t-1} \mathrm{E}[\mathbb{I}_{\{\mathcal{T}=n\}} X_n \mid \mathcal{F}_{t-1}] + \mathrm{E}[\mathbb{I}_{\{\mathcal{T}\geqslant t\}} X_t \mid \mathcal{F}_{t-1}] \\
&= \sum_{n=1}^{t-1} \mathbb{I}_{\{\mathcal{T}=n\}} X_n + \mathrm{E}[\mathbb{I}_{\{\mathcal{T}\geqslant t\}} X_t \mid \mathcal{F}_{t-1}]
\end{aligned}$$

(since $\mathbb{I}_{\{\mathcal{T}=n\}} X_n$ is $\mathcal{F}_{t-1}$-measurable for $0 \leqslant n \leqslant t-1$)

$$= \sum_{n=1}^{t-1} \mathbb{I}_{\{\mathcal{T}=n\}} X_n + \mathbb{I}_{\{\mathcal{T}\geqslant t\}} \mathrm{E}[X_t \mid \mathcal{F}_{t-1}]$$

(since $\mathbb{I}_{\{\mathcal{T}\geqslant t\}} = 1 - \mathbb{I}_{\{\mathcal{T}\leqslant t-1\}}$ is $\mathcal{F}_{t-1}$-measurable)

$$= \sum_{n=1}^{t-1} \mathbb{I}_{\{\mathcal{T}=n\}} X_n + \mathbb{I}_{\{\mathcal{T}\geqslant t\}} X_{t-1}$$

(by the above martingale property)

$$\begin{aligned}
&= \sum_{n=1}^{t-2} \mathbb{I}_{\{\mathcal{T}=n\}} X_n + \mathbb{I}_{\{\mathcal{T}=t-1\}} X_{t-1} + \mathbb{I}_{\{\mathcal{T}\geqslant t\}} X_{t-1} \\
&= \sum_{n=1}^{t-2} \mathbb{I}_{\{\mathcal{T}=n\}} X_n + \mathbb{I}_{\{\mathcal{T}\geqslant t-1\}} X_{t-1} = X_{(t-1)\wedge\mathcal{T}}.
\end{aligned}$$

That is, the stopped process is a martingale. The proof for a supermartingale (or submartingale) is very similar and left to the reader. □

Let us see how this theorem can be used in practice. In Game Theory, a martingale is a betting strategy such that the gambler doubles the bet after every loss. By doing this, the gambler guarantees that the first win would recover all previous losses plus it would give a profit equal to the original stake. Let us consider the following game of chance. Flip a coin repeatedly to generate a sequence $\omega \in \Omega_\infty := \prod_{i=1}^{\infty}\{H, T\}$. Win an amount equal to the stake on heads (with probability $p > 0$) and lose on tails (with probability $1 - p > 0$). Consider the "martingale strategy": if the gambler loses, then he/she doubles the stake and plays again; if he/she wins, then he/she takes the winnings and quits playing. Suppose that the initial stake is $\alpha_1 = \$1$. The stake at time $t \geqslant 2$ is

$$\alpha_t(\omega) = \alpha_t(\omega_1, \ldots, \omega_{t-1}) = \begin{cases} 2^{t-1} & \text{if } \omega_1 = \cdots = \omega_{t-1} = H, \\ 0 & \text{otherwise.} \end{cases}$$

For a given outcome ω, the Profit & Loss, or wealth, at time t is

$$V_t(\omega) = \sum_{k=1}^{t} \alpha_k(\omega) X_k(\omega), \text{ where } X_k(\omega) = \begin{cases} 1 & \text{if } \omega_k = H, \\ -1 & \text{if } \omega_k = T. \end{cases}$$

The time to quit playing, $\mathcal{T} = \min\{t : \omega_t = H\}$, is a stopping time w.r.t. the natural filtration generated by the coin flips. Note that $\mathbb{P}(\mathcal{T} < \infty) = 1$ since, by the continuity of the probability measure,

$$\mathbb{P}(\mathcal{T} = \infty) = \lim_{n\to\infty} \mathbb{P}(\mathcal{T} > n) = \lim_{n\to\infty} \mathbb{P}(\omega_1 = T, \ldots, \omega_n = T) = \lim_{n\to\infty} (1-p)^n = 0.$$

At time $\mathcal{T}$, the total winning is \$1:

$$V_{\mathcal{T}} = -1 - 2 - 4 - \cdots - 2^{\mathcal{T}-2} + 2^{\mathcal{T}-1} = 1.$$

Let us find the average total loss at time $\mathcal{T} - 1$, i.e., one step before winning:

$$\begin{aligned} \mathrm{E}[V_{\mathcal{T}-1}] &= \mathrm{E}[1 - 2^{\mathcal{T}-1}] = 1 - \sum_{n=1}^{\infty} 2^{n-1}\mathbb{P}(\mathcal{T} = n) \\ &\quad \left(\text{using } \mathbb{P}(\mathcal{T} = n) = \mathbb{P}(\omega_1 = T, \ldots, \omega_{n-1} = T, \omega_n = H) = p(1-p)^{n-1}\right) \\ &= 1 - \sum_{n=1}^{\infty} 2^{n-1} p(1-p)^{n-1} = 1 - p\sum_{m=0}^{\infty}(2-2p)^m = \begin{cases} 1 - \frac{p}{2p-1} & \text{if } p > \frac{1}{2} \\ -\infty & \text{if } p \leqslant \frac{1}{2} \end{cases} \end{aligned}$$

This game of chance creates a paradox. On the one hand, with probability 1 we stop playing after a finite number of games and our total winning is \$1. On the other hand, the average total loss is infinite (when $p \leqslant \frac{1}{2}$) if we quit playing one step before winning. Note that the process $\{V_t\}_{t\geqslant 0}$ is a martingale when $p = \frac{1}{2}$, so it is a fair game but with unbounded loss.

6.4 Exercises

Exercise 6.1. Define events (a)–(c) as subsets of the coin toss sample space

$$\Omega_3 = \{TTT, TTH, THT, THH, HTT, HTH, HHT, HHH\}.$$

(a) The second toss results in a head.

(b) A tail comes before a head.

(c) No heads.

Assuming that $\mathbb{P}(\text{Head}) = 1/3$ find the probabilities of these events.

Exercise 6.2. Define an indicator function $\mathbb{I}_A$ of event A as follows:

$$\mathbb{I}_A(\omega) = \begin{cases} 0 & \text{if } \omega \notin A, \\ 1 & \text{if } \omega \in A. \end{cases}$$

Prove the following properties:

(a) $\mathbb{I}_{A\cap B} = \mathbb{I}_A \cdot \mathbb{I}_B$;

(b) $\mathbb{I}_{A\cup B} = \mathbb{I}_A + \mathbb{I}_B - \mathbb{I}_A \cdot \mathbb{I}_B$;

(c) $\mathbb{I}_{A^{\complement}} = 1 - \mathbb{I}_A$.

Exercise 6.3. Express the indicator function of $(A \backslash B) \cup (B \backslash A)$ in terms of the indicator functions of $\mathbb{I}_A$ and $\mathbb{I}_B$.

Exercise 6.4. Consider a three-period binomial model with $S_0 = 4$, $u = 2$, $d = 0.5$, and $p = \frac{1}{4}$.

(a) Construct a binomial tree.

(b) Determine the probability distribution of the stock price S_3. Find its expected value.

(c) Determine the probability distributions of the geometric average $\sqrt[3]{S_1 S_2 S_3}$. Find its expected value.

Exercise 6.5. Consider a binomial model with $S_0 = 4$, $u = 2$, $d = 0.5$, and $p = \frac{1}{4}$. Let $\mathcal{T} = \mathcal{T}_b^+ := \min\{n = 0, 1, 2, \ldots : S_n \geqslant 6\}$ be the first passage time above level $b = 6$. Construct the following events as subsets of the sample space Ω_3 and find their probabilities:

$$\text{(a) } \{\mathcal{T} \leqslant 3\}; \quad \text{(b) } \{\mathcal{T} = 3\}; \quad \text{(c) } \{\mathcal{T} > 3\}.$$

Exercise 6.6. Consider the six-period model with parameters given in Figure 6.4.

(a) Determine the events $\{\mathcal{T}_b^+ = k\}$ for $k = 0, 1, 2, 3, 4, 5, 6$ and $k = \infty$ as subsets of Ω_6.

(b) Assume $p = 1/4$ is the probability of an upward move and determine the probability mass function (i.e., distribution) of $\mathcal{T}_b^+$.

(c) Let $\hat{S}_t := S_{t \wedge \mathcal{T}}$, where $\mathcal{T} = \mathcal{T}_b^+$. Determine the probability mass function of $\hat{S}_t$ for $t = 2, 3, 4, 5, 6$.

Exercise 6.7. Throw two dice. Suppose that we only know that the face values, which have been turned up, are either even or odd. Define the sample space Ω of all outcomes and find a disjoint partition of Ω and the corresponding σ-algebra that represents this information.

Exercise 6.8. Throw two dice. Suppose that we only know the sum of the face values, which have been turned up. Define the sample space Ω that includes all the possible outcomes and find a disjoint partition of Ω that represents this information.

Exercise 6.9. Find a σ-algebra over Ω_3 generated by:

(a) the number of D's in ω, D_3;

(b) the log-returns $\ln\left(\frac{S_2}{S_1}\right)$ and $\ln\left(\frac{S_3}{S_2}\right)$.

Exercise 6.10. Let $\mathcal{F}_n =$ be a σ-algebra generated by n coin tosses. Find the smallest n such that the following event belongs to $\mathcal{F}_n$:

(a) $A = \{$the first occurrence of heads is preceded by no more than 10 tails$\}$;

(b) $B = \{$there is at least 1 head in the sequence $\omega_1, \omega_2, \ldots\}$;

(c) $C = \{$the first 100 tosses produce the same outcome$\}$;

(d) $D = \{$there are no more than 2 heads and 2 tails among the first 5 tosses$\}$.

Exercise 6.11. Show that in the three-period binomial model the smallest σ-algebras respectively generated by the random vector (S_1, S_2, S_3) and by the product $S_1 \cdot S_2 \cdot S_3$ are not the same. What can you say about $\sigma(S_1, S_2)$ and $\sigma(S_1 \cdot S_2)$?

Exercise 6.12. Let $\{M_n\}_{n\geqslant 0}$ be a symmetric simple random walk on Ω_3. Show that the sequence $(\sigma(M_1), \sigma(M_1 + M_2), \sigma(M_1 + M_2 + M_3))$ is not a filtration.

Exercise 6.13. Consider the N-period binomial model. Let $\{F_n\}_{0\leqslant n\leqslant N}$ be the natural filtration. Show that the time-k value M_k of the random walk process, $\{M_n, 0 \leqslant n \leqslant N\}$, is $\mathcal{F}_n$-measurable iff $n \geqslant k$.

Exercise 6.14. Consider the binomial probability space $(\Omega_4, \mathcal{F}, \mathbb{P})$ for any $\mathbb{P}(\mathsf{U}) = p \in (0,1)$. Compute the conditional expectation

$$\mathrm{E}[\mathsf{D}_4 \mid \{\text{at least two U's}\}].$$

Exercise 6.15. Let $\mathbb{P} = \{A_1, A_2, \ldots, A_M\}$ be a disjoint partition of a sample space Ω. Prove the law of total probability

$$\mathbb{P}(B) = \sum_{m=1}^{M} \mathbb{P}(B \mid A_m)\,\mathbb{P}(A_m), \quad B \subset \Omega.$$

Exercise 6.16. Prove (6.25) by using the fact that (since it is $\mathcal{G}$-measurable) Y is a simple random variable of the form $Y = \sum_{A'\in\mathcal{P}} \mathrm{E}[Y \mid A']\,\mathbb{I}_{A'}$. Combine this with the identity $\mathbb{I}_A\,\mathbb{I}_{A'} = \mathbb{I}_A$ if $A = A'$, and $\mathbb{I}_A \cdot \mathbb{I}_{A'} = 0$ if $A \neq A'$, for any two atoms $A, A' \in \mathcal{P}$.

Exercise 6.17. Consider the binomial probability space $(\Omega_N, \mathcal{F}, \mathbb{P}_N)$. Let $\mathcal{F} = \sigma(\mathsf{U}_N)$ be the smallest σ-algebra generated by the number of U's. Find the following conditional expectations w.r.t. a σ-algebra $\mathcal{F}$:

(a) $\mathrm{E}[\mathsf{D}_N \mid \mathcal{F}]$;

(b) $\mathrm{E}[\mathsf{UbD} \mid \mathcal{F}]$, where $\mathsf{UbD}(\omega)$ is the number of U's before the first D in $\omega \in \Omega_N$.

Exercise 6.18. Consider the filtered binomial probability space $(\Omega_N, \mathcal{F}, \mathbb{P}_N, \{\mathcal{F}_n\}_{0\leqslant n\leqslant N})$. Find $\mathrm{E}_n[\mathsf{U}_m]$ and $\mathrm{E}_n[\mathsf{D}_m]$ for arbitrary n and m, $0 \leqslant n, m \leqslant N$.

Exercise 6.19. Let X be any integrable random variable on a probability space $(\Omega, \mathcal{F}, \mathbb{P})$ (that is, $\mathrm{E}[|X|] < \infty$). Let $\mathbb{F} = \{\mathcal{F}_n\}_{n\geqslant 0} \subset \mathcal{F}$ be a filtration. Prove that the stochastic process $\{X_n\}_{n\geqslant 0}$ defined by $X_n := \mathrm{E}[X \mid \mathcal{F}_n]$ for all $n \geqslant 0$ is a martingale w.r.t. the filtration $\mathbb{F}$.

Exercise 6.20. Let $X_0, X_1, X_2, \ldots$ be a sequence of i.i.d. random variables with common moments $\mathrm{E}[X_1] = 0$ and $\mathrm{E}[X_1^2] = b^2$. Let $\mathbb{F} = \{\mathcal{F}_n\}_{n\geqslant 0}$ be the natural filtration, i.e., $\mathcal{F}_n = \sigma(X_0, X_1, \ldots, X_n)$. Prove that the following processes are martingales w.r.t. the filtration $\mathbb{F}$:

(a) $Y_n = \sum\limits_{k=0}^{n} X_k$, $n = 0, 1, 2, \ldots$;

(b) $Z_n = \left(\sum\limits_{k=0}^{n} X_k\right)^2 - b^2\,n$, $n = 0, 1, 2, \ldots$.

Exercise 6.21. Let $\{M_n\}_{n\geqslant 0}$ be a symmetric simple random walk. Show that the process $\{Y_n\}_{n\geqslant 1}$ defined by

$$Y_n = (-1)^n \cos(\pi M_n)$$

is a martingale w.r.t. the natural filtration $\{\mathcal{F}_n\}_{n\geqslant 0}$. [Hint: Make use of the identity $\cos(a + b) = \cos a \cos b - \sin a \sin b$ when proving the martingale expectation property.]

Exercise 6.22. Let $\{M_n\}_{n\geqslant 0}$ be a symmetric simple random walk. Fix a real b. Prove that the process $S_n = \mathrm{e}^{bM_n}\left(\frac{2}{\mathrm{e}^b+\mathrm{e}^{-b}}\right)^n$, $n = 0, 1, 2, \ldots$, is a martingale w.r.t. the natural filtration $\{\mathcal{F}_n\}_{n\geqslant 0}$.

Exercise 6.23. Let $\{M_n\}_{n\geqslant 0}$ be a symmetric simple random walk. Prove that the process $M_n^2 - n$, $n = 0, 1, 2, \ldots$ is a martingale w.r.t. the natural filtration.

Exercise 6.24. Using a similar procedure as in Example 6.16, show that the random walk $\{M_n\}_{n\geqslant 0}$ is a Markov process with stationary and independent increments.

Exercise 6.25. Let $\{Z_n\}_{n\geqslant 0}$ be any sequence of square integrable random variables, i.e., assume $\mathrm{E}[Z_n^2] < \infty$ for all $n \geqslant 0$. Show that if the process $\{Z_n\}_{n\geqslant 0}$ is a martingale w.r.t. a filtration $\{\mathcal{F}_n\}_{n\geqslant 0}$, then the (squared) process $\{Z_n^2\}_{n\geqslant 0}$ is a submartingale w.r.t. the same filtration.

Exercise 6.26. Let $\{M_n\}_{n\geqslant 0}$ be an *asymmetric* simple random walk, i.e., $M_0 = 0$ and $M_n = \sum_{k=1}^{n} X_k$, $n \geqslant 1$, where $X_1, X_2, \ldots$ is a sequence of i.i.d. Bernoulli random variables such that $X_1 = 1$ with probability $p \neq \frac{1}{2}$ and $X_1 = -1$ with probability $q = 1 - p$ for some $p \in (0, 1)$. Show that $\{Z_n := M_n - n(p - q);\ n = 0, 1, 2, \ldots\}$ is a martingale w.r.t. its natural filtration.

Exercise 6.27. Show that if $\mathcal{T}$ is a stopping time w.r.t. some filtration then so is $\mathcal{T} \wedge m = \min\{\mathcal{T}, m\}$ for any fixed integer $m \geqslant 0$.

Exercise 6.28. Show that if $\mathcal{T}_1$ and $\mathcal{T}_2$ are stopping times w.r.t. some filtration then so are $\mathcal{T}_1 \wedge \mathcal{T}_2 = \min\{\mathcal{T}_1, \mathcal{T}_2\}$ and $\mathcal{T}_1 \vee \mathcal{T}_2 = \max\{\mathcal{T}_1, \mathcal{T}_2\}$.

Exercise 6.29. Consider the multi-period binomial model. Let $\overline{S}_n := \dfrac{S_n}{(1+r)^n}$, $n \geqslant 0$, be a discounted stock price process with the interest rate $r \geqslant 0$. Find the up-move probability p such that $\{\overline{S}_n\}_{n\geqslant 0}$ is a martingale w.r.t. the natural filtration generated by the stock price process.

Chapter 7

Replication and Pricing in the Binomial Tree Model

7.1 The Standard Binomial Tree Model

By combining the probabilistic framework in Chapter 6 with the main formal concepts of derivative asset pricing presented for the single-period model in Chapter 5, we are now ready to formally discuss derivative asset pricing within the multi-period binomial tree model. Let us begin by recalling the salient features of the standard T-period (recombining) binomial tree model on the space $(\Omega, \mathbb{P}, \mathcal{F}, \mathbb{F})$ with two assets, namely, a risky stock S and a risk-free asset B, such as a bank account or zero-coupon bond. The model is specified as follows.

- The time is discrete: $t \in \{0, 1, 2, \ldots, T\}$.
- There are 2^T possible market scenarios:

  $$\Omega \equiv \Omega_T = \{\omega = \omega_1\omega_2 \cdots \omega_T \ : \ \omega_t \in \{\mathsf{D}, \mathsf{U}\}, \, t = 1, 2, \ldots, T\},$$

  where each scenario can be represented by a path in a multi-period recombining binomial tree.
- The set of events is the power set $\mathcal{F} = 2^\Omega$.
- The probability function $\mathbb{P} \colon \mathcal{F} \to [0, 1]$ is given by

  $$\mathbb{P}(E) = \sum_{\omega \in E} \mathbb{P}(\omega), \ E \in \mathcal{F}, \text{ where } \mathbb{P}(\omega) \equiv \mathbb{P}(\{\omega\}) = p^{\#\mathsf{U}(\omega)}(1-p)^{\#\mathsf{D}(\omega)}, \tag{7.1}$$

  and $p \in (0, 1)$ is a probability of the event $\{\omega_t = \mathsf{U}\} = \{\omega \in \Omega \ : \ \omega_t = \mathsf{U}\}$ for every $t = 1, 2, \ldots, T$.
- The flow of information is described by the filtration $\mathbb{F} = \{\mathcal{F}_t\}_{0 \leqslant t \leqslant T}$, where $\mathcal{F}_0 = \emptyset$ and $\mathcal{F}_t$ is generated by the first t market moves $\omega_1, \ldots, \omega_t$ for every $t = 1, 2, \ldots, T$, i.e., $\mathcal{F}_t = \sigma(\mathcal{P}_t)$, where the partition $\mathcal{P}_t$ is a collection of atoms of the form

  $$A_{\omega_1^*, \omega_2^*, \ldots, \omega_t^*} = \{\omega \in \Omega \ : \ \omega_n = \omega_n^* \text{ for all } n = 1, 2, \ldots, t\}, \quad \omega_1^*, \omega_2^*, \ldots, \omega_t^* \in \{\mathsf{D}, \mathsf{U}\}$$

  (in particular, $\mathcal{F}_T \equiv \mathcal{F} = 2^\Omega$).
- The stochastic stock price process, $\{S_t\}_{0 \leqslant t \leqslant T}$, which is adapted to the filtration $\mathbb{F}$, is given by the recurrence

  $$S_t(\omega) = S_{t-1}(\omega) u^{\#\mathsf{U}(\omega_t)} d^{\#\mathsf{D}(\omega_t)}, \quad t = 1, 2, \ldots, T,$$

  or, equivalently, by the relationship

  $$S_t(\omega) = S_0 u^{\mathsf{U}_t(\omega)} d^{\mathsf{D}_t(\omega)}, \quad t = 0, 1, 2, \ldots, T,$$

where $\mathsf{U}_t(\omega) = \#\mathsf{U}(\omega_1, \omega_2, \ldots, \omega_t)$ and $\mathsf{D}_t(\omega) = \#\mathsf{D}(\omega_1, \omega_2, \ldots, \omega_t)$ count, respectively, the number of downward and upward market moves; d and u are, respectively, downward and upward market movement factors which satisfy $0 < d < u$; and $\omega \in \Omega_T$ is a market scenario. The initial price of the stock, $S_0 > 0$, is known.

- The deterministic price process, $\{B_t\}_{0 \leqslant t \leqslant T}$, for the risk-free asset is given by

$$B_t = B_0(1+r)^t, \quad t = 0, 1, 2, \ldots, T,$$

where $r > 0$ is a one-period return. With loss of generality, we assume that we deal with a bank account such that $B_0 = 1$. Note that for a unit zero-coupon bond paying \$1 at time T, the initial value is $B_0 = (1+r)^{-T}$.

Recall that a single-period binomial-tree model admits no (static) arbitrage portfolios in base assets iff there exists an equivalent martingale measure (EMM) $\widetilde{\mathbb{P}}^{(g)}$ for numéraire $g \in \{B, S\}$. An EMM $\widetilde{\mathbb{P}}^{(g)}$ is defined so that it is equivalent to the real-world measure $\mathbb{P}$ (which is also called an actual or physical probability measure) and the base asset price processes discounted by g are $\widetilde{\mathbb{P}}^{(g)}$-martingales. Let us extend this idea to the multi-period case. The probability function in (7.1) is specified by a single probability $p = \mathbb{P}(\mathsf{U})$ of an upward move over a single period. The respective risk-neutral probability of an upward move, $\tilde{p}^{(g)} = \widetilde{\mathbb{P}}^{(g)}(\mathsf{U})$, for either choice $g = B$ or $g = S$, is

$$\tilde{p} \equiv \tilde{p}^{(B)} = \frac{1+r-d}{u-d} \quad \text{or} \quad \tilde{q} \equiv \tilde{p}^{(S)} = \tilde{p} \cdot \frac{u}{1+r} = \frac{1+r-d}{u-d} \cdot \frac{u}{1+r}.$$

In either case, the probability $\tilde{p}^{(g)} \in (0,1)$ exists iff

$$d < 1 + r < u. \tag{7.2}$$

By replacing in (7.1) the probability p by a risk-neutral probability $\tilde{p}^{(g)}$, we can construct a risk-neutral probability measure $\widetilde{\mathbb{P}}^{(g)}$ defined on the σ-algebra $\mathcal{F} = 2^{\Omega}$. As is proved below, the base assets process discounted by numéraire g are $\widetilde{\mathbb{P}}^{(g)}$-martingales. In other words, according to Definition 5.9, the probability measures $\widetilde{\mathbb{P}}^{(B)}$ and $\widetilde{\mathbb{P}}^{(S)}$ are equivalent martingale measures for the multi-period binomial model.

Theorem 7.1. *The discounted base asset price processes*

$$\left\{\overline{S}_t := \frac{S_t}{g_t}\right\}_{0 \leqslant t \leqslant T} \quad \text{and} \quad \left\{\overline{B}_t := \frac{B_t}{g_t}\right\}_{0 \leqslant t \leqslant T}$$

are $\widetilde{\mathbb{P}}^{(g)}$-martingales for $g \in \{B, S\}$, i.e.,

$$\widetilde{\mathrm{E}}_t^{(g)}[\overline{S}_{t+1}] = \overline{S}_t \text{ and } \widetilde{\mathrm{E}}_t^{(g)}[\overline{B}_{t+1}] = \overline{B}_t$$

holds for all $t = 0, 1, \ldots, T-1$, iff the condition (7.2) holds.

Proof. Let us consider the case with $g = B$ (the other case is treated similarly). The discounted risk-free asset price process $\overline{B}_t$ is equal to 1 for all times t and hence the process $\{\overline{B}_t\}$ is a martingale. We now show that the process $\{\overline{S}_t\}$ is a martingale relative to $\widetilde{\mathbb{P}} \equiv \widetilde{\mathbb{P}}^{(B)}$. Fix arbitrarily $t \in \{0, 1, \ldots, T-1\}$. We have

$$\begin{aligned}
\widetilde{\mathrm{E}}_t\left[\frac{S_{t+1}}{B_{t+1}}\right] &= \widetilde{\mathrm{E}}_t\left[\frac{S_t u^{\#\mathsf{U}(\omega_{t+1})} d^{\#\mathsf{D}(\omega_{t+1})}}{(1+r)B_t}\right] = \frac{S_t}{B_t}\widetilde{\mathrm{E}}\left[\frac{1}{1+r} u^{\#\mathsf{U}(\omega_{t+1})} d^{\#\mathsf{D}(\omega_{t+1})}\right] \\
&= \frac{S_t}{B_t}\left(\frac{u}{1+r}\tilde{p} + \frac{d}{1+r}(1-\tilde{p})\right).
\end{aligned}$$

The last expression is equal to $\frac{S_t}{B_t}$ and hence the martingale condition for the discounted stock price process is fulfilled iff $\tilde{p} = \frac{1+r-d}{u-d}$. The condition $\tilde{p} \in (0,1)$ is equivalent to (7.2). □

The EMM $\widetilde{\mathbb{P}}^{(g)}$ for $g \in \{B, S\}$ exists iff (7.2) holds. Thus, according to the fundamental theorem of asset pricing (proved for the one-period case), there are no arbitrage portfolios iff $d < 1+r < u$, i.e., the return on the risk-free asset is strictly between the downward and upward returns on the stock: $d-1 < r < u-1$. In the next sections, we will introduce the notion of an arbitrage portfolio strategy and will prove the fundamental theorems in the multi-period case. Apparently, the condition (7.2) guarantees the absence of arbitrage strategy in the binomial tree model as well.

7.2 Self-Financing Strategies and Their Value Processes

Consider an investor who begins with an initial capital to be invested in base securities. Suppose that injecting or withdrawing funds is not allowed in the future time, although the investor can modify the investment portfolio by changing the positions in base assets. For example, the investor may sell some stock shares and invest the proceeds without risk. As a result, a sequence of investment portfolios in the base assets indexed by time is constructed. Recall from Section 2.2.4 in Chapter 2 that such a sequence of portfolios that does not allow for injecting or withdrawing funds is called a *self-financing strategy.* A self-financing strategy allows the investor to create a portfolio with a target probability distribution or hedge a cash flow during a period of time. Self-financing strategies are important for the no-arbitrage pricing of derivative securities when combined with replication in the multi-period setting where trading (i.e., portfolio re-balancing) in the base assets is allowed at times $t = 0, 1, \ldots, T$. The simplest example of a self-financing strategy is a static portfolio in the base assets that does not change in time.

For the binomial model there are only two base assets, namely, a risky stock S and a risk-free bank account B. Thus, any investment (or trading) strategy $\mathbf{\Phi}$ is a sequence of portfolios in the two base assets:

$$\mathbf{\Phi} = \{\varphi_t\}_{0 \leqslant t \leqslant T-1}, \text{ where } \varphi_t = (\beta_t, \Delta_t).$$

Throughout we shall use β_t and Δ_t to denote the respective positions in assets B and S at time t. For each $t = 0, 1, \ldots, T-1$, the portfolio φ_t is formed at time t and held until time $t+1$, i.e., $\varphi_t = (\beta_t, \Delta_t)$ is the portfolio held in the time period $[t, t+1)$. At time $t+1$ the investor can re-balance the portfolio to form the new portfolio $\varphi_{t+1} = (\beta_{t+1}, \Delta_{t+1})$, which is held in the time period $[t+1, t+2)$, and so on. At each trading time, we will insist that the re-balancing of the positions must satisfy the self-financing condition.

Let us begin with time $t = 0$. The investor begins with a given initial capital or wealth Π_0 which completely finances the initial portfolio with positions $\varphi_0 = (\beta_0, \Delta_0)$ in the base securities, i.e., with acquisition value

$$\Pi_0 = \Pi_0[\varphi_0] := \Delta_0 S_0 + \beta_0 B_0.$$

We use the notation $\Pi_t[\varphi]$ to denote the time-t value of a portfolio $\varphi = (\beta, \Delta)$ in the base assets B and S:

$$\Pi_t[\varphi] \equiv \Pi_t[(\beta, \Delta)] := \Delta S_t + \beta B_t.$$

By the above self-financing of the initial portfolio, the initial position Δ_0 in the stock determines the initial investment in the risk-free asset:

$$\beta_0 = \frac{\Pi_0 - \Delta_0 S_0}{B_0} \implies \Pi_0 = \Delta_0 S_0 + \left(\frac{\Pi_0 - \Delta_0 S_0}{B_0}\right) B_0.$$

The investor holds this portfolio until a time just prior to time $t = 1$, and at time $t = 1$ the investor liquidates it to form a new one. The liquidation value is the value of the portfolio with the positions being those at time 0 but with the prices of the base assets being those at the present time $t = 1$:

$$\Pi_1 := \Pi_1[\varphi_0] = \Delta_0 S_1 + \beta_0 B_1 = \Delta_0 S_1 + (\Pi_0 - \Delta_0 S_0)\frac{B_1}{B_0} = \Delta_0 S_1 + (\Pi_0 - \Delta_0 S_0)(1 + r).$$

These proceeds are used entirely to finance the formation of a new portfolio $\varphi_1 = (\beta_1, \Delta_1)$ with the same acquisition value $\Pi_1 = \Pi_1[\varphi_1] = \Delta_1 S_1 + \beta_1 B_1$. This is the self-financing condition applied at time $t = 1$, giving

$$\beta_1 = \frac{\Pi_1 - \Delta_1 S_1}{B_1} \implies \Pi_1 = \Delta_1 S_1 + \left(\frac{\Pi_1 - \Delta_1 S_1}{B_1}\right) B_1.$$

The position Δ_1 is determined based on the information available at time 1, i.e., $\Delta_1 = \Delta_1(\omega_1)$. By repeating the same procedure at every time step, we obtain general formulae for the equivalent liquidation and acquisition values Π_t, $t = 1, \ldots, T$, for any self-financing strategy:

$$\Pi_t := \Pi_t[\varphi_{t-1}] = \Delta_{t-1} S_t + (\Pi_{t-1} - \Delta_{t-1} S_{t-1})(1 + r), \tag{7.3}$$

$$= \Pi_t[\varphi_t] = \Delta_t S_t + \underbrace{\left(\frac{\Pi_t - \Delta_t S_t}{B_t}\right)}_{=\beta_t} B_t, \tag{7.4}$$

where $\varphi_t = (\beta_t, \Delta_t)$ is a portfolio in the base assets B and S formed at time $t = 0, 1, \ldots, T-1$. Since the liquidation value and the acquisition value of a self-financing strategy are the same at every date $t \geqslant 0$, we will only speak of the value Π_t of a self-financing strategy.

At every time step $t \geqslant 0$, the positions Δ_t and β_t are determined based on the market information available at time t. In other words, Δ_t and β_t depend on the first t market moves and we express this as

$$\Delta_t(\omega) = \Delta_t(\omega_1 \ldots \omega_t), \quad \beta_t(\omega) = \beta_t(\omega_1 \ldots \omega_t).$$

Therefore, the portfolio process $\{\varphi_t\}_{0 \leqslant t \leqslant T-1}$ is adapted to the natural filtration $\mathbb{F}$, i.e. Δ_t and β_t are $\mathcal{F}_t$-measurable random variables for all $t \geqslant 0$.

Clearly, any self-financing strategy in the binomial tree model is fully described by the process $\{\Delta_t\}_{0 \leqslant t \leqslant T-1}$ of stock positions (i.e., the delta positions) and the initial value Π_0. The positions β_t can be calculated with the use of the self-financing condition:

$$\beta_t = \frac{\Pi_t - \Delta_t S_t}{B_t}. \tag{7.5}$$

Since the delta process is adapted to the natural filtration $\mathbb{F}$, the value process is expected to be adapted to $\mathbb{F}$, as is proved in the next proposition.

Proposition 7.2. *Let $\{\Delta_t\}_{0 \leqslant t \leqslant T-1}$ be a process adapted to the natural filtration $\mathbb{F} = \{\mathcal{F}_t\}_{0 \leqslant t \leqslant T}$ of a binomial tree model, and let Π_0 be the initial known capital. Then the value process $\{\Pi_t\}_{0 \leqslant t \leqslant T}$ defined recursively by the* ***wealth equation***

$$\Pi_t = \Delta_{t-1} S_t + (\Pi_{t-1} - \Delta_{t-1} S_{t-1})(1 + r), \quad 1 \leqslant t \leqslant T, \tag{7.6}$$

is adapted to the natural filtration as well, i.e., $\Pi_t = \Pi_t(\omega_1 \omega_2 \ldots \omega_t)$ for all $0 \leqslant t \leqslant T$.

Proof. Let us prove the assertion by induction. The initial value Π_0 is an $\mathcal{F}_0$-measurable constant. For any $1 \leqslant t \leqslant T$, the value $\Pi_t = \Delta_{t-1}S_t + (\Pi_{t-1} - \Delta_{t-1}S_{t-1})(1+r)$ is $\mathcal{F}_t$-measurable since Π_t is a linear combination of $\mathcal{F}_t$-measurable variables S_t, S_{t-1}, and Π_{t-1}. The latter is measurable w.r.t. $\mathcal{F}_{t-1} \subset \mathcal{F}_t$. □

Example 7.1. Consider a three-period recombining binomial tree model with $S_0 = 8$, $B_0 = 1$, $u = \frac{3}{2}$, $d = \frac{1}{2}$, and $r = \frac{1}{4}$. Find the terminal value of a self-financing strategy with the initial value $\Pi_0 = 10$ and stock positions Δ_t given as the number of upward moves in the first t market movements for each $t = 0, 1, 2$.

Solution. The recombining binomial tree for the stock price process is given in Figure 7.1. Now construct the self-financing strategy and calculate its value step by step going forward

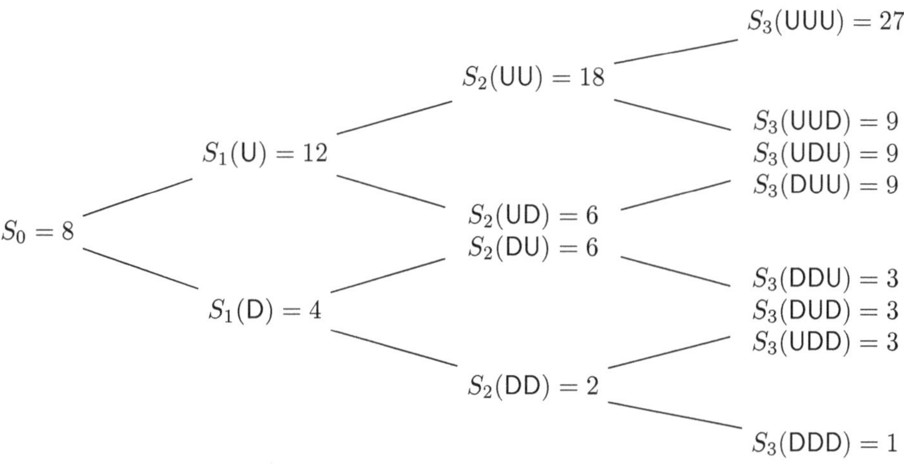

FIGURE 7.1: A three-period recombining binomial tree.

in time. Note that the positions $\Delta_0 = 0$, $\Delta_1 = \#\mathsf{U}(\omega_1)$, $\Delta_2 = \#\mathsf{U}(\omega_1\omega_2)$ are $\mathcal{F}_0$-, $\mathcal{F}_1$-, and $\mathcal{F}_2$-measurable, respectively. The self-financing investment strategy is therefore adapted to the natural filtration. Construct the strategy for each $t = 0, 1, 2$ as follows.

$t = 0$: Since $\Delta_0 = 0$, at time 0 all the capital is invested in the bank account with $\beta_0 = \frac{\Pi_0}{B_0} = 10$.

$t = 1$: There are no shares of stock in our investment portfolio so far. Therefore, its value is independent of ω_1 and by the above wealth equation:

$$\Pi_1 = \beta_0 B_1 = \Pi_0 \cdot (1+r) = 12.5.$$

If $\omega_1 = \mathsf{D}$, then $\Delta_1(\mathsf{D}) = 0$ and $\beta_1(\mathsf{D}) = \beta_0 = 10$. If $\omega_1 = \mathsf{U}$, then $\Delta_1(\mathsf{U}) = 1$ and

$$\beta_1(\mathsf{U}) = \frac{\Pi_1(\mathsf{U}) - S_1(\mathsf{U})\Delta_1(\mathsf{U})}{B_1} = \frac{12.5 - 12 \cdot 1}{1.25} = 0.4.$$

$t = 2$: The self-financed portfolio has value $\Pi_2(\omega_1\omega_2) = \beta_1(\omega_1)B_2 + \Delta_1(\omega_1)S_2(\omega_1\omega_2)$ for each of four scenarios: $\omega_1\omega_2 \in \{\mathsf{DD}, \mathsf{DU}, \mathsf{UD}, \mathsf{UU}\}$. Let us calculate the respective liquidation values of Δ_2 and β_2 in each case.

- If $\omega_1\omega_2 = \mathsf{DD}$, then

$$\Pi_2(\mathsf{DD}) = \beta_1(\mathsf{D})B_2 + \Delta_1(\mathsf{D})S_2(\mathsf{DD}) = 10 \cdot 1.25^2 + 0 \cdot 2 = 15.625,$$
$$\Delta_2(\mathsf{DD}) = \#\mathsf{U}(\mathsf{DD}) = 0,$$
$$\beta_2(\mathsf{DD}) = \frac{\Pi_2(\mathsf{DD}) - S_2(\mathsf{DD})\Delta_2(\mathsf{DD})}{B_2} = \frac{15.625 - 2 \cdot 0}{1.25^2} = 10.$$

- If $\omega_1\omega_2 = \mathsf{DU}$, then

$$\Pi_2(\mathsf{DU}) = \beta_1(\mathsf{D})B_2 + \Delta_1(\mathsf{D})S_2(\mathsf{DU}) = 10 \cdot 1.25^2 + 0 \cdot 6 = 15.625,$$
$$\Delta_2(\mathsf{DU}) = \#\mathsf{U}(\mathsf{DU}) = 1,$$
$$\beta_2(\mathsf{DU}) = \frac{\Pi_2(\mathsf{DU}) - S_2(\mathsf{DU})\Delta_2(\mathsf{DU})}{B_2} = \frac{15.625 - 6 \cdot 1}{1.25^2} = 6.16.$$

- If $\omega_1\omega_2 = \mathsf{UD}$, then

$$\Pi_2(\mathsf{UD}) = \beta_1(\mathsf{U})B_2 + \Delta_1(\mathsf{U})S_2(\mathsf{UD}) = 0.4 \cdot 1.25^2 + 1 \cdot 6 = 6.625,$$
$$\Delta_2(\mathsf{UD}) = \#\mathsf{U}(\mathsf{UD}) = 1,$$
$$\beta_2(\mathsf{UD}) = \frac{\Pi_2(\mathsf{UD}) - S_2(\mathsf{UD})\Delta_2(\mathsf{UD})}{B_2} = \frac{6.625 - 6 \cdot 1}{1.25^2} = 0.4.$$

- If $\omega_1\omega_2 = \mathsf{UU}$, then

$$\Pi_2(\mathsf{UU}) = \beta_1(\mathsf{U})B_2 + \Delta_1(\mathsf{U})S_2(\mathsf{UU}) = 0.4 \cdot 1.25^2 + 1 \cdot 18 = 18.625,$$
$$\Delta_2(\mathsf{UU}) = \#\mathsf{U}(\mathsf{UU}) = 2,$$
$$\beta_2(\mathsf{UU}) = \frac{\Pi_2(\mathsf{UU}) - S_2(\mathsf{UU})\Delta_2(\mathsf{UU})}{B_2} = \frac{18.625 - 18 \cdot 2}{1.25^2} = -11.12.$$

The terminal value Π_3 of the strategy is calculated by

$$\Pi_3(\omega_1\omega_2\omega_3) = \beta_2(\omega_1\omega_2)B_3 + \Delta_2(\omega_1\omega_2)S_3(\omega_1\omega_2\omega_3), \quad \omega_1, \omega_2, \omega_3 \in \{\mathsf{D}, \mathsf{U}\}.$$

The full details of this three-period strategy, for every scenario, are summarized in Figure 7.2. □

7.2.1 Equivalent Martingale Measures for the Binomial Model

As mentioned at the start of this chapter, we fix the filtration to be the one generated by the market moves (i.e., the scenarios in the binomial tree); hence, the filtration also corresponds to the natural filtration generated by the stock price process. Since our main interest is in pricing derivative assets, the martingale property will be relevant under risk-neutral probability measures. We already referred to these measures as so-called equivalent martingale measures (EMMs). In the next chapter we shall define an EMM for the general case of N basic securities within the multi-period setting. At this point we need only to deal with the binomial multi-period model with two basic securities: the risk-free bank account B (or money market account, or zero-coupon bond) and the risky stock S. Hence, given a numéraire asset g, a corresponding EMM $\widetilde{\mathbb{P}} \equiv \widetilde{\mathbb{P}}^{(g)}$ is defined such that the discounted value processes of the risk-free asset and the stock are both $\widetilde{\mathbb{P}}^{(g)}$-martingales (see Theorem 7.1).

Recall Chapter 5, where we proved that for a single-period (Arrow–Debreu) model the value process of any static portfolio in the base securities, discounted by a numéraire asset $g \in \{B, S\}$, is a $\widetilde{\mathbb{P}}^{(g)}$-martingale, i.e., the discounted portfolio value process is a martingale under the measure $\widetilde{\mathbb{P}}^{(g)}$. We are now ready to extend such a result for the binomial tree model to the multi-period setting in the case of dynamic self-financing portfolio strategies.

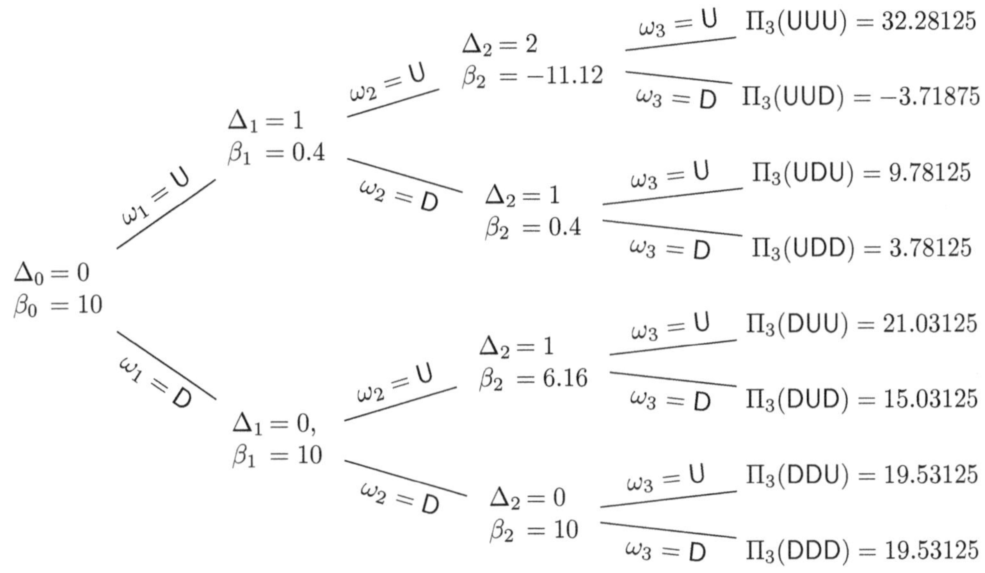

$\Delta_0 = 0$, $\beta_0 = 10$

- $\omega_1 = \mathsf{U}$: $\Delta_1 = 1$, $\beta_1 = 0.4$
  - $\omega_2 = \mathsf{U}$: $\Delta_2 = 2$, $\beta_2 = -11.12$
    - $\omega_3 = \mathsf{U}$: $\Pi_3(\mathsf{UUU}) = 32.28125$
    - $\omega_3 = \mathsf{D}$: $\Pi_3(\mathsf{UUD}) = -3.71875$
  - $\omega_2 = \mathsf{D}$: $\Delta_2 = 1$, $\beta_2 = 0.4$
    - $\omega_3 = \mathsf{U}$: $\Pi_3(\mathsf{UDU}) = 9.78125$
    - $\omega_3 = \mathsf{D}$: $\Pi_3(\mathsf{UDD}) = 3.78125$
- $\omega_1 = \mathsf{D}$: $\Delta_1 = 0$, $\beta_1 = 10$
  - $\omega_2 = \mathsf{U}$: $\Delta_2 = 1$, $\beta_2 = 6.16$
    - $\omega_3 = \mathsf{U}$: $\Pi_3(\mathsf{DUU}) = 21.03125$
    - $\omega_3 = \mathsf{D}$: $\Pi_3(\mathsf{DUD}) = 15.03125$
  - $\omega_2 = \mathsf{D}$: $\Delta_2 = 0$, $\beta_2 = 10$
    - $\omega_3 = \mathsf{U}$: $\Pi_3(\mathsf{DDU}) = 19.53125$
    - $\omega_3 = \mathsf{D}$: $\Pi_3(\mathsf{DDD}) = 19.53125$

FIGURE 7.2: A self-financing strategy and its value process constructed in Example 7.1.

Theorem 7.3. *Suppose that an equivalent martingale measure $\widetilde{\mathbb{P}} \equiv \widetilde{\mathbb{P}}^{(g)}$ for a numéraire asset g (e.g., the bank account B or stock S) exists. Let $\{\Delta_t\}_{0\leqslant t\leqslant T-1}$ be a process adapted to the natural filtration of the binomial tree model. Let $\{\Pi_t\}_{0\leqslant t\leqslant T}$ be generated recursively by the wealth equation (7.6). Then the value process discounted by g, $\left\{\overline{\Pi}_t := \frac{\Pi_t}{g_t}\right\}_{0\leqslant t\leqslant T}$, is a $\widetilde{\mathbb{P}}$-martingale.*

Proof. Note that, for every $0 \leqslant t \leqslant T$, the value $\overline{\Pi}_t$ is clearly $\mathcal{F}_t$-measurable and integrable, i.e., $\widetilde{\mathrm{E}}[|\overline{\Pi}_t|] < \infty$. It then suffices to prove the single-step martingale expectation property under the $\widetilde{\mathbb{P}}$-measure: $\widetilde{\mathrm{E}}[\overline{\Pi}_{t+1} \mid \mathcal{F}_t] \equiv \widetilde{\mathrm{E}}_t[\overline{\Pi}_{t+1}] = \overline{\Pi}_t$, for every $0 \leqslant t \leqslant T-1$. We now use the fact that the discounted stock price and risk-free asset price processes, $\overline{S}_t := \frac{S_t}{g_t}$ and $\overline{B}_t := \frac{B_t}{g_t}$, are $\widetilde{\mathbb{P}}$-martingales and that $\Delta_t, \overline{S}_t, \overline{B}_t, \overline{\Pi}_t$ are $\mathcal{F}_t$-measurable. From the wealth equation (7.6) with $B_{t+1}/B_t = 1 + r$ we have:

$$
\begin{aligned}
\widetilde{\mathrm{E}}_t[\overline{\Pi}_{t+1}] \equiv \widetilde{\mathrm{E}}_t\left[\frac{\Pi_{t+1}}{g_{t+1}}\right] &= \widetilde{\mathrm{E}}_t\left[\frac{\Delta_t S_{t+1} + (\Pi_t - \Delta_t S_t)(B_{t+1}/B_t)}{g_{t+1}}\right] \\
&= \widetilde{\mathrm{E}}_t\left[\Delta_t \frac{S_{t+1}}{g_{t+1}} + \left(\frac{\Pi_t}{g_t} - \Delta_t \frac{S_t}{g_t}\right) \frac{B_{t+1}/g_{t+1}}{B_t/g_t}\right] \\
&= \widetilde{\mathrm{E}}_t\left[\Delta_t \overline{S}_{t+1} + (\overline{\Pi}_t - \Delta_t \overline{S}_t) \frac{\overline{B}_{t+1}}{\overline{B}_t}\right] \\
&= \Delta_t \underbrace{\widetilde{\mathrm{E}}_t\left[\overline{S}_{t+1}\right]}_{=\overline{S}_t} + \left(\frac{\overline{\Pi}_t - \Delta_t \overline{S}_t}{\overline{B}_t}\right) \underbrace{\widetilde{\mathrm{E}}_t\left[\overline{B}_{t+1}\right]}_{=\overline{B}_t} \\
&= \Delta_t \overline{S}_t + \overline{\Pi}_t - \Delta_t \overline{S}_t = \overline{\Pi}_t.
\end{aligned}
$$

Hence, the process $\{\overline{\Pi}_t\}$ is a $\widetilde{\mathbb{P}}$-martingale. □

Suppose that the interest rate r is fixed and that the risk-free bank account (or zero-coupon bond) is used as numéraire. Then we obtain the following well-known result for the risk-neutral expectation (and risk-neutral growth rate) of the self-financing portfolio value process.

Corollary 7.4. *Let* $\widetilde{\mathbb{P}} \equiv \widetilde{\mathbb{P}}^{(B)}$ *be the EMM for the numéraire* $g = B$, *i.e.,* $\widetilde{\mathrm{E}}_t[\frac{\Pi_{t+1}}{B_{t+1}}] = \frac{\Pi_t}{B_t}$ *and* $\widetilde{\mathrm{E}}_t[\Pi_{t+1}] = (1+r)\Pi_t$ *since* $B_{t+1} = (1+r)B_t$ *for all* $t \geqslant 0$. *Then,*

$$\widetilde{\mathrm{E}}_s[\Pi_t] = (1+r)^{(t-s)}\Pi_s \text{ for all } 0 \leqslant s \leqslant t \leqslant T,$$

and hence $\widetilde{\mathrm{E}}_0[\Pi_t] = (1+r)^t\Pi_0$ *for all* $t \geqslant 0$.

Example 7.2. Verify that the self-financed portfolio value process in Example 7.1, discounted by the risk-free asset price process, is a martingale under the measure $\widetilde{\mathbb{P}} \equiv \widetilde{\mathbb{P}}^{(B)}$.

Solution. The risk-neutral probabilities for the up and down moves are

$$\tilde{p} = \widetilde{\mathbb{P}}(\omega_i = \mathsf{U}) = \frac{1.25 - 0.5}{1.5 - 0.5} = 0.75 \text{ and } \widetilde{\mathbb{P}}(\omega_i = \mathsf{D}) = 1 - \tilde{p} = 0.25.$$

It is sufficient to verify that $\Pi_t = \frac{1}{1+r}\widetilde{\mathrm{E}}_t[\Pi_{t+1}]$ for $t = 0, 1, 2$.

$$t=0: \quad \Pi_0 = 10 \stackrel{?}{=} \frac{1}{1+r}\widetilde{\mathrm{E}}_0[\Pi_1] = \frac{\Pi_1(\mathsf{U})\cdot\tilde{p} + \Pi_1(\mathsf{D})\cdot(1-\tilde{p})}{1+r} = \frac{12.5\cdot 0.75 + 12.5\cdot 0.25}{1.25} = 10 \quad \checkmark$$

$$t=1: \quad \Pi_1(\mathsf{D}) = 12.5 \stackrel{?}{=} \frac{1}{1+r}\widetilde{\mathrm{E}}_1[\Pi_2](\mathsf{D}) = \frac{\Pi_2(\mathsf{DU})\cdot\tilde{p} + \Pi_2(\mathsf{DD})\cdot(1-\tilde{p})}{1+r} = \frac{15.625\cdot 0.75 + 15.625\cdot 0.25}{1.25} = 12.5 \quad \checkmark$$

$$\Pi_1(\mathsf{U}) = 12.5 \stackrel{?}{=} \frac{1}{1+r}\widetilde{\mathrm{E}}_1[\Pi_2](\mathsf{U}) = \frac{\Pi_2(\mathsf{UU})\cdot\tilde{p} + \Pi_2(\mathsf{UD})\cdot(1-\tilde{p})}{1+r} = \frac{18.625\cdot 0.75 + 6.625\cdot 0.25}{1.25} = 12.5 \quad \checkmark$$

$$t=2: \quad \Pi_2(\mathsf{DD}) = 15.625 \stackrel{?}{=} \frac{1}{1+r}\widetilde{\mathrm{E}}_2[\Pi_3](\mathsf{DD}) = \frac{\Pi_3(\mathsf{DDU})\cdot\tilde{p} + \Pi_3(\mathsf{DDD})\cdot(1-\tilde{p})}{1+r} = \frac{19.53125\cdot 0.75 + 19.53125\cdot 0.25}{1.25} = 15.625 \quad \checkmark$$

$$\Pi_2(\mathsf{DU}) = 15.625 \stackrel{?}{=} \frac{1}{1+r}\widetilde{\mathrm{E}}_2[\Pi_3](\mathsf{DU}) = \frac{\Pi_3(\mathsf{DUU})\cdot\tilde{p} + \Pi_3(\mathsf{DUD})\cdot(1-\tilde{p})}{1+r} = \frac{21.03125\cdot 0.75 + 15.03125\cdot 0.25}{1.25} = 15.625 \quad \checkmark$$

$$\Pi_2(\mathsf{UD}) = 6.625 \stackrel{?}{=} \frac{1}{1+r}\widetilde{\mathrm{E}}_2[\Pi_3](\mathsf{UD}) = \frac{\Pi_3(\mathsf{UDU})\cdot\tilde{p} + \Pi_3(\mathsf{UDD})\cdot(1-\tilde{p})}{1+r} = \frac{9.78125\cdot 0.75 + 3.78125\cdot 0.25}{1.25} = 6.625 \quad \checkmark$$

$$\Pi_2(\mathsf{UU}) = 18.625 \stackrel{?}{=} \frac{1}{1+r}\widetilde{\mathrm{E}}_2[\Pi_3](\mathsf{UU}) = \frac{\Pi_3(\mathsf{UUU})\cdot\tilde{p} + \Pi_3(\mathsf{UUD})\cdot(1-\tilde{p})}{1+r} = \frac{32.28125\cdot 0.75 - 3.78125\cdot 0.25}{1.25} = 18.625 \quad \checkmark$$

□

Another corollary of Theorem 7.3 is the first fundamental theorem of asset pricing (FTAP). It states that there are no arbitrage opportunities iff there exists an EMM $\widetilde{\mathbb{P}}^{(g)}$ for a numéraire g. So far, such an assertion has been proved for static portfolios in the one-period or multi-period setting. However, one can create an arbitrage opportunity by manipulating with a portfolio in base assets without injecting or withdrawing funds. In other words, a specially constructed self-financing strategy can be an arbitrage. Let us prove that there are no self-financing arbitrage strategies in a binomial tree model iff there exists an EMM. First, we need to have a formal definition of an arbitrage in a multi-period trading model as follows.

Definition 7.1. An *arbitrage strategy* in a multi-period model is a self-financing strategy with nonnegative value process and zero initial value Π_0 such that $\mathbb{P}(\Pi_t > 0) > 0$ holds at some time $t \in \{1, 2, \ldots, T\}$.

Suppose that a binomial tree model admits no arbitrage. In particular, there are no static (i.e., with constant positions in base assets as in any single-period setting) arbitrage portfolios. Then, as was proved before, there exists an EMM. To complete the proof of the FTAP for the binomial tree model we only need to prove the following converse statement.

Lemma 7.5. *Suppose that an EMM $\widetilde{\mathbb{P}} = \widetilde{\mathbb{P}}^{(g)}$ for a numéraire g exists. Then there are no self-financing arbitrage strategies.*

Proof. Suppose that a self-financed arbitrage strategy with initial value zero and delta positions $\{\Delta_t\}_{t\geqslant 0}$ exists. The value process for such a strategy must satisfy the wealth equation in (7.6) such that (by the arbitrage assumption) $\Pi_t(\omega) \geqslant 0$ for all $\omega \in \Omega$ and $t \geqslant 0$. Moreover, arbitrage implies that there exists a time $m \in \{1, 2, \ldots, T\}$ and market scenario $\omega^* \in \Omega$ such that $\Pi_m(\omega^*) > 0$. Therefore,

$$\widetilde{\mathrm{E}}_0[\overline{\Pi}_m] = \sum_{\omega\in\Omega} \overline{\Pi}_m(\omega)\,\widetilde{\mathbb{P}}(\omega) \geqslant \frac{\Pi_m(\omega^*)}{g(\omega^*)}\,\widetilde{\mathbb{P}}(\omega^*) > 0.$$

This contradicts Theorem 7.3, since $\widetilde{\mathrm{E}}[\overline{\Pi}_m] = \overline{\Pi}_0 = \Pi_0/g_0 = 0$. □

7.3 Dynamic Replication in the Binomial Tree Model

7.3.1 Dynamic Replication of Payoffs

A key idea in modern finance is the replication of a financial claim with the use of portfolios in other (base) assets. The two most important applications of replication is the no-arbitrage pricing of derivative securities and hedging their liabilities. Consider a derivative security with maturity at time T and payoff function $X : \Omega_T \to \mathbb{R}$. Recall that $\mathcal{L}(\Omega_T)$ denotes the collection of all payoff functions on the sample space Ω_T. The structure of a payoff can be quite general. It can depend on the whole path ω, or on a quantity calculated along the stock price path such as an average of the stock prices over some time window, or only on the terminal stock price S_T. For example, the payoff of a non-path-dependent derivative, such as a standard European call or put option, with exercise only at maturity T, is a function of only the terminal stock price S_T and is given by $X(\omega) = \Lambda(S_T(\omega))$ with function $\Lambda\colon \mathbb{R}_+ \to \mathbb{R}$, where $\mathbb{R}_+ := [0, \infty)$. Hence, in what follows we allow for generally path-dependent payoffs where the random variable X is $\mathcal{F}_T$-measurable, i.e., the payoff is determined by possibly the entire sequence of market moves $\omega \equiv \omega_1 \ldots \omega_T \in \Omega_T$.

Let $\{V_t\}_{0 \leqslant t \leqslant T}$ denote the *price process of the derivative security*, i.e., V_t is the price of the derivative security at time t. At maturity time, the derivative price is given by the payoff function $V_T = X$, where $X\colon \Omega_T \to \mathbb{R}$. The writer of a derivative needs to calculate the no-arbitrage current price, V_0, of the contract. As we saw in great detail in Chapters 2 and 5, in the single-period model, the price of a derivative security must equal the initial value of a portfolio replicating the derivative payoff at maturity. In the multi-period setting, such a replication can only be done dynamically with the use of self-financing strategies. It is possible to construct such a strategy that replicates the whole derivative price process from time 0 until maturity T. Knowledge of the derivative price process is required for hedging the writer's liabilities.

Definition 7.2. A self-financing strategy $\{\varphi_t\}_{0 \leqslant t \leqslant T-1}$ is said to *replicate* the payoff X at maturity T if its value at maturity T, given by $\Pi_T = \Pi_T[\varphi_{T-1}]$, equals the payoff value for all possible scenarios, i.e.,

$$\Pi_T(\omega) = X(\omega) \text{ for all } \omega \in \Omega.$$

According to the law of one price, the initial value Π_0 of a strategy that replicates the payoff X of a derivative maturing at time T must equal the initial value V_0 of the derivative, or else an arbitrage opportunity exists. Moreover, we shall prove that in the absence of arbitrage opportunities, the price of the derivative, V_t is equal to the value $\Pi_t \equiv \Pi_t[\varphi_t]$ of the replicating strategy at every time t.

To construct a self-financing strategy that replicates a derivative with payoff X, we proceed as follows. First, we construct a no-arbitrage derivative price process $\{V_t\}_{0 \leqslant t \leqslant T}$ recursively backward in time starting from maturity time T. Second, we obtain a sequence of (delta) positions in the stock, $\{\Delta_t\}_{0 \leqslant t \leqslant T-1}$, corresponding to the price process. Finally, we show that the process $\{\Delta_t\}$ is nothing more than a replicating strategy for the derivative price process so that the value process $\{\Pi_t\}_{0 \leqslant t \leqslant T}$ generated by the strategy coincides with the derivative price process at *all intermediate dates and at maturity*, i.e., $\Pi_t(\omega) = V_t(\omega)$ for all $0 \leqslant t \leqslant T$ and all scenarios $\omega \in \Omega$. Note that $\Pi_T(\omega) = V_T(\omega) = X(\omega)$ holds at maturity by the definition of a replicating strategy. Before we prove a general result, let us study how this procedure works for the simple one- and two-period cases.

Example 7.3 (The cases with $T = 1$ and $T = 2$). Assume that the binomial tree model admits no-arbitrage, i.e. $d < 1 + r < u$ holds. Let $\widetilde{\mathbb{P}}$ be the usual risk-neutral measure (for the numéraire asset $g = B$). Construct the replicating strategy and find the no-arbitrage prices for an arbitrary derivative with maturity time (a) $T = 1$ and (b) $T = 2$.

Solution.

Case with $T = 1$. In the one-period case, the replicating portfolio is static and formed at time 0, as we already saw in Chapter 2. The initial value Π_0 is invested in Δ_0 shares of stock, leaving us with the risk-free asset position with value $\Pi_0 - \Delta_0 S_0$. At time 1, the portfolio value is $\Pi_1(\omega) = \Delta_0 S_1(\omega) + (\Pi_0 - \Delta_0 S_0)(1 + r)$. Let us choose $\Pi_0 = V_0$ and Δ_0 such that $\Pi_1(\omega) = V_1(\omega)$ for $\omega \in \Omega_1 = \{\mathsf{D}, \mathsf{U}\}$. Since there are two possibilities for ω_1, we obtain the linear system of two equations in two unknowns Δ_0 and V_0:

$$\begin{cases} \Pi_1(\mathsf{U}) = V_1(\mathsf{U}) \\ \Pi_1(\mathsf{D}) = V_1(\mathsf{D}) \end{cases} \iff \begin{cases} \Delta_0 S_1(\mathsf{U}) + (V_0 - \Delta_0 S_0)(1 + r) = V_1(\mathsf{U}) \\ \Delta_0 S_1(\mathsf{D}) + (V_0 - \Delta_0 S_0)(1 + r) = V_1(\mathsf{D}) \end{cases}$$

$$\iff \begin{cases} (S_1(\mathsf{U}) - (1 + r)S_0)\Delta_0 + (1 + r)V_0 = V_1(\mathsf{U}) \\ (S_1(\mathsf{D}) - (1 + r)S_0)\Delta_0 + (1 + r)V_0 = V_1(\mathsf{D}) \end{cases} \tag{7.7}$$

This system has a unique solution with Δ_0 obtained by simply subtracting the first and second equations in (7.7):

$$\Delta_0 = \frac{V_1(\mathsf{U}) - V_1(\mathsf{D})}{S_1(\mathsf{U}) - S_1(\mathsf{D})}. \tag{7.8}$$

Substituting this expression for Δ_0 into either of the equations in (7.7) gives the current price of the derivative in terms of known (payoff) values $V_1(\mathsf{U})$ and $V_1(\mathsf{D})$:

$$\begin{aligned} V_0 &= (1+r)^{-1}\left[V_1(\mathsf{U}) - (S_1(\mathsf{U}) - (1+r)S_0)\Delta_0\right] \\ &= (1+r)^{-1}\left[\frac{V_1(\mathsf{U})(S_1(\mathsf{U}) - S_1(\mathsf{D})) - (S_1(\mathsf{U}) - (1+r)S_0)(V_1(\mathsf{U}) - V_1(\mathsf{D}))}{S_1(\mathsf{U}) - S_1(\mathsf{D})}\right] \\ &= (1+r)^{-1}\left[\frac{(1+r)S_0 - S_1(\mathsf{D})}{S_1(\mathsf{U}) - S_1(\mathsf{D})}V_1(\mathsf{U}) + \frac{S_1(\mathsf{U}) - (1+r)S_0}{S_1(\mathsf{U}) - S_1(\mathsf{D})}V_1(\mathsf{D})\right] \\ &= (1+r)^{-1}\left[\frac{(1+r)S_0 - S_0 d}{S_0 u - S_0 d}V_1(\mathsf{U}) + \frac{S_0 u - (1+r)S_0}{S_0 u - S_0 d}V_1(\mathsf{D})\right] \\ &= (1+r)^{-1}\left[\tilde{p}V_1(\mathsf{U}) + (1-\tilde{p})V_1(\mathsf{D})\right] \\ &= (1+r)^{-1}\widetilde{\mathrm{E}}_0[V_1]. \end{aligned} \tag{7.9}$$

Note that the last expression corresponds to the discounted expected value of the derivative payoff with risk-neutral probabilities with asset B as numéraire:

$$\tilde{p} = \frac{(1+r)S_0 - S_0 d}{S_0 u - S_0 d} = \frac{(1+r) - d}{u - d}, \quad 1 - \tilde{p} = \frac{S_0\, u - (1+r)S_0}{S_0\, u - S_0\, d} = \frac{u - (1+r)}{u - d}.$$

Case with $T = 2$. Let us construct a replicating strategy $\{\Delta_t\}_{t=0,1}$ with the time-2 value satisfying $\Pi_2(\omega) = V_2(\omega)$ for all $\omega \in \Omega_2 = \{\mathsf{DD}, \mathsf{DU}, \mathsf{UD}, \mathsf{UU}\}$. At time 0, the strategy is already specified above, i.e., take a position Δ_0 in shares of stock and a position in the risk-free bank account with value $\Pi_0 - \Delta_0 S_0$. This gives Δ_0 and V_0, respectively, as in (7.8) and (7.9). At time 1, we liquidate the old portfolio and form a new portfolio having value $\Pi_1(\omega_1) = V_1(\omega_1)$ and with new position $\Delta_1(\omega_1)$ in the stock for any given market scenario with first move ω_1. At time 2, the value of this new portfolio, as given by the wealth equation for $t = 2$, becomes

$$\Pi_2(\omega_1\omega_2) = \Delta_1(\omega_1)S_2(\omega_1\omega_2) + (V_1(\omega_1) - \Delta_1(\omega_1)S_1(\omega_1))(1+r).$$

By replication, $\Pi_2(\omega_1\omega_2)$ must equal $V_2(\omega_1\omega_2)$ for all four possible market scenarios $\omega_1\omega_2$. Hence, we obtain a system of four linear equations which are grouped into two pairs of equations. The first pair (for $\omega_1 = \mathsf{U}$) gives two equations in two unknowns $\Delta_1(\mathsf{U}), V_1(\mathsf{U})$:

$$\begin{cases} V_2(\mathsf{UU}) = \Delta_1(\mathsf{U})S_2(\mathsf{UU}) + (V_1(\mathsf{U}) - \Delta_1(\mathsf{U})S_1(\mathsf{U}))(1+r) \\ V_2(\mathsf{UD}) = \Delta_1(\mathsf{U})S_2(\mathsf{UD}) + (V_1(\mathsf{U}) - \Delta_1(\mathsf{U})S_1(\mathsf{U}))(1+r) \end{cases} \tag{7.10}$$

and the second pair (for $\omega_1 = \mathsf{D}$) in the two unknowns $\Delta_1(\mathsf{D}), V_1(\mathsf{D})$:

$$\begin{cases} V_2(\mathsf{DU}) = \Delta_1(\mathsf{D})S_2(\mathsf{DU}) + (V_1(\mathsf{D}) - \Delta_1(\mathsf{D})S_1(\mathsf{D}))(1+r) \\ V_2(\mathsf{DD}) = \Delta_1(\mathsf{D})S_2(\mathsf{DD}) + (V_1(\mathsf{D}) - \Delta_1(\mathsf{D})S_1(\mathsf{D}))(1+r) \end{cases} \tag{7.11}$$

At this point it is important to observe that both of these pairs of equations are of the same form as (7.7) and hence are solved in the same manner. In particular, solving (7.10) gives

$$\Delta_1(\mathsf{U}) = \frac{V_2(\mathsf{UU}) - V_2(\mathsf{UD})}{S_2(\mathsf{UU}) - S_2(\mathsf{UD})} \tag{7.12}$$

and

$$\begin{aligned} V_1(\mathsf{U}) &= (1+r)^{-1}\left[\frac{(1+r)S_1(\mathsf{U}) - S_2(\mathsf{UD})}{S_2(\mathsf{UU}) - S_2(\mathsf{UD})}V_2(\mathsf{UU}) + \frac{S_2(\mathsf{UU}) - (1+r)S_1(\mathsf{U})}{S_2(\mathsf{UU}) - S_2(\mathsf{UD})}V_2(\mathsf{UD})\right] \\ V_1(\mathsf{U}) &= (1+r)^{-1}\left[\frac{(1+r)S_1(\mathsf{U}) - S_1(\mathsf{U})d}{S_1(\mathsf{U})u - S_1(\mathsf{U})d}V_2(\mathsf{UU}) + \frac{S_1(\mathsf{U})u - (1+r)S_1(\mathsf{U})}{S_1(\mathsf{U})u - S_1(\mathsf{U})d}V_2(\mathsf{UD})\right] \\ &= (1+r)^{-1}\left[\tilde{p}V_2(\mathsf{UU})) + (1-\tilde{p})V_2(\mathsf{UD})\right] \\ &= (1+r)^{-1}\widetilde{\mathrm{E}}_1[V_2](\mathsf{U}). \end{aligned} \tag{7.13}$$

Similarly, solving (7.11) gives

$$\Delta_1(\mathsf{D}) = \frac{V_2(\mathsf{DU}) - V_2(\mathsf{DD})}{S_2(\mathsf{DU}) - S_2(\mathsf{DD})} \tag{7.14}$$

and

$$\begin{aligned} V_1(\mathsf{D}) &= (1+r)^{-1}\left[\frac{(1+r)S_1(\mathsf{D}) - S_2(\mathsf{DD})}{S_2(\mathsf{DU}) - S_2(\mathsf{DD})}V_2(\mathsf{DU}) + \frac{S_2(\mathsf{DU}) - (1+r)S_1(\mathsf{D})}{S_2(\mathsf{DU}) - S_2(\mathsf{DD})}V_2(\mathsf{DD})\right] \\ &= (1+r)^{-1}\left[\tilde{p}V_2(\mathsf{DU})) + (1-\tilde{p})V_2(\mathsf{DD})\right] \\ &= (1+r)^{-1}\widetilde{\mathrm{E}}_1[V_2](\mathsf{D}). \end{aligned} \tag{7.15}$$

Combining both expressions for the derivative prices at $t = 1$ gives

$$\begin{aligned} &V_1(\omega_1) = (1+r)^{-1}\left[\tilde{p}V_2(\omega_1\mathsf{U}) + (1-\tilde{p})V_2(\omega_1\mathsf{D})\right] \\ &\Longrightarrow V_1 = (1+r)^{-1}\widetilde{\mathrm{E}}_1[V_2]. \end{aligned}$$

We hence see from this last equation, and Equation (7.9), that the derivative prices at times $t = 0$ and $t = 1$ are given by the discounted (risk-neutral) expected value of the derivative price at time $t + 1$. Hence, by the tower property, the initial price of the derivative is expressible as an expectation of the payoff at time $t = T = 2$, discounted back by two periods:

$$V_0 = (1+r)^{-1}\widetilde{\mathrm{E}}_0\left[V_1\right] = (1+r)^{-2}\,\widetilde{\mathrm{E}}_0\left[\widetilde{\mathrm{E}}_1[V_2]\right] = (1+r)^{-2}\,\widetilde{\mathrm{E}}_0[V_2].$$

□

Now, we consider the general case for any finite number of $T \geqslant 1$ periods. Assume the absence of arbitrage opportunities. Let us define recursively backward in time the price process $\{V_t\}_{0\leqslant t\leqslant T}$ of a derivative with maturity time T and given payoff function $X \in \mathcal{L}(\Omega_T)$. Fix arbitrarily time $t \in \{0, 1, \dots, T-1\}$ and market moves $\omega_1, \omega_2, \dots, \omega_t \in \{\mathsf{D}, \mathsf{U}\}$. The stock price S_{t+1} conditional on $\omega_1, \omega_2, \dots, \omega_t$ follows a binomial single-period sub-tree: $S_{t+1} = S_t u$ or $S_{t+1} = S_t d$ if $\omega_{t+1} = \mathsf{U}$ or $\omega_{t+1} = \mathsf{D}$, respectively. Applying the single-step discounted expectation formula (e.g., as in Equation (7.9)) to the sub-tree originated at $S_t(\omega_1\omega_2\dots\omega_t)$ gives

$$V_t(\omega_1\omega_2\dots\omega_t) = \frac{1}{1+r}\left(\tilde{p}V_{t+1}(\omega_1\omega_2\dots\omega_t\mathsf{U}) + (1-\tilde{p})V_{t+1}(\omega_1\omega_2\dots\omega_t\mathsf{D})\right) \tag{7.16}$$

$$\equiv \frac{1}{1+r}\widetilde{\mathrm{E}}_t[V_{t+1}](\omega_1\omega_2\dots\omega_t), \quad t = 0, 1, \dots, T-1, \tag{7.17}$$

where $\tilde{p}$ and $1 - \tilde{p}$ are risk-neutral probabilities given above. At maturity, we set

$$V_T(\omega_1\omega_2\dots\omega_T) = X(\omega_1\omega_2\dots\omega_T). \tag{7.18}$$

Define the strategy $\{\Delta_t\}_{0\leqslant t\leqslant T-1}$ as follows:

$$\Delta_t(\omega_1\omega_2\dots\omega_t) = \frac{V_{t+1}(\omega_1\omega_2\dots\omega_t\mathsf{U}) - V_{t+1}(\omega_1\omega_2\dots\omega_t\mathsf{D})}{S_{t+1}(\omega_1\omega_2\dots\omega_t\mathsf{U}) - S_{t+1}(\omega_1\omega_2\dots\omega_t\mathsf{D})}, \quad t = 0, 1, \dots, T-1. \tag{7.19}$$

Now, let us prove that the derivative price process $\{V_t\}_{0\leqslant t\leqslant T}$ and the value process for the self-financing strategy $\{\Delta_t\}_{0\leqslant t\leqslant T-1}$ have the same value at all times $t = 0, 1, \dots, T$, i.e., the portfolio value process with (delta hedging) strategy $\{\Delta_t\}_{0\leqslant t\leqslant T-1}$ replicates the derivative price process.

Theorem 7.6. *Consider a derivative security with $\mathcal{F}_T$-measurable payoff X, at maturity T. Define the derivative price process $\{V_t\}_{0\leqslant t\leqslant T}$ by (7.16)–(7.18) and the self-financing portfolio strategy $\{\Delta_t\}_{0\leqslant t\leqslant T-1}$ by (7.19). Set $\Pi_0 = V_0$ and construct recursively forward in time the value process $\{\Pi_t\}_{0\leqslant t\leqslant T}$ via the wealth equation (7.6). Then, the strategy replicates the derivative price process at every time, i.e.*

$$\Pi_t(\omega_1\omega_2\ldots\omega_t) = V_t(\omega_1\omega_2\ldots\omega_t) \tag{7.20}$$

holds for all $t = 0, 1, \ldots, T$ and all market moves $\omega_1, \omega_2, \ldots, \omega_t \in \{\mathsf{D}, \mathsf{U}\}$.

Proof. We prove the assertion by induction. For $t = 0$, the equality $\Pi_0 = V_0$ follows trivially by the definition of the initial price. Now assume that (7.20) holds for some time t and show that it holds for time $t+1$. Fix an arbitrary sequence of moves $\omega_1\omega_2\ldots\omega_t$. By the induction assumption, $\Pi_t(\omega_1\omega_2\ldots\omega_t) = V_t(\omega_1\omega_2\ldots\omega_t)$, and since $\omega_{t+1} = \mathsf{U}$ or D, we need to prove that

$$\Pi_{t+1}(\omega_1\ldots\omega_t\mathsf{D}) = V_{t+1}(\omega_1\ldots\omega_t\mathsf{D}) \text{ and } \Pi_{t+1}(\omega_1\ldots\omega_t\mathsf{U}) = V_{t+1}(\omega_1\ldots\omega_t\mathsf{U}).$$

Let us only consider the case with $\omega_{t+1} = \mathsf{D}$ and $S_{t+1} = S_t d$, since the case with $\omega_{t+1} = \mathsf{U}$ is treated similarly. The wealth equation (7.6) then gives

$$\begin{aligned}\Pi_{t+1}(\omega_1\ldots\omega_t\mathsf{D}) &= \Delta_t(\omega_1\ldots\omega_t)S_t(\omega_1\ldots\omega_t)d\\ &\quad + (\Pi_t(\omega_1\ldots\omega_t) - \Delta_t(\omega_1\ldots\omega_t)S_t(\omega_1\ldots\omega_t))(1+r)\\ &= \Delta_t(\omega_1\ldots\omega_t)(d-(1+r))S_t(\omega_1\ldots\omega_t) + (1+r)\Pi_t(\omega_1\ldots\omega_t)\end{aligned} \tag{7.21}$$

where Δ_t is given by (7.19):

$$\Delta_t(\omega_1\ldots\omega_t) = \frac{V_{t+1}(\omega_1\ldots\omega_t\mathsf{U}) - V_{t+1}(\omega_1\ldots\omega_t\mathsf{D})}{(u-d)S_t(\omega_1\ldots\omega_t)}.$$

Substituting this into (7.21), cancelling out S_t, and using $\tilde{p} = (1+r-d)/(u-d)$ and $\Pi_t = V_t$ gives

$$\Pi_{t+1}(\omega_1\ldots\omega_t\mathsf{D}) = -\tilde{p}\,(V_{t+1}(\omega_1\ldots\omega_t\mathsf{U}) - V_{t+1}(\omega_1\ldots\omega_t\mathsf{D})) + (1+r)V_t(\omega_1\ldots\omega_t)$$

(and by using Equation (7.16))

$$\begin{aligned}&= \tilde{p}V_{t+1}(\omega_1\ldots\omega_t\mathsf{D}) - \tilde{p}V_{t+1}(\omega_1\ldots\omega_t\mathsf{U})\\ &\quad + \tilde{p}V_{t+1}(\omega_1\ldots\omega_t\mathsf{U}) + (1-\tilde{p})V_{t+1}(\omega_1\ldots\omega_t\mathsf{D})\\ &= V_{t+1}(\omega_1\ldots\omega_t\mathsf{D}).\end{aligned}$$

□

Remarks.

1. Applying the tower property of conditional expectations to (7.17) gives us the (multi-step ahead) risk-neutral pricing formula for any derivative with payoff $V_T = X \in \mathcal{L}(\Omega)$ at maturity T:

$$V_t(\omega_1\omega_2\ldots\omega_t) = \frac{1}{(1+r)^{T-t}}\widetilde{\mathrm{E}}_t[V_T](\omega_1\omega_2\ldots\omega_t), \quad 0 \leqslant t \leqslant T, \tag{7.22}$$

or, in short,

$$V_t = \frac{1}{(1+r)^{T-t}}\widetilde{\mathrm{E}}_t[V_T], \quad 0 \leqslant t \leqslant T, \tag{7.23}$$

In particular, the initial derivative price is

$$V_0 = \frac{1}{(1+r)^T}\widetilde{\mathrm{E}}_0[V_T] = \frac{1}{(1+r)^T}\sum_{\omega\in\Omega_T} V_T(\omega)\,\widetilde{\mathbb{P}}(\omega). \tag{7.24}$$

2. Theorem 7.6 is a particular case of a more general result presented below in Section 7.3.2, where here we take $X_t = 0$ for all $t = 0, 1, \ldots, T-1$ and $X_T = X \neq 0$, i.e., the derivative has a payoff (i.e., a cash flow) only at maturity T.

3. A corollary of Theorems 7.3 and 7.6 is the property that, in the measure $\widetilde{\mathbb{P}} = \widetilde{\mathbb{P}}^{(g)}$ where $g \in \{B, S\}$, the discounted derivative price process $\left\{\overline{V}_t := \frac{V_t}{g_t}\right\}_{0 \leqslant t \leqslant T}$ is a martingale:

$$\widetilde{\mathrm{E}}_s\left[\overline{V}_t\right] = \overline{V}_s \text{ for all } 0 \leqslant s \leqslant t \leqslant T.$$

Note that this is consistent with the fact that, within the measure $\widetilde{\mathbb{P}} = \widetilde{\mathbb{P}}^{(B)}$, the growth rate of the portfolio value is given by the interest rate, i.e., $B_t = (1+r)^{t-s} B_s$ gives $\widetilde{\mathrm{E}}_s\left[V_t\right] = (1+r)^{t-s} V_s$.

Theorem 7.6 can be generalized to the case with an arbitrary numéraire asset g. Equation (7.17) is rewritten as follows:

$$V_t(\omega_1\omega_2\ldots\omega_t) = g_t(\omega_1\omega_2\ldots\omega_t)\widetilde{\mathrm{E}}_t^{(g)}\left[\frac{V_{t+1}}{g_{t+1}}\right](\omega_1\omega_2\ldots\omega_t), \tag{7.25}$$

or, equivalently,

$$\frac{V_t}{g_t} = \widetilde{\mathrm{E}}_t^{(g)}\left[\frac{V_{t+1}}{g_{t+1}}\right], \tag{7.26}$$

where $\widetilde{\mathrm{E}}_t^{(g)}$ denotes the expectation conditional on $\mathcal{F}_t$ and w.r.t. the risk-neutral probability measure $\widetilde{\mathbb{P}}^{(g)}$. Applying the tower property to (7.26) gives

$$\frac{V_t}{g_t} = \widetilde{\mathrm{E}}_t^{(g)}\left[\frac{V_T}{g_T}\right] \implies V_t = g_t\widetilde{\mathrm{E}}_t^{(g)}\left[\frac{V_T}{g_T}\right] \text{ for all } t \in \{0, 1, \ldots, T\}. \tag{7.27}$$

That is, the time-t derivative price V_t discounted by the numéraire g_t is given by the mathematical expectation (in the risk-neutral measure $\widetilde{\mathbb{P}}^{(g)}$) of the discounted payoff V_T/g_T. Note how this more general result recovers Equation (7.22) when we choose the bank account as numéraire, i.e., when $g_t = B_t$ we obtain the usual discount factor $\frac{g_t}{g_T} = \frac{B_t}{B_T} = (1+r)^{-(T-t)}$.

In a model with a finite number of states, the conditional mathematical expectation in (7.22) or (7.27) can be calculated as a sum over $\omega \in \Omega$. In particular, (7.22) can be written using Equation (6.33) of Chapter 6 where $t = n$, $T = N$, and $X \equiv V_T$:

$$V_t(\omega_1\ldots\omega_t) = (1+r)^{-(T-t)} \sum_{\omega_{t+1},\ldots,\omega_T \in \{\mathsf{D},\mathsf{U}\}} \cdots \sum \tilde{p}^{\#\mathsf{U}(\omega_{t+1}\ldots\omega_T)}(1-\tilde{p})^{\#\mathsf{D}(\omega_{t+1}\ldots\omega_T)} V_T(\omega), \tag{7.28}$$

where $\tilde{p} \equiv \widetilde{\mathbb{P}}^{(B)}(\mathsf{U}) = \frac{1+r-d}{u-d}$ and $\omega = \omega_1\ldots\omega_T$, $0 \leqslant t \leqslant T$. If, instead, we choose the stock as numéraire, i.e., $g_t = S_t$, then (7.27) gives us yet another equivalent formula for the derivative price:

$$V_t(\omega_1\ldots\omega_t) = S_t(\omega_1\ldots\omega_t) \sum_{\omega_{t+1},\ldots,\omega_T \in \{\mathsf{D},\mathsf{U}\}} \cdots \sum \tilde{q}^{\#\mathsf{U}(\omega_{t+1}\ldots\omega_T)}(1-\tilde{q})^{\#\mathsf{D}(\omega_{t+1}\ldots\omega_T)} \frac{V_T(\omega)}{S_T(\omega)}, \tag{7.29}$$

where now $\tilde{q} \equiv \widetilde{\mathbb{P}}^{(S)}(\mathsf{U}) = \frac{1+r-d}{u-d} \cdot \frac{u}{1+r}$.

In summary, we have derived two methods for calculating derivative prices. The first method involves the use of a single-step recurrence formula as in (7.16) or (7.26), which is used to calculate the prices one by one recursively backward in time. The second method employs a multi-step pricing formulas as in (7.27), (7.28), or (7.29). It allows us to calculate the derivative price at any intermediate date by averaging the values of the payoff function weighted by the risk-neutral probabilities of all the $(T-t)$-step paths $\omega_{t+1}\ldots\omega_T$.

The replicating strategy $\{\Delta_t\}_{t\geqslant 0}$ is also called a *delta hedging strategy.* It allows the writer of a derivative contract to hedge perfectly the contract until the expiration date. Suppose that the writer sells one derivative contract for V_0 dollars and uses the proceeds to form a replicating portfolio, (β_0, Δ_0), in the bank account and the stock. As a result, the total value of the investment portfolio of one short derivative, Δ_0 shares of stock, and β_0 units of the risk-free asset is zero. At the end of each period, the investor changes (i.e., re-adjusts) the positions in the stock using (7.19) and (7.5). According to Theorem 7.6, the total value of the portfolio remains equal to zero. At maturity, the investor closes the positions in the stock and bank account to pay out the premium (if any) to the holder of the derivative contract. The proceeds cover the payoff in full. The whole situation is a zero sum game without any risk involved.

The next question arising naturally is whether the binomial tree model is *complete*, that is, every claim can be replicated. Recall that a single-period model is said to be complete if every payoff can be replicated by a portfolio in base assets.

Definition 7.3. A multi-period model is said to be *complete* if every $\mathcal{F}_T$-measurable payoff $X\colon \Omega_T \to \mathbb{R}$ can be replicated by a self-financing trading strategy in base assets.

Theorem 7.6 states that every payoff can be replicated by a self-financing strategy. Thus, the binomial tree model is indeed *complete.* To illustrate this important result, we will find the price process and replication strategy for a path-dependent derivative in the following example.

Example 7.4. Consider a binomial tree model from Example 7.1. Determine the prices and replicating strategy for a *lookback call option* with payoff

$$X = S_3 - \min_{0\leqslant t\leqslant 3} S_t.$$

Solution. First, we shall construct a tree of possible scenarios. To calculate the payoff function at maturity $T=3$ we need to know the minimum price

$$m_3(\omega) = \min_{0\leqslant t\leqslant 3} S_t(\omega)$$

for all market scenarios $\omega \in \Omega_3$. The sampled minimum $m_t = \min_{0\leqslant n\leqslant t} S_n$, $t=0,1,2,3$, can be calculated simultaneously with stock prices using the formula $m_t = \min\{m_{t-1}, S_t\}$, $t=1,2,3$, where $m_0 = S_0$. Since the value of m_t depends on the historical path of the stock price process (i.e., a market scenario ω), the result can be represented using a nonrecombining scenario tree, as given in Figure 7.3. As is clearly seen from Figure 7.3, the tree cannot be reduced to a recombining one since $m_2(\mathsf{DU}) \neq m_2(\mathsf{UD})$, $m_3(\mathsf{UUD}) \neq m_3(\mathsf{UDU}) \neq m_3(\mathsf{DUU})$, etc.

To calculate the derivative prices V_t, $t=0,1,2$, we first evaluate the payoff function

$$V_3(\omega) = S_3(\omega) - m_3(\omega)$$

for all eight possible scenarios $\omega \in \Omega_3$. After that, using the backward recursion (7.16) with

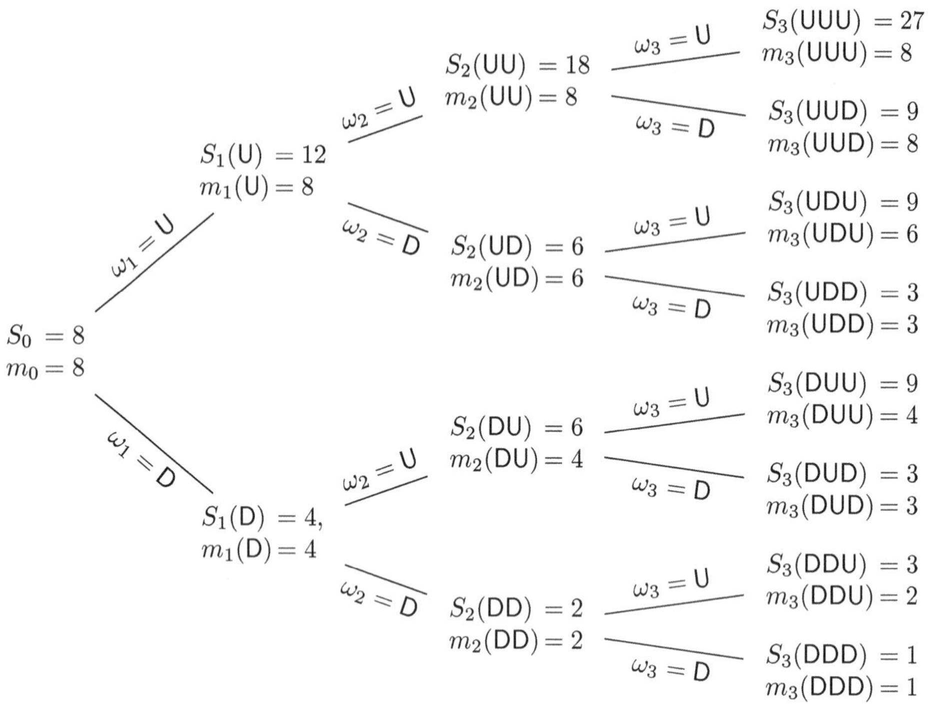

$S_3(\mathsf{UUU}) = 27$, $m_3(\mathsf{UUU}) = 8$
$S_2(\mathsf{UU}) = 18$, $m_2(\mathsf{UU}) = 8$
$\omega_3 = \mathsf{U}$
$\omega_3 = \mathsf{D}$
$S_3(\mathsf{UUD}) = 9$, $m_3(\mathsf{UUD}) = 8$
$\omega_2 = \mathsf{U}$
$S_1(\mathsf{U}) = 12$, $m_1(\mathsf{U}) = 8$
$S_3(\mathsf{UDU}) = 9$, $m_3(\mathsf{UDU}) = 6$
$\omega_1 = \mathsf{U}$
$\omega_2 = \mathsf{D}$
$S_2(\mathsf{UD}) = 6$, $m_2(\mathsf{UD}) = 6$
$\omega_3 = \mathsf{U}$
$\omega_3 = \mathsf{D}$
$S_3(\mathsf{UDD}) = 3$, $m_3(\mathsf{UDD}) = 3$
$S_0 = 8$, $m_0 = 8$
$S_3(\mathsf{DUU}) = 9$, $m_3(\mathsf{DUU}) = 4$
$\omega_1 = \mathsf{D}$
$S_2(\mathsf{DU}) = 6$, $m_2(\mathsf{DU}) = 4$
$\omega_3 = \mathsf{U}$
$\omega_3 = \mathsf{D}$
$\omega_2 = \mathsf{U}$
$S_3(\mathsf{DUD}) = 3$, $m_3(\mathsf{DUD}) = 3$
$S_1(\mathsf{D}) = 4$, $m_1(\mathsf{D}) = 4$
$S_3(\mathsf{DDU}) = 3$, $m_3(\mathsf{DDU}) = 2$
$\omega_2 = \mathsf{D}$
$S_2(\mathsf{DD}) = 2$, $m_2(\mathsf{DD}) = 2$
$\omega_3 = \mathsf{U}$
$\omega_3 = \mathsf{D}$
$S_3(\mathsf{DDD}) = 1$, $m_3(\mathsf{DDD}) = 1$

FIGURE 7.3: A binomial tree with stock prices S and sampled minimum values m calculated for each node. There are $|\Omega_3| = 8$ distinct paths from time 0 to 3.

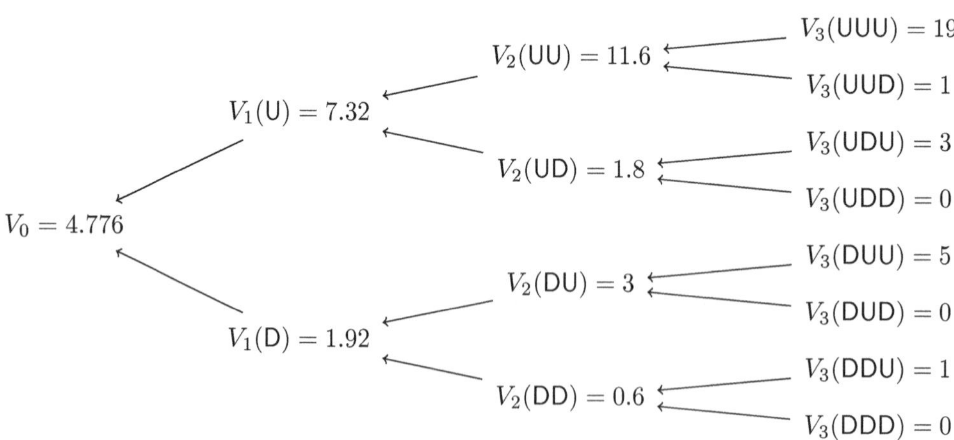

FIGURE 7.4: A scenario tree with derivative prices calculated for each node of a three-period nonrecombining tree.

$\tilde{p} = \frac{3}{4}$ and $r = \frac{1}{4}$, we compute derivative prices as follows:

$$
\begin{aligned}
V_2(\omega_1\omega_2) &= \frac{1}{1+r}\left(\tilde{p}V_3(\omega_1\omega_2\mathsf{U}) + (1-\tilde{p})V_3(\omega_1\omega_2\mathsf{D})\right) = \frac{3V_3(\omega_1\omega_2\mathsf{U}) + V_3(\omega_1\omega_2\mathsf{D})}{5},\\
V_1(\omega_1) &= \frac{1}{1+r}\left(\tilde{p}V_2(\omega_1\mathsf{U}) + (1-\tilde{p})V_2(\omega_1\mathsf{D})\right) = \frac{3V_2(\omega_1\mathsf{U}) + V_2(\omega_1\mathsf{D})}{5},\\
V_0 &= \frac{1}{1+r}\left(\tilde{p}V_1(\mathsf{U}) + (1-\tilde{p})V_1(\mathsf{D})\right) = \frac{3V_1(\mathsf{U}) + V_1(\mathsf{D})}{5},
\end{aligned}
$$

where $\omega_1, \omega_2 \in \{\mathsf{D}, \mathsf{U}\}$. The results of our computations are summarized in Figure 7.4.

We can now also obtain the replicating strategy using the delta hedging formula in (7.19) for Δ_t, $t = 0, 1, 2$:

$$
\begin{aligned}
\Delta_0 &= \frac{V_1(\mathsf{U}) - V_1(\mathsf{D})}{S_1(\mathsf{U}) - S_1(\mathsf{D})} = \frac{7.32 - 1.92}{12 - 4} = \frac{27}{40} = 0.675,\\
\Delta_1(\mathsf{U}) &= \frac{V_1(\mathsf{UU}) - V_1(\mathsf{UD})}{S_1(\mathsf{UU}) - S_1(\mathsf{UD})} = \frac{11.6 - 1.8}{18 - 6} = \frac{49}{60} \cong 0.816,\\
\Delta_1(\mathsf{D}) &= \frac{V_1(\mathsf{DU}) - V_1(\mathsf{DD})}{S_1(\mathsf{DU}) - S_1(\mathsf{DD})} = \frac{3 - 0.6}{6 - 2} = \frac{3}{5} = 0.6,\\
\Delta_2(\mathsf{UU}) &= \frac{V_3(\mathsf{UUU}) - V_3(\mathsf{UUD})}{S_3(\mathsf{UUU}) - S_3(\mathsf{UUD})} = \frac{19 - 1}{27 - 9} = 1,\\
\Delta_2(\mathsf{UD}) &= \frac{V_3(\mathsf{UDU}) - V_3(\mathsf{UDD})}{S_3(\mathsf{UDU}) - S_3(\mathsf{UDD})} = \frac{3 - 0}{9 - 3} = 0.5,\\
\Delta_2(\mathsf{DU}) &= \frac{V_3(\mathsf{DUU}) - V_3(\mathsf{DUD})}{S_3(\mathsf{DUU}) - S_3(\mathsf{DUD})} = \frac{5 - 0}{9 - 3} = \frac{5}{6},\\
\Delta_2(\mathsf{DD}) &= \frac{V_3(\mathsf{DDU}) - V_3(\mathsf{DDD})}{S_3(\mathsf{DDU}) - S_3(\mathsf{DDD})} = \frac{1 - 0}{3 - 1} = 0.5.
\end{aligned}
$$

The alternative method for computing the derivative prices is to use (7.28). For example, the initial price can be computed by the following discounted expectation:

$$
\begin{aligned}
V_0 &= (1+r)^{-3}\widetilde{\mathrm{E}}[V_3] = (1+r)^{-3}\sum_{\omega\in\Omega_3}\widetilde{\mathbb{P}}(\omega)V_3(\omega)\\
&= \left(\frac{5}{4}\right)^{-3}\sum_{\omega_1,\omega_2,\omega_3\in\{\mathsf{D},\mathsf{U}\}}\left(\frac{3}{4}\right)^{\#\mathsf{U}(\omega_1\omega_2\omega_3)}\left(\frac{1}{4}\right)^{\#\mathsf{D}(\omega_1\omega_2\omega_3)} V_3(\omega_1\omega_2\omega_3)\\
&= \left(\frac{4}{5}\right)^3\left\{19\cdot\left(\frac{3}{4}\right)^3 + (5+3)\cdot\left(\frac{3}{4}\right)^2\frac{1}{4} + 1\cdot\left[\left(\frac{3}{4}\right)^2\frac{1}{4} + \left(\frac{1}{4}\right)^2\frac{3}{4}\right]\right\}\\
&= 4.776,
\end{aligned}
$$

□

7.3.2 Replication and Valuation of Random Cash Flows

Consider a financial contract whose payoff is spread over all T periods. Let the time-t payment be X_t for all $t \in \{0, 1, \ldots, T\}$. In general X_t is an $\mathcal{F}_t$-measurable random variable. In other words, X_t is a function of the first t market moves $\omega_1, \omega_2, \ldots, \omega_t$. So we deal with a stream of random cash flows $\{X_t\}_{0\leqslant t\leqslant T}$ adapted to the filtration $\mathbb{F}$. We allow these payments to be negative as well as positive. If a payment is positive (negative), then the holder, who is long on the contract, receives (makes) the payment.

Such a stream of cash flows can be replicated by a self-financing strategy in the base securities S and B. Start with the capital $\Pi_0 = V_0$. At time $t = 0$, we make the payment X_0 and form a portfolio (β_0, Δ_0) with the total cost of $\Pi_0 - X_0$. At time $t = 1$ its value is

$$\Pi_1 = \Delta_0 S_1 + \beta_0 B_1 = \Delta_0 S_1 + (\Pi_0 - X_0 - \Delta_0 S_0)(1 + r).$$

At every time $t \in \{1, \ldots, T\}$, we liquidate the portfolio $(\beta_{t-1}, \Delta_{t-1})$ whose total value (before the payment X_t is made) is governed by the following wealth equation:

$$\Pi_t = \Delta_{t-1} S_t + (\Pi_{t-1} - X_{t-1} - \Delta_{t-1} S_{t-1})(1 + r). \tag{7.30}$$

Theorem 7.7. *Consider a derivative security with the payoff process $\{X_t\}_{0 \leqslant t \leqslant T}$ adapted to the filtration $\mathbb{F}$. Define the derivative price process $\{V_t\}_{0 \leqslant t \leqslant T}$ by the following backward-in-time recursion:*

$$V_T(\omega_1, \ldots, \omega_T) = X_T(\omega_1, \ldots, \omega_T), \tag{7.31}$$

$$\begin{aligned} V_t(\omega_1, \ldots, \omega_t) &= X_t(\omega_1, \ldots, \omega_t) \\ &\quad + \frac{1}{1+r}\left(\tilde{p} V_{t+1}(\omega_1, \ldots, \omega_t, \mathsf{U}) + (1 - \tilde{p}) V_{t+1}(\omega_1, \ldots, \omega_t, \mathsf{D})\right), \\ &\quad t = 0, 1, \ldots, T - 1; \end{aligned} \tag{7.32}$$

and the strategy $\{\Delta_t\}_{0 \leqslant t \leqslant T-1}$ by (7.19). Then the value process $\{\Pi_t\}_{0 \leqslant t \leqslant T}$ given by the wealth equation (7.30) coincides with the derivative price process:

$$\Pi_t(\omega_1 \omega_2 \ldots \omega_t) = V_t(\omega_1 \omega_2 \ldots \omega_t)$$

for all $0 \leqslant t \leqslant T$ and all $\omega_1, \omega_2, \ldots, \omega_t \in \{\mathsf{D}, \mathsf{U}\}$. In other words, $\{\Delta_t\}_{0 \leqslant t \leqslant T-1}$ is a replicating strategy for the derivative security with the payoff process $\{X_t\}_{0 \leqslant t \leqslant T}$.

Proof. The proof is analogous to that of Theorem 7.6 and is left as an exercise for the reader. □

Equation (7.32) can be written in a compact form using the risk-neutral mathematical expectation:

$$V_t = X_t + (1 + r)^{-1} \widetilde{\mathrm{E}}_t[V_{t+1}], \quad t = 0, 1, \ldots, T - 1.$$

Applying this formula successively and using the tower property gives

$$V_t = X_t + \widetilde{\mathrm{E}}_t \left[\sum_{s=t+1}^{T} (1 + r)^{-(s-t)} X_s \right]. \tag{7.33}$$

In other words, the value V_t of the stream of cash flows at time t is given by the sum of the time-t risk-neutral values of all present and future payments. Here, we assume the risk-neutral measure $\widetilde{\mathbb{P}} \equiv \widetilde{\mathbb{P}}^{(B)}$.

7.4 Pricing and Hedging Non-Path-Dependent Derivatives

Equation (7.28) allows us to calculate the current derivative price by averaging the discounted payoff values calculated for all market scenarios. However, it is time and memory consuming to tabulate the payoff function for 2^T scenarios. For example, a naive implementation of (7.28) for a 100-period binomial tree leads to a large amount of computation with $2^{100} \cong 1.268 \times 10^{30}$ market scenarios!

We know that in a recombining binomial tree many market scenarios lead to the same prices of S. For example, there are only $T+1$ different time-T stock prices: $S_T = S_0 u^n d^{T-n}$ for $n = 0, 1, \ldots, T$. Consider a non-path-dependent derivative. At maturity, the payoff is a function of the stock price, $X = \Lambda(S_T)$ (recall that Λ is used here to denote a known payoff function of the price S_T of the underlying asset). Thus, we expect that the derivative price is a function of the spot price for any time t preceding the maturity: $V_t = V_t(S_t)$. The number of possible values of S_t is $t+1$, which is significantly less than 2^t. Therefore, we may organize the computation in a more efficient manner by computing derivative values on the recombining binomial tree.

To prove that the time-t derivative price V_t is only a function of the current spot S_t, we use the Markov property of the stock price process within the risk-neutral expectation pricing formula. Recall the definition of a discrete-time Markov process from Chapter 6. From the tower property, we have that if the Markov property holds for one period of time then it holds for any multiple time periods. Clearly the stock price process $\{S_t\}_{t \geqslant 0}$ is a Markov process since, given any (Borel) function g,

$$\widetilde{\mathrm{E}}_t[g(S_{t+1})] = f(S_t), \text{ where } f(x) := \tilde{p}g(xu) + (1-\tilde{p})g(xd).$$

[Note that the Markov property holds in any equivalent probability measure.] Therefore, from the multi-step Markov property of the stock price process applied to the risk-neutral expectation of the discounted payoff function $\Lambda(S_T)$, the time-t derivative price must be given by a function of t, S_t, say $f(t, S_t)$:

$$V_t = \widetilde{\mathrm{E}}_t\left[(1+r)^{-(T-t)}\Lambda(S_T)\right] = f(t, S_t).$$

This is an equation stating the equivalence of two random variables. On any given $\omega \in \Omega$ we have $V_t(\omega) = f(t, S_t(\omega))$. Given a numerical value S for the stock price for a given outcome ω, i.e., $S_t(\omega) = S$, then the derivative price function $f(t, S)$ is an ordinary function of ordinary variables t, S. Note that below we shall also write $V_t(S)$ to denote the time-t derivative price for $S_t = S$.

The possible values of S correspond to the nodes of the binomial tree and the derivative value can be calculated recursively backward in time on all these nodes:

$$V_t(S_{t,n}) = \frac{1}{1+r}\left[\tilde{p}V_{t+1}(S_{t,n}u) + (1-\tilde{p})V_{t+1}(S_{t,n}d)\right], \quad 0 \leqslant t \leqslant T-1, \tag{7.34}$$

and $V_T(S_{T,n}) \equiv \Lambda(S_{T,n})$, where $S_{t,n} := S_0 u^n d^{t-n}$ for all $n = 0, 1, \ldots, t$ and $t = 0, 1, \ldots, T$. Hence, this recovers the backward recurrence formula in Equation (4.12) of Chapter 4.

On the other hand, we can derive a direct formula for $V_t(S)$ at $t = 0, 1, \ldots, T-1$, given as the discounted mathematical expectation of the payoff function. Applying (7.22) to a European-style derivative with payoff $\Lambda(S_T)$ gives

$$\begin{aligned} V_t(S_t) &= (1+r)^{-(T-t)}\widetilde{\mathrm{E}}_t\left[V_T(S_T)\right] \\ &= (1+r)^{-(T-t)}\widetilde{\mathrm{E}}_t\left[\Lambda\left(S_t\frac{S_T}{S_t}\right)\right]. \end{aligned} \tag{7.35}$$

The price S_t is $\mathcal{F}_t$-measurable. The ratio $\frac{S_T}{S_t}$ is independent of $\mathcal{F}_t$, since it is a function of $\omega_{t+1}, \ldots, \omega_T$, i.e. for all $\omega = \omega_1 \ldots \omega_T$ we have

$$\left(\frac{S_T}{S_t}\right)(\omega) = u^{\#\mathsf{U}(\omega_{t+1},\ldots,\omega_T)}\, d^{(T-t)-\#\mathsf{U}(\omega_{t+1},\ldots,\omega_T)} = u^{H_{T-t}(\omega)}d^{(T-t)-H_{T-t}(\omega)}.$$

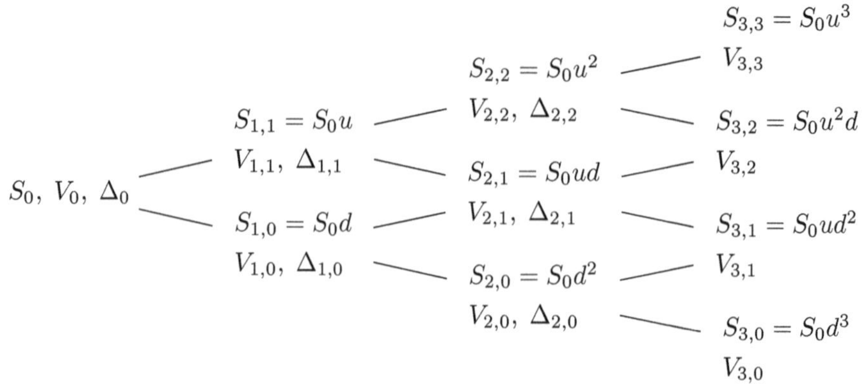

$t=0$	$t=1$	$t=2$	$t=3$
			$S_{3,3} = S_0u^3$, $V_{3,3}$
		$S_{2,2} = S_0u^2$, $V_{2,2}$, $\Delta_{2,2}$	
	$S_{1,1} = S_0u$, $V_{1,1}$, $\Delta_{1,1}$		$S_{3,2} = S_0u^2d$, $V_{3,2}$
S_0, V_0, Δ_0		$S_{2,1} = S_0ud$, $V_{2,1}$, $\Delta_{2,1}$	
	$S_{1,0} = S_0d$, $V_{1,0}$, $\Delta_{1,0}$		$S_{3,1} = S_0ud^2$, $V_{3,1}$
		$S_{2,0} = S_0d^2$, $V_{2,0}$, $\Delta_{2,0}$	
			$S_{3,0} = S_0d^3$, $V_{3,0}$

FIGURE 7.5: A recombining binomial tree with stock prices S, derivative prices V, and replication stock positions Δ given for each node.

Here, $H_{T-t}(\omega) := \#\mathsf{U}(\omega_{t+1}, \ldots, \omega_T) = \mathsf{U}_T(\omega) - \mathsf{U}_t(\omega)$, where U_t is defined in Equation (6.3) of Chapter 6. The variable H_{T-t} counts the number of upward moves in the sequence $\omega_{t+1}, \ldots, \omega_T$. Hence, $H_{T-t} \sim Bin(T-t, \tilde{p})$ under $\widetilde{\mathbb{P}}$. The random variable H_{T-t} is independent of $\mathcal{F}_t$; the variable S_t is $\mathcal{F}_t$-measurable. Therefore, by using Proposition 6.7, we obtain

$$\begin{aligned} V_t &= \frac{1}{(1+r)^{T-t}} \widetilde{\mathrm{E}}_t \left[\Lambda \left(S_t u^{H_{T-t}} d^{(T-t)-H_{T-t}} \right) \right] \\ &= \frac{1}{(1+r)^{T-t}} \widetilde{\mathrm{E}} \left[\Lambda \left(S u^{H_{T-t}} d^{(T-t)-H_{T-t}} \right) \right] \bigg|_{S=S_t} \\ &= \frac{1}{(1+r)^{T-t}} \sum_{n=0}^{T-t} \Lambda \left(S_t u^n d^{(T-t)-n} \right) \binom{T-t}{n} \tilde{p}^n (1-\tilde{p})^{(T-t)-n} := V_t(S_t). \end{aligned} \tag{7.36}$$

As is seen from the above equation, V_t is a function of S_t. By setting $t = 0$, we can recover the risk-neutral pricing formula (4.13) from Chapter 4 for the initial price V_0 of a European-style derivative.

Using (7.34), we can compute derivative prices for each node of the recombining binomial tree. As a result, we obtain the *derivative pricing function*, $V_t(S)$,, which is a function of calendar (actual) time $t \geqslant 0$ and stock price (spot) $S > 0$. It is defined for every node of the binomial tree and hence can be represented in a tree form as well (see Figure 7.5).

The stock positions of the replication strategy $\{\Delta_t\}_{0 \leqslant t \leqslant T-1}$ given by (7.19) are functions of derivative and stock prices. Therefore, the position Δ_t does not explicitly depend on $\omega_1, \ldots, \omega_t$, but it is a function of only $S_t = S_t(\omega_1, \ldots, \omega_t)$ and given by

$$\Delta_t(S_t) = \frac{V_{t+1}(S_t u) - V_{t+1}(S_t d)}{(u-d)S_t}. \tag{7.37}$$

Equation (7.5) takes the following form:

$$\beta_t(S_t) = \frac{V_t(S_t) - \Delta_t(S_t)S_t}{B_t} = \frac{V_t(S_t) - (V_{t+1}(S_t u) - V_{t+1}(S_t d))/(u-d)}{B_t}. \tag{7.38}$$

Hence the computation of the replication strategy reduces to the calculation of stock positions for each node of the recombining binomial tree. As a result we obtain a binomial tree where three quantities are computed for each node (t, n), where $t = 0, 1, \ldots, T$ and $n = 0, 1, \ldots, t$: the stock price $S_{t,n} := S_0 u^n d^{t-n}$, the derivative price $V_{t,n} := V_t(S_{t,n})$, and the replication position $\Delta_{t,n} := \Delta_t(S_{t,n})$ (see Figure 7.5).

The recurrence formulae (7.34) and (7.37) can now be written in a more compact form:

$$\begin{aligned} V_{t,n} &= \frac{1}{1+r}\left[\tilde{p}V_{t+1,n+1} + (1-\tilde{p})V_{t+1,n}\right], \\ \Delta_{t,n} &= \frac{V_{t+1,n+1} - V_{t+1,n}}{S_{t,n}\,(u-d)} \quad \text{for } 0 \leqslant t \leqslant T-1 \text{ and } 0 \leqslant n \leqslant t. \end{aligned}$$

The terminal condition is $V_{T,n} = \Lambda(S_{T,n})$ for $0 \leqslant n \leqslant T$.

Example 7.5. Consider the three-period recombining binomial tree model from Example 7.1 with $S_0 = 8$, $B_0 = 1$, $u = \frac{3}{2}$, $d = \frac{1}{2}$, and $r = \frac{1}{4}$. Determine the option prices and replication strategy for a standard European call with maturity time $T = 3$ and strike price $K = 8$ (i.e., the call option is at the money).

Solution. We first calculate the payoff values and then the call option prices $C_t(S_{t,n})$ at $t = 0, 1, 2$ using the backward-in-time recursion (7.34), which takes the following form:

$$C_t(S_{t,n}) = \frac{1}{1+r}\left(\tilde{p}C_{t+1}(S_{t,n}u) + (1-\tilde{p})C_{t+1}(S_{t,n}d)\right) = \frac{3C_{t+1}(S_{t,n}u) + C_{t+1}(S_{t,n}d)}{5},$$

where $S_{t,n} = S_0 u^n d^{t-n} = 8 \cdot 1.5^n \cdot 0.5^{t-n}$ for $n = 0, 1, \ldots, t$.

$t = 3$: At maturity $T = 3$, calculate the payoff values $C_3(S_3) = (S_3 - 8)^+$, where $S_3 \in \{1, 3, 9, 27\}$:

$$C_3(1) = C_3(3) = 0, \quad C_3(9) = 1, \quad C_3(27) = 19.$$

$t = 2$: Calculate the option prices at time $t = 2$:

$$\begin{aligned} C_2(2) &= \frac{3C_3(3) + C_3(1)}{5} = 0, \quad C_2(6) = \frac{3C_3(9) + C_3(3)}{5} = 0.6, \\ C_2(18) &= \frac{3C_3(27) + C_3(9)}{5} = 11.6. \end{aligned}$$

$t = 1$: Calculate the option prices at time $t = 1$:

$$C_1(4) = \frac{3C_2(6) + C_2(2)}{5} = 0.36, \quad C_1(12) = \frac{3C_2(18) + C_2(6)}{5} = 7.08.$$

$t = 0$: The initial option price is

$$C_0(8) = \frac{3C_1(12) + C_1(4)}{5} = 4.32.$$

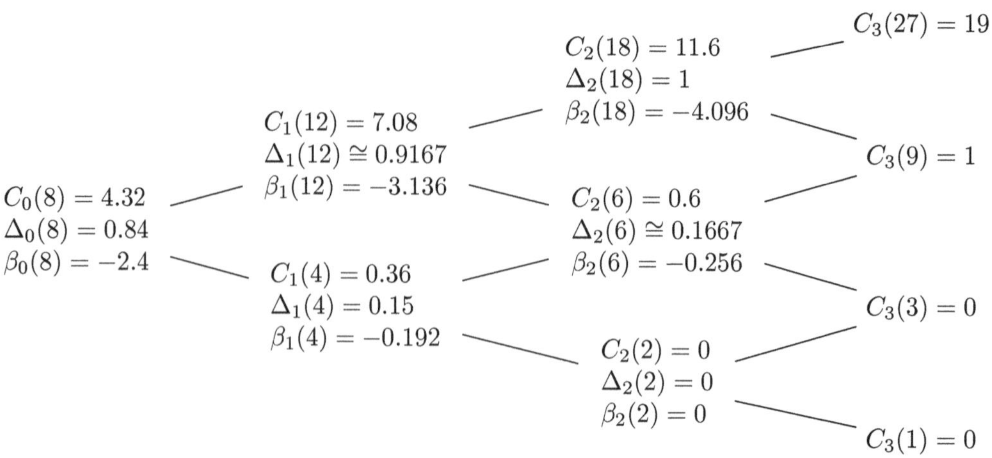

FIGURE 7.6: Call option prices and replication portfolio positions β and Δ calculated for each node of the recombining binomial tree.

Now, compute the replication strategy $\{(\beta_t, \Delta_t)\}_{t=0,1,2}$ using the formulae (7.37) and (7.38):

$$\Delta_0(8) = \frac{C_1(12) - C_1(4)}{12 - 4} = 0.84, \qquad \beta_0(8) = \frac{C_0(8) - 8\Delta_0(8)}{1} = -2.4,$$
$$\Delta_1(4) = \frac{C_2(6) - C_2(2)}{6 - 2} = 0.15, \qquad \beta_1(4) = \frac{C_1(4) - 4\Delta_1(4)}{1.25} = -0.192,$$
$$\Delta_1(12) = \frac{C_2(18) - C_2(6)}{18 - 6} = \frac{11}{12} \cong 0.9167, \qquad \beta_1(12) = \frac{C_1(12) - 12\Delta_1(12)}{1.25} = -3.136,$$
$$\Delta_2(2) = \frac{C_3(3) - C_3(1)}{3 - 1} = 0, \qquad \beta_2(2) = \frac{C_2(2) - 2\Delta_2(2)}{1.25^2} = 0,$$
$$\Delta_2(6) = \frac{C_3(9) - C_3(3)}{9 - 3} = \frac{1}{6} \cong 0.1667, \qquad \beta_2(6) = \frac{C_2(6) - 6\Delta_2(6)}{1.25^2} = -0.256,$$
$$\Delta_2(18) = \frac{C_3(27) - C_3(9)}{27 - 9} = 1, \qquad \beta_2(18) = \frac{C_2(18) - 18\Delta_2(18)}{1.25^2} = -4.096.$$

The results are summarized in Figure 7.6. □

7.5 Pricing Formulae for Standard European Options

The formula (7.36) allows us to price any non-path-dependent derivative with a given payoff function Λ including nonlinear functions of the stock price. However, the option price formula (7.36) can be written in a simpler form for the standard European call and put payoff functions. In Chapter 4 we present the binomial pricing formulae (4.17) and (4.18) for initial (time-0) prices of the European call and put options. Those formulae are given in terms of the binomial CDF $\mathcal{B}$. Let us generalize that result and derive a closed-form formula of the time-t derivative price (with arbitrary t) for a class of derivatives with piecewise linear payoffs.

According to the pricing formula (7.35), the derivative price is given by the risk-neutral expectation of the payoff function. First, we observe that the European call, put, and chooser option payoffs, which are, respectively, given by $(S-K)^+$, $(K-S)^+$, and $S \wedge K$, can be represented as a linear combination of functions from the list

$$\{S,\ K,\ S\,\mathbb{I}_{\{S\leqslant K\}},\ K\,\mathbb{I}_{\{S\leqslant K\}}\}.$$

Indeed, the call and put payoffs admit the following respective representations:

$$(S-K)^+ = (S-K)\,(1-\mathbb{I}_{\{S\leqslant K\}}) = S-K-S\,\mathbb{I}_{\{S\leqslant K\}} + K\,\mathbb{I}_{\{S\leqslant K\}},$$
$$(K-S)^+ = (K-S)\,\mathbb{I}_{\{S\leqslant K\}} = K\,\mathbb{I}_{\{S\leqslant K\}} - S\,\mathbb{I}_{\{S\leqslant K\}}.$$

Using the linearity of the mathematical expectation in (7.35) and the martingale property for the discounted stock price process gives the following formulae for time-t prices of the standard European call and put options:

$$\begin{aligned} C_t(S_t) &= \frac{1}{(1+r)^{T-t}}\widetilde{\mathrm{E}}_t[(S_T-K)^+] \\ &= S_t - \frac{K}{(1+r)^{T-t}} - \frac{1}{(1+r)^{T-t}}\widetilde{\mathrm{E}}_t[S_T\,\mathbb{I}_{\{S_T\leqslant K\}}] + \frac{K}{(1+r)^{T-t}}\widetilde{\mathrm{E}}_t[\mathbb{I}_{\{S_T\leqslant K\}}], \quad (7.39) \\ P_t(S_t) &= \frac{1}{(1+r)^{T-t}}\widetilde{\mathrm{E}}_t[(K-S_T)^+] \\ &= \frac{K}{(1+r)^{T-t}}\widetilde{\mathrm{E}}_t[\mathbb{I}_{\{S_T\leqslant K\}}] - \frac{1}{(1+r)^{T-t}}\widetilde{\mathrm{E}}_t[S_T\,\mathbb{I}_{\{S_T\leqslant K\}}], \quad 0\leqslant t\leqslant T. \quad (7.40) \end{aligned}$$

As is seen, the evaluation of European call and put options reduces to computation of the conditional expectations $\widetilde{\mathrm{E}}_t[\mathbb{I}_{\{S_T\leqslant K\}}]$ and $\widetilde{\mathrm{E}}_t[S_T\,\mathbb{I}_{\{S_T\leqslant K\}}]$. The expectation of the indicator function, $\mathbb{I}_{\{S_T\leqslant K\}}$, is equal to the conditional probability of the event $\{S_T\leqslant K\}$:

$$\widetilde{\mathrm{E}}_t[\mathbb{I}_{\{S_T\leqslant K\}}] = \widetilde{\mathbb{P}}(S_T\leqslant K \mid S_t). \quad (7.41)$$

Recall that the ratio S_T/S_t can be expressed in terms of a binomial random variable:

$$\frac{S_T}{S_t} = u^{H_{T-t}}\,d^{T-t-H_{T-t}}, \quad (7.42)$$

where $H_{T-t}\sim Bin(T-t,\tilde{p})$ is independent of S_t. Dividing both parts of the inequality $S_T\leqslant K$ by S_t and using (7.42) allows for reducing the event $\{S_T\leqslant K\}$ to an equivalent event for the binomial random variable H_{T-t}:

$$\frac{S_T}{S_t}\leqslant\frac{K}{S_t} \iff u^{H_{T-t}}\,d^{T-t-H_{T-t}}\leqslant\frac{K}{S_t} \iff \left(\frac{u}{d}\right)^{H_{T-t}}\leqslant\frac{K}{S_t}\,d^{-(T-t)} \iff$$
$$H_{T-t}\,\ln\left(\frac{u}{d}\right)\leqslant\ln\left(\frac{K}{S_t}\right)-(T-t)\ln d \iff H_{T-t}\leqslant m_{T-t},$$

where $m_{T-t}(S_t)\equiv m(K,S_t,u,d,T-t)$ is given by

$$m_{T-t}(S_t) = \max\left\{m\ :\ 0\leqslant m\leqslant T-t;\ S_t\,u^m\,d^{(T-t)-m}\leqslant K\right\} = \left\lfloor\frac{\ln\left(\frac{K}{S_t}\right)-(T-t)\ln d}{\ln\left(\frac{u}{d}\right)}\right\rfloor.$$

Here we assume that the strike price K lies in the interval $[S_t d^{T-t}, S_t u^{T-t}]$, which includes all possible stock prices S_T conditional on S_t. If $K < S_t d^{T-t}$ holds, i.e., the strike price is less than the minimum value of S_T conditional on S_t, then we set $m_{T-t} = -\infty$. If $K > S_t u^{T-t}$ holds, i.e., the strike price is greater than the maximum value of S_T conditional on S_t, then we set $m_{T-t} = T - t$. Now, the conditional expectation in (7.41) is written as

$$\widetilde{\mathrm{E}}_t[\mathbb{I}_{\{S_T \leqslant K\}}] = \widetilde{\mathbb{P}}(H_{T-t} \leqslant m_{T-t}(S))|_{S=S_t} = \mathcal{B}(m_{T-t}(S_t); T-t, \tilde{p}), \tag{7.43}$$

where $\mathcal{B}(m; n, p)$ is the CDF of $\mathsf{X} \sim Bin(n, p)$ given by

$$\mathcal{B}(m; n, p) = \mathbb{P}(\mathsf{X} \leqslant m) = \sum_{k=0}^{m} \binom{n}{k} p^k (1-p)^{n-k}.$$

Note that if $K < S_t d^{T-t}$ holds, then $\widetilde{\mathrm{E}}_t[\mathbb{I}_{\{S_T \leqslant K\}}] = 0$; if $K > S_t u^{T-t}$ holds, then $\widetilde{\mathrm{E}}_t[\mathbb{I}_{\{S_T \leqslant K\}}] = 1$.

Similarly, we calculate the conditional expectation of a product of the stock price S_T and indicator function $\mathbb{I}_{\{S_T \leqslant K\}}$ as follows:

$$\begin{aligned}
\widetilde{\mathrm{E}}_t[S_T \, \mathbb{I}_{\{S_T \leqslant K\}}] &= \widetilde{\mathrm{E}}_t[S_t \, u^{H_{T-t}} \, d^{T-t-H_{T-t}} \, \mathbb{I}_{\{H_{T-t} \leqslant m_{T-t}\}}] \\
&= S_t \, \widetilde{\mathrm{E}}_t[u^{H_{T-t}} \, d^{T-t-H_{T-t}} \, \mathbb{I}_{\{H_{T-t} \leqslant m_{T-t}\}}] \\
&= S_t \sum_{k=0}^{m_{T-t}} \binom{T-t}{k} u^k \, \tilde{p}^k \, d^{(T-t)-k} \, (1-\tilde{p})^{(T-t)-k}.
\end{aligned} \tag{7.44}$$

The summation in (7.44) can also be written as a binomial CDF. Consider the probability

$$\tilde{q} = \frac{u}{1+r} \cdot \tilde{p},$$

which is the risk-neutral probability of an upward market move in the EMM $\widetilde{\mathbb{P}}^{(S)}$. The probability of a downward market move is

$$1 - \tilde{q} = \frac{d}{1+r}(1-\tilde{p}).$$

Thus, we have $u\tilde{p} = (1+r)\tilde{q}$ and $d(1-\tilde{p}) = (1+r)(1-\tilde{q})$. Hence,

$$u^k \, \tilde{p}^k \, d^{(T-t)-k} \, (1-\tilde{p})^{(T-t)-k} = (1+r)^{T-t} \, \tilde{q}^k \, (1-\tilde{q})^{(T-t)-k}.$$

Substituting this into (7.44) gives

$$\begin{aligned}
\widetilde{\mathrm{E}}_t[S_T \, \mathbb{I}_{\{S_T \leqslant K\}}] &= S_t \, (1+r)^{T-t} \sum_{k=0}^{m_{T-t}} \binom{T-t}{k} \tilde{q}^k \, (1-\tilde{q})^{(T-t)-k} \\
&= S_t \, (1+r)^{T-t} \mathcal{B}(m_{T-t}; T-t, \tilde{q}).
\end{aligned} \tag{7.45}$$

By combining (7.39) or (7.40) with (7.43) and (7.45), we obtain a binomial pricing formula for the European call or put option:

$$C_t(S_t) = S_t - \frac{K}{(1+r)^{T-t}} - \frac{1}{(1+r)^{T-t}} S_t (1+r)^{T-t} \mathcal{B}(m_{T-t}; T-t, \tilde{q}) \tag{7.46}$$
$$+ \frac{K}{(1+r)^{T-t}} \mathcal{B}(m_{T-t}; T-t, \tilde{p})$$
$$= S_t (1 - \mathcal{B}(m_{T-t}; T-t, \tilde{q})) - \frac{K}{(1+r)^{T-t}} (1 - \mathcal{B}(m_{T-t}; T-t, \tilde{p})), \tag{7.47}$$
$$P_t(S_t) = \frac{K}{(1+r)^{T-t}} \mathcal{B}(m_{T-t}; T-t, \tilde{p}) - \frac{1}{(1+r)^{T-t}} S_t (1+r)^{T-t} \mathcal{B}(m_{T-t}; T-t, \tilde{q})$$
$$= \frac{K}{(1+r)^{T-t}} \mathcal{B}(m_{T-t}; T-t, \tilde{p}) - S_t \mathcal{B}(m_{T-t}; T-t, \tilde{q}). \tag{7.48}$$

Note the option price in (7.47) and (7.48) satisfy the *put-call parity*:

$$C_t(S_t) - P_t(S_t) = S_t - \frac{K}{(1+r)^{T-t}}, \quad 0 \leqslant t \leqslant T.$$

That is, a portfolio of a long call and a short put is equivalent to a long forward contract with the same strike price and expiry date. As a result, the put (call) option value can be computed by combining the put-call parity and the call (put) pricing formula.

7.6 Pricing and Hedging Path-Dependent Derivatives

As the name implies, the payoff function of a path-dependent derivative depends on some quantity calculated along a trajectory (sample path) of the underlying security price process. The options or derivatives with path-dependent payoffs are called *exotic* since their features are more complex than commonly traded "vanilla" derivatives such as standard European and American put and call options. Examples of such path-dependent quantities include the observed maximum and minimum values of a security price process, the arithmetic and geometric averages, etc. By themselves, such quantities are generally not Markovian. However, when coupled with the underlying security price process, the combination forms a Markov vector process. By the Markov property, it then follows that we can price such path-dependent European-style derivatives by a backward recurrence method involving only derivative prices at the relevant nodes corresponding to the joint values of the underlying (stock) and whatever path-dependent quantities make up the payoff. The backward recurrence pricing formula involves derivative prices on a lattice with nodes specified by the stock price and path-dependent values.

7.6.1 Average Asset Prices and Asian Options

Asian options are typical examples of exotic options where the payoff functions depend on some form of averaging of the underlying asset price over the life of the option. By considering different types of averaging, such as arithmetic or geometric, one can generate different types of options. So the payoff of an Asian option maturing at time T depends on the arithmetic average, A_T, or the geometric average, G_T, of the prices of the underlying asset S. The averages A_t and G_t, observed from time 0 to t, are, respectively, defined by

$$A_t = \frac{1}{t+1}\sum_{n=0}^{t} S_n, \quad G_t = \left(\prod_{n=0}^{t} S_n\right)^{\frac{1}{t+1}}, \quad 0 \leqslant t \leqslant T. \tag{7.49}$$

As follows from (7.49), at time 0 we have $A_0 = G_0 = S_0$. Let us only consider the case with the arithmetic average. The results for the geometric average can be obtained by replacing A with G in the payoff functions and formulae below.

Clearly, the average price process $\{A_t\}_{t\geqslant 0}$ is, by itself, not Markovian since we cannot calculate A_t by using only A_{t-1}. However, the vector process $\{(S_t,\, A_t)\}_{t\geqslant 0}$ is Markovian. The evolution of this process from time $t-1$ to time t for $t \geqslant 1$ is given by the following equation:

$$\begin{bmatrix} S_{t-1} \\ A_{t-1} \end{bmatrix} \to \begin{bmatrix} S_t \\ A_t \end{bmatrix} = \begin{bmatrix} Y_t S_{t-1} \\ \frac{1}{t+1}\left(tA_{t-1} + S_{t-1}Y_t\right) \end{bmatrix}, \tag{7.50}$$

where the random variable $Y_t := S_t/S_{t-1} = u\mathbb{I}_{\{\omega_t = \mathsf{U}\}} + d\mathbb{I}_{\{\omega_t=\mathsf{D}\}}$ depends only on ω_t. Hence, given knowledge of the vector process $(S_{t-1},\, A_{t-1})$ at time $t-1$, we obtain the vector process $(S_t,\, A_t)$ at time t given only the information of the outcome ω_t. That is, the pair S_{t-1}, A_{t-1} is $\mathcal{F}_{t-1}$-measurable and Y_t is independent of $\mathcal{F}_{t-1}$ so that we may use the appropriate multidimensional version of Proposition 6.7 (with random vector (S_{t-1}, A_{t-1}) and random variable Y_t). Hence, given any (Borel) function $f : \mathbb{R}^2 \to \mathbb{R}$,

$$\mathrm{E}_{t-1}[f(S_t,\, A_t)] = \mathrm{E}_{t-1}\left[f\big(Y_t S_{t-1},\, \frac{1}{t+1}(tA_{t-1} + S_{t-1}Y_t)\big)\right] = g(S_{t-1}, A_{t-1}). \tag{7.51}$$

We note that this same result also follows directly by applying Equation (6.33) of Chapter 6, giving

$$g(x,y) = \mathrm{E}\big[f\big(Y_t x,\, \frac{1}{t+1}(ty+xY_t)\big)\big] = pf\big(ux,\, \frac{1}{t+1}(ty+xu)\big)+(1-p)f\big(dx,\, \frac{1}{t+1}(ty+xd)\big).$$

This proves that the above vector process is Markovian.

For the geometric average we have

$$\begin{aligned} G_t &= ((S_0 \cdots S_{t-1}) \cdot S_t)^{\frac{1}{t+1}} = \left((G_{t-1})^t \cdot S_t\right)^{\frac{1}{t+1}} \\ &= \left((G_{t-1})^t \cdot S_{t-1}Y_t\right)^{\frac{1}{t+1}} = \left((G_{t-1})^t S_{t-1}\right)^{\frac{1}{t+1}} Y_t^{\frac{1}{t+1}}. \end{aligned} \tag{7.52}$$

By itself, the process $\{G_t\}_{t\geqslant 0}$ is not Markovian. However, the vector (G_t, S_t) is determined by (G_{t-1}, S_{t-1}) and Y_t where G_{t-1} and S_{t-1} are $\mathcal{F}_{t-1}$-measurable and Y_t is independent of $\mathcal{F}_{t-1}$. Hence, by the same steps as above, it follows that the vector process $\{(S_t,\, G_t)\}_{t\geqslant 0}$ is Markovian.

In general, the payoff function of an Asian-style option exercised at maturity time $t = T$ is a function of the terminal price and the average price of the underlying asset: $\Lambda(S_T, A_T)$ (or $\Lambda(S_T, G_T)$). There are two main examples of Asian options: a floating price (denoted *AFP*) and a floating strike (denoted *AFS*). Sometimes, these options are also called an average price option and an average strike option, respectively. The payoff functions of Asian calls (denoted C) and puts (denoted P) at maturity time T are as follows:

$$\begin{aligned} \Lambda_C^{AFP}(S_T, A_T) &= (A_T - K)^+, \quad \Lambda_P^{AFP}(S_T, A_T) = (K - A_T)^+, \\ \Lambda_C^{AFS}(S_T, A_T) &= (S_T - A_T)^+, \quad \Lambda_P^{AFS}(S_T, A_T) = (A_T - S_T)^+. \end{aligned} \tag{7.53}$$

Here K is a fixed strike price for the average price options. Note that the payoff of a floating price option is obtained by replacing the terminal asset price S_T with the average value A_T in the payoff function of the respective standard European option. The average strike options do not have a fixed strike price. Their payoffs can also be deduced from the standard European options by replacing the fixed strike with the average value A_T.

7.6.2 Extreme Asset Prices and Lookback Options

Lookback options are another example of path-dependent options that we consider. Their payoffs depend on the maximum or minimum value of the underlying asset price attained during the life of the option. The option allows the holder to "look back" over time to determine the payoff. Recall from Chapter 6 that, in the discrete time setting, the maximum and minimum prices of asset S are, respectively, defined by:

$$M_t = \max_{n=0,\ldots,t} S_n, \quad m_t = \min_{n=0,\ldots,t} S_n, \quad t = 0, 1, \ldots, T. \tag{7.54}$$

As in the case with the average process, the sampled maximum process, $\{M_t\}_{0\leqslant t\leqslant T}$, and sampled minimum process, $\{m_t\}_{0\leqslant t\leqslant T}$, are not Markovian by themselves. To update the extreme value for a single-period transition, we use the following recurrences:

$$M_t = \max\{M_{t-1}, S_t\} = \max\{M_{t-1}, Y_t S_{t-1}\}, \tag{7.55}$$
$$m_t = \min\{m_{t-1}, S_t\} = \min\{m_{t-1}, Y_t S_{t-1}\}, \tag{7.56}$$

for $t = 1, 2, \ldots, T$, where $S_{t-1}, M_{t-1}, m_{t-1}$ are $\mathcal{F}_{t-1}$-measurable random variables and Y_t is independent of $\mathcal{F}_{t-1}$. Hence, the pair of vector processes $\{(S_t,\, M_t)\}_{t\geqslant 0}$ and $\{(S_t,\, m_t)\}_{t\geqslant 0}$ are Markovian.

There exist two kinds of lookback options: with floating strike and with floating price. For the *floating strike lookback* (denoted LFS), the option's strike price is floating and determined at maturity. The payoff of an LFS option is the maximum difference between the market asset's price at maturity and the floating strike. An LFS call gives its holder the right to buy at the lowest price recorded during the option's life. An LFS put gives the right to sell at the highest price recorded during the option's life. The payoffs to the holder are given by

$$\Lambda_C^{LFS} = (S_T - m_T)^+ = S_T - m_T, \quad \Lambda_P^{LFS} = (M_T - S_T)^+ = M_T - S_T, \tag{7.57}$$

respectively, for the LFS call and the LFS put. At maturity, the LFS options are never out of the money since $m_T \leqslant S_T \leqslant M_T$ by definition of the extreme values.

As for the floating price lookback (denoted LFP) options, their payoffs are the maximum differences between the optimal (maximum or minimum) underlying asset price and fixed strike. The payoff functions are given by

$$\Lambda_C^{LFP} = (M_T - K)^+, \quad \Lambda_P^{LFP} = (K - m_T)^+, \tag{7.58}$$

respectively, for the lookback call and the lookback put: In other words, the LFP options are structured so that the call (put) option has payoffs given by the underlying asset price at its highest (lowest) realized level during the lifetime of the option. Note that LFP options have the possibility of expiring worthless.

7.6.3 Recursive Evaluation of Path-Dependent Options

The risk-neutral pricing formula (7.22) allows us to compute the price of any path-dependent derivative. First, we simply need to calculate the path-dependent payoff for each possible path in the nonrecombing binomial tree (recall that there are 2^T possible paths). Second, we compute the sum of payoff values multiplied by respective risk-neutral probabilities. Third, we multiply the result obtained by a discounting factor.

As in the case with non-path-dependent European-style options, the computational cost may be reduced by calculating the derivative prices recursively backward in time for every possible value of the spot price and path-dependent quantity. Consider a lookback option where the payoff $\Lambda(S_T, m_T)$ is a function of only S_T and m_T or only m_T. This includes any of the above mentioned lookback options on the minimum. Since $\{(S_t, m_t)\}_{t\geqslant 0}$ is a Markov process, we have that the option price at any time t is given as a function of only random variables S_t and m_t, i.e., $V_t = V_t(S_t, m_t)$ where $V_t(\omega_1, \ldots, \omega_t) = V_t(S_t(\omega_1, \ldots, \omega_t), m_t(\omega_1, \ldots, \omega_t))$. Then, by using (7.56) within (7.17) we have

$$\begin{aligned} V_t(S_t, m_t) &= \frac{1}{1+r}\widetilde{E}_t\left[V_{t+1}(S_{t+1}, m_{t+1})\right] . \\ &= \frac{1}{1+r}\widetilde{\mathrm{E}}_t\left[V_{t+1}(S_tY_{t+1}, m_t \wedge (S_tY_{t+1}))\right] \\ &= \frac{1}{1+r}\left[\tilde{p}V_{t+1}(S_tu, m_t \wedge (S_tu)) + (1-\tilde{p})V_{t+1}(S_td, m_t \wedge (S_td))\right] . \end{aligned}$$

We use the standard notation $a\wedge b := \min\{a, b\}$ and $a\vee b := \max\{a, b\}$. Hence, the derivative price at each node $(S_t, m_t) = (S, m)$ can be computed by employing the backward recurrence relation

$$V_t(S, m) = \frac{1}{1+r}\left[\tilde{p}V_{t+1}(Su, m \wedge (Su)) + (1-\tilde{p})V_{t+1}(Sd, m \wedge (Sd))\right] , \tag{7.59}$$

for all $t = 1, \ldots, T-1$ and for $t = T$: $V_T(S, m) = \Lambda(S, m)$. Note that if $u > 1$, then we may simply replace the argument $m \wedge (Su)$ by m.

By a very similar analysis, due to the fact that $\{(S_t, M_t)\}_{t\geqslant 0}$ is Markov, we can derive a backward recurrence pricing formula for a lookback option whose payoff $\Lambda(S_T, M_T)$ is a function of only S_T and the maximum M_T (or only M_T). In particular, by using (7.55) within (7.17), the derivative price at each node $(S_t, M_t) = (S, M)$ satisfies

$$V_t(S, M) = \frac{1}{1+r}\left[\tilde{p}V_{t+1}(Su, M \vee (Su)) + (1-\tilde{p})V_{t+1}(Sd, M \vee (Sd))\right] , \tag{7.60}$$

for all $t = 1, \ldots, T-1$ and $V_T(S, M) = \Lambda(S, M)$. If $d < 1$, then the argument $M\vee(Sd) = M$.

In the more general case of a lookback option, the payoff function can depend on both the terminal maximum and the minimum values of the stock, and possibly the terminal stock price, i.e., $\Lambda(S_T, m_T, M_T)$. Then, since the triplet $\{(S_t, m_t, M_t)\}_{t\geqslant 0}$ is a vector Markov process, it follows that the option price process is a function $V_t = V_t(S_t, m_t, M_t)$. That is, the option price at any time $t = 0, 1, \ldots, T$ is given by the (ordinary) function $V_t(S, m, M)$ at each node $S_t = S$, $m_t = m$, and $M_t = M$. The above analysis leads to the backward recurrence formula:

$$\begin{aligned} V_t(S, m, M) = \frac{1}{1+r}\big[&\tilde{p}V_{t+1}(Su, m \wedge (Su), M \vee (Su)) \\ &+ (1-\tilde{p})V_{t+1}(Sd, m \wedge (Sd), M \vee (Sd))\big], \end{aligned} \tag{7.61}$$

for $t = 0, 1, \ldots, T-1$ and $V_T(S, m, M) = \Lambda(S, m, M)$. Again, if $d < 1$, then $M \vee (Sd) = M$ and if $u > 1$, then $m \wedge (Su) = m$.

Example 7.6. Consider the three-period model in Example 7.1 with $S_0 = 8$, $B_0 = 1$, $u = \frac{3}{2}$, $d = \frac{1}{2}$, $r = \frac{1}{4}$, $T = 3$. Find prices for the floating strike lookback call option with payoff as in Example 7.4.

Solution. The nodes (S, m) at each time $t = 0, 1, 2, 3$ are displayed in Figure 7.3. Beginning with time $t = 3$, we have seven distinct pairs of values:

$$(S_3, m_3) = (S, m) \in \{(27, 8),\ (9, 8),\ (9, 6),\ (9, 4),\ (3, 3),\ (3, 2),\ (1, 1)\} .$$

The derivative price at those nodes is simply the payoff value, $V_3(S, m) = \Lambda(S, m) = S - m$:

$$V_3(27, 8) = 19,\ V_3(9, 8) = 1,\ V_3(9, 6) = 3,\ V_3(9, 4) = 5$$
$$V_3(3, 3) = 0,\ V_3(3, 2) = 1,\ V_3(1, 1) = 0\,.$$

The nodes at time $t = 2$ are $(S_2, m_2) = (S, m) \in \{(18, 8),\ (6, 6),\ (6, 4),\ (2, 2)\}$. Equation (7.59) with $\tilde{p} = \frac{3}{4}$ now reads

$$V_t(S, m) = \frac{3}{5} V_{t+1}\left(\frac{3S}{2}, m \wedge \frac{3S}{2}\right) + \frac{1}{5} V_{t+1}\left(\frac{S}{2}, m \wedge \frac{S}{2}\right).$$

Applying this equation for $t = 2$ to all four nodes and using the above payoff values gives

$$V_2(18, 8) = \frac{3}{5} V_3(27, 8) + \frac{1}{5} V_3(9, 8) = 11.60;\ V_2(6, 6) = \frac{3}{5} V_3(9, 6) + \frac{1}{5} V_3(3, 3) = 1.80;$$
$$V_2(6, 4) = \frac{3}{5} V_3(9, 4) + \frac{1}{5} V_3(3, 3) = 3.00;\ V_2(2, 2) = \frac{3}{5} V_3(3, 2) + \frac{1}{5} V_3(1, 1) = 0.60\,.$$

Applying again the recurrence formula for $t = 1$ to the two nodes $(S_1, m_1) = (S, m) \in \{(12, 8),\ (4, 4)\}$:

$$V_1(12, 8) = \frac{3}{5} V_2(18, 8) + \frac{1}{5} V_2(6, 6) = 7.32;\ V_1(4, 4) = \frac{3}{5} V_2(6, 4) + \frac{1}{5} V_2(2, 2) = 1.92\,.$$

Finally, the price at current time $t = 0$ is calculated for $(S_0, m_0) = (S_0, S_0) = (8, 8)$:

$$V_0(8, 8) = \frac{3}{5} V_1(12, 8) + \frac{1}{5} V_1(4, 4) = 4.776.$$

Note that this agrees with the price V_0 in Example 7.4 which was computed by two other methods. ☐

The delta hedging strategy for the above path-dependent European-style options can also be computed based on the fact that option prices are given as functions whose arguments correspond to the nodal values. Consider the above first type of lookback options where $V_t = V_t(S_t, m_t)$. Then, $\Delta_t = \Delta_t(S_t, m_t)$ in (7.19). Hence, the corresponding hedging position in the stock at time t, given $S_t = S, m_t = m$, is given by

$$\Delta_t(S, m) = \frac{V_{t+1}(Su, m \wedge (Su)) - V_{t+1}(Sd, m \wedge (Sd))}{S\,(u - d)}\,. \tag{7.62}$$

Similarly, the delta hedging position for a lookback option with price $V_t = V_t(S_t, M_t)$ at time t, given $S_t = S, M_t = M$, is

$$\Delta_t(S, M) = \frac{V_{t+1}(Su, M \vee (Su)) - V_{t+1}(Sd, M \vee (Sd))}{S\,(u - d)}\,. \tag{7.63}$$

For the more general lookback options where $V_t = V_t(S_t, m_t, M_t)$, the hedging position at time t, given $S_t = S, m_t = m, M_t = M$, is

$$\Delta_t(S, m, M) = \frac{V_{t+1}(Su, m \wedge (Su), M \vee (Su)) - V_{t+1}(Sd, m \wedge (Sd), M \vee (Sd))}{S\,(u - d)}\,. \tag{7.64}$$

7.6.3.1 Pricing Lookback Options on a Two-Dimensional Lattice

Let us construct a complete computational scheme for pricing a lookback derivative whose payoff $\Lambda(S, m)$ depends on the terminal and minimum asset prices. The case when the payoff is a function of the maximum asset price is considered similarly. For simplicity, assume that $ud = 1$ (i.e., we deal with a symmetric stock lattice) to reduce the range of possible values of S_t and m_t. At time $t \in \{0, 1, \ldots, T-1\}$, the stock price can take one of $t+1$ values:

$$S_t \in \{S_{t,n} = S_0 u^{2n-t} \;:\; n = 0, 1, \ldots, t\}.$$

The minimum price process $\{m_t\}_{t \geqslant 0}$ is nonincreasing. Thus, the value of m_t does not exceed S_0 at any time t. Let us find the range of values of m_t. At time $t = 0$, there is only one value of m_0, namely, $m_0 = S_0$. At time $t = 1$, there are two possible values: $m_1 \in \{S_0, S_0 u^{-1}\}$. At time $t = 2$, there are three possible values: $m_2 \in \{S_0, S_0 u^{-1}, S_0 u^{-2}\}$. By induction, we can show that m_t has $t+1$ possible values:

$$m_t \in \{S_0 u^{k-t} \;:\; k = 0, 1, \ldots, t\}.$$

Thus, at time t, the random vector $[S_t, m_t]$ may have at most $(t+1)^2$ values:

$$[S_t, m_t] \in \left\{[S_0 u^{2n-t}, S_0 u^{k-t}] \;:\; n = 0, 1, \ldots, t,\; k = 0, 1, \ldots, t\right\}. \tag{7.65}$$

The time-t lookback derivative value is a function of S_t and m_t, i.e., $V_t = V_t(S_t, m_t)$. Let $V_{t,n,k}$ denote the time-t derivative value at the node $[S_t, m_t] = [S_0 u^{2n-t}, S_0 u^{k-t}]$. Using the backward recurrence formula (7.61), we construct a scheme for computing the lookback derivative prices on a two-dimensional lattice of values of S_t and m_t. Since not all combinations in (7.65) are possible, we first find the range for the minimum price m_t conditional on $S_t = S_{t,n} = S_0 u^{2n-t}$. The node $(t, S_{t,n})$ of a binomial tree is attained by a path $\omega_1 \omega_2 \ldots \omega_t$ with n upward moves and $t-n$ downward moves. Considering all such paths, we find that the lowest value of m_t is attained on the path with $\omega_1 = \cdots = \omega_{t-n} = \mathsf{D}$ and $\omega_{t-n+1} = \cdots = \omega_t = \mathsf{U}$. It is equal to $S_0 u^{n-t}$. The largest value of m_t attained on a path with n upward moves is equal to $\min\{S_0, S_0 u^{2n-t}\} = S_0 u^{\min\{t,2n\}-t}$. It is achieved on the path with $\omega_1 = \cdots = \omega_n = \mathsf{U}$ and $\omega_{n+1} = \cdots = \omega_t = \mathsf{D}$. Thus, there are $\min\{t-n, n\}+1$ possible values of m_t given that $S_t = S_{t,n}$:

$$m_t \in \left\{S_0 u^{k-t} \;:\; n \leqslant k \leqslant \min\{t, 2n\}\right\}.$$

We now derive a backward-in-time recursion scheme for evaluation of the lookback derivative prices for every attainable value of $[S_t, m_t]$. At maturity, the derivative values are equal to the respective values of the payoff function. For $t < T$, using the one-period risk-neutral pricing formula (7.59) with $S_t = S_0 u^{2n-t}$ and $m_t = S_0 u^{k-t}$ (where $0 \leqslant n \leqslant t$ and $n \leqslant k \leqslant \min\{t, 2n\}$) gives

$$\begin{aligned}
V_t(S_0 u^{2n-t}, S_0 u^{k-t}) &= \frac{1}{1+r}\left[\tilde{p}\, V_{t+1}(S_0 u^{2n-t+1}, S_0 u^{k-t} \wedge S_0 u^{2n-t+1})\right. \\
&\qquad \left. + (1-\tilde{p})\, V_{t+1}(S_0 u^{2n-t-1}, S_0 u^{k-t} \wedge S_0 u^{2n-t-1})\right] \\
&= \frac{1}{1+r}\left[\tilde{p}\, V_{t+1}(S_0 u^{2(n+1)-(t+1)}, S_0 u^{(k+1)-(t+1)})\right. \\
&\qquad \left. + (1-\tilde{p})\, V_{t+1}(S_0 u^{2n-(t+1)}, S_0 u^{\min\{k+1,2n\}-(t+1)})\right].
\end{aligned}$$

Therefore, we obtain the following recursion scheme:

$$V_{T,n,k} = \Lambda\left(S_0 u^{2n-T}, S_0 u^{k-T}\right), \quad 0 \leqslant n \leqslant T,\ n \leqslant k \leqslant \min\{T, 2n\}, \tag{7.66}$$

$$V_{t,n,k} = \frac{1}{1+r}\left[\tilde{p}\, V_{t+1,n+1,k+1} + (1-\tilde{p})\, V_{t+1,n,\min\{k+1,2n\}}\right], \quad 0 \leqslant t \leqslant T, \tag{7.67}$$
$$0 \leqslant n \leqslant t,\ n \leqslant k \leqslant \min\{t, 2n\}.$$

To compare the scheme (7.66)–(7.67) with the general approach (7.17)–(7.18), we find and compare the total number of derivative values to be calculated by using each method. The general method is implemented on a nonrecombining binomial tree with $1+2+\cdots+2^T = \mathcal{O}(2^{T+1})$ nodes. The scheme (7.66)–(7.67) needs to compute derivative prices for $n_t := \sum_{n=0}^{t}(\min\{n, t-n\}+1)$ distinct pairs of values of $[S_t, m_t]$ for each $t = 0, 1, \ldots, T$. If t is even, then $n_t = \frac{(t+2)^2}{4}$. If t is odd, then $n_t = \frac{(t+1)(t+3)}{4}$. Therefore, there are $\mathcal{O}(\frac{1}{12}T^3)$ values to be calculated. For large values of T, the scheme (7.66)–(7.67) with a polynomial computational cost is much more efficient than the general approach whose cost is an exponential function of T. Note that pricing of a non-path-dependent option on a recombining binomial tree with T periods requires $\mathcal{O}(\frac{1}{2}T^2)$ arithmetic operations.

7.7 American Options

Recall that an American call (put) option gives the right to buy (to sell) the underlying asset for the strike price agreed in advance at any time from the time the option is written (which is time $t = 0$) to the expiry time $t = T$. The holder of an American option may exercise the option at any time up to and including the expiry date. In the discrete time setting, the option can only be exercised at times $0, 1, \ldots, T$.

7.7.1 Writer's Perspective: Pricing and Hedging

Let us review the pricing of an American derivative in a recombining binomial tree model. Let $\Lambda(S_t)$ be a non-path-dependent European-style payoff to the holder of an American derivative at time $t \in \{0, 1, \ldots, T\}$. For example, the call and put payoffs are $\Lambda(S_t) = (S_t - K)^+$ and $\Lambda(S_t) = (K - S_t)^+$, respectively, if the option is exercised at time $0 \leqslant t \leqslant T$. Let $V_t(S_t)$ denote the time-t value of the non-path-dependent American derivative for spot S_t that has not been exercised yet. At expiry time T, the value of the derivative is

$$V_T(S_T) = \max\{\Lambda(S_T), 0\}. \tag{7.68}$$

Given the American derivative has not been exercised before time $t \in \{0, \ldots, T-2, T-1\}$, the holder has the choice to exercise the derivative immediately at time t with payoff $\Lambda(S_t)$ or wait until the next time moment $t+1$ when the derivative will be worth $V_{t+1}(S_{t+1})$. The value $\Lambda(S_t)$ is called the *intrinsic value* (at time t). The time-t value of the latter alternative, called the *continuation value*, is given by a one-step derivative pricing formula:

$$\frac{1}{1+r}\widetilde{\mathrm{E}}_t\left[V_{t+1}(S_{t+1})\right] = \frac{1}{1+r}\left(\tilde{p}V_{t+1}(S_t u) + (1-\tilde{p})V_{t+1}(S_t d)\right)$$

where $\tilde{p} = \frac{1+r-d}{u-d}$. Since the holder of the option may choose either alternative, the time-t value of the derivative is the maximum of the intrinsic value and continuation value:

$$V_t(S_t) = \max\left\{\Lambda(S_t), \frac{1}{1+r}\left(\tilde{p}V_{t+1}(S_t u) + (1-\tilde{p})V_{t+1}(S_t d)\right)\right\}, \tag{7.69}$$

for $t = 0, 1, 2, \ldots, T-1$. Equations (7.68)–(7.69) allow us to evaluate American derivative prices on a recombining binomial tree starting from the expiry date and then proceeding backward in time. As we have seen in the European case, option prices can be used to construct a delta hedging (self-financing) portfolio process that perfectly replicates the European option prices. Let us study whether an American option can be replicated by a self-financing strategy.

Example 7.7. Consider the three-period recombining binomial tree model in Example 7.1 with $S_0 = 8$, $u = \frac{3}{2}$, $d = \frac{1}{2}$, and $r = \frac{1}{4}$. Find and compare the prices of the standard European and American put options with common strike $K = 8$ and expiration $T = 3$.

Solution. The risk-neutral probabilities are $\tilde{p} = 0.75$ and $1-\tilde{p} = 0.25$. For every stock price node, we apply recurrence formulae (7.34) and (7.69) to evaluate the respective prices of the European and American put options. Let P_t^E and P_t^A, respectively, denote the time-t prices of the European and American put options for $t = 0, 1, 2, 3$. At the expiration date, the option value must equal the payoff function:

$$P_3^E(S_3) = P_3^A(S_3) = (8 - S_3)^+, \quad S_3 \in \{1, 3, 9, 27\}.$$

To determine the price for the times before expiration, use the following recurrences:

$$\begin{aligned}
P_t^E(S_t) &= \frac{1}{1.25} \cdot \left(\frac{3}{4} \cdot P_{t+1}^E\left(S_t \cdot {}^3\!/_2\right) + \frac{1}{4} \cdot P_{t+1}^E\left(S_t \cdot {}^1\!/_2\right) \right) \\
&= \frac{3 \cdot P_{t+1}^E(1.5 \cdot S_t) + P_{t+1}^E(0.5 \cdot S_t)}{5}, \\
P_t^A(S_t) &= \max\left\{ (8 - S_t)^+, \frac{3 \cdot P_{t+1}^A(1.5 \cdot S_t) + P_{t+1}^A(0.5 \cdot S_t)}{5} \right\}, \\
&\text{where } S_t = 8 \cdot 2^n \cdot 0.5^{t-n}, \quad n = 0, 1, \ldots, t, \quad t = 2, 1, 0.
\end{aligned}$$

As a result, we obtain the option prices given in Figure 7.7. The values of the American put are boxed at those nodes of the tree where the intrinsic value is greater than or equal to the continuation value:

$$\begin{aligned}
P_2^A(6) &= \max\left\{ 8 - 6, \frac{3 \cdot 0 + 5}{5} \right\} = 2 \vee 1 = 2, \\
P_2^A(2) &= \max\left\{ 8 - 2, \frac{3 \cdot 5 + 7}{5} \right\} = 6 \vee 4.4 = 6, \\
P_1^A(4) &= \max\left\{ 8 - 4, \frac{3 \cdot 2 + 6}{5} \right\} = 4 \vee 2.4 = 4.
\end{aligned}$$

So the holder of the American put should exercise the option early when the continuation value is strictly less than the intrinsic value, which happens when $S_1 = 4$, or $S_2 = 6$, or $S_2 = 2$. The price of the American option is strictly larger than that of the European put for those cases. Therefore, the initial price of the American put is strictly larger than that of the European put. □

Example 7.8. Construct a hedging strategy for the American put in Example 7.7.

Solution. $t = 0$**:** The initial value, Π_0, of the hedging strategy is the same as that of the option price: $\Pi_0 = P_0^A(8) = 1.04$. We calculate the number of shares of stock, Δ_0,

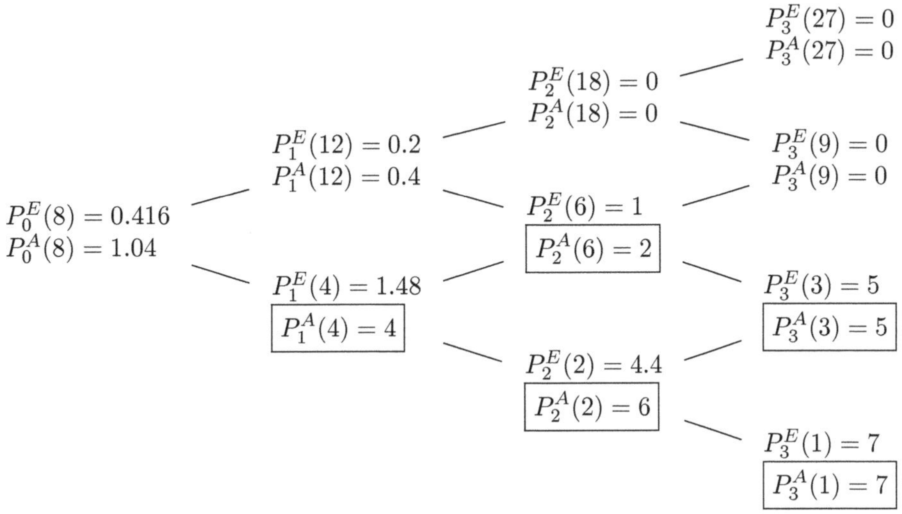

$P_3^E(27) = 0$
$P_3^A(27) = 0$

$P_2^E(18) = 0$
$P_2^A(18) = 0$

$P_1^E(12) = 0.2$
$P_1^A(12) = 0.4$

$P_3^E(9) = 0$
$P_3^A(9) = 0$

$P_0^E(8) = 0.416$
$P_0^A(8) = 1.04$

$P_2^E(6) = 1$
$\boxed{P_2^A(6) = 2}$

$P_1^E(4) = 1.48$
$\boxed{P_1^A(4) = 4}$

$P_3^E(3) = 5$
$\boxed{P_3^A(3) = 5}$

$P_2^E(2) = 4.4$
$\boxed{P_2^A(2) = 6}$

$P_3^E(1) = 7$
$\boxed{P_3^A(1) = 7}$

FIGURE 7.7: Prices of the standard European put (P^E) and American put (P^A) options.

and the position in the risk-free account, $\beta_0 B_0$, so that the hedging portfolio replicates the option prices at time 1. From (7.19), obtain the hedging position in the stock:

$$\Delta_0 = \frac{P_1^A(12) - P_1^A(4)}{12 - 4} = \frac{0.4 - 4}{8} = -0.45.$$

From the self-financing condition, the amount in the bank account is given by $\beta_0 B_0 = \Pi_0 - \Delta_0 S_0 = 1.04 + 0.45 \cdot 8 = 4.64$.

$t = 1$: At time $t = 1$, the stock price is either at $S_1 = S_1(\mathsf{U}) = 12$ ($\omega_1 = \mathsf{U}$) or at $S_1 = S_1(\mathsf{D}) = 4$ ($\omega_1 = \mathsf{D}$). Consider the case with $\omega_1 = \mathsf{U}$. There are two possible values of $P_2^A(S_2(\mathsf{U}\omega_2))$. The portfolio that hedges against these two possibilities has

$$\Delta_1(\mathsf{U}) = \frac{P_2^A(S_2(\mathsf{UU})) - P_2^A(S_2(\mathsf{UD}))}{S_2(\mathsf{UU}) - S_2(\mathsf{UD})} = \frac{P_2^A(18) - P_2^A(6)}{18 - 6} = \frac{0 - 2}{12} = -\frac{1}{6}$$

shares of stock. In case $\omega_1 = \mathsf{D}$, the value of the hedging portfolio is

$$\Pi_1(\mathsf{D}) = (1 + r)\beta_0 B_0 + \Delta_0 S_1(\mathsf{D}) = 4.64 \cdot 1.25 - 0.45 \cdot 4 = 4.$$

The holder may exercise the option at time 1. If this is the case, then as writer we liquidate the hedging portfolio and deliver the premium $(K - S_1(\mathsf{D}))^+ = (8 - 4)^+ = 4$ to the holder. If the option is not exercised (i.e., is kept alive) then we continue hedging. At time $t = 2$, the option may be worth $P_2^A(6) = 2$ if $\omega_2 = \mathsf{U}$ or $P_2^A(2) = 6$ if $\omega_2 = \mathsf{D}$. To hedge against these two possibilities, we need a portfolio whose value is given by the continuation value a time $t = 1$:

$$\frac{4}{5}\left(0.75 \cdot P_2^A(6) + 0.25 \cdot P_2^A(2)\right) = 2.4\,.$$

So we can consume the surplus \$4 − \$2.4 = \$1.6 and continue hedging with the remaining \$2.4. This means that, under scenario $\omega_1 = \mathsf{D}$, the holder of the American put missed out on what would have been an optimal exercise opportunity at time 1. The writer's hedging position in this case is

$$\Delta_1(\mathsf{D}) = \frac{P_2^A(6) - P_2^A(2)}{6-2} = \frac{2-6}{4} = -1.$$

$t = 2$: There are three possibilities to consider. First, assume that $S_2 = 18$. The option is out of the money and the payoff value is zero regardless of the value of ω_3. The value of the hedging portfolio is zero. The position in stock changes to $\Delta_2(18) = 0$. Second, consider the case that $S_2 = 6$. Again, if the holder exercises the option, then we liquidate the hedging portfolio, which is worth \$2, and use the proceeds to pay out the premium. Otherwise, we consume \$2 − \$1 = \$1 and use the remaining \$1 to hedge against two possibilities: $P_3^A(9) = 0$ and $P_3^A(3) = 5$. We set

$$\Delta_2(S_2 = 6) = \frac{P_3^A(9) - P_3^A(3)}{9-3} = \frac{0-5}{6} = -\frac{5}{6}.$$

The last case is when $S_2 = 2$. The value of the hedging portfolio is \$6, which is sufficient to cover our liabilities if the holder decides to exercise. If the holder declines to exercise the option, then we consume \$6 − \$4.4 = \$1.6 and use the remaining \$4.4 to construct a hedging portfolio with $\Delta_2(S_2 = 2) = \frac{5-7}{3-1} = -1$ shares of stock. □

As we can see, an American derivative can be replicated by means of a delta hedging portfolio strategy and a consumption process. If the holder of the derivative does not exercise at the optimal time, then all excess value due to delayed optimal exercise is consumed by the writer. The above example is a special case of a general algorithm for no-arbitrage pricing and hedging of a (path-independent) American derivative for a binomial model with parameters $0 < d < 1+r < u$. That is, assuming a given payoff function $\Lambda(S)$, let the prices $V_t(S), t = T, T-1, \ldots, 1, 0$ at every node $(S_t, t) = (s, t)$ satisfy the recurrence relations in (7.68) and (7.69). Moreover, let the hedging position $\Delta_t = \Delta_t(S_t)$ in the stock be given by

$$\Delta_t = \frac{V_{t+1}(S_t u) - V_{t+1}(S_t d)}{S_t(u-d)}, \tag{7.70}$$

and the corresponding consumption be given by the difference of the derivative value and the continuation value,

$$C_t = V_t(S_t) - \frac{1}{1+r}\left(\tilde{p}V_{t+1}(S_t u) + (1-\tilde{p})V_{t+1}(S_t d)\right), \tag{7.71}$$

for $t = 0, 1, 2, \ldots, T-1$. If we begin with a portfolio having initial capital $\Pi_0 = V_0(S_0)$ and having time-t value Π_t given recursively forward in time by the (modified wealth equation due to consumption),

$$\Pi_{t+1} = \Delta_t S_{t+1} + (1+r)(\Pi_t - C_t - \Delta_t S_t), \tag{7.72}$$

then the portfolio process will replicate the American derivative value for any scenario, i.e.,

$$\Pi_t(\omega) = V_t(S_t(\omega))$$

for every $\omega \in \Omega_T$ and $t = 0, \ldots, T$. It should be noted that Equation (7.69) guarantees that the consumption is nonnegative, i.e., $C_t \geqslant 0$ for every $t \geqslant 0$. Moreover, the derivative prices are always as valuable as the intrinsic (early payoff) value, $V_t^A(S_t) \geqslant \Lambda(S_t)$, and by replication this also implies $\Pi_t \geqslant \Lambda(S_t)$, for all $t \geqslant 0$.

The above assertion of replication can be proven using similar steps as in the proof of Theorem 7.6. By assumption, the claim is obviously true for $t = 0$. We now give a compact version of a proof of replication that may be instructive. By induction, we assume $\Pi_t = V_t^A(S_t)$ and show $\Pi_{t+1} = V_{t+1}^A(S_{t+1})$ for $t = 0, 1, \ldots, T-1$, as follows. Equation (7.71) gives

$$(1+r)(\Pi_t - C_t) = (1+r)(V_t(S_t) - C_t) = \tilde{p}V_{t+1}(S_t u) + (1-\tilde{p})V_{t+1}(S_t d)\,.$$

Then, using this expression and (7.70) within (7.72) gives

$$\begin{aligned}
\Pi_{t+1} &= \Delta_t(S_{t+1} - (1+r)S_t) + (1+r)(\Pi_t - C_t)\\
&= (V_{t+1}(S_t u) - V_{t+1}(S_t d)) \cdot \frac{S_{t+1} - (1+r)S_t}{S_t(u-d)} + (1+r)(V_t(S_t) - C_t)\\
&= (V_{t+1}(S_t u) - V_{t+1}(S_t d)) \cdot \left((1-\tilde{p})\mathbb{I}_{\{\omega_{t+1}=\mathsf{U}\}} - \tilde{p}\,\mathbb{I}_{\{\omega_{t+1}=\mathsf{D}\}} \right)\\
&\quad + \tilde{p}V_{t+1}(S_t u) + (1-\tilde{p})V_{t+1}(S_t d)\\
&= V_{t+1}(S_t u)\,\mathbb{I}_{\{\omega_{t+1}=\mathsf{U}\}} + V_{t+1}(S_t d)\,\mathbb{I}_{\{\omega_{t+1}=\mathsf{D}\}}\\
&= V_{t+1}(S_{t+1})\,.
\end{aligned}$$

Here we made use of the random variable representation for the stock price at time $t+1$ in terms the stock price at time t: $S_{t+1} = S_t u\,\mathbb{I}_{\{\omega_{t+1}=\mathsf{U}\}} + S_t d\,\mathbb{I}_{\{\omega_{t+1}=\mathsf{D}\}}$.

Although we do not consider any applications of *path-dependent American derivatives*, we point out that the above equations for the pricing algorithm and replication strategy (by delta hedging and consumption) are the same in structure within the standard binomial model. The payoff $\Lambda(S_T)$ is replaced by Λ_T, which is any $\mathcal{F}_T$-measurable random variable (not necessarily given as a function of S_T) that evaluates to a number $\Lambda_T(\omega)$ for every $\omega \in \Omega_T$. There is a sequence of F_t-measurable random variables, $\{\Lambda_t\}_{t=0,1,\ldots,T}$, representing the intrinsic values at times $t = 0, 1, \ldots, T$. The derivative prices V_t, at time t, are $\mathcal{F}_t$-measurable random variables that evaluate to a number $V_t(\omega) = V_t(\omega_1 \ldots \omega_t)$ for every sequence $\omega_1 \ldots \omega_t$, $t = 0, 1, \ldots, T$. The backward recurrence relation in Equation (7.69) is then a special case of the following relation:

$$V_t(\omega_1 \ldots \omega_t) = \max\left\{\Lambda_t(\omega_1 \ldots \omega_t), \frac{1}{1+r}\left[\tilde{p}V_{t+1}(\omega_1 \ldots \omega_t\mathsf{U}) + (1-\tilde{p})V_{t+1}(\omega_1 \ldots \omega_t\mathsf{D})\right]\right\}, \tag{7.73}$$

for $t = 0, 1, \ldots, T-1$ and $V_T(\omega_1 \ldots \omega_T) = \max\{\Lambda_T(\omega_1 \ldots \omega_T), 0\}$. The above delta hedging and consumption relations, (7.70) and (7.71), are replaced in the obvious manner by

$$\Delta_t(\omega_1 \ldots \omega_t) = \frac{V_{t+1}(\omega_1 \ldots \omega_t\mathsf{U}) - V_{t+1}(\omega_1 \ldots \omega_t\mathsf{D})}{S_{t+1}(\omega_1 \ldots \omega_t\mathsf{U}) - S_{t+1}(\omega_1 \ldots \omega_t\mathsf{D})}, \tag{7.74}$$

and

$$C_t(\omega_1 \ldots \omega_t) = V_t(\omega_1 \ldots \omega_t) - \frac{1}{1+r}\left[\tilde{p}V_{t+1}(\omega_1 \ldots \omega_t\mathsf{U}) + (1-\tilde{p})V_{t+1}(\omega_1 \ldots \omega_t\mathsf{D})\right]. \tag{7.75}$$

Finally, the wealth equation is defined exactly as in (7.72), which leads to replication, i.e., $\Pi_t(\omega_1 \ldots \omega_t) = V_t(\omega_1 \ldots \omega_t)$ for all scenarios $\omega_1 \ldots \omega_t$ and times $t = 0, 1, \ldots, T$.

7.7.2 Buyer's Perspective: Optimal Exercise

Let us assume we are dealing with a non-path-dependent American derivative. The analysis and equations extend in the obvious manner, as pointed out in the previous section.

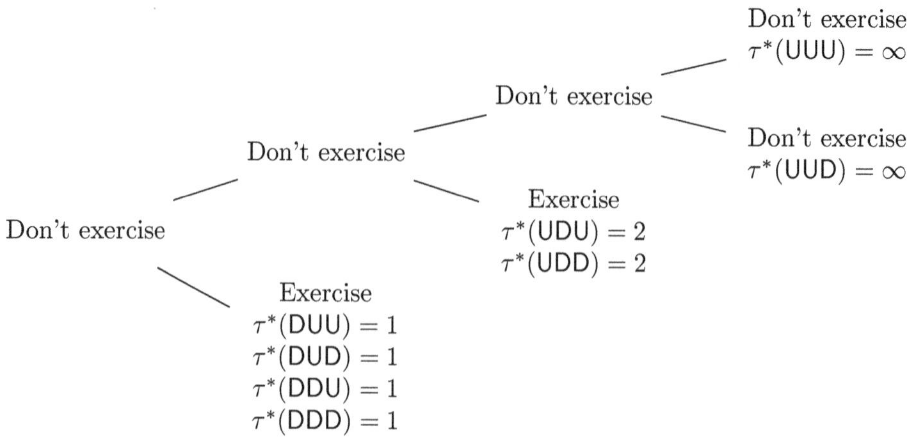

FIGURE 7.8: Exercise rule τ^* of Example 7.7.

An American derivative (with assumed payoff function $\Lambda(S)$) can be exercised by its holder at any time $t \in \{0, 1, 2 \ldots, T\}$ before the expiry date T. To decide when to exercise the derivative, the holder needs to follow a certain strategy, which is simply a rule that tells if the derivative should be exercised at a particular time t based on the information revealed up to that point. An exercise strategy can be described by a function that maps the set of scenarios Ω to the set of dates. Given the state of the world $\omega \in \Omega$, the holder exercises at time $t = \tau(\omega)$ and receives the payoff $\Lambda\left(S_{\tau(\omega)}(\omega)\right)$. In a discrete-time model, we have $\tau\colon \Omega \to \{0, 1, \ldots, T, \infty\}$. If $\tau = \infty$, then the derivative should not be exercised (for example, the option is out of the money). Since the decision to exercise at time t (i.e., $\tau(\omega) = t$) only depends on the information revealed up to that time, the function τ is adapted to the filtration $\mathbb{F} = \{\mathcal{F}_t\}_{0 \leqslant t \leqslant T}$. That is, for all $0 \leqslant t \leqslant T$ we have $\{\omega \in \Omega \ : \ \tau(\omega) \leqslant t\} \in \mathcal{F}_t$. In other words, τ is a stopping time. Let $\mathcal{S}_{t,T}$ be the set of all stopping times τ such that $\tau\colon \Omega \to \{t, t+1, \ldots, T, \infty\}$, where $t = 0, 1, \ldots, T$. In particular, $\mathcal{S}_{0,T}$ contains every stopping time in the T-period model.

Let us come back to Example 7.7. As follows from the solution, the holder should exercise the option as soon as the intrinsic value exceeds the continuation value. In particular, if the stock price goes down at time $t = 1$ (i.e., $\omega_1 = \mathsf{D}$), then the option should be exercised at time $t = 1$. If the stock price goes up at time $t = 1$ (i.e., $\omega_1 = \mathsf{U}$), then there is no advantage to exercising at time $t = 1$ and the holder should wait until time $t = 2$. If the stock price goes down at time $t = 2$ (i.e., the scenario observed so far is $\omega_1\omega_2 = \mathsf{UD}$), then it is beneficial to exercise the option. If the stock price goes up, then the option is out of the money and there is no advantage to an early exercise of this American put. The exercising rule obtained is summarized in Figure 7.8, i.e., $\tau^*(A_{\mathsf{D}}) = 1, \tau^*(A_{\mathsf{UD}}) = 2, \tau^*(A_{\mathsf{UU}}) = \infty$ where $A_{\mathsf{D}} = \{\mathsf{DUU}, \mathsf{DUD}, \mathsf{DDU}, \mathsf{DDD}\}, A_{\mathsf{UD}} = \{\mathsf{UDU}, \mathsf{UDD}\}, A_{\mathsf{UU}} = \{\mathsf{UUU}, \mathsf{UUD}\}$.

Recall from Chapter 6 that, given a stochastic process $\{X_t\}_{t \geqslant 0}$ and stopping time τ, we can define a stopped process $Y_t(\omega) := X_{t \wedge \tau}$, i.e.,

$$Y_t(\omega) := X_{t \wedge \tau(\omega)}(\omega) \equiv X_{\min\{t, \tau(\omega)\}}(\omega), \quad t \geqslant 0,\ \omega \in \Omega. \tag{7.76}$$

For example, consider the three-period binomial model with choice $\tau = \tau^*$. Then, for $t = 0$

we have $Y_0 = X_0$. For $t \geqslant 1$ we have the following:

$$\begin{aligned}
\omega \in A_{\mathsf{D}} &: \; Y_t(\omega) = X_{t\wedge 1}(\omega) = X_1(\omega)\,,\\
\omega \in A_{\mathsf{UD}} &: Y_t(\omega) = X_{t\wedge 2}(\omega) = X_1(\omega)\mathbb{I}_{\{t=1\}} + X_2(\omega)\mathbb{I}_{\{t\geqslant 2\}}\,,\\
\omega \in A_{\mathsf{UU}} &: Y_t(\omega) = X_{t\wedge\infty}(\omega) = X_t(\omega)\,.
\end{aligned}$$

Example 7.9. Consider the American price process $\{P_t^A\}_{0\leqslant t\leqslant 3}$ constructed in Example 7.7. Show that (a) the discounted process $\left\{\overline{P}_t^A := \frac{P_t^A}{(1+r)^t}\right\}_{0\leqslant t\leqslant 3}$ is a supermartingale under the risk-neutral measure $\widetilde{\mathbb{P}}$; (b) the discounted stopped process $\left\{\overline{P}_{t\wedge\tau^*}^A\right\}_{0\leqslant t\leqslant 3}$, where the stopping time τ^* is the optimal exercising strategy presented in Figure 7.8, is a $\widetilde{\mathbb{P}}$-martingale .

Solution. First, let us construct the discounted derivative price process $\left\{\overline{P}_t^A\right\}_{0\leqslant t\leqslant 3}$. Each American put time-t value in Figure 7.7 is divided by $(1+r)^t = 1.25^t$. The result is presented in Figure 7.9.

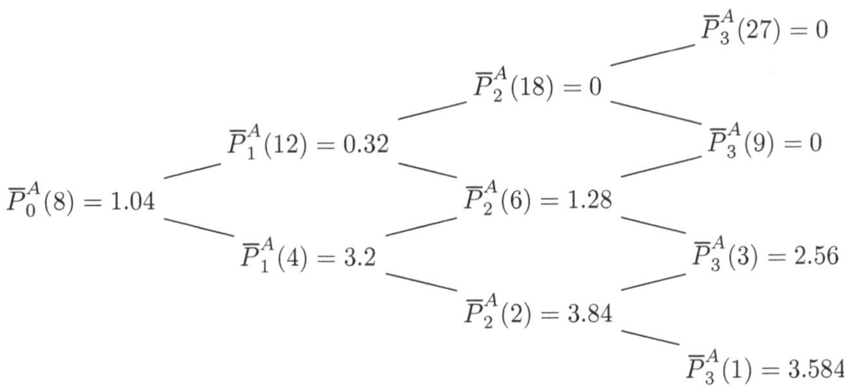

FIGURE 7.9: Values of the discounted price process $\{\overline{P}_t^A\}$ for the standard American put.

To verify whether the American put price process is a $\widetilde{\mathbb{P}}$-supermartingale, we only need to check if the inequality $\overline{P}_t^A(S_t) \geqslant \widetilde{\mathrm{E}}[\overline{P}_{t+1}^A(S_{t+1}) \mid S_t] \equiv 0.75 \cdot \overline{P}_{t+1}^A(\frac{3}{2}S_t) + 0.25 \cdot \overline{P}_{t+1}^A(\frac{1}{2}S_t)$ holds for each $t = 0, 1, 2$ and every possible price S_t. If we have a strict inequality in at least one case, then the process is a supermartingale but not a martingale. Checking this condition for each time step we have:

$$\begin{aligned}
t=0:&\quad \overline{P}_0^A(8) = 1.04 \overset{?}{\geqslant} \widetilde{\mathrm{E}}[\overline{P}_1^A] = 0.32 \cdot 0.75 + 3.2 \cdot 0.25 = 1.04 && \checkmark\\
t=1:&\quad \overline{P}_1^A(4) = 3.2 \overset{?}{\geqslant} \widetilde{\mathrm{E}}[\overline{P}_2^A \mid S_1 = 4] = 1.28 \cdot 0.75 + 3.84 \cdot 0.25 = 1.92 && \checkmark\\
&\quad \overline{P}_1^A(12) = 0.32 \overset{?}{\geqslant} \widetilde{\mathrm{E}}[\overline{P}_2^A \mid S_1 = 12] = 0 \cdot 0.75 + 1.28 \cdot 0.25 = 0.32 && \checkmark\\
t=2:&\quad \overline{P}_2^A(2) = 3.84 \overset{?}{\geqslant} \widetilde{\mathrm{E}}[\overline{P}_3^A \mid S_2 = 2] = 2.56 \cdot 0.75 + 3.584 \cdot 0.25 = 2.816 && \checkmark\\
&\quad \overline{P}_2^A(6) = 1.28 \overset{?}{\geqslant} \widetilde{\mathrm{E}}[\overline{P}_3^A \mid S_2 = 6] = 0 \cdot 0.75 + 2.56 \cdot 0.25 = 0.64 && \checkmark\\
&\quad \overline{P}_2^A(18) = 0 \overset{?}{\geqslant} \widetilde{\mathrm{E}}[\overline{P}_3^A \mid S_2 = 18] = 0 \cdot 0.75 + 0 \cdot 0.25 = 0 && \checkmark
\end{aligned}$$

Since every conditional expectation satisfies the above inequality, the discounted American put price process is indeed a supermartingale. Finally, consider the stopped process defined by $\overline{P}^*_t := \overline{P}^A_{t\wedge\tau^*}$ where the optimal stopping time τ^* is given in Figure 7.8. The dynamics of the stopped process can be represented by a scenario tree of all market moves, as in Figure 7.10. It is not difficult to verify that this stopped process is a $\widetilde{\mathbb{P}}$-martingale. Indeed, for every choice of market moves $\omega_1, \omega_2 \in \{\mathsf{D}, \mathsf{U}\}$ we have that

$$\begin{aligned}\overline{P}^*_0 &= \widetilde{\mathrm{E}}_0[\overline{P}^*_1] = \tilde{p}\overline{P}^*_1(\mathsf{U}) + (1-\tilde{p})\overline{P}^*_1(\mathsf{D}) = 0.75\overline{P}^*_1(\mathsf{U}) + 0.25\overline{P}^*_1(\mathsf{D}),\\ \overline{P}^*_1(\omega_1) &= \widetilde{\mathrm{E}}_1[\overline{P}^*_2](\omega_1) = \tilde{p}\overline{P}^*_2(\omega_1\mathsf{U}) + (1-\tilde{p})\overline{P}^*_2(\omega_1\mathsf{D}),\\ \overline{P}^*_2(\omega_1\omega_2) &= \widetilde{\mathrm{E}}_2[\overline{P}^*_3](\omega_1\omega_2) = \tilde{p}\overline{P}^*_3(\omega_1\omega_2\mathsf{U}) + (1-\tilde{p})\overline{P}^*_3(\omega_1\omega_2\mathsf{D}).\end{aligned}$$ □

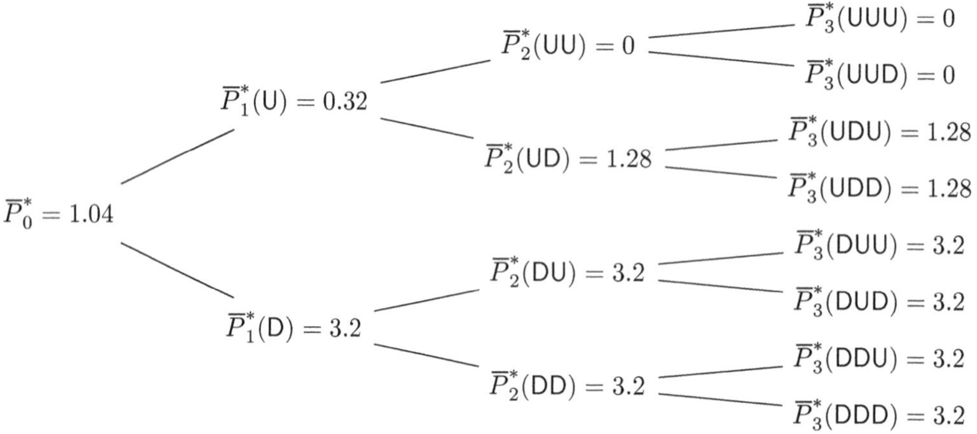

FIGURE 7.10: A full scenario tree for the three-period binomial model with values of the stopped derivative price process $\{\overline{P}^*_t := \overline{P}^A_{t\wedge\tau^*}\}$ for the standard American put.

As is seen from the above example, not exercising an American option at the optimal time is equivalent to the option being an unfavorable game for its holder. An interesting fact is that stopping at the optimal time may turn a supermartingale into a martingale. In general, a stopped (super-)martingale is a (super-)martingale, as was proved in the Optional Sampling Theorem from Chapter 6.

Now let us come back to the problem of pricing of an American derivative at time $t = 0$ based on an optimal exercise strategy. Suppose that the buyer follows a certain exercise strategy $\tau \in \mathcal{S}_{0,T}$, which is not necessarily an optimal one. If $\tau(\omega) \leqslant T$ for scenario $\omega \in \Omega$, then the holder exercises the derivative at time $\tau(\omega)$ to realize the payoff $\Lambda_{\tau(\omega)}(\omega)$ or $\Lambda(S_{\tau(\omega)}(\omega))$ for the non-path-dependent case. If $\tau(\omega) = \infty$, then the derivative is not exercised at all and the terminal payoff is zero. Therefore, the holder's payoff at time $t = T\wedge\tau$ is a product of the payoff function and the indicator function of the event $\{\tau \leqslant T\}$. Generally, for any $\mathcal{F}_t$-measurable intrinsic value Λ_t the payoff at time τ is $\mathbb{I}_{\{\tau\leqslant T\}}\Lambda_\tau$. If the payoff is only dependent on the stock price (i.e., non-path-dependent American options where $\Lambda_t = \Lambda(S_t)$), then the time-τ payoff will be given by $\mathbb{I}_{\{\tau\leqslant T\}}\Lambda(S_\tau)$. To find the time-0 risk-neutral value, denoted by $v_0(\tau)$, of this payoff under the exercise rule τ, we can apply the risk-neutral pricing formula for the random cash flow $\{\mathbb{I}_{\{\tau=t\}}\Lambda_t\}_{0\leqslant t\leqslant T}$ (see

Section 7.3.2):

$$v_0(\tau) = \widetilde{\mathrm{E}}_0\left[\sum_{t=0}^{T} \mathbb{I}_{\{\tau=t\}} \frac{\Lambda_t}{(1+r)^t}\right] = \widetilde{\mathrm{E}}_0\left[\mathbb{I}_{\{\tau\leqslant T\}} \frac{\Lambda_\tau}{(1+r)^\tau}\right].$$

Here, we use the fact that $\mathbb{I}_{\{\tau\leqslant T\}} = \mathbb{I}_{\{\tau=0\}} + \mathbb{I}_{\{\tau=1\}} + \cdots + \mathbb{I}_{\{\tau=T\}}$.

At time $t = 0$ it is unknown what strategy is the optimal one. Moreover, the writer of the derivative has no control on what exercise strategy will be used by the buyer, who can choose any one from the set $\mathcal{S}_{0,T}$. Therefore, it is reasonable to set the fair price V_0 of the American derivative equal to the maximum over all values of $v_0(\tau)$ for all possible exercise strategies:

$$V_0 = \max_{\tau\in\mathcal{S}_{0,T}} v_0(\tau) = \max_{\tau\in\mathcal{S}_{0,T}} \widetilde{\mathrm{E}}_0\left[\mathbb{I}_{\{\tau\leqslant T\}} \frac{\Lambda_\tau}{(1+r)^\tau}\right]. \tag{7.77}$$

The optimal exercise strategy (the optimal stopping time random variable) is defined as the stopping time $\tau^* \in \mathcal{S}_{0,T}$ which maximizes the expected final payoff under the risk-neutral measure. Hence, the price of the American claim is given in terms of an optimal stopping time τ^* by

$$V_0 = \widetilde{\mathrm{E}}_0\left[\mathbb{I}_{\{\tau^*\leqslant T\}} \frac{\Lambda_{\tau^*}}{(1+r)^{\tau^*}}\right]. \tag{7.78}$$

Let us see that selling the American derivative for an amount other than V_0 leads to arbitrage. Suppose that the holder of the derivative does not withdraw the proceeds at time $t = \tau^*$ from the market, but allows them to grow at the risk-free rate r from time τ^* to time T. The final payoff is

$$\frac{B_T}{B_{\tau^*}} \mathbb{I}_{\{\tau^*\leqslant T\}} \Lambda_{\tau^*} = (1+r)^{T-\tau} \mathbb{I}_{\{\tau^*\leqslant T\}} \Lambda_{\tau^*}.$$

Let a self-financing strategy $\mathbf{\Phi}^*$ replicate this payoff at time T. Hence the initial value $\Pi_0[\mathbf{\Phi}^*]$ is equal to V_0 in (7.77). Note that in reality neither the seller of the American derivative nor the buyer knows in advance what strategy will be optimal and what the final payoff will be. Hence, at time $t = 0$ such a replicating strategy $\mathbf{\Phi}^*$ is unknown. As we saw in Example 7.8, an American derivative can be replicated by two processes, namely, a portfolio process and a consumption process. The worst case scenario for the seller of the derivative is when the buyer follows the optimal stopping strategy and he or she will exercise at the optimal time, leaving nothing to the writer to consume for free.

Now, suppose that the American derivative can be purchased for an amount less than $V_0 = \Pi_0[\mathbf{\Phi}^*]$. Then an arbitrage opportunity can be created by purchasing the cheaper derivative and selling short the more expensive strategy $\mathbf{\Phi}^*$. If the initial price of the American derivative is larger than V_0, then an arbitrage opportunity is available to the seller of the option, who can invest the proceeds in $\mathbf{\Phi}^*$. So we may conclude that the no-arbitrage price of the American derivative is equal to $\Pi_0[\mathbf{\Phi}^*] = V_0$.

The next example gives an explicit implementation of Equations (7.77) and (7.78).

Example 7.10. Find values $v_0(\tau)$ for each exercise rule $\tau \in \mathcal{S}_{0,3}$ for the American put option from Example 7.7. Find an optimal stopping time τ^* such that $v_0(\tau^*)$ is a maximum value. Compare $v_0(\tau^*)$ with the no-arbitrage price P_0^A calculated in Example 7.7. Moreover, compare the optimal stopping time τ^* with the exercise rule given in Figure 7.8.

Solution. Without loss of generality we can assume that the option has to be exercised before or at the expiry time but with a possibly zero payoff to the holder if the option is out of the money or it is not optimal to exercise the option. In this case it is sufficient to only consider finite stopping times of the form $\tau\colon \Omega \to \{0,1,2,3\}$, i.e., $\tau \in \widehat{\mathcal{S}}_{0,3}$, while calculating the maximum in (7.77). Then, (7.77) simplifies as follows:

$$P_0^A = \max_{\tau \in \widehat{\mathcal{S}}_{0,3}} \widetilde{\mathrm{E}}_0\left[\frac{\Lambda(S_\tau)}{(1+r)^\tau}\right] = \max_{\tau \in \widehat{\mathcal{S}}_{0,3}} \widetilde{\mathrm{E}}\left[\frac{(8-S_\tau)^+}{(1+r)^\tau}\right].$$

Using the definition of a stopping time, we can find all elements of $\widehat{\mathcal{S}}_{0,T}$. For example, Table 7.11 defines all 26 finite stopping times of $\widehat{\mathcal{S}}_{0,3}$ for a three-period model. We denote these random variables by $\hat{\tau}_k, k = 1,\ldots,26$.

TABLE 7.11: Stopping times of $\widehat{\mathcal{S}}_{0,3}$.

ω	$\hat{\tau}_1$	$\hat{\tau}_2$	$\hat{\tau}_3$	$\hat{\tau}_4$	$\hat{\tau}_5$	$\hat{\tau}_6$	$\hat{\tau}_7$	$\hat{\tau}_8$	$\hat{\tau}_9$	$\hat{\tau}_{10}$	$\hat{\tau}_{11}$	$\hat{\tau}_{12}$	$\hat{\tau}_{13}$	$\hat{\tau}_{14}$	$\hat{\tau}_{15}$	$\hat{\tau}_{16}$	$\hat{\tau}_{17}$	$\hat{\tau}_{18}$	$\hat{\tau}_{19}$	$\hat{\tau}_{20}$	$\hat{\tau}_{21}$	$\hat{\tau}_{22}$	$\hat{\tau}_{23}$	$\hat{\tau}_{24}$	$\hat{\tau}_{25}$	$\hat{\tau}_{26}$
UUU	0	1	1	1	1	1	2	2	3	3	2	2	2	2	3	2	2	2	3	3	3	3	3	3	2	3
UUD	0	1	1	1	1	1	2	2	3	3	2	2	2	2	3	2	2	2	3	3	3	3	3	3	2	3
UDU	0	1	1	1	1	1	2	3	2	3	2	2	2	3	2	2	3	3	2	2	3	3	3	2	3	3
UDD	0	1	1	1	1	1	2	3	2	3	2	2	2	3	2	2	3	3	2	2	3	3	3	2	3	3
DUU	0	1	2	2	3	3	1	1	1	1	2	2	3	2	2	3	3	2	2	3	2	3	2	3	3	3
DUD	0	1	2	2	3	3	1	1	1	1	2	2	3	2	2	3	3	2	2	3	2	3	2	3	3	3
DDU	0	1	2	3	2	3	1	1	1	1	2	3	2	2	2	3	2	3	3	2	2	2	3	3	3	3
DDD	0	1	2	3	2	3	1	1	1	1	2	3	2	2	2	3	2	3	3	2	2	2	3	3	3	3

TABLE 7.12: The mathematical expectation $\widetilde{\mathrm{E}}\left[(1+r)^{-\tau}\Lambda(S_\tau)\right]$ is calculated for each stopping time $\tau = \hat{\tau}_k$ of Table 7.11.

$\hat{\tau}_1$	$\hat{\tau}_2$	$\hat{\tau}_3$	$\hat{\tau}_4$	$\hat{\tau}_5$	$\hat{\tau}_6$	$\hat{\tau}_7$	$\hat{\tau}_8$	$\hat{\tau}_9$	$\hat{\tau}_{10}$	$\hat{\tau}_{11}$	$\hat{\tau}_{12}$	$\hat{\tau}_{13}$
0	0.8	0.48	0.416	0.36	0.296	1.04	0.92	1.04	0.92	0.72	0.656	0.6
$\hat{\tau}_{14}$	$\hat{\tau}_{15}$	$\hat{\tau}_{16}$	$\hat{\tau}_{17}$	$\hat{\tau}_{18}$	$\hat{\tau}_{19}$	$\hat{\tau}_{20}$	$\hat{\tau}_{21}$	$\hat{\tau}_{22}$	$\hat{\tau}_{23}$	$\hat{\tau}_{24}$	$\hat{\tau}_{25}$	$\hat{\tau}_{26}$
0.6	0.72	0.536	0.48	0.536	0.656	0.6	0.6	0.48	0.536	0.536	0.416	0.416

For each stopping time $\tau = \hat{\tau}_k$ of Table 7.11, we can calculate the risk-neutral expectation of the discounted payoff as follows:

$$\begin{aligned}
\widetilde{\mathrm{E}}_0\left[\frac{\Lambda(S_\tau)}{(1+r)^\tau}\right] &= \sum_{\omega\in\Omega} \frac{\Lambda(S_{\tau(\omega)}(\omega))}{(1+r)^{\tau(\omega)}}\tilde{p}^{\#\mathsf{U}(\omega)}(1-\tilde{p})^{\#\mathsf{D}(\omega)} \\
&= \sum_{\omega\in\Omega_3} 1.25^{-\tau(\omega)}\cdot\left(8 - S_{\tau(\omega)}(\omega)\right)^+ \cdot 0.75^{\#\mathsf{U}(\omega)}\cdot 0.25^{\#\mathsf{D}(\omega)}.
\end{aligned}$$

For example, for $\tau = \hat{\tau}_8$:

$$\begin{aligned}
\widetilde{\mathrm{E}}_0\left[\frac{(8-S_\tau)^+}{1.25^\tau}\right] &= \sum_{\omega\in\Omega_3} 1.25^{-\hat{\tau}_8(\omega)}\left(8 - S_0 u^{\#\mathsf{U}_{\hat{\tau}_8(\omega)}(\omega)} d^{\#\mathsf{D}_{\hat{\tau}_8(\omega)}(\omega)}\right)\cdot 0.75^{\#\mathsf{U}(\omega)}\cdot 0.25^{\#\mathsf{D}(\omega)} \\
&= \underbrace{1.25^{-2}\cdot(8-18)^+\cdot\left(0.75^3 + 0.75^2\cdot 0.25\right)}_{\hat{\tau}_8(\omega)=2 \text{ for } \omega\in A_{\mathsf{UU}}} \\
&\quad + \underbrace{1.25^{-3}\cdot\left((8-9)^+\cdot 0.75^2\cdot 0.25 + (8-3)^+\cdot 0.75\cdot 0.25^2\right)}_{\hat{\tau}_8(\omega)=3 \text{ for } \omega\in A_{\mathsf{UD}}} \\
&\quad + \underbrace{1.25^{-1}\cdot(8-4)^+\cdot\left(0.75^2\cdot 0.25 + 2\cdot 0.75\cdot 0.25^2 + 0.25^3\right)}_{\hat{\tau}_8(\omega)=1 \text{ for } \omega\in A_{\mathsf{D}}} \\
&= 0+0+0+0.12+0.45+0.15+0.15+0.05 = 0.92.
\end{aligned}$$

The results are presented in Table 7.12. As is seen, the maximum is equal to $P_0^A = 1.04$ and it is attained for stopping times $\tau = \hat{\tau}_7$ and $\hat{\tau}_9$. The exercise rules $\hat{\tau}_7$ and $\hat{\tau}_9$ of Table 7.11 and the optimal exercise rule τ^* of Figure 7.8 are all equivalent since

$$\mathbb{I}_{\{\tau^*(\omega)\leqslant 3\}}(8 - S_{\tau^*(\omega)}(\omega))^+ = (8 - \Lambda(S_{\hat{\tau}_7(\omega)}(\omega)))^+ = (8 - S_{\hat{\tau}_9(\omega)}(\omega))^+$$

holds for all $\omega \in \Omega_3$. □

In the above three-period example, we considered all conceivable stopping times (i.e., stopping rules) and respectively computed the set of all discounted expected values of the put payoff. The fair value of the American put was then given by taking the maximum over all such computed values. Underlying all of this is a method (i.e., a rule) for choosing the optimal exercise time. As depicted in Figure 7.8, the rule corresponds to defining an optimal stopping time, τ^*. It turns out that this is given by the random variable:

$$\tau^* = \min\{t \in \{0,1,2,3\} \;:\; P_t^A = \Lambda_t\} = \min\{t \in \{0,1,2,3\} \;:\; P_t^A(S_t) = (8 - S_t)^+\}.$$

We note that, in the case that this set is empty, we define the minimum to be ∞. That is, we put $\tau^* = \infty$ if $P_t^A(S_t(\omega)) > (8 - S_t(\omega))^+$ for all $t \geqslant 0, \omega \in \Omega_3$. Given a scenario ω, then $\tau^*(\omega)$ is the nonnegative integer corresponding to the first time at which the American price equals the intrinsic value, i.e., the smallest integer $t \in \{0,1,2,3\}$ such that $P_t^A(S_t(\omega)) = (8 - S_t(\omega))^+$:

$$\tau^*(\omega) = \min\{t \in \{0,1,2,3\} \;:\; P_t^A(S_t(\omega)) = (8 - S_t(\omega))^+\}.$$

It is simple to verify that this produces the optimal stopping time. For instance, take the outcome $\omega \in A_{\mathsf{UU}}$. For $t = 0$, $P_0^A(S_0) = P_0^A(8) = 1.04 > (8-8)^+ = 0$. For $t = 1$, $P_1^A(S_1(\omega)) = P_1^A(S_1(\mathsf{U})) = P_1^A(12) = 0.4 > (8 - 12)^+ = 0$. For $t = 2$, $P_2^A(S_2(\omega)) = P_2^A(S_2(\mathsf{UU})) = P_2^A(18) = 0 = (8-18)^+ = (8 - S_2(\mathsf{UU}))^+$. Hence, $\tau^*(\omega) = 2$. Similarly, the reader can verify that $\tau^*(\omega) = 2$, for $\omega \in A_{\mathsf{UD}}$, and $\tau^*(\omega) = 1$, for $\omega \in A_{\mathsf{D}}$. Therefore, the above method generates an optimal stopping time where $\tau^* = \hat{\tau}_7$.

The following proposition now justifies the above for a generally path-dependent American option on the standard T-period binomial model. Here, it suffices to consider an optimal stopping $\tau^* \in \mathcal{S}_{0,T}$ which corresponds to the optimal stopping time in (7.78) for the time-0 no-arbitrage price the Amercian option. Again, we note that in Equation (7.79) below, if $\min\{t \in \{0,1,\ldots,T\} \;:\; V_t(\omega) = \Lambda_t(\omega)\} = \emptyset$ then we put $\tau^*(\omega) = \infty$ for any such $\omega \in \Omega_T$.

Proposition 7.8. *The stopping time defined by*

$$\tau^* = \min\{t \in \{0,1,\ldots,T\}\;;\; V_t = \Lambda_t\} \tag{7.79}$$

maximizes the expectation on the right-hand side of (7.77). That is, the time-0 American option price is given by V_0^A in (7.78) with optimal stopping time τ^* given by (7.79).

Proof. The proof is rather immediate once we use the fact that the stopped discounted American option value process defined by $\left\{\frac{V_{t\wedge\tau^*}}{(1+r)^{t\wedge\tau^*}}\right\}_{t=0,1,\ldots,T}$ is a $\widetilde{\mathbb{P}}$-martingale. This is shown by considering the backward recurrence relation in (7.73). In particular, arbitrarily fix the first t market moves and consider the two possible mutually exclusive events whose union is $A_{\omega_1,\ldots,\omega_t}$: (i) $\{\omega \in A_{\omega_1,\ldots,\omega_t} \;:\; \tau^*(\omega) > t\}$ and (ii) $\{\omega \in A_{\omega_1,\ldots,\omega_t} \;:\; \tau^*(\omega) \leqslant t\}$.

For case (i): $\tau^*(\omega) \wedge t = t$ and hence the American option price is given by the continuation value at time t. The latter is greater than the intrinsic value, i.e., $V_t^A(\omega) > \Lambda_t(\omega)$. Therefore, Equation (7.73) implies

$$V_{\tau^*\wedge t}(\omega) \equiv V_{\tau^*(\omega)\wedge t}(\omega) = V_t(\omega) = \frac{1}{1+r}\widetilde{\mathrm{E}}_t\left[V_{t+1}\right](\omega) = \frac{1}{1+r}\widetilde{\mathrm{E}}_t\left[V_{\tau^*\wedge(t+1)}\right](\omega)$$

where we used $\tau^*(\omega) \wedge (t+1) = t+1$ since $\tau^*(\omega) \geqslant t+1$. Moreover, we then have $(1+r)^{\tau^*(\omega)\wedge(t+1)}/(1+r)^{\tau^*(\omega)\wedge t} = (1+r)^{t+1}/(1+r)^t = 1+r$, which implies the martingale expectation property:

$$\frac{V_{\tau^*\wedge t}}{(1+r)^{\tau^*\wedge t}}(\omega) = \widetilde{\mathrm{E}}_t\left[\frac{V_{\tau^*\wedge(t+1)}}{(1+r)^{\tau^*\wedge(t+1)}}\right](\omega).$$

For case (i) we have $\tau^*(\omega) \wedge t = \tau^*(\omega)$, giving

$$\frac{V_{\tau^*(\omega)\wedge t}(\omega)}{(1+r)^{\tau^*(\omega)\wedge t}} = \frac{V_{\tau^*(\omega)}(\omega)}{(1+r)^{\tau^*(\omega)}} = \frac{\Lambda_{\tau^*(\omega)}(\omega)}{(1+r)^{\tau^*(\omega)}} = \frac{\Lambda_{\tau^*}}{(1+r)^{\tau^*}}(\omega)$$

for all $t > \tau^*(\omega)$. Hence, for $t > \tau^*(\omega)$ the discounted process is a constant and therefore satisfies the martingale property.

By using the $\widetilde{\mathbb{P}}$-martingale property of the discounted stopped price process T-steps forward, and the fact that $V_{\tau^*} = \Lambda_{\tau^*}$, we finally obtain the required pricing formula:

$$V_0 = \widetilde{\mathrm{E}}_0\left[\frac{V_{\tau^*\wedge T}}{(1+r)^{\tau^*\wedge T}}\right] = \widetilde{\mathrm{E}}_0\left[\mathbb{I}_{\{\tau^*\leqslant T\}}\frac{V_{\tau^*}}{(1+r)^{\tau^*}}\right] = \widetilde{\mathrm{E}}_0\left[\mathbb{I}_{\{\tau^*\leqslant T\}}\frac{\Lambda_{\tau^*}}{(1+r)^{\tau^*}}\right].$$

Here we also used the fact that $1 = \mathbb{I}_{\{\tau^*\leqslant T\}} + \mathbb{I}_{\{\tau^*=\infty\}}$ where $\mathbb{I}_{\{\tau^*=\infty\}}V_{\tau^*\wedge T} = \mathbb{I}_{\{\tau^*=\infty\}}V_T = 0$ since the event of never exercising the option amounts to zero payoff. □

In a similar way, we can define the value process $\{v_t\}_{0\leqslant t\leqslant T}$. Fix $t \in \{0, 1, \ldots, T\}$. Assuming that the derivative has not been exercised before time t, the holder may exercise it any time from t until T. Let the holder follow an exercise strategy $\tau \in \mathcal{S}_{t,T}$. The no-arbitrage time-$t$ value (relative to the rule τ) is given by the risk-neutral expectation of the discounted payoff conditional on the information available at time t: $v_t(\tau) = \widetilde{\mathrm{E}}_t\left[\mathbb{I}_{\{\tau\leqslant T\}}\frac{\Lambda_\tau}{(1+r)^{\tau-t}}\right]$. The fair price of the American derivative, V_t, which is independent of the exercise rule used by the holder of the derivative, is given by the largest possible value $V_t(\tau)$:

$$V_t = \max_{\tau\in\mathcal{S}_{t,T}} \widetilde{\mathrm{E}}_t\left[\mathbb{I}_{\{\tau\leqslant T\}}\frac{\Lambda_\tau}{(1+r)^{\tau-t}}\right] = \widetilde{\mathrm{E}}_t\left[\mathbb{I}_{\{\tau^*\leqslant T\}}\frac{\Lambda_{\tau^*}}{(1+r)^{\tau^*}}\right], \tag{7.80}$$

where $\tau^* \in \mathcal{S}_{t,T}$ is an optimal exercise time, which maximizes the expected final payoff in (7.80).

7.7.3 Early-Exercise Boundary

An American option should be exercised according to the optimal stopping rule τ^* given in (7.79), i.e., as soon as the value $V_t(S_t)$ becomes equal to $\Lambda_t(S_t)$, for maximal gain. As a result, the binomial lattice is divided into two parts: a continuation domain for which the option is not exercised,

$$\mathcal{D}_C = \{(t, S_{t,n}) \;:\; 0\leqslant t\leqslant T,\ 0\leqslant n\leqslant t,\ V_t(S_{t,n}) > \Lambda_t(S_{t,n})\}, \tag{7.81}$$

and a stopping domain whereby the option is exercised early,

$$\mathcal{D}_S = \{(t, S_{t,n}) \;:\; 0\leqslant t\leqslant T,\ 0\leqslant n\leqslant t,\ V_t(S_{t,n}) = \Lambda_t(S_{t,n})\}. \tag{7.82}$$

The structure of the stopping domain may be quite complicated, and it is defined by the payoff function Λ and whether the underlying asset pays dividends. However, for the standard monotonic piecewise call and put payoffs, the stopping domain turns out to be simply connected. The early-exercise boundary, which separates the continuation and stopping domains, is a connected path in the binomial tree given by

$$\{(t, S) \; : \; 0 \leqslant t \leqslant T, \; S = S_t^*\},$$

where

$$S_t^* = \min\{S_{t,n} \; : \; C_t^A(S_{t,n}) = (S_{t,n} - K)^+, \; 0 \leqslant n \leqslant t\} \tag{7.83}$$

for a call and

$$S_t^* = \max\{S_{t,n} \; : \; P_t^A(S_{t,n}) = (K - S_{t,n})^+, \; 0 \leqslant n \leqslant t\} \tag{7.84}$$

for a put. Note that since the American option value is always nonnegative, the superscript + signs are omitted in (7.83) and (7.84).

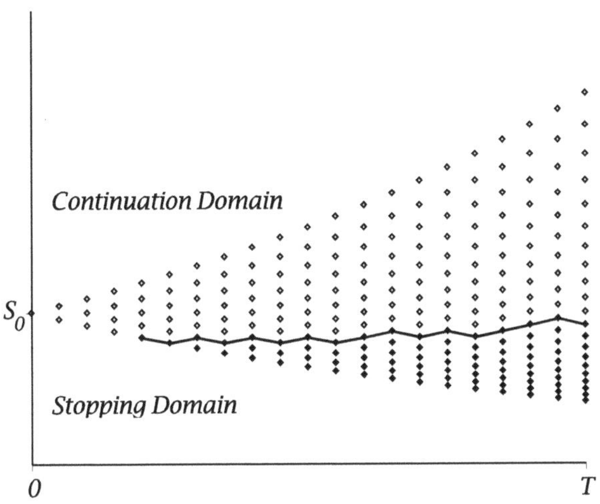

FIGURE 7.13: The early-exercise boundary for an American put option. The black nodes correspond to the stopping domain.

A typical early exercise boundary for an American put option is shown in Figure 7.13. The stopping domain is marked with black nodes. As follows from the definition of the stopping domain, the optimal exercise time τ^* defined in (7.79) is also the first hitting time of the stopping domain $\mathcal{D}_S$. That is, τ^* is the first time when the stock price process reaches the early-exercise boundary S^*:

$$\tau^* = \min\{t \; : \; 0 \leqslant t \leqslant T, \; S_t = S_t^*\}.$$

7.7.4 Pricing American Options: The Case with Dividends

Using a nonarbitrage argument, we have proved earlier that the price of a standard American call (with intrinsic value $\Lambda_t = (S_t - K)^+$, $0 \leqslant t \leqslant T$) is the same as that of the standard European call paying $(S_T - K)^+$ at expiration T. Note that the situation is different for a dividend-paying asset. If this is the case, then, as is demonstrated below, the American call can have a larger initial value than the European call.

Consider a dividend-paying asset that follows the binomial model between the dividend payment times. We assume that at each time $t = 1, 2, \ldots, T$, the dividend payment at time t denoted D_t is equal to $q_t S_t^*$ for some nonrandom $q_t \in [0, 1)$, where S_t^* denotes the

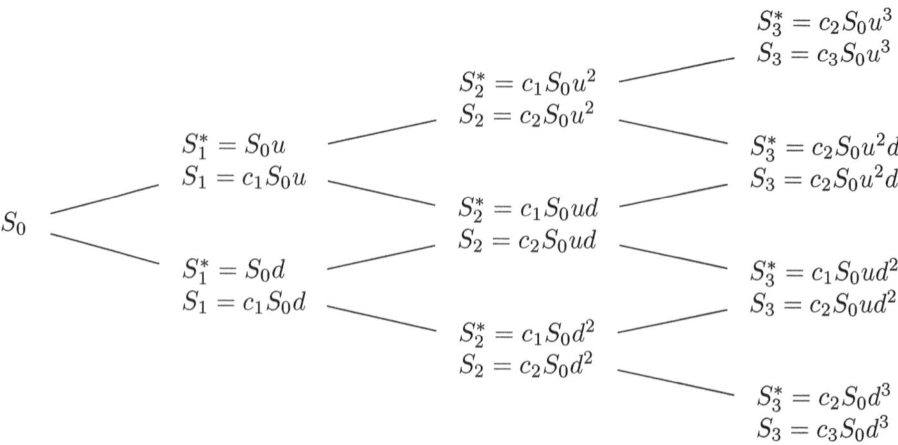

FIGURE 7.14: A three-period recombining tree for the asset price process with dividend payments $D_t = q_t S_t^*$. Here $c_t = (1-q_1)(1-q_2)\cdots(1-q_t)$ is the cumulative after-dividend factor.

asset price just prior to the dividend payment. The no-arbitrage asset price just after the dividend payment is

$$S_t = S_t^* - D_t = S_t^* - q_t S_t^* = (1-q_t)S_t^*.$$

In the binomial model, the asset prices satisfy the following recurrence relation:

$$S_t(\omega) = (1-q_t)S_t^*(\omega) = (1-q_t)S_{t-1}(\omega)Y_t(\omega_t), \quad 1 \leqslant t \leqslant T, \quad \omega \in \Omega_T,$$

where the factor $Y_t(\omega_t) := u^{\#\mathsf{U}(\omega_t)} d^{\#\mathsf{D}(\omega_t)}$ equals u with probability p or d with probability $1-p$, and it is independent of S_{t-1}. Applying the relation sequentially at times $t, t-1, \ldots, 1$ gives

$$S_t(\omega) = \prod_{n=1}^{t}(1-q_n)Y_n(\omega_n)\, S_0 = \prod_{n=1}^{t}(1-q_n)\, S_0 u^{\mathsf{U}_t(\omega)} d^{\mathsf{D}_t(\omega)} := c_t S_0 u^{\mathsf{U}_t(\omega)} d^{\mathsf{D}_t(\omega)},$$

where c_t is the cumulative after-dividend factor defined by

$$c_t = (1-q_1)(1-q_2)\cdots(1-q_t).$$

The evolution of the asset price process can be represented by a recombining binomial tree (see a three-period tree presented in Figure 7.14).

In the general case, where the dividend payments, $\{D_t\}_{1\leqslant t\leqslant T}$, are not given as a percentage of the asset price, the recurrence formula for the asset price takes the form:

$$S_t(\omega) = S_t^*(\omega) - D_t = S_{t-1}u^{\#\mathsf{U}(\omega_t)} d^{\#\mathsf{D}(\omega_t)} - D_t, \quad 1 \leqslant t \leqslant T, \quad \omega \in \Omega_T.$$

In general, the price dynamics of such an asset cannot be represented by a recombining binomial tree. In the case of rare dividend payments, we have a partly recombining tree.

Example 7.11. Consider the three-period binomial tree model in Example 7.1 with $S_0 = 8$, $u = \frac{3}{2}$, $d = \frac{1}{2}$, and $r = \frac{1}{4}$. Assume the stock pays dividends at times $t = 2$ and $t = 3$ and each dividend payment is 10% of the stock price. That is, $q_1 = 0$ and $q_2 = q_3 = 0.1$.

(a) Construct the binomial tree with prior- and post-dividend stock prices.

(b) Find and compare the prices of the standard European and American call options with common strike $K = 6$ and expiration $T = 3$.

Solution. The recombining binomial tree for the stock price process is given in Figure 7.1. The risk-neutral probabilities are $\tilde{p} = 0.75$ and $1 - \tilde{p} = 0.25$. At each node of the tree, we apply recurrence formulae (7.34) and (7.69) to evaluate the respective prices of the European and American call options. Let C_t^E and C_t^A, respectively, denote the time-t prices of the European and American put options for $t = 0, 1, 2, 3$. At the expiration date, the option value must equal the payoff function: $C_3^E(S) = C_3^A(S) = (S - 6)^+$ for $S \in \{0.9, 2.7, 8.1, 24.3\}$. To determine the price for the times before expiration, use the following recurrences:

$$C_t^E(S) = \frac{3 \cdot C_{t+1}^E(1.5 \cdot S) + C_{t+1}^E(0.5 \cdot S)}{5},$$
$$C_t^A(S) = \max\left\{(S-6)^+, \frac{3 \cdot C_{t+1}^A(1.5 \cdot S) + C_{t+1}^A(0.5 \cdot S)}{5}\right\}, \quad t = 2, 1, 0.$$

As a result, we obtain the option prices given in Figure 7.16. The values of the American put are boxed at those nodes of the tree where the intrinsic value is greater than or equal to the continuation value. Therefore, the American call option should be exercised at time $t = 2$ when the stock price is $S_2 = 16.2$, or at time $t = 3$ when the price S_3 exceeds the strike $K = 6$. Note that the initial price of the American call, C_0^A, is strictly larger than that of the European call, C_0^E. □

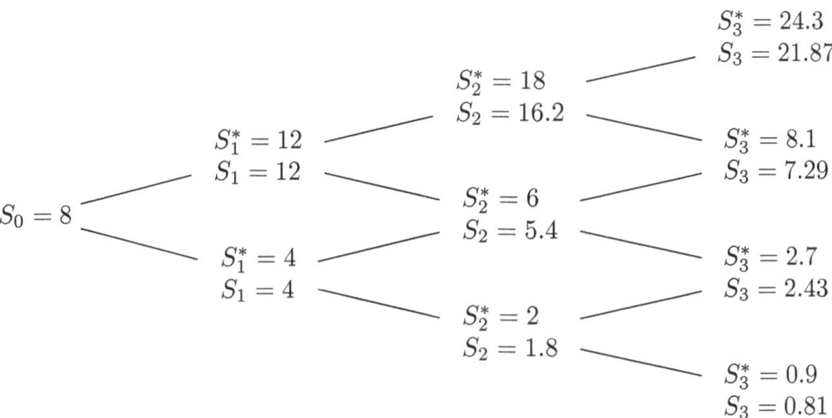

FIGURE 7.15: A three-period recombining binomial tree for the stock paying the dividend of $0.1S_t^*$ at times $t = 1, 2$.

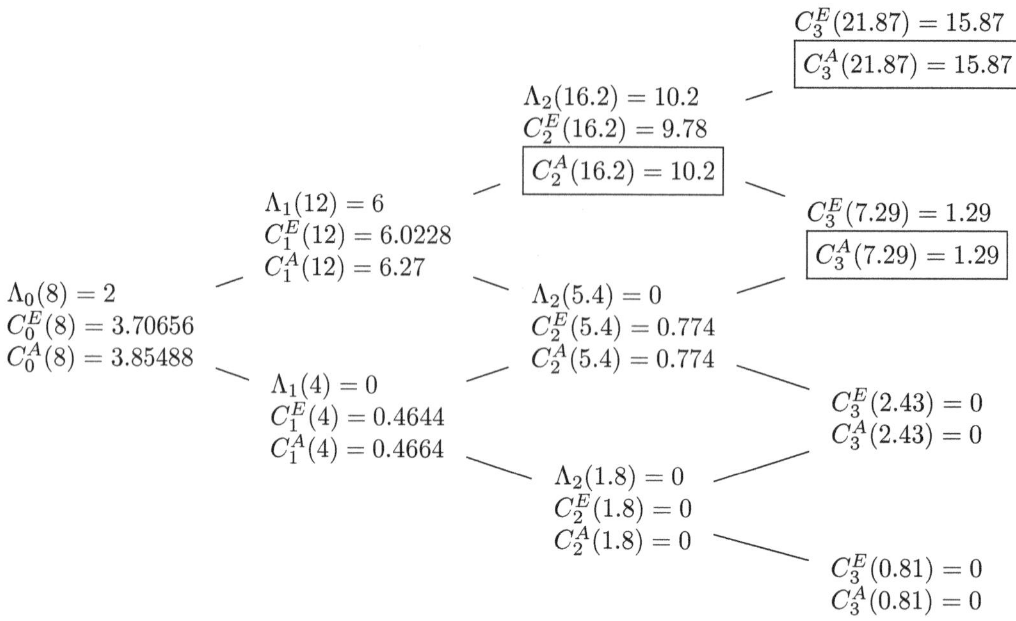

FIGURE 7.16: Prices of the standard European call, C_t^E, and American call, C_t^A, on the dividend-paying stock.

7.8 Exercises

Exercise 7.1. Consider a T-period binomial tree model with stock price $S_{t,n} = S_0 u^n d^{t-n}$ at each node (t, n) of the binomial tree for every $n = 0, 1, \ldots, t$ and every $t = 0, 1, \ldots, T$.

(a) Let $v, t \in \{0, 1, \ldots, T\}$ be two dates such that $v < t$. Find the joint probability $\mathbb{P}(S_t = S_{t,n}, S_v = S_{v,m})$, $n = 0, 1, \ldots, t$, $m = 0, 1, \ldots, v$.

(b) Find the conditional probability $\mathbb{P}(S_t = S_{t,n} \mid S_v = S_{v,m})$ using the result of (a).

Exercise 7.2. Consider a three-period binomial model with $S_0 = 100$, $u = 1.15$, and $d = 0.9$.

(a) Find the distribution of S_3 conditional on $S_1 = 115$.

(b) Find the distribution of S_3 conditional on $S_1 = 90$.

Exercise 7.3. Consider a three-period binomial tree model with $S_0 = 8$, $u = 1.5$, $d = 0.5$, and $r = 0.25$. Consider a European put option that expires at $T = 3$ and provides payoff $(10 - S_3)^+$. Let $P_{t,n}$ be the price of this option at time t given that $S_t = S_{t,n}$.

(a) Compute recursively backward in time the prices $P_{t,n}$, $t = 3, 2, 1, 0$, $n = 0, 1, \ldots, t$. Present the prices calculated in a recombining binomial tree diagram.

(b) Construct a hedging strategy $\{\Delta_t\}_{t=0,1,2}$. Represent it using a recombining binomial tree diagram.

Exercise 7.4. Complete the proof of Theorem 7.6 by showing that

$$\Pi_{t+1}(\omega_1 \dots \omega_t \mathsf{U}) = V_{t+1}(\omega_1 \dots \omega_t \mathsf{U}).$$

Exercise 7.5. Prove Theorem 7.7.

Exercise 7.6. Find the initial risk-neutral value of a contract that pays \$1 at the first time t when $\#\mathsf{U}(\omega_1, \omega_2, \dots, \omega_t) = n$ for some nonrandom $n \in \{1, 2, \dots, T\}$.

Exercise 7.7. Consider a T-period binomial tree model. For $t \geqslant 0$, define $G_t = (S_0 \cdot S_1 \cdots S_t)^{\frac{1}{t+1}}$ to be the geometric average of stock prices between times zero and t. Consider an Asian call option that expires at time T and provides payoff $(G_T - K)^+$ with strike price K. Let $V_t(S, G)$ denote the price of this option at time $t \geqslant 0$ given that $S_t = S$ and $G_t = G$. In particular, $V_T(S, G) = (G - K)^+$.

(a) Develop a one-step pricing formula for $V_t = V_t(S, G)$ in terms of V_{t+1}, S, and G.

(b) Develop a formula for $\Delta_t(S, G)$, the number of shares of stock held in a replicating portfolio during the period $[t, t+1]$, given that $S_t = S$ and $G_t = G$.

Exercise 7.8. Consider a three-period binomial model with $S_0 = 8$, $u = 1.5$, $d = 0.5$, and $r = 0.25$. Consider an Asian call option that expires at time three and provides payoff $(G_3 - 8)^+$.

(a) Compute the prices V_t for each $t = 3, 2, 1, 0$, using a backward-time recursion. Write these prices on a nonrecombining binomial tree diagram.

(b) Construct a hedging strategy $\{\Delta_t\}_{t=0,1,2}$. Represent it on a nonrecombining binomial tree diagram.

Exercise 7.9. Consider a binomial-tree arbitrage-free model with T periods and arithmetic floating-price (Asian) call and put options with respective payoff functions

$$C_T = (A_T - K)^+ \quad \text{and} \quad P_T = (K - A_T)^+,$$

where $K > 0$ is a strike price and $A_T = \frac{1}{T+1} \sum_{n=0}^{T} S_n$.

(a) Find the risk-neutral conditional expectation $\widetilde{\mathrm{E}}_0[A_T]$.

(b) Using the result of (a), find the put-call parity for the Asian options at time 0. That is, express the difference of initial prices $C_0 - P_0$ in terms of S_0, K, T, and r.

Exercise 7.10. Consider a T-period arbitrage-free binomial model and floating-price look-back call and put options with fixed strike $K > 0$ defined by the respective payoffs

$$C_T = \left(M_T - K\right)^+ \quad \text{and} \quad P_T = \left(K - M_T\right)^+,$$

where $M_t := \max_{0 \leqslant n \leqslant t} S_n$ for $0 \leqslant t \leqslant T$.

(a) Prove (or argue) whether the floating-price lookback call and put option prices C_t and P_t at arbitrary time $0 \leqslant t \leqslant T$ are functions of the pair of values (M_t, S_t) or whether they are just functions of M_t.

(b) Assume $ud = 1$ (i.e., a symmetric stock lattice) and $T = 3$ (i.e., a three-period model). Determine the arbitrage-free present values C_0 and P_0. Leave your answers as expressions in terms of the parameters S_0, r, u, d.

Exercise 7.11. Consider the T-period binomial model with dividend-paying stock where the stock price process $\{S_t\}_{0\leqslant t\leqslant T}$ follows

$$S_{t+1} = (1 - q_{t+1})Y_{t+1}S_t$$

with $Y_{t+1}(\omega) = u\mathbb{I}_{\{\omega_{t+1}=\mathsf{U}\}} + d\mathbb{I}_{\{\omega_{t+1}=\mathsf{D}\}}$ and $q_{t+1}(\omega_1\ldots\omega_t\omega_{t+1}) \in (0,1)$. In this case the wealth equation takes the form

$$\begin{aligned}\Pi_{t+1} &= \Delta_t S_{t+1} + (1+r)(\Pi_t - \Delta_t S_t) + \Delta_t q_{t+1}Y_{t+1}S_t \\ &= \Delta_t Y_{t+1}S_t + (1+r)(\Pi_t - \Delta_t S_t)\end{aligned}$$

where r is a constant interest rate and $B_t = (1+r)^t$ is the value at time $t = 0,1,\ldots,T$ of one dollar initially invested in the bank account at time 0.

(a) Show that the discounted wealth process $\{\frac{\Pi_t}{B_t}\}_{0\leqslant t\leqslant T}$ is a martingale under the risk-neutral measure $\widetilde{\mathbb{P}} \equiv \widetilde{\mathbb{P}}^{(B)}$. Hence, w.r.t. this measure, state the risk-neutral pricing formula for the price of a derivative claim at time n with payoff Π_T at terminal time T.

(b) Show that the discounted stock price process $\{\frac{S_t}{B_t}\}_{0\leqslant t\leqslant T}$ is not a martingale but a strict supermartingale under the $\widetilde{\mathbb{P}}$-measure, i.e., show that $\widetilde{\mathrm{E}}_t\left[\frac{S_{t+1}}{B_{t+1}}\right] < \frac{S_t}{B_t}$.

(c) Denote $c_t = (1-q_1)(1-q_2)\cdots(1-q_t)$. Clearly, c_t is $\mathcal{F}_t$-measurable. Show that the process $\left(\frac{S_t}{c_tB_t}\right)_{0\leqslant t\leqslant T}$ is a $\widetilde{\mathbb{P}}$-martingale.

Exercise 7.12. Consider the T-period binomial model with stock price $S_{t,n} = S_0u^nd^{t-n}$ at each node (t,n) with $t = 0,1,\ldots,T$ and $n = 0,1,\ldots,t$. Let $V_{t,n} = V_t(S_{t,n})$ denote the (no-arbitrage) price of any standard (non-path-dependent) European option at the node (t,n); let the relative derivative prices be defined by $\overline{V}_{t,n} := \frac{V_{t,n}}{S_{t,n}}$. Determine α and β (explicitly in terms of the model parameters) in the recurrence relation:

$$\overline{V}_{t,n} = \alpha\overline{V}_{t+1,n+1} + \beta\overline{V}_{t+1,n}.$$

Exercise 7.13. Consider a standard T-period binomial model and an Arrow–Debreu derivative security $\mathcal{E} \equiv \mathcal{E}^{\bar{\omega}}$ that pays one dollar if the *single particular* sequence $\bar{\omega} := \bar{\omega}_1\cdots\bar{\omega}_T$ of T market moves occurs and pays zero otherwise. Hence, this is a derivative with payoff given by the indicator function random variable $\mathcal{E}_T = \mathbb{I}_{\{\omega=\bar{\omega}\}}$:

$$\mathcal{E}_T(\omega_1\omega_2\cdots\omega_T) = \begin{cases}1 & \text{if } \omega_1 = \bar{\omega}_1, \omega_2 = \bar{\omega}_2, \ldots, \omega_T = \bar{\omega}_T, \\ 0 & \text{otherwise.}\end{cases}$$

(a) Derive a formula for the no-arbitrage derivative value, $\mathcal{E}_t$, for time $t = 0,1,\ldots,T$.

(b) Consider the self-financed replicating portfolio strategy for this derivative. Derive a formula for the delta hedging position in the stock, $\Delta_t(\omega_1\omega_2\cdots\omega_t)$, for time $t = 0,1,\ldots,T-1$.

Exercise 7.14 (The Law of One Price for the Binomial Tree Model)**.** Consider two securities with respective discrete-time price processes $\{X_t\}_{0\leqslant t\leqslant T}$ and $\{Y_t\}_{0\leqslant t\leqslant T}$ adapted to the binomial filtration $\mathbb{F} = \{\mathcal{F}_t\}_{0\leqslant t\leqslant T}$. Suppose that $X_T(\omega) = Y_T(\omega)$ for all $\omega \in \Omega_T$. Prove that $X_t(\omega_1\ldots\omega_t) = Y_t(\omega_1\ldots\omega_t)$ for all $0 \leqslant t \leqslant T$ and $\omega_1,\ldots,\omega_t \in \{\mathsf{D},\mathsf{U}\}$, or else an arbitrage opportunity exists.

Exercise 7.15. Consider a T-period binomial arbitrage-free model and a chooser option expiring at time T with payoff

$$(a)\ \Lambda(S) = S \vee K, \quad (b)\ \Lambda(S) = S \wedge K.$$

Derive the risk-neutral pricing formula in terms of the cumulative probability mass function of the binomial distribution $\mathsf{Bin}(n,p)$ given by

$$\mathcal{B}(m;n,p) = \sum_{k=0}^{m} \binom{n}{k} p^k (1-p)^{n-k}, \quad 0 \leqslant m \leqslant n.$$

Exercise 7.16. Consider a T-period binomial arbitrage-free model and a European-style derivative expiring at time T, namely,

(a) an asset-or-nothing call option with payoff

$$\Lambda(S) = \begin{cases} S & \text{if } S > K, \\ 0 & \text{if } S \leqslant K, \end{cases}$$

(b) a bear spread with payoff

$$\Lambda(S) = \begin{cases} K_2 - K_1 & S \leqslant K_1, \\ K_2 - S & K_1 < S \leqslant K_2, \\ 0 & S \geqslant K_2, \end{cases}$$

(c) a strangle with payoff

$$\Lambda(S) = \begin{cases} K_1 - S & S \leqslant K_1, \\ 0 & K_1 < S < K_2, \\ S - K_2 & S \geqslant K_2. \end{cases}$$

Assume that $K_1 < S_0 < K_2$ in (b) and (c). Derive the risk-neutral pricing formula in terms of the cumulative probability mass function of the binomial distribution.

Exercise 7.17. Let $\sigma > 0$ and $r_\infty \geqslant 0$ be, respectively, an annual volatility and annual rate of interest compounded continuously. Consider the following parameterization of an N-period binomial model with the length of each period equal to $\delta t = \frac{1}{N}$:

$$u = \mathrm{e}^{\sigma\sqrt{\delta t}}, \quad d = \frac{1}{u} = \mathrm{e}^{-\sigma\sqrt{\delta t}}, \quad 1 + r = \mathrm{e}^{r_\infty \delta t}.$$

Show that the binomial model is arbitrage free iff $N > \left(\frac{r_\infty}{\sigma}\right)^2$ holds.

Exercise 7.18. Consider a 3-month call option on a dividend-paying stock with spot price $S_0 = \$50$ and strike price $K = \$51$. The dividend is due at the end of month 1 and is equal to \$2.50. Use a three-period binomial model with annual interest rate $r_\infty = 5\%$ and annual volatility $\sigma = 12\%$ parametrized as in Exercise 7.17. Find the no-arbitrage initial values of

(a) the European derivative,

(b) the American derivative.

Exercise 7.19. In the three-period binomial model with $S_0 = 8$, $u = 1.5$, $d = 0.5$, and $1+r = 1.25$, determine the derivative prices, the delta hedging strategy, the consumption process, and the optimal stopping time for the American straddle that expiries at $T = 3$ and has intrinsic value $\Lambda(S) = (8 - S)^+ + (S - 8)^+$.

Exercise 7.20. In the three-period binomial model with $S_0 = 8$, $u = 1.5$, $d = 0.5$, and $1+r = 1.25$, determine the derivative prices, the delta hedging strategy, the consumption process, and the optimal stopping time for the American-Asian put that expiries at time three and whose intrinsic value at each time $t = 0, 1, 2, 3$ is

$$\Lambda_t = \left(4 - \frac{1}{t+1}\sum_{n=0}^{t} S_n\right)^+ .$$

Exercise 7.21. Consider the American derivative of Example 7.20. Find the optimal exercise policy (optimal stopping time) τ^*. Verify that the initial derivative price is equal to

$$\widetilde{\mathrm{E}}\left[\frac{\Lambda_\tau \mathbb{I}_{\{\tau^* \leqslant 3\}}}{(1+r)^{\tau^*}}\right].$$

Chapter 8

General Multi-Asset Multi-Period Model

In this chapter we construct a general multi-asset discrete-time model with a finite state space. Most of the results of Chapter 5 and 7 will be generalized so that a single-period model and a binomial tree model become special cases of a more general framework.

8.1 Main Elements of the Model

Let us describe all main components of a general multi-period model defined on a filtered probability space $(\Omega, \mathcal{F}, \mathbb{P}, \mathbb{F})$.

- There are $T + 1$ trading dates, $t \in \{0, 1, 2, \ldots, T\}$. Denote the collection of trading dates by $\mathbf{T}$.

- A state space $\Omega = \{\omega^1, \omega^2, \ldots, \omega^M\}$ represents all possible final states (or scenarios) of the world at time T.

- A filtration $\mathbb{F} = \{\mathcal{F}_t\}_{t\in\mathbf{T}}$ describes the arrival of information about the market with the passage of time:

$$\Omega = \mathcal{F}_0 \subseteq \mathcal{F}_1 \subseteq \mathcal{F}_2 \subseteq \ldots \subseteq \mathcal{F}_T \equiv \mathcal{F} = 2^\Omega.$$

  For each date $t \in \mathbf{T}$, the respective partition $\mathcal{P}_t$ of Ω that generates $\mathcal{F}_t$ is given by

$$\mathcal{P}_t = \{A_t^1, A_t^2, \ldots, A_t^{k_t}\}. \tag{8.1}$$

  Here A_t^j, $j = 1, 2, \ldots, k_t$, are atoms, and k_t is the number of atoms of $\mathcal{P}_t$. The partitions form an information structure, i.e., $\mathcal{P}_{t-1} \preceq \mathcal{P}_t$ and for all $t = 1, 2, \ldots, T$. Therefore, the numbers k_t form an increasing sequence so that

$$1 = k_0 \leqslant k_1 \leqslant \cdots \leqslant k_T = M.$$

- A probability measure $\mathbb{P}$ on Ω describes the "real-world" probabilities for the possible states of the world, i.e., $p_j = \mathbb{P}(\omega^j) > 0$ is the probability that the economy will be revealed to be in state $\omega^j \in \Omega$ at time T.

The information structure $\{\mathcal{P}_t\}_{t\in\mathbf{T}}$ can be represented by a nonrecombining tree, where each node is an atom and two atoms are connected by an edge if one of the atoms is a subset of the other. We call it the *information tree*. Any final outcome $\omega^j \in \Omega$ is contained in the unique sequence of atoms, one from each partition $\mathcal{P}_t$:

$$\{\omega^j\} = A_T^j \subseteq A_{T-1}^{j_{T-1}} \subseteq \cdots \subseteq A_1^{j_1} \subseteq A_0^1 = \Omega,$$

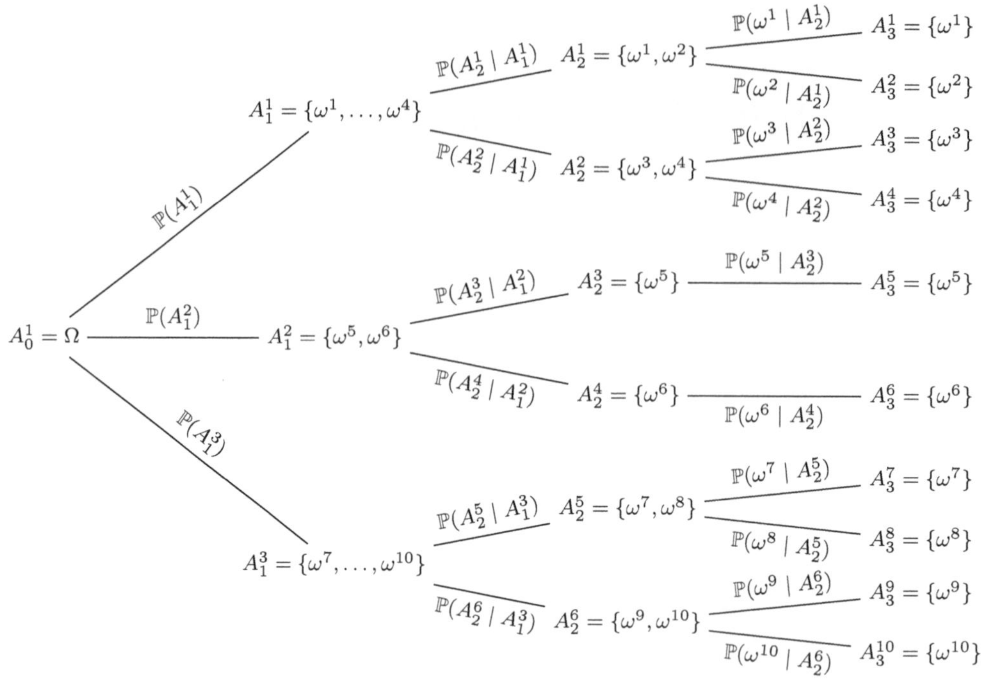

FIGURE 8.1: An information tree with conditional probabilities of going from one atom to another.

where the indices $j_t \in \{1, 2, \ldots, k_t\}$ depend on j. The sequence of atoms

$$A_0^1 \to A_1^{j_1} \to \cdots \to A_{T-1}^{j_{T-1}} \to A_T^j$$

form a path in the information tree that goes from atom $A_0^1 = \Omega$ to atom $A_T^j = \{\omega^j\}$. Such a path uniquely represents ω^j. A particular example of a three-period information tree is given in Figure 8.1.

The probability $\mathbb{P}(\omega^j)$ can be computed as a product of conditional probabilities:

$$\begin{aligned}
\mathbb{P}(\omega^j) &= \mathbb{P}(\{\omega^j\} \mid A_{T-1}^{j_{T-1}}) \cdot \mathbb{P}(A_{T-1}^{j_{T-1}}) \\
&= \mathbb{P}(\{\omega^j\} \mid A_{T-1}^{j_{T-1}}) \cdot \mathbb{P}(A_{T-1}^{j_{T-1}} \mid A_{T-2}^{j_{T-2}}) \cdot \mathbb{P}(A_{T-2}^{j_{T-2}}) \\
&\vdots \\
&= \mathbb{P}(\{\omega^j\} \mid A_{T-1}^{j_{T-1}}) \cdot \mathbb{P}(A_{T-1}^{j_{T-1}} \mid A_{T-2}^{j_{T-2}}) \cdots \mathbb{P}(A_2^{j_2} \mid A_1^{j_1}) \cdot \mathbb{P}(A_1^{j_1}).
\end{aligned} \tag{8.2}$$

Therefore, we may refer to $\mathbb{P}(\omega^j)$ as a *path probability*. Equation (8.2) gives us another method of constructing the probability measure $\mathbb{P}$. We only need to define the conditional probability $\mathbb{P}(A_{t+1} \mid A_t)$ for every pair of atoms $A_t \in \mathcal{P}_t$ and $A_{t+1} \in \mathcal{P}_{t+1}$ with the property $A_{t+1} \subseteq A_t$.

For calculating derivative prices at an intermediate time $t \in \mathbf{T}$ we need conditional expectations of the form $\mathrm{E}_t[X] \equiv \mathrm{E}[X \mid \mathcal{F}_t]$ (where $X\colon \Omega \to \mathbb{R}$ is a random variable). The σ-algebra $\mathcal{F}_t$ is generated by the partition $\mathcal{P}_t$ with k_t atoms, as given in (8.1). Therefore, the conditional expectation $\mathrm{E}_t[X]$ is a random variable that is constant on atoms of $\mathcal{F}_t$; it is given by

$$\mathrm{E}_t[X](\omega) = \sum_{i=1}^{k_t} \mathrm{E}[X \mid A_t^i]\, \mathbb{I}_{A_t^i}(\omega)\,.$$

The conditional expectation of X given atom A_t^i is calculated as follows:

$$\mathrm{E}[X \mid A_t^i] = \frac{\mathrm{E}[X\, \mathbb{I}_{A_t^i}]}{\mathbb{P}(A_t^i)} = \frac{1}{\mathbb{P}(A_t^i)} \sum_{\omega \in A_t^i} X(\omega)\, \mathbb{P}(\omega)\,, \quad i = 1, 2, \ldots, k_t\,.$$

For each state $\omega^j \in A_t^i$, we can find a unique sequence of atoms

$$\{\omega^j\} = A_T^j \subseteq A_{T-1}^{j_{T-1}} \subseteq \cdots \subseteq A_{t+1}^{j_{t+1}} \subseteq A_t^i.$$

Thus, the probability of state ω^j can be calculated as follows:

$$\mathbb{P}(\omega^j) = \mathbb{P}(\{\omega^j\} \mid A_{T-1}^{j_{T-1}})\, \mathbb{P}(A_{T-1}^{j_{T-1}} \mid A_{T-2}^{j_{T-2}}) \cdots \mathbb{P}(A_{t+1}^{j_{t+1}} \mid A_t^i)\, \mathbb{P}(A_t^i)\,.$$

As a result, we can simplify the formula of the conditional expectation to obtain

$$\mathrm{E}[X \mid A_t^i] = \sum_{\omega \in A_t^i} X(\omega)\, \mathbb{P}(\{\omega^j\} \mid A_{T-1}^{j_{T-1}})\, \mathbb{P}(A_{T-1}^{j_{T-1}} \mid A_{T-2}^{j_{T-2}}) \cdots \mathbb{P}(A_{t+1}^{j_{t+1}} \mid A_t^i)\,.$$

In conclusion, let us consider several examples of multi-asset multi-period models.

Example 8.1. Clearly, the standard binomial lattice model fits the definition of a general multi-period model. The state space Ω is the collection of $M = 2^T$ distinct paths in the recombining binomial tree with T periods, where each path is coded by a sequence of the letters D's and U's (of length T). That is, $\omega = \omega_1\omega_2 \ldots \omega_T$, where each $\omega_t \in \{\mathsf{D}, \mathsf{U}\}$ is the time-t market move. The partition $\mathcal{P}_t$ consists of 2^t atoms. Each atom of $\mathcal{P}_t$ contains 2^{T-t} paths, all having identical first t market moves. The probability measure $\mathbb{P}$ on Ω is defined by the probability $p \in (0, 1)$ of a single upward move U. The probability of scenario $\omega \in \Omega$ is found to be

$$\mathbb{P}(\omega) = p^{\#\mathsf{U}(\omega)}(1-p)^{\#\mathsf{D}(\omega)}.$$

Select an atom $A_t \in \mathcal{P}_t$ so that $\omega \in A_t$ iff $\omega_1 = \bar{\omega}_1, \ldots, \omega_t = \bar{\omega}_t$ for some $\bar{\omega}_1, \ldots, \bar{\omega}_t \in \{\mathsf{D}, \mathsf{U}\}$. That is, $A_t = A_{\bar{\omega}_1 \ldots \bar{\omega}_t}$. This atom is a subset of $A_{\bar{\omega}_1 \ldots \bar{\omega}_{t-1}} \in \mathcal{P}_{t-1}$. The conditional probability of $A_{\bar{\omega}_1 \ldots \bar{\omega}_t}$ given $A_{\bar{\omega}_1 \ldots \bar{\omega}_{t-1}}$ is

$$\mathbb{P}(A_{\bar{\omega}_1 \ldots \bar{\omega}_t} \mid A_{\bar{\omega}_1 \ldots \bar{\omega}_{t-1}}) = p^{\#\mathsf{U}(\bar{\omega}_t)}(1-p)^{\#\mathsf{D}(\bar{\omega}_t)}.$$

Therefore, the probability of an individual scenario $\omega \in \Omega$ is computed as

$$\mathbb{P}(\omega) = p^{\#\mathsf{U}(\omega)}(1-p)^{\#\mathsf{D}(\omega)}.$$

Example 8.2. A generalization of the binomial tree model is a *multinomial tree model*, where at each time moment there exist $n \geqslant 2$ possible continuations denoted by M_j for $j = 1, 2, \ldots, n$. Note that for a binomial model we use $\mathsf{D} \equiv \mathsf{M}_1$ and $\mathsf{U} \equiv \mathsf{M}_2$. A scenario in this model is a path in a nonrecombining n-nomial tree with T periods. Each path can be coded by a sequence of length T formed from the symbols $\mathsf{M}_1, \mathsf{M}_2, \ldots, \mathsf{M}_n$:

$$\Omega = \big\{\omega_1\omega_2 \ldots \omega_T \;:\; \omega_1, \omega_2, \ldots, \omega_T \in \{\mathsf{M}_1, \mathsf{M}_2, \ldots, \mathsf{M}_n\}\big\}.$$

Clearly, the state space Ω has $M = n^T$ possible scenarios. The probability measure $\mathbb{P}$ is defined by the collection of probabilities $p_i \in (0, 1)$, $i = 1, 2, \ldots, n$, in which p_i is the probability of the single move M_i. Let $\#\mathsf{M}_i(\omega)$ give the number of M_i in the sequence ω. The probability of $\omega \in \Omega$ is computed as

$$\mathbb{P}(\omega) = p_1^{\#\mathsf{M}_1(\omega)} p_2^{\#\mathsf{M}_2(\omega)} \cdots p_n^{\#\mathsf{M}_n(\omega)}.$$

Example 8.3. Consider a two-period trinomial model. Assume that by the end of period one the economy can end up in one of three possible scenarios. Moreover, for each of these possibilities, there are three alternative continuations throughout period two. Therefore, in total there are nine possible states of the world ($M = 9$) at time $T = 2$. The filtration $\mathbb{F} = \{\mathcal{F}_t\}_{t\in\{0,1,2\}}$ is generated by the information structure $\mathcal{P}_0 \preceq \mathcal{P}_1 \preceq \mathcal{P}_2$ with the following partitions:

$$\begin{aligned}
\mathcal{P}_0 &= \{A_0^1 = \{\omega^1, \omega^2, \ldots, \omega^9\} = \Omega\},\\
\mathcal{P}_1 &= \{A_1^1 = \{\omega^1, \omega^2, \omega^3\}, A_1^2 = \{\omega^4, \omega^5, \omega^6\}, A_1^3 = \{\omega^7, \omega^8, \omega^9\}\},\\
\mathcal{P}_2 &= \{A_2^1 = \{\omega^1\}, A_2^2 = \{\omega^2\}, \ldots, A_2^9 = \{\omega^9\}\}.
\end{aligned}$$

8.2 Assets, Portfolios, and Strategies

8.2.1 Payoffs and Assets

Consider a financial contract maturing at time $T_m \in \{1, 2, \ldots, T\}$. The *payoff* of such a contract (to its holder) is a real-valued function defined on the set of states of the world. Such a function is contingent on the information available at time T_m and is therefore represented by an $\mathcal{F}_{T_m}$-measurable random variable $X\colon \Omega \to \mathbb{R}$. Trivial examples include constant payoffs.

Since the σ-algebra $\mathcal{F}_{T_m}$ is generated by the partition $\mathcal{P}_{T_m}$, the $\mathcal{F}_{T_m}$-measurability of X means that the payoff X is constant on the atoms $A_{T_m}^1, A_{T_m}^2, \ldots, A_{T_m}^{k_{T_m}}$ of $\mathcal{P}_{T_m}$. Thus, payoff X can be represented by the vector

$$X(\mathcal{P}_{T_m}) = \left[X(A_{T_m}^1),\ X(A_{T_m}^2),\ \ldots,\ X(A_{T_m}^{k_{T_m}})\right] \in \mathbb{R}^{k_{T_m}}.$$

A linear combination of $\mathcal{F}_{T_m}$-measurable random variables (i.e., payoffs maturing at the same time T_m) is again an $\mathcal{F}_{T_m}$-measurable random variable (i.e., a payoff maturing at time T_m). Thus, the set of payoffs maturing at time T_m is a vector subspace of the space $\mathcal{L}(\Omega)$. Recall that $\mathcal{L}(\Omega)$ consists of all payoff functions defined on Ω. We denote such a subspace by $\mathcal{L}_{T_m}(\Omega)$.

A European-style asset (also called a security) S maturing at time $T_m \leqslant T$ is described by

- a nonnegative terminal payoff function $S_{T_m}\colon \Omega \to \mathbb{R}_+$ at time T_m,
- nonnegative prices $S_t\colon \Omega \to \mathbb{R}_+$ with $t \in \{0, 1, \ldots, T_m - 1\}$.

Since the asset price S_t is contingent on the information available at time t, it is an $\mathcal{F}_t$-measurable random variable for all t. In other words, the asset S is described by a nonnegative price process $\{S_t\}_{0\leqslant t\leqslant T_m}$ adapted to the filtration $\{\mathcal{F}_t\}_{0\leqslant t\leqslant T_m}$.

The main building blocks of our model are $N+1$ base assets $S^0, S^1, S^2, \ldots, S^N$ all maturing at time T. Here, S^0 is an asset with a strongly positive price process. It usually refers to a bank account or to a money market account. We will adopt a dual notation for such an asset: $S^0 \equiv B$. The other N assets with nonnegative price processes $\{S_t^i\}_{0\leqslant t\leqslant T}$, $i = 1, 2, \ldots, N$, usually refer to equity stocks.

Let us define the short rate process $\{r_t\}_{1\leqslant t\leqslant T}$ as a sequence of one-period returns on B:

$$r_t := \frac{B_t - B_{t-1}}{B_{t-1}} \text{ for all } t \in \{1, 2, \ldots, T\}.$$

The condition that $r_t(\omega) > -1$ for all $\omega \in \Omega$ and for all $t \in \{1, 2, \dots, T\}$ guarantees that the price process $\{B_t\}_{0 \leqslant t \leqslant T}$ is strictly positive (provided that $B_0 > 0$). Additionally, we assume that the short rate process is $\mathbb{F}$-*predictable*, that is, r_t is $\mathcal{F}_{t-1}$-measurable for all $t \in \{1, 2, \dots, T\}$. So the interest rate r_t is known at time $t-1$. Therefore, the bank account process is $\mathbb{F}$-predictable as well.

The bank account B is used here as a numéraire asset. The discounted base asset prices are obtained by dividing the nondiscounted (original) prices by the price of the numéraire at the same time:

$$\overline{S}_t^i := \frac{S_t^i}{B_t} \text{ for } i \in \{0, 1, 2, \dots, N\} \text{ and } t \in \{0, 1, \dots, T\}.$$

Notice that $\overline{S}_t^0 \equiv 1$ for all t.

Example 8.4. Consider the two-period trinomial model from Example 8.3. Let us add three assets to the model, namely, a risk-free cash account B with initial value $B_0 = 1$ and interest rate $r = 0$ and two stocks S^1 and S^2 with respective initial values $S_0^1 = 50$ and $S_0^2 = 100$. The time-1 and time-2 values of the stocks are specified in the following table.

ω	$S_1^1(\omega)$	$S_1^2(\omega)$	$S_2^1(\omega)$	$S_2^2(\omega)$
ω_1	50	115	45	130
ω_2	50	115	45	120
ω_3	50	115	60	95
ω_4	40	115	45	135
ω_5	40	115	35	120
ω_6	40	115	40	105
ω_7	60	70	60	100
ω_8	60	70	55	40
ω_9	60	70	70	70

The asset price processes $\{S_t^1\}_{t \in \{0,1,2\}}$ and $\{S_t^2\}_{t \in \{0,1,2\}}$ are adapted to the filtration since for every $t \in \{0, 1, 2\}$ and every $i \in \{1, 2\}$ the price S_t^i is constant on atoms of the partition $\mathcal{P}_t$. For example, $S_1^1(\omega) = 50$ and $S_2^1(\omega) = 115$ for all $\omega \in \{\omega^1, \omega^2, \omega^3\} = A_1^1 \in \mathcal{P}_1$.

8.2.2 Static and Dynamic Portfolios

A portfolio is a combination of positions in several (or all) base assets. If the positions do not change as time passes by, then we speak of a static portfolio. Let β denote the position in B, where $\beta < 0$ corresponds to a loan and $\beta > 0$ is an investment; let Δ^i be the position in S^i for $i = 1, 2, \dots, N$, where $\Delta^i < 0$ corresponds to shorting the asset and $\Delta^i > 0$ is a long position. Such a static portfolio is represented by the vector

$$\varphi = \left[\beta, \Delta^1, \Delta^2, \cdots, \Delta^N\right] \in \mathbb{R}^{N+1},$$

As opposed to static portfolios in a single-period model, a portfolio held in a multi-period environment can be re-balanced at intermediate dates. The portfolio holder can change or even liquidate some positions and open others. A sequence of re-balanced portfolios in the base assets indexed by time, $\{\varphi_t\}_{0 \leqslant t \leqslant T_m - 1}$, is called a *portfolio strategy* maturing at time $T_m \in \{1, 2, \dots, T\}$. The portfolio φ_t is formed at time t and held from time t to $t+1$. At time $t+1$, φ_t is liquidated and a new portfolio φ_{t+1} is set up. This procedure is repeated for every $t \in \{0, 1, \dots, T_m - 1\}$. At time T_m, the final portfolio $\varphi_{T_m - 1}$ is liquidated and the proceeds are consumed.

The re-balancing of a portfolio at any date is contingent on the information available at that time. Therefore, the re-balancing is done in a way such that φ_t is $\mathcal{F}_t$-measurable for each $t \in \{0, 1, \ldots, T_m - 1\}$. Mathematically speaking, a portfolio strategy is a stochastic vector process adapted to the filtration $\{\mathcal{F}_t\}_{0 \leqslant t \leqslant T_m - 1}$.

Recall the notion of the value (wealth) of a portfolio. The time-t value of a static portfolio φ in the base assets is defined as

$$\Pi_t[\varphi] = \beta B_t + \Delta^1 S_t^1 + \Delta^2 S_t^2 + \cdots + \Delta^N S_t^N.$$

For a portfolio strategy $\mathbf{\Phi} = \{\varphi_t\}_{0 \leqslant t \leqslant T_m - 1}$ maturing at time T_m, in which the portfolio $\varphi_t = \left[\beta_t, \Delta_t^1, \Delta_t^2, \cdots, \Delta_t^N\right]$ is held during the period $[t, t+1]$, we define its *acquisition value* $\Pi_t[\varphi_t]$ and its *liquidation value* $\Pi_{t+1}[\varphi_t]$, respectively, by

$$\Pi_t[\varphi_t] = \beta_t B_t + \Delta_t^1 S_t^1 + \Delta_t^2 S_t^2 + \cdots + \Delta_t^N S_t^N,$$
$$\Pi_{t+1}[\varphi_t] = \beta_t B_{t+1} + \Delta_t^1 S_{t+1}^1 + \Delta_t^2 S_{t+1}^2 + \cdots + \Delta_t^N S_{t+1}^N, \quad t \in \{0, 1, \ldots, T_m - 1\}.$$

At each date $t \in \{1, 2, \ldots, T_m - 1\}$, the portfolio φ_{t-1} is liquidated with proceeds of $\Pi_t[\varphi_{t-1}]$ just before the new portfolio φ_t is acquired for $\Pi_t[\varphi_t]$. This process is referred to as *portfolio re-balancing.*

8.2.3 Self-Financing Strategies

The difference $\Pi_t[\varphi_t] - \Pi_t[\varphi_{t-1}]$ may be positive, meaning that some funds are withdrawn, or negative, meaning that some funds are injected. *Self-financing strategies* do not allow withdrawing or injecting funds at intermediate dates. At each time $t \in \{1, 2, \ldots, T_m - 1\}$, the portfolio value just before re-balancing is exactly the same as the value after re-balancing: $\Pi_t[\varphi_t] = \Pi_t[\varphi_{t-1}]$. A self-financing strategy finances each re-balancing on its own without withdrawing or injecting funds. A trivial example of a self-financing strategy is a static portfolio.

Since for a self-financing strategy there is no need to distinguish between its acquisition and liquidation values, we will only speak of the portfolio value of a self-financing strategy $\mathbf{\Phi}$ given by

$$\Pi_t \equiv \Pi_t^{\mathbf{\Phi}} = \beta_{t-1} B_t + \Delta_{t-1}^1 S_t^1 + \Delta_{t-1}^2 S_t^2 + \cdots + \Delta_{t-1}^N S_t^N, \quad t \in \{1, 2, \ldots, T_m\}.$$

The initial value of a self-financing strategy $\mathbf{\Phi}$ given by

$$\Pi_0 = \beta_0 B_0 + \Delta_0^1 S_0^1 + \Delta_0^2 S_0^2 + \cdots + \Delta_0^N S_0^N$$

is referred to as the *initial cost* (of setting up the strategy). The terminal value Π_{T_m} at maturity is referred to as the *payoff* of $\mathbf{\Phi}$.

The self-financing condition can be written as follows:

$$\begin{aligned}\Pi_{t+1}[\varphi_{t+1}] - \Pi_{t+1}[\varphi_t] &= B_{t+1}(\beta_{t+1} - \beta_t) + \sum_{i=1}^N S_{t+1}^i (\Delta_{t+1}^i - \Delta_t^i) \\ &= \delta\beta_t \cdot B_{t+1} + \sum_{i=1}^N \delta\Delta_t^i \cdot S_{t+1}^i = 0 \qquad (8.3)\end{aligned}$$

for $t \in \{0, 1, \ldots, T_m - 1\}$, where $\delta\beta_t := \beta_{t+1} - \beta_t$ and $\delta\Delta_t^i := \Delta_{t+1}^i - \Delta_t^i$ are single-period changes in the strategy positions. Let the one-period changes in the portfolio value and in the asset prices be given by

$$\delta\Pi_t := \Pi_{t+1} - \Pi_t, \quad \delta B_t := B_{t+1} - B_t, \quad \delta S_t^i := S_{t+1}^i - S_t^i,$$

respectively. The change in the re-balanced portfolio value from time t to time $t+1$ is

$$\begin{aligned}\delta\Pi_t &= \left(B_{t+1}\beta_{t+1} + \sum_{i=1}^{N} S^i_{t+1}\Delta^i_{t+1}\right) - \left(B_t\beta_t + \sum_{i=1}^{N} S^i_t\Delta^i_t\right)\\ &= B_{t+1}(\beta_t + \delta\beta_t) + \sum_{i=1}^{N} S^i_{t+1}(\Delta^i_t + \delta\Delta^i_t) - \left(B_t\beta_t + \sum_{i=1}^{N} S^i_t\Delta^i_t\right)\\ &= \beta_t\,\delta B_t + \sum_{i=1}^{N}\Delta^i_t\,\delta S^i_t.\end{aligned}$$

As is seen, the self-financing condition is equivalent to the statement that the changes in portfolio values are due only to changes in base asset prices.

The self-financing condition imposes restrictions on the positions in the base assets. At any time t, the portfolio value Π_t and the delta-positions of a self-financing strategy determine the position β in the bank account B:

$$\Pi_t = \beta_t B_t + \sum_{i=1}^{N}\Delta^i_t S^i_t \implies \beta_t = \frac{\Pi_t - \sum_{i=1}^{N}\Delta^i_t S^i_t}{B_t}.$$

Moreover, we can show that the value process $\{\Pi_t\}_{0\leqslant t\leqslant T_m}$ is governed by a wealth equation which is similar to (7.6) derived in the case with only two base assets:

$$\Pi_t = \sum_{i=1}^{N}\Delta^i_{t-1}S^i_t + \left(\Pi_{t-1} - \sum_{i=1}^{N}\Delta^i_{t-1}S^i_{t-1}\right)(1+r_t), \quad 1\leqslant t\leqslant T_m. \tag{8.4}$$

As a result, a self-financing strategy, whose value dynamics follow (8.4), can be described by the delta strategy $\{\Delta_t\}_{0\leqslant t\leqslant T_m-1}$ and the initial cost Π_0.

Recall that a linear combination of two portfolios, which is merely a linear combination of two vectors, is a portfolio again. The linear combination of two strategies $\mathbf{\Phi} = \{\varphi_t\}_{t\geqslant 0}$ and $\mathbf{\Psi} = \{\psi_t\}_{t\geqslant 0}$ maturing at the same time T_m is defined by taking a linear combination of the portfolios φ_t and ψ_t at each time $t \in \{0, 1, \ldots, T_m - 1\}$. Any linear combination $a\mathbf{\Phi} + b\mathbf{\Psi}$ with $a, b \in \mathbb{R}$, of two self-financing strategies $\mathbf{\Phi}$ and $\mathbf{\Psi}$ maturing at the same time T_m is again a self-financing strategy with the same maturity. The proof follows from the fact the acquisition and liquidation values are linear functions of the positions. For example, the liquidation time-t value of the strategy $a\mathbf{\Phi} + b\mathbf{\Psi}$ is

$$\begin{aligned}\Pi_t[a\varphi_{t-1} + b\psi_{t-1}] &= B_t\left(a\beta^{\mathbf{\Phi}}_{t-1} + b\beta^{\mathbf{\Psi}}_{t-1}\right) - \sum_{i=1}^{N} S^i_t\left(a\Delta^{\mathbf{\Phi}}_{t-1} + b\Delta^{\mathbf{\Psi}}_{t-1}\right)\\ &= a\left(B_t\beta^{\mathbf{\Phi}}_{t-1} + \sum_{i=1}^{N} S^i_t\Delta^{\mathbf{\Phi}}_{t-1}\right) + b\left(B_t\beta^{\mathbf{\Psi}}_{t-1} + \sum_{i=1}^{N} S^i_t\Delta^{\mathbf{\Psi}}_{t-1}\right)\\ &= a\Pi_t[\varphi_{t-1}] + b\Pi_t[\psi_{t-1}].\end{aligned}$$

Applying the self-financing condition (8.3) to the strategy $a\mathbf{\Phi} + b\mathbf{\Psi}$ gives

$$\begin{aligned}\Pi_t[a\varphi_t + b\psi_t] - \Pi_t[a\varphi_{t-1} + b\psi_{t-1}] &= a\left(\Pi_{t+}[\varphi_t] - \Pi_t[\varphi_{t-1}]\right) + b\left(\Pi_{t+}[\psi_t] - \Pi_t[\psi_{t-1}]\right)\\ &= a\cdot 0 + b\cdot 0 = 0,\end{aligned}$$

that is, $a\mathbf{\Phi}+b\mathbf{\Psi}$ is a self-financing strategy. In particular, the sum $\mathbf{\Phi}+\mathbf{\Psi}$ and the multipliers $a\mathbf{\Phi}$ and $b\mathbf{\Psi}$ are all self-financing strategies.

Suppose that we are given a self-financing strategy $\mathbf{\Phi} = \{\varphi_t\}_{0\leqslant t\leqslant T_m-1}$ maturing at time T_m. The terminal value (at time T_m) of this strategy is $\Pi^{\mathbf{\Phi}}_{T_m}$. We can lock in this value by liquidating the portfolio φ_{T_m-1} at time T_m and investing all proceeds in the bank account with strictly positive returns. The new strategy $\mathbf{\Phi}' = \{\varphi'_t\}_{0\leqslant t\leqslant T-1}$ is thus defined by

$$\varphi'_t := \begin{cases} \varphi_t & \text{if } 0 \leqslant t \leqslant T_m - 1, \\ \left[\overline{\Pi}, 0, \cdots, 0\right] & \text{if } t \geqslant T_m, \end{cases} \tag{8.5}$$

where $\overline{\Pi} := \overline{\Pi}^{\mathbf{\Phi}}_{T_m} = \frac{\Pi^{\mathbf{\Phi}}_{T_m}}{B_{T_m}}$ is the discounted value of the terminal value of the strategy $\mathbf{\Phi}$. We refer to the trading strategy $\mathbf{\Phi}'$ as the strategy obtained by *locking the portfolio value* of $\mathbf{\Phi}$ at maturity. This new strategy is now maturing at time T. Thus, any trading strategy can be converted into a strategy maturing at the terminal time T.

8.2.4 Replication of Payoffs

Consider a financial contract maturing at time $T_m \leqslant T$ with payoff V_{T_m}. As usual, our objective is to find a fair initial price of such a contract. In the multi-period world, this goal can be achieved by constructing a dynamic portfolio strategy that replicates the payoff function. The value of the contract is then given by the cost of setting up the replication strategy.

Definition 8.1. A payoff V_{T_m} maturing at time T_m is said to be *attainable* if there exists a self-financing strategy $\mathbf{\Phi} = \{\varphi_t\}_{0\leqslant t\leqslant T_m-1}$ maturing at time T_m such that its terminal value coincides with the payoff for all possible market scenarios: $\Pi^{\mathbf{\Phi}}_{T_m} = V_{T_m}$. Any trading strategy with this property is called a *replicating* or *hedging strategy* for the payoff V_{T_m}. A market model is said to be *complete* if every payoff for every maturity is attainable.

The initial fair value V_0 of a contract with payoff V_{T_m} maturing at time T_m is given by the cost of setting up a strategy $\mathbf{\Phi}$ that replicates V_{T_m}:

$$V_0 = \Pi^{\mathbf{\Phi}}_0, \text{ where } \Pi^{\mathbf{\Phi}}_{T_m} = X_{T_m}. \tag{8.6}$$

If there exist several replicating strategies (all having the same terminal value), then their initial values are expected to be the same. In other words, the Law of One Price is expected to hold for each maturity T_m. Otherwise, the definition (8.6) is meaningless.

For a multi-period model, the Law of One Price is stated as follows: any two self-financing strategies maturing at the same time and having the same terminal value have the same initial value. If the Law of One Price does not hold, then for each attainable payoff there exists a replicating strategy with an arbitrary pre-specified initial value. The proof of this statement is identical to that for single-period models considered in Chapter 5 (see Proposition 5.7). As in the single-period case, we can show that the Law of One Price holds iff each strategy replicating a zero payoff has initial value zero (see Lemma 5.6).

8.3 Fundamental Theorems of Asset Pricing

8.3.1 Arbitrage Strategies

Recall that an *admissible strategy* is a self-financing strategy with nonnegative values for all dates from time zero until the maturity. An *arbitrage strategy* is an admissible strategy with zero initial value and positive terminal value. In other words, a self-financing strategy $\mathbf{\Phi} = \{\varphi_t\}_{0 \leqslant t \leqslant T_m - 1}$ maturing at time T_m is an arbitrage opportunity if the conditions

$$\Pi_0^{\mathbf{\Phi}} = 0, \quad \Pi_t^{\mathbf{\Phi}} \geqslant 0 \text{ for all } t \in \{1, 2, \ldots, T_m\}, \quad \text{and } \mathbb{P}(\Pi_{T_m}^{\mathbf{\Phi}} > 0) > 0$$

all hold.

When we say that the market model admits no arbitrage opportunity, we mean that the model admits no arbitrage strategies for any maturity $T_m \in \{1, 2, \ldots, T\}$. The same convention is assumed when we say that the Law of One Price holds. However, it is sufficient to only consider strategies maturing at time T, as is shown in the following lemma.

Lemma 8.1. *Consider a multi-period, finite-state model.*

(a) The Law of One price holds for all maturities iff it holds for strategies maturing at time T.

(b) There are no arbitrage strategies for all maturities iff the market admits no arbitrage opportunities for the maturity T.

Proof. The proof of the sufficiency part is trivial. Let us prove the necessity part. Being given a strategy with maturity $T_m \leqslant T$, we can construct another strategy with maturity T by locking the portfolio value at time T_m.

(a) As was noticed above, the Law of One Price holds for maturity $T_m \leqslant T$ iff each strategy replicating the zero claim at time T_m has initial value zero. Let $\mathbf{\Phi} = \{\varphi_t\}_{0 \leqslant t \leqslant T_m - 1}$ be a strategy replicating the zero paryoff $X_{T_m} \equiv 0$. Then the strategy $\mathbf{\Phi}' = \{\varphi_t'\}_{0 \leqslant t \leqslant T-1}$ defined by

$$\varphi_t' := \begin{cases} \varphi_t & \text{if } 0 \leqslant t \leqslant T_m - 1 \\ \mathbf{0} & \text{if } t \geqslant T_m \end{cases}$$

is self-financing and replicates the zero claim with maturity T. If the Law of One Price holds for maturity T, then $\Pi_0[\mathbf{\Phi}] = \Pi_0[\mathbf{\Phi}'] = 0$. Therefore, the Law of One Price holds for maturity T_m.

(b) Assume that there are no arbitrage strategies with maturity T. Let us prove that there are no arbitrage opportunities for any time $T_m \leqslant T$. Indeed, if $\mathbf{\Phi} = \{\varphi_t\}_{0 \leqslant t \leqslant T_m - 1}$ is an arbitrage strategy for time T_m, then the strategy $\mathbf{\Phi}'$ constructed by (8.5) is an arbitrage opportunity for time T, which contradicts the assumption. □

8.3.2 Enhancing the Law of One Price

In Chapter 5, we proved that the no-arbitrage condition implies that the Law of One Price holds for static portfolios. Let us show that if no arbitrage strategy exists then the Law of One Price holds in a stronger sense: the strategies replicating the same payoff have the same value process. In other words, if two self-financing strategies $\mathbf{\Phi}$ and $\mathbf{\Psi}$ have the same terminal value, i.e., $\Pi_{T_m}^{\mathbf{\Phi}} = \Pi_{T_m}^{\mathbf{\Psi}}$, then not only their initial values $\Pi_0^{\mathbf{\Phi}}$ and $\Pi_0^{\mathbf{\Psi}}$ are the same, but at every intermediate date $t \in \{0, 1, \ldots, T_m - 1\}$ the values $\Pi_t^{\mathbf{\Phi}}$ and $\Pi_t^{\mathbf{\Psi}}$ are the same. First, we shall prove that any self-financing strategy with a nonnegative payoff is admissible provided that there are no arbitrage opportunities.

Lemma 8.2. *Assume the absence of arbitrage. Let $\mathbf{\Phi} = \{\varphi_t\}_{0 \leqslant t \leqslant T_m - 1}$ be a self-financing strategy maturing at time T_m.*

(i) If $\Pi_{T_m}^{\mathbf{\Phi}} \geqslant 0$ holds, then $\Pi_t^{\mathbf{\Phi}} \geqslant 0$ holds for all $t \in \{0, 1, 2, \ldots, T_m\}$.

(ii) If $\Pi^{\Phi}_{T_m} = 0$ *holds (i.e.,* Φ *replicates the zero claim), then* $\Pi^{\Phi}_t = 0$ *holds for all* $t \in \{0, 1, 2, \ldots, T_m\}$.

Proof. Suppose that statement (i) is not true. Let t_* be the largest time $t \in \{0, 1, \ldots, T_m - 1\}$ such that $\Pi^{\Phi}_t \geqslant 0$ does not hold. That is, there exists a state of the world $\omega_* \in \Omega$ for which $\Pi^{\Phi}_{t_*}(\omega_*) < 0$. We also have that $\Pi^{\Phi}_t \geqslant 0$ for all $t_* < t \leqslant T_m$. Let $A_* \in \mathcal{P}_{t_*}$ be the atom that contains the state ω_*. Since $\Pi^{\Phi}_{t_*}$ is $\mathcal{F}_{t_*}$-measurable, it is constant (and negative) on the atom A_*. Define a new self-financing strategy $\Psi = \{\psi_t\}_{0 \leqslant t \leqslant T_m - 1}$ as follows:

$$\psi_t(\omega) = \begin{cases} 0 & \text{if } 0 \leqslant t < t_* \text{ or } \omega \notin A_*, \\ \varphi_t(\omega) - \varphi_*(\omega) & \text{if } t_* \leqslant t < T_m \text{ and } \omega \in A_*, \end{cases}$$

where $\varphi_* = \left[\frac{\Pi^{\Phi}_{t_*}}{B_{t_*}}, 0, \cdots, 0\right]$. The value process of the strategy Ψ is zero for $t < t_*$. For $t > t_*$, the value of the strategy is

$$\Pi_t[\psi_t] = \mathbb{I}_{A_*} \cdot \left(\Pi^{\Phi}_t - \frac{\Pi^{\Phi}_{t_*}}{B_{t_*}} B_t\right).$$

At time t_*, the self-financing condition holds:

$$\Pi_{t_*}[\psi_{t_*-1}] = 0 = \Pi_{t_*}[\psi_{t_*}] = \mathbb{I}_{A_*} \cdot \left(\Pi^{\Phi}_{t_*} - \frac{\Pi^{\Phi}_{t_*}}{B_{t_*}} B_{t_*}\right).$$

The value Π^{Ψ}_t is nonnegative for all $t > t_*$ since $\Pi^{\Phi}_t \geqslant 0$ and $-\frac{\Pi^{\Phi}_{t_*}}{B_{t_*}} B_t = \frac{|\Pi^{\Phi}_{t_*}|}{B_{t_*}} B_t > 0$. Moreover, $\Pi^{\Psi}_t(\omega) > 0$ for all $\omega \in A_*$ and all $t > t_*$. Therefore, Ψ is an arbitrage strategy for maturity T_m, contradicting the no-arbitrage assumption. Hence, our supposition is wrong and statement (i) holds.

Now, assume that $\Pi^{\Phi}_{T_m} = 0$. By statement (i), which has been just proved, we have that

$$\Pi^{\Phi}_{T_m} \geqslant 0 \implies \Pi^{\Phi}_t \geqslant 0 \text{ for all } t \in \mathbf{T}_m,$$

where $\mathbf{T}_m := \{0, 1, \ldots, T_m\}$. On the other hand $\Pi^{-\Phi}_{T_m} = 0$, hence

$$\Pi^{-\Phi}_{T_m} \geqslant 0 \implies \Pi^{-\Phi}_t \geqslant 0 \text{ for all } t \in \mathbf{T}_m \implies \Pi^{\Phi}_t \leqslant 0 \text{ for all } t \in \mathbf{T}_m.$$

Therefore, $\Pi^{\Phi}_t = 0$ for all $t \in \mathbf{T}_m$. □

As follows from Lemma 8.2, the initial no-arbitrage price of a positive payoff is strictly positive. Moreover, any intermediate price Π_t of such a payoff is strictly positive at some atoms of the partition $\mathcal{P}_t$, otherwise an arbitrage opportunity exists. Recall that $X \in \mathrm{L}(\Omega)$ is said to be positive if $X(\omega) \geqslant 0$ for all $\omega \in \Omega$ and $X(\omega^*) > 0$ for at least one $\omega^* \in \Omega$.

A stronger version of the Law of One Price follows from Lemma 8.2.

Corollary 8.3. *Assume the absence of arbitrage. Suppose that two self-financing strategies* Φ *and* Ψ *both maturing at* T_m *have the same terminal value, i.e.,* $\Pi^{\Phi}_{T_m} = \Pi^{\Psi}_{T_m}$*. Then* $\Pi^{\Phi}_t = \Pi^{\Psi}_t$ *holds for all* $0 \leqslant t \leqslant T_m$.

Proof. The new self-financing strategy $\Phi - \Psi$ replicates the zero claim maturing at T_m. As follows from statement (ii) of Lemma 8.2, all intermediate-time values of $\Phi - \Psi$ are zero. By using the linearity of the value function, we obtain

$$\Pi_t[\Phi - \Psi] = 0 \implies \Pi_t[\Phi] - \Pi_t[\Psi] = 0 \implies \Pi_t[\Phi] = \Pi_t[\Psi],$$

for all $0 \leqslant t \leqslant T_m$. □

8.3.3 Equivalent Martingale Measures

Recall the concept of a *numéraire asset.* In a discrete-time model, a *numéraire* is any trading asset g with a strictly positive price process, that is, $g_0 > 0$ and $\mathbb{P}(g_t > 0) = 1$ for all $t \in \{1, 2, \ldots, T\}$. The bank account B satisfies the above definition and it will be our primary choice for a numéraire asset.

Definition 8.2. A probability distribution $\widetilde{\mathbb{P}}^{(g)}$ on the state space Ω is called an equivalent martingale measure (EMM) or a risk-neutral measure relative to numéraire g if

(1) $\widetilde{\mathbb{P}}^{(g)}$ is equivalent to the real-world probability distribution $\mathbb{P}$ (that is, $\widetilde{\mathbb{P}}^{(g)}(\omega) > 0$ for all ω from a finite state space Ω);

(2) for each base asset the discounted price process is a $\widetilde{\mathbb{P}}^{(g)}$-martingale, that is,

$$\widetilde{\mathrm{E}}^{(g)}\left[\frac{S^i_{t+1}}{g_{t+1}} \,\Big|\, \mathcal{F}_t\right] = \frac{S^i_t}{g_t},$$

for all $i \in \{0, 1, 2, \ldots, N\}$ and all $t \in \{0, 1, \ldots, T-1\}$.

Our primary choice for a numéraire asset will be the bank account B. Since B_t discounted by B_t is identically one, a probability measure $\widetilde{\mathbb{P}} \equiv \widetilde{\mathbb{P}}^{(g=B)}$ is a martingale measure if the discounted base asset price processes $\{\overline{S}^i_t := \frac{S^i_t}{B_t}\}_{0 \leqslant t \leqslant T}$, $i = 1, 2, \ldots, N$, are all $\widetilde{\mathbb{P}}$-martingales.

Let us review some properties of an EMM which were first discussed in Chapter 7.

Lemma 8.4. *A probability measure $\widetilde{\mathbb{P}}$ is an EMM with a numéraire asset g iff for any self-financing strategy, the value process discounted by g is a $\widetilde{\mathbb{P}}$-martingale as well.*

Proof. Consider a self-financing strategy $\{\varphi_t = [\beta_t, \Delta^1_t, \cdots, \Delta^N_t]\,; 0 \leqslant t \leqslant T-1\}$, and let $\widetilde{\mathbb{P}} \equiv \widetilde{\mathbb{P}}^{(g)}$ be an EMM. Define the discounted portfolio value process:

$$\overline{\Pi}_t := \frac{\Pi_t}{g_t}, \quad 0 \leqslant t \leqslant T.$$

Now, compute the conditional risk-neutral expectation of $\overline{\Pi}_{t+1}$ given $\mathcal{F}_t$:

$$\widetilde{\mathrm{E}}_t[\overline{\Pi}_{t+1}] = \widetilde{\mathrm{E}}_t\left[\beta_t \overline{B}_{t+1} + \Delta^1_t \overline{S}^1_{t+1} + \cdots + \Delta^N_t \overline{S}^N_{t+1}\right]$$

(use the linearity and the fact that φ_t is adapted to $\mathcal{F}_t$)

$$\begin{aligned} &= \beta_t \widetilde{\mathrm{E}}_t\left[\overline{B}_{t+1}\right] + \Delta^1_t \widetilde{\mathrm{E}}_t\left[\overline{S}^1_{t+1}\right] + \cdots + \Delta^N_t \widetilde{\mathrm{E}}_t\left[\overline{S}^N_{t+1}\right] \\ &= \beta_t \overline{B}_t + \Delta^1_t \overline{S}^1_t + \cdots + \Delta^N_t \overline{S}^N_t \end{aligned}$$

(use the self-financing condition)

$$= \overline{\Pi}_t.$$

That is, $\widetilde{\mathrm{E}}_t[\overline{\Pi}_{t+1}] = \overline{\Pi}_t$ for any $t \in \{0, 1, \ldots, T-1\}$, which is the martingale condition for the discounted value process. The converse is obvious, since any base asset forms a static portfolio with only one nonzero position. Since a static portfolio is a particular case of a self-financing strategy, for each base asset the discounted price process is a $\widetilde{\mathbb{P}}$-martingale. □

The martingale property of self-financing strategies is the key to pricing derivatives. Suppose that the EMM $\widetilde{\mathbb{P}}^{(g)}$ exists. The time-t no-arbitrage value V_t of an attainable payoff V_{T_m} has to be equal to the time-t cost of a strategy $\mathbf{\Phi}$ that replicates V_{T_m}:

$$V_t = \Pi_t^{\mathbf{\Phi}} \text{ for any time } t \text{ with } 0 \leqslant t \leqslant T_m \leqslant T \tag{8.7}$$

Then, thanks to the fact that the discounted portfolio value process $\left\{ \overline{\Pi}_t^{\mathbf{\Phi}} \right\}$ is a $\widetilde{\mathbb{P}}^{(g)}$-martingale and $\Pi_{T_m}^{\mathbf{\Phi}} = V_{T_m}$ holds at maturity T_m, the price V_t is given by the risk-neutral mathematical expectation of the payoff function V_{T_m}:

$$V_t = g_t \widetilde{\mathrm{E}}_t^{(g)} \left[\frac{\Pi_{T_m}^{\mathbf{\Phi}}}{g_{T_m}} \right] = g_t \widetilde{\mathrm{E}}_t^{(g)} \left[\frac{V_{T_m}}{g_{T_m}} \right] \text{ or, equivalently, } \frac{V_t}{g_t} = \widetilde{\mathrm{E}}_t^{(g)} \left[\frac{V_{T_m}}{g_{T_m}} \right]. \tag{8.8}$$

That is, the discounted derivative price process $\left\{ \overline{V}_t := \frac{V_t}{g_t} \right\}$ is a $\widetilde{\mathbb{P}}^{(g)}$-martingale.

To apply the risk-neutral pricing formula (8.8) we only need to construct the EMM $\widetilde{\mathbb{P}}^{(g)}$. In the next section we show how this problem is reduced to solution of a system of linear equations.

8.3.4 Calculation of Martingale Measures

Since the state space Ω is finite, the computation of the martingale measure $\widetilde{\mathbb{P}}$ (relative to the numéraire $g = B$) reduces to computing $\widetilde{\mathbb{P}}(\omega)$ for each $\omega \in \Omega$. As was pointed out in the beginning of this chapter, any state of the world $\omega^j \in \Omega$ lies in a unique sequence of atoms, one from each partition $\mathcal{P}_t$:

$$\{\omega^j\} = A_T^j \subseteq A_{T-1}^{j_{T-1}} \subseteq \cdots \subseteq A_1^{j_1} \subseteq A_0^1 = \Omega.$$

By using (8.2), we can calculate the risk-neutral probability $\widetilde{\mathbb{P}}(\omega^j)$ as a product of conditional probabilities:

$$\widetilde{\mathbb{P}}(\omega^j) = \widetilde{\mathbb{P}}(\{\omega^j\} \mid A_{T-1}^{j_{T-1}}) \cdot \widetilde{\mathbb{P}}(A_{T-1}^{j_{T-1}} \mid A_{T-2}^{j_{T-2}}) \cdots \widetilde{\mathbb{P}}(A_2^{j_2} \mid A_1^{j_1}) \cdot \widetilde{\mathbb{P}}(A_1^{j_1}).$$

Thus, our goal is to calculate the conditional probability $\widetilde{\mathbb{P}}(A_{t+1} \mid A_t)$ for every pair of atoms $A_t \in \mathcal{P}_t$ and $A_{t+1} \in \mathcal{P}_{t+1}$ such that $A_{t+1} \subseteq A_t$. Fix arbitrarily time $t \in \{0, 1, \ldots, T-1\}$ and atom $A_t \in \mathcal{P}_t$. The partition $\mathcal{P}_{t+1}$ is obtained by refining the partition $\mathcal{P}_t$. Suppose that the atom A_t is a union of $\ell \geqslant 1$ atoms

$$A_{t+1}^1, A_{t+1}^2, \ldots, A_{t+1}^\ell \in \mathcal{P}_{t+1}. \tag{8.9}$$

Let us find ℓ conditional probabilities $\widetilde{\mathbb{P}}(A_{t+1}^j \mid A_t)$, $j = 1, 2, \ldots, \ell$, such that for each base asset S^i the discounted price process satisfies the martingale condition

$$\widetilde{\mathrm{E}}_t \left[\overline{S}_{t+1}^i \right] = \overline{S}_t^i, \; t \geqslant 0.$$

Since $\overline{S}_t^i$ and $\widetilde{\mathrm{E}}_t \left[\overline{S}_{t+1}^i \right]$ are both $\mathcal{F}_t$-measurable random variables, they are constant on atoms of $\mathcal{P}_t$. Therefore, for the atom A_t, we can write $N+1$ equations (one for each $i = 0, 1, 2, \ldots, N$) of the form

$$\overline{S}_t^i(A_t) = \widetilde{\mathrm{E}}_t \left[\overline{S}_{t+1}^i \right](A_t) = \widetilde{\mathrm{E}} \left[\overline{S}_{t+1}^i \mid A_t \right] = \sum_{j=1}^{\ell} \overline{S}_{t+1}^i(A_{t+1}^j) \widetilde{\mathbb{P}}(A_{t+1}^j \mid A_t). \tag{8.10}$$

When $i = 0$, Equation (8.10) reduces to

$$1 = \sum_{j=1}^{\ell} \widetilde{\mathbb{P}}(A_{t+1}^j \mid A_t),$$

since $\overline{S}^0 \equiv 1$. The above equation means that $\widetilde{\mathbb{P}}(\,\cdot \mid A_t)$ is a probability measure on the state space $\widehat{\Omega} = \left\{\widehat{\omega}^j := A_{t+1}^j;\ j = 1, 2, \ldots \ell\right\}$. Thus, we obtain a system of $N+1$ equations with ℓ unknowns $q_j = \widetilde{\mathbb{P}}(A_{t+1}^j \mid A_t)$, $j = 1, 2, \ldots, \ell$,

$$\begin{cases} \sum_{j=1}^{\ell} \overline{S}_{t+1}^1(A_{t+1}^j) q_j = \overline{S}_t^1(A_t), \\ \vdots \\ \sum_{j=1}^{\ell} \overline{S}_{t+1}^N(A_{t+1}^j) q_j = \overline{S}_t^N(A_t), \\ \sum_{j=1}^{\ell} q_j = 1. \end{cases} \tag{8.11}$$

Such a system needs to be solved for every atom of the partitions $\mathcal{P}_0, \mathcal{P}_1, \ldots, \mathcal{P}_{T-1}$. In total, there are $1 + k_1 + k_2 + \cdots + k_{T-1}$ systems like (8.11). Solving all these systems gives us all the conditional probabilities of the form $\widetilde{\mathbb{P}}(A_{t+1} \mid A_t)$. As a result, we can calculate the probability of every state ω (and hence the probability of any event) using Equation (8.2). Let us illustrate this method of computation of a martingale measure by the following example.

Example 8.5. Consider the two-period trinomial model with three base assets as given in Examples 8.3 and 8.4. Find the EMM $\widetilde{\mathbb{P}}^{(B)}$.

Solution. Since the interest rate is zero, discounting does not change the stock prices. First, let us find the probabilities $q_k := \widetilde{\mathbb{P}}(A_1^k)$, $k = 1, 2, 3$, for atoms of $\mathcal{P}_1$ by solving the following system of linear equations:

$$\begin{cases} 50q_1 + 40q_2 + 60q_3 = 50 \\ 115q_1 + 115q_2 + 70q_3 = 100 \\ q_1 + q_2 + q_3 = 1 \end{cases} \iff \begin{bmatrix} q_1 \\ q_2 \\ q_3 \end{bmatrix} = \begin{bmatrix} 1/3 \\ 1/3 \\ 1/3 \end{bmatrix}.$$

At time $t = 1$, there are three trinomial sub-models, hence we need to solve three systems of linear equations to find the conditional probabilities $q_j^k := \widetilde{\mathbb{P}}(A_2^j \mid A_1^k)$ for $j = 1, 2, \ldots, 9$ and $k = 1, 2, 3$ as follows:

$$\begin{cases} 45q_1^1 + 45q_2^1 + 60q_3^1 = 50 \\ 130q_1^1 + 120q_2^1 + 95q_3^1 = 115 \\ q_1^1 + q_2^1 + q_3^1 = 1 \end{cases} \iff \begin{bmatrix} q_1^1 \\ q_2^1 \\ q_3^1 \end{bmatrix} = \begin{bmatrix} 1/3 \\ 1/3 \\ 1/3 \end{bmatrix},$$

$$\begin{cases} 45q_4^2 + 35q_5^2 + 40q_6^2 = 40 \\ 135q_4^2 + 120q_5^2 + 105q_6^2 = 115 \\ q_4^2 + q_5^2 + q_6^2 = 1 \end{cases} \iff \begin{bmatrix} q_4^2 \\ q_5^2 \\ q_6^2 \end{bmatrix} = \begin{bmatrix} 2/9 \\ 2/9 \\ 5/9 \end{bmatrix},$$

$$\begin{cases} 60q_7^3 + 55q_8^3 + 70q_9^3 = 60 \\ 100q_7^3 + 40q_8^3 + 70q_9^3 = 70 \\ q_7^3 + q_8^3 + q_9^3 = 1 \end{cases} \iff \begin{bmatrix} q_7^3 \\ q_8^3 \\ q_9^3 \end{bmatrix} = \begin{bmatrix} 2/5 \\ 2/5 \\ 1/5 \end{bmatrix}.$$

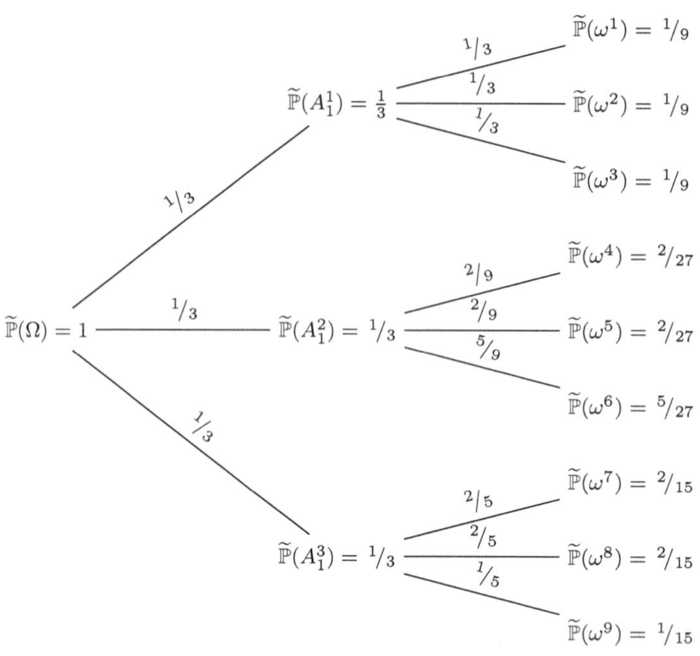

FIGURE 8.2: The tree with state probabilities.

All the above systems have positive solutions, thus the EMM exists and the model is arbitrage-free. The state probabilities can be computed as follows:

$$\tilde{p}_j \equiv \widetilde{\mathbb{P}}(\omega^j) = \widetilde{\mathbb{P}}(A_2^j \mid A_1^k) \cdot \widetilde{\mathbb{P}}(A_1^k) = q_j^k \cdot q_k,$$

where $k = 1$ if $j = 1, 2, 3$, $k = 2$ if $j = 4, 5, 6$, and $k = 3$ if $j = 7, 8, 9$. Hence, we obtain

$$\tilde{p}_1 = \frac{1}{9},\ \tilde{p}_2 = \frac{1}{9},\ \tilde{p}_3 = \frac{1}{9},\ \tilde{p}_4 = \frac{2}{27},\ \tilde{p}_5 = \frac{2}{27},\ \tilde{p}_6 = \frac{5}{27},\ \tilde{p}_7 = \frac{2}{15},\ \tilde{p}_8 = \frac{2}{15},\ \tilde{p}_9 = \frac{1}{15}.$$

The final solution is illustrated in Figure 8.2. □

Example 8.6. Consider the two-period trinomial model with three base assets as given in Examples 8.3 and 8.4. Find the no-arbitrage prices V_0 and V_1 of a basket call option maturing at time $T = 2$ with the payoff $V_2(\omega) = (\max\{2S_2^1(\omega), S_2^2(\omega)\} - 120)^+$, $\omega \in \Omega$.

Solution. First, calculate the payoff function $V_2(\omega)$ for all possible scenarios ω as follows:

ω	ω^1	ω^2	ω^3	ω^4	ω^5	ω^6	ω^7	ω^8	ω^9
$2S_2^1(\omega)$	90	90	120	90	70	80	120	110	140
$S_2^1(\omega)$	130	120	95	135	120	105	100	40	70
$\max\{2S_2^1(\omega), S_2^2(\omega)\}$	130	120	120	135	120	105	120	110	140
$V_2(\omega)$	10	20	20	15	10	0	10	0	20

Applying the pricing formula (8.8) to the payoff V_2 gives

$$
\begin{aligned}
V_0 &= B_0 \sum_{j=1}^{9} \frac{V_2(\omega^j)}{B_2(\omega^j)} \tilde{p}_j = \sum_{j=1}^{9} V_2(\omega^j)\tilde{p}_j \quad (\text{since } B_t \equiv 1) \\
&= \frac{1}{9} \cdot 10 + \frac{1}{9} \cdot 20 + \frac{1}{9} \cdot 20 + \frac{2}{27} \cdot 15 + \frac{2}{27} \cdot 10 + \frac{5}{27} \cdot 0 + \frac{2}{15} \cdot 10 + \frac{2}{15} \cdot 0 + \frac{1}{15} \cdot 20 \\
&= \frac{272}{27} \cong 10.074.
\end{aligned}
$$

The time-1 price V_1 is an $\mathcal{F}_1$-measurable random variable. It is constant on atoms of $\mathcal{P}_1$ and is given by

$$
V_1(A_1^i) = B_1(A_1^i) \sum_{j=1}^{9} \frac{V_2(\omega^j)}{B_2(\omega^j)} \widetilde{\mathbb{P}}(\{\omega^j\} \mid A_1^i) = \sum_{j=1}^{9} V_2(\omega^j) \widetilde{\mathbb{P}}(\{\omega^j\} \mid A_1^i) \text{ for } i = 1, 2, 3 .
$$

Calculate the value of V_1 on each atom:

$$
\begin{aligned}
V_1(A_1^1) &= 10 \cdot \frac{1}{3} + 20 \cdot \frac{1}{3} + 20 \cdot \frac{1}{3} = \frac{50}{3} \cong 17.333 , \\
V_1(A_1^2) &= 15 \cdot \frac{5}{9} + 20 \cdot \frac{2}{9} + 0 \cdot \frac{2}{9} = \frac{115}{9} \cong 12.778 , \\
V_1(A_1^3) &= 10 \cdot \frac{2}{5} + 0 \cdot \frac{2}{5} + 20 \cdot \frac{1}{5} = \frac{40}{5} = 8 .
\end{aligned}
$$

□

8.3.5 The First and Second FTAPs

The first and second fundamental theorems of asset pricing (FTAPs) for a multi-period model are formulated in exactly the same way as those for a single-period model. The absence of arbitrage is equivalent to the existence of an equivalent martingale measure. Moreover, under the no-arbitrage assumption, the martingale measure is unique iff the model is complete, i.e., every payoff can be replicated by a self-financing strategy. For simplicity of presentation, we assume that the bank account B serves as a numéraire asset, and the (equivalent) martingale measure $\widetilde{\mathbb{P}}$ is constructed relative to such a numéraire.

In Chapter 5, we proved the first and second FTAPs for a single-period model. Let us prove the theorems by reducing the multi-period model to a single-period case. As was discussed in the beginning of this section, a multi-period model with a finite number of states can be viewed as a union of a finite number of single-period sub-models, where the states of the world are atoms of the partitions $\mathcal{P}_0, \mathcal{P}_1, \ldots, \mathcal{P}_T$. Let us consider a sub-model originated at some atom $A_t \in \mathcal{P}_t$ with $t \in \{0, 1, \ldots, T-1\}$. Suppose that the atom A_t is a union of ℓ atoms as listed in (8.9). The multi-period model can be represented by a multinomial tree (see Figure 8.1), where every scenario $\omega \in A_t$ can be viewed as a path that goes through the atom A_t to one of the atoms from the list in (8.9). Such a single-step transition from time t to $t+1$ can be described by a single-period model with the state space

$$
\widehat{\Omega} = \left\{ \widehat{\omega}^1 := A_{t+1}^1, \, \widehat{\omega}^2 := A_{t+1}^2, \, \ldots, \, \widehat{\omega}^\ell := A_{t+1}^\ell \right\} . \tag{8.12}
$$

The state probabilities are given by the conditional probabilities of going from A_t to one of its ℓ successors $A_{t+1}^1, A_{t+1}^2, \ldots, A_{t+1}^\ell$:

$$
\widehat{\mathbb{P}}\left(\widehat{\omega}^j\right) := \mathbb{P}\left(A_{t+1}^j \mid A_t\right) = \frac{\mathbb{P}(A_{t+1}^j)}{\mathbb{P}(A_t)}, \quad j = 1, 2, \ldots, \ell. \tag{8.13}
$$

The single-period sub-model obtained is illustrated in Figure 8.3.

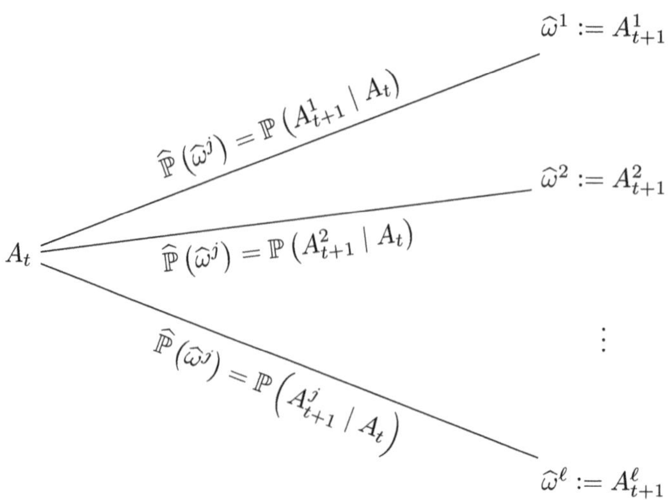

FIGURE 8.3: A single-period sub-model.

To complete the construction of a single-period model, we construct initial price vector $\widehat{\mathbf{S}}_0$ and payoff matrix $\widehat{\mathbf{D}}$. Since the base asset price process is adapted to the filtration $\mathbb{F}$, the prices B_t, $S^1_t, \ldots, S^N_t$ are constant on $A_t \in \mathcal{P}_t$, and the prices B_{t+1}, $S^1_{t+1}, \ldots, S^N_{t+1}$ are constant on each of $A^1_{t+1}, \ldots, A^\ell_{t+1}$. Therefore, the initial price vector and payoff matrix are given by values of the base assets respectively calculated on A_t and atoms of (8.9):

$$\widehat{\mathbf{S}}_0 = \left[B_t(A_t),\, S^1_t(A_t),\, \cdots,\, S^N_t(A_t)\right]^\top, \tag{8.14}$$

$$\widehat{\mathbf{D}} = \begin{bmatrix} B_{t+1}(A^1_{t+1}) & B_{t+1}(A^2_{t+1}) & \cdots & B_{t+1}(A^\ell_{t+1}) \\ S^1_{t+1}(A^1_{t+1}) & S^1_{t+1}(A^2_{t+1}) & \cdots & S^1_{t+1}(A^\ell_{t+1}) \\ \vdots & \vdots & \ddots & \vdots \\ S^N_{t+1}(A^1_{t+1}) & S^N_{t+1}(A^2_{t+1}) & \cdots & S^N_{t+1}(A^\ell_{t+1}) \end{bmatrix}. \tag{8.15}$$

As a result, we obtain a single-period model with the state space $\widehat{\Omega}$, probability distribution function $\widehat{\mathbb{P}}$, initial price vector $\widehat{\mathbf{S}}_0$, and payoff matrix $\widehat{\mathbf{D}}$. Recall that there are $k_0 = 1$ sub-models at time 0, k_1 sub-models at time 1, and so on. In total, we have $k_0 + k_1 + \cdots + k_{T-1}$ single-period sub-models.

The proof of the FTAPs is split into several steps. First, we prove that there are no arbitrage strategies in the multi-period model iff there are no arbitrage portfolios in each single-period sub-model (Lemma 8.5). We also prove a similar result about the completeness of models (Lemma 8.6). Second, we prove that there exists a (unique) martingale measure for the multi-period model iff each sub-model has a (unique) martingale measure (Lemma 8.7).

Lemma 8.5. *There are no arbitrage opportunities in a multi-period model iff each single-period sub-model is arbitrage-free.*

Proof. Let us prove that if each single-period sub-model is arbitrage-free, then there are no arbitrage strategies. Suppose there exists an admissible arbitrage strategy $\boldsymbol{\Phi}$ maturing at time T_m. Let $t \geqslant 0$ be the largest time such that $\Pi^{\boldsymbol{\Phi}}_t \equiv 0$ and $\Pi^{\boldsymbol{\Phi}}_{t+1} > 0$. Such time t exists since the initial value $\Pi^{\boldsymbol{\Phi}}_0$ is zero and the terminal value $\Pi^{\boldsymbol{\Phi}}_{T_m}$ is positive. By construction, the $(t+1)$-time value of the strategy is positive for some scenario $\omega^j \in \Omega$, that is, $\Pi^{\boldsymbol{\Phi}}_{t+1}(\omega^j) > 0$.

Let ω^j be contained in a chain of atoms:

$$\{\omega^j\} \subseteq A_{T-1}^{j_{T-1}} \subseteq \cdots \subseteq A_{t+1}^{j_{t+1}} \subseteq A_t^{j_t} \subseteq \cdots \subseteq A_0^1 = \Omega.$$

The value process is adapted to the filtration $\mathbb{F}$. Therefore, the strategy values $\Pi_t^{\boldsymbol{\Phi}}$ and $\Pi_{t+1}^{\boldsymbol{\Phi}}$ are constant on $A_t^{j_t}$ and $A_{t+1}^{j_{t+1}}$, respectively. Consider the single-period sub-model originated at atom $A_t^{j_t}$ and the portfolio $\boldsymbol{\psi} := \boldsymbol{\varphi}_t(\omega^j)$. Since $\Pi_t[\boldsymbol{\varphi}_t] \equiv 0$, the initial value of $\boldsymbol{\psi}$ is zero. As $\Pi_{t+1}[\boldsymbol{\varphi}_t](\omega^j) > 0$, the terminal value of $\boldsymbol{\psi}$ is strictly positive on the atom $A_{t+1}^{j_{t+1}}$. Therefore, $\boldsymbol{\psi}$ is an arbitrage portfolio in the sub-model originated at $A_t^{j_t}$. We arrive at a contradiction. Hence, the multi-period model is arbitrage-free. The converse is obvious since every static portfolio in a single-period sub-model is a special case of a dynamic strategy. Indeed, consider a single-period sub-model originated at some atom $A_t^j \in \mathcal{P}_t$ and a portfolio $\boldsymbol{\psi}$ in base assets with initial value zero. Construct a strategy maturing at time $t+1$ as follows: $\boldsymbol{\varphi}_s \equiv \mathbf{0}$ for all $s \in \{0, 1, \ldots, t-1\}$, $\boldsymbol{\varphi}_t(\omega) = \boldsymbol{\psi}$ if $\omega \in A_t^j$ and $\boldsymbol{\varphi}_t(\omega) = \mathbf{0}$ otherwise. The strategy is self-financing. If the portfolio $\boldsymbol{\psi}$ is an arbitrage opportunity, then $\Pi_{t+1}[\boldsymbol{\varphi}_t] > 0$ or, in other words, it is an arbitrage strategy that contradicts the assumption. □

Lemma 8.6. *A multi-period model is complete iff each single-period sub-model is complete.*

Proof. The proof is left as an exercise for the reader. □

Lemma 8.7. *There exists a (unique) martingale measure for a multi-period model iff each single-period sub-model has a (unique) martingale measure.*

Proof. Being given the risk-neutral probability distribution function for each one-period sub-model, we have the conditional risk-neutral probabilities of the form $\widetilde{\mathbb{P}}(A_{t+1} \mid A_t)$ for all $t \in \{0, 1, \ldots, T-1\}$ and any two atoms $A_t \in \mathcal{P}_t$, $A_{t+1} \in \mathcal{P}_{t+1}$ such that $A_{t+1} \subseteq A_t$. Using the path probability formula (8.2), we can then calculate the risk-neutral probability $\widetilde{\mathbb{P}}(\omega)$ for each outcome $\omega \in \Omega$. If the risk-neutral probability measure $\widetilde{\mathbb{P}}$ is known, the state probabilities for each sub-model can be computed by (8.13) where $\widetilde{\mathbb{P}}$ is used in place of $\mathbb{P}$. As usual, in the case with a finite probability space $(\Omega, \mathbb{P})$, the probability of event $E \subseteq \Omega$ is calculated as $\mathbb{P}(E) = \sum_{\omega \in E} \mathbb{P}(\omega)$. Clearly, the probability distribution is unique for each sub-model iff the probability distribution $\widetilde{\mathbb{P}}$ defined on Ω is unique. Moreover, the base asset price processes discounted by the numéraire asset B, $\{\overline{S}_t^i = \frac{S_t^i}{B_t}\}_{0 \leqslant t \leqslant T}$, are all martingales iff

$$\widetilde{\mathrm{E}}_t\left[\overline{S}_{t+1}^i\right](A_t) = \widetilde{\mathrm{E}}\left[\overline{S}_{t+1}^i \mid \overline{S}_t^i(A_t)\right] = \overline{S}_t^i \tag{8.16}$$

holds for every time $t \in \{0, 1, \ldots, T-1\}$, all $i = 1, 2, \ldots, N$, and each atom $A_t \in \mathcal{P}_t$. Equation (8.16) is nothing but a single-period martingale condition. □

8.3.6 Pricing and Hedging Derivatives

The pricing formula (8.8) allows us to calculate the price process for any derivative with an attainable payoff V_{T_m} at maturity T_m. Since for every time t the derivative price V_t is equal to the conditional expectation of V_{T_m} given the σ-algebra $\mathcal{F}_t$, this price is $\mathcal{F}_t$-measurable (hence, the derivative price process $\{V_t\}_{0 \leqslant t \leqslant T_m}$ is $\mathbb{F}$-adapted). By construction, any $\mathcal{F}_t$-measurable random variable is constant on atoms of the partition $\mathcal{P}_t$. So we only need to calculate the derivative price $V_t(A_t)$ for every atom $A_t \in \mathcal{P}_t$. Suppose that A_t is a union of ℓ atoms from $\mathcal{P}_{t+1}$ as listed in (8.9). Since the discounted derivative price process is a $\widetilde{\mathbb{P}}$-martingale, $\overline{V}_t(A_t)$ is equal to the risk-neutral mathematical expectation of $\overline{V}_{t+1}$,

$$\overline{V}_t(A_t) = \widetilde{\mathrm{E}}_t\left[\overline{V}_{t+1}\right](A_t) = \widetilde{\mathrm{E}}\left[\overline{V}_{t+1} \mid A_t\right] = \sum_{j=1}^{\ell} \overline{V}_{t+1}(A^j_{t+1})\,\widetilde{\mathbb{P}}(A^j_{t+1} \mid A_t).$$

Therefore, the derivative prices can be computed using a backward-in-time recursion as follows:

$$V_t(A_t) = \frac{1}{1+r_{t+1}} \sum_{j=1}^{\ell} V_{t+1}(A^j_{t+1})\,\widetilde{\mathbb{P}}(A^j_{t+1} \mid A_t) \text{ for all } A_t \in \mathcal{P}_t \text{ and all } t = 0, 1, \ldots, T_m - 1.$$

At maturity T_m, the terminal payoff $V_{T_m}(A_{T_m})$ is known for every atom $A_{T_m} \in \mathcal{P}_{T_m}$. Using the above equation, we calculate the derivative prices $V_t(A_t)$ for all $A_t \in \mathcal{P}_t$ and for all $t = T_m - 1, T_m - 2, \ldots, 0$. In the end, we find the initial price V_0. In total, to obtain a complete derivative price process, we need to calculate $1 + k_1 + \cdots + k_{T_m - 1}$ conditional expectations. One can notice that the same approach is used to price derivatives in the binomial tree model.

Knowing the derivative prices $V_t(A_t)$ for all $t \in \{0, 1, \ldots, T_m\}$ and all $A_t \in \mathcal{P}_t$, we can find the self-financing portfolio strategy $\boldsymbol{\Phi} = \{\varphi_t\}_{0 \leqslant t \leqslant T_m - 1}$ hedging the derivative. Suppose that the market model is arbitrage free and complete. Fix arbitrarily time $t \in \{0, 1, \ldots, T_m - 1\}$ and atom $A_t \in \mathcal{P}_t$. The portfolio $\varphi_t(A_t)$ can be found by solving a single-period replication problem. Consider the one-period sub-model originated at the atom A_t. The state space $\widehat{\Omega}$, initial price vector $\widehat{\mathbf{S}}_0$, and payoff matrix $\widehat{\mathbf{D}}$ are given in (8.12), (8.14), and (8.15), respectively. Find a portfolio in the base assets replicating the payoff vector $V_{t+1}(\widehat{\Omega}) = \left[V_{t+1}(A^1_{t+1}),\, V_{t+1}(A^2_{t+1}),\, \ldots,\, V_{t+1}(A^\ell_{t+1})\right]$. The portfolio vector is a solution to the matrix-vector equation $\psi\widehat{\mathbf{D}} = V_{t+1}(\widehat{\Omega})$. Set $\varphi_t(A_t)$ equal to the replicating portfolio obtained. By construction, the initial cost to set up this portfolio is equal to $V_t(A_t)$. Repeat these steps for all t and all $A_t \in \mathcal{P}_t$ to obtain a self-financing portfolio strategy $\boldsymbol{\Phi}$ that replicates the derivative process, i.e., $\Pi^{\boldsymbol{\Phi}}_t = V_t$ for all $t = 0, 1, \ldots, T_m$.

8.3.7 Radon–Nikodym Derivative Process and Change of Numéraire

In the previous sections, all main results were derived using the risk-neutral probability measure $\widetilde{\mathbb{P}} \equiv \widetilde{\mathbb{P}}^{(B)}$ (also called an EMM) relative to the numéraire asset $g = B$. Suppose the EMM $\widetilde{\mathbb{P}}^{(B)}$ is known and there exists another asset S with a strictly positive price process. How can we find the EMM $\widetilde{\mathbb{P}}^{(S)}$ relative to the numéraire asset $g = S$ without recomputing all risk-neutral probabilities from scratch? As we learned in Chapter 5, it is the Radon–Nikodym derivative that connects two equivalent probability measures. Recall its definition. Let $\mathbb{P}$ and $\widehat{\mathbb{P}}$ be two probability measures on the same finite sample space Ω so that $\mathbb{P}(\omega) > 0$ and $\widehat{\mathbb{P}}(\omega) > 0$ for all scenarios $\omega \in \Omega$. That is, $\mathbb{P}$ and $\widehat{\mathbb{P}}$ are equivalent probability measures. The Radon–Nikodym derivative of $\widehat{\mathbb{P}}$ w.r.t. $\mathbb{P}$ is the random variable ϱ defined by

$$\varrho(\omega) = \frac{\widehat{\mathbb{P}}(\omega)}{\mathbb{P}(\omega)} \text{ for } \omega \in \Omega\,.$$

The mathematical expectation of a random variable X on Ω under $\widehat{\mathbb{P}}$ can also be computed under the probability measure $\mathbb{P}$ with the use of ϱ as follows:

$$\widehat{\mathrm{E}}[X] = \sum_{\omega} X(\omega)\widehat{\mathbb{P}}(\omega) = \sum_{\omega} X(\omega) \underbrace{\frac{\widehat{\mathbb{P}}(\omega)}{\mathbb{P}(\omega)}}_{=\varrho(\omega)} \mathbb{P}(\omega) = \sum_{\omega} \varrho(\omega) Y(\omega) \mathbb{P}(\omega) = \mathrm{E}[\varrho X]\,. \tag{8.17}$$

In particular, we can calculate the risk-neutral expectation $\widetilde{\mathrm{E}}[X]$ by computing $\mathrm{E}[\varrho\, X]$ under the actual (real-world) probability measure, where ϱ is the Radon–Nikodym derivative of $\widetilde{\mathbb{P}}$ w.r.t. $\mathbb{P}$.

Consider an $\mathcal{F}_t$-measurable random variable X with $0 \leqslant t \leqslant T$. For example, X is a payoff maturing at time n. The computation of expected values of X while switching the probability measure can be simplified further with the use of the *Radon–Nikodym derivative process*, which is defined just below. Moreover, to price derivatives at any intermediate time $t \in \{0, 1, \ldots, T\}$, we need a conditional expectation under $\widetilde{\mathbb{P}}$ and the Radon–Nikodym process handles it as well.

Definition 8.3. Let $\mathbb{P}$ and $\widehat{\mathbb{P}}$ be two probability measures on a finite sample space Ω so that $\mathbb{P}(\omega) > 0$ and $\widehat{\mathbb{P}}(\omega) > 0$ for every scenario ω (hence, $\mathbb{P}$ and $\widehat{\mathbb{P}}$ are equivalent). Let ϱ be the Radon–Nikodym derivative of $\widehat{\mathbb{P}}$ w.r.t. $\mathbb{P}$ given by $\varrho(\omega) = \frac{\widehat{\mathbb{P}}(\omega)}{\mathbb{P}(\omega)}$. The Radon–Nikodym derivative process is defined as

$$\varrho_t = \mathrm{E}_t[\varrho], \quad t \in \{0, 1, \ldots, T\}. \tag{8.18}$$

In particular, $\varrho_T = \varrho$ for all scenarios. Since $\mathrm{E}[\varrho] = 1$, the initial value ϱ_0 is 1.

As we proved in Chapter 6, any random process $\{X_t\}_{0 \leqslant t \leqslant T}$ defined by

$$X_t = \mathrm{E}[X \mid \mathcal{F}_t], \quad t \in \{0, 1, \ldots, T\},$$

is adapted to the filtration $\{\mathcal{F}_t\}_{0 \leqslant t \leqslant T}$ and is a martingale. Hence, the Radon–Nikodym derivative process $\{\varrho_t\}_{0 \leqslant t \leqslant T}$ is a $\mathbb{P}$-martingale.

Suppose that X_t is an $\mathcal{F}_t$-measurable random variable for some $t \in \{0, 1, \ldots, T\}$. In this case, Equation (8.17) simplifies further. Applying the tower property gives

$$\widehat{\mathrm{E}}[X_t] = \mathrm{E}[\varrho\, X_t] = \mathrm{E}[\mathrm{E}_t[\varrho\, X_t]] = \mathrm{E}[X_t\, \mathrm{E}_t[\varrho]] = \mathrm{E}[X_t\, \varrho_t]. \tag{8.19}$$

The σ-algebra $\mathcal{F}_t$ is generated from a partition $\mathcal{P}_t$ that consists of atoms $A_t^1, \ldots, A_t^{k_t}$. Hence, an $\mathcal{F}_t$ -measurable random variable is constant on the atoms of $\mathcal{P}_t$. Fix an atom $A_t^i \in \mathcal{P}_t$ and consider the random variable $X_t = \mathbb{I}_{A_t^i}$. On the one hand, we have

$$\widehat{\mathrm{E}}[X_t] = \widehat{\mathbb{P}}(A_t^i). \tag{8.20}$$

On the other hand, we have

$$\mathrm{E}[\varrho_t\, X_t] = \sum_{\omega} \varrho_t(\omega)\, X_t(\omega)\, \mathbb{P}(\omega) = \sum_{\omega \in A_t^i} \varrho_t(\omega) \mathbb{P}(\omega).$$

Since ϱ_t is $\mathcal{F}_t$-measurable, it is constant on A_t^i. Therefore, we have

$$\mathrm{E}[\varrho_t\, X_t] = \varrho_t(A_t^i)\, \mathbb{P}(A_t^i), \tag{8.21}$$

where $\varrho_t(A_t^i)$ is the value of ϱ_t on the atom A_t^i. According to (8.19), the quantities in (8.20) and (8.21) are equal, and hence

$$\varrho_t(A_t^i) = \frac{\widehat{\mathbb{P}}(A_t^i)}{\mathbb{P}(A_t^i)} \tag{8.22}$$

for any $A_t^i \in \mathcal{P}_t$. The partition $\mathcal{P}_t$ consists of k_t atoms. So, the Radon–Nikodym derivative process can be written as a weighted sum of k_t indicator functions:

$$\varrho_t(\omega) = \sum_{i=1}^{k_t} \varrho_t(A_t^i)\, \mathbb{I}_{A_t^i}(\omega) = \sum_{i=1}^{k_t} \frac{\widehat{\mathbb{P}}(A_t^i)}{\mathbb{P}(A_t^i)}\, \mathbb{I}_{A_t^i}(\omega), \quad t = 0, 1, \ldots, T. \tag{8.23}$$

Example 8.7. Find the Radon–Nikodym derivative process for the binomial tree model.

Solution. Let p and q be the probability of an upward move under probability measures $\mathbb{P}$ and $\mathbb{Q}$, respectively. Every atom $A_t^i \in \mathcal{P}_t$ is of the form

$$A_t^i = A_{\bar{\omega}_1\bar{\omega}_2\ldots\bar{\omega}_t} = \{\omega \in \Omega_T \; : \; \omega_1 = \bar{\omega}_1, \omega_2 = \bar{\omega}_2, \ldots, \omega_t = \bar{\omega}_t\}$$

for some $\bar{\omega}_1, \bar{\omega}_2, \ldots, \bar{\omega}_t \in \{\mathsf{U}, \mathsf{D}\}$. The probability of A_t^i under $\mathbb{P}$ is

$$\mathbb{P}(A_t^i) = p^{\#\mathsf{U}(\bar{\omega}_1,\bar{\omega}_2,\ldots,\bar{\omega}_t)}(1-p)^{\#\mathsf{D}(\bar{\omega}_1,\bar{\omega}_2,\ldots,\bar{\omega}_t)}.$$

The same probability computed under the measure $\mathbb{Q}$ is

$$\mathbb{Q}(A_t^i) = q^{\#\mathsf{U}(\bar{\omega}_1,\bar{\omega}_2,\ldots,\bar{\omega}_t)}(1-q)^{\#\mathsf{D}(\bar{\omega}_1,\bar{\omega}_2,\ldots,\bar{\omega}_t)}.$$

So, the random variable ϱ_t is a function of the first t market moves and is given by

$$\varrho_t(\omega_1, \omega_2, \ldots, \omega_t) = \left(\frac{q}{p}\right)^{\#\mathsf{U}(\omega_1,\omega_2,\ldots,\omega_t)} \left(\frac{1-q}{1-p}\right)^{\#\mathsf{D}(\omega_1,\omega_2,\ldots,\omega_t)},$$

for any $\omega_1, \omega_2, \ldots, \omega_t \in \{\mathsf{U}, \mathsf{D}\}$. □

Now we derive an analogue of (8.17) for conditional expectations. Fix arbitrarily $t \in \{0, 1, \ldots, T\}$. The conditional expectation of a random variable X given an atom $A_t^i \in \mathcal{P}_t$ under the probability measure $\widehat{\mathbb{P}}$ is

$$\widehat{\mathrm{E}}[X \mid A_t^i] = \frac{1}{\widehat{\mathbb{P}}(A_t^i)} \widehat{\mathrm{E}}[X\,\mathbb{I}_{A_t^i}]$$

(apply (8.17) and divide and multiply by $\mathbb{P}(A_t^i)$)

$$= \frac{\mathbb{P}(A_t^i)}{\widehat{\mathbb{P}}(A_t^i)} \frac{\mathrm{E}[\varrho X\,\mathbb{I}_{A_t^i}]}{\mathbb{P}(A_t^i)} = \underbrace{\frac{\mathbb{P}(A_t^i)}{\widehat{\mathbb{P}}(A_t^i)}}_{=1/\varrho_t(A_t^i)} \mathrm{E}[\varrho X \mid A_t^i]$$

$$= \frac{1}{\varrho_t(A_t^i)} \mathrm{E}[\varrho X \mid A_t^i] = \mathrm{E}\left[(\varrho/\varrho_t)\, X \mid A_t^i\right].$$

According to the definition of a conditional mathematical expectation given a σ-algebra, which is generated by a finite partition, we have

$$\begin{aligned}\widehat{\mathrm{E}}_t[X] &= \sum_{i=1}^{k_t} \widehat{\mathrm{E}}[X \mid A_t^i]\,\mathbb{I}_{A_t^i} = \sum_{i=1}^{k_t} \mathrm{E}\left[(\varrho/\varrho_t)X \mid A_t^i\right]\mathbb{I}_{A_t^i} \\ &= \mathrm{E}_t\left[(\varrho/\varrho_t)X\right] = \mathrm{E}_t\left[(\varrho_T/\varrho_t)X\right] = \frac{1}{\varrho_t}\mathrm{E}_t\left[\varrho_T X\right]. \end{aligned} \tag{8.24}$$

Note that if we set $t = 0$, then (8.24) reduces to (8.17) since $\varrho_0 = 1$.

Suppose we deal with two equivalent martingale measures for respective numéraire assets f and g, where, for example, one numéraire is a risk-free bond and the other is some base stock. In this case, we can obtain a simple explicit formula of the Radon–Nikodym derivative process. According to the pricing formula (8.8), we have the following equivalence of conditional expectations under $\widetilde{\mathbb{P}}^{(g)}$ and $\widetilde{\mathbb{P}}^{(f)}$:

$$\widetilde{\mathrm{E}}_t^{(g)}\left[\frac{g_t}{g_T}X\right] = \widetilde{\mathrm{E}}_t^{(f)}\left[\frac{f_t}{f_T}X\right] \implies \widetilde{\mathrm{E}}_t^{(g)}[X] = \widetilde{\mathrm{E}}_t^{(f)}\left[\frac{g_T/g_t}{f_T/f_t}X\right] \tag{8.25}$$

for all $t \in \{0, 1, \dots, T\}$ and any payoff X. On the other hand, applying (8.24) with $\widehat{\mathbb{P}} = \widetilde{\mathbb{P}}^{(g)}$ and $\mathbb{P} = \widetilde{\mathbb{P}}^{(f)}$ gives

$$\widetilde{\mathrm{E}}_t^{(g)}[X] = \widetilde{\mathrm{E}}_t^{(f)}\left[\frac{\varrho_T}{\varrho_t} X\right] \tag{8.26}$$

where $\{\varrho_t\}_{0 \leqslant t \leqslant T}$ is the Radon–Nikodym derivative process of $\widetilde{\mathbb{P}}^{(g)}$ w.r.t. $\widetilde{\mathbb{P}}^{(f)}$. Combining (8.25) and (8.26), we have

$$\widetilde{\mathrm{E}}_t^{(f)}\left[\frac{\varrho_T}{\varrho_t} X\right] = \widetilde{\mathrm{E}}_t^{(f)}\left[\frac{g_T/g_t}{f_T/f_t} X\right] = \widetilde{\mathrm{E}}_t^{(f)}\left[\frac{(g_T/g_0)/(f_T/f_0)}{(g_t/g_0)/(f_t/f_0)} X\right].$$

Let us fix some state $\bar{\omega}$ and take $X = \mathbb{I}_{\{\bar{\omega}\}}$. The above identity becomes

$$\left(\frac{\varrho_T}{\varrho_t}\right)(\bar{\omega}) = \left(\frac{(g_T/g_0)/(f_T/f_0)}{(g_t/g_0)/(f_t/f_0)}\right)(\bar{\omega}).$$

Since, as follows from (5.53), the Radon–Nikodym derivative $\varrho \equiv \varrho_T$ is equal to $\frac{g_T/g_0}{f_T/f_0}$, we obtain the formula of the Radon–Nikodym derivative process:

$$\varrho_t = \frac{g_t/g_0}{f_t/f_0}, \quad t \in \{0, 1, \dots, T\}. \tag{8.27}$$

Combining (8.22) and (8.27) allows us to express probabilities of atoms under $\widetilde{\mathbb{P}}^{(g)}$ in terms of probabilities under $\widetilde{\mathbb{P}}^{(f)}$:

$$\widetilde{\mathbb{P}}^{(g)}(A_t^i) = \frac{g_t(A_t^i)/g_0}{f_t(A_t^i)/f_0}\, \widetilde{\mathbb{P}}^{(f)}(A_t^i) \text{ for } A_t^i \in \mathcal{P}_t.$$

In particular, by setting $t = T$, we obtain the known relation from Chapter 5,

$$\widetilde{\mathbb{P}}^{(g)}(\omega) = \frac{g_T(\omega)/g_0}{f_T(\omega)/f_0}\, \widetilde{\mathbb{P}}^{(f)}(\omega).$$

8.4 Examples of Discrete-Time Models

8.4.1 Binomial Tree Model with Stochastic Volatility

Consider a binomial tree model where the upward and downward stock returns, u and d, and the interest rate r are stochastic processes adapted to the natural filtration. For each $n \geqslant 1$, the factors and the interest rate depend on the first $n-1$ market moves:

$$u = u_n(\omega_1\omega_2 \dots \omega_{n-1}), \quad d = d_n(\omega_1\omega_2 \dots \omega_{n-1}), \quad r = r_n(\omega_1\omega_2 \dots \omega_{n-1}).$$

That is, the processes $\{u_n\}_{n \geqslant 1}$, $\{d_n\}_{n \geqslant 1}$, $\{r_n\}_{n \geqslant 1}$ are all $\mathbb{F}$-predictable. At time zero, the initial factors u_1 and d_1 and the initial rate r_1 are constant. The initial stock price S_0 is positive. Assuming that $0 < d_n(\omega_1\omega_2 \dots \omega_{n-1}) \leqslant u_n(\omega_1\omega_2 \dots \omega_{n-1})$ holds for every $n \geqslant 1$ and all $\omega_1, \omega_2, \dots, \omega_{n-1} \in \{\mathsf{D}, \mathsf{U}\}$ we guarantee the positiveness of future stock prices. The dynamics of the stock price at time $n \geqslant 1$ is given by

$$S_n(\omega_1\omega_2 \dots \omega_{n-1}\omega_n) = \begin{cases} u_n(\omega_1\omega_2 \dots \omega_{n-1}) S_{n-1}(\omega_1\omega_2 \dots \omega_{n-1}) & \text{if } \omega_n = \mathsf{U}, \\ d_n(\omega_1\omega_2 \dots \omega_{n-1}) S_{n-1}(\omega_1\omega_2 \dots \omega_{n-1}) & \text{if } \omega_n = \mathsf{D}. \end{cases}$$

The bank account process is stochastic and its value at time $n \geqslant 2$ is given by

$$B_n(\omega_1\omega_2 \dots \omega_{n-1}\omega_n) = B_{n-1}(\omega_1\omega_2 \dots \omega_{n-2})(1 + r_n(\omega_1\omega_2 \dots \omega_{n-1})).$$

At time 1, we have

$$S_1(\omega_1) = \begin{cases} u_1 S_0 & \text{if } \omega_1 = \mathsf{U}, \\ d_1 S_0 & \text{if } \omega_1 = \mathsf{D} \end{cases} \quad \text{and} \quad B_1 = B_0(1 + r_1).$$

The initial value B_0 is positive. Typically, we assume that $B_0 = 1$. As is seen from the above equations, the bank account value B_n depends on $\omega_1\omega_2 \dots \omega_{n-1}$ and is independent of ω_n. That is, the process $\{B_n\}_{n\geqslant 0}$ is $\mathbb{F}$-predictable.

Let us discuss the main characteristics of this model.

- Every scenario $\omega = \omega_1\omega_2 \dots \omega_T$ can be considered as a path in a binomial tree, which is not necessarily a recombining one. The recombination of a binomial tree is essential since the nodes of one time step increase linearly with the number of time steps in a recombining tree. In a nonrecombining tree, in contrast, the nodes increase exponentially, so that the tree can only be built for a few steps, even using modern computers. A two-period binomial tree is recombining iff $u_1 d_2(\mathsf{U}) = d_1 u_2(\mathsf{D})$ holds. A general multi-period binomial tree is recombining iff every two-period sub-tree is recombining. So, we have the following necessary and sufficient condition:

$$u_n(\omega_1 \dots \omega_{n-1})\, d_{n+1}(\omega_1 \dots \omega_{n-1}\mathsf{U}) = d_n(\omega_1 \dots \omega_{n-1})\, u_{n+1}(\omega_1 \dots \omega_{n-1}\mathsf{D}) \tag{8.28}$$

  for all $n \in \{1, 2, \dots, T-1\}$ and all $\omega_1, \omega_2, \dots, \omega_{n-1} \in \{\mathsf{U}, \mathsf{D}\}$. For example, a three-period binomial tree, which contains three two-period sub-trees, is recombining iff the following conditions hold:

$$\begin{aligned} u_1\, d_2(\mathsf{U}) &= d_1\, u_2(\mathsf{D}), \\ u_2(\mathsf{U})\, d_3(\mathsf{UU}) &= d_2(\mathsf{U})\, u_3(\mathsf{UD}), \\ u_2(\mathsf{D})\, d_3(\mathsf{DU}) &= d_2(\mathsf{D})\, u_3(\mathsf{DD}). \end{aligned}$$

- This model is arbitrage-free iff every single-period binomial sub-model is arbitrage-free. Therefore, there is no arbitrage iff

$$d_n(\omega_1\omega_2 \dots \omega_{n-1}) < 1 + r_n(\omega_1\omega_2 \dots \omega_{n-1}) < u_n(\omega_1\omega_2 \dots \omega_{n-1}) \tag{8.29}$$

  holds for all $n \in \{1, 2, \dots, T\}$ and all market moves $\omega_1, \omega_2, \dots, \omega_{n-1} \in \{\mathsf{D}, \mathsf{U}\}$.

- To find the risk-neutral state probabilities $\widetilde{\mathbb{P}}(\omega_1\omega_2 \dots \omega_T)$ for a binomial model with T periods, we need to compute the risk-neutral probabilities for each single-period sub-model. At time zero, there is only one binomial sub-model and the two risk-neutral probabilities are

$$\tilde{p}_1(\mathsf{D}) = \frac{u_1 - 1 - r_1}{u_1 - d_1}, \quad \tilde{p}_1(\mathsf{U}) = \frac{1 + r_1 - d_1}{u_1 - d_1}.$$

  Fix arbitrarily $n \geqslant 1$ and $\omega_1, \omega_2, \dots, \omega_n \in \{\mathsf{D}, \mathsf{U}\}$. Consider a binomial sub-tree originated from the path $\omega_1, \omega_2, \dots, \omega_n$ in a binomial tree. The risk-neutral probabilities of the events $\{\omega_{n+1} = \mathsf{D}\}$ and $\{\omega_{n+1} = \mathsf{U}\}$ conditional on $\omega_1, \omega_2, \dots, \omega_n$ are, respectively, given by

$$\tilde{p}_{n+1}(\mathsf{D} \mid \omega_1, \omega_2, \dots, \omega_n) = \frac{u_{n+1}(\omega_1\omega_2 \dots \omega_n) - 1 - r_{n+1}(\omega_1\omega_2 \dots \omega_n)}{u_{n+1}(\omega_1\omega_2 \dots \omega_n) - d_{n+1}(\omega_1\omega_2 \dots \omega_n)},$$

$$\tilde{p}_{n+1}(\mathsf{U} \mid \omega_1, \omega_2, \dots, \omega_n) = \frac{1 + r_{n+1}(\omega_1\omega_2 \dots \omega_n) - d_{n+1}(\omega_1\omega_2 \dots \omega_n)}{u_{n+1}(\omega_1\omega_2 \dots \omega_n) - d_{n+1}(\omega_1\omega_2 \dots \omega_n)}.$$

Finally, the risk-neutral state probability of scenario $\omega = \omega_1\omega_2\ldots\omega_T \in \Omega_T$ is

$$\widetilde{\mathbb{P}}(\omega_1\omega_2\ldots\omega_T) = \tilde{p}_1(\omega_1)\tilde{p}_2(\omega_2 \mid \omega_1)\cdots\tilde{p}_n(\omega_n \mid \omega_1\omega_2\ldots\omega_{n-1}).$$

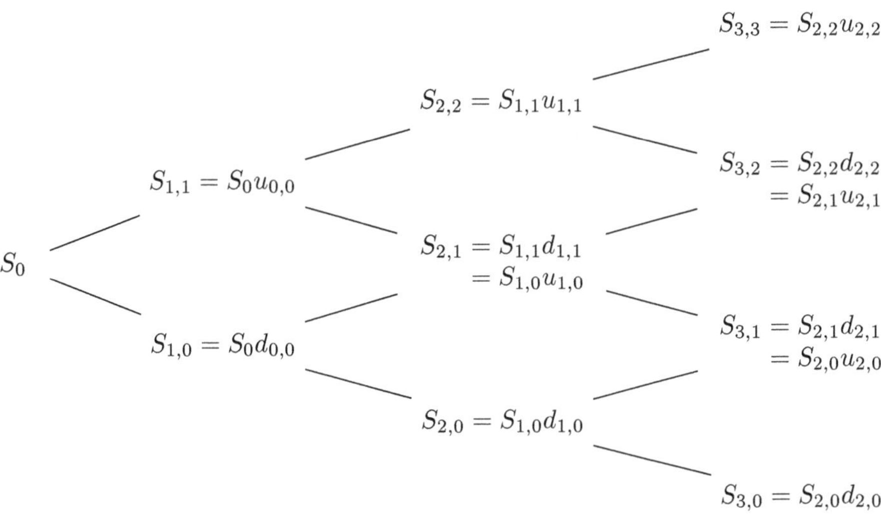

FIGURE 8.4: A schematic representation of a three-period binomial lattice with state-dependent market move factors.

A useful version of a stochastic binomial model is a model with state-dependent returns (see Figure 8.4). Recall some facts about binomial lattices. In a recombining binomial lattice, all paths with the same number of upward and downward moves lead to the same node. The node (n, m) of a binomial lattice (with integer coordinates n and m so that $0 \leqslant m \leqslant n$) is reached from the root node $(0, 0)$ by making m upward moves and $n - m$ downward moves. There are $\binom{n}{m}$ of such paths. The stock price at the node (n, m) is denoted by $S_{n,m}$. Let the factors u and d and the risk-free rate r be functions of n and m: $d = d_{n,m}$, $u = u_{n,m}$, $r = r_{n,m}$. The recombining condition (8.28) takes the form

$$u_{n,m}\, d_{n+1,m+1} = d_{n,m}\, u_{n+1,m}$$

for all integers n and m with the property $0 \leqslant m \leqslant n \leqslant T - 2$. The stock prices are given by the following iterative formulae:

$$\begin{aligned} S_{n+1,0} &= S_{n,0}d_{n,0}, \\ S_{n+1,m} &= S_{n,m}d_{n,m} = S_{n,m-1}u_{n,m-1} \text{ for } 1 \leqslant m \leqslant n, \\ S_{n+1,n+1} &= S_{n,n}u_{n,n}, \end{aligned}$$

for any $n \in \{0, 1, \ldots, T - 1\}$. Although we do not discuss here practical examples, a state-dependent binomial tree can be constructed as an approximation of a continuous-time stock price process with time- and state-dependent volatility $\sigma(t, S)$. The tree consistent with a given volatility structure is called the *implied tree.*

8.4.2 Binomial Tree Model for Interest Rates

The goal of this sub-section is to develop a binomial-tree model for stochastic interest rates. This model can be used for pricing (zero-)coupon bonds and other interest rate derivatives. Consider a discrete-time model with the finite state space $\Omega = 2^T$, filtration $\{\mathcal{F}_n\}_{n\in\{0,1,\ldots,T\}}$, and risk-free *rate of interest* $\{r_n\}_{n\in\{1,2,\ldots,T\}}$. Let us assume that the rate r_n is $\mathcal{F}_{n-1}$ measurable for every $n \in \{1, 2, \ldots, T\}$. In particular, the rate r_1 is deterministic. We assume that we deal with a binomial model. Hence, r_n is a function of the first $n-1$ market moves: $r_n = r_n(\omega_1, \ldots, \omega_{n-1})$. We also assume that the rates r_n are all positive for all possible market scenarios.

The rate r_n is valid for the nth period, i.e., one dollar invested in the bank account at time $n-1$ grows to $1 + r_n$ dollars at time n. So, V_0 invested at time 0 grows to $V_n = V_0(1+r_1)(1+r_2)\cdots(1+r_n)$ at time n. Since the rates $r_1, r_2, \ldots, r_n$ are all $\mathcal{F}_{n-1}$-measurable, the accumulated value V_n is a function of $\omega_1, \omega_2, \ldots, \omega_{n-1}$.

Let us introduce a bank account process $\{B_n\}_{n=0,1,\ldots,T}$ defined by

$$B_0 = 1 \text{ and } B_n = (1+r_1)(1+r_2)\cdots(1+r_n) \text{ for all } n \in \{1, 2, \ldots, T\}.$$

In addition, define the discount factor (at time n) as

$$D_n = \frac{B_0}{B_n} = \frac{1}{(1+r_1)(1+r_2)\cdots(1+r_n)}.$$

The risk-neutral pricing formula (8.8) says that the time-n value of a payment V_m received at time $m \leqslant T$ (where the payoff V_m is $\mathcal{F}_m$-measurable) is

$$V_n = B_n \widetilde{\mathrm{E}}_n \left[\frac{V_m}{B_m}\right] \text{ for } n \in \{0, 1, \ldots, m-1\},$$

where the risk-neutral expectation is taken under the EMM $\widetilde{\mathbb{P}}^{(B)}$. Therefore the prices of a zero-coupon bond (ZCB) that pays one dollar at maturity time m are

$$Z_{n,m} = B_n \widetilde{\mathrm{E}}_n \left[\frac{1}{B_m}\right] = \widetilde{\mathrm{E}}_n \left[\frac{1}{(1+r_{n+1})(1+r_{n+2})\cdots(1+r_m)}\right] \text{ for } n \in \{0, 1, \ldots, m-1\}. \tag{8.30}$$

At the maturity time m, the price of the ZCB is $Z_{m,m} = 1$. We can also find the yield rate $y_{n,m}$ with $0 \leqslant n \leqslant m$, which is a constant rate of interest for the period from time n to m defined so that it is equivalent to the rate of return of the ZCB maturing at time m. Solving the respective time-value equation gives the yield rate:

$$Z_{n,m} = (1 + y_{n,m})^{-(m-n)} \implies y_{n,m} = Z_{n,m}^{-1/(m-n)} - 1.$$

A coupon bond with maturity m can be modelled as a stream of payments $C_1, C_2, \ldots, C_m$. For each $n \in \{1, 2, \ldots, m-1\}$, the value C_n is the coupon payment made at time n. The last payment C_m includes the redemption value as well as any coupon due at time m. In the case of the zero-coupon bond maturing at time m, we have $C_1 = C_2 = \ldots = C_{m-1} = 0$ and $C_m = 1$. The no-arbitrage value of a coupon bond is equal to the no-arbitrage value of a portfolio of zero-coupon bonds with C_n bonds maturing at time $n = 1, 2, \ldots, m$. Thus, the initial price P_0 of a coupon-paying bond is calculated as follows:

$$P_0 = \sum_{n=1}^{m} C_n Z_{0,n}.$$

In general, the initial no-arbitrage value V_0 of a stream of random payments $C_1, C_2, \ldots, C_n$,

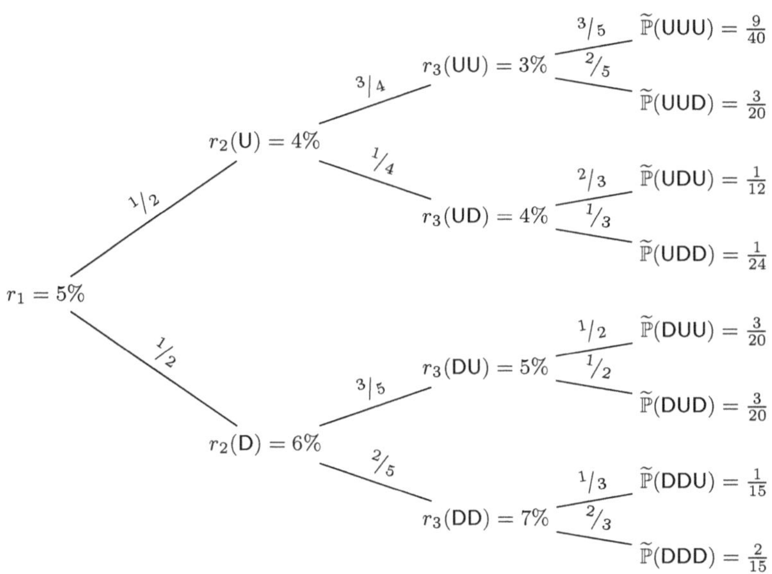

FIGURE 8.5: A three-period binomial model of stochastic interest rates.

where C_m is $\mathcal{F}_m$-measurable for each $m = 1, 2, \ldots, n$, is a sum of no-arbitrage values of individual payments:

$$V_0 = \sum_{n=1}^{m} \widetilde{\mathrm{E}}_0 \left[\frac{C_n}{(1+r_1)(1+r_2)\cdots(1+r_n)} \right] . \tag{8.31}$$

Example 8.8. Consider a nonrecombining binomial tree model whose risk-neutral probabilities and interest rates are as shown in Figure 8.5. Compute the time-0 price of (i) a zero-coupon bond maturing at time 3 and (ii) a coupon bond with payments $C_1 = C_2 = C_3 = \frac{1}{3}$.

Solution. First, we compute the time-0 prices $Z_{0,n}$ of zero-coupon bonds maturing at times $n = 1, 2, 3$, respectively:

$$\begin{aligned}
Z_{0,3} &= \sum_{\omega_1,\omega_2,\omega_3 \in \{\mathsf{D},\mathsf{U}\}} \frac{1}{(1+r_1)(1+r_2(\omega_1))(1+r_3(\omega_1\omega_2))} \widetilde{\mathbb{P}}(\omega_1\omega_2\omega_3) \\
&= \sum_{\omega_1,\omega_2 \in \{\mathsf{D},\mathsf{U}\}} \frac{1}{(1+r_1)(1+r_2(\omega_1))(1+r_3(\omega_1\omega_2))} \widetilde{\mathbb{P}}(A_{\omega_1\omega_2}) \\
&= \frac{1/5}{1.05 \cdot 1.06 \cdot 1.07} + \frac{3/10}{1.05 \cdot 1.06 \cdot 1.05} + \frac{1/8}{1.05 \cdot 1.04 \cdot 1.04} + \frac{3/8}{1.05 \cdot 1.04 \cdot 1.03} \\
&\cong 0.868116\,,
\end{aligned}$$

$$\begin{aligned}
Z_{0,2} &= \sum_{\omega_1,\omega_2 \in \{\mathsf{D},\mathsf{U}\}} \frac{1}{(1+r_1)(1+r_2(\omega_1))} \widetilde{\mathbb{P}}(A_{\omega_1\omega_2}) = \sum_{\omega_1 \in \{\mathsf{D},\mathsf{U}\}} \frac{1}{(1+r_1)(1+r_2(\omega_1))} \widetilde{\mathbb{P}}(A_{\omega_1}) \\
&= \frac{1/2}{1.05 \cdot 1.06} + \frac{1/2}{1.05 \cdot 1.04} \cong 0.907112\,, \\
Z_{0,1} &= \sum_{\omega_1 \in \{\mathsf{D},\mathsf{U}\}} \frac{1}{1+r_1} \widetilde{\mathbb{P}}(A_{\omega_1}) = \frac{1}{1+r_1} = \frac{1}{1.05} \cong 0.952381\,.
\end{aligned}$$

In the above calculations, we used the risk-neutral probabilities $\widetilde{\mathbb{P}}(A_{\omega_1\omega_2})$ computed from transition probabilities, which are given as weights in Figure 8.5:

$$\begin{aligned}
\widetilde{\mathbb{P}}(A_{\mathsf{DD}}) &= \widetilde{\mathbb{P}}(A_{\mathsf{D}})\,\widetilde{\mathbb{P}}(A_{\mathsf{DD}} \mid A_{\mathsf{D}}) = \frac{1}{2}\cdot\frac{2}{5} = \frac{1}{5}, \\
\widetilde{\mathbb{P}}(A_{\mathsf{DU}}) &= \widetilde{\mathbb{P}}(A_{\mathsf{D}})\,\widetilde{\mathbb{P}}(A_{\mathsf{DU}} \mid A_{\mathsf{D}}) = \frac{1}{2}\cdot\frac{3}{5} = \frac{3}{10}, \\
\widetilde{\mathbb{P}}(A_{\mathsf{UD}}) &= \widetilde{\mathbb{P}}(A_{\mathsf{U}})\,\widetilde{\mathbb{P}}(A_{\mathsf{UD}} \mid A_{\mathsf{U}}) = \frac{1}{2}\cdot\frac{1}{4} = \frac{1}{8}, \\
\widetilde{\mathbb{P}}(A_{\mathsf{UU}}) &= \widetilde{\mathbb{P}}(A_{\mathsf{U}})\,\widetilde{\mathbb{P}}(A_{\mathsf{UU}} \mid A_{\mathsf{U}}) = \frac{1}{2}\cdot\frac{3}{4} = \frac{3}{8}.
\end{aligned}$$

So, the price of the zero-coupon bond maturing at time 3 is $Z_{0,3} \cong 0.868116$. The price of the coupon bond is

$$\frac{1}{3}\cdot Z_{0,1} + \frac{1}{3}\cdot Z_{0,2} + \frac{1}{3}\cdot Z_{0,2} = \frac{0.952381 + 0.907112 + 0.868116}{3} \cong 0.909203. \qquad \square$$

In Example 8.8, the bond prices are calculated using the risk-neutral probabilities given us a priori. In practice, we deal with the reverse situation: the market prices of bonds with different term structures are known and then the risk-neutral probability measure $\widetilde{\mathbb{P}}$ needs to be determined. Recall that the discounted bond price processes $\{\overline{Z}_{n,m} = Z_{n,m}/B_n\}_{0\leqslant n\leqslant m}$ are $\widetilde{\mathbb{P}}$-martingales for all $m = 1, 2, \ldots, T$. Therefore, the risk-neutral probabilities can be found by solving

$$\begin{cases}
(1 + r_{n+1}(\omega_1\ldots\omega_n))Z_{n,m}(\omega_1\ldots\omega_n) = Z_{n+1,m}(\omega_1\ldots\omega_n\mathsf{D})\,\widetilde{\mathbb{P}}(A_{\omega_1\ldots\omega_n\mathsf{D}} \mid A_{\omega_1\ldots\omega_n}) \\
\qquad\qquad + Z_{n+1,m}(\omega_1\ldots\omega_n\mathsf{U})\,\widetilde{\mathbb{P}}(A_{\omega_1\ldots\omega_n\mathsf{U}} \mid A_{\omega_1\ldots\omega_n}) \\
\widetilde{\mathbb{P}}(A_{\omega_1\ldots\omega_n\mathsf{D}} \mid A_{\omega_1\ldots\omega_n}) + \widetilde{\mathbb{P}}(A_{\omega_1\ldots\omega_n\mathsf{U}} \mid A_{\omega_1\ldots\omega_n}) = 1
\end{cases} \tag{8.32}$$

for all $\omega_1, \ldots, \omega_n \in \{\mathsf{D}, \mathsf{U}\}$ and all n and m with $1 \leqslant n+1 < m \leqslant T$. Clearly, under the no-arbitrage assumption, all probabilities have to be between 0 and 1 and have to be independent of maturity.

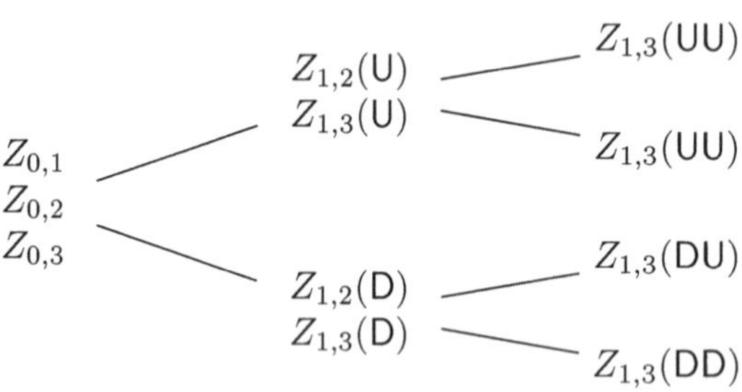

FIGURE 8.6: A binomial tree with bond prices.

For example, in the three-period case, we have three zero-coupon bonds with maturities $m = 1$, $m = 2$, and $m = 3$, respectively. The bond prices can be organized in a binomial tree, given in Figure 8.6. Since the face value of any ZCB is one, we omit the values $Z_{m,m}$ for all $m = 1, 2, 3$ in the above figure and truncate the tree to two periods. To find the state probabilities $\widetilde{\mathbb{P}}(A_{\mathsf{D}})$ and $\widetilde{\mathbb{P}}(A_{\mathsf{U}})$, we solve either

$$Z_{0,2} = \frac{1}{1+r_1}\left[Z_{1,2}(\mathsf{D})\widetilde{\mathbb{P}}(A_{\mathsf{D}}) + Z_{1,2}(\mathsf{U})\widetilde{\mathbb{P}}(A_{\mathsf{U}})\right]$$

or

$$Z_{0,3} = \frac{1}{1+r_1}\left[Z_{1,3}(\mathsf{D})\widetilde{\mathbb{P}}(A_{\mathsf{D}}) + Z_{1,3}(\mathsf{U})\widetilde{\mathbb{P}}(A_{\mathsf{U}})\right].$$

The result has to be independent of maturity or there is an arbitrage opportunity otherwise. The proof of this fact is left as an exercise for the reader. To find the conditional probabilities $\widetilde{\mathbb{P}}(A_{\omega_1\omega_2} \mid A_{\omega_1})$ for $\omega_1, \omega_2 \in \{\mathsf{D}, \mathsf{U}\}$, we solve the following equations:

$$Z_{1,3}(\mathsf{D}) = \frac{1}{1+r_2(\mathsf{D})}\left[Z_{2,3}(\mathsf{DD})\widetilde{\mathbb{P}}(A_{\mathsf{DD}} \mid A_{\mathsf{D}}) + Z_{2,3}(\mathsf{DU})\widetilde{\mathbb{P}}(A_{\mathsf{DU}} \mid A_{\mathsf{D}})\right]$$
$$Z_{1,3}(\mathsf{U}) = \frac{1}{1+r_2(\mathsf{U})}\left[Z_{2,3}(\mathsf{UD})\widetilde{\mathbb{P}}(A_{\mathsf{UD}} \mid A_{\mathsf{U}}) + Z_{2,3}(\mathsf{UU})\widetilde{\mathbb{P}}(A_{\mathsf{UU}} \mid A_{\mathsf{U}})\right].$$

Note that we also use the fact that $\widetilde{\mathbb{P}}$ is a probability measure. Since the value of a ZCB at maturity is always one, the above process cannot be used to determine the probabilities $\widetilde{\mathbb{P}}(A_{\omega_1\omega_2\omega_3} \mid A_{\omega_1\omega_2})$ for the last period. However, these probabilities are not required for pricing bonds and other interest rate derivatives.

8.5 Exercises

Exercise 8.1. Prove that a multi-period model is complete iff every single-period sub-model is complete (Lemma 8.6).

Exercise 8.2. Consider a binomial model with stochastic volatility and stochastic interest rates presented in Section 8.4.1. Find a formula for the number of shares of stock, Δ_t, held at each time $t \in \{0, 1, 2 \ldots, T-1\}$ by a self-financing portfolio strategy that replicates a derivative with a given payoff V_T.

Exercise 8.3. Consider a two-period market model with four scenarios and two base assets: a risk-free asset B with the prices $B_0 = 50$, $B_1 = 55$, and $B_2 = 60$ and a risky stock S whose prices are given by

Scenario	$S_0(\omega)$	$S_1(\omega)$	$S_2(\omega)$
ω_1	50	60	70
ω_2	50	60	50
ω_3	50	45	50
ω_4	50	45	40

(a) Find the EMM relative to the numéraire $g = B$.

(b) If instead $S_2(\omega_1) = 65$, show how to construct an arbitrage strategy.

Exercise 8.4. For the model described in Examples 8.3 and 8.4, construct a self-financing strategy replicating the payoff vector

$$V_2(\Omega) = [10,\ 20,\ 30,\ 40,\ 50,\ 60,\ 70,\ 80,\ 90]\,.$$

Exercise 8.5. Consider the two-period trinomial model with three base assets as given in Examples 8.3 and 8.4.

(a) Construct the Radon–Nikodym process $\{\varrho_t\}_{t\in\{0,1,2\}}$ of $\widetilde{\mathbb{P}}^{(S^1)}$ w.r.t. $\widetilde{\mathbb{P}}^{(B)}$.

(b) Using the result of Example 8.5, find the EMM $\widetilde{\mathbb{P}}^{(S^1)}$.

Exercise 8.6. Consider the following game. There is a set of 3 coins labelled 1, 2, and 3. The first coin has probability of heads 0.4, the second coin has probability of heads 0.6, and the third coin is fair. Your initial capital is \$1. You toss the three coins in turn. If coin n results in heads, then your wealth is increased by factor of $n+1$; otherwise, the wealth is decreased by factor of $n+1$, where $n = 1, 2, 3$. Draw a scenario tree indicating possible outcomes along with their probabilities. Find the distribution of your final wealth.

Exercise 8.7. For a given integer m with $1 \leqslant m \leqslant T$, consider a contract that pays r_m at time m. Show that the time-zero no-arbitrage price of this contract is equal to $Z_{0,m-1} - Z_{0,m}$.

Exercise 8.8. For a given integer m with $1 \leqslant m \leqslant T$, an m-period *interest rate swap* is a contract that makes payments $S_1, S_2, \ldots, S_m$ at times $1, 2, \ldots, m$, respectively, where $S_n = K - r_n$ for $n = 1, 2, \ldots, m$, and K is a fixed rate.

(a) Show that the time-zero no-arbitrage value of the swap contract is equal to

$$K\sum_{n=1}^{m} Z_{0,n} - (1 - Z_{0,m})\,.$$

(b) Find the time-zero no-arbitrage value of a three-period swap with $K = 4\%$ for the model of Example 8.8.

Exercise 8.9. For a given integer m with $1 \leqslant m \leqslant T$, an m-period *interest rate cap* is a contract that makes payments $C_1, C_2, \ldots, C_m$ at times $1, 2, \ldots, m$, respectively, where $C_n = (r_n - K)^+$ for $n = 1, 2, \ldots, m$, and K is a fixed rate. An m-period *interest rate floor* is a contract that makes payments $F_1, F_2, \ldots, F_m$ at times $1, 2, \ldots, m$, respectively, where $F_n = (K - r_n)^+$ for $n = 1, 2, \ldots, m$, and K is a fixed rate.

(a) Consider interest rate swap, cap, and floor with the same period m. Show that the sum of the initial no-arbitrage values of the swap and cap gives the initial no-arbitrage value of the floor:

$$\text{Swap}_m + \text{Cap}_m = \text{Floor}_m\,.$$

(b) Find the initial no-arbitrage values of the three-period floor and cap with $K = 4\%$ for the model in Example 8.8.

Exercise 8.10. Consider a binomial tree model for interest rates. Let the risk-free interest rates and no-arbitrage prices of bonds with all possible maturities be given. Show that the lack of arbitrage implies that the risk-neutral probabilities defined by (8.32) are independent of maturity.

Part III

Continuous-Time Modelling

Chapter 9

Essentials of General Probability Theory

In this chapter we present some measure-theoretic foundations of probability theory and random variables. This then leads us into the main tools and formulas for computing expectations and conditional expectations of random variables (more importantly, continuous random variables) under different probability measures. The main formulas that are provided here are used in further chapters for the understanding and quantitative modelling of continuous-time stochastic financial models.

9.1 Random Variables and Lebesgue Integration

Within the foundation of general probability theory, the mathematical expectation of any real-valued random variable X defined on a probability space $(\Omega, \mathcal{F}, \mathbb{P})$ is a so-called *Lebesgue integral w.r.t. a given probability measure* $\mathbb{P}$ over the sample space Ω. In order to define an integral of a random variable (i.e., a measurable set function) w.r.t. some given measure, we need a measurable space and a measure. The measurable space here is the pair $(\Omega, \mathcal{F})$, where $\mathcal{F}$ is a σ-algebra of events in Ω, and the measure is the probability measure function $\mathbb{P}\colon \Omega \to [0,1]$. Before providing a precise definition of such an integral, we make a couple of remarks. Namely, Ω is any abstract set having either a finite, infinitely countable, or uncountable number of elements. So here we are dealing with a general probability space that includes the finite probability spaces studied in previous chapters as special cases. As usual, we denote every element in Ω by ω, where $\{\omega\}$ is a singleton set in Ω. Any random variable X has a *positive part* $X_+ := \max\{X, 0\} \geqslant 0$ and a *negative part* $X_- := \max\{-X, 0\} \geqslant 0$, where $X = X_+ - X_-$ and $|X| = X_+ + X_-$. Note that both random variables X_+ and X_- are *nonnegative*. The expectation of X w.r.t. the measure $\mathbb{P}$ is defined as the difference of two nonnegative expectations:

$$\mathrm{E}[X] = \mathrm{E}[X_+] - \mathrm{E}[X_-] .$$

We write this, term by term, in the notation of a Lebesgue integral w.r.t. the measure $\mathbb{P}$ as follows:

$$\underbrace{\int_\Omega X(\omega)\,\mathrm{d}\mathbb{P}(\omega)}_{\mathrm{E}[X]} = \underbrace{\int_\Omega X_+(\omega)\,\mathrm{d}\mathbb{P}(\omega)}_{\mathrm{E}[X_+]} - \underbrace{\int_\Omega X_-(\omega)\,\mathrm{d}\mathbb{P}(\omega)}_{\mathrm{E}[X_-]} . \tag{9.1}$$

The absolute value of X has expectation $\mathrm{E}[\,|X|\,] = \mathrm{E}[X_+] + \mathrm{E}[X_-]$. X is said to be integrable w.r.t. measure $\mathbb{P}$ iff $\mathrm{E}[\,|X|\,] < \infty$, i.e., $\mathrm{E}[X_+] < \infty$ and $\mathrm{E}[X_-] < \infty$, hence $\mathrm{E}[X] < \infty$. A common notation used to state this is to write $X \in L^1(\Omega, \mathcal{F}, \mathbb{P})$. If $\mathrm{E}[X_+] = \infty$ and $\mathrm{E}[X_-] < \infty$, then we set $\mathrm{E}[X] = \infty$; if $\mathrm{E}[X_+] < \infty$ and $\mathrm{E}[X_-] = \infty$, then we set $\mathrm{E}[X] = -\infty$; if $\mathrm{E}[X_+] = \mathrm{E}[X_-] = \infty$, then $\mathrm{E}[X]$ is not defined. Note that for strictly positive or nonnegative X, $X_+ \equiv X$ ($X_- \equiv 0$). For strictly negative or nonpositive X we have

$X_- \equiv -X$ ($X_+ \equiv 0$). The Lebesgue integral of a random variable X over any subset $A \in \mathcal{F}$ is defined as the Lebesgue integral (over Ω) of the random variable $\mathbb{I}_A X$:

$$\int_A X(\omega)\, \mathrm{d}\mathbb{P}(\omega) \equiv \int_\Omega \mathbb{I}_A(\omega) X(\omega)\, \mathrm{d}\mathbb{P}(\omega) \equiv \mathrm{E}[\mathbb{I}_A X] \tag{9.2}$$

where $\mathbb{I}_A(\omega) = 1$ if $\omega \in A$ and $\mathbb{I}_A(\omega) = 0$ if $\omega \notin A$. Moreover, putting $X \equiv 1$ gives the probability under measure $\mathbb{P}$ of any event $A \in \mathcal{F}$ as a Lebesgue integral w.r.t. $\mathbb{P}$ over A:

$$\mathrm{E}[\mathbb{I}_A] = \mathbb{P}(A) = \int_A \mathrm{d}\mathbb{P}(\omega) = \int_\Omega \mathbb{I}_A(\omega)\, \mathrm{d}\mathbb{P}(\omega)\,. \tag{9.3}$$

Since $\mathbb{P}$ is an assumed probability measure, we must have $\mathbb{P}(\Omega) = \mathrm{E}[\mathbb{I}_\Omega] = 1$.

We now develop the Lebesgue integral of X w.r.t. measure $\mathbb{P}$ by first considering its definition for any ($\mathbb{P}$-a.s.) nonnegative real-valued[1] random variable $X\colon \Omega \to [0, \infty)$. The symbol $\mathbb{P}$-a.s. stands for almost surely w.r.t. to probability measure $\mathbb{P}$, i.e., a relation holds $\mathbb{P}$-a.s. means that it holds with probability one. So $X \geqslant 0$ ($\mathbb{P}$-a.s.) means that the set for which $X \geqslant 0$ has probability one: $\mathbb{P}(\{\omega \in \Omega : X(\omega) \geqslant 0\}) = 1$. We form a partition of the positive real line: $0 = y_0 < y_1 < y_2 < \ldots < y_k < y_{k+1} < \ldots < y_n$, where $y_n \to \infty$ as $n \to \infty$. A partition is defined such that the maximum sub-interval spacing approaches zero, i.e., $\lim_{n\to\infty} \max_{0\leqslant k\leqslant n-1} (y_{k+1} - y_k) = 0$ is implied. Then, the sets in $\mathcal{F}$ defined by $A_k := X^{-1}([y_k, y_{k+1})) \equiv \{\omega \in \Omega : y_k \leqslant X(\omega) < y_{k+1}\}, k = 0, 1, \ldots, n-1$, and $A_n := X^{-1}([y_n, \infty)) = \{\omega \in \Omega : X(\omega) \geqslant y_n\}$ are mutually exclusive and form a partition of the sample space Ω for every $n \geqslant 0$. The Lebesgue integral is defined as the limit of a partial sum:

$$\int_\Omega X(\omega)\, \mathrm{d}\mathbb{P}(\omega) := \lim_{n\to\infty} \sum_{k=0}^{n} y_k \mathbb{P}(A_k) \equiv \sum_{k=0}^{\infty} y_k \mathbb{P}(A_k)\,. \tag{9.4}$$

Assuming it exists, this sum gives a unique value (including cases where it may be infinite). This definition uses the lower value y_k of X on every set A_k. An equivalent definition replaces the value y_k with the upper value y_{k+1} for every $k \geqslant 1$. For any real $X = X_+ - X_-$, we simply use (9.4) for each respective nonnegative Lebesgue integral of X_+ and X_- and subtract to obtain the Lebesgue integral of X according to (9.1).

Assume $\{A_i\}$ is a *countable collection of disjoint sets* in $\mathcal{F}$. Then, applying the identity $\mathbb{I}_{\cup_i A_i} = \sum_i \mathbb{I}_{A_i}$ to (9.2) and using the linearity property of the Lebesgue integral gives

$$\int_{\cup_i A_i} X(\omega)\, \mathrm{d}\mathbb{P}(\omega) = \sum_i \int_\Omega X(\omega)\mathbb{I}_{A_i}(\omega)\, \mathrm{d}\mathbb{P}(\omega) = \sum_i \int_{A_i} X(\omega)\, \mathrm{d}\mathbb{P}(\omega)\,, \tag{9.5}$$

i.e., $\mathrm{E}[X\, \mathbb{I}_{\cup_i A_i}] = \sum_i \mathrm{E}[X\, \mathbb{I}_{A_i}]$. For $X \equiv 1$ we have $\mathrm{E}[\mathbb{I}_{\cup_i A_i}] = \sum_i \mathrm{E}[\mathbb{I}_{A_i}]$:

$$\mathbb{P}\Big(\bigcup_i A_i\Big) = \sum_i \mathbb{P}(A_i).$$

This recovers the countable additivity property of $\mathbb{P}$, which must hold for $\mathbb{P}$ to be a measure.

To see how the Lebesgue integral in (9.4) works, consider the class of random variables having the form of a finite sum of indicator random variables $X = \sum_{j=0}^{N} x_j \mathbb{I}_{C_j}$ with each $x_j \in [0, \infty)$, $C_j = \{\omega \in \Omega : X(\omega) = x_j\} \subset \mathcal{F}$, and where $\{C_j\}_{j=0}^{N}$ forms a partition of Ω.

[1]The definition also extends to the case $X\colon \Omega \to [0, \infty]$ where X can equal ∞, i.e., $X = X \cdot \mathbb{I}_{\{0\leqslant X<\infty\}} + \infty \cdot \mathbb{I}_{\{X=\infty\}}$ where the usual convention $0 \cdot \infty = 0$ is adopted. If $\mathbb{P}(X = \infty) = 0$, then we simply take $X = X \cdot \mathbb{I}_{\{0\leqslant X<\infty\}}$. If $\mathbb{P}(X = \infty) > 0$, i.e., the set $\{\omega \in \Omega\colon X(\omega) = \infty\}$ has positive $\mathbb{P}$-measure, then $\mathrm{E}[X] = \infty$.

This is called a *simple function* or *simple random variable.* Any random variable that can take only finitely many different values, $x_0, \ldots, x_N$, (including possibly a zero value) has this form. Note also that any simple random variable can be represented as the difference of two *nonnegative simple random variables.* For each sufficiently small interval $[y_k, y_{k+1})$ we have $A_k = X^{-1}([y_k, y_{k+1})) = X^{-1}(x_j) = C_j$ if there is a value $x_j \in [y_k, y_{k+1})$ for some $0 \leqslant j \leqslant N$, or otherwise $A_k = X^{-1}([y_k, y_{k+1})) = \emptyset$. For sufficiently large n, each x_j value will be contained in exactly one sub-interval and the nth partial sum in (9.4) does not change value as n is increased indefinitely since only N intervals will contain an x_j value (with probability measure $\mathbb{P}(C_j)$) and the rest of the intervals yield $A_k = \emptyset$ (i.e., $\mathbb{P}(A_k) = 0$). Hence, the Lebesgue integral in (9.4) is given as the finite sum

$$\int_\Omega X(\omega)\, \mathrm{d}\mathbb{P}(\omega) = \sum_{j=0}^{N} x_j \mathbb{P}(C_j) = \sum_{j=0}^{N} x_j \mathbb{P}(X = x_j)\,. \tag{9.6}$$

This corresponds to the formula in (6.17) for a discrete random variable on a finite state space Ω. However, here the simple random variable is discrete and defined more generally for uncountably infinite Ω. For a discrete random variable X taking on a countably infinite number of values $\{x_0, x_1, \ldots\}$, with infinitely countable or uncountable Ω, it readily follows that the Lebesgue integral recovers the familiar expectation formula:

$$\mathrm{E}[X] = \int_\Omega X(\omega)\, \mathrm{d}\mathbb{P}(\omega) = \sum_{j=0}^{\infty} x_j \mathbb{P}(X = x_j)$$

where the probabilities $p_j \equiv \mathbb{P}(X = x_j)$ define the probability mass function (PMF) of X in the measure $\mathbb{P}$.

Simple random variables are particularly useful and provide an alternative equivalent way of defining the Lebesgue integral. In standard measure and integration theory, one begins by taking (9.6) as the definition for the Lebesgue integral of any simple random variable. Then, for any $X \geqslant 0$ (a.s.) the Lebesgue integral in (9.4) is equivalently defined as the supremum of Lebesgue integral values over the set of all nonnegative simple random variables having value not greater than X:

$$\mathrm{E}[X] = \int_\Omega X(\omega)\, \mathrm{d}\mathbb{P}(\omega) := \sup\left\{\int_\Omega X^*(\omega)\, \mathrm{d}\mathbb{P}(\omega) : X^* \text{ is simple } 0 \leqslant X^* \leqslant X\right\}. \tag{9.7}$$

This last definition is rather abstract but it gives rise to other more explicit representations for the Lebesgue integral upon using the fact that a nonnegative X can be expressed (a.s.) as a limiting sequence of nonnegative simple random variables. Then, the Lebesgue integral of any nonnegative X can be computed as the limit of the Lebesgue integrals corresponding to the sequence of nonnegative simple random variables that converges to X. To see how this arises, we need to first recall the concept of pointwise convergence of a sequence of random variables. We recall that a sequence of random variables, $X_n, n = 1, 2, \ldots,$ is said to converge pointwise almost surely to some random variable X when $X_n \to X$ (a.s.) as $n \to \infty$, i.e., $\mathbb{P}\left(\left\{\omega \in \Omega : \lim_{n\to\infty} X_n(\omega) = X(\omega)\right\}\right) = 1$. A sequence of nonnegative random variables $X_n, n = 1, 2, \ldots,$ is said to *converge pointwise monotonically* to X (a.s.) if $X_n \to X$ and $0 \leqslant X_1 \leqslant X_2 \leqslant \ldots \leqslant X$ (a.s.), i.e., $\mathbb{P}(\{\omega \in \Omega : X_n(\omega) \nearrow X(\omega)\}) = 1$. If we have such a sequence, then each successive random variable in the sequence will approximate X better than the previous one and the approximation becomes exact in the limit $n \to \infty$. It stands to reason that the corresponding sequence of expected values (Lebesgue integrals for each X_n) will be increasingly better approximations and will give the expected value (Lebesgue integral) of X in the limit $n \to \infty$. This is summarized (without proof) in the following well-known theorem, appropriately called the Monotone Convergence Theorem (MCT) for random variables.

Theorem 9.1 (Monotone Convergence for random variables). *If $X_n, n = 1, 2, \ldots,$ is a sequence of nonnegative random variables converging pointwise monotonically to X (a.s.), then*

$$\lim_{n\to\infty} \mathrm{E}[X_n] = \mathrm{E}[X], \quad i.e., \quad \lim_{n\to\infty} \int_\Omega X_n(\omega)\, \mathrm{d}\mathbb{P}(\omega) = \int_\Omega X(\omega)\, \mathrm{d}\mathbb{P}(\omega).$$

In fact, we have monotonic convergence, i.e., $\mathrm{E}[X_n] \nearrow \mathrm{E}[X]$ as $n \to \infty$.

The MCT has many applications. For instance, we recover the known continuity properties of a probability measure $\mathbb{P}$. That is, let $A_1, A_2, \ldots, A_n, \ldots$ be subsets (events) in Ω. If $A_1 \subset A_2 \subset \cdots$, then $\lim_{n\to\infty} A_n = \bigcup_{n=1}^{\infty} A_n$ and we have (monotone continuity from below):

$$\mathbb{P}\left(\bigcup_{n=1}^{\infty} A_n\right) = \lim_{n\to\infty} \mathbb{P}(A_n).$$

If $A_1 \supset A_2 \supset \cdots$, then $\lim_{n\to\infty} A_n = \bigcap_{n=1}^{\infty} A_n$ and we have (monotone continuity from above):

$$\mathbb{P}\left(\bigcap_{n=1}^{\infty} A_n\right) = \lim_{n\to\infty} \mathbb{P}(A_n).$$

These follow as a corollary of Theorem 9.1. For example, monotone continuity from above is obtained by defining the sequence of monotonically increasing random variables $X_n = 1 - \mathbb{I}_{A_n}$, where $A_{n+1} \subset A_n$ for all $n \geqslant 1$. Hence, $X_n \nearrow X \equiv 1 - \mathbb{I}_A$, $A = \bigcap_{n=1}^{\infty} A_n$. By MCT we have $\lim_{n\to\infty} \mathrm{E}[X_n] = \mathrm{E}[X]$, i.e.,

$$\lim_{n\to\infty} \mathrm{E}[1 - \mathbb{I}_{A_n}] = \mathrm{E}[1 - \mathbb{I}_A] \implies \lim_{n\to\infty} \mathbb{P}(A_n) = \mathbb{P}(A) = \mathbb{P}\left(\bigcap_{n=1}^{\infty} A_n\right).$$

We now apply MCT to obtain a more explicit formula for $\mathrm{E}[X]$ when $X \geqslant 0$ (a.s.) by producing a monotonically increasing sequence of nonnegative simple random variables. There are many ways to do so. An explicit example is the sequence defined by

$$X_n(\omega) := \begin{cases} k/2^n & \text{if } \omega \in A_n^{(k)} \\ 0 & \text{otherwise} \end{cases} \tag{9.8}$$

with sets $A_n^{(k)} := \{\omega \in \Omega : k/2^n \leqslant X(\omega) < (k+1)/2^n\}$, $k = 0, \ldots, 2^{2n}$. For every $n \geqslant 1$, X_n is a simple random variable:

$$X_n(\omega) = \sum_{k=0}^{2^{2n}} \frac{k}{2^n} \mathbb{I}_{A_n^{(k)}}(\omega) = \sum_{k=0}^{2^{2n}} \frac{k}{2^n} \mathbb{I}_{\{X \in [\frac{k}{2^n}, \frac{k+1}{2^n})\}}(\omega). \tag{9.9}$$

We leave it as an exercise for the reader to verify that $X_n(\omega) \nearrow X(\omega)$ for any $X \geqslant 0$. Therefore, by using (9.6) as the expected value for every simple X_n and passing to the limit with the use of MCT, the Lebesgue integral (expectation) of any nonnegative random variable X is given by

$$\mathrm{E}[X] = \lim_{n\to\infty} \mathrm{E}[X_n] = \lim_{n\to\infty} \sum_{k=0}^{2^{2n}} \frac{k}{2^n} \mathbb{P}\left(\frac{k}{2^n} \leqslant X < \frac{k+1}{2^n}\right). \tag{9.10}$$

This is equivalent to (9.4) but here the partitions are chosen such that the convergence is monotonic, whereas the series in (9.4) is generally not monotonic. For arbitrary $X = X_+ - X_-$ we use the above construction for the two separate nonnegative random variables X_+ and X_- and subtract the two expectations, giving $\mathrm{E}[X]$, assuming we don't have $\infty - \infty$.

The Lebesgue integral above was presented in the general context of an abstract state space Ω with generally uncountable numbers of abstract elements ω and random variables X (i.e., measurable set functions) on a probability space $(\Omega, \mathcal{F}, \mathbb{P})$. It is important to consider the case where $\Omega \subset \mathbb{R}$, i.e., every ω is a real number. Recall our discussion of the Borel σ-algebra $\mathcal{B}(\mathbb{R})$. The pair $(\mathbb{R}, \mathcal{B}(\mathbb{R}))$, i.e., $\Omega = \mathbb{R}$, $\mathcal{F} = \mathcal{B}(\mathbb{R})$, is a measurable space. The *Lebesgue measure*[2] on $\mathbb{R}$, which we denote by m, is a measure that assigns a nonnegative real value or infinity to each Borel set $B \in \mathcal{B}(\mathbb{R})$, i.e., $m : B \to [0, \infty) \cup \infty$, such that all intervals $[a, b], a \leqslant b$, have measure equal to their length: $m([a, b]) = b - a$. All semi-infinite intervals $(-\infty, a], (-\infty, a), [b, \infty)$, or (b, ∞) have infinite Lebesgue measure. A single point has Lebesgue measure zero, $m(\{x\}) = 0$ for any point $x \in \mathbb{R}$. The empty set also has zero measure. Any set B that is a countable union of points also has zero measure since m is countably additive, i.e.,

$$m\left(\bigcup_{n=1}^{\infty} B_n\right) = \sum_{n=1}^{\infty} m(B_n)$$

for all disjoint Borel sets $B_n, n \geqslant 1$. This also implies finite additivity holds where $m(\cup_{n=1}^{N} B_n) = \sum_{n=1}^{N} m(B_n)$ for any $N \geqslant 1$. The set of rational numbers $\mathbb{Q}$ is countable with Lebesgue measure zero. The irrationals $\mathbb{R} \cap \mathbb{Q}^{\complement} = \mathbb{R} \backslash \mathbb{Q}$ are uncountable. Since $(\mathbb{R} \backslash \mathbb{Q}) \cup \mathbb{Q} = \mathbb{R}$, the Lebesgue measure of any Borel set B is unchanged if we remove all the rational numbers from it, i.e., as a disjoint union, $B = (B \backslash \mathbb{Q}) \cup (B \cap \mathbb{Q})$, hence $m(B) = m(B \backslash \mathbb{Q}) + m(B \cap \mathbb{Q}) = m(B \backslash \mathbb{Q})$. Note that $B \cap \mathbb{Q} \subset \mathbb{Q}$ so $m(B \cap \mathbb{Q}) \leqslant m(\mathbb{Q}) = 0 \implies m(B \cap \mathbb{Q}) = 0$.

There are also other more peculiar Borel sets that are uncountable but yet have Lebesgue measure zero. The well-known Cantor ternary set on $[0, 1]$, which is discussed in detail in many textbooks on real analysis, is such an example. In the interest of space, we don't discuss this set in any detail here. It suffices to note that the points in this set are too sparsely distributed over the unit interval $[0, 1]$ and hence do not accumulate any length measure. On the other hand, the points in the Cantor set cannot be counted (i.e., listed in a sequence in one-to-one with the integers). The Cantor set is constructed by starting with $[0, 1]$ and removing the middle third interval $(\frac{1}{3}, \frac{2}{3})$ and then repeating this process of removing the middle third for all remaining intervals in succession. One can then make a simple argument to show that the set is not countable, in essentially the same manner that the total number of branches in a binomial tree having an infinite number of time steps cannot be counted either.

By the above discussion, we see that the triplet $(\Omega, \mathcal{F}, \mathbb{P}) := ([0, 1], \mathcal{B}([0, 1]), m)$ serves as an example of a probability space where m acts as a *uniform probability measure* on the set $\Omega = [0, 1] \equiv \{x \in \mathbb{R} : 0 \leqslant x \leqslant 1\}$. By our previous notation, we may also write this as $\Omega = \{\omega \in \mathbb{R} : 0 \leqslant \omega \leqslant 1\}$. Any real-valued random variable $X : [0, 1] \to \mathbb{R}$ is a Borel (measurable) function where $X^{-1}(B) \in \mathcal{B}([0, 1])$ for every $B \in \mathcal{B}(\mathbb{R})$. Its expected value is then the Lebesgue integral w.r.t. the Lebesgue (uniform probability) measure m:

$$\mathrm{E}[X] = \int_{[0,1]} X(\omega) \, \mathrm{d}m(\omega) . \tag{9.11}$$

[2]The Lebesgue measure m of an interval $I_n = [a_n, b_n]$, or $(a_n, b_n]$, or $[a_n, b_n)$, or (a_n, b_n), is its length, $m(I_n) = \ell(I_n) := b_n - a_n, a_n \leqslant b_n$. The measure m of a Lebesgue-measurable set (for our purposes a Borel set) B is defined precisely as the smallest total length among all countable unions of intervals in $\mathbb{R}$ that (cover) contain B, i.e., for any $B \in \mathcal{B}(\mathbb{R})$, $m(B) := \inf\{\sum_{n=1}^{\infty} \ell(I_n) : B \subset \cup_{n=1}^{\infty} I_n\}$.

For example, the outcome of the experiment of picking a real number in $[0,1]$ uniformly at random is captured by the value of the random variable $X(\omega) = \omega$. The probability that a number is chosen within some arbitrary subinterval $[a,b] \subset [0,1]$ must therefore be the length of the interval. In this case the event is represented as the set $\{X \in [a,b]\} \equiv \{a \leqslant \omega \leqslant b\}$. Its probability is the Lebesgue integral of the indicator random variable $\mathbb{I}_{\{X\in[a,b]\}}(\omega) = \mathbb{I}_{\{a\leqslant\omega\leqslant b\}}$:

$$\mathbb{P}(X \in [a,b]) = \mathrm{E}[\mathbb{I}_{\{X\in[a,b]\}}] = \int_{[0,1]} \mathbb{I}_{[a,b]}(\omega)\, dm(\omega) \equiv \int_{[a,b]} dm(\omega) = m([a,b]) = b - a\,.$$

Note that $\mathbb{P}(\Omega) = \mathbb{P}(X \in [0,1]) = m([0,1]) = 1$. Combining this with the countable additivity property of m and the fact that $m : B \to [0,1]$ for any $B \in \mathcal{B}([0,1]) = \mathcal{B}(\mathbb{R}) \cap [0,1]$ shows that the Lebesgue measure m restricted to the unit interval $[0,1]$ is a proper probability measure. Note that we also get the probability of picking any finite or countable set of numbers in $[0,1]$ is zero since the Lebesgue measure of such a set is zero. In fact, the probability of picking a rational number is zero:

$$\mathbb{P}(X \in [0,1] \cap \mathbb{Q}) = \int_{[0,1]\cap\mathbb{Q}} dm(\omega) = m([0,1] \cap \mathbb{Q}) = 0\,.$$

The probability of picking an irrational number is 1, $\mathbb{P}(X \in [0,1]\backslash\mathbb{Q}) = m([0,1]\backslash\mathbb{Q}) = 1$. As well, the probability that we pick a number to be in the Cantor set $\mathcal{C}$ is zero since $\mathbb{P}(X \in \mathcal{C}) = m(\mathcal{C}) = 0$.

If we have a Lebesgue-measurable set $\Omega \subset \mathbb{R}$ with finite nonzero measure, $0 < m(\Omega) < \infty$, then the Lebesgue measure restricted to Ω, denoted by m_Ω, gives $m_\Omega(B) = m(B)$ for any set $B \in \mathcal{B}(\Omega)$. [Note: We assume that Ω is a Borel set since any Lebesgue-measurable set is either a Borel set or close to it in the sense that all sets that are Lebesgue-measurable and not Borel are null sets of Lebesgue measure zero.] The set Ω has "nonzero finite length." Typically, Ω is an interval or a combination of intervals, but need not be. In the above example, $\Omega = [0,1]$ (i.e., the unit interval) and $m_{[0,1]}(B) = m(B)$ for any $B \in \mathcal{B}([0,1])$. Then, $m_\Omega : B \to [0,c]$, $c \equiv m(\Omega)$, for any $B \in \mathcal{B}(\Omega)$. So m_Ω is a uniform measure on the space $(\Omega, \mathcal{B}(\Omega))$. We simply normalize this measure to obtain a *uniform probability measure* defined by $\mathbb{P}(B) := \frac{1}{c} \cdot m_\Omega(B)$, for all $B \in \mathcal{B}(\Omega)$, with $\mathbb{P}(\Omega) = \frac{1}{c} \cdot m_\Omega(\Omega) = \frac{c}{c} = 1$. Hence, $(\Omega, \mathcal{B}(\Omega), \mathbb{P})$ is a probability space where $\mathcal{B}(\Omega)$ contains all the events.

At this point we recall (from real analysis) the Lebesgue integral over $\mathbb{R}$ w.r.t. the Lebesgue measure m, which is defined regardless of any association to a probability space. The Lebesgue integral, w.r.t. m over $\mathbb{R}$, is defined for any Lebesgue-measurable function f, i.e., if $f^{-1}(I)$ is a Lebesgue-measurable set for any interval $I \in \mathbb{R}$. For our purposes, it suffices to assume that f is any Borel-measurable real-valued function, i.e., the set $f^{-1}(B) \equiv \{x \in \mathbb{R} : f(x) \in B\} \in \mathcal{B}(\mathbb{R})$ for any $B \in \mathcal{B}(\mathbb{R})$. The Lebesgue integral (w.r.t. m) over $\mathbb{R}$ is first defined for a nonnegative Borel function f. By this we mean f is nonnegative *almost everywhere* (abbreviated a.e. or m-a.e.), i.e., the set of points where $f < 0$ has Lebesgue measure zero: $m(\{x \in \mathbb{R} : f(x) < 0\}) = 0$. Essentially, by putting $\Omega = \mathbb{R}$, $\mathcal{F} = \mathcal{B}(\mathbb{R})$, and replacing $\mathbb{P}(\omega)$ by $m(x)$ and $X(\omega)$ by $f(x)$, equivalent definitions follow as those displayed above for the probability (expectation) Lebesgue integrals on the generally abstract measurable sets $(\Omega, \mathcal{F})$ with measure $\mathbb{P}$. In the Lebesgue integral the measurable space is $(\mathbb{R}, \mathcal{B}(\mathbb{R}))$ and the measure is the Lebesgue measure m. The Lebesgue integral of a nonnegative Borel function f (w.r.t. m over $\mathbb{R}$) can be defined by

$$\int_{\mathbb{R}} f(x)\, dm(x) := \lim_{n\to\infty} \sum_{k=0}^{n} y_k\, m(B_k)\,, \tag{9.12}$$

assuming the sum converges in $\mathbb{R}$ or equals ∞. This is the analogue of the definition in

(9.4). Here, the partitions $y_k, k \geqslant 0$ are defined above (9.4) and $B_k := f^{-1}([y_k, y_{k+1})) \equiv \{x \in \mathbb{R} : y_k \leqslant f(x) < y_{k+1}\}, k = 0, 1, \ldots, n-1$, $B_n := f^{-1}([y_n, \infty)) = \{x \in \mathbb{R} : f(x) \geqslant y_n\}$.

An equivalent (and more standard definition) is to first define the integral for any simple function $\varphi(x) = \sum_{k=1}^{n} a_k \mathbb{I}_{A_k}(x)$, with a_k's as real numbers and A_k's as Lebesgue-measurable sets (for our purposes these are Borel sets in $\mathbb{R}$):

$$\int_{\mathbb{R}} \varphi(x)\,\mathrm{d}m(x) := \sum_{k=1}^{n} a_k\, m(A_k)\,. \tag{9.13}$$

[Note that the usual convention $0 \cdot \infty = 0$ is used throughout.] Then, we define the Lebesgue integral for any nonnegative f as the supremum over the set of Lebesgue integral values of all nonnegative simple functions that are less than or equal to f:

$$\int_{\mathbb{R}} f(x)\,\mathrm{d}m(x) := \sup\left\{\int_{\mathbb{R}} \varphi(x)\,\mathrm{d}m(x) : \varphi \text{ is a simple function, } 0 \leqslant \varphi \leqslant f\right\}. \tag{9.14}$$

The Lebesgue integral, w.r.t. m over $\mathbb{R}$, of any Lebesgue-measurable (or Borel function) $f = f_+ - f_-$, with $f_\pm \geqslant 0$, is then

$$\int_{\mathbb{R}} f(x)\,\mathrm{d}m(x) = \int_{\mathbb{R}} f_+(x)\,\mathrm{d}m(x) - \int_{\mathbb{R}} f_-(x)\,\mathrm{d}m(x)\,. \tag{9.15}$$

As in the above case of the expectation of a random variable, f is integrable iff $\int_{\mathbb{R}} f_+\,\mathrm{d}m < \infty$ and $\int_{\mathbb{R}} f_-\,\mathrm{d}m < \infty$, i.e., $|f| = f_+ + f_-$ has a finite integral. If $\int_{\mathbb{R}} f_+\,\mathrm{d}m = \infty$ and $\int_{\mathbb{R}} f_-\,\mathrm{d}m < \infty$, then $\int_{\mathbb{R}} f\,\mathrm{d}m = \infty$; if $\int_{\mathbb{R}} f_+\,\mathrm{d}m < \infty$ and $\int_{\mathbb{R}} f_-\,\mathrm{d}m = \infty$, then $\int_{\mathbb{R}} f\,\mathrm{d}m = -\infty$; if $\int_{\mathbb{R}} f_+\,\mathrm{d}m = \int_{\mathbb{R}} f_-\,\mathrm{d}m = \infty$, then $\int_{\mathbb{R}} f\,\mathrm{d}m = \infty - \infty$ is not defined. Note that the Lebesgue integral of f over any measurable (Borel) set $A \in \mathcal{B}(\mathbb{R})$ is the integral of the (Borel) measurable function $\mathbb{I}_A f$ over $\mathbb{R}$:

$$\int_{A} f(x)\,\mathrm{d}m(x) = \int_{\mathbb{R}} \mathbb{I}_A(x) f(x)\,\mathrm{d}m(x)\,. \tag{9.16}$$

[Note: f is assumed (Borel) measurable so that the function $\mathbb{I}_A f$ is also measurable for any measurable set A.] The positive and negative parts of $\mathbb{I}_A f$ are $(\mathbb{I}_A f)_\pm = \mathbb{I}_A f_\pm$, giving

$$\begin{aligned}\int_{A} f(x)\,\mathrm{d}m(x) &= \int_{\mathbb{R}} \mathbb{I}_A(x) f_+(x)\,\mathrm{d}m(x) - \int_{\mathbb{R}} \mathbb{I}_A(x) f_-(x)\,\mathrm{d}m(x) \\ &= \int_{A} f_+(x)\,\mathrm{d}m(x) - \int_{A} f_-(x)\,\mathrm{d}m(x)\,. \end{aligned} \tag{9.17}$$

The MCT for random variables (Theorem 9.1) has an obvious analogue for sequences of Lebesgue-measurable functions, which we now state for Borel-measurable functions. Recall that a sequence of functions, $f_n, n = 1, 2, \ldots$, converges pointwise almost everywhere (a.e.) to a function f when $f_n(x) \to f(x)$ (a.e.) as $n \to \infty$. That is, the set for which convergence does not hold is a null set with Lebesgue measure zero: $m(\{x \in \mathbb{R} : \lim_{n\to\infty} f_n(x) \neq f(x)\}) = 0$. A sequence of nonnegative functions $f_n, n = 1, 2, \ldots$, is said to *converge pointwise monotonically* to f (a.e.) if $f_n(x) \to f(x)$ and $0 \leqslant f_1(x) \leqslant f_2(x) \leqslant \ldots \leqslant f(x)$ (a.e.), i.e., the set of values of $x \in \mathbb{R}$ for which these relations do not hold has Lebesgue measure zero. If we have such a sequence, then each successive function will better approximate f and the approximation becomes exact in the limit $n \to \infty$. Moreover, if the sequence of functions are Borel functions, then their corresponding Lebesgue integrals converge to the Lebesgue integral of the limiting function f. This is summarized in the MCT (Monotone Convergence Theorem) for functions. Its proof is given in any standard textbook on real analysis.

Theorem 9.2 (Monotone Convergence for functions). *If $f_n(x), n = 1, 2, \ldots, x \in \mathbb{R}$, is a sequence of nonnegative Borel functions converging pointwise monotonically (a.e.) to some function f, then*

$$\lim_{n\to\infty} \int_{\mathbb{R}} f_n(x)\,\mathrm{d}m(x) = \int_{\mathbb{R}} f(x)\,\mathrm{d}m(x)\,.$$

In fact, we have monotonic convergence, i.e., $\int_{\mathbb{R}} f_n(x)\,\mathrm{d}m(x) \nearrow \int_{\mathbb{R}} f(x)\,\mathrm{d}m(x)$, as $n \to \infty$.

For every nonnegative measurable (or Borel) function f there exists a sequence of nonnegative simple functions, $f_n, n \geqslant 1$, converging monotonically to f, i.e., $f_n(x) \nearrow f(x)$ (a.e.) as $n \to \infty$. In fact, the analogue of the sequence in (9.9) is the sequence of simple functions:

$$f_n(x) = \sum_{k=0}^{2^{2n}} \frac{k}{2^n} \mathbb{I}_{\{f^{-1}\left([\frac{k}{2^n}, \frac{k+1}{2^n})\right)\}}(x) \equiv \sum_{k=0}^{2^{2n}} \frac{k}{2^n} \mathbb{I}_{\{f(x)\in[\frac{k}{2^n}, \frac{k+1}{2^n})\}}\,. \tag{9.18}$$

Using MCT (Theorem 9.2) and the simple formula in (9.13) gives us the following representation for the Lebesgue integral, w.r.t. m over $\mathbb{R}$, of any nonnegative function f:

$$\int_{\mathbb{R}} f(x)\,\mathrm{d}m(x) = \lim_{n\to\infty} \int_{\mathbb{R}} f_n(x)\,\mathrm{d}m(x) = \lim_{n\to\infty} \sum_{k=0}^{2^{2n}} \frac{k}{2^n}\, m(A_k^n)\,, \tag{9.19}$$

where $A_k^n := f^{-1}([\frac{k}{2^n}, \frac{k+1}{2^n})) \equiv \{x \in \mathbb{R} : \frac{k}{2^n} \leqslant f(x) < \frac{k+1}{2^n}\}$. Such a series can be obtained for both f_+ and f_- and then the Lebesgue integral of $f = f_+ - f_-$ w.r.t. m over $\mathbb{R}$ is given by the difference, provided the result is defined (not $\infty - \infty$). Note also that the Lebesgue integral of a nonnegative f over any measurable (Borel) set B, using (9.16), has the representation

$$\int_B f(x)\,\mathrm{d}m(x) = \lim_{n\to\infty} \int_{\mathbb{R}} \mathbb{I}_B(x) f_n(x)\,\mathrm{d}m(x) = \lim_{n\to\infty} \sum_{k=0}^{2^{2n}} \frac{k}{2^n}\, m(A_k^n \cap B)\,. \tag{9.20}$$

The theory of Lebesgue integration w.r.t. m provides a framework for integrating a general class of real-valued functions. Lebesgue integration also forms the foundation for probability theory. The actual computation of many specific integrals is, in practice, a difficult task, as there are no known techniques for integrating particular functions. However, most Lebesgue integrals that we encounter are equivalent to a corresponding Riemann integral. This then allows us to use all the known powerful techniques of elementary calculus to compute Riemann integrals. Consider a continuous function $f : [a, b] \to \mathbb{R}$. Then, f is integrable and the function $F(x) := \int_{[a,x]} f(y)\,\mathrm{d}m(y) \equiv \int_a^x f(y)\,\mathrm{d}m(y)$ is differentiable for $x \in (a, b)$, i.e., $F'(x) = f(x)$. This is the fundamental theorem of calculus. The theorem just below relates the Lebesgue integral w.r.t. m and the Riemann integral of a bounded function over a finite interval. This theorem is proven in most standard textbooks on real analysis. Note that the statement "f is continuous (a.e.)" means that the set of points for which f is not continuous has Lebesgue measure zero.

Theorem 9.3 (Lebesgue versus Riemann Integration). *Let $f : [a, b] \to \mathbb{R}$ be bounded. Then:*

(i) f is Riemann-integrable, i.e., $\int_a^b f(x)\,\mathrm{d}x$ is defined, if and only if f is continuous (a.e.).

(ii) If f is Riemann-integrable, then the Lebesgue integral over $[a, b]$ is also defined and the two integrals are the same, i.e., $\int_a^b f(x)\,\mathrm{d}x = \int_{[a,b]} f(x)\,\mathrm{d}m(x)$.

Hence, when computing the Lebesgue integral of a function over an interval with a well-defined Riemann integral, we can simply equate the Lebesgue integral with the corresponding Riemann integral. For example, we write $\int_a^b f(x)\,\mathrm{d}x \equiv \int_{[a,b]} f(x)\,\mathrm{d}m(x)$. As well, assuming the existence of Riemann integrals for semi-infinite or infinite intervals, we write $\int_a^\infty f(x)\,\mathrm{d}x \equiv \int_{[a,\infty)} f(x)\,\mathrm{d}m(x)$, $\int_{-\infty}^b f(x)\,\mathrm{d}x \equiv \int_{(-\infty,b]} f(x)\,\mathrm{d}m(x)$, $\int_{-\infty}^\infty f(x)\,\mathrm{d}x \equiv \int_{\mathbb{R}} f(x)\,\mathrm{d}m(x)$. The same goes for other improper well-defined Riemann integrals. More generally, if the integral is over some arbitrary Borel set B, it is also customary to use short-hand notation for the Lebesgue integral w.r.t. m over B as $\int_B f(x)\,\mathrm{d}x \equiv \int_B f(x)\,\mathrm{d}m(x)$. If the integrand function f (or $f \cdot \mathbb{I}_B$) is not continuous (a.e.), then the Riemann integral is not defined and the integral is understood to be the corresponding Lebesgue integral, assuming it exists.

The expected value of a general real-valued random variable X defined on a probability space $(\Omega, \mathcal{F}, \mathbb{P})$ is a Lebesgue integral w.r.t. $\mathbb{P}$ over Ω. This is a construction that is general, yet not always practical when dealing with a generally abstract sample space Ω. For discrete random variables, the expectation reduces to a sum involving the probability mass function. We also saw that, for a continuous uniform random variable, its expectation (Lebesgue integral w.r.t. $\mathbb{P}$) reduces to a Lebesgue integral w.r.t. m and hence the latter can also be expressed as a Riemann integral. The transformation from a Lebesgue integral w.r.t. measure $\mathbb{P}$ over Ω into a Lebesgue integral (or Riemann) over $\mathbb{R}$ makes the theory more practical. This allows $\mathrm{E}[X]$ to be expressed in terms of integrals over $\mathbb{R}$, rather than over Ω, as follows.

We begin by recalling what a distribution measure is for a random variable X. Let B be any Borel set in $\mathbb{R}$. The *distribution measure* of X w.r.t. a probability measure $\mathbb{P}$ is defined by the set function

$$\mu_X(B) := \mathbb{P}(X^{-1}(B)) \equiv \mathbb{P}(X \in B). \tag{9.21}$$

It is important to note that μ_X measures subsets in $\mathbb{R}$, whereas $\mathbb{P}$ measures subsets in Ω. Namely, the probability of the event $\{X \in B\}$ is computed as a μ_X-measure of B. The cumulative distribution function (CDF), F_X, of the random variable X, w.r.t. $\mathbb{P}$, is then given in terms of this measure by

$$F_X(x) := \mu_X((-\infty, x]) = \mathbb{P}(X \in (-\infty, x]) \equiv \mathbb{P}(X \leqslant x),\ x \in \mathbb{R}. \tag{9.22}$$

Recall that any CDF is generally a right-continuous monotone nondecreasing function with limiting values $\lim_{x\to-\infty} F_X(x) \equiv F_X(-\infty) = 0$ and $\lim_{x\to\infty} F_X(x) \equiv F_X(\infty) = 1$.

Since the measure $\mathbb{P}$ is countably additive, μ_X is countably additive. Indeed, for any countable collection of pairwise disjoint Borel sets $\{B_i\}$ the corresponding pre-images $\{X^{-1}(B_i)\}$ are pairwise disjoint sets in Ω. Hence,

$$\mu_X\Big(\bigcup_i B_i\Big) = \mathbb{P}\Big(X^{-1}\Big(\bigcup_i B_i\Big)\Big) = \mathbb{P}\Big(\bigcup_i X^{-1}(B_i)\Big) = \sum_i \mathbb{P}(X^{-1}(B_i)) = \sum_i \mu_X(B_i).$$

Moreover, the measure is normalized, $\mu_X(\mathbb{R}) = \mathbb{P}(X \in \mathbb{R}) = 1$, so $(\mathbb{R}, \mathcal{B}, \mu_X)$ is in fact a probability space.

Since $(\mathbb{R}, \mathcal{B}, \mu_X)$ is a measure space, we can define a Lebesgue integral of a Borel function $f(x)$ w.r.t. $\mu_X(x)$ over $\mathbb{R}$ in a similar manner as the Lebesgue integral w.r.t. the measure m. For a simple function $\varphi(x) = \sum_{k=1}^n a_k \mathbb{I}_{A_k}(x)$, with A_k's as Borel sets in $\mathbb{R}$:

$$\int_{\mathbb{R}} \varphi(x)\,\mathrm{d}\mu_X(x) := \sum_{k=1}^n a_k\, \mu_X(A_k)\,. \tag{9.23}$$

This is the analogue of (9.13). Then, the Lebesgue integral w.r.t. $\mu_X(x)$ over $\mathbb{R}$ for any nonnegative f is defined as the supremum over Lebesgue integral values of all nonnegative simple functions that are less than or equal to f:

$$\int_{\mathbb{R}} f(x)\,\mathrm{d}\mu_X(x) := \sup\left\{\int_{\mathbb{R}} \varphi(x)\,\mathrm{d}\mu_X(x) : \varphi \text{ is a simple function, } 0 \leqslant \varphi \leqslant f\right\}. \tag{9.24}$$

The Lebesgue integral, w.r.t. μ_X over $\mathbb{R}$, of any Borel function $f = f_+ - f_-$, is then given by the difference of the two nonnegative Lebesgue integrals:

$$\int_{\mathbb{R}} f(x)\,\mathrm{d}\mu_X(x) = \int_{\mathbb{R}} f_+(x)\,\mathrm{d}\mu_X(x) - \int_{\mathbb{R}} f_-(x)\,\mathrm{d}\mu_X(x)\,, \tag{9.25}$$

and for any Borel set B in $\mathbb{R}$ we have

$$\int_B f(x)\,\mathrm{d}\mu_X(x) = \int_{\mathbb{R}} \mathbb{I}_B(x) f(x)\,\mathrm{d}\mu_X(x) = \int_{\mathbb{R}} \mathbb{I}_B(x) f_+(x)\,\mathrm{d}\mu_X(x) - \int_{\mathbb{R}} \mathbb{I}_B(x) f_-(x)\,\mathrm{d}\mu_X(x)\,. \tag{9.26}$$

Using MCT and the sequence of simple functions in (9.18), the Lebesgue integral of any nonnegative function f, w.r.t. μ_X over $\mathbb{R}$, is given by

$$\int_{\mathbb{R}} f(x)\,\mathrm{d}\mu_X(x) = \lim_{n\to\infty} \int_{\mathbb{R}} f_n(x)\,\mathrm{d}\mu_X(x) = \lim_{n\to\infty} \sum_{k=0}^{2^{2n}} \frac{k}{2^n}\,\mu_X\big(A_k^n\big)\,, \tag{9.27}$$

where $A_k^n := f^{-1}\big([\frac{k}{2^n}, \frac{k+1}{2^n})\big)$.

Based on the above construction, we have the following result, which gives the expected value of a random variable $g(X)$ as a Lebesgue integral of the (ordinary) function $g(x)$ w.r.t. the distribution measure of X over $\mathbb{R}$. Here we assume that $g(X)$ is integrable, i.e., $\mathrm{E}[\,|g(X)|\,] < \infty$.

Theorem 9.4. *Given a random variable X on $(\Omega, \mathcal{F}, \mathbb{P})$ and a Borel function $g : \mathbb{R} \to \mathbb{R}$,*

$$\mathrm{E}[g(X)] \equiv \int_{\Omega} g(X(\omega))\,\mathrm{d}\mathbb{P}(\omega) = \int_{\mathbb{R}} g(x)\,\mathrm{d}\mu_X(x)\,. \tag{9.28}$$

Proof. In many analysis textbooks we find a standard way to prove this by first showing that (9.28) follows trivially for the simplest case of a Boolean indicator function $g(x) = \mathbb{I}_A(x)$ and then by the linearity property of integrals the result is shown to hold for any simple function $g(x) = \sum_{k=1}^n a_k \mathbb{I}_{A_k}(x)$. Equation (9.28) is then shown to hold for any nonnegative function by using MCT and finally it follows for any $g(x) = g_+(x) - g_-(x)$. As an alternate proof, it is now instructive to see how (9.28) follows directly using (9.27) for nonnegative g with the sequence g_n defined as in (9.18) with f replaced by g:

$$\begin{aligned}
\int_{\mathbb{R}} g(x)\,\mathrm{d}\mu_X(x) = \lim_{n\to\infty} \int_{\mathbb{R}} g_n(x)\,\mathrm{d}\mu_X(x) &= \lim_{n\to\infty} \sum_{k=0}^{2^{2n}} \frac{k}{2^n}\,\mu_X(A_k^n) \\
&= \lim_{n\to\infty} \sum_{k=0}^{2^{2n}} \frac{k}{2^n}\,\mathbb{P}(X^{-1}(A_k^n)) \\
&= \lim_{n\to\infty} \sum_{k=0}^{2^{2n}} \frac{k}{2^n}\,\mathbb{P}\Big(g(X) \in [\frac{k}{2^n}, \frac{k+1}{2^n})\Big) \\
&\equiv \int_{\Omega} g(X(\omega))\,\mathrm{d}\mathbb{P}(\omega)\,.
\end{aligned}$$

Here we used the definition in (9.21) for each set $A_k^n = g^{-1}([\frac{k}{2^n}, \frac{k+1}{2^n}))$ and manipulated the set $X^{-1}(A_k^n) \equiv \{\omega \in \Omega : X(\omega) \in A_k^n\} = \{\omega \in \Omega : g(X(\omega)) \in g(A_k^n)\} = \{\omega \in \Omega : g(X(\omega)) \in [\frac{k}{2^n}, \frac{k+1}{2^n})\}$. Hence, (9.28) holds for both nonnegative parts g_+ and g_- of g, i.e., it must hold for any Borel function g. □

The above expectation formula is useful if the integral on the right-hand side of (9.28) can be computed more explicitly. This is still in the form of a Lebesgue integral w.r.t. the measure μ_X. However, we can reduce this to more familiar forms depending on the type of random variable X.

In the simplest case of a constant random variable $X \equiv a$, the distribution measure is the *Dirac measure*, $\mu_X(\cdot) = \delta_a(\cdot)$, i.e., for any Borel set B,

$$\delta_a(B) := \begin{cases} 1 & \text{if } a \in B \\ 0 & \text{if } a \notin B. \end{cases} \tag{9.29}$$

In particular, $\delta_a(\{a\}) = 1$ and $\delta_a(\{x\}) = 0$ for $x \neq a$. By (9.22), the CDF is simply $F_X(x) := \mu_X((-\infty, x]) = \delta_a((-\infty, x]) = \mathbb{I}_{\{x \geqslant a\}}$. The expected value as a Lebesgue integral w.r.t. $\mathbb{P}$ is trivially given since $\mathrm{E}[g(X)] = g(a)\mathbb{P}(\Omega) = g(a)$. According to (9.28), this value must equal

$$\mathrm{E}[g(X)] = \int_{\mathbb{R}} g(x)\,\mathrm{d}\delta_a(x) = g(a). \tag{9.30}$$

This gives us the formula for computing an integral w.r.t. the Dirac measure and is known as the sifting property since the Dirac measure picks out only the integrand value for $x = a$. Extending this to any purely discrete random variable that can take on distinct values a_i with probability $p_i > 0, i = 1, 2, \ldots, \sum_i p_i = 1$, the distribution measure is a linear combination of Dirac measures at each point in the range of X: $\mu_X(B) = \sum_i p_i \delta_{a_i}(B)$. In particular, $\mu_X(\{a_i\}) = p_i$. Then, using (9.30) and the fact that the integral w.r.t. a linear combination of measures is the linear combination of integrals w.r.t. each measure,

$$\mathrm{E}[g(X)] = \int_{\mathbb{R}} g(x)\,\mathrm{d}\mu_X(x) = \sum_i p_i \int_{\mathbb{R}} g(x)\,\mathrm{d}\delta_{a_i}(x) = \sum_i p_i g(a_i). \tag{9.31}$$

Using (9.22), the CDF is given as the piecewise constant (staircase function),

$$F_X(x) = \sum_i p_i \delta_{a_i}((-\infty, x]) = \sum_i p_i \mathbb{I}_{\{x \geqslant a_i\}}, \tag{9.32}$$

with jump discontinuities only at points $x = a_i$, i.e., $F_X(a_i) - F_X(a_i-) = p_i$ and $F_X(x) - F_X(x-) = 0$ for all x values not equal to any a_i. This is the familiar form for the CDF of a purely discrete random variable. Observe that, if g is continuous at the points a_i, the expectation of $g(X)$ is equal to the Riemann–Stieltjes integral of g with F_X as integrator:

$$\mathrm{E}[g(X)] = \int_{\mathbb{R}} g(x)\,\mathrm{d}F_X(x) = \sum_i g(a_i)(F_X(a_i) - F_X(a_i-)) = \sum_i g(a_i)p_i. \tag{9.33}$$

We can also define a distribution measure for a random variable X given as a so-called mixture of random variables with each having its own distribution measure, i.e., $\mu_X(B) = \sum_i p_i \mu_i(B)$ where $p_i \geqslant 0, i = 1, 2, \ldots, \sum_i p_i = 1$, and each μ_i is a distribution measure on $(\mathbb{R}, \mathcal{B})$. Then, the expected value of $g(X)$ is given as a linear combination of Lebesgue integrals w.r.t. each measure μ_i:

$$\mathrm{E}[g(X)] = \int_{\mathbb{R}} g(x)\,\mathrm{d}\mu_X(x) = \sum_i p_i \int_{\mathbb{R}} g(x)\,\mathrm{d}\mu_i(x). \tag{9.34}$$

Let us now consider the application of Theorem 9.4 to the most common important case of a continuous random variable. This is the case where the random variable has a probability density function (PDF) $f_X(x)$. In this case there is a nonnegative integrable Borel function f_X such that

$$\mu_X(B) = \int_B f_X(x)\,\mathrm{d}m(x)\,, \text{ for all Borel sets } B, \tag{9.35}$$

with CDF

$$F_X(b) := \int_{(-\infty,b]} f_X(x)\,\mathrm{d}m(x) \tag{9.36}$$

for all $b \in \mathbb{R}$. In this case, the CDF F_X is continuous and hence has no jumps, i.e., $F_X(x) = F_X(x-) = F_X(x+)$ for all x. Moreover, F_X is not just continuous but in fact an *absolutely continuous function.* The reader may wish to consult a textbook on real analysis to learn more about this technical detail. It suffices here to point out that F_X is differentiable and its derivative is the PDF: $F'_X \equiv f_X$. The distribution measure is said to be *absolutely continuous w.r.t. the Lebesgue measure m.* In this case, X is an absolutely continuous random variable but we simply say that it is a continuous random variable.[3] Based on (9.35) it is easy to prove, using similar steps as in the above proof of (9.28) combined with the linearity property of the Lebesgue integral w.r.t. m and where $\mu_X(A_k^n) = \int_{A_k^n} f_X(x)\,\mathrm{d}m(x)$, that the expectation is a Lebesgue integral of gf_X w.r.t. m:

$$\mathrm{E}[g(X)] = \int_{-\infty}^{\infty} g(x)\,f_X(x)\,\mathrm{d}m(x)\,. \tag{9.37}$$

In most (and for our purposes essentially all) applications, the density f_X (when it exists) is bounded and continuous (a.e.) on $\mathbb{R}$, i.e., the CDF is the Riemann integral of the PDF,

$$F_X(b) := \int_{-\infty}^{b} f_X(x)\,\mathrm{d}x\,,\ b \in \mathbb{R}. \tag{9.38}$$

Moreover, if g is continuous (a.e.) on $\mathbb{R}$, then (9.37) reduces to the familiar well-known formula for the expected value of $g(X)$ as a Riemann integral:

$$\mathrm{E}[g(X)] = \int_{-\infty}^{\infty} g(x)\,f_X(x)\,\mathrm{d}x\,, \tag{9.39}$$

assuming both $g_\pm$ are Riemann-integrable, $\mathrm{E}[\,|g(X)|\,] \equiv \int_{-\infty}^{\infty} |g(x)|\,f_X(x)\,\mathrm{d}x < \infty$, i.e., $\mathrm{E}[g_+] < \infty$ and $\mathrm{E}[g_-] < \infty$.

The standard normal $X \sim Norm(0,1)$ is an important example of a continuous random variable having positive Gaussian density $f_X(x) = n(x) := \frac{1}{\sqrt{2\pi}}\mathrm{e}^{-x^2/2}$, for all real x. The distribution μ_X is absolutely continuous w.r.t. the Lebesgue measure. In fact, f_X is bounded and continuous on $\mathbb{R}$ with the CDF given by the Riemann integral:

$$F_X(b) := \mathbb{P}(X \leqslant b) = \int_{-\infty}^{b} n(x)\,\mathrm{d}x \equiv \mathcal{N}(b)\,,\ b \in \mathbb{R}.$$

[3]There also exist very special types of random variables X where F_X is not an absolutely continuous function, yet it has zero derivative $F'_X(x) \equiv 0$ on a set of Lebesgue measure zero (a.e.). In this case X is said to be *singularly continuous* where the CDF is a nondecreasing monotone continuous function with zero derivative (a.e.) and hence there does not exist a PDF f_X, i.e., (9.36) (and (9.38)) does not hold. The expectation of $g(X)$ is still defined as a Lebesgue integral in (9.28). The so-called Cantor function on $[0,1]$ is a well-known textbook example of a CDF of a singularly continuous random variable that is uniformly distributed on the Cantor set $\mathcal{C}$. The Cantor CDF is constant on the complement of the Cantor set, i.e., $F'_X(x) \equiv 0$ for $x \in [0,1]\backslash\mathcal{C}$. There are other known interesting properties of such random variables. However, throughout this text we will have no need for such singular cases so that all continuous random variables are also absolutely continuous.

Clearly, $\mathcal{N}'(x) = n(x)$ where $F_X(x) = \mathcal{N}(x)$ is a proper CDF since it is monotonically increasing from $\mathcal{N}(-\infty) = 0$ to $\mathcal{N}(\infty) = 1$.

We now wish to make one last connection of the expectation in (9.28) to the so-called *Lebesgue–Stieltjes integral* encountered in real analysis. For any type of random variable X, we then realize that $\mathrm{E}[g(X)]$, when g is continuous (a.e.), is simply a Riemann–Stieltjes integral of g with CDF F_X as integrator. Beginning with (9.22), observe that for any *semi-open interval* $(a, b]$ the probability $\mathbb{P}(X \in (a, b])$ is given equivalently by the distribution measure $\mu_X((a, b])$ or the difference of the CDF values at the interval endpoints:

$$\mu_X((a,b]) = \mu_X((-\infty, b]) - \mu_X((-\infty, a]) = F_X(b) - F_X(a). \tag{9.40}$$

For any semi-open interval, $\ell_{F_X}((a,b]) := F_X(b) - F_X(a)$ defines its "length relative to F_X." Since $\ell_{F_X}((a,c]) = \ell_{F_X}((a,b]) + \ell_{F_X}((b,c])$ for $a < b < c$, this length is additive (cumulative) for adjoining intervals. In the special case that $F_X(x) = x$ we recover the usual length $b - a$. In contrast to the usual length, an infinitesimal interval does not necessarily have length zero relative to F_X since F_X is a CDF that may have jump discontinuities. In fact, a singleton set has length equal to the size of the jump discontinuity of the CDF:

$$\ell_{F_X}(\{x\}) = \lim_{\epsilon \to 0} \ell_{F_X}((x - \epsilon, x]) = F_X(x) - \lim_{\epsilon \to 0} F_X(x - \epsilon) = F_X(x) - F_X(x-).$$

For a purely continuous random variable X there are no jumps so all points have zero such length, but for a random variable having a discrete part there is a nonzero length given by the PMF values $p_i = F_X(x_i) - F_X(x_i-)$ at the points corresponding to the countable set of discrete values $\{x_i\}$ of X. For the other types of intervals we have $\ell_{F_X}([a,b]) = F_X(b) - F_X(a-), \ell_{F_X}((a,b)) = F_X(b-) - F_X(a), \ell_{F_X}([a,b)) = F_X(b-) - F_X(a-)$.

Based on the above definition of the length ℓ_F, for a given CDF F, the definition of the Lebesgue measure m is generalized to the *Lebesgue–Stieltjes measure generated by* F. This measure is denoted by m_F. The measure $m_F : \mathbb{R} \to [0,1]$, defined for any Borel set $B \in \mathcal{B}(\mathbb{R})$, is the smallest total length relative to F of all countable unions of semi-open intervals in $\mathbb{R}$ that contain B:

$$m_F(B) := \inf\left\{\sum_{n=1}^{\infty} \ell_F(I_n) : I_n = (a_n, b_n],\, a_n \leqslant b_n,\, B \subset \bigcup_{n=1}^{\infty} I_n\right\}. \tag{9.41}$$

Hence, m_F is a measure that assigns a value in $[0,1]$ for each Borel set B, such that all semi-open intervals $I_n = (a_n, b_n]$ have measure $m_F((a_n, b_n]) = \ell_F((a_n, b_n]) = F(b_n) - F(a_n)$. All intervals, including semi-infinite intervals, have finite measure; in particular, $m_F(\mathbb{R}) = F(\infty) - F(-\infty) = 1 - 0 = 1$. This measure is countably additive on $\mathcal{B}(\mathbb{R})$:

$$m_F\left(\bigcup_{n=1}^{\infty} B_n\right) = \sum_{n=1}^{\infty} m_F(B_n)$$

for all disjoint Borel sets $B_n, n \geqslant 1$. Because of the equivalence relation in (9.40) and the fact that the σ-algebra generated by all semi-open intervals in $\mathbb{R}$ is $\mathcal{B}(\mathbb{R})$, the distribution measure is the same as the Lebesgue–Stieltjes measure generated by the CDF, i.e., $\mu_X(B) = m_{F_X}(B)$ for every Borel set B. For example, $X \equiv a$ has CDF $F_X(x) = \mathbb{I}_{\{x \geqslant a\}}$. So, the measure $m_{F_X} = \delta_a$ is the Dirac measure concentrated at a. For any purely discrete random variable, with CDF in (9.32), the measure $m_{F_X} = \sum_i p_i \delta_{a_i}$ is a weighted sum of the Dirac measures for each point with its corresponding PMF value, i.e., $m_{F_X}(B) = \sum_i p_i \delta_{a_i}(B)$.

The Lebesgue–Stieltjes integral of a function w.r.t. m_{F_X} is defined in the same manner as in (9.23), (9.24), (9.25), and (9.26) with the notation that μ_X is replaced by m_{F_X}. In particular, the expectation in (9.28) is now recognized as the Lebesgue-Stieltjes integral of g w.r.t. m_{F_X} over $\mathbb{R}$, denoted by $\int_{\mathbb{R}} g(x)\,\mathrm{d}m_{F_X}(x)$, where we write equivalently

$$\mathrm{E}[g(X)] = \int_{\mathbb{R}} g(x)\,\mathrm{d}\mu_X(x) \equiv \int_{\mathbb{R}} g(x)\,\mathrm{d}m_{F_X}(x)\,. \tag{9.42}$$

If g is continuous (a.e.) then it can be shown that the Lebesgue-Stieltjes integral is the same as the corresponding Riemann-Stieltjes integral of g with CDF F_X as integrator function:

$$\mathrm{E}[g(X)] = \int_{\mathbb{R}} g(x)\,\mathrm{d}m_{F_X}(x) = \int_{\mathbb{R}} g(x)\,\mathrm{d}F_X(x)\,. \tag{9.43}$$

Since any CDF is a monotone (nondecreasing) bounded function, then it is of bounded variation and hence can always be used as integrator. If X is absolutely continuous, then $\mathrm{d}F_X(x) = F_X'(x)\,\mathrm{d}x = f_X(x)\,\mathrm{d}x$ and the Riemann-Stieltjes integral is the same as the usual Riemann integral in (9.39). For a purely discrete random variable with CDF in (9.32), the Riemann-Stieltjes integral in (9.43) is given by (9.33). The Riemann-Stieltjes integral in (9.43) also gives $\mathrm{E}[g(X)]$ for all other types of mixture random variables, as shown in Section 11.2.1, for example. Such random variables can have a discrete and continuous part (the continuous part being either absolutely continuous and/or singularly continuous).

Based on the above expectation formulas, one can also proceed to compute various quantities such as the moments $\mathrm{E}[X^n], n \geqslant 1$, of a real-valued random variable X (assuming $\mathrm{E}[|X|^n] < \infty$); the moment generating function $M_X(t) := \mathrm{E}[\mathrm{e}^{tX}]$, which is either infinite or a function of the real parameter t on some interval of convergence about $t = 0$; the characteristic function $\phi_X(t) := \mathrm{E}[\mathrm{e}^{itX}]$, $i \equiv \sqrt{-1}$, which is bounded for all $t \in \mathbb{R}$, i.e., $|\phi_X(t)| \leqslant \mathrm{E}[\,|\mathrm{e}^{itX}|\,] = 1$. The characteristic function (or the moment generating function) is useful for computing the mean and variance as well as various moments of a random variable. The relevant formulas and theorems related to these functions are part of standard material that is covered in most textbooks on probability theory and are hence (in the interest of space) simply omitted here.

In closing this section we mention one other general result, which is the *change of variable formula for an expectation* given in (9.44) just below. This allows us to compute the expectation $\mathrm{E}[g(X)]$ as an integral w.r.t. the distribution measure μ_Y of the random variable defined by $Y := g(X)$. Note that, since X is a random variable on $(\Omega, \mathcal{F}, \mathbb{P})$, then Y is also a random variable on $(\Omega, \mathcal{F}, \mathbb{P})$. That is, g is a Borel function, $g^{-1}(B) \in \mathcal{B}$, giving $Y^{-1}(B) = X^{-1}(g^{-1}(B)) \in \mathcal{F}$ for every $B \in \mathcal{B}$. Assuming Y is integrable, i.e., $\mathrm{E}[\,|Y|\,] < \infty$, then

$$\mathrm{E}[Y] := \int_{\Omega} Y(\omega)\,\mathrm{d}\mathbb{P}(\omega) = \int_{\mathbb{R}} g(x)\,\mathrm{d}\mu_X(x) = \int_{\mathbb{R}} y\,\mathrm{d}\mu_Y(y)\,, \tag{9.44}$$

where $\mu_Y(B) := \mathbb{P}(Y^{-1}(B)) \equiv \mathbb{P}(Y \in B)$, for every $B \in \mathcal{B}$. The proof of this formula follows readily from the relation between the two distribution measures: $\mu_Y(B) := \mathbb{P}(Y^{-1}(B)) = \mathbb{P}(X^{-1}(g^{-1}(B))) = \mu_X(g^{-1}(B))$. Hence, in the first equation line in the proof of Theorem 9.4 we have $\mu_X(A_k^n) = \mu_X(g^{-1}([\frac{k}{2^n}, \frac{k+1}{2^n}))) = \mu_Y([\frac{k}{2^n}, \frac{k+1}{2^n}))$, i.e.,

$$\int_{\mathbb{R}} g(x)\,\mathrm{d}\mu_X(x) = \lim_{n\to\infty} \sum_{k=0}^{2^{2n}} \frac{k}{2^n}\,\mu_X(A_k^n) = \lim_{n\to\infty} \sum_{k=0}^{2^{2n}} \frac{k}{2^n}\,\mu_Y\big(\big[\frac{k}{2^n}, \frac{k+1}{2^n}\big)\big) = \int_{\mathbb{R}} y\,\mathrm{d}\mu_Y(y)$$

which proves the formula. In summary, we see that $\mathrm{E}[g(X)]$ can be evaluated in three different ways: (i) by integrating the random variable $Y \equiv g(X)$ w.r.t. $\mathbb{P}$ over Ω, (ii) by integrating the function $g(x)$ w.r.t. distribution measure $\mu_X(x)$ over $\mathbb{R}$, or (iii) by integrating the function $f(y) = y$ w.r.t. distribution measure $\mu_Y(y)$ over $\mathbb{R}$.

9.2 Multidimensional Lebesgue Integration

The above integration theory for Borel functions of a single variable (and random variables defined as functions of a single random variable) extends into the general multidimensional case. The construction of Lebesgue integrals mirrors the above single-variable case. We recall the Borel sets in $\mathbb{R}^n$, $\mathcal{B}_n \equiv \mathcal{B}(\mathbb{R}^n)$ and Borel functions defined over $\mathbb{R}^n$ in Section 6.2.2. In $\mathbb{R}^2$, we denote the Lebesgue measure by $m_2 \colon \mathcal{B}_2 \to [0, \infty]$. It can be defined formally as an extension of the definition given above for the Lebesgue measure m. Given any two intervals I_1, I_2, then m_2 measures the area of the rectangle $I_1 \times I_2$: $m_2(I_1 \times I_2) = \ell(I_1)\ell(I_2)$. In terms of the Lebesgue measure in one dimension we have the *product measure* $m_2(I_1 \times I_2) = m(I_1)m(I_2)$. Note that the null sets (having zero measure w.r.t. m_2) include any countable union of points in $\mathbb{R}^2$ as well as some uncountable sets of the form $A \times \{b\}$, $A \subset \mathbb{R}$, $b \in \mathbb{R}$ or $\{a\} \times B$, $a \in \mathbb{R}$, $B \subset \mathbb{R}$. Also, any graph or curve in $\mathbb{R}^2$ has zero m_2 measure. The Lebesgue integral of a Borel function over $\mathbb{R}^2$ w.r.t. measure m_2 is defined in similar fashion to what we have above for the single variable case. A simple Borel function $\varphi(x,y) = \sum_{k=1}^{n} a_k \mathbb{I}_{A_k}(x,y)$, $a_k \in \mathbb{R}$, with all $A_k \in \mathcal{B}_2$, has Lebesgue integral

$$\int_{\mathbb{R}^2} \varphi(x,y)\,\mathrm{d}m_2(x,y) := \sum_{k=1}^{n} a_k\, m_2(A_k)\,. \tag{9.45}$$

The Lebesgue integral of any nonnegative Borel function f is defined by

$$\int_{\mathbb{R}^2} f(x,y)\,\mathrm{d}m_2(x,y) := \sup\left\{\int_{\mathbb{R}^2} \varphi(x,y)\,\mathrm{d}m_2(x,y) : \varphi \text{ is a simple function, } 0 \leqslant \varphi \leqslant f\right\}. \tag{9.46}$$

The Lebesgue integral, w.r.t. m_2 over $\mathbb{R}^2$, of any Borel function $f = f_+ - f_-$, with $f_\pm \geqslant 0$, is then

$$\int_{\mathbb{R}^2} f\,\mathrm{d}m_2 \equiv \int_{\mathbb{R}^2} f(x,y)\,\mathrm{d}m_2(x,y) = \int_{\mathbb{R}^2} f_+(x,y)\,\mathrm{d}m_2(x,y) - \int_{\mathbb{R}^2} f_-(x,y)\,\mathrm{d}m_2(x,y)\,. \tag{9.47}$$

f is integrable iff $\int_{\mathbb{R}^2} |f|\,\mathrm{d}m_2 < \infty$. We denote this by writing $f \in L^1(\mathbb{R}^2, \mathcal{B}_2, m_2)$. The Lebesgue integral of f over any Borel set $B \subset \mathbb{R}^2$ is the Lebesgue integral of $\mathbb{I}_B f$ over $\mathbb{R}^2$:

$$\int_B f\,\mathrm{d}m_2 = \int_{\mathbb{R}^2} \mathbb{I}_B(x,y) f(x,y)\,\mathrm{d}m_2(x,y)\,. \tag{9.48}$$

Assuming an integrable function, $f \in L^1(\mathbb{R}^2, \mathcal{B}_2, m_2)$, then Fubini's Theorem can be applied for interchanging the order of integration:

$$\int_{\mathbb{R}^2} f\,\mathrm{d}m_2 = \int_{\mathbb{R}} \left(\int_{\mathbb{R}} f(x,y)\,\mathrm{d}m(x)\right) \mathrm{d}m(y) = \int_{\mathbb{R}} \left(\int_{\mathbb{R}} f(x,y)\,\mathrm{d}m(y)\right) \mathrm{d}m(x)\,. \tag{9.49}$$

What is important for us is when f is continuous (m_2–a.e.) in (9.47) or $\mathbb{I}_B f$ is continuous in (9.48). The Lebesgue integral in (9.47) is then equal to the Riemann (double) integral over $\mathbb{R}^2$:

$$\int_{\mathbb{R}^2} f\,\mathrm{d}m_2 = \iint_{\mathbb{R}^2} f(x,y)\,\mathrm{d}x\,\mathrm{d}y = \int_{\mathbb{R}} \left(\int_{\mathbb{R}} f(x,y)\,\mathrm{d}x\right) \mathrm{d}y = \int_{\mathbb{R}} \left(\int_{\mathbb{R}} f(x,y)\,\mathrm{d}y\right) \mathrm{d}x \tag{9.50}$$

where we assume $f \in L^1(\mathbb{R}^2, \mathcal{B}_2, m_2)$, i.e., the function is integrable. In all our applications this will be the case where, for fixed $x \in \mathbb{R}$, f is continuous in y and for fixed $y \in \mathbb{R}$, f is continuous in x. The set B in (9.48) is usually a rectangular region $[a,b] \times [c,d]$, or of type $a \leqslant x \leqslant b, h_1(x) \leqslant y \leqslant h_2(x)$, or $g_1(y) \leqslant x \leqslant g_2(y), c \leqslant y \leqslant d$, etc. Given a set $B = B_1 \times B_2 = \{(x,y) \in \mathbb{R}^2 \colon x \in B_1, y \in B_2\}$, and assuming $\mathbb{I}_B f$ is continuous and integrable on $\mathbb{R}^2$, the Lebesgue integral in (9.48) is then

$$\int_{B_1 \times B_2} f \, \mathrm{d}m_2 = \int_{B_2} \left(\int_{B_1} f(x,y) \, \mathrm{d}x \right) \mathrm{d}y = \int_{B_1} \left(\int_{B_2} f(x,y) \, \mathrm{d}y \right) \mathrm{d}x \,. \tag{9.51}$$

We note that when the integrals only have meaning as Lebesgue integrals, we interpret the Riemann integrals as convenient shorthand notation for the corresponding Lebesgue integrals.

In $\mathbb{R}^3$, the Lebesgue measure, $m_3 \colon \mathcal{B}_3 \to [0,\infty]$, measures the volume of $I_1 \times I_2 \times I_3$, where $m_3(I_1 \times I_2 \times I_3) = \ell(I_1)\ell(I_2)\ell(I_3)$. For any $B = B_1 \times B_2 \times B_3 \in \mathcal{B}_3$, we have the product measure $m_3(B) = m(B_1)m(B_2)m(B_3)$. The null sets of m_3 include any countable union of points in $\mathbb{R}^3$ as well as uncountable sets of the form $A \times B \times \{c\}$, $A, B \subset \mathbb{R}$, $c \in \mathbb{R}$, or $A \times \{b\} \times C$, $A, C \subset \mathbb{R}$, $b \in \mathbb{R}$, or $\{a\} \times B \times C$, $B, C \subset \mathbb{R}$, $a \in \mathbb{R}$, all surfaces and lines, etc.. The Lebesgue integral of a Borel function $f \colon \mathbb{R}^3 \to \mathbb{R}$ w.r.t. measure m_3 over $\mathbb{R}^3$ is defined in analogy with the above construction in $\mathbb{R}^2$. If $f(x_1, x_2, x_3)$ is integrable w.r.t. m_3 over $\mathbb{R}^3$ (denoted as $f \in L^1(\mathbb{R}^3, \mathcal{B}_3, m_3)$) and is furthermore a continuous function (m_3–a.e.) of the three variables, then its Lebesgue integral is a Riemann (triple) integral over $\mathbb{R}^3$:

$$\int_{\mathbb{R}^3} f \, \mathrm{d}m_3 = \int_{\mathbb{R}} \int_{\mathbb{R}} \int_{\mathbb{R}} f(x_1, x_2, x_3) \, \mathrm{d}x_1 \, \mathrm{d}x_2 \, \mathrm{d}x_3 \,, \tag{9.52}$$

where we can also change the order of integration by successive application of Fubini's Theorem. For a Borel set $B = B_1 \times B_2 \times B_3 = \{(x_1, x_2, x_3) \in \mathbb{R}^3 \colon x_1 \in B_1, x_2 \in B_2, x_3 \in B_3\}$ we have

$$\begin{aligned}
\int_{B_1 \times B_2 \times B_3} f \, \mathrm{d}m_3 &= \int_{B_3} \int_{B_2} \int_{B_1} f(x_1, x_2, x_3) \, \mathrm{d}m(x_1) \, \mathrm{d}m(x_2) \, \mathrm{d}m(x_3) \\
&= \int_{B_3} \int_{B_2} \int_{B_1} f(x_1, x_2, x_3) \, \mathrm{d}x_1 \, \mathrm{d}x_2 \, \mathrm{d}x_3 \\
&= \int_{\mathbb{R}^3} f(x_1, x_2, x_3) \, \mathbb{I}_B(x_1, x_2, x_3) \, \mathrm{d}x_1 \, \mathrm{d}x_2 \, \mathrm{d}x_3 \,,
\end{aligned} \tag{9.53}$$

where the Riemann integral is used as shorthand for the Lebesgue integral and is equivalent to it when $f(x_1, x_2, x_3) \, \mathbb{I}_B(x_1, x_2, x_3)$ is a continuous function of x_1, x_2, x_3.

More generally, in $\mathbb{R}^n$ the Lebesgue measure, $m_n \colon \mathcal{B}_n \to [0,\infty]$, gives the n-dimensional volume of any n-dimensional cube: $m_n(I_1 \times \ldots \times I_n) = \ell(I_1) \times \ldots \times \ell(I_n)$. For every cartesian n-tuple $B = B_1 \times \ldots \times B_n \in \mathcal{B}_n$, $m_n(B) = m(B_1)m(B_2) \cdots m(B_n)$ is a product measure. Null sets of m_n are sets having zero n-dimensional volume and these include any countable union of points in $\mathbb{R}^n$, hyperplanes, lines, etc. The Lebesgue integral of a Borel function $f \colon \mathbb{R}^n \to \mathbb{R}$ w.r.t. m_n over $\mathbb{R}^n$ for all $n \geqslant 2$ is constructed as in the above case of $n = 2$. If $f \in L^1(\mathbb{R}^n, \mathcal{B}_n, m_n)$, i.e., $f(\mathbf{x}) \equiv f(x_1, \ldots, x_n)$ is integrable w.r.t. m_n over $\mathbb{R}^n$, and is furthermore a continuous function (m_n–a.e.) of the n variables $\mathbf{x}$, then its Lebesgue integral is equal to its Riemann integral over $\mathbb{R}^n$:

$$\int_{\mathbb{R}^n} f \, \mathrm{d}m_n = \int_{\mathbb{R}} \cdots \int_{\mathbb{R}} f(x_1, \ldots, x_n) \, \mathrm{d}x_1 \ldots \mathrm{d}x_n \,. \tag{9.54}$$

For a Borel set $B = B_1 \times \ldots \times B_n = \{\mathbf{x} \in \mathbb{R}^n : x_1 \in B_1, \ldots, x_n \in B_n\}$ we have

$$\begin{aligned}\int_{B_1 \times \ldots \times B_n} f \, \mathrm{d}m_n &= \int_{B_n} \cdots \int_{B_1} f(x_1, \ldots, x_n) \, \mathrm{d}m(x_1) \ldots \mathrm{d}m(x_n) \\ &= \int_{B_n} \cdots \int_{B_1} f(x_1, \ldots, x_n) \, \mathrm{d}x_1 \ldots \mathrm{d}x_n \\ &= \int_{\mathbb{R}^n} f(\mathbf{x}) \, \mathbb{I}_B(\mathbf{x}) \, \mathrm{d}^n\mathbf{x}, \end{aligned} \tag{9.55}$$

where the Riemann integral is used as shorthand for the Lebesgue integral and is equivalent to it when $f(\mathbf{x}) \, \mathbb{I}_B(\mathbf{x})$ is continuous on $\mathbb{R}^n$.

9.3 Multiple Random Variables and Joint Distributions

Let us now see how distributions and expectations are formulated for multiple random variables (i.e., random vectors) by first considering a pair of random variables $(X, Y) : \Omega \to \mathbb{R}^2$ defined on the same probability space $(\Omega, \mathcal{F}, \mathbb{P})$. The *joint distribution measure*

$$\mu_{X,Y} : \mathcal{B}_2 \to [0, 1]$$

is the measure induced by the pair (X, Y) and defined by

$$\mu_{X,Y}(B) := \mathbb{P}((X, Y) \in B), \;\; B \in \mathcal{B}_2. \tag{9.56}$$

This measure is countably additive and assigns a number in $[0, 1]$ to a Borel set B in $\mathbb{R}^2$ which corresponds to the probability of the event $\{(X, Y) \in B\} \equiv \{\omega \in \Omega : (X(\omega), Y(\omega)) \in B\}$. This measure is normalized so that $\mu_{X,Y}(\mathbb{R}^2) = \mathbb{P}((X, Y) \in \mathbb{R}^2) = 1$ and so the measure space $(\mathbb{R}^2, \mathcal{B}_2, \mu_{X,Y})$ is also a probability space. Writing $B = B_1 \times B_2$, $B_1, B_2 \in \mathcal{B}$, then we see that

$$\mu_{X,Y}(B_1 \times B_2) = \mathbb{P}(X^{-1}(B_1) \cap Y^{-1}(B_2)) = \mathbb{P}(X \in B_1, Y \in B_2) \tag{9.57}$$

gives the probability of the joint event $\{X \in B_1\} \cap \{Y \in B_2\} \equiv \{X \in B_1, Y \in B_2\}$. This joint measure determines the univariate (marginal distribution) measures of X and Y by letting $B_1 = \mathbb{R}$ or $B_2 = \mathbb{R}$:

$$\mu_{X,Y}(B \times \mathbb{R}) = \mathbb{P}(X \in B, Y \in \mathbb{R}) = \mathbb{P}(X \in B) = \mu_X(B), \tag{9.58}$$

$$\mu_{X,Y}(\mathbb{R} \times B) = \mathbb{P}(X \in \mathbb{R}, Y \in B) = \mathbb{P}(Y \in B) = \mu_Y(B), \tag{9.59}$$

for all $B \in \mathcal{B}$.

Letting $B_1 = (-\infty, x], B_2 = (-\infty, y]$ in (9.57) gives the joint CDF of (X, Y):

$$F_{X,Y}(x, y) := \mu_{X,Y}((-\infty, x] \times (-\infty, y]) = \mathbb{P}(X \leqslant x, Y \leqslant y), \;\; x, y \in \mathbb{R}. \tag{9.60}$$

This CDF is right-continuous on $\mathbb{R}^2$, monotone in both x and y, and recovers the univariate (marginal) CDF of X or Y in the respective limits:

$$\lim_{y \to \infty} F_{X,Y}(x, y) \equiv F_{X,Y}(x, \infty) = \mathbb{P}(X \leqslant x) = \mu_X((-\infty, x]) = F_X(x), \;\; x \in \mathbb{R}, \tag{9.61}$$

$$\lim_{x \to \infty} F_{X,Y}(x, y) \equiv F_{X,Y}(\infty, y) = \mathbb{P}(Y \leqslant y) = \mu_Y((-\infty, y]) = F_Y(y), \;\; y \in \mathbb{R}. \tag{9.62}$$

Taking the limit of infinite argument in the marginal CDF (in either case) gives

$$\lim_{x\to\infty,y\to\infty} F_{X,Y}(x,y) \equiv F_{X,Y}(\infty,\infty) = F_X(\infty) = F_Y(\infty) = 1.$$

Taking a decreasing sequence of numbers $x_n \searrow -\infty$ gives a decreasing sequence of sets approaching the empty set: $(-\infty, x_n] \times (-\infty, y] \searrow \emptyset$. By monotone continuity (from above) of the measure, $\mu_{X,Y}((-\infty, x_n] \times (-\infty, y]) \searrow \mu_{X,Y}(\emptyset) = 0$, i.e., for any $y \in \mathbb{R}$,

$$F_{X,Y}(-\infty, y) \equiv \lim_{x\to-\infty} F_{X,Y}(x,y) = \lim_{x_n \searrow -\infty} \mu_{X,Y}((-\infty, x_n] \times (-\infty, y]) = \mu_{X,Y}(\emptyset) = 0\,.$$

Similarly, $\lim_{y\to-\infty} F_{X,Y}(x,y) \equiv F_{X,Y}(x,-\infty) = 0\,,\ x \in \mathbb{R}$. These two relations must clearly hold since X and Y are in $\mathbb{R}$ and so $\mathbb{P}(X < -\infty, Y \leqslant y) = 0$ and $\mathbb{P}(X \leqslant x, Y < -\infty) = 0$.

Let x_1, x_2, y_1, y_2 be real numbers such that $x_1 < x_2$ and $y_1 < y_2$; then the joint measure of the semi-open rectangle $(x_1, x_2] \times (y_1, y_2]$ is given by

$$\begin{aligned}\mu_{X,Y}((x_1,x_2] \times (y_1,y_2]) &= \mathbb{P}(x_1 < X \leqslant x_2,\ y_1 < Y \leqslant y_2) \\ &= F_{X,Y}(x_2,y_2) - F_{X,Y}(x_1,y_2) - F_{X,Y}(x_2,y_1) + F_{X,Y}(x_1,y_1)\,. \end{aligned} \tag{9.63}$$

Based on this relation, the definition of a Lebesgue–Stieltjes measure for a single random variable, defined in (9.41), can be extended to a Lebesgue–Stieltjes measure generated by the joint CDF $F_{X,Y}$ of (X,Y). The quantity in (9.63) can be viewed as a measure of an "area relative to the joint CDF" $F_{X,Y}$ for any semi-open rectangle in $\mathbb{R}^2$. The Lebesgue–Stieltjes measure generated by $F_{X,Y}$, which we denote by $m_F^{X,Y}$, is the measure function $m_F^{X,Y} : \mathbb{R}^2 \to [0,1]$, defined for any Borel set $B = B_1 \times B_2 \in \mathcal{B}_2$, that assigns the smallest total area relative to $F_{X,Y}$ (using (9.63)) of all countable unions of semi-open rectangles $I_k \times J_l \equiv (a_k, b_k] \times (c_l, d_l]$, $a_k \leqslant b_k$, $c_l \leqslant d_l$ in $\mathbb{R}^2$ that contain B:

$$m_F^{X,Y}(B) := \inf\left\{ \sum_{k=1}^{\infty}\sum_{l=1}^{\infty} \mu_{X,Y}(I_k \times J_l) : B_1 \subset \bigcup_{k=1}^{\infty} I_k,\ B_2 \subset \bigcup_{l=1}^{\infty} J_l \right\}. \tag{9.64}$$

This measure is equivalent to the joint distribution measure, i.e., $m_F^{X,Y}(B) = \mu_{X,Y}(B)$.

Since $(\mathbb{R}^2, \mathcal{B}_2, \mu_{X,Y})$ is a measure, we can define the Lebesgue integral w.r.t. $\mu_{X,Y}$ over $\mathbb{R}^2$ (i.e., the Lebesgue–Stieltjes integral w.r.t. $\mu_{X,Y}$ or equivalently w.r.t. $m_F^{X,Y}$) in very similar manner as was done above for the Lebesgue–Stieltjes integral w.r.t. μ_X in (9.23) - (9.68). For any simple function

$$\varphi(x,y) = \sum_{k=1}^{K}\sum_{l=1}^{L} a_{k,l} \mathbb{I}_{B_1^k \times B_2^l}(x)$$

with $B_1^k \times B_2^l \in \mathcal{B}_2$, $a_{k,l} \in \mathbb{R}$, its Lebesgue integral w.r.t. $\mu_{X,Y}$ over $\mathbb{R}^2$ is defined by

$$\int_{\mathbb{R}^2} \varphi(x,y)\, \mathrm{d}\mu_{X,Y}(x,y) := \sum_{k=1}^{K}\sum_{l=1}^{L} a_{k,l}\, \mu_{X,Y}(B_1^k \times B_2^l)\,. \tag{9.65}$$

Based on this definition, the Lebesgue–Stieltjes integral w.r.t. $\mu_{X,Y}$ over $\mathbb{R}^2$ for any nonnegative Borel function $f\colon \mathbb{R}^2 \to \mathbb{R}$ is defined as the supremum over integral values of all nonnegative simple functions $\varphi \leqslant f$:

$$\int_{\mathbb{R}^2} f(x,y)\, \mathrm{d}\mu_{X,Y}(x,y) := \sup\left\{ \int_{\mathbb{R}^2} \varphi(x,y)\, \mathrm{d}\mu_{X,Y}(x,y) : \varphi \text{ is simple},\ 0 \leqslant \varphi \leqslant f \right\}. \tag{9.66}$$

For any Borel function $f = f_+ - f_-$, the Lebesgue–Stieltjes integral is given by the difference of the two nonnegative integrals:

$$\int_{\mathbb{R}^2} f(x,y)\,\mathrm{d}\mu_{X,Y}(x,y) = \int_{\mathbb{R}^2} f_+(x,y)\,\mathrm{d}\mu_{X,Y}(x,y) - \int_{\mathbb{R}^2} f_-(x,y)\,\mathrm{d}\mu_{X,Y}(x,y)\,, \tag{9.67}$$

and for any Borel set B in $\mathbb{R}^2$ we have

$$\int_B f(x,y)\,\mathrm{d}\mu_{X,Y}(x,y) = \int_{\mathbb{R}^2} \mathbb{I}_B(x,y) f(x,y)\,\mathrm{d}\mu_{X,Y}(x,y)\,. \tag{9.68}$$

We note that MCT is a general property that also applies to all Lebesgue–Stieltjes integrals.

Based on the above construction, we have the following result for the expected value of $h(X,Y)(\omega) \equiv h(X(\omega),Y(\omega))$, defined as a Borel function of two random variables (X,Y). Here we assume that $h(X,Y)$ is integrable, i.e., $\mathrm{E}[\,|h(X,Y)|\,] < \infty$.

Theorem 9.5. *Given a pair of random variables (X,Y) on $(\Omega,\mathcal{F},\mathbb{P})$ and a Borel function $h:\mathbb{R}^2 \to \mathbb{R}$,*

$$\mathrm{E}[h(X,Y)] \equiv \int_\Omega h(X(\omega),Y(\omega))\,\mathrm{d}\mathbb{P}(\omega) = \int_{\mathbb{R}^2} h(x,y)\,\mathrm{d}\mu_{X,Y}(x,y)\,. \tag{9.69}$$

The proof of (9.69) is very similar to the proof of (9.28). In the special case $h(x,y) = g(x)$, (9.69) recovers (9.28). The Lebesgue–Stieltjes integral in (9.69) is a very general representation for $\mathrm{E}[h(X,Y)]$ where h is a Borel function. That is, X or Y can be any type of random variable, i.e., any combination of discrete, absolutely continuous, or singularly continuous. The expectation in (9.69) reduces to various useful and familiar formulas for $\mathrm{E}[h(X,Y)]$ that a student learns in a standard course in probability theory. For our purpose, there are two main cases: discrete or continuous (we simply say continuous to mean absolutely continuous).

Assume that h is a continuous function (m_2–a.e.). This is virtually always the case in practice and certainly the case for all our applications in this text. The Lebesgue–Stieltjes is then a Riemann–Stieltjes integral over $\mathbb{R}^2$. Let us consider the simple case where both X and Y are discrete random variables and $h(x,y)$ is continuous at all values $(x,y) = (x_i,y_j)$ in the range of (X,Y); then the Riemann–Stieltjes integral simply recovers the summation formula in the joint PMF $p_{X,Y}(x,y) \equiv \mathbb{P}(X = x, Y = y)$ of (X,Y) at the support values:

$$\mathrm{E}[h(X,Y)] = \sum_{\text{all } x_i}\sum_{\text{all } y_j} p_{X,Y}(x_i,y_j)\,h(x_i,y_j)$$

assuming $\mathrm{E}[\,|h(X,Y)|\,] < \infty$ (i.e., summation converging for both negative and positive parts of h). Letting $h(X,Y) = \mathbb{I}_{\{X\leqslant x, Y\leqslant y\}}$, for any fixed real values (x,y), recovers the joint CDF as a (two-dimensional piecewise constant) staircase function in (x,y):

$$F_{X,Y}(x,y) = \mathbb{P}(X \leqslant x, Y \leqslant y) = \mathrm{E}[\mathbb{I}_{\{X\leqslant x, Y\leqslant y\}}] = \sum_{x_i\leqslant x}\sum_{y_j\leqslant y} p_{X,Y}(x_i,y_j)$$

with jump discontinuities at only the support values $(x,y) = (x_i,y_j)$ of the PMF. This recovers the formula in (9.32) when $y \to \infty$.

Let us now consider the case where (X,Y) are continuous with joint density denoted by $f_{X,Y}$. In this case, every Borel set $B \in \mathcal{B}_2$ has joint measure given by the Lebesgue integral of a nonnegative integrable Borel function $f_{X,Y}:\mathbb{R}^2 \to \mathbb{R}$ (namely, the joint PDF) over B:

$$\mu_{X,Y}(B) = \int_B f_{X,Y}(x,y)\,\mathrm{d}m_2(x,y) \equiv \int_{-\infty}^{\infty}\int_{-\infty}^{\infty} \mathbb{I}_B(x,y) f_{X,Y}(x,y)\,\mathrm{d}x\,\mathrm{d}y\,. \tag{9.70}$$

Recall that we sometimes simply write the Lebesgue integral as a Riemann integral using the convention we adopted in the previous section. Of course, the two are equal if $\mathbb{I}_B f_{X,Y}$ is a continuous function in (x, y). Since (9.70) holds for all $B \in \mathcal{B}_2$, then it holds for all sets of the form $B = (-\infty, a] \times (-\infty, b]$. Using $\mathbb{I}_B(x, y) = \mathbb{I}_{\{x \leqslant a, y \leqslant b\}}$ and the definition (9.60) we have the joint CDF

$$F_{X,Y}(a, b) = \int_{-\infty}^{b} \int_{-\infty}^{a} f_{X,Y}(x, y)\, dx\, dy\,, \quad a, b \in \mathbb{R}. \tag{9.71}$$

Since $F_{X,Y}(\infty, \infty) = 1$, then $f_{X,Y}$ integrates to unity on all of $\mathbb{R}^2$. In fact, (9.71) holds iff (9.70) holds for all Borel sets $B \in \mathcal{B}_2$. The joint CDF is continuous on $\mathbb{R}^2$ and related to the joint PDF by differentiating (9.71),

$$f_{X,Y}(x, y) = \frac{\partial^2}{\partial x \partial y} F_{X,Y}(x, y)\,, \quad x, y \in \mathbb{R}.$$

The marginal CDF of X and Y are given by (9.61)–(9.62) and taking either limit $a \to \infty$ or $b \to \infty$ in (9.71) gives

$$F_X(a) = \int_{-\infty}^{a} \left(\int_{-\infty}^{\infty} f_{X,Y}(x, y)\, dy \right) dx \text{ and } F_Y(b) = \int_{-\infty}^{b} \left(\int_{-\infty}^{\infty} f_{X,Y}(x, y)\, dx \right) dy,$$

for all $a, b \in \mathbb{R}$. Hence, the existence of the joint PDF $f_{X,Y}$ implies the existence of the respective marginal densities of X and Y:

$$f_X(x) = \int_{-\infty}^{\infty} f_{X,Y}(x, y)\, dy \quad \text{and} \quad f_Y(y) = \int_{-\infty}^{\infty} f_{X,Y}(x, y)\, dx. \tag{9.72}$$

We note that the converse is generally not true. Recall from our discussion of a single random variable, the marginal densities are nonnegative Borel functions that exist whenever (see (9.35) and (9.36))

$$\mu_X(B) = \int_B f_X(x)\, dx \quad \text{and} \quad \mu_Y(B) = \int_B f_Y(y)\, dy$$

for all Borel sets $B \subset \mathbb{R}$, or equivalently whenever

$$F_X(a) = \int_{-\infty}^{a} f_X(x)\, dx \quad \text{and} \quad F_Y(b) = \int_{-\infty}^{b} f_Y(y)\, dy$$

for all $a, b \in \mathbb{R}$. The expectations $\mathrm{E}[h_1(X)]$ and $\mathrm{E}[h_2(Y)]$, for single-variable Borel functions h_1 and h_2, are therefore given by the respective Riemann (Lebesgue) integrals over $\mathbb{R}$:

$$\mathrm{E}[h_1(X)] = \int_{\mathbb{R}} h_1(x) f_X(x)\, dx \quad \text{and} \quad \mathrm{E}[h_2(Y)] = \int_{\mathbb{R}} h_2(y) f_Y(y)\, dy\,. \tag{9.73}$$

For jointly continuous (X, Y), (9.70) holds, and it is readily proven (in the same manner that (9.37) or (9.39) is proven) that the expected value in (9.69) is given by the integral over $\mathbb{R}^2$ of the joint PDF multiplied by h, i.e.,

$$\mathrm{E}[h(X, Y)] = \int_{-\infty}^{\infty} \int_{-\infty}^{\infty} h(x, y) f_{X,Y}(x, y)\, dx\, dy\,. \tag{9.74}$$

Recall from Definition 6.13 that X and Y are mutually independent if

$$\mathbb{P}(X \in B_1, Y \in B_2) = \mathbb{P}(X \in B_1)\, \mathbb{P}(Y \in B_2) \tag{9.75}$$

for all $B_1, B_2 \in \mathcal{B}(\mathbb{R})$. That is, for a Borel rectangle $B = B_1 \times B_2$ the joint distribution measure given by (9.57) is now a product of the marginal distribution measures:

$$\mu_{X,Y}(B_1 \times B_2) = \mathbb{P}(X \in B_1)\,\mathbb{P}(Y \in B_2) = \mu_X(B_1)\,\mu_Y(B_2) := \mu_{X\times Y}(B_1 \times B_2)\,. \quad (9.76)$$

Hence, (9.75) and (9.76) are equivalent. From (9.60) we also have that independence is equivalent to

$$F_{X,Y}(x,y) = F_X(x)\,F_Y(y), \quad x, y \in \mathbb{R}. \quad (9.77)$$

Moreover, two continuous random variables (X,Y) are independent if and only if their joint PDF is the product of the marginal PDFs,

$$f_{X,Y}(x,y) = f_X(x)\,f_Y(y), \quad x, y \in \mathbb{R}. \quad (9.78)$$

This is easily proven. In particular, assuming (X,Y) are independent, then (9.71) gives

$$\int_{-\infty}^{b}\int_{-\infty}^{a} f_{X,Y}(x,y)\,\mathrm{d}x\,\mathrm{d}y = F_{X,Y}(a,b) = F_X(a)\,F_Y(b)$$
$$= \int_{-\infty}^{a} f_X(x)\,\mathrm{d}x \int_{-\infty}^{b} f_Y(y)\,\mathrm{d}y = \int_{-\infty}^{b}\int_{-\infty}^{a} f_X(x) f_Y(y)\,\mathrm{d}x\,\mathrm{d}y$$

for all $a, b \in \mathbb{R}$. This implies $f_{X,Y}(x,y) = f_X(x) f_Y(y)$. We leave the proof of the converse as an exercise for the reader. When (X,Y) are independent, the general expectation formula for all types of random variables, as given by (9.69), is now a Lebesgue–Stieltjes integral w.r.t. the above product measure $\mu_{X\times Y}$:

$$\mathrm{E}[h(X,Y)] = \int_{\mathbb{R}^2} h(x,y)\,\mathrm{d}\mu_{X\times Y}(x,y) = \int_{\mathbb{R}}\int_{\mathbb{R}} h(x,y)\,\mathrm{d}\mu_X(x)\,\mathrm{d}\mu_Y(y)\,, \quad (9.79)$$

where the order of integration in μ_X and μ_Y is interchangeable according to Fubini's Theorem. In the case that $h(X,Y) = h_1(X)h_2(Y)$,

$$\mathrm{E}[h(X,Y)] = \left(\int_{\mathbb{R}} h_1(x)\,\mathrm{d}\mu_X(x)\right)\left(\int_{\mathbb{R}} h_2(y)\,\mathrm{d}\mu_Y(y)\right) = \mathrm{E}[h_1(X)]\,\mathrm{E}[h_2(Y)]\,. \quad (9.80)$$

Of course, for continuous (X,Y) this product of expectations is given by (9.73). Taking $h_1(x) = x, h_2(y) = y, h(x,y) = xy$ shows that two mutually independent random variables have zero covariance

$$\mathrm{Cov}(X,Y) \equiv \mathrm{E}[XY] - \mathrm{E}[X]\,\mathrm{E}[Y] = 0\,. \quad (9.81)$$

The converse is generally not true.

An important example of a jointly continuous random vector (X,Y) is the standard bivariate normal distribution where $\mathrm{E}[X] = \mathrm{E}[Y] = 0$, $\mathrm{Cov}(X,Y) = \rho$, $|\rho| < 1$. The well-known joint PDF is

$$f_{X,Y}(x,y) = n_2(x,y;\rho) := \frac{1}{2\pi\sqrt{1-\rho^2}} \exp\left(-\frac{x^2+y^2-2\rho xy}{2(1-\rho^2)}\right), \quad x, y \in \mathbb{R}\,. \quad (9.82)$$

The joint distribution measure $\mu_{X,Y}$ is absolutely continuous w.r.t. Lebesgue measure m_2; i.e., for all $B \in \mathcal{B}_2$ we have

$$\mu_{X,Y}(B) = \int_B n_2(x,y;\rho)\,\mathrm{d}m_2(x,y) \equiv \int_{-\infty}^{\infty}\int_{-\infty}^{\infty} \mathbb{I}_B(x,y) n_2(x,y;\rho)\,\mathrm{d}x\,\mathrm{d}y\,. \quad (9.83)$$

The joint CDF is

$$F_{X,Y}(a,b) = \mathcal{N}_2(x,y;\rho) := \int_{-\infty}^{b}\int_{-\infty}^{a} n_2(x,y;\rho)\,\mathrm{d}x\,\mathrm{d}y, \quad a,b \in \mathbb{R}. \tag{9.84}$$

The functions n_2 and $\mathcal{N}_2$ denote the standard bivariate normal PDF and CDF, respectively, where $n_2(x,y;\rho) = \frac{\partial^2}{\partial x \partial y}\mathcal{N}_2(x,y;\rho)$. We note also the symmetry: $n_2(x,y,\rho) = n_2(y,x,\rho)$ and $\mathcal{N}_2(x,y,\rho) = \mathcal{N}_2(y,x,\rho)$. The marginal CDFs of X and Y are the standard normal CDF and follow simply from the limiting values of the joint CDF (see (9.61)–(9.62)) :

$$F_X(x) = F_{X,Y}(x,\infty) = \mathcal{N}_2(x,\infty;\rho) = \mathcal{N}(x)\,, \quad F_Y(y) = F_{X,Y}(\infty,y) = \mathcal{N}_2(\infty,y;\rho) = \mathcal{N}(y)\,.$$

Here we used the integral definition of $\mathcal{N}_2$ in (9.84). Hence, X and Y are identically distributed *Norm*$(0,1)$ random variables with standard normal (marginal) PDF

$$f_X(x) = \mathcal{N}'(x) = n(x)\,, \quad f_Y(y) = \mathcal{N}'(y) = n(y)\,,$$

$n(z) := \frac{1}{\sqrt{2\pi}}\mathrm{e}^{-z^2/2}$, $z \in \mathbb{R}$. [Note that these marginal PDFs also follow by integrating the joint PDF according to (9.72).] We observe that the pair (X,Y) is mutually independent if and only if the correlation coefficient $\rho = 0$, i.e.,

$$f_{X,Y}(x,y) = n_2(x,y;0) = \frac{1}{2\pi}\mathrm{e}^{-(x^2+y^2)/2} = n(x)n(y) = f_X(x)\,f_Y(y), \quad x,y \in \mathbb{R},$$

and the integral in (9.84) factors into

$$F_{X,Y}(a,b) = \mathcal{N}_2(a,b;0) = \mathcal{N}(a)\,\mathcal{N}(b) = F_X(a)\,F_Y(b), \quad a,b \in \mathbb{R}.$$

Substituting the above joint PDF into (9.74) gives the expectation of a Borel function of the normal pair (X,Y) as

$$\mathrm{E}[h(X,Y)] = \int_{-\infty}^{\infty}\int_{-\infty}^{\infty} h(x,y) n_2(x,y;\rho)\,\mathrm{d}x\,\mathrm{d}y\,. \tag{9.85}$$

For $\rho = 0$, $n_2(x,y;0) = n(x)n(y)$, and for $h(x,y) = h_1(x)h_2(y)$ this expectation reduces to (9.80), where

$$\mathrm{E}[h_1(X)] = \int_{\mathbb{R}} h_1(x)\,\mathrm{d}\mu_X(x) = \int_{\mathbb{R}} h_1(x)n(x)\,\mathrm{d}x\,,$$

$$\mathrm{E}[h_2(Y)] = \int_{\mathbb{R}} h_2(x)\,\mathrm{d}\mu_Y(y) = \int_{\mathbb{R}} h_2(y)n(y)\,\mathrm{d}y\,.$$

The above formulation extends to the more general case of an *n-dimensional real-valued random vector* $\mathbf{X} = (X_1,\ldots,X_n) \in \mathbb{R}^n$, for all integers $n \geqslant 1$. Each X_i is a random variable on $(\Omega,\mathcal{F},\mathbb{P})$ where $X_i^{-1}(B_i) \in \mathcal{F}$ for every $B_i \in \mathcal{B}(\mathbb{R})$, $i = 1,\ldots,n$. As a random vector $\mathbf{X}\colon \Omega \to \mathbb{R}^n$, for every Borel set $B = B_1 \times \ldots \times B_n \in \mathcal{B}_n$,

$$\mathbf{X}^{-1}(B) \equiv \{\mathbf{X} \in B\} \equiv \{X_1 \in B_1,\ldots,X_n \in B_n\} \in \mathcal{F}\,.$$

The joint distribution measure of $\mathbf{X} = (X_1,\ldots,X_n)$, which generalizes (9.56), is defined by

$$\mu_{\mathbf{X}}(B) \equiv \mu_{X_1,\ldots,X_n}(B) := \mathbb{P}(\mathbf{X} \in B) \equiv \mathbb{P}(X_1 \in B_1,\ldots,X_n \in B_n)\,. \tag{9.86}$$

This measure assigns a probability to a Borel set B in $\mathbb{R}^n$ which corresponds to the probability of the joint event $\{\omega \in \Omega\colon X_1(\omega) \in B_1,\ldots,X_n(\omega) \in B_n\} = \{X_1 \in B_1\}\cap\ldots\cap\{X_n \in B_n\}$. It is normalized, $\mu_{\mathbf{X}}(\mathbb{R}^n) = \mathbb{P}(\mathbf{X} \in \mathbb{R}^n) = 1$, so $(\mathbb{R}^n,\mathcal{B}_n,\mu_{\mathbf{X}})$ is a probability space.

The joint (n-dimensional) measure $\mu_{\mathbf{X}}$ determines all the univariate, bivariate, trivariate, etc., distribution measures for all single random variables X_i, pairs (X_i, X_j), triples (X_i, X_j, X_k), etc. This follows by setting some of the appropriate sets among $B_1, \ldots, B_n$ equal to $\mathbb{R}$. For example, setting all sets $B_j = \mathbb{R}$ for all $j \neq i$, and $B_i = A \in \mathcal{B}$, gives the univariate marginal distribution measures

$$\mu_{\mathbf{X}}(B_1 \times \ldots \times B_n) = \mathbb{P}(X_i \in A) = \mu_{X_i}(A)\,, \quad i = 1, \ldots, n.$$

The bivariate (marginal) distribution measure of a pair (X_i, X_j), $i < j$, is obtained by setting all sets $B_k = \mathbb{R}$ for all $k \neq i, k \neq j$, and $B_i = A \in \mathcal{B}, B_j = B \in \mathcal{B}$:

$$\mu_{\mathbf{X}}(B_1 \times \ldots \times B_n) = \mathbb{P}(X_i \in A, X_j \in B) = \mu_{X_i, X_j}(A \times B)\,, \quad i < j = 1, \ldots, n.$$

Letting $B_1 = (-\infty, x_1], B_2 = (-\infty, x_2], \ldots, B_n = (-\infty, x_n]$ in (9.86) gives the multivariate joint CDF of $(X_1, \ldots, X_n)$:

$$F_{\mathbf{X}}(\mathbf{x}) \equiv F_{(X_1,\ldots,X_n)}(x_1, \ldots, x_n) = \mathbb{P}(X_1 \leqslant x_1, \ldots, X_n \leqslant x_n)\,, \tag{9.87}$$

$\mathbf{x} = (x_1 \ldots, x_n) \in \mathbb{R}^n$. This CDF is right-continuous on $\mathbb{R}^n$ and is a nondecreasing monotonic function in all variables $x_1, \ldots, x_n$. The marginal CDFs of each X_i are recovered in the limit that $x_j \to \infty$, for all $j \neq i$ in (9.87):

$$F_{X_i}(x_i) = \mathbb{P}(X_i \leqslant x_i) = F_{(X_1,\ldots,X_n)}(\infty, \ldots, \infty, x_i, \infty, \ldots, \infty), \; x_i \in \mathbb{R}. \tag{9.88}$$

Similarly, the (marginal) joint CDF for each random vector pair (X_i, X_j), $i < j$, is obtained by letting $x_k \to \infty$ for all $k \neq i, k \neq j$:

$$F_{X_i,X_j}(x_i, x_j) = \mathbb{P}(X_i \leqslant x_i, X_j \leqslant x_j) = \lim_{\textit{all } x_k \to \infty; k \neq i, k \neq j} F_{(X_1,\ldots,X_n)}(x_1, \ldots, x_n)\,.$$

All other (marginal) joint CDFs are obtained in the appropriate limits. For example, we can consider any k-dimensional random vector such as $(X_1, \ldots, X_k)$, for any $1 \leqslant k \leqslant n$, having joint CDF

$$F_{X_1,\ldots,X_k}(x_1, \ldots, x_k) = F_{(X_1,\ldots,X_n)}(x_1, \ldots, x_k, \infty, \ldots, \infty)\,.$$

More generally, the joint CDF $F_{X_{i_1},\ldots,X_{i_k}}(y_1, \ldots, y_k)$, where $(y_1, \ldots, y_k) \in \mathbb{R}^k$, of any k-dimensional random vector $(X_{i_1}, \ldots, X_{i_k})$, $1 \leqslant i_1 < \ldots < i_k \leqslant n$, $1 \leqslant k \leqslant n$, taken from $(X_1, \ldots, X_n)$ is obtained by setting $x_j = \infty$ for all $j \notin \{i_1, i_2, \ldots, i_k\}$ in the n-dimensional joint CDF $F_{(X_1,\ldots,X_n)}(x_1, \ldots, x_n)$. This corresponds to the probability

$$F_{X_{i_1},\ldots,X_{i_k}}(y_1, \ldots, y_k) = \mathbb{P}(X_{i_1} \leqslant y_1, \ldots, X_{i_k} \leqslant y_k). \tag{9.89}$$

As in the two-dimensional case, the joint CDF evaluates to zero when setting any one of its arguments to $-\infty$. Setting all $x_i = \infty$ gives unity: $F_{\mathbf{X}}(\infty, \ldots, \infty) = \mathbb{P}(\mathbf{X} \in \mathbb{R}^n) = 1$.

The random vector $\mathbf{Y} = (Y_1, \ldots, Y_k) := (X_{i_1}, \ldots, X_{i_k})$, for any $1 \leqslant k \geqslant n$, has a joint distribution measure defined by

$$\mu_{\mathbf{Y}}(B_1 \times \ldots \times B_k) = \mathbb{P}(X_{i_1} \in B_1, \ldots, X_{i_k} \in B_k)$$

for any Borel set $B_1 \times \ldots \times B_k \in \mathcal{B}_k \equiv \mathcal{B}(\mathbb{R}^k)$. Hence, $(\mathbb{R}^k, \mathcal{B}_k, \mu_{\mathbf{Y}})$ is a probability space for each $k = 1, \ldots, n$.

The relation in (9.63) can be extended to any n-dimensional semi-open rectangle with the use of the n-dimensional joint CDF $F_{\mathbf{X}}(\mathbf{x})$. Moreover, the Lebesgue–Stieltjes measure defined in (9.64) can be extended into n dimensions accordingly, as generated by $F_{\mathbf{X}}$. In fact, the n-dimensional joint distribution measure $\mu_{\mathbf{X}}$ in (9.86) is the same Lebesgue–Stieltjes measure on $\mathbb{R}^n$. The above construction of the Lebesgue–Stieltjes integral w.r.t. $\mu_{X,Y}$ (for dimension $n = 2$), provided by (9.65),(9.66), (9.67), and (9.68), extends in the obvious manner into dimension $n \geqslant 2$. We write the Lebesgue–Stieltjes integral of a Borel function $f\colon \mathbb{R}^n \to \mathbb{R}$ w.r.t. the joint distribution measure $\mu_{\mathbf{X}}$ as the difference of two nonnegative integrals

$$\int_{\mathbb{R}^n} f(\mathbf{x})\,\mathrm{d}\mu_{\mathbf{X}}(\mathbf{x}) = \int_{\mathbb{R}^n} f_+(\mathbf{x})\,\mathrm{d}\mu_{\mathbf{X}}(\mathbf{x}) - \int_{\mathbb{R}^n} f_-(\mathbf{x})\,\mathrm{d}\mu_{\mathbf{X}}(\mathbf{x})\,, \tag{9.90}$$

and for any Borel set B in $\mathbb{R}^n$ we have

$$\int_B f(\mathbf{x})\,\mathrm{d}\mu_{\mathbf{X}}(\mathbf{x}) = \int_{\mathbb{R}^n} \mathbb{I}_B(\mathbf{x}) f(\mathbf{x})\,\mathrm{d}\mu_{\mathbf{X}}(\mathbf{x})\,. \tag{9.91}$$

Here, $f(\mathbf{x}) \equiv f(x_1, \ldots, x_n)$, $\mathrm{d}\mu_{\mathbf{X}}(\mathbf{x}) \equiv \mathrm{d}\mu_{X_1,\ldots,X_n}(x_1, \ldots, x_n)$ is shorthand vector notation.

Given a Borel function, $h : \mathbb{R}^n \to \mathbb{R}$, of a random vector $\mathbf{X} = (X_1, \ldots, X_n)$ on a probability space $(\Omega, \mathcal{F}, \mathbb{P})$, Theorem 9.5 is generalized to give the expected value of the random variable $h(\mathbf{X}) \equiv h(X_1, \ldots, X_n)$ as an integral of $h(\mathbf{x}) = h(x_1, \ldots, x_n)$ w.r.t. the joint distribution measure $\mu_{\mathbf{X}}$ over $\mathbb{R}^n$:

$$\mathrm{E}[h(\mathbf{X})] \equiv \int_\Omega h(\mathbf{X}(\omega))\,\mathrm{d}\mathbb{P}(\omega) = \int_{\mathbb{R}^n} h(\mathbf{x})\,\mathrm{d}\mu_{\mathbf{X}}(\mathbf{x})\,. \tag{9.92}$$

This formula can be proven in the same manner as the proof of (9.69). This Lebesgue–Stieltjes integral is a general representation for the expected value $\mathrm{E}[h(\mathbf{X})]$ where h is a Borel function on $\mathbb{R}^n$. So, each component of the vector $(X_1, \ldots, X_n)$ can be any type of random variable, i.e., any combination of discrete, absolutely continuous, or singularly continuous random variables. The two main types of random variables of interest to us are either discrete or continuous (i.e., absolutely continuous).

The case where all X_i are discrete (as in the binomial and multinomial financial models considered in previous chapters) simply generalizes the above double summation formulas in the case of two variables to multiple (n-fold) summation formulas involving the joint PMF $p_{X_1,\ldots,X_n}(x_1, \ldots, x_n) \equiv \mathbb{P}(X_1 = x_1, \ldots, X_n = x_n)$ at the support values:

$$\mathrm{E}[h(X_1, \ldots, X_n)] = \sum_{\text{all } x_1} \cdots \sum_{\text{all } x_n} p_{X_1,\ldots,X_n}(x_1, \ldots, x_n)\, h(x_1, \ldots, x_n)\,.$$

Here we assume $\mathrm{E}[\,|h(X_1, \ldots, X_n)|\,] < \infty$ (i.e., we assume the sums converge for both negative and positive parts of h). Choosing the indicator function $h(\mathbf{X}) = \mathbb{I}_{\{X_1 \leqslant a_1, \ldots, X_n \leqslant a_n\}}$ recovers the joint CDF:

$$F_{(X_1,\ldots,X_n)}(a_1, \ldots, a_n) = \mathrm{E}[\mathbb{I}_{\{X_1 \leqslant a_1, \ldots, X_n \leqslant a_n\}}] = \sum_{x_1 \leqslant a_1} \cdots \sum_{x_n \leqslant a_n} p_{X_1,\ldots,X_n}(x_1, \ldots, x_n)\,.$$

This is a (n-dimensional piecewise constant) staircase function in the variables $(a_1, \ldots, a_n)$ with jump discontinuities occurring at only the support values of the PMF.

The most important case for continuous random variables is when $(X_1, \ldots, X_n)$ are continuous with joint density $f_{X_1,\ldots,X_n}(x_1, \ldots, x_n) \equiv f_{\mathbf{X}}(\mathbf{x})$ as a nonnegative integrable Borel function $f_{\mathbf{X}}\colon \mathbb{R}^n \to \mathbb{R}$, i.e., when every Borel set $B \in \mathcal{B}_n$ has joint measure given by the Lebesgue integral of the joint density over B:

$$\mu_{\mathbf{X}}(B) = \int_B f_{\mathbf{X}}\,\mathrm{d}m_n \equiv \int_{\mathbb{R}^n} \mathbb{I}_B(\mathbf{x}) f_{\mathbf{X}}(\mathbf{x})\,\mathrm{d}^n\mathbf{x}\,. \tag{9.93}$$

This is the generalization of (9.70), where the Lebesgue integral is written as a Riemann integral on $\mathbb{R}^n$. The Lebesgue and Riemann integrals are equal if $\mathbb{I}_B f_{\mathbf{X}}$ is (m_n-a.e.) continuous on $\mathbb{R}^n$. The joint CDF is obtained by setting $B = (-\infty, x_1] \times \ldots \times (-\infty, x_n]$, where $\mathbb{I}_B(y_1, \ldots, y_n) = \mathbb{I}_{\{y_1 \leqslant x_1, y_n \leqslant x_n\}}$:

$$F_{X_1,\ldots,X_n}(x_1, \ldots, x_n) = \int_{-\infty}^{x_n} \cdots \int_{-\infty}^{x_1} f_{X_1,\ldots,X_n}(y_1, \ldots, y_n)\, \mathrm{d}y_1 \ldots \mathrm{d}y_n\,, \tag{9.94}$$

for all $x_1, \ldots, x_n \in \mathbb{R}$. Note that the joint PDF $f_{\mathbf{X}}$ integrates to unity since $\mu_{\mathbf{X}}(\mathbb{R}^n) = \mathbb{P}(\mathbf{X} \in \mathbb{R}^n) = 1$. As we proved for $n = 2$, the relation in (9.94) is equivalent to (9.93). The joint CDF is continuous on $\mathbb{R}^n$ and related to the joint PDF by differentiation,

$$f_{X_1,\ldots,X_n}(x_1, \ldots, x_n) = \frac{\partial^n}{\partial x_1 \ldots \partial x_n} F_{X_1,\ldots,X_n}(x_1, \ldots, x_n), \quad x_1, \ldots, x_n \in \mathbb{R}. \tag{9.95}$$

Note that (9.95) implies the existence of all marginal PDFs (densities) for all univariate X_i, bivariate (X_i, X_j), etc. In particular, all k-dimensional random vectors $(X_{i_1}, \ldots, X_{i_k})$, $1 \leqslant i_1 < \ldots < i_k \leqslant n$, $1 \leqslant k \leqslant n$, have CDF as in (9.89). Using this relation in (9.94) gives us the joint (marginal) PDF of $(X_{i_1}, \ldots, X_{i_k})$ as an $(n-k)$-dimensional integral of the joint PDF over $\mathbb{R}^{n-k}$ in the integration variables y_j, for all $j \notin \{i_1, i_2, \ldots, i_k\}$. For example, the CDF of the random vector consisting of the first k variables, $(X_1, \ldots, X_k)$, is

$$F_{X_1,\ldots,X_k}(x_1, \ldots, x_k) = F_{X_1,\ldots,X_k,X_{k+1},\ldots,X_n}(x_1, \ldots, x_k, \infty, \ldots, \infty) \tag{9.96}$$

$$= \int_{-\infty}^{x_k} \cdots \int_{-\infty}^{x_1} \left(\int_{-\infty}^{\infty} \cdots \int_{-\infty}^{\infty} f_{X_1,\ldots,X_n}(y_1, \ldots, y_n)\, \mathrm{d}y_{k+1} \ldots \mathrm{d}y_n \right) \mathrm{d}y_1 \ldots \mathrm{d}y_k\,. \tag{9.97}$$

The $(n-k)$-dimensional (inner) integral is the joint PDF of $(X_1, \ldots, X_k)$. This is an integrable Borel function $f_{X_1,\ldots,X_k} : \mathbb{R}^k \to R$, given by

$$f_{X_1,\ldots,X_k}(x_1, \ldots, x_k) = \int_{-\infty}^{\infty} \cdots \int_{-\infty}^{\infty} f_{X_1,\ldots,X_n}(x_1, \ldots, x_k, x_{k+1}, \ldots, x_n)\, \mathrm{d}x_{k+1} \ldots \mathrm{d}x_n\,.$$

Hence, for every $k = 1, \ldots, n$, we have the marginal CDF and PDF relations:

$$F_{X_1,\ldots,X_k}(x_1, \ldots, x_k) = \int_{-\infty}^{x_k} \cdots \int_{-\infty}^{x_1} f_{X_1,\ldots,X_k}(y_1 \ldots y_k)\, \mathrm{d}y_1 \ldots \mathrm{d}y_k\,, \tag{9.98}$$

and

$$f_{X_1,\ldots,X_k}(x_1, \ldots, x_k) = \frac{\partial^k}{\partial x_1 \ldots \partial x_k} F_{X_1,\ldots,X_k}(x_1, \ldots, x_k), \quad x_1, \ldots, x_k \in \mathbb{R}. \tag{9.99}$$

Based on (9.93), it can be proven that the expectation formula in (9.92) takes the form of an integral over $\mathbb{R}^n$ involving the joint PDF:

$$\mathrm{E}[h(\mathbf{X})] = \int_{\mathbb{R}^n} h(\mathbf{x}) f_{\mathbf{X}}(\mathbf{x})\, \mathrm{d}^n\mathbf{x}$$

$$\equiv \int_{-\infty}^{\infty} \cdots \int_{-\infty}^{\infty} h(x_1, \ldots, x_n)\, f_{X_1,\ldots,X_n}(x_1, \ldots, x_n)\, \mathrm{d}x_1 \ldots \mathrm{d}x_n\,. \tag{9.100}$$

Note that (9.74) is a special case of this formula for $n = 2$ dimensions. All marginal CDFs in (9.89) are also conveniently expressed as expectations of indicator functions where

$$F_{X_{i_1},\ldots,X_{i_k}}(y_1, \ldots, y_k) = \mathbb{P}(X_{i_1} \leqslant y_1, \ldots, X_{i_k} \leqslant y_k) = \mathrm{E}[\mathbb{I}_{X_{i_1} \leqslant y_1, \ldots, X_{i_k} \leqslant y_k}]\,.$$

It is convenient to define $\mathbf{Y} = (Y_1, \ldots, Y_k) := (X_{i_1}, \ldots, X_{i_k})$. Now, differentiating all k

arguments of the (marginal) joint CDF gives the (marginal) joint PDF for a continuous random vector $\mathbf{Y}$,

$$f_{Y_1,\ldots,Y_k}(y_1,\ldots,y_k) = \frac{\partial^k}{\partial y_1 \ldots \partial y_k} F_{Y_1,\ldots,Y_k}(y_1,\ldots,y_k)\,.$$

Hence, if $h\colon \mathbb{R}^k \to \mathbb{R}$ is a Borel function of only $(X_{i_1},\ldots,X_{i_k}) \equiv (Y_1,\ldots,Y_k)$, with $1 \leqslant k \leqslant n$ components from $\mathbf{X}$, (9.100) reduces to a k-dimensional integral involving the (marginal) joint PDF of $\mathbf{Y}$:

$$\mathrm{E}[h(Y_1,\ldots,Y_k)] = \int_{-\infty}^{\infty} \cdots \int_{-\infty}^{\infty} h(y_1,\ldots,y_k)\, f_{Y_1,\ldots,Y_k}(y_1,\ldots,y_k)\,\mathrm{d}y_1 \ldots \mathrm{d}y_k\,. \tag{9.101}$$

Based on (9.101), and choosing appropriate functions for h, we can in principle compute several quantities of interest, such as moments, product moments, joint moment generating functions, joint characteristic functions, etc., as long as the integrals exist. In particular, the covariance between any two continuous random variables in $\mathbf{X}$, say X_i and X_j, is computed by making use of the joint PDF of the pair (X_i, X_j) and the marginal densities of X_i and X_j:

$$\begin{aligned}\mathrm{Cov}(X_i,X_j) &:= \mathrm{E}[X_i X_j] - \mathrm{E}[X_i]\,\mathrm{E}[X_j] \\ &= \int_{\mathbb{R}}\int_{\mathbb{R}} xy f_{X_i,X_j}(x,y)\,\mathrm{d}x\,\mathrm{d}y - \left(\int_{\mathbb{R}} x f_{X_i}(x)\,\mathrm{d}x\right)\left(\int_{\mathbb{R}} y f_{X_j}(y)\,\mathrm{d}y\right). \end{aligned} \tag{9.102}$$

Let us now consider the case where $X_1,\ldots,X_n$ are independent. By property 1 of Definition 6.14, it follows that the joint distribution measure in (9.86) is now a product measure on $\mathbb{R}^n$:

$$\mu_{X_1,\ldots,X_n}(B) = \prod_{i=1}^{n} \mathbb{P}(X_i \in B_i) = \prod_{i=1}^{n} \mu_{X_i}(B_i) := \mu_{X_1 \times \ldots \times X_n}(B) \tag{9.103}$$

for all Borel sets $B = B_1 \times \ldots \times B_n$ in $\mathbb{R}^n$. The joint CDF is then the product of marginal CDFs,

$$F_{X_1,\ldots,X_n}(x_1,\ldots,x_n) = \prod_{i=1}^{n} \mathbb{P}(X_i \leqslant x_i) = \prod_{i=1}^{n} F_{X_i}(x_i)\,. \tag{9.104}$$

For continuous random variables then, by differentiating (9.104) according to (9.95), the joint PDF is the product of marginal densities

$$f_{X_1,\ldots,X_n}(x_1,\ldots,x_n) = \prod_{i=1}^{n} f_{X_i}(x_i)\,. \tag{9.105}$$

In fact, it can be shown that (9.105) and (9.104) are equivalent in the case of continuous random variables.

The expectation formula in (9.79) extends to n dimensions,

$$\mathrm{E}[h(\mathbf{X})] = \int_{\mathbb{R}} \cdots \int_{\mathbb{R}} h(\mathbf{x})\,\mathrm{d}\mu_{X_1}(x_1) \ldots \mathrm{d}\mu_{X_n}(x_n)\,, \tag{9.106}$$

where the order of integration is interchangeable according to Fubini's Theorem. Similarly, in the case where $h(\mathbf{X}) = h_1(X_1)h_2(X_2)\cdots h(X_n)$, (9.80) extends to a product of n expectations:

$$\mathrm{E}[h_1(X_1)h_2(X_2)\cdots h(X_n)] = \prod_{i=1}^{n}\int_{\mathbb{R}} h_i(x_i)\,\mathrm{d}\mu_{X_i}(x_i) = \prod_{i=1}^{n}\mathrm{E}[h_i(X_i)]\,, \tag{9.107}$$

where we assume that all product functions are integrable, $\mathrm{E}[\,|h_i(X_i)|\,] < \infty$, $i = 1,\ldots,n$. For continuous random variables we have the usual formula for the expectation involving the marginal densities,

$$\mathrm{E}[h_1(X_1)h_2(X_2)\cdots h(X_n)] = \prod_{i=1}^{n}\mathrm{E}[h_i(X_i)] = \prod_{i=1}^{n}\int_{-\infty}^{\infty} h_i(x)f_{X_i}(x)\,\mathrm{d}x\,. \tag{9.108}$$

The above formulas in the case of independence have analogues for any sub-collection of random variables, i.e., for any random vector $\mathbf{Y} = (Y_1,\ldots,Y_k) := (X_{i_1},\ldots,X_{i_k})$, $1 \leqslant i_1 < i_2 < \ldots < i_k \leqslant n$, $1 \leqslant k \leqslant n$, as discussed above. If all components are independent, then the joint distribution measure of $\mathbf{Y}$ is simply the product measure on $\mathbb{R}^k$:

$$\mu_{Y_1,\ldots,Y_k}(B) = \prod_{i=1}^{k}\mu_{Y_i}(B_i) := \mu_{Y_1\times\ldots\times Y_k}(B)$$

for all Borel sets $B = B_1\times\ldots\times B_k$ in $\mathbb{R}^k$. The joint CDF of $\mathbf{Y}$ is the product of the marginal CDFs

$$F_{Y_1,\ldots,Y_k}(y_1,\ldots,y_n) = \prod_{i=1}^{k}F_{Y_i}(y_i)\,,$$

with joint PDF (for the case of a continuous random vector) as a product of the marginal densities

$$f_{Y_1,\ldots,Y_k}(y_1,\ldots,y_n) = \prod_{i=1}^{k}f_{Y_i}(y_i)\,.$$

Note that in the case that X_i and X_j are independent, $f_{X_i,X_j}(x,y) = f_{X_i}(x)f_{X_j}(y)$, $\mathrm{E}[X_iX_j] = \mathrm{E}[X_i]\mathrm{E}[X_j]$, so (9.102) gives zero covariance, as required.

9.4 Conditioning

Now that we are equipped with general probability theory, we revisit the subject of conditioning and conditional expectations of random variables. We basically already covered the main topics in Section 6.3.2 of Chapter 6. We recall Definition 6.12 for the expectation of a random variable conditional on a σ-algebra. The definition was stated very generally using expectations. Since any expectation is in fact a Lebesgue integral w.r.t. a given probability measure $\mathbb{P}$, we can also state Definition 6.12 in the equivalent manner using Lebesgue integral notation. In particular, property (ii) in Definition 6.12 reads

$$\int_B X(\omega)\,\mathrm{d}\mathbb{P}(\omega) = \int_B \mathrm{E}[X\mid\mathcal{G}](\omega)\,\mathrm{d}\mathbb{P}(\omega)$$

for every $B \in \mathcal{G}$.

It is instructive to see how Proposition 6.7 on independence for two random variables follows in the case of continuous random variables. Let X and Y be jointly (absolutely) continuous and independent random variables possessing a joint PDF $f_{X,Y}(x,y) = f_X(x)f_Y(y)$ with marginal PDFs $f_X(x)$ and $f_Y(y)$. The conditional PDF of Y given $X = x$ is hence the marginal PDF of Y: $f_{Y|X}(y|x) = f_{X,Y}(x,y)/f_X(x) = f_Y(y)$. Evaluating the conditional expectation in the usual manner gives

$$\begin{aligned}\mathrm{E}[h(X,Y) \mid X = x] &= \mathrm{E}[h(x,Y) \mid X = x] \\ &= \int_{\mathbb{R}} h(x,y) f_{Y|X}(y|x)\,\mathrm{d}y = \int_{\mathbb{R}} h(x,y) f_Y(y)\,\mathrm{d}y = \mathrm{E}[h(x,Y)]\,.\end{aligned}$$

Hence, as a random variable we have $\mathrm{E}[h(X,Y) \mid X] = \int_{\mathbb{R}} h(X,y) f_Y(y)\,\mathrm{d}y := g(X)$, i.e., $\mathrm{E}[h(X,Y) \mid X](\omega) = \int_{\mathbb{R}} h(X(\omega),y) f_Y(y)\,\mathrm{d}y := g(X(\omega))$, for each $\omega \in \Omega$.

Now, to formally show that the above expectation formula is in fact the correct one we need to verify properties (i) and (ii) of Definition 6.12 with $\mathcal{G} = \sigma(X)$. Property (i) holds since g defined by the above integral is a Borel function and hence $\sigma(g(X)) \subset \sigma(X)$, i.e., $g(X)$ is $\sigma(X)$-measurable. For property (ii), we need to show that

$$\mathrm{E}\big[\mathbb{I}_{X\in B} \cdot \mathrm{E}[h(X,Y) \mid X]\big] \equiv \mathrm{E}\big[\mathbb{I}_{X\in B} \cdot g(X)\big] = \mathrm{E}\big[\mathbb{I}_{X\in B} \cdot h(X,Y)\big]$$

for every Borel set B in $\sigma(X)$, i.e., B is any Borel set in the range of X and so we take any $B \in \mathcal{B}(\mathbb{R})$. Expressing these expectations using the joint PDF gives

$$\mathrm{E}\big[\mathbb{I}_{X\in B} \cdot g(X)\big] = \int_B g(x)\left(\int_{\mathbb{R}} f_{X,Y}(x,y)\,\mathrm{d}y\right)\mathrm{d}x = \int_B g(x) f_X(x)\,\mathrm{d}x$$

and

$$\mathrm{E}\big[\mathbb{I}_{X\in B} \cdot h(X,Y)\big] = \int_B \left(\int_{\mathbb{R}} h(x,y) f_{X,Y}(x,y)\,\mathrm{d}y\right)\mathrm{d}x\,.$$

For these two quantities to be equal, for every Borel set B, we necessarily must have the equivalence of the x-integrands, i.e.,

$$g(x) = \int_{\mathbb{R}} h(x,y)\frac{f_{X,Y}(x,y)}{f_X(x)}\,\mathrm{d}y = \int_{\mathbb{R}} h(x,y) f_{Y|X}(y|x)\,\mathrm{d}y = \int_{\mathbb{R}} h(x,y) f_Y(y)\,\mathrm{d}y\,.$$

This proves the above expression for $g(x)$ and hence for $\mathrm{E}[h(X,Y) \mid X]$.

If $\mathbf{X}$ and $\mathbf{Y}$ are jointly continuous and independent random vectors, then their joint PDF is $f_{\mathbf{X},\mathbf{Y}}(\mathbf{x},\mathbf{y}) = f_{\mathbf{X}}(\mathbf{x})f_{\mathbf{Y}}(\mathbf{y})$ with marginal PDFs $f_{\mathbf{X}}(\mathbf{x})$ and $f_{\mathbf{Y}}(\mathbf{y})$. The conditional PDF of $\mathbf{Y}$ given $\mathbf{X} = \mathbf{x}$ is the marginal PDF of $\mathbf{Y}$: $f_{\mathbf{Y}|\mathbf{X}}(\mathbf{y}|\mathbf{x}) = f_{\mathbf{Y}}(\mathbf{y})$. By basic probability theory, the conditional expectation is given by the n-dimensional integral:

$$\mathrm{E}[h(\mathbf{X},\mathbf{Y}) \mid \mathbf{X} = \mathbf{x}] = \int_{\mathbb{R}^n} h(\mathbf{x},\mathbf{y}) f_{\mathbf{Y}|\mathbf{X}}(\mathbf{y}|\mathbf{x})\,\mathrm{d}^n\mathbf{y} = \int_{\mathbb{R}^n} h(\mathbf{x},\mathbf{y}) f_{\mathbf{Y}}(\mathbf{y})\,\mathrm{d}^n\mathbf{y} = \mathrm{E}[h(\mathbf{x},\mathbf{Y})]\,.$$

A similar analysis as given above (for the case of two random variables) formally shows that this is the correct formula satisfying properties (i) and (ii) of Definition 6.12 with $\mathcal{G} = \sigma(\mathbf{X})$. In this case the Borel sets $B \subset \mathbb{R}^n$.

Although we have already covered many of the important properties of conditioning in Section 6.3.3 of Chapter 6, it is still useful to summarize them in the following theorem since they pertain to the more general theory of random variables. Many of the proofs are rather straightforward and are standard in real analysis so we don't repeat them here. We note that we have also proven some of the properties in the discrete setting in Chapter 6 and that we have already used some of the properties stated below.

Theorem 9.6. *Let X and Y be random variables on $(\Omega, \mathcal{F}, \mathbb{P})$. Then, the following hold.*

1. *(Linearity) For any real constants a, b:*
$$\mathrm{E}[a\,X + b\,Y \mid \mathcal{G}] = a\,\mathrm{E}[X \mid \mathcal{G}] + b\,\mathrm{E}[Y \mid \mathcal{G}]\,.$$
2. *(Nested Expectation)*
$$\mathrm{E}[\,\mathrm{E}[X \mid \mathcal{G}]\,] = \mathrm{E}[X]\,.$$
3. *(Tower Property) For any $\mathcal{H} \subset \mathcal{G} \subset \mathcal{F}$,*
$$\mathrm{E}[\mathrm{E}[X \mid \mathcal{G}] \mid \mathcal{H}] = \mathrm{E}[X \mid \mathcal{H}]\,.$$
4. *(Independence) If X is independent of $\mathcal{G}$ then*
$$\mathrm{E}[X \mid \mathcal{G}] = \mathrm{E}[X]\,.$$
5. *(Measurability) If X is $\mathcal{G}$-measurable then*
$$\mathrm{E}[X \mid \mathcal{G}] = X\,.$$
6. *(Positivity) If $X \geqslant 0$ (a.s.) then $\mathrm{E}[X \mid \mathcal{G}] \geqslant 0$ (a.s.).*
7. *(Monotone Convergence) If $X_n, n \geqslant 1$, is a nonnegative sequence of random variables on $(\Omega, \mathcal{F}, \mathbb{P})$ and increases (a.s.) to X, then the sequence $\mathrm{E}[X_n \mid \mathcal{G}], n \geqslant 1$, increases (a.s.) to $\mathrm{E}[X \mid \mathcal{G}]$.*
8. *(Pulling out what is known) If Y is $\mathcal{G}$-measurable and XY is integrable then*
$$\mathrm{E}[XY \mid \mathcal{G}] = Y\mathrm{E}[X \mid \mathcal{G}]\,.$$
9. *(Conditional Jensen's Inequality) If $\phi : \mathbb{R} \to \mathbb{R}$ is a convex function and X is integrable then*
$$\mathrm{E}[\phi(X) \mid \mathcal{G}] \geqslant \phi\,(\mathrm{E}[X \mid \mathcal{G}])\,.$$

9.5 Changing Probability Measures

Here we shall keep the discussion very succinct. Let us begin by defining a new probability measure $\widehat{\mathbb{P}} \equiv \widehat{\mathbb{P}}^{(\varrho)}$ by

$$\widehat{\mathbb{P}}(A) \equiv \int_A \mathrm{d}\widehat{\mathbb{P}}(\omega) := \int_A \varrho(\omega)\,\mathrm{d}\mathbb{P}(\omega)\,, \tag{9.109}$$

for all $A \in \mathcal{F}$, i.e., $\widehat{\mathbb{P}}(A) \equiv \widehat{\mathrm{E}}[\mathbb{I}_A] := \mathrm{E}[\varrho\,\mathbb{I}_A]$, such that ϱ is chosen to be a nonnegative (a.s.) random variable on $(\Omega, \mathcal{F})$ having unit expectation under measure $\mathbb{P}$:

$$\mathrm{E}[\varrho] \equiv \int_\Omega \varrho(\omega)\,\mathrm{d}\mathbb{P}(\omega) = 1.$$

Note that in order for $\widehat{\mathbb{P}}$ to be a probability measure we necessarily have $1 = \widehat{\mathbb{P}}(\Omega) = \mathrm{E}[\varrho\mathbb{I}_\Omega] = \mathrm{E}[\varrho]$. The measure $\widehat{\mathbb{P}}$ is also countably additive from the countable additivity property of the Lebesgue integral w.r.t. $\mathbb{P}$, i.e., for any countable collection of pairwise disjoint sets $\{A_i\} \in \mathcal{F}$ we have, setting $A \equiv \cup_i A_i$ in (9.109),

$$\widehat{\mathbb{P}}(\cup_i A_i) = \mathrm{E}[\varrho \mathbb{I}_{\cup_i A_i}] = \mathrm{E}[\varrho \sum_i \mathbb{I}_{A_i}] = \sum_i \mathrm{E}[\varrho \mathbb{I}_{A_i}] = \sum_i \widehat{\mathbb{P}}(A_i).$$

Hence, $\widehat{\mathbb{P}}$ is a probability measure and $(\Omega, \mathcal{F}, \widehat{\mathbb{P}})$ is a probability space.

The random variable ϱ corresponds to the so-called *Radon–Nikodym derivative* of $\widehat{\mathbb{P}}$ w.r.t. $\mathbb{P}$. Note that $\mathrm{d}\widehat{\mathbb{P}}(\omega) = \varrho(\omega)\,\mathrm{d}\mathbb{P}(\omega)$. It is customary notation to denote ϱ by $\frac{\mathrm{d}\widehat{\mathbb{P}}}{\mathrm{d}\mathbb{P}}$, i.e., $\mathrm{d}\widehat{\mathbb{P}}(\omega) = \frac{\mathrm{d}\widehat{\mathbb{P}}}{\mathrm{d}\mathbb{P}}(\omega)\,\mathrm{d}\mathbb{P}(\omega)$. The notation arises naturally when we compute the expectation of a random variable in the two different measures. The expectation of a random variable X in the original $\mathbb{P}$-measure is denoted by $\mathrm{E}[X]$ and we let $\widehat{\mathrm{E}}[X]$ denote the expectation in the new $\widehat{\mathbb{P}}$-measure. Using the definition in (9.109), the expectation of X under measure $\widehat{\mathbb{P}}$ equals the expectation of $X\varrho \equiv X\frac{\mathrm{d}\widehat{\mathbb{P}}}{\mathrm{d}\mathbb{P}}$ under measure $\mathbb{P}$:

$$\widehat{\mathrm{E}}[X] = \int_\Omega X(\omega)\,\mathrm{d}\widehat{\mathbb{P}}(\omega) = \int_\Omega X(\omega)\varrho(\omega)\,\mathrm{d}\mathbb{P}(\omega) = \mathrm{E}[X\varrho] \equiv \mathrm{E}\left[X\frac{\mathrm{d}\widehat{\mathbb{P}}}{\mathrm{d}\mathbb{P}}\right]. \tag{9.110}$$

Moreover, if ϱ is strictly positive (a.s.) and Y is integrable under measure $\mathbb{P}$, then its expectation under measure $\mathbb{P}$ equals the expectation of $\frac{Y}{\varrho} \equiv Y\frac{\mathrm{d}\mathbb{P}}{\mathrm{d}\widehat{\mathbb{P}}}$ under measure $\widehat{\mathbb{P}}$:

$$\mathrm{E}[Y] = \int_\Omega Y(\omega)\,\mathrm{d}\mathbb{P}(\omega) = \int_\Omega Y(\omega)\frac{1}{\varrho(\omega)}\,\mathrm{d}\widehat{\mathbb{P}}(\omega) = \widehat{\mathrm{E}}\left[\frac{Y}{\varrho}\right] = \widehat{\mathrm{E}}\left[Y\frac{\mathrm{d}\mathbb{P}}{\mathrm{d}\widehat{\mathbb{P}}}\right]. \tag{9.111}$$

Note that $\frac{\mathrm{d}\mathbb{P}}{\mathrm{d}\widehat{\mathbb{P}}} = \left(\frac{\mathrm{d}\widehat{\mathbb{P}}}{\mathrm{d}\mathbb{P}}\right)^{-1} = \frac{1}{\varrho}$ for any strictly positive ϱ.

Remark: Here we don't state the formal Radon-Nikodym theorem. There are different versions of it in measure theory where measures can be more general than probability measures. One version of the Radon-Nikodym theorem goes as follows. Given two finite measures μ and ν on a space $(\Omega, \mathcal{F})$ where ν is absolutely continuous w.r.t. μ (i.e., all sets of μ-measure zero are ν-measure zero), then there is a nonnegative $\mathcal{F}$-measurable function $h \equiv \frac{\mathrm{d}\nu}{\mathrm{d}\mu}$ such that the ν-measure of any set $A \in \mathcal{F}$ is given by a Lebesgue integral w.r.t. μ: $\nu(A) = \int_A h\,\mathrm{d}\mu$. Moreover, this function is unique w.r.t. measure μ. For our purposes, the Radon–Nikodym theorem guarantees that, given two equivalent probability measures $\widehat{\mathbb{P}}$ and $\mathbb{P}$, there is nonnegative random variable ϱ satisfying the above relations. Moreover, ϱ is unique (a.s.). The Radon–Nikodym theorem also applies to distribution measures of random variables and joint random variables.

We have already seen how measure changes are applied for discrete random variables. Let us briefly see how measure changes can be applied in the case of absolutely continuous random variables. Assume $X \in \mathbb{R}$ has a PDF $f(x)$ under measure $\mathbb{P}$ and a PDF $\widehat{f}(x)$ under measure $\widehat{\mathbb{P}}$. Although we can further generalize, we shall also assume that these densities are (a.e.) positive on $\mathbb{R}$. Then, for any $b \in \mathbb{R}$ we have the CDF of X under measure $\widehat{\mathbb{P}}$, $\widehat{F}_X(b) \equiv \widehat{\mu}_X((-\infty, b])$ as

$$\widehat{F}_X(b) = \widehat{\mathrm{E}}[\mathbb{I}_{\{X\leqslant b\}}] \equiv \int_{-\infty}^b \widehat{f}(x)\,\mathrm{d}x = \int_{-\infty}^b \frac{\widehat{f}(x)}{f(x)} f(x)\,\mathrm{d}x \equiv \mathrm{E}\left[\frac{\widehat{f}(X)}{f(X)}\mathbb{I}_{\{X\leqslant b\}}\right]. \tag{9.112}$$

By the Radon–Nikodym derivative, we see that the ratio of densities gives the Radon–Nikodym derivative for changing distribution measures of X. In this case, $\frac{\mathrm{d}\widehat{\mu}_X}{\mathrm{d}\mu_X}(x) = \frac{\widehat{f}(x)}{f(x)}$. This is known as a likelihood ratio. As a random variable we have the Radon–Nikodym derivative $\varrho = \varrho(X) = \frac{\widehat{f}(X)}{f(X)}$. This also generalizes to the multidimensional case in the obvious manner as the ratio of the joint PDFs.

An example of the likelihood ratio in measure changes is to consider a normal random variable. Say that $X \sim Norm(a, \sigma^2)$, i.e., that it has mean a and variance σ^2 under measure $\mathbb{P}$. Then, defining the change of measure $\mathbb{P} \to \widehat{\mathbb{P}}$, i.e., change of distribution measure $\mu_X \to \widehat{\mu}_X$, by the Radon–Nikodym random variable

$$\varrho = \frac{\mathrm{d}\widehat{\mathbb{P}}}{\mathrm{d}\mathbb{P}} = \frac{\sigma}{\widehat{\sigma}} \exp\left[\frac{(X-a)^2}{2\sigma^2} - \frac{(X-\widehat{a})^2}{2\widehat{\sigma}^2}\right] \equiv h(X)$$

we have that $X \sim Norm(\widehat{a}, \widehat{\sigma}^2)$ under measure $\widehat{\mathbb{P}}$. This follows from the above likelihood ratio of densities. The PDF of X under measure $\widehat{\mathbb{P}}$ equals its PDF under measure $\mathbb{P}$ times the likelihood ratio $\frac{\mathrm{d}\widehat{\mu}_X}{\mathrm{d}\mu_X}(x) = h(x) = \frac{\widehat{f}(x)}{f(x)}$. Hence,

$$h(X) = \frac{\widehat{f}(X)}{f(X)} = \frac{\frac{1}{\widehat{\sigma}\sqrt{2\pi}} \mathrm{e}^{\frac{-(X-\widehat{a})^2}{2\widehat{\sigma}^2}}}{\frac{1}{\sigma\sqrt{2\pi}} \mathrm{e}^{\frac{-(X-a)^2}{2\sigma^2}}}$$

which gives the above result. We note that this change of measure changes both the mean and variance of a normal random variable.

The next (and last) result of this chapter gives us a formula for computing the expectation (under a given measure $\mathbb{P}$) of a random variable conditional on any sub-σ-algebra $\mathcal{G} \subset \mathcal{F}$ via the corresponding conditional expectation of ϱX under another equivalent measure $\widehat{\mathbb{P}}$. The new measure $\widehat{\mathbb{P}}$ is defined in (9.109) with (Radon–Nikodym derivative) random variable ϱ. The theorem is useful when considering measure changes while calculating conditional expectations involving stochastic processes such as those driven by Brownian motions. The sub-σ-algebras are part of a filtration for Brownian motion. The conditioning on the filtration simplifies even further when we are dealing with practical applications involving Markov processes.

Theorem 9.7 (General Bayes Formula). *Let $(\Omega, \mathcal{F}, \mathbb{P})$ and $(\Omega, \mathcal{F}, \widehat{\mathbb{P}})$ be two probability spaces with $\mathbb{P} \sim \widehat{\mathbb{P}}$ (i.e., $\mathbb{P}$ and $\widehat{\mathbb{P}}$ are equivalent probability measures). Let the random variable X be integrable w.r.t. $\widehat{\mathbb{P}}$ and set $\varrho \equiv \frac{\mathrm{d}\widehat{\mathbb{P}}}{\mathrm{d}\mathbb{P}}$. Then, the random variable ϱX is integrable w.r.t. $\mathbb{P}$ and its expectation under measure $\widehat{\mathbb{P}}$ conditional on a sub-σ-algebra $\mathcal{G} \subset \mathcal{F}$ is given by (a.s.)*

$$\widehat{\mathrm{E}}[X \mid \mathcal{G}] = \frac{\mathrm{E}[\varrho X \mid \mathcal{G}]}{\mathrm{E}[\varrho \mid \mathcal{G}]}. \tag{9.113}$$

Proof. All we need to verify is that the right-hand side of (9.113), $Y := \frac{\mathrm{E}[\varrho X|\mathcal{G}]}{\mathrm{E}[\varrho|\mathcal{G}]}$, is (almost surely) the random variable $\widehat{\mathrm{E}}[X \mid \mathcal{G}]$, i.e., the expectation of X under $\widehat{\mathbb{P}}$ conditional on $\mathcal{G}$. Note that Y must satisfy the two properties in Definition 6.12 with $\widehat{\mathbb{P}}$ as measure. Hence, we need to show: (i) Y is $\mathcal{G}$-measurable; (ii) $\widehat{\mathrm{E}}[\mathbb{I}_A Y] = \widehat{\mathrm{E}}[\mathbb{I}_A X]$, for every event $A \in \mathcal{G}$. Property (i) is obviously satisfied since Y is a ratio of two $\mathcal{G}$-measurable random variables and hence is $\mathcal{G}$-measurable. Property (ii) is shown by first applying the change of measure $\widehat{\mathbb{P}} \to \mathbb{P}$ for an unconditional expectation via (9.110), then making use of the tower property $\mathrm{E}[\mathrm{E}[\cdot \mid \mathcal{G}]] = \mathrm{E}[\cdot]$ in reverse order, pulling out the $\mathcal{G}$-measurable random variable $\mathbb{I}_A Y$ in the inner conditional expectation, cancelling out the $\mathrm{E}[\varrho \mid \mathcal{G}]$ term, re-applying the tower property and changing back measures $\mathbb{P} \to \widehat{\mathbb{P}}$ in the final expectation:

$$\begin{aligned}\widehat{\mathrm{E}}[\mathbb{I}_A Y] = \mathrm{E}[\varrho \mathbb{I}_A Y] = \mathrm{E}[\mathrm{E}[\varrho \mathbb{I}_A Y \mid \mathcal{G}]] = \mathrm{E}[\mathbb{I}_A Y \, \mathrm{E}[\varrho \mid \mathcal{G}]] &= \mathrm{E}[\mathbb{I}_A \mathrm{E}[\varrho X \mid \mathcal{G}]] \\ &= \mathrm{E}[\mathrm{E}[\varrho \mathbb{I}_A X \mid \mathcal{G}]] \\ &= \mathrm{E}[\varrho \mathbb{I}_A X] = \widehat{\mathrm{E}}[\mathbb{I}_A X] .\end{aligned}$$

Finally, the assumption that X is integrable w.r.t. $\widehat{\mathbb{P}}$, i.e., $\widehat{\mathrm{E}}[\,|X|\,] < \infty$, implies that ϱX is integrable w.r.t. $\mathbb{P}$ by changing measures:

$$\widehat{\mathrm{E}}[\,|X|\,] = \mathrm{E}[\varrho\,|X|\,] = \mathrm{E}[\,|\varrho X|\,] < \infty . \qquad \square$$

Chapter 10

One-Dimensional Brownian Motion and Related Processes

Brownian motion is a keystone in the foundation of mathematical finance. Brownian motion is a continuous-time stochastic process but it can be constructed as a limiting case of symmetric random walks. Recall that a similar approach was used to obtain the log-normal price model as a limiting case of binomial models. Since the probability distribution of Brownian motion is normal, it is reasonable to begin with reviewing some of the important basic facts about the multivariate normal distribution.

10.1 Multivariate Normal Distributions

10.1.1 Multivariate Normal Distribution

The n-variate normal distribution is determined by an $n \times 1$ mean vector $\boldsymbol{\mu}$ and an $n \times n$ positive definite covariance matrix $\mathbf{C} = \mathbf{C}^\top$:

$$\boldsymbol{\mu} = \left[\mu_1, \mu_2, \ldots, \mu_n\right]^\top, \quad \mathbf{C} = \left[C_{ij}\right]_{i,j=1,\ldots,n}.$$

Note: throughout we use superscript $\top$ to denote the transpose. We denote the n-variate normal distribution with mean vector $\boldsymbol{\mu}$ and covariance matrix $\mathbf{C}$ by $Norm_n(\boldsymbol{\mu}, \mathbf{C})$. Assuming the nondegenerate case with a positive definite covariance matrix, the joint probability density function (PDF) is an n-variate real-valued function

$$f(\mathbf{x}) = \frac{1}{(2\pi)^{n/2}\left[\det \mathbf{C}\right]^{n/2}} \mathrm{e}^{-\frac{1}{2}(\mathbf{x}-\boldsymbol{\mu})^\top \mathbf{C}^{-1}(\mathbf{x}-\boldsymbol{\mu})}, \quad \mathbf{x} \in \mathbb{R}^n. \tag{10.1}$$

This PDF is that of random vector $\mathbf{X} = \left[X_1, X_2, \ldots, X_n\right]^\top \sim Norm_n(\boldsymbol{\mu}, \mathbf{C})$, i.e., $\mathbf{X}$ has n-variate normal distribution with mean $\boldsymbol{\mu} = \mathrm{E}[\mathbf{X}]$ and covariance $\mathbf{C} = \mathrm{E}[\mathbf{X}\mathbf{X}^\top] - \boldsymbol{\mu}\boldsymbol{\mu}^\top$. As components, $\mu_i = \mathrm{E}[X_i]$ and $C_{ij} = \mathrm{Cov}(X_i, X_j) = \mathrm{E}[X_i X_j] - \mu_i \mu_j, i, j = 1, 2 \ldots, n$. It is a well-known fact that normal random variables are independent iff they are uncorrelated with zero covariances $C_{ij} = 0$ for all $i \neq j$. Therefore, if the covariance is a diagonal matrix, $\mathbf{C} = \mathrm{diag}(\sigma_1^2, \sigma_2^2, \ldots, \sigma_n^2)$, where $\sigma_i^2 = \mathrm{E}[(X_i - \mu_i)^2]$, then the components of $\mathbf{X}$ are (mutually) independent normally distributed random variables. In this case, the joint PDF is a product of one-dimensional marginal densities:

$$f(\mathbf{x}) = \prod_{i=1}^{n} \frac{1}{\sigma_i} n\left(\frac{x_i - \mu_i}{\sigma_i}\right) = \prod_{i=1}^{n} \frac{1}{\sqrt{2\pi}\sigma_i} \mathrm{e}^{-\frac{(x_i-\mu_i)^2}{2\sigma_i^2}}.$$

Here we used the notation $n(z) = \mathcal{N}'(z) = \frac{\mathrm{e}^{-z^2/2}}{\sqrt{2\pi}}$ for the PDF of the standard normal. Another well-known fact is that a sum of normal random variables is again normally distributed:

$$X_i \sim Norm\left(\mu_i, \sigma_i^2\right),\ i = 1, 2 \implies X_1 + X_2 \sim Norm\left(\mu_1 + \mu_2, \mathrm{Var}\left(X_1 + X_2\right)\right), \tag{10.2}$$

where $\mathrm{Var}\left(X_1 + X_2\right) = \sigma_1^2 + 2\,\mathrm{Cov}(X_1, X_2) + \sigma_2^2$. This property extends to the multivariate case as follows. Let $\mathbf{a}$ and $\mathbf{B}$ be an $m \times 1$ constant vector and an $m \times n$ constant matrix, respectively, with $1 \leqslant m \leqslant n$. Suppose that $\mathbf{X} \sim Norm_n(\boldsymbol{\mu}, \mathbf{C})$; then the m-variate random vector $\mathbf{a} + \mathbf{B}\mathbf{X}$ is normally distributed with mean vector $\mathbf{a} + \mathbf{B}\,\boldsymbol{\mu}$. Its covariance matrix is given by $\mathrm{E}[\mathbf{B}(\mathbf{X} - \boldsymbol{\mu})(\mathbf{B}(\mathbf{X} - \boldsymbol{\mu}))^\top] = \mathbf{B}\,\mathrm{E}[(\mathbf{X} - \boldsymbol{\mu})(\mathbf{X} - \boldsymbol{\mu})^\top]\,\mathbf{B}^\top = \mathbf{B}\,\mathbf{C}\,\mathbf{B}^\top$. Hence, provided that $\det(\mathbf{B}\,\mathbf{C}\,\mathbf{B}^\top) \neq 0$, we have:

$$\mathbf{X} \sim Norm_n(\boldsymbol{\mu}, \mathbf{C}) \implies \mathbf{a} + \mathbf{B}\,\mathbf{X} \sim Norm_m(\mathbf{a} + \mathbf{B}\,\boldsymbol{\mu}, \mathbf{B}\,\mathbf{C}\,\mathbf{B}^\top). \tag{10.3}$$

This property allows us to express an arbitrary multivariate normal vector in terms of independent standard normal variables. Since the covariance matrix $\mathbf{C}$ is positive definite, it admits the Cholesky factorization: $\mathbf{C} = \mathbf{L}\,\mathbf{L}^\top$ with a lower-triangular $n \times n$ matrix $\mathbf{L}$. Let $Z_1, Z_2, \ldots, Z_n$ be i.i.d. standard normals. The vector $\mathbf{Z} = \left[Z_1, Z_2, \ldots, Z_n\right]^\top$ is then an n-variate normal with mean vector zero and having the identity covariance matrix $\mathbf{I}$. Applying (10.3) to $\mathbf{Z} \sim Norm_n(\mathbf{0}, \mathbf{I})$ gives $\boldsymbol{\mu} + \mathbf{L}\,\mathbf{Z} \sim Norm_n(\boldsymbol{\mu}, \mathbf{C})$, i.e., in this case the vector $\boldsymbol{\mu} + \mathbf{L}\,\mathbf{Z}$ has mean $\boldsymbol{\mu} + \mathbf{L}\,\mathbf{0} = \boldsymbol{\mu}$ and covariance matrix $\mathbf{L}\,\mathbf{I}\,\mathbf{L}^\top = \mathbf{L}\,\mathbf{L}^\top = \mathbf{C}$.

10.1.2 Conditional Normal Distributions

Suppose that $\mathbf{X} = \left[X_1, X_2, \ldots, X_n\right]^\top \sim Norm_n(\boldsymbol{\mu}, \mathbf{C})$, $n \geqslant 2$. Let us split (partition) the vector $\mathbf{X}$ into two parts:

$$\mathbf{X} = \begin{bmatrix} \mathbf{X}_1 \\ \mathbf{X}_2 \end{bmatrix}, \text{ where } \mathbf{X}_1 \in \mathbb{R}^m \text{ and } \mathbf{X}_2 \in \mathbb{R}^{n-m}$$

for some m with $1 < m < n$. Correspondingly, we split the vector $\boldsymbol{\mu}$ and matrix $\mathbf{C}$ to represent them in a block form:

$$\boldsymbol{\mu} = \begin{bmatrix} \boldsymbol{\mu}_1 \\ \boldsymbol{\mu}_2 \end{bmatrix} \text{ and } \mathbf{C} = \begin{bmatrix} \mathbf{C}_{11} & \mathbf{C}_{12} \\ \mathbf{C}_{21} & \mathbf{C}_{22} \end{bmatrix},$$

where $\boldsymbol{\mu}_1 \in \mathbb{R}^m$ and $\boldsymbol{\mu}_2 \in \mathbb{R}^{n-m}$ are the respective mean vectors of $\mathbf{X}_1$ and $\mathbf{X}_2$; $\mathbf{C}_{11} = \mathrm{E}[\mathbf{X}_1\mathbf{X}_1^\top]$ is $m \times m$, $\mathbf{C}_{12} = \mathrm{E}[\mathbf{X}_1\mathbf{X}_2^\top]$ is $m \times (n-m)$, $\mathbf{C}_{21} = \mathbf{C}_{12}^\top$ is $(n-m) \times m$, and $\mathbf{C}_{22} = \mathrm{E}[\mathbf{X}_2\mathbf{X}_2^\top]$ is $(n-m) \times (n-m)$. Then, the *conditional distribution of* $\mathbf{X}_1$ *given the value of* $\mathbf{X}_2 = \mathbf{x}_2$ *is normal:*

$$\mathbf{X}_1 | \{\mathbf{X}_2 = \mathbf{x}_2\} \sim Norm_m\left(\boldsymbol{\mu}_1 + \mathbf{C}_{12}\,\mathbf{C}_{22}^{-1}(\mathbf{x}_2 - \boldsymbol{\mu}_2), \mathbf{C}_{11} - \mathbf{C}_{12}\,\mathbf{C}_{22}^{-1}\mathbf{C}_{21}\right). \tag{10.4}$$

For example, consider a two-dimensional (bivariate) normal random vector

$$\mathbf{X} = \begin{bmatrix} X_1 \\ X_2 \end{bmatrix} \sim Norm_2\left(\begin{bmatrix} \mu_1 \\ \mu_2 \end{bmatrix}, \begin{bmatrix} \sigma_1^2 & \rho\sigma_1\sigma_2 \\ \rho\sigma_1\sigma_2 & \sigma_2^2 \end{bmatrix}\right).$$

In conformity with (10.4), X_1 conditional on X_2 (and vice versa) is normally distributed. In this case, $m = n - m = 1$ so the block matrices are simply numbers: $\mathbf{C}_{11} = \sigma_1^2$, $\mathbf{C}_{12} = \mathbf{C}_{21} = \rho\sigma_1\sigma_2$, $\mathbf{C}_{22} = \sigma_2^2$, and $\mathbf{x}_2 - \boldsymbol{\mu}_2 = x_2 - \mu_2$. In particular,

$$X_1 \mid \{X_2 = x_2\} \sim Norm\left(\mu_1 + \rho\frac{\sigma_1}{\sigma_2}(x_2 - \mu_2), \sigma_1^2(1 - \rho^2)\right). \tag{10.5}$$

10.2 Standard Brownian Motion

10.2.1 One-Dimensional Symmetric Random Walk

Consider the binomial sample space $\Omega \equiv \Omega_\infty$:

$$\Omega_\infty = \prod_{i=1}^{\infty}\{\mathsf{D},\mathsf{U}\} = \{\omega_1\omega_2\ldots \mid \omega_i \in \{\mathsf{D},\mathsf{U}\}, \text{ for all } i \geqslant 1\}$$

generated by a Bernoulli experiment using any number of repeated up or down moves. We assume a probability measure $\mathbb{P}$ such that the up and down moves are equally probable. That is, each outcome $\boldsymbol{\omega} = \omega_1\omega_2\ldots \in \Omega_\infty$ is a sequence of U's and D's where $\mathbb{P}(\omega_i = \mathsf{D}) = \mathbb{P}(\omega_i = \mathsf{U}) = \frac{1}{2}$ for all $i \geqslant 1$. We note that, in contrast to the binomial model where we had a finite number of moves $N \geqslant 1$ with $\Omega = \Omega_N$, the number of moves is now infinite and the sample space Ω_∞ is *uncountable.* Nevertheless, the discrete-valued random variables that were previously defined on Ω_N (e.g., $\mathsf{U}_n, \mathsf{D}_n, \mathsf{B}_n$, etc.) are still defined in the same way on Ω_∞. The sample space now also admits uncountable partitions. Obviously, we still have all the countable partitions in the same manner as we discussed for Ω_N. For example, the event that the first move is up is the atom $A_\mathsf{U} = \{\omega_1 = \mathsf{U}\} = \{\mathsf{U}\omega_2\omega_3\ldots \mid \omega_i \in \{\mathsf{D},\mathsf{U}\}, \text{ for all } i \geqslant 2\}$ and its complement is $A_\mathsf{D} = \{\omega_1 = \mathsf{D}\} = \{\mathsf{D}\omega_2\omega_3\ldots \mid \omega_i \in \{\mathsf{D},\mathsf{U}\}, \text{ for all } i \geqslant 2\}$ where $\Omega_\infty = A_\mathsf{U} \cup A_\mathsf{D}$. In particular, the atoms that correspond to fixing (resolving) the first $n \geqslant 1$ moves are given similarly,

$$A_{\omega_1^*\ldots\omega_n^*} := \{\omega_1 = \omega_1^*, \ldots, \omega_n = \omega_n^*\} = \{\omega_1^*\ldots\omega_n^*\omega_{n+1}\ldots \mid \omega_i \in \{\mathsf{D},\mathsf{U}\}, i \geqslant n+1\}.$$

The union of these 2^n atoms is a partition of Ω_∞, for every choice of $n \geqslant 1$. We denote by $\mathcal{F}_\infty$ the σ-algebra generated by all the above atoms corresponding to any number of moves $n \geqslant 1$. Recall that for every $N \geqslant 1$, $\{\mathcal{F}_n\}_{1\leqslant n\leqslant N}$ is the natural filtration generated by the moves up to time N. We saw that the natural filtration can be equivalently generated by any of the above sets of random variables that contain the information on all the moves, i.e., $\mathcal{F}_0 = \{\emptyset, \Omega\}$, $\mathcal{F}_N = \sigma(\omega_1, \ldots, \omega_N) = \sigma(M_1, \ldots, M_N) = \sigma(\{\mathsf{U}_k : 1 \leqslant n \leqslant N\})$, etc. For every N, $\{\mathcal{F}_n\}_{1\leqslant n\leqslant N}$ is the natural filtration for the binomial model up to time N, where $\mathcal{F}_n \subset \mathcal{F}_{n+1}$ for all $n \geqslant 0$. In the limiting case of $N \to \infty$ we have $\mathcal{F}_\infty = \cup_{n=1}^{\infty}\mathcal{F}_n$, or equivalently: $\mathcal{F}_\infty = \sigma(\{\omega_n : n \geqslant 1\}) = \sigma(\{M_n : n \geqslant 1\}) = \sigma(\{\mathsf{U}_n : n \geqslant 1\})$. Hence, $\mathcal{F}_\infty$ contains all the possible events in the binomial model with infinite moves for which we can compute probabilities of occurrence. For every given $N \geqslant 1$, $(\Omega_N, \mathcal{F}_N, \mathbb{P})$ is a probability space on which we can define any random variable that may depend on up to N numbers of moves (i.e., random variables that are $\mathcal{F}_n$-measurable for $1 \leqslant n \leqslant N$). Passing to the limiting case, the triplet $(\Omega_\infty, \mathcal{F}_\infty, \mathbb{P})$ is then a probability space for any random variable that depends on any number of moves (i.e., random variables that are $\mathcal{F}_n$-measurable for all $n \geqslant 1$), i.e., $(\Omega_\infty, \mathcal{F}_\infty, \mathbb{P}, \{\mathcal{F}_n\}_{n\geqslant 1})$ is a filtered probability space.

Recall that the symmetric random walk $\{M_n\}_{n\geqslant 0}$ is defined by $M_0 = 0$ and

$$M_n(\boldsymbol{\omega}) = \sum_{k=1}^{n} X_k(\boldsymbol{\omega}), \text{ where } X_k(\boldsymbol{\omega}) = \begin{cases} 1 & \text{if } \omega_k = \mathsf{U}, \\ -1 & \text{if } \omega_k = \mathsf{D}, \end{cases}$$

for $n, k \geqslant 1$. Let us summarize the properties of the symmetric random walk.

1. It is a martingale w.r.t. $\{F_n\}_{n\geqslant 0}$ and has zero mean, $\mathrm{E}[M_n] \equiv 0$ for all times $n \geqslant 0$.
2. It is a square-integrable process, i.e., $\mathrm{E}[M_n^2] < \infty$, such that $\mathrm{Var}(M_n) = n$ and $\mathrm{Cov}(M_n, M_k) = n \wedge k$, for $n, k \geqslant 0$.

Proof. Calculate the variance and covariance of this process as follows:

$$\begin{aligned}
\mathrm{Var}(M_n) &= \sum_{k=1}^{n} \mathrm{Var}(X_k) = n \quad (\text{since } \{X_k\}_{k\geqslant 1} \text{ are i.i.d.}),\\
\mathrm{Cov}(M_n, M_k) &= \mathrm{E}[M_n\, M_k] - \underbrace{\mathrm{E}[M_n]\,\mathrm{E}[M_k]}_{=0} = \mathrm{E}[\,\mathrm{E}[M_n\, M_k \mid \mathcal{F}_{n\wedge k}]\,]\\
&= \mathrm{E}[\,\mathrm{E}[M_{n\wedge k}\, M_{n\vee k} \mid \mathcal{F}_{n\wedge k}]\,] = \mathrm{E}[M_{n\wedge k}\,\mathrm{E}[M_{n\vee k} \mid \mathcal{F}_{n\wedge k}]]\\
&= \mathrm{E}[M_{n\wedge k}^2] = \mathrm{Var}(M_{n\wedge k}) = n \wedge k.
\end{aligned}$$

□

We note that the above result is also easily proven by using the independence of $M_n - M_k$ and M_k, for $n \geqslant k$.

3. It is a process with independent and stationary increments such that $\mathrm{E}[M_n - M_k] = 0$ and $\mathrm{Var}(M_n - M_k) = |n - k|$, for $n, k \geqslant 0$.

Proof. The independence and stationarity of increments are easily proved (as shown in a previous example). The expected value and variance are

$$\begin{aligned}
\mathrm{E}[M_n - M_k] &= \mathrm{E}[M_n] - \mathrm{E}[M_k] = 0 - 0 = 0,\\
\mathrm{Var}(M_n - M_k) &= \mathrm{Var}(M_n) - 2\,\mathrm{Cov}(M_n, M_k) + \mathrm{Var}(M_k)\\
&= n - 2(n \wedge k) + k = n \vee k - n \wedge k = |n - k|.
\end{aligned}$$

□

The next step toward Brownian motion (BM) is the construction of a *scaled symmetric random walk*. We denote this process by $W^{(n)}$ and it will be derived from a symmetric random walk. Fix $n \in \mathbb{N}$. For $t \geqslant 0$, define $W^{(n)}(t) := \frac{1}{\sqrt{n}} M_{nt}$ provided that nt is an integer. Suppose that nt is noninteger. Find s such that ns is an integer and $ns < nt < ns + 1$ holds. Since ns is the largest integer less than or equal to nt, it is equal to $\lfloor nt \rfloor$. We hence define $W^{(n)}(t) := W^{(n)}(s) = \frac{1}{\sqrt{n}} M_{ns}$ to obtain the following general formula of the scaled symmetric random walk:

$$W^{(n)}(t) = \frac{1}{\sqrt{n}} M_{\lfloor nt \rfloor}, \quad t \geqslant 0. \tag{10.6}$$

As is seen from (10.6), $W^{(n)}$ is a *continuous-time* stochastic process with piecewise-constant right-continuous sample paths. At every time index $t = \frac{k}{n}$, $k = 1, 2, \ldots$, the process has a jump of size $\frac{1}{\sqrt{n}}$. Figure 10.1 depicts a sample path of this process for $n = 100$.

Clearly, the time-t realization $W^{(n)}(t)$ depends on the first $\lfloor nt \rfloor$ up or down moves $\omega_1, \ldots, \omega_k$, with $k = \lfloor nt \rfloor$. Note that since the process $W^{(n)}$ is constant on the time interval, we have $\lfloor nt \rfloor \leqslant nt < \lfloor nt \rfloor + 1$. Thus, $W^{(n)}(t)$ is measurable w.r.t. $\mathcal{F}_{\lfloor nt \rfloor} = \sigma(M_1, \ldots, M_{\lfloor nt \rfloor}) = \sigma(W^{(n)}(u) : 0 \leqslant u \leqslant t)$. Hence, the filtration defined by $\mathbb{F}^{(n)} := \{\mathcal{F}^{(n)}(t) := \mathcal{F}_{\lfloor nt \rfloor}; t \geqslant 0\}$ is in fact the natural filtration for the process $\{W^{(n)}(t)\}_{t\geqslant 0}$. The process is hence adapted to $\mathbb{F}^{(n)}$. Since the scaled symmetric random walk is obtained by re-scaling (and constant interpolation) of the symmetric random walk, it inherits the properties of the original process. For all $t, s \geqslant 0$ *such that ns and nt are integers*, we have

1. $\mathrm{E}\left[W^{(n)}(t) \mid \mathcal{F}^{(n)}(s)\right] = W^{(n)}(s)$ (martingale property) for $0 \leqslant s \leqslant t$, and zero time-$t$ expectation $\mathrm{E}[W^{(n)}(t)] = 0$;

2. $\mathrm{Var}(W^{(n)}(t)) = t$ and $\mathrm{Cov}(W^{(n)}(t), W^{(n)}(s)) = t \wedge s$.

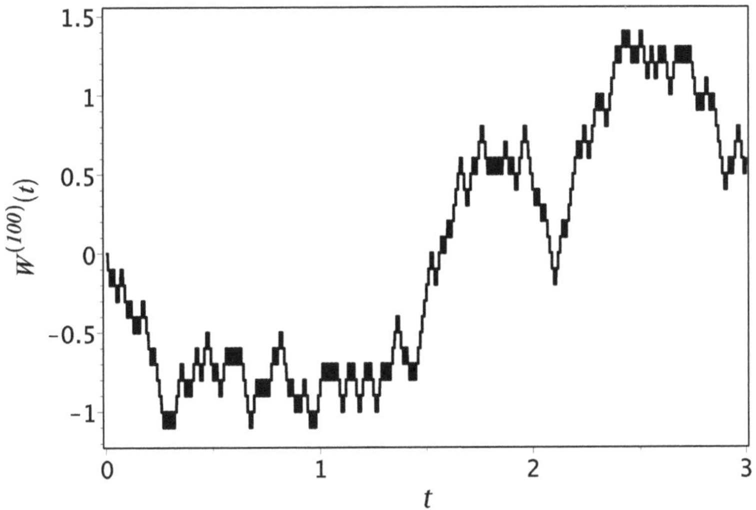

FIGURE 10.1: A sample path of $W^{(100)}$.

3. Let $0 \leqslant t_0 < t_1 < \cdots < t_n$ be such that nt_j is an integer for all $0 \leqslant j \leqslant n$, then the increments $W^{(n)}(t_1) - W^{(n)}(t_0), \ldots, W^{(n)}(t_n) - W^{(n)}(t_{n-1})$ are independent and stationary with mean zero and variance $\operatorname{Var}(W^{(n)}(t_j) - W^{(n)}(t_{j-1})) = t_j - t_{j-1}$ for $1 \leqslant j \leqslant n$.

The proof of these properties is left as an exercise for the reader (see Exercise 10.5).

An important characteristic of a stochastic process is the so-called *quadratic variation.* This quantity is a random variable that characterizes the variability of a stochastic process. For $t \geqslant 0$, the quadratic variation of the scaled random walk $W^{(n)}$ on $[0, t]$, denoted by $[W^{(n)}, W^{(n)}](t)$, is calculated as follows on any given path $\omega \in \Omega$. We go from time 0 to time t along the path, calculating the increments over each time step of size $\frac{1}{n}$. All these increments are then squared and summed up. For $t \geqslant 0$, such that $\lfloor nt \rfloor = nt$ is an integer,

$$[W^{(n)}, W^{(n)}](t) = \sum_{k=1}^{nt} \left[W^{(n)}\left(\frac{k}{n}\right) - W^{(n)}\left(\frac{k-1}{n}\right) \right]^2$$
$$= \sum_{k=1}^{nt} \left[\frac{1}{\sqrt{n}} X_k \right]^2 = \sum_{k=1}^{nt} \frac{1}{n} = t \quad (\text{since } X_k^2 = 1,\ 1 \leqslant k \leqslant n). \tag{10.7}$$

The above quadratic variation was calculated to be time t for any path or outcome ω. For every path of the scaled random walk, the quadratic variation $[W^{(n)}, W^{(n)}](t)$ is hence a constant and equal to t, i.e., $[W^{(n)}, W^{(n)}](t) = t$ is a constant random variable. Recall that the variance of $W^{(n)}(t)$ also has the same value.

The probability distribution of M_m is that of a shifted and scaled binomial random variable: $M_m = 2\,\mathsf{U}_m - m$ with $\mathsf{U}_m \sim Bin(m, \frac{1}{2})$. Therefore, the mass probabilities of the scaled random walk are also binomial ones:

$$\mathbb{P}\left(W^{(n)}(t) = \frac{2k - m}{\sqrt{n}} \right) = \mathbb{P}(\mathsf{U}_m = k) = \frac{m!}{k!\,(m-k)!} \left(\frac{1}{2} \right)^m, \quad k = 0, 1, \ldots, m, \quad m = \lfloor nt \rfloor.$$

For example, $W^{(100)}(0.1)$ (with $n = 100, t = 0.1, nt = 10$) takes its value on the set

$$\{-1,\ -0.8,\ -0.6,\ -0.4,\ -0.2,\ 0,\ 0.2,\ 0.4,\ 0.6,\ 0.8,\ 1\}.$$

The probability mass function is plotted in Figure 10.2.

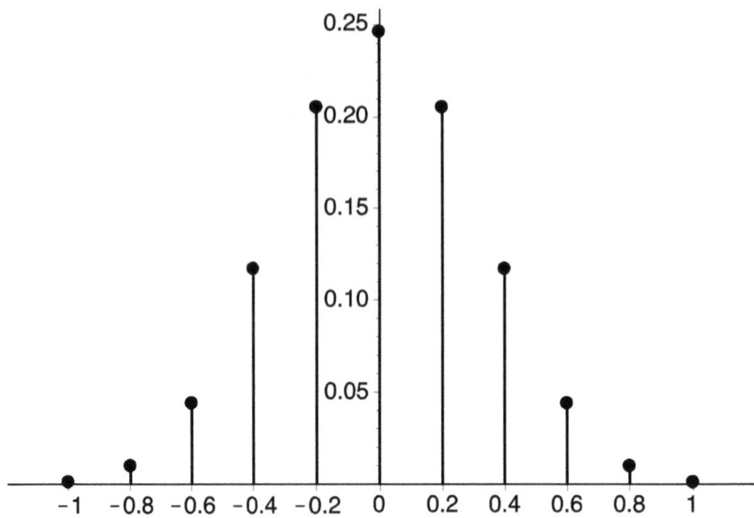

FIGURE 10.2: The probability mass function of $W^{(100)}(0.1)$.

Let us find the limiting distribution of $W^{(n)}(t)$, as $n \to \infty$, for fixed $t > 0$. Since $W^{(n)}(t)$ is equal to a scaled sum of i.i.d. random variables X_k, $k \geqslant 1$, with $\mathrm{E}[X_1] = 0$ and $\mathrm{Var}(X_1) = 1$, the Central Limit Theorem can be applied to yield

$$W^{(n)}(t) = \frac{1}{\sqrt{n}} \sum_{k=1}^{\lfloor nt \rfloor} X_k = \sqrt{\frac{\lfloor nt \rfloor}{n}} \left(\frac{1}{\sqrt{\lfloor nt \rfloor}} \sum_{k=1}^{\lfloor nt \rfloor} X_k \right) \xrightarrow{d} \sqrt{t}\, Z, \text{ as } n \to \infty,$$

where $Z \sim \mathit{Norm}(0,1)$. In the above derivation, we use the property $nt - 1 \leqslant \lfloor nt \rfloor \leqslant nt$, so $\lim_{n\to\infty} \frac{\lfloor nt \rfloor}{n} = t$. Thus,

$$W^{(n)}(t) \xrightarrow{d} \mathit{Norm}(0,t), \text{ as } n \to \infty.$$

To demonstrate the convergence of the probability distribution of $W^{(n)}(t)$ to the normal distribution, we plot and compare a histogram for $W^{(n)}(t)$ and a normal density curve for $\mathit{Norm}(0,t)$ in Figure 10.3. The histogram is constructed for $W^{(100)}(0.1)$ by replacing each mass probability in Figure 10.2 by a bar with width 0.2 (since the distance between mass points equals 0.2) and height chosen so that the area of the bar is equal to the respective mass probability. For example, $\mathbb{P}(W^{(100)}(0.1) = 0.2) = \mathbb{P}(\mathsf{U}_{10} = 6) \cong 0.20508$, hence we draw a histogram bar centered at 0.2 with width 0.2 and height $\frac{0.20508}{0.2} \cong 1.02539$. As a result, the total area of the histogram is one (as well as the area under the normal curve). Note the close agreement between the histogram and the normal density in Figure 10.3.

The limit of scaled random walks $\{W^{(n)}(t)\}_{t\geqslant 0}$ taken simultaneously, as $n \to \infty$, for all $t \in [0,T]$, $T > 0$, gives us a new continuous-time stochastic process called *Brownian motion.* It is denoted by $\{W(t)\}_{t\geqslant 0}$. Brownian motion can be viewed as a scaled random walk with infinitesimally small steps so that the process moves equally likely upward or downward by $\sqrt{\mathrm{d}t}$ in each infinitesimal time interval $\mathrm{d}t$. To govern the behaviour of Brownian motion, the number of (up or down) moves becomes infinite in any time interval. Equivalently, if each move is like a coin toss, then the coin needs to be tossed "infinitely fast." Thus,

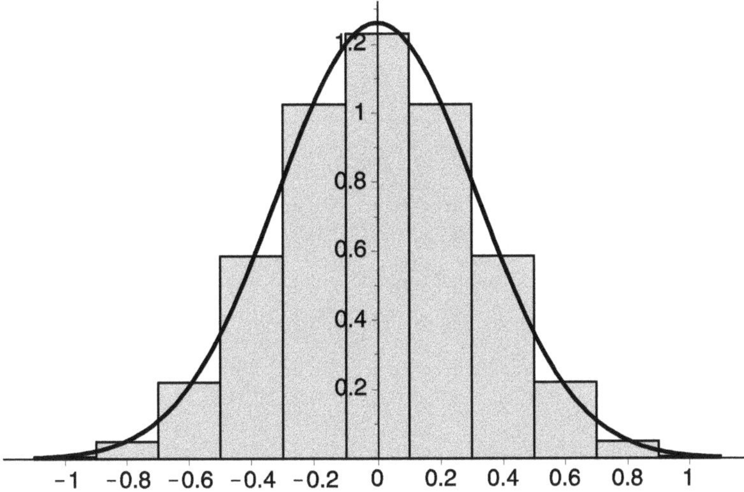

FIGURE 10.3: Comparison of the histogram constructed for $W^{(100)}(0.1)$ and the density curve for $Norm(0, 0.1)$.

$W(t)$, for all $t > 0$, can be viewed as a function of an uncountably infinite sequence of elementary moves (or coin tosses). Different outcomes (sequences of ω's) produce different sample paths. Any particular outcome $\boldsymbol{\omega}^* \in \Omega_\infty$ produces a particular path $\{W(t, \boldsymbol{\omega}^*)\}_{t \geqslant 0}$. Although sample paths of $W^{(n)}$ are piecewise-constant functions of t that are discontinuous at $t = k/n$, $k \geqslant 1$, the size of jumps at the points of discontinuity goes to zero, as $n \to \infty$. Moreover, the length of intervals where $W^{(n)}$ is constant goes to zero as $n \to \infty$, as well. In the limiting case, the Brownian paths become continuous everywhere and nonconstant on any interval no matter how small.

Let us summarize the differences between a scaled random walk $\{W^{(n)}(t)\}_{t \in [0,T]}$ and Brownian motion $\{W(t)\}_{t \in [0,T]}$, in the following table.

	$W^{(n)}(t)$	$W(t)$
RANGE:	discrete and bounded	continuous on $(-\infty, \infty)$
DISTRIBUTION:	approximately normal	normal
SAMPLE PATHS:	piecewise-constant	continuous and nonconstant on any time interval

10.2.2 Formal Definition and Basic Properties of Brownian Motion

Definition 10.1. A real-valued continuous-time stochastic process $\{W(t)\}_{t \geqslant 0}$ defined on a probability space $(\Omega, \mathcal{F}, \mathbb{P})$ is called a *standard Brownian motion* if it satisfies the following.

1. (Almost every path starts at the origin.) $W(0) = 0$ with probability one.
2. (Independence of nonoverlapping increments) For all $n \in \mathbb{N}$ and every choice of time partition $0 \leqslant t_0 < t_1 < \cdots < t_n$, the increments

$$W(t_1) \equiv W(t_1) - W(t_0), W(t_2) - W(t_1), \ldots, W(t_n) - W(t_{n-1})$$

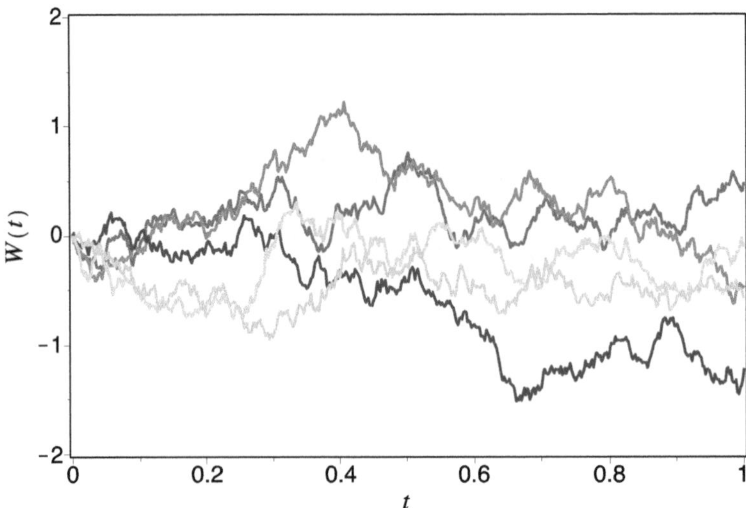

FIGURE 10.4: Sample paths of a standard Brownian motion.

are jointly independent.

3. (Normality of any increment) For all $0 \leqslant s < t$, the increment $W(t)-W(s)$ is normally distributed with mean 0 and variance $t-s$:

$$\begin{aligned} &\mathrm{E}[W(t)-W(s)] = 0, \quad \mathrm{Var}(W(t)-W(s)) = t-s, \\ &W(t)-W(s) \sim Norm(0, t-s). \end{aligned} \tag{10.8}$$

4. (Continuity of paths) For almost all $\omega \in \Omega$ (i.e., with probability one), the sample path $W(t,\omega)$ is a continuous function of time t.

Brownian motion (which we also abbreviate as BM) is named after the botanist Robert Brown who was the first to observe and describe the motion of a pollen particle suspended in fluid as an irregular random motion. In 1900, Louis Bachelier used Brownian motion as a model for movements of stock prices in his mathematical theory of speculations. In 1905, Albert Einstein obtained differential equations for the distribution function of Brownian motion. He argued that the random movement observed by Robert Brown is due to bombardment of the particle by molecules of the fluid. However, it was Norbert Wiener who constructed the mathematical foundation of BM as a stochastic process in 1931. Sometimes, this process is also called the Wiener process and this fact explains the notation used to denote Brownian motion by the letter W.

Although the processes $W^{(n)}$ and W have different probability distributions and sample paths, some of the properties of a scaled random walk are inherited by Brownian motion. For example, by making use of the independence of increments for adjacent (or nonoverlapping) time intervals $W(t)-W(s)$ and $W(s)-W(0)=W(s)$ and the fact that $\mathrm{E}[W(t)]=\mathrm{E}[W(s)]=0$, the covariance of $W(s)$ and $W(t)$, $0 \leqslant s \leqslant t$, is given by

$$\begin{aligned} \mathrm{Cov}(W(s),W(t)) = \mathrm{E}[W(s)\,W(t)] &= \mathrm{E}[W(s)\,(W(t)-W(s)) + W^2(s)] \\ &= \mathrm{E}[W(s)]\,\mathrm{E}[W(t)-W(s)] + \mathrm{Var}(W(s)) = 0 + s = s. \end{aligned}$$

In general, we have

$$\mathrm{Cov}(W(s),W(t)) = s \wedge t, \ \text{ for } s,t \geqslant 0. \tag{10.9}$$

Another feature that BM and symmetric random walks have in common is the martingale property. The *natural filtration for Brownian motion* is the σ-algebra generated by the Brownian motion observed up to time $t \geqslant 0$:

$$\mathcal{F}_t^W = \sigma(\{W(s) : 0 \leqslant s \leqslant t\}).$$

BM is hence automatically adapted to the filtration $\mathbb{F}^W \equiv \{\mathcal{F}_t^W\}_{t \geqslant 0}$, i.e., $W(t)$ is obviously $\mathcal{F}_t^W$-measurable for every $t \geqslant 0$. In what follows we will assume any appropriate filtration $\mathbb{F} = \{\mathcal{F}_t\}_{t \geqslant 0}$ for BM (of which $\mathbb{F}^W$ is one such filtration) such that all of the defining properties of Brownian motion hold. For all $0 \leqslant u \leqslant s < t$, $W(t) - W(s)$ is independent of $W(u)$ and hence it follows that the increment $W(t) - W(s)$ is independent of the σ-algebra $\mathcal{F}_s^W$ generated by all Brownian paths up to time s. So, for any appropriate filtration $\mathbb{F}$ for BM, we must have that $W(t) - W(s)$ is independent of $\mathcal{F}_s$. In what follows, we shall also use the shorthand notation for the conditional expectation of a random variable X w.r.t. a given σ-algebra $\mathcal{F}_t$, i.e., we shall often interchangeably write $\mathrm{E}[X \mid \mathcal{F}_t] \equiv \mathrm{E}_t[X]$ (for shorthand) wherever it is convenient.

Below we consider some examples of processes that are martingales w.r.t. any assumed filtration $\{\mathcal{F}_t\}_{t \geqslant 0}$ for BM. One such process is Brownian motion. For any finite time $t \geqslant 0$, the BM process is obviously integrable since $\mathrm{E}[|W(t)|] \leqslant \sqrt{\mathrm{E}[W^2(t)]} = \sqrt{t} < \infty$. In fact, this expectation is computed exactly since $W(t) \stackrel{d}{=} \sqrt{t}Z$, where $Z \sim \mathit{Norm}(0,1)$, giving $\mathrm{E}[|W(t)|] = \sqrt{t}\mathrm{E}[|Z|] = 2\sqrt{t}\int_0^\infty \frac{\mathrm{e}^{-z^2/2}}{\sqrt{2\pi}} z dz = \sqrt{\frac{2t}{\pi}}$.

Theorem 10.1. *Brownian motion is a martingale w.r.t.* $\{\mathcal{F}_t\}_{t \geqslant 0}$.

Proof. BM is adapted to its natural filtration and hence assumed adapted to $\{\mathcal{F}_t\}_{t \geqslant 0}$. Since we already showed that $W(t)$ is integrable, we need only show that $\mathrm{E}[W(t) \mid \mathcal{F}_s] \equiv \mathrm{E}_s[W(t)] = W(s)$, for $0 \leqslant s \leqslant t$. Now, $W(s)$ is $\mathcal{F}_s$-measurable and $W(t) - W(s)$ is independent of $\mathcal{F}_s$, hence

$$\begin{aligned}\mathrm{E}_s[W(t)] &= \mathrm{E}_s[(W(t) - W(s)) + W(s)] = \mathrm{E}_s[W(t) - W(s)] + \mathrm{E}_s[W(s)] \\ &= \mathrm{E}[W(t) - W(s)] + W(s) = 0 + W(s) = W(s).\end{aligned}$$

□

The next example shows that squaring BM and subtracting by time gives a martingale.

Example 10.1. Prove that $\{W^2(t) - t\}_{t \geqslant 0}$ is a martingale w.r.t. any filtration $\{\mathcal{F}_t\}_{t \geqslant 0}$ for BM.

Solution. Since $W(t)$ is $\mathcal{F}_t$-measurable, then $X_t := W^2(t) - t$ is $\mathcal{F}_t$-measurable. Moreover, $\mathrm{E}[|X_t|] \leqslant \mathrm{E}[W^2(t)] + t = 2t < \infty$. It remains to show that $\mathrm{E}_s[X_t] = X_s$, for $0 \leqslant s \leqslant t$. We now calculate the conditional time-s expectation of $W^2(t)$ for $s \leqslant t$, by using the fact that $W(s)$ and $W^2(s)$ are $\mathcal{F}_s$-measurable and that $W(t) - W(s)$ and $(W(t) - W(s))^2$ are independent of $\mathcal{F}_s$:

$$\begin{aligned}\mathrm{E}_s\left[W^2(t)\right] &= \mathrm{E}_s\left[(W(t) - W(s) + W(s))^2\right] \\ &= \mathrm{E}_s\left[(W(t) - W(s))^2 + 2W(s)(W(t) - W(s)) + W^2(s)\right] \\ &= \mathrm{E}_s\left[(W(t) - W(s))^2\right] + 2\mathrm{E}_s\left[W(s)(W(t) - W(s))\right] + \mathrm{E}_s\left[W^2(s)\right] \\ &= \mathrm{E}\left[(W(t) - W(s))^2\right] + 2W(s)\,\mathrm{E}[W(t) - W(s)] + W^2(s) \\ &= \mathrm{Var}(W(t) - W(s)) + 2W(s) \cdot 0 + W^2(s) = (t - s) + W^2(s).\end{aligned}$$

Therefore, $\mathrm{E}_s\left[W^2(t) - t\right] = (t - s) + W^2(s) - t = W^2(s) - s$. □

We note that an alternative way to compute $\mathrm{E}_s\left[W^2(t)\right] \equiv \mathrm{E}\left[W^2(t) \mid \mathcal{F}_s\right]$ is to use the fact that $W(s)$ is $\mathcal{F}_s$-measurable and $Y \equiv W(t) - W(s)$ is independent of $\mathcal{F}_s$. Hence, $\mathrm{E}\left[W^2(t) \mid \mathcal{F}_s\right] = \mathrm{E}\left[(Y + W(s))^2 \mid \mathcal{F}_s\right] = g(W(s)) = t - s + W^2(s)$, where Proposition 6.7 gives $g(x) = \mathrm{E}\left[(Y+x)^2\right] = \mathrm{E}[Y^2] + 2x\mathrm{E}[Y] + x^2 = t - s + 2x \cdot 0 + x^2 = t - s + x^2$. [Note that in Proposition 6.7 we have set $\mathcal{G} \equiv \mathcal{F}_s$, $X \equiv W(s)$, $Y \equiv W(t) - W(s)$.]

The following is an example of a so-called *exponential martingale*. This type of martingale will turn out to be very useful when pricing options. Recall $\mathrm{E}[\mathrm{e}^{\alpha X}] = \mathrm{e}^{\alpha\mu + \alpha^2\sigma^2/2}$ is the moment generating function (MGF) of a normal random variable $X \sim Norm(\mu, \sigma^2)$. Hence, the MGF of $W(t)$ is given by $\mathrm{E}[\mathrm{e}^{\alpha W(t)}] = \mathrm{e}^{\alpha^2 t/2}$ since $W(t) \sim Norm(0, t)$.

Example 10.2. Prove that $\{\mathrm{e}^{\alpha W(t) - \alpha^2 t/2}\}_{t \geqslant 0}$, for all $\alpha \in \mathbb{R}$, is a martingale w.r.t. any filtration $\{\mathcal{F}_t\}_{t \geqslant 0}$ for BM.

Solution. Since $W(t)$ is $\mathcal{F}_t$-measurable, then $X_t := \mathrm{e}^{\alpha W(t) - \alpha^2 t/2}$ is $\mathcal{F}_t$-measurable. The process is integrable since $\mathrm{E}[|X_t|] = \mathrm{e}^{-\alpha^2 t/2}\mathrm{E}[\mathrm{e}^{\alpha W(t)}] = 1$. We now show that $\mathrm{E}_s[X_t] = X_s$, for $0 \leqslant s \leqslant t$. Indeed, since $\mathrm{e}^{\alpha W(s)}$ is $\mathcal{F}_s$-measurable and $W(t) - W(s) \sim Norm(0, t-s)$ is independent of $\mathcal{F}_s$:

$$\mathrm{E}_s\left[\mathrm{e}^{\alpha W(t)}\right] = \mathrm{E}_s\left[\mathrm{e}^{\alpha(W(t)-W(s))}\mathrm{e}^{\alpha W(s)}\right] = \mathrm{e}^{\alpha W(s)}\mathrm{E}\left[\mathrm{e}^{\alpha(W(t)-W(s))}\right] = \mathrm{e}^{\alpha W(s)}\mathrm{e}^{\alpha^2(t-s)/2}.$$

Therefore, $\mathrm{E}[\mathrm{e}^{\alpha W(t) - \alpha^2 t/2} \mid \mathcal{F}_s] = \mathrm{e}^{\alpha W(s) - \alpha^2 s/2}$. □

10.2.3 Multivariate Distribution of Brownian Motion

The random variables $W(t_1), W(t_2), \ldots, W(t_n)$ are jointly normally distributed for all time points $0 \leqslant t_1 < t_2 < \cdots < t_n$. Therefore, their joint n-variate distribution is determined by the mean vector and covariance matrix. In accordance with (10.8) and (10.9), we have

$$\mu_i = \mathrm{E}[W(t_i)] = 0 \text{ and } C_{ij} = \mathrm{Cov}\left(W(t_i), W(t_j)\right) = \mathrm{E}\left[W(t_i)\,W(t_j)\right] = t_i \wedge t_j,$$

for $1 \leqslant i, j \leqslant n$. Therefore, $\mathbf{W} := \left[W(t_1), W(t_2), \cdots, W(t_n)\right]^\top$, where each component random variable corresponds to the Brownian path at the respective time points, has the n-variate normal distribution with mean vector zero and covariance matrix

$$\begin{bmatrix} \mathrm{E}[W^2(t_1)] & \mathrm{E}[W(t_1)\,W(t_2)] & \cdots & \mathrm{E}[W(t_1)\,W(t_n)] \\ \mathrm{E}[W(t_2)\,W(t_1)] & \mathrm{E}[W^2(t_2)] & \cdots & \mathrm{E}[W(t_2)\,W(t_n)] \\ \vdots & \vdots & \ddots & \vdots \\ \mathrm{E}[W(t_n)\,W(t_1)] & \mathrm{E}[W(t_n)\,W(t_2)] & \cdots & \mathrm{E}[W^2(t_n)] \end{bmatrix} = \begin{bmatrix} t_1 & t_1 & \cdots & t_1 \\ t_1 & t_2 & \cdots & t_2 \\ \vdots & \vdots & \ddots & \vdots \\ t_1 & t_2 & \cdots & t_n \end{bmatrix} \tag{10.10}$$

Thus, the joint PDF is given by (10.1) with $\boldsymbol{\mu} = \mathbf{0}$ and covaraince matrix $\mathbf{C}$ in (10.10):

$$f_{\mathbf{W}}(\mathbf{x}) = \frac{1}{(2\pi)^{n/2}\left[\det \mathbf{C}\right]^{n/2}} \mathrm{e}^{-\frac{1}{2}\mathbf{x}^\top \mathbf{C}^{-1}\mathbf{x}}, \quad \mathbf{x} \in \mathbb{R}^n. \tag{10.11}$$

Example 10.3. Let $0 < t_1 < t_2 < t_3$. Find the probability distribution of

$$W(t_1) - 2W(t_2) + 3W(t_3).$$

Solution. The vector $\mathbf{W} = \left[W(t_1), W(t_2), W(t_3)\right]^\top$ has the trivariate normal distribution:

$$\mathbf{W} \sim Norm_3\left(\mathbf{0} \equiv \begin{bmatrix} 0 \\ 0 \\ 0 \end{bmatrix}, \mathbf{C} = \begin{bmatrix} t_1 & t_1 & t_1 \\ t_1 & t_2 & t_2 \\ t_1 & t_2 & t_3 \end{bmatrix}\right).$$

Set $\mathbf{b} = [1, -2, 3]^\top$. Then, in accordance with (10.3), $W(t_1) - 2W(t_2) + 3W(t_3) = \mathbf{b}^\top \mathbf{W}$ is a scalar normal random variable with mean $\mathbf{b}^\top \mathbf{0} \equiv 0$ and variance

$$\mathbf{b}^\top \mathbf{C}\,\mathbf{b} = [1, -2, 3] \begin{bmatrix} t_1 & t_1 & t_1 \\ t_1 & t_2 & t_2 \\ t_1 & t_2 & t_3 \end{bmatrix} \begin{bmatrix} 1 \\ -2 \\ 3 \end{bmatrix} = 3t_1 - 8t_2 + 9t_3.$$

Alternatively, by using the property that a linear combination of normal random variables is again a normal variate, we have that $W(t_1){-}2W(t_2){+}3W(t_3)$ is normally distributed with mean

$$\mathrm{E}[W(t_1) - 2W(t_2) + 3W(t_3)] = \mathrm{E}[W(t_1)] - 2\mathrm{E}[W(t_2)] + 3\mathrm{E}[W(t_3)] = 0$$

and variance

$$\begin{aligned} &\mathrm{Var}(W(t_1) - 2W(t_2) + 3W(t_3)) = \mathrm{Var}(W(t_1)) + \mathrm{Var}(-2W(t_2)) + \mathrm{Var}(3W(t_3)) \\ &\quad + 2\,\mathrm{Cov}(W(t_1), -2W(t_2)) + 2\,\mathrm{Cov}(-2W(t_2), 3W(t_3)) + 2\,\mathrm{Cov}(W(t_1), 3W(t_3)) \\ &= t_1 + 4t_2 + 9t_3 - 4t_1 - 12t_2 + 6t_1 = 3t_1 - 8t_2 + 9t_3. \end{aligned}$$ □

Example 10.4. Consider a standard BM. Calculate the following probabilities:

(a) $\mathbb{P}(W(t) \leqslant 0)$ for $t > 0$;

(b) $\mathbb{P}(W(1) \leqslant 0, W(2) \leqslant 0)$.

Solution.
(a) Since for all $t > 0$, $W(t) \stackrel{d}{=} \sqrt{t}Z$, where $Z \sim \mathit{Norm}(0,1)$, we have

$$\mathbb{P}(W(t) \leqslant 0) = \mathbb{P}(\sqrt{t}Z \leqslant 0) = \mathbb{P}(Z \leqslant 0) = \mathcal{N}(0) = \frac{1}{2}.$$

(b) Using the independence of the increments $W(2) - W(1) \stackrel{d}{=} Z_1$ and $W(1) - W(0) \stackrel{d}{=} Z_2$, where Z_1 and Z_2 are independent standard normal random variables with joint PDF $f_{Z_1,Z_2}(z_1, z_2) = f_{Z_1}(z_1) f_{Z_2}(z_2) = n(z_1)\, n(z_2)$, we obtain

$$\begin{aligned} \mathbb{P}(W(1) \leqslant 0, W(2) \leqslant 0) &= \mathbb{P}(W(1) \leqslant 0, W(1) + (W(2) - W(1)) \leqslant 0) \\ &= \mathbb{P}(Z_1 \leqslant 0, Z_1 + Z_2 \leqslant 0) \\ &= \iint_D n(z_1)\, n(z_2)\, \mathrm{d}z_1\, \mathrm{d}z_2 = \int_{-\infty}^{0} \left(\int_{-\infty}^{-z_1} n(z_2)\, \mathrm{d}z_2 \right) n(z_1)\, \mathrm{d}z_1 \\ &\quad \big(\text{where } D = \{(z_1, z_2) \in \mathbb{R}^2 \,:\, z_1 + z_2 \leqslant 0 \text{ and } z_1 \leqslant 0\}\big) \\ &= \int_{-\infty}^{0} \mathcal{N}(-z_1)\, n(z_1)\, \mathrm{d}z_1 = \int_0^\infty \mathcal{N}(x)\, n(x)\, \mathrm{d}x \\ &\quad \big(\text{by change of variables } x = -z_1 \text{ and where } n(-x) = n(x)\big) \\ &= \int_0^\infty \mathcal{N}(x) \mathcal{N}'(x)\, \mathrm{d}x = \frac{1}{2}\mathcal{N}^2(x)\Big|_0^\infty = \frac{1}{2}(1^2 - (1/2)^2) = \frac{3}{8}. \end{aligned}$$

Not surprisingly, this example verifies that $W(1)$ and $W(2)$ are dependent random variables since the joint probability

$$\mathbb{P}(W(1) \leqslant 0, W(2) \leqslant 0) = \frac{3}{8} \neq \frac{1}{4} = \mathbb{P}(W(1) \leqslant 0)\, \mathbb{P}(W(2) \leqslant 0).$$ □

10.2.4 The Markov Property and the Transition PDF

Our goal is the derivation of the transition probability distribution of Brownian motion. In particular, given a filtration $\{\mathcal{F}_t\}_{t\geqslant 0}$ for BM, we want to derive a formula for calculating conditional probabilities of the form

$$\mathbb{P}(W(s) \in A \mid \mathcal{F}_t) \text{ for } 0 \leqslant t < s \text{ and Borel set } A \in \mathcal{B}(\mathbb{R}).$$

Our approach is based on two properties of Brownian motion, namely, time and space homogeneity and the Markov property.

Let $\{W(t)\}_{t\geqslant 0}$ be a standard Brownian motion and let $x \in \mathbb{R}$. The process $W^{(x)}$ defined by

$$W^{(x)}(t) := x + W(t), \quad x \in \mathbb{R}$$

is a Brownian motion. Standard BM W and $W^{(x)}$ have identical increments: $W^{(x)}(t+s) - W^{(x)}(t) = W(t+s) - W(t)$, $s > 0, t \geqslant 0$. Hence, $W^{(x)}$ is a continuous-time stochastic process with the same independent normal increments as W and continuous sample paths. The only difference w.r.t. standard BM is that this process starts at an arbitrary real value x: $W^{(x)}(0) = x$. The property that $x+W(t)$ is Brownian motion for all x is called the *space homogeneity*. Another property of BM is its *time homogeneity*, meaning that the process defined by $\widehat{W}(t) := W(t+T) - W(T)$, $t \geqslant 0$, is Brownian motion for any fixed $T \geqslant 0$. In other words, translation of a Brownian sample path along both space and time axes gives us again a Brownian path.

Theorem 10.2. *Brownian motion is a Markov process.*

Proof. Fix any (Borel) function $h\colon \mathbb{R} \to \mathbb{R}$ and take $\{\mathcal{F}_t\}_{t\geqslant 0}$ as a filtration for BM. We need to show that there exists a (Borel) function $g\colon \mathbb{R} \to \mathbb{R}$ such that

$$\mathrm{E}[h(W(t)) \mid \mathcal{F}_s] \equiv \mathrm{E}_s[h(W(t))] = g(W(s)), \text{ for all } 0 \leqslant s \leqslant t,$$

i.e., this means that $\mathrm{E}[h(W(t)) \mid \mathcal{F}_s] = \mathrm{E}[h(W(t))|W(s)]$ for all $0 \leqslant s \leqslant t$. Indeed, since $W(s)$ is $\mathcal{F}_s$-measurable and $Y := W(t) - W(s)$ is independent of $\mathcal{F}_s$, we can apply Proposition 6.7, giving

$$\mathrm{E}_s[h(W(t))] = \mathrm{E}_s[h(Y + W(s))] = g(W(s))$$

where $g(x) = \mathrm{E}[h(Y+x)] = \mathrm{E}[h(\sqrt{t-s}\, Z + x)]$, $Z \sim Norm(0,1)$. In fact, using the density of Z we have $g(x) = \int_{\mathbb{R}} h(\sqrt{t-s}\, z + x) n(z)\, dz$. □

We remark that, for $s = t$, this result simply gives $g(x) = h(x)$ and recovers the trivial case that $\mathrm{E}_s[h(W(s))] = h(W(s))$, i.e., $W(s)$, and hence $h(W(s))$, is $\mathcal{F}_s$-measurable. For $s < t$, we can also re-express $g(x)$ by using a linear change of integration variables $z \to y = \sqrt{t-s}\, z + x$, $z = (y-x)/\sqrt{t-s}$, giving:

$$g(x) = \int_{\mathbb{R}} h(y) p(s,t;x,y)\, \mathrm{d}y \tag{10.12}$$

where $p(s,t;x,y) := \frac{\mathrm{e}^{-\frac{(y-x)^2}{2(t-s)}}}{\sqrt{2\pi(t-s)}}$. Below, we shall see that this function has a very special role and is the so-called transition PDF of BM. The above Markov property means that when taking an expectation of a function of BM at a future time t conditional on all the information (path history) of the BM up to an earlier time $s < t$, it is the same as only conditioning on knowledge of the BM at time s. In particular,

$$\mathrm{E}_s[h(W(t))] = \mathrm{E}[h(W(t)) \mid W(s)] = \int_{\mathbb{R}} h(y) p(s,t;W(s),y)\, \mathrm{d}y.$$

Based on this formula, we can then compute the expected value of $h(W(t))$ conditional on any value $W(s) = x$ of BM at time $s < t$ as

$$\mathrm{E}[h(W(t)) \mid W(s) = x] = \int_{\mathbb{R}} h(y) p(s,t;x,y)\,\mathrm{d}y. \tag{10.13}$$

By this formula it is then clear that $p(s,t;x,y)$ is in fact the conditional PDF of random variable $W(t)$, given $W(s)$, i.e.,

$$\mathrm{E}[h(W(t)) \mid W(s) = x] = \int_{\mathbb{R}} h(y) f_{W(t)|W(s)}(y|x)\,\mathrm{d}y \implies f_{W(t)|W(s)}(y|x) = p(s,t;x,y).$$

By the Markov property of W we have, for all $0 \leqslant s \leqslant t$ and Borel set A,

$$\mathbb{P}(W(t) \in A \mid \mathcal{F}_s) = \mathrm{E}_s[\mathbb{I}_{\{W(t) \in A\}}] = \mathrm{E}[\mathbb{I}_{\{W(t)\in A\}} \mid W(s)] = \mathbb{P}(W(t) \in A \mid W(s)), \tag{10.14}$$

i.e., this probability is a function of $W(s)$ and A. Hence, the evaluation of probabilities of the form (10.14) and other expectations of functions of BM reduces to calculation of integrals involving $p(s,t;x,y)$. The transition PDF, and its associated transition probability function, plays a pivotal role in the theory of continuous-time Markov processes.

Definition 10.2. A *transition probability function* for a continuous-time Markov process $\{X(t)\}_{t \geqslant 0}$ is a real-valued function P given by the conditional probability

$$P(s,t;x,y) := \mathbb{P}(X(t) \leqslant y \mid X(s) = x), \quad 0 \leqslant s \leqslant t, \quad x, y \in \mathbb{R}.$$

Suppose that the function P is absolutely continuous (w.r.t. the Lebesgue measure) and there exists a real-valued nonnegative function p such that

$$P(s,t;x,y) = \int_{-\infty}^{y} p(s,t;x,z)\,\mathrm{d}z. \tag{10.15}$$

This function p is called a *transition PDF* of the process X. By differentiating (10.15), we obtain

$$p(s,t;x,y) = \frac{\partial}{\partial y} P(s,t;x,y).$$

By definition, the transition probability function P and transition PDF p are, respectively, the conditional CDF and PDF of random variable $X(t)$ given a value for $X(s)$:

$$F_{X(t)|X(s)}(y \mid x) = P(s,t;x,y) \text{ and } f_{X(t)|X(s)}(y \mid x) = p(s,t;x,y). \tag{10.16}$$

The CDF $P(s,t;x,y)$ hence represents the probability that the value, $X(t)$, of the process at time t is at most y given that it has known value $X(s) = x$ at an earlier time $s < t$. In other words, in terms of a path-wise description, it represents the probability of the event $\{\omega \in \Omega \mid X(t,\omega) \leqslant y,\ X(s,\omega) = x,\ 0 \leqslant s < t\}$. The transition PDF represents the density of such paths in an infinitesimal interval $\mathrm{d}y$ about the value y, i.e., we can formally write

$$P(s,t;x,y+\mathrm{d}y) - P(s,t;x,y) \equiv \mathbb{P}\big(X(t) \in (y, y+\mathrm{d}y] \mid X(s) = x\big) = p(s,t;x,y)\,\mathrm{d}y.$$

An important class of processes occurring in many models of finance and other fields is one in which the transition CDF (and PDF) can be written as a function of the difference $t - s$ of the time variables s and t. This property is stated formally below.

Definition 10.3. A stochastic process $\{X(t)\}_{t \geqslant 0}$ is called *time-homogeneous* if, for all $s \geqslant 0$, $t > 0$, $x \in \mathbb{R}$, and $A \in \mathcal{B}(\mathbb{R})$,

$$\mathbb{P}(X(s+t) \in A \mid X(s) = x) = \mathbb{P}(X(t) \in A \mid X(0) = x). \tag{10.17}$$

If we deal with a time-homogeneous process X, then the probability distribution function for the transition $X(s) = x \to X(t) = y$ is a function of the time increment $t-s$. That is, the time variables t and s occur as the difference of the two. Thus, the transition PDF can be written as a function of three arguments: $p(t; x, y)$. We think of the value x as the starting value of the process and y as the endpoint value of the process after a time lapse $t > 0$. The two versions of the transition PDF relate to each other as follows (since $(s+t) - s = t$):

$$p(t; x, y) = p(s, s+t; x, y) \text{ for } s \geqslant 0,\ t > 0,\ x, y \in \mathbb{R}.$$

Transition probabilities of a time-homogeneous process X are calculated by integrating the transition PDF $p(t; x, y)$:

$$\mathbb{P}(X(s+t) \in A \mid X(s) = x) = \int_A p(t; x, y)\, \mathrm{d}y, \text{ for } s \geqslant 0,\ t > 0,\ A \in \mathcal{B}(\mathbb{R}).$$

Brownian motion W is a Markov process, hence it is possible to define its transition probability function. Since we can write $W(t)$ as a sum $W(s) + (W(t) - W(s))$, where the increment $W(t) - W(s) \sim Norm(0, t-s)$ is independent of $W(s)$, for $s < t$, the conditional distribution of $W(t)$ given $W(s)$ (i.e., conditional on a given value $W(s) = x$) is normal. This can be seen by a straightforward application of (10.5), where we identify $x_2 \equiv x$, $X_1 \equiv W(t)$, $X_2 \equiv W(s)$, $\mu_1 = \mu_2 = \mathrm{E}[W(t)] = \mathrm{E}[W(s)] = 0$, $\sigma_1^2 = \mathrm{Var}(W(t)) = t$, $\sigma_2^2 = \mathrm{Var}(W(s)) = s$, $\rho = \mathrm{Cov}(W(t), W(s))/\sqrt{st} = \sqrt{s/t}$, and hence

$$W(t) \mid \{W(s) = x\} \sim Norm(x, t-s),\ 0 \leqslant s < t.$$

So the conditional CDF is given by $F_{W(t)|W(s)}(y, x) = \mathcal{N}\left(\frac{y-x}{\sqrt{t-s}}\right)$, where $\mathcal{N}$ is the standard normal CDF. This gives the transition probability function P, which can also be derived as follows:

$$\begin{aligned} P(s, t; x, y) &= \mathbb{P}(W(t) \leqslant y \mid W(s) = x) = \mathbb{P}(W(s) + (W(t) - W(s)) \leqslant y \mid W(s) = x) \\ &= \mathbb{P}\left(x + \sqrt{t-s}\, Z \leqslant y\right) = \mathbb{P}\left(Z \leqslant \frac{y-x}{\sqrt{t-s}}\right) = \mathcal{N}\left(\frac{y-x}{\sqrt{t-s}}\right), \end{aligned}$$

where $Z \sim Norm(0, 1)$. The transition PDF for BM is hence:

$$p(s, t; x, y) = \frac{\partial}{\partial y}\mathcal{N}\left(\frac{y-x}{\sqrt{t-s}}\right) = \frac{1}{\sqrt{t-s}}\, n\left(\frac{y-x}{\sqrt{t-s}}\right) = \frac{\exp\left(-\frac{(y-x)^2}{2(t-s)}\right)}{\sqrt{2\pi(t-s)}}. \tag{10.18}$$

Note that this is entirely consistent with the expression we obtained above just after the proof of the Markov property of Brownian motion. As is seen from (10.18), the transition PDF of Brownian motion is a function of the spatial point difference $x - y$ (thanks to the space-homogeneity property) and time difference $t - s$ (thanks to the time-homogeneity property). The transition distribution of BM does not change with any shift in time and in space. This property is seen by letting $p_0(t; x)$ be the PDF of $W(t)$, i.e.,

$$p_0(t; x) := \frac{\partial}{\partial x}\mathbb{P}(W(t) \leqslant x) = \frac{\partial}{\partial x}\mathcal{N}\left(\frac{x}{\sqrt{t}}\right) = \frac{1}{\sqrt{t}}\, n\left(\frac{x}{\sqrt{t}}\right) = \frac{\mathrm{e}^{-x^2/2t}}{\sqrt{2\pi t}}. \tag{10.19}$$

Then, the transition PDF p is given by

$$p(s, t; y, x) = p_0(t-s; x-y), \quad 0 \leqslant s < t, \quad x, y, \in \mathbb{R}. \tag{10.20}$$

In the previous section, we obtained the joint PDF of the path skeleton, $W(t_1), \ldots, W(t_n)$, of Brownian motion evaluated at n arbitrary time points $0 = t_0 < t_1 < t_2 < \cdots < t_n$. The resulting n-variate Gaussian distribution formula in (10.11) looks somewhat complicated since it involves the evaluation of the inverse of the covariance matrix $\mathbf{C}$ given in (10.10). However, by using the Markov property of Brownian motion, we can derive a simpler expression for the joint density. As is well known, a joint PDF of a random vector can be expressed as a product of marginal and conditional univariate densities. So, we have

$$\begin{aligned} f_{\mathbf{W}}(\mathbf{x}) &\equiv f_{W(t_1),\ldots,W(t_n)}(x_1,\ldots,x_n) \\ &= f_{W(t_1)}(x_1)\, f_{W(t_2)|W(t_1)}(x_2 \mid x_1) \times \cdots \times f_{W(t_n)|W(t_1),\ldots,W(t_{n-1})}(x_n \mid x_1,\ldots,x_{n-1}). \end{aligned}$$

Applying the Markov property gives us the following equivalences in distribution:

$$W(t_k) \mid \{W(t_1),\ldots,W(t_{k-1})\} \stackrel{d}{=} W(t_k) \mid W(t_{k-1}), \quad 2 \leqslant k \leqslant n.$$

Thus, we obtain

$$\begin{aligned} f_{\mathbf{W}}(\mathbf{x}) &= \prod_{k=1}^{n} f_{W(t_k)|W(t_{k-1})}(x_k \mid x_{k-1}) = \prod_{k=1}^{n} p(t_{k-1},t_k;x_{k-1},x_k) \\ &= \prod_{k=1}^{n} p_0(t_k - t_{k-1}; x_k - x_{k-1}) = \prod_{k=1}^{n} \frac{1}{\sqrt{t_k - t_{k-1}}} \cdot n\left(\frac{x_k - x_{k-1}}{\sqrt{t_k - t_{k-1}}}\right) \\ &= \frac{1}{(2\pi)^{n/2}\sqrt{(t_1 - t_0)\cdots(t_n - t_{n-1})}} \cdot \exp\left(-\frac{1}{2}\sum_{k=1}^{n}\frac{(x_k - x_{k-1})^2}{t_k - t_{k-1}}\right), \end{aligned} \tag{10.21}$$

where $f_{W(t_1)|W(t_0)}(x_1 \mid x_0) \equiv f_{W(t_1)}(x_1) = p_0(t_1;x_1)$ since $W(t_0) \equiv W(0) = x_0 \equiv 0$ for a standard Brownian motion. So now we have two formulas, (10.11) and (10.21), of the joint PDF $f_{\mathbf{W}}$ of Brownian motion on n time points. However, one can show that they are equivalent (see Exercise 10.17). Equation (10.21) clearly shows that the joint PDF of a Brownian path along any discrete set of time points is the product of the transition PDF for BM for each time step.

10.2.5 Quadratic Variation and Nondifferentiability of Paths

Although sample paths of Brownian motion $W(t,\omega)$, $\omega \in \Omega$, are almost surely, i.e., with probability one, continuous functions of time index t, they do not look like regular functions that appear in a textbook on calculus. First, Brownian sample paths are fractals, meaning that any zoomed-in part of a Brownian path looks very much the same as the original trajectory. Second, Brownian sample paths are (almost surely) not monotone or linear in any finite interval $(t, t+\delta t)$ and are not differentiable (w.r.t. t) at any point. A formal proof of the second statement is based on the concept of the variation of a function, but let us first provide some probabilistic arguments against the differentiability of $W(t)$. Consider a finite difference

$$\frac{W(t+\delta t) - W(t)}{\delta t} \stackrel{d}{=} \frac{\sqrt{\delta t}\, Z}{\delta t} = \frac{Z}{\sqrt{\delta t}}, \quad Z \sim Norm(0,1). \tag{10.22}$$

Clearly, for any $\kappa > 0$,

$$\mathbb{P}\left(\left|\frac{Z}{\sqrt{\delta t}}\right| > \kappa\right) = \mathbb{P}(|Z| > \kappa\sqrt{\delta t}) \to \mathbb{P}(|Z| > 0) = 1, \text{ as } \delta t \to 0.$$

Therefore, the ratio $\frac{W(t+\delta t)-W(t)}{\delta t}$ is unbounded, as $\delta t \to 0$, with probability one. So the ratio in (10.22) cannot (almost surely) converge to some finite limit. Recall that for a differentiable function we have the convergence $\frac{f(t+\delta t)-f(t)}{\delta t} \to f'(t)$, as $\delta t \to 0$.

Definition 10.4. The nonnegative quantity $V_{[a,b]}^{(p)}(f)$, defined by

$$V_{[a,b]}^{(p)}(f) = \limsup_{\delta t^{(n)} \to 0} \sum_{i=1}^{n} |f(t_i) - f(t_{i-1})|^p,$$

where the limit is taken over all possible partitions $a = t_0 < t_1 < \cdots < t_n = b$ of $[a,b]$ shrinking as $n \to \infty$ and $\delta t^{(n)} := \max_{i=1}^{n}(t_i - t_{i-1}) \to 0$, is called a *p-variation* (for $p > 0$) of a function $f\colon \mathbb{R} \to \mathbb{R}$ on $[a,b]$, $a < b$. If $V_{[a,b]}^{(p)}(f) < \infty$, then f is said to be a *function of bounded p-variation* on $[a,b]$.

Of particular interest are the first ($p = 1$) and quadratic (second) ($p = 2$) variations. Let us consider regular functions such as monotone and differentiable functions and let us find their variations.

Proposition 10.3.

1. *Bounded monotone functions have bounded first variations.*
2. *Differentiable functions with a bounded derivative have bounded first variations.*
3. *The quadratic variation of a differentiable function with a bounded derivative is zero.*

Proof.

1. Consider a function f that is monotone on $[a,b]$. Assume that f is nondecreasing. Then, for any partition $a = t_0 < t_1 < \cdots < t_n = b$,

$$\sum_{i=1}^{n} |f(t_i) - f(t_{i-1})| = \sum_{i=1}^{n} (f(t_i) - f(t_{i-1})) = f(b) - f(a),$$

since $f(t_i) \geqslant f(t_{i-1})$. Therefore, the first variation of a nondecreasing (increasing) function f is equal to $f(b) - f(a)$. For a nonincreasing (decreasing) function, $V_{[a,b]}^{(1)}(f) = f(a) - f(b)$. By combining these two cases, we obtain $V_{[a,b]}^{(1)}(f) = |f(b) - f(a)| < \infty$.

2. Now, consider a differentiable function $f \in D(a,b)$ with a bounded derivative $|f'(t)| \leqslant M < \infty$, for all $t \in (a,b)$. By the mean value theorem, for $u, v \in (a,b)$ with $u < v$, there exists $t \in [u,v]$ such that $\frac{f(v)-f(u)}{v-u} = f'(t)$. Therefore, for every partition of $[a,b]$, we obtain

$$\sum_{i=1}^{n} |f(t_i) - f(t_{i-1})| = \sum_{i=1}^{n} |f'(t_i^*)|\,(t_i - t_{i-1}) \leqslant M \sum_{i=1}^{n} (t_i - t_{i-1}) = M\,(b-a),$$

where $t_i^* \in [t_{i-1}, t_i]$, $1 \leqslant i \leqslant n$. Therefore,

$$V_{[a,b]}^{(1)}(f) \leqslant M\,(b-a) < \infty.$$

Moreover, assuming that $|f'|$ is integrable on $[a,b]$, we obtain that

$$V_{[a,b]}^{(1)}(f) = \lim_{n\to\infty} \sum_{i=1}^{n} |f'(t_i^*)|\,(t_i - t_{i-1}) = \int_a^b |f'(t)|\,\mathrm{d}t.$$

3. First, let us find the upper bound of the sum of squared increments:

$$\sum_{i=1}^{n} (f(t_i) - f(t_{i-1}))^2 = \sum_{i=1}^{n} (f'(t_i^*))^2 \, (t_i - t_{i-1})^2$$
$$\leqslant \delta t^{(n)} \, M^2 \sum_{i=1}^{n} (t_i - t_{i-1}) = \delta t^{(n)} \, M^2 \, (b-a).$$

Since $\delta t^{(n)} \to 0$ as $n \to \infty$, the upper bound converges to zero, hence the quadratic variation is zero. □

Now, we turn our attention to Brownian motion. In the theorem that follows, we prove that the first variation of BM is infinite and the second variation is nonzero (almost surely). Therefore, almost surely, a Brownian path cannot be monotone on any time interval since otherwise its first variation on that interval would be finite. Moreover, a Brownian sample path is not differentiable w.r.t. t, since otherwise a differentiable path would have zero quadratic variation. Since Brownian motion is time and space homogeneous, it is sufficient to consider the case with a standard BM on the interval $[0, t]$. Note that it is customary to denote the quadratic variation of the BM process W on $[0, t]$ by $[W, W](t)$.

Theorem 10.4. $V^{(1)}_{[0,t]}(W) = \infty$ *and* $[W, W](t) = t$ *for all* $t > 0$.

Proof. Let us first prove that the quadratic variation of BM on $[0, t]$ is finite and equals t. Consider a partition $0 = t_0 < t_1 < \cdots < t_n = t$ of $[0, t]$. Find the expected value and variance of $V_n = \sum_{i=1}^{n}(W(t_i) - W(t_{i-1}))^2$:

$$\mathrm{E}[V_n] = \sum_{i=1}^{n} \mathrm{E}\left[(W(t_i) - W(t_{i-1}))^2\right] = \sum_{i=1}^{n} (t_i - t_{i-1}) = t$$
$$\mathrm{Var}(V_n) = \sum_{i=1}^{n} \mathrm{Var}\left((W(t_i) - W(t_{i-1}))^2\right) = 2 \sum_{i=1}^{n} (t_i - t_{i-1})^2 \leqslant 2\, t\, \delta t^{(n)},$$

where we use the property $W(t_i) - W(t_{i-1}) \stackrel{d}{=} \sqrt{t_i - t_{i-1}}\, Z$ with $Z \sim Norm(0, 1)$ and hence

$$\mathrm{Var}\left((W(t_i) - W(t_{i-1}))^2\right) = (t_i - t_{i-1})^2 \; \mathrm{Var}(Z^2) = (t_i - t_{i-1})^2 \; \{\mathrm{E}[Z^4] - (\mathrm{E}[Z^2])^2\}$$
$$= (t_i - t_{i-1})^2 \, (3 - 1^2) = 2(t_i - t_{i-1})^2.$$

Therefore, $\mathrm{Var}(V_n) \to 0$, as $n \to \infty$ (i.e., $\delta t^{(n)} \to 0$). The variance of a random variable is zero iff it is (a.s.) a constant. Thus, $[W, W](t) = \lim_{n\to\infty} V_n = t$ (a.s.).

Now, consider the first variation of BM. We find a lower bound of the partial sum of the absolute value of Brownian increments:

$$s_n := \sum_{i=1}^{n} |W(t_i) - W(t_{i-1})| = \sum_{i=1}^{n} \frac{(W(t_i) - W(t_{i-1}))^2}{|W(t_i) - W(t_{i-1})|} \geqslant \frac{\sum_{i=1}^{n}(W(t_i) - W(t_{i-1}))^2}{\sup_{1\leqslant i\leqslant n} |W(t_i) - W(t_{i-1})|}.$$

In the limiting case, $\sum_{i=1}^{n}(W(t_i) - W(t_{i-1}))^2 \to [W, W](t) = t < \infty$, as $n \to \infty$. It can be shown that almost every path of BM is absolutely continuous, therefore (a.s.)

$$\sup_{1\leqslant i\leqslant n} |W(t_i) - W(t_{i-1})| \to 0, \text{ as } n \to \infty.$$

[Note that we are not proving the known absolute continuity of Brownian paths.] Thus, the first variation of BM, given by $\lim_{n\to\infty} s_n$, is infinite since s_n is bounded from below by a ratio converging to ∞ as $n \to \infty$. □

The results of Theorem 10.4 are in agreement with similar properties of the scaled random walk. In (10.7), we proved that $[W^{(n)}, W^{(n)}](t) = t$. So the quadratic variations of the processes W and $W^{(n)}$ are the same. Moreover, one can show that the first variation of $W^{(n)}$ on $[0, t]$ is equal to $\frac{\lfloor nt \rfloor}{\sqrt{n}} = \sqrt{nt}$ (given that nt is an integer). As $n \to \infty$, scaled random walks $W^{(n)}$ converge to Brownian motion and $V^{(1)}_{[0,t]}\left(W^{(n)}\right) = \frac{\lfloor nt \rfloor}{\sqrt{n}} \to \infty = V^{(1)}_{[0,t]}(W)$.

10.3 Some Processes Derived from Brownian Motion

10.3.1 Drifted Brownian Motion

Let μ and $\sigma > 0$ be real constants. A scaled Brownian motion with linear drift, denoted by $W^{(\mu,\sigma)}$, is defined by

$$W^{(\mu,\sigma)}(t) := \mu t + \sigma W(t). \tag{10.23}$$

We call this process (and its extensions described below) *drifted Brownian motion*. The expected value and variance of $W^{(\mu,\sigma)}(t)$ are:

$$\begin{aligned}
\mathrm{E}\left[W^{(\mu,\sigma)}(t)\right] &= \mathrm{E}[\mu t + \sigma W(t)] = \mu t + \sigma \mathrm{E}[W(t)] = \mu t, \\
\mathrm{Var}\left(W^{(\mu,\sigma)}(t)\right) &= \mathrm{Var}(\mu t + \sigma W(t)) = \sigma^2 \,\mathrm{Var}(W(t)) = \sigma^2 t.
\end{aligned}$$

Since $W(t)$ is normally distributed, the sum $\mu t + \sigma W(t)$ is a normal random variable as well. Moreover, any increment of drifted Brownian motion, $W^{(\mu,\sigma)}(t) - W^{(\mu,\sigma)}(s)$, with $0 \leqslant s < t$, is normally distributed:

$$\begin{aligned}
W^{(\mu,\sigma)}(t) - W^{(\mu,\sigma)}(s) &= \mu\,(t-s) + \sigma\,(W(t) - W(s)) \\
&\stackrel{d}{=} \mu\,(t-s) + \sigma\,\sqrt{t-s}\,Z \sim Norm(\mu\,(t-s), \sigma^2\,(t-s)),
\end{aligned}$$

where $Z \sim Norm(0,1)$. Let us find the transition probability law of the drifted BM. Since Brownian motion is a homogeneous process, it is sufficient to find the probability distribution of the time-t value of a BM with drift:

$$\mathbb{P}\left(W^{(\mu,\sigma)}(t) \leqslant x\right) = \mathbb{P}\left(\mu t + \sigma W(t) \leqslant x\right) = \mathbb{P}\left(\frac{W(t)}{\sqrt{t}} \leqslant \frac{x - \mu t}{\sigma\sqrt{t}}\right) = \mathcal{N}\left(\frac{x - \mu t}{\sigma\sqrt{t}}\right).$$

The PDF of $W^{(\mu,\sigma)}(t)$ is then

$$p_0^{(\mu,\sigma)}(t;x) := \frac{\partial}{\partial x}\mathbb{P}\left(W^{(\mu,\sigma)}(t) \leqslant x\right) = \frac{\partial}{\partial x}\mathcal{N}\left(\frac{x-\mu t}{\sigma\sqrt{t}}\right) = \frac{1}{\sigma\sqrt{t}}\,n\left(\frac{x-\mu t}{\sigma\sqrt{t}}\right).$$

Therefore, the transition probability distribution function and transition PDF for the process $\{W^{(\mu,\sigma)}(t), t \geqslant 0\}$ are, respectively,

$$\begin{aligned}
P(s,t;x,y) &= \mathbb{P}\left(W^{(\mu,\sigma)}(t) \leqslant y \mid W^{(\mu,\sigma)}(s) = x\right) = \mathbb{P}\left(W^{(\mu,\sigma)}(t-s) \leqslant y - x\right) \\
&= \mathcal{N}\left(\frac{y - x - \mu(t-s)}{\sigma\sqrt{t-s}}\right), \\
p(s,t;x,y) &= \frac{\partial}{\partial y}\mathcal{N}\left(\frac{y-x-\mu(t-s)}{\sigma\sqrt{t-s}}\right) = \frac{1}{\sigma\sqrt{t-s}}\,n\left(\frac{y-x-\mu(t-s)}{\sigma\sqrt{t-s}}\right) \\
&= p_0^{(\mu,\sigma)}(t-s; y-x),
\end{aligned}$$

for $0 \leqslant s < t$ and $x, y \in \mathbb{R}$. Clearly, the transition probability distribution is the same if the underlying BM is not a standard one and is starting at some nonzero point. A drifted BM, $\{X(t), t \geqslant 0\}$, starting at $x_0 \in \mathbb{R}$, is

$$X(t) := x_0 + W^{(\mu,\sigma)}(t) = x_0 + \mu t + \sigma W(t). \tag{10.24}$$

Another possible extension is the case where the drift and scale parameters are generally time-dependent functions, $\mu = \mu(t)$ and $\sigma = \sigma(t)$. The drifted BM starting at x_0 takes the following form:

$$X(t) := x_0 + \mu(t)t + \sigma(t)W(t). \tag{10.25}$$

10.3.2 Geometric Brownian Motion

Geometric Brownian motion (GBM) is the most well-known stochastic model for modelling positive asset price processes and pricing contingent claims. It is a keystone of the Black–Scholes–Merton theory of option pricing. However, from the mathematical point of view, the GBM is just an exponential function of a drifted Brownian motion. For constant drift μ and volatility parameter σ, GBM is defined by

$$S(t) := \mathrm{e}^{x_0 + W^{(\mu,\sigma)}(t)} = \mathrm{e}^{x_0 + \mu t + \sigma W(t)} = S_0\, \mathrm{e}^{\mu t + \sigma W(t)}, \quad t \geqslant 0, \quad S_0 > 0. \tag{10.26}$$

The process starts at $S_0 = \mathrm{e}^{x_0}$, since $W^{(\mu,\sigma)}(0) = W(0) = 0$. This process is hence strictly positive for all $t \geqslant 0$. Since $S(t)$ is an exponential function of a normal variable, the probability distribution of the time-t realization of GBM is log-normal. The CDF of $S(t)$ is

$$\mathbb{P}(S(t) \leqslant x) = \mathbb{P}\left(S_0\, \mathrm{e}^{\mu t + \sigma W(t)} \leqslant x\right) = \mathbb{P}\left(W(t) \leqslant \frac{\ln(x/S_0) - \mu t}{\sigma}\right) = \mathcal{N}\left(\frac{\ln(x/S_0) - \mu t}{\sigma\sqrt{t}}\right),$$

for $x > 0$. The PDF of $S(t)$ can be obtained by differentiating the above CDF:

$$f_{S(t)}(x) = \frac{\partial}{\partial x}\mathcal{N}\left(\frac{\ln(x/S_0) - \mu t}{\sigma\sqrt{t}}\right) = \frac{1}{x\sigma\sqrt{t}}\, n\left(\frac{\ln(x/S_0) - \mu t}{\sigma\sqrt{t}}\right) = \frac{1}{x\sigma\sqrt{2\pi t}}\mathrm{e}^{-\frac{(\ln(x/S_0) - \mu t)^2}{2\sigma^2 t}}.$$

By combining the expression (10.26) written for $S(t_1)$ and $S(t_0)$ with $0 \leqslant t_0 < t_1$, we can express $S(t_1)$ in terms of $S(t_0)$ as follows:

$$S(t_1) = S(t_0)\, \mathrm{e}^{\mu\,(t_1 - t_0) + \sigma\,(W(t_1) - W(t_0))} \stackrel{d}{=} S(t_0)\, \mathrm{e}^{\mu\,(t_1 - t_0) + \sigma\, W(t_1 - t_0)},$$

where $W(t_1 - t_0) \stackrel{d}{=} \sqrt{t_1 - t_0}\, Z$, $Z \sim \mathit{Norm}(0, 1)$. This proves that GBM is a time-homogeneous process and its transition PDF is, for any $0 \leqslant s < t$, $x, y > 0$,

$$p(s, t; x, y) = f_{S(t)|S(s)}(y|x) = \frac{1}{y\sigma\sqrt{2\pi(t-s)}} \exp\left(-\frac{[\ln(y/x) - \mu(t-s)]^2}{2\sigma^2(t-s)}\right). \tag{10.27}$$

10.3.3 Brownian Bridge

A bridge process is obtained by conditioning on the value of the original process at some future time. For example, consider a standard Brownian motion *pinned at the origin* at some time $T > 0$, i.e., $W(0) = W(T) = 0$. As a result, we obtain a continuous time process defined for time $t \in [0, T]$ such that its probability distribution is the distribution of a standard BM conditional on $W(T) = 0$. This process is called a *standard Brownian bridge*. We shall denote it by $B^{(0,0)}_{[0,T]}$ or simply B. Figure 10.5 depicts some sample paths of this process for $T = 1$. The process can be expressed in terms of a standard BM as follows:

$$B(t) = W(t) - \frac{t}{T}W(T), \quad 0 \leqslant t \leqslant T. \tag{10.28}$$

First, we show that it satisfies the boundary condition $B(0) = B(T) = 0$:

$$B(0) = W(0) - \frac{0}{T}W(T) = 0, \quad B(T) = W(T) - \frac{T}{T}W(T) = W(T) - W(T) = 0.$$

Second, we show that realizations of the Brownian bridge have the same probability distribution as those of the Brownian motion $W(t)$ conditional on $W(T) = 0$. For all $t \in (0, T)$, the time-t realization $B(t)$ is normally distributed as a linear combination of normal random variables. The mean and variance are, respectively,

$$\mathrm{E}\left[B(t)\right] = \mathrm{E}\left[W(t) - \frac{t}{T}W(T)\right] = 0,$$

$$\mathrm{Var}\left(B(t)\right) = \mathrm{E}\left[\left(W(t) - \frac{t}{T}W(T)\right)^2\right] = \frac{t(T-t)}{T}.$$

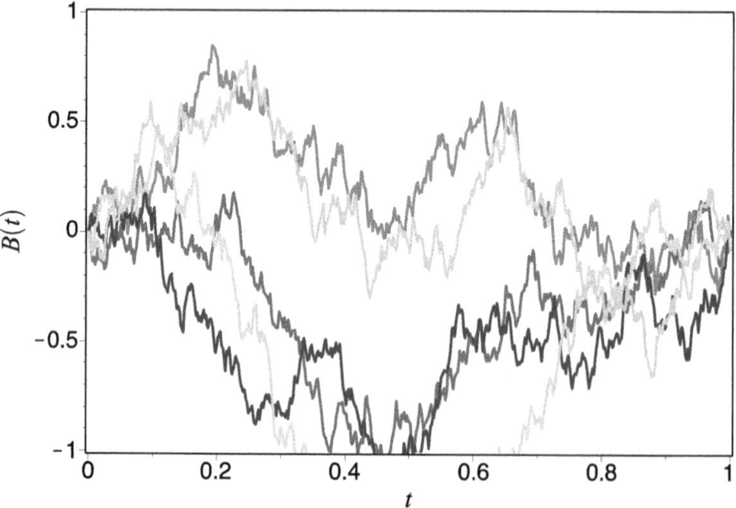

FIGURE 10.5: Sample paths of a standard Brownian bridge for $t \in [0, 1]$. Note that all paths begin and end at value zero.

Let us find the conditional distribution of $W(t) \mid W(T)$. Since $[W(t), W(T)]$ is jointly normally distributed with zero mean vector and covariance matrix $\begin{bmatrix} t & t \\ t & T \end{bmatrix}$, the conditional distribution is again normal. Applying (10.5) gives the probability distribution of $W(t)$, given $W(T) = y$, as normal with mean $y\frac{t}{T}$ and variance $\frac{t(T-t)}{T}$, i.e., $B(t) \sim N(y\frac{t}{T}, \frac{t(T-t)}{T})$. Therefore, for the standard Brownian bridge (pinned at the origin: $W(0) = W(T) = y = 0$),

$$B(t) \sim Norm\left(0, \frac{t(T-t)}{T}\right), \quad 0 < t < T. \tag{10.29}$$

On the other hand, the conditional PDF for $W(t) \mid W(T)$ can be expressed in terms of the transition PDF of BM given in (10.20) as follows:

$$\begin{aligned}
f_{W(t)|W(T)}(x \mid y) &= \frac{f_{W(t),W(T)}(x,y)}{f_{W(T)}(y)} = \frac{f_{W(t)}(x)\, f_{W(T)|W(t)}(y \mid x)}{f_{W(T)}(y)} \\
&= \frac{p_0(t;x)\, p_0(T-t;y-x)}{p_0(T;y)} \\
&= \frac{\frac{1}{\sqrt{2\pi t}} \exp\left(-\frac{x^2}{2t}\right) \frac{1}{\sqrt{2\pi(T-t)}} \exp\left(-\frac{(y-x)^2}{2(T-t)}\right)}{\frac{1}{\sqrt{2\pi T}} \exp\left(-\frac{y^2}{2T}\right)} \\
&= \frac{1}{\sqrt{2\pi \frac{t(T-t)}{T}}} \exp\left(-\frac{1}{2}\left(\frac{x^2}{t} + \frac{(y-x)^2}{T-t} - \frac{y^2}{T}\right)\right).
\end{aligned}$$

Simplifying the exponent in the above expression gives

$$\begin{aligned}
\frac{x^2}{t} + \frac{(y-x)^2}{T-t} - \frac{y^2}{T} &= \frac{x^2T(T-t) + (y-x)^2tT - y^2t(T-t)}{tT(T-t)} \\
&= \frac{x^2T^2 - 2xytT + y^2t^2}{tT(T-t)} = \frac{(xT-yt)^2}{tT(T-t)} \\
&= \frac{(x-yt/T)^2}{t(T-t)/T}.
\end{aligned}$$

Finally, we obtain

$$f_{W(t)|W(T)}(x \mid y) = \frac{1}{\sqrt{2\pi \frac{t(T-t)}{T}}} \exp\left(-\frac{1}{2}\frac{\left(x - y\frac{t}{T}\right)^2}{\frac{t(T-t)}{T}}\right) = \frac{1}{\sqrt{\frac{t(T-t)}{T}}}\, n\left(\frac{x - y\frac{t}{T}}{\sqrt{\frac{t(T-t)}{T}}}\right). \tag{10.30}$$

Again, we conclude that $W(t) \mid \{W(T) = y\} \sim Norm\left(y\frac{t}{T}, \frac{t(T-t)}{T}\right)$ for $0 < t < T$.

In general, the Brownian bridge from a to b on $[0,T]$, denoted by $B^{(a,b)}_{[0,T]}(t)$, is obtained by adding a linear drift function to a standard Brownian bridge (from 0 to 0):

$$B^{(a,b)}_{[0,T]}(t) = a + \frac{(b-a)t}{T} + B^{(0,0)}_{[0,T]}(t) = a + \frac{(b-a)t}{T} + W(t) - \frac{t}{T}W(T), \quad t \in [0,T]. \tag{10.31}$$

Since adding a nonrandom function to a normally distributed random variable again gives a normal random variable, the realizations of a Brownian bridge are normally distributed:

$$B^{(a,b)}_{[0,T]}(t) \sim Norm\left(a + \frac{(b-a)t}{T}, \frac{t(T-t)}{T}\right), \quad t \in (0,T).$$

10.3.4 Gaussian Processes

Most of the continuous-time processes we have considered so far belong to the class of Gaussian processes. The most well-known representative of such a class is Brownian motion. Other examples of Gaussian processes include BM with drift and the Brownian bridge process. Gaussian processes are so named because their realizations have a normal probability distribution (which is also called a Gaussian distribution).

Definition 10.5. A continuous-time process $\{X(t)\}_{t \geqslant 0}$ is called a *Gaussian process*, if for every partition $0 \leqslant t_1 < t_2 < \cdots < t_n$, the random variables $X(t_1), X(t_2), \ldots, X(t_n)$ are jointly normally distributed.

The probability distribution of a normal vector $\mathbf{X} = \left[X(t_1), X(t_2), \ldots, X(t_n)\right]^\top$ is determined by the mean vector and covariance matrix. Thus, a Gaussian process is determined by the *mean value function* defined by $m_X(t) := \mathrm{E}[X(t)]$ and *covariance function* defined by $c_X(t,s) := \mathrm{Cov}(X(t), X(s)) = \mathrm{E}[(X(t) - m_X(t))(X(s) - m_X(s))]$, for $t, s \geqslant 0$. The probability distribution of the time-t realization is normal: $X(t) \sim Norm\,(m_X(t), c_X(t,t))$ for all $t > 0$. The probability distribution of the vector $\mathbf{X}$ is $Norm_n(\boldsymbol{\mu}, \mathbf{C})$ with

$$\boldsymbol{\mu} = \left[m_X(t_1), m_X(t_2), \ldots, m_X(t_n)\right]^\top \text{ and } \mathbf{C}_{ij} = c_X(t_i, t_j),\ 1 \leqslant i, j \leqslant n.$$

The function $m_X(t)$ defines a curve on the time-space plane where the sample paths of $\{X(t)\}$ concentrate around it.

Examples of Gaussian processes include the following processes:

1. A constant distribution process $X(t) \equiv Z$, where $Z \sim Norm(a, b^2)$, with $m_X(t) = a$ and $c_X(t,s) = b^2$;

2. A piecewise constant process $Y(t) = Z_{\lfloor t \rfloor}$, where $Z_k \sim Norm(a_k, b_k^2)$, $k \in \mathbb{N}_0$, are independent random variables, with $m_Y(t) = a_{\lfloor t \rfloor}$ and $c_Y(t,s) = b_{\lfloor t \rfloor} b_{\lfloor s \rfloor} \mathbb{I}_{\{\lfloor t \rfloor = \lfloor s \rfloor\}}$, i.e., the covariance function is zero unless $\lfloor s \rfloor = \lfloor t \rfloor$, in which case $c_Y(t,s) = b_{\lfloor t \rfloor}^2$.

Example 10.5. Let $\{Z_k\}_{k \in \mathbb{N}}$ be i.i.d. standard normal random variables and define $X(t) = \sum_{k=1}^{\lfloor t \rfloor} Z_k$, $t \geqslant 0$. Show that $\{X(t)\}_{t \geqslant 0}$ is a Gaussian process. Find its mean function and covariance function.

Solution. For any time $t \geqslant 0$, $X(t)$ is a sum of standard normals, hence $X(t)$ is normal. Therefore, any finite-dimensional distribution of the process X is multivariate normal as well. The expected value and variance of $X(t)$ are

$$\mathrm{E}[X(t)] = \sum_{k=1}^{\lfloor t \rfloor} \mathrm{E}[Z_k] = 0 \text{ and } \mathrm{Var}(X(t)) = \sum_{k=1}^{\lfloor t \rfloor} \mathrm{Var}(Z_k) = \lfloor t \rfloor.$$

For $0 \leqslant s < t$ we have

$$X(t) = X(s) + \sum_{k=\lfloor s \rfloor + 1}^{\lfloor t \rfloor} Z_k.$$

Thus, the covariance of $X(s)$ and $X(t)$ is

$$\mathrm{Cov}(X(s), X(t)) = \mathrm{Cov}(X(s), X(s)) + \mathrm{Cov}\left(X(s), \sum_{k=\lfloor s \rfloor + 1}^{\lfloor t \rfloor} Z_k\right) = \mathrm{Var}(X(s)) + 0 = \lfloor s \rfloor.$$

The above argument applies similarly if we assume $s \geqslant t$, giving $\mathrm{Cov}(X(s), X(t)) = \lfloor t \rfloor$. Thus, $m_X(t) = 0$ and $c_X(s,t) = \lfloor s \wedge t \rfloor$. □

Clearly, Brownian motion and some other processes derived from it are Gaussian processes.

- Standard Brownian motion is a Gaussian process with $m(t) = 0$ and $c(t,s) = t \wedge s$, for $s, t \geqslant 0$. Moreover, a Gaussian process having such covariance and mean functions and continuous sample paths is a standard Brownian motion (see Exercise 10.18).

- Brownian motion starting at $x \in \mathbb{R}$ is a Gaussian process with $m(t) = x$ and $c(s,t) = s \wedge t$, $s, t \geqslant 0$.

- Clearly, adding a deterministic function to a Gaussian process produces another Gaussian process. Therefore, the drifted Brownian motion starting at $x \in \mathbb{R}$, $\{W^{(\mu,\sigma)}(t) = x+\mu t+\sigma W(t) \,:\, t \geqslant 0\}$, is a Gaussian process with $m(t) = x+\mu t$ and $c(s,t) = \sigma^2\,(s \wedge t)$, $s,t \geqslant 0$. The drifted BM with time-dependent drift $\mu(t)$ and volatility $\sigma(t)$ is a Gaussian process where $m(t) = x + \mu(t)t$ and $c(s,t) = \sigma(s)\sigma(t)\,(s \wedge t)$.

- The standard Brownian bridge from 0 to 0 on $[0,T]$ is a Gaussian process with mean $m(t) = 0$ and $c(s,t) = s \wedge t - \frac{st}{T}$, $s,t \in [0,T]$.

Proof. The standard Brownian bridge is the process $B^{(0,0)}_{[0,T]}(t) = W(t) - \frac{t}{T}W(T)$. For $s,t \in [0,T]$, we simply calculate the mean and covariance functions as follows:

$$
\begin{aligned}
m(t) &= \mathrm{E}\left[W(t) - \frac{t}{T}W(T)\right] = \mathrm{E}[W(t)] - \frac{t}{T}\mathrm{E}[W(T)] = 0\,, \\
c(s,t) &= \mathrm{E}\left[\left(W(s) - \frac{s}{T}W(T)\right)\left(W(t) - \frac{t}{T}W(T)\right)\right] \\
&= \mathrm{E}[W(s)\,W(t)] - \frac{t}{T}\mathrm{E}[W(s)\,W(T)] - \frac{s}{T}\mathrm{E}[W(t)\,W(T)] + \frac{st}{T^2}\mathrm{E}[W^2(T)] \\
&= s \wedge t - \frac{st}{T} - \frac{st}{T} + \frac{stT}{T^2} = s \wedge t - \frac{st}{T}\,.
\end{aligned}
$$

□

- The general Brownian bridge from a to b on $[0,T]$ is a Gaussian process with $m(t) = a + \frac{(b-a)t}{T}$ and $c(s,t) = s \wedge t - \frac{st}{T}$, $s,t \in [0,T]$. The proof is left as an exercise for the reader (see Exercise 10.19).

- Geometric Brownian motion is **not** a Gaussian process, but the logarithm of GBM is so (see Exercise 10.20).

10.4 First Hitting Times and Maximum and Minimum of Brownian Motion

From Section 6.3.7, we recall the discussion and basic definitions of first passage times for a stochastic process. We are now interested in developing various formulae for the distribution of first passage times, the distribution of the sampled maximum (or minimum) up to a given time $t > 0$, as well as the joint distribution of the sampled maximum (or minimum) and the process value at a given time $t > 0$ for standard BM as well as translated and scaled BM with drift.

10.4.1 The Reflection Principle: Standard Brownian Motion

Let us recall some basic definitions from Section 6.3.7 that we now specialize to standard BM, $W = \{W(t)\}_{t\geqslant 0}$, under a given (fixed) measure $\mathbb{P}$. The first hitting time, to a given level $m \in \mathbb{R}$, for standard BM is defined by

$$
\mathcal{T}_m := \inf\{t \geqslant 0 \,:\, W(t) = m\}\,. \tag{10.32}
$$

We recall that either terminology, first hitting time or first passage time, is equivalent since the paths of W are continuous (a.s.). In what follows we can equally say that $\mathcal{T}_m$ is a first hitting time or a first passage time to level m. The sampled maximum and minimum of W, from time 0 to time $t \geqslant 0$, are respectively denoted by

$$M(t) := \sup_{0 \leqslant u \leqslant t} W(u) \tag{10.33}$$

and

$$m(t) := \inf_{0 \leqslant u \leqslant t} W(u) . \tag{10.34}$$

Note: $W(0) = M(0) = m(0) = 0$, $M(t) \geqslant 0$ and $M(t)$ is increasing in t. Similarly, $m(t) \leqslant 0$ and $m(t)$ is decreasing in t. If $m > 0$, then $\mathcal{T}_m$ is the same as the first hitting time up to level m. If $m < 0$, then $\mathcal{T}_m$ is the same as the first hitting time down to level m. Hence, we have the equivalence of events:

$$\{\mathcal{T}_m \leqslant t\} = \{M(t) \geqslant m\}, \text{ if } m > 0; \quad \{\mathcal{T}_m \leqslant t\} = \{m(t) \leqslant m\}, \text{ if } m < 0. \tag{10.35}$$

The path symmetries and Markov (memoryless) properties of standard BM are key ingredients in what follows. We already know that BM is time and space homogeneous. In particular, $\{W(t+s) - W(s)\}_{t \geqslant 0}$ is a standard BM for any fixed time $s \geqslant 0$. A more general version of this is the property that $\{W(t + \mathcal{T}) - W(\mathcal{T})\}_{t \geqslant 0}$ is a standard BM, independent of $\{W(u) : 0 \leqslant u \leqslant \mathcal{T}\}$, where $\mathcal{T}$ is a stopping time w.r.t. a filtration for BM. We don't prove this latter property but it is a consequence of what is known as the *strong Markov property* of BM. Based on this property, the distribution of $\mathcal{T}_m$ is now easily derived.

Proposition 10.5 (First Hitting Time Distribution for Standard BM). *The cumulative distribution function of the first hitting time in (10.32), for a level $m \neq 0$, is given by*

$$\mathbb{P}(\mathcal{T}_m \leqslant t) = 2\,\mathbb{P}(W(t) \geqslant |m|) = 2\left[1 - \mathcal{N}\left(\frac{|m|}{\sqrt{t}}\right)\right], \quad 0 < t < \infty, \tag{10.36}$$

and zero for all $t \leqslant 0$.

Proof. We prove the result for $|m| = m > 0$, as the formula follows for $m < 0$ ($-m = |m|$) by simple reflection symmetry of BM where $-W$ is also a standard BM. If a BM path lies above m at time t, this implies that the path has already hit level m by time t. By the above strong Markov property of BM, $\{W(t + \mathcal{T}_m) - W(\mathcal{T}_m)\}_{t \geqslant 0}$ is a standard BM. By spatial symmetry of BM paths, given $t > \mathcal{T}_m$, then $W(t) - W(\mathcal{T}_m) \sim Norm(0, t - \mathcal{T}_m)$, i.e., we have the conditional probability $\mathbb{P}(W(t) - W(\mathcal{T}_m) > 0 \mid \mathcal{T}_m < t) = \frac{1}{2}$. Now, by continuity of BM, note that the event that BM is greater than level m at time t, i.e., $W(t) > m$, is the same as the joint event that it already hit level m, $\mathcal{T}_m < t$, and that $W(t) > W(\mathcal{T}_m)$. So, we have the equivalence of probabilities:

$$\begin{aligned} \mathbb{P}(W(t) > m) &= \mathbb{P}(W(t) - W(\mathcal{T}_m) > 0, \mathcal{T}_m < t) \\ &= \mathbb{P}(W(t) - W(\mathcal{T}_m) > 0 \mid \mathcal{T}_m < t)\,\mathbb{P}(\mathcal{T}_m < t) \\ &= \frac{1}{2}\mathbb{P}(\mathcal{T}_m < t) . \end{aligned}$$

This gives (10.36) for $m > 0$ since $\mathbb{P}(W(t) > m) = 1 - \mathbb{P}(W(t) \leqslant m) = 1 - \mathcal{N}(m/\sqrt{t})$. □

Note: $W(t)$ and $\mathcal{T}_m$ are continuous random variables, i.e., $\mathbb{P}(W(t) > m) = \mathbb{P}(W(t) \geqslant m)$ and $\mathbb{P}(\mathcal{T}_m < t) = \mathbb{P}(\mathcal{T}_m \leqslant t)$.

From (10.36) we see that, given an infinite time, BM will eventually hit any finite fixed level with probability one, i.e.,

$$\mathbb{P}(\mathcal{T}_m < \infty) = \lim_{t\to\infty} \mathbb{P}(\mathcal{T}_m \leqslant t) = 2\left[1 - \lim_{t\to\infty} \mathcal{N}\left(\frac{|m|}{\sqrt{t}}\right)\right] = 2\left[1 - \mathcal{N}(0)\right] = 2\left[1 - \frac{1}{2}\right] = 1\,.$$

In the opposing limit, given any $m \neq 0$,

$$\mathbb{P}(\mathcal{T}_m = 0) = \lim_{t\searrow 0} \mathbb{P}(\mathcal{T}_m \leqslant t) = 2\left[1 - \lim_{t\to 0} \mathcal{N}\left(\frac{|m|}{\sqrt{t}}\right)\right] = 2[1 - \mathcal{N}(\infty)] = 2[1-1] = 0.$$

Hence, the function defined by $F_{\mathcal{T}_m}(t) := \mathbb{P}(\mathcal{T}_m \leqslant t)$ is a proper CDF. The PDF (density) of the first hitting time is given by differentiating (10.36) w.r.t. t, $f_{\mathcal{T}_m}(t) := \frac{\partial}{\partial t} F_{\mathcal{T}_m}(t) \equiv \frac{\partial}{\partial t}\mathbb{P}(\mathcal{T}_m \leqslant t)$:

$$f_{\mathcal{T}_m}(t) = \frac{|m|}{t\sqrt{t}}\, n\left(\frac{|m|}{\sqrt{t}}\right) = \frac{|m|}{t\sqrt{2\pi t}} \mathrm{e}^{-m^2/2t}\,,\quad 0 < t < \infty. \tag{10.37}$$

An important symmetry, which we state without any proof, is called the *reflection principle* for standard BM. This states that, given a stopping time $\mathcal{T}$ (w.r.t. a filtration for BM), the process $\mathcal{W} = \{\mathcal{W}(t)\}_{t\geqslant 0}$ defined by

$$\mathcal{W}(t) := \begin{cases} W(t) & \text{for } t \leqslant \mathcal{T}, \\ 2W(\mathcal{T}) - W(t) & \text{for } t > \mathcal{T}, \end{cases}$$

is also a standard BM. In particular, if we let $\mathcal{T} = \mathcal{T}_m$, then $W(\mathcal{T}) = W(\mathcal{T}_m) = m$ and every path of $\mathcal{W}$ is a path of W up to the first hitting time $\mathcal{T}_m$ to level m and the reflection of a path of W for time $t > \mathcal{T}_m$. Since $\mathcal{W}$ and W are equivalent realizations of standard BM, this means that every path of a standard Brownian has a reflected path. This is depicted in Figure 10.6.

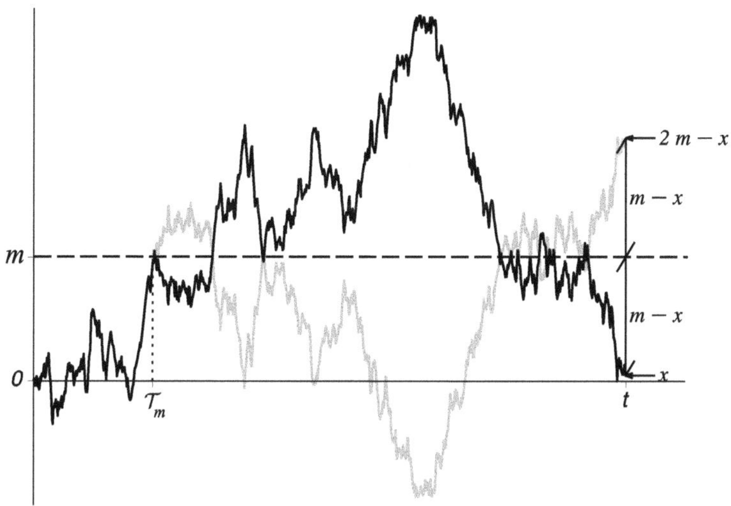

FIGURE 10.6: A sample path of standard BM reaching level $m > 0$ at the first hitting time $\mathcal{T}_m$ and its reflected path about level m for times past $\mathcal{T}_m$.

Based on the reflection principle, we can now obtain the probability of the joint event that BM at time t, $W(t)$, is below a value $x \in \mathbb{R}$ and the sampled maximum of BM in the interval $[0,t]$, $M(t)$, is above level $m > 0$, where $x \leqslant m$. This then leads to the joint distribution for the pair of random variables $M(t), W(t)$.

Proposition 10.6 (Joint Distribution of $M(t), W(t)$). *For all $m > 0$, $-\infty < x \leqslant m$, $t \in (0,\infty)$, we have*

$$\mathbb{P}(M(t) \geqslant m, W(t) \leqslant x) = \mathbb{P}(W(t) \geqslant 2m - x) = 1 - \mathcal{N}\left(\frac{2m-x}{\sqrt{t}}\right). \tag{10.38}$$

Hence, for all $t \in (0,\infty)$, the joint PDF of $(M(t), W(t))$ is

$$f_{M(t),W(t)}(m,x) = \frac{2(2m-x)}{t\sqrt{2\pi t}} \mathrm{e}^{-(2m-x)^2/2t}, \quad \text{for } x \leqslant m,\ m > 0, \tag{10.39}$$

and zero for all other values of m, x.

Proof. As depicted in Figure 10.6, observe that for every Brownian path that hits level m before $t > 0$ and has value $x \leqslant m$ at time t there is a (reflected) Brownian path that has level $2m - x$. Indeed, by the reflection principle (taking $\mathcal{T} = \mathcal{T}_m$, so $\mathcal{W}(t) = 2m - W(t)$ for $t > \mathcal{T}_m$) and the equivalence in (10.35) for $m > 0$:

$$\begin{aligned}
\mathbb{P}(M(t) \geqslant m, W(t) \leqslant x) &= \mathbb{P}(\mathcal{T}_m \leqslant t, W(t) \leqslant x) \\
&= \mathbb{P}(\mathcal{T}_m \leqslant t, 2m - W(t) \geqslant 2m - x) \\
&= \mathbb{P}(\mathcal{T}_m \leqslant t, \mathcal{W}(t) \geqslant 2m - x) \\
&= \mathbb{P}(\mathcal{W}(t) \geqslant 2m - x).
\end{aligned}$$

The last equality obtains as follows. Since $x \leqslant m$, then $2m - x \geqslant 2m - m = m$, which implies $\mathcal{W}(t) \geqslant m$, i.e., the reflected BM and hence the BM has already reached level m. This implies that the joint probability is just the probability of the event $\{\mathcal{W}(t) \geqslant 2m - x\}$ and so (10.38) obtains since $\mathcal{W}$ and W are both standard BM. The formula in (10.39) follows by the standard definition of the joint PDF as the second mixed partial derivative of the joint CDF. Note the minus sign here since we are directly differentiating the joint probability in (10.38), which involves $\{M(t) \geqslant m\}$ instead of $\{M(t) \leqslant m\}$:

$$\begin{aligned}
f_{M(t),W(t)}(m,x) &= -\frac{\partial^2}{\partial m \partial x} \mathbb{P}(M(t) \geqslant m, W(t) \leqslant x) \\
&= \frac{\partial}{\partial m} \frac{\partial}{\partial x} \mathcal{N}\left(\frac{2m-x}{\sqrt{t}}\right) \\
&= \frac{-1}{\sqrt{t}} \frac{\partial}{\partial m} n\left(\frac{2m-x}{\sqrt{t}}\right) \\
&= \frac{-1}{\sqrt{t}} \left[\frac{-2(2m-x)}{t}\right] n\left(\frac{2m-x}{\sqrt{t}}\right),
\end{aligned}$$

$n(z) \equiv \frac{\mathrm{e}^{-z^2/2}}{\sqrt{2\pi}}$, which is the right-hand side of (10.39). □

The joint CDF of $(M(t), W(t))$ now follows easily by writing the event $\{W(t) \leqslant x\}$ as a union of two mutually exclusive events,

$$\{W(t) \leqslant x\} = \{M(t) > m, W(t) \leqslant x\} \cup \{M(t) \leqslant m, W(t) \leqslant x\},$$

and equating probabilities on both sides,

$$\mathbb{P}(W(t) \leqslant x) = \mathbb{P}(M(t) > m, W(t) \leqslant x) + \mathbb{P}(M(t) \leqslant m, W(t) \leqslant x).$$

Isolating the second term on the right-hand side and using (10.38) for the first term on the right-hand side (where $\mathbb{P}(M(t) > m, W(t) \leqslant x) = \mathbb{P}(M(t) \geqslant m, W(t) \leqslant x)$ since $M(t)$ is continuous) gives the joint CDF for $m > 0, x \leqslant m$,

$$F_{M(t),W(t)}(m,x) := \mathbb{P}(M(t) \leqslant m, W(t) \leqslant x) = \mathbb{P}(W(t) \leqslant x) - \mathbb{P}(W(t) \geqslant 2m - x)$$
$$= \mathcal{N}\left(\frac{x}{\sqrt{t}}\right) - \mathcal{N}\left(\frac{x-2m}{\sqrt{t}}\right), \qquad (10.40)$$

and $F_{M(t),W(t)}(m,x) = F_{M(t)}(m)$ for $x > m > 0$, $F_{M(t),W(t)}(m,x) \equiv 0$ for $m \leqslant 0$, for all $t \in (0,\infty)$. We also see that this joint CDF recovers the joint PDF in (10.39). For $m > 0$, $x \leqslant m$, it is given by simply differentiating (10.40),

$$f_{M(t),W(t)}(m,x) := \frac{\partial^2}{\partial m \partial x} F_{M(t),W(t)}(m,x) = \frac{\partial^2}{\partial m \partial x} \mathcal{N}\left(\frac{2m-x}{\sqrt{t}}\right),$$

and $f_{M(t),W(t)}(m,x) \equiv 0$ for all other values of (m,x).

The (marginal) CDF of $M(t)$ follows by setting $x = m$ in (10.40),

$$F_{M(t)}(m) := \mathbb{P}(M(t) \leqslant m) = \mathbb{P}(M(t) \leqslant m, W(t) \leqslant m) = 2\mathcal{N}\left(\frac{m}{\sqrt{t}}\right) - 1, \qquad (10.41)$$

for $m > 0$ and zero for $m \leqslant 0$. Note that this is a CDF for the continuous random variable $M(t)$ where $F_{M(t)}$ is continuous for $m \in \mathbb{R}$, monotonically increasing for $m > 0$, $F_{M(t)}(-\infty) = 0$ and $F_{M(t)}(\infty) = 1$. For $m > 0$, $\{\mathcal{T}_m \leqslant t\} \equiv \{M(t) \geqslant m\}$,

$$F_{\mathcal{T}_m}(t) = \mathbb{P}(M(t) \geqslant m) = 1 - \mathbb{P}(M(t) \leqslant m) = 1 - F_{M(t)}(m) = 2 - 2\mathcal{N}\left(\frac{m}{\sqrt{t}}\right), \ t > 0.$$

This therefore recovers the CDF of $\mathcal{T}_m$ in (10.36) in the case that $m > 0$.

The next proposition follows simply from Proposition 10.6 and leads to the joint distribution of the pair $(m(t), W(t))$, i.e., the sampled minimum of BM in the interval $[0,t]$, $m(t)$, and the value of standard BM at time t.

Proposition 10.7 (Joint Distribution of $m(t), W(t)$)**.** *For all $m < 0$, $m \leqslant x < \infty$, $t \in (0,\infty)$, we have*

$$\mathbb{P}(m(t) \leqslant m, W(t) \geqslant x) = \mathbb{P}(W(t) \geqslant x - 2m) = 1 - \mathcal{N}\left(\frac{x-2m}{\sqrt{t}}\right). \qquad (10.42)$$

Hence, for all $t \in (0,\infty)$, the joint PDF of $(m(t), W(t))$ is

$$f_{m(t),W(t)}(m,x) = \frac{2(x-2m)}{t\sqrt{2\pi t}} \mathrm{e}^{-(x-2m)^2/2t}, \ \textit{for } x \geqslant m, \ m < 0, \qquad (10.43)$$

and zero otherwise.

Proof. One way to prove the result is to apply the same steps as in the proof of Proposition 10.6, but with arguments applying to a picture that is the reflection (about the vertical axis) of the picture in Figure 10.6. Here we simply make use of the fact that reflected BM defined by $\{B(t) := -W(t)\}_{t\geqslant 0}$ is also a standard BM. Let $M_B(t) := \sup_{0\leqslant u\leqslant t} B(u)$. Note that $-M_B(t) = -\sup_{0\leqslant u\leqslant t}(-W(u)) = \inf_{0\leqslant u\leqslant t} W(u) \equiv m(t)$. Since $\{B(t)\}_{t\geqslant 0}$ is a standard BM, by Proposition 10.6, with values $m' > 0$, $x' \leqslant m'$, $t > 0$, we have

$$\mathbb{P}\left(M_B(t) \geqslant m', B(t) \leqslant x'\right) = \mathbb{P}(B(t) \geqslant 2m' - x') = 1 - \mathcal{N}\left(\frac{2m'-x'}{\sqrt{t}}\right).$$

The left-hand side is re-expressed as:

$$\begin{aligned}\mathbb{P}\left(M_B(t) \geqslant m', B(t) \leqslant x'\right) &= \mathbb{P}\left(-M_B(t) \leqslant -m', -W(t) \leqslant x'\right) \\ &= \mathbb{P}\left(m(t) \leqslant -m', W(t) \geqslant -x'\right)\end{aligned}$$

Substituting $m = -m'$, $x = -x'$, where $m < 0$, $x \geqslant m$, gives the result in (10.42). The joint density in (10.43) follows by differentiating as in the above proof of (10.39). □

The (marginal) CDF of $m(t)$, $F_{m(t)}$, follows by expressing $\{W(t) \geqslant m\}$ as a union of two mutually exclusive events (note: $\{m(t) > m\} = \{m(t) > m, W(t) > m\}$ since $\{m(t) > m, W(t) \leqslant m\} = \emptyset$):

$$\begin{aligned}\{W(t) > m\} &= \{m(t) \leqslant m, W(t) > m\} \cup \{m(t) > m, W(t) > m\} \\ &= \{m(t) \leqslant m, W(t) > m\} \cup \{m(t) > m\}.\end{aligned}$$

Computing probabilities and using (10.42) for $x = m > 0$ gives

$$\begin{aligned}\mathbb{P}(m(t) > m) &= \mathbb{P}(W(t) > m) - \mathbb{P}(m(t) \leqslant m, W(t) \geqslant m) \\ &= 2\mathcal{N}\left(-\frac{m}{\sqrt{t}}\right) - 1 \\ &= 1 - 2\mathcal{N}\left(\frac{m}{\sqrt{t}}\right)\end{aligned}$$

where we used $\mathcal{N}(z) + \mathcal{N}(-z) = 1$. Hence,

$$F_{m(t)}(m) := \mathbb{P}(m(t) \leqslant m) = 1 - \mathbb{P}(m(t) > m) = 2\mathcal{N}\left(\frac{m}{\sqrt{t}}\right) \tag{10.44}$$

for $-\infty < m < 0$ and $F_{m(t)}(m) \equiv 1$ for $m \geqslant 0$. The reader can easily check that this is a proper CDF for the continuous random variable $m(t)$. Moreover, this also recovers the CDF of the first hitting time $\mathcal{T}_m$ in (10.36) for $m = -|m| < 0$, i.e., we have the equivalence of the CDFs, $F_{m(t)}(m) = F_{\mathcal{T}_m}(t)$, for all $t > 0, m < 0$.

By writing $\{W(t) \geqslant x\} = \{m(t) \leqslant m, W(t) \geqslant x\} \cup \{m(t) > m, W(t) \geqslant x\}$ and taking probabilities of this disjoint union, while using (10.42), gives another useful relation:

$$\begin{aligned}\mathbb{P}(m(t) > m, W(t) \geqslant x) &= \mathbb{P}(W(t) \geqslant x) - \mathbb{P}(W(t) \geqslant x - 2m) \\ &= \mathcal{N}\left(\frac{x - 2m}{\sqrt{t}}\right) - \mathcal{N}\left(\frac{x}{\sqrt{t}}\right)\end{aligned} \tag{10.45}$$

for $m < 0$, $m \leqslant x < \infty$, $t > 0$. For given $t > 0$, the joint CDF of $m(t), W(t)$ follows by using (10.42) and (10.44):

$$\begin{aligned}F_{m(t),W(t)}(m, x) := \mathbb{P}(m(t) \leqslant m, W(t) \leqslant x) &= \mathbb{P}(m(t) \leqslant m) - \mathbb{P}(m(t) \leqslant m, W(t) \geqslant x) \\ &= 2\mathcal{N}\left(\frac{m}{\sqrt{t}}\right) - \mathcal{N}\left(\frac{2m - x}{\sqrt{t}}\right),\end{aligned} \tag{10.46}$$

for $m < 0$, $x \geqslant m$, $F_{m(t),W(t)}(m, x) = F_{m(t),W(t)}(m, m) = \mathcal{N}\left(\frac{m}{\sqrt{t}}\right)$ for $x < m < 0$, and $F_{m(t),W(t)}(m, x) \equiv 0$ for $m \geqslant 0$.

Before closing this section we make another important connection to BM that is killed at a given level m. Given $m \neq 0$, standard *BM killed at level* m is standard BM up to the first hitting time $\mathcal{T}_m$ at which time the process is "killed and sent to the so-called cemetery state $\partial^\dagger$." [Remark: We can also define a similar BM that is "frozen or absorbed" at the level m.] Let $\{W_{(m)}(t)\}_{t\geqslant 0}$ denote the standard BM killed at level m, then

$$W_{(m)}(t) := \begin{cases} W(t) & \text{for } t < \mathcal{T}_m, \\ \partial^\dagger & \text{for } t \geqslant \mathcal{T}_m. \end{cases} \tag{10.47}$$

For $m > 0$, according to (10.35), $\{t < \mathcal{T}_m\} = \{M(t) < m\}$. The event that the process $W_{(m)}$ has not been killed and hence lies below or at any $x < m$ at time t is the joint event that W lies below or at $x < m$ at time t and the maximum of W up to time t is less than m:

$$\begin{aligned}\mathbb{P}(W_{(m)}(t) \leqslant x) &= \mathbb{P}(M(t) < m, W(t) \leqslant x) = F_{M(t),W(t)}(m,x) \\ &= \mathcal{N}\left(\frac{2m-x}{\sqrt{t}}\right) - \mathcal{N}\left(-\frac{x}{\sqrt{t}}\right).\end{aligned}$$

Here we used (10.40). Differentiating this expression w.r.t. x gives the density $p^{W_{(m)}}$ for killed standard BM (starting at $W_{(m)}(0) = 0$) on its state space $x \in (-\infty, m)$, $m > 0$:

$$p^{W_{(m)}}(t;0,x)\,\mathrm{d}x \equiv \mathbb{P}(W_{(m)}(t) \in \mathrm{d}x) = \mathbb{P}(M(t) \leqslant m, W(t) \in \mathrm{d}x) \tag{10.48}$$

where

$$\begin{aligned}p^{W_{(m)}}(t;0,x) &= \frac{\partial}{\partial x}\mathbb{P}(W_{(m)}(t) \leqslant x) = \frac{\partial}{\partial x}F_{M(t),W(t)}(m,x) \\ &= \frac{1}{\sqrt{t}}\left[n\left(\frac{x}{\sqrt{t}}\right) - n\left(\frac{x-2m}{\sqrt{t}}\right)\right] \\ &= p_0(t;x) - p_0(t;2m-x),\end{aligned} \tag{10.49}$$

for $x < m$, and $p^{W_{(m)}}(t;0,x) \equiv 0$ for $x \geqslant m$. Here $p_0(t;x)$, defined in (10.19), is the Gaussian density for standard BM, W. Note that we also readily obtain the transition PDF for the killed BM by the time and space homogeneous property (see (10.20)) with $m \to m - x$,

$$\begin{aligned}p^{W_{(m)}}(s,t;y,x) &= p_0(t-s;y-x) - p_0(t;2(m-x)-(y-x)) \\ &= p_0(t-s;y-x) - p_0(t;2m-x-y),\end{aligned} \tag{10.50}$$

$0 \leqslant s < t$, $-\infty < x, y < m$, $m > 0$.

For $m < 0$, $\{t < \mathcal{T}_m\} = \{m(t) > m\}$. The event that $W_{(m)}$ has not been killed and hence lies at or above a value $x > m$ at time t is the joint event that W lies above or at $x > m$ at time t and the minimum of W up to time t is greater than m:

$$\begin{aligned}\mathbb{P}(W_{(m)}(t) \geqslant x) &= \mathbb{P}(m(t) > m, W(t) \geqslant x) \\ &= \mathcal{N}\left(\frac{x-2m}{\sqrt{t}}\right) - \mathcal{N}\left(\frac{x}{\sqrt{t}}\right),\end{aligned}$$

where we used (10.45). Following the same steps as above gives the density $p^{W_{(m)}}$ for killed standard BM on its state space $x \in (m, \infty)$, $m < 0$:

$$p^{W_{(m)}}(t;0,x)\,\mathrm{d}x \equiv \mathbb{P}(W_{(m)}(t) \in \mathrm{d}x) = \mathbb{P}(m(t) \geqslant m, W(t) \in \mathrm{d}x) \tag{10.51}$$

where

$$p^{W_{(m)}}(t;0,x) = -\frac{\partial}{\partial x}\mathbb{P}(W_{(m)}(t) \geqslant x) = p_0(t;x) - p_0(t;x-2m), \tag{10.52}$$

for $x > m$ and $p^{W_{(m)}}(t;0,x) \equiv 0$ for $x \leqslant m$. Note that this expression is the same as that in (10.49), but now $x > m$, $m < 0$.

For any finite time $t > 0$, it is simple to show that the densities in (10.49) and (10.52) are strictly positive on their respective domains of definition. However, they do not integrate to unity, as is clear from their definitions. In fact, they can be used to obtain the distribution of $M(t)$ and $m(t)$, respectively, and hence to obtain the distribution of the first hitting time $\mathcal{T}_m$. For example, let's take the case with $m > 0$. Then, the probability of event $\{\mathcal{T}_m \leqslant t\} \doteq \{M(t) \geqslant m\}$ is the probability that the process has been killed and is hence in the state $\partial^\dagger$ at time t. Hence, the CDF of the first hitting time to level $m > 0$ is given by

$$
\begin{aligned}
F_{\mathcal{T}_m}(t) = 1 - \mathbb{P}(M(t) < m) &= 1 - \int_{-\infty}^{m} \mathbb{P}(M(t) < m, W(t) \in \mathrm{d}x) \\
&= 1 - \int_{-\infty}^{m} p^{W_{(m)}}(t;0,x)\,\mathrm{d}x \\
&= \int_{m}^{\infty} p_0(t;x)\,\mathrm{d}x + \int_{-\infty}^{m} p_0(t;2m-x)\,\mathrm{d}x \\
&= 2 - 2\mathcal{N}\left(\frac{m}{\sqrt{t}}\right), \; t > 0, \qquad (10.53)
\end{aligned}
$$

which recovers our previously derived formula in (10.36) for $m > 0$. We leave it to the reader to verify that the CDF in (10.36) for the first hitting time down to level $m < 0$ is also recovered by integrating the density in (10.52):

$$
\begin{aligned}
F_{\mathcal{T}_m}(t) = 1 - \mathbb{P}(m(t) > m) &= 1 - \int_{m}^{\infty} \mathbb{P}(m(t) > m, W(t) \in \mathrm{d}x) \\
&= 1 - \int_{m}^{\infty} p^{W_{(m)}}(t;0,x)\,\mathrm{d}x\,. \qquad (10.54)
\end{aligned}
$$

10.4.2 Translated and Scaled Driftless Brownian Motion

Let us now consider the process $X = \{X(t)\}_{t\geqslant 0}$ as a scaled and translated BM defined as in (10.24), $x_0 \in \mathbb{R}, \sigma > 0$, but with *zero drift* $\mu = 0$:

$$
X(t) := x_0 + W^{(0,\sigma)}(t) = x_0 + \sigma W(t)\,. \qquad (10.55)
$$

Hence, all paths of X start at the point $X(0) = x_0 \in \mathbb{R}$. The relationship between $X(t)$ and $W(t)$ is simply a monotonically increasing mapping, i.e., $X(t) = f(W(t))$ where $f(w) := x_0 + \sigma w$, $f^{-1}(x) := (x - x_0)/\sigma$. As a consequence, all formulae derived in Section 10.4.1 for the CDF (PDF) of the first hitting time, the maximum, as shown just below.

The first hitting time of the X process to a level $m \in \mathbb{R}$, denoted by $\mathcal{T}_m^X$, is given by the first hitting time of standard BM, W, to the level $f^{-1}(m) = (m - x_0)/\sigma$:

$$
\begin{aligned}
\mathcal{T}_m^X &:= \inf\{t \geqslant 0 \,:\, X(t) = m\} \\
&= \inf\{t \geqslant 0 \,:\, x_0 + \sigma W(t) = m\} \\
&= \inf\{t \geqslant 0 \,:\, W(t) = (m - x_0)/\sigma\} \\
&\equiv \mathcal{T}_{(m-x_0)/\sigma}\,. \qquad (10.56)
\end{aligned}
$$

That is, sending $m \to (m - x_0)/\sigma$ in $\mathcal{T}_m$ for W gives $\mathcal{T}_m^X$. Similarly, for the sampled maximum and minimum of X, from time 0 to time $t \geqslant 0$, we have the simple relations

$$M^X(t) := \sup_{0 \leqslant u \leqslant t} X(u) = \sup_{0 \leqslant u \leqslant t} [x_0 + \sigma W(u)] \\ = x_0 + \sigma \sup_{0 \leqslant u \leqslant t} W(u) = x_0 + \sigma M(t) \tag{10.57}$$

and similarly

$$m^X(t) := \inf_{0 \leqslant u \leqslant t} X(u) = x_0 + \sigma\, m(t)\,. \tag{10.58}$$

Note that $x_0 = M^X(0) = m^X(0)$, $m^X(t) \leqslant X(t) \leqslant M^X(t)$, $M^X(t)$ is increasing and $m^X(t)$ is decreasing.

Based on (10.56), the formula in (10.36) directly gives us the CDF of $\mathcal{T}_m^X$, for all $m \in \mathbb{R}$:

$$\mathbb{P}(\mathcal{T}_m^X \leqslant t) = \mathbb{P}(\mathcal{T}_{\frac{(m-x_0)}{\sigma}} \leqslant t) = 2\left[1 - \mathcal{N}\left(\frac{|m - x_0|}{\sigma\sqrt{t}}\right)\right], \quad 0 < t < \infty. \tag{10.59}$$

Clearly, we have the equivalence of events:

$$\{M^X(t) \geqslant m\} = \{M(t) \geqslant (m - x_0)/\sigma\}\,, \quad \text{for all } m \geqslant x_0,$$
$$\{m^X(t) \leqslant m\} = \{m(t) \leqslant (m - x_0)/\sigma\}\,, \quad \text{for all } m \leqslant x_0,$$
$$\{X(t) \leqslant x\} = \{W(t) \leqslant (x - x_0)/\sigma\}\,, \quad \text{for all } x \in \mathbb{R}.$$

Based on these relations, all relevant formulae for the marginal and joint CDFs immediately follow from those for standard BM upon sending $m \to (m - x_0)/\sigma$ and $x \to (x - x_0)/\sigma$ in the CDF formulae of Section 10.4.1. For example, using (10.38), for $m > x_0$, $x \leqslant m$:

$$\mathbb{P}(M^X(t) \geqslant m, X(t) \leqslant x) = \mathbb{P}(M(t) \geqslant (m - x_0)/\sigma, W(t) \leqslant (x - x_0)/\sigma) \\ = 1 - \mathcal{N}\left(\frac{2m - (x + x_0)}{\sigma\sqrt{t}}\right). \tag{10.60}$$

For all $t > 0$, the joint CDF of $M^X(t), X(t)$ is obtained by sending $m \to (m - x_0)/\sigma$ and $x \to (x - x_0)/\sigma$ in (10.40):

$$F_{M^X(t),X(t)}(m, x) = \mathcal{N}\left(\frac{2m - (x + x_0)}{\sigma\sqrt{t}}\right) - \mathcal{N}\left(\frac{x_0 - x}{\sigma\sqrt{t}}\right), \quad m > x_0, m \geqslant x, \tag{10.61}$$

$F_{M^X(t),X(t)}(m, x) \equiv 0$ for $m \leqslant x_0$ and $F_{M^X(t),X(t)}(m, x) = F_{M^X(t)}(m)$ for $x > m > x_0$. Differentiating gives the joint PDF,

$$f_{M^X(t),X(t)}(m, x) = \frac{2(2m - (x + x_0))}{\sigma^3 t\sqrt{2\pi t}} \mathrm{e}^{-(2m-(x+x_0))^2/2\sigma^2 t}, \tag{10.62}$$

for $x < m, x_0 < m$ and zero otherwise.

The (marginal) CDF of $M^X(t)$, $t > 0$, follows from (10.41), upon sending $m \to (m - x_0)/\sigma$,

$$F_{M^X(t)}(m) := \mathbb{P}(M^X(t) \leqslant m) = 2\mathcal{N}\left(\frac{m - x_0}{\sigma\sqrt{t}}\right) - 1\,, \tag{10.63}$$

for $m > x_0$ and zero for $m \leqslant x_0$.

Similarly, for $t > 0$, the joint variables $m^X(t), X(t)$ satisfy (from (10.42)):

$$\mathbb{P}(m^X(t) \leqslant m, X(t) \geqslant x) = 1 - \mathcal{N}\left(\frac{x + x_0 - 2m}{\sigma\sqrt{t}}\right) \tag{10.64}$$

for $x_0 > m, x \geqslant m$ and with joint PDF

$$f_{m^X(t),X(t)}(m, x) = \frac{2(x + x_0 - 2m)}{\sigma^3 t\sqrt{2\pi t}} \mathrm{e}^{-(2m-(x+x_0))^2/2\sigma^2 t}, \tag{10.65}$$

for $x > m, x_0 > m$ and zero otherwise.

Upon sending $m \to (m - x_0)/\sigma$ and $x \to (x - x_0)/\sigma$, the relation in (10.45) gives

$$\mathbb{P}(m^X(t) > m, X(t) \geqslant x) = \mathcal{N}\left(\frac{x + x_0 - 2m}{\sigma\sqrt{t}}\right) - \mathcal{N}\left(\frac{x - x_0}{\sigma\sqrt{t}}\right) \tag{10.66}$$

for $m < x_0$, $m \leqslant x < \infty$, $t > 0$ and (10.46) gives the joint CDF of $m^X(t), X(t)$:

$$F_{m^X(t),X(t)}(m, x) = 2\mathcal{N}\left(\frac{m - x_0}{\sigma\sqrt{t}}\right) - \mathcal{N}\left(\frac{2m - x - x_0}{\sigma\sqrt{t}}\right), \tag{10.67}$$

for $m < x_0$, $x \geqslant m$, $F_{m^X(t),X(t)}(m, x) = F_{m^X(t),X(t)}(m, m) = \mathcal{N}\left(\frac{m-x_0}{\sigma\sqrt{t}}\right)$ for $x < m < x_0$, and $F_{m^X(t),X(t)}(m, x) \equiv 0$ for $m \geqslant x_0$.

Formulae for the transition PDF of driftless BM killed at level m and started at $x_0 < m$, or $x_0 > m$, are given by (10.49) or (10.52) upon making the substitution $m \to (m - x_0)/\sigma$ and $x \to (x - x_0)/\sigma$.

10.4.3 Brownian Motion with Drift

We now consider BM with a constant drift, i.e., let the process $\{X(t)\}_{t\geqslant 0}$ be defined by (10.23) for $\sigma = 1$:

$$X(t) := W^{(\mu,1)}(t) \equiv \mu t + W(t). \tag{10.68}$$

We remark that it suffices to consider this case, as it also leads directly to the respective formulae for the process defined in (10.24), where the BM is started at an arbitrary point x_0 and has arbitrary volatility parameter $\sigma > 0$. This is due to the relation

$$W^{(\mu,\sigma)}(t) \equiv \mu t + \sigma W(t) = \sigma[\frac{\mu}{\sigma}t + W(t)] = \sigma W^{(\frac{\mu}{\sigma},1)}(t). \tag{10.69}$$

So, the process in (10.24) is obtained by scaling with σ and translating by x_0 the process X in (10.68), as in the previous section, and now by applying the additional scaling of the drift, i.e., send $\mu \to \frac{\mu}{\sigma}$. For clarity, we shall discuss this scaling and translation in Section 10.4.3.1.

The sampled maximum and minimum of the process $X \equiv W^{(\mu,1)}$ are defined by

$$M^X(t) := \sup_{0\leqslant s\leqslant t} X(s) = \sup_{0\leqslant s\leqslant t} [\mu s + W(s)] \tag{10.70}$$

and

$$m^X(t) := \inf_{0\leqslant s\leqslant t} X(s) = \inf_{0\leqslant s\leqslant t} [\mu s + W(s)]. \tag{10.71}$$

In contrast to our previous expressions for driftless or standard BM, the drift term is a *time-dependent function* that is added to the standard (zero drift) BM. Hence, we cannot use the reflection principle in this case. In Chapter 11 we shall see how to implement methods in probability measure changes in order to calculate expectations of random variables that are functionals of standard Brownian motion. A very powerful tool is Girsanov's Theorem. Section 11.8.1 contains some applications of Girsanov's Theorem. Here we simply borrow the key results derived in Section 11.8.1 that allow us to derive formulae for the CDF of the first hitting time of drifted BM and other joint event probabilities, PDFs and CDFs for the above sampled maximum and minimum random variables.

Our two main formulae are given by (11.104) and (11.105). The joint CDF in (11.106) and (11.107) are important cases. These were used in Section 11.8.1 to derive the joint PDF of $M^X(t), X(t)$ and of $m^X(t), X(t)$, given in (11.109) and (11.110). The joint CDF in (11.106) also leads to the useful relation

$$\mathbb{P}(M^X(t) \leqslant m, X(t) \in \mathrm{d}x) = \mathrm{e}^{-\frac{1}{2}\mu^2 t + \mu x}\, \mathbb{P}(M(t) \leqslant m, W(t) \in \mathrm{d}x)\,, \tag{10.72}$$

for $x < m, m > 0$. To see how this identity arises, we use the definition of the joint PDF of $M^X(t), X(t)$ as the second partial derivative of the CDF, i.e., $f_{M^X(t),X(t)}(w, y) = \frac{\partial}{\partial y}\frac{\partial}{\partial w} F_{M^X(t),X(t)}(w, y)$. Hence, for $x < m, m > 0$, the joint CDF of $M^X(t), X(t)$ is expressed equivalently as

$$\begin{aligned}
\mathbb{P}(M^X(t) \leqslant m, X(t) \leqslant x) &\equiv F_{M^X(t),X(t)}(m, x) \\
&= \int_{-\infty}^{x}\int_{0}^{m} f_{M^X(t),X(t)}(w, y)\,\mathrm{d}w\,\mathrm{d}y \\
&= \int_{-\infty}^{x}\int_{0}^{m} \frac{\partial}{\partial y}\frac{\partial}{\partial w} F_{M^X(t),X(t)}(w, y)\,\mathrm{d}w\,\mathrm{d}y \\
&= \int_{-\infty}^{x} \frac{\partial}{\partial y}\left[\int_{0}^{m} \frac{\partial}{\partial w} F_{M^X(t),X(t)}(w, y)\,\mathrm{d}w\right]\mathrm{d}y \\
&= \int_{-\infty}^{x} \frac{\partial}{\partial y}\left[F_{M^X(t),X(t)}(m, y) - F_{M^X(t),X(t)}(0, y)\right]\mathrm{d}y \\
&= \int_{-\infty}^{x} \frac{\partial}{\partial y} F_{M^X(t),X(t)}(m, y)\,\mathrm{d}y \\
&\equiv \int_{-\infty}^{x} \mathbb{P}(M^X(t) \leqslant m, X(t) \in \mathrm{d}y)\,.
\end{aligned} \tag{10.73}$$

Note that $F_{M^X(t),X(t)}(0, y) = \mathbb{P}(M^X(t) \leqslant 0, X(t) \leqslant y) \equiv 0$. On the other hand, by substituting the expression in (11.109), i.e., $f_{M^X(t),X(t)}(w, y) = \mathrm{e}^{-\frac{1}{2}\mu^2 t + \mu y} f_{M(t),W(t)}(w, y)$, into the first integral above and then repeating the same steps (but now with $F_{M(t),W(t)}$ in the place of $F_{M^X(t),X(t)}$) gives

$$\begin{aligned}
\mathbb{P}(M^X(t) \leqslant m, X(t) \leqslant x) &= \int_{-\infty}^{x} \mathrm{e}^{-\frac{1}{2}\mu^2 t + \mu y}\left[\int_{0}^{m} f_{M(t),W(t)}(w, y)\,\mathrm{d}w\right]\mathrm{d}y \\
&= \int_{-\infty}^{x} \mathrm{e}^{-\frac{1}{2}\mu^2 t + \mu y}\frac{\partial}{\partial y} F_{M(t),W(t)}(m, y)\,\mathrm{d}y \\
&\equiv \int_{-\infty}^{x} \mathrm{e}^{-\frac{1}{2}\mu^2 t + \mu y}\mathbb{P}(M(t) \leqslant m, W(t) \in \mathrm{d}y)\,.
\end{aligned} \tag{10.74}$$

Hence, (10.72) holds by equating integrands in (10.73) and (10.74).

Note that the probabilities in (10.72) represent the probability for $X(t)$ (or respectively $W(t)$) to take on a value in an infinitesimal interval around the point x, assuming that the sampled maximum $M^X(t)$ (or respectively $M(t)$) is less than m, i.e., the respective probabilities correspond to a density in x (obtained by differentiating the joint CDF w.r.t. x) times the infinitesimal $\mathrm{d}x$,

$$\mathbb{P}(M^X(t) \leqslant m, X(t) \in \mathrm{d}x) = \frac{\partial}{\partial x} F_{M^X(t),X(t)}(m,x)\,\mathrm{d}x$$

and

$$\mathbb{P}(M(t) \leqslant m, W(t) \in \mathrm{d}x) = \frac{\partial}{\partial x} F_{M(t),W(t)}(m,x)\,\mathrm{d}x\,.$$

We have already seen that this last probability is related to the density for standard BM killed at $m > 0$. This is given by (10.48) and (10.49). The previous probability is similarly related to the density for drifted BM killed at $m > 0$.

For any $m \neq 0$, *drifted BM killed at level* m is defined as

$$X_{(m)}(t) := \begin{cases} X(t) & \text{for } t < \mathcal{T}_m^X, \\ \partial^\dagger & \text{for } t \geqslant \mathcal{T}_m^X, \end{cases} \tag{10.75}$$

where $\mathcal{T}_m^X$ is the first hitting time of the drifted BM, $X = W^{(\mu,1)}$, to level m. For $m > 0$, the process $X_{(m)}$ has state space $(-\infty, m)$ and starts at zero, as does $W_{(m)}$. Hence, taking any value $x < m$, the joint CDF $F_{M^X(t),X(t)}(m,x)$ is equivalent to the probability that the drifted BM with killing at m lies below x, i.e.,

$$\mathbb{P}(X_{(m)}(t) \leqslant x) = \mathbb{P}(M^X(t) < m, X(t) \leqslant x) = F_{M^X(t),X(t)}(m,x) \tag{10.76}$$

and the analogue of (10.48) is

$$p^{X_{(m)}}(t;0,x)\,\mathrm{d}x \equiv \mathbb{P}(X_{(m)}(t) \in \mathrm{d}x) = \mathbb{P}(M^X(t) \leqslant m, X(t) \in \mathrm{d}x)\,. \tag{10.77}$$

Hence, (10.72) is equivalent to

$$\mathbb{P}(X_{(m)}(t) \in \mathrm{d}x) = \mathrm{e}^{-\frac{1}{2}\mu^2 t + \mu x}\,\mathbb{P}(W_{(m)}(t) \in \mathrm{d}x)\,. \tag{10.78}$$

In particular, the density $p^{X_{(m)}}$ for the drifted killed BM, started at $X_{(m)}(0) = 0$, is related to the density $p^{W_{(m)}}$ in (10.49) by

$$\begin{aligned} p^{X_{(m)}}(t;0,x) &= \mathrm{e}^{-\frac{1}{2}\mu^2 t + \mu x} p^{W_{(m)}}(t;0,x) \\ &= \mathrm{e}^{-\frac{1}{2}\mu^2 t + \mu x}\left[p_0(t;x) - p_0(t;2m - x)\right], \end{aligned} \tag{10.79}$$

for $x < m$ and zero otherwise. We can rewrite this density in a more convenient form by multiplying and completing the squares in the two exponents,

$$\begin{aligned} p^{X_{(m)}}(t;0,x) &= \mathrm{e}^{-\frac{1}{2}\mu^2 t + \mu x}\left(\frac{\mathrm{e}^{-x^2/2t}}{\sqrt{2\pi t}} - \frac{\mathrm{e}^{-(2m-x)^2/2t}}{\sqrt{2\pi t}}\right) \\ &= \frac{\mathrm{e}^{-(x-\mu t)^2/2t}}{\sqrt{2\pi t}} - \mathrm{e}^{2\mu m}\,\frac{\mathrm{e}^{-(x-\mu t-2m)^2/2t}}{\sqrt{2\pi t}} \\ &= p_0(t;x-\mu t) - \mathrm{e}^{2\mu m}\,p_0(t;x-\mu t-2m)\,, \end{aligned} \tag{10.80}$$

for $x < m$ and zero otherwise. This expression involves a linear combination of densities for standard BM at any point $x \in (-\infty, m)$. Note that setting $\mu = 0$ recovers $p^{W_{(m)}}$ in (10.49). In the limit $m \to \infty$, this density recovers the density $p^X(t;0,x) \equiv p_0(t;x-\mu t)$ for drifted BM, $X \equiv W^{(\mu,1)} \in \mathbb{R}$, with no killing.

The joint CDF of $M^X(t), X(t)$ can be obtained by evaluating a double integral of the joint PDF. Alternatively, since we now have the density $p^{X_{(m)}}$ at hand, we can simply use it to compute a single integral to obtain the joint CDF. Indeed, using (10.80) within the last integral in (10.73):

$$\begin{aligned}
F_{M^X(t),X(t)}(m,x) &= \int_{-\infty}^{x} \mathbb{P}(M^X(t) \leqslant m, X(t) \in \mathrm{d}y) = \int_{-\infty}^{x} \mathbb{P}(X_{(m)}(t) \in \mathrm{d}y) \\
&= \int_{-\infty}^{x} p^{X_{(m)}}(t;0,y)\,\mathrm{d}y \\
&= \int_{-\infty}^{x} p_0(t; y-\mu t)\,\mathrm{d}y - \mathrm{e}^{2\mu m}\int_{-\infty}^{x} p_0(t; y-\mu t-2m)\,\mathrm{d}y \\
&= \int_{-\infty}^{x-\mu t} p_0(t;y)\,\mathrm{d}y - \mathrm{e}^{2\mu m}\int_{-\infty}^{x-\mu t-2m} p_0(t;y)\,\mathrm{d}y \\
&= \mathbb{P}(W(t) \leqslant x-\mu t) - \mathrm{e}^{2\mu m}\,\mathbb{P}(W(t) \leqslant x-\mu t-2m) \\
&= \mathcal{N}\left(\frac{x-\mu t}{\sqrt{t}}\right) - \mathrm{e}^{2\mu m}\,\mathcal{N}\left(\frac{x-\mu t-2m}{\sqrt{t}}\right), \qquad (10.81)
\end{aligned}$$

for $x \leqslant m, m > 0$. For $x > m > 0$, $F_{M^X(t),X(t)}(m,x) = F_{M^X(t),X(t)}(m,m) = F_{M^X(t)}(m)$ and $F_{M^X(t),X(t)}(m,x) \equiv 0$ for $m \leqslant 0$.

The (marginal) CDF of $M^X(t)$ is given by setting $x = m$ in (10.81),

$$\begin{aligned}
F_{M^X(t)}(m) &= F_{M^X(t),X(t)}(m,m) = \mathbb{P}(X_{(m)}(t) \leqslant m) \\
&= \mathcal{N}\left(\frac{m-\mu t}{\sqrt{t}}\right) - \mathrm{e}^{2\mu m}\,\mathcal{N}\left(\frac{-m-\mu t}{\sqrt{t}}\right), \quad m > 0, \qquad (10.82)
\end{aligned}$$

and $F_{M^X(t)}(m) \equiv 0$ for $m \leqslant 0$. The first hitting time up to level $m > 0$ has CDF

$$F_{\mathcal{T}_m^X}(t) = 1 - F_{M^X(t)}(m) = \mathcal{N}\left(\frac{\mu t-m}{\sqrt{t}}\right) + \mathrm{e}^{2\mu m}\,\mathcal{N}\left(\frac{-m-\mu t}{\sqrt{t}}\right), \qquad (10.83)$$

for $t > 0$, and zero otherwise.

We now consider the case where $m < 0$. The analogue of (10.72) for the joint pair $m^X(t), X(t)$ is

$$\mathbb{P}(m^X(t) \geqslant m, X(t) \in \mathrm{d}x) = \mathrm{e}^{-\frac{1}{2}\mu^2 t + \mu x}\,\mathbb{P}(m(t) \geqslant m, W(t) \in \mathrm{d}x), \qquad (10.84)$$

for $x > m, m < 0$. The derivation is very similar to that given in (10.73) and (10.74). We can relate this to the first hitting time $\mathcal{T}_m^X$, which is now the first hitting time of the drifted BM down to level $m < 0$. The process $X_{(m)}$ has state space (m, ∞). The identity in (10.78) is now valid for $x > m$, $m < 0$. The analogues of (10.76) and (10.77) are

$$\mathbb{P}(X_{(m)}(t) \geqslant x) = \mathbb{P}(m^X(t) > m, X(t) \geqslant x) \qquad (10.85)$$

and

$$p^{X_{(m)}}(t;0,x)\,\mathrm{d}x \equiv \mathbb{P}(X_{(m)}(t) \in \mathrm{d}x) = \mathbb{P}(m^X(t) \geqslant m, X(t) \in \mathrm{d}x), \qquad (10.86)$$

$x > m, m < 0$. By (10.78), the density $p^{X_{(m)}}(t;0,x)$ for the process $X_{(m)} \in (m, \infty)$ is still given by the expression in (10.79), or equivalently in (10.80), but now for $x > m$, $m < 0$. The density $p^{X_{(m)}}(t;0,x)$ is identically zero for $x \leqslant m$.

The derivation of the joint CDF of $m^X(t), X(t)$ is left as an exercise for the reader. In what follows it is useful to compute the following joint probability (using similar steps as in (10.81)):

$$\begin{aligned}\mathbb{P}(m^X(t) \geqslant m, X(t) \geqslant x) &= \int_x^\infty p^{X_{(m)}}(t;0,y)\,\mathrm{d}y \\ &= \mathbb{P}(W(t) \geqslant x - \mu t) - \mathrm{e}^{2\mu m}\,\mathbb{P}(W(t) \geqslant x - \mu t - 2m) \\ &= \mathcal{N}\left(\frac{-x+\mu t}{\sqrt{t}}\right) - \mathrm{e}^{2\mu m}\,\mathcal{N}\left(\frac{-x+\mu t+2m}{\sqrt{t}}\right), \end{aligned} \tag{10.87}$$

for $x \geqslant m, m < 0$. The CDF of the first hitting time down to level $m < 0$ follows from this expression for $x = m$:

$$\begin{aligned}F_{\mathcal{T}_m^X}(t) &= 1 - \mathbb{P}(m^X(t) \geqslant m) = 1 - \mathbb{P}(m^X(t) \geqslant m, X(t) \geqslant m) \\ &= 1 - \int_m^\infty p^{X_{(m)}}(t;0,y)\,\mathrm{d}y \\ &= 1 - \mathcal{N}\left(\frac{-m+\mu t}{\sqrt{t}}\right) + \mathrm{e}^{2\mu m}\,\mathcal{N}\left(\frac{-m+\mu t+2m}{\sqrt{t}}\right) \\ &= \mathcal{N}\left(\frac{m-\mu t}{\sqrt{t}}\right) + \mathrm{e}^{2\mu m}\,\mathcal{N}\left(\frac{m+\mu t}{\sqrt{t}}\right)\end{aligned} \tag{10.88}$$

for $t > 0$, and zero otherwise. For given $t > 0$, this expression is also the CDF of $m^X(t)$, $F_{m^X(t)}(m) = F_{\mathcal{T}_m}(t)$ for $-\infty < m \leqslant 0$, and $F_{m^X(t)}(m) \equiv 1$ for $m \geqslant 0$.

Other probabilities related to (10.72) and (10.84) also follow. For example, we can compute the probability that drifted BM process X is within an infinitesimal interval containing $x \in \mathbb{R}$ jointly with the condition that the process has already hit level $m > 0$ (i.e., its sampled maximum $M^X(t) \geqslant m$). This probability is given by

$$\begin{aligned}\mathbb{P}(M^X(t) \geqslant m, X(t) \in \mathrm{d}x) &= \mathbb{P}(X(t) \in \mathrm{d}x) - \mathbb{P}(M^X(t) \leqslant m, X(t) \in \mathrm{d}x) \\ &= \mathbb{P}(X(t) \in \mathrm{d}x) - \mathbb{P}(X_{(m)}(t) \in \mathrm{d}x) \\ &= [p^X(t;0,x) - p^{X_{(m)}}(t;0,x)]\,\mathrm{d}x\end{aligned} \tag{10.89}$$

where we used (10.77) and (10.78). We can write down the explicit expression by combining $p^X(t;0,x) = p_0(t;x-\mu t)$ with the density in (10.80), i.e., for $m > 0$:

$$\begin{aligned}&\mathbb{P}(M^X(t) \geqslant m, X(t) \in \mathrm{d}x) \\ &= \{p_0(t;x-\mu t) - [p_0(t;x-\mu t) - \mathrm{e}^{2\mu m}\,p_0(t;x-\mu t-2m)]\mathbb{I}_{\{x<m\}}\}\,\mathrm{d}x \\ &= \begin{cases}\mathrm{e}^{2\mu m}\,p_0(t;x-\mu t-2m)\,\mathrm{d}x & \text{for } x < m, \\ p_0(t;x-\mu t)\,\mathrm{d}x & \text{for } x \geqslant m.\end{cases}\end{aligned} \tag{10.90}$$

For $m < 0$, a similar derivation gives the probability that the drifted BM process is within an infinitesimal interval jointly with the condition that the process has already hit level $m < 0$ (i.e., its sampled minimum $m^X(t) \leqslant m$):

$$\mathbb{P}(m^X(t) \leqslant m, X(t) \in \mathrm{d}x) = \begin{cases}\mathrm{e}^{2\mu m}\,p_0(t;x-\mu t-2m)\,\mathrm{d}x & \text{for } x > m, \\ p_0(t;x-\mu t)\,\mathrm{d}x & \text{for } x \leqslant m.\end{cases} \tag{10.91}$$

10.4.3.1 Translated and Scaled Brownian Motion with Drift

All the formulae derived for the process $W^{(\mu,1)}$ in the previous section are trivially extended to generate all the corresponding formulae for drifted BM starting at arbitrary

$x_0 \in \mathbb{R}$ and with any volatility parameter $\sigma > 0$. To see this, consider the process defined by (10.24). According to (10.69),

$$X(t) = x_0 + \mu t + \sigma W(t) = x_0 + \sigma W^{(\frac{\mu}{\sigma},1)}(t) . \tag{10.92}$$

The sampled maximum of this process is given by

$$M^X(t) := \sup_{0\leqslant u\leqslant t} X(u) = x_0 + \sigma \sup_{0\leqslant u\leqslant t} W^{(\frac{\mu}{\sigma},1)}(u) \equiv x_0 + \sigma M^{(\frac{\mu}{\sigma},1)}(t) . \tag{10.93}$$

Here we are using $M^{(\mu,\sigma)}(t)$ to denote the sampled maximum of process $W^{(\mu,\sigma)}$. The analogous relation for the sampled minimum $m^X(t)$ follows in the obvious manner.

Computing the joint CDF of $M^X(t), X(t)$,

$$\begin{aligned} F_{M^X(t),X(t)}(m,x) &= \mathbb{P}(M^X(t) \leqslant m, X(t) \leqslant x) \\ &= \mathbb{P}\left(x_0 + \sigma M^{(\frac{\mu}{\sigma},1)}(t) \leqslant m,\, x_0 + \sigma W^{(\frac{\mu}{\sigma},1)}(t) \leqslant x\right) \\ &= \mathbb{P}\left(M^{(\frac{\mu}{\sigma},1)}(t) \leqslant \frac{m-x_0}{\sigma},\, W^{(\frac{\mu}{\sigma},1)}(t) \leqslant \frac{x-x_0}{\sigma}\right) \\ &= \mathbb{P}\left(M^{(\mu',1)}(t) \leqslant m',\, W^{(\mu',1)}(t) \leqslant x'\right) \end{aligned} \tag{10.94}$$

where $\mu' = \frac{\mu}{\sigma}$, $m' = \frac{m-x_0}{\sigma}$, and $x' = \frac{x-x_0}{\sigma}$. Hence, this is the joint CDF of the random variable pair $M^{(\mu',1)}(t), W^{(\mu',1)}(t)$, where the function is evaluated at arguments m', x'. By definition, this is given by the formula for the joint CDF in (10.81) with the *variable replacements*:

$$m \to \frac{m-x_0}{\sigma}, \quad x \to \frac{x-x_0}{\sigma}, \quad \mu \to \frac{\mu}{\sigma} . \tag{10.95}$$

These are precisely the same variable replacements that we saw in Section 10.4.2 for driftless BM. In the case of drifted BM we see that the volatility parameter σ also enters into the adjusted drift. By simply making the above variable replacements in (10.81), we obtain the joint CDF,

$$F_{M^X(t),X(t)}(m,x) = \mathcal{N}\left(\frac{x-x_0-\mu t}{\sigma\sqrt{t}}\right) - \mathrm{e}^{\frac{2\mu}{\sigma^2}(m-x_0)} \mathcal{N}\left(\frac{x+x_0-\mu t - 2m}{\sigma\sqrt{t}}\right), \tag{10.96}$$

for $x, x_0 < m$. For $x \geqslant m > x_0$, $F_{M^X(t),X(t)}(m,x) = F_{M^X(t),X(t)}(m,m) = F_{M^X(t)}(m)$ and $F_{M^X(t),X(t)}(m,x) \equiv 0$ for $m \leqslant x_0$. The CDF of $M^X(t)$ is given by

$$F_{M^X(t)}(m) = \mathcal{N}\left(\frac{m-x_0-\mu t}{\sigma\sqrt{t}}\right) - \mathrm{e}^{\frac{2\mu}{\sigma^2}(m-x_0)} \mathcal{N}\left(\frac{x_0-m-\mu t}{\sigma\sqrt{t}}\right), \quad x_0 < m . \tag{10.97}$$

By the above analysis it follows that all formulae for any marginal CDF or joint CDF of the random variables $M^X(t), m^X(t), X(t)$, as well as the CDF of the first hitting time, derived in the previous section for the process $X(t) = W^{(\mu,1)}(t)$, now give rise to the respective formulae for the process defined by (10.92) upon making the variable replacements in (10.95). Note that for every differentiation of a (joint) CDF w.r.t. variables x, m, or t, there is an extra factor of $\frac{1}{\sigma}$ that multiplies the expression for the resulting PDF. For example, let $X_{(m)}$ be the process X in (10.92) killed at an upper level $m > x_0$. The density for $X_{(m)}(t)$ is then given by applying (10.95) to (10.80) and multiplying by $\frac{1}{\sigma}$:

$$p^{X_{(m)}}(t; x_0, x) = \frac{\mathrm{e}^{-(x-x_0-\mu t)^2/2\sigma^2 t}}{\sigma\sqrt{2\pi t}} - \mathrm{e}^{\frac{2\mu}{\sigma^2}(m-x_0)} \frac{\mathrm{e}^{-(x+x_0-\mu t-2m)^2/2\sigma^2 t}}{\sigma\sqrt{2\pi t}} \tag{10.98}$$

for $x_0, x < m$ and zero otherwise. The same expression for the density is also valid for the domain $x_0, x > m$ where m is a lower killing level. Similarly, the corresponding expressions for the probability densities in (10.90) and (10.91) follow upon making the variable replacements in (10.95) and multiplying by $\frac{1}{\sigma}$.

For completeness, we now re-state the formulae in (10.83), (10.87), and (10.88) for the drifted BM in (10.92):

$$F_{\mathcal{T}_m^X}(t) = \mathcal{N}\left(\frac{x_0 + \mu t - m}{\sigma\sqrt{t}}\right) + \mathrm{e}^{\frac{2\mu}{\sigma^2}(m-x_0)}\mathcal{N}\left(\frac{x_0 - \mu t - m}{\sigma\sqrt{t}}\right), \quad x_0 < m, t > 0, \tag{10.99}$$

$$\mathbb{P}(m^X(t) \geqslant m, X(t) \geqslant x) = \mathcal{N}\left(\frac{x_0 + \mu t - x}{\sigma\sqrt{t}}\right) - \mathrm{e}^{\frac{2\mu}{\sigma^2}(m-x_0)}\mathcal{N}\left(\frac{2m - x - x_0 + \mu t}{\sigma\sqrt{t}}\right), \tag{10.100}$$

for $x, x_0 \geqslant m, t > 0$, and

$$F_{\mathcal{T}_m^X}(t) = \mathcal{N}\left(\frac{m - x_0 - \mu t}{\sigma\sqrt{t}}\right) + \mathrm{e}^{\frac{2\mu}{\sigma^2}(m-x_0)}\mathcal{N}\left(\frac{m - x_0 + \mu t}{\sigma\sqrt{t}}\right), \quad x_0 > m, t > 0. \tag{10.101}$$

10.5 Exercises

Exercise 10.1. The PDF of a sum of two continuous random variables X and Y is given by the convolution of the PDF's f_X and f_Y:

$$f_{X+Y}(z) = \int_{-\infty}^{\infty} f_X(x) f_Y(z - x)\,\mathrm{d}x. \tag{10.102}$$

(a) Use (10.102) to show that a sum of two independent standard normal variables results in a normal variable. Find the PDF of such a sum.

(b) Assuming that $X_1 \sim Norm(\mu_1, \sigma_1^2)$ and $X_2 \sim Norm(\mu_2, \sigma_2^2)$ are correlated with $\mathrm{Corr}(X_1, X_2) = \rho$, find the mean and variance of $a_1 X_1 + a_2 X_2$ for $a_1, a_2 \in \mathbb{R}$.

Exercise 10.2. Consider a log-normal random variable $S = \mathrm{e}^{a+bZ}$ with $Z \sim Norm(0, 1)$. Find formulae for the following expected values:

(a) $\mathrm{E}[\,\mathbb{I}_{\{S>K\}}\,]$, where K is a real positive constant;

(b) $\mathrm{E}[\,S\,\mathbb{I}_{\{S>K\}}\,]$;

(c) $\mathrm{E}[\,\max(S, K)\,]$, where you may use the property

$$\max(S, K) = S\,\mathbb{I}_{\{S>K\}} + K\,\mathbb{I}_{\{K \geqslant S\}} = S\,\mathbb{I}_{\{S>K\}} - K\,\mathbb{I}_{\{S>K\}} + K.$$

Express your answers in terms of the standard normal CDF function $\mathcal{N}$ and constants a, b, and K. [Note: This calculation is part of the Black–Scholes–Merton theory of pricing vanilla European style options. K is the constant strike price and S is the random stock price.]

Exercise 10.3. Suppose $\mathbf{X} = [X_1, X_2, X_3]^\top$ is a three-dimensional Gaussian (normal) random vector with mean vector zero and covariance matrix

$$\mathbf{C} = \mathrm{E}\left[(\mathbf{X} - \mathrm{E}[\mathbf{X}])(\mathbf{X} - \mathrm{E}[\mathbf{X}])^\top\right] = \mathrm{E}\left[\mathbf{X}\mathbf{X}^\top\right] = \begin{bmatrix} 4 & 1 & 2 \\ 1 & 2 & 1 \\ 2 & 1 & 3 \end{bmatrix}.$$

Set $Y = 1 + X_1 - 2X_2 + X_3$ and $Z = X_1 - 2X_3$.

(a) Find the probability distribution of Y.

(b) Find the probability distribution of the vector $[Y, Z]^\top$.

(c) Find a linear combination $W = aY + bZ$ that is independent of X_1.

Exercise 10.4. Take $X_0 = 0$ and define $X_k = X_{k-1} + Z_k$, for $k = 1, 2, \ldots, n$, where Z_k are i.i.d. standard normals. So we have

$$X_k = \sum_{j=1}^{k} Z_j \,. \tag{10.103}$$

Let $\mathbf{X} \in R_n$ be the vector $\mathbf{X} = [X_1, X_2, \ldots, X_n]^\top$.

(a) Show that $\mathbf{X}$ is a multivariate normal.

(b) Use the formula (10.103) to calculate the covariances $\mathrm{Cov}(X_k, X_j)$, for $1 \leqslant k, j \leqslant n$.

(c) Use the answers to part (b) to write a formula for the elements of $\mathbf{C} = \mathrm{Cov}(\mathbf{X}, \mathbf{X}) = \mathrm{E}\left[(\mathbf{X} - \mathrm{E}[\mathbf{X}])(\mathbf{X} - \mathrm{E}[\mathbf{X}])^\top\right]$.

(d) Write the joint PDF of $\mathbf{X}$.

Exercise 10.5. Prove the following properties of a scaled symmetric random walk $W^{(n)}(t)$, $t \geqslant 0$. For all $0 \leqslant s \leqslant t$ such that ns and nt are integers, we have

(a) $\mathrm{E}\left[W^{(n)}(t) \mid \mathcal{F}^{(n)}(s)\right] = W^{(n)}(s)$ (the martingale property) and $\mathrm{E}[W^{(n)}(t)] = 0$;

(b) $\mathrm{Var}(W^{(n)}(t)) = t$ and $\mathrm{Cov}(W^{(n)}(t), W^{(n)}(s)) = s$.

(c) Let $0 \leqslant t_0 < t_1 < \cdots < t_n$ be such that nt_j is an integer for all $0 \leqslant j \leqslant n$, then the increments $W^{(n)}(t_1) - W^{(n)}(t_0), \ldots, W^{(n)}(t_n) - W^{(n)}(t_{n-1})$ are independent and stationary with mean zero and variance $\mathrm{Var}(W^{(n)}(t_j) - W^{(n)}(t_{j-1})) = t_j - t_{j-1}$, $1 \leqslant j \leqslant n$.

Exercise 10.6. Show that the following processes are standard Brownian motions.

(a) $X(t) := -W(t)$.

(b) $X(t) := W(T + t) - W(T)$, where $T \in (0, \infty)$.

(c) $X(t) := cW(t/c^2)$, where $c \neq 0$ is a real constant.

(d) $X(t) := tW(1/t)$, $t > 0$, and $X(0) := 0$.

Exercise 10.7. Let $\{B(t)\}_{t \geqslant 0}$ and $\{W(t)\}_{t \geqslant 0}$ be two independent Brownian motions. Show that $X(t) = (B(t) + W(t))/\sqrt{2}$, $t \geqslant 0$, is also a Brownian motion. Find the coefficient of correlation between $B(t)$ and $X(t)$.

Exercise 10.8. For Brownian motion $\{W(t)\}_{t\geqslant 0}$ and its natural filtration $\{\mathcal{F}(t)\}_{t\geqslant 0}$, calculate $\mathrm{E}_s[W^3(t)]$ for $0 \leqslant s < t$.

Exercise 10.9. Find the distribution of $W(1) + W(2) + \cdots + W(n)$ for $n = 1, 2, \ldots$

Exercise 10.10. Let $a_1, a_2, \ldots, a_n \in \mathbb{R}$ and $0 < t_1 < t_2 < \cdots < t_n$. Find the distribution of $\sum_{k=1}^{n} a_k\, W(t_k)$. Note that the choice with $a_k = \frac{1}{n}$, $1 \leqslant k \leqslant n$, leads to Asian options.

Exercise 10.11. Suppose that the processes $\{X(t)\}_{t\geqslant 0}$ and $\{Y(t)\}_{t\geqslant 0}$ are respectively given by $X(t) = x_0 + \mu_x t + \sigma_x W(t)$ and $Y(t) = y_0 + \mu_y t + \sigma_y W(t)$, where x_0, y_0, μ_x, μ_y, $\sigma_x > 0$, and $\sigma_y > 0$ are real constants. Find the covariance $\mathrm{Cov}(X(t), Y(s))$ for $s, t \geqslant 0$.

Exercise 10.12. Consider the process $X(t) := x_0 + \mu t + \sigma W(t)\, t \geqslant 0$, where x_0, μ, and σ are real constants. Show that

$$\mathrm{E}[\max(X(t) - K, 0)] = (x_0 + \mu t - K)\,\mathcal{N}\left(\frac{x_0 + \mu t - K}{\sigma\sqrt{t}}\right) + \sigma\sqrt{t}\, n\left(\frac{x_0 + \mu t - K}{\sigma\sqrt{t}}\right).$$

Exercise 10.13. By directly calculating partial derivatives, verify that the transition PDF

$$p_0(t; x) = \frac{1}{\sqrt{2\pi t}}\mathrm{e}^{-\frac{x^2}{2t}}$$

of standard Brownian motion satisfies the *heat equation*, also called the *diffusion equation*:

$$\frac{\partial u(t,x)}{\partial t} = \frac{1}{2}\frac{\partial^2 u(t,x)}{\partial x^2}.$$

Exercise 10.14. Find the transition PDF of $\{W^n(t)\}_{t\geqslant 0}$ for $n = 1, 2, \ldots$

Exercise 10.15. Find the transition PDF of $\{S^n(t)\}_{t\geqslant 0}$, where $S(t) = S_0\,\mathrm{e}^{\mu t + \sigma W(t)}, t \geqslant 0$, $S_0 > 0$, for $n = 1, 2, \ldots$.

Exercise 10.16. Find the transition probability function and transition PDF of the process $X(t) := |W(t)|$, $t \geqslant 0$. [Hint: make use of $\{|W(t)| < x\} = \{-x < W(t) < x\}$.] We note that this process corresponds to nonnegative standard BM reflected at the origin.

Exercise 10.17. Show that formulae (10.11) and (10.21) are equivalent by considering successive Brownian increments and applying (10.3).

Exercise 10.18. Prove that a Gaussian process with the covariance function $c(s,t) = s \wedge t$, mean function $m(t) \equiv 0$, and continuous sample paths is a standard Brownian motion.

Exercise 10.19. Show that the mean and covariance functions of the Brownian bridge from a to b on $[0, T]$ are, respectively, $m(t) = a + \frac{(b-a)t}{T}$ and $c(s,t) = s \wedge t - \frac{st}{T}$, for $s, t \in [0, T]$.

Exercise 10.20. Find the mean and covariance functions of the GBM process defined by (10.26).

Exercise 10.21. Consider the GBM process $S(t) = S_0\,\mathrm{e}^{\mu t + \sigma W(t)}, t \geqslant 0$, $S_0 > 0$. The respective sampled maximum and minimum of this process are defined by

$$M^S(t) := \sup_{0\leqslant u\leqslant t} S(u) \quad \text{and} \quad m^S(t) := \inf_{0\leqslant u\leqslant t} S(u),$$

and the first hitting time to a level $B > 0$ is defined by $\mathcal{T}_B^S = \inf\{t \geqslant 0 : S(t) = B\}$. Derive expressions for the following:

(a) $F_{M^S(t),S(t)}(y,x) := \mathbb{P}(M^S(t) \leqslant y, S(t) \leqslant x)$, for all $t > 0$, $0 < x \leqslant y < \infty$, $S_0 \leqslant y$;

(b) $F_{M^S(t)}(y) := \mathbb{P}(M^S(t) \leqslant y)$, for all $t > 0$, $S_0 \leqslant y < \infty$;

(c) $F_{\mathcal{T}_B^S}(t) := \mathbb{P}(\mathcal{T}_B^S \leqslant t)$, for all $t > 0$, $S_0 < B$;

(d) $\mathbb{P}(M^S(t) \leqslant y, S(t) \in \mathrm{d}x)$, for all $t > 0$, $x \leqslant y < \infty$, $S_0 \leqslant y$;

(e) $\mathbb{P}(m^S(t) \geqslant y, S(t) \geqslant x)$, for all $t > 0$, $0 < y \leqslant x < \infty$, $S_0 \geqslant y$;

(f) $F_{\mathcal{T}_B^S}(t) := \mathbb{P}(\mathcal{T}_B^S \leqslant t)$, for all $t > 0$, $S_0 > B$;

(g) $\mathbb{P}(m^S(t) \geqslant y, S(t) \in \mathrm{d}x)$, for all $t > 0$, $0 < y \leqslant x < \infty$, $S_0 > y$.

Exercise 10.22. Consider the drifted BM defined by $X(t) := \mu t + W(t)$, $t \geqslant 0$, with $M^X(t)$ and $m^X(t)$ defined by (10.70) and (10.71). For any $T > 0$, show that

$$\mathbb{P}\left(M^X(T) \leqslant m \,|\, X(T) = x\right) = 1 - \mathrm{e}^{-2m(m-x)/T}, \quad x \leqslant m,\ m \geqslant 0,$$

and

$$\mathbb{P}\left(m^X(T) \geqslant m \,|\, X(T) = x\right) = 1 - \mathrm{e}^{-2m(m-x)/T}, \quad x \geqslant m,\ m \leqslant 0.$$

Hint: Use an appropriate conditioning.

Chapter 11

Introduction to Continuous-Time Stochastic Calculus

11.1 The Riemann Integral of Brownian Motion

11.1.1 The Riemann Integral

Let f be a real-valued function defined on $[0, T]$. We now recall the precise definition of the Riemann integral of f on $[0, T]$ as follows.

- For $n \in \mathbb{N}$, consider a partition P_n of the interval $[0, T]$:

  $$P_n = \{t_0, t_1, \ldots, t_n\}, \quad 0 = t_0 < t_1 < \cdots < t_n = T.$$

  Define $\Delta t_i = t_i - t_{i-1}$, $i = 1, 2, \ldots, n$.

- Introduce an intermediate partition Q_n for the partition P_n:

  $$Q_n = \{s_1, s_2, \ldots, s_n\}, \quad t_{i-1} \leqslant s_i \leqslant t_i, \quad i = 1, 2, \ldots, n.$$

- Define the Riemann (nth partial) sum as a weighted average of the values of f:

  $$S_n = S_n(f, P_n, Q_n) = \sum_{i=1}^{n} f(s_i)\Delta t_i \,.$$

- Suppose that the mesh size $\delta(P_n) := \max_{1\leqslant i\leqslant n} \Delta t_i$ goes to zero as $n \to \infty$. If the limit $\lim_{n\to\infty} S_n$ exists and does not depend on the choice of partitions P_n and Q_n, then this limit is called the *Riemann integral* of f on $[0, T]$, denoted as usual by $\int_0^T f(t)\,\mathrm{d}t$. The function f is called the *integrand*. If the Riemann integral exists, then f is said to be *Riemann integrable* on $[0, T]$. For instance, if f is continuous on $[0, T]$ (or the set of discontinuities of f is finite), then f is Riemann integrable on $[0, T]$. In fact, the Riemann integral exists if f is m–a.e. (i.e., Lebesgue almost everywhere) continuous on $[0, T]$.

11.1.2 The Integral of a Brownian Path

Our goal is now to consider computing the Riemann integral of a Brownian sample path w.r.t. time $t \in [0, T]$, i.e., for an outcome $\omega \in \Omega$:

$$I(T, \omega) := \int_0^T W(t, \omega)\,\mathrm{d}t \,. \tag{11.1}$$

Recall that, with probability one (i.e., for almost all $\omega \in \Omega$), Brownian paths are continuous functions of time t. Hence, almost any sample path $(t, W(t,\omega))$, $0 \leqslant t \leqslant T$, is continuous. The Riemann integral (11.1) of such a sample path hence exists and is given by

$$\int_0^T W(t,\omega)\,\mathrm{d}t = \lim_{\delta(P_n)\to 0} \sum_{i=1}^{n} W(s_i,\omega)(t_i - t_{i-1})\,.$$

It hence suffices to consider a uniform partition P_n of $[0,T]$ with step size $\Delta t = \frac{T}{n}$ and time points $t_i = i\,\Delta t$ for $0 \leqslant i \leqslant n$. Let Q_n be chosen so that $s_i = t_i$, $1 \leqslant i \leqslant n$. We then have the nth Riemann sum $S_n(\omega) := \Delta t \cdot \sum_{i=1}^{n} W(i\Delta t, \omega)$ and its limit converging to the Riemann integral of the Brownian path:

$$I(T,\omega) = \lim_{n\to\infty} S_n(\omega) = \lim_{n\to\infty} \frac{T}{n} \sum_{i=1}^{n} W(i\frac{T}{n},\omega)$$

for almost all $\omega \in \Omega$. Hence, as a random variable, the Riemann integral of Brownian motion is (a.s.) uniquely given by

$$I(T) \equiv \int_0^T W(t)\,\mathrm{d}t = \lim_{n\to\infty} S_n = \lim_{n\to\infty} \frac{T}{n} \sum_{i=1}^{n} W(i\frac{T}{n}).$$

We now show that the nth Riemann sum S_n is a normally distributed random variable.

Proposition 11.1. *The Riemann sum S_n is normally distributed with mean and variance*

$$\mathrm{E}[S_n] = 0 \quad \textit{and} \quad \mathrm{E}[(S_n)^2] = \frac{T(T+\Delta t)(T+\Delta t/2)}{3}\,.$$

Proof. Since S_n is a linear combination of jointly normal random variables, it is normally distributed. The expected value is

$$\mathrm{E}[S_n] = \mathrm{E}\left[\Delta t \cdot \sum_{i=1}^{n} W(t_i)\right] = \Delta t \sum_{i=1}^{n} \mathrm{E}[W(t_i)] = 0.$$

Then, $\mathrm{Var}(S_n) = \mathrm{E}[(S_n)^2]$ is given by [note: $t_i \wedge t_j = (\Delta t)(i \wedge j)$]

$$\begin{aligned}
\mathrm{E}[(S_n)^2] &= \mathrm{E}\left[\left(\Delta t \cdot \sum_{i=1}^{n} W(t_i)\right)^2\right] = (\Delta t)^2\,\mathrm{E}\left[\left(\sum_{i=1}^{n} W(t_i)\right)\left(\sum_{j=1}^{n} W(t_j)\right)\right] \\
&= (\Delta t)^2 \sum_{i=1}^{n}\sum_{j=1}^{n} \mathrm{E}[W(t_i)W(t_j)] = (\Delta t)^2 \sum_{i=1}^{n}\sum_{j=1}^{n} t_i \wedge t_j \\
&= (\Delta t)^3 \sum_{i=1}^{n}\sum_{j=1}^{n} i \wedge j = (\Delta t)^3 \sum_{k=1}^{n} k^2 = (\Delta t)^3\,\frac{n(n+1)(2n+1)}{6} \\
&= \frac{(\Delta t \cdot n)(\Delta t \cdot n + \Delta t)(\Delta t \cdot n + \Delta t/2)}{3} = \frac{T(T+\Delta t)(T+\Delta t/2)}{3}
\end{aligned}$$

since $\Delta t \cdot n = T$. □

The Reimann sums $\{S_n\}_{n\geqslant 1}$ hence form a sequence of normally distributed random variables. The limit of such a sequence is hence normally distributed and this gives us that

$I(T) = \lim_{n\to\infty} S_n$ is a normal random variable. This is true for all time values $T > 0$. As $n \to \infty$ (and hence $\Delta t = T/n \to 0$), we obtain the mean and variance of $I(T)$:

$$\mathrm{E}[S_n] \to \mathrm{E}[I(T)] = 0, \quad \mathrm{E}[S_n^2] \to \mathrm{E}[I^2(T)] = \frac{T^3}{3}.$$

Thus, the stochastic process $\{I(t)\}_{t>0}$ is a Gaussian process where $I(t) \sim Norm(0, t^3/3)$.

Alternatively, to obtain the moments of $I(t)$ it is instructive to apply the Fubini Theorem that allows for changing the order of the time integral and expectation integral. For all t and s with $0 \leqslant s \leqslant t$, we have

$$\begin{aligned}
\mathrm{E}[I(s)I(t)] &= \mathrm{E}\left[\left(\int_0^s W(u)\,\mathrm{d}u\right)\left(\int_0^t W(v)\,\mathrm{d}v\right)\right] = \int_0^t\int_0^s \mathrm{E}[W(u)W(v)]\,\mathrm{d}u\,\mathrm{d}v \\
&= \int_0^t\int_0^s \min\{u,v\}\,\mathrm{d}u\,\mathrm{d}v = \int_0^s\int_0^s \min\{u,v\}\,\mathrm{d}u\,\mathrm{d}v + \int_s^t\left(\int_0^s \min\{u,v\}\,\mathrm{d}u\right)\mathrm{d}v \\
&= \int_0^s\int_0^s \min\{u,v\}\,\mathrm{d}u\,\mathrm{d}v + \int_s^t\left(\int_0^s u\,\mathrm{d}u\right)\mathrm{d}v \\
&= \frac{s^3}{3} + (t-s)\frac{s^2}{2}.
\end{aligned}$$

The last line follows by computing each integral separately. The second integral follows trivially. The first integral is readily computed by writing $\min\{u,v\} = \min\{u,v\}\mathbb{I}_{u\leqslant v} + \min\{u,v\}\mathbb{I}_{v\leqslant u}$ and by symmetry:

$$\begin{aligned}
\int_0^s\int_0^s \min\{u,v\}\,\mathrm{d}u\,\mathrm{d}v &= \int_0^s\int_0^s \min\{u,v\}\mathbb{I}_{u\leqslant v}\,\mathrm{d}u\,\mathrm{d}v + \int_0^s\int_0^s \min\{u,v\}\mathbb{I}_{v\leqslant u}\,\mathrm{d}v\,\mathrm{d}u \\
&= \int_0^s\left(\int_0^v u\,\mathrm{d}u\right)\mathrm{d}v + \int_0^s\left(\int_0^u v\,\mathrm{d}v\right)\mathrm{d}u \\
&= 2\int_0^s\left(\int_0^v u\,\mathrm{d}u\right)\mathrm{d}v = \int_0^s v^2\,\mathrm{d}v = s^3/3.
\end{aligned}$$

Hence, applying the above formula for $s = t$ gives the variance $\mathrm{E}[I^2(t)] = \mathrm{E}[I(t)I(t)] = \frac{t^3}{3}$. The mean function of the integral process I is zero, $m_I(t) = \mathrm{E}[I(t)] = 0$ and the covariance function is

$$c_I(s,t) = \mathrm{E}[I(s)I(t)] = \frac{(s\wedge t)^3}{3} + |t-s|\frac{(s\wedge t)^2}{2}.$$

Example 11.1. Show that $Y(t) := W^3(t) - 3\int_0^t W(u)\,\mathrm{d}u$, $t \geqslant 0$, is a martingale w.r.t. any filtration $\{\mathcal{F}_t\}_{t\geqslant 0}$ for Brownian motion.

Solution. First we note that $Y(t) := W^3(t) - 3I(t)$, where the integral $I(t) \equiv \int_0^t W(u)\,\mathrm{d}u$ is a function of the history of the Brownian motion up to time t and is hence $\mathcal{F}_t$–measurable. That is, the integral process $\{I(t)\}_{t\geqslant 0}$ is adapted to the filtration and hence so is the process $\{Y(t)\}_{t\geqslant 0}$. The process is also integrable since $\mathrm{E}[|Y(t)|] \leqslant \mathrm{E}[|W^3(t)|] + 3\mathrm{E}[|I(t)|] = \mathrm{E}[|W^3(t)|] + 3\int_0^t \mathrm{E}[|W(u)|]\,\mathrm{d}u < \infty$. Now, for times $t, s \geqslant 0$ we consider $\mathrm{E}[I(t+s) \mid \mathcal{F}_t] \equiv \mathrm{E}_t[I(t+s)]$:

$$\begin{aligned}
\mathrm{E}_t[I(t+s)] &= \mathrm{E}_t\left[I(t) + \int_t^{t+s} W(u)\,\mathrm{d}u\right] = I(t) + \mathrm{E}_t\left[\int_t^{t+s} W(u)\,\mathrm{d}u\right] \\
&= I(t) + \int_t^{t+s} \mathrm{E}_t[W(u)]\,\mathrm{d}u = I(t) + \int_t^{t+s} W(t)\,\mathrm{d}u \\
&= I(t) + W(t)\int_t^{t+s} \mathrm{d}u = I(t) + sW(t),
\end{aligned}$$

where we used the martingale property of Brownian motion and Fubini's theorem in one of the terms. Note that in explicit integral form we have shown

$$\mathrm{E}_t\left[\int_0^{t+s} W(u)\,\mathrm{d}u\right] = \int_0^t W(u)\,\mathrm{d}u + sW(t).$$

Using the fact that $\{W^3(t) - 3tW(t)\}_{t\geqslant 0}$ and $\{W(t)\}_{t\geqslant 0}$ are martingales, we obtain

$$\begin{aligned}\mathrm{E}_t\left[W^3(t+s)\right] &= \mathrm{E}_t\left[W^3(t+s) - 3(t+s)W(t+s)\right] + \mathrm{E}_t\left[3(t+s)W(t+s)\right]\\ &= W^3(t) - 3tW(t) + 3(t+s)W(t) = W^3(t) + 3sW(t).\end{aligned}$$

Therefore,

$$\begin{aligned}\mathrm{E}_t[Y(t+s)] &= \mathrm{E}_t\left[W^3(t+s)\right] - 3\mathrm{E}_t[I(t+s)]\\ &= W^3(t) + 3sW(t) - 3I(t) - 3sW(t) = W^3(t) - 3I(t) = Y(t).\end{aligned}$$ □

11.2 The Riemann–Stieltjes Integral of Brownian Motion

Since it is possible to integrate Brownian paths (and functions of Brownian motion) w.r.t. time, it is interesting to find out what other integrals can be calculated for Brownian motion. The Riemann–Stieltjes integral generalizes the Riemann integral. It provides an integral of one function w.r.t. another appropriate one. So our goal is to define the integral of one stochastic process w.r.t. another one (say, w.r.t. Brownian motion).

11.2.1 The Riemann–Stieltjes Integral

The construction of the Riemann–Stieltjes integral goes as follows. Let f be a bounded function and g be a monotonically increasing function, both defined on $[0, T]$.

- For $n \in \mathbb{N}$, introduce partitions P_n and Q_n in the same manner as is done for the Riemann integral.
- If the limit of the partial sum over any (shrinking) partition,

$$\lim_{\substack{n\to\infty \\ \delta(P_n)\to 0}} \sum_{i=1}^{n} f(s_i)(g(t_i) - g(t_{i-1})),$$

  exists and is independent of the choice of P_n and Q_n, then it is called the *Riemann–Stieltjes integral* of f w.r.t. g on $[0, T]$ and is denoted by

$$\int_0^T f(t)\,\mathrm{d}g(t).$$

  The function g is called the *integrator.*
- By taking $g(x) = x$, the Riemann integral is simply seen to be a special case of the Riemann–Stieltjes integral.

Let us take a look at some important examples, as follows.

(1) Consider the *Heaviside* (unit) step function, H, defined by

$$H(x) = \mathbb{I}_{[0,\infty)}(x) = \begin{cases} 0 & \text{if } x < 0, \\ 1 & \text{if } x \geqslant 0. \end{cases}$$

Let f be continuous at an interior point $s \in (0, T)$, c be a nonnegative constant, and let $g(x) = c\,H(x - s)$. Then,

$$\int_0^T f(t)\,\mathrm{d}g(t) = c\,f(s).$$

Hence, when integrator g is a simple step function, the integral simply picks out one value of the continuous function f and this value corresponds to the value of f at the point of discontinuity of g. This is a *sifting property* of the step function integrator g.

(2) The first example extends into the more general case of a step function $g(x)$ assumed as a mixture of Heaviside unit step functions: $g(x) = \sum_{n=1}^{\infty} c_n H(x - s_n)$, where $c_n \geqslant 0$ for $n = 1, 2, 3, \ldots$ are chosen such that $\sum_{n=1}^{\infty} c_n$ converges and $\{s_n\}_{n\geqslant 1}$ is a sequence of distinct points in $(0, T)$. If f is continuous on $[0, T]$, then

$$\int_0^T f(t)\,\mathrm{d}g(t) = \sum_{n=1}^{\infty} c_n f(s_n).$$

We see that the integral is a sum over f evaluated at all points of discontinuity of g within the integration interval $[0, T]$. This extends the sifting property in the above first example.

(3) Suppose that f and g' are Riemann integrable on $[0, T]$. In that case

$$\int_0^T f(t)\,\mathrm{d}g(t) = \int_0^T f(t)g'(t)\,\mathrm{d}t.$$

Hence, when the integrator is differentiable, the Riemann–Stieltjes integral is simply the Riemann integral of fg', i.e., we formally have the differential $\mathrm{d}g(t) = g'(t)\,\mathrm{d}t$.

(4) Consider a CDF F that is a mixture of a discrete CDF F_1 and a continuous CDF F_2:

$$F(x) = w_1 F_1(x) + w_2 F_2(x), \quad F_1(x) = \sum_{n=1}^{\infty} p_n\, H(x - x_n), \quad F_2(x) = \int_{-\infty}^{x} p(t)\,\mathrm{d}t,$$

where w_1 and w_2 are nonnegative weights summing to one, $\{p_n\}_{n\geqslant 1}$ and $\{x_n\}_{n\geqslant 1}$ are, respectively, the mass probabilities and mass points of the discrete distribution, and $p(x) = F_2'(x)$ is the PDF of the continuous distribution. Then, for a bounded f:

$$\begin{aligned}\int_{-\infty}^{\infty} f(x)\,\mathrm{d}F(x) &= w_1 \int_{-\infty}^{\infty} f(x)\,\mathrm{d}F_1(x) + w_2 \int_{-\infty}^{\infty} f(x)\,\mathrm{d}F_2(x) \\ &= w_1 \sum_{n=1}^{\infty} p_n f(x_n) + w_2 \int_{-\infty}^{\infty} f(x)\,p(x)\,\mathrm{d}x.\end{aligned}$$

In the first integral, with F_1 as integrator, we used the result in example (2). This is an example in which the Riemann–Stieltjes integral gives us the expected value of a function $f(X)$ of a random variable X having a mixture distribution given by F_1 and F_2 with respective mixture probabilities (weights) w_1 and w_2. In particular, the above equation can be read as $\mathrm{E}[f(X)] = w_1\mathrm{E}^{(1)}[f(X)] + w_2\mathrm{E}^{(2)}[f(X)]$.

The Riemann–Stieltjes integral $\int_0^T f(t)\,\mathrm{d}g(t)$ can be extended on a larger class of functions. Recall that the p–variation of $f\colon [0,T] \to \mathbb{R}$ is

$$V_{[0,T]}^{(p)}(f) = \limsup_{\delta(P_n)\to 0} \sum_{i=1}^{n} |f(t_i) - f(t_{i-1})|^p,$$

where the limit is taken over all possible partitions $0 = t_0 < t_1 < \cdots < t_n = T$, shrinking as $n \to \infty$. The following result (stated without proof) shows that we can consider the Riemann–Stieltjes integral on a fairly extensive combination of functions f and integrator g whose combined variational properties satisfy a certain condition.

Proposition 11.2. *Assume that f and g do not have discontinuities at the same points within the integration interval $[0,T]$. Let the p-variation of f and the q-variation of g be finite for some $p, q > 0$, such that $\frac{1}{p} + \frac{1}{q} > 1$. Then, the Riemann–Stieltjes integral $\int_0^T f(t)\,\mathrm{d}g(t)$ exists and is finite.*

For example, if the integrator g is a function of bounded variation on $[0,T]$, and f is a continuous function, then both functions have finite (first) variation. Hence, we may use $p = q = 1$ in the above proposition and this confirms that the Riemann–Stieltjes integral of f w.r.t. g is defined.

11.2.2 Integrals w.r.t. Brownian Motion

It is known that (a.s.) the p-variation of a Brownian sample path on $[0,T]$ is finite for $p \geqslant 2$ and infinite for $p < 2$. In particular, we proved that the quadratic variation of Brownian motion is bounded but the first variation is unbounded. Applying Proposition 11.2 for $q = 2$ to the Riemann–Stieltjes integral $\int_0^T f(t)\,\mathrm{d}W(t)$ gives us that such an integral w.r.t. Brownian motion is well-defined if the p-variation of f is finite for some $p \in (0,2)$. For example, the integral exists if f is a function of bounded variation ($p = 1$) such as a monotone function or a continuously differentiable function. Thus, for example, the integrals

$$\int_0^T \mathrm{e}^t\,\mathrm{d}W(t), \quad \int_0^T t^\alpha\,\mathrm{d}W(t) \ \ (\alpha \geqslant 1)$$

exist as Riemann–Stieltjes integrals. However, the integral

$$\int_0^T W(t)\,\mathrm{d}W(t) \tag{11.2}$$

does *not* (a.s.) exist as a Riemann–Stieltjes integral. First, note that Proposition 11.2 is not applicable to the integral in (11.2) since $V_{[0,T]}^{(p)}(W)$ is finite iff $p \geqslant 2$. Hence, for $f(t) = g(t) = W(t)$ we have $p = q$ and $\frac{1}{p} + \frac{1}{p} \leqslant 1$ for $p \geqslant 2$. Second, let us show we can obtain different values of the integral in (11.2) for different intermediate partitions Q_n.

Consider the Riemann–Stieltjes sum $S_n = \sum_{i=1}^n W(t_{i-1})(W(t_i) - W(t_{i-1}))$, i.e., with the intermediate nodes $s_i = t_{i-1}$ for $i = 1, 2, \ldots, n$. We rewrite S_n as follows, upon using the algebraic identity $a(b-a) = -\frac{1}{2}(a-b)^2 + \frac{1}{2}(b^2 - a^2)$:

$$\begin{aligned} S_n &= -\frac{1}{2}\sum_{i=1}^{n}\{(W(t_i) - W(t_{i-1}))^2 - (W^2(t_i) - W^2(t_{i-1}))\} \\ &= -\frac{1}{2}\sum_{i=1}^{n}(W(t_i) - W(t_{i-1}))^2 + \frac{1}{2}\underbrace{\sum_{i=1}^{n}(W^2(t_i) - W^2(t_{i-1}))}_{=W^2(t_n)-W^2(t_0)=W^2(T)-W^2(0)}. \end{aligned}$$

As $n \to \infty$ and $\delta(P_n) \to 0$, we have $\sum_{i=1}^{n}(W(t_i) - W(t_{i-1}))^2 \to [W,W](T) = T$, i.e., the quadratic variation of Brownian motion. Therefore,

$$\lim_{n\to\infty} S_n = \frac{1}{2}\left(W^2(T) - W^2(0)\right) - \frac{T}{2}.$$

This limit is called the *Itô integral* of BM:

$$\int_0^T W(t)\,\mathrm{d}W(t) = \frac{1}{2}\left(W^2(T) - W^2(0)\right) - \frac{T}{2} = \frac{1}{2}(W^2(T) - T).$$

Recall that for a differentiable function f we would have obtained (the ordinary calculus result)

$$\int_0^T f(t)\,\mathrm{d}f(t) = \frac{f^2(T) - f^2(0)}{2}.$$

Consider another choice of the intermediate partition with upper endpoint $s_i = t_i$ for every $i = 1, 2, \ldots, n$. The Riemann–Stieltjes sum is then

$$\begin{aligned} S_n^* &= \sum_{i=1}^{n} W(t_i)(W(t_i) - W(t_{i-1})) \\ &= \underbrace{\frac{1}{2}\sum_{i=1}^{n}(W(t_i) - W(t_{i-1}))^2}_{\to[W,W](T)=T,\text{ as } n\to\infty} + \underbrace{\frac{1}{2}\sum_{i=1}^{n}(W^2(t_i) - W^2(t_{i-1}))}_{=W^2(T)-W^2(0)} \\ &\to \frac{1}{2}\left(W^2(T) - W^2(0)\right) + \frac{T}{2}, \text{ as } n \to \infty. \end{aligned}$$

For $0 \leqslant \alpha \leqslant 1$, consider a weighted average of S_n and S_n^*:

$$\begin{aligned} \alpha S_n + (1-\alpha)S_n^* &= \sum_{i=1}^{n}(\alpha W(t_{i-1}) + (1-\alpha)W(t_i))(W(t_i) - W(t_{i-1})) \\ &\to \frac{1}{2}\left(W^2(T) - W^2(0)\right) + \frac{T}{2} - \alpha T, \text{ as } n \to \infty. \end{aligned}$$

An interesting case is when the midpoint is used, i.e., $\alpha = \frac{1}{2}$. The respective limit is called the *Stratonovich integral* of BM:

$$\int_0^T W(t) \circ \mathrm{d}W(t) = \frac{1}{2}\left(W^2(T) - W^2(0)\right).$$

The Stratonovich integral satisfies the usual rules of (nonstochastic) ordinary calculus such as the chain rule and integration by parts. The two types of stochastic integrals are related as

$$\int_0^T W(t) \circ \mathrm{d}W(t) = \int_0^T W(t)\,\mathrm{d}W(t) + \frac{T}{2}.$$

For a continuously differentiable function $f\colon \mathbb{R} \to \mathbb{R}$, it can be shown that the following conversion formula applies:

$$\int_0^T f(W(t)) \circ \mathrm{d}W(t) = \int_0^T f(W(t))\,\mathrm{d}W(t) + \frac{1}{2}\int_0^T f'(W(t))\,\mathrm{d}t,$$

where the respective integrals correspond to the Stratonovich and Itô integrals of a differentiable function $f(W(t))$ of BM. We note, however, that we have yet to give a precise general definition of such stochastic integrals. This is the topic of the next section.

11.3 The Itô Integral and Its Basic Properties

11.3.1 The Itô Integral for Simple Processes

Our goal is to give a construction of the Itô stochastic integral w.r.t. standard Brownian motion. Generally, we shall assume that the integrand is some other stochastic process that is adapted to a chosen filtration $\mathbb{F} = \{\mathcal{F}_t\}_{t\geqslant 0}$ for Brownian motion. We can, for instance, choose $\mathbb{F}$ as the natural filtration for Brownian motion, $\mathbb{F}^W = \{\mathcal{F}_t^W\}_{t\geqslant 0}$. In what follows, we shall consider all processes defined on some time interval $[0, T]$, for any $T > 0$, and we begin by considering a simple case with a piecewise-constant integrand.

Definition 11.1. A continuous-time stochastic process $\{C(t)\}_{0\leqslant t\leqslant T}$ defined on the filtered probability space $(\Omega, \mathcal{F}, \mathbb{P}, \mathbb{F})$ is said to be a *simple process* (or *step-stochastic process*) if there exists a time partition $P_n = \{t_0, t_1, \ldots, t_n\}$ of $[0, T]$, where $t_0 = 0$ and $t_n = T$, such that the process C is constant on each subinterval $[t_i, t_{i+1})$, $0 \leqslant i \leqslant n-1$. In other words, there exists random variables $\xi_0, \xi_1, \ldots, \xi_{n-1}$ such that:

1. ξ_i is $\mathcal{F}_{t_i}$-measurable for $i = 0, 1, \ldots, n-1$ (i.e., the process C is adapted to $\mathbb{F}$);
2. $C(t) = \sum_{i=0}^{n-1} \xi_i \mathbb{I}_{[t_i, t_{i+1})}(t)$, i.e., $C(t) = \xi_i$ for $t \in [t_i, t_{i+1})$.

The simple process C is said to be *square integrable* if

$$\mathrm{E}\left[\int_0^T C^2(s)\,\mathrm{d}s\right] < \infty \iff \mathrm{E}[\xi_i^2] < \infty \text{ for } i = 0, 1, \ldots, n-1.$$

The process $\{C(t)\}_{0\leqslant t\leqslant T}$ is defined as a right-continuous step process. For instance, a piecewise-constant approximation of Brownian motion is the simple process:

$$C(t) = \xi_i \equiv W(t_i) \text{ for } t \in [t_i, t_{i+1}).$$

Note that the process $C(t)$ on the interval $t \in [t_i, t_{i+1})$ is fixed to BM at time t_i (it is a $Norm(0, t_i)$ random variable). For any path ω, the graph of $C(t, \omega)$ as function of time $t \in [0, T]$ is piecewise constant (step function) with fixed value $C(t, \omega) = W(t_i, \omega) \equiv \xi_i(\omega)$ on every time interval $t \in [t_i, t_{i+1})$. Figure 11.1 depicts a sample path of BM and an approximation to it by the path of a simple process on the interval $[0, 1]$.

The Itô integral of a simple process can be defined as a Riemann–Stieltjes sum evaluated at the *left endpoint* of subintervals $[t_i, t_{i+1})$. Hence, the simplest case of an Itô integral of an indicator function $\mathbb{I}_{[a,b]}(t)$, $0 \leqslant a < b \leqslant T$, is the Riemann–Stieltjes integral w.r.t. W:

$$\int_0^T \mathbb{I}_{[a,b]}(t)\,\mathrm{d}W(t) = \int_a^b \mathrm{d}W(t) = W(b) - W(a).$$

The Riemann–Stieltjes integral of a step function (i.e., a linear combination of indicator functions) gives us the working definition of the Itô integral of a simple process as follows.

Definition 11.2. The Itô integral $I(t)$ of a simple process $C(s) = \sum_{i=0}^{n-1} \xi_i \mathbb{I}_{[t_i,t_{i+1})}(s)$ on any interval $[0,t]$, $0 \leqslant t \leqslant T$, is

$$I(t) \equiv \int_0^t C(s)\,\mathrm{d}W(s) = \sum_{i=0}^{k-1} \xi_i (W(t_{i+1}) - W(t_i)) + \xi_k (W(t) - W(t_k)), \tag{11.3}$$

for $t_k \leqslant t \leqslant t_{k+1}$. For the integration interval $[0,T]$, we obtain

$$I(T) = \int_0^T C(s)\,\mathrm{d}W(s) = \sum_{i=0}^{n-1} \xi_i (W(t_{i+1}) - W(t_i)).$$

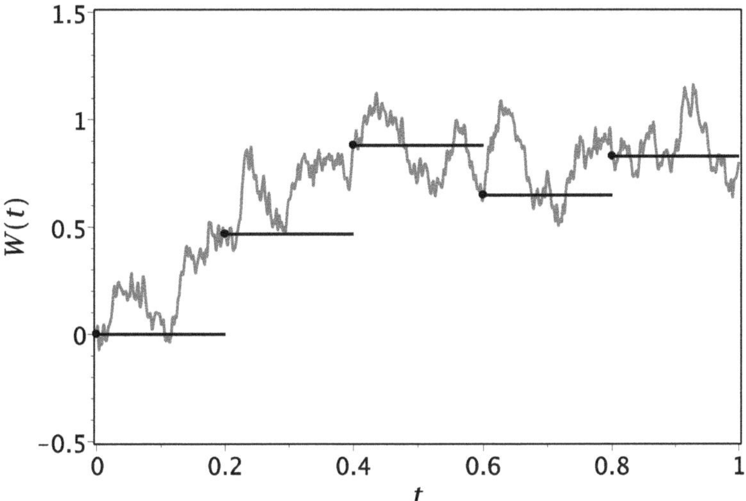

FIGURE 11.1: A Brownian sample path and its approximation by a simple process.

The Itô integral of a general process $X \equiv \{X(t), t \geqslant 0\}$ adapted to a filtration $\mathbb{F}$ for BM is defined as the mean-square limit of integrals of simple processes that approximate X. Consider a *square-integrable* [1] continuous-time process $\{X(t)\}_{0 \leqslant t \leqslant T}$ adapted to $\mathbb{F}$:

$$\mathrm{E}\left[\int_0^T X^2(t)\,\mathrm{d}t\right] < \infty \text{ and } X(t) \text{ is } \mathcal{F}_t\text{-measurable for } 0 \leqslant t \leqslant T.$$

[1] The square integrability condition is also denoted by writing $X \in L^2([0,T],\Omega)$. Throughout we assume that the integrand process X is measurable. That is, for every Borel set $B \in \mathcal{B}(\mathbb{R})$, the sets $\{(t,\omega) : X(t,\omega) \in B\} \in \mathcal{B}(\mathbb{R}_+) \times \mathcal{F}$. By Fubini's Theorem, assuming $\mathrm{E}[X^2(t)] < \infty$ for all $t \geqslant 0$, then this expectation is a Lebesgue-measurable function of time t and we may interchange the expectation integral with the time integral: $\mathrm{E}\left[\int_0^T X^2(t)\,\mathrm{d}t\right] = \int_0^T \mathrm{E}\left[X^2(t)\right]\,\mathrm{d}t$.

The process X on $[0,T]$ can be approximated by a sequence of simple processes as follows:

- select a partition $P_n = \{t_0, t_1, \dots, t_n\}$ of $[0,T]$, e.g., $t_i = i\frac{T}{n}$ for $i = 0, 1, \dots, n$;
- set $\xi_i = X(t_i)$ for $i = 0, 1, \dots, n-1$;
- set $C^{(n)}(t) = \sum_{i=0}^{n-1} \xi_i \mathbb{I}_{[t_i, t_{i+1})}(t)$.

As the maximum step size $\delta(P_n)$ goes to 0 as $n \to \infty$, the sequence of simple processes $\{C^{(n)}(t)\}_{n\geqslant 1}$ gives a better and better approximation of the continuously varying process X. The precise convergence condition is specified by requiring that

$$\lim_{n\to\infty} \mathrm{E}\left[\int_0^T \left(X(t) - C^{(n)}(t)\right)^2 \mathrm{d}t\right] = 0. \tag{11.4}$$

Given an adapted process X satisfying the above square integrability condition, it can be proven that there exists such a sequence of square-integrable and adapted simple processes such that (11.4) holds. The corresponding sequence is said to approximate the process X. Then, the *Itô integral* $I(t)$, $0 \leqslant t \leqslant T$, of a general process X is defined as the mean-square limit of integrals of an approximating sequence of simple processes:

$$I(t) \equiv \int_0^t X(s)\,\mathrm{d}W(s) := \lim_{n\to\infty} \int_0^t C^{(n)}(s)\,\mathrm{d}W(s) \equiv \lim_{n\to\infty} I^{(n)}(t). \tag{11.5}$$

The above mean-square limit really means that the sequence of Itô integral random variables $\{I^{(n)}(t)\}_{n\geqslant 1}$ converges to the random variable $I(t)$ in the sense of $L^2(\Omega)$, i.e., for each $t \geqslant 0$ we have

$$\lim_{n\to\infty} \mathrm{E}\left[\left(I(t) - I^{(n)}(t)\right)^2\right] = 0. \tag{11.6}$$

The assumed square integrability condition on X ensures that $I(t)$ exists and is given uniquely (a.s.). That is, for *any approximating sequence* satisfying the condition in (11.4), it can be shown that $\mathrm{E}\left[\left(I^{(m)}(t) - I^{(n)}(t)\right)^2\right] \to 0$, as $m, n \to \infty$. This implies that $\{I^{(n)}(t)\}_{n\geqslant 1}$ is a Cauchy sequence in $L^2(\Omega)$ and therefore has a unique limit in the L^2 sense.

11.3.2 Properties of the Itô Integral

The Itô integral $I(t) \equiv I_X(t) := \int_0^t X(s)\,\mathrm{d}W(s)$ of a continuous-time stochastic process $\{X(t)\}_{t\geqslant 0}$, which is adapted to a filtration $\mathbb{F} = \{\mathcal{F}_t\}_{t\geqslant 0}$ for Brownian motion and assumed to satisfy the *square-integrability condition* $\mathrm{E}\left[\int_0^T X^2(t)\,\mathrm{d}t\right] < \infty$, has the following properties.

(1) **Continuity.** Sample paths $\{I(t;\omega)\}_{0\leqslant t\leqslant T}$ are continuous functions of time t (a.s.).

(2) **Adaptivity.** $I(t)$ is $\mathcal{F}_t$-measurable for all $t \in [0,T]$.

(3) **Linearity.** Let $I_1(t) = \int_0^t X_1(s)\,\mathrm{d}W(s)$ and $I_2(t) = \int_0^T X_2(s)\,\mathrm{d}W(s)$ and assume that processes X_1 and X_2 meet the same requirements as those specified for process X above. Then, $c_1 I_1(t) + c_2 I_2(t) = \int_0^t (c_1 X_1(s) + c_2 X_2(s))\,\mathrm{d}W(s)$ for constants $c_1, c_2 \in \mathbb{R}$.

(4) **Martingale.** $\{I(t)\}_{0\leqslant t\leqslant T}$ is a martingale w.r.t. filtration $\mathbb{F}$.

(5) **Zero mean.** $\mathrm{E}[I(t)] = 0$ for $0 \leqslant t \leqslant T$.

(6) **Itô isometry.** $\mathrm{Var}(I(t)) = \mathrm{E}[I^2(t)] = \mathrm{E}\left[\left(\int_0^t X(s)\,\mathrm{d}W(s)\right)^2\right] = \int_0^t \mathrm{E}[X^2(s)]\,\mathrm{d}s$, i.e., the variance of an Itô integral on $[0,t]$ is equal to a Riemann integral (w.r.t. time variable s) of the second moment of the integrand process as a function of time $s \in [0,t]$.

Proof. To simplify the proof, we suppose that $\{X(t), t \geqslant 0\}$ is a simple process having the form $X(t) = \sum_{i=0}^{n-1} \xi_i \mathbb{I}_{[t_i,t_{i+1})}(t)$ with each ξ_i as $\mathcal{F}_{t_i}$-measurable.

(1) Fix $\omega \in \Omega$. Then $I(t, \omega)$ is a Riemann–Stieltjes integral of a piecewise-constant step function $X(s, \omega)$ with respect to the continuous integrator function $W(s, \omega)$ on the interval $s \in [0, t]$. Such an integral is a continuous function of the upper limit t.

(2) Let $t \in [t_k, t_{k+1}]$. Then, it is clear from the expression in (11.3) that $I(t)$ is $\mathcal{F}_t$-measurable since it is a function of only Brownian motions up to time t and all $\xi_i, 0 \leqslant i \leqslant k$, random variables are $\mathcal{F}_{t_i}$-measurable and hence $\mathcal{F}_t$-measurable.

(3) Suppose that X_1 and X_2 are defined on the same partition P_n (otherwise we combine the partitions for X_1 and X_2) and are given by:

$$X_j(t) = \sum_{i=0}^{n-1} \xi_i^{(j)} \mathbb{I}_{[t_i,t_{i+1})}(t), \quad j = 1, 2.$$

Then, $c_1 X_1(t) + c_2 X_2(t) = \sum_{i=0}^{n-1} (c_1 \xi_i^{(1)} + c_2 \xi_i^{(2)}) \mathbb{I}_{[t_i,t_{i+1})}(t)$ is a simple process. Integrate it on $[0, T]$ to obtain

$$\int_0^T (c_1 X_1(s) + c_2 X_2(s))\, \mathrm{d}W(s) = \sum_{i=1}^{n} (c_1 \xi_{i-1}^{(1)} + c_2 \xi_{i-1}^{(2)})(W(t_i) - W(t_{i-1}))$$
$$= c_1 \sum_{i=1}^{n} \xi_{i-1}^{(1)} (W(t_i) - W(t_{i-1})) + c_2 \sum_{i=1}^{n} \xi_{i-1}^{(2)} (W(t_i) - W(t_{i-1}))$$
$$= c_1 \int_0^T X_1(s)\, \mathrm{d}W(s) + c_2 \int_0^T X_2(s)\, \mathrm{d}W(s).$$

(4) Fix s and t such that $0 \leqslant s \leqslant t \leqslant T$. Let us show that $\mathrm{E}[I(t) \mid \mathcal{F}_s] = I(s)$. Suppose that $s \in [t_{m-1}, t_m]$ and $t \in [t_{k-1}, t_k]$ for some $1 \leqslant m \leqslant k \leqslant n$. Represent the integral of X on $[0, t]$ as follows:

$$I(t) = \underbrace{\sum_{i=0}^{m-2} \xi_i (W(t_{i+1}) - W(t_i)) + \xi_{m-1}(W(s) - W(t_{m-1}))}_{=I(s) \text{ is } \mathcal{F}_s\text{-measurable}}$$
$$+ \xi_{m-1}(W(t_m) - W(s)) + \sum_{j=m}^{k-2} \xi_j (W(t_{j+1}) - W(t_j)) + \xi_{k-1}(W(t) - W(t_{k-1})).$$

By taking the expectation of $I(t)$ conditional on $\mathcal{F}_s$ and applying properties of conditional expectations, we obtain

$$\mathrm{E}_s[I(t)] = I(s) + \underbrace{\xi_{m-1} \mathrm{E}_s[W(t_m) - W(s)]}_{(\text{since } \xi_{m-1} \text{ is } \mathcal{F}_s\text{-measurable})}$$
$$+ \sum_{j=m}^{k-2} \mathrm{E}_s \left[\xi_j (W(t_{j+1}) - W(t_j))\right] + \mathrm{E}_s \left[\xi_{k-1}(W(t) - W(t_{k-1}))\right].$$

Since $W(t_m) - W(s)$ is independent of $\mathcal{F}_s$, we have

$$\mathrm{E}_s[W(t_m) - W(s)] = \mathrm{E}[W(t_m) - W(s)] = 0.$$

By using the tower property and the independence property, we obtain

$$\begin{aligned}\mathrm{E}_s\left[\xi_j\left(W(t_{j+1})-W(t_j)\right)\right] &= \mathrm{E}_s\left[\xi_j\,\mathrm{E}_{t_j}[W(t_{j+1})-W(t_j)]\right]\\ &= \mathrm{E}_s\left[\xi_j\,\mathrm{E}[W(t_{j+1})-W(t_j)]\right]=0\end{aligned}$$

for $j=m+1,\ldots,k-2$. A similar step can be applied to the last expectation $\mathrm{E}_s\left[\xi_{k-1}\,\mathrm{E}_{t_{k-1}}[W(t)-W(t_{k-1})]\right]=\mathrm{E}_s\left[\xi_{k-1}\,\mathrm{E}[W(t)-W(t_{k-1})]\right]=0$. Therefore, we have $\mathrm{E}_s[I(t)]=I(s)$ for $0\leqslant s\leqslant t\leqslant T$. Since the Itô integral is adapted to $\mathbb{F}$ and is assumed integrable, $\mathrm{E}[|I(t)|]<\infty$, then it is a martingale w.r.t. $\mathbb{F}$.

(5) Since the integral process I is a martingale, the expected value $\mathrm{E}[I(t)]$ is constant and equal to $\mathrm{E}[I(0)]=0$ for all $0\leqslant t\leqslant T$. We note that the martingale property also gives us the identity

$$\mathrm{E}\left[\int_s^t X(u)\,\mathrm{d}W(u)\,\middle|\,\mathcal{F}_s\right]=\mathrm{E}[I(t)-I(s)\mid\mathcal{F}_s]=\mathrm{E}[I(t)\mid\mathcal{F}_s]-I(s)=I(s)-I(s)=0.$$

(6) For ease of presentation, we assume that $t=t_k$, for some $k=1,\ldots,n$, since the proof follows similarly for any value of $t\geqslant 0$. Then,

$$I(t_k)=\sum_{i=0}^{k-1}\xi_i\left(W(t_{i+1})-W(t_i)\right)=\sum_{i=0}^{k-1}\xi_i\,Z_i,$$

where $Z_i\equiv W(t_{i+1})-W(t_i)\sim Norm(0,t_{i+1}-t_i), i=0,\ldots,k-1$, are i.i.d. random variables. Note that each Z_i is independent of ξ_i since ξ_i is $\mathcal{F}_{t_i}$-measurable and the Brownian increment $W(t_{i+1})-W(t_i)$ is independent of $\mathcal{F}_{t_i}$. Squaring $I(t_k)$ and taking its expectation gives

$$\mathrm{E}\left[I^2(t_k)\right]=\sum_{i=0}^{k-1}\mathrm{E}\left[\xi_i^2Z_i^2\right]+2\sum\sum_{0\leqslant i<j\leqslant k-1}\mathrm{E}\left[\xi_i\xi_jZ_iZ_j\right].$$

The second summation involves expectations of products with $i<j$, i.e., $j\geqslant i+1$, and hence ξ_i,ξ_j and Z_i are $\mathcal{F}_{t_j}$-measurable. Applying the tower property by conditioning on $\mathcal{F}_{t_j}$ then gives

$$\begin{aligned}\mathrm{E}\left[\xi_i\xi_jZ_iZ_j\right] &= \mathrm{E}\left[\mathrm{E}[\xi_i\xi_jZ_iZ_j\mid\mathcal{F}_{t_j}]\right]\quad(\xi_i\xi_jZ_i\text{ is }\mathcal{F}_{t_j}\text{-measurable})\\ &= \mathrm{E}\left[\xi_i\xi_jZ_i\,\mathrm{E}[Z_j\mid\mathcal{F}_{t_j}]\right]=\mathrm{E}\left[\xi_i\xi_jZ_i\,\mathrm{E}[Z_j]\right]=0.\end{aligned}$$

Here we used the fact that $Z_j\equiv W(t_{j+1})-W(t_j)$ is independent of $\mathcal{F}_{t_j}$ and $\mathrm{E}[Z_j]=0$. [We note that the result is also more simply derived by using the independence of Z_j and $\xi_i\xi_jZ_i$.] Therefore, the above second summation is zero and we have from the first sum (upon using the independence of ξ_i and Z_i):

$$\begin{aligned}\mathrm{E}[I^2(t)] &= \sum_{i=0}^{k-1}\mathrm{E}\left[\xi_i^2Z_i^2\right]=\sum_{i=0}^{k-1}\mathrm{E}\left[\xi_i^2\right]\mathrm{E}\left[Z_i^2\right]\\ &= \sum_{i=0}^{k-1}\mathrm{E}\left[\xi_i^2\right](t_{i+1}-t_i)=\int_0^{t_k}\mathrm{E}\left[X^2(s)\right]\,\mathrm{d}s=\mathrm{E}\left[\int_0^{t_k}X^2(s)\,\mathrm{d}s\right],\end{aligned}$$

where we used the step function $\mathrm{E}[X^2(s)]=\sum_{i=0}^{k-1}\mathrm{E}[\xi_i^2]\,\mathbb{I}_{[t_i,t_{i+1})}(s)$, for $0\leqslant s\leqslant t_k$.

□

It is important to note that not all properties that are valid for Riemann integrals are necessarily true for Itô integrals. For example, suppose that two processes X and Y satisfy $X(t) \leqslant Y(t)$ (a.s.), i.e., $\mathbb{P}(X(t) \leqslant Y(t)) = 1$, for $0 \leqslant t \leqslant T$. Then, it is true that $\int_0^t X(s)\,\mathrm{d}s \leqslant \int_0^t Y(s)\,\mathrm{d}s$ (a.s.) for $0 \leqslant t \leqslant T$. However, this type of integral inequality property is *not* generally valid for Itô integrals $I_X(t) = \int_0^t X(s)\,\mathrm{d}W(s)$ and $I_Y(t) = \int_0^t Y(s)\,\mathrm{d}W(s)$. For example, consider the trivial case of constant processes $X(t) \equiv 0$ and $Y(t) \equiv 1$. Clearly, $\mathbb{P}(X(t) \leqslant Y(t)) = \mathbb{P}(0 \leqslant 1) = 1$. However, $I_X(t) \equiv 0$ and $I_Y(t) = \int_0^t \mathrm{d}W(s) = W(t)$ so that $\mathbb{P}(I_X(t) \leqslant I_Y(t)) = \mathbb{P}(0 \leqslant W(t)) = 1/2 \neq 1$.

Example 11.2. Show whether or not the following integrals are well-defined.

(a) $\int_0^1 \mathrm{e}^{W(t)}\,\mathrm{d}W(t)$.

(b) $\int_0^1 W(t+1)\,\mathrm{d}W(t)$.

(c) $\int_0^t \mathrm{e}^{W^2(s)}\,\mathrm{d}W(s)$ for $t \geqslant 0$.

(d) $\int_0^1 (1-t)^{-a}\,\mathrm{d}W(t)$ for $a \in \mathbb{R}$.

Solution.

(a) For $0 \leqslant t \leqslant 1$, the integrand $\mathrm{e}^{W(t)}$ is $\mathcal{F}_t$-measurable. So the integrand process, $X(t) := \mathrm{e}^{W(t)}$, is adapted to a filtration for Brownian motion. Now, we check if the integrand is square integrable:

$$\int_0^1 \mathrm{E}[\mathrm{e}^{2W(t)}]\,\mathrm{d}t = \int_0^1 \mathrm{e}^{2t}\,\mathrm{d}t = \frac{\mathrm{e}^2 - 1}{2} < \infty.$$

Therefore, the integral is defined.

(b) Note, in this case the integrand process $X(t) := W(t+1)$ is $\mathcal{F}_{t+1}$-measurable, but not $\mathcal{F}_t$-measurable and hence the Itô integral is not defined.

(c) First, find the second moment of the integrand:

$$\mathrm{E}\big[\mathrm{e}^{2W^2(s)}\big] = \int_{-\infty}^{\infty} \mathrm{e}^{2sz^2}\, n(z)\,\mathrm{d}z = \int_{-\infty}^{\infty} \frac{1}{\sqrt{2\pi}} \mathrm{e}^{-(\frac{1}{2}-2s)z^2}\,\mathrm{d}z$$

and this has finite value $\frac{1}{\sqrt{1-4s}}$ iff $s < \frac{1}{4}$. Now, the integral of $\mathrm{E}\big[\mathrm{e}^{2W^2(s)}\big]$ on $s \in [0,t]$ is finite iff $0 \leqslant t \leqslant \frac{1}{4}$. So the Itô integral is defined for $t \in [0, \frac{1}{4}]$.

(d) Note that the integrand $X(t) = (1-t)^{-a}$ is just an ordinary function of time t, and $\int_0^1 \mathrm{E}[X^2(t)]\,\mathrm{d}t = \int_0^1 (1-t)^{-2a}\,\mathrm{d}t < \infty$ iff $a < \frac{1}{2}$. So the Itô integral is defined iff $a < \frac{1}{2}$.

□

Before discussing further properties of the Itô integral, we now present a useful formula for computing the covariance between two Itô integrals (w.r.t. the same Brownian motion) as follows by Itô isometry. In particular, let X and Y be two adapted processes such that each satisfies the square integrability condition, i.e., assume $\int_0^t \mathrm{E}[X^2(s)]\,\mathrm{d}s < \infty$ and $\int_0^t \mathrm{E}[Y^2(s)]\,\mathrm{d}s < \infty$. Then, $I_X(t) := \int_0^t X(s)\,\mathrm{d}W(s)$ and $I_Y(t) := \int_0^t Y(s)\,\mathrm{d}W(s)$ have covariance

$$\mathrm{E}[I_X(t)I_Y(t)] \equiv \mathrm{E}\left[\int_0^t X(s)\,\mathrm{d}W(s) \int_0^t Y(s)\,\mathrm{d}W(s)\right] = \int_0^t \mathrm{E}[X(s)Y(s)]\,\mathrm{d}s. \tag{11.7}$$

Note that the Itô integrals have zero mean, $\mathrm{E}[I_X(t)] = \mathrm{E}[I_Y(t)] = 0$. Hence, their covariance $\mathrm{Cov}(I_X(t), I_Y(t)) = \mathrm{E}[I_X(t)I_Y(t)]$. The formula in (11.7) is readily proven by writing the product $I_X I_Y = \frac{1}{2}(I_X + I_Y)^2 - \frac{1}{2}I_X^2 - \frac{1}{2}I_Y^2 = \frac{1}{2}I_{X+Y}^2 - \frac{1}{2}I_X^2 - \frac{1}{2}I_Y^2$. Using linearity of expectations and applying Itô isometry three times gives the result:

$$\begin{aligned}
\mathrm{E}[I_X(t)I_Y(t)] &= \frac{1}{2}\left(\mathrm{E}[I_{X+Y}^2(t)] - \mathrm{E}[I_X^2(t)] - \mathrm{E}[I_Y^2(t)]\right) \\
&= \frac{1}{2}\left(\int_0^t \mathrm{E}[(X(s)+Y(s))^2]\,\mathrm{d}s - \int_0^t \mathrm{E}[X^2(s)]\,\mathrm{d}s - \int_0^t \mathrm{E}[Y^2(s)]\,\mathrm{d}s\right) \\
&= \int_0^t \mathrm{E}\left[\frac{1}{2}(X(s)+Y(s))^2 - \frac{1}{2}X^2(s) - \frac{1}{2}Y^2(s)\right]\mathrm{d}s = \int_0^t \mathrm{E}[X(s)Y(s)]\,\mathrm{d}s.
\end{aligned}$$

The result in (11.7) also leads to a formula for the covariance, $\mathrm{Cov}(I_X(t), I_Y(u)) = \mathrm{E}[I_X(t)I_Y(u)]$, between two Itô integrals at different times, $0 \leqslant t \leqslant u$:

$$\mathrm{E}[I_X(t)I_Y(u)] \equiv \mathrm{E}\left[\int_0^t X(s)\,\mathrm{d}W(s) \int_0^u Y(s)\,\mathrm{d}W(s)\right] = \int_0^t \mathrm{E}[X(s)Y(s)]\,\mathrm{d}s. \tag{11.8}$$

This follows from the martingale property of an Itô integral and by conditioning on $\mathcal{F}_t$, with $I_X(t)$ as $\mathcal{F}_t$-measurable, while using the tower property:

$$\mathrm{E}[I_X(t)I_Y(u)] = \mathrm{E}\big[\,\mathrm{E}[I_X(t)I_Y(u) \mid \mathcal{F}_t]\,\big] = \mathrm{E}\big[\,I_X(t)\,\mathrm{E}[I_Y(u) \mid \mathcal{F}_t]\,\big] = \mathrm{E}[I_X(t)I_Y(t)].$$

11.4 Itô Processes and Their Properties

11.4.1 Gaussian Processes Generated by Itô Integrals

The Itô integral of a nonrandom (ordinary) differentiable function f can be considered as a Riemann–Stieltjes integral with any path of Brownian motion acting as the integrator function w.r.t. time. Thus it can be reduced to a Riemann integral by using the integration by parts formula:

$$I(t) = \int_0^t f(s)\,\mathrm{d}W(s) = f(t)W(t) - f(0)W(0) - \int_0^t f'(s)W(s)\,\mathrm{d}s.$$

We know that the Riemann integral of Brownian motion is a Gaussian process. In a similar way, one can prove that the integral $\int_0^t f'(s)W(s)\,\mathrm{d}s$ is a Gaussian process as well (being considered as a function of the upper limit t). Thus, $I(t)$ is a Gaussian process as well. This result is proved below for a more general case.

Theorem 11.3. *Let $f\colon [0,\infty) \to \mathbb{R}$ be a nonrandom function such that $\int_0^T f^2(t)\,\mathrm{d}t < \infty$ for some $T > 0$. Then, the Itô integral $I(t) = \int_0^t f(s)\,\mathrm{d}W(s)$, $0 \leqslant t \leqslant T$, is a Gaussian process with mean zero and covariance function given by*

$$c_I(t,s) := \mathrm{Cov}(I(t), I(s)) = \int_0^{t\wedge s} f^2(u)\,\mathrm{d}u, \quad 0 \leqslant t, s \leqslant T.$$

Proof. It suffices to show the property for $0 \leqslant s \leqslant t \leqslant T$ where $t \wedge s = s$. By the zero mean property of Itô integrals, we have $m_I(t) = \mathrm{E}[I(t)] \equiv 0$. For the covariance function $c_I(t, s)$ we have, upon using the tower property and martingale property of the Itô integral,

$$c_I(t,s) \equiv \mathrm{E}\left[\int_0^t f(u)\,\mathrm{d}W(u) \int_0^s f(u)\,\mathrm{d}W(u)\right] \\ = \mathrm{E}[I(t)I(s)] = \mathrm{E}\big[I(s)\,\mathrm{E}[I(t) \mid F_s]\big] = \mathrm{E}[I^2(s)].$$

This expectation is evaluated by the Itô isometry formula:

$$\mathrm{E}[I^2(s)] \equiv \mathrm{E}\left[\left(\int_0^s f(u)\,\mathrm{d}W(u)\right)^2\right] = \int_0^s \mathrm{E}\left[f^2(u)\right]\,\mathrm{d}u = \int_0^s f^2(u)\,\mathrm{d}u$$

where $f(u)$ is nonrandom. Finally, by using the Itô formula presented in the next subsection, we can obtain the moment generating function of $I(t)$:

$$M_{I(t)}(\alpha) := \mathrm{E}\left[\mathrm{e}^{\alpha I(t)}\right] = \mathrm{e}^{\frac{1}{2}\alpha^2 \int_0^t f^2(s)\,\mathrm{d}s}, \quad \forall \alpha \in \mathbb{R}.$$

This is the unique moment generating function of a normal random variable with mean zero and variance $\int_0^t f^2(s)\,\mathrm{d}s$. Therefore, $I(t) \sim \mathit{Norm}(0, \int_0^t f^2(s)\,\mathrm{d}s)$. ☐

It should be remarked that the above result can be stated as

$$\int_0^t f(s)\,\mathrm{d}W(s) \sim \mathit{Norm}\left(0, \int_0^t f^2(s)\,\mathrm{d}s\right) \stackrel{d}{=} W(g(t))$$

where g is a function of time t as defined by $g(t) := \int_0^t f^2(s)\,\mathrm{d}s$. That is, the Itô integral of the ordinary function f on $[0, t]$ has the same distribution as standard Brownian motion at a time given by $g(t)$. This is a simple type of *time-changed Brownian motion* where in this case the time change, $g(t)$, is an ordinary function of time t.

Example 11.3. The process $X(t) = \int_0^t s\,\mathrm{d}W(s)$ is a Gaussian process with mean zero and variance $\mathrm{Var}(X(t)) = \int_0^t s^2\,\mathrm{d}s = t^3/3$, i.e., $X(t) \stackrel{d}{=} W(g(t))$ where $g(t) = t^3/3$.

11.4.2 Itô Processes

The sum of an Itô integral of a stochastic process and an ordinary (Riemann) integral generates another stochastic process called an Itô process.

Definition 11.3. Let $\{\mu(t)\}_{t \geqslant 0}$ and $\{\sigma(t)\}_{t \geqslant 0}$ be adapted to a filtration $\{\mathcal{F}_t\}_{t \geqslant 0}$ for standard Brownian motion and satisfying

$$\int_0^T \mathrm{E}[|\mu(t)|]\,\mathrm{d}t < \infty \text{ and } \int_0^T \mathrm{E}[\sigma^2(t)]\,\mathrm{d}t < \infty.$$

Then, the process

$$X(t) = X_0 + \int_0^t \mu(s)\,\mathrm{d}s + \int_0^t \sigma(s)\,\mathrm{d}W(s) \tag{11.9}$$

is well-defined for $0 \leqslant t \leqslant T$. It is called an *Itô process*. The processes $\{\mu(t)\}_{t \geqslant 0}$ and $\{\sigma(t)\}_{t \geqslant 0}$ are respectively called the *drift coefficient* process and the *diffusion or volatility coefficient* process.

The Itô process X can also be described by its so-called *stochastic differential equation (SDE)* which is obtained by "formally differentiating" (11.9) w.r.t. the time parameter t:

$$\mathrm{d}X(t) = \mu(t)\,\mathrm{d}t + \sigma(t)\,\mathrm{d}W(t). \tag{11.10}$$

We note that this SDE, along with the initial condition $X(0) = X_0$, is a shorthand way of writing the stochastic integral equation in (11.9). We interpret (11.10) through (11.9), where the latter has proper mathematical meaning as a sum of a Riemann integral and an Itô stochastic integral. That is, the Itô process $X \equiv \{X(t)\}_{t\geqslant 0}$ can be viewed as a solution to the SDE in (11.10) with the initial condition $X(0) = X_0$. The differential representation in (11.10) only has rigorous mathematical meaning by way of the respective integral representations in (11.9).

Some examples of Itô processes are as follows.

(a) Let $X(0) = x_0$, $\mu(t) \equiv \mu$ and $\sigma(t) \equiv \sigma$ be constants. Then, we obtain a drifted BM (i.e., BM with constant drift μ):

$$X(t) = x_0 + \int_0^t \mu\,\mathrm{d}s + \int_0^t \sigma\,\mathrm{d}W(s) = x_0 + \mu t + \sigma W(t).$$

(b) Let $\mu = \mu(t, X(t))$ and $\sigma = \sigma(t, X(t))$ be functions of both time t and the process value $X(t)$ at time t. The Itô process implicitly defined by the stochastic integral equation

$$X(t) = X_0 + \int_0^t \mu(s, X(s))\,\mathrm{d}s + \int_0^t \sigma(s, X(s))\,\mathrm{d}W(s)$$

is called a *diffusion process.*

(c) Let $\mu(t)$ and $\sigma(t)$ be nonrandom (ordinary) functions of time t. Then,

$$X(t) = X_0 + \int_0^t \mu(s)\,\mathrm{d}s + \int_0^t \sigma(s)\,\mathrm{d}W(s)\,,\quad t \geqslant 0,$$

with constant $X_0 \in \mathbb{R}$, is a Gaussian process with mean and covariance functions

$$m_X(t) = X_0 + \int_0^t \mu(u)\,du \quad\text{and}\quad c_X(t,s) = \int_0^{t\wedge s} \sigma^2(u)\,du.$$

The Itô process defined in (11.9) is given by a sum of a Riemann integral of μ and an Itô integral of σ. Both integrals being considered as functions of the upper limit t have continuous sample paths. Therefore, the Itô process has continuous sample paths as well.

So far, we have defined the Itô process as a stochastic integral w.r.t. Brownian motion. More generally, we can also define a stochastic integral w.r.t. an Itô process. Let the process $\{Y(t)\}_{t\geqslant 0}$ be adapted to a filtration for BM. We define the stochastic integral of Y *w.r.t. the Itô process* X, defined in (11.9), as follows:

$$\int_0^t Y(s)\,\mathrm{d}X(s) := \int_0^t Y(s)\mu(s)\,\mathrm{d}s + \int_0^t Y(s)\sigma(s)\,\mathrm{d}W(s), \quad t \geqslant 0.$$

Note that this is like substituting the stochastic differential $\mathrm{d}X(s) = \mu(s)\,\mathrm{d}s + \sigma(s)\,\mathrm{d}W(s)$ (given by (11.10)) into the left-hand integral and writing it as a sum of a Riemann and Itô integral. Note that in case the process is standard Brownian motion, i.e., $X(t) = W(t)$ with $\mu \equiv 0, \sigma \equiv 1$, we simply recover $\int_0^t Y(s)\,\mathrm{d}W(s)$, the stochastic integral of Y w.r.t. Brownian motion.

11.4.3 Quadratic (Co-)Variation

An important characteristic of a stochastic process is the *quadratic variation* that measures the accumulated variability of the process along its path. The quadratic variation is a path-dependent quantity. Recall that for Brownian motion we derived its quadratic variation on a time interval $[0,t]$ as $[W,W](t) = t$. So Brownian motion *accumulates quadratic variation at rate one per unit time.* This gives us a simple differential "rule":

$$\mathrm{d}[W,W](t) \equiv \mathrm{d}W(t)\,\mathrm{d}W(t) \equiv (\,\mathrm{d}W(t))^2 = \mathrm{d}t.$$

A practical way of thinking about this result is to say that a Brownian increment is of order $\mathcal{O}((\,\mathrm{d}t)^{1/2})$ as $\mathrm{d}t \to 0$. We essentially already used this fact in showing the non-differentiability of Brownian paths. We also saw that, formally, the quadratic variation of a continuously differentiable function f is zero. This fact is also realized by noting that $\mathrm{d}[f,f](t) = (\,\mathrm{d}f(t))^2 = (f'(t))^2(\,\mathrm{d}t)^2 = \mathcal{O}((\,\mathrm{d}t)^2)$ is negligible as $\mathrm{d}t \to 0$.

One can prove that the quadratic variation of the Itô integral $I_X = \int_0^t X(s)\,\mathrm{d}W(s)$ is

$$[I_X, I_X](t) = \int_0^t X^2(s)\,\mathrm{d}s, \quad t \geqslant 0. \tag{11.11}$$

So, the integral $I_X(t)$ *accumulates quadratic variation at the (generally random) rate of $X^2(t)$ per unit time at every time $t \geqslant 0$.* That is, in differential form (11.11) gives us the "rule":

$$\mathrm{d}[I_X, I_X](t) \equiv \mathrm{d}I_X(t)\,\mathrm{d}I_X(t) \equiv (\,\mathrm{d}I_X(t))^2 = X^2(t)\,\mathrm{d}t.$$

Similarly, we can define the *quadratic covariation* of two processes:

$$[X,Y](t) = \lim_{\delta(P_n)\to 0} \sum_{i=1}^{n} (X(t_i) - X(t_{i-1}))(Y(t_i) - Y(t_{i-1})). \tag{11.12}$$

Clearly, the quadratic covariation is a bilinear functional. Let us consider several examples.

1. Let $X(t)$ be a continuously differentiable ($C^1(\mathbb{R})$) function that satisfies $\mathrm{d}X(t) = \mu_X(t)\,\mathrm{d}t$ and let $Y(t)$ be an Itô process. Then, $[X,Y](t) = 0$ for $t \geqslant 0$. In differential form, this fact reads as $\mathrm{d}X(t)\,\mathrm{d}Y(t) = 0$. Since Brownian motion is itself an Itô process and the function $X(t) = t$ belongs to $C^1(\mathbb{R})$, we have $[t,W](t) = 0$. This last fact is recorded in differential form as the "rule": $\mathrm{d}t\,\mathrm{d}W(t) = 0$, i.e., an infinitesimal time increment times an infinitesimal Brownian increment gives zero.

Proof. For $t \geqslant 0$,

$$\begin{aligned} \big|[X,Y](t)\big| &\leqslant \lim_{\delta(P_n)\to 0} \left| \sum_{i=1}^{n} (X(t_i) - X(t_{i-1}))(Y(t_i) - Y(t_{i-1})) \right| \\ &\leqslant \underbrace{\lim_{\delta(P_n)\to 0} \max_{1\leqslant i\leqslant n} \big|Y(t_i) - Y(t_{i-1})\big|}_{=0 \text{ a.s.}} \cdot \underbrace{\lim_{\delta(P_n)\to 0} \sum_{i=1}^{n} \big|X(t_i) - X(t_{i-1})\big|}_{=V_X^{(1)}(t)<\infty} = 0 \end{aligned}$$

Here we applied the Heine–Cantor theorem, which states that a continuous function (in this case Y is a.s. continuous) on a finite interval is uniformly continuous and the fact that sample paths of X have finite first variation. ☐

2. The covariation of two Itô processes X and Y defined by

$$X(t) = X(0) + \int_0^t \mu_X(s)\, ds + \int_0^t \sigma_X(s)\, dW(s)$$

and

$$Y(t) = Y(0) + \int_0^t \mu_Y(s)\, ds + \int_0^t \sigma_Y(s)\, dW(s)$$

is given by

$$[X, Y](t) = \int_0^t \sigma_X(s)\sigma_Y(s)\, ds. \tag{11.13}$$

A simple (heuristic) way to arrive at this result is to make recourse to the simple differential rules. In particular, the two processes satisfy

$$dX(t) = \mu_X(t)\, dt + \sigma_X(t)\, dW(t) \quad \text{and} \quad dY(t) = \mu_Y(t)\, dt + \sigma_Y(t)\, dW(t).$$

Hence, by multiplying out all terms in the differentials, we have

$$\begin{aligned} dX(t)\, dY(t) &= \{\mu_X(t)\, dt + \sigma_X(t)\, dW(t)\}\{\mu_Y(t)\, dt + \sigma_Y(t)\, dW(t)\} \\ &= \mu_X(t)\mu_Y(t)(dt)^2 + (\mu_X(t)\sigma_Y(t) + \mu_Y(t)\sigma_X(t))\, dt\, dW(t) \\ &\quad + \sigma_X(t)\sigma_Y(t)(dW(t))^2 . \end{aligned}$$

Now, using the rules $(dt)^2 \equiv 0$, $dt\, dW(t) \equiv 0$ and $(dW(t))^2 = dt$ gives the differential form of the quadratic covariation in (11.13):

$$d[X, Y](t) \equiv dX(t)\, dY(t) = \sigma_X(t)\sigma_Y(t)\, dt.$$

An important application of quadratic covariation is the integration by parts formula given just below. Consider a sequence of partitions $\{P_n\}_{n\geqslant 1}$ of $[0, t]$ (such that $\delta(P_n) \to 0$, as $n \to \infty$) and rewrite the sum of products of increments of X and Y on the partition P_n as follows:

$$\begin{aligned} \sum_{i=1}^{n}(X(t_i) - X(t_{i-1}))(Y(t_i) - Y(t_{i-1})) &= \underbrace{\sum_{i=1}^{n} \left(X(t_i)Y(t_i) - X(t_{i-1})Y(t_{i-1})\right)}_{=X(t)Y(t)-X(0)Y(0)} \\ - \underbrace{\sum_{i=1}^{n} X(t_{i-1})\big(Y(t_i) - Y(t_{i-1})\big)}_{\to\int_0^t X(s)\, dY(s),\ \text{as } n\to\infty} &- \underbrace{\sum_{i=1}^{n} Y(t_{i-1})\big(X(t_i) - X(t_{i-1})\big)}_{\to\int_0^t Y(s)\, dX(s),\ \text{as } n\to\infty} . \end{aligned}$$

Thus, the quadratic covariation of two Itô processes is

$$[X, Y](t) = X(t)Y(t) - X(0)Y(0) - \int_0^t X(s)\, dY(s) - \int_0^t Y(s)\, dX(s).$$

Alternatively, we write

$$X(t)Y(t) - X(0)Y(0) = \int_0^t X(s)\, dY(s) + \int_0^t Y(s)\, dX(s) + [X, Y](t). \tag{11.14}$$

In differential form this gives us the important *Itô product rule*:

$$d(X(t)Y(t)) = X(t)\, dY(t) + Y(t)\, dX(t) + dX(t)\, dY(t). \tag{11.15}$$

The reader should observe that the stochastic differential of a product of two processes *does not* obey the same differential product rule as in ordinary calculus. The extra term $\mathrm{d}X(t)\,\mathrm{d}Y(t) \equiv \mathrm{d}[X,Y](t)$ is the product of the two differentials, which is generally nonzero. In particular, if both processes X and Y are driven by a Brownian increment $\mathrm{d}W(t)$ then their paths are nondifferentiable and hence the quadratic covariation $[X,Y](t)$ is nonzero. Later we shall see that the above product rule also follows as a special case of the Itô formula derived for smooth functions of two processes.

11.5 Itô's Formula for Functions of BM and Itô Processes

11.5.1 Itô's Formula for Functions of BM

The Itô formula is a stochastic chain rule that allows us to find stochastic differentials of functions of Brownian motion as well as functions of an Itô process. The ordinary chain rule written for two differentiable functions f and g is as follows:

- $\frac{\mathrm{d}}{\mathrm{d}t} f(g(t)) = f'(g(t))\, g'(t)$ (derivative form);
- $\mathrm{d}f(g(t)) = f'(g(t))\, \mathrm{d}g(t)$ (differential form);
- $f(g(t)) - f(g(0)) = \int_0^t f'(g(s))\, \mathrm{d}g(s)$ (integral form).

However, we cannot immediately apply this rule to $f(W(t))$ since Brownian motion W has nondifferentiable sample paths. Assume that f has continuous derivatives of first, second, and higher orders. Consider the Taylor series expansion for a smooth function f about the value $W(t)$:

$$f(W(t+\delta t)) - f(W(t)) = f'(W(t)) \underbrace{\left(W(t+\delta t) - W(t)\right)}_{\text{of order } (\delta t)^{\frac{1}{2}}} + \frac{1}{2} f''(W(t)) \underbrace{\left(W(t+\delta t) - W(t)\right)^2}_{\text{of order } \delta t}$$

$$+ \frac{1}{6} f'''(W(t)) \underbrace{\left(W(t+\delta t) - W(t)\right)^3}_{\text{of order } (\delta t)^{\frac{3}{2}}} + \cdots,$$

where δt is a small time increment. A heuristic argument that leads us to the simplest version of the Itô formula goes as follows. In the infinitesimal limit, we take $\delta t \to \mathrm{d}t$ and $W(t+\delta t) - W(t) \to \mathrm{d}W(t)$, and we neglect all terms of order $(\delta t)^{3/2}$ and smaller (of higher power than $3/2$ in δt) to obtain

$$\mathrm{d}f(W(t)) = f'(W(t))\, \mathrm{d}W(t) + \frac{1}{2} f''(W(t))(\mathrm{d}W(t))^2.$$

By applying the simple rule $(\mathrm{d}W(t))^2 = \mathrm{d}t$, we obtain the *Itô formula* for $f(W(t))$, which can be stated in the respective differential and integral forms:

$$\mathrm{d}f(W(t)) = \frac{1}{2} f''(W(t))\, \mathrm{d}t + f'(W(t))\, \mathrm{d}W(t), \tag{11.16}$$

$$\int_0^t \mathrm{d}f(W(s)) := f(W(t)) - f(W(0)) = \frac{1}{2} \int_0^t f''(W(s))\, \mathrm{d}s + \int_0^t f'(W(s))\, \mathrm{d}W(s). \tag{11.17}$$

This formula holds for any twice continuously differentiable function $f \in C^2(\mathbb{R})$. The expression (11.17) tells us that $f(W) := \{f(W(t))\}_{t \geqslant 0}$ is an Itô process.

Only the skeleton of a proof of the Itô formula (11.17) is outlined below:

Proof. Let $P_n = \{t_i\}_{1\leqslant i\leqslant n}$ be a partition of $[0,t]$. Write $f(W(t)) - f(W(0))$ as a telescopic sum over points of P_n:

$$f(W(t)) - f(W(0)) = \sum_{i=1}^{n} \big(f(W(t_i)) - f(W(t_{i-1}))\big).$$

Now, apply Taylor's expansion formula to each term of the above sum:

$$f(W(t_i)) - f(W(t_{i-1})) = f'(W(t_{i-1}))\big(W(t_i) - W(t_{i-1})\big) + \frac{1}{2}f''(\theta_i)\big(W(t_i) - W(t_{i-1})\big)^2,$$

where θ_i lies between $W(t_{i-1})$ and $W(t_i)$ for $i = 1, 2, \ldots, n$. By taking limits as $n \to \infty$ and $\delta(P_n) \to 0$, the partial sums converge (in $L^2(\Omega)$) to the respective integrals:

$$\sum_{i=1}^{n} f'(W(t_{i-1}))\big(W(t_i) - W(t_{i-1})\big) \to \int_0^t f'(W(s))\,\mathrm{d}W(s) \quad \text{(an Itô integral)},$$

$$\sum_{i=1}^{n} f''(\theta_i)\big(W(t_i) - W(t_{i-1})\big)^2 \to \int_0^t f''(W(s))\,\mathrm{d}s \quad \text{(a Riemann integral)}. \qquad \square$$

Example 11.4. Find the stochastic differential $\mathrm{d}f(W(t))$ for functions:

(a) $f(x) = x^n$, $n \in \mathbb{N}$;

(b) $f(x) = \mathrm{e}^{\alpha x}$, $\alpha \in \mathbb{R}$.

Solution.
(a) Differentiating gives $f'(x) = nx^{n-1}$, $f''(x) = n(n-1)x^{n-2}$. Thus, (11.17) with $f(W(t)) = W^n(t), f(W(0)) = W^n(0) = 0$ reads

$$W^n(t) = \frac{n(n-1)}{2}\int_0^t W^{n-2}(s)\,\mathrm{d}s + n\int_0^t W^{n-1}(s)\,\mathrm{d}W(s).$$

The differential form of the above representation is

$$\mathrm{d}W^n(t) = \frac{n(n-1)}{2}W^{n-2}(t)\,\mathrm{d}t + nW^{n-1}(t)\,\mathrm{d}W(t).$$

By taking $n = 2$, we now also recover the well-known formula of the Itô integral of Brownian motion that we derived previously:

$$W^2(t) = \int_0^t \mathrm{d}s + 2\int_0^t W(s)\,\mathrm{d}W(s) \implies \int_0^t W(s)\,\mathrm{d}W(s) = \frac{1}{2}W^2(t) - \frac{t}{2}.$$

For $n = 3$, we have

$$W^3(t) = 3\int_0^t W(s)\,\mathrm{d}s + 3\int_0^t W^2(s)\,\mathrm{d}W(s).$$

Note that the Itô integral $\int_0^t W^2(s)\,\mathrm{d}W(s)$ is a (square-integrable) martingale with the property $\int_0^t \mathrm{E}[W^4(s)]\,\mathrm{d}s < \infty$. Hence, the process defined by $Y(t) := W^3(t) - 3\int_0^t W(s)\,\mathrm{d}s$, $t \geqslant 0$, is a martingale w.r.t. any Brownian filtration. We have proven this fact earlier in Example 11.1, but now it follows simply by applying an appropriate Itô formula and from the martingale property of the Itô integral.

(b) Differentiating gives $f'(x) = \alpha f(x)$ and $f''(x) = \alpha^2 f(x)$. Denote $X(t) = f(W(t)) =$

$e^{\alpha W(t)}$. Recall that X is a geometric Brownian motion (GBM). Now, by applying the Itô formula in (11.16) we have the stochastic differential of GBM:

$$\mathrm{d}X(t) = \frac{\alpha^2}{2} X(t)dt + \alpha X(t)\,\mathrm{d}W(t). \qquad \square$$

There are various important extensions of the Itô formula. In particular, consider the case of a stochastic process defined by $X(t) := f(t, W(t))$, $t \geqslant 0$, where the function $f(t,x) \in C^{1,2}$, i.e., we assume that the functions $f_t(t,x) := \frac{\partial f}{\partial t}(t,x)$, $f_x(t,x) := \frac{\partial f}{\partial x}(t,x)$, $f_{tx}(t,x) := \frac{\partial^2 f}{\partial t \partial x}(t,x)$, and $f_{xx}(t,x) := \frac{\partial^2 f}{\partial x^2}(t,x)$ are continuous. Let us heuristically apply a Taylor expansion to the differential $\mathrm{d}f(t,W(t)) = f(t+\mathrm{d}t, W(t)+\mathrm{d}W(t)) - f(t,W(t))$ and keep only terms up to second order in the Brownian increment $\mathrm{d}W(t)$ and first order in the time increment $\mathrm{d}t$:

$$\begin{aligned}\mathrm{d}f(t,W(t)) &= f_t(t,W(t))\,\mathrm{d}t + f_x(t,W(t))\,\mathrm{d}W(t) \\ &\quad + f_{tx}(t,W(t))\,\mathrm{d}t\,\mathrm{d}W(t) + \frac{1}{2} f_{xx}(t,W(t))(\mathrm{d}W(t))^2 + \cdots\end{aligned}$$

By the simple rules we have

$$(\mathrm{d}W(t))^2 \equiv \mathrm{d}t,\ (\mathrm{d}t)^2 \equiv 0,\ \ \mathrm{d}t\,\mathrm{d}W(t) \equiv 0.$$

Collecting the coefficient terms in $\mathrm{d}t$ and $\mathrm{d}W(t)$, the differential and integral forms of the *Itô formula* for $f(t,W(t))$ are then respectively given by

$$\mathrm{d}f(t,W(t)) = \left(f_t(t,W(t)) + \frac{1}{2} f_{xx}(t,W(t))\right)\mathrm{d}t + f_x(t,W(t))\,\mathrm{d}W(t), \tag{11.18}$$

$$\begin{aligned}f(t,W(t)) - f(0,W(0)) &= \int_0^t \left(f_u(u,W(u)) + \frac{1}{2} f_{xx}(u,W(u))\right)\mathrm{d}u \\ &\quad + \int_0^t f_x(u,W(u))\,\mathrm{d}W(u),\end{aligned} \tag{11.19}$$

for all $0 \leqslant t \leqslant T$.

Example 11.5. Find the stochastic differential of the GBM process $S(t) = S_0 e^{\alpha t + \sigma W(t)}$, $t \geqslant 0$, with constants $S_0 > 0$, $\alpha, \sigma \in \mathbb{R}$.

Solution. We represent $S(t) = f(t,W(t))$, where $f(t,x) := S_0 e^{\alpha t + \sigma x}$. Hence,

$$f_t(t,x) = \alpha f(t,x),\ \ f_x(t,x) = \sigma f(t,x),\ \ f_{xx}(t,x) = \sigma^2 f(t,x).$$

Substituting these partial derivatives into the Itô formula (11.18) gives

$$\begin{aligned}\mathrm{d}S(t) &= \left(\alpha f(t,W(t)) + \frac{\sigma^2}{2} f(t,W(t))\right)\mathrm{d}t + \sigma f(t,W(t))\,\mathrm{d}W(t) \\ &= (\alpha + \frac{\sigma^2}{2})S(t)\,\mathrm{d}t + \sigma S(t)\,\mathrm{d}W(t).\end{aligned} \qquad \square$$

Note that, in the above example, if we put $\alpha = \mu - \sigma^2/2$, with parameter $\mu \in \mathbb{R}$, then $S(t) = S_0 e^{(\mu - \sigma^2/2)t + \sigma W(t)}$ is a GBM satisfying the SDE

$$\mathrm{d}S(t) = \mu S(t)\,\mathrm{d}t + \sigma S(t)\,\mathrm{d}W(t)$$

with initial condition $S(0) = S_0$ and it is an Itô process where

$$S(t) = S_0 + \mu \int_0^t S(u)\,\mathrm{d}u + \sigma \int_0^t S(u)\,\mathrm{d}W(u).$$

Hence, $S(t) = S_0 \mathrm{e}^{(\mu-\sigma^2/2)t+\sigma W(t)}$, for $t \geqslant 0$, is an explicit solution to the above stochastic integral (or differential) equation whereby $S(t)$ is explicitly given as (an exponential) function in the BM $W(t)$ at time t. It is a martingale iff the drift coefficient $\mu = 0$. In fact, in the previous chapter, we already proved (using different methods) that $M(t) := \mathrm{e}^{-\frac{\sigma^2}{2}t+\sigma W(t)}$ is a martingale. It follows that the discounted process by $\mathrm{e}^{-\mu t}S(t) = M(t), t \geqslant 0$, is a martingale w.r.t. a filtration for BM.

11.5.2 Itô's Formula for Itô Processes

We are now ready to extend the Itô formula to the case of a process defined in terms of a smooth function of an Itô process and time t. Consider an Itô process $\{X(t)\}_{0\leqslant t\leqslant T}$ with the stochastic differential

$$\mathrm{d}X(t) = \mu(t)\,\mathrm{d}t + \sigma(t)\,\mathrm{d}W(t).$$

As in the previous version of the Itô formula obtained above, assume that $f(t,x) \in C^{1,2}$. Then, $Y(t) := f(t, X(t))$, $0 \leqslant t \leqslant T$, is also an Itô process. To obtain its stochastic differential we apply a Taylor expansion and keep only terms up to second order in the increment $\mathrm{d}X(t)$ and first order in the time increment $\mathrm{d}t$:

$$\mathrm{d}Y(t) = f_t(t, X(t))\,\mathrm{d}t + f_x(t, X(t))\,\mathrm{d}X(t) + \frac{1}{2} f_{xx}(t, X(t))(\mathrm{d}X(t))^2.$$

Note that the mixed partial derivative term $f_{tx}(t, X(t))\,\mathrm{d}t\,\mathrm{d}X(t) \equiv 0$ since $\mathrm{d}t\,\mathrm{d}X(t) = \mu(t)(\mathrm{d}t)^2 + \sigma(t)\,\mathrm{d}t\,\mathrm{d}W(t) \equiv 0$ upon using the simple rules $(\mathrm{d}t)^2 \equiv 0$, $\mathrm{d}t\,\mathrm{d}W(t) \equiv 0$. Also, $(\mathrm{d}X(t))^2 = \sigma^2(t)\,\mathrm{d}t$, and inserting the differential for $\mathrm{d}X(t)$ into the above equation and combining all coefficients multiplying $\mathrm{d}t$ and $\mathrm{d}W(t)$ finally gives us the *Itô formula* in differential form:

$$\begin{aligned} \mathrm{d}Y(t) \equiv \mathrm{d}f(t, X(t)) = \Big(f_t(t, X(t)) + \mu(t) f_x(t, X(t)) + \frac{1}{2}\sigma^2(t) f_{xx}(t, X(t)) \Big)\,\mathrm{d}t \\ + \sigma(t) f_x(t, X(t))\,\mathrm{d}W(t). \end{aligned} \tag{11.20}$$

The integral form of this is

$$\begin{aligned} f(t, X(t)) - f(0, X(0)) = \int_0^t \Big(f_s(s, X(s)) + \mu(s) f_x(s, X(s)) + \frac{1}{2}\sigma^2(s) f_{xx}(s, X(s)) \Big)\,\mathrm{d}s \\ + \int_0^t \sigma(s) f_x(s, X(s))\,\mathrm{d}W(s), \end{aligned} \tag{11.21}$$

for $0 \leqslant t \leqslant T$.

Note that in case $Y(t) = f(X(t))$, i.e., $f(t,x) = f(x)$ is not an explicit function of the time variable, then $f_t(t,x) \equiv 0$ and all partial derivatives are simply ordinary derivatives: $f_x(t,x) = f'(x), f_{xx}(t,x) = f''(x)$. The differential form of the Itô formula is $\mathrm{d}f(X(t)) = f'(X(t))\,\mathrm{d}X(t) + \frac{1}{2} f''(X(t))(\mathrm{d}X(t))^2$, i.e.,

$$\mathrm{d}f(X(t)) = \Big(\mu(t) f'(X(t)) + \frac{1}{2}\sigma^2(t) f''(X(t)) \Big)\,\mathrm{d}t + \sigma(t) f'(X(t))\,\mathrm{d}W(t) \tag{11.22}$$

and in integral form

$$f(X(t)) - f(X(0)) = \int_0^t \left(\mu(s)f'(X(s)) + \frac{1}{2}\sigma^2(s)f''(X(s))\right) \mathrm{d}s + \int_0^t \sigma(s)f'(X(s))\,\mathrm{d}W(s). \tag{11.23}$$

Observe that (11.16) and (11.17) are recovered by (11.22) and (11.23) in the special case where the Itô process is Brownian motion: $X = W$ where $\mu \equiv 0$, $\sigma \equiv 1$. Similarly, (11.18) and (11.19) are special cases of (11.20) and (11.21).

Example 11.6. Let $Y(t) := \ln X(t)$, $t \geqslant 0$, where $\{X(t)\}_{t\geqslant 0}$ is an Itô process with stochastic differential

$$\mathrm{d}X(t) = aX(t)\,\mathrm{d}t + bX(t)\,\mathrm{d}W(t).$$

Find the SDE for the process Y and then find explicit representations for $Y(t)$ and $X(t)$ in terms of $W(t)$.

Solution. In this case we define $f(x) := \ln x$, where $Y(t) = f(X(t))$. Differentiating gives $f'(x) = \frac{1}{x}, f''(x) = -\frac{1}{x^2}$. Applying (11.22) with $\mu(t) = aX(t), \sigma(t) = bX(t)$ gives

$$\begin{aligned}\mathrm{d}Y(t) &= \left(aX(t)\frac{1}{X(t)} + \frac{1}{2}(bX(t))^2\left(\frac{-1}{X^2(t)}\right)\right) \mathrm{d}t + bX(t)\frac{1}{X(t)}\,\mathrm{d}W(t)\\ &= \left(a - \frac{b^2}{2}\right) \mathrm{d}t + b\,\mathrm{d}W(t).\end{aligned}$$

Integrating this equation therefore shows that the process Y is a drifted Brownian motion starting at $Y(0) = \ln X(0)$:

$$Y(t) = \ln X(0) + \left(a - \frac{b^2}{2}\right) t + bW(t).$$

By inverting the transformation, we find the original process $X(t)$ as a closed-form expression in $W(t)$:

$$X(t) = \mathrm{e}^{Y(t)} = \mathrm{e}^{\ln X(0)+\left(a-\frac{b^2}{2}\right)t+bW(t)} = X(0)\mathrm{e}^{\left(a-\frac{b^2}{2}\right)t+bW(t)}. \qquad \square$$

An Itô integral $I(t) = \int_0^t \sigma(s)\,\mathrm{d}W(s)$, $t \geqslant 0$, is a martingale (provided that the stochastic integral is well-defined). However, a time integral $\int_0^t \mu(s)\,\mathrm{d}s$ is generally not a martingale. Thus, the Itô formula can be used to verify whether or not a stochastic process that is a function of an Itô process is a martingale.

Example 11.7. Verify whether or not the following processes are martingales w.r.t. a filtration for BM:

(a) $X(t) = Z^2(t) - \int_0^t f^2(u)\,\mathrm{d}u$, where $Z(t) = \int_0^t f(u)\,\mathrm{d}W(u)$ and f is an ordinary continuous function for $t \geqslant 0$;

(b) $Y(t) = V^2(t) - \frac{t^2}{2}$, where $V(t) = \int_0^t W(u)\,\mathrm{d}W(u)$.

Solution.
(a) The process Z is Gaussian with the stochastic differential $\mathrm{d}Z(t) = f(t)\,\mathrm{d}W(t) =$

$\mu(t)\,\mathrm{d}t + \sigma(t)\,\mathrm{d}W(t)$ where $\mu(t) \equiv 0$ and $\sigma(t) \equiv f(t)$. The process X is given by $X(t) = g(t, Z(t))$ with $g(t,x) := x^2 - \int_0^t f^2(u)\,\mathrm{d}u$. Taking derivatives of g:

$$g_t(t,x) = -f^2(t), \quad g_x(t,x) = 2x, \quad g_{xx}(t,x) \equiv 2.$$

Applying (11.20) gives a stochastic differential with zero drift,

$$\begin{aligned}\mathrm{d}X(t) &= \left[g_t(t,Z(t)) + 0 \cdot g_x(t,Z(t)) + \frac{1}{2}f^2(t)g_{xx}(t,Z(t))\right]\mathrm{d}t + f(t)g_x(t,Z(t))\,\mathrm{d}W(t) \\ &= \left(-f^2(t) + \frac{1}{2}(2f^2(t))\right)\mathrm{d}t + 2f(t)Z(t)\,\mathrm{d}W(t) = 2f(t)Z(t)\,\mathrm{d}W(t).\end{aligned}$$

In integral form, $X(t) = 2\int_0^t f(u)Z(u)\,\mathrm{d}W(u)$ since $X(0) = 0$. Thus, $X(t)$ is an Itô integral. It satisfies the square-integrability condition

$$\mathrm{E}\left[\int_0^t (f(u)Z(u))^2\,\mathrm{d}u\right] = \int_0^t f^2(u)\mathrm{E}[Z^2(u)]\,\mathrm{d}u < \infty$$

since $\mathrm{E}[Z^2(u)] = \int_0^u f^2(s)\,\mathrm{d}s$ is a continuous function of $u \geqslant 0$. Hence the process X is a martingale.

(b) First, find the stochastic differential of Y:

$$\mathrm{d}Y(t) = 2V(t)\,\mathrm{d}V(t) + (\,\mathrm{d}V(t))^2 - t\,\mathrm{d}t = (W^2(t) - t)\,\mathrm{d}t + 2V(t)W(t)\,\mathrm{d}W(t).$$

Thus, $Y(t)$ is a sum of an Itô integral (which is a martingale) and a Riemann integral of a function of Brownian motion:

$$Y(t) = 2\int_0^t V(u)W(u)\,\mathrm{d}W(u) + \int_0^t (W^2(u) - u)\,\mathrm{d}u.$$

Note that $Y(0) = 0$. Let us show that the Riemann integral above is not a martingale. As a first simple check, we can try to verify whether the expected value of $I(t) := \int_0^t (W^2(u) - u)\,\mathrm{d}u$ is nonconstant over time:

$$\mathrm{E}[I(t)] = \int_0^t (\underbrace{\mathrm{E}\left[W^2(u)\right]}_{u} - u)\,\mathrm{d}u = 0$$

for $t \geqslant 0$. So the expectation is constant and we cannot yet conclude whether or not the process is a martingale. We hence need to necessarily calculate the conditional expectation to verify whether the process satisfies the martingale property. For $t, s > 0$, we have, upon using the martingale property of $\{W^2(t) - t\}_{t \geqslant 0}$:

$$\begin{aligned}\mathrm{E}_t\left[I(t+s)\right] &= I(t) + \mathrm{E}_t\left[\int_t^{t+s} (W^2(u) - u)\,\mathrm{d}u\right] = I(t) + \int_t^{t+s} \mathrm{E}_t\left[(W^2(u) - u)\right]\mathrm{d}u \\ &= I(t) + \int_t^{t+s} (W^2(t) - t)\,\mathrm{d}u \\ &= I(t) + (W^2(t) - t)\int_t^{t+s} \mathrm{d}u = I(t) + s(W^2(t) - t) \neq I(t).\end{aligned}$$

In conclusion, $\mathrm{E}_t\left[Y(t+s)\right] \neq Y(t)$ and hence the process Y is not a martingale. □

11.6 Stochastic Differential Equations

An equation of the form of an Itô stochastic differential

$$\mathrm{d}X(t) = \mu(t, X(t))\,\mathrm{d}t + \sigma(t, X(t))\,\mathrm{d}W(t) \tag{11.24}$$

where the coefficient drift $\mu(t,x)$ and volatility $\sigma(t,x)$ are given (known) functions and $X(t)$ is the unknown process is called a *stochastic differential equation (SDE).* Equations of this form are of great importance in financial modelling. In practice, (11.24) is subject to an initial condition $X(0) = X_0$ where X_0 is either a random variable or simply a constant $X_0 = x \in \mathbb{R}$. As was mentioned in a previous section, an SDE of the type in (11.24), with constant X_0, is also called a diffusion. We will study diffusions in some depth a little later in the text.

A process X is a so-called *strong solution to the SDE* in (11.24) if, for all $t \geqslant 0$ (or $t \in [0,T]$ if time is restricted to some finite interval $[0,T]$), the process satisfies

$$X(t) = X(0) + \int_0^t \mu(s, X(s))\,\mathrm{d}s + \int_0^t \sigma(s, X(s))\,\mathrm{d}W(s)$$

where both integrals are assumed to exist. The randomness is completely driven by the underlying Brownian motion. So, in case $\sigma \equiv 0$ the equation is simply an ordinary first order ODE. It is important to note that a solution $X(t)$ is an adapted process that is some *representation or functional* written in terms of the Brownian motion up to time t, i.e., $X(t) = F(t, \{W(s); 0 \leqslant s \leqslant t\})$. We have in fact already seen some cases (see Examples 11.5 and 11.6) where the solution $X(t) = F(t, W(t))$ is just a function of the Brownian motion at the endpoint time t. A strong solution hence also gives a path-wise representation of the process $\{X(t)\}_{t\geqslant 0}$. In most cases strong solutions to SDEs cannot be found explicitly, although we can still compute a number of important properties of the process. An alternative and important type of solution is a so-called *weak solution*, which is a solution in distribution. We now turn our attention to so-called linear SDEs, as these form the simplest class of SDEs that have some applications in finance and for which a unique strong solution can be found explicitly.

11.6.1 Solutions to Linear SDEs

A linear SDE is an equation of the form

$$\mathrm{d}X(t) = (\alpha(t) + \beta(t)X(t))\,\mathrm{d}t + (\gamma(t) + \delta(t)X(t))\,\mathrm{d}W(t) \tag{11.25}$$

where the coefficients $\alpha(t), \beta(t), \gamma(t), \delta(t)$ are given adapted processes. These are assumed to be continuous functions of time t. We note that they can simply be ordinary (nonrandom) functions of t or may also be random but not functions of the process $X(t)$. When $\alpha(t), \beta(t), \gamma(t), \delta(t)$ are non-random functions of time, then the process is a diffusion with linear SDE of the form $\mathrm{d}X(t) = a(t, X(t))\,\mathrm{d}t + b(t, X(t))\,\mathrm{d}W(t)$, with both coefficient functions being linear in the state variable: $a(t,x) = \alpha(t) + \beta(t)x$ and $b(t,x) = \gamma(t) + \delta(t)x$. The stochastic equations considered in Examples 11.5 and 11.6 are simple linear SDEs. The nice thing about an SDE of the form (11.25) is that we have explicit solutions, as we now derive.

Equation (11.25) is readily solved by first considering the simpler case when $\alpha(t) \equiv \gamma(t) \equiv 0$. Denoting the simpler process by U, the SDE in (11.25) takes the form

$$\mathrm{d}U(t) = \beta(t)U(t)\,\mathrm{d}t + \delta(t)U(t)\,\mathrm{d}W(t). \tag{11.26}$$

This SDE is now solved by considering the logarithm of the process, $Y(t) := \ln U(t)$, and applying Itô's formula (see Example 11.6):

$$\mathrm{d}Y(t) = \mathrm{d}\ln U(t) = \frac{\mathrm{d}U(t)}{U(t)} - \frac{1}{2}\left(\frac{\mathrm{d}U(t)}{U(t)}\right)^2 = \left(\beta(t) - \frac{1}{2}\delta^2(t)\right)\mathrm{d}t + \delta(t)\,\mathrm{d}W(t).$$

Putting this SDE in integral form and using $U(t) = \mathrm{e}^{Y(t)}$ gives

$$U(t) = U(0)\exp\left[\int_0^t \left(\beta(s) - \frac{1}{2}\delta^2(s)\right)\mathrm{d}s + \int_0^t \delta(s)\,\mathrm{d}W(s)\right]. \tag{11.27}$$

This solution is compactly written as a product: $U(t) = U(0)\mathrm{e}^{\int_0^t \beta(s)\,\mathrm{d}s}\mathcal{E}_t(\delta\cdot W)$, where we denote the *stochastic exponential* of an adapted process $\{\delta(s), 0 \leqslant s \leqslant t\}$, w.r.t. BM on the time interval $[0,t]$, by

$$\mathcal{E}_t(\delta\cdot W) := \exp\left[-\frac{1}{2}\int_0^t \delta^2(s)\,\mathrm{d}s + \int_0^t \delta(s)\,\mathrm{d}W(s)\right]. \tag{11.28}$$

Note that, by setting $\beta(t) \equiv 0$ in (11.26), the solution to the SDE

$$\mathrm{d}U(t) = \delta(t)U(t)\,\mathrm{d}W(t) \quad \text{subject to } U(0) = 1$$

is the stochastic exponential in (11.28). This type of process plays an important role in derivative pricing theory so we shall revisit it in Section 11.8, as well as its multidimensional version in Section 11.10. Note that when the coefficients $\beta(t)$ and $\delta(t)$ are nonrandom functions of time, and $U(0)$ is taken as a positive constant, the Itô integral $\int_0^t \delta(s)\,\mathrm{d}W(s) \sim Norm(0, \int_0^t \delta^2(s)\,\mathrm{d}s)$, i.e., is a Gaussian process, and $\ln\frac{U(t)}{U(0)} \sim Norm(\mu_t, v_t)$ with mean $\mu_t = \int_0^t (\beta(s) - \frac{1}{2}\delta^2(s))\,\mathrm{d}s$ and variance $v_t = \int_0^t \delta^2(s)\,\mathrm{d}s$. That is, $U(t)$ is a lognormal random variable and hence $\{U(t)\}_{t\geqslant 0}$ is a GBM process. In particular, for the case of constant coefficients we recover a GBM process as in Example 11.6. In more complicated general cases where $\delta(t)$ and $\beta(t)$ are random variables (for example, functionals of BM up to time t) then the exponent in (11.27) is not a normal random variable and hence the process $\{U(t)\}_{t\geqslant 0}$ is not a GBM.

Finally, the solution $X(t)$ for the general linear SDE in (11.25) is now readily derived based on the solution to (11.26). The trick is to write it as a product, $X(t) = U(t)V(t)$ where $U(t)$ is given by (11.27) and hence satisfies the SDE in (11.26), and where $V(t)$ satisfies the SDE:

$$\mathrm{d}V(t) = \left[\frac{\alpha(t) - \gamma(t)\delta(t)}{U(t)}\right]\mathrm{d}t + \frac{\gamma(t)}{U(t)}\,\mathrm{d}W(t) \tag{11.29}$$

with initial conditions chosen as $U(0) = 1$ and $V(0) = X(0)$. By the Itô product formula (11.15) $X(t) = U(t)V(t)$ satisfies $\mathrm{d}X(t) = U(t)\,\mathrm{d}V(t) + V(t)\,\mathrm{d}U(t) + \mathrm{d}U(t)\,\mathrm{d}V(t)$. Using (11.26) and (11.29) and by the usual rules we have that $X(t)$ satisfies the SDE in (11.25) with initial condition $U(0)V(0) = X(0)$, i.e., $X(t)$ solves (11.25) with arbitrary initial condition $X(0)$. An explicit representation for $V(t)$ is obtained simply from the integral form of (11.29) with $V(0) = X(0)$:

$$V(t) = X(0) + \int_0^t \frac{\alpha(s) - \gamma(s)\delta(s)}{U(s)}\,\mathrm{d}s + \int_0^t \frac{\gamma(s)}{U(s)}\,\mathrm{d}W(s). \tag{11.30}$$

Hence, the solution to the general linear SDE (11.25) is given by $X(t) = U(t)V(t)$ where $U(t)$ and $V(t)$ are respectively given by (11.27) and (11.30) with $U(0) = 1$. That is,

$$X(t) = U(t)\left(X(0) + \int_0^t [\alpha(s) - \gamma(s)\delta(s)]U^{-1}(s)\,\mathrm{d}s + \int_0^t \gamma(s)U^{-1}(s)\,\mathrm{d}W(s)\right) \tag{11.31}$$

where $U(t) = e^{\int_0^t (\beta(s) - \frac{1}{2}\delta^2(s))\,ds + \int_0^t \delta(s)\,dW(s)} = e^{\int_0^t \beta(s)\,ds}\mathcal{E}_t(\delta \cdot W)$ and $U^{-1}(s) \equiv 1/U(s) = e^{-\int_0^s \beta(u)\,du}\mathcal{E}_s^{-1}(\delta \cdot W)$, $0 \leqslant s \leqslant t$.

Example 11.8. Solve the SDE

$$dX(t) = (\alpha - \beta X(t))\,dt + \sigma\,dW(t)$$

for all $t \geqslant 0$, subject to $X(0) = x$ with constants $x, \alpha, \beta, \sigma \in \mathbb{R}$.

Solution. Note that the SDE is of the form in (11.25) with constant coefficients $\alpha(t) = \alpha, \beta(t) = -\beta, \gamma(t) = \sigma, \delta(t) \equiv 0$. The expression in (11.28) simplifies to $\mathcal{E}_t(\delta \cdot W) = \mathcal{E}_t(0) = 1$ and $U(t) = e^{-\beta t}, U^{-1}(s) = e^{\beta s}$. Substituting into (11.31) gives the solution

$$\begin{aligned} X(t) &= e^{-\beta t}\left(x + \alpha \int_0^t e^{\beta s}\,ds + \sigma \int_0^t e^{\beta s}\,dW(s)\right) \\ &= e^{-\beta t}x + \frac{\alpha}{\beta}(1 - e^{-\beta t}) + \sigma e^{-\beta t}\int_0^t e^{\beta s}\,dW(s). \end{aligned} \tag{11.32}$$

□

By letting $X(t) := f(t, Y(t))$, $Y(t) = \int_0^t e^{\beta s}\,dW(s)$, where $f(t, y) := e^{-\beta t}x + \frac{\alpha}{\beta}(1 - e^{-\beta t}) + \sigma e^{-\beta t}y$, and applying an Itô formula, the reader should verify that $X(t)$ satisfies the above SDE. We note that another simple alternative way to arrive at this solution (without the use of (11.25)) is to define $\tilde{X}(t) := e^{\beta t}X(t)$, which has the effect of eliminating the state variable dependence in the SDE. Indeed, by applying an Itô formula we have $d\tilde{X}(t) = \alpha e^{\beta t}\,dt + \sigma e^{\beta t}\,dW(t)$, with coefficients independent of $\tilde{X}(t)$. Integrating, with $\tilde{X}(0) = X(0) = x$, gives the solution to $\tilde{X}(t)$:

$$\tilde{X}(t) = x + \frac{\alpha}{\beta}(e^{\beta t} - 1) + \sigma \int_0^t e^{\beta s}\,dW(s). \tag{11.33}$$

The solution in (11.32) follows by $X(t) = e^{-\beta t}\tilde{X}(t)$.

In Example 11.8, the solution represented in (11.32) is a Gaussian process involving an Itô integral that is a normal random variable, i.e., $X(t) = \mu_t + \sigma e^{-\beta t}Y(t)$ where $\mu_t = E[X(t)] = e^{-\beta t}x + \frac{\alpha}{\beta}(1 - e^{-\beta t})$ and $Y(t) := \int_0^t e^{\beta s}\,dW(s)$ is a zero-mean normal random variable:

$$Y(t) \sim Norm(0, \text{Var}(Y(t))) = Norm\left(0, \int_0^t e^{2\beta s}\,ds\right) = Norm\left(0, \frac{1}{2\beta}(e^{2\beta t} - 1)\right).$$

Hence, $X(t) \sim Norm(\mu_t, \sigma^2 e^{-2\beta t}\,\text{Var}(Y(t))) = Norm(e^{-\beta t}x + \frac{\alpha}{\beta}(1 - e^{-\beta t}), \frac{\sigma^2}{2\beta}(1 - e^{-2\beta t}))$. Applying the formulae in (11.8) to (11.32) also gives us the covariance function $c_X(t, v) := \text{Cov}(X(t), X(v)), 0 \leqslant t \leqslant v$:

$$\begin{aligned} c_X(t, v) &= \sigma^2 e^{-\beta(t+v)}\,\text{Cov}\left(\int_0^t e^{\beta s}\,dW(s), \int_0^v e^{\beta s}\,dW(s)\right) \\ &= \sigma^2 e^{-\beta(t+v)}\int_0^t e^{2\beta s}\,ds = \frac{\sigma^2}{2\beta}e^{-\beta(t+v)}(e^{2\beta t} - 1) = \frac{\sigma^2}{2\beta}(e^{-\beta(v-t)} - e^{-\beta(v+t)}). \end{aligned}$$

We can also represent the solution as a functional of the Brownian motion up to time t by making use of the Itô product rule: $d(e^{\beta t}W(t)) = \beta e^{\beta t}W(t)\,dt + e^{\beta t}\,dW(t) \implies e^{\beta t}W(t) = \beta\int_0^t e^{\beta s}W(s)\,ds + \int_0^t e^{\beta s}\,dW(s) \implies \int_0^t e^{\beta s}\,dW(s) = e^{\beta t}W(t) - \beta\int_0^t e^{\beta s}W(s)\,ds$. Putting this last expression into (11.32) gives $X(t) = F(t, \{W(s); 0 \leqslant s \leqslant t\})$ with functional F of the Brownian path defined by

$$F(t, \{W(s); 0 \leqslant s \leqslant t\}) := \mathrm{e}^{-\beta t}x + \frac{\alpha}{\beta}(1 - \mathrm{e}^{-\beta t}) + \sigma W(t) - \sigma\beta \int_0^t \mathrm{e}^{-\beta(t-s)} W(s)\,\mathrm{d}s. \tag{11.34}$$

In the chapter on interest rate modelling we shall see that the above process, referred to as the Vasicek model for $\alpha, \beta, \sigma > 0$, is among the simplest one used to model the instantaneous (short) interest rate. A natural extension of this model is to allow the coefficients to be time dependent functions. The resulting linear SDE is explicitly solved in the following example.

Example 11.9. Solve the SDE

$$\mathrm{d}X(t) = (a(t) - b(t)X(t))\,\mathrm{d}t + \sigma(t)\,\mathrm{d}W(t)$$

subject to $X(0) = x \in \mathbb{R}$, where $a(t), b(t), \sigma(t)$ are nonrandom continuous functions of time $t \geqslant 0$.

Solution. The SDE is of the form in (11.25) with coefficient functions $\alpha(t) \equiv a(t)$, $\beta(t) \equiv -b(t)$, $\gamma(t) \equiv \sigma(t)$, $\delta(t) \equiv 0$. Hence, $\mathcal{E}_t(\delta \cdot W) = \mathcal{E}_t(0) = 1$ and $U(t) = \mathrm{e}^{-\int_0^t b(s)\,\mathrm{d}s}, U^{-1}(s) = \mathrm{e}^{\int_0^s b(u)\,\mathrm{d}u}$. Substituting into (11.31) gives us the explicit solution

$$\begin{aligned} X(t) &= \mathrm{e}^{-\int_0^t b(u)\,\mathrm{d}u}\left(x + \int_0^t \mathrm{e}^{\int_0^s b(u)\,\mathrm{d}u} a(s)\,\mathrm{d}s + \int_0^t \mathrm{e}^{\int_0^s b(u)\,\mathrm{d}u} \sigma(s)\,\mathrm{d}W(s)\right) \\ &= x\mathrm{e}^{-\int_0^t b(u)\,\mathrm{d}u} + \int_0^t \mathrm{e}^{-\int_s^t b(u)\,\mathrm{d}u} a(s)\,\mathrm{d}s + \int_0^t \mathrm{e}^{-\int_s^t b(u)\,\mathrm{d}u} \sigma(s)\,\mathrm{d}W(s). \end{aligned} \tag{11.35}$$

□

The above process is Gaussian since the integrand in the Itô integral is an ordinary (nonrandom) function of time. By applying Itô isometry on the Itô integral in (11.35) we have that $X(t)$ is normally distributed with mean and variance:

$$\mathrm{E}[X(t)] = x\mathrm{e}^{-\int_0^t b(u)\,\mathrm{d}u} + \int_0^t \mathrm{e}^{-\int_s^t b(u)\,\mathrm{d}u} a(s)\,\mathrm{d}s\,, \tag{11.36}$$

$$\mathrm{Var}(X(t)) = \mathrm{E}\left[\left(\int_0^t \mathrm{e}^{-\int_s^t b(u)\,\mathrm{d}u} \sigma(s)\,\mathrm{d}W(s)\right)^2\right] = \int_0^t \mathrm{e}^{-2\int_s^t b(u)\,\mathrm{d}u} \sigma^2(s)\,\mathrm{d}s\,. \tag{11.37}$$

The covariance function for this process follows by using (11.8) on the first line in (11.35) after subtracting the mean (for $t \leqslant v$):

$$\begin{aligned} c_X(t, v) &= \mathrm{e}^{-\int_0^t b(u)\,\mathrm{d}u} \mathrm{e}^{-\int_0^v b(u)\,\mathrm{d}u}\,\mathrm{Cov}\left(\int_0^t \mathrm{e}^{\int_0^s b(u)\,\mathrm{d}u} \sigma(s)\,\mathrm{d}W(s), \int_0^v \mathrm{e}^{\int_0^s b(u)\,\mathrm{d}u} \sigma(s)\,\mathrm{d}W(s)\right) \\ &= \mathrm{e}^{-2\int_0^t b(u)\,\mathrm{d}u} \mathrm{e}^{-\int_t^v b(u)\,\mathrm{d}u} \int_0^t \mathrm{e}^{2\int_0^s b(u)\,\mathrm{d}u} \sigma^2(s)\,\mathrm{d}s \\ &= \mathrm{e}^{-\int_t^v b(u)\,\mathrm{d}u} \int_0^t \mathrm{e}^{-2\int_s^t b(u)\,\mathrm{d}u} \sigma^2(s)\,\mathrm{d}s. \end{aligned}$$

11.6.2 Existence and Uniqueness of a Strong Solution of an SDE

An important question when finding a strong solution to an SDE of the form given by (11.24) is whether such a solution exists and, if so, whether the solution is unique. The following theorem gives sufficient conditions for the existence and uniqueness of a strong solution to the SDE in (11.24). We omit the proof of this theorem as the technical details can be found in other more specialized textbooks on stochastic analysis. The conditions in the theorem are not necessary, but are rather mild sufficient conditions that guarantee the existence a unique strong solution, i.e., that there is a unique process $\{X(t)\}_{t\geqslant 0}$ satisfying (11.24).

Theorem 11.4. *Assume the following conditions are satisfied:*

- *The coefficient functions $\mu(t,x)$ and $\sigma(t,x)$ are locally Lipschitz in x, uniformly in t. That is, for arbitrary positive constants T and N, there exists a constant K depending possibly only on T and N such that*

$$|\mu(t,x)-\mu(t,y)|+|\sigma(t,x)-\sigma(t,y)|<K|x-y|,$$

  whenever $|x|,|y|\leqslant N$ and $0\leqslant t\leqslant T$.

- *The coefficient functions $\mu(t,x)$ and $\sigma(t,x)$ satisfy the linear growth condition in the variable x, i.e.,*

$$|\mu(t,x)|+|\sigma(t,x)|\leqslant K(1+|x|).$$

- *The initial value $X(0)$ is independent of the Brownian motion up to arbitrary time T and has a finite second moment, i.e., $X(0)$ is independent of $\mathcal{F}_T^W\equiv\sigma(W(t);0\leqslant t\leqslant T)$ and $\mathrm{E}[X^2(0)]<\infty$.*

Then, the SDE in (11.24) has a unique strong solution $\{X(t)\}_{t\geqslant 0}$ with continuous paths $X(t,\omega),t\geqslant 0$.

For a given SDE, the conditions in the above theorem can be readily checked. For example, the first (Lipschitz) condition holds if the coefficient functions have continuous first partial derivatives $\frac{\partial\mu}{\partial x}(t,x)$ and $\frac{\partial\sigma}{\partial x}(t,x)$. The second condition is satisfied when the coefficient functions $\mu(t,x)$ and $\sigma(t,x)$ have at most a linear growth in x for large values of x and are also bounded for arbitrarily small values of x. In most cases the SDE is subject to a constant initial condition $X(0)\in\mathbb{R}$ so that the above third condition is automatically satisfied. In the case of a general linear SDE we have already shown that the solution is given by (11.31). This includes cases when the coefficients $\alpha(t),\beta(t),\gamma(t),\delta(t)$ are bounded nonrandom functions of time. In these cases, $\mu(t,x)$ and $\sigma(t,x)$ are linear functions of x, for all t, and hence the conditions in the above theorem are indeed satisfied and so a unique strong solution exists. In Examples 11.8 and 11.9 we have solved the SDE and found the unique strong solution as given by (11.32) and (11.35), respectively. We note that unique strong solutions also exist when the coefficient functions in the SDE satisfy milder conditions than those listed in the above theorem. For instance, for a time-homogeneous SDE with (time-independent) drift and volatility coefficients $\mu(x)$ and $\sigma(x)$, it can be shown that there exists a unique strong solution when $\mu(x)$ satisfies a Lipschitz condition, $|\mu(x)-\mu(y)|<K|x-y|$, and $\sigma(x)$ satisfies a Hölder condition, $|\sigma(x)-\sigma(y)|<K|x-y|^{\alpha}$, with order $\alpha\geqslant 1/2$, for some constant K.

11.7 The Markov Property, Feynman–Kac Formulae, and Transition CDFs and PDFs

We have already shown that Brownian motion is a Markov process. Generally, a process $\{X(t)\}_{t\geqslant 0}$ has the Markov property if the probability of the event $\{X(t) \leqslant y\}$, $y \in \mathbb{R}$, conditional on all the past information $\mathcal{F}_s$ (i.e., the information about the complete history of the process up to a prior time s) is the same as its probability conditional on only knowing the process endpoint value $X(s)$, for all $0 \leqslant s \leqslant t$. That is, the Markov property can be formally stated equivalently as

$$\mathbb{P}(X(t) \leqslant y \mid \mathcal{F}_s) = \mathbb{P}(X(t) \leqslant y \mid X(s)) \tag{11.38}$$

or, for any Borel function $h : \mathbb{R} \to \mathbb{R}$,

$$\mathrm{E}[h(X(t)) \mid \mathcal{F}_s] = \mathrm{E}[h(X(t)) \mid X(s)], \tag{11.39}$$

for all $0 \leqslant s \leqslant t$. Note that the specific choice of $h(x) := \mathbb{I}_{x\leqslant y}$ recovers (11.38). Hence, when the Markov property holds, conditioning on natural filtration $\mathcal{F}_s = \sigma(X(u); 0 \leqslant u \leqslant s)$ is equal to conditioning on $\sigma(X(s))$. In particular, for the case of a discrete-time process $X(t_0), X(t_1), \ldots, X(t_{n-1}), X(t_n), \ldots$, with times $t_0 < t_1 < \ldots < t_{n-1} < t_n < \ldots$, the above property takes the form that is familiar in the theory of discrete-time Markov chains:

$$\mathbb{P}(X(t_m) \leqslant y \mid X(t_0), X(t_1), \ldots, X(t_{n-1}), X(t_n)) = \mathbb{P}(X(t_m) \leqslant y \mid X(t_n)) \tag{11.40}$$

for all $m \geqslant n$, i.e., when conditioning on the values of the process for a set $\{t_i\}_{0\leqslant i\leqslant n}$ of previous times, the only conditioning that is relevant is the value of the process at the most recent time t_n. We note that this follows from (11.38) by setting $s = t_n, t = t_m$, i.e., $\mathcal{F}_s \equiv \mathcal{F}_{t_n} = \sigma(X(t_0), X(t_1), \ldots, X(t_{n-1}), X(t_n))$ and $\sigma(X(s)) = \sigma(X(t_n))$, and then using the usual shorthand notation $\mathbb{P}(A \mid \sigma(Y_1, \ldots, Y_n)) \equiv \mathbb{P}(A \mid Y_1, \ldots, Y_n)$ for expressing the probability of an event A conditional on a σ-algebra generated by a set of random variables $Y_1, \ldots, Y_n$.

We are interested in computing conditional expectations involving functions of a process $X = \{X(t)\}_{t\geqslant 0} \in \mathbb{R}$ that solves a given SDE as in (11.24) subject to some initial condition $X(0) = x$. In particular, we will need to compute expectations as in (11.39) for the case that the process X is Markov. Let us now fix some time $T \geqslant 0$. Then, we shall denote the conditional expectation of a function $h(X(T))$, conditioned on the process having a given value $X(t) = x \in \mathbb{R}$ at a time $t \leqslant T$, by

$$\mathrm{E}_{t,x}[h(X(T))] := \mathrm{E}[h(X(T)) \mid X(t) = x].$$

So the subscript x, t is shorthand notation for conditioning on a given value $X(t) = x$ of the process at time t. It should be clear that this conditional expectation is an ordinary (nonrandom) function of the ordinary variables x and t, i.e., $\mathrm{E}_{t,x}[h(X(T))] = g(t, x)$ for any fixed T. Therefore, the Markov property in (11.39) is expressible as $\mathrm{E}[h(X(T)) \mid \mathcal{F}_t] = \mathrm{E}_{t,X(t)}[h(X(T))] = g(t, X(t))$, for all $0 \leqslant t \leqslant T$. This expectation is now the *random variable* given by the function $g(t, X(t))$ of the random variable $X(t)$ and evaluates to $g(t, x)$ upon setting $X(t) = x$. Hence, if we know the conditional probability distribution of random variable $X(T)$, given $X(t) = x$, then we could compute $g(t, x)$. That is, assume the conditional probability density function (PDF) of $X(T)$, given $X(t)$, exists. Recall from the previous chapter that this PDF is the transition PDF for the process X. From Definition 10.2, we have

$$\mathrm{E}_{t,x}[h(X(T))] = \int_{\mathbb{R}} h(y) f_{X(T)|X(t)}(y \mid x)\, \mathrm{d}y = \int_{\mathbb{R}} h(y) p(t, T; x, y)\, \mathrm{d}y. \tag{11.41}$$

In some cases this integral can be computed analytically. Otherwise, we need to employ a numerical method. For example, if the expression for the PDF is known, then we can compute the integral using an appropriate numerical quadrature algorithm. Monte Carlo methods can generally be used to compute the above integral by sampling (i.e., simulating) the paths of the process at time T given their fixed value x at time t. Different simulation approaches may be applied. One approach is to use a time-stepping algorithm for simulating the paths according to the SDE. Alternatively, if the transition PDF is known, then we can sample the (path endpoint) value $X(T)$ according to its distribution. For details on these techniques, we refer the reader to the chapter on Monte Carlo methods.

The next theorem tells us that solutions to an SDE are Markov processes.

Theorem 11.5. *Let $\{X(t)\}_{t\geqslant 0}$ be a solution to the SDE in (11.24) with some given initial condition. Then,*

$$\mathrm{E}[h(X(T)) \mid \mathcal{F}_t] = g(t, X(t)) \tag{11.42}$$

where $g(t,x) = \mathrm{E}_{t,x}[h(X(T))]$, for all $0 \leqslant t \leqslant T$ and Borel function h.

A rigorous proof of this result is beyond the scope of this text. The important content of this result is that the expectation of any function of the process at a future time $T \geqslant t$, conditional on the filtration (or path history) at time t, is given simply by its expectation conditional only on the path value at time t. In practice, we can apply this theorem by first computing $\mathrm{E}_{t,x}[h(X(T))]$, and then putting $X(t)$ in the place of variable x.

A simple nonrigorous, yet instructive, argument that leads us to the fact that the solution to an SDE has the Markov property is to let $T = t + \Delta t$, for a small time step $\Delta t \approx 0$. Then, by the integral form of the SDE:

$$X(t+\Delta t) = X(t) + \int_t^{t+\Delta t} \mu(s, X(s))\,\mathrm{d}s + \int_t^{t+\Delta t} \sigma(s, X(s))\,\mathrm{d}W(s).$$

This expresses the value of the process at future time $t + \Delta t$ in terms of its value at any current time t plus an ordinary integral and an Itô integral. For small Δt, the integrals are well approximated by holding the integrand coefficient functions constant and evaluated at the left endpoint $s = t$ of the time interval $[t, t + \Delta t]$. This gives us the approximation $X(t + \Delta t) \approx X(t) + \mu(t, X(t))\Delta t + \sigma(t, X(t))\Delta W(t)$, where $\Delta W(t) \equiv W(t + \Delta t) - W(t)$. Hence, the left-hand side of (11.38), for $T = t + \Delta t$, is approximated by

$$\mathbb{P}(X(t+\Delta t) \leqslant y \mid \mathcal{F}_t) \approx \mathbb{P}(X(t) + \mu(t, X(t))\Delta t + \sigma(t, X(t))\Delta W(t) \leqslant y \mid \mathcal{F}_t) = g(X(t))$$

where $g(x) = \mathbb{P}(x + \mu(t,x)\Delta t + \sigma(t,x)\Delta W(t) \leqslant y)$. Here we used the independence proposition where the Brownian increment $\Delta W(t)$ is independent of $\mathcal{F}_t$ and $X(t)$ is $\mathcal{F}_t$-measurable. We can now put back the conditioning on $X(t)$ in the unconditional probability since $\Delta W(t)$ is independent of $X(t)$, so the function $g(x)$ is equally given by the conditional probability: $g(x) = \mathbb{P}(x + \mu(t,x)\Delta t + \sigma(t,x)\Delta W(t) \leqslant y \mid X(t)) \approx \mathbb{P}(X(t+\Delta t) \leqslant y \mid X(t))$. Hence, we recover (approximately) the Markov property in (11.38), $\mathbb{P}(X(t + \Delta t) \leqslant y \mid \mathcal{F}_t) \approx \mathbb{P}(X(t+\Delta t) \leqslant y \mid X(t))$. In the above theorem this relation holds exactly.

By the Markov property of the process X, we also have the following martingale property for a process defined via an expectation of some function of $X(T)$ conditional on $X(t)$.

Proposition 11.6. *Let $\{X(t)\}_{t\geqslant 0}$ satisfy the SDE in (11.24) subject to some initial condition. Let $\phi : \mathbb{R} \to \mathbb{R}$ be a Borel function and define $f(t,x) := \mathrm{E}_{t,x}[\phi(X(T))]$ for fixed $T > 0$, assuming $\mathrm{E}_{t,x}[|\phi(X(T))|] < \infty$. Then, the stochastic process*

$$Y_t := f(t, X(t)),\ 0 \leqslant t \leqslant T,$$

is a martingale w.r.t. any filtration $\{\mathcal{F}_t\}_{t\geqslant 0}$ for Brownian motion.

Proof. Let $0 \leqslant s \leqslant t \leqslant T$. Note that, based on Theorem 11.5,

$$\mathrm{E}[\phi(X(T)) \mid \mathcal{F}_t] = \mathrm{E}_{t,X(t)}[\phi(X(T))] = f(t, X(t)) = Y_t$$

for any $0 \leqslant t \leqslant T$. Using this relation and applying the tower property gives the martingale expectation property:

$$\mathrm{E}[Y_t \mid \mathcal{F}_s] = \mathrm{E}[\,\mathrm{E}[\phi(X(T)) \mid \mathcal{F}_t] \mid \mathcal{F}_s] = \mathrm{E}[\phi(X(T)) \mid \mathcal{F}_s] = f(s, X(s)) = Y_s. \qquad \square$$

Based on the Markov property of any solution to an SDE, we are now ready to discuss the very important connection that exists between an SDE and a PDE. In what follows we will find it very convenient to make use of the differential operator $\mathcal{G}$ defined by

$$\mathcal{G}_{t,x} f(t,x) := \frac{1}{2}\sigma^2(t,x)\frac{\partial^2 f}{\partial x^2}(t,x) + \mu(t,x)\frac{\partial f}{\partial x}(t,x). \tag{11.43}$$

This differential operator in the variables (t, x) is the so-called *generator* for the process X and it acts on all functions $f \in C^{1,2}$, i.e., having continuous partial derivatives $\frac{\partial f}{\partial t}$, $\frac{\partial f}{\partial x}$ and $\frac{\partial^2 f}{\partial x^2}$. Using this operator we can now rewrite the differential and integral forms of the Itô formula in (11.20) and (11.21) as

$$\mathrm{d}f(t, X(t)) = \left(\frac{\partial}{\partial t} + \mathcal{G}_{t,x}\right) f(t, X(t))\,\mathrm{d}t + \sigma(t, X(t))\frac{\partial f}{\partial x}(t, X(t))\,\mathrm{d}W(t). \tag{11.44}$$

and

$$f(t, X(t)) = f(0, X(0)) + \int_0^t \left(\frac{\partial}{\partial s} + \mathcal{G}_{s,x}\right) f(s, X(s))\,\mathrm{d}s + \int_0^t \sigma(s, X(s))\frac{\partial f}{\partial x}(s, X(s))\,\mathrm{d}W(s)\,. \tag{11.45}$$

Fix a time $T > 0$. If we assume that the Itô integral in (11.45) is a martingale, for $0 \leqslant t \leqslant T$, whereby the square integrability condition holds, i.e., assuming

$$\int_0^T \mathrm{E}\left[\left(\sigma(s, X(s))\frac{\partial f}{\partial x}(s, X(s))\right)^2\right] \mathrm{d}s < \infty\,, \tag{11.46}$$

we have a useful representation of a martingale Markov process. In particular, the process $\{M_f(t), 0 \leqslant t \leqslant T\}$ defined by

$$M_f(t) := f(t, X(t)) - \int_0^t \left(\frac{\partial}{\partial s} + \mathcal{G}_{s,x}\right) f(s, X(s))\,\mathrm{d}s \tag{11.47}$$

is a martingale. To see this, first note that

$$\mathrm{E}\left[\int_t^T \sigma(s, X(s))\frac{\partial f}{\partial x}(s, X(s))\,\mathrm{d}W(s)\,\middle|\,\mathcal{F}_t\right] = 0\,,$$

since the Itô integral is a martingale. Using the Itô formula (11.45) for times t and T:

$$f(T, X(T)) = f(t, X(t)) + \int_t^T \left(\frac{\partial}{\partial s} + \mathcal{G}_{s,x}\right) f(s, X(s))\,\mathrm{d}s + \int_t^T \sigma(s, X(s))\frac{\partial f}{\partial x}(s, X(s))\,\mathrm{d}W(s)\,.$$

Taking expectations, conditional on $\mathcal{F}_t$, on both sides of this equation while using the above (zero expectation) relation gives

$$\mathrm{E}[f(T, X(T))\,|\,\mathcal{F}_t] = f(t, X(t)) + \mathrm{E}\left[\int_t^T \left(\frac{\partial}{\partial s} + \mathcal{G}_{s,x}\right) f(s, X(s))\,\mathrm{d}s\,\middle|\,\mathcal{F}_t\right].$$

Now, writing the integral appearing inside the last expectation as the difference

$$\int_0^T \left(\frac{\partial}{\partial s} + \mathcal{G}_{s,x}\right) f(s, X(s))\, \mathrm{d}s - \int_0^t \left(\frac{\partial}{\partial s} + \mathcal{G}_{s,x}\right) f(s, X(s))\, \mathrm{d}s$$

and using the fact that the $[0,t]$-integral is $\mathcal{F}_t$-measurable and rearranging terms we obtain

$$\mathrm{E}\left[f(T, X(T)) - \int_0^T \left(\frac{\partial}{\partial s} + \mathcal{G}_{s,x}\right) f(s, X(s))\, \mathrm{d}s \,\middle|\, \mathcal{F}_t\right]$$
$$= f(t, X(t)) - \int_0^t \left(\frac{\partial}{\partial s} + \mathcal{G}_{s,x}\right) f(s, X(s))\, \mathrm{d}s.$$

By the definition in (11.47) we therefore have shown the martingale property:

$$\mathrm{E}[M_f(T) \mid \mathcal{F}_t] = M_f(t)\,,\ 0 \leqslant t \leqslant T. \tag{11.48}$$

One important consequence of this martingale property is the following theorem, which shows that a solution to certain parabolic PDEs (which we will later see are closely related to the Black–Scholes PDE) can be represented as a conditional expectation.

Theorem 11.7 (Feynman–Kac). *Given a fixed $T > 0$, let $\{X(t)\}_{t\geqslant 0}$ satisfy the SDE in (11.24) and let $\phi : \mathbb{R} \to \mathbb{R}$ be a Borel function. Moreover, assume the square integrability condition (11.46) holds. Let $f(t,x)$ be a $C^{1,2}$ function solving the PDE $\frac{\partial f}{\partial t} + \mathcal{G}_{t,x} f = 0$, i.e.,*

$$\frac{\partial f}{\partial t}(t,x) + \frac{1}{2}\sigma^2(t,x)\frac{\partial^2 f}{\partial x^2}(t,x) + \mu(t,x)\frac{\partial f}{\partial x}(t,x) = 0\,, \tag{11.49}$$

for all x, $0 < t < T$, subject to the condition $f(T,x) = \phi(x)$. Then, assuming that $\mathrm{E}_{t,x}[|\phi(X(T))|] < \infty$, $f(t,x)$ has the representation

$$f(t,x) = \mathrm{E}_{t,x}[\phi(X(T))] \equiv \mathrm{E}[\phi(X(T)) \mid X(t) = x] \tag{11.50}$$

for all x, $0 \leqslant t \leqslant T$.

We note that if $\phi(x)$ is continuous, then $f(T-,x) \equiv \lim_{t \nearrow T} f(t,x) = f(T,x) = \phi(x)$, i.e., we have continuity of the solution at $t = T$, for all x.

Proof. Assuming the square integrability condition (11.46), then according to the above discussion we have that the process defined in (11.47) satisfies the martingale property in (11.48). Now, let f satisfy the PDE in (11.49). This implies that the integral in (11.47) vanishes since the integrand function is identically zero, i.e., $\frac{\partial}{\partial s} f(s,x) + \mathcal{G}_{s,x} f(s,x) = 0$, for all $s > 0$ and all values of x. Hence, the process $M_f(t) := f(t, X(t)), 0 \leqslant t \leqslant T$, is a martingale. Combining this with the Markov property of the process, and finally substituting the terminal condition for the random variable $f(T, X(T)) = \phi(X(T))$, we have

$$f(t, X(t)) = \mathrm{E}\big[f(T, X(T)) \,\big|\, \mathcal{F}_t\big] = \mathrm{E}_{t,X(t)}\big[f(T, X(T))\big] = \mathrm{E}_{t,X(t)}\big[\phi(X(T))\big]$$

so that $f(t,x) = \mathrm{E}_{t,x}\big[\phi(X(T))\big]$ for all $x, 0 \leqslant t \leqslant T$. □

This theorem hence shows that the solution at current time t, given by (11.50), is in the form of a conditional expectation of the random variable $\phi(X(T))$, where ϕ is the given boundary value function and $X(T)$ is the random variable corresponding to the endpoint value of the process at future (terminal) time T, where the process solves the SDE in

(11.24) subject to it having current time-t value $X(t) = x$. From our previous discussion surrounding (11.41), we see that this theorem gives us a probabilistic representation of the solution to the parabolic PDE in (11.49) subject to the (terminal time) boundary value function $f(T,x) = \phi(x)$. Alternatively, the theorem can be used in the opposite sense; that is, it provides a PDE approach for evaluating a conditional expectation of the form in (11.50).

We now consider the simplest example of how this theorem is used in the case where the underlying Itô process is just standard Brownian motion. In particular, we solve the simple heat equation on the real line and thereby obtain a probabilistic representation of the solution as an expectation involving the endpoint value of Brownian motion.

Example 11.10. Solve the boundary value problem:

$$\frac{\partial f}{\partial t} + \frac{1}{2}\frac{\partial^2 f}{\partial x^2} = 0,$$

for $x \in \mathbb{R}, 0 \leqslant t \leqslant T$, subject to $f(x,T) = \phi(x)$ where ϕ is an arbitrary function. Give the explicit solution for $\phi(x) = x^2$.

Solution. Observe that this PDE is of the same form as in (11.49) with coefficient functions $\sigma(t,x) \equiv 1, \mu(t,x) \equiv 0$. The corresponding SDE in (11.24) is then

$$\mathrm{d}X(t) = \mathrm{d}W(t).$$

This trivial linear SDE has the solution $X(t) = x_0 + W(t)$. As seen below, the end result does not depend on x_0 since $X(T) - X(t) = W(T) - W(t)$. Using (11.50) and the fact that $W(T) - W(t)$ and $W(t)$ are independent, the solution takes the equivalent forms

$$\begin{aligned} f(t,x) &= \mathrm{E}[\phi(X(T)) \mid X(t) = x] = \mathrm{E}[\phi(W(T) - W(t) + X(t)) \mid X(t) = x] \\ &= \mathrm{E}[\phi(W(T) - W(t) + x)] \\ &= \mathrm{E}[\phi(W(T)) \mid W(t) = x] \\ &= \int_{-\infty}^{\infty} \phi(y) p(t,T;x,y)\, dy\,, \quad \text{for } t < T, \end{aligned}$$

where $p(t,T;x,y) := \frac{e^{-\frac{(y-x)^2}{2(T-t)}}}{\sqrt{2\pi(T-t)}}$ is the transition PDF of standard Brownian motion. The last line follows immediately from (10.13). Note that, for $t = T$, the boundary value condition is satisfied where $f(T,x) = \mathrm{E}[\phi(W(T) - W(T) + x)] = \mathrm{E}[\phi(x)] = \phi(x)$.

For $\phi(x) = x^2$, we simply have

$$\begin{aligned} f(t,x) &= \mathrm{E}[(W(T) - W(t) + x)^2] \\ &= \mathrm{E}[(W(T) - W(t))^2] + 2x\mathrm{E}[W(T) - W(t)] + x^2 = (T - t) + x^2\,, \end{aligned}$$

for all $0 \leqslant t \leqslant T$. Note that the terminal condition is satisfied, $f(T,x) = x^2$, and this function satisfies the above PDE since $\frac{\partial f}{\partial t} + \frac{1}{2}\frac{\partial^2 f}{\partial x^2} = -1 + \frac{1}{2}(2) = 0$. □

We note that the square integrability condition in Theorem 11.7 can be shown to hold in the above example. Moreover, $f(t,x)$ is a $C^{1,2}$ function since $p(t,T;x,y)$ is a $C^{1,2}$ function in the (t,x) variables for $t < T$. In the case $\phi(x) = x^2$, we see that this follows trivially. More generally, assuming an arbitrary ϕ function such that the above y-integral exists, we can verify that the above integral represents a solution to the PDE. Indeed, the corresponding linear differential operator in (11.43) is now $\mathcal{G}_{t,x}f := \frac{1}{2}\frac{\partial^2 f}{\partial x^2}$, and reversing the order of differentiation and integration (w.r.t. the y variable) we have, for all $t < T$:

$$\left(\frac{\partial}{\partial t} + \mathcal{G}_{t,x}\right) f(t,x) = \int_{-\infty}^{\infty} \phi(y) \left(\frac{\partial}{\partial t} + \mathcal{G}_{t,x}\right) p(t,T;x,y)\, dy = 0\,,$$

since (as can be verified explicitly and directly) the above PDF $p = p(t,T;x,y)$ solves the PDE $\frac{\partial p}{\partial t} + \mathcal{G}_{t,x}p = 0$. For continuous $\phi(x)$, the solution is also continuous w.r.t. time $t \in [0,T]$ where $f(T-,x) = f(T,x) = \phi(x)$. This is the case as the transition PDF approaches the *Dirac delta function* centred at zero, denoted by $\delta(\cdot)$, as $t \nearrow T$, i.e.,

$$p(T-,T;x,y) \equiv \lim_{t\nearrow T} p(t,T;x,y) = \lim_{\tau\searrow 0} \frac{\mathrm{e}^{-\frac{(y-x)^2}{2\tau}}}{\sqrt{2\pi\tau}} = \delta(x-y).$$

Here we have used one representation of the Dirac delta function as the limit of an infinitesimally narrow Gaussian PDF. The Dirac delta function is even, $\delta(x-y) = \delta(y-x)$, and has the defining (sifting) property:

$$f(T-,x) = \int_{\mathbb{R}} p(T-,T;x,y)\phi(y)\, dy = \int_{\mathbb{R}} \delta(x-y)\phi(y)\, dy = \phi(x)\,,$$

for any function ϕ that is continuous at x. The above delta function terminal condition is a general property of any transition PDF, as shown in Proposition 11.8 below. The above sifting property arises naturally by the Dirac measure defined in (9.29). Viewed as a distribution over $\mathbb{R}$, the only outcome that occurs with probability one is the single point x. The Dirac delta function can then be related to the Dirac (singular) measure $\delta_x(y)$ for a given point $x \in \mathbb{R}$, $\mathrm{d}\delta_x(y) = \delta(y-x)\,\mathrm{d}y$. Formally, the Dirac delta function $\delta(x)$ is also related to the Heaviside unit step function $H(x)$, where $H(x)$ equals 1 for $x > 0$, equals 0 for $x < 0$ and equals $1/2$ at $x = 0$. In particular, its derivative is the delta function, $H'(x) = \delta(x)$. As a function of y, we write the differential $\mathrm{d}H(y-x) = H'(y-x)\,\mathrm{d}y = \delta(y-x)\,\mathrm{d}y$. Hence, when considered as a Riemann–Stieltjes integral with integrator $H(y-x)$, a function $\phi(y)$ that is continuous at the point $y = x$ will exhibit the sifting property:

$$\int_I \phi(y)\delta(y-x)\,\mathrm{d}y \equiv \int_I \phi(y)\,\mathrm{d}H(y-x) = \phi(x)$$

if x is in any interval $I \subset \mathbb{R}$ and the integral is zero if $x \notin I$. Note that when $I = \mathbb{R}$ the integral equals $\phi(x)$. The reader will note that exactly the same property is satisfied if we use the unit indicator function $\mathbb{I}_{\{y \geqslant x\}}$ as integrator in the place of $H(y-x)$. The two are equivalent for all $y \neq x$ and they are used as alternate definitions of the unit step function.

We can now use Theorem 11.7 to obtain the *backward Kolmogorov PDE* (in the so-called backward-time variables t, x) that is solved by any transition CDF, $P(t,T;x,y) := \mathbb{P}(X(T) \leqslant y \mid X(t) = x)$, and hence, its corresponding transition PDF for a diffusion process with SDE in (11.24).

Proposition 11.8. *Assume the square-integrability condition in Theorem 11.7 holds. Then, a transition PDF, $p = p(t,T;x,y)$, for the process $\{X(t)\}_{t\geqslant 0}$ with the generator in (11.43) solves the backward Kolmogorov PDE:*

$$\left(\frac{\partial}{\partial t} + \mathcal{G}_{t,x}\right) p = 0\,, \tag{11.51}$$

where $\lim_{t\nearrow T} p(t,T;x,y) \equiv p(T-,T;x,y) = \delta(x-y)$.

Proof. We begin by writing the transition probability function as a conditional expectation:

$$P(t,T;x,y) = \mathbb{P}(X(T) \leqslant y \mid X(t) = x) = \mathrm{E}[\mathbb{I}_{\{X(T)\leqslant y\}} \mid X(t) = x] \,.$$

By the above Feynman–Kac theorem, then P (for fixed T, y) solves the PDE

$$\left(\frac{\partial}{\partial t} + \mathcal{G}_{t,x}\right) P(t,T;x,y) = 0 \,,$$

with terminal condition $P(T,T;x,y) = \mathbb{P}(y \geqslant x) = \mathbb{I}_{\{y\geqslant x\}} \equiv \phi(x)$. Taking partial derivatives w.r.t. y on both sides of the above PDE, and using the fact that the order of the differential operators $\left(\frac{\partial}{\partial t} + \mathcal{G}_{t,x}\right)$ and $\frac{\partial}{\partial y}$ can be reversed, gives

$$\left(\frac{\partial}{\partial t} + \mathcal{G}_{t,x}\right) \frac{\partial}{\partial y} P(t,T;x,y) = 0.$$

This is exactly (11.51) since the transition PDF $p(t,T;x,y) := \frac{\partial}{\partial y} P(t,T;x,y)$. The delta function terminal condition is seen to arise as follows, since the transition function approaches the unit step function as $t \nearrow T$,

$$\begin{aligned} \mathrm{d}P(T-,T;x,y) &= \mathrm{d}H(y-x) = \delta(y-x)\,\mathrm{d}y \\ &= p(T-,T;x,y)\,\mathrm{d}y \,. \end{aligned}$$

□

We remark that any function p (or P) that is a solution to the backward Kolmogorov PDE and is a conditional density (or distribution) function of some Markov process, as a diffusion or Itô process, is a transition PDF (or CDF). In fact, a transition PDF $p = p(t,T;x,y)$ is called a *fundamental solution* to the Kolmogorov PDE in (11.51) and its defining properties are that: (i) p is nonnegative, jointly continuous in the variables $t, T; x, y$, twice continuously differentiable in the spatial variables and continuously differentiable in the time variables; (ii) for any bounded Borel function ϕ then the function defined by $u(t,x) := \int_{\mathbb{R}} \phi(y) p(t,T;x,y)\,\mathrm{d}y$ is bounded and also satisfies the same Kolmogorov PDE; (iii) for continuous ϕ, $\lim_{t\nearrow T} u(t,x) \equiv u(T-,x) = \phi(x)$ for all x. Property (iii) is equivalent to the Dirac delta function limit, $\lim_{t\nearrow T} p(t,T;x,y) = p(T-,T;x,y) = \delta(x-y)$. Hence, generally, if given a transition PDF p, the conditional expectation in (11.50), i.e.,

$$f(t,x) = \int_{\mathbb{R}} \phi(y)\, p(t,T;x,y) dy \,, \tag{11.52}$$

solves the backward Kolmogorov PDE in (11.49) with terminal condition $f(T,x) = \phi(x)$. We showed this specifically for the simple case of Brownian motion in the above example.

Example 11.11. Consider a GBM process $\{S(t)\}_{t\geqslant 0} \in \mathbb{R}_+$ with SDE

$$\mathrm{d}S(t) = \mu S(t)\,\mathrm{d}t + \sigma S(t)\,\mathrm{d}W(t) \,,$$

where $\mu, \sigma > 0$ are constants.

(a) Provide the corresponding backward Kolmogorov PDE and obtain the transition CDF and PDF.

(b) Solve the PDE

$$\frac{\partial f}{\partial t} + \frac{1}{2}\sigma^2 x^2 \frac{\partial^2 f}{\partial x^2} + \mu x \frac{\partial f}{\partial x} = 0 \,,$$

for $x > 0, t \leqslant T$, subject to $f(T,x) = \phi(x)$ where ϕ is an arbitrary function. Give the explicit solution for $\phi(x) = x\mathbb{I}_{\{x>a\}}$, with constant $a > 0$.

Solution.
(a) The drift and diffusion coefficient functions are time independent linear functions: $\mu(t,x) = \mu x$ and $\sigma(t,x) = \sigma x$. According to (11.43), the generator $\mathcal{G}_{t,x} \equiv \mathcal{G}_x$ is the differential operator

$$\mathcal{G}_x := \frac{1}{2}\sigma^2 x^2 \frac{\partial^2}{\partial x^2} + \mu x \frac{\partial}{\partial x}.$$

The transition CDF $P = P(t,T;x,y)$ hence solves the PDE in (11.51):

$$\frac{\partial P}{\partial t} + \frac{1}{2}\sigma^2 x^2 \frac{\partial^2 P}{\partial x^2} + \mu x \frac{\partial P}{\partial x} = 0\,,$$

for all $t < T, x, y > 0$ with $P(T,T;x,y) = \mathbb{I}_{\{x \leqslant y\}}$. The transition CDF is given by the conditional expectation:

$$P(t,T;x,y) = \mathbb{P}(S(T) \leqslant y \mid S(t) = x) = \mathrm{E}[\mathbb{I}_{\{S(T)\leqslant y\}} \mid S(t) = x]\,.$$

We have already computed this in the previous chapter by substituting the strong solution for GBM in the form

$$S(T) = S(t)\mathrm{e}^{(\mu - \frac{1}{2}\sigma^2)(T-t) + \sigma(W(T)-W(t))},$$

giving

$$\begin{aligned} P(t,T;x,y) &= \mathrm{E}[\mathbb{I}_{\{\ln(S(T)/y)\leqslant 0\}} \mid S(t) = x] \\ &= \mathbb{P}\left(\frac{W(T)-W(t)}{\sqrt{T-t}} \leqslant -\frac{\ln(x/y) + (\mu - \frac{1}{2}\sigma^2)(T-t)}{\sigma\sqrt{T-t}}\right) \\ &= \mathcal{N}\left(\frac{\ln(y/x) - (\mu - \frac{1}{2}\sigma^2)(T-t)}{\sigma\sqrt{T-t}}\right). \end{aligned} \tag{11.53}$$

Here we used the fact that $W(T) - W(t)$ is independent of $W(t)$, and hence independent of $S(t)$, where $W(T) - W(t) \stackrel{d}{=} \sqrt{T-t}Z$, $Z \sim Norm(0,1)$. Differentiating the above CDF with respect to y gives the known lognormal density (see (10.27) with the drift replacement $\mu \to \mu - \frac{1}{2}\sigma^2$):

$$p(t,T;x,y) = \frac{1}{y\sigma\sqrt{2\pi(T-t)}} \exp\left(-\frac{[\ln(y/x) - (\mu - \frac{1}{2}\sigma^2)(T-t)]^2}{2\sigma^2(T-t)}\right), \tag{11.54}$$

for all $x, y > 0, t < T$, and zero otherwise. The reader can verify that the transition CDF in (11.53) has limit $P(T-,T;x,y) = H(y-x)$.

(b) The solution $f(t,x)$ can be obtained by the Feynman–Kac Theorem 11.7 or, alternatively, directly from (11.52). Let's solve for $f(t,x)$ using both equivalent approaches. In the first approach, we use the above strong solution to the SDE. By (11.50), and the independence of $W(T) - W(t)$ and $S(t)$, we have

$$\begin{aligned} f(t,x) &= \mathrm{E}[\phi(S(T)) \mid S(t) = x] = \mathrm{E}[\phi(S(t)\mathrm{e}^{(\mu-\frac{1}{2}\sigma^2)(T-t)+\sigma(W(T)-W(t))}) \mid S(t) = x] \\ &= \mathrm{E}[\phi(x\mathrm{e}^{(\mu-\frac{1}{2}\sigma^2)(T-t)+\sigma(W(T)-W(t))})] \\ &= \mathrm{E}[\phi(x\mathrm{e}^{(\mu-\frac{1}{2}\sigma^2)(T-t)+\sigma\sqrt{T-t}Z})] \\ &= \int_{-\infty}^{\infty} \phi(x\mathrm{e}^{(\mu-\frac{1}{2}\sigma^2)(T-t)+\sigma\sqrt{T-t}z})n(z)\,\mathrm{d}z\,. \end{aligned}$$

This integral, assuming it exists, represents the solution to the PDE for arbitrary function ϕ. In particular, for $\phi(y) = y\mathbb{I}_{\{y>a\}} = y\mathbb{I}_{\{\ln(y/a)>0\}}$ we have

$$\begin{aligned} f(t,x) &= x\mathrm{e}^{(\mu-\frac{1}{2}\sigma^2)(T-t)}\mathrm{E}\left[\mathrm{e}^{\sigma\sqrt{T-t}Z}\mathbb{I}_{\{\ln(x/a)+(\mu-\frac{1}{2}\sigma^2)(T-t)+\sigma\sqrt{T-t}Z>0\}}\right] \\ &= x\mathrm{e}^{(\mu-\frac{1}{2}\sigma^2)(T-t)}\mathrm{E}\left[\mathrm{e}^{\sigma\sqrt{T-t}Z}\mathbb{I}_{\{Z>A\}}\right] \end{aligned}$$

with constant $A \equiv -\frac{\ln(x/a)+(\mu-\frac{1}{2}\sigma^2)(T-t)}{\sigma\sqrt{T-t}}$, for all $t < T$. For $t = T$, we simply have $f(T,x) = x\mathbb{I}_{\{x>a\}}$. This expectation is evaluated (see identity (A.1)) using $\mathrm{E}[\mathrm{e}^{BZ}\mathbb{I}_{\{Z>A\}}] = \mathrm{e}^{B^2/2}\mathcal{N}(B-A)$, with constant $B \equiv \sigma\sqrt{T-t}$, giving

$$\begin{aligned} f(t,x) &= x\mathrm{e}^{(\mu-\frac{1}{2}\sigma^2)(T-t)} \cdot \mathrm{e}^{\frac{1}{2}\sigma^2(T-t)}\mathcal{N}\left(\sigma\sqrt{T-t} + \frac{\ln(x/a)+(\mu-\frac{1}{2}\sigma^2)(T-t)}{\sigma\sqrt{T-t}}\right) \\ &= x\mathrm{e}^{\mu(T-t)}\mathcal{N}\left(\frac{\ln(x/a)+(\mu+\frac{1}{2}\sigma^2)(T-t)}{\sigma\sqrt{T-t}}\right). \end{aligned}$$

The reader can check that this expression solves the above PDE by computing the partial derivatives $\frac{\partial f}{\partial t}$, $\frac{\partial^2 f}{\partial x^2}$, $\frac{\partial f}{\partial x}$. Moreover, in the limit $t \nearrow T$ (defining $\tau = T - t$):

$$f(T-,x) = \lim_{\tau\searrow 0} x\mathrm{e}^{\mu\tau}\mathcal{N}\left(\frac{\ln(x/a)+(\mu-\frac{1}{2}\sigma^2)\tau}{\sigma\sqrt{\tau}}\right) = \lim_{\tau\searrow 0} x\,\mathcal{N}\left(\frac{\ln(x/a)}{\sigma\sqrt{\tau}}\right) = x\,H(x-a)\,.$$

[Note that this equals $f(T,x) = \phi(x) = x\,\mathbb{I}_{\{x>a\}}$ for all x, except at the point of discontinuity $x = a$ of $\phi(x)$, i.e., $\phi(a) = 0$ and $f(T-,a) = aH(0) = a/2$.]

In the second approach we use (11.52) and insert the above transition PDF to obtain

$$\begin{aligned} f(t,x) &= \int_0^\infty \phi(y)\,p(t,T;x,y)\,\mathrm{d}y = \frac{1}{\sigma\sqrt{2\pi(T-t)}}\int_0^\infty \phi(y)\mathrm{e}^{-\frac{1}{2}\left[\frac{\ln(y/x)-\mu(T-t)}{\sigma\sqrt{T-t}}\right]^2}\frac{\mathrm{d}y}{y} \\ &\left(\text{let } z = \frac{\ln(y/x)-\mu(T-t)}{\sigma\sqrt{T-t}},\ y = x\mathrm{e}^{(\mu-\frac{1}{2}\sigma^2)(T-t)+\sigma\sqrt{T-t}z},\ \frac{\mathrm{d}y}{y} = \sigma\sqrt{T-t}\,\mathrm{d}z\right) \\ &= \int_0^\infty \phi(x\mathrm{e}^{(\mu-\frac{1}{2}\sigma^2)(T-t)+\sigma\sqrt{T-t}z})\frac{\mathrm{e}^{-\frac{1}{2}z^2}}{\sqrt{2\pi}}\,\mathrm{d}z\,. \end{aligned}$$

As required, this produces exactly the same solution as we have above by the first approach. Of course, we should not be surprised by this fact since, by definition, $p(t,T;x,y)$ is the conditional density of $S(T)$ at y, given $S(t) = x$, and hence $f(t,x) = \mathrm{E}[\phi(S(T)) \mid S(t) = x] = \int_0^\infty \phi(y)\,p(t,T;x,y)\,\mathrm{d}y$. □

In the next example we obtain the transition CDF/PDF for the GBM process with time-dependent drift and diffusion coefficients.

Example 11.12. Consider a GBM process $\{S(t)\}_{t\geqslant 0} \in \mathbb{R}_+$ with SDE

$$\mathrm{d}S(t) = \mu(t)S(t)\,\mathrm{d}t + \sigma(t)S(t)\,\mathrm{d}W(t)\,, \tag{11.55}$$

where $\mu(t), \sigma(t) > 0$ are continuous (ordinary) functions of time $t \geqslant 0$. State the corresponding backward Kolmogorov PDE and obtain the transition CDF and PDF.

Solution. We have a linear SDE with coefficient functions $\mu(t,x) = \mu(t)x$ and $\sigma(t,x) = \sigma(t)x$. The corresponding generator is the differential operator

$$\mathcal{G}_{t,x} := \frac{1}{2}\sigma^2(t)x^2\frac{\partial^2}{\partial x^2} + \mu(t)x\frac{\partial}{\partial x}.$$

The transition CDF $P = P(t,T;x,y)$ solves the PDE in (11.51):

$$\frac{\partial P}{\partial t} + \frac{1}{2}\sigma^2(t)x^2\frac{\partial^2 P}{\partial x^2} + \mu(t)x\frac{\partial P}{\partial x} = 0\,,$$

for all $t < T, x, y > 0$ with $P(T,T;x,y) = \mathbb{I}_{\{x \leqslant y\}}$. By the Feynman–Kac Theorem 11.7, the transition CDF is obtained by evaluating the conditional expectation

$$P(t,T;x,y) = \mathbb{P}(S(T) \leqslant y \mid S(t) = x) = \mathbb{P}(X(T) \leqslant \ln y \mid X(t) = \ln x)\,,$$

where we define the process $X(t) := \ln S(t), t \geqslant 0$. The SDE (11.55) has unique strong solution given by (11.31) with $\alpha(t) \equiv \gamma(t) \equiv 0$, $\beta(t) \equiv \mu(t)$, $\delta(t) \equiv \sigma(t)$, i.e.,

$$S(t) = S(0)\mathrm{e}^{\int_0^t \mu(s)\,\mathrm{d}s - \frac{1}{2}\int_0^t \sigma^2(s)\,\mathrm{d}s + \int_0^t \sigma(s)\,\mathrm{d}W(s)}\,,$$

hence

$$S(T) = S(t)\,\mathrm{e}^{\int_t^T \mu(s)\,\mathrm{d}s - \frac{1}{2}\int_t^T \sigma^2(s)\,\mathrm{d}s + \int_t^T \sigma(s)\,\mathrm{d}W(s)}$$

and

$$X(T) = X(t) + \int_t^T \mu(s)\,\mathrm{d}s - \frac{1}{2}\int_t^T \sigma^2(s)\,\mathrm{d}s + \int_t^T \sigma(s)\,\mathrm{d}W(s)\,.$$

It is convenient to define the *time-averaged* drift and volatility functions:

$$\bar{\mu}(t,T) := \frac{1}{T-t}\int_t^T \mu(s)\,\mathrm{d}s\,, \quad \bar{\sigma}(t,T) := \sqrt{\frac{1}{T-t}\int_t^T \sigma^2(s)\,\mathrm{d}s}\,. \tag{11.56}$$

Since $\int_t^T \sigma(s)\,\mathrm{d}W(s) \stackrel{d}{=} W\left(\bar{\sigma}^2(t,T)(T-t)\right) \stackrel{d}{=} \bar{\sigma}(t,T)\sqrt{T-t}Z$, $Z \sim Norm(0,1)$,

$$X(T) \stackrel{d}{=} X(t) + [\bar{\mu}(t,T) - \frac{1}{2}\bar{\sigma}^2(t,T)](T-t) + \bar{\sigma}(t,T)\sqrt{T-t}Z\,,$$

where $X(T) - X(t)$, and hence Z, is independent of $X(t)$. Combining these facts into the above gives:

$$\begin{aligned} P(t,T;x,y) &= \mathbb{P}\left(X(T) - X(t) \leqslant \ln y - X(t) \mid X(t) = \ln x\right) = \mathbb{P}\left(X(T) - X(t) \leqslant \ln(y/x)\right) \\ &= \mathbb{P}\left([\bar{\mu}(t,T) - \frac{1}{2}\bar{\sigma}^2(t,T)](T-t) + \bar{\sigma}(t,T)\sqrt{T-t}Z \leqslant \ln(y/x)\right) \\ &= \mathcal{N}\left(\frac{\ln(y/x) - [\bar{\mu}(t,T) - \frac{1}{2}\bar{\sigma}^2(t,T)](T-t)}{\bar{\sigma}(t,T)\sqrt{T-t}}\right). \end{aligned} \tag{11.57}$$

We note that this is the form of the transition CDF for standard GBM in Example 11.11, wherein the drift and volatility coefficients in (11.53) are now replaced by the time-averaged ones: $\mu \to \bar{\mu}(t,T)$ and $\sigma \to \bar{\sigma}(t,T)$. Observe, however, that the CDF in (11.57) is not a function of only $T - t$, i.e., the GBM process with time-dependent coefficients is an example of a time-inhomogeneous process (i.e., not time-homogeneous as in Example 11.11). Differentiating (11.53) with respect to y gives the lognormal density (analogous to 11.54):

$$p(t,T;x,y) = \frac{1}{y\bar{\sigma}\sqrt{2\pi(T-t)}}\exp\left(-\frac{[\ln(y/x) - (\bar{\mu} - \frac{1}{2}\bar{\sigma}^2)(T-t)]^2}{2\bar{\sigma}^2(T-t)}\right), \tag{11.58}$$

for all $x, y > 0, t < T$, and zero otherwise, where $\bar{\mu} \equiv \bar{\mu}(t,T)$, $\bar{\sigma} \equiv \bar{\sigma}(t,T)$. □

Example 11.13. Consider the process $\{X(t)\}_{t\geqslant 0} \in \mathbb{R}$ in Example 11.8, i.e.,

$$\mathrm{d}X(t) = (\alpha - \beta X(t))\,\mathrm{d}t + \sigma\,\mathrm{d}W(t)\,. \tag{11.59}$$

For $\alpha = 0$, this process is specifically called the *Ornstein–Uhlenbeck process* or OU process for short. Derive the corresponding transition CDF and PDF.

Solution. From the analysis in Example 11.8, we have the strong solution for $X(T)$ in terms of $X(t) = x$, which we can write in equivalent forms using time-changed BM:

$$\begin{aligned}
X(T) &= \mathrm{e}^{-\beta(T-t)}x + \frac{\alpha}{\beta}(1 - \mathrm{e}^{-\beta(T-t)}) + \sigma\int_t^T \mathrm{e}^{-\beta(T-s)}\,\mathrm{d}W(s) \\
&\stackrel{d}{=} \mathrm{e}^{-\beta(T-t)}x + \frac{\alpha}{\beta}(1 - \mathrm{e}^{-\beta(T-t)}) + \sigma W\left(\frac{1-\mathrm{e}^{-2\beta(T-t)}}{2\beta}\right) \\
&\stackrel{d}{=} \mathrm{e}^{-\beta(T-t)}x + \frac{\alpha}{\beta}(1 - \mathrm{e}^{-\beta(T-t)}) + \sigma \mathrm{e}^{-\beta(T-t)} W\left(\frac{\mathrm{e}^{2\beta(T-t)}-1}{2\beta}\right) \\
&\stackrel{d}{=} \mathrm{e}^{-\beta(T-t)}x + \frac{\alpha}{\beta}(1 - \mathrm{e}^{-\beta(T-t)}) + \sigma\sqrt{\frac{1-\mathrm{e}^{-2\beta(T-t)}}{2\beta}}\,Z\,.
\end{aligned}$$

The last line displays $X(T)$ as a normal random variable, where $Z \sim$ *Norm*$(0,1)$ and independent of $X(t)$. Hence, the transition CDF is a normal CDF:

$$P(t,T;x,y) = \mathbb{P}\left(X(T) \leqslant y \mid X(t) = x\right) = \mathcal{N}\left(\frac{y - [\mathrm{e}^{-\beta(T-t)}x + \frac{\alpha}{\beta}(1-\mathrm{e}^{-\beta(T-t)})]}{\sigma\sqrt{(1-\mathrm{e}^{-2\beta(T-t)})/2\beta}}\right), \tag{11.60}$$

and the transition PDF is the Gaussian function

$$p(t,T;x,y) = \frac{1}{\sigma}\sqrt{\frac{2\beta}{1-\mathrm{e}^{-2\beta(T-t)}}}\; n\left(\frac{y - [\mathrm{e}^{-\beta(T-t)}x + \frac{\alpha}{\beta}(1-\mathrm{e}^{-\beta(T-t)})]}{\sigma\sqrt{(1-\mathrm{e}^{-2\beta(T-t)})/2\beta}}\right), \tag{11.61}$$

for all $x, y \in \mathbb{R}$, $t < T$. □

Note that a transition PDF p (or CDF P) solving a given Kolmogorov PDE as in (11.51), subject to $p(T-,T;x,y) = \delta(x-y)$, is in general cases not necessarily a unique solution. This is the case even if we require p (or P) to be a PDF (or CDF). If a diffusion has one or both of its endpoints (left or right endpoint) as a regular boundary, then the behaviour of the process at the endpoint can be specified differently. An example of this is the specification of a regular reflecting boundary versus a regular killing (absorbing) boundary as in the case of Brownian motion (BM) that is either reflected or killed at an upper or lower finite boundary point. The known transition PDFs for both respective cases are of course different, yet both solve the same Kolmogorov PDE for Brownian motion and have limit $p(T-,T;x,y) = \delta(x-y)$. The key point is that the Kolmogorov PDE and terminal time condition make no mention of the boundary conditions imposed on the solution as a function of the spatial variable x. To obtain a unique fundamental solution that corresponds to a transition PDF (assuming of course that such a solution exists) one generally needs to also specify the spatial boundary conditions at both endpoints of the process. In some cases, such as for BM or drifted BM on $\mathbb{R}$, both endpoints $\pm\infty$ of the process are natural boundaries (not regular) and there is then a unique transition PDF on $\mathbb{R}$, i.e., the Gaussian PDF we have already derived. Similarly, for GBM the two endpoints of the state space $(0,\infty)$ are natural boundaries and hence the process has a unique transition PDF on $\mathbb{R}_+$, i.e., the known lognormal PDF.

The following result extends Theorem 11.7 and, as we shall see in later chapters, is used for pricing (single-asset) financial derivatives via a PDE based approach.

Theorem 11.9 ("Discounted" Feynman–Kac). *Fix $T > 0$ and let $\{X(t)\}_{t\geqslant 0}$ satisfy the SDE in (11.24). Let the same assumptions stated in Theorem (11.7) hold and assume $r(t,x) : [0,T] \times \mathbb{R} \to \mathbb{R}$ is a lower-bounded continuous function. Then, the function defined by the conditional expectation*

$$f(t,x) := \mathrm{E}_{t,x}[\mathrm{e}^{-\int_t^T r(u,X(u))\,\mathrm{d}u}\phi(X(T))] \equiv \mathrm{E}[\mathrm{e}^{-\int_t^T r(u,X(u))\,\mathrm{d}u}\phi(X(T)) \mid X(t) = x] \tag{11.62}$$

solves the PDE $\frac{\partial f}{\partial t} + \mathcal{G}_{t,x} f - r(t,x) f = 0$, i.e.,

$$\frac{\partial f}{\partial t}(t,x) + \frac{1}{2}\sigma^2(t,x)\frac{\partial^2 f}{\partial x^2}(t,x) + \mu(t,x)\frac{\partial f}{\partial x}(t,x) - r(t,x)f(t,x) = 0\,, \tag{11.63}$$

for all x, $0 < t < T$, subject to the terminal condition $f(T,x) = \phi(x)$.

Proof. This result follows by first rewriting the exponential factor as

$$\mathrm{e}^{-\int_t^T r(u,X(u))\,\mathrm{d}u} = \mathrm{e}^{-\int_0^T r(u,X(u))\,\mathrm{d}u} \cdot \mathrm{e}^{\int_0^t r(u,X(u))\,\mathrm{d}u}\,.$$

The process defined by $g_t := \mathrm{e}^{-\int_0^t r(u,X(u))\,\mathrm{d}u} f(t,X(t))$ is a martingale since $g_t = \mathrm{E}[g_T \mid \mathcal{F}_t]$, where $g_T = \mathrm{e}^{-\int_0^T r(u,X(u))\,\mathrm{d}u} f(T,X(T)) = \mathrm{e}^{-\int_0^T r(u,X(u))\,\mathrm{d}u}\phi(X(T))$:

$$\begin{aligned} g_t = \mathrm{e}^{-\int_0^t r(u,X(u))\,\mathrm{d}u} f(t,X(t)) &= \mathrm{E}_{t,X(t)}[\mathrm{e}^{-\int_0^T r(u,X(u))\,\mathrm{d}u}\phi(X(T))] \\ &= \mathrm{E}[\mathrm{e}^{-\int_0^T r(u,X(u))\,\mathrm{d}u}\phi(X(T)) \mid \mathcal{F}_t]\,. \end{aligned}$$

Note that g_T is $\mathcal{F}_T$-measurable and assumed integrable. The last step consists of computing the stochastic differential of g_t via the Itô product formula. To do so, define $I(t) := \int_0^t r(u,X(u))\,\mathrm{d}u$ giving $\mathrm{d}I(t) = r(t,X(t))\,\mathrm{d}t$, $(\,\mathrm{d}I(t))^2 \equiv 0$ and

$$\mathrm{d}[\mathrm{e}^{-\int_0^t r(u,X(u))\,\mathrm{d}u}] = \mathrm{d}\mathrm{e}^{-I(t)} = -\mathrm{e}^{-I(t)}\,\mathrm{d}I(t) = -\mathrm{e}^{-\int_0^t r(u,X(u))\,\mathrm{d}u} r(t,X(t))\,\mathrm{d}t\,.$$

Hence, using this and (11.44) within the Itô product formula gives

$$\begin{aligned} \mathrm{d}g_t &= \mathrm{e}^{-I(t)} \cdot \mathrm{d}f(t,X(t)) + f(t,X(t)) \cdot \mathrm{d}\mathrm{e}^{-I(t)} + \mathrm{d}\mathrm{e}^{-I(t)} \cdot \mathrm{d}f(t,X(t)) \\ &= \mathrm{e}^{-I(t)} \cdot \mathrm{d}f(t,X(t)) + f(t,X(t)) \cdot \mathrm{d}\mathrm{e}^{-I(t)} \\ &= \mathrm{e}^{-I(t)}\left[\mathrm{d}f(t,X(t)) - f(t,X(t)\,\mathrm{d}I(t)\right] \\ &= \mathrm{e}^{-I(t)}\Bigg[\left(\frac{\partial}{\partial t} f(t,X(t)) + \mathcal{G}_{t,x} f(t,X(t)) - r(t,X(t)) f(t,X(t))\right)\mathrm{d}t \\ &\qquad + \sigma(t,X(t))\frac{\partial f}{\partial x}(t,X(t))\,\mathrm{d}W(t)\Bigg]\,. \end{aligned}$$

By the martingale condition, the drift coefficient (i.e., the expression multiplying $\mathrm{d}t$) must vanish for all values $X(t) = x$ and time t; namely, $\left(\frac{\partial}{\partial t} + \mathcal{G}_{t,x} - r(t,x)\right) f(t,x) = 0$. This is precisely the PDE in (11.63). Finally, the terminal condition follows trivially from (11.62) for $t = T$: $f(T,x) = \mathrm{E}[\mathrm{e}^{-\int_T^T r(u,X(u))\,\mathrm{d}u}\phi(X(T)) \mid X(T) = x] = \mathrm{E}[\phi(X(T)) \mid X(T) = x] = \phi(x)$. □

An important special case is when $r(t,x) = r$ is a constant. Then, the function defined by the conditional expectation, $f(t,x) := \mathrm{e}^{-r(T-t)}\mathrm{E}_{t,x}[\phi(X(T))]$, solves

$$\frac{\partial f}{\partial t}(t,x) + \frac{1}{2}\sigma^2(t,x)\frac{\partial^2 f}{\partial x^2}(t,x) + \mu(t,x)\frac{\partial f}{\partial x}(t,x) - rf(t,x) = 0\,, \tag{11.64}$$

with terminal condition $f(T,x) = \phi(x)$.

11.7.1 Forward Kolmogorov PDE

Proposition 11.8 states that a transition PDF solves the Kolmogorov PDE (11.51) in the *backward variables* (t, x). We can also define the differential operator $\tilde{\mathcal{G}} \equiv \tilde{\mathcal{G}}_{T,y}$ acting on the *forward variables* (T, y):

$$\tilde{\mathcal{G}} f(T, y) := \frac{1}{2} \frac{\partial^2}{\partial y^2} \left(\sigma^2(T, y) f(T, y) \right) - \frac{\partial}{\partial y} \left(\mu(T, y) f(T, y) \right). \tag{11.65}$$

This is also referred to as the differential adjoint to the generator $\mathcal{G}$. It can be shown that under fairly general conditions the transition PDF $p = p(t, T; x, y)$ (considered as a function of T, y, for any fixed t, x) satisfies the so-called *forward Kolmogorov or Fokker–Planck* PDE

$$\frac{\partial p}{\partial T} = \tilde{\mathcal{G}} p, \tag{11.66}$$

with $\lim_{T \searrow t} p(t, T; x, y) = \delta(y - x)$. The name forward derives from the fact that y refers to the value of the process at future time $T > t$.

The formal proof of (11.66), and under what conditions it holds true, requires a rather technical discussion that is beyond our scope. However, it is instructive to see how (11.66) arises from the backward PDE. Let the interval $\mathcal{I}$ denote the state space of process X. For example, $\mathcal{I} = \mathbb{R}$ for standard BM, $\mathcal{I} = \mathbb{R}_+$ for GBM, $\mathcal{I} = (L, \infty)$ for GBM killed at a lower level $L > 0$, etc. In our heuristic justification of (11.66) we shall now make use of the Chapman–Kolmogorov relation:

$$p(t, T; x, y) = \int_{\mathcal{I}} p(t, t'; x, x') p(t', T; x', y) \, \mathrm{d}x' \tag{11.67}$$

for any $t < t' < T$, $x, y \in \mathcal{I}$. Equation (11.67) is an important general property that follows from the Markov property. To derive this relation, consider the joint PDF of the triplet $(X(T), X(t'), X(t))$ and applying conditioning gives

$$f_{X(T),X(t'),X(t)}(y, x', x) = f_{X(t)}(x) \cdot f_{X(t')|X(t)}(x'|x) \cdot f_{X(T)|X(t')}(y|x')$$

where $f_{X(T)|X(t'),X(t)}(y|x', x) = f_{X(T)|X(t')}(y|x')$ by the Markov property. Dividing both sides by the PDF of $X(t)$, $f_{X(t)}(x)$, and using the definition of the transition PDF in (10.16), gives the joint PDF of the pair $X(T), X(t')$ conditional on $X(t) = x$:

$$f_{X(T),X(t')|X(t)}(y, x'|x) \equiv \frac{f_{X(T),X(t'),X(t)}(y, x', x)}{f_{X(t)}(x)} = p(t, t'; x, x') p(t', T; x', y) .$$

Integrating out the x' variable gives the PDF of $X(T)$ conditional on $X(t) = x$:

$$f_{X(T)|X(t)}(y|x) = \int_{\mathcal{I}} f_{X(T),X(t')|X(t)}(y, x'|x) \, \mathrm{d}x' = \int_{\mathcal{I}} p(t, t'; x, x') p(t', T; x', y) \, \mathrm{d}x' .$$

By definition, $f_{X(T)|X(t)}(y|x) = p(t, T; x, y)$ and therefore we obtain (11.67).

To arrive at (11.66) we begin by differentiating both sides of (11.67) w.r.t. t' and note that $\frac{\partial}{\partial t'} p(t, T; x, y) \equiv 0$, giving

$$\int_{\mathcal{I}} \left[p(t', T; x', y) \frac{\partial}{\partial t'} p(t, t'; x, x') + p(t, t'; x, x') \frac{\partial}{\partial t'} p(t', T; x', y) \right] \mathrm{d}x' \equiv 0 . \tag{11.68}$$

We leave the first integral term as is, but re-express the second part of the integral by using the backward PDE, $\frac{\partial}{\partial t'} p(t', T; x', y) = -\mathcal{G}_{t',x'} p(t', T; x', y)$, to obtain

$$\int_{\mathcal{I}} p(t, t'; x, x') \frac{\partial}{\partial t'} p(t', T; x', y) \, \mathrm{d}x' = - \int_{\mathcal{I}} p(t, t'; x, x') \, \mathcal{G}_{t',x'} p(t', T; x', y) \, \mathrm{d}x' .$$

The next step consists of using the differential operator $\mathcal{G}_{t',x'}$, applying integration by parts on the above right-hand integral and assuming that contributions from the boundaries of $\mathcal{I}$ vanish (see Exercise 11.35) to obtain

$$\int_{\mathcal{I}} p(t,t';x,x')\,\mathcal{G}_{t',x'}p(t',T;x',y)\,\mathrm{d}x' = \int_{\mathcal{I}} p(t',T;x',y)\,\tilde{\mathcal{G}}_{t',x'}p(t,t';x,x')\,\mathrm{d}x' . \tag{11.69}$$

This shows that $\tilde{\mathcal{G}}$ indeed acts as the corresponding adjoint operator to $\mathcal{G}$. Using this relation into the second term in the integrand of (11.68) gives

$$\int_{\mathcal{I}} p(t',T;x',y)\left[\frac{\partial}{\partial t'}p(t,t';x,x') - \tilde{\mathcal{G}}_{t',x'}p(t,t';x,x')\right]\mathrm{d}x' \equiv 0 . \tag{11.70}$$

Since this integral is identically zero for arbitrary given values $t' > t$, $x' \in \mathcal{I}$, then (assuming a large enough family of positive transition PDFs $p(t',T;x',y)$ as functions of x') the integrand must be zero for all $x' \in \mathcal{I}$. This implies that the term in brackets in the integrand must equal zero, i.e., for fixed backward variables t, x we have the forward Kolmogorov PDE, $\frac{\partial p}{\partial t'} = \tilde{\mathcal{G}}_{t',x'}p$, in the forward variables t', x' for an arbitrary transition PDF $p = p(t,t';x,x')$.

11.7.2 Transition CDF/PDF for Time-Homogeneous Diffusions

In many applications, including derivative pricing, the stochastic process is assumed to be time-homogeneous. We recall the definition of a time-homogeneous process from the previous chapter, i.e., the relation in (10.17). For a time-homogeneous diffusion process, this means that the drift and diffusion coefficient functions are only functions of the "spatial variable" and are not functions of time t: $\mu(x,t) = \mu(x)$ and $\sigma(x,t) = \sigma(x)$. The generator $\mathcal{G}_{t,x} \equiv \mathcal{G}_x$ for such a process is then of the form

$$\mathcal{G}_x := \frac{1}{2}\sigma^2(x)\frac{\partial^2}{\partial x^2} + \mu(x)\frac{\partial}{\partial x} . \tag{11.71}$$

Since the transition PDF (or CDF) satisfies a *time-homogeneous PDE*, it is then a *function of the time difference*: $\tau \equiv T - t$, i.e., we write it as $p(\tau;x,y)$ and the transition CDF as $P(\tau;x,y)$. This time dependence on $\tau = T - t$ can also be realized from the conditional expectation definition of the transition CDF. Indeed, the defining relation in (10.17) implies

$$\begin{aligned} P(t,T;x,y) \equiv \mathbb{P}(X(T) \leqslant y \mid X(t) = x) &= \mathbb{P}(X(t+\tau) \leqslant y \mid X(t) = x) \\ &= \mathbb{P}(X(\tau) \leqslant y \mid X(0) = x) \\ &= P(0,\tau;x,y) \equiv P(\tau;x,y) , \end{aligned}$$

and $p(\tau;x,y) = \frac{\partial}{\partial y}P(\tau;x,y)$. Writing $p(t,T;x,y) = p(\tau;x,y)$ and using the fact that $\frac{\partial \tau}{\partial T} = 1$ and $\frac{\partial \tau}{\partial t} = -1$ gives

$$\frac{\partial p(t,T;x,y)}{\partial T} = \frac{\partial p(\tau;x,y)}{\partial \tau} \quad\text{and}\quad \frac{\partial p(t,T;x,y)}{\partial t} = -\frac{\partial p(\tau;x,y)}{\partial \tau} .$$

The backward and forward Kolmogorov PDEs are then given by

$$\frac{\partial p}{\partial \tau} = \frac{1}{2}\sigma^2(x)\frac{\partial^2 p}{\partial x^2} + \mu(x)\frac{\partial p}{\partial x} \qquad \text{(backward)} \tag{11.72}$$

$$\frac{\partial p}{\partial \tau} = \frac{1}{2}\frac{\partial^2}{\partial y^2}\left(\sigma^2(y)p\right) - \frac{\partial}{\partial y}\left(\mu(y)p\right) \qquad \text{(forward)} \tag{11.73}$$

for a transition PDF $p = p(\tau; x, y)$ and the same PDEs for the corresponding CDF $P(\tau; x, y)$. The previous terminal condition is now an *initial condition* where

$$\lim_{\tau \searrow 0} p(\tau, x, y) \equiv p(0+, x, y) = \delta(x - y) \text{ and } P(0, x, y) = \mathbb{I}_{\{x \leqslant y\}} . \tag{11.74}$$

Note that, by time homogeneity, the conditional expectation in (11.50) in the above Feynman–Kac Theorem gives

$$\mathrm{E}[\phi(X(T)) \mid X(t) = x] = \mathrm{E}[\phi(X(t + \tau)) \mid X(t) = x] = \mathrm{E}[\phi(X(\tau)) \mid X(0) = x] := f(\tau, x) .$$

That is, (11.52) now reads

$$f(\tau, x) = \int_{\mathbb{R}} \phi(y)\, p(\tau; x, y) dy , \tag{11.75}$$

where f solves the backward Kolmogorov PDE

$$\frac{\partial f}{\partial \tau} = \frac{1}{2}\sigma^2(x)\frac{\partial^2 f}{\partial x^2} + \mu(x)\frac{\partial f}{\partial x} \tag{11.76}$$

with initial condition $f(0, x) = \phi(x)$. Note: $f(0+, x) \equiv f(0, x)$ for continuous $\phi(x)$.

Assuming a constant discount function $r(t, x) = r$, we observe that the discounted expectation is also a function of variables τ, x, i.e., we have the function

$$v(\tau, x) = \mathrm{e}^{-r(T-t)}\mathrm{E}_{t,x}[\phi(X(T))] = \mathrm{e}^{-r\tau} f(\tau, x) = \mathrm{e}^{-r\tau} \int_{\mathbb{R}} \phi(y)\, p(\tau; x, y) dy$$

satisfying the PDE

$$\frac{\partial v}{\partial \tau} = \frac{1}{2}\sigma^2(x)\frac{\partial^2 v}{\partial x^2} + \mu(x)\frac{\partial v}{\partial x} - rv \tag{11.77}$$

with initial condition $v(0, x) = \phi(x)$. This is the time-homogeneous version of (11.64).

We have already seen several specific examples of time-homogeneous processes such as standard BM, GBM in Example 11.11, and the OU process in Example 11.13. In Example 11.11, the GBM process is time homogeneous with coefficient functions $\mu(x) = \mu x$ and $\sigma(x) = \sigma x$, and having respective transition CDF and PDF:

$$P(\tau; x, y) = \mathcal{N}\left(\frac{\ln(y/x) - (\mu - \frac{1}{2}\sigma^2)\tau}{\sigma\sqrt{\tau}}\right) \tag{11.78}$$

and

$$p(\tau; x, y) = \frac{1}{y\sigma\sqrt{2\pi\tau}} \exp\left(-\frac{[\ln(y/x) - (\mu - \frac{1}{2}\sigma^2)\tau]^2}{2\sigma^2\tau}\right), \tag{11.79}$$

$x, y > 0, \tau > 0$. The reader can verify by direct differentiation that both functions satisfy (11.72) and (11.73) with the appropriate initial condition in (11.74). This is also the case for the time-homogeneous OU process, where setting $\tau = T - t$ in (11.60) and (11.61) gives the transition CDF and PDF that satisfy the above time-homogeneous Kolmogorov PDEs with $\mu(x) = \alpha - \beta x$, $\sigma(x) = \sigma$. In contrast, for nonconstant $\mu(t)$ and (or) nonconstant $\sigma(t)$, the GBM process in Example 11.11 is time inhomogeneous , i.e., the transition functions in (11.57) and (11.58) cannot be written as functions of only $\tau = T - t$ in the time variables, but rather depend on both t and T, separately, via the time-averaged quantities in (11.56).

11.8 Radon–Nikodym Derivative Process and Girsanov's Theorem

Our main goal in this section is to use and build upon the basic tools and ideas developed in Section 9.5 of Chapter 9 in order to understand how to construct a certain type of probability measure change which introduces a drift in the BM. In particular, we are interested in a measure change, say $\mathbb{P} \to \widehat{\mathbb{P}}$, whereby we begin with $W := \{W(t)\}_{t\geqslant 0}$ as a standard BM under measure $\mathbb{P}$ and then define a new process, which we denote by $\widehat{W} := \{\widehat{W}(t)\}_{t\geqslant 0}$, such that $\widehat{W}$ is a standard BM under the new measure $\widehat{\mathbb{P}}$. We will see that the measure change from $\mathbb{P} \to \widehat{\mathbb{P}}$ is constructed by using a positive random variable that is an exponential $\mathbb{P}$-martingale and that there is a precise relationship between the two Brownian motions W and $\widehat{W}$ which will differ only by a drift component. This is the essence of Girsanov's Theorem, whose statement and proof are given later. The change of measure has many useful applications and will also allow us to compute conditional expectations of processes or random variables that are functionals of Brownian motion under two different probability measures $\mathbb{P}$ and $\widehat{\mathbb{P}}$. These two measures will be equivalent in the sense that (as we recall from our previous discussion on equivalent probability measures) all events having zero probability under one measure also have zero probability under the other measure.

Let us begin by fixing a filtered probability space $(\Omega, \mathcal{F}, \mathbb{P}, \mathbb{F})$, where $\mathbb{F} = \{\mathcal{F}_t\}_{t\geqslant 0}$ is any filtration for standard Brownian motion and recall Definition 10.1 of a $(\mathbb{P}, \mathbb{F})$-BM. This is shorthand for a standard Brownian motion $W := \{W(t)\}_{t\geqslant 0}$ w.r.t. a given filtration $\mathbb{F}$ and a measure $\mathbb{P}$. That is, W has (a.s.) continuous paths started at $W(0) = 0$ and, under given measure $\mathbb{P}$, it has normally distributed increments $W(t) - W(s) \sim Norm(0, t-s)$ that are independent of $\mathcal{F}_s$, for all $0 \leqslant s < t$. From these original defining properties we then showed that W has quadratic variation $[W, W](t) = t$ and that it is a $(\mathbb{P}, \mathbb{F})$-martingale (shorthand for a martingale w.r.t. filtration $\mathbb{F}$ and measure $\mathbb{P}$). Now, let's assume that we have a process that we know is a continuous martingale started at zero and with the same quadratic variation formula as standard Brownian motion. The question is whether or not this process is a standard Brownian motion. It turns out that the answer is yes, the process is a standard Brownian motion as stated in the following theorem, originally due to Lévy. In what follows, this will give us a useful way to recognize when a martingale process is in fact a standard Brownian motion. The characterization makes no assumption of the normality and independence of increments! Rather, these properties are implied. Besides the martingale property, the requirement of continuity of all paths and the fact that they must start at zero, the recognition that we have a BM follows from the assumption of the above quadratic variation formula.

[*Technical Remark*: We note that the proof of Theorem 11.10 below makes use of a general version of the Itô formula in (11.18). Although we do not prove it, it turns out that we have the same Itô formula as in (11.18) if W is replaced by a continuous martingale process $M := \{M(t)\}_{t\geqslant 0}$, that starts at zero and has quadratic variation $[M, M](t) = t$, i.e., $\mathrm{d}M(t)\,\mathrm{d}M(t) = \mathrm{d}[M, M](t) = \mathrm{d}t$:

$$\mathrm{d}f(t, M(t)) = \left(f_t(t, M(t)) + \frac{1}{2} f_{xx}(t, M(t))\right) \mathrm{d}t + f_x(t, M(t))\,\mathrm{d}M(t). \tag{11.80}$$

Essentially one can think of this as the Itô formula in (11.20) where $X \equiv M$ with zero drift

$\mu \equiv 0$ and unit diffusion function $\sigma \equiv 1$. The integral form of (11.80) is (see (11.19))

$$f(t, M(t)) = f(0, M(0)) + \int_0^t \left(f_u(u, M(u)) + \frac{1}{2} f_{xx}(u, M(u))\right) \mathrm{d}u + \int_0^t f_x(u, M(u))\, \mathrm{d}M(u). \tag{11.81}$$

Assuming the usual square integrability condition as we did for any Itô integral w.r.t. BM, the above stochastic integral w.r.t. the increment $\mathrm{d}M(u)$ is defined in a similar fashion and is a martingale having zero expected value. Note that if $f(t,x)$ is a $C^{1,2}$ function satisfying the PDE $f_t(t,x) + \frac{1}{2} f_{xx}(t,x) = 0$, then the process defined by $Y(t) := f(t, M(t))$ is a martingale.]

Theorem 11.10 (Lévy's characterization of standard BM). *Let the process* $\{M(t)\}_{t \geqslant 0}$ *be a continuous* $(\mathbb{P}, \mathbb{F})$*-martingale started at* $M(0) = 0$ *(a.s.) and with quadratic variation* $[M, M](t) = t$ *for all* $t \geqslant 0$. *Then,* $\{M(t)\}_{t \geqslant 0}$ *is a standard* $(\mathbb{P}, \mathbb{F})$*-BM.*

Proof. Since we have already assumed that M has continuous paths all starting at zero, from the definition of a $(\mathbb{P}, \mathbb{F})$-BM we have left to show that $M(t) - M(s) \sim Norm(0, t-s)$ and that these increments are independent of $\mathcal{F}_s$ for all $0 \leqslant s < t$. For this purpose, consider the function $f(t,x) = \mathrm{e}^{-\frac{1}{2}\theta^2 t + \theta x}$ with arbitrary real parameter θ. Since $f(t,x)$ satisfies the PDE $f_t(t,x) + \frac{1}{2} f_{xx}(t,x) = 0$, from the above discussion we have that the process $f(t, M(t)) = \mathrm{e}^{-\frac{1}{2}\theta^2 t + \theta M(t)}, t \geqslant 0$, is a martingale. In fact, we recognize this as an example of an exponential martingale. Taking the expectation of the process at time t, conditional on filtration $\mathcal{F}_s$, $s \leqslant t$, and using the martingale property gives the *conditional* moment-generating function (m.g.f.) of $M(t) - M(s)$ as a function of θ:

$$\mathrm{E}\left[\mathrm{e}^{-\frac{1}{2}\theta^2 t + \theta M(t)} \mid \mathcal{F}_s\right] = \mathrm{e}^{-\frac{1}{2}\theta^2 s + \theta M(s)} \implies \mathrm{E}\left[\mathrm{e}^{\theta(M(t) - M(s))} \mid \mathcal{F}_s\right] = \mathrm{e}^{\frac{1}{2}\theta^2 (t-s)}.$$

This is equivalent to the m.g.f. of $M(t) - M(s)$, which is the m.g.f. of a $Norm(0, t-s)$ random variable (by the tower property):

$$\mathrm{E}\left[\mathrm{e}^{\theta(M(t) - M(s))}\right] = \mathrm{E}\left[\mathrm{E}\left[\mathrm{e}^{\theta(M(t) - M(s))} \mid \mathcal{F}_s\right]\right] = \mathrm{e}^{\frac{1}{2}\theta^2 (t-s)}.$$

Hence, as function of θ, the m.g.f. and the m.g.f. conditional on $\mathcal{F}_s$ are the same and correspond to that of a $Norm(0, t-s)$ random variable, i.e., $M(t) - M(s) \sim Norm(0, t-s)$ and $M(t) - M(s)$ is independent of $\mathcal{F}_s$, for all $s \leqslant t$. □

[*Remark*: In what follows, we will only distinguish between different probability measures while *fixing a filtration* $\mathbb{F}$ *for BM*. Hence, we shall also write $\mathbb{P}$-BM to mean standard $(\mathbb{P}, \mathbb{F})$-BM, i.e., standard BM w.r.t. filtration $\mathbb{F}$ and under measure $\mathbb{P}$. Equivalently, we shall also say that W is a BM under the measure $\mathbb{P}$. We shall also sometimes loosely say BM (or Brownian motion) where we clearly really mean *standard* BM. Also, we simply say $\mathbb{P}$-martingale to mean a $(\mathbb{P}, \mathbb{F})$-martingale and $\widehat{\mathbb{P}}$-martingale to mean a $(\widehat{\mathbb{P}}, \mathbb{F})$-martingale when $\mathbb{F}$ is fixed.]

In what follows we let the probability measure $\widehat{\mathbb{P}} \equiv \widehat{\mathbb{P}}^{(\varrho)}$ be defined by (9.109) of Section 9.5, i.e., $\widehat{\mathbb{P}}(A) := \int_A \varrho(\omega)\, \mathrm{d}\mathbb{P}(\omega)$, $A \in \mathcal{F}$, with Radon–Nikodym random variable $\varrho \equiv \frac{\mathrm{d}\widehat{\mathbb{P}}}{\mathrm{d}\mathbb{P}}$ assumed positive (almost surely) with unit expectation under measure $\mathbb{P}$, $\mathrm{E}[\varrho] = 1$. We recall how ϱ is used in (9.110) and (9.111) for computing the expectation of any integrable random variable under measures $\widehat{\mathbb{P}}$ and $\mathbb{P}$, respectively. Shortly we shall explicitly specify this random variable and, in fact, its precise specification is a key ingredient in Girsanov's Theorem. However, for the moment we can keep our assumptions on ϱ as is (which are as general as possible). In preparation for our main result, we will need to define and discuss some basic properties of a so-called *Radon–Nikodym derivative process of* $\widehat{\mathbb{P}}$ *w.r.t.* $\mathbb{P}$. In a previous chapter on discrete-time financial models, we defined a similar process but in a discrete-time stochastic setting. In continuous time we shall fix some terminal time $T > 0$ and define the Radon–Nikodym derivative process $\{\varrho_t\}_{0 \leqslant t \leqslant T}$ (of measure $\widehat{\mathbb{P}} \equiv \widehat{\mathbb{P}}^{(\varrho)}$ w.r.t. measure $\mathbb{P}$ for a given filtration $\mathbb{F}$) by

$$\varrho_t := \mathrm{E}[\,\varrho \mid \mathcal{F}_t],\ \ 0 \leqslant t \leqslant T. \tag{11.82}$$

We remark that it is customary to also use the following *more explicit equivalent notations* for the random variable ϱ_t:

$$\varrho_t \equiv \left(\frac{\mathrm{d}\widehat{\mathbb{P}}}{\mathrm{d}\mathbb{P}}\right)_t \overset{\text{or}}{\equiv} \left(\frac{\mathrm{d}\widehat{\mathbb{P}}^{(\varrho)}}{\mathrm{d}\mathbb{P}}\right)_t \overset{\text{or}}{\equiv} \left(\frac{\mathrm{d}\widehat{\mathbb{P}}^{(\varrho)}}{\mathrm{d}\mathbb{P}}\right)_{\mathcal{F}_t}.$$

Hence (11.82) is also written as $\left(\frac{\mathrm{d}\widehat{\mathbb{P}}}{\mathrm{d}\mathbb{P}}\right)_t := \mathrm{E}[\frac{\mathrm{d}\widehat{\mathbb{P}}}{\mathrm{d}\mathbb{P}} \mid \mathcal{F}_t]$. These notations really spell out the definition in (11.82) and also visually remind us of the "direction of the measure change," e.g., $\mathbb{P} \to \widehat{\mathbb{P}}$. In what follows we shall try to keep our notation less cumbersome as long as there is no ambiguity.

Clearly ϱ_t is $\mathcal{F}_t$-measurable and integrable, $\mathrm{E}[|\varrho_t|] \leqslant \mathrm{E}[|\varrho|] = \mathrm{E}[\varrho] = 1 < \infty$, for all $t \in [0, T]$. By the tower property and the definition in (11.82), we immediately we see that the process $\{\varrho_t\}_{0 \leqslant t \leqslant T}$ is a $\mathbb{P}$-martingale (recall the Doob-Lévy martingale):

$$\mathrm{E}[\,\varrho_t \mid \mathcal{F}_s] = \mathrm{E}\big[\mathrm{E}[\,\varrho \mid \mathcal{F}_t] \mid \mathcal{F}_s\big] = \mathrm{E}[\,\varrho \mid \mathcal{F}_s] = \varrho_s\,, \quad 0 \leqslant s \leqslant t \leqslant T.$$

By definition, the process also starts with unit value: $\varrho_0 = \mathrm{E}[\,\varrho \mid \mathcal{F}_0] \equiv \mathrm{E}[\,\varrho\,] = 1$. Hence, by the martingale property, the process has unit expectation, $\mathrm{E}[\varrho_t\,] = \varrho_0 = 1$, for all $t \in [0, T]$.

The next proposition gives a useful formula for computing the $\widehat{\mathbb{P}}$-measure expectation of an $\mathcal{F}_t$-measurable random variable X, conditional on information up to a time s prior to time t, as a $\mathbb{P}$-measure conditional expectation of $X \cdot (\varrho_t/\varrho_s)$. The ratio ϱ_t/ϱ_s of the Radon–Nikodym derivative process at times s and t adjusts for the change of measure in the conditional expectation.

Proposition 11.11. *Let* $\widehat{\mathbb{P}}$ *be defined by* $\widehat{\mathbb{P}}(A) := \int_A \varrho(\omega)\,\mathrm{d}\mathbb{P}(\omega)$, $A \in \mathcal{F}$, *with process* $\varrho_t := \mathrm{E}[\,\varrho \mid \mathcal{F}_t], 0 \leqslant t \leqslant T$. *Assume the random variable* X *is integrable w.r.t.* $\widehat{\mathbb{P}}$ *and* $\mathcal{F}_t$*-measurable for a given time* $t \in [0, T]$. *Then, for all* $0 \leqslant s \leqslant t$,

$$\widehat{\mathrm{E}}\big[X \mid \mathcal{F}_s\big] = \varrho_s^{-1}\mathrm{E}\big[\varrho_t\, X \mid \mathcal{F}_s\big]\,. \tag{11.83}$$

Proof. This result follows as a simple application of Theorem 9.7 where we set $\mathcal{G} \equiv \mathcal{F}_s$, and $\mathcal{F}_s \subset \mathcal{F}_t \subset \mathcal{F}$ implies $\mathcal{G} \subset \mathcal{F}$. Then, upon using the definition in (11.82) for time s, the formula in (9.113) gives

$$\widehat{\mathrm{E}}\big[X \mid \mathcal{F}_s\big] = \frac{\mathrm{E}[\,\varrho X \mid \mathcal{F}_s]}{\mathrm{E}[\,\varrho \mid \mathcal{F}_s]} = \varrho_s^{-1}\mathrm{E}[\,\varrho X \mid \mathcal{F}_s]\,.$$

The last expectation on the right is now recast by reversing the tower property, by conditioning on $\mathcal{F}_t$, and using the fact that X is $\mathcal{F}_t$-measurable (so it is pulled out of the inner expectation conditional on $\mathcal{F}_t$ below):

$$\mathrm{E}[\,\varrho X \mid \mathcal{F}_s] = \mathrm{E}\big[\mathrm{E}[\,\varrho X \mid \mathcal{F}_t] \mid \mathcal{F}_s\big] = \mathrm{E}\big[X\,\mathrm{E}[\,\varrho \mid \mathcal{F}_t] \mid \mathcal{F}_s\big] = \mathrm{E}\big[X\,\varrho_t \mid \mathcal{F}_s\big]\,.$$

In the last step we used the definition $\mathrm{E}[\,\varrho \mid \mathcal{F}_t] = \varrho_t$. ☐

Note that a special case of (11.83) is when $s = 0$. Since $\varrho_0 = 1$, $\widehat{\mathrm{E}}[X \mid \mathcal{F}_0] = \widehat{\mathrm{E}}[X]$ and $\mathrm{E}[\varrho_t X \mid \mathcal{F}_0] = \mathrm{E}[\varrho_t X]$, we have

$$\widehat{\mathrm{E}}[X] = \mathrm{E}[\varrho_t X] \text{ for } \mathcal{F}_t\text{-measurable } X. \tag{11.84}$$

Consider a continuous-time stochastic process $\{X(t)\}_{t \geqslant 0}$ adapted to the filtration $\mathbb{F}$. Since $X(t)$ is $\mathcal{F}_t$-measurable for every $t \geqslant 0$, we may put $X = X(t)$ in (11.83) to obtain

$$\widehat{\mathrm{E}}[X(t) \mid \mathcal{F}_s] = \varrho_s^{-1}\mathrm{E}[\varrho_t X(t) \mid \mathcal{F}_s], \; 0 \leqslant s \leqslant t \leqslant T. \tag{11.85}$$

As a consequence of this property we have the following result.

Proposition 11.12. *A continuous-time adapted stochastic process $\{M(t)\}_{0 \leqslant t \leqslant T}$ is a $\widehat{\mathbb{P}}$-martingale if and only if $\{\varrho_t M(t)\}_{0 \leqslant t \leqslant T}$ is a $\mathbb{P}$-martingale.*

Proof. Assume $\{M(t)\}_{0 \leqslant t \leqslant T}$ is a $\widehat{\mathbb{P}}$-martingale. Then, using (11.85) with $X(t) \equiv M(t)$,

$$M(s) = \widehat{\mathrm{E}}[M(t) \mid \mathcal{F}_s] = \varrho_s^{-1}\mathrm{E}[\varrho_t M(t) \mid \mathcal{F}_s] \implies \varrho_s M(s) = \mathrm{E}[\varrho_t M(t) \mid \mathcal{F}_s]$$

for $0 \leqslant s \leqslant t \leqslant T$, where the last relation is the $\mathbb{P}$-martingale property of $\{\varrho_t M(t)\}_{0 \leqslant t \leqslant T}$. The converse follows since all the above steps may be reversed. Moreover, $\{M(t)\}_{0 \leqslant t \leqslant T}$ is adapted to $\mathbb{F}$ and integrable w.r.t. $\widehat{\mathbb{P}}$ if and only if $\{\varrho_t M(t)\}_{0 \leqslant t \leqslant T}$ is adapted to $\mathbb{F}$ and integrable w.r.t. $\mathbb{P}$. □

We are now finally ready to state and prove Girsanov's Theorem for the case of standard Brownian motion.

Theorem 11.13 (Girsanov's Theorem for BM)**.** *Let $\{W(t)\}_{0 \leqslant t \leqslant T}$ be a standard $\mathbb{P}$-BM w.r.t. a filtration $\mathbb{F} = \{\mathcal{F}_t\}_{0 \leqslant t \leqslant T}$ and assume the process $\{\gamma(t)\}_{0 \leqslant t \leqslant T}$ is adapted to $\mathbb{F}$, for a given $T > 0$. Define*

$$\varrho_t := \exp\left(-\frac{1}{2}\int_0^t \gamma^2(s)\,\mathrm{d}s + \int_0^t \gamma(s)\,\mathrm{d}W(s)\right), 0 \leqslant t \leqslant T, \tag{11.86}$$

and the probability measure $\widehat{\mathbb{P}} \equiv \widehat{\mathbb{P}}^{(\varrho)}$ by the Radon–Nikodym derivative $\frac{\mathrm{d}\widehat{\mathbb{P}}}{\mathrm{d}\mathbb{P}} = \left(\frac{\mathrm{d}\widehat{\mathbb{P}}}{\mathrm{d}\mathbb{P}}\right)_T \equiv \varrho_T$. Furthermore, assume the square-integrability condition holds:

$$\mathrm{E}\left[\int_0^T \varrho_s^2\,\gamma^2(s)\,\mathrm{d}s\right] < \infty. \tag{11.87}$$

Then, the process $\{\widehat{W}(t)\}_{0 \leqslant t \leqslant T}$ defined by

$$\widehat{W}(t) := W(t) - \int_0^t \gamma(s)\,\mathrm{d}s \tag{11.88}$$

is a standard $\widehat{\mathbb{P}}$-BM w.r.t. filtration $\mathbb{F}$.

Some clarifying remarks on Theorem 11.13 before its proof:

1. The condition in (11.87) is required to ensure that $\{\varrho_t\}_{0 \leqslant t \leqslant T}$ is a $\mathbb{P}$-martingale with $\mathrm{E}[\varrho_t] = 1$, i.e., this corresponds to the Itô process $\int_0^t \varrho_s\,\gamma(s)\,\mathrm{d}W(s), 0 \leqslant t \leqslant T$, being a martingale. An equivalent and more practically verified condition that guarantees the process $\{\varrho_t\}_{0 \leqslant t \leqslant T}$ is a $\mathbb{P}$-martingale is the so-called *Novikov condition*:

$$\mathrm{E}\left[\exp\left(\frac{1}{2}\int_0^T \gamma^2(s)\,\mathrm{d}s\right)\right] < \infty. \tag{11.89}$$

2. The differential increments of the two Brownian motions are simply related: $\mathrm{d}W(t) = \mathrm{d}\widehat{W}(t) + \gamma(t)\,\mathrm{d}t$ and $\mathrm{d}\widehat{W}(t) = \mathrm{d}W(t) - \gamma(t)\,\mathrm{d}t$.

3. Pay attention to the consistent and correct use of the $\pm$ signs. In this regard, we note that the Radon–Nikodym derivative random variable in (11.86) can *equivalently* be written as
$$\varrho_t = \exp\left(-\frac{1}{2}\int_0^t \theta^2(s)\,\mathrm{d}s - \int_0^t \theta(s)\,\mathrm{d}W(s)\right).$$
Note the $-$ sign instead of the $+$ sign in front of the Itô integral. Then, (11.88) is replaced by $\widehat{W}(t) := W(t) + \int_0^t \theta(s)\,\mathrm{d}s$, i.e., $\mathrm{d}\widehat{W}(t) = \mathrm{d}W(t) + \theta(t)\,\mathrm{d}t$. This is obtained simply by setting $\gamma(t) = -\theta(t)$ in the original definition where $\gamma^2(t) = \theta^2(t)$.

4. In general, γ is an adapted process so that ϱ_t is a functional of BM from time 0 to t. In particular, the Radon–Nikodym derivative process has the form of an exponential $\mathbb{P}$-martingale in the process γ w.r.t. the $\mathbb{P}$-BM, i.e., by the definition in (11.28) we have $\varrho_t \equiv \varrho_t^{(\gamma)} = \mathcal{E}_t(\gamma \cdot W)$. Dividing the process value at any two times $0 \leqslant s < t \leqslant T$ gives
$$\frac{\varrho_t}{\varrho_s} \equiv \frac{\left(\frac{\mathrm{d}\widehat{\mathbb{P}}}{\mathrm{d}\mathbb{P}}\right)_t}{\left(\frac{\mathrm{d}\widehat{\mathbb{P}}}{\mathrm{d}\mathbb{P}}\right)_s} = \frac{\mathcal{E}_t(\gamma \cdot W)}{\mathcal{E}_s(\gamma \cdot W)} = \exp\left(-\frac{1}{2}\int_s^t \gamma^2(u)\,\mathrm{d}u + \int_s^t \gamma(u)\,\mathrm{d}W(u)\right).$$

5. Note that $\mathbb{F}$ is any filtration for BM. It can, but need not be, the natural filtration $\mathbb{F}^W$ generated by W.

6. In the simplest case we can choose a constant process, $\gamma(t) = \gamma =$ constant, where
$$\varrho_t \equiv \varrho_t^{(\gamma)} = \mathrm{e}^{-\frac{1}{2}\gamma^2 t + \gamma W(t)} \tag{11.90}$$
and $\widehat{W}(t) \equiv \widehat{W}^{(\gamma)}(t) := W(t) - \gamma t, 0 \leqslant t \leqslant T$, is a $\widehat{\mathbb{P}}$-BM.

Proof. First let us verify that $\{\varrho_t\}_{0 \leqslant t \leqslant T}$ is a Radon–Nikodym derivative process. By the assumption in (11.87) (or the Novikov condition) we have that $\{\varrho_t\}_{0 \leqslant t \leqslant T}$ is a $\mathbb{P}$-martingale; in fact it is an exponential $\mathbb{P}$-martingale. This can be seen by applying Itô's formula to the stochastic exponential in (11.86), giving
$$\mathrm{d}\varrho_t = \varrho_t \gamma(t)\,\mathrm{d}W(t) \implies \varrho_t = \varrho_0 + \int_0^t \varrho_s \gamma(s)\,\mathrm{d}W(s)$$
where the Itô integral is a martingale (under measure $\mathbb{P}$) by the condition in (11.87). Because of the $\mathbb{P}$-martingale property, $\mathrm{E}[\varrho_t] = \varrho_0 = \mathrm{e}^0 = 1$, $0 \leqslant t \leqslant T$. In particular, $\mathrm{E}[\varrho_T] = 1$ and ϱ_T is also nonnegative. Hence, $\varrho \equiv \frac{\mathrm{d}\widehat{\mathbb{P}}}{\mathrm{d}\mathbb{P}} = \varrho_T$ is a proper Radon–Nikodym derivative and by the $\mathbb{P}$-martingale property the process in (11.86) satisfies the definition in (11.82), i.e., it is indeed a Radon–Nikodym derivative process.

We now show that the process $\widehat{W}$ defined by (11.88) is a standard $\widehat{\mathbb{P}}$-BM by verifying all the defining properties in Theorem 11.10 with measure $\widehat{\mathbb{P}}$ (filtration $\mathbb{F}$ fixed):

(i) The process starts at zero, $\widehat{W}(0) = W(0) = 0$, and is continuous in time since $\widehat{W}(t) := W(t) - \int_0^t \gamma(s)\,\mathrm{d}s$ where $W(t)$ and the integral $\int_0^t \gamma(s)\,\mathrm{d}s$ are both continuous in $t \geqslant 0$.

(ii) $\mathrm{d}[\widehat{W}, \widehat{W}](t) = \mathrm{d}\widehat{W}(t)\,\mathrm{d}\widehat{W}(t) = (\mathrm{d}W(t) - \gamma(t)\,\mathrm{d}t)(\mathrm{d}W(t) - \gamma(t)\,\mathrm{d}t) = \mathrm{d}W(t)\,\mathrm{d}W(t) = \mathrm{d}t$, i.e., the process has quadratic variation $[\widehat{W}, \widehat{W}](t) = t$.

(iii) $\{\widehat{W}(t)\}_{0\leqslant t\leqslant T}$ is a $\widehat{\mathbb{P}}$-martingale. By Proposition 11.12, this follows if we can show that the process $\{\varrho_t\widehat{W}(t)\}_{0\leqslant t\leqslant T}$ is a $\mathbb{P}$-martingale. To show the latter, we compute the stochastic differential by Itô's product rule (using $\mathrm{d}\varrho_t = \varrho_t\gamma(t)\,\mathrm{d}W(t)$ and $\mathrm{d}\widehat{W}(t) = \mathrm{d}W(t) - \gamma(t)\,\mathrm{d}t$ and setting $\mathrm{d}W(t)\,\mathrm{d}W(t) = \mathrm{d}t$, $\mathrm{d}W(t)\,\mathrm{d}t = 0$):

$$\begin{aligned}\mathrm{d}\big(\varrho_t\widehat{W}(t)\big) &= \varrho_t\ \mathrm{d}\widehat{W}(t) + \widehat{W}(t)\ \mathrm{d}\varrho_t + \mathrm{d}\varrho_t\ \mathrm{d}\widehat{W}(t)\\ &= \varrho_t[\,\mathrm{d}W(t) - \gamma(t)\,\mathrm{d}t] + \varrho_t\gamma(t)\widehat{W}(t)\,\mathrm{d}W(t) + \varrho_t\gamma(t)\,\mathrm{d}W(t)[\,\mathrm{d}W(t) - \gamma(t)\,\mathrm{d}t]\\ &= \varrho_t[1 + \gamma(t)\widehat{W}(t)]\,\mathrm{d}W(t)\,.\end{aligned}$$

This is a stochastic differential with a zero drift term (i.e., the coefficient in $\mathrm{d}t$ is zero). In integral form, where $\varrho_0\widehat{W}(0) = 0$, we have

$$\varrho_t\widehat{W}(t) = \int_0^t \varrho_s\,[1 + \gamma(s)\widehat{W}(s)]\,\mathrm{d}W(s)\,,\ \ 0 \leqslant t \leqslant T.$$

By the assumed boundedness of $\int_0^t \gamma(s)\,\mathrm{d}s$, and the fact that the BM $W(t)$ is bounded (a.s.), $\widehat{W}(t)$ is bounded (a.s.) for all $0 \leqslant t \leqslant T$. Combining this fact with the square-integrability condition (11.87), it follows that the above Itô integral is defined as it satisfies the square-integrability condition, $\mathrm{E}\left[\int_0^T \varrho_s^2\,[1+\gamma(s)\widehat{W}(s)]^2\,\mathrm{d}s\right] < \infty$, and is hence a $\mathbb{P}$-martingale, i.e., $\{\varrho_t\widehat{W}(t)\}_{0\leqslant t\leqslant T}$ is a $\mathbb{P}$-martingale. □

11.8.1 Some Applications of Girsanov's Theorem

Let's begin by considering a simple example of how Girsanov's Theorem can be applied to change probability measures so as to eliminate the drift in a drifted Brownian process.

Example 11.14. Let $X(t) \equiv W^{(\mu,\sigma)}(t)$ be a drifted BM process (recall (10.23))

$$X(t) := \mu t + \sigma W(t),$$

where $\{W(t)\}_{t\geqslant 0}$ is a standard $\mathbb{P}$-BM. Find a measure under which $\{X(t)\}_{0\leqslant t\leqslant T}$, for any $T > 0$, is a scaled BM with zero drift.

Solution. We note that the drift μ and volatility parameter $\sigma > 0$ are constants. Hence, by using Girsanov's Theorem we define a measure $\widehat{\mathbb{P}}$, $\frac{\mathrm{d}\widehat{\mathbb{P}}}{\mathrm{d}\mathbb{P}} = \varrho_T$, where ϱ_t is given by (11.90). Now, $\widehat{W}(t) := W(t) - \gamma t$ is a standard $\widehat{\mathbb{P}}$-BM and writing $X(t)$ in terms of $\widehat{W}(t)$ gives

$$X(t) = \mu t + \sigma W(t) = \mu t + \sigma(\widehat{W}(t) + \gamma t) = (\mu + \sigma\gamma)t + \sigma\widehat{W}(t).$$

So the drift coefficient of $X(t)$ is now $\mu + \sigma\gamma$, while the volatility parameter multiplying the standard $\widehat{\mathbb{P}}$-BM is still σ. Note that we can also see this in stochastic differential form:

$$\mathrm{d}X(t) = \mu\,\mathrm{d}t + \sigma\,\mathrm{d}W(t) = \mu\,\mathrm{d}t + \sigma(\,\mathrm{d}\widehat{W}(t) + \gamma\,\mathrm{d}t) = (\mu + \sigma\gamma)\,\mathrm{d}t + \sigma\,\mathrm{d}\widehat{W}(t).$$

Hence, choosing $\gamma = -\mu/\sigma$ gives zero drift, $\mu + \sigma\gamma = 0$, and the process $X(t) = \sigma\widehat{W}(t)$ is a zero-drift scaled BM under measure $\widehat{\mathbb{P}}$. The measure change $\mathbb{P} \to \widehat{\mathbb{P}}$, $\frac{\mathrm{d}\widehat{\mathbb{P}}}{\mathrm{d}\mathbb{P}} = \varrho_T$, is defined explicitly by the Radon–Nikodym derivative process

$$\varrho_t \equiv \left(\frac{\mathrm{d}\widehat{\mathbb{P}}}{\mathrm{d}\mathbb{P}}\right)_t = \exp\left(-\frac{\mu^2 t}{2\sigma^2} - \frac{\mu}{\sigma}W(t)\right),\ \ 0 \leqslant t \leqslant T. \tag{11.91}$$

□

For the above example, we can also find the CDF of $X(t)$ in the $\widehat{\mathbb{P}}$-measure, denoted by $\widehat{F}_{X(t)}$. It is instructive to see the two ways to obtain this CDF. One way is to simply use $X(t) = \sigma\widehat{W}(t) \stackrel{d}{=} \sigma\sqrt{t}\widehat{Z}$, $\widehat{Z} \sim Norm(0,1)$ under measure $\widehat{\mathbb{P}}$:

$$\widehat{F}_{X(t)}(x) \equiv \widehat{\mathbb{P}}(X(t) \leqslant x) = \widehat{\mathbb{P}}\left(\widehat{Z} \leqslant \frac{x}{\sigma\sqrt{t}}\right) = \mathcal{N}\left(\frac{x}{\sigma\sqrt{t}}\right).$$

The other way is to compute a $\mathbb{P}$-measure expectation using ϱ_t and apply the identity in (11.84) since $\mathbb{I}_{\{X(t)\leqslant x\}} = \mathbb{I}_{\{\mu t+\sigma W(t)\leqslant x\}}$ is an $\mathcal{F}_t$-measurable random variable:

$$\begin{aligned}\widehat{F}_{X(t)}(x) = \widehat{\mathrm{E}}\big[\mathbb{I}_{\{X(t)\leqslant x\}}\big] = \mathrm{E}\big[\varrho_t\mathbb{I}_{\{X(t)\leqslant x\}}\big] &= \mathrm{E}\big[\mathrm{e}^{-\frac{1}{2}\gamma^2 t+\gamma W(t)}\mathbb{I}_{\{\mu t+\sigma W(t)\leqslant x\}}\big] \\ &= \mathrm{e}^{-\frac{1}{2}\gamma^2 t}\mathrm{E}\big[\mathrm{e}^{\gamma W(t)}\mathbb{I}_{\{W(t)\leqslant (x-\mu t)/\sigma\}}\big] \\ &= \mathrm{e}^{-\frac{1}{2}\gamma^2 t}\cdot \mathrm{e}^{\frac{1}{2}\gamma^2 t}\mathcal{N}\left(\frac{x-\mu t}{\sigma\sqrt{t}} - \gamma\sqrt{t}\right) \\ &= \mathcal{N}\left(\frac{x-(\mu+\sigma\gamma)t}{\sigma\sqrt{t}}\right) = \mathcal{N}\left(\frac{x}{\sigma\sqrt{t}}\right)\end{aligned}$$

where $\mu + \sigma\gamma = 0$. Note that here we used the expectation identity (A.2) in the Appendix where $W(t) \sim Norm(0,t)$ under measure $\mathbb{P}$.

The CDF of $X(t)$ in the $\mathbb{P}$-measure was already computed in Section 10.3.1, i.e.,

$$F_{X(t)}(x) \equiv \mathbb{P}(X(t) \leqslant x) \equiv \mathbb{P}(W^{(\mu,\sigma)}(t) \leqslant x) = \mathcal{N}\left(\frac{x-\mu t}{\sigma\sqrt{t}}\right).$$

We therefore see from the above two expressions for the CDF of the process at time t (in the two different measures) that the measure change $\mathbb{P} \to \widehat{\mathbb{P}}$ eliminates the drift μt when $\gamma = -\mu/\sigma$. Observe that $X(t) \sim Norm(0, \sigma^2 t)$ under the $\widehat{\mathbb{P}}$-measure:

$$\widehat{\mathrm{E}}\big[X(t)\big] = \sigma\widehat{\mathrm{E}}\big[\widehat{W}(t)\big] = 0\,, \quad \widehat{\mathrm{E}}\big[X^2(t)\big] = \sigma^2\widehat{\mathrm{E}}\big[\widehat{W}^2(t)\big] = \sigma^2 t\,.$$

In contrast, $X(t) \sim Norm(\mu t, \sigma^2 t)$ under the $\mathbb{P}$-measure.

In previous chapters we saw how measure changes are employed in discrete-time asset price models such as the binomial model. In particular, we discussed various risk-neutral measures. By using Girsanov's Theorem, we can now consider our first example of how to construct a risk-neutral measure for a single stock GBM price process in *continuous time.*

Example 11.15. (Changing the drift in GBM) Assume a non-dividend-paying stock price process with SDE

$$\mathrm{d}S(t) = S(t)\big[\mu\,\mathrm{d}t + \sigma\,\mathrm{d}W(t)\big]\,,$$

where $\{W(t)\}_{t\geqslant 0}$ is a standard BM under the physical (real-world) measure $\mathbb{P}$, μ is a constant physical (i.e., historical) growth rate, and $\sigma > 0$ is a constant volatility. Find the risk-neutral probability measure $\widetilde{\mathbb{P}}$ defined such that the discounted stock price process $\{\overline{S}(t) := \mathrm{e}^{-rt}S(t)\}_{0\leqslant t\leqslant T}$, for any $T > 0$, is a $\widetilde{\mathbb{P}}$-martingale, where r is a constant interest rate.

Solution. By the strong solution of the SDE

$$\begin{aligned}S(t) = S(0)\,\mathrm{e}^{(\mu-\sigma^2/2)t+\sigma W(t)} \implies \overline{S}(t) = \mathrm{e}^{-rt}S(t) &= S(0)\,\mathrm{e}^{(\mu-r)t}\cdot\mathrm{e}^{-\sigma^2 t/2+\sigma W(t)} \\ &\equiv S(0)\,\mathrm{e}^{(\mu-r)t}\cdot\mathcal{E}_t(\sigma\cdot W)\,.\end{aligned}$$

We recognize $\{\mathcal{E}_t(\sigma \cdot W) := \mathrm{e}^{-\sigma^2 t/2+\sigma W(t)}\}_{t\geqslant 0}$ as a (exponential) $\mathbb{P}$-martingale with unit expectation, $\mathrm{E}[\mathcal{E}_t(\sigma \cdot W)] = 1$ (see Example 10.2 in Chapter 10). So we now proceed to eliminate the drift $\mu - r$ by expressing W in terms of a new BM, $\widetilde{W}$, in the new measure $\widetilde{\mathbb{P}}$. Since $\mu - r$ and σ are constants, we can accomplish this by employing a measure change as in the above example:

$$\varrho_t \equiv \left(\frac{\mathrm{d}\widetilde{\mathbb{P}}}{\mathrm{d}\mathbb{P}}\right)_t = \mathrm{e}^{-\frac{1}{2}\gamma^2 t+\gamma W(t)} \equiv \mathcal{E}_t(\gamma \cdot W), \tag{11.92}$$

where $\frac{\mathrm{d}\widetilde{\mathbb{P}}}{\mathrm{d}\mathbb{P}} = \varrho_T$ and $\widetilde{W}(t) = W(t) - \gamma t$ is a standard $\widetilde{\mathbb{P}}$-BM. Substituting $W(t) = \widetilde{W}(t) + \gamma t$ into the above exponential expression gives

$$\overline{S}(t) = S(0)\,\mathrm{e}^{(\mu-r)t} \cdot \mathrm{e}^{-\sigma^2 t/2+\sigma(\widetilde{W}(t)+\gamma t)} = \overline{S}(0)\,\mathrm{e}^{(\mu-r+\sigma\gamma)t} \cdot \mathcal{E}_t(\sigma \cdot \widetilde{W}) \tag{11.93}$$

where $\overline{S}(0) = S(0)$. Note that $\{\mathcal{E}_t(\sigma \cdot \widetilde{W}) := \mathrm{e}^{-\sigma^2 t/2+\sigma\widetilde{W}(t)}\}_{t\geqslant 0}$ is a $\widetilde{\mathbb{P}}$-martingale where:

$$\widetilde{\mathrm{E}}[\,\mathcal{E}_t(\sigma \cdot \widetilde{W}) \mid \mathcal{F}_u] = \mathcal{E}_u(\sigma \cdot \widetilde{W}),\ u \leqslant t.$$

Clearly, by setting $\gamma = (r - \mu)/\sigma$, we have $\mu - r + \sigma\gamma = 0$ and this gives the unique measure change for eliminating the drift in (11.93), giving the discounted stock price process as a $\widetilde{\mathbb{P}}$-martingale, i.e.,

$$\overline{S}(t) = \overline{S}(0) \cdot \mathcal{E}_t(\sigma \cdot \widetilde{W}),\ \ 0 \leqslant t \leqslant T, \tag{11.94}$$

where

$$\widetilde{\mathrm{E}}[\overline{S}(t) \mid \mathcal{F}_u] = \overline{S}(u),\ \ 0 \leqslant u \leqslant t \leqslant T. \tag{11.95}$$

In summary, the risk-neutral measure is the unique measure obtained with the Radon–Nikodym derivative process and measure change defined by (11.92) with $\gamma = (r - \mu)/\sigma$:

$$\left(\frac{\mathrm{d}\widetilde{\mathbb{P}}}{\mathrm{d}\mathbb{P}}\right)_t = \mathcal{E}_t\left(\frac{(r-\mu)}{\sigma} \cdot W\right),\ 0 \leqslant t \leqslant T;\quad \frac{\mathrm{d}\widetilde{\mathbb{P}}}{\mathrm{d}\mathbb{P}} = \mathcal{E}_T\left(\frac{(r-\mu)}{\sigma} \cdot W\right). \tag{11.96}$$

□

Note that the measure $\widetilde{\mathbb{P}}$ is uniquely specified by (11.96), where $\gamma = (r - \mu)/\sigma$ always exists since $\sigma > 0$. We can also see directly how to choose the above measure change by working with the SDE where the Brownian increment $\mathrm{d}W(t) = \mathrm{d}\widetilde{W}(t) + \gamma\,\mathrm{d}t$ is used within the original SDE:

$$\mathrm{d}S(t) = S(t)\big[\mu\,\mathrm{d}t + \sigma(\,\mathrm{d}\widetilde{W}(t) + \gamma\,\mathrm{d}t\,)\big] = S(t)\big[(\mu + \sigma\gamma)\,\mathrm{d}t + \sigma\,\mathrm{d}\widetilde{W}(t)\big]. \tag{11.97}$$

Taking the stochastic differential of $\overline{S}(t) \equiv \mathrm{e}^{-rt}S(t)$ and using the above $\mathrm{d}S(t)$ term:

$$\begin{aligned} \mathrm{d}\overline{S}(t) = \mathrm{d}(\mathrm{e}^{-rt}S(t)) &= \mathrm{e}^{-rt}\big[\mathrm{d}S(t) - rS(t)\,\mathrm{d}t\big] \\ &= \mathrm{e}^{-rt}S(t)\big[(\mu - r + \sigma\gamma)\,\mathrm{d}t + \sigma\,\mathrm{d}\widetilde{W}(t)\big] \\ &= \overline{S}(t)\big[(\mu - r + \sigma\gamma)\,\mathrm{d}t + \sigma\,\mathrm{d}\widetilde{W}(t)\big] & (11.98) \\ \implies \mathrm{d}\overline{S}(t) &= \sigma\overline{S}(t)\,\mathrm{d}\widetilde{W}(t) & (11.99) \end{aligned}$$

where the last expression with zero drift is obtained by choosing $\gamma = (r - \mu)/\sigma$, i.e., by employing the measure change defined in (11.96). Note that the SDE in (11.99) with initial condition $\overline{S}(0)$ is equivalent to (11.94), which is its unique solution. For an arbitrary choice of γ the SDE with drift in (11.98) subject to initial condition $\overline{S}(0)$ is equivalent to (11.93), which is its unique solution. Finally, note that choosing $\gamma = (r - \mu)/\sigma$ in (11.97) gives the stock price *drifting at the risk-free rate within the risk-neutral measure*:

$$\mathrm{d}S(t) = S(t)\big[r\,\mathrm{d}t + \sigma\,\mathrm{d}\widetilde{W}(t)\big] \tag{11.100}$$

with unique solution

$$S(t) = S(0)\,\mathrm{e}^{rt} \cdot \mathcal{E}_t(\sigma \cdot \widetilde{W}) = S(0)\,\mathrm{e}^{(r-\sigma^2/2)t+\sigma\widetilde{W}(t)} \tag{11.101}$$

equivalent to (11.94). The $\widetilde{\mathbb{P}}$-martingale property in (11.95) is equivalently expressed as

$$\widetilde{\mathrm{E}}[\,S(t) \mid \mathcal{F}_u] = \mathrm{e}^{r(t-u)}S(u)\,,\ 0 \leqslant u \leqslant t \leqslant T. \tag{11.102}$$

In Example 11.14 we used Girsanov's Theorem to obtain a new measure $\widehat{\mathbb{P}}$, defined by the Radon–Nikodym process in (11.91), such that the process $X(t) \equiv W^{(\mu,\sigma)}(t)$ is a scaled standard $\widehat{\mathbb{P}}$-BM. We now employ the same measure change and thereby compute expectations and joint probabilities of events associated with the sampled maximum or minimum of BM with drift. In particular, let's simply set $\sigma = 1$ and consider the process defined by (10.68) in Section 10.4.3, i.e.,

$$X(t) = \mu t + W(t) = \widehat{W}(t).$$

The expression in (11.91), for $\sigma = 1$, gives the Radon–Nikodym derivative for the change of measure $\mathbb{P} \to \widehat{\mathbb{P}}$, $\varrho_t = \left(\frac{\mathrm{d}\widehat{\mathbb{P}}}{\mathrm{d}\mathbb{P}}\right)_t = \mathrm{e}^{-\frac{1}{2}\mu^2 t-\mu W(t)}$. Hence, the Radon–Nikodym derivative for the change of measure $\widehat{\mathbb{P}} \to \mathbb{P}$ is expressed in terms of the $\widehat{\mathbb{P}}$-BM, $\widehat{W}(t) = W(t) + \mu t$, as

$$\frac{1}{\varrho_t} = \left(\frac{\mathrm{d}\mathbb{P}}{\mathrm{d}\widehat{\mathbb{P}}}\right)_t = \mathrm{e}^{\frac{1}{2}\mu^2 t+\mu W(t)} = \mathrm{e}^{-\frac{1}{2}\mu^2 t+\mu\widehat{W}(t)}\,. \tag{11.103}$$

Let A, B be any two Borel sets in $\mathbb{R}$ and consider the $\mathcal{F}_t$-measurable indicator random variables $\mathbb{I}_{\{M^X(t)\in A, X(t)\in B\}}$ and $\mathbb{I}_{\{m^X(t)\in A, X(t)\in B\}}$ where the respective sampled maximum, $M^X(t)$, and minimum, $m^X(t)$, of the drifted BM process X are defined in (10.70) and (10.71). That is,

$$M^X(t) = \sup_{0\leqslant u\leqslant t} X(u) = \sup_{0\leqslant u\leqslant t} \widehat{W}(u) \equiv M^{\widehat{W}}(t)$$

and

$$m^X(t) = \inf_{0\leqslant u\leqslant t} X(u) = \inf_{0\leqslant u\leqslant t} \widehat{W}(u) \equiv m^{\widehat{W}}(t)\,.$$

The sampled maximum $M(t) \equiv M^W(t)$ and minimum $m(t) \equiv m^W(t)$ of the standard $\mathbb{P}$-BM, W, are defined in (10.33) and (10.34). Applying the change of measure while using (11.103) within (11.84) gives

$$\begin{aligned}
\mathbb{P}(M^X(t) \in A, X(t) \in B) &\equiv \mathrm{E}\big[\mathbb{I}_{\{M^X(t)\in A, X(t)\in B\}}\big] \\
&= \widehat{\mathrm{E}}\big[\varrho_t^{-1}\mathbb{I}_{\{M^X(t)\in A, X(t)\in B\}}\big] \\
&= \mathrm{e}^{-\frac{1}{2}\mu^2 t}\widehat{\mathrm{E}}\big[\mathrm{e}^{\mu\widehat{W}(t)}\mathbb{I}_{\{M^{\widehat{W}}(t)\in A, \widehat{W}(t)\in B\}}\big] \\
&= \mathrm{e}^{-\frac{1}{2}\mu^2 t}\mathrm{E}\big[\mathrm{e}^{\mu W(t)}\mathbb{I}_{\{M(t)\in A, W(t)\in B\}}\big]\,.
\end{aligned} \tag{11.104}$$

In the last equation line we simply removed all "hats" since the random variables $M^{\widehat{W}}(t)$

and $\widehat{W}(t)$ under measure $\widehat{\mathbb{P}}$ are the same as $M(t)$ and $W(t)$ under measure $\mathbb{P}$. By the same steps as in (11.104) we have

$$\mathbb{P}(m^X(t) \in A, X(t) \in B) = \mathrm{e}^{-\frac{1}{2}\mu^2 t}\mathrm{E}\left[\mathrm{e}^{\mu W(t)}\mathbb{I}_{\{m(t)\in A, W(t)\in B\}}\right]. \tag{11.105}$$

Equations (11.104) and (11.105) can be used to compute the probability of any joint event involving either pair $M^X(t), X(t)$ or $m^X(t), X(t)$. For example, taking intervals $A = (-\infty, m]$, $B = (-\infty, x]$ gives the respective joint CDFs

$$\begin{aligned} F_{M^X(t),X(t)}(m,x) &:= \mathbb{P}(M^X(t) \leqslant m, X(t) \leqslant x) \\ &= \mathrm{e}^{-\frac{1}{2}\mu^2 t}\mathrm{E}\left[\mathrm{e}^{\mu W(t)}\mathbb{I}_{\{M(t)\leqslant m, W(t)\leqslant x\}}\right] \end{aligned} \tag{11.106}$$

and

$$\begin{aligned} F_{m^X(t),X(t)}(m,x) &:= \mathbb{P}(m^X(t) \leqslant m, X(t) \leqslant x) \\ &= \mathrm{e}^{-\frac{1}{2}\mu^2 t}\mathrm{E}\left[\mathrm{e}^{\mu W(t)}\mathbb{I}_{\{m(t)\leqslant m, W(t)\leqslant x\}}\right]. \end{aligned} \tag{11.107}$$

Expressing the expectation in (11.106) as an integral over the joint density of $M(t), W(t)$:

$$F_{M^X(t),X(t)}(m,x) = \mathrm{e}^{-\frac{1}{2}\mu^2 t}\int_0^m \int_{-\infty}^x \mathrm{e}^{\mu y} f_{M(t),W(t)}(w,y)\,\mathrm{d}y\,\mathrm{d}w. \tag{11.108}$$

Differentiating, and making use of the known joint PDF of $M(t), W(t)$ in (10.39), gives the joint PDF of $M^X(t), X(t)$

$$\begin{aligned} f_{M^X(t),X(t)}(m,x) &= \mathrm{e}^{-\frac{1}{2}\mu^2 t + \mu x} f_{M(t),W(t)}(m,x) \\ &= \frac{2(2m-x)}{t\sqrt{2\pi t}}\mathrm{e}^{-\frac{1}{2}\mu^2 t + \mu x - (2m-x)^2/2t}, \end{aligned} \tag{11.109}$$

for $x \leqslant m, m > 0$ and zero otherwise. Similarly, the joint PDF of $m^X(t), X(t)$ follows from (11.107) and the joint PDF in (10.43),

$$\begin{aligned} f_{m^X(t),X(t)}(m,x) &= \mathrm{e}^{-\frac{1}{2}\mu^2 t + \mu x} f_{m(t),W(t)}(m,x) \\ &= \frac{2(x-2m)}{t\sqrt{2\pi t}}\mathrm{e}^{-\frac{1}{2}\mu^2 t + \mu x - (x-2m)^2/2t}, \end{aligned} \tag{11.110}$$

for $x \geqslant m, m < 0$, and zero otherwise. Other applications of (11.104) and (11.105) are given in Section 10.4.3.

11.9 Brownian Martingale Representation Theorem

Before moving on to the next section on multidimensional (vector) BM we state a result that we will later see has some theoretical importance in replication (hedging) and pricing derivative contracts within a continuous-time financial model driven by a single BM. We have already learned that, given an adapted process $\{X(t)\}_{0\leqslant t\leqslant T}$ with $\int_0^T \mathrm{E}[X^2(t)]\,\mathrm{d}t < \infty$, the Itô process $\{I(t) := \int_0^t X(s)\,\mathrm{d}W(s)\}_{0\leqslant t\leqslant T}$ is a $(\mathbb{P}, \mathbb{F})$-martingale where $\{W(t)\}_{t\geqslant 0}$ is a $(\mathbb{P}, \mathbb{F})$-BM. A question that one may ask is: Are all $(\mathbb{P}, \mathbb{F})$-martingales expressible as an

Itô process? It turns out that this is the case if we consider martingales that are *square integrable* and we also restrict the filtration to be the *natural filtration generated by the BM*, i.e., if $\mathbb{F} = \mathbb{F}^W = \{F_t^W\}_{t\geqslant 0} := \{\sigma(W(s) : 0 \leqslant s \leqslant t)\}_{t\geqslant 0}$. We summarize this in the following known theorem without proof.

Theorem 11.14 (Brownian Martingale Representation Theorem). *Assume $\{M(t)\}_{0\leqslant t\leqslant T}$ is a $(\mathbb{P}, \mathbb{F}^W)$-martingale and that it is square integrable, i.e.,*

$$\mathrm{E}[M^2(t)] < \infty\,, \quad \textit{for every } t \in [0,T].$$

Then, there exists an $\mathbb{F}^W$-adapted process $\{\theta(t)\}_{0\leqslant t\leqslant T}$ such that (a.s.)

$$M(t) = M(0) + \int_0^t \theta(u)\,\mathrm{d}W(u)\,. \tag{11.111}$$

This theorem tells us that if a process is a square-integrable martingale, w.r.t. a given measure $\mathbb{P}$ and natural filtration $\mathbb{F}^W$ generated by the standard $\mathbb{P}$-BM W, then it can be expressed as a sum of its initial value and an Itô integral in the $\mathbb{P}$-BM. The integrand of the Itô integral is a process that is adapted to $\mathbb{F}^W$. Note that the Itô integral itself is a square-integrable $(\mathbb{P}, \mathbb{F}^W)$-martingale and also continuous in time. So the martingale having this representation is also continuous in time (i.e., the process has no jumps).

We are now ready to state a closely related result that is a consequence of the above theorem and will later be applicable to our discussion of derivative replication in Chapter 12. Let us consider what happens when we change measures $\mathbb{P} \to \widehat{\mathbb{P}}$ as defined in Girsanov's Theorem 11.13. As we already noted, $\mathbb{F}$ could be any filtration for the $\mathbb{P}$-BM, W. Now set $\mathbb{F} = \mathbb{F}^W$ where $\{\gamma(t)\}_{0\leqslant t\leqslant T}$ is assumed to be $\mathbb{F}^W$-adapted and clearly the time integral of this process occurring in (11.88) is $\mathbb{F}^W$-adapted. In particular, the σ-algebra $\sigma\left(\int_0^t \gamma(s)\,\mathrm{d}s\right) \subset \mathcal{F}_t^W$ for every $t \geqslant 0$. Then, by the definition in (11.88), the σ-algebra $\mathcal{F}_t^{\widehat{W}} := \sigma(\widehat{W}(u) : 0 \leqslant u \leqslant t) = \mathcal{F}_t^W$. Hence, if $\{\gamma(t)\}_{0\leqslant t\leqslant T}$ is chosen as an $\mathbb{F}^W$-adapted process, then the natural filtration $\mathbb{F}^{\widehat{W}} = \{\mathcal{F}_t^{\widehat{W}}\}_{0\leqslant t\leqslant T}$, generated by $\widehat{W}$ in (11.88), is equal to the natural filtration $\mathbb{F}^W$, generated by W, i.e., $\mathbb{F}^W = \mathbb{F}^{\widehat{W}}$. In summary, by combining these facts with Theorem 11.14 we have the result below. This states that, if the change of measure $\mathbb{P} \to \widehat{\mathbb{P}}$ is defined via Girsanov's Theorem with an $\mathbb{F}^W$-adapted process, then we can always express a square-integrable $(\widehat{\mathbb{P}}, \mathbb{F}^W)$-martingale as its initial value plus an Itô integral in the $\widehat{\mathbb{P}}$-BM.

Proposition 11.15. *Let the measure $\widehat{\mathbb{P}}$ be defined as in Girsanov's Theorem 11.13 with the assumption that the process $\{\gamma(t)\}_{0\leqslant t\leqslant T}$ is $\mathbb{F}^W$-adapted. If $\{M(t)\}_{0\leqslant t\leqslant T}$ is a square-integrable $(\widehat{\mathbb{P}}, \mathbb{F}^W)$-martingale, then there exists an adapted process, say $\{\widehat{\theta}(t)\}_{0\leqslant t\leqslant T}$, such that (a.s.)*

$$M(t) = M(0) + \int_0^t \widehat{\theta}(u)\,\mathrm{d}\widehat{W}(u)\,. \tag{11.112}$$

Proof. By the above argument we have $\mathbb{F}^W = \mathbb{F}^{\widehat{W}}$. Hence, $\{M(t)\}_{0\leqslant t\leqslant T}$ is a square-integrable $(\widehat{\mathbb{P}}, \mathbb{F}^{\widehat{W}})$-martingale where $\widehat{\mathrm{E}}[M^2(t)] = \mathrm{E}[\varrho_t M^2(t)] \leqslant \mathrm{E}[M^2(t)] < \infty$. It now follows trivially by Theorem 11.14 that there exists an $\mathbb{F}^{\widehat{W}}$-adapted (and hence $\mathbb{F}^W$-adapted) process $\{\widehat{\theta}(t)\}_{0\leqslant t\leqslant T}$ such that (11.112) holds (a.s.). □

11.10 Stochastic Calculus for Multidimensional BM

11.10.1 The Itô Integral and Itô's Formula for Multiple Processes on Multidimensional BM

We now extend the definition of one-dimensional standard BM $\{W(t)\}_{t\geqslant 0}$ into d dimensions for any finite integer $d \geqslant 1$. As seen below, the extension to multiple dimensions is fairly straightforward as we take each component as an independent one-dimensional standard BM. Notation needs to be introduced to precisely denote each component BM and boldface is used for a vector BM.

Definition 11.4. A standard BM in $\mathbb{R}^d$ (or standard d-dimensional BM) is a vector process

$$\mathbf{W}(t) \equiv (W_1(t), W_2(t), \ldots, W_d(t)), \, t \geqslant 0,$$

where each component process $\{W_i(t)\}_{t\geqslant 0}$, $1 \leqslant i \leqslant d$, is an *independent one-dimensional standard BM* in $\mathbb{R}$.

Hence, each component is *i.i.d.* where $W_i(t) \sim Norm(0,t)$ and $W_i(t) - W_i(s) \sim Norm(0, t-s)$, $1 \leqslant i \leqslant d$, $0 \leqslant s \leqslant t$. We call this a standard vector BM since, by construction, each component is an identical and independent copy of a one-dimensional standard BM. That is, $W_i(t)$ and $W_j(t)$ are independent if $i \neq j$. A filtration $\mathbb{F} = \{\mathcal{F}_t\}_{t\geqslant 0}$ is a filtration for standard d-dimensional BM if it is a filtration for each component BM, $\{W_i(t)\}_{t\geqslant 0}$. The natural filtration for $\{\mathbf{W}(t)\}_{t\geqslant 0}$, denoted by $\mathbb{F}^{\mathbf{W}}$, is the filtration generated by all components of the standard d-dimensional BM. Given any filtration $\mathbb{F}$ for $\{\mathbf{W}(t)\}_{t\geqslant 0}$, we must have that $\{\mathbf{W}(t)\}_{t\geqslant 0}$ is $\mathbb{F}$-adapted, i.e., $\mathbf{W}(t)$ is $\mathcal{F}_t$-measurable, and that each Brownian *vector increment* $\mathbf{W}(t+s) - \mathbf{W}(t)$ is independent of $\mathcal{F}_t$ for $s, t \geqslant 0$.

Since each component is a standard BM, then we have the usual properties such as the quadratic variation formula for each $1 \leqslant i \leqslant d$:

$$[W_i, W_i](t) = t \implies \mathrm{d}[W_i, W_i](t) \equiv \mathrm{d}W_i(t)\,\mathrm{d}W_i(t) = \mathrm{d}t\,. \tag{11.113}$$

Moreover, $[f, W_i](t)$ has zero covariation for any continuously differentiable function $f(t)$. In particular, for each $1 \leqslant i \leqslant d$,

$$[t, W_i](t) = 0 \implies \mathrm{d}W_i(t)\,\mathrm{d}t = 0\,. \tag{11.114}$$

The covariation of two independent Brownian motions is zero, i.e.,

$$[W_i, W_j](t) = 0 \implies \mathrm{d}[W_i, W_j](t) \equiv \mathrm{d}W_i(t)\,\mathrm{d}W_j(t) = 0\,, \quad \text{for } i \neq j. \tag{11.115}$$

It is simple to see how this arises by considering a time partition $\{0 = t_0, t_1, \ldots, t_n = t\}$ and forming the partial sum of products of individual Brownian increments:

$$Q_n^{i,j}(t) := \sum_{k=1}^{n} (W_i(t_k) - W_i(t_{k-1}))\,(W_j(t_k) - W_j(t_{k-1}))$$

for $i \neq j$. Using the fact that the increments are all mutually independent with mean zero, $\mathrm{E}[W_i(t_k) - W_i(t_{k-1})] = 0$, for every k, then $\mathrm{E}[Q_n^{i,j}(t)] = 0$. Since all n terms in the sum are mutually independent, the variance of the sum is the sum of the individual variances. Using the independence of the product terms, where $\mathrm{E}[(W_i(t_k) - W_i(t_{k-1}))^2] = \mathrm{E}[(W_j(t_k) - W_j(t_{k-1}))^2] = t_k - t_{k-1}$, gives

$$\begin{aligned}
\operatorname{Var}\left(Q_n^{i,j}(t)\right) &= \sum_{k=1}^{n} \mathrm{E}[(W_i(t_k) - W_i(t_{k-1}))^2]\, \mathrm{E}[(W_j(t_k) - W_j(t_{k-1}))^2] \\
&= \sum_{k=1}^{n} (t_k - t_{k-1})^2 \\
&\leqslant \Delta_n \sum_{k=1}^{n} (t_k - t_{k-1}) = \Delta_n (t_n - t_0) = \Delta_n t
\end{aligned}$$

where $\Delta_n := \max\limits_{k=1,\dots n} (t_k - t_{k-1})$ is the maximum time increment over the partition. Clearly, $Var(Q_n^{i,j}(t))) \to 0$ as $\Delta_n \to 0$, i.e., this implies that, for all $t \geqslant 0$, the random variable $Q_n^{i,j}(t)$ converges to its expected value $\mathrm{E}[Q_n^{i,j}(t)] = 0$ as $\Delta_n \to 0$ and hence the co-variation $[W_i, W_j](t) := \lim\limits_{\Delta_n \to 0} Q_n^{i,j}(t)$ must be zero for $i \neq j$.

For convenience we summarize the above "basic rules" for the stochastic increments as follows:

$$\mathrm{d}W_i(t)\,\mathrm{d}W_j(t) = \delta_{ij}\,\mathrm{d}t\,, \quad \mathrm{d}W_i(t)\,\mathrm{d}t = 0\,, \quad (\mathrm{d}t)^2 = 0, \tag{11.116}$$

where $\delta_{ij} = 1$ if $i = j$, and 0 if $i \neq j$.

As in the case of standard BM in one dimension, there is a similar useful characterization of a standard d-dimensional BM due to Lévy which we state in the following lemma. The result can be proven based on multidimensional extensions of the Itô formula.

Theorem 11.16 (Lévy's Characterization of a Standard Multidimensional BM). *Consider the vector-valued process $\{\mathbf{M}(t) := M_1(t), \dots, M_d(t)\}_{t \geqslant 0}$ where each component process $\{M_i(t)\}_{t \geqslant 0}$, $1 \leqslant i \leqslant d$, is a continuous $(\mathbb{P}, \mathbb{F})$-martingale starting at $M_i(0) = 0$ (a.s.) and having quadratic variation $[M_i, M_i](t) = t$, for all $t \geqslant 0$. Also, assume $[M_i, M_j](t) = 0$ for $i \neq j$. Then, $\{\mathbf{M}(t)\}_{t \geqslant 0}$ is a standard d-multidimensional $(\mathbb{P}, \mathbb{F})$-BM.*

According to this result, a vector process is a standard vector BM (in a given measure and filtration) if we can verify that *every component process* is a martingale with continuous paths starting at zero, has the same quadratic variation as a standard BM, and all covariations among different components are zero. Basically, this means that each component is an i.i.d. standard one-dimensional BM.

Let us fix a filtration $\mathbb{F} = \{\mathcal{F}_t\}_{t \geqslant 0}$ for BM in $\mathbb{R}^d$ for a given integer $d \geqslant 1$. The formulae and concepts we developed in previous sections on the Itô integral, Itô's formula for a function of an Itô process and SDEs can be generalized to a multiple (vector) BM and multiple Itô processes that are driven by the vector BM in $\mathbb{R}^d$. Let us first discuss this extension for the case of BM in $\mathbb{R}^2$, i.e., $d = 2$ where $\mathbf{W}(t) = (W_1(t), W_2(t))$. We can have any number of Itô processes that can be represented as an Itô integral w.r.t. $\mathbf{W}(t)$ plus a drift term which is a Riemann (or Lebesgue) integral. Consider two Itô processes $X \equiv \{X(t)\}_{t \geqslant 0}$ and $Y \equiv \{Y(t)\}_{t \geqslant 0}$ which form a vector process, $(X(t), Y(t))_{t \geqslant 0}$. Let $\mu_X(t)$ and $\mu_Y(t)$ be $\mathcal{F}_t$-adapted drift coefficients of processes X and Y, respectively. The diffusion or volatility coefficient vectors are $\mathcal{F}_t$-adapted vectors in $\mathbb{R}^2$ denoted by

$$\boldsymbol{\sigma}_X(t) = (\sigma_{X,1}(t), \sigma_{X,2}(t)) \quad \text{and} \quad \boldsymbol{\sigma}_Y(t) = (\sigma_{Y,1}(t), \sigma_{Y,2}(t))$$

for processes X and Y, respectively. The two processes have the representations:

$$X(t) = X(0) + \int_0^t \mu_X(u)\,\mathrm{d}u + \int_0^t \boldsymbol{\sigma}_X(u)\cdot \mathrm{d}\mathbf{W}(u)\,, \tag{11.117}$$

$$Y(t) = Y(0) + \int_0^t \mu_Y(u)\,\mathrm{d}u + \int_0^t \boldsymbol{\sigma}_Y(u)\cdot \mathrm{d}\mathbf{W}(u)\,. \tag{11.118}$$

In each case, the first (Riemann or Lebesgue) integral is the drift term and the second integral is a sum of two Itô integrals; one is w.r.t. the first component of the volatility vector and the first BM and the second is w.r.t. the second component of the volatility vector and the second BM. That is, we define

$$\int_0^t \boldsymbol{\sigma}_X(u)\cdot \mathrm{d}\mathbf{W}(u) := \int_0^t \sigma_{X,1}(u)\,\mathrm{d}W_1(u) + \int_0^t \sigma_{X,2}(u)\,\mathrm{d}W_2(u)\,, \tag{11.119}$$

$$\int_0^t \boldsymbol{\sigma}_Y(u)\cdot \mathrm{d}\mathbf{W}(u) := \int_0^t \sigma_{Y,1}(u)\,\mathrm{d}W_1(u) + \int_0^t \sigma_{Y,2}(u)\,\mathrm{d}W_2(u)\,. \tag{11.120}$$

Given a time $T > 0$, throughout we shall assume the square integrability condition holds for the Itô integrals on all time intervals $[0, t]$, $0 \leqslant t \leqslant T$, i.e., given an adapted vector process $\{\boldsymbol{\sigma}(t) = (\sigma_1(t), \sigma_2(t))\}_{t\geqslant 0}$ then we assume

$$\mathrm{E}\left[\left(\int_0^T \boldsymbol{\sigma}(t)\cdot \mathrm{d}\mathbf{W}(t)\right)^2\right] = \int_0^T \mathrm{E}\left[\|\boldsymbol{\sigma}(t)\|^2\right]\,\mathrm{d}t < \infty\,. \tag{11.121}$$

where $\|\boldsymbol{\sigma}(t)\|^2 \equiv \sum_{i=1}^d \sigma_i^2(t)$ is the square magnitude of the volatility vector, e.g., for $d = 2$ then $\|\boldsymbol{\sigma}(t)\|^2 = \sigma_1^2(t)+\sigma_2^2(t)$. This condition is equivalent to requiring $\int_0^T \mathrm{E}\left[\sigma_i^2(t)\right]\,\mathrm{d}t < \infty$, for every component i, and it guarantees the martingale property,

$$\mathrm{E}\left[\int_0^T \boldsymbol{\sigma}(s)\cdot \mathrm{d}\mathbf{W}(s)\,\middle|\,\mathcal{F}_t\right] = \int_0^t \boldsymbol{\sigma}(s)\cdot \mathrm{d}\mathbf{W}(s)\,,\ \ 0 \leqslant t \leqslant T. \tag{11.122}$$

Hence, the d-dimensional Itô integrals have zero expectation. The equality in (11.121) is the Itô isometry formula for vector BM, which is a special case of the covariance formula,

$$\mathrm{Cov}\left(\int_0^t \boldsymbol{\sigma}(s)\cdot \mathrm{d}\mathbf{W}(s), \int_0^t \boldsymbol{\gamma}(s)\cdot \mathrm{d}\mathbf{W}(s)\right) = \int_0^t \mathrm{E}\left[\boldsymbol{\sigma}(s)\cdot\boldsymbol{\gamma}(s)\right]\,\mathrm{d}s\,, \tag{11.123}$$

where $\boldsymbol{\sigma}(t)$ and $\boldsymbol{\gamma}(t)$ are $\mathcal{F}_t$-adapted d-dimensional vectors. This is readily derived by writing out the two Itô integrals as sums of (one-dimensional) Itô integrals, as in (11.119), and then using the covariance relation for each pair of Itô integrals. Also, we assume that any drift coefficient $\mu(t)$ is integrable,

$$\mathrm{E}\left[\int_0^T |\mu(t)|\,\mathrm{d}t\right] < \infty\,. \tag{11.124}$$

The Itô integrals in (11.119)–(11.120) are the one-dimensional Itô integrals w.r.t. a single standard BM which is taken as either W_1 or W_2. The Riemann (Lebesgue) integrals in (11.117) and (11.118) are continuous functions of time and therefore have zero quadratic variation. To obtain the quadratic variation of the X process, note that the quadratic variation of each Itô integral in (11.119),

$$I_{X,1}(t) := \int_0^t \sigma_{X,1}(u)\,\mathrm{d}W_1(u) \quad\text{and}\quad I_{X,2}(t) := \int_0^t \sigma_{X,2}(u)\,\mathrm{d}W_2(u),$$

is computed according to (11.11) (where W_1 and W_2 individually act as W):

$$[I_{X,1}, I_{X,1}](t) \equiv \int_0^t \sigma_{X,1}^2(u)\,\mathrm{d}u \quad \text{and} \quad [I_{X,2}, I_{X,2}](t) \equiv \int_0^t \sigma_{X,2}^2(u)\,\mathrm{d}u\,. \tag{11.125}$$

Since $[W_1, W_2](t) = 0$, i.e., $\mathrm{d}W_1(t)\,\mathrm{d}W_2(t) = 0$, the covariation of the two integrals is zero: $[I_{X,1}, I_{X,2}](t) = 0$. Hence, the quadratic variation of the X process is the quadratic variation of the Itô integral in (11.119), which, in turn, is the sum of the two quadratic variations in (11.125):

$$\begin{aligned}[X,X](t) = [I_{X,1}, I_{X,1}](t) + [I_{X,2}, I_{X,2}](t) &= \int_0^t \left(\sigma_{X,1}^2(u) + \sigma_{X,2}^2(u)\right)\,\mathrm{d}u \\ &= \int_0^t \|\boldsymbol{\sigma}_X(u)\|^2\,\mathrm{d}u\,. \end{aligned} \tag{11.126}$$

Similarly, the Y process has quadratic variation

$$[Y,Y](t) = \int_0^t \left(\sigma_{Y,1}^2(u) + \sigma_{Y,2}^2(u)\right)\,\mathrm{d}u = \int_0^t \|\boldsymbol{\sigma}_Y(u)\|^2\,\mathrm{d}u\,. \tag{11.127}$$

The stochastic differential forms of (11.126) and (11.127) are

$$\mathrm{d}[X,X](t) = \mathrm{d}X(t)\,\mathrm{d}X(t) = \left(\sigma_{X,1}^2(t) + \sigma_{X,2}^2(t)\right)\,\mathrm{d}t = \|\boldsymbol{\sigma}_X(t)\|^2\,\mathrm{d}t\,, \tag{11.128}$$

$$\mathrm{d}[Y,Y](t) = \mathrm{d}Y(t)\,\mathrm{d}Y(t) = \left(\sigma_{Y,1}^2(t) + \sigma_{Y,2}^2(t)\right)\,\mathrm{d}t = \|\boldsymbol{\sigma}_Y(t)\|^2\,\mathrm{d}t\,. \tag{11.129}$$

It is easier to obtain (11.128) and (11.129) by working directly with the stochastic differential forms of (11.117) and (11.118),

$$\begin{aligned}\mathrm{d}X(t) &= \mu_X(t)\,\mathrm{d}t + \boldsymbol{\sigma}_X(t)\cdot\mathrm{d}\mathbf{W}(t) \equiv \mu_X(t)\,\mathrm{d}t + \sigma_{X,1}(t)\,\mathrm{d}W_1(t) + \sigma_{X,2}(t)\,\mathrm{d}W_2(t)\,,\\ \mathrm{d}Y(t) &= \mu_Y(t)\,\mathrm{d}t + \boldsymbol{\sigma}_Y(t)\cdot\mathrm{d}\mathbf{W}(t) \equiv \mu_Y(t)\,\mathrm{d}t + \sigma_{Y,1}(t)\,\mathrm{d}W_1(t) + \sigma_{Y,2}(t)\,\mathrm{d}W_2(t)\,,\end{aligned}$$

and then applying the rules in (11.116). For example, by squaring the differential $\mathrm{d}X(t)$ and setting the terms $\mathrm{d}t\,\mathrm{d}W_1(t) = \mathrm{d}t\,\mathrm{d}W_2(t) = 0$, $\mathrm{d}W_1(t)\,\mathrm{d}W_2(t) = 0$, $(\mathrm{d}t)^2 = 0$, and $(\mathrm{d}W_1(t))^2 = (\mathrm{d}W_2(t))^2 = \mathrm{d}t$, we obtain

$$\begin{aligned}\mathrm{d}X(t)\,\mathrm{d}X(t) \equiv (\mathrm{d}X(t))^2 &= (\mu_X(t)\,\mathrm{d}t + \boldsymbol{\sigma}_X(t)\cdot\mathrm{d}\mathbf{W}(t))^2\\ &= \boldsymbol{\sigma}_X(t)\cdot\boldsymbol{\sigma}_X(t)\,\mathrm{d}t = \|\boldsymbol{\sigma}_X(t)\|^2\,\mathrm{d}t\,.\end{aligned}$$

This recovers the result in (11.128). A similar derivation based on squaring $\mathrm{d}Y(t)$ gives (11.129). The covariation is also simpler to compute based on this differential approach. By multiplying the two stochastic differentials and applying the simple rules in (11.116),

$$\begin{aligned}\mathrm{d}[X,Y](t) = \mathrm{d}X(t)\,\mathrm{d}Y(t) &= (\mu_X(t)\,\mathrm{d}t + \boldsymbol{\sigma}_X(t)\cdot\mathrm{d}\mathbf{W}(t))\,(\mu_Y(t)\,\mathrm{d}t + \boldsymbol{\sigma}_Y(t)\cdot\mathrm{d}\mathbf{W}(t))\\ &= (\boldsymbol{\sigma}_X(t)\cdot\mathrm{d}\mathbf{W}(t))\,(\boldsymbol{\sigma}_Y(t)\cdot\mathrm{d}\mathbf{W}(t))\\ &= \boldsymbol{\sigma}_X(t)\cdot\boldsymbol{\sigma}_Y(t)\,\mathrm{d}t\,.\end{aligned} \tag{11.130}$$

The last equation line is obtained as follows:

$$\begin{aligned}(\boldsymbol{\sigma}_X(t)\cdot\mathrm{d}\mathbf{W}(t))\,(\boldsymbol{\sigma}_Y(t)\cdot\mathrm{d}\mathbf{W}(t)) &= \sum_{i=1}^{d=2}\sum_{j=1}^{d=2}\sigma_{X,i}(t)\sigma_{Y,j}(t)\,\underbrace{\mathrm{d}W_i(t)\,\mathrm{d}W_j(t)}_{=\delta_{ij}\,\mathrm{d}t}\\ &= \left(\sum_{i=1}^{d=2}\sigma_{X,i}(t)\sigma_{Y,i}(t)\right)\,\mathrm{d}t = \boldsymbol{\sigma}_X(t)\cdot\boldsymbol{\sigma}_Y(t)\,\mathrm{d}t\,.\end{aligned}$$

The integral form of (11.130) gives the covariation of the two Itô processes,

$$[X,Y](t) = \int_0^t \boldsymbol{\sigma}_X(u)\cdot\boldsymbol{\sigma}_Y(u)\,\mathrm{d}u = \int_0^t (\sigma_{X,1}(u)\sigma_{Y,1}(u) + \sigma_{X,2}(u)\sigma_{Y,2}(u))\,\mathrm{d}u\,.$$

The Itô formula in (11.20) and (11.21) for a function of one Itô process, and time t, extends further to the slightly more general case of a *function of two Itô processes and time* t. We simply state this important result as a lemma (without proof). The main idea, and a simple way to remember the formula in (11.131), is to Taylor expand $f(t,x,y)$ up to terms of order $\mathrm{d}t$, $(\mathrm{d}x)^2, (\mathrm{d}y)^2$ and then replace ordinary variables $x \to X(t)$, $y \to Y(t)$ and ordinary differentials by their respective stochastic differentials: $\mathrm{d}x \to \mathrm{d}X(t)$, $\mathrm{d}y \to \mathrm{d}Y(t)$, $(\mathrm{d}x)^2 \to (\mathrm{d}X(t))^2 \equiv \mathrm{d}[X,X](t)$, $(\mathrm{d}y)^2 \to (\mathrm{d}Y(t))^2 \equiv \mathrm{d}[Y,Y](t)$, and $\mathrm{d}x\,\mathrm{d}y \to \mathrm{d}X(t)\,\mathrm{d}Y(t) \equiv \mathrm{d}[X,Y](t)$.

Lemma 11.17 (Itô Formula for a Function of Two Processes). *Assume $f(t,x,y)$ is a $C^{1,2,2}$ function on $\mathbb{R}_+ \times \mathbb{R}^2$, i.e., having continuous derivatives $f_t \equiv \frac{\partial f}{\partial t}$, $f_x \equiv \frac{\partial f}{\partial x}$, $f_y \equiv \frac{\partial f}{\partial y}$, $f_{xx} \equiv \frac{\partial^2 f}{\partial x^2}$, $f_{xy} \equiv \frac{\partial^2 f}{\partial x \partial y}$ and $f_{yy} \equiv \frac{\partial^2 f}{\partial y^2}$. Let the processes X and Y be Itô processes as given in (11.117) and (11.118). Then, the process defined by $F(t) := f(t,X(t),Y(t)), t \geqslant 0$, has stochastic differential $\mathrm{d}F(t) \equiv \mathrm{d}f(t,X(t),Y(t))$ given by*

$$\begin{aligned}\mathrm{d}f(t,X(t),Y(t)) &= f_t(t,X(t),Y(t))\,\mathrm{d}t + f_x(t,X(t),Y(t))\,\mathrm{d}X(t) + f_y(t,X(t),Y(t))\,\mathrm{d}Y(t)\\ &\quad + \frac{1}{2}f_{xx}(t,X(t),Y(t))\;\mathrm{d}[X,X](t) + \frac{1}{2}f_{yy}(t,X(t),Y(t))\;\mathrm{d}[Y,Y](t)\\ &\quad + f_{xy}(t,X(t),Y(t))\;\mathrm{d}[X,Y](t)\,. \end{aligned} \tag{11.131}$$

In integral form,

$$f(t,X(t),Y(t)) = f(0,X(0),Y(0)) \tag{11.132}$$

$$\begin{aligned}&+ \int_0^t \Big[f_u(u,X(u),Y(u)) + \frac{1}{2}\|\boldsymbol{\sigma}_X(u)\|^2 f_{xx}(u,X(u),Y(u))\\ &\qquad + \frac{1}{2}\|\boldsymbol{\sigma}_Y(u)\|^2 f_{yy}(u,X(u),Y(u)) + \boldsymbol{\sigma}_X(u)\cdot\boldsymbol{\sigma}_Y(u)\, f_{xy}(u,X(u),Y(u))\Big]\,\mathrm{d}u\\ &+ \int_0^t f_x(u,X(u),Y(u))\,\mathrm{d}X(u) + \int_0^t f_y(u,X(u),Y(u))\,\mathrm{d}Y(u)\,. \end{aligned} \tag{11.133}$$

It should be remarked (and we shall see later when we present the general form of the Itô formula for functions of multiple processes driven by multiple Brownian motions) that this lemma is generally valid for any number $d \geqslant 1$ of underlying Brownian motions, although we have focused our present discussion on taking $d = 2$ as the base case. For $d \geqslant 2$ the volatilities are d-dimensional vectors and the standard BM is a d-dimensional vector (standard) BM. For the case that $d = 1$ we simply have the vectors becoming scalars, e.g., $\boldsymbol{\sigma}_X(t) \to \sigma_X(t)$, $\boldsymbol{\sigma}_Y(t) \to \sigma_Y(t)$, and $\mathbf{W}(t) \to W(t)$.

Observe that the first integral in (11.133) is a Riemann (or Lebesgue) integral on the time interval $[0,t]$, whereas the second and third integrals are stochastic integrals w.r.t. the Itô processes X in (11.117) and Y in (11.118). The representation of $\mathrm{d}f(t,X(t),Y(t))$ in (11.131) and its corresponding integral form in (11.133) is written in terms of the stochastic differentials of X and Y. The Itô formula is also equivalently rewritten by substituting the above stochastic differentials for $\mathrm{d}X(t)$ and $\mathrm{d}Y(t)$. Then, (11.131) takes the form

$$\begin{aligned}\mathrm{d}f &= \left(f_t + \mu_X(t)f_x + \mu_Y(t)f_y + \frac{1}{2}\|\boldsymbol{\sigma}_X(t)\|^2 f_{xx} + \frac{1}{2}\|\boldsymbol{\sigma}_Y(t)\|^2 f_{yy} + \boldsymbol{\sigma}_X(t)\cdot\boldsymbol{\sigma}_Y(t) f_{xy}\right)\mathrm{d}t \\ &\quad + (f_x\boldsymbol{\sigma}_X(t) + f_y\boldsymbol{\sigma}_Y(t))\cdot \mathrm{d}\mathbf{W}(t) \\ &\equiv \mu_f(t)\,\mathrm{d}t + \boldsymbol{\sigma}_f(t)\cdot\mathrm{d}\mathbf{W}(t) \end{aligned}\tag{11.134}$$

where $f \equiv f(t, X(t), Y(t)), f_x \equiv f_x(t, X(t), Y(t))$, etc., is used to compact the expressions. In the second equation line we simply identified the drift $\mu_f(t)$ and volatility vector $\boldsymbol{\sigma}_f(t)$ for the process $\{f(t, X(t), Y(t))\}_{t\geqslant 0}$. We see that $\mu_f(t)$ and $\boldsymbol{\sigma}_f(t)$ are adapted processes defined explicitly as functions of $f(t, X(t), Y(t))$ and its partial derivatives, as well as functions of linear combinations of the drift and volatility vector coefficients of processes X and Y. In particular, the volatility vector $\boldsymbol{\sigma}_f(t) := f_x\,\boldsymbol{\sigma}_X(t) + f_y\,\boldsymbol{\sigma}_Y(t) = (\sigma_{f,1}(t), \sigma_{f,2}(t))$ has components

$$\sigma_{f,1}(t) = f_x\sigma_{X,1}(t) + f_y\sigma_{Y,1}(t)\,, \quad \sigma_{f,2}(t) = f_x\sigma_{X,2}(t) + f_y\sigma_{Y,2}(t)\,. \tag{11.135}$$

Hence, $\{F(t)\}_{t\geqslant 0} \equiv \{f(t, X(t), Y(t))\}_{t\geqslant 0}$ is an Itô process satisfying the stochastic integral equation

$$\begin{aligned}F(t) &= F(0) + \int_0^t \mu_f(u)\,\mathrm{d}u + \int_0^t \boldsymbol{\sigma}_f(u)\cdot\mathrm{d}\mathbf{W}(u) \\ &\equiv F(0) + \int_0^t \mu_f(u)\,\mathrm{d}u + \int_0^t \sigma_{f,1}(u)\,\mathrm{d}W_1(u) + \int_0^t \sigma_{f,2}(u)\,\mathrm{d}W_2(u)\,. \end{aligned}\tag{11.136}$$

The following example shows that the Itô Product Rule, derived previously, now follows simply by applying the Itô formula in (11.131).

Example 11.16. Let $\{X(t)\}_{t\geqslant 0}$ and $\{Y(t)\}_{t\geqslant 0}$ be Itô processes. Obtain the stochastic differential of their product.

Solution. Defining the function $f(t, x, y) := xy$ gives the product $F(t) := f(t, X(t), Y(t)) = X(t)Y(t)$ as an Itô process whose stochastic differential is given according to (11.131). In this case the function is independent of t and has derivatives:

$$f_t = 0,\ f_x = y,\ f_y = x,\ f_{xy} = 1,\ f_{xx} = f_{yy} = 0\,.$$

Substituting these terms into (11.131) (with $x = X(t), y = Y(t)$) gives

$$\begin{aligned}\mathrm{d}(X(t)Y(t)) &\equiv \mathrm{d}f(t, X(t), Y(t)) = 0\cdot\mathrm{d}t + Y(t)\,\mathrm{d}X(t) + X(t)\,\mathrm{d}Y(t) + \frac{1}{2}\cdot 0\cdot\mathrm{d}X(t)\,\mathrm{d}X(t) \\ &\quad + \frac{1}{2}\cdot 0\cdot\mathrm{d}Y(t)\,\mathrm{d}Y(t) + 1\cdot\mathrm{d}X(t)\,\mathrm{d}Y(t) \\ &= Y(t)\,\mathrm{d}X(t) + X(t)\,\mathrm{d}Y(t) + \mathrm{d}X(t)\,\mathrm{d}Y(t)\,. \end{aligned}\tag{11.137}$$

Assuming $X(t)Y(t) \neq 0$, we note that a useful way to represent this is to divide by $X(t)Y(t)$ (i.e., factor out the product), giving the relative differential

$$\frac{\mathrm{d}F(t)}{F(t)} \equiv \frac{\mathrm{d}(X(t)Y(t))}{X(t)Y(t)} = \frac{\mathrm{d}X(t)}{X(t)} + \frac{\mathrm{d}Y(t)}{Y(t)} + \frac{\mathrm{d}X(t)}{X(t)}\frac{\mathrm{d}Y(t)}{Y(t)}\,. \tag{11.138}$$

We can also write this in the form of (11.134):

$$\begin{aligned}\frac{\mathrm{d}(X(t)Y(t))}{X(t)Y(t)} &= \left(\frac{\mu_X(t)}{X(t)} + \frac{\mu_Y(t)}{Y(t)} + \frac{\boldsymbol{\sigma}_X(t)}{X(t)}\cdot\frac{\boldsymbol{\sigma}_Y(t)}{Y(t)}\right)\mathrm{d}t + \left(\frac{\boldsymbol{\sigma}_X(t)}{X(t)} + \frac{\boldsymbol{\sigma}_Y(t)}{Y(t)}\right)\cdot\mathrm{d}\mathbf{W}(t) \\ &\equiv \mu_{XY}(t)\,\mathrm{d}t + \boldsymbol{\sigma}_{XY}(t)\cdot\mathrm{d}\mathbf{W}(t)\,. \end{aligned}\tag{11.139}$$

This shows how the drift $\mu_{XY}(t)$ and volatility vector $\boldsymbol{\sigma}_{XY}(t)$ for the product process $F = XY$ are related to the drifts and volatility vectors of the processes X and Y. □

Another important example is the *Quotient Rule* for the stochastic differential of a ratio of two Itô processes. This rule is useful when pricing derivatives where we need to compute the drift and volatility of a process defined by a ratio of two asset price processes.

Example 11.17. Let $\{X(t)\}_{t\geqslant 0}$ and $\{Y(t)\}_{t\geqslant 0} \neq 0$ be Itô processes. Obtain the stochastic differential of their ratio $F(t) := \frac{X(t)}{Y(t)}$.

Solution. Let $f(t,x,y) := x/y$, i.e., $f(t,X(t),Y(t)) = X(t)/Y(t)$ is an Itô process with its stochastic differential given by (11.131). The relevant partial derivatives are

$$f_t = 0,\ f_x = \frac{1}{y},\ f_y = -\frac{x}{y^2},\ f_{xy} = -\frac{1}{y^2},\ f_{yy} = \frac{2x}{y^3},\ f_{xx} = 0\,.$$

Substituting these terms into (11.131) (with $x = X(t), y = Y(t)$) gives

$$\begin{aligned}
\mathrm{d}\left(\frac{X(t)}{Y(t)}\right) &\equiv \mathrm{d}F(t) \equiv \mathrm{d}f(t,X(t),Y(t)) \\
&= \frac{1}{Y(t)}\,\mathrm{d}X(t) - \frac{X(t)}{Y^2(t)}\,\mathrm{d}Y(t) + \frac{X(t)}{Y^3(t)}(\,\mathrm{d}Y(t))^2 - \frac{1}{Y^2(t)}\,\mathrm{d}X(t)\,\mathrm{d}Y(t) \\
&= \left(\frac{\mathrm{d}X(t)}{Y(t)} - \frac{X(t)}{Y(t)}\frac{\mathrm{d}Y(t)}{Y(t)}\right)\left(1 - \frac{\mathrm{d}Y(t)}{Y(t)}\right). \qquad (11.140)
\end{aligned}$$

This is written in a more convenient form (for later use) by dividing through by $X(t)/Y(t)$:

$$\frac{\mathrm{d}F(t)}{F(t)} \equiv \frac{\mathrm{d}\frac{X(t)}{Y(t)}}{\frac{X(t)}{Y(t)}} = \left(\frac{\mathrm{d}X(t)}{X(t)} - \frac{\mathrm{d}Y(t)}{Y(t)}\right)\left(1 - \frac{\mathrm{d}Y(t)}{Y(t)}\right). \qquad (11.141)$$

Substituting the expressions for $\mathrm{d}X(t)$ and $\mathrm{d}Y(t)$, applying the basic rules, and combining terms in $\mathrm{d}t$ and $\mathrm{d}\mathbf{W}(t)$ gives the form in (11.134) as

$$\begin{aligned}
\frac{\mathrm{d}\frac{X(t)}{Y(t)}}{\frac{X(t)}{Y(t)}} &= \left(\frac{\mu_X(t)}{X(t)} - \frac{\mu_Y(t)}{Y(t)} + \frac{\boldsymbol{\sigma}_Y(t)}{Y(t)}\cdot\left(\frac{\boldsymbol{\sigma}_Y(t)}{Y(t)} - \frac{\boldsymbol{\sigma}_X(t)}{X(t)}\right)\right)\mathrm{d}t + \left(\frac{\boldsymbol{\sigma}_X(t)}{X(t)} - \frac{\boldsymbol{\sigma}_Y(t)}{Y(t)}\right)\cdot\mathrm{d}\mathbf{W}(t) \\
&\equiv \mu_{\frac{X}{Y}}(t)\,\mathrm{d}t + \boldsymbol{\sigma}_{\frac{X}{Y}}(t)\cdot\mathrm{d}\mathbf{W}(t)\,. \qquad (11.142)
\end{aligned}$$

This gives the drift $\mu_{\frac{X}{Y}}(t)$ and volatility vector $\boldsymbol{\sigma}_{\frac{X}{Y}}(t)$ for the quotient process $F = \frac{X}{Y}$ in terms of the drifts and volatility vectors of the individual processes X and Y. □

The Itô product and quotient rules in (11.139) and (11.142) take on a more compact form if the processes X and Y can be represented in terms of the so-called *log-drifts and log-volatility vectors* (sometimes also referred to as local drift and local volatility). It will turn out to be particularly convenient when we later model the asset (e.g., stock) price processes. That is, assume processes X and Y satisfy the SDEs

$$\frac{\mathrm{d}X(t)}{X(t)} = \mu_X(t)\,\mathrm{d}t + \boldsymbol{\sigma}_X(t)\cdot\mathrm{d}\mathbf{W}(t)\,, \qquad (11.143)$$

$$\frac{\mathrm{d}Y(t)}{Y(t)} = \mu_Y(t)\,\mathrm{d}t + \boldsymbol{\sigma}_Y(t)\cdot\mathrm{d}\mathbf{W}(t)\,, \qquad (11.144)$$

where $\mu_X(t), \mu_Y(t), \boldsymbol{\sigma}_X(t), \boldsymbol{\sigma}_Y(t)$ are $\mathcal{F}_t$-adapted log-drifts and log-volatility vectors. Note that these SDEs are quite general. The difference is that the previous coefficients are related to these "log-coefficients" by sending the previous coefficients $\mu_X(t) \to \mu_X(t)X(t)$, $\mu_Y(t) \to \mu_Y(t)Y(t)$, $\boldsymbol{\sigma}_X(t) \to \boldsymbol{\sigma}_X(t)X(t)$, $\boldsymbol{\sigma}_Y(t) \to \boldsymbol{\sigma}_Y(t)Y(t)$. The Itô formula applied to (11.143) and (11.144) still gives (11.138) and (11.140). However, now the terms occurring in (11.139) and (11.142) simplify, where we replace the previous ratios $\frac{\mu_X(t)}{X(t)} \to \mu_X(t)$, $\frac{\mu_Y(t)}{Y(t)} \to \mu_Y(t)$, $\frac{\boldsymbol{\sigma}_X(t)}{X(t)} \to \boldsymbol{\sigma}_X(t)$, $\frac{\boldsymbol{\sigma}_Y(t)}{Y(t)} \to \boldsymbol{\sigma}_Y(t)$, giving

$$\begin{aligned}\frac{\mathrm{d}(X(t)Y(t))}{X(t)Y(t)} &= \big(\mu_X(t) + \mu_Y(t) + \boldsymbol{\sigma}_X(t) \cdot \boldsymbol{\sigma}_Y(t)\big)\,\mathrm{d}t + \big(\boldsymbol{\sigma}_X(t) + \boldsymbol{\sigma}_Y(t)\big) \cdot \mathrm{d}\mathbf{W}(t) \qquad (11.145)\\ &\equiv \mu_{XY}(t)\,\mathrm{d}t + \boldsymbol{\sigma}_{XY}(t) \cdot \mathrm{d}\mathbf{W}(t)\,.\end{aligned}$$

Here, $\mu_{XY}(t)$ and $\boldsymbol{\sigma}_{XY}(t)$ denote the log-drift and log-volatility vector of process XY and

$$\begin{aligned}\frac{\mathrm{d}\frac{X(t)}{Y(t)}}{\frac{X(t)}{Y(t)}} &= \big(\mu_X(t) - \mu_Y(t) + \boldsymbol{\sigma}_Y(t) \cdot (\boldsymbol{\sigma}_Y(t) - \boldsymbol{\sigma}_X(t))\big)\,\mathrm{d}t + \big(\boldsymbol{\sigma}_X(t) - \boldsymbol{\sigma}_Y(t)\big) \cdot \mathrm{d}\mathbf{W}(t)\\ &\equiv \mu_{\frac{X}{Y}}(t)\,\mathrm{d}t + \boldsymbol{\sigma}_{\frac{X}{Y}}(t) \cdot \mathrm{d}\mathbf{W}(t) \qquad (11.146)\end{aligned}$$

where $\mu_{\frac{X}{Y}}(t)$ and $\boldsymbol{\sigma}_{\frac{X}{Y}}(t)$ denote the log-drift and log-volatility vector of process $\frac{X}{Y}$.

For $d = 1$, the SDEs in (11.143)–(11.146) are all of the form in (11.26) where all volatility coefficients are scalars. The processes can therefore be represented as in (11.27). Given initial values $X(0)$ and $Y(0)$:

$$\begin{aligned}X(t) &= X(0)\exp\left[\int_0^t (\mu_X(s) - \frac{1}{2}\sigma_X^2(s))\,\mathrm{d}s + \int_0^t \sigma_X(s)\,\mathrm{d}W(s)\right],\\ Y(t) &= Y(0)\exp\left[\int_0^t (\mu_Y(s) - \frac{1}{2}\sigma_Y^2(s))\,\mathrm{d}s + \int_0^t \sigma_Y(s)\,\mathrm{d}W(s)\right],\\ X(t)Y(t) &= X(0)Y(0)\exp\left[\int_0^t (\mu_{XY}(s) - \frac{1}{2}\sigma_{XY}^2(s))\,\mathrm{d}s + \int_0^t \sigma_{XY}(s)\,\mathrm{d}W(s)\right],\\ \frac{X(t)}{Y(t)} &= \frac{X(0)}{Y(0)}\exp\left[\int_0^t (\mu_{\frac{X}{Y}}(s) - \frac{1}{2}\sigma_{\frac{X}{Y}}^2(s))\,\mathrm{d}s + \int_0^t \sigma_{\frac{X}{Y}}(s)\,\mathrm{d}W(s)\right].\end{aligned}$$

The reader can verify that the above third equation obtains by multiplying the expressions in the first and second equations, while the fourth equation obtains by dividing the expressions in the first and second equations. In the special case that the log-drift and log-volatility vectors are nonrandom (constants or ordinary functions of time t) the above processes are all GBM processes.

The above representations readily extend to the general vector case of $d \geqslant 1$. Consider the X process. Its natural logarithm has SDE:

$$\begin{aligned}\mathrm{d}\ln X(t) &= \frac{\mathrm{d}X(t)}{X(t)} - \frac{1}{2}\left(\frac{\mathrm{d}X(t)}{X(t)}\right)^2\\ &= \mu_X(t)\,\mathrm{d}t + \boldsymbol{\sigma}_X(t) \cdot \mathrm{d}\mathbf{W}(t) - \frac{1}{2}\big(\boldsymbol{\sigma}_X(t) \cdot \mathrm{d}\mathbf{W}(t)\big)^2\\ &= \left(\mu_X(t) - \frac{1}{2}\|\boldsymbol{\sigma}_X(t)\|^2\right)\mathrm{d}t + \boldsymbol{\sigma}_X(t) \cdot \mathrm{d}\mathbf{W}(t)\,.\end{aligned}$$

In integral form,

$$\ln\frac{X(t)}{X(0)} = \int_0^t (\mu_X(s) - \frac{1}{2}\|\boldsymbol{\sigma}_X(s)\|^2)\,\mathrm{d}s + \int_0^t \boldsymbol{\sigma}_X(s) \cdot \mathrm{d}\mathbf{W}(s)\,.$$

By exponentiating, $\frac{X(t)}{X(0)} = \exp(\ln \frac{X(t)}{X(0)})$,

$$X(t) = X(0) \exp\left[\int_0^t (\mu_X(s) - \frac{1}{2}\|\boldsymbol{\sigma}_X(s)\|^2)\,\mathrm{d}s + \int_0^t \boldsymbol{\sigma}_X(s)\cdot \mathrm{d}\mathbf{W}(s)\right]. \tag{11.147}$$

This expresses $X(t)$ in the general case of d-dimensional BM and reduces to the above expression in case $d = 1$. Similar expressions hold for the other processes. In fact, given an adapted drift $\mu(t)$ and volatility vector $\boldsymbol{\sigma}(t)$ (satisfying the above integrability assumptions), the SDE

$$\frac{\mathrm{d}U(t)}{U(t)} = \mu(t)\,\mathrm{d}t + \boldsymbol{\sigma}(t)\cdot \mathrm{d}\mathbf{W}(t) \tag{11.148}$$

with initial value $U(0)$ is equivalent to the representation

$$\begin{aligned} U(t) &= U(0) \exp\left[\int_0^t (\mu(s) - \frac{1}{2}\|\boldsymbol{\sigma}(s)\|^2)\,\mathrm{d}s + \int_0^t \boldsymbol{\sigma}(s)\cdot \mathrm{d}\mathbf{W}(s)\right] \\ &= U(0)\mathrm{e}^{\int_0^t \mu(s)\,\mathrm{d}s} \cdot \mathcal{E}_t(\boldsymbol{\sigma}\cdot \mathbf{W}) \end{aligned} \tag{11.149}$$

where the vector BM version of the stochastic exponential in (11.28) is defined by

$$\mathcal{E}_t(\boldsymbol{\sigma}\cdot \mathbf{W}) := \exp\left[-\frac{1}{2}\int_0^t \|\boldsymbol{\sigma}(s)\|^2\,\mathrm{d}s + \int_0^t \boldsymbol{\sigma}(s)\cdot \mathrm{d}\mathbf{W}(s)\right]. \tag{11.150}$$

Hence, each process satisfying (11.143)–(11.146) has an equivalent representation as in (11.149). For the X process we see that (11.147) has precisely the form in (11.149). For processes Y, XY, and X/Y the same form obtains in the obvious manner where the corresponding drifts and volatility vectors, for the respective processes, are substituted into (11.149).

We now further extend our discussion to arbitrary dimensions $d \geqslant 1$. As already noted, all the formulae presented so far are valid for $d \geqslant 1$. Given a d-dimensional $\mathcal{F}_t$-adapted vector $\boldsymbol{\gamma}(t) = (\gamma_1(t), \ldots, \gamma_d(t))$, the Itô integral w.r.t. d-dimensional BM is defined by the sum of one-dimensional Itô integrals,

$$\int_0^t \boldsymbol{\gamma}(s)\cdot \mathrm{d}\mathbf{W}(s) := \sum_{i=1}^d \int_0^t \gamma_i(s)\,\mathrm{d}W_i(s). \tag{11.151}$$

Throughout we shall assume that all such Itô integrals are square-integrable martingales for all times $0 \leqslant t \leqslant T$, given some $T > 0$, where (11.121)–(11.123) hold.

Let $\{\mathbf{X}(t) \equiv (X_1(t), X_2(t), \ldots, X_n(t))\}_{t\geqslant 0} \in \mathbb{R}^n$, $n \geqslant 1$, be an n-dimensional Itô vector process where each component is a real-valued Itô process driven by a d-dimensional BM:

$$\begin{aligned} \mathrm{d}X_i(t) &= \mu_i(t)\,\mathrm{d}t + \sum_{j=1}^d \sigma_{ij}(t)\,\mathrm{d}W_j(t) \\ &\equiv \mu_i(t)\,\mathrm{d}t + \boldsymbol{\sigma}_i(t)\cdot \mathrm{d}\mathbf{W}(t) \end{aligned} \tag{11.152}$$

with corresponding integral form

$$\begin{aligned} X_i(t) &= \int_0^t \mu_i(s)\,\mathrm{d}s + \sum_{j=1}^d \int_0^t \sigma_{ij}(s)\,\mathrm{d}W_j(s) \\ &\equiv \int_0^t \mu_i(s)\,\mathrm{d}s + \int_0^t \boldsymbol{\sigma}_i(s)\cdot \mathrm{d}\mathbf{W}(s) \end{aligned} \tag{11.153}$$

for $i = 1, \ldots, n$. Each $\{\mu_i(t)\}_{t \geqslant 0}$ is an integrable adapted process. The coefficients $\sigma_{ij}(t)$ are adapted and satisfy the square-integrability condition, $\int_0^T \mathrm{E}[\sigma_{ij}^2(s)]\,\mathrm{d}s < \infty$. The $n \times d$ matrix of coefficients $\boldsymbol{\sigma}(t) := [\sigma_{ij}(t)]_{i=1,\ldots,n;\, j=1,\ldots,d}$ is the *matrix of volatilities where the* i*th row gives the volatility vector* $\boldsymbol{\sigma}_i(t)$ *of the* i*th process* X_i:

$$\boldsymbol{\sigma}_i(t) = (\sigma_{i1}(t), \sigma_{i2}(t), \ldots, \sigma_{id}(t)) \,, \quad i = 1, \ldots, n \,.$$

The jth component of $\boldsymbol{\sigma}_i(t)$ is the volatility coefficient $\sigma_{ij}(t)$.

Being F_t-adapted, the coefficients $\mu_i(t)$ and $\sigma_{ij}(t)$ can generally depend on the entire path of the vector process $\mathbf{X}$ up to time t, e.g., they can be functionals of the Brownian path $\{\mathbf{W}(s) : 0 \leqslant s \leqslant t\}$. Generally this can be an intractable situation. However, an important case (and one that leads to some tractable models) is when these coefficients are known (defined) functions of the endpoint value of the process: $\mu_i(t) := \mu_i(t, \mathbf{X}(t))$ and $\boldsymbol{\sigma}_i(t) := \boldsymbol{\sigma}_i(t, \mathbf{X}(t))$. In this case, we say that the coefficients are *state dependent* and the vector process $\{\mathbf{X}(t)\}_{t \geqslant 0}$ is a *vector-valued diffusion process* with each component process solving an SDE of the form

$$\mathrm{d}X_i(t) = \mu_i(t, \mathbf{X}(t))\,\mathrm{d}t + \boldsymbol{\sigma}_i(t, \mathbf{X}(t)) \cdot \mathrm{d}\mathbf{W}(t) \,. \tag{11.154}$$

We will return to this case later.

We now turn to the more general multidimensional version of the Itô formula by extending Lemma 11.17 to the case of a function of time and any $n \geqslant 1$ processes. In preparation, we already computed the quadratic variation and the covariation of two Itô processes driven by a vector BM (see (11.126), (11.127), and (11.130)). In particular, any component process X_i has quadratic variation

$$[X_i, X_i](t) = \int_0^t \|\boldsymbol{\sigma}_i(u)\|^2\,\mathrm{d}u = \sum_{j=1}^d \int_0^t \sigma_{ij}^2(u)\,\mathrm{d}u \tag{11.155}$$

which is written equivalently in differential form as

$$\mathrm{d}[X_i, X_i](t) \equiv \mathrm{d}X_i(t)\,\mathrm{d}X_i(t) \equiv (\,\mathrm{d}X_i(t))^2 = \|\boldsymbol{\sigma}_i(t)\|^2\,\mathrm{d}t \,. \tag{11.156}$$

Any pair of processes X_i, X_j has covariation

$$[X_i, X_j](t) = \int_0^t \boldsymbol{\sigma}_i(u) \cdot \boldsymbol{\sigma}_j(u)\,\mathrm{d}u \,, \tag{11.157}$$

or in differential form,

$$\mathrm{d}[X_i, X_j](t) \equiv \mathrm{d}X_i(t)\,\mathrm{d}X_j(t) = \boldsymbol{\sigma}_i(t) \cdot \boldsymbol{\sigma}_j(t)\,\mathrm{d}t \,. \tag{11.158}$$

It is also useful to write the above as $\mathrm{d}[X_i, X_j](t) = C_{ij}(t)\,\mathrm{d}t$ by defining the coefficients

$$C_{ij}(t) := \boldsymbol{\sigma}_i(t) \cdot \boldsymbol{\sigma}_j(t) = \sum_{k=1}^n \sigma_{ik}(t)\,\sigma_{jk}(t) \,. \tag{11.159}$$

These are the elements of an $n \times n$ matrix $\mathbf{C}(t) := [C_{ij}(t)]_{i,j=1,\ldots,n}$, where $C_{ij}(t)$ are related to the *instantaneous covariances* between the differential increments of the two processes X_i and X_j. In terms of the matrix $\boldsymbol{\sigma}(t)$, the instantaneous covariances are given by $C_{ij}(t) = (\boldsymbol{\sigma}(t)\,\boldsymbol{\sigma}(t)^\top)_{ij}$ where $\boldsymbol{\sigma}(t)^\top$ is the $d \times n$ transpose of the matrix $\boldsymbol{\sigma}(t)$. Given the instantaneous covariances, we also define the *instantaneous correlations*

$$\rho_{ij}(t) := \frac{\boldsymbol{\sigma}_i(t) \cdot \boldsymbol{\sigma}_j(t)}{\|\boldsymbol{\sigma}_i(t)\|\,\|\boldsymbol{\sigma}_j(t)\|} = \frac{C_{ij}(t)}{\|\boldsymbol{\sigma}_i(t)\|\,\|\boldsymbol{\sigma}_j(t)\|} \,. \tag{11.160}$$

[Remark: We can see, at least heuristically, that these coefficients are a measure of the instantaneous correlations between the increments of a pair of processes X_i and X_j when we divide the differential of the covariation with the square root of the product of the differentials of the variations, $\frac{\mathrm{d}[X_i,X_j](t)}{\sqrt{\mathrm{d}[X_i,X_i](t)\cdot \mathrm{d}[X_j,X_j](t)}} = \frac{C_{ij}(t)}{\|\boldsymbol{\sigma}_i(t)\|\,\|\boldsymbol{\sigma}_j(t)\|} = \rho_{ij}(t)$.]

As in the case of Lemma 11.17, the main idea, which also gives us a simple way to remember the Itô formula for smooth functions $f(t,x_1,\ldots,x_n)$ of n variables and time t, is to apply a Taylor expansion of f up to terms of first order in the time increment $\mathrm{d}t$, and up to second order in the increments $\mathrm{d}x_1,\ldots,\mathrm{d}x_n$. The stochastic differential form of the Itô formula is then obtained upon replacing the ordinary variables $x_i \to X_i(t)$, and replacing all ordinary differentials by the respective stochastic differentials, $\mathrm{d}x_i \to \mathrm{d}X_i(t)$, $\mathrm{d}x_i\,\mathrm{d}x_j \to \mathrm{d}[X_i,X_j](t)$, $i,j=1,\ldots,n$. We then also write the formula more explicitly by applying the basic rules in (11.116). The formal proof (which we omit) follows in the same manner as the proof for the two variable case in Lemma 11.17. Here we simply state this important result as a lemma.

Lemma 11.18 (Itô Formula for a Function of Several Processes). *Let the vector-valued process $\{\mathbf{X}(t) \equiv (X_1(t),X_2(t),\ldots,X_n(t))\}_{t\geqslant 0}$ satisfy the SDE in (11.152) and assume $f(t,\mathbf{x}) \equiv f(t,x_1,\ldots,x_n)$ is a real-valued function on $\mathbb{R}_+\times\mathbb{R}^n$ that is continuously differentiable with respect to time t and twice continuously differentiable with respect to the n variables $x_1,\ldots,x_n$. Then, the process defined by $F(t) := f(t,\mathbf{X}(t)) \equiv f(t,X_1(t),\ldots,X_n(t))$, $t\geqslant 0$, is an Itô process with stochastic differential $\mathrm{d}F(t) \equiv \mathrm{d}f(t,\mathbf{X}(t))$ given by*

$$\begin{aligned}\mathrm{d}F(t) &= \frac{\partial f}{\partial t}(t,\mathbf{X}(t))\,\mathrm{d}t + \sum_{i=1}^{n}\frac{\partial f}{\partial x_i}(t,\mathbf{X}(t))\,\mathrm{d}X_i(t)\\ &\quad + \frac{1}{2}\sum_{i=1}^{n}\sum_{j=1}^{n}\frac{\partial^2 f}{\partial x_i\partial x_j}(t,\mathbf{X}(t))\,\mathrm{d}[X_i,X_j](t)\end{aligned} \tag{11.161}$$

Note that the formula in (11.131) is recovered in the special case that $n=2$ by setting $X_1(t)\equiv X(t), X_2(t)\equiv Y(t)$. The stochastic differential in (11.161) has meaning when written as an Itô process in integral form. Using (11.158), (11.161) takes the form

$$\begin{aligned}\mathrm{d}f(t,\mathbf{X}(t)) &= \left[\frac{\partial f}{\partial t}(t,\mathbf{X}(t)) + \frac{1}{2}\sum_{i=1}^{n}\sum_{j=1}^{n}C_{ij}(t)\frac{\partial^2 f}{\partial x_i\partial x_j}(t,\mathbf{X}(t))\right]\mathrm{d}t\\ &\quad + \sum_{i=1}^{n}\frac{\partial f}{\partial x_i}(t,\mathbf{X}(t))\,\mathrm{d}X_i(t)\,.\end{aligned} \tag{11.162}$$

This expression involves the time differential and a linear combination of stochastic differentials in the component processes. By further making use of (11.152), we obtain Itô's formula, extending (11.134), where the stochastic differential is written in terms of the vector BM increment:

$$\begin{aligned}\mathrm{d}f(t,\mathbf{X}(t)) &= \left[\frac{\partial f}{\partial t}(t,\mathbf{X}(t)) + \frac{1}{2}\sum_{i=1}^{n}\sum_{j=1}^{n}C_{ij}(t)\frac{\partial^2 f}{\partial x_i\partial x_j}(t,\mathbf{X}(t)) + \sum_{i=1}^{n}\mu_i(t)\frac{\partial f}{\partial x_i}(t,\mathbf{X}(t))\right]\mathrm{d}t\\ &\quad + \sum_{i=1}^{n}\frac{\partial f}{\partial x_i}(t,\mathbf{X}(t))\,\boldsymbol{\sigma}_i(t)\cdot\mathrm{d}\mathbf{W}(t)\\ &\equiv \mu_f(t)\,\mathrm{d}t + \boldsymbol{\sigma}_f(t)\cdot\mathrm{d}\mathbf{W}(t)\,.\end{aligned} \tag{11.163}$$

In the second equation line we identified the drift and volatility vector of the process $\{f(t,\mathbf{X}(t))\}_{t\geqslant 0}$.

11.10.2 Multidimensional SDEs, Feynman–Kac Formulae, and Transition CDFs and PDFs

The main concepts, theorems, and formulae that we established in Section 11.7 for the case of a single process driven by a one-dimensional BM also carry over into the multidimensional case with appropriate assumptions in place. Here we only give a very brief account of some of the relevant results. Our main starting point is the n-dimensional diffusion process solving the system of SDEs in (11.154), i.e.,

$$\mathrm{d}X_i(t) = \mu_i(t, \mathbf{X}(t))\,\mathrm{d}t + \sum_{j=1}^{d} \sigma_{ij}(t, \mathbf{X}(t))\,\mathrm{d}W_j(t)\,, \quad i = 1, \ldots, n\,. \tag{11.164}$$

In integral form,

$$X_i(t) = X_i(0) + \int_0^t \mu_i(s, \mathbf{X}(s))\,\mathrm{d}s + \sum_{j=1}^{d} \int_0^t \sigma_{ij}(s, \mathbf{X}(s))\,\mathrm{d}W_j(s)\,, \quad i = 1, \ldots, n\,. \tag{11.165}$$

The drift and volatility coefficients, $\mu_i(t, \mathbf{x})$ and $\sigma_{ij}(t, \mathbf{x})$, are given functions of time and variables $\mathbf{x} = (x_1, \ldots, x_n)$.

Theorem 11.4, which provides sufficient conditions on the existence and uniqueness of a strong solution to the one-dimensional SDE (11.24), extends in a similar manner to the above system of SDEs. The absolute values for the Lipschitz condition and the linear growth condition on the coefficients are now replaced by appropriate vector and matrix norms. We denote the drift vector by $\boldsymbol{\mu}(t, \mathbf{x}) = (\mu_1(t, \mathbf{x}), \ldots, \mu_n(t, \mathbf{x}))$. The norm of a vector $\mathbf{v} \in \mathbb{R}^n$ is $\|\mathbf{v}\| := \sqrt{\sum_{i=1}^n v_i^2}$ and the norm of a matrix A with elements a_{ij} is defined similarly as $\|A\| := \sqrt{\sum_{i,j} a_{ij}^2}$. If there is a constant $K > 0$ such that the Lipschitz condition

$$\|\boldsymbol{\mu}(t, \mathbf{x}) - \boldsymbol{\mu}(t, \mathbf{y})\| + \|\boldsymbol{\sigma}(t, \mathbf{x}) - \boldsymbol{\sigma}(t, \mathbf{y})\| \leqslant K\|\mathbf{x} - \mathbf{y}\|$$

and the linear growth condition

$$\|\boldsymbol{\mu}(t, \mathbf{x})\| + \|\boldsymbol{\sigma}(t, \mathbf{x})\| \leqslant K(1 + \|\mathbf{x}\|)$$

are both satisfied, for $\mathbf{x}, \mathbf{y} \in \mathbb{R}^n$, then this ensures that there is a unique vector process $\{\mathbf{X}(t)\}_{t\geqslant 0}$ solving (11.165) and that the paths of the vector process are continuous in time. As in the one-dimensional case, these conditions are not necessary, but are sufficient conditions to guarantee the existence of a unique strong solution.

The solution to (11.165) is also a vector Markov process, i.e.,

$$\mathbb{P}(\mathbf{X}(t) \leqslant \mathbf{y} \mid \mathcal{F}_s) = \mathbb{P}(\mathbf{X}(t) \leqslant \mathbf{y} \mid \mathbf{X}(s)) \tag{11.166}$$

for all $\mathbf{y} \in \mathbb{R}^n$ or, for Borel function $h : \mathbb{R}^n \to \mathbb{R}$,

$$\mathrm{E}[h(\mathbf{X}(t)) \mid \mathcal{F}_s] = \mathrm{E}[h(\mathbf{X}(t)) \mid X(s)]\,, \quad 0 \leqslant s \leqslant t. \tag{11.167}$$

The conditional expectation of $h(\mathbf{X}(T))$, given the vector value $\mathbf{X}(t) = \mathbf{x} \in \mathbb{R}^n$ at a time $t \leqslant T$, is denoted by

$$\mathrm{E}_{t,\mathbf{x}}[h(\mathbf{X}(T))] := \mathrm{E}[h(\mathbf{X}(T)) \mid \mathbf{X}(t) = \mathbf{x}].$$

As in the scalar case, the subscripts $t, \mathbf{x}$ are shorthand for conditioning on a given vector value $\mathbf{X}(t) = \mathbf{x}$ at time t. The conditional expectation is a function of the ordinary variables $\mathbf{x}$ and t, i.e., $\mathrm{E}_{t,\mathbf{x}}[h(\mathbf{X}(T))] = g(t, \mathbf{x})$ for fixed T. The Markov property is expressible as $\mathrm{E}[h(\mathbf{X}(T)) \mid \mathcal{F}_t] = \mathrm{E}_{t,\mathbf{X}(t)}[h(\mathbf{X}(T))] = g(t, \mathbf{X}(t))$, for all $0 \leqslant t \leqslant T$. Hence, if we know the conditional probability distribution of the random vector $\mathbf{X}(T)$, given $\mathbf{X}(t) = x$, then we can compute the function $g(t, \mathbf{x})$.

As in the scalar case, the conditional PDF of $\mathbf{X}(T)$, given $\mathbf{X}(t)$ is the (joint) transition PDF, $p(t,T;\mathbf{x},\mathbf{y}) \equiv p(t,T;x_1,\ldots,x_n,y_1,\ldots,y_n)$, for the vector process $\mathbf{X}$ obtained by differentiating the corresponding (joint) transition CDF, $P(t,T;\mathbf{x},\mathbf{y})$:

$$p(t,T;\mathbf{x},\mathbf{y}) = \frac{\partial^n}{\partial y_1 \ldots \partial y_n} P(t,T;\mathbf{x},\mathbf{y}), \tag{11.168}$$

$$\begin{aligned} P(t,T;\mathbf{x},\mathbf{y}) &:= \mathbb{P}(\mathbf{X}(T) \leqslant \mathbf{y} \mid \mathbf{X}(t) = \mathbf{x}) \\ &\equiv \mathbb{P}(X_1(T) \leqslant y_1, \ldots, X_n(T) \leqslant y_n \mid X_1(t) = x_1, \ldots, X_n(t) = x_n) \\ &= \int_{-\infty}^{y_1} \cdots \int_{-\infty}^{y_n} p(t,T;\mathbf{x},\mathbf{z})\,\mathrm{d}\mathbf{z}. \end{aligned} \tag{11.169}$$

As in the one-dimensional case, the Markov and tower property lead to the multidimensional version of the Chapman–Kolmogorov relation:

$$p(s,t;\mathbf{x},\mathbf{y}) = \int_{\mathbb{R}^n} p(s,u;\mathbf{x},\mathbf{z}) p(u,t;\mathbf{z},\mathbf{y})\,\mathrm{d}\mathbf{z}, \quad s < u < t. \tag{11.170}$$

Given any Borel set $B \in \mathcal{B}(\mathbb{R}^n)$, the probability that the time-t vector process has value in B, given that it has value $\mathbf{x} \in \mathbb{R}^n$ at some earlier time $s < t$, is given by integrating the transition PDF over B:

$$P(\mathbf{X}(t) \in B \mid \mathbf{X}(s) = \mathbf{x}) = \int_B p(s,t;\mathbf{x},\mathbf{y})\,\mathrm{d}\mathbf{y}. \tag{11.171}$$

The multidimensional analogue of (11.41) for computing a conditional expectation is an integral over $\mathbb{R}^n$:

$$\mathrm{E}_{t,\mathbf{x}}[h(\mathbf{X}(T))] = \int_{\mathbb{R}^n} h(\mathbf{y}) p(t,T;\mathbf{x},\mathbf{y})\,\mathrm{d}\mathbf{y}, \quad t < T. \tag{11.172}$$

As in the case of a scalar diffusion on $\mathbb{R}$ with the generator in (11.43), the generator for the above vector diffusion process $\{\mathbf{X}(t)\}_{t\geqslant 0}$ on $\mathbb{R}^n$ is defined by the differential operator $\mathcal{G}_{t,\mathbf{x}}$ acting on a smooth function $f = f(t,\mathbf{x})$,

$$\mathcal{G}_{t,\mathbf{x}} f := \frac{1}{2} \sum_{i=1}^{n} \sum_{j=1}^{n} C_{ij}(t,\mathbf{x}) \frac{\partial^2 f}{\partial x_i \partial x_j} + \sum_{i=1}^{n} \mu_i(t,\mathbf{x}) \frac{\partial f}{\partial x_i} \tag{11.173}$$

where $[C_{ij}(t,\mathbf{x}) = \sum_{k=1}^{d} \sigma_{ik}(t,\mathbf{x}) \sigma_{jk}(t,\mathbf{x})]_{i,j=1,\ldots,n}$ is the *diffusion matrix*. We shall assume that this matrix is positive definite where $\mathbf{v}^\top \mathbf{C} \mathbf{v} > 0$ for nonzero $\mathbf{v} \in \mathbb{R}^n$. The differential and integral forms of the Itô formula are now written compactly (extending (11.44) and (11.45) to the multidimensional case):

$$\mathrm{d}f(t,\mathbf{X}(t)) = \left(\frac{\partial}{\partial t} + \mathcal{G}_{t,\mathbf{x}}\right) f(t,\mathbf{X}(t))\,\mathrm{d}t + \sum_{j=1}^{d} \left(\sum_{i=1}^{n} \frac{\partial f}{\partial x_i}(t,\mathbf{X}(t))\,\sigma_{ij}(t,\mathbf{X}(t)) \right) \mathrm{d}W_j(t) \tag{11.174}$$

and

$$\begin{aligned} f(t,\mathbf{X}(t)) &= f(0,\mathbf{X}(0)) + \int_0^t \left(\frac{\partial}{\partial s} + \mathcal{G}_{s,\mathbf{x}}\right) f(s,\mathbf{X}(s))\,\mathrm{d}s \\ &\quad + \sum_{j=1}^{d} \int_0^t \left(\sum_{i=1}^{n} \frac{\partial f}{\partial x_i}(s,\mathbf{X}(s))\,\sigma_{ij}(s,\mathbf{X}(s)) \right) \mathrm{d}W_j(s). \end{aligned} \tag{11.175}$$

The analogues of (11.46)–(11.48) also follow if we fix a time $T > 0$ and assume the square-integrability condition,

$$\int_0^T \mathrm{E}\left[\left(\sum_{i=1}^{n} \frac{\partial f}{\partial x_i}(s, \mathbf{X}(s))\sigma_{ij}(s, \mathbf{X}(s))\right)^2\right] \mathrm{d}s < \infty\,, \quad j = 1, \ldots, n, \tag{11.176}$$

which ensures that all Itô integrals (w.r.t. each BM, W_j) in (11.175) are martingales. By using a similar argument as in the one-dimensional case, the process $\{M_f(t)\}_{0 \leqslant t \leqslant T}$ defined by

$$M_f(t) := f(t, \mathbf{X}(t)) - \int_0^t \left(\frac{\partial}{\partial s} + \mathcal{G}_{s,\mathbf{x}}\right) f(s, \mathbf{X}(s))\, \mathrm{d}s \tag{11.177}$$

is a martingale, i.e., (11.48) holds. As a particular application of (11.177), we obtain a martingale defined by $M_f(t) := f(t, \mathbf{X}(t))$ if the function f solves the PDE: $\frac{\partial f}{\partial t} + \mathcal{G}_{t,\mathbf{x}} f = 0$.

The martingale property of the process defined in (11.177) allows us to extend Theorems 11.7 and 11.9 to the multidimensional case. Here, we simply state useful versions of the multidimensional extensions. Their proofs involve some similar steps as in the one-dimensional case.

Theorem 11.19 (Multidimensional Feynman–Kac). *Let $\{\mathbf{X}(t) := (X_1(t), \ldots, X_n(t))\}_{0 \leqslant t \leqslant T}$ solve the system of SDEs in (11.164) and let $\phi : \mathbb{R}^n \to \mathbb{R}$ be a Borel function. Also, assume the square-integrability condition (11.176) holds. Suppose the function $f(t, \mathbf{x})$ is a solution to the backward Kolmogorov PDE $\frac{\partial f}{\partial t} + \mathcal{G}_{t,\mathbf{x}} f = 0$, i.e.,*

$$\frac{\partial f}{\partial t}(t, \mathbf{x}) + \frac{1}{2}\sum_{i=1}^{n}\sum_{j=1}^{n} C_{ij}(t, \mathbf{x}) \frac{\partial^2 f}{\partial x_i \partial x_j}(t, \mathbf{x}) + \sum_{i=1}^{n} \mu_i(t, \mathbf{x}) \frac{\partial f}{\partial x_i}(t, \mathbf{x}) = 0\,, \tag{11.178}$$

for all $\mathbf{x} \in \mathbb{R}^n$, $t < T$, subject to the terminal condition $f(T, \mathbf{x}) = \phi(\mathbf{x})$. Then, assuming $\mathrm{E}_{t,\mathbf{x}}[|\phi(\mathbf{X}(T))|] < \infty$, $f(t, \mathbf{x})$ has the representation

$$f(t, \mathbf{x}) = \mathrm{E}_{t,\mathbf{x}}[\phi(\mathbf{X}(T))] \equiv \mathrm{E}[\phi(\mathbf{X}(T)) \mid \mathbf{X}(t) = \mathbf{x}] \tag{11.179}$$

for all $\mathbf{x} \in \mathbb{R}^n$, $0 \leqslant t \leqslant T$.

The slightly more general result below includes an additional exponential discount factor via a discounting function $r(t, \mathbf{x})$. Theorem 11.19 is then a particular case of this theorem by simply setting $r(t, \mathbf{x}) \equiv 0$.

Theorem 11.20 ("Discounted" Feynman–Kac). *Let $\{\mathbf{X}(t) := (X_1(t), \ldots, X_n(t))\}_{0 \leqslant t \leqslant T}$ solve the system of SDEs in (11.164) and assume the square integrability condition (11.176) holds. Let $\phi : \mathbb{R}^n \to \mathbb{R}$ be a Borel function and $r(t, \mathbf{x}) : [0, T] \times \mathbb{R}^n \to \mathbb{R}$ be a lower-bounded continuous function. Then, the function defined by the conditional expectation*

$$f(t, \mathbf{x}) := \mathrm{E}_{t,\mathbf{x}}[\mathrm{e}^{-\int_t^T r(u, \mathbf{X}(u))\, \mathrm{d}u} \phi(\mathbf{X}(T))] \equiv \mathrm{E}[\mathrm{e}^{-\int_t^T r(u, \mathbf{X}(u))\, \mathrm{d}u} \phi(\mathbf{X}(T)) \mid \mathbf{X}(t) = x] \tag{11.180}$$

solves the PDE $\frac{\partial f}{\partial t} + \mathcal{G}_{t,\mathbf{x}} f - r(t, \mathbf{x}) f = 0$, i.e.,

$$\frac{\partial f}{\partial t}(t, \mathbf{x}) + \frac{1}{2}\sum_{i=1}^{n}\sum_{j=1}^{n} C_{ij}(t, \mathbf{x}) \frac{\partial^2 f}{\partial x_i \partial x_j}(t, \mathbf{x}) + \sum_{i=1}^{n} \mu_i(t, \mathbf{x}) \frac{\partial f}{\partial x_i}(t, \mathbf{x}) - r(t, \mathbf{x}) f(t, \mathbf{x}) = 0\,, \tag{11.181}$$

for all $\mathbf{x}$, $0 < t < T$, subject to the terminal condition $f(T, \mathbf{x}) = \phi(\mathbf{x})$.

In the special case that the discount function is a constant, $r(t,\mathbf{x}) \equiv r$, then (11.180) simplifies since the discount factor is simply $\mathrm{e}^{-r(T-t)}$ where $f(t,\mathbf{x}) = \mathrm{e}^{-r(T-t)}\mathrm{E}_{t,\mathbf{x}}[\phi(\mathbf{X}(T))]$.

The multidimensional version of Proposition 11.8 also follows where both the transition PDF and CDF solve the backward Kolmogorov PDE in (11.178) in the (backward) variables $(t,\mathbf{x})$. In particular, fixing a time $T > 0$ and a vector $\mathbf{y} \in \mathbb{R}^n$, and setting $\phi(\mathbf{x}) = \mathbb{I}_{\{x_1 \leqslant y_1,\ldots,x_n \leqslant y_n\}}$ in Proposition 11.19 implies that the transition CDF

$$P(t,T;\mathbf{x},\mathbf{y}) \equiv \mathrm{E}[\mathbb{I}_{\{X_1(T)\leqslant y_1,\ldots,X_n(T)\leqslant y_n\}} \mid X_1(t) = x_1,\ldots,X_n(t) = x_n]$$

solves the PDE in (11.178) with terminal condition as the indicator function, $P(T,T;\mathbf{x},\mathbf{y}) = \mathbb{P}(x_1 \leqslant y_1,\ldots,x_n \leqslant y_n) = \mathbb{I}_{\{x_1\leqslant y_1,\ldots,x_n\leqslant y_n\}}$. The transition PDF, $p = p(t,T;\mathbf{x},\mathbf{y})$, is obtained from the CDF by differentiating in the $\mathbf{y}$ variables, according to (11.168). Hence, p also solves (11.178) and the terminal condition is given by

$$\lim_{t\nearrow T} p(t,T;\mathbf{x},\mathbf{y}) = \delta(\mathbf{y}-\mathbf{x})$$

where $\delta(\mathbf{y}-\mathbf{x}) = \delta(y_1 - x_1)\cdots\delta(y_n - x_n)$ is the n-dimensional Dirac delta function as a product of univariate delta functions.

The transition PDF $p(t,T;\mathbf{x},\mathbf{y})$ is the conditional PDF of the random vector $\mathbf{X}(T)$ at $\mathbf{y}$, given $\mathbf{X}(t) = \mathbf{x}$. Hence, according to (11.179) the solution to the PDE problem takes the form of an integral of the product of the transition PDF and the function $\phi(\mathbf{y})$:

$$f(t,\mathbf{x}) = \mathrm{E}[\phi(\mathbf{X}(T)) \mid \mathbf{X}(t) = \mathbf{x}] = \int_{\mathbb{R}^n} \phi(\mathbf{y})p(t,T;\mathbf{x},\mathbf{y})\,\mathrm{d}\mathbf{y}\,. \tag{11.182}$$

That is, the transition PDF is the fundamental solution to the PDE problem stated in Theorem 11.19.

For many applications the vector diffusion process is time homogeneous where the drift and diffusion matrix are time independent, $\mu_i(t,\mathbf{x}) \equiv \mu_i(\mathbf{x})$ and $C_{ij}(t,\mathbf{x}) = C_{ij}(\mathbf{x}) = \sum_{k=1}^{d} \sigma_{ik}(\mathbf{x})\sigma_{jk}(\mathbf{x})$. The generator is then a differential operator only in the $\mathbf{x}$ variables,

$$\mathcal{G}_{t,\mathbf{x}} = \mathcal{G}_{\mathbf{x}} := \frac{1}{2}\sum_{i=1}^{n}\sum_{j=1}^{n} C_{ij}(\mathbf{x})\frac{\partial^2}{\partial x_i \partial x_j} + \sum_{i=1}^{n}\mu_i(\mathbf{x})\frac{\partial}{\partial x_i}.$$

Defining $\tau := T - t$, the solution in (11.179) is a function of $(\tau,\mathbf{x})$, i.e., $f = f(\tau,\mathbf{x})$, and the backward PDE in (11.178) takes the form

$$\frac{\partial f}{\partial \tau} = \frac{1}{2}\sum_{i=1}^{n}\sum_{j=1}^{n} C_{ij}(\mathbf{x})\frac{\partial^2 f}{\partial x_i \partial x_j} + \sum_{i=1}^{n}\mu_i(\mathbf{x})\frac{\partial f}{\partial x_i} \tag{11.183}$$

subject to the initial condition $f(0,\mathbf{x}) = \phi(\mathbf{x})$. Moreover, if the discount function in Theorem 11.20 is time independent, $r(t,\mathbf{x}) \equiv r(\mathbf{x})$, then the operator $\mathcal{G}_{t,\mathbf{x}} - r(t,\mathbf{x}) \equiv \mathcal{G}_{\mathbf{x}} - r(\mathbf{x})$ is time independent, i.e., the PDE in (11.181) is time-homogeneous:

$$\frac{\partial f}{\partial \tau} = \frac{1}{2}\sum_{i=1}^{n}\sum_{j=1}^{n} C_{ij}(\mathbf{x})\frac{\partial^2 f}{\partial x_i \partial x_j} + \sum_{i=1}^{n}\mu_i(\mathbf{x})\frac{\partial f}{\partial x_i} - r(\mathbf{x})f\,, \tag{11.184}$$

with the solution represented in (11.180) as a function $f = f(\tau,\mathbf{x})$ having initial condition $f(0,\mathbf{x}) = \phi(\mathbf{x})$.

For the time-homogeneous case we hence have the transition CDF and PDF as functions of $\tau,\mathbf{x},\mathbf{y}$ where we equivalently write $P(t,T;\mathbf{x},\mathbf{y})$ as $P(\tau;\mathbf{x},\mathbf{y})$ and $p(t,T;\mathbf{x},\mathbf{y})$ as $p(\tau;\mathbf{x},\mathbf{y})$.

Both $P(\tau;\mathbf{x},\mathbf{y})$ and $p(\tau;\mathbf{x},\mathbf{y})$ solve the PDE in (11.183) where $p(0+;\mathbf{x},\mathbf{y}) = \delta(\mathbf{x}-\mathbf{y})$ and $P(0;\mathbf{x},\mathbf{y})$ given by the n-dimensional unit step function $\mathbb{I}_{\{\mathbf{y}\geqslant\mathbf{x}\}} = \mathbb{I}_{\{y_1\geqslant x_1,\dots,y_n\geqslant x_n\}}$.

As a first example of how Theorem 11.19 can be applied in practice, consider a simple 2-dimensional process ($n=2$) driven by a 2-dimensional BM ($d=2$). That is, let $\{\mathbf{X}(t) = [X_1(t), X_2(t)]^\top\}_{t\geqslant 0} \in \mathbb{R}^2$ be two scaled and drifted Brownian motions satisfying the system of SDEs:

$$\begin{aligned}
\mathrm{d}X_1(t) &= \mu_1\,\mathrm{d}t + \sigma_1\,\mathrm{d}W_1(t) \equiv \mu_1\,\mathrm{d}t + \boldsymbol{\sigma}_1\cdot\mathrm{d}\mathbf{W}(t)\,,\\
\mathrm{d}X_2(t) &= \mu_2\,\mathrm{d}t + \sigma_2\rho\,\mathrm{d}W_1(t) + \sigma_2\sqrt{1-\rho^2}\,\mathrm{d}W_2(t) \equiv \mu_2\,\mathrm{d}t + \boldsymbol{\sigma}_2\cdot\mathrm{d}\mathbf{W}(t)\,,
\end{aligned} \tag{11.185}$$

where $\rho\in(-1,1)$ is a constant correlation coefficient, $\boldsymbol{\sigma}_1 = [\sigma_1, 0]$ and $\boldsymbol{\sigma}_2 = [\sigma_2\rho, \sigma_2\sqrt{1-\rho^2}]$ are volatility vectors with magnitudes $\|\boldsymbol{\sigma}_1\| = \sigma_1$, $\|\boldsymbol{\sigma}_2\| = \sigma_2$. Note that $\boldsymbol{\sigma}_1\cdot\boldsymbol{\sigma}_2 = \rho\sigma_1\sigma_2$. We can also represent this system of SDEs in vector-matrix notation, $\mathrm{d}\mathbf{X}(t) = \boldsymbol{\mu}\,\mathrm{d}t + \boldsymbol{\sigma}\,\mathrm{d}\mathbf{W}(t)$:

$$\begin{bmatrix}\mathrm{d}X_1(t)\\ \mathrm{d}X_2(t)\end{bmatrix} = \begin{bmatrix}\mu_1\\ \mu_2\end{bmatrix}\mathrm{d}t + \begin{bmatrix}\sigma_1 & 0\\ \sigma_2\rho & \sigma_2\sqrt{1-\rho^2}\end{bmatrix}\begin{bmatrix}\mathrm{d}W_1(t)\\ \mathrm{d}W_2(t)\end{bmatrix}.$$

The above 2×2 is the $\boldsymbol{\sigma}$-matrix whose rows correspond to the volatility vectors $\boldsymbol{\sigma}_1$ and $\boldsymbol{\sigma}_2$. The diffusion matrix $\mathbf{C} = \boldsymbol{\sigma}\boldsymbol{\sigma}^\top$ is then

$$\mathbf{C} = \begin{bmatrix}\sigma_1 & 0\\ \sigma_2\rho & \sigma_2\sqrt{1-\rho^2}\end{bmatrix}\begin{bmatrix}\sigma_1 & \sigma_2\rho\\ 0 & \sigma_2\sqrt{1-\rho^2}\end{bmatrix} = \begin{bmatrix}\sigma_1^2 & \rho\sigma_1\sigma_2\\ \rho\sigma_1\sigma_2 & \sigma_2^2\end{bmatrix},$$

where the 2×2 matrix $\boldsymbol{\sigma}$ is the lower Cholesky factorization of $\mathbf{C}$.

The SDEs in (11.185), subject to arbitrary initial conditions $X_1(t) = x_1, X_2(t) = x_2$, are solved by simply integrating from time t to T:

$$\begin{aligned}
X_1(T) &= x_1 + \mu_1(T-t) + \boldsymbol{\sigma}_1\cdot(\mathbf{W}(T)-\mathbf{W}(t))\,,\\
X_2(T) &= x_2 + \mu_2(T-t) + \boldsymbol{\sigma}_2\cdot(\mathbf{W}(T)-\mathbf{W}(t))\,.
\end{aligned} \tag{11.186}$$

We can express these random variables in terms of two standard normal random variables:

$$\begin{aligned}
X_1(T) &= x_1 + \mu_1\tau + \sigma_1\sqrt{\tau}Z_1\,,\\
X_2(T) &= x_2 + \mu_2\tau + \sigma_2\sqrt{\tau}Z_2\,.
\end{aligned} \tag{11.187}$$

where

$$Z_1 := \frac{\boldsymbol{\sigma}_1\cdot(\mathbf{W}(T)-\mathbf{W}(t))}{\sigma_1\sqrt{\tau}}\,,\quad Z_2 := \frac{\boldsymbol{\sigma}_2\cdot(\mathbf{W}(T)-\mathbf{W}(t))}{\sigma_2\sqrt{\tau}}\,, \tag{11.188}$$

$\tau := T-t$. The correlation (and covariance) between these two standard normals equals ρ. This follows from the fact that $W_i(T) - W_i(t), i = 1,2$, are i.i.d. $Norm(0,\tau)$:

$$\begin{aligned}
\mathrm{Cov}(Z_1,Z_2) &= \frac{1}{\sigma_1\sigma_2\tau}\,\mathrm{Cov}\left(\boldsymbol{\sigma}_1\cdot(\mathbf{W}(T)-\mathbf{W}(t)), \boldsymbol{\sigma}_2\cdot(\mathbf{W}(T)-\mathbf{W}(t))\right)\\
&= \frac{1}{\sigma_1\sigma_2\tau}\sum_{i=1}^{2}\sum_{j=1}^{2}\sigma_{1i}\sigma_{2j}\,\mathrm{Cov}\left(W_i(T)-W_i(t), W_j(T)-W_j(t)\right)\\
&= \frac{1}{\sigma_1\sigma_2\tau}\sum_{i=1}^{2}(\sigma_{1i}\sigma_{2i})\tau = \frac{\boldsymbol{\sigma}_1\cdot\boldsymbol{\sigma}_2}{\sigma_1\sigma_2} = \rho\,.
\end{aligned}$$

Hence, by (11.187), the matrix of correlations, ρ_{ij}, of the component processes is the 2×2 correlation matrix given by $\rho_{11} = \rho_{22} = 1$, $\rho_{12} = \rho_{21} = \rho$, i.e.,

$$\rho_{12} := \text{Corr}\,(X_1(T), X_2(T)) = \frac{\text{Cov}((X_1(T), X_2(T))}{\sqrt{\text{Var}(X_1(T))\,\text{Var}(X_2(T))}} = \text{Cov}\,(Z_1, Z_2) = \rho\,.$$

The covariance matrix of $\mathbf{X}(T)$ is given by $\text{Cov}(X_i(T), X_j(T)) = C_{ij}\tau = (\boldsymbol{\sigma}_i \cdot \boldsymbol{\sigma}_j)\tau = \rho_{ij}\sigma_i\sigma_j\tau$; $i, j = 1, 2$. The time-scaled solution vector $\frac{1}{\sqrt{\tau}}\mathbf{X}(T)$ is a bivariate normal,

$$\left[\frac{X_1(T)}{\sqrt{\tau}}, \frac{X_2(T)}{\sqrt{\tau}}\right]^\top \sim Norm_2\left(\left[\frac{x_1 + \mu_1\tau}{\sqrt{\tau}}, \frac{x_2 + \mu_2\tau}{\sqrt{\tau}}\right]^\top, \mathbf{C}\right).$$

The time-homogeneous transition CDF for the vector process is obtained by computing a joint conditional probability while using (11.186), or (11.187), and the fact that $\mathbf{W}(T) - \mathbf{W}(t)$ is independent of $\mathbf{X}(t)$, i.e., the pair Z_1, Z_2 is independent of the pair $X_1(t), X_2(t)$:

$$\begin{aligned}
P(\tau; x_1, x_2, y_1, y_2) &:= \mathbb{P}\left(X_1(T) \leqslant y_1, X_2(T) \leqslant y_2 \mid X_1(t) = x_1, X_2(t) = x_2\right) \\
&= \mathbb{P}\left(x_1 + \mu_1\tau + \sigma_1\sqrt{\tau}Z_1 \leqslant y_1, x_2 + \mu_2\tau + \sigma_2\sqrt{\tau}Z_2 \leqslant y_2\right) \\
&= \mathbb{P}\left(Z_1 \leqslant \frac{y_1 - x_1 - \mu_1\tau}{\sigma_1\sqrt{\tau}}, Z_2 \leqslant \frac{y_2 - x_2 - \mu_2\tau}{\sigma_2\sqrt{\tau}}\right) \\
&= \mathcal{N}_2\left(\frac{y_1 - x_1 - \mu_1\tau}{\sigma_1\sqrt{\tau}}, \frac{y_2 - x_2 - \mu_2\tau}{\sigma_2\sqrt{\tau}}; \rho\right). \qquad (11.189)
\end{aligned}$$

This is a bivariate normal CDF and differentiating (using the chain rule) gives the transition PDF as a bivariate normal density,

$$\begin{aligned}
p(\tau; x_1, x_2, y_1, y_2) &\equiv \frac{\partial^2}{\partial y_1 \partial y_2} P(\tau; x_1, x_2, y_1, y_2) \\
&= \frac{1}{\sigma_1\sigma_2\tau} n_2\left(\frac{y_1 - x_1 - \mu_1\tau}{\sigma_1\sqrt{\tau}}, \frac{y_2 - x_2 - \mu_2\tau}{\sigma_2\sqrt{\tau}}; \rho\right) \\
&= \frac{1}{2\pi\tau\sigma_1\sigma_2\sqrt{1-\rho^2}} \exp\left(-\frac{z_1^2 + z_2^2 - 2\rho z_1 z_2}{2(1-\rho^2)}\right), \qquad (11.190)
\end{aligned}$$

where $z_1 = \frac{y_1 - x_1 - \mu_1\tau}{\sigma_1\sqrt{\tau}}$, $z_2 = \frac{y_2 - x_2 - \mu_2\tau}{\sigma_2\sqrt{\tau}}$. According to Theorem 11.19, both transition CDF and PDF, P and p, solve the time-homogeneous PDE in (11.183) in the variables τ, x_1, x_2, for fixed arbitrary real values of y_1, y_2. Using the above explicit constant expressions for $C_{11} = \sigma_1^2, C_{22} = \sigma_2^2, C_{12} = C_{21} = \rho\sigma_1\sigma_2$, and the constant drift coefficients μ_1 and μ_2, p and P solve the backward PDE, i.e.,

$$\frac{\partial p}{\partial \tau} = \frac{1}{2}\sigma_1^2\frac{\partial^2 p}{\partial x_1^2} + \frac{1}{2}\sigma_2^2\frac{\partial^2 p}{\partial x_2^2} + \rho\sigma_1\sigma_2\frac{\partial^2 p}{\partial x_1 \partial x_2} + \mu_1\frac{\partial p}{\partial x_1} + \mu_2\frac{\partial p}{\partial x_2}$$

and similarly for P. We leave it as an exercise for the reader to show that the transition CDF in (11.190) has limit $P(0+; x_1, x_2, y_1, y_2) = \mathbb{I}_{\{y_1 \geqslant x_1, y_2 \geqslant x_2\}}$ for all $\mathbf{x} \neq \mathbf{y}$. The Dirac delta function initial condition for the transition PDF then follows by formally differentiating the step function to obtain $p(0+; x_1, x_2, y_1, y_2) = \delta(y_1 - x_1)\delta(y_2 - x_2)$. The reader can verify by direct differentiation that the transition PDF in (11.190) satisfies the above PDE and is hence the fundamental solution. As a density in the variables y_1, y_2, the transition PDF should also integrate to unity over $\mathbb{R}^2$. That is, the event $\{(X_1(T), X_2(T)) \in \mathbb{R}^2\}$, conditional on $X_1(t) = x_1, X_2(t) = x_2$, must have unit probability. This is directly verified as follows:

$$\begin{aligned}
&\mathbb{P}\left(X_1(T) < \infty, X_2(T) < \infty \mid X_1(t) = x_1, X_2(t) = x_2\right) \\
&= \lim_{y_1\to\infty, y_2\to\infty} P(\tau; x_1, x_2, y_1, y_2) \\
&= \lim_{y_1\to\infty, y_2\to\infty} \mathcal{N}_2\left(\frac{y_1 - x_1 - \mu_1\tau}{\sigma_1\sqrt{\tau}}, \frac{y_2 - x_2 - \mu_2\tau}{\sigma_2\sqrt{\tau}}; \rho\right) \\
&= \mathcal{N}_2\left(\infty, \infty; \rho\right) = 1\,.
\end{aligned}$$

From (11.169), this implies that the PDF integrates to unity for all $x_1, x_2 \in \mathbb{R}$,

$$\int_{-\infty}^{\infty}\int_{-\infty}^{\infty} p(\tau; x_1, x_2, y_1, y_2)\, \mathrm{d}y_1\, \mathrm{d}y_2 = 1\,.$$

Alternatively, this is easily shown by directly integrating the bivariate density in (11.190). Note that in the special case when $\rho = 0$, the two processes are independent (uncorrelated) drifted and scaled BM. The joint transition PDF and CDF are simply products of the one-dimensional PDFs and CDFs of the component processes. This is consistent with the fact that the above PDE is separable in the variables x_1 and x_2 and hence admits a solution as a product of individual functions of x_1 and x_2.

Let us now consider a *multidimensional GBM process.* This is an important example that arises in Chapter 13 where we consider derivative pricing within a standard economic model containing multiple stocks whose price processes are correlated geometric Brownian motions. In particular, consider $n \geqslant 1$ strictly positive stock price processes $\mathbf{S}(t) := (S_1(t), \ldots, S_n(t))$, $t \geqslant 0$, that are driven by a standard $d \geqslant 1$ dimensional BM, $\mathbf{W}(t) = (W_1(t), \ldots, W_d(t))$:

$$\begin{aligned}
\mathrm{d}S_i(t) &= S_i(t)\left[\mu_i\, \mathrm{d}t + \sum_{j=1}^{d} \sigma_{ij}\, \mathrm{d}W_j(t)\right] \\
&\equiv S_i(t)\left[\mu_i\, \mathrm{d}t + \boldsymbol{\sigma}_i \cdot \mathrm{d}\mathbf{W}(t)\right], \quad i = 1, \ldots, n\,,
\end{aligned} \tag{11.191}$$

where the log-drifts μ_i and log-volatilities σ_{ij} are assumed to be constant parameters. It is important to stress the distinction that these are *log-coefficients*, although by standard convention we are still using similar symbols for the drift and volatility coefficient functions! To be precise, by identifying the SDE in (11.191) with that in (11.164) (where $\mathbf{X}(t) \to \mathbf{S}(t)$) we see that the drift and volatility *coefficient functions* for the i-th stock price in (11.191) are state-dependent (time-independent) linear functions of the i-th variable

$$\mu_i(t, \mathbf{x}) \equiv \mu_i(\mathbf{x}) = \mu_i x_i \quad \text{and} \quad \sigma_{ij}(t, \mathbf{x}) \equiv \sigma_{ij}(\mathbf{x}) = \sigma_{ij} x_i \tag{11.192}$$

where on the right of each equality are the log-drift and log-volatility parameters μ_i and σ_{ij}. [We note that the symbols μ_i and σ_{ij} are constant parameters when denoted without arguments and they are the drift and volatility coefficient functions when denoted with arguments.] So the above SDEs are time homogeneous with linear functions $\mu_i(t, \mathbf{S}(t)) \equiv \mu_i(\mathbf{S}(t)) = \mu_i S_i(t)$ and $\sigma_{ij}(t, \mathbf{S}(t)) \equiv \sigma_{ij}(\mathbf{S}(t)) = S_i(t)\sigma_{ij}$.

The *log-volatility coefficient matrix* $\boldsymbol{\sigma} = [\sigma_{ij}]_{i=1,\ldots,n;j=1,\ldots,d}$ is an $n \times d$ constant matrix with the i-th row being the $1 \times d$ volatility vector $\boldsymbol{\sigma}_i = [\sigma_{i1} \ldots, \sigma_{id}]$ for the i-th stock price process. The system of SDEs in (11.191) has matrix-vector form:

$$\begin{bmatrix} \frac{\mathrm{d}S_1(t)}{S_1(t)} \\ \vdots \\ \frac{\mathrm{d}S_n(t)}{S_n(t)} \end{bmatrix} = \begin{bmatrix} \mu_1 \\ \vdots \\ \mu_n \end{bmatrix} \mathrm{d}t + \begin{bmatrix} \sigma_{11} & \cdots & \sigma_{1d} \\ \vdots & & \vdots \\ \sigma_{n1} & \cdots & \sigma_{nd} \end{bmatrix} \begin{bmatrix} \mathrm{d}W_1(t) \\ \vdots \\ \mathrm{d}W_d(t) \end{bmatrix}. \tag{11.193}$$

As shown just below, the *log-diffusion matrix* $\mathbf{C} = \boldsymbol{\sigma}\boldsymbol{\sigma}^\top$ is proportional to the $n \times n$ matrix of covariances among the log-returns of the stocks. We assume that $\mathbf{C}$ is nonsingular. As usual, we define the $n \times n$ matrix of correlations, $\boldsymbol{\rho} := [\rho_{ij}]_{i,j=1,\ldots,n}$ where $C_{ij} = \boldsymbol{\sigma}_i \cdot \boldsymbol{\sigma}_j = \rho_{ij}\sigma_i\sigma_j$, where the i-th volatility vector has magnitude denoted by $\sigma_i > 0$,

$$C_{ii} = \sigma_i^2 \equiv \|\boldsymbol{\sigma}_i\|^2 = \sigma_{i1}^2 + \ldots + \sigma_{id}^2 .$$

The system in (11.191) is readily solved by considering the log-prices defined by $X_i(t) := \ln S_i(t)$. In fact, we have already solved this problem. See (11.148)–(11.150), where each SDE in (11.191) is of the form in (11.148) with solution of the form in (11.149). For the sake of clarity, we repeat the same steps here by using Itô's formula where (upon substituting the expression in (11.191)):

$$\begin{aligned} \mathrm{d}X_i(t) = \mathrm{d}\ln S_i(t) = \frac{\mathrm{d}S_i(t)}{S_i(t)} - \frac{1}{2}\left(\frac{\mathrm{d}S_i(t)}{S_i(t)}\right)^2 &= \left(\mu_i - \frac{1}{2}\|\boldsymbol{\sigma}_i\|^2\right)\mathrm{d}t + \boldsymbol{\sigma}_i \cdot \mathrm{d}\mathbf{W}(t) \\ &= \left(\mu_i - \frac{1}{2}\sigma_i^2\right)\mathrm{d}t + \boldsymbol{\sigma}_i \cdot \mathrm{d}\mathbf{W}(t) \end{aligned}$$

with initial condition $X_i(0) = \ln S_i(0)$, where $S_i(0), i = 1, \ldots, n$, are the initial stock prices. Integrating and exponentiating gives the stock prices $S_i(t) = \mathrm{e}^{X_i(t)}$ for all $t \geqslant 0$:

$$S_i(t) = S_i(0)\,\mathrm{e}^{(\mu_i - \frac{1}{2}\sigma_i^2)t + \boldsymbol{\sigma}_i \cdot \mathbf{W}(t)} = S_i(0)\,\mathrm{e}^{\mu_i t}\mathcal{E}_t(\boldsymbol{\sigma}_i \cdot \mathbf{W}) . \tag{11.194}$$

It is easy to verify by computing the stochastic differential of this expression, upon directly applying Itô's formula, that each $S_i(t)$ solves (11.191). The solution in (11.194) is in fact the unique strong solution subject to the initial price vector $[S_1(0), \ldots, S_n(0)]^\top$.

The second expression in (11.194) involves an exponential $\mathbb{P}$-martingale,

$$\mathcal{E}_t(\boldsymbol{\sigma}_i \cdot \mathbf{W}) = \exp\left[-\frac{1}{2}\sigma_i^2 t + \boldsymbol{\sigma}_i \cdot \mathbf{W}(t)\right] = \exp\left[-\frac{1}{2}\sigma_i^2 t + \sum_{j=1}^{d}\sigma_{ij}W_j(t)\right].$$

To see that this is a $\mathbb{P}$-martingale with expectation one, note that $\boldsymbol{\sigma}_i \cdot \mathbf{W}(t)$ is normal with mean $\mathrm{E}[\boldsymbol{\sigma}_i \cdot \mathbf{W}(t)] = \sum_{j=1}^{d}\sigma_{ij}\mathrm{E}[W_j(t)] = 0$ and variance

$$\mathrm{Var}\left(\boldsymbol{\sigma}_i \cdot \mathbf{W}(t)\right) = \mathrm{Var}\left(\sum_{j=1}^{d}\sigma_{ij}W_j(t)\right) = \sum_{j=1}^{d}\sigma_{ij}^2\,\mathrm{Var}(W_j(t)) = \|\boldsymbol{\sigma}_i\|^2 t = \sigma_i^2 t .$$

Here we used the fact that all $W_j(t)$ BMs are i.i.d. *Norm*$(0, t)$. Hence, by the expression for the m.g.f. of a normal random variable, $\mathrm{E}[\mathrm{e}^{\alpha\boldsymbol{\sigma}_i \cdot \mathbf{W}(t)}] = \mathrm{e}^{\frac{1}{2}\alpha^2\sigma_i^2 t}$ for any α, i.e., $\mathrm{E}[\mathrm{e}^{\boldsymbol{\sigma}_i \cdot \mathbf{W}(t)}] = \mathrm{e}^{\frac{1}{2}\sigma_i^2 t}$ and $\mathrm{E}[\mathcal{E}_t(\boldsymbol{\sigma}_i \cdot \mathbf{W})] = 1$, for all $t \geqslant 0$. So the mean of the price in (11.194) is

$$\mathrm{E}\left[S_i(t)\right] = S_i(0)\mathrm{e}^{\mu_i t}\mathrm{E}\left[\mathcal{E}_t(\boldsymbol{\sigma}_i \cdot \mathbf{W})\right] = S_i(0)\,\mathrm{e}^{\mu_i t} . \tag{11.195}$$

As in the one-dimensional GBM stock model, the log-normal drift parameter μ_i is therefore the (physical) growth rate of the i-th price process in the (physical) measure $\mathbb{P}$.

From the strong solution in (11.194) we have

$$S_i(T) = S_i(t)\,\mathrm{e}^{(\mu_i - \frac{1}{2}\sigma_i^2)(T-t) + \boldsymbol{\sigma}_i \cdot (\mathbf{W}(T) - \mathbf{W}(t))} . \tag{11.196}$$

It is convenient to define the log-return random variables over a time interval $\tau := T - t$,

$$X_i := \ln\frac{S_i(T)}{S_i(t)} = \alpha_i + \sigma_i\sqrt{\tau}Z_i, \tag{11.197}$$

$\alpha_i \equiv (\mu_i - \frac{1}{2}\sigma_i^2)\tau$, where

$$Z_i := \frac{\boldsymbol{\sigma}_i \cdot (\mathbf{W}(T) - \mathbf{W}(t))}{\sigma_i\sqrt{\tau}} = \frac{1}{\sigma_i\sqrt{\tau}}\sum_{j=1}^{d} \sigma_{ij}(W_j(T) - W_j(t)),$$

$i = 1, \ldots, n$. The Z_i's are normal random variables since they are linear combinations of Brownian increments. The vector $\mathbf{Z} = [Z_1, \ldots, Z_n]^\top$ has multivariate normal distribution:

$$[Z_1, \ldots, Z_n]^\top \sim Norm_n(\mathbf{0}, \boldsymbol{\rho}) \,. \tag{11.198}$$

To verify this, the covariances are computed using the same steps as in the calculation of the covariance of Z_1 and Z_2 in (11.188):

$$\begin{aligned} \operatorname{Cov}(Z_i, Z_j) &= \frac{1}{\sigma_i\sqrt{\tau}}\frac{1}{\sigma_j\sqrt{\tau}} \operatorname{Cov}(\boldsymbol{\sigma}_i \cdot (\mathbf{W}(T) - \mathbf{W}(t)), \boldsymbol{\sigma}_j \cdot (\mathbf{W}(T) - \mathbf{W}(t))) \\ &= \frac{1}{\sigma_i\sqrt{\tau}}\frac{1}{\sigma_j\sqrt{\tau}} \boldsymbol{\sigma}_i \cdot \boldsymbol{\sigma}_j \tau = \frac{\boldsymbol{\sigma}_i \cdot \boldsymbol{\sigma}_j}{\sigma_i\sigma_j} = \rho_{ij} \,. \end{aligned} \tag{11.199}$$

Hence, $\operatorname{Cov}(X_i, X_j) = \tau\sigma_i\sigma_j \operatorname{Cov}(Z_i, Z_j) = \tau\sigma_i\sigma_j\rho_{ij} = \tau C_{ij}$. The log-returns are therefore jointly normally distributed:

$$[X_1, \ldots, X_n]^\top \sim Norm_n\left([\alpha_1, \ldots, \alpha_n]^\top, \tau\mathbf{C}\right) \,. \tag{11.200}$$

In particular, the matrix $\boldsymbol{\rho}$ is in fact the *matrix of correlations of the stock price log-returns*:

$$\operatorname{Corr}(X_i, X_j) = \frac{\operatorname{Cov}(X_i, X_j)}{\sqrt{\operatorname{Var}(X_i)\operatorname{Var}(X_j)}} = \frac{\tau C_{ij}}{\sqrt{(\sigma_i^2\tau)(\sigma_j^2\tau)}} = \frac{C_{ij}}{\sigma_i\sigma_j} = \rho_{ij} \,. \tag{11.201}$$

The above GBM process is time homogeneous. Let $\mathbf{x} = (x_1, \ldots, x_n)$ and $\mathbf{y} = (y_1, \ldots, y_n)$ be strictly positive vectors in $\mathbb{R}_+^n$. The (joint) transition CDF of the stock price process $\mathbf{S}(t)$ is given by the conditional probability below which is calculated by using independence among all log-returns $\{X_i \equiv \frac{S_i(T)}{S_i(t)}\}_{i=1,\ldots,n}$ and time-t stock prices $\{S_i(t)\}_{i=1,\ldots,n}$:

$$\begin{aligned} P(\tau; \mathbf{x}, \mathbf{y}) &:= \mathbb{P}\left(S_1(T) \leqslant y_1, \ldots, S_n(T) \leqslant y_n \mid S_1(t) = x_1, \ldots, S_n(t) = x_n\right) \\ &= \mathbb{P}\left(\ln\frac{S_1(T)}{S_1(t)} \leqslant \ln\frac{y_1}{x_1}, \ldots, \ln\frac{S_n(T)}{S_n(t)} \leqslant \ln\frac{y_n}{x_n} \,\middle|\, S_1(t) = x_1, \ldots, S_n(t) = x_n\right) \\ &= \mathbb{P}\left(X_1 \leqslant \ln\frac{y_1}{x_1}, \ldots, X_n \leqslant \ln\frac{y_n}{x_n}\right) \\ &= \mathbb{P}\left(Z_1 \leqslant a_1, \ldots, Z_n \leqslant a_n\right) \\ &= \mathcal{N}_n\left(a_1 \ldots, a_n; \boldsymbol{\rho}\right) \end{aligned} \tag{11.202}$$

where $a_i \equiv \frac{\ln\frac{y_i}{x_i} - \alpha_i}{\sigma_i\sqrt{\tau}} = \frac{\ln\frac{y_i}{x_i} - (\mu_i - \sigma_i^2/2)\tau}{\sigma_i\sqrt{\tau}}$. The function $\mathcal{N}_n(a_1, \ldots, a_n; \boldsymbol{\rho})$ is the n-variate standard normal CDF of $\mathbf{Z}$ with given correlation matrix $\boldsymbol{\rho}$:

$$\mathcal{N}_n(a_1, \ldots, a_n; \boldsymbol{\rho}) = \int_{-\infty}^{a_n} \cdots \int_{-\infty}^{a_1} n_n(z_1, \ldots, z_n; \boldsymbol{\rho})\, \mathrm{d}z_1 \ldots \mathrm{d}z_n \,.$$

The standard normal PDF of $\mathbf{Z}$, $n_n(z_1, \ldots, z_n; \boldsymbol{\rho}) = \frac{\partial^n}{\partial z_1 \ldots \partial z_n}\mathcal{N}_n(z_1 \ldots, z_n; \boldsymbol{\rho})$, is given by the n-variate Gaussian density

$$n_n(z_1, \ldots, z_n; \boldsymbol{\rho}) = \frac{1}{\sqrt{(2\pi)^n \det\boldsymbol{\rho}}} \exp\left(-\frac{1}{2}\mathbf{z} \cdot \boldsymbol{\rho}^{-1} \cdot \mathbf{z}^\top\right), \tag{11.203}$$

$\mathbf{z} = [z_1, \ldots, z_n] \in \mathbb{R}^n$. Differentiating according to (11.168), and applying the chain rule, gives the (joint) transition PDF for the time-homogeneous GBM stock price process:

$$\begin{aligned} p(\tau; \mathbf{x}, \mathbf{y}) &= \frac{\partial^n}{\partial y_1 \ldots \partial y_n} P(\tau; \mathbf{x}, \mathbf{y}) = \left(\prod_{i=1}^{n} \frac{\partial a_i}{\partial y_i} \right) \frac{\partial^n}{\partial a_1 \ldots \partial a_n} \mathcal{N}_n (a_1 \ldots, a_n; \boldsymbol{\rho}) \\ &= \left(\prod_{i=1}^{n} \frac{1}{y_i \sigma_i \sqrt{\tau}} \right) n_n (a_1 \ldots, a_n; \boldsymbol{\rho}) \\ &= \frac{1}{y_1 \cdots y_n \sigma_1 \cdots \sigma_n \sqrt{(2\pi\tau)^n \det \boldsymbol{\rho}}} \exp\left(-\frac{1}{2} \mathbf{a} \cdot \boldsymbol{\rho}^{-1} \cdot \mathbf{a}^\top \right), \end{aligned} \tag{11.204}$$

$\mathbf{a} = [a_1, \ldots, a_n]$ and a_i's defined above. This is of the form of a *multivariate log-normal density*. Note that this density can also be written in terms of the covariance matrix $\mathbf{C} = \mathbf{D}\boldsymbol{\rho}\mathbf{D}$, $\mathbf{D} = \text{diag}(\sigma_1, \ldots, \sigma_n)$, $\boldsymbol{\rho}^{-1} = \mathbf{D}\mathbf{C}^{-1}\mathbf{D}$.

By (11.192), the diffusion matrix function has elements

$$C_{ij}(\mathbf{x}) = \sum_{k=1}^{d} \sigma_{ik}(\mathbf{x})\, \sigma_{jk}(\mathbf{x}) = \sum_{k=1}^{d} x_i \sigma_{ik}\, x_j \sigma_{jk} = x_i x_j \sum_{k=1}^{d} \sigma_{ik} \sigma_{jk} = x_i x_j C_{ij}$$

with constants $C_{ij} = \boldsymbol{\sigma}_i \cdot \boldsymbol{\sigma}_j = \rho_{ij} \sigma_i \sigma_j$. Hence, the time-homogeneous PDE in (11.183) takes the equivalent form:

$$\begin{aligned} \frac{\partial f}{\partial \tau} &= \frac{1}{2} \sum_{i=1}^{n} \sum_{j=1}^{n} \rho_{ij} \sigma_i \sigma_j x_i x_j \frac{\partial^2 f}{\partial x_i \partial x_j} + \sum_{i=1}^{n} \mu_i x_i \frac{\partial f}{\partial x_i} \\ &= \frac{1}{2} \sum_{i=1}^{n} \sigma_i^2 x_i^2 \frac{\partial^2 f}{\partial x_i^2} + \sum_{j=1}^{n} \sum_{i<j} \rho_{ij} \sigma_i \sigma_j x_i x_j \frac{\partial^2 f}{\partial x_i \partial x_j} + \sum_{i=1}^{n} \mu_i x_i \frac{\partial f}{\partial x_i}. \end{aligned} \tag{11.205}$$

The Feynman–Kac Theorem 11.19 assures us that the transition CDF in (11.202) and PDF in (11.204) both solve the PDE (11.205) in the variables $\tau > 0, \mathbf{x} \in \mathbb{R}_+^n$, for fixed $\mathbf{y} \in \mathbb{R}_+^n$. The initial condition $P(0; \mathbf{x}, \mathbf{y}) = \mathbb{I}_{\{\mathbf{x} \leqslant \mathbf{y}\}} \equiv \mathbb{I}_{\{x_1 \leqslant y_1, \ldots, x_n \leqslant y_n\}}$ follows from the basic limit properties of the multivariate normal CDF. We leave it to the reader to verify. Then, by multiple differentiation of the step functions, the initial condition $p(0+; \mathbf{x}, \mathbf{y}) = \delta(\mathbf{x} - \mathbf{y})$ is obtained for the transition PDF.

A common case is when $n = d = 2$. The above formulation simplifies since we have only one correlation coefficient ρ for the log-returns of stocks 1 and 2, where $\rho_{12} = \rho_{21} \equiv \rho$, $\rho_{11} = \rho_{22} = 1$. The log-diffusion matrix of covariances has elements $C_{11} = \sigma_1^2, C_{22} = \sigma_2^2, C_{12} = C_{21} = \rho\sigma_1\sigma_2$. The log-volatility vectors are $\boldsymbol{\sigma}_1 = [\sigma_{11}, \sigma_{12}] = [\sigma_1, 0]$ and $\boldsymbol{\sigma}_2 = [\sigma_{21}, \sigma_{22}] = [\rho\sigma_2, \sigma_2\sqrt{1-\rho^2}]$. In this case, the system of SDEs in (11.191) simplifies for two stock prices driven by two BMs:

$$\begin{aligned} \mathrm{d}S_1(t) &= S_1(t)[\mu_1 \,\mathrm{d}t + \sigma_1 \,\mathrm{d}W_1(t)] \\ \mathrm{d}S_2(t) &= S_2(t)[\mu_2 \,\mathrm{d}t + \sigma_2 \rho \,\mathrm{d}W_1(t) + \sigma_2 \sqrt{1-\rho^2} \,\mathrm{d}W_2(t)]. \end{aligned} \tag{11.206}$$

The unique solution is

$$\begin{aligned} S_1(t) &= S_1(0)\, \mathrm{e}^{(\mu_1 - \frac{1}{2}\sigma_1^2)t + \sigma_1 W_1(t)}, \\ S_2(t) &= S_2(0)\, \mathrm{e}^{(\mu_2 - \frac{1}{2}\sigma_2^2)t + \sigma_2(\rho W_1(t) + \sqrt{1-\rho^2} W_2(t))}. \end{aligned}$$

The transition CDF and PDF obtain as a special case of (11.202) and (11.204) for $n = 2$:

$$P(\tau; x_1, x_2, y_1, y_2) = \mathcal{N}_2\left(\frac{\ln\frac{y_1}{x_1} - (\mu_1 - \frac{1}{2}\sigma_1^2)\tau}{\sigma_1\sqrt{\tau}}, \frac{\ln\frac{y_2}{x_2} - (\mu_2 - \frac{1}{2}\sigma_2^2)\tau}{\sigma_2\sqrt{\tau}}; \rho\right) \tag{11.207}$$

and

$$p(\tau; x_1, x_2, y_1, y_2) = \frac{1}{y_1 y_2 \sigma_1 \sigma_2 \tau} n_2\left(\frac{\ln\frac{y_1}{x_1} - (\mu_1 - \frac{1}{2}\sigma_1^2)\tau}{\sigma_1\sqrt{\tau}}, \frac{\ln\frac{y_2}{x_2} - (\mu_2 - \frac{1}{2}\sigma_2^2)\tau}{\sigma_2\sqrt{\tau}}; \rho\right). \tag{11.208}$$

By the Feynman–Kac Theorem 11.19, these functions solve the time-homogeneous PDE in (11.205) for $n = 2$:

$$\frac{\partial f}{\partial \tau} = \frac{1}{2}\sigma_1^2 x_1^2 \frac{\partial^2 f}{\partial x_1^2} + \frac{1}{2}\sigma_2^2 x_2^2 \frac{\partial^2 f}{\partial x_2^2} + \rho\sigma_1\sigma_2 x_1 x_2 \frac{\partial^2 f}{\partial x_1 \partial x_2} + \mu_1 x_1 \frac{\partial f}{\partial x_1} + \mu_2 x_2 \frac{\partial f}{\partial x_2}. \tag{11.209}$$

The reader can verify that the transition CDF has the limiting form $P(0+; x_1, x_2, y_1, y_2) = \mathbb{I}_{\{y_1 \geqslant x_1, y_2 \geqslant x_2\}}$, for all $\mathbf{x} \neq \mathbf{y}$, and $p(0+; x_1, x_2, y_1, y_2) = \delta(y_1 - x_1)\delta(y_2 - x_2)$.

11.10.3 Girsanov's Theorem for Multidimensional BM

We recall Girsanov's Theorem 11.13 where the measure change was constructed in terms of a Radon–Nikodym process which has the form of an exponential martingale involving a single standard BM in the original measure $\mathbb{P}$. Based on our knowledge of multidimensional BM and Itô integrals on multidimensional BM we can now consider the multidimensional version of Girsanov's Theorem. The main ingredients are as in Theorem 11.13 where the single BM is now a multidimensional BM. As usual, we fix some filtered probability space $(\Omega, \mathcal{F}, \mathbb{P}, \mathbb{F})$, where $\mathbb{F} = \{\mathcal{F}_t\}_{t\geqslant 0}$ is any filtration for standard Brownian motion.

Theorem 11.21 (Girsanov's Theorem for Multidimensional BM). *Fix a time $T > 0$ and let $\mathbf{W}(t) = (W_1(t), \ldots, W_d(t)), 0 \leqslant t \leqslant T$, be a standard d-dimensional $\mathbb{P}$-BM with respect to a filtration $\mathbb{F} = \{\mathcal{F}_t\}_{0\leqslant t\leqslant T}$. Assume the vector process $\gamma(t) = (\gamma_1(t), \ldots, \gamma_d(t)), 0 \leqslant t \leqslant T$, is adapted to $\mathbb{F}$ such that*

$$\mathrm{E}\left[\exp\left(\frac{1}{2}\int_0^T \|\gamma(s)\|^2\,\mathrm{d}s\right)\right] < \infty. \tag{11.210}$$

Define

$$\varrho_t := \exp\left(-\frac{1}{2}\int_0^t \|\gamma(s)\|^2\,\mathrm{d}s + \int_0^t \gamma(s)\cdot\mathrm{d}\mathbf{W}(s)\right), 0 \leqslant t \leqslant T, \tag{11.211}$$

and the probability measure $\widehat{\mathbb{P}} \equiv \widehat{\mathbb{P}}^{(\varrho)}$ by the Radon–Nikodym derivative $\frac{\mathrm{d}\widehat{\mathbb{P}}}{\mathrm{d}\mathbb{P}} = \left(\frac{\mathrm{d}\widehat{\mathbb{P}}}{\mathrm{d}\mathbb{P}}\right)_T \equiv \varrho_T$. Then, the process $\widehat{\mathbf{W}}(t) = (\widehat{W}_1(t), \ldots, \widehat{W}_d(t)), 0 \leqslant t \leqslant T$, defined by

$$\widehat{\mathbf{W}}(t) := \mathbf{W}(t) - \int_0^t \gamma(s)\,\mathrm{d}s \tag{11.212}$$

is a standard d-dimensional $\widehat{\mathbb{P}}$-BM w.r.t. filtration $\mathbb{F}$.

We don't provide the proof of this result here. We leave it as an exercise where one can apply similar steps as in (i)–(iii) in the proof of Theorem 11.13. In the multidimensional case we have d-dimensional BM and Lévy's characterization in Theorem 11.16 can be applied.

The same remarks as were stated for Theorem 11.13 also apply to Theorem 11.21, where the adapted process is now the vector $\boldsymbol{\gamma}$ rather than the scalar γ. Of course, this multidimensional version generalizes Theorem 11.13, which obtains in the simplest case with $d = 1$. The Radon–Nikodym derivative process that defines the change of measure $\mathbb{P} \to \widehat{\mathbb{P}}$, such that the new d-dimensional BM, $\widehat{\mathbf{W}}$, is a $\widehat{\mathbb{P}}$-BM, is given by the exponential $\mathbb{P}$-martingale in (11.211). It can be proven that the Novikov condition in (11.210) guarantees that the process $\{\varrho_t\}_{0\leqslant t\leqslant T}$ is indeed a proper Radon–Nikodym derivative process, i.e., it is a $\mathbb{P}$-martingale with constant unit expectation, $\mathrm{E}[\varrho_t] = 1$ for all $t \in [0, T]$. By Itô's formula, the stochastic exponential in (11.211) is equivalent to the stochastic differential (using $\varrho_0 = 1$):

$$\mathrm{d}\varrho_t = \varrho_t \boldsymbol{\gamma}(t) \cdot \mathrm{d}\mathbf{W}(t) \implies \varrho_t = 1 + \int_0^t \varrho_s \boldsymbol{\gamma}(s) \cdot \mathrm{d}\mathbf{W}(s)\,.$$

The Novikov condition assures us that the Itô integral satisfies the integrability condition as in (11.121) and is therefore a $\mathbb{P}$-martingale with zero expectation. By the definition in (11.150) we have $\varrho_t \equiv \varrho_t^{(\boldsymbol{\gamma})} = \mathcal{E}_t(\boldsymbol{\gamma} \cdot \mathbf{W})$. Dividing the process value at any two times $0 \leqslant s < t \leqslant T$ gives

$$\frac{\varrho_t}{\varrho_s} \equiv \frac{(\frac{\mathrm{d}\widehat{\mathbb{P}}}{\mathrm{d}\mathbb{P}})_t}{(\frac{\mathrm{d}\widehat{\mathbb{P}}}{\mathrm{d}\mathbb{P}})_s} = \frac{\mathcal{E}_t(\boldsymbol{\gamma} \cdot \mathbf{W})}{\mathcal{E}_s(\boldsymbol{\gamma} \cdot \mathbf{W})} = \exp\left(-\frac{1}{2}\int_s^t \|\boldsymbol{\gamma}(u)\|^2\, \mathrm{d}u + \int_s^t \boldsymbol{\gamma}(u) \cdot \mathrm{d}\mathbf{W}(u)\right).$$

In general, $\boldsymbol{\gamma}$ is an adapted vector process so that ϱ_t is a functional of d-dimensional BM from time 0 to t. By choosing a constant vector, $\boldsymbol{\gamma}(t) = \boldsymbol{\gamma}$, we have the simple and important special case where the Radon–Nikodym derivative process is also a GBM expressed equivalently as

$$\varrho_t \equiv \varrho_t^{(\boldsymbol{\gamma})} = \mathrm{e}^{-\frac{1}{2}\|\boldsymbol{\gamma}\|^2 t + \boldsymbol{\gamma} \cdot \mathbf{W}(t)} = \mathrm{e}^{\frac{1}{2}\|\boldsymbol{\gamma}\|^2 t + \boldsymbol{\gamma} \cdot \widehat{\mathbf{W}}(t)} \tag{11.213}$$

where $\widehat{\mathbf{W}}(t) \equiv \widehat{\mathbf{W}}^{(\boldsymbol{\gamma})}(t) := \mathbf{W}(t) - \boldsymbol{\gamma} t$, $0 \leqslant t \leqslant T$, is a $\widehat{\mathbb{P}}$-BM. We recall that the random variable $\boldsymbol{\gamma} \cdot \mathbf{W}(t) \sim Norm(0, \|\boldsymbol{\gamma}\|^2 t)$. Hence, from its m.g.f. (under measure $\mathbb{P}$) we have $\mathrm{E}[\mathrm{e}^{\boldsymbol{\gamma} \cdot \mathbf{W}(t)}] = \mathrm{e}^{\frac{1}{2}\|\boldsymbol{\gamma}\|^2 t}$, i.e., $\mathrm{E}[\varrho_t] = 1$ for all $t \geqslant 0$. This also follows trivially from the fact that the process is easily verified to be a $\mathbb{P}$-martingale and therefore must have constant expectation under measure $\mathbb{P}$, i.e., $\mathrm{E}[\varrho_t] = \mathrm{E}[\varrho_0] = \mathrm{E}[1] = 1$.

The multidimensional version of Girsanov's Theorem 11.21 has many far-reaching applications. We now give one of its applications that is particularly important for financial derivative pricing theory. Namely, we shall use Girsanov's Theorem to find a risk-neutral measure such that all stock prices $S_1(t), \ldots, S_n(t)$ in the multidimensional GBM model have a *common drift rate equal to a constant* r. Here, r is again the fixed (continuously compounded) interest rate. This problem is equivalent to applying Girsanov's Theorem to construct an equivalent martingale measure $\widehat{\mathbb{P}} \equiv \widetilde{\mathbb{P}}$, such that all discounted stock price processes defined by $\{\overline{S}_i(t) := \mathrm{e}^{-rt} S_i(t)\}_{t\geqslant 0}$, $i = 1, \ldots, n$, are $\widetilde{\mathbb{P}}$-martingales. For the case of a single stock ($n = 1$) driven by one BM ($d = 1$), we have already solved this problem in Example 11.15.

For each i-th stock, the (log-)drift μ_i and all components of the (log-)volatility vector $\boldsymbol{\sigma}_i$ in (11.191) are constants. It follows that the measure change that we will need to employ uses (11.213), i.e., with constant d-dimensional vector $\boldsymbol{\gamma} = [\gamma_1, \ldots, \gamma_d]$:

$$\varrho_t \equiv \left(\frac{\mathrm{d}\widetilde{\mathbb{P}}}{\mathrm{d}\mathbb{P}}\right)_t = \mathrm{e}^{-\frac{1}{2}\|\boldsymbol{\gamma}\|^2 t + \boldsymbol{\gamma} \cdot \mathbf{W}(t)} \tag{11.214}$$

with d-dimensional $\widetilde{\mathbb{P}}$-BM given by $\widetilde{\mathbf{W}}(t) \equiv \widehat{\mathbf{W}}^{(\gamma)}(t) := \mathbf{W}(t) - \gamma t$. In terms of stochastic differentials, $\mathrm{d}\widetilde{\mathbf{W}}(t) = \mathrm{d}\mathbf{W}(t) - \gamma\,\mathrm{d}t$. A quick method of arriving at the change of measure is to consider the SDE satisfied by the stock prices (with respect to the physical $\mathbb{P}$-BM) in (11.191) and write $\mathrm{d}\mathbf{W}(t) = \mathrm{d}\widetilde{\mathbf{W}}(t) + \gamma\,\mathrm{d}t$:

$$\begin{aligned}\mathrm{d}S_i(t) &= S_i(t)\big[\mu_i\,\mathrm{d}t + \boldsymbol{\sigma}_i\cdot(\,\mathrm{d}\widetilde{\mathbf{W}}(t) + \gamma\,\mathrm{d}t)\big]\\ &= S_i(t)\big[(\mu_i + \boldsymbol{\sigma}_i\cdot\gamma)\,\mathrm{d}t + \boldsymbol{\sigma}_i\cdot\mathrm{d}\widetilde{\mathbf{W}}(t)\big]\,,\ i = 1,\ldots,n. \end{aligned} \tag{11.215}$$

Hence, a risk-neutral measure exists if we can find a vector γ such that the log-drift coefficient equals r for every $i = 1,\ldots,n$. That is, we have

$$\mathrm{d}S_i(t) = S_i(t)\big[r\,\mathrm{d}t + \boldsymbol{\sigma}_i\cdot\mathrm{d}\widetilde{\mathbf{W}}(t)\big]\,,\ i = 1,\ldots,n, \tag{11.216}$$

which is equivalent to $\{\overline{S}_i(t)\}_{t\geqslant 0}$, $i = 1,\ldots,n$, being $\widetilde{\mathbb{P}}$-martingales, if and only if γ solves $\mu_i + \boldsymbol{\sigma}_i\cdot\gamma = r$, for each $i = 1,\ldots,n$. This is a linear system of n-equations in d unknowns $\gamma_1,\ldots,\gamma_d$. Using the components of the (log-)volatility vectors, $\boldsymbol{\sigma}_i = [\sigma_{i1},\ldots,\sigma_{id}]$, we then have an $n\times d$ linear system

$$\begin{bmatrix}\sigma_{11} & \cdots & \sigma_{1d}\\ \vdots & & \vdots\\ \sigma_{n1} & \cdots & \sigma_{nd}\end{bmatrix}\begin{bmatrix}\gamma_1\\ \vdots\\ \gamma_d\end{bmatrix} = \begin{bmatrix} r-\mu_1\\ \vdots\\ r-\mu_n\end{bmatrix}. \tag{11.217}$$

In compact notation this reads $\boldsymbol{\sigma}\gamma^\top = \mathbf{b}$, where $\boldsymbol{\sigma}$ is $n\times d$, $\gamma^\top$ is $d\times 1$, and $\mathbf{b} := r\mathbf{1} - \boldsymbol{\mu}$ is $n\times 1$, where $\boldsymbol{\mu} = [\mu_1,\ldots,\mu_n]^\top$, $\mathbf{1} = [1,\ldots,1]^\top$.

Hence, the question of the existence of a risk-neutral measure is answered quite simply by applying standard linear algebra. Generally a solution vector γ exists if and only if $\mathbf{b}$ is spanned by the d column vectors of $\boldsymbol{\sigma}$. Here we should point out that we are seeking a solution for *arbitrary physical drift vector* $\boldsymbol{\mu}\in\mathbb{R}^n$. A solution vector γ exists for any $\mathbf{b}\in\mathbb{R}^n$, and hence for any $\boldsymbol{\mu}\in\mathbb{R}^n$, if the d column vectors of $\boldsymbol{\sigma}$ span $\mathbb{R}^n$, i.e., if $\operatorname{rank}(\boldsymbol{\sigma}) = n$. In the case when $\operatorname{rank}(\boldsymbol{\sigma}) = n < d$, we have an infinite (continuum) number of solution vectors γ and each corresponds to a (different) risk-neutral measure $\widetilde{\mathbb{P}} \equiv \widetilde{\mathbb{P}}^{(\gamma)}$. This is therefore the case where the risk-neutral measure exists and is not unique. If $\operatorname{rank}(\boldsymbol{\sigma}) = n = d$, which is the case where the *number of stocks equals the number of independent BMs and the $n\times n$ matrix $\boldsymbol{\sigma}$ has an inverse $\boldsymbol{\sigma}^{-1}$*, then the risk-neutral measure $\widetilde{\mathbb{P}}$ exists and is uniquely given by $\gamma^\top = \boldsymbol{\sigma}^{-1}(r\mathbf{1} - \boldsymbol{\mu})$, i.e., $\gamma_j = \sum_{i=1}^n(\boldsymbol{\sigma}^{-1})_{ji}(r-\mu_i)$, $j = 1,\ldots,d$. Finally, if $d < n$, then $\operatorname{rank}(\boldsymbol{\sigma}) \leqslant d < n$ (the d column vectors of $\boldsymbol{\sigma}$ do not span all of $\mathbb{R}^n$) and hence a solution vector γ exists only for $\boldsymbol{\mu}$ vectors such that $\mathbf{b}$ is in the span of the column vectors of $\boldsymbol{\sigma}$. In this case, there does not exist a risk-neutral measure $\widetilde{\mathbb{P}}$ for arbitrary $\boldsymbol{\mu}$.

11.10.4 Martingale Representation Theorem for Multidimensional BM

In closing this chapter, we simply state (without proof) the multidimensional version of Theorem 11.14. This is of importance when discussing hedging financial derivatives in an economy with multiple assets that are driven by a multidimensional BM. The theorem is quite similar and extends Theorem 11.14 in a rather obvious fashion whereby the Itô integrals, and hence the Itô processes, are defined with respect to a d-dimensional BM where $d \geqslant 1$. The theorem basically states that a square-integrable $(\mathbb{P},\mathbb{F}^W)$-martingale is expressible as its initial value plus an Itô integral in the d-dimensional BM and some $\mathbb{F}^W$-adapted vector process as integrand. In the result below, we combine Theorem 11.14 and Proposition 11.15 into one Theorem for the more general case of multidimensional BM.

Theorem 11.22 (Multidimensional Brownian Martingale Representation Theorem). *Let $\mathbf{W}(t) = (W_1(t), \ldots, W_d(t))$ be a d-dimensional standard BM where $\mathbb{F}^W$ denotes its natural filtration. Assume $\{M(t)\}_{0 \leqslant t \leqslant T}$ is a square-integrable $(\mathbb{P}, \mathbb{F}^W)$-martingale. Then, there exists an $\mathbb{F}^W$-adapted d-dimensional process (a.s.) $\boldsymbol{\theta}(t) = (\theta_1(t), \ldots, \theta_d(t)), 0 \leqslant t \leqslant T$, such that*

$$M(t) = M(0) + \int_0^t \boldsymbol{\theta}(u) \cdot \mathrm{d}\mathbf{W}(u) \equiv M(0) + \sum_{j=1}^{d} \int_0^t \theta_j(u) \cdot \mathrm{d}W_j(u)\,, \tag{11.218}$$

for all $t \in [0, T]$. Moreover, let $\widehat{\mathbb{P}}$ be a measure constructed using Girsanov's Theorem 11.21 with the assumption that the d-dimensional process $\{\gamma(t)\}_{0 \leqslant t \leqslant T}$ is $\mathbb{F}^W$-adapted. If the process $\{\widehat{M}(t)\}_{0 \leqslant t \leqslant T}$ is a square-integrable $(\widehat{\mathbb{P}}, \mathbb{F}^W)$-martingale, then there exists an adapted d-dimensional process $\{\widehat{\boldsymbol{\theta}}(t) = (\widehat{\theta}_1(t), \ldots, \widehat{\theta}_d(t))\}_{0 \leqslant t \leqslant T}$, such that (a.s.)

$$\widehat{M}(t) = \widehat{M}(0) + \int_0^t \widehat{\boldsymbol{\theta}}(u) \cdot \mathrm{d}\widehat{\mathbf{W}}(u)\,. \tag{11.219}$$

Again we stress that the martingale having this representation is (a.s.) continuous in time (i.e., the process has no jumps) since it is an Itô process.

11.11 Exercises

Exercise 11.1. In each case show whether or not the stochastic integral is well-defined. Note: *you do not need to compute the values of the integrals.*

$$\text{(a)} \int_0^1 (W^2(t) + W(t) + 1)\,\mathrm{d}W(t); \quad \text{(b)} \int_0^1 |W(t)|^{-1/2}\,\mathrm{d}W(t);$$

$$\text{(c)} \int_0^1 W(\tfrac{1}{t})\,\mathrm{d}W(t); \quad \text{(d)} \int_0^1 |tW(t)|^{-1/4}\,\mathrm{d}W(t);$$

$$\text{(e)} \int_0^1 W^a(t)\,\mathrm{d}W(t).$$

For part (e) find all values of the parameter $a \in \mathbb{R}$ for which the integral is well-defined. [Hint for singular integrands: Recall that if $t = 0$ is a singular point of a function $f(t)$, where $f(t) = \mathcal{O}(|t|^p)$ with $p < 0$, as $t \to 0$, then $\int_0^1 f(t)\,\mathrm{d}t < \infty$ iff $p > -1$.]

Exercise 11.2. In each case show whether or not the stochastic integral is well-defined. Note: *you do not need to compute the values of the integrals.*

$$\text{(a)} \int_0^1 W(2t)\,\mathrm{d}W(t); \quad \text{(b)} \int_0^1 W(\tfrac{t}{2})\,\mathrm{d}W(t);$$

$$\text{(c)} \int_1^\infty W(\tfrac{1}{t})\,\mathrm{d}W(t); \quad \text{(d)} \int_0^1 |W(t)|^{1/2}\,\mathrm{d}W(t).$$

Exercise 11.3. For any $\alpha \in (0, 1)$, define the stochastic integral

$$\int_0^T W(t) \diamond \mathrm{d}_\alpha W(t) := \lim_{\delta(P_n)\to 0} \sum_{i=1}^{n} [\alpha W(t_{i-1}) + (1-\alpha)W(t_i)](W(t_i) - W(t_{i-1})),$$

where $P_n = \{0 = t_0 < t_1 < \ldots < t_n = T\}$ is a finite partition of $[0, T]$, and $\delta(P_n)$ is the mesh size. Write $\int_0^T W(t) \diamond \mathrm{d}_\alpha W(t)$ as a linear combination of the Itô integral $\int_0^T W(t)\,\mathrm{d}W(t)$ and the Stratonovich integral $\int_0^T W(t) \circ \mathrm{d}W(t)$.

Exercise 11.4. Evaluate the following repeated (double) stochastic integral:

$$\int_0^t \left(\int_0^s \mathrm{d}W(u) \right) \mathrm{d}W(s).$$

Exercise 11.5. Show that $\int_0^t W^2(s)\,\mathrm{d}W(s) = \frac{1}{3}W^3(t) - \int_0^t W(s)\,\mathrm{d}s$.

[*Hint*: You may use an appropriate Itô formula.]

Exercise 11.6. Use the Itô isometry property to calculate the variances of the Itô integrals

$$\text{(a)} \int_0^t |W(s)|^{1/2}\,\mathrm{d}W(s), \quad \text{(b)} \int_0^t |W(s) + s|^2\,\mathrm{d}W(s), \quad \text{(c)} \int_0^t (W(s) + s)^{3/2}\,\mathrm{d}W(s).$$

Explain why the above integrals are well-defined.

Exercise 11.7. Calculate

$$\lim_{n\to\infty,\ \delta(P_n)\to 0} \sum_{i=1}^{n} (2W(t_{i-1}) + W(t_i))(W(t_i) - W(t_{i-1})),$$

where $P_n = \{0 = t_0 < t_1 < t_2 < \ldots < t_n = T\}$ is a partition of $[0, T]$ with mesh size $\delta(P_n) = \max_{i=1,\ldots,n} |t_i - t_{i-1}|$.

Exercise 11.8. Using Itô's formula, show that the process defined by

$$X(t) := W^4(t) - 6\int_0^t W^2(u)\,\mathrm{d}u,$$

$t \geqslant 0$, is a martingale w.r.t. a filtration for Brownian motion.

Exercise 11.9. Use Itô's formula to show that for any integer $k \geqslant 2$,

$$\mathrm{E}[W^k(t)] = \frac{k(k-1)}{2} \int_0^t \mathrm{E}[W^{k-2}(s)]\,\mathrm{d}s,$$

and use this to derive a formula for all the moments of the standard normal distribution.

Exercise 11.10. Show that $M(t) := \mathrm{e}^{t/2}\sin(W(t)), t \geqslant 0$, is a martingale w.r.t. a filtration for Brownian motion.

Exercise 11.11. Use Itô's formula to show that for any nonrandom, continuously differentiable function $f(t)$, the following formula of integration by parts is true:

$$\int_0^t f(s)\,\mathrm{d}W(s) = f(t)W(t) - \int_0^t f'(s)W(s)\,\mathrm{d}s.$$

Exercise 11.12. Use Itô's formula to find the stochastic differentials for the following functions of Brownian motion:

(a) $\mathrm{e}^{W(t)}$; (b) $W^k(t)$, $k \geqslant 0$; (c) $\cos(tW(t))$; (d) $\mathrm{e}^{W^2(t)}$; (e) $\arctan(t + W(t))$.

Exercise 11.13. Using Itô's formula, show that $Y(t) := W^3(t) - 3tW(t), t \geqslant 0$, is a martingale w.r.t. a filtration for Brownian motion.

Exercise 11.14. Define $Z(t) = \exp(\sigma W(t))$. Use Itô's formula to write down a stochastic differential for $Z(t)$. Then, by taking the mathematical expectation, find an ordinary (deterministic) first order linear differential equation for $m(t) := \mathrm{E}[Z(t)]$ and solve it to show that

$$\mathrm{E}[\exp(\sigma W(t))] = \exp\left(\frac{\sigma^2}{2}t\right).$$

Exercise 11.15. Let $\mathcal{N}(x)$ be the standard normal CDF and consider the process

$$X(t) := \mathcal{N}\left(\frac{W(t)}{\sqrt{T-t}}\right), \quad 0 \leqslant t < T.$$

Express this process as an Itô process, i.e. you need to *determine the explicit expressions for the adapted drift* $\mu(t)$ *and diffusion* $\sigma(t)$ *and provide the explicit form for* $X(t)$ as:

$$X(t) = X(0) + \int_0^t \mu(s)\,ds + \int_0^t \sigma(s)\,dW(s).$$

Show that the process is a martingale w.r.t. any filtration for BM. Find the limiting value $X(T-) = \lim_{t \nearrow T} X(t)$. What is the state space for the X process.

Exercise 11.16. Suppose that the processes $X := \{X(t)\}_{t \geqslant 0}$ and $Y := \{Y(t)\}_{t \geqslant 0}$ have the log-normal dynamics:

$$\begin{aligned} dX(t) &= X(t)(\mu_X\,\mathrm{d}t + \sigma_X\,\mathrm{d}W(t)) \\ dY(t) &= Y(t)(\mu_Y\,\mathrm{d}t + \sigma_Y\,\mathrm{d}W(t)). \end{aligned}$$

Show that the process $Z(t) := \frac{Y(t)}{X(t)}$ is also log-normal, with dynamics

$$dZ(t) = Z(t)(\mu_Z\,\mathrm{d}t + \sigma_Z\,\mathrm{d}W(t)),$$

and determine the coefficients μ_Z and σ_Z in terms of those of X and Y. Solve the same problem now assuming that X and Y are governed by two correlated Brownian motions W^X and W^Y, respectively, where $\mathrm{Corr}(W^X(t), W^Y(t)) = \rho t$, i.e., $\mathrm{d}W^X(t)\,\mathrm{d}W^Y(t) = \rho\,\mathrm{d}t$, for a given correlation coefficient $-1 \leqslant \rho \leqslant 1$.

Exercise 11.17. Let a time-homogeneous diffusion $X(t)$ have a stochastic differential with drift coefficient function $\mu(x) = 3x - 1$ and diffusion coefficient function $\sigma(x) = 2\sqrt{x}$. Assuming that $X(t) \geqslant 0$, find the stochastic differential for the process $Y(t) := \sqrt{X(t)}$. Find the generator for $Y(t)$.

Exercise 11.18. Let $X(t) = tW^2(t)$ and $Y(t) = e^{W(t)}$. Find the stochastic differential of $Z(t) := \frac{X(t)}{Y(t)}$. Compute the mean and variance of $Z(t)$.

Exercise 11.19. Let $X(t)$ be a time-homogeneous diffusion process solving an SDE with drift and diffusion coefficient functions $\mu(x) = cx$ and $\sigma(x) = \sigma$, respectively, where c, σ are constants and with initial condition $X(0) = x \in \mathbb{R}$. Consider the process defined by $Y(t) := X^2(t) - 2c\int_0^t X^2(s)\,\mathrm{d}s - \sigma^2 t$, $t \geqslant 0$.

(a) Represent the Y process as an Itô process and show that is a martingale w.r.t. any filtration for Brownian motion.

(b) Compute the mean and variance of $Y(t)$ for all $t \geqslant 0$.

Exercise 11.20. Consider the linear SDE in (11.25) in the case where $\delta(t) \equiv 0$ and where $\alpha(t), \beta(t), \gamma(t)$ are continuous nonrandom functions of time $t \geqslant 0$. Assume a constant initial condition $X(0) = x_0$. Show that the process $\{X(t)\}_{t\geqslant 0}$ is a Gaussian process and compute its mean and covariance functions.

Exercise 11.21. Use the Itô formula to write down stochastic differentials for the following processes:

(a) $Y(t) = \exp\left(\sigma W(t) - \frac{1}{2}\sigma^2 t\right)$,

(b) $Z(t) = f(t)W(t)$ where f is a continuously differentiable function.

Exercise 11.22. A time-homogeneous diffusion process X has a stochastic differential with respective drift and diffusion coefficient functions $\mu(x) = 0$ and $\sigma(x) = x(1-x)$. Assuming $0 < X(t) < 1$, show that the process $Y(t) := \ln\left(\frac{X(t)}{1-X(t)}\right)$ has a constant diffusion coefficient.

Exercise 11.23. Let $X(t) := (1-t)\int_0^t \frac{\mathrm{d}W(s)}{1-s}$, where $0 \leqslant t < 1$. Provide the stochastic differential equation for $X(t)$ in the form $\mathrm{d}X(t) = a(t, X(t))\,\mathrm{d}t + b(t, X(t))\,\mathrm{d}W(t)$. Check your answer by solving the SDE obtained subject to the initial condition $X(0) = 0$.

Exercise 11.24. Solve the following linear SDEs:

(a) $\mathrm{d}X(t) = W(t)X(t)\,\mathrm{d}t + W(t)X(t)\,\mathrm{d}W(t)\,, \quad X(0) = 1;$

(b) $\mathrm{d}X(t) = \alpha(\theta - X(t))\,\mathrm{d}t + \sigma X(t)\,\mathrm{d}W(t)\,, \quad X(0) = x \in \mathbb{R};$

(c) $\mathrm{d}X(t) = a(t)X(t)\,\mathrm{d}t + \sigma X(t)\,\mathrm{d}W(t)\,, \quad X(0) = x \in \mathbb{R}.$

(d) $\mathrm{d}X(t) = X(t)\,\mathrm{d}t + X(t)\,\mathrm{d}W(t)$, $X(0) = 1$.

Assume α, θ, σ are positive constants.

Exercise 11.25. Let $g(y)$ be a given function of y, and suppose that $y = f(x)$ is a solution of the ODE $\mathrm{d}y = g(y)\,\mathrm{d}x$, that is, $f'(x) = g(f(x))$. Show that $X(t) = f(W(t))$ is a solution of the SDE

$$\mathrm{d}X(t) = \frac{1}{2}g(X(t))g'(X(t))\,\mathrm{d}t + g(X(t))\,\mathrm{d}W(t).$$

Exercise 11.26. Use Exercise 11.25 to solve the following nonlinear SDEs, subject to $X(0) = x_0 \in \mathbb{R}$.

(a) $\mathrm{d}X(t) = \frac{\sigma^2}{4}\,\mathrm{d}t + \sigma\sqrt{X(t)}\,\mathrm{d}W(t);$

(b) $\mathrm{d}X(t) = X^3(t)\,\mathrm{d}t + X^2(t)\,\mathrm{d}W(t);$

(c) $\mathrm{d}X(t) = \frac{1}{2}\mathrm{e}^{2X(t)}\,\mathrm{d}t + \mathrm{e}^{X(t)}\,\mathrm{d}W(t).$

In each case, find the time interval for which the solution exists. [Hint: In each case, $f(x)$ of Exercise 11.25 is determined by solving a first order separable ODE. For parts (b) and (c) the solution exists up to an "explosion time" when the solution becomes singular.]

Exercise 11.27. Show that for any $u \in \mathbb{R}$, the function $f(t,x) = \exp(ux - u^2t/2)$ solves the backward PDE for Brownian motion. Take the first, second, and third derivatives of $\exp(ux - u^2t/2)$ w.r.t. u, and set $u = 0$, to show that functions x, $x^2 - t$, $x^3 - 3tx$ also solve the backward equation for Brownian motion. Deduce that $W^2(t) - t$ and $W^3(t) - 3tW(t)$ are martingales.

Exercise 11.28. Consider the drifted BM, $X(t) = x_0 + \mu t + \sigma W(t)$, and define the process $Y(t) := f(X(t))$, $t \geqslant 0$. By applying an appropriate Itô formula, obtain a general expression for a twice continuously differentiable function $f(x), x \in \mathbb{R}$, such that $\{Y(t)\}_{t \geqslant 0}$ is a martingale w.r.t. any filtration for BM.

Exercise 11.29. Derive a system of diffusion-type SDEs for the coupled processes $X(t) = \cos(W(t))$ and $Y(t) = \sin(W(t))$.

Exercise 11.30. Consider the process defined by $X(t) = \sinh(C + t + W(t))$, $t \geqslant 0$, where $C = \sinh^{-1} x_0$ with initial condition $X(0) = x_0$. This process is a diffusion on $\mathbb{R}$ and it satisfies an SDE of the form

$$\mathrm{d}X(t) = \mu(X(t))dt + \sigma(X(t))\,\mathrm{d}W(t)\,.$$

(i) Find the coefficient functions $\mu(x)$ and $\sigma(x)$.
(ii) Provide the backward Kolmogorov PDE and the terminal condition for the transition PDF $p(t, T; x, y)$.
(iii) Derive analytical expressions for the transition CDF and PDF of the process X.

Exercise 11.31. Give the probabilistic representation of the solution $f(t, x)$ of the PDE

$$\frac{\partial f}{\partial t} + \frac{x^2}{2}\frac{\partial^2 f}{\partial x^2} = 0,\ 0 \leqslant t \leqslant T, \quad f(T, x) = x^2\,.$$

Solve this PDE using the solution of the respective SDE.

Exercise 11.32. Consider the boundary value problem for the heat equation:

$$\frac{\partial V}{\partial t} + \frac{1}{2}\frac{\partial^2 V}{\partial x^2} = 0, \quad V(1, x) = f(x)$$

where f, the boundary value for time $t = 1$, is given, and where we are looking for a solution $V = V(t, x)$ defined for $0 \leqslant t \leqslant 1$ and $x \in \mathbb{R}$. Show that the solution is

$$V(t, x) = \int_{-\infty}^{\infty} f(y) e^{-\frac{(x-y)^2}{2(1-t)}} \frac{dy}{\sqrt{2\pi(1-t)}}\,.$$

Can you think of a function f for which the solution formula would not make sense?

Exercise 11.33. Consider the boundary value problem for the heat equation with a drift term:

$$\frac{\partial V}{\partial t} + \frac{1}{2}\frac{\partial^2 V}{\partial x^2} + a\frac{\partial V}{\partial x} = 0, \quad V(1, x) = f(x)$$

where f, the boundary value for time $t = 1$, is given, and a is a real constant. Derive an explicit (integral) formula for the solution $V = V(t, x)$.

Exercise 11.34. Let $f(t, x)$ satisfy the PDE

$$\frac{\partial f}{\partial t} + \frac{1}{2}\sigma^2 x^2 \frac{\partial^2 f}{\partial x^2} + \mu x \frac{\partial f}{\partial x} = 0, \quad 0 \leqslant t \leqslant T,\ x \in \mathbb{R}_+,$$

for fixed $T > 0$, with real constants $\sigma > 0$, μ. Solve for $f(t, x)$ subject to the terminal condition $f(T, x) = \mathbb{I}_{\{K_1 < x < K_2\}}$, where $K_2 > K_1 > 0$ are constants.

Exercise 11.35. To compact notation, we suppress all other variables except x' and denote $f(x') \equiv p(t,t';x,x')$ and $g(x') \equiv p(t',T;x',y)$. Using the definition of $\mathcal{G}_{t',x'}$, the left-hand integral in (11.69) becomes

$$\int_{\mathcal{I}} f(x')\, \mathcal{G}_{t',x'} g(x')\, \mathrm{d}x' = \frac{1}{2}\int_{\mathcal{I}} f(x')\sigma^2(t',x')\frac{\partial^2 g}{\partial x'^2}\, \mathrm{d}x' + \int_{\mathcal{I}} f(x')\mu(t',x')\frac{\partial g}{\partial x'}\, \mathrm{d}x' .$$

Now apply integration by parts to both integrals (in the first integral note that $\frac{\partial^2 g}{\partial x'^2} = \frac{\partial}{\partial x'}\left(\frac{\partial g}{\partial x'}\right)$). State appropriate assumptions that allow you to set the boundary terms to zero. Then, apply integration by parts again on the remaining integral containing $\sigma^2(t',x')$. Again, state appropriate assumptions that allow you to set the boundary terms to zero. In the end, obtain

$$\int_{\mathcal{I}} f(x')\, \mathcal{G}_{t',x'} g(x')\, \mathrm{d}x' = \int_{\mathcal{I}} g(x')\, \tilde{\mathcal{G}}_{t',x'} f(x')\, \mathrm{d}x'$$

where $\tilde{\mathcal{G}}_{t',x'} f = \frac{1}{2}\frac{\partial^2}{\partial x'^2}\left(\sigma^2(t',x')f\right) - \frac{\partial}{\partial x'}\left(\mu(t',x')f\right)$.

Exercise 11.36. Assume that a stock price process $\{S(t)\}_{t\geqslant 0}$ satisfies the SDE

$$\mathrm{d}S(t) = rS(t)\,\mathrm{d}t + \sigma S(t)\,\mathrm{d}\widetilde{W}(t)\,,$$

with constants $r, \sigma > 0$, and where $\{\widetilde{W}(t)\}_{t\geqslant 0}$ is a standard $\widetilde{\mathbb{P}}$-BM. By using Girsanov's Theorem, find the explicit expression for the Radon–Nikodym derivative process

$$\varrho_t := \left(\frac{\mathrm{d}\widehat{\mathbb{P}}}{\mathrm{d}\widetilde{\mathbb{P}}}\right)_t$$

such that the process defined by $\widehat{S}(t) := \frac{e^{rt}}{S(t)}, t \geqslant 0$, is a $\widehat{\mathbb{P}}$-martingale. Give the SDE satisfied by the stock price $S(t)$ w.r.t. the $\widehat{\mathbb{P}}$-BM.

Exercise 11.37. Consider a stock price process $\{S(t)\}_{t\geqslant 0}$ that obeys the SDE

$$\mathrm{d}S(t) = \mu S(t)dt + \sigma(S(t))^{1+\beta}\,\mathrm{d}W(t), \quad S(0) = S > 0,$$

with constant parameters $\sigma \neq 0$, β, μ.

(a) Assume the process $\int_0^t (S(u))^{1+\beta}\,\mathrm{d}W(u)$ is a $\mathbb{P}$-martingale w.r.t. the filtration generated by the standard $\mathbb{P}$-Brownian motion $\{W(t)\}_{t\geqslant 0}$. Determine an exact expression for the mean of the process $\mathrm{E}[S(t)]$. [Hint: You may write $S(t)$ in terms of an exponential martingale by considering the process defined by $X(t) := \ln S(t)$.]

(b) Fix $T > 0$ and assume the existence of a Radon–Nikodym derivative process $\varrho_t := \left(\frac{\mathrm{d}\widetilde{\mathbb{P}}}{\mathrm{d}\mathbb{P}}\right)_t, 0 \leqslant t \leqslant T$, such that $\{e^{-rt}S(t)\}_{0\leqslant t\leqslant T}$ is a $\widetilde{\mathbb{P}}$-martingale. Give the form of ϱ_t as an exponential $\mathbb{P}$-martingale.

Exercise 11.38. Consider a one-dimensional general diffusion process $\{X(t)\}_{t\geqslant 0}$ having a transition PDF $p(s,t;x,y)$, $s < t$, w.r.t. a given probability measure $\mathbb{P}$, for all x, y in the state space of the process. Assume a change of measure $\mathbb{P} \to \widehat{\mathbb{P}}$ is defined by a Radon–Nikodym derivative process

$$\varrho_t := \left(\frac{\mathrm{d}\widehat{\mathbb{P}}}{\mathrm{d}\mathbb{P}}\right)_t = h(t, X(t))$$

for all $t \geqslant 0$ and where $h(t,x)$ is some given Borel function of t,x. Let $\widehat{p}(s,t;x,y)$ be the transition PDF w.r.t. the measure $\widehat{\mathbb{P}}$. Show that the two transition PDFs are related by

$$\widehat{p}(s,t;x,y) = \frac{h(t,y)}{h(s,x)}\, p(s,t;x,y)\,.$$

[Hint: Consider the definition of the transition CDF (w.r.t. the measure $\widehat{\mathbb{P}}$):

$$\widehat{P}(s,t;x,y) = \widehat{\mathbb{P}}(X(t) \leqslant y \mid X(s) = x) = \widehat{\mathrm{E}}[\mathbb{I}_{\{X(t)\leqslant y\}} \mid X(s) = x]$$

and make use of the Markov property and the change of measure for computing expectations.]

Chapter 12

Risk-Neutral Pricing in the (B, S) Economy: One Underlying Stock

At this point we have the necessary tools in stochastic analysis for developing the theory of derivative pricing and hedging in continuous time in an economy where risky assets are modelled as Itô processes. In this chapter we consider an economy with two securities: a single tradable risky asset, namely, a stock, and a money market (bank) account or a bond. We refer to this as a (B, S) economy with only two tradable assets: B stands for the bank account or bond and S stands for the stock. More specifically, this chapter is devoted to presenting the theoretical framework solely within the classical case of an economy where the interest rate is fixed and the stock price is modelled as a standard geometric Brownian motion (GBM) with constant growth rate and constant volatility. The case in which the stock pays a dividend yield is also included later. The only source of randomness driving the stock price is a single Brownian motion. This is also referred to as the standard Black–Scholes framework. This can be thought of as the continuous-time analogue of the standard binomial tree model which was formally discussed in great detail in Chapter 7. The analogues of the up and down market moves in discrete time are now random movements of the underlying (driving) Brownian motion. The same important underlying concepts of self-financing, replication, hedging, arbitrage, and no-arbitrage pricing of derivative contracts in discrete time now carry over into the continuous-time setting.

All appropriately discounted tradable assets are martingales under an equivalent martingale (risk-neutral) measure. In particular, by using a self-financing replication strategy, we will arrive at the risk-neutral pricing formula that expresses the current price of any attainable European-style derivative, including contracts with path-dependent payoffs, as a conditional expectation (under the risk-neutral or equivalent martingale measure) of the discounted payoff. The class of derivative securities that are attainable is quite general. This chapter also includes a discussion of this and its relation to hedging and pricing. When pricing a non-path-dependent European-style derivative, the inherent Markov property reduces the risk-neutral pricing formula to a conditional expectation where the (discounted) Feynman-Kac formula allows us to arrive at the infamous Black–Scholes–Merton PDE for the current price of the derivative (option) contract. Recall that in Chapter 11 we made the important connection between the SDE of an Itô process, the (discounted) conditional expectation of a (payoff) function of the terminal value of the process, and the corresponding terminal (or initial) value PDE problem. This is the essence of the Feynman–Kac representation. Our discussion on the theoretical framework of the (B, S) economy culminates in the risk-neutral pricing formulation. We then apply this formulation to derivative pricing problems, where we explicitly derive pricing and hedging formulae for various European contracts, such as standard calls and puts, as well as more complex options such as compound options. We finally also apply the risk-neutral pricing formulation to path-dependent derivatives such as barrier options and lookback options.

12.1 Replication (Hedging) and Derivative Pricing in the Simplest Black–Scholes Economy

Following Section 11.8 of Chapter 11, we fix a filtered probability space $(\Omega, \mathcal{F}, \mathbb{P}, \mathbb{F})$, where $\mathbb{F} = \{\mathcal{F}_t\}_{0 \leqslant t \leqslant T}$ is a filtration for standard Brownian motion, i.e., $\{W(t)\}_{t \geqslant 0}$ is a standard $(\mathbb{P}, \mathbb{F})$-BM where $\mathbb{P}$ is the physical (real-world) measure. The first base security, B, in this market is the bank account whose price process is denoted by $\{B(t)\}_{0 \leqslant t \leqslant T}$. In this section we assume a constant interest rate r where $B(t) = \mathrm{e}^{rt}$, i.e., $B(0) = 1$ and $B(t)$ is one unit (dollar) of investment compounded continuously with fixed rate r over time $[0, t]$. The second base security is the stock whose price process is assumed to be a standard GBM satisfying the SDE

$$\mathrm{d}S(t) = \mu S(t)\,\mathrm{d}t + \sigma S(t)\,\mathrm{d}W(t)\,,$$

with constant drift μ and constant volatility $\sigma > 0$. We recall Example 11.15 in Section11.8 of Chapter 11. There we showed, by a simple application of Girsanov's Theorem, that there is a unique risk-neutral measure $\widetilde{\mathbb{P}}$, defined by (11.96), where $\{\widetilde{W}(t) := W(t) + \frac{(\mu - r)}{\sigma} t\}_{t \geqslant 0}$ is a standard $\widetilde{\mathbb{P}}$-BM. The discounted stock price process is a $\widetilde{\mathbb{P}}$-martingale. For convenience, we repeat some of the important equations here. In particular, the stock satisfies the SDE

$$\mathrm{d}S(t) = rS(t)\,\mathrm{d}t + \sigma S(t)\,\mathrm{d}\widetilde{W}(t)\,, \tag{12.1}$$

with solution

$$S(t) = S(0)\,\mathrm{e}^{(r - \frac{1}{2}\sigma^2)t + \sigma \widetilde{W}(t)}\,. \tag{12.2}$$

The discounted stock price process, $\overline{S}(t) \equiv \mathrm{e}^{-rt} S(t) = D(t)S(t)$ with discount factor $D(t) := 1/B(t) = \mathrm{e}^{-rt}$, is a $\widetilde{\mathbb{P}}$-martingale:

$$\mathrm{d}\overline{S}(t) = \sigma \overline{S}(t)\,\mathrm{d}\widetilde{W}(t)\,, \quad \text{i.e.,} \quad \overline{S}(t) = S(0) + \sigma \int_0^t \overline{S}(u)\,\mathrm{d}\widetilde{W}(u)\,. \tag{12.3}$$

Note also that $\mathrm{d}B(t) = rB(t)\,\mathrm{d}t$ has an identically zero coefficient in $\mathrm{d}\widetilde{W}(t)$ and that $\overline{B}(t) := B(t)/B(t) = 1$ is trivially a (constant) martingale under any measure. Hence, $\widetilde{\mathbb{P}}$ is an equivalent martingale measure (EMM) with the bank account as a numéraire asset.

As in the discrete-time models, we assume no arbitrage in the market and will, below, set up a *self-financing replicating portfolio strategy* in the two base securities such that the value of the portfolio replicates the value of the financial derivative at some future (maturity) time T. In the absence of arbitrage, the time-t value of the self-financing replicating portfolio must therefore equal the no-arbitrage price of the derivative. Let $\{V_t\}_{0 \leqslant t \leqslant T}$ denote the *price process of the derivative security* where V_t is the price of the derivative at time $t \leqslant T$. At maturity T, the derivative price is given by the payoff value V_T, which is an $\mathcal{F}_T$-measurable random variable. In the present model, V_T is generally some functional of the Brownian motion (BM) up to time T or equivalently some functional of the underlying stock price process $\{S(t)\}_{0 \leqslant t \leqslant T}$. This functional can be quite complex for a general path-dependent payoff, although for a non-path-dependent derivative (as, for example, a standard call or put) the payoff is only a function of the terminal stock value $S(T)$.

In this (B, S) economy any portfolio (trading) strategy is a continuous sequence of portfolios in the two base assets: (β_t, Δ_t), $0 \leqslant t \leqslant T$, with each process $\{\beta_t\}_{0 \leqslant t \leqslant T}$ and $\{\Delta_t\}_{0 \leqslant t \leqslant T}$ assumed to be adapted to the filtration $\mathbb{F}$. As in the binomial model, β_t represents the time-t position in the bank account where $\beta_t < 0$ corresponds to a loan and $\beta_t > 0$ is an investment. The hedge position Δ_t is the number of shares held in the stock at time

t where $\Delta_t < 0$ corresponds to shorting the stock and $\Delta_t > 0$ is a long position. Since we are in continuous time, trading (i.e., portfolio re-balancing) is allowed at any moment in time. The investor begins with a given initial wealth Π_0, which completely finances the initial portfolio with positions (β_0, Δ_0) and subsequently trades at every time $t \in [0, T]$ while holding Δ_t shares in the stock and β_t units in the bank account, i.e., this is represented by the portfolio value process for $t \in [0, T]$:

$$\Pi_t = \Delta_t S(t) + \beta_t B(t). \tag{12.4}$$

In what follows, we will only consider *self-financing* portfolio strategies. In analogy with the binomial model, a self-financing portfolio strategy is one in which the differential change in portfolio value is due only to differential changes in the prices of the base assets. Essentially this means that the investor holds the positions (β_t, Δ_t) during the infinitesimal time window $[t, t + \mathrm{d}t)$. At time $t + \mathrm{d}t$ the BM will have changed by a differential amount $\mathrm{d}W(t)$, the stock will have changed its share price by a differential amount $\mathrm{d}S(t)$, and the investment in the bank account will have either accrued interest (if $\beta_t > 0$) or will have decreased in value if $\beta_t < 0$. Formally, a self-financing portfolio strategy is then a portfolio strategy $(\beta_t, \Delta_t), 0 \leqslant t \leqslant T$, such that the cumulative gain in portfolio value is given by

$$\Pi_t = \Pi_0 + \int_0^t \beta_u \,\mathrm{d}B(u) + \int_0^t \Delta_u \,\mathrm{d}S(u)\,, \quad \text{for all } t \in [0, T], \tag{12.5}$$

with probability one (a.s.). It is more convenient to work with the differential form of (12.5),

$$\mathrm{d}\Pi_t = \beta_t \,\mathrm{d}B(t) + \Delta_t \,\mathrm{d}S(t)\,.$$

From (12.4) we can express the bank account investment (or loan) as $\beta_t B(t) = \Pi_t - \Delta_t S(t)$ and substituting this into the above differential, where $\beta_t \,\mathrm{d}B(t) = r\beta_t B(t)\,\mathrm{d}t$, gives

$$\mathrm{d}\Pi_t = r\beta_t B(t)\,\mathrm{d}t + \Delta_t \,\mathrm{d}S(t) = r(\Pi_t - \Delta_t S(t))\,\mathrm{d}t + \Delta_t \,\mathrm{d}S(t)\,. \tag{12.6}$$

We can recognize this as the differential form of the continuous-time analogue of the wealth equation we encountered in the binomial model.

As in the binomial model, the discounted self-financing portfolio value process, defined by $\{\overline{\Pi}_t := \mathrm{e}^{-rt}\Pi_t\}_{0 \leqslant t \leqslant T}$, is a $\widetilde{\mathbb{P}}$-martingale. This is readily shown by applying the Itô product rule to the process $\mathrm{e}^{-rt}\Pi_t \equiv D(t)\Pi_t$, where $\mathrm{d}D(t)\,\mathrm{d}\Pi_t = -rD(t)\,\mathrm{d}t\,\mathrm{d}\Pi_t \equiv 0$, and using (12.6):

$$\begin{aligned}
\mathrm{d}\overline{\Pi}_t \equiv \mathrm{d}(D(t)\Pi_t) &= \Pi_t \,\mathrm{d}D(t) + D(t)\,\mathrm{d}\Pi_t \\
&= -rD(t)\Pi_t \,\mathrm{d}t + D(t)\,[\,r(\Pi_t - \Delta_t S(t))\,\mathrm{d}t + \Delta_t \,\mathrm{d}S(t)\,] \\
&= \Delta_t \,D(t)\,[\,-rS(t)\,\mathrm{d}t + \mathrm{d}S(t)\,] \\
&= \Delta_t \,\mathrm{d}\overline{S}(t) \\
&= \Delta_t \,\sigma \overline{S}(t)\,\mathrm{d}\widetilde{W}(t)\,.
\end{aligned} \tag{12.7}$$

In the last equation line we made use of (12.3). This therefore shows that $\{\overline{\Pi}_t\}_{0 \leqslant t \leqslant T}$ is a $\widetilde{\mathbb{P}}$-martingale and that changes in this discounted self-financing portfolio are due only to changes in the discounted stock price. In integral form we have

$$\overline{\Pi}_t = \overline{\Pi}_0 + \int_0^t \Delta_u \,\mathrm{d}\overline{S}(u) = \Pi_0 + \int_0^t \sigma \overline{S}(u)\Delta_u \,\mathrm{d}\widetilde{W}(u)\,, \quad 0 \leqslant t \leqslant T. \tag{12.8}$$

Note: we are assuming that the above Itô integral is square integrable. This guarantees that the process defined by (12.8) is a $\widetilde{\mathbb{P}}$-martingale. This technical detail can be verified later once the option price, and hence the delta position, is obtained.

As in the binomial model, we wish to price derivative contracts that can be replicated by a self-financing portfolio strategy. Let us therefore give a definition of this for the above continuous-time model.

Definition 12.1. A self-financing strategy $(\beta_t, \Delta_t), 0 \leqslant t \leqslant T$, is said to *replicate* the $\mathcal{F}_T$-measurable payoff V_T at maturity T if $\Pi_T = V_T$ (a.s.), i.e., $\mathbb{P}(\Pi_T = V_T) = 1$. We also say that the payoff V_T is attainable.

Let us suppose that a self-financing portfolio strategy that replicates the derivative payoff V_T exists. Later we give some discussion on this existence. Then, the cost at any time $t \leqslant T$ to set up such a strategy, i.e., the portfolio value Π_t, must equal the *time-t price of the derivative* V_t in order for the investor to hedge (at time t) the short position in the derivative security that has the given attainable payoff V_T at future time T, and hence avoid arbitrage. Therefore, we set $V_t = \Pi_t$ for all $0 \leqslant t \leqslant T$ for any attainable payoff V_T.

The key step now is to make use of the $\widetilde{\mathbb{P}}$-martingale property in (12.8). In particular, we have

$$D(t)\Pi_t = \widetilde{\mathrm{E}}[\, D(T)\Pi_T \mid \mathcal{F}_t]$$

and, since $V_t = \Pi_t$, then $D(t)V_t = \widetilde{\mathrm{E}}[\, D(T)V_T \mid \mathcal{F}_t]$. The discounted derivative price process, $\{D(t)V_t \equiv \mathrm{e}^{-rt}V_t\}_{0\leqslant t\leqslant T}$, is therefore a $\widetilde{\mathbb{P}}$-martingale. We can combine the discount factors, where $D(T)/D(t) = B(t)/B(T) = \mathrm{e}^{-r(T-t)}$, giving

$$V_t = B(t)\,\widetilde{\mathrm{E}}\left[\frac{V_T}{B(T)}\,\middle|\,\mathcal{F}_t\right] = \mathrm{e}^{-r(T-t)}\widetilde{\mathrm{E}}[\, V_T \mid \mathcal{F}_t],\quad 0 \leqslant t \leqslant T. \tag{12.9}$$

This is the *risk-neutral pricing formula* for the above (B, S) model with a constant interest rate. It is the continuous-time analogue of (7.23) for the binomial model in Chapter 7. We remark that (12.9) was derived for the simplest case where the stock is a standard GBM, but in Chapter 13 we will arrive at the same formula for more general continuous-time stock price processes as long as the payoff is attainable and there exists a risk-neutral measure $\widetilde{\mathbb{P}}$ where the discounted stock (base asset) price process is a $\widetilde{\mathbb{P}}$-martingale. In Chapter 13 we shall also define an arbitrage strategy in the multi-asset continuous-time framework where it is shown that the existence of a risk-neutral measure implies that no arbitrage strategies are possible within the market model.

The formula in (12.9) may seem very simple; however, its practical use rests upon our ability to calculate the expectation of the payoff conditional on the filtration at time t. The payoff is an $\mathcal{F}_T$-measurable random variable and it can, in some cases, be quite complex, as it may have a complicated dependence on the path of the stock price process. Hence, the conditional expectation can be quite challenging to compute for complex path-dependent payoffs. The main idea is to simplify the $\mathcal{F}_t$-conditional expectation to an expectation that can be readily computed. In most practical situations the derivative has a payoff structure that is not too complex. In the binomial model, we have already used the risk-neutral pricing framework to value derivatives having commonly encountered payoffs, such as the standard European call and put options as well as path-dependent options such as lookback and Asian options. Our main tools for analytically pricing such options were the Markov property and the Independence Proposition 6.7 or 6.8. For continuous-time models these tools will also be used within the risk-neutral framework. Moreover, we shall also have other tools at our disposal, such as the PDE approach that is a result of the Feyman–Kac theorems.

Let's now consider the case of a standard (non-path-dependent) European option with payoff $V_T = \Lambda(S(T))$ as a $\mathcal{F}_T$-measurable random variable that is a function of only the *terminal (maturity time T) value of the stock price.* The important simplification that now follows is due to the Markov property where the conditioning on $\mathcal{F}_t$ is replaced with a conditioning on $S(t)$. We remind the reader of our discussions in Section 11.7 on conditional expectations and the Markov property. In particular, recall that a solution to an SDE is a Markov process and hence (11.42) in Theorem 11.5 applies. The time-t derivative value (expressed as a $\sigma(S(t))$-measurable random variable) then takes the form

$$\begin{aligned} V_t &= \mathrm{e}^{-r(T-t)}\widetilde{\mathrm{E}}[\Lambda(S(T)) \mid \mathcal{F}_t] \\ &= \mathrm{e}^{-r(T-t)}\widetilde{\mathrm{E}}[\Lambda(S(T)) \mid S(t)] = V(t, S(t)) \,. \end{aligned} \tag{12.10}$$

The *derivative pricing function*, $V(t, S)$, is a function of calendar (actual) time t and the spot[1] (ordinary variable) $S > 0$ and is given by the discounted expectation of the payoff conditional on $S(t) = S$:

$$V(t, S) = \mathrm{e}^{-r(T-t)}\widetilde{\mathrm{E}}_{t,S}[\Lambda(S(T))] := \mathrm{e}^{-r(T-t)}\widetilde{\mathrm{E}}[\Lambda(S(T)) \mid S(t) = S] \,. \tag{12.11}$$

Given the share price for the stock at time $t \leqslant T$, which is the spot $S > 0$, then (12.11) gives us the no-arbitrage time-t price of a European option having non-path-dependent attainable payoff $\Lambda(S(T))$ at maturity T, where $\Lambda : \mathbb{R}_+ \to \mathbb{R}$ is the payoff function.

We can now generally express the pricing function in (12.11) as an integral over the standard normal density $n(z)$ or equivalently as an integral over the risk-neutral transition PDF. This has essentially already been done in Example 11.11 of Chapter 11 where related formulas were derived under measure $\mathbb{P}$ with drift μ (instead of $\widetilde{\mathbb{P}}$ with drift r). Using (12.2), the time-T stock price is given in terms of the time-t price and the $\widetilde{\mathbb{P}}$-BM increment:

$$S(T) = S(t)\,\mathrm{e}^{(r-\frac{1}{2}\sigma^2)(T-t)+\sigma(\widetilde{W}(T)-\widetilde{W}(t))} \equiv S(t)\,\mathrm{e}^{(r-\frac{1}{2}\sigma^2)\tau+\sigma\sqrt{\tau}\widetilde{Z}} \,. \tag{12.12}$$

Throughout, we conveniently define $\tau := T - t$ as the time to maturity and

$$\widetilde{Z} := \frac{\widetilde{W}(T) - \widetilde{W}(t)}{\sqrt{\tau}} \tag{12.13}$$

which is a standard normal random variable, i.e., $\widetilde{Z} \sim Norm(0, 1)$ under measure $\widetilde{\mathbb{P}}$. Note that $\widetilde{Z}$ (or $\frac{S(T)}{S(t)}$) is independent of $S(t)$ since $\widetilde{W}(T) - \widetilde{W}(t)$ is independent of $\widetilde{W}(t)$. Combining these facts into (12.11) gives

$$\begin{aligned} V(t, S) &= \mathrm{e}^{-r\tau}\widetilde{\mathrm{E}}\big[\Lambda\big(S(t)\,\mathrm{e}^{(r-\frac{1}{2}\sigma^2)\tau+\sigma\sqrt{\tau}\widetilde{Z}}\big) \mid S(t) = S\big] \\ &= \mathrm{e}^{-r\tau}\widetilde{\mathrm{E}}\big[\Lambda\big(S\,\mathrm{e}^{(r-\frac{1}{2}\sigma^2)\tau+\sigma\sqrt{\tau}\widetilde{Z}}\big)\big] \\ &= \mathrm{e}^{-r\tau}\int_{-\infty}^{\infty} \Lambda(S\,\mathrm{e}^{(r-\frac{1}{2}\sigma^2)\tau+\sigma\sqrt{\tau}z})n(z)\,\mathrm{d}z \,. \end{aligned} \tag{12.14}$$

By changing integration variables in (12.14), i.e., letting $y = S\,\mathrm{e}^{(r-\frac{1}{2}\sigma^2)\tau+\sigma\sqrt{\tau}z}$, gives the equivalent pricing formula

$$V(t, S) = \mathrm{e}^{-r\tau}\int_0^{\infty} \Lambda(y)\,\widetilde{p}(\tau; S, y)\,\mathrm{d}y \,, \tag{12.15}$$

[1] Assuming there is no confusion, when discussing pricing formulas we prefer to choose more appropriate letters for some of the ordinary variables. In this case we denote the spot value by using the dummy variable S, instead of using some other letter like x, in the context of a pricing formula.

where $\widetilde{p}(\tau; S, y)$ denotes the risk-neutral transition PDF of the stock price process in (12.2):

$$\widetilde{p}(\tau; S, y) = \frac{1}{y\sigma\sqrt{2\pi\tau}} \exp\left(-\frac{[\ln(y/S) - (r - \frac{1}{2}\sigma^2)\tau]^2}{2\sigma^2\tau}\right) ; \quad S, y > 0, \, \tau > 0. \tag{12.16}$$

We recall that this is the time-homogeneous log-normal density in y given by (11.79) with drift parameter μ set to the risk-free rate r. Note that when the stock price process is time homogeneous, as in the present case, the pricing formula is a function of the time to maturity $\tau = T - t$ and we shall also denote it by $v(\tau, S) := V(t, S)$, i.e., $v(\tau, S) := V(T - \tau, S)$.

We see that (12.14) and (12.15) provide two equivalent expectation (integral) approaches for pricing standard European options on a stock. The transition PDF $\widetilde{p}(\tau; S, y)$ is the fundamental solution to the PDE in (11.72) with the spot value as a backward variable, $x \equiv S$, linear diffusion coefficient function $\sigma(x) \equiv \sigma(S) = \sigma S$, and linear drift coefficient function $\mu(x) \equiv \mu(S) = rS$. The discounted transition PDF, $\mathrm{e}^{-r\tau}\widetilde{p}(\tau; S, y)$, and hence the pricing function $v = v(\tau, S)$ in (12.15), satisfies the PDE in the variables (S, τ) (see (11.77) in Chapter 11):

$$\frac{\partial v}{\partial \tau} = \frac{1}{2}\sigma^2 S^2 \frac{\partial^2 v}{\partial S^2} + rS\frac{\partial v}{\partial S} - rv \,, \tag{12.17}$$

subject to the payoff condition $v(0+, S) = \Lambda(S)$, which is an initial condition in the time to maturity $\tau \searrow 0$. This is the Black–Scholes partial differential equation (BSPDE) for the pricing function *expressed as function of the spot and time to maturity variables* (S, τ). Since $\frac{\partial V}{\partial t} = \frac{\partial \tau}{\partial t} \cdot \frac{\partial v}{\partial \tau} = -\frac{\partial v}{\partial \tau}$, then (12.17) is equivalent to

$$\frac{\partial V}{\partial t} + \frac{1}{2}\sigma^2 S^2 \frac{\partial^2 V}{\partial S^2} + rS\frac{\partial V}{\partial S} - rV = 0 \,, \tag{12.18}$$

subject to the payoff condition $V(T-, S) = \Lambda(S)$, which is a terminal condition in time $t \nearrow T$. This is the usual BSPDE satisfied by the pricing function $V = V(t, S)$ in the variables (t, S). In fact, the BSPDE in (12.18) arises by direct application of the discounted Feynman–Kac Theorem 11.9 to the conditional expectation in (12.11). [Note that Theorem 11.9 is the same when we replace $\mathbb{P}$, E, and W everywhere by $\widetilde{\mathbb{P}}$, $\widetilde{\mathrm{E}}$, and $\widetilde{W}$, respectively. All this means is that the probability measure is now the risk-neutral measure $\widetilde{\mathbb{P}}$. Here, we have the dummy variable $x \equiv S$ and the process $X(t) \equiv S(t)$ is GBM with coefficient functions defined by $\mu(t, S) := rS, \sigma(t, S) := \sigma S$ in the SDE (12.1). In particular, for constant interest rate r the pricing function $V(t, S)$ satisfies the PDE (11.64) in the variables t and S, which is the BSPDE in (12.18).]

Let us now turn our attention to the problem of replicating a derivative claim, i.e., *the hedging problem* in the simple (B, S) model. We first see how this problem is solved in the case of non-path-dependent payoffs where $V_T = \Lambda(S(T))$ and $V_t = V(t, S(t))$. By applying the Itô formula with the differential in (12.1) we obtain:

$$\begin{aligned} \mathrm{d}[\mathrm{e}^{-rt}V(t, S(t))] &= \mathrm{e}^{-rt}\left[\mathrm{d}V(t, S(t)) - rV(t, S(t))\,\mathrm{d}t\right] \\ &= \mathrm{e}^{-rt}\left(\frac{\partial V}{\partial t}(t, S(t)) + \mathcal{G}\,V(t, S(t)) - rV(t, S(t))\right)\mathrm{d}t \\ &\quad + \sigma\mathrm{e}^{-rt}S(t)\frac{\partial V}{\partial S}(t, S(t))\,\mathrm{d}\widetilde{W}(t)\,, \end{aligned}$$

where $\mathcal{G}\,V(t, S) := \frac{1}{2}\sigma^2 S^2 \frac{\partial^2}{\partial S^2}V(t, S) + rS\frac{\partial}{\partial S}V(t, S)$ is the differential generator corresponding to the GBM process with SDE in (12.1). Now using the fact that the pricing function $V(t, S)$ satisfies the BSPDE in (12.18), i.e., $\left(\frac{\partial}{\partial t} + \mathcal{G} - r\right)V(t, S) = 0$ for all (t, S), then

$$\mathrm{d}[\mathrm{e}^{-rt}V(t, S(t))] = \sigma\overline{S}(t)\frac{\partial V}{\partial S}(t, S(t))\,\mathrm{d}\widetilde{W}(t)\,.$$

In integral form:

$$e^{-rt}V(t,S(t)) = V(0,S(0)) + \int_0^t \sigma\overline{S}(u)\frac{\partial V}{\partial S}(u,S(u))\,d\widetilde{W}(u) \tag{12.19}$$

for all $t \in [0,T]$. Assuming the square-integrability condition on the integrand of this Itô integral, $\{e^{-rt}V(t,S(t))\}_{0\leqslant t\leqslant T}$ is a $\widetilde{\mathbb{P}}$-martingale. For portfolio replication we require $\Pi_t = V(t,S(t))$, or $\overline{\Pi}_t = e^{-rt}V(t,S(t))$, for all $t \in [0,T]$. We see from (12.8) that this can be achieved by setting the initial value of the portfolio $\Pi_0 = V_0 = V(0,S(0))$ and by choosing the delta position such that the respective Itô integrals in (12.8) and (12.19) are equal, i.e., the delta hedge is achieved by choosing

$$\Delta_t = \frac{\partial V}{\partial S}(t,S(t)) \equiv \frac{\partial V}{\partial S}(t,S)\bigg|_{S=S(t)}, \quad \text{for all } t \in [0,T). \tag{12.20}$$

Hence, (12.20) gives the *time-t (delta) position in the stock required to dynamically replicate the derivative claim in a self-financing portfolio strategy.* Clearly, Δ_t is $\mathcal{F}_t$-measurable (in fact it is $\sigma(S(t))$-measurable) and is given uniquely by the first derivative (w.r.t. the spot variable S) of the pricing function evaluated at $S = S(t)$. Defining the function $\Delta(t,S) := \frac{\partial V}{\partial S}(t,S)$, then $\Delta(t,S)$ gives the time-t position in the stock given the spot value S. Once the pricing function $V(t,S)$ is known (and hence its derivative computed) the self-financing replicating portfolio in (12.4) is given by $\Delta_t = \Delta(t,S(t))$ positions in the stock and $\beta_t = e^{-rt}[V(t,S(t)) - S(t)\Delta(t,S(t))]$ units in the bank account, i.e., the value of the bank account portion of the portfolio at time t is $V(t,S(t)) - S(t)\Delta(t,S(t))$.

Consider the general case where V_T is any $\mathcal{F}_T$-measurable payoff (i.e., generally path dependent). Now, we generally have $V_t \neq V(t,S(t))$ (i.e., we do not generally have V_t as a function of t and $S(t)$) and hence the above BSPDE and Feynman–Kac results are not generally applicable. However, what is important is that the discounted derivative value process $\{D(t)V_t\}_{0\leqslant t\leqslant T}$ is a $(\widetilde{\mathbb{P}},\mathbb{F})$-martingale. We now show that we can guarantee that V_T is attainable if we make two general assumptions on the payoff:

1. V_T is square integrable, i.e., $E[V_T^2] < \infty$.

2. V_T is $\mathcal{F}_T^W$-measurable.

[Note that these two conditions are already implicit in the above non-path-dependent case.] The discounted price process $\{D(t)V_t\}_{0\leqslant t\leqslant T}$ is then a square-integrable $(\widetilde{\mathbb{P}},\mathbb{F}^W)$-martingale. We can now make use of Theorem 11.14, in particular Proposition 11.15 of Chapter 11, where we identify $M(t) \equiv D(t)V_t$ and identify the measure $\widehat{\mathbb{P}} = \widetilde{\mathbb{P}}$, with constant $\gamma(t) = \gamma \equiv (r-\mu)/\sigma$ being $\mathcal{F}_t^W$-adapted. This implies[2] $\mathbb{F}^W = \mathbb{F}^{\widetilde{W}}$ and hence $\{D(t)V_t\}_{0\leqslant t\leqslant T}$ is a square-integrable $(\widetilde{\mathbb{P}},\mathbb{F}^{\widetilde{W}})$-martingale. Hence, there exists an adapted process, we now denote by $\{\widetilde{\theta}(t)\}_{0\leqslant t\leqslant T}$, such that (note $D(0)V_0 = V_0$):

$$D(t)V_t = V_0 + \int_0^t \widetilde{\theta}(t)\,d\widetilde{W}(u)\,,\ \ 0 \leqslant t \leqslant T. \tag{12.21}$$

For portfolio replication we require $\Pi_t = V_t$, for all $t \in [0,T]$. By (12.21) we see that this is

[2]It is also trivial to see in this case that the natural filtration $\mathbb{F}^W = \{\mathcal{F}_t^W\}_{t\geqslant 0}$ generated by W ($\mathbb{P}$-measure BM) is the same as the natural filtration $\mathbb{F}^{\widetilde{W}} = \{\mathcal{F}_t^{\widetilde{W}}\}_{t\geqslant 0}$ generated by $\widetilde{W}$ ($\widetilde{\mathbb{P}}$-measure BM) since the two Brownian motions differ only by a constant: $\widetilde{W}(t) = W(t) + \frac{(\mu-r)}{\sigma}t$, so $\mathcal{F}_t^{\widetilde{W}} = \mathcal{F}_t^W$ for every $t \geqslant 0$.

achieved by setting $\Pi_0 = V_0$ and by choosing the delta position such that the integrands in (12.8) and (12.21) are equal, i.e., the delta hedge is achieved by setting [note $\overline{S}(t) = D(t)S(t)$]

$$\widetilde{\theta}(t) = \sigma D(t)S(t)\Delta_t\,, \quad \text{i.e.,} \quad \Delta_t = \frac{\widetilde{\theta}(t)}{\sigma D(t)S(t)} \tag{12.22}$$

for all $t \in [0,T]$. This solution for Δ_t exists, for every $\widetilde{\theta}(t)$, since we are assuming that the volatility parameter $\sigma > 0$. Note also that the stock price is a GBM with $S(0) > 0$ and therefore cannot hit zero in any finite time. Of course, implicit in this existence are also the above assumptions on the derivative payoff. Namely, V_T must be $\mathcal{F}_T^W$-measurable and this means that we can only guarantee replication of payoffs whose only source of randomness is the BM driving the stock itself.

We have therefore shown that the model can replicate (hedge) any complex arbitrary path-dependent derivatives with a payoff satisfying the above two assumptions. In this sense we can say that the (B, S) model is a *complete market model.* The formula in (12.22) asserts the existence of a hedging strategy and therefore justifies the use of the risk-neutral pricing formula in (12.9). We remark that generally (12.22) does not give a practical (or explicit) construction of the hedging strategy. However, there are important examples of common payoffs for which we do have an explicit formula for the delta hedge. For example, in the particular case of a non-path-dependent standard European option we already showed that $\widetilde{\theta}(t)$ is given explicitly in terms of the derivative of the pricing function:

$$\widetilde{\theta}(t) = \sigma D(t)S(t)\frac{\partial V}{\partial S}(t, S(t))$$

with the delta position given by (12.20).

12.1.1 Pricing Standard European Calls and Puts

We now use the risk-neutral pricing formulation to derive the well-known Black–Scholes–Merton formulae for the prices of a standard call and put option in the simplest (B, S) model of Section 12.1, where the stock price is a GBM with constant volatility $\sigma > 0$ and the bank account has constant interest rate r. Here we are assuming a zero dividend on the stock, but the inclusion of a stock dividend is quite simple, as shown later.

Consider a call option with payoff $V_T \equiv C_T = (S(T) - K)^+$. This is an example of a non-path-dependent option with payoff function $\Lambda(x) := (x - K)^+$. Let $S(t) = S > 0$ be the spot price of the stock at time $t \leqslant T$. Since the payoff depends only on the terminal stock price $S(T)$, the (current) time-t price of this call with strike $K > 0$ can be found by directly evaluating the integral in (12.14) using the identity (A.1) in the Appendix. We leave this as an exercise for the reader. We recall that in Chapter 4 a brute force integration of (12.14) was also used in deriving the time-0 price of the put option with payoff $\Lambda(x) = (K - x)^+$. Here we carry out the derivation in an explicit manner that is instructive since it clearly displays the use of the conditional expectation approach and the connection between the different random variables and independence. In particular, using (12.11), the time-t price of the call, $C(t, S)$, is the discounted risk-neutral expectation of the payoff at time $T > t$, conditional on $S(t) = S$:

$$\begin{aligned} C(t,S) &= \mathrm{e}^{-r(T-t)}\,\widetilde{\mathrm{E}}_{t,S}\big[(S(T)-K)^+\big] \\ &= \mathrm{e}^{-r\tau}\,\widetilde{\mathrm{E}}_{t,S}\big[(S(T)-K)\,\mathbb{I}_{\{S(T)>K\}}\big] \\ &= \mathrm{e}^{-r\tau}\,\widetilde{\mathrm{E}}_{t,S}\big[S(T)\,\mathbb{I}_{\{S(T)>K\}}\big] - \mathrm{e}^{-r\tau}\,K\widetilde{\mathrm{E}}_{t,S}\big[\mathbb{I}_{\{S(T)>K\}}\big] \\ &= \mathrm{e}^{-r\tau}\,\widetilde{\mathrm{E}}_{t,S}\big[S(T)\,\mathbb{I}_{\{S(T)>K\}}\big] - \mathrm{e}^{-r\tau}\,K\,\widetilde{\mathbb{P}}_{t,S}\big(S(T)>K\big)\,, \end{aligned} \tag{12.23}$$

with $S(T)$ given by (12.12). Two conditional expectations need to be computed (where the last term has been expressed as a conditional probability). In both cases we use the fact that $\widetilde{Z}$ in (12.13) is independent of $S(t)$ and, by the Independence Proposition 6.7, this allows us to remove the conditioning upon setting $S(t) = S$. First, let us rewrite the event $\{S(T) > K\}$ in terms of $\widetilde{Z}$ and $S(t)$ by simply dividing $S(T)$ in (12.12) by $S(t)$, dividing K by $S(t)$, and taking natural logarithms:

$$\{S(T) > K\} = \left\{ \ln \frac{S(T)}{S(t)} > \ln \frac{K}{S(t)} \right\} = \left\{ \widetilde{Z} > -d_-\left(\frac{S(t)}{K}, \tau\right) \right\}$$

where we define

$$d_-(x, \tau) := \frac{\ln x + (r - \frac{1}{2}\sigma^2)\tau}{\sigma\sqrt{\tau}}, \quad d_+(x, \tau) := \frac{\ln x + (r + \frac{1}{2}\sigma^2)\tau}{\sigma\sqrt{\tau}} = d_-(x, \tau) + \sigma\sqrt{\tau} \quad (12.24)$$

for all $x > 0$, $\tau := T - t > 0$. We use the above representation for the event and substitute the expression for $S(T)$, given in (12.12), into the first expectation in (12.23), which is now readily computed by setting $S(t) = S$, using the independence of $\widetilde{Z}$ and $S(t)$, and then applying the identity[3] in (A.1) of the Appendix to evaluate the (unconditional) expectation:

$$\begin{aligned}
\widetilde{\mathrm{E}}_{t,S}\left[S(T)\,\mathbb{I}_{\{S(T)>K\}}\right] &= \widetilde{\mathrm{E}}\left[S(t)\,\mathrm{e}^{(r-\frac{1}{2}\sigma^2)\tau+\sigma\sqrt{\tau}\widetilde{Z}}\,\mathbb{I}_{\{\widetilde{Z}>-d_-(\frac{S(t)}{K},\tau)\}} \mid S(t) = S\right] \\
&= S\,\mathrm{e}^{(r-\frac{1}{2}\sigma^2)\tau}\,\widetilde{\mathrm{E}}\left[\mathrm{e}^{\sigma\sqrt{\tau}\widetilde{Z}}\,\mathbb{I}_{\{\widetilde{Z}>-d_-(\frac{S}{K},\tau)\}}\right] \\
&= S\,\mathrm{e}^{(r-\frac{1}{2}\sigma^2)\tau}\,\mathrm{e}^{\frac{1}{2}(\sigma\sqrt{\tau})^2}\,\mathcal{N}\left(\sigma\sqrt{\tau} + d_-(\frac{S}{K}, \tau)\right) \\
&= \mathrm{e}^{r\tau} S\,\mathcal{N}\left(d_+(\frac{S}{K}, \tau)\right). \qquad (12.25)
\end{aligned}$$

The conditional probability in (12.23) now follows quite simply:

$$\begin{aligned}
\widetilde{\mathbb{P}}_{t,S}\big(S(T) > K\big) = \widetilde{\mathbb{P}}\left(\widetilde{Z} > -d_-(\frac{S(t)}{K}, \tau) \,\middle|\, S(t) = S\right) &= \widetilde{\mathbb{P}}\left(\widetilde{Z} > -d_-(\frac{S}{K}, \tau)\right) \\
&= \mathcal{N}\left(d_-(\frac{S}{K}, \tau)\right). \qquad (12.26)
\end{aligned}$$

Note that this is also obtained directly from the transition CDF, $\widetilde{P}$, of the GBM process under measure $\widetilde{\mathbb{P}}$, where

$$\widetilde{P}(t, T; S, K) \equiv \widetilde{\mathbb{P}}_{t,S}\big(0 < S(T) \leqslant K\big) = 1 - \widetilde{\mathbb{P}}_{t,S}\big(S(T) > K\big) = \mathcal{N}\left(-d_-(\frac{S}{K}, \tau)\right).$$

Substituting (12.25) and (12.26) into (12.23) completes our derivation of the pricing formula for the standard call:

$$C(t, S) = S\mathcal{N}\left(d_+(\frac{S}{K}, \tau)\right) - \mathrm{e}^{-r\tau} K\mathcal{N}\left(d_-(\frac{S}{K}, \tau)\right); \quad \tau = T - t. \qquad (12.27)$$

Having priced the call, we can now easily derive the pricing formula for the put price,

[3]In applying the expectation identities in the Appendix we need to simply identify the parameters A, B and the mean μ and variance σ^2 of the appropriate normal random variable X in the given measure. Note that the parameter σ used in the above equations is obviously the symbol for the volatility of the stock. Of course, this σ is not the same σ used throughout the Appendix! For the expectation in (12.25), we employ the formula in (A.1) by identifying $X \equiv \widetilde{Z}$ as *Norm*$(\mu = 0, \sigma^2 = 1)$ and $A \equiv -d_-(\frac{S}{K}, \tau)$, $B \equiv \sigma\sqrt{\tau}$.

$P(t,S)$, by recalling the simple symmetry between the call and put payoffs, i.e., a portfolio in one long call and one short put is equivalent to a portfolio in one long forward contract:

$$(S(T)-K)^+ - (K-S(T))^+ = S(T)-K\,. \tag{12.28}$$

Taking discounted risk-neutral expectations on both sides of (12.28) gives

$$\underbrace{\mathrm{e}^{-r\tau}\,\widetilde{\mathrm{E}}_{t,S}\big[(S(T)-K)^+\big]}_{C(t,S)} - \underbrace{\mathrm{e}^{-r\tau}\,\widetilde{\mathrm{E}}_{t,S}\big[(K-S(T))^+\big]}_{P(t,S)} = \underbrace{\mathrm{e}^{-r\tau}\,\widetilde{\mathrm{E}}_{t,S}\big[S(T)\big]}_{S} - \mathrm{e}^{-r\tau}K \tag{12.29}$$

where we recognize the left-hand side as the difference in the time-t price of the call and put, at strike K. The right-hand side is the time-t price of a forward contract with payoff $S(T)-K$. This is the *put-call parity relation* that we have already encountered in our primer chapter (Chapter 4) on derivative securities. It is important to note that this relation is valid for quite general models (i.e., beyond the GBM model). The only assumption is that the discounted stock price is a $\widetilde{\mathbb{P}}$-martingale. Substituting the call price in (12.27) into (12.29) gives the pricing formula for the standard put:

$$\begin{aligned}
P(t,S) &= C(t,S) + \mathrm{e}^{-r\tau}K - S \\
&= \mathrm{e}^{-r\tau}K\left[1-\mathcal{N}\left(d_-(\tfrac{S}{K},\tau)\right)\right] - S\left[1-\mathcal{N}\left(d_+(\tfrac{S}{K},\tau)\right)\right] \\
&= \mathrm{e}^{-r\tau}K\mathcal{N}\left(-d_-(\tfrac{S}{K},\tau)\right) - S\mathcal{N}\left(-d_+(\tfrac{S}{K},\tau)\right)\ ;\ \tau = T-t.
\end{aligned} \tag{12.30}$$

We leave it as an exercise for the reader to show by direct differentiation that the above call and put pricing functions in (12.27) and (12.30) solve the BSPDE. The initial conditions $\tau \searrow 0$ (equivalently $t \nearrow T$) on the above pricing functions are readily shown to be satisfied by working out the limiting forms (see also Example 11.11):

$$\begin{aligned}
\lim_{\tau\searrow 0}\mathcal{N}\left(d_\pm(\tfrac{S}{K},\tau)\right) &= \lim_{\tau\searrow 0}\mathcal{N}\left(\frac{\ln(S/K)}{\sigma\sqrt{\tau}} + (r\pm\frac{1}{2}\sigma^2)\frac{\sqrt{\tau}}{\sigma}\right) = \lim_{\tau\searrow 0}\mathcal{N}\left(\frac{\ln(S/K)}{\sigma\sqrt{\tau}}\right) \\
&= \begin{cases} \mathcal{N}(\infty)=1 & \text{if } S>K \\ \mathcal{N}(0)=1/2 & \text{if } S=K \\ \mathcal{N}(-\infty)=0 & \text{if } S<K \end{cases} \\
&= H(S-K)\,.
\end{aligned} \tag{12.31}$$

Hence, both limits equal the unit step function centered at $S=K$. Using both these limits in (12.27) verifies the payoff condition for the call:

$$\begin{aligned}
\lim_{t\nearrow T} C(t,S) &= S\lim_{\tau\searrow 0}\mathcal{N}\left(d_+(\tfrac{S}{K},\tau)\right) - K\lim_{\tau\searrow 0}\mathrm{e}^{-r\tau}\mathcal{N}\left(d_-(\tfrac{S}{K},\tau)\right) \\
&= (S-K)H(S-K) \\
&= (S-K)\mathbb{I}_{\{S>K\}} = (S-K)^+.
\end{aligned}$$

By the same steps applied to (12.30), we verify the payoff condition:

$$\lim_{t\nearrow T} P(t,S) = (K-S)H(K-S) = (K-S)^+\,.$$

The asymptotic values for the call and put pricing functions for small and large spot values are readily obtained. We leave it as an exercise to show, by directly computing the limits, that

$$P(t,S) \sim \mathrm{e}^{-r\tau} K\,, \ \text{as } S \searrow 0, \quad \text{and} \quad P(t,S) \sim 0\,, \ \text{as } S \nearrow \infty\,, \tag{12.32}$$

and

$$C(t,S) \sim 0\,, \ \text{as } S \searrow 0, \quad \text{and} \quad C(t,S) \sim S - \mathrm{e}^{-r\tau} K \sim \infty\,, \ \text{as } S \nearrow \infty\,. \tag{12.33}$$

The financial interpretation of the above limits is clear. In the limit that the time-t stock price (spot S) is very close to zero, it will remain close to zero within a finite time to maturity $\tau = T - t$. This means that the call will certainly expire out of the money (hence is worthless) and the put will expire completely in the money with payoff K and time-t value $\mathrm{e}^{-r\tau} K$. In the limit of arbitrarily large spot value, the stock price will remain arbitrarily large, i.e., the put will certainly expire out of the money (hence is worthless) and the call will expire in the money with an arbitrarily large payoff.

12.1.2 Hedging Standard European Calls and Puts

In Chapter 4 we computed the "delta" of a standard call and put by differentiating the pricing functions. Let us denote the delta of the call at calendar time t by $\Delta_c(t,S) = \frac{\partial C}{\partial S}(t,S)$ and the delta of the put by $\Delta_p(t,S) = \frac{\partial P}{\partial S}(t,S)$. Note that Δ_c and Δ_p are also functions of (τ, S), since the pricing functions in (12.27) and (12.30) are expressible as functions of (τ, S). We recall a previous method to derive Δ_c by simply differentiating (12.27) w.r.t. S, where $\mathcal{N}'(x) = n(x) = \mathrm{e}^{-\frac{1}{2}x^2}/\sqrt{2\pi}$, while using $\frac{\partial}{\partial S} d_\pm(\frac{S}{K}, \tau) = \frac{1}{S\sigma\sqrt{\tau}}$,

$$\begin{aligned}\Delta_c(t,S) &= \mathcal{N}\left(d_+(\frac{S}{K}, \tau)\right) + \frac{K}{S\sigma\sqrt{2\pi\tau}}\left[\frac{S}{K}\,\mathrm{e}^{-\frac{1}{2}d_+^2(\frac{S}{K},\tau)} - \mathrm{e}^{-r\tau - \frac{1}{2}d_-^2(\frac{S}{K},\tau)}\right] \\ &= \mathcal{N}\left(d_+(\frac{S}{K}, \tau)\right)\,. \end{aligned} \tag{12.34}$$

Here we used the following identity with $x = S/K$ (which we leave as a somewhat tedious exercise in algebra for the reader to show):

$$x\,\mathrm{e}^{-\frac{1}{2}d_+^2(x,\tau)} \equiv \mathrm{e}^{\ln x - \frac{1}{2}d_+^2(x,\tau)} = \mathrm{e}^{-r\tau - \frac{1}{2}d_-^2(x,\tau)}\,. \tag{12.35}$$

The above derivation of Δ_c is correct but perhaps not the most instructive. We now give an alternate derivation of Δ_c by directly connecting it to the discounted risk-neutral conditional expectation in (12.25). From the risk-neutral pricing formula in (12.14),

$$C(t,S) = \mathrm{e}^{-r\tau}\widetilde{\mathrm{E}}\big[\big(S\,\mathrm{e}^{(r-\frac{1}{2}\sigma^2)\tau + \sigma\sqrt{\tau}\widetilde{Z}} - K\big)^+\big] = \mathrm{e}^{-r\tau}\int_{-\infty}^{\infty}\big(S\,\mathrm{e}^{(r-\frac{1}{2}\sigma^2)\tau+\sigma\sqrt{\tau}z} - K\big)^+ n(z)\,\mathrm{d}z\,.$$

We now differentiate w.r.t. S (as a parameter) inside the expectation or inside the integral and use the property $\frac{\partial}{\partial S}(aS - K)^+ = aH(aS-K) = a\mathbb{I}_{\{aS \geqslant K\}}$. In the integral we have $a \equiv a(z) \equiv \mathrm{e}^{(r-\frac{1}{2}\sigma^2)\tau+\sigma\sqrt{\tau}z}$ or in the expectation $a \equiv a(\widetilde{Z}) \equiv \mathrm{e}^{(r-\frac{1}{2}\sigma^2)\tau+\sigma\sqrt{\tau}\widetilde{Z}}$. We can write the steps compactly as follows by differentiating inside the expectation (note: $S(T) = Sa(\widetilde{Z})$),

$$\begin{aligned}\Delta_c(t,S) = \mathrm{e}^{-r\tau}\widetilde{\mathrm{E}}\left[\frac{\partial}{\partial S}(S\,a(\widetilde{Z}) - K)^+\right] &= \mathrm{e}^{-r\tau}\widetilde{\mathrm{E}}\big[a(\widetilde{Z})\,\mathbb{I}_{\{S\,a(\widetilde{Z})>K\}}\big] \\ &= \frac{1}{S}\mathrm{e}^{-r\tau}\widetilde{\mathrm{E}}_{t,S}\big[S(T)\,\mathbb{I}_{\{S(T)>K\}}\big] \\ &= \mathcal{N}\left(d_+(\frac{S}{K}, \tau)\right)\,.\end{aligned}$$

Here we identified the conditional expectation in (12.25).

It is important to note that the delta of a call is strictly positive, i.e., $\Delta_c(t, S) > 0$ for all $t < T$, $S > 0$, since the CDF $\mathcal{N}(x)$ is strictly positive for all $x \in \mathbb{R}$. Hence, according to (12.20), to replicate a call the writer is continuously re-balancing the self-financing portfolio while always maintaining a *delta positive (long) position in the stock* given by $\Delta_t \equiv \Delta_c(t, S(t)) = \mathcal{N}\left(d_+(\frac{S(t)}{K}, \tau)\right)$ for $\tau = T - t > 0$. In particular, the time-t investment in the stock, given spot $S(t) = S$, is

$$\Delta_t S(t) = S\,\Delta_c(t, S) = S\mathcal{N}\left(d_+(\frac{S}{K}, \tau)\right).$$

Since $\Pi_t = C(t, S(t)) = C(t, S)$, the time-$t$ value of the bank account (cash) portion of the self-financing portfolio is, upon using (12.27),

$$\beta_t B(t) = \Pi_t - \Delta_t S(t) = C(t, S) - S\,\Delta_c(t, S) = -\mathrm{e}^{-r\tau}\, K\mathcal{N}\left(d_-(\frac{S}{K}, \tau)\right).$$

This quantity is negative for all $t < T$, $S > 0$. Hence, the writer of the call is always maintaining a *negative position (loan) in the bank account.*

The delta position at maturity $t = T$ is given by the limit $t \nearrow T$ of the function in (12.34). From (12.31) we see that, in the limit of zero time to maturity, Δ_c approaches the unit step function with discontinuity at $S = K$. Moreover, for any fixed $\tau > 0$, we observe that Δ_c is a strictly increasing function of S with limiting values of zero and unity:

$$\lim_{S \searrow 0} \Delta_c(t, S) = \mathcal{N}(-\infty) = 0 \quad \text{and} \quad \lim_{S \nearrow \infty} \Delta_c(t, S) \to \mathcal{N}(\infty) = 1\,.$$

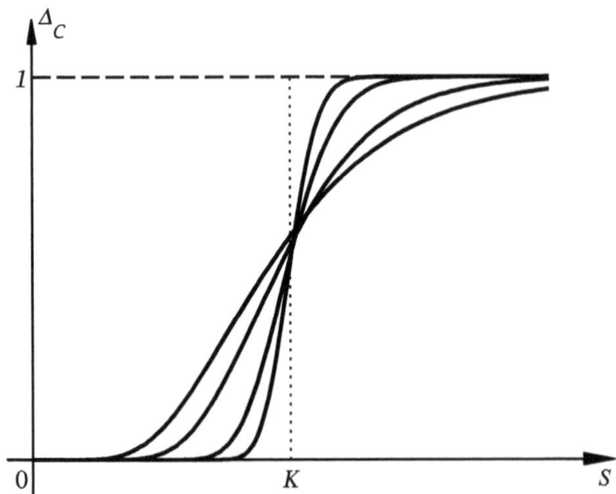

FIGURE 12.1: Typical plots of the call delta as function of spot for four different values of time to maturity τ.

Figure 12.1 contains typical plots of Δ_c for different values of τ. We clearly see that Δ_c approaches the unit step function as $\tau \searrow 0$. As $\tau \searrow 0$, (12.34) gives $\Delta_c \to 0$ for $S < K$, $\Delta_c \to 1$ for $S > K$, and $\Delta_c \to \frac{1}{2}$ for $S = K$. We see that Δ_c gets progressively steeper for smaller and smaller values of τ and eventually becomes the unit step function. At a time t

just before expiry (i.e., $\tau \approx 0$), if the stock price lies above K, then $\Delta_c \approx 1$ and the hedging replicates the positive payoff of the call after paying off the loan in the amount of K (see (12.31) and the above expression for $\beta_t B(t)$). On the other hand, if the stock price is below K (out of the money), then $\Delta_c \approx 0$ and $\beta_t B(t) \approx 0$, so the hedge replicates the zero payoff of the call.

There are also scenarios where the stock price $S(t)$ just before expiry ($t \approx T$) stays very close to the strike K until time T. There is a fairly high probability (which is easily computed) that the stock price will fluctuate between values below K and above K. These are scenarios where the stock is said to be "pinning the strike." For values of τ close to zero this would mean that the (hedge) position in the stock would have to be re-balanced, in a short time, between a very small long position ($\Delta_c \approx 0$) and a large long position where the replicating portfolio is almost all stock ($\Delta_c \approx 1$). This re-balancing requires either selling off a large portion of the underlying stock or buying up a large portion of stock so as to maintain the correct hedge. In particular, if the stock price suddenly moves just before expiry from one side of the strike to the other, then the writer or trader must rapidly trade enough of the underlying stock before expiration in order to hedge the loss against such a movement. In the (B, S) theoretical (idealized) model such transactions are assumed to occur instantaneously (as efficiently as is required!) and without any liquidity issues in trading, i.e., all scenarios are, in theory, hedged. In the real world these kinds of scenarios cannot be hedged effectively in time since instantaneous re-balancing is obviously not possible. Moreover, there are transaction fees associated with each trade. The risk associated with options trading whereby the market price of the stock is pinning the strike is referred to as *pin risk*. Later we consider the pricing of a so-called soft-strike call. This contract differs from the standard call, as its payoff is everywhere differentiable and it has a continuous range of strikes. This type of option avoids the problem of pin risk when delta hedging.

The delta of a put option, Δ_p, follows trivially by put-call parity,

$$\Delta_p(t, S) = \frac{\partial}{\partial S}\left(C(t, S) - S + \mathrm{e}^{-r\tau}K\right) = \Delta_c(t, S) - 1 = \mathcal{N}\left(d_+(\frac{S}{K}, \tau)\right) - 1$$
$$= -\mathcal{N}\left(-d_+(\frac{S}{K}, \tau)\right). \qquad (12.36)$$

The delta of a put is therefore strictly negative and is simply related to the call delta by $\Delta_p = \Delta_c - 1$. A put option is replicated by continuously re-balancing with a *delta negative (short) position in the stock* given by $\Delta_t \equiv \Delta_p(t, S(t)) = -\mathcal{N}\left(-d_+(\frac{S(t)}{K}, \tau)\right)$. Given spot $S(t) = S$, the time-t value of the stock portion of the replicating portfolio for the put is given by

$$\Delta_t S(t) = S\,\Delta_p(t, S) = -S\mathcal{N}\left(-d_+(\frac{S}{K}, \tau)\right).$$

The self-financing replicating portfolio value for the put is $\Pi_t = P(t, S(t)) = P(t, S)$, so the corresponding time-t investment in the bank account is

$$\beta_t B(t) = P(t, S) - S\,\Delta_p(t, S) = \mathrm{e}^{-r\tau} K\mathcal{N}\left(-d_-(\frac{S}{K}, \tau)\right).$$

In direct contrast to the call, the put is replicated by always maintaining a *positive investment in the bank account.*

At maturity $t = T$, $\Delta_p(T, S) = \Delta_c(T, S) - 1 = H(S - K) - 1$. For $\tau > 0$, Δ_p is a strictly increasing function of S with limiting values:

$$\lim_{S \searrow 0} \Delta_p(t, S) = -1 \quad \text{and} \quad \lim_{S \nearrow \infty} \Delta_p(t, S) \to 0.$$

The above discussion on hedging and pin risk associated with the call also applies in an obviously similar manner to the put with given strike K. We note that for the put delta the corresponding plots of Δ_p as a function of S are as in Figure 12.1, where the origin of the vertical axis is simply shifted up by unity.

In closing this section we recall that the delta is one among other so-called "Greeks" of an option that are of interest to practitioners such as options traders. Recall that in Section 4.3.4.4 of Chapter 4 we provided some discussion of these quantities. This chapter is focused on risk-neutral pricing and hedging of options. We leave it as an exercise for the reader to derive the corresponding formulas for the gamma, theta, vega, and rho of a standard call and put expressed as a function of S, τ, K, r, σ.

12.1.3 Europeans with Piecewise Linear Payoffs

Equations (12.25) and (12.26) are useful for pricing any European option having a *piecewise linear payoff* with possibly a finite (or countable) number of discontinuities. Actually, the standard call and put options are important special cases of such payoffs with no discontinuity and a discontinuity in their first derivatives. Let's now consider a simple example of a piecewise constant payoff with a single jump discontinuity.

Example 12.1. (Asset-or-Nothing Binary Call) Consider an option that pays the holder the value of the underlying share price of the stock if the stock price at expiry is above a given strike K and is otherwise worthless, i.e., the holder gets the asset (stock) or nothing with payoff function

$$\Lambda(S) = S\,\mathbb{I}_{\{S\geqslant K\}} = \begin{cases} S & \text{if } S \geqslant K, \\ 0 & \text{if } S < K. \end{cases}$$

Derive the risk-neutral pricing formula and the hedging position in the stock for the European option with this payoff.

Solution. Observe that the payoff is also the first term in the standard call option. The risk-neutral pricing formula immediately follows from (12.25):

$$V(t,S) = \mathrm{e}^{-r\tau}\,\widetilde{\mathrm{E}}_{t,S}\left[S(T)\,\mathbb{I}_{\{S(T)\geqslant K\}}\right] = S\mathcal{N}\left(d_+(\frac{S}{K},\tau)\right),$$

where $\tau = T - t$. The delta hedge is obtained by straightforward differentiation,

$$\Delta(t,S) = \mathcal{N}\left(d_+(\frac{S}{K},\tau)\right) + \frac{1}{\sigma\sqrt{\tau}}n\left(d_+(\frac{S}{K},\tau)\right).$$

Note that the time-t replicating portfolio is always long in the stock with a bank loan in the amount of $|V(t,S) - S\Delta(t,S)| = \frac{S}{\sigma\sqrt{\tau}}n\left(d_+(\frac{S}{K},\tau)\right)$. □

Recall that a cash-or-nothing binary call has payoff $\Lambda(S) = \mathbb{I}_{\{S\geqslant K\}}$. Hence, a standard call struck at K is a portfolio consisting of K short positions in a cash-or-nothing call and one long position in an asset-or-nothing call, both struck at K.

We now consider a derivation of the risk-neutral pricing formula for a European option with arbitrary piecewise linear payoff, assuming the standard GBM model for the stock. As a general form for the payoff we consider the sum of linear functions restricted to any number $n \geqslant 1$ of nonoverlapping intervals:

$$\Lambda(S) = \sum_{i=1}^{n}(A_iS + B_i)\,\mathbb{I}_{\{a_i\leqslant S<b_i\}} \tag{12.37}$$

with any real constants A_i, B_i and where $0 \leqslant a_1 < b_1 \leqslant a_2 < b_2 \leqslant \ldots \leqslant a_n < b_n \leqslant \infty$. This function can account for a number of jump discontinuities (including no discontinuities), piecewise constant payoffs, and piecewise linear payoffs with any combination of positive and negative slopes.

As a concrete example, let $n = 1$. A call with $\Lambda(S) = (S - K)^+$ is then given by (12.37) with parameter choice $A_1 = 1$, $B_1 = -K$, $a_1 = K$, $b_1 = \infty$. A cash-or-nothing binary call with $\Lambda(S) = \mathbb{I}_{\{S \geqslant K\}}$ obtains with $A_1 = 0$, $B_1 = 1$, $a_1 = K$, $b_1 = \infty$. The payoff of an asset-or-nothing call, $\Lambda(S) = S\,\mathbb{I}_{\{S \geqslant K\}}$, corresponds to $A_1 = 1$, $B_1 = 0$, $a_1 = K$, $b_1 = \infty$. For $n = 2$, an example is the butterfly spread with strikes $K_1 < K_2 < K_3$, $K_2 = (K_1 + K_3)/2$, which is representable as a linear combination of piecewise linear functions with payoff $\Lambda(S) = (S - K_1)\,\mathbb{I}_{\{K_1 \leqslant S < K_2\}} + (K_3 - S)\,\mathbb{I}_{\{K_2 \leqslant S < K_3\}}$. This corresponds to setting the parameters in (12.37) to $A_1 = 1, B_1 = -K_1, a_1 = K_1, b_1 = K_2$ and $A_2 = -1, B_2 = K_3, a_2 = K_2, b_2 = K_3$.

In order to make use of the conditional expectation identities in (12.25) and (12.26) we write $\mathbb{I}_{\{a_i \leqslant S < b_i\}} = \mathbb{I}_{\{S \geqslant a_i\}} - \mathbb{I}_{\{S \geqslant b_i\}}$ and express the payoff in (12.37) as

$$\Lambda(S) = \sum_{i=1}^{n} \left\{ A_i S \big(\mathbb{I}_{\{S \geqslant a_i\}} - \mathbb{I}_{\{S \geqslant b_i\}}\big) + B_i \big(\mathbb{I}_{\{S \geqslant a_i\}} - \mathbb{I}_{\{S \geqslant b_i\}}\big) \right\} .$$

Applying the identities in (12.25) and (12.26) and the linearity property of expectations, we arrive at an analytical expression for the risk-neutral pricing formula of a European option with payoff in (12.37):

$$\begin{aligned}
V(t, S) &= \mathrm{e}^{-r\tau}\, \widetilde{\mathrm{E}}_{t,S}\big[\Lambda(S(T))\big] \\
&= \mathrm{e}^{-r\tau} \sum_{i=1}^{n} \Big\{ A_i \left(\widetilde{\mathrm{E}}_{t,S}\big[S(T)\mathbb{I}_{\{S(T) \geqslant a_i\}}\big] - \widetilde{\mathrm{E}}_{t,S}\big[S(T)\mathbb{I}_{\{S(T) \geqslant b_i\}}\big] \right) \\
&\qquad + B_i \left(\widetilde{\mathbb{P}}_{t,S}(S(T) \geqslant a_i) - \widetilde{\mathbb{P}}_{t,S}(S(T) \geqslant b_i) \right) \Big\} \\
&= \sum_{i=1}^{n} \Big\{ A_i S \left[\mathcal{N}\Big(d_+\big(\frac{S}{a_i}, \tau\big)\Big) - \mathcal{N}\Big(d_+\big(\frac{S}{b_i}, \tau\big)\Big) \right] \\
&\qquad + \mathrm{e}^{-r\tau} B_i \left[\mathcal{N}\Big(d_-\big(\frac{S}{a_i}, \tau\big)\Big) - \mathcal{N}\Big(d_-\big(\frac{S}{b_i}, \tau\big)\Big) \right] \Big\}
\end{aligned} \tag{12.38}$$

where $\tau = T - t$ and $d_\pm(x, \tau)$ functions defined in (12.24). We leave as an exercise the derivation of a general formula for the delta hedging position $\Delta(t, S)$ for this option. The pricing formula in (12.38) is applicable to several payoff forms that occur in practice and we leave some as assigned exercises at the end of this chapter.

12.1.4 Power Options

Power options differ from vanilla European options in that the payoff function is not linear but raised to some power in the underlying spot. We now show how to analytically value such options under the GBM model. Typically, the payoff of a power option is a quadratic function of the stock price. The widest possible application of power options is for addressing the nonlinear risk of option sellers. There was proposed a class of soft-strike options which do not have a single fixed strike price but a continuous range of strikes spread over an interval. As was mentioned earlier, such options allow for addressing limitations of a standard delta hedging when the underlying asset is pinning the strike at the expiration of the option.

More generally, the payoff of a power option may involve the terminal stock price raised to some power, e.g., $S^\alpha(T) \equiv (S(T))^\alpha$ with either positive or negative exponent $\alpha \neq 0$, as well as other terms involving $S^\alpha(T)$ times an indicator function restricting the value of $S(T)$ on some interval. Let's assume that the payoff is some linear combination of elemental payoffs having any of the three forms

$$S^\alpha(T) \quad \text{or} \quad S^\alpha(T)\,\mathbb{I}_{\{A_1<S(T)\leqslant A_2\}} \quad \text{or} \quad S^\alpha(T)\,\mathbb{I}_{\{S(T)>A\}}$$

where $A_1 < A_2$ and A are nonnegative constants. Hence, by the risk-neutral derivative pricing formulation the present value of a power option at time $t < T$ will involve expectations of the above payoffs under the risk-neutral measure $\widetilde{\mathbb{P}}$. We can handle all of the above three payoffs by simply deriving a formula for the conditional expectation of the payoff $S^\alpha(T)\,\mathbb{I}_{\{S(T)>A\}}$, for any $A \geqslant 0$, since

$$S^\alpha(T)\,\mathbb{I}_{\{A_1<S(T)\leqslant A_2\}} = S^\alpha(T)\,\mathbb{I}_{\{S(T)>A_1\}} - S^\alpha(T)\,\mathbb{I}_{\{S(T)>A_2\}}\,. \tag{12.39}$$

Note also that for $A_1 = 0$ we have

$$S^\alpha(T)\,\mathbb{I}_{\{0<S(T)\leqslant A_2\}} = S^\alpha(T) - S^\alpha(T)\,\mathbb{I}_{\{S(T)>A\}}$$

where $\mathbb{I}_{\{S(T)>0\}} = 1$ since the stock price is always positive. The expectation of $S^\alpha(T)$, conditional on a given spot value $S(t) = S > 0$, for any $t < T$, is easily calculated using (12.12) raised to the exponent α. Again we use the fact that $\widetilde{Z}$ in (12.13) is independent of $S(t)$ (under measure $\widetilde{\mathbb{P}}$), which allows us to remove the conditioning upon setting $S(t) = S$:

$$\begin{aligned}
\widetilde{\mathrm{E}}_{t,S}\big[S^\alpha(T)\big] \equiv \widetilde{\mathrm{E}}\big[S^\alpha(T)\,|\,S(t)=S\big] &= \mathrm{e}^{\alpha(r-\sigma^2/2)(T-t)}\,\widetilde{\mathrm{E}}\big[S^\alpha(t)\,\mathrm{e}^{\alpha\sigma\sqrt{T-t}\widetilde{Z}}\,|\,S(t)=S\big] \\
&= S^\alpha\,\mathrm{e}^{\alpha(r-\sigma^2/2)(T-t)}\,\widetilde{\mathrm{E}}\big[\mathrm{e}^{\alpha\sigma\sqrt{T-t}\widetilde{Z}}\big] \\
&= S^\alpha\,\mathrm{e}^{\alpha(r-\sigma^2/2)(T-t)}\,\mathrm{e}^{\frac{1}{2}\alpha^2\sigma^2(T-t)} \\
&= S^\alpha\,\mathrm{e}^{\alpha(r+\frac{1}{2}\sigma^2(\alpha-1))\tau}
\end{aligned} \tag{12.40}$$

where $\tau := T - t$. The only difference between the conditional expectation in (12.40) and that of $S^\alpha(T)\,\mathbb{I}_{\{S(T)>A\}}$ is the indicator function term. By the exact same step as in our derivation of the standard call and put options, the indicator random variable term simplifies,

$$\mathbb{I}_{\left\{\ln\frac{S(T)}{S(t)}>\ln\frac{A}{S}\right\}} = \mathbb{I}_{\{\widetilde{Z}>-d_-(\frac{S}{A},\tau)\}}\,,$$

with $d_\pm$ defined in (12.24), i.e., $d_\pm(\frac{S}{A},\tau) \equiv \frac{\ln(S/A)+(r\pm\frac{1}{2}\sigma^2)\tau}{\sigma\sqrt{\tau}}$. Then, conditioning on $S(t) = S$ and using independence,

$$\begin{aligned}
\widetilde{\mathrm{E}}_{t,S}\big[S^\alpha(T)\,\mathbb{I}_{\{S(T)>A\}}\big] &= \mathrm{e}^{\alpha(r-\sigma^2/2)\tau}\,\widetilde{\mathrm{E}}\big[S^\alpha(t)\,\mathbb{I}_{\{\widetilde{Z}>-d_-(\frac{S(t)}{A},\tau)\}}\,\mathrm{e}^{\alpha\sigma\sqrt{\tau}\widetilde{Z}}\,|\,S(t)=S\big] \\
&= S^\alpha\,\mathrm{e}^{\alpha(r-\sigma^2/2)\tau}\,\widetilde{\mathrm{E}}\big[\mathbb{I}_{\{\widetilde{Z}>-d_-(\frac{S}{A},\tau)\}}\,\mathrm{e}^{\alpha\sigma\sqrt{\tau}\widetilde{Z}}\big] \\
&= S^\alpha\,\mathrm{e}^{\alpha(r+\frac{1}{2}\sigma^2(\alpha-1))\tau}\,\mathcal{N}\left(d_+(\frac{S}{A},\tau) + (\alpha-1)\sigma\sqrt{\tau}\right).
\end{aligned} \tag{12.41}$$

The last expectation was computed using the identity in (A.1) of the Appendix and noting that $d_-(x,\tau) = d_+(x,\tau) - \sigma\sqrt{\tau}$. Note that this formula also recovers (12.40) in the limit $A \searrow 0$. This follows by monotone convergence of the expectations where $\mathbb{I}_{\{S(T)>A\}} \nearrow \mathbb{I}_{\{S(T)>0\}} = 1$, as $A \searrow 0$. Based on (12.39) and the linearity property of the expectation, using (12.41) for $A = A_1$ and for $A = A_2$ leads to the formula

$$\begin{aligned}\widetilde{\mathrm{E}}_{t,S}\left[S^\alpha(T)\,\mathbb{I}_{\{A_1<S(T)\leqslant A_2\}}\right] &= \widetilde{\mathrm{E}}_{t,S}\left[S^\alpha(T)\,\mathbb{I}_{\{S(T)>A_1\}}\right] - \widetilde{\mathrm{E}}_{t,S}\left[S^\alpha(T)\,\mathbb{I}_{\{S(T)>A_2\}}\right] \\ &= S^\alpha\,\mathrm{e}^{\alpha(r+\frac{1}{2}\sigma^2(\alpha-1))\tau}\left[\mathcal{N}\left(d_-(\frac{S}{A_1},\tau)+\alpha\sigma\sqrt{\tau}\right) - \mathcal{N}\left(d_-(\frac{S}{A_2},\tau)+\alpha\sigma\sqrt{\tau}\right)\right]. \quad (12.42)\end{aligned}$$

We now use the formulae in (12.40)–(12.42) to price a soft-strike call option in the following example.

Example 12.2. (Soft-Strike Call Option) Consider the soft-strike European call option with payoff function

$$\Lambda_a(S) = \begin{cases} 0 & \text{if } S < K - a, \\ \frac{1}{4a}(S-K+a)^2 & \text{if } K-a \leqslant S \leqslant K+a, \\ S-K & \text{if } S > K+a, \end{cases} \quad (12.43)$$

where the constant $a \in [0, K]$ and K is a central strike value.

(a) Describe the main features of the graph of $\Lambda_a(S)$ for all $S > 0$.

(b) Assume the stock price process $\{S(t)\}_{t\geqslant 0}$ is a GBM with constant interest rate r and volatility σ. Let $S(t) = S$ be the spot price of the stock at current time $t < T$. Derive the no-arbitrage pricing formula for a European-style option with the above payoff function.

Solution. (a) (see Figure 12.2) Note that $\Lambda_a(S) \geqslant (S-K)^+$, where $\Lambda_a(S) \searrow (S-K)^+$ as $a \searrow 0$, i.e., as a function of the parameter a, the soft-strike payoff decreases monotonically to the standard call payoff as $a \searrow 0$. In contrast to the standard call payoff $(S-K)^+$ whose derivative w.r.t. S has a unit jump discontinuity at $S = K$, the payoff $\Lambda_a(S)$ has a continuous derivative for all S. The left and right derivatives at $S = K - a$ are the same, $\Lambda_a'(K-a) = 0$, and the left and right derivatives at $S = K + a$ are the same, $\Lambda_a'(K+a) = 1$. Combining this with the derivative for $S \in (K-a, K+a)$, the payoff has a continuous derivative w.r.t. S given by

$$\Lambda_a'(S) = \begin{cases} 0 & \text{if } S < K - a, \\ \frac{1}{2a}(S-K+a) & \text{if } K-a \leqslant S \leqslant K+a, \\ 1 & \text{if } S > K+a. \end{cases} \quad (12.44)$$

Moreover, the payoff has a piecewise constant second derivative $\Lambda_a''(S)$. In fact, the payoff can be expressed as an integral over the standard call payoff function $(S-k)^+$ by employing a continuum of strikes $k \in (K-a, K+a)$ (see Exercise 12.14).

(b) Express $\Lambda_a(S(T))$ as a linear combination of payoffs involving the different powers of $S(T)$ multiplying indicator random variables:

$$\begin{aligned}\Lambda_a(S(T)) = \frac{1}{4a}&\left(S^2(T) - 2(K-a)S(T) + (K-a)^2\right)\mathbb{I}_{\{K-a\leqslant S(T)\leqslant K+a\}} \\ &+ S(T)\,\mathbb{I}_{\{S(T)>K+a\}} - K\,\mathbb{I}_{\{S(T)>K+a\}}\,.\end{aligned}$$

The option value at current time t is then the discounted conditional expectation of this payoff. Let $V(t, S) \equiv C(t, S; K, a)$ denote the time-t value of the soft-strike call

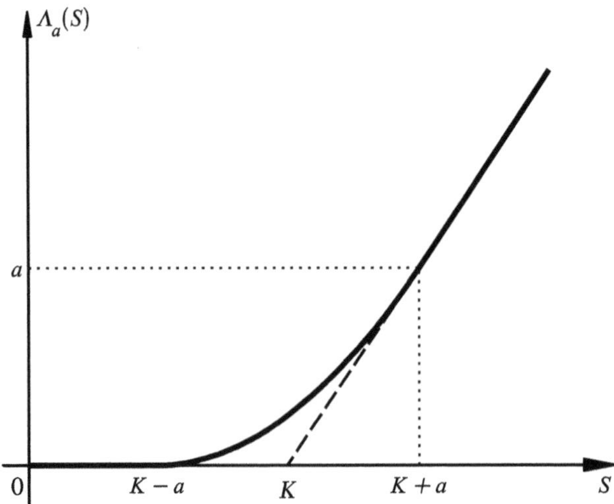

FIGURE 12.2: The payoff of a soft-strike call centred at strike K.

where we include the dependence on the parameters a, K defining the payoff. By linearity of expectations:

$$\begin{aligned}
C(t,S;K,a) &= \mathrm{e}^{-r\tau}\,\widetilde{\mathrm{E}}_{t,S}\big[\Lambda_a(S(T))\big] \\
&= \mathrm{e}^{-r\tau}\bigg(\frac{1}{4a}\widetilde{\mathrm{E}}_{t,S}\big[S^2(T)\,\mathbb{I}_{\{K-a\leqslant S(T)\leqslant K+a\}}\big] \\
&\quad - \frac{(K-a)}{2a}\widetilde{\mathrm{E}}_{t,S}\big[S(T)\,\mathbb{I}_{\{K-a\leqslant S(T)\leqslant K+a\}}\big] + \widetilde{\mathrm{E}}_{t,S}\big[S(T)\,\mathbb{I}_{\{S(T)>K+a\}}\big] \\
&\quad + \frac{(K-a)^2}{4a}\widetilde{\mathbb{P}}_{t,S}\big(K-a\leqslant S(T)\leqslant K+a\big) - K\,\widetilde{\mathbb{P}}_{t,S}\big(S(T)>K+a\big)\bigg).
\end{aligned}$$

We now use the formulas given in (12.41) and (12.42) for respective powers of $\alpha = 1, 2$, as well as our previously derived formulas for the risk-neutral conditional probability that $S(T)$ lies above a strike level or within an interval. Combining all terms gives

$$\begin{aligned}
C(t,S;K,a) \;&= S^2\tfrac{\mathrm{e}^{(r+\sigma^2)\tau}}{4a}\left[\mathcal{N}\left(d_+(\tfrac{S}{K-a},\tau)+\sigma\sqrt{\tau}\right)-\mathcal{N}\left(d_+(\tfrac{S}{K+a},\tau)+\sigma\sqrt{\tau}\right)\right] \\
&\quad -\tfrac{(K-a)}{2a}S\left[\mathcal{N}\left(d_+(\tfrac{S}{K-a},\tau)\right)-\mathcal{N}\left(d_+(\tfrac{S}{K+a},\tau)\right)\right] \\
&\quad +\tfrac{(K-a)^2}{4a}\mathrm{e}^{-r\tau}\left[\mathcal{N}\left(d_-(\tfrac{S}{K-a},\tau)\right)-\mathcal{N}\left(d_-(\tfrac{S}{K+a},\tau)\right)\right] \\
&\quad +S\mathcal{N}\left(d_+(\tfrac{S}{K+a},\tau)\right)-K\mathrm{e}^{-r\tau}\mathcal{N}\left(d_-(\tfrac{S}{K+a},\tau)\right) \qquad (12.45)
\end{aligned}$$

where $\tau = T - t$ is the time to maturity and $d_\pm(x,\tau)$ are defined in (12.24). □

Based on the pricing formula in (12.45) we can compute the delta hedging position in the stock. Moreover, we can readily price the corresponding *soft-strike put* option with given center strike K and width a. We leave this as an exercise for the reader (see Exercise 12.15). There are also other power options that are readily priced with the use of the formulae in (12.40)–(12.42) and these are assigned as exercises at the end of this chapter.

12.1.5 Dividend Paying Stock

12.1.5.1 The Case of Continuous Dividend Paying Stock

Let us now consider the above (B, S) model where the stock is a standard GBM which also pays a dividend with a constant continuous yield $q > 0$ per unit of time. During an infinitesimal time $\mathrm{d}t$ the holder of the stock receives a dividend payment of $qS(t)\,\mathrm{d}t$ that is in proportion to the stock price at time t. We can see this from the return due only to the dividend, $\frac{S(t+\delta t)-S(t)}{S(t)} = \mathrm{e}^{q\delta t} - 1 \approx q\delta t$ for small time interval $\delta t \approx 0$. By no-arbitrage this dividend payment to the stock holder must be exactly balanced by a decrease in the share price of the stock. Hence, in the physical $\mathbb{P}$-measure there is the additional negative drift term, $-qS(t)\,\mathrm{d}t$, due to the dividend. The stock price process is a GBM with SDE

$$\mathrm{d}S(t) = (\mu - q)S(t)\,\mathrm{d}t + \sigma S(t)\,\mathrm{d}W(t)\,.$$

By the same unique risk-neutral measure $\widetilde{\mathbb{P}}$ as above, where $\widetilde{W}(t) := W(t) + \frac{(\mu-r)}{\sigma}t$ is a standard $\widetilde{\mathbb{P}}$-BM, the above SDE takes the form

$$\mathrm{d}S(t) = S(t)\big[(r-q)\,\mathrm{d}t + \sigma\,\mathrm{d}\widetilde{W}(t)\big]\,. \tag{12.46}$$

For a nonzero dividend, the risk-neutral drift of the stock price is $(r - q)$ in the place of r. Given an arbitrary initial stock value $S(0) > 0$, this SDE has a unique solution

$$S(t) = S(0)\,\mathrm{e}^{(r-q-\sigma^2/2)t+\sigma\widetilde{W}(t)}\,. \tag{12.47}$$

The time-T stock price is given in terms of the time-t price and a $\widetilde{\mathbb{P}}$-BM increment:

$$S(T) = S(t)\,\mathrm{e}^{(r-q-\sigma^2/2)(T-t)+\sigma(\widetilde{W}(T)-\widetilde{W}(t))}\,. \tag{12.48}$$

Note that the process $\{\widehat{S}(t) := \mathrm{e}^{qt}S(t)\}_{t\geqslant 0}$ acts as a nondividend stock, i.e., it has the same risk-neutral drift, r, as any non-dividend-paying asset. Discounting the $\widehat{S}$ price process with the bank account gives $\left\{\overline{\widehat{S}}(t) := D(t)\widehat{S}(t) \equiv \mathrm{e}^{-rt}\widehat{S}(t) \equiv \mathrm{e}^{-(r-q)t}S(t)\right\}_{t\geqslant 0}$ as a $\widetilde{\mathbb{P}}$-martingale.

The replicating portfolio is the same as given in (12.4). The portfolio is invested in the amount of $\Delta_t S(t)$ in the stock, so the dividend payment is $q\Delta_t S(t)\,\mathrm{d}t$ over time $\mathrm{d}t$. This term is now added to the differential change in the self-financing portfolio in (12.6), giving

$$\begin{aligned}
\mathrm{d}\Pi_t &= r\beta_t B(t)\,\mathrm{d}t + \Delta_t\,\mathrm{d}S(t) + q\Delta_t S(t)\,\mathrm{d}t \\
&= [r\Pi_t - (r-q)\Delta_t S(t)]\,\mathrm{d}t + \Delta_t\,\mathrm{d}S(t) \\
&= r\Pi_t\,\mathrm{d}t + \sigma\Delta_t S(t)\,\mathrm{d}\widetilde{W}(t)\,.
\end{aligned} \tag{12.49}$$

In the last line we used (12.46). This is of the same form as in (12.7) and hence (12.8) still holds where the discounted self-financing portfolio value process $\{\overline{\Pi}_t\}_{0\leqslant t\leqslant T}$ is a $\widetilde{\mathbb{P}}$-martingale. The risk-neutral pricing formula in (12.9) as well as (12.10) and (12.11) still hold. The conditions and arguments given in Section 12.1 that guarantee that a claim can be replicated (hedged) are the same as in the case of zero dividend on the stock.

The only difference is that the stock has drift parameter $(r - q)$ instead of r, i.e., $S(t)$ is given by (12.47) instead of (12.2) and (12.48) replaces (12.12). So the question is, in terms of pricing, what does this change? The general answer is actually very simple given the solution in the case that $q = 0$. Let's take a look at how (12.14)–(12.20) change. Using (12.48) in the place of (12.12), the expectations in (12.14) and (12.15) now become

$$V(t, S) = \mathrm{e}^{-r\tau}\int_{-\infty}^{\infty}\Lambda(S\,\mathrm{e}^{(r-q-\frac{1}{2}\sigma^2)\tau+\sigma\sqrt{\tau}z})n(z)\,\mathrm{d}z = \mathrm{e}^{-r\tau}\int_0^{\infty}\Lambda(y)\,\widetilde{p}(\tau; S, y)\,\mathrm{d}y\,, \tag{12.50}$$

where $\widetilde{p}(\tau; S, y)$ is now the risk-neutral transition PDF of the stock price GBM process with drift $r - q$, i.e., with r replaced by $r - q$ in (12.16). The generator for the stock price process with SDE in (12.46) is defined by $\mathcal{G}V := \frac{1}{2}\sigma^2 S^2 \frac{\partial^2}{\partial S^2} V + (r-q)S\frac{\partial}{\partial S}V$. Hence, the respective Black–Scholes partial differential equations in (12.17) and (12.18) are now

$$\frac{\partial v}{\partial \tau} = \frac{1}{2}\sigma^2 S^2 \frac{\partial^2 v}{\partial S^2} + (r-q)S\frac{\partial v}{\partial S} - rv\,, \tag{12.51}$$

subject to $v(0, S) = \Lambda(S)$ and

$$\frac{\partial V}{\partial t} + \frac{1}{2}\sigma^2 S^2 \frac{\partial^2 V}{\partial S^2} + (r-q)S\frac{\partial V}{\partial S} - rV = 0\,, \tag{12.52}$$

subject to $V(T, S) = \Lambda(S)$. Note that the dividend only changes the drift term where $rS\frac{\partial}{\partial S}$ has been replaced by $(r-q)S\frac{\partial}{\partial S}$. Everything else is the same, including the discount factor. Equations (12.19) and (12.20) are then also the same, i.e., the delta hedge position is the same. Of course, the derivative value $V(t, S)$ for $q \neq 0$ differs from the value when $q = 0$.

We now show that the derivative pricing formula for $q \neq 0$ (nonzero stock dividend) obtains trivially from the corresponding pricing formula for $q = 0$ (zero dividend), given arbitrary payoff $V_T = \Lambda(S(T))$. To precisely describe this, let $V(t, S; r, q)$ be the time-t price of the derivative for given interest rate r and stock dividend q. This function has a dependence on parameters r and q, as well as other parameters that we simply suppress. The corresponding price when $q = 0$ is then $V(t, S; r, 0)$. The function $V(t, S; r, q)$ is given by (12.50) and $V(t, S; r, 0)$ is given by (12.14). Multiplying out the discount factor in both cases gives $e^{r\tau}V(t, S; r, q) = [e^{r\tau}V(t, S; r, 0)]|_{r \to r-q}$, which is equivalent to

$$V(t, S; r, q) = e^{-q(T-t)}V(t, S; r-q, 0)\,. \tag{12.53}$$

From this relation we see that the pricing function (on the left) for $q \neq 0$ is given by the corresponding pricing function (on the right) for $q = 0$ after replacing the interest rate r by $r - q$ and multiplying the function by $e^{-q\tau}$, $\tau = T - t$. The simple example below shows how (12.53) is very easily applied to a standard call and put.

Example 12.3. Apply the symmetry in (12.53) to (12.27) and obtain pricing formulae for the standard European call for a stock with continuous constant dividend yield q.

Solution. Let $C(t, S; r, q) \equiv C(\tau, S, K, \sigma, r, q)$ denote the pricing formula for the call with stock dividend q. For $q = 0$, $C(t, S; r, 0) \equiv C(\tau, S, K, \sigma, r, 0)$ is given by (12.27):

$$C(\tau, S, K, \sigma, r, 0) = S\mathcal{N}\left(\frac{\ln\frac{S}{K} + (r + \frac{1}{2}\sigma^2)\tau}{\sigma\sqrt{\tau}}\right) - e^{-r\tau}K\mathcal{N}\left(\frac{\ln\frac{S}{K} + (r - \frac{1}{2}\sigma^2)\tau}{\sigma\sqrt{\tau}}\right).$$

Applying (12.53), $C(t, S; r, q) = e^{-q\tau}C(t, S; r-q, 0) = e^{-q\tau}C(\tau, S, K, \sigma, r-q, 0)$:

$$\begin{aligned} C(t, S; r, q) &= e^{-q\tau}S\mathcal{N}\left(\frac{\ln\frac{S}{K} + (r - q + \frac{1}{2}\sigma^2)\tau}{\sigma\sqrt{\tau}}\right) - e^{-r\tau}K\mathcal{N}\left(\frac{\ln\frac{S}{K} + (r - q - \frac{1}{2}\sigma^2)\tau}{\sigma\sqrt{\tau}}\right) \\ &= e^{-q\tau}S\mathcal{N}\left(d_+(\frac{e^{-q\tau}S}{K}, \tau)\right) - e^{-r\tau}K\mathcal{N}\left(d_-(\frac{e^{-q\tau}S}{K}, \tau)\right) \end{aligned} \tag{12.54}$$

where $\tau = T - t$ and $d_\pm(x, \tau)$ are defined in (12.24). □

This example also points out another very useful simple symmetry where the standard option pricing function, for given dividend q, is given by the original (zero dividend) pricing function with spot value S replaced by the "effective spot" value $\mathrm{e}^{-q\tau}S$ (keeping everything else the same). This is an alternatively useful symmetry that we can apply to immediately obtain the pricing formula for $q \neq 0$ by substituting $\mathrm{e}^{-q\tau}S$ for S within the pricing formula for $q = 0$. We can express this additional symmetry as

$$V(t, S; r, q) = V(t, \mathrm{e}^{-q\tau}S; r, 0)\,. \tag{12.55}$$

We see quite trivially how this symmetry works in the above example of the call where setting $\mathrm{e}^{-q\tau}S$ for S in (12.27) gives (12.54). Applying either (12.53) or (12.55) to (12.30) gives the corresponding pricing formula for the put option on a dividend paying stock,

$$P(t, S; r, q) = \mathrm{e}^{-r\tau} K\mathcal{N}\left(-d_-(\frac{\mathrm{e}^{-q\tau}S}{K}, \tau)\right) - \mathrm{e}^{-q\tau}S\mathcal{N}\left(-d_+(\frac{\mathrm{e}^{-q\tau}S}{K}, \tau)\right). \tag{12.56}$$

The put-call parity relation in (12.29) now takes the more general form,

$$C(t, S) - P(t, S) = \mathrm{e}^{-q\tau}S - \mathrm{e}^{-r\tau} K \tag{12.57}$$

where we simply write $C(t, S) = C(t, S; r, q)$ and $P(t, S) = P(t, S; r, q)$.

Corresponding symmetry relations for the delta hedging position also follow. Let us denote $\Delta(t, S) \equiv \Delta(t, S; r, q) = \frac{\partial}{\partial S}V(t, S; r, q)$. Then,

$$\Delta(t, S; r, q) = \mathrm{e}^{-q\tau}\frac{\partial}{\partial S}V(t, S; r - q, 0) = \mathrm{e}^{-q\tau}\Delta(t, S; r - q, 0)\,. \tag{12.58}$$

By the chain rule, differentiating (12.55) gives us an alternative symmetry:

$$\Delta(t, S; r, q) = \mathrm{e}^{-q\tau}\frac{\partial}{\partial x}V(t, x; r, 0)\bigg|_{x=\mathrm{e}^{-q\tau}S} = \mathrm{e}^{-q\tau}\Delta(t, \mathrm{e}^{-q\tau}S; r, 0)\,. \tag{12.59}$$

For example, consider a call where (12.34) gives $\Delta_c(t, S; r, 0)$ and by either (12.58) or (12.59):

$$\Delta_c(t, S; r, q) = \mathrm{e}^{-q\tau}\mathcal{N}\left(d_+(\frac{\mathrm{e}^{-q\tau}S}{K}, \tau)\right) \equiv \mathrm{e}^{-q\tau}\mathcal{N}\left(\frac{\ln\frac{S}{K} + (r - q + \frac{1}{2}\sigma^2)\tau}{\sigma\sqrt{\tau}}\right). \tag{12.60}$$

We point out that this formula can also be obtained using our previous derivations, i.e., without use of the symmetry relation in (12.58).

If we assume no knowledge of the above symmetry and no prior pricing formula for $q = 0$, then we can of course derive the pricing formula in (12.54) from first principle, by employing the same steps that lead to (12.27) where we had $q = 0$. Namely, we substitute the expression for $S(T)$ in (12.48) into the discounted expectation in (12.23) and evaluate by using the identities in (12.25) and (12.26) where now the drift $r - q$ replaces r, i.e., (12.25) and (12.26) become

$$\widetilde{\mathrm{E}}_{t,S}\left[S(T)\,\mathbb{I}_{\{S(T)>K\}}\right] = \mathrm{e}^{(r-q)\tau}S\mathcal{N}\left(d_+(\frac{\mathrm{e}^{-q\tau}S}{K}, \tau)\right) \tag{12.61}$$

and

$$\widetilde{\mathbb{P}}_{t,S}(S(T) > K) = \mathcal{N}\left(d_-(\frac{\mathrm{e}^{-q\tau}S}{K}, \tau)\right). \tag{12.62}$$

Combining these into (12.23) gives the call pricing formula in (12.54). Using similar steps, or by put-call parity, we obtain the above put pricing formula. The pricing formulae for

European derivatives on a dividend paying stock for all other types of payoffs, including those we considered in Sections 12.1.3 and 12.1.4, also follow by the above symmetry relation. Alternatively the formulae can be derived from first principles based on the identities in (12.61) and (12.62). In the case of power options on a dividend payoff stock, we have the identities in (12.40), (12.41), and (12.42) where r is replaced by $r - q$ (or equivalently do not replace r but replace S by $\mathrm{e}^{-q\tau}S$).

[*Technical Remark*: We now give a more technical argument showing that the above symmetry relation in (12.53) holds as a particular case of a similar symmetry for more complex path-dependent payoffs. Consider an arbitrary European derivative where the payoff V_T has a path dependence on the stock, i.e., is a functional of the stock price process from time t to T. Examples of these derivatives are barrier options, Asian options, lookback options, etc. In general, the derivative price V_t is an $\mathcal{F}_t$-measurable random variable where (12.9) holds for any attainable claim V_T. More specifically, denote by $\{S^{(r-q)}(u) : t \leqslant u \leqslant T\}$ the path of the stock price process defined by the risk-neutral drift parameter $r - q$ (and volatility σ) and let $V_t = V_t^{(r,q)}$ be the corresponding derivative price at time t, for any constant interest rate r and constant dividend yield q. Now, let V_T depend on any segment of the stock price path, including the entire path history from time t to T. We write this as a functional, $V_T^{(r,q)} = F\left(\{S^{(r-q)}(u) : t \leqslant u \leqslant T\}\right)$. Note that $V_T^{(r-q,0)} = F\left(\{S^{(r-q-0)}(u) : t \leqslant u \leqslant T\}\right) = V_T^{(r,q)}$. Hence, by (12.9),

$$V_t^{(r,q)} = \mathrm{e}^{-r(T-t)}\widetilde{\mathrm{E}}\left[V_T^{(r,q)} \mid \mathcal{F}_t\right] = \mathrm{e}^{-q(T-t)} \cdot \mathrm{e}^{-(r-q)(T-t)}\widetilde{\mathrm{E}}\left[V_T^{(r-q,0)} \mid \mathcal{F}_t\right]$$
$$= \mathrm{e}^{-q(T-t)} V_t^{(r-q,0)} . \tag{12.63}$$

So (12.53) is recovered for standard non-path-dependent payoffs with spot $S(t) = S$ giving $V_t^{(r,q)} = V(t, S; r, q)$. However, we note that the relation in (12.55) does not generally hold for path-dependent derivative pricing.]

12.1.5.2 The Case of Discrete-Time Dividends

The continuous-time dividend payment model of a stock in the previous section led to analytical pricing formulae for European derivatives that have the same form as the formulae with zero dividend. We shall mostly adopt this model in further applications. In practice, however, stocks in the market pay dividends at discrete regular intervals of time. Let's suppose that at present time $t_0 < T$ we know the fixed future dividend dates, $T_i, i = 1, \ldots, N$, $t_0 < T_1 < \ldots < T_N \leqslant T$. At each time T_i the dividend payment $\mathrm{div}(T_i)$ is a proportion of the stock value at time T_i, i.e., $\mathrm{div}(T_i) = d_i S(T_i)$, where $0 \leqslant d_i \leqslant 1$ is the dividend percentage. When no dividend is paid at time T_i we have $d_i = 0$ and when the full share value of the stock is paid at time T_i then $d_i = 1$ and the stock becomes worthless for all time after T_i. Typically, we can assume $0 < d_i < 1$ for $i = 1, \ldots, N$. We remark that by writing $d_i = q_i\,(T_i - T_{i-1})$ then q_i is a dividend yield (rate) that is fixed within the time interval $(T_{i-1}, T_i]$.

The model for the stock is then as follows. Between the ith dividend date and just before the $(i+1)$th dividend date, i.e., within *any time interval between dividend payments*, the stock price evolves simply as a GBM according to the SDE in (12.1). Hence the stock price at time $t_0 \leqslant t < T_1$ is given in terms of the spot $S(t_0) \equiv S_0$ as

$$S(t) = S_0\, \mathrm{e}^{(r-\frac{1}{2}\sigma^2)(t-t_0)+\sigma(\widetilde{W}(t)-\widetilde{W}(t_0))} .$$

At time $t = T_1^-$:

$$S(T_1^-) = S_0\, \mathrm{e}^{(r-\frac{1}{2}\sigma^2)(T_1-t_0)+\sigma(\widetilde{W}_1-\widetilde{W}(t_0))} ,$$

where we use the notation $\widetilde{W}_i \equiv \widetilde{W}(T_i)$ for $i = 1, \ldots, N$. For times $t \in [T_{i-1}, T_i)$, $i = 2, \ldots, N$, we have

$$S(t) = S(T_{i-1})\, \mathrm{e}^{(r-\frac{1}{2}\sigma^2)(t-T_{i-1})+\sigma(\widetilde{W}(t)-\widetilde{W}_{i-1})},$$

and in particular for $t = T_i^-$, just before the ith dividend date,

$$S(T_i^-) = S(T_{i-1})\, \mathrm{e}^{(r-\frac{1}{2}\sigma^2)(T_i-T_{i-1})+\sigma(\widetilde{W}_i-\widetilde{W}_{i-1})}. \tag{12.64}$$

Finally, for the last interval $[T_N, T]$,

$$S(T) = S(T_N)\, \mathrm{e}^{(r-\frac{1}{2}\sigma^2)(T-T_N)+\sigma(\widetilde{W}(T)-\widetilde{W}_N)}.$$

Now, at each dividend payment date T_i, $i = 1, \ldots, N$, the stock price instantaneously decreases by a fraction d_i due to the dividend payment. This is expressed as

$$S(T_i) = (1 - d_i) S(T_i^-). \tag{12.65}$$

Hence, substituting the expression for $S(T_i^-)$ in (12.64) into this last equation gives the evolution of the stock price from one discrete dividend date to the next as

$$\frac{S(T_i)}{S(T_{i-1})} = (1 - d_i)\, \mathrm{e}^{(r-\frac{1}{2}\sigma^2)(T_i-T_{i-1})+\sigma(\widetilde{W}_i-\widetilde{W}_{i-1})}, \quad i = 2, \ldots, N. \tag{12.66}$$

For time t_0 to just after the first dividend payment time T_1 we have

$$\frac{S(T_1)}{S(t_0)} = (1 - d_1)\, \mathrm{e}^{(r-\frac{1}{2}\sigma^2)(T_1-t_0)+\sigma(\widetilde{W}_1-\widetilde{W}(t_0))}. \tag{12.67}$$

Multiplying out the stock price ratios for all adjoining time intervals including the first and last interval (note: this is a telescoping product) gives

$$\begin{aligned}
S(T) = S_0 \frac{S(T)}{S(t_0)} &= \frac{S(T_1)}{S(t_0)} \left(\prod_{i=2}^{N} \frac{S(T_i)}{S(T_{i-1})} \right) \frac{S(T)}{S(T_N)} \\
&= S_0 (1 - d_1)\, \mathrm{e}^{(r-\frac{1}{2}\sigma^2)(T_1-t_0)+\sigma(\widetilde{W}_1-\widetilde{W}(t_0))} \cdot \prod_{i=2}^{N} (1 - d_i)\, \mathrm{e}^{(r-\frac{1}{2}\sigma^2)(T_i-T_{i-1})+\sigma(\widetilde{W}_i-\widetilde{W}_{i-1})} \\
&\quad \cdot \mathrm{e}^{(r-\frac{1}{2}\sigma^2)(T-T_N)+\sigma(\widetilde{W}(T)-\widetilde{W}_N)} \\
&= S_0 \Big[\prod_{i=1}^{N} (1 - d_i) \Big] \mathrm{e}^{(r-\frac{1}{2}\sigma^2)(T-t_0)+\sigma(\widetilde{W}(T)-\widetilde{W}(t_0))} \\
&= \widetilde{S}_0 \mathrm{e}^{(r-\frac{1}{2}\sigma^2)(T-t_0)+\sigma(\widetilde{W}(T)-\widetilde{W}(t_0))},
\end{aligned} \tag{12.68}$$

where $\widetilde{S}_0 \equiv S_0 \prod_{i=1}^{N} (1 - d_i) = S_0 (1 - d_1) \cdots (1 - d_N)$. In the last line we simplified the sum in the exponents where $\widetilde{W}_1 - \widetilde{W}(t_0) + \widetilde{W}_2 - \widetilde{W}_1 + \ldots + \widetilde{W}_N - \widetilde{W}_{N-1} + \widetilde{W}(T) - \widetilde{W}_N = \widetilde{W}(T) - \widetilde{W}(t_0)$.

From (12.68) we see that the stock price $S(T)$ is a GBM random variable with drift r and volatility σ. The overall discount factor due to all the dividends is multiplying the actual initial stock price S_0 giving the initial value $\widetilde{S}_0$ *now acting as effective initial stock price* at time-t_0. We recall that in the case of a continuous dividend the quantity $\mathrm{e}^{-q\tau} S_0$ acts in the place of $\widetilde{S}_0$ when pricing a non-path-dependent European derivatives. Hence, all the time-t_0 pricing formulae for non-path-dependent (standard) European derivatives on the stock are obtained by using the initial value $\widetilde{S}_0$ in the place of S_0. This is clear by substituting the time-T stock price expression in (12.68) into the risk-neutral pricing (expectation) formula with (non-path-dependent) payoff $V_T = \Lambda(S(T))$. In particular, let $V(t_0, S; \mathbf{d})$ represent the European pricing formula for the above stock model with discrete dividends, within the time interval (t_0, T), grouped in a vector $\mathbf{d} = (d_1, \ldots, d_N)$. Accordingly, let $V(t_0, S) \equiv V(t_0, S; \mathbf{0})$ be the pricing function for zero dividends. Then, the analogue of (12.55) is the relation

$$V(t_0, S; \mathbf{d}) = V(t_0, \widetilde{S}_0), \quad \text{where } \widetilde{S}_0 \equiv S_0 \prod_{i=1}^{N} (1 - d_i). \tag{12.69}$$

The pricing functions for a standard call option, put option, power option, etc., follow immediately based on the formulae for zero dividends. For example, simply setting the spot value to $\widetilde{S}_0 \equiv S_0 \prod_{i=1}^{N}(1-d_i)$ in (12.27) gives the pricing formula for the time-t_0 call option with discrete dividends $d_1, \ldots, d_N$ within the time to maturity $\tau = T - t_0$:

$$C(t_0, S; \mathbf{d}) = \widetilde{S}_0 \mathcal{N}(\widetilde{d}_+) - \mathrm{e}^{-r\tau} K \mathcal{N}(\widetilde{d}_-) \tag{12.70}$$

where

$$\widetilde{d}_\pm := d_\pm\left(\frac{\widetilde{S}_0}{K}, \tau\right) = \frac{\ln \frac{S_0}{K} + \sum_{i=1}^{N} \ln(1 - d_i) + (r \pm \frac{1}{2}\sigma^2)\tau}{\sigma\sqrt{\tau}}.$$

12.2 Forward Starting and Compound Options

We are now equipped to readily develop pricing and hedging formulae for other classes of options besides the standard European options considered so far where we assumed a payoff $V_T = \Lambda(S(T))$. Rather than having a payoff that depends solely on the terminal stock price $S(T)$ at maturity T, there are European-style options that can have stipulations at a finite number of intermediate dates within the lifetime of the option. These types of options are examples of what can be termed *multistage options*. As in most options, there are simpler and more complex versions of these contracts.

Examples of simpler contracts are *forward starting* options with one intermediate date $T_1 < T$. The holder enters the contract at a time $t < T_1$ such that at time T_1 the contract has the value of an option (say a European call) on an underlying stock which expires at T. Generally we can represent the option value at time T_1 as a function of $S(T_1)$: $V_{T_1} = V_{T_1}(T_1, S(T_1))$. The time-$T$ payoff, $V_T = \Lambda(S(T_1), S(T))$, of the option can depend upon $S(T_1)$ and $S(T)$. By risk-neutral pricing we have (discounting from T back to T_1)

$$V_{T_1} = \mathrm{e}^{-r(T-T_1)} \widetilde{\mathrm{E}}_{T_1, S(T_1)}[\Lambda(S(T_1), S(T))]. \tag{12.71}$$

Let $V(t, S) \equiv V(t, S; T_1, T)$ denote the time-t value of the forward starting option with given spot $S(t) = S$. By the Markov property of the stock price process, conditioning on $\mathcal{F}_t$ reduces to conditioning on $S(t)$. Applying again the risk-neutral pricing formula (discounting from time T_1 to t) and the tower property[4] finally gives us the time-t price as a single conditional expectation discounted from T to t:

$$\begin{aligned} V(t, S; T_1, T) &= \mathrm{e}^{-r(T_1-t)} \widetilde{\mathrm{E}}_{t,S}[V_{T_1}] \\ &= \mathrm{e}^{-r(T_1-t)} \cdot \mathrm{e}^{-r(T-T_1)} \widetilde{\mathrm{E}}_{t,S}\left[\widetilde{\mathrm{E}}_{T_1,S(T_1)}[\Lambda(S(T_1), S(T))]\right] \\ &= \mathrm{e}^{-r(T-t)} \widetilde{\mathrm{E}}_{t,S}\left[\Lambda(S(T_1), S(T))\right]. \end{aligned} \tag{12.72}$$

[4]We remind the reader of the shorthand notation we have been adopting for conditional expectations. In particular, $\widetilde{\mathrm{E}}_{t,S}[(\cdot)] \equiv \widetilde{\mathrm{E}}[(\cdot) \mid S(t) = S]$ is a number where $S > 0$ is a spot value and $\widetilde{\mathrm{E}}_{T_1,S(T_1)}[(\cdot)] \equiv \widetilde{\mathrm{E}}[(\cdot) \mid S(T_1)]$ is a $\sigma(S(T_1))$-measurable (and hence $\mathcal{F}_{T_1}$-measurable) random variable where we are conditioning on the σ-algebra, $\sigma(S(T_1))$, generated by the stock at time T_1.

For instance, in a *forward starting call* the holder enters the contract at a time $t < T_1$, prior to an intermediate date T_1, whose value at time T_1 is a call on an underlying stock initiated at the "forward" time T_1. The call initiated at time T_1 expires at date $T > T_1$ and has some strike specification, which is generally a function of the stock price $S(T_1)$ (i.e., the strike is not a constant specified value). Specifically, the forward starting call can be specified as having strike $K_{T_1} = S(T_1)$, then $\Lambda(S(T_1), S(T)) = (S(T) - S(T_1))^+$. Hence, as viewed at present time t, the strike is a random variable corresponding to the price of the stock at time T_1. Inserting the payoff into (12.72) gives the time-t price of this forward starting call

$$C(t, S; T_1, T) = \mathrm{e}^{-r(T-t)} \widetilde{\mathrm{E}}_{t,S}\left[\,(S(T) - S(T_1))^+\,\right].$$

Let's assume the stock price is a GBM given by (12.2) with zero dividend. The addition of a continuous constant dividend yield q on the stock can be done trivially by applying the symmetry relation in (12.53) to the resulting pricing formula. The above expectation is readily evaluated by writing the payoff as

$$(S(T) - S(T_1))^+ = S(T_1)\left(\frac{S(T)}{S(T_1)} - 1\right)^+ = S(T_1)\,(Y-1)^+$$

where random variable $Y := \frac{S(T)}{S(T_1)} = \mathrm{e}^{(r-\frac{1}{2}\sigma^2)(T-T_1)+\sigma(\widetilde{W}(T)-\widetilde{W}(T_1))}$ is independent of $S(T_1)$. Hence, using the tower property in reverse gives a nested expectation with an inner expectation conditional on $S(T_1)$ as in (12.72):

$$\begin{aligned} C(t, S; T_1, T) &= \mathrm{e}^{-r(T-t)} \widetilde{\mathrm{E}}_{t,S}\left[\,\widetilde{\mathrm{E}}_{T_1,S(T_1)}[\,S(T_1)\,(Y-1)^+\,]\,\right] \\ &= \mathrm{e}^{-r(T-t)} \widetilde{\mathrm{E}}_{t,S}\left[\,S(T_1)\,\widetilde{\mathrm{E}}[(Y-1)^+\,]\,\right] \\ &= \mathrm{e}^{-r(T-T_1)} \widetilde{\mathrm{E}}[(Y-1)^+\,] \cdot \mathrm{e}^{-r(T_1-t)} \widetilde{\mathrm{E}}_{t,S}\left[\,S(T_1)\,\right]. \end{aligned} \tag{12.73}$$

Here we pulled $S(T_1)$ out of the inner expectation as it is $\mathcal{F}_{T_1}$-measurable; then the inner condition is dropped since Y is independent of $S(T_1)$, and finally the unconditional expectation is a constant that is factored out of the expectation conditional on $S(t) = S$. Note that we have also factored the discount term into two parts. The first term on the right in (12.73) is recognized as the Black–Scholes price of a call with no dividend, time to maturity $T - T_1$, effective strike and spot of unity:

$$\mathrm{e}^{-r(T-T_1)} \widetilde{\mathrm{E}}[(Y-1)^+\,] = \mathcal{N}(d_+) - \mathrm{e}^{-r(T-T_1)} \mathcal{N}(d_-)\,, \tag{12.74}$$

where

$$d_\pm \equiv d_\pm(1, T-T_1) = \frac{\ln(1) + (r \pm \frac{1}{2}\sigma^2)(T-T_1)}{\sigma\sqrt{T-T_1}} = \left(\frac{r}{\sigma} \pm \frac{1}{2}\sigma\right)\sqrt{T-T_1}\,.$$

The second term in (12.73) gives the spot S since $\widetilde{\mathrm{E}}_{t,S}\left[\,S(T_1)\,\right] = \mathrm{e}^{r(T_1-t)}S$. Multiplying the expression in (12.74) by S produces the pricing formula:

$$C(t, S; T_1, T) = S[\mathcal{N}(d_+) - \mathrm{e}^{-r(T-T_1)} \mathcal{N}(d_-)]\,. \tag{12.75}$$

We observe that this pricing function is linear in the spot S since the term in square brackets depends only on parameters r, σ, T_1, T. The delta hedge is then trivially given by the term in brackets, $\Delta(t, S) = \frac{\partial}{\partial S} C(t, S; T_1, T) = [\mathcal{N}(d_+) - \mathrm{e}^{-r(T-T_1)} \mathcal{N}(d_-)]$. This is a constant hedge position having no dependence on time t and spot S. What is then interesting is that this forward starting call can be hedged statically in time! Other related forward starting option problems are left as exercises for the reader (see Exercises 12.19 and 12.20).

We now consider the problem of pricing a more complex class of options known as *compound options*. As the name implies, such contracts are options on options. Here we shall assume the simplest types of compound options that involve an (outer) option to buy or sell another (inner or embedded) option at some future time. For standard European-style options the payoff is simply a function of the underlying stock price at some expiry time $T > t$ where t is current calendar time. In a compound option the essential difference is that its value at some future time, say $T_1 > t$, is a specified function of an option price on an underlying stock whereby the latter option is initiated at time T_1 and matures at a future time $T_2 > T_1$ with a specified payoff function. In a compound option, the role of an underlying asset is not played by the stock but rather *the embedded option on the stock plays the role of "underlying asset" for the (outer) option.*

Generally, a European compound option is defined by an outer payoff function $\phi^{(1)}$ and inner payoff function $\phi^{(2)}$. We are interested in pricing the compound option at time $t < T_1$. Given $t < T_1 < T_2$, let $V_t = V(S(t), t, T_1, T_2)$, $t \leqslant T_1$, be the *value process* of the compound option. Note that we have also denoted this as a function of given exercise times T_1, T_2. At time T_1, with stock price $S(T_1)$, the value of the compound option is given by $V_{T_1} = \phi^{(1)}(V^{(2)}(S(T_1), T_1, T_2))$, where $V^{(2)}(S(T_1), T_1, T_2)$ is the value of the underlying (inner) option at time T_1 with time to expiry $T_2 - T_1$ and payoff value $V^{(2)}(S(T_2), T_2, T_2) = \phi^{(2)}(S(T_2))$. Note that $V^{(2)}(S(T_1), T_1, T_2)$ is an ordinary function of the random variable $S(T_1)$ and can be viewed as a random asset value at time T_1. Given spot $S(t) = S$, then by the risk-neutral pricing formulation the arbitrage-free price of the compound option is given by the conditional expectation (under the risk-neutral measure $\widetilde{\mathbb{P}}$) of the discounted value, $\mathrm{e}^{-r(T_1-t)} V_{T_1}$, of the payoff of the outer option at time T_1:

$$V(t, S; T_1, T_2) = \mathrm{e}^{-r(T_1-t)} \widetilde{\mathrm{E}}_{t,S}\left[\phi^{(1)}(V^{(2)}(S(T_1), T_1, T_2))\right] \tag{12.76}$$

where

$$V^{(2)}(S(T_1), T_1, T_2) = \mathrm{e}^{-r(T_2-T_1)} \widetilde{\mathrm{E}}_{T_1, S(T_1)}\left[\phi^{(2)}(S(T_2))\right]. \tag{12.77}$$

The order of the steps for obtaining the price $V(t, S; T_1, T_2)$ by the above expectation approach is as follows.

1. Determine the time-T_1 price $V^{(2)}(S_1, T_1, T_2)$ of the embedded option on the stock having expiry $T_2 > T_1$ and spot variable S_1.

2. Set $S_1 = S(T_1)$ to obtain the payoff of the outer option at time T_1; as a random variable this payoff is $V_{T_1} = \phi^{(1)}(V_{T_1}^{(2)})$ where $V_{T_1}^{(2)} \equiv V^{(2)}(S(T_1), T_1, T_2)$.

3. Compute the discounted risk-neutral expectation in (12.76), i.e., the time-t price is $V(t, S) \equiv V(t, S; T_1, T_2) = \mathrm{e}^{-r(T_1-t)} \widetilde{\mathrm{E}}_{t,S}\left[V_{T_1}\right]$.

The most common examples of European compound options are a *call-on-a-call*, *put-on-a-call*, *put-on-a-put* and *call-on-a-put*. These four options are characterized by two expiration dates T_1 and T_2 and two strike values K_1 and K_2 and with respective payoff functions $\phi^{(1)}(x) = (x - K_1)^+$ and $\phi^{(2)}(x) = (x - K_2)^+$; $\phi^{(1)}(x) = (K_1 - x)^+$ and $\phi^{(2)}(x) = (x - K_2)^+$; $\phi^{(1)}(x) = (K_1 - x)^+$ and $\phi^{(2)}(x) = (K_2 - x)^+$; $\phi^{(1)}(x) = (x - K_1)^+$ and $\phi^{(2)}(x) = (K_2 - x)^+$. For example, the call-on-a-call contract gives the holder the right (but not the obligation) to buy an underlying call option for a fixed strike price K_1 at calendar time T_1 and where the underlying call is specified by strike K_2 and time to expiry $T_2 - T_1$.

As a concrete example, let us specifically value the call-on-a-call option by implementing (12.76) within the usual GBM process for the stock price process having dividend q in an economy with constant interest rate r. Denote the value of the underlying call at time T_1 by $C_{T_1} \equiv C_{T_1}(S(T_1), K_2, T_2)$. Hence, $V^{(2)}(S(T_1), T_1, T_2) = C_{T_1}(S(T_1), K_2, T_2)$ in (12.76) is given explicitly by the standard call price formula, i.e., for time-T_1 spot value $S(T_1) = S_1 > 0$ and time to maturity $T_2 - T_1$:

$$C_{T_1}(S_1, K_2, T_2) = \mathrm{e}^{-q(T_2-T_1)} S_1 \mathcal{N}(d_+) - K_2 \mathrm{e}^{-r(T_2-T_1)} \mathcal{N}(d_-), \tag{12.78}$$

where $d_\pm \equiv d_\pm(\frac{S_1}{K_2}, T_2 - T_1)$. Throughout this section, we define

$$d_\pm(x, \tau) := \frac{\ln x + (r - q \pm \frac{\sigma^2}{2})\tau}{\sigma\sqrt{\tau}}, \quad x, \tau > 0. \tag{12.79}$$

From (12.76), the call-on-a-call option value, denoted by $V^{cc}(S, t)$, is given by

$$V^{cc}(t, S) = \mathrm{e}^{-r(T_1-t)} \widetilde{\mathrm{E}}_{t,S}\left[(C_{T_1}(S(T_1), K_2, T_2) - K_1)^+\right]. \tag{12.80}$$

Note that the random variable within this expectation is nonzero only when $C_{T_1} > K_1$. Recall that the call pricing function $C_{T_1}(S_1, K_2, T_2)$ is a strictly increasing function of the spot variable S_1 where $C_{T_1}(S_1, K_2, T_2) \to 0$, as $S_1 \to 0+$, and $C_{T_1}(S_1, K_2, T_2) \to \infty$, as $S_1 \to \infty$. The graph of $C_{T_1}(S_1, K_2, T_2)$ versus S_1 must therefore cross the level $K_1 > 0$ at exactly one (critical) point, i.e., at $S_1 = S_1^*$. This point is the root of the equation

$$C_{T_1}(S_1^*, K_2, T_2) = K_1.$$

Note that by (12.78) we see that this is a nonlinear algebraic equation so that S_1^*, being a function of K_1, K_2 and $T_2 - T_1$, is in practice obtained numerically. Given the point S_1^*, and since $C_{T_1}(S_1, K_2, T_2)$ is strictly increasing in S_1, we have the equivalence $\mathbb{I}_{\{C_{T_1}(S(T_1),K_2,T_2) > K_1\}} = \mathbb{I}_{\{S(T_1)>S_1^*\}}$, hence

$$(C_{T_1}(S(T_1), K_2, T_2) - K_1)^+ = (C_{T_1}(S(T_1), K_2, T_2) - K_1)\mathbb{I}_{\{S(T_1)>S_1^*\}}.$$

So (12.80) now reads

$$V^{cc}(t, S) = \mathrm{e}^{-r\tau_1} \widetilde{\mathrm{E}}_{t,S}\left[C_{T_1}(S(T_1), K_2, T_2)\mathbb{I}_{\{S(T_1)>S_1^*\}}\right] - K_1 \mathrm{e}^{-r\tau_1} \widetilde{\mathrm{E}}_{t,S}\left[\mathbb{I}_{\{S(T_1)>S_1^*\}}\right] \tag{12.81}$$

where we define $\tau_1 := T_1 - t$ and $\tau_2 := T_2 - t$ in what follows.

The two conditional expectations in (12.81) are readily evaluated by using the strong solution representation of the stock price process. In particular, the second expectation is evaluated using

$$S(T_1) = S(t)\mathrm{e}^{(r-q-\sigma^2/2)\tau_1 + \sigma(\widetilde{W}(T_1) - \widetilde{W}(t))} \tag{12.82}$$

and the fact that $\widetilde{W}(T_1) - \widetilde{W}(t)$ and $\widetilde{W}(t)$ are independent. Hence, $\frac{S(T_1)}{S(t)}$ and $S(t)$ are independent, giving

$$\begin{aligned}
\widetilde{\mathrm{E}}_{t,S}\left[\mathbb{I}_{\{S(T_1)>S_1^*\}}\right] &= \widetilde{\mathrm{E}}\left[\mathbb{I}_{\{\frac{S(T_1)}{S(t)} > \frac{S_1^*}{S}\}} \mid S(t) = S\right] = \widetilde{\mathrm{E}}\left[\mathbb{I}_{\{\frac{S(T_1)}{S(t)} > \frac{S_1^*}{S}\}}\right] \\
&= \widetilde{\mathbb{P}}\left(\ln\frac{S(T_1)}{S(t)} > \ln\frac{S_1^*}{S}\right) \\
&= \widetilde{\mathbb{P}}\left(\frac{\widetilde{W}(T_1) - \widetilde{W}(t)}{\sqrt{\tau_1}} < a_-\right) = \mathcal{N}(a_-)
\end{aligned} \tag{12.83}$$

where we denote $a_\pm \equiv d_\pm(\frac{S}{S_1^*}, \tau_1)$. The last equality follows since $\frac{\widetilde{W}(T_1)-\widetilde{W}(t)}{\sqrt{\tau_1}} \sim Norm(0, 1)$ under measure $\widetilde{\mathbb{P}}$.

The first conditional expectation in (12.81) is re-expressed as follows:

$$\begin{aligned}
&\widetilde{\mathrm{E}}_{t,S}\left[\mathbb{I}_{\{S(T_1)>S_1^*\}}C_{T_1}(S(T_1),K_2,T_2)\right] \\
&= \mathrm{e}^{-r(T_2-T_1)}\widetilde{\mathrm{E}}_{t,S}\left[\mathbb{I}_{\{S(T_1)>S_1^*\}}\widetilde{\mathrm{E}}_{T_1,S(T_1)}\left[(S(T_2)-K_2)^+\right]\right] \\
&= \mathrm{e}^{-r(\tau_2-\tau_1)}\widetilde{\mathrm{E}}_{t,S}\left[\widetilde{\mathrm{E}}_{T_1,S(T_1)}\left[\mathbb{I}_{\{S(T_1)>S_1^*,S(T_2)>K_2\}}(S(T_2)-K_2)\right]\right] \\
&= \mathrm{e}^{-r(\tau_2-\tau_1)}\widetilde{\mathrm{E}}_{t,S}\left[\mathbb{I}_{\{S(T_1)>S_1^*,\,S(T_2)>K_2\}}(S(T_2)-K_2)\right] \\
&= \mathrm{e}^{-r(\tau_2-\tau_1)}\widetilde{\mathrm{E}}_{t,S}\left[\mathbb{I}_{\{S(T_1)>S_1^*,\,S(T_2)>K_2\}}S(T_2)\right] \\
&\qquad - K_2\,\mathrm{e}^{-r(\tau_2-\tau_1)}\widetilde{\mathrm{E}}_{t,S}\left[\mathbb{I}_{\{S(T_1)>S_1^*,\,S(T_2)>K_2\}}\right].
\end{aligned} \tag{12.84}$$

Note that in the third line from the top we have moved the indicator random variable $\mathbb{I}_{\{S(T_1)>S_1^*\}}$ to the inside of the inner expectation since it is known at time T_1 (i.e., it is $\sigma(S(T_1))$-measurable). In the third line we have altogether eliminated the inner conditional expectation (i.e., the conditioning on $S(T_1)$) simply by using iterated conditioning (i.e., the tower property). The two conditional expectations in the last equation line of (12.84) are evaluated as follows. The last expectation is a joint probability. Upon using the condition $S(t) = S$, the fact that $S(T_1)/S(t)$ and $S(T_2)/S(t)$ are both independent of $S(t)$, and using (12.82) and

$$S(T_2) = S(t)\mathrm{e}^{(r-q-\sigma^2/2)\tau_2+\sigma(\widetilde{W}(T_2)-\widetilde{W}(t))} \tag{12.85}$$

we have

$$\begin{aligned}
\widetilde{\mathrm{E}}_{t,S}\left[\mathbb{I}_{\{S(T_1)>S_1^*,\,S(T_2)>K_2\}}\right] &= \widetilde{\mathrm{E}}\left[\mathbb{I}_{\left\{\frac{S(T_1)}{S(t)}>\frac{S_1^*}{S},\,\frac{S(T_2)}{S(t)}>\frac{K_2}{S}\right\}}\right] \\
&= \widetilde{\mathbb{P}}\left(\ln\frac{S(T_1)}{S(t)} > \ln\frac{S_1^*}{S},\ \ln\frac{S(T_2)}{S(t)} > \ln\frac{K_2}{S}\right) \\
&= \widetilde{\mathbb{P}}\left(\frac{\widetilde{W}(T_1)-\widetilde{W}(t)}{\sqrt{\tau_1}} > -a_-,\ \frac{\widetilde{W}(T_2)-\widetilde{W}(t)}{\sqrt{\tau_2}} > -b_-\right)
\end{aligned} \tag{12.86}$$

where $b_\pm \equiv d_\pm(\frac{S}{K_2},\tau_2)$. Since the increments $\widetilde{W}(T_1)-\widetilde{W}(t)$ and $\widetilde{W}(T_2)-\widetilde{W}(t)$ are $Norm(0,\tau_1)$ and $Norm(0,\tau_2)$, respectively, the random variables $Z_1 := \frac{\widetilde{W}(T_1)-\widetilde{W}(t)}{\sqrt{\tau_1}}$ and $Z_2 := \frac{\widetilde{W}(T_2)-\widetilde{W}(t)}{\sqrt{\tau_2}}$ are both standard normals under the risk-neutral measure $\widetilde{\mathbb{P}}$. Moreover, using the independence of nonoverlapping Brownian increments, their covariance (in the $\widetilde{\mathbb{P}}$-measure) is given by

$$\begin{aligned}
\widetilde{\mathrm{Cov}}(Z_1,Z_2) &= \frac{1}{\sqrt{\tau_1\tau_2}}\widetilde{\mathrm{Cov}}(\widetilde{W}(T_1)-\widetilde{W}(t),\widetilde{W}(T_2)-\widetilde{W}(t)) \\
&= \frac{1}{\sqrt{\tau_1\tau_2}}\widetilde{\mathrm{Cov}}(\widetilde{W}(T_1)-\widetilde{W}(t),\widetilde{W}(T_2)-\widetilde{W}(T_1)+\widetilde{W}(T_1)-\widetilde{W}(t)) \\
&= \frac{1}{\sqrt{\tau_1\tau_2}}\widetilde{\mathrm{Var}}(\widetilde{W}(T_1)-\widetilde{W}(t)) = \frac{1}{\sqrt{\tau_1\tau_2}}\tau_1 = \sqrt{\frac{\tau_1}{\tau_2}} = \sqrt{\frac{T_1-t}{T_2-t}}.
\end{aligned}$$

Hence the vector (Z_1,Z_2) has standard normal bivariate distribution with correlation coefficient $\rho \equiv \sqrt{\tau_1/\tau_2}$. By symmetry, $(-Z_1,-Z_2)$ has the same bivariate distribution. Hence, (12.86) gives

$$\widetilde{\mathrm{E}}_{t,S}\left[\mathbb{I}_{\{S(T_1)>S_1^*,\,S(T_2)>K_2\}}\right] = \widetilde{\mathbb{P}}(Z_1 > -a_-, Z_2 > -b_-) \tag{12.87}$$

$$= \widetilde{\mathbb{P}}(Z_1 < a_-, Z_2 < b_-) = \mathcal{N}_2(a_-,b_-;\rho)\,. \tag{12.88}$$

Following similar steps as led to (12.86) above, and inserting the exponential form in (12.85) for $S(T_2)$ where $S(t) = S$, the second to last conditional expectation in (12.84) is now conveniently rewritten in terms of Z_1 and Z_2 and evaluated:

$$\begin{aligned}
\widetilde{\mathrm{E}}_{t,S}\left[\mathbb{I}_{\{S(T_1)>S_1^*,\, S(T_2)>K_2\}} S(T_2)\right] &= S\,\widetilde{\mathrm{E}}\left[\mathbb{I}_{\{\ln\frac{S(T_1)}{S(t)}>\ln\frac{S_1^*}{S},\, \ln\frac{S(T_2)}{S(t)}>\ln\frac{K_2}{S}\}} S(T_2)/S(t)\right] \\
&= S\mathrm{e}^{(r-q-\sigma^2/2)\tau_2}\,\widetilde{\mathrm{E}}\left[\mathbb{I}_{\{Z_1>-a_-,\, Z_2>-b_-\}}\mathrm{e}^{\sigma\sqrt{\tau_2}Z_2}\right] \\
&= S\mathrm{e}^{(r-q-\sigma^2/2)\tau_2}\,\widetilde{\mathrm{E}}\left[\mathbb{I}_{\{Z_1<a_-,\, Z_2<b_-\}}\mathrm{e}^{-\sigma\sqrt{\tau_2}Z_2}\right] \\
&= S\mathrm{e}^{(r-q-\sigma^2/2)\tau_2}\,\mathrm{e}^{\frac{1}{2}\sigma^2\tau_2}\,\mathcal{N}_2\big(a_- + \rho\sigma\sqrt{\tau_2},\, b_- + \sigma\sqrt{\tau_2}\,;\rho\big) \\
&= S\mathrm{e}^{(r-q)\tau_2}\,\mathcal{N}_2\big(a_- + \sigma\sqrt{\tau_1},\, b_- + \sigma\sqrt{\tau_2}\,;\rho\big) \\
&= S\mathrm{e}^{(r-q)\tau_2}\,\mathcal{N}_2\big(a_+,\, b_+\,;\rho\big)\,.
\end{aligned} \tag{12.89}$$

We note that in evaluating the last expectation we used the identity in (A.13) of the Appendix.

Finally, by combining the expressions in (12.89), (12.88), (12.84), and (12.83) into (12.81) gives the explicit formula for the compound call-on-a-call:

$$V^{cc}(t, S) = S\mathrm{e}^{-q\tau_2}\mathcal{N}_2\big(a_+,\, b_+\,;\rho\big) - K_2\mathrm{e}^{-r\tau_2}\mathcal{N}_2\big(a_-,\, b_-\,;\rho\big) - K_1\mathrm{e}^{-r\tau_1}\mathcal{N}(a_-) \tag{12.90}$$

where $a_\pm, b_\pm, \rho$ are defined above. Note that the option value is a function of the spot S, the two strike values K_1, K_2, and the two time to expiration values τ_1, τ_2.

The other types of compound options can be valued in similar fashion. For example, we leave the valuation of the put-on-a-put as an exercise. There also exists a form of put-call parity among some pairs of compound options. In particular, the call-on-a-call option value and the corresponding put-on-a-call option value $V^{pc}(S, t)$ are related by

$$V^{cc}(t, S) - V^{pc}(t, S) = C_t(S, K_2, T_2) - \mathrm{e}^{-r\tau_1}K_1. \tag{12.91}$$

Namely, the difference in the time-t value of the call-on-a-call and put-on-a-call (with spot $S(t) = S$ and given inner and outer strike and maturity pairs K_1, T_1 and K_2, T_2) is simply the time-t value of a standard call (with spot $S(t) = S$, strike and maturity K_2, T_2) minus the discounted inner strike value $\mathrm{e}^{-r\tau_1}K_1$. This is shown as follows. According to (12.76), the put-on-a-call option has value

$$V^{pc}(t, S) = \mathrm{e}^{-r\tau_1}\widetilde{\mathrm{E}}_{t,S}\left[(K_1 - C_{T_1})^+\right]$$

where $C_{T_1} \equiv C_{T_1}(S(T_1), K_2, T_2)$ is the value of the (inner) call option initiated at T_1 and maturing at $T_2 > T_1$. Using the simple identity $(K_1 - C_{T_1})^+ = (C_{T_1} - K_1)^+ - C_{T_1} + K_1$ within the above expectation gives

$$V^{pc}(t, S) = \mathrm{e}^{-r\tau_1}\widetilde{\mathrm{E}}_{t,S}\left[(C_{T_1} - K_1)^+\right] - \mathrm{e}^{-r\tau_1}\widetilde{\mathrm{E}}_{t,S}\left[C_{T_1}\right] + \mathrm{e}^{-r\tau_1}K_1\,. \tag{12.92}$$

The first expectation is $V^{cc}(S, t)$. By the tower property, we now show that the second expectation reduces to the value $C_t(S, K_2, T_2)$ of a standard call with spot S, strike K_2, and maturity $T_2 > t$. Note that the call value C_{T_1} is a random variable expressed here as function of the time-T_1 spot random variable $S(T_1)$, and hence its value is given by the discounted expected value of the payoff $(S(T_2) - K_2)^+$ at time T_2, conditional on $S(T_1)$:

$$C_{T_1} \equiv C_{T_1}(S(T_1), K_2, T_2) = \widetilde{\mathrm{E}}_{T_1, S(T_1)}\left[\mathrm{e}^{-r(T_2-T_1)}(S(T_2) - K_2)^+\right].$$

Substituting this representation for C_{T_1} into the second expectation in (12.92) and invoking the tower property, while combining the discount factors, gives

$$\begin{aligned}
e^{-rT_1}\widetilde{E}_{t,S}\left[C_{T_1}\right] &= e^{-r(T_1-t)}\widetilde{E}_{t,S}\left[\widetilde{E}_{T_1,S(T_1)}\left[e^{-r(T_2-T_1)}(S(T_2)-K_2)^+\right]\right] \\
&= e^{-r(T_2-t)}\widetilde{E}_{t,S}\left[(S(T_2)-K_2)^+\right] = C_t(S,K_2,T_2). \qquad (12.93)
\end{aligned}$$

Hence (12.91) is recovered from (12.92).

Another way to obtain (12.93) is simply to note that the discounted call price process $e^{-rt}C_t \equiv e^{-rt}C(S(t),K_2,T_2)$, for $t < T_2$ and fixed K_2, T_2, is a martingale under the risk-neutral measure $\widetilde{\mathbb{P}}$. Hence, combining the $\widetilde{\mathbb{P}}$-martingale and Markov properties:

$$e^{-rt}C_t = \widetilde{E}\left[e^{-rT_1}C_{T_1}|\mathcal{F}_t\right] = e^{-rT_1}\widetilde{E}_{t,S(t)}\left[C_{T_1}\right] = e^{-rT_1}\widetilde{E}_{t,S(t)}\left[C(S(T_1),K_2,T_2)\right].$$

Then, setting $S(t) = S$ gives (12.93). We remark that the above put-call parity type relation is valid for quite general models of the stock price process, i.e., it holds for GBM and other models where we assume the discounted stock price process is a martingale under the risk-neutral measure $\widetilde{\mathbb{P}}$ and the stock is not allowed to default. Of course, under more general models the pricing formulas for the compound options will not involve univariate and bi-variate standard normal CDFs as we derived above for the GBM model. With the exception of some families of alternative models, one has to resort to numerical methods for pricing compound options under more complex stochastic models for the stock.

12.3 Some European-Style Path-Dependent Derivatives

We now consider the application of risk-neutral pricing to path-dependent European options whose payoff is dependent on the underlying stock price history over the lifetime of the contract. We shall specialize to the pricing of two types of path-dependent options, namely, barrier options and lookback options. As seen in the examples below, these classes of options have a payoff that is a function of a combination of the stock price $S(T)$ at maturity $T > 0$ and the realized (sampled) maximum $M^S(T)$ or the realized minimum $m^S(T)$ of the stock price process $\{S(t)\}_{t\geqslant 0}$ where

$$M^S(t) := \sup_{0\leqslant u\leqslant t} S(u) \quad \text{and} \quad m^S(t) := \inf_{0\leqslant u\leqslant t} S(u) \qquad (12.94)$$

for all $0 \leqslant t \leqslant T$. There are many variations of the payoff for these options. Some payoffs, such as in the case of a so-called double barrier option, are functions of the triplet $M^S(T), m^S(T), S(T)$. Here we shall focus our attention on developing analytical pricing formulae for *single-barrier* options and lookback options whose payoff is *either* a function of the pair $M^S(T), S(T)$ *or* a function of the pair $m^S(T), S(T)$, separately. Given the joint distribution of either pair of random variables, there are two main types of payoffs for which we can in principle derive pricing formulae. In the first case, the payoff is assumed to be a (Borel) function, $\phi : R^2_+ \to \mathbb{R}$, of the terminal stock price and its realized maximum and in the second case it is a function of the terminal stock price and its realized minimum:

$$(i)\; V_T = \phi(M^S(T), S(T)) \quad \text{and} \quad (ii)\; V_T = \phi(m^S(T), S(T)). \qquad (12.95)$$

Let's first take a look at the payoffs that define some single-barrier option contracts. For such contracts the payoff simplifies into a product of a function of the terminal stock price $\Lambda(S(T))$ and an indicator function involving either the realized maximum $M^S(T)$ or minimum $m^S(T)$ of the stock price during the option's lifetime. There are two basic types of single-barrier options: (i) *knock-out* options that have a nonzero payoff only if a level $B > 0$ *is not attained* and (ii) *knock-in* options that have a nonzero payoff only if level B *is attained* during the option's lifetime. The different versions of these correspond to whether level B is a *lower barrier or an upper barrier.* Letting $\Lambda(S(T))$ be the effective payoff of a standard (non-path-dependent) European option, e.g., $\Lambda(x) = (x - K)^+$ for a call and $\Lambda(x) = (K - x)^+$ for a put struck at $K > 0$, we have the following four different payoffs for a single barrier at level B:

(a) Up-and-out: $V_T^{UO} = \Lambda(S(T))\,\mathbb{I}_{\{M^S(T)<B\}}$, where $\phi(M, S) := \Lambda(S)\,\mathbb{I}_{\{M<B\}}$;

(b) Down-and-out: $V_T^{DO} = \Lambda(S(T))\,\mathbb{I}_{\{m^S(T)>B\}}$, where $\phi(m, S) := \Lambda(S)\,\mathbb{I}_{\{m>B\}}$;

(c) Up-and-in: $V_T^{UI} = \Lambda(S(T))\,\mathbb{I}_{\{M^S(T)\geqslant B\}}$, where $\phi(M, S) := \Lambda(S)\,\mathbb{I}_{\{M\geqslant B\}}$;

(d) Down-and-in: $V_T^{DI} = \Lambda(S(T))\,\mathbb{I}_{\{m^S(T)\leqslant B\}}$, where $\phi(m, S) := \Lambda(S)\,\mathbb{I}_{\{m\leqslant B\}}$. (12.96)

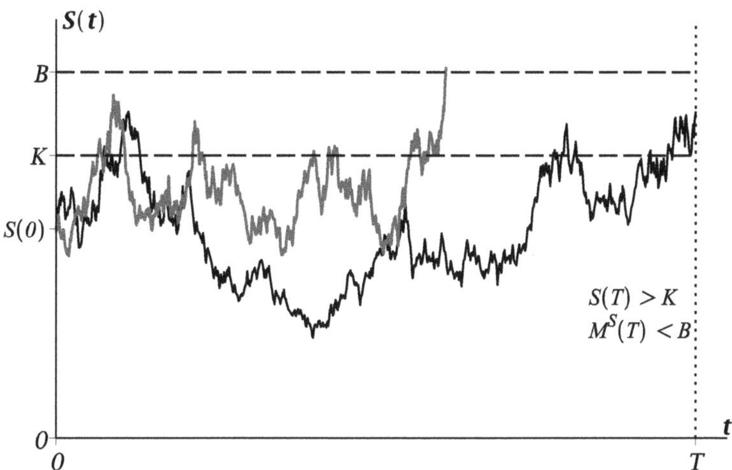

FIGURE 12.3: Two types of stock price paths starting at $S(0) < B$ are depicted for an up-and-out call with strike K. Only paths in the set $\{M^S(T) < B, S(T) > K\}$, i.e., paths that do not surpass level B and also end up above the strike at terminal time T give a positive payoff for an up-and-out call struck at K.

For example, an up-and-out call with strike K is defined as having the payoff of a call, $(S(T) - K)^+$, if the realized maximum value of the underlying stock price stays below the barrier level B and has otherwise zero payoff if the stock price attains or goes above level B at any time until T. See Figure 12.3. We write this as $C_T^{UO} = (S(T) - K)^+\,\mathbb{I}_{\{M^S(T)<B\}}$, where $\phi(M, S) = (S - K)^+\,\mathbb{I}_{\{M<B\}}$. The down-and-out call has payoff $C_T^{DO} = (S(T) - K)^+\,\mathbb{I}_{\{m^S(T)>B\}}$. In the case of a knock-out put, the up-and-out put has payoff $P_T^{UO} = (K - S(T))^+\,\mathbb{I}_{\{M^S(T)<B\}}$ and the down-and-out put has payoff $P_T^{DO} = (K - S(T))^+\,\mathbb{I}_{\{m^S(T)>B\}}$. See Figure 12.4.

On the other hand, an up-and-in call with strike K is defined as having a call payoff if the stock price has attained or has gone above B at any time until T and has otherwise zero payoff, i.e., the payoff is $C_T^{UI} = (S(T) - K)^+\,\mathbb{I}_{\{M^S(T)\geqslant B\}}$. Similarly, a down-and-in call has a payoff that is nonzero only if the stock price has fallen below or at level B,

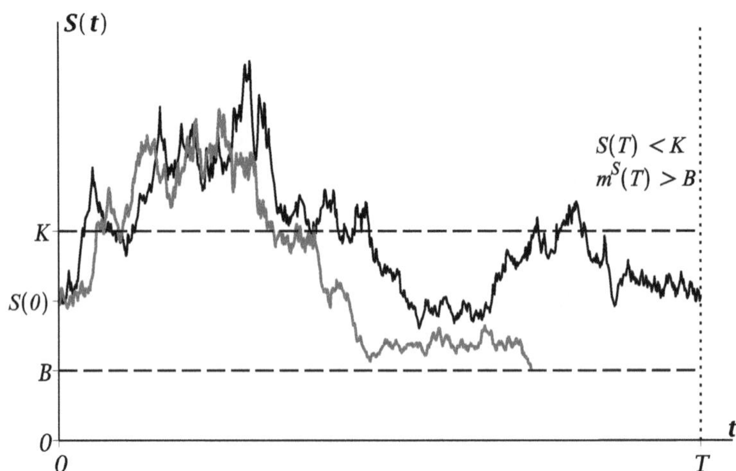

FIGURE 12.4: Two types of stock price paths starting at $S(0) > B$ are depicted for a down-and-out put with strike K. Only paths in the set $\{m^S(T) > B, S(T) < K\}$, i.e., paths that do not fall below level B and also end up below the strike at terminal time T give a positive payoff for a down-and-out put struck at K.

$C_T^{DI} = (S(T)-K)^+ \mathbb{I}_{\{m^S(T) \leqslant B\}}$. For the up-and-in put and down-and-in put, with strike K, we have $P_T^{UI} = (K - S(T))^+ \mathbb{I}_{\{M^S(T) \geqslant B\}}$ and $P_T^{DI} = (K - S(T))^+ \mathbb{I}_{\{m^S(T) \leqslant B\}}$, respectively.

There is a very simple and useful symmetry relation between the knock-in and knock-out payoffs. Since we have the obvious relations

$$\mathbb{I}_{\{M^S(T)<B\}} + \mathbb{I}_{\{M^S(T) \geqslant B\}} = \mathbb{I}_{\{m^S(T)>B\}} + \mathbb{I}_{\{m^S(T) \leqslant B\}} = 1$$

then

$$V_T^{UO} + V_T^{UI} = V_T^{DO} + V_T^{DI} = \Lambda(S(T)) \,. \tag{12.97}$$

This is known as "knock-in-knock-out" symmetry. By computing the pricing formula for the knock-out (or knock-in) option then the pricing formula for the corresponding knock-in (or knock-out) follows simply by subtracting the former from the price of the standard European option having payoff $\Lambda(S(T))$. That is, letting $V_t^{UO}, V_t^{UI}, V_t^{DO}, V_t^{DI}$ represent the respective time-t barrier option prices, $t \leqslant T$, then by risk-neutral pricing we have

$$V_t^{UO} + V_t^{UI} = V_t^{DO} + V_t^{DI} = V_t \,, \tag{12.98}$$

where V_t is the time-t price of the standard European option with payoff $\Lambda(S(T))$.

We now turn to the definition of lookback options in the continuous time setting. We recall from the discrete-time setting (see Section 7.6.2 of Chapter 7) that there are two main kinds of lookback options, with either floating strike (LFS) or floating price (LFP). We list the four common lookback option payoffs:

(a) Floating strike call (LFS call): $C_T^{LFS} = (S(T) - m^S(T))^+ = S(T) - m^S(T)$;

(b) Floating strike put (LFS put): $P_T^{LFS} = (M^S(T) - S(T))^+ = M^S(T) - S(T)$;

(c) Floating price call (LFP call): $C_T^{LFP} = (M^S(T) - K)^+$;

(d) Floating price put (LFP put): $P_T^{LFP} = (K - m^S(T))^+$. (12.99)

Note that the LFS options are never out of the money since $m^S(T) \leqslant S(T) \leqslant M^S(T)$. For an LFS option the strike price is floating as it is not preassigned but rather determined by the realized maximum or minimum value of the stock during the lifetime of the option. The payoff of an LFS option is the maximum difference between the stock's price at maturity and the floating strike. The LFS call gives its holder the right to buy at the lowest stock price realized during the option's lifetime, whereas the LFS put gives the right to sell at the highest realized stock price. For the LFP options, the payoffs are the maximum differences between the optimal (maximum or minimum) stock price and the fixed strike K. LFP options are designed so that the call (or put) has a payoff given by the stock price at its highest (or lowest) realized value during the option's lifetime.

12.3.1 Risk-Neutral Pricing under GBM

Before specializing and thereby simplifying the problem to the pricing of barrier options and lookback options, covered in Sections 12.3.2 and 12.3.3, we now present the risk-neutral pricing formulation for the two general types of payoffs in (12.95) above. We assume $\{S(t)\}_{t\geqslant 0}$ to be a GBM given by (12.2) if $q = 0$ or (12.47) if $q \neq 0$. The stock price in (12.47) is given by a strictly increasing exponential mapping

$$S(t) = S_0 e^{\sigma X(t)}, \; S(0) = S_0 \tag{12.100}$$

where the drifted $\widetilde{\mathbb{P}}$-BM process X is defined by (see (10.68) of Section 10.4.3)

$$X(t) := \widetilde{W}^{(\nu,1)}(t) \equiv \nu t + \widetilde{W}(t)\,, \;\; \nu := \frac{(r - q - \frac{1}{2}\sigma^2)}{\sigma}\,. \tag{12.101}$$

Note that $\nu := \frac{(r-\frac{1}{2}\sigma^2)}{\sigma}$ for the zero dividend case. Hence, the realized maximum and minimum of the stock price in (12.94) are related trivially to the maximum and minimum of the drifted BM:

$$M^S(t) = S_0 e^{\sigma M^X(t)} \quad \text{and} \quad m^S(t) = S_0 e^{\sigma m^X(t)} \tag{12.102}$$

where (see (10.70))

$$M^X(t) := \sup_{0\leqslant u\leqslant t} X(u) \quad \text{and} \quad m^X(t) := \inf_{0\leqslant u\leqslant t} X(u)\,. \tag{12.103}$$

In what follows we will be conditioning on $\mathcal{F}_t$ for any fixed current time t, $0 \leqslant t \leqslant T$. By the time homogeneity property of the stock price process, it is convenient to define $\tau := T - t$. Let us first consider expressing the joint random variables $M^S(T), S(T)$ as functions of the $\mathcal{F}_t$-measurable random variables $M^S(t), S(t)$. Using (12.100) the stock price at time T is

$$S(T) = S(t) \exp\left(\sigma[X(T) - X(t)]\right) = S(t) \exp\left(\sigma \mathcal{X}(\tau)\right) \tag{12.104}$$

where we define $\mathcal{X}(s) := X(s+t) - X(t) = \nu s + \widetilde{W}(s+t) - \widetilde{W}(t)$. Note that the process $\{\widetilde{W}(s+t) - \widetilde{W}(t)\}_{s\geqslant 0}$, for fixed t, is a standard $\widetilde{\mathbb{P}}$-BM $\{\widetilde{W}(s)\}_{s\geqslant 0}$; hence $\mathcal{X}(s)$ is the drifted BM, $\{\widetilde{W}^{(\nu,1)}(s)\}_{s\geqslant 0}$. The realized maximum of the stock price up to time T is the larger of the realized maximum up to time t and the realized maximum from time t to T:

$$M^S(T) = \max\left\{M^S(t), \sup_{t\leqslant u\leqslant T} S(u)\right\} = \max\left\{M^S(t), S(t)\, e^{\sigma M^{\mathcal{X}}(\tau)}\right\}, \tag{12.105}$$

where $M^{\mathcal{X}}(\tau) := \sup_{0 \leqslant s \leqslant \tau} \mathcal{X}(s)$. To arrive at the last term, we employed the steps:

$$\begin{aligned}
\sup_{t \leqslant u \leqslant T} S(u) &= S(t) \cdot \exp\left(\sigma \sup_{t \leqslant u \leqslant T} [X(u) - X(t)]\right) \\
&= S(t) \cdot \exp\left(\sigma \sup_{0 \leqslant s \leqslant \tau} [X(s+t) - X(t)]\right) \\
&= S(t) \cdot \exp\left(\sigma \sup_{0 \leqslant s \leqslant \tau} \mathcal{X}(s)\right). \qquad (12.106)
\end{aligned}$$

By similar steps, the sampled minimum of the stock price takes the form

$$m^S(T) = \min\left\{m^S(t), S(t)\,\mathrm{e}^{\sigma m^{\mathcal{X}}(\tau)}\right\}, \qquad (12.107)$$

where $m^{\mathcal{X}}(\tau) := \inf_{0 \leqslant s \leqslant \tau} \mathcal{X}(s)$.

Based on (12.105) and (12.107), $\{(M^X(t), X(t))\}_{t \geqslant 0}$ and $\{(m^X(t), X(t))\}_{t \geqslant 0}$ are both (vector) Markov processes. Observe that both pairs of random variables $M^{\mathcal{X}}(\tau), \mathcal{X}(\tau)$ and $m^{\mathcal{X}}(\tau), \mathcal{X}(\tau)$ are $\mathcal{F}_t$-independent (and hence also independent of the random variables $S(t), m^S(t)$, and $M^S(t)$). Moreover, both pairs of random variables have the same joint distribution as $(M^X(\tau), X(\tau))$, and $(m^X(\tau), X(\tau))$, respectively. In particular, the joint PDF of $M^{\mathcal{X}}(\tau), \mathcal{X}(\tau)$ (in the risk-neutral measure $\widetilde{\mathbb{P}}$) is given by (sending $t \to \tau$ and $\mu \to \nu$ in (11.109) of Chapter 11)

$$\begin{aligned}
\widetilde{f}_{M^{\mathcal{X}}(\tau),\mathcal{X}(\tau)}(w,x) \equiv \widetilde{f}_{M^X(\tau),X(\tau)} &:= \frac{\partial^2}{\partial w \partial x}\widetilde{\mathbb{P}}(M^X(\tau) \leqslant w, X(\tau) \leqslant x) \\
&= \frac{2(2w-x)}{\tau\sqrt{2\pi\tau}}\mathrm{e}^{-\frac{1}{2}\nu^2\tau + \nu x - (2w-x)^2/2\tau}, \qquad (12.108)
\end{aligned}$$

for $-\infty < x < w, w > 0$ and zero otherwise. The risk-neutral joint PDF of $m^{\mathcal{X}}(\tau), \mathcal{X}(\tau)$ is given by (sending $t \to \tau$ and $\mu \to \nu$ in (11.110) of Chapter 11)

$$\begin{aligned}
\widetilde{f}_{m^{\mathcal{X}}(\tau),\mathcal{X}(\tau)}(w,x) \equiv \widetilde{f}_{m^X(\tau),X(\tau)}(w,x) &:= \frac{\partial^2}{\partial w \partial x}\widetilde{\mathbb{P}}(m^X(\tau) \leqslant w, X(\tau) \leqslant x) \\
&= \frac{2(x-2w)}{\tau\sqrt{2\pi\tau}}\mathrm{e}^{-\frac{1}{2}\nu^2\tau + \nu x - (x-2w)^2/2\tau}, \qquad (12.109)
\end{aligned}$$

for $x > w, w < 0$ and zero otherwise.

By the joint Markov property and using (12.104) and (12.105), a European option with payoff (i) in (12.95) has time-t no-arbitrage price (expressed as an $\mathcal{F}_t$-measurable random variable) given by

$$\begin{aligned}
V_t &= V(t, S(t), M^S(t)) \\
&= \mathrm{e}^{-r(T-t)}\widetilde{\mathrm{E}}[\,\phi(M^S(T), S(T))\,|\mathcal{F}_t] \\
&= \mathrm{e}^{-r(T-t)}\widetilde{\mathrm{E}}[\,\phi(\max\{M^S(t), S(t)\,\mathrm{e}^{\sigma M^{\mathcal{X}}(\tau)}\}, S(t)\,\mathrm{e}^{\sigma \mathcal{X}(\tau)})\,|S(t), M^S(t)]. \qquad (12.110)
\end{aligned}$$

For any positive real values $M^S(t) = M$, $S(t) = S > 0$, $M \geqslant S$, i.e., the spot values of the sampled maximum up to calendar time t and the stock price at calendar time t, the general pricing formula is obtained by computing this expectation while using the fact that

$M^{\mathcal{X}}(\tau), \mathcal{X}(\tau)$ are $\mathcal{F}_t$-independent:

$$\begin{aligned} V(t,S,M) &= \mathrm{e}^{-r\tau}\widetilde{\mathrm{E}}[\phi(\max\{M^S(t), S(t)\,\mathrm{e}^{\sigma M^{\mathcal{X}}(\tau)}\}, S(t)\,\mathrm{e}^{\sigma\mathcal{X}(\tau)})\,|S(t)=S, M^S(t)=M] \\ &= \mathrm{e}^{-r\tau}\widetilde{\mathrm{E}}[\phi(\max\{M, S\,\mathrm{e}^{\sigma M^{\mathcal{X}}(\tau)}\}, S\,\mathrm{e}^{\sigma\mathcal{X}(\tau)})] \\ &= \mathrm{e}^{-r\tau}\int_0^\infty\int_{-\infty}^w \phi(\max\{M, S\,\mathrm{e}^{\sigma w}\}, S\,\mathrm{e}^{\sigma x})\widetilde{f}_{M^X(\tau),X(\tau)}(w,x)\,\mathrm{d}x\,\mathrm{d}w\,, \end{aligned} \tag{12.111}$$

where $\tau = T - t$ is the time to maturity. This is a double integral of the joint density in (12.108) multiplied by the *effective payoff*, $h(w, x) := \phi(\max\{M, S\,\mathrm{e}^{\sigma w}\}, S\,\mathrm{e}^{\sigma x})$, which is a function of w, x.

Applying the same steps by using (12.104) and (12.107), the time-t no-arbitrage pricing formula for the European option with payoff (ii) in (12.95) for given real positive spot values, $0 < m^S(t) = m \leqslant S = S(t)$, is given by

$$\begin{aligned} V(t,S,m) &= \mathrm{e}^{-r\tau}\widetilde{\mathrm{E}}[\phi(\min\{m^S(t), S(t)\,\mathrm{e}^{\sigma m^{\mathcal{X}}(\tau)}\}, S(t)\,\mathrm{e}^{\sigma\mathcal{X}(\tau)})\,|S(t)=S, m^S(t)=m] \\ &= \mathrm{e}^{-r\tau}\widetilde{\mathrm{E}}[\phi(\min\{m, S\,\mathrm{e}^{\sigma m^{\mathcal{X}}(\tau)}\}, S\,\mathrm{e}^{\sigma\mathcal{X}(\tau)})] \\ &= \mathrm{e}^{-r\tau}\int_{-\infty}^0\int_w^\infty \phi(\min\{m, S\,\mathrm{e}^{\sigma w}\}, S\,\mathrm{e}^{\sigma x})\widetilde{f}_{m^X(\tau),X(\tau)}(w,x)\,\mathrm{d}x\,\mathrm{d}w\,, \end{aligned} \tag{12.112}$$

$\tau = T - t$. This is now a double integral involving the joint density in (12.109) and the effective payoff given by $g(w, x) := \phi(\min\{m, S\,\mathrm{e}^{\sigma w}\}, S\,\mathrm{e}^{\sigma x})$.

[We remark that one can always generally write a stock price as in (12.100), using an exponential (monotonic) function of a Markov process X which is specified as a more complex process. Then, the pricing formulae in (12.111) and (12.112) can be used for more general stock price models as long as there exist joint densities $\widetilde{f}_{M^X(\tau),X(\tau)}$ and $\widetilde{f}_{m^X(\tau),X(\tau)}$ and also that the discounted stock price process is a $\widetilde{\mathbb{P}}$-martingale. Of course, for a more general stock price model that is not a GBM process, the process X is not specified simply as a drifted BM but as a more complex process. The joint densities for such processes will also be more complex than those for drifted BM given in (12.108) and (12.109).]

Note that both pricing formulae in (12.111) and (12.112) are functions of $\tau = T - t$. For example, we can write the price in (12.111) as a function $v(\tau, S, M)$ where $v(\tau, S, M) = V(t, S, M) = V(T-\tau, S, M)$ and similarly for the pricing function in (12.112). Note that the above pricing formulae are generally valid for any intermediate time and that the spot values $S(t) = S, M^S(t) = M, m^S(t) = m$ are known at intermediate time t. However, the payoff is generally a function of the realized maximum $M^S(T)$ (or minimum $m^S(T)$) involving the continuous sampling of the stock price *starting at a prior time* $t_0 = 0$. These are therefore referred to as "seasoned" contracts. This general situation is depicted in Figure 12.5.

Let's now specialize to the case where the realized maximum and minimum are computed *starting from current time* t. Then $S(t) = M^S(t) = M^S(t)$, i.e., with spot values $S = M = m$, where in the integrands of (12.111) and (12.112) we have, respectively,

$$\max\{M, S\,\mathrm{e}^{\sigma w}\} = \max\{S, S\,\mathrm{e}^{\sigma w}\} = S\,\mathrm{e}^{\sigma w}\,,\ \text{since } w > 0\,,$$

$$\min\{m, S\,\mathrm{e}^{\sigma w}\} = \min\{S, S\,\mathrm{e}^{\sigma w}\} = S\,\mathrm{e}^{\sigma w}\,,\ \text{since } w < 0\,.$$

Hence both payoff functions in (12.111) and (12.112) have the form $\phi(S\,\mathrm{e}^{\sigma w}, S\,\mathrm{e}^{\sigma x})$ and the option pricing formulae are functions of only the spot S and $\tau = T - t$. In particular, the pricing formula in (12.111) is reduced to $V(t, S, M) = V(t, S) = v(\tau, S)$:

$$v(\tau, S) = \mathrm{e}^{-r\tau}\int_0^\infty\int_{-\infty}^w \phi(S\,\mathrm{e}^{\sigma w}, S\,\mathrm{e}^{\sigma x})\widetilde{f}_{M^X(\tau),X(\tau)}(w,x)\,\mathrm{d}x\,\mathrm{d}w\,. \tag{12.113}$$

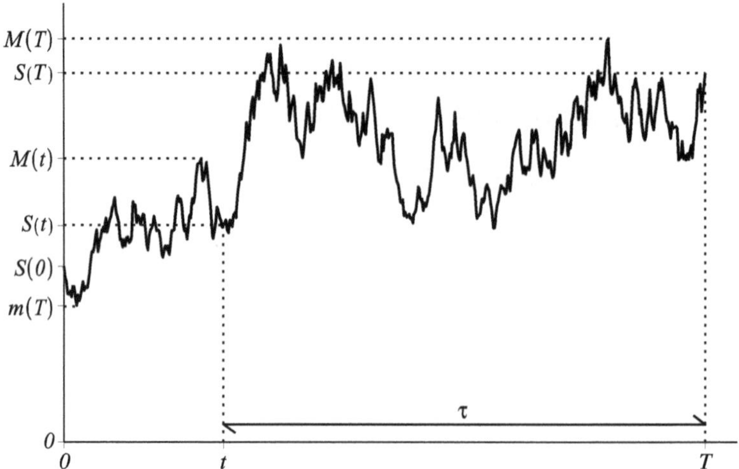

FIGURE 12.5: A sample stock price path is shown with its initial value, its value and realized maximum and minimum at both the intermediate (current) time t and at terminal time T.

Setting the current calendar time $t = 0$, $S = S(0) = S_0$, $\tau = T$ gives the price expressed as function of spot S_0, and the time to maturity, which is now represented by the variable T, $V(0, S_0) = v(T, S_0)$:

$$v(T, S_0) = \mathrm{e}^{-rT} \int_0^{\infty} \int_{-\infty}^{w} \phi(S_0\, \mathrm{e}^{\sigma w}, S_0\, \mathrm{e}^{\sigma x}) \widetilde{f}_{M^X(T),X(T)}(w, x)\, \mathrm{d}x\, \mathrm{d}w\,. \tag{12.114}$$

Of course, we need only compute one of these as (12.114) obtains trivially from (12.113) and vice versa. For options involving the realized minimum, (12.112) gives $V(t, S, m) = V(t, S) = v(\tau, S)$:

$$v(\tau, S) = \mathrm{e}^{-r\tau} \int_{-\infty}^{0} \int_{w}^{\infty} \phi(S\, \mathrm{e}^{\sigma w}, S\, \mathrm{e}^{\sigma x}) \widetilde{f}_{m^X(\tau),X(\tau)}(w, x)\, \mathrm{d}x\, \mathrm{d}w \tag{12.115}$$

or expressed as a function of T and $S(0) = S_0$, where we simply make the variable replacements $S \to S_0$ and $\tau \to T$ in the derived pricing function $v(\tau, S)$.

12.3.2 Pricing Single Barrier Options

For barrier options the contacts are specified such that the sampling of the maximum and minimum of the stock price starts at current time t. So we have the case discussed above where $S(t) = M^S(t) = m^S(t)$ (or $S_0 = S(0) = M^S(0) = m^S(0)$ for current time $t = 0$). Hence the pricing formulae in (12.113)–(12.115) are our general starting point. Given a spot value $S(t) = S$, we denote the respective time-t pricing functions for cases (a)–(d) defined in (12.96) by $V^{UO}(t, S; B)$, $V^{DO}(t, S; B)$, $V^{UI}(t, S; B)$, and $V^{DI}(t, S; B)$. As functions of time to maturity we write these pricing functions equally as $v^{UO}(\tau, S; B)$, $v^{DO}(\tau, S; B)$, $v^{UI}(\tau, S; B)$, and $v^{DI}(\tau, S; B)$. By knock-in-knock-out symmetry in (12.98), we need only derive a pricing formula for either knock-out or knock-in options as we can use the pricing formula for the standard (vanilla) option to obtain one pricing formula from the other:

$$V^{UO}(t, S; B) + V^{UI}(t, S; B) = V^{DO}(t, S; B) + V^{DI}(t, S; B) = V(t, S), \qquad (12.116)$$

where $V(t, S)$ is the time-t pricing formula for the standard European option with payoff function Λ.

For barrier options we see that the overall payoff function ϕ in all cases (a)–(d) in (12.96) is a product of an indicator function in the first argument and the effective payoff function Λ in the second argument. Hence, in the integrand of (12.113)–(12.115) we have in the respective cases (a)–(d) in (12.96):

$$\begin{aligned}
&\text{(a) } \phi(S\,\mathrm{e}^{\sigma w}, S\,\mathrm{e}^{\sigma x}) = \mathbb{I}_{\{S\,\mathrm{e}^{\sigma w} < B\}}\Lambda(S\,\mathrm{e}^{\sigma x}) = \mathbb{I}_{\{w<b\}}\Lambda(S\,\mathrm{e}^{\sigma x}),\ -\infty < x < w, w > 0;\\
&\text{(b) } \phi(S\,\mathrm{e}^{\sigma w}, S\,\mathrm{e}^{\sigma x}) = \mathbb{I}_{\{S\,\mathrm{e}^{\sigma w} > B\}}\Lambda(S\,\mathrm{e}^{\sigma x}) = \mathbb{I}_{\{w>b\}}\Lambda(S\,\mathrm{e}^{\sigma x}),\ w < x < \infty, w < 0;\\
&\text{(c) } \phi(S\,\mathrm{e}^{\sigma w}, S\,\mathrm{e}^{\sigma x}) = \mathbb{I}_{\{S\,\mathrm{e}^{\sigma w} \geqslant B\}}\Lambda(S\,\mathrm{e}^{\sigma x}) = \mathbb{I}_{\{w\geqslant b\}}\Lambda(S\,\mathrm{e}^{\sigma x}),\ -\infty < x < w, w > 0;\\
&\text{(d) } \phi(S\,\mathrm{e}^{\sigma w}, S\,\mathrm{e}^{\sigma x}) = \mathbb{I}_{\{S\,\mathrm{e}^{\sigma w} \leqslant B\}}\Lambda(S\,\mathrm{e}^{\sigma x}) = \mathbb{I}_{\{w\leqslant b\}}\Lambda(S\,\mathrm{e}^{\sigma x}),\ w < x < \infty, w < 0; \quad (12.117)
\end{aligned}$$

where $b := \frac{1}{\sigma}\ln\frac{B}{S}$. These are all product functions in the integrand variables x and w. This leads to an important simplification in (12.113–(12.115) which reduce to single integrals, as given in the following result where the pricing formulae for knock-out barrier options are single integrals (in x) involving the effective payoff $\Lambda(S\,\mathrm{e}^{\sigma x})$ and the risk-neutral probability density for the drifted BM in (12.101) that is killed at the effective barrier level b.

Proposition 12.1 (Pricing Formulae for Single-Barrier Knock-Out Options). *Assume a constant interest rate r and constant continuous dividend yield q on a stock whose price process is a GBM with constant volatility σ. Let $B > 0$ be an arbitrary knock-out barrier level, $S(t) = S > 0$ be the stock spot price, and $\Lambda(\cdot)$ be the effective payoff function. Then, for $S < B$ the up-and-out option has value*

$$V^{UO}(t, S; B) = \mathrm{e}^{-r\tau}\int_{-\infty}^{b}\Lambda(S\mathrm{e}^{\sigma x})\,\widetilde{p}^{X^{(b)}}(\tau; 0, x)\,\mathrm{d}x \qquad (12.118)$$

and $V^{UO}(t, S; B) \equiv 0$ for $S \geqslant B$. For $S > B$, the down-and-out option has the value

$$V^{DO}(t, S; B) = \mathrm{e}^{-r\tau}\int_{b}^{\infty}\Lambda(S\mathrm{e}^{\sigma x})\,\widetilde{p}^{X^{(b)}}(\tau; 0, x)\,\mathrm{d}x \qquad (12.119)$$

and $V^{DO}(t, S; B) \equiv 0$ for $S \leqslant B$, where $\tau = T - t > 0$ is the time to maturity, $b := \frac{1}{\sigma}\ln\frac{B}{S}$, $\nu := \frac{(r-q-\frac{1}{2}\sigma^2)}{\sigma}$, and $\widetilde{p}^{X^{(b)}}$ is the (risk-neutral) density,

$$\begin{aligned}
\widetilde{p}^{X^{(b)}}(\tau; 0, x) &= p_0(\tau; x - \nu\tau) - \mathrm{e}^{2\nu b}\,p_0(\tau; x - \nu\tau - 2b)\\
&\equiv \frac{1}{\sqrt{\tau}}n\left(\frac{x - \nu\tau}{\sqrt{\tau}}\right) - \left(\frac{B}{S}\right)^{\frac{2(r-q)}{\sigma^2}-1}\frac{1}{\sqrt{\tau}}n\left(\frac{x - \nu\tau - 2b}{\sqrt{\tau}}\right), \qquad (12.120)
\end{aligned}$$

defined on the respective domains $(-\infty, b)$ and (b, ∞).

Proof. We prove (12.118), as (12.119) follows similarly. Using (a) in (12.117) within (12.113), changing the order of integration and evaluating the inner integral (as was done in (10.73) of Section 10.4.3 of Chapter 10):

$$
\begin{aligned}
V^{UO}(t,S;B) &= \mathrm{e}^{-r\tau}\int_{-\infty}^{b}\Lambda(S\,\mathrm{e}^{\sigma x})\left(\int_0^b \widetilde{f}_{M^X(\tau),X(\tau)}(w,x)\,\mathrm{d}w\right)\mathrm{d}x \\
&= \mathrm{e}^{-r\tau}\int_{-\infty}^{b}\Lambda(S\,\mathrm{e}^{\sigma x})\frac{\partial}{\partial x}\widetilde{F}_{M^X(\tau),X(\tau)}(b,x)\,\mathrm{d}x \\
&= \mathrm{e}^{-r\tau}\int_{-\infty}^{b}\Lambda(S\,\mathrm{e}^{\sigma x})\,\widetilde{\mathbb{P}}(M^X(\tau)\leqslant b, X(\tau)\in\mathrm{d}x) \\
&= \mathrm{e}^{-r\tau}\int_{-\infty}^{b}\Lambda(S\,\mathrm{e}^{\sigma x})\,\widetilde{p}^{X_{(b)}}(\tau;0,x)\,\mathrm{d}x\,,
\end{aligned}
$$

for $b>0$ and is identically zero for $b\leqslant 0$, i.e., $V^{UO}(t,S;B)\equiv 0$ for $S\geqslant B$. Here we made use of (10.77) and (10.80) of Section 10.4.3 of Chapter 10, with the variable replacements for the drift $\mu\to\nu$ and level $m\to b$. Note that

$$
2\nu b = \left(\frac{2(r-q)}{\sigma^2}-1\right)\ln\frac{B}{S} \implies \mathrm{e}^{2\nu b} = \mathrm{e}^{\left(\frac{2(r-q)}{\sigma^2}-1\right)\ln\frac{B}{S}} = \left(\frac{B}{S}\right)^{\frac{2(r-q)}{\sigma^2}-1}. \tag{12.121}
$$

□

[Remark: The prices $v^{UO}(T,S_0;B)=V^{UO}(0,S_0;B)$ and $v^{DO}(T,S_0;B)=V^{DO}(0,S_0;B)$, expressing the current time-0 price with maturity T, follow in the obvious manner by setting $t=0$, i.e., replacing $\tau\to T$ and $S\to S_0$ in the above formulae.]

Note that the density function in (12.120) is a linear combination of two normal densities. Hence, to apply (12.118) or (12.119) we need to compute an integral of the function $g(x):=\Lambda(S\,\mathrm{e}^{\sigma x})$, times $\mathbb{I}_{\{x<b\}}$ or $\mathbb{I}_{\{x>b\}}$, against a normal PDF in x. Let's now consider pricing an up-and-out call option where

$$
\Lambda(S\,\mathrm{e}^{\sigma x}) = (S\,\mathrm{e}^{\sigma x}-K)\mathbb{I}_{\{S\,\mathrm{e}^{\sigma x}>K\}} = (S\,\mathrm{e}^{\sigma x}-K)\mathbb{I}_{\{x>\kappa\}} = S\,\mathrm{e}^{\sigma x}\mathbb{I}_{\{x>\kappa\}} - K\,\mathbb{I}_{\{x>\kappa\}}\,,
$$

$\kappa:=\frac{1}{\sigma}\ln\frac{K}{S}$ and we assume the nontrivial case with $S<B$. Substituting this expression into the integrand in (12.118) gives the price of the up-and-out call as a difference of two integrals:

$$
C^{UO}(t,S,K;B) = \mathrm{e}^{-r\tau}S\int_\kappa^b \mathrm{e}^{\sigma x}\,\widetilde{p}^{X_{(b)}}(\tau;0,x)\,\mathrm{d}x - \mathrm{e}^{-r\tau}K\int_\kappa^b \widetilde{p}^{X_{(b)}}(\tau;0,x)\,\mathrm{d}x \tag{12.122}
$$

if $\kappa<b\equiv\frac{1}{\sigma}\ln\frac{B}{S}$, i.e., $K<B$. Note that $C^{UO}(t,S,K;B)\equiv 0$ if $\kappa\geqslant b$ (i.e., $K\geqslant B$). It is also clear from Figure 12.3 that paths which are in the money (above the strike) are necessarily above or at level B. Since all paths give zero payoff, the price of the up-and-out call must be identically zero when $K\geqslant B$. For $K<B$ the price is given by computing the two integrals in (12.122) upon substituting the density in (12.120). The second integral in (12.122) is a combination of two integrals involving the standard normal PDF which are readily evaluated by changing variables or simply using either identity (A.1) or (A.2) in the Appendix:

$$
\begin{aligned}
&\int_\kappa^b \widetilde{p}^{X_{(b)}}(\tau;0,x)\,\mathrm{d}x \\
&= \int_\kappa^b \frac{\mathrm{e}^{-(x-\nu\tau)^2/2\tau}}{\sqrt{2\pi\tau}}\,\mathrm{d}x - \mathrm{e}^{2\nu b}\int_\kappa^b \frac{\mathrm{e}^{-(x-(\nu\tau+2b))^2/2\tau}}{\sqrt{2\pi\tau}}\,\mathrm{d}x \\
&= \mathcal{N}\left(\frac{b-\nu\tau}{\sqrt{\tau}}\right) - \mathcal{N}\left(\frac{\kappa-\nu\tau}{\sqrt{\tau}}\right) - \mathrm{e}^{2\nu b}\left[\mathcal{N}\left(-\frac{b+\nu\tau}{\sqrt{\tau}}\right) - \mathcal{N}\left(\frac{\kappa-2b-\nu\tau}{\sqrt{\tau}}\right)\right].
\end{aligned} \tag{12.123}
$$

We can now express this in terms of the original parameters B, K, S, r, q, σ using (12.121) and the algebraic relations

$$\frac{b+\nu\tau}{\sqrt{\tau}} = d_-\big(\frac{B}{S},\tau\big)\,; \;\; \frac{b-\nu\tau}{\sqrt{\tau}} = -d_-\big(\frac{S}{B},\tau\big); \;\; \frac{\kappa-2b-\nu\tau}{\sqrt{\tau}} = -d_-\big(\frac{B^2}{KS},\tau\big)$$
$$\frac{\kappa+\nu\tau}{\sqrt{\tau}} = d_-\big(\frac{K}{S},\tau\big)\,; \;\; \frac{\kappa-\nu\tau}{\sqrt{\tau}} = -d_-\big(\frac{S}{K},\tau\big)\,,$$

where we define $d_+(x,\tau) := \dfrac{\ln x + (r-q+\frac{1}{2}\sigma^2)\tau}{\sigma\sqrt{\tau}}, d_-(x,\tau) = d_+(x,\tau) - \sigma\sqrt{\tau}$. Substituting these expressions into (12.123) and using the identity $\mathcal{N}(x) + \mathcal{N}(-x) = 1$ gives the exact integral:

$$\begin{aligned}\int_{\kappa=\frac{1}{\sigma}\ln\frac{K}{S}}^{b=\frac{1}{\sigma}\ln\frac{B}{S}} \widetilde{p}^{X_{(b)}}(\tau;0,x)\,\mathrm{d}x &= \mathcal{N}\left(d_-\big(\frac{S}{K},\tau\big)\right) - \mathcal{N}\left(d_-\big(\frac{S}{B},\tau\big)\right) \\ &- \left(\frac{B}{S}\right)^{\frac{2(r-q)}{\sigma^2}-1}\left[\mathcal{N}\left(d_-\big(\frac{B^2}{KS},\tau\big)\right) - \mathcal{N}\left(d_-\big(\frac{B}{S},\tau\big)\right)\right].\end{aligned} \tag{12.124}$$

We leave it as an exercise for the reader to apply similar steps to show that the (discounted) first integral in (12.122) is given by

$$\begin{aligned}\mathrm{e}^{-(r-q)\tau}\int_{\kappa=\frac{1}{\sigma}\ln\frac{K}{S}}^{b=\frac{1}{\sigma}\ln\frac{B}{S}} \mathrm{e}^{\sigma x}\,\widetilde{p}^{X_{(b)}}(\tau;0,x)\,\mathrm{d}x &= \mathcal{N}\left(d_+\big(\frac{S}{K},\tau\big)\right) - \mathcal{N}\left(d_+\big(\frac{S}{B},\tau\big)\right) \\ &- \left(\frac{B}{S}\right)^{\frac{2(r-q)}{\sigma^2}+1}\left[\mathcal{N}\left(d_+\big(\frac{B^2}{KS},\tau\big)\right) - \mathcal{N}\left(d_+\big(\frac{B}{S},\tau\big)\right)\right].\end{aligned} \tag{12.125}$$

Substituting the integral expressions in (12.124) and (12.125) into (12.122) and combining terms gives the analytically exact pricing formula for the up-and-out call for $K < B$:

$$C^{UO}(t,S,K;B) = C(t,S,K) - C^{UI}(t,S,K;B) \tag{12.126}$$

where

$$C(t,S,K) = \mathrm{e}^{-q\tau} S\,\mathcal{N}\left(d_+\big(\frac{S}{K},\tau\big)\right) - \mathrm{e}^{-r\tau} K\,\mathcal{N}\left(d_-\big(\frac{S}{K},\tau\big)\right)$$

is the Black–Scholes pricing formula for a standard call on a dividend paying stock and C^{UI} is the *up-and-in call pricing formula* for $K < B$:

$$\begin{aligned}C^{UI}(t,S,K;B) &= e^{-q\tau} S\mathcal{N}\Big(d_+\big(\frac{S}{B},\tau\big)\Big) - e^{-r\tau} K\mathcal{N}\Big(d_-\big(\frac{S}{B},\tau\big)\Big) \\ &+ e^{-q\tau} S\Big(\frac{B}{S}\Big)^{\frac{2(r-q)}{\sigma^2}+1}\Big[\mathcal{N}\Big(d_+\big(\frac{B^2}{KS},\tau\big)\Big) - \mathcal{N}\Big(d_+\big(\frac{B}{S},\tau\big)\Big)\Big] \\ &- e^{-r\tau} K\Big(\frac{B}{S}\Big)^{\frac{2(r-q)}{\sigma^2}-1}\Big[\mathcal{N}\Big(d_-\big(\frac{B^2}{KS},\tau\big)\Big) - \mathcal{N}\Big(d_-\big(\frac{B}{S},\tau\big)\Big)\Big],\end{aligned} \tag{12.127}$$

where $\tau = T - t$ is time to maturity. Note that for $K \geqslant B$, $C^{UI}(t,S,K;B) = C(t,S,K)$.

Pricing formulae for other up-and-out (and up-and-in) options are readily derived using similar steps and by combining the above integral identities in (12.124) and (12.125) within

(12.118). For down-and-out (and down-and-in) we use (12.119) and develop similar identities to (12.124) and (12.125) for evaluating the pricing integrals. The derivations of pricing formulae for down-and-out (and down-and-in) call and put options are left as exercises at the end of this chapter.

Example 12.4. (Up-and-Out Put Price) Derive the time-t, $t < T$, no-arbitrage pricing formula of an up-and-out put option with payoff

$$P_T^{UO} = (K - S(T))^+ \, \mathbb{I}_{\{M^S(T) < B\}}$$

where $B > 0$ is the knock-out barrier and $K > 0$ the strike. Assume the stock is a GBM with constant interest rate and continuous dividend yield q.

Solution. We take spot $S < B$. For an up-and-out option we use (12.118) with put payoff

$$\Lambda(S\,\mathrm{e}^{\sigma x}) = K\,\mathbb{I}_{\{x<\kappa\}} - S\,\mathrm{e}^{\sigma x}\mathbb{I}_{\{x<\kappa\}}\,, \quad \kappa \equiv \frac{1}{\sigma}\ln\frac{K}{S}\,.$$

The integral over the density is restricted to $x < b$, $b \equiv \frac{1}{\sigma}\ln\frac{B}{S}$. Since $\mathbb{I}_{\{x<\kappa\}}\mathbb{I}_{\{x<b\}} = \mathbb{I}_{\{x<b\wedge\kappa\}}$,

$$\Lambda(S\,\mathrm{e}^{\sigma x})\mathbb{I}_{\{x<b\}} = K\,\mathbb{I}_{\{x<b\wedge\kappa\}} - S\,\mathrm{e}^{\sigma x}\mathbb{I}_{\{x<b\wedge\kappa\}}$$

where $b \wedge \kappa \equiv \min(b, \kappa) = \frac{1}{\sigma}\ln\frac{K\wedge B}{S}$. The price of the up-and-out put is then given by

$$P^{UO}(t, S, K; B) = \mathrm{e}^{-r\tau}K\int_{-\infty}^{b\wedge\kappa} \widetilde{p}^{X_{(b)}}(\tau; 0, x)\,\mathrm{d}x - \mathrm{e}^{-r\tau}S\int_{-\infty}^{b\wedge\kappa} \mathrm{e}^{\sigma x}\widetilde{p}^{X_{(b)}}(\tau; 0, x)\,\mathrm{d}x\,.$$

There are two cases: (i) $B < K$ or (ii) $B \geqslant K$. For $B < K$, $b \wedge \kappa = b$ and the price is

$$P^{UO}(t, S, K; B) = \mathrm{e}^{-r\tau}K\int_{-\infty}^{b} \widetilde{p}^{X_{(b)}}(\tau; 0, x)\,\mathrm{d}x - \mathrm{e}^{-r\tau}S\int_{-\infty}^{b} \mathrm{e}^{\sigma x}\widetilde{p}^{X_{(b)}}(\tau; 0, x)\,\mathrm{d}x\,.$$

The two integrals can be computed using the same steps and identities used for the up-and-out call above. However, there is a shortcut based on (12.124) and (12.125) in the limit that the lower point of integration goes to $-\infty$. That is, the above two integrals correspond to taking the limit $K \to 0+$, $\ln\frac{K}{S} \to -\infty$, in the expressions in (12.124) and (12.125). Since $d_\pm(\frac{B^2}{KS}, \tau) \to \infty$, $d_\pm(\frac{S}{K}, \tau) \to \infty$, all $\mathcal{N}(\cdot)$ terms with these arguments approach $\mathcal{N}(\infty) = 1$, as $K \to 0+$. Upon using the symmetry $1 - \mathcal{N}(z) = \mathcal{N}(-z)$ in the resulting expressions we obtain:

$$\int_{-\infty}^{b=\frac{1}{\sigma}\ln\frac{B}{S}} \widetilde{p}^{X_{(b)}}(\tau; 0, x)\,\mathrm{d}x = \mathcal{N}\left(-d_-\left(\frac{S}{B}, \tau\right)\right) - \left(\frac{B}{S}\right)^{\frac{2(r-q)}{\sigma^2}-1}\mathcal{N}\left(-d_-\left(\frac{B}{S}, \tau\right)\right) \tag{12.128}$$

and

$$\mathrm{e}^{-(r-q)\tau}\int_{-\infty}^{b=\frac{1}{\sigma}\ln\frac{B}{S}} \mathrm{e}^{\sigma x}\,\widetilde{p}^{X_{(b)}}(\tau; 0, x)\,\mathrm{d}x = \mathcal{N}\left(-d_+\left(\frac{S}{B}, \tau\right)\right) - \left(\frac{B}{S}\right)^{\frac{2(r-q)}{\sigma^2}+1}\mathcal{N}\left(-d_+\left(\frac{B}{S}, \tau\right)\right). \tag{12.129}$$

Substituting these integrals gives the explicit pricing function for $B < K$:

$$\begin{aligned} P^{UO}(t, S, K; B) = \mathrm{e}^{-r\tau}K&\left[\mathcal{N}\left(-d_-\left(\frac{S}{B}, \tau\right)\right) - \left(\frac{B}{S}\right)^{\frac{2(r-q)}{\sigma^2}-1}\mathcal{N}\left(-d_-\left(\frac{B}{S}, \tau\right)\right)\right] \\ -\mathrm{e}^{-q\tau}S&\left[\mathcal{N}\left(-d_+\left(\frac{S}{B}, \tau\right)\right) - \left(\frac{B}{S}\right)^{\frac{2(r-q)}{\sigma^2}+1}\mathcal{N}\left(-d_+\left(\frac{B}{S}, \tau\right)\right)\right], \end{aligned} \tag{12.130}$$

$\tau = T - t$. For $B \geqslant K$, $\kappa \leqslant b$, $b \wedge \kappa = \kappa$ and the price is given by

$$P^{UO}(t, S, K; B) = \mathrm{e}^{-r\tau} K \int_{-\infty}^{\kappa} \widetilde{p}^{X_{(b)}}(\tau; 0, x)\, \mathrm{d}x - \mathrm{e}^{-r\tau} S \int_{-\infty}^{\kappa} \mathrm{e}^{\sigma x} \widetilde{p}^{X_{(b)}}(\tau; 0, x)\, \mathrm{d}x\,.$$

In this case we express each integral on $(-\infty, \kappa)$ as the integral on $(-\infty, b)$ minus the integral on (κ, b). Then, we can use the difference of (12.128) and (12.124) to obtain the first integral on $(-\infty, \kappa)$ and the difference of (12.129) and (12.125). Combining terms and simplifying, we have the explicit pricing function for $B \geqslant K$:

$$\begin{aligned} P^{UO}(t, S, K; B) = P(t, S, K) &+ \mathrm{e}^{-q\tau} S \left(\frac{B}{S}\right)^{\frac{2(r-q)}{\sigma^2}+1} \mathcal{N}\left(-d_+\left(\frac{B^2}{KS}, \tau\right)\right) \\ &- \mathrm{e}^{-r\tau} K \left(\frac{B}{S}\right)^{\frac{2(r-q)}{\sigma^2}-1} \mathcal{N}\left(-d_-\left(\frac{B^2}{KS}, \tau\right)\right), \end{aligned} \tag{12.131}$$

$\tau = T - t$, where $P(t, S, K) = \mathrm{e}^{-r\tau} K \mathcal{N}\left(-d_-(\frac{S}{K}, \tau)\right) - \mathrm{e}^{-q\tau} S \mathcal{N}\left(-d_+(\frac{S}{K}, \tau)\right)$ is the Black–Scholes pricing formula for a standard put on a dividend paying stock. □

In closing this section we show that barrier options can also be "delta hedged" and we also make the connection between the risk-neutral pricing approach and the corresponding Black–Scholes PDE (BSPDE) for pricing single barrier options. We focus our discussion on the up-and-out and down-and-out options. The analysis for knock-in barrier options follows from knock-in-knock-out symmetry. As we have shown above, the general pricing formulae for the knock-out options are given by (12.118) and (12.119) of Proposition 12.1. Assuming that the integrals in (12.118) and (12.119) exist, and that we can evaluate them, we have completely solved the pricing problem for single barrier options. Alternatively, we now show that the pricing function is a solution to a BSPDE subject to appropriate boundary conditions. We have already seen how the risk-neutral pricing formulation is related to the BSPDE for the case of a standard (no barrier) European option. In particular, the pricing function $V(t, S)$ in (12.15) is expressed as an (expectation) integral of the payoff against the risk-neutral transition PDF in (12.16) for the stock price process on the domain $(0, \infty)$. The discounted risk-neutral transition PDF, and therefore $V(t, S)$, solves the BSPDE (12.51) in the variables (S, τ) and (12.52) in the variables (t, S).

To see how the BSPDE arises for an up-and-out option, we apply a change of integration variables by letting $y = S\mathrm{e}^{\sigma x}$ ($x = \frac{1}{\sigma} \ln \frac{y}{S}$) in (12.118), which then takes the form

$$V^{UO}(t, S; B) = \mathrm{e}^{-r\tau} \int_0^B \Lambda(y)\, \widetilde{p}^{S_{(B)}}(\tau; S, y)\, \mathrm{d}y \tag{12.132}$$

where $\widetilde{p}^{S_{(B)}}(\tau; S, y)$ is defined for all S, y values on the interval $(0, B)$:

$$\begin{aligned} \widetilde{p}^{S_{(B)}}(\tau; S, y) \equiv & \frac{1}{\sigma y \sqrt{\tau}} n\left(\frac{\ln \frac{y}{S} - (r - q - \frac{1}{2}\sigma^2)\tau}{\sigma\sqrt{\tau}}\right) \\ & - \left(\frac{B}{S}\right)^{\frac{2(r-q)}{\sigma^2}-1} \frac{1}{\sigma y \sqrt{\tau}} n\left(\frac{\ln \frac{Sy}{B^2} - (r - q - \frac{1}{2}\sigma^2)\tau}{\sigma\sqrt{\tau}}\right). \end{aligned} \tag{12.133}$$

As shown in Exercise 12.22, this is the risk-neutral transition PDF for the stock price process killed at the first-hitting time to level B on *either* interval $(0, B)$ or (B, ∞). The latter interval is used for the down-and-out option where (12.119) takes the same form as in (12.132) but with (B, ∞) as the integration interval in place of $(0, B)$.

For any fixed y, the discounted transition PDF, $v(\tau, S, y) := \mathrm{e}^{-r\tau} \widetilde{p}^{S_{(B)}}(\tau; S, y)$, solves the time-homogeneous BSPDE in (12.51) subject to the initial condition $v(0+,S,y) = \delta(S-y)$ (with the Dirac delta function δ). In fact, it is a fundamental solution on the interval $(0,B)$ (as well as on the interval (B,∞)) with zero boundary conditions at the barrier level B and at either endpoint $S \to 0+$ or $S \to \infty$. Assuming the integral in (12.132) exists and the resulting pricing function is $C^{1,2}$ (continuously differentiable in t (or τ) and twice differentiable in S), we can apply the differential operator $(\frac{\partial}{\partial t} + \mathcal{L}^{BS})$ (acting on variables t and S) on both sides of (12.132). Note that $\mathcal{L}^{BS}$ is the Black–Scholes operator as in (12.52). By interchanging the order of differentiation and integration (in the dummy variable y), and using the fact that $v(\tau,S,y)$ solves the BSPDE, gives

$$\frac{\partial}{\partial t}V^{UO} + \mathcal{L}^{BS}V^{UO} = \int_0^B \underbrace{\left(\frac{\partial}{\partial t}v(\tau,S,y) + \mathcal{L}^{BS}v(\tau,S,y)\right)}_{\equiv 0} \Lambda(y)\, dy = 0\,. \tag{12.134}$$

Hence, $V^{UO} \equiv V^{UO}(t,S;B)$ is a solution to the BSPDE in (12.52) on the rectangular domain $0 < S < B, 0 \leqslant t < T$ or equivalently the BSPDE in (12.51) for $0 < S < B, \tau \in (0,T]$. The terminal condition (or initial condition $\tau \to 0+$) is given by the payoff function where $V^{UO}(T,S;B) \equiv V^{UO}(T-,S;B) = \Lambda(S)$, for any continuous Λ and for $0 \leqslant S \leqslant B$. The boundary conditions at the endpoints of $(0,B)$ are given by (note: $S=0$ is the limit $S \searrow 0$)

$$\begin{aligned} &V^{UO}(t,S=0;B) = \mathrm{e}^{-r(T-t)}\Lambda(0)\,, && 0 \leqslant t \leqslant T\,,\\ &V^{UO}(t,S=B;B) = 0\,, && 0 \leqslant t < T\,. \end{aligned} \tag{12.135}$$

The boundary condition at $S=0$ is due to the stock price staying at zero if it is set to zero and hence the payoff will be $\Lambda(0)$, which is discounted by $\mathrm{e}^{-r(T-t)}$ to obtain its time-t value. The second condition corresponds to the option being worthless if the spot is at the barrier level any time before maturity. As an example, for an up-and-out call its value at the lower boundary $S=0$ is $C^{UO}(t,0,K;B) = \mathrm{e}^{-r(T-t)}\Lambda(0) = \mathrm{e}^{-r(T-t)}(0-K)^+ = 0$ and at $S=B$ we have $C^{UO}(t,B,K;B) = 0$. For an up-and-out put, $P^{UO}(t,0,K;B) = \mathrm{e}^{-r(T-t)}(K-0)^+ = \mathrm{e}^{-r(T-t)}K$ and $P^{UO}(t,B,K;B) = 0$ at $S=B$.

For a down-and-out option the analysis is similar, leading to the same BSPDE in (12.52) for $V^{DO} \equiv V^{DO}(t,S;B)$ on the rectangular domain $B < S < \infty, 0 \leqslant t < T$ or equivalently the BSPDE in (12.51) for $B < S < \infty, \tau \in (0,T]$. The terminal (or initial) time condition is again the payoff function, $V^{DO}(T,S;B) = \Lambda(S)$, for $B \leqslant S < \infty$, and with boundary endpoint conditions:

$$\begin{aligned} &\lim_{S\to\infty} V^{DO}(t,S;B) = \lim_{S\to\infty} V(t,S)\,, && 0 \leqslant t \leqslant T\,,\\ &V^{DO}(t,S=B;B) = 0\,, && 0 \leqslant t < T\,. \end{aligned} \tag{12.136}$$

The first condition states that the value of the down-and-out option and the corresponding standard option value $V(t,S)$ should be the same in the limit of infinite stock value. This is due to the stock price staying close to infinity and not hitting the lower knock-out barrier (in finite time) if it starts close to infinity. For the GBM model this is the case where the boundary at infinity is a natural boundary. The second boundary condition is again due to the option expiring worthless if the spot is at the barrier level before maturity. For example, a down-and-out call has value $C^{DO}(t,B,K;B) = 0$ at $S=B$ and $C^{DO}(t,S,K;B) \sim C(t,S,K) \sim \mathrm{e}^{-q(T-t)}S - \mathrm{e}^{-r(T-t)}K$, as $S \to \infty$. For a down-and-out put, $P^{DO}(t,B,K;B) = 0$ and $P^{DO}(t,S,K;B) \sim P(t,S,K) \sim 0$, as $S \to \infty$.

Let $V(t,S;B)$ denote either pricing function $V^{UO}(t,S;B)$ or $V^{DO}(t,S;B)$. For any given $B > 0$, we argued above that $V(t,S;B)$ is a $C^{1,2}$ function that solves the BSPDE in the (dummy) variables t, S. We can therefore apply Itô's formula to the discounted process defined via the function $V(t, S; B)$, i.e., $\{e^{-rt}V(t, S(t); B)\}_{t\geqslant 0}$. Taking the stochastic differential and using the fact that $V(t, S; B)$ solves the BSPDE gives

$$
\begin{aligned}
\mathrm{d}\left[e^{-rt}V(t, S(t); B)\right] &= e^{-rt}\left(\frac{\partial}{\partial t} + \mathcal{L}^{BS}\right)V(t, S(t); B)\,\mathrm{d}t + \sigma\overline{S}(t)\frac{\partial}{\partial S}V(t, S(t); B)\,\mathrm{d}\widetilde{W}(t)\,, \\
&= \sigma\overline{S}(t)\frac{\partial}{\partial S}V(t, S(t); B)\,\mathrm{d}\widetilde{W}(t)\,.
\end{aligned}
$$

Note that this stochastic differential and that of the discounted price process for the knock-out barrier option are the same for all times before the stock price hits the barrier level B. Equating this with the expression in (12.7) gives the hedging position $\Delta_t = \frac{\partial}{\partial S}V(t, S(t); B)$. For a given realization of the stock price process, this is then the hedging position held in the stock for all times t up to the first hitting time to the (knock-out) level B, or otherwise up to maturity time T if the stock price does not attain the level B during the option's lifetime. In particular, for every spot value $S(t) = S < B$, the hedging formula for an up-and-out option is the delta of the pricing function, $\Delta^{UO}(t, S; B) = \frac{\partial}{\partial S}V^{UO}(t, S; B)$. Similarly, for $S(t) = S > B$, a down-and-out option is hedged using $\Delta^{DO}(t, S; B) = \frac{\partial}{\partial S}V^{DO}(t, S; B)$.

12.3.3 Pricing Lookback Options

We can now proceed to derive pricing formulae for generally "seasoned" lookback options of types (a)–(d) with payoffs defined in (12.99) where conditioning is on knowledge of the sampled stock price maximum, $M^S(t) = M \geqslant S$, or minimum $m^S(t) = m \leqslant S$, i.e., we are entering the contract at time t where the realized maximum or minimum up to time t generally differs from the stock (spot) price $S(t) = S$. Our main pricing formulae are (12.111) and (12.112). We therefore need the effective payoff functions in the integrand of either case. For example, consider the floating strike (LFS) call with payoff $C_T^{LFS} = \phi(m^S(T), S(T)) = S(T) - m^S(T)$ in (a) of (12.99), i.e. $\phi(x, y) := y - x$. Hence, the effective payoff for this option is the integrand function in (12.112) given by

$$
\begin{aligned}
g(w, x) &:= \phi\big(\min\{m, S\,e^{\sigma w}\}, S\,e^{\sigma x}\big) \\
&= S\,e^{\sigma x} - \min\{m, S\,e^{\sigma w}\} \\
&= S\,e^{\sigma x} - \left[S\,e^{\sigma w}\,\mathbb{I}_{\{S\,e^{\sigma w} < m\}} + m\,\mathbb{I}_{\{S\,e^{\sigma w} \geqslant m\}}\right] \\
&= S\,e^{\sigma x} - S\,e^{\sigma w}\,\mathbb{I}_{\{w < \widehat{m}\}} - m\,\mathbb{I}_{\{w \geqslant \widehat{m}\}}
\end{aligned}
$$

where $\widehat{m} := \frac{1}{\sigma}\ln\frac{m}{S} \leqslant 0$. For case (b) in (12.99) we have $\phi(M^S(T), S(T)) = M^S(T) - S(T)$, i.e., $\phi(x, y) := x - y$. Hence, the effective payoff in (12.111) is

$$
\begin{aligned}
h(w, x) := \phi\big(\max\{M, S\,e^{\sigma w}\}, S\,e^{\sigma x}\big) &= \max\{M, S\,e^{\sigma w}\} - S\,e^{\sigma x} \\
&= M\,\mathbb{I}_{\{w < \widehat{M}\}} + S\,e^{\sigma w}\,\mathbb{I}_{\{w \geqslant \widehat{M}\}} - S\,e^{\sigma x}\,,
\end{aligned}
$$

where $\widehat{M} := \frac{1}{\sigma}\ln\frac{M}{S} \geqslant 0$.

The reader can verify that the effective payoffs for cases (c) and (d) are as given below where we summarize the effective payoffs for lookbacks (a)–(d) in (12.99):

$$\text{(a) } g(w,x) = S\,\mathrm{e}^{\sigma x} - S\,\mathrm{e}^{\sigma w}\,\mathbb{I}_{\{w<\widehat{m}\}} - m\,\mathbb{I}_{\{w\geqslant\widehat{m}\}}; \tag{12.137}$$

$$\text{(b) } h(w,x) = M\,\mathbb{I}_{\{w<\widehat{M}\}} + S\,\mathrm{e}^{\sigma w}\,\mathbb{I}_{\{w\geqslant\widehat{M}\}} - S\,\mathrm{e}^{\sigma x}; \tag{12.138}$$

$$\text{(c) } h(w,x) = \begin{cases} (S\,\mathrm{e}^{\sigma w} - K)\mathbb{I}_{\{w>\kappa\}} & \text{for } M < K, \\ M\,\mathbb{I}_{\{w<\widehat{M}\}} + S\,\mathrm{e}^{\sigma w}\,\mathbb{I}_{\{w\geqslant\widehat{M}\}} - K & \text{for } M \geqslant K; \end{cases} \tag{12.139}$$

$$\text{(d) } g(w,x) = \begin{cases} K - m\,\mathbb{I}_{\{w\geqslant\widehat{m}\}} - S\,\mathrm{e}^{\sigma w}\mathbb{I}_{\{w<\widehat{m}\}} & \text{for } m < K, \\ (K - S\,\mathrm{e}^{\sigma w})\,\mathbb{I}_{\{w<\kappa\}} & \text{for } m \geqslant K, \end{cases} \tag{12.140}$$

where $\kappa := \frac{1}{\sigma}\ln\frac{K}{S}$.

It is important to note that the functions in (c) and (d) depend only on w (not x). Moreover, the functions in (a) and (b) are simply sums of functions that depend on only one of the variables, either x or w and not both. This therefore simplifies the pricing integrals in (12.111) and (12.112), which are then sums of single integrals involving either (risk-neutral) marginal density in $m^X(\tau)$ or $M^X(\tau)$. Recall that integrating a joint PDF in one of its arguments (over $\mathbb{R}$) produces the corresponding marginal PDF. This simplification is given explicitly below for the above cases (a)–(d) where the pricing formulae are reduced to single integrals involving the marginal CDF or PDF of $m^X(\tau)$ and $M^X(\tau)$ and other more trivial integrals for the expected value of the drifted BM.

We now state these CDFs and PDFs for further use below when computing expectation integrals within the $\widetilde{\mathbb{P}}$-measure. The CDFs of $m^X(\tau)$ and $M^X(\tau)$ were derived in Section 10.4.3 of Chapter 10. Under the risk-neutral measure $\widetilde{\mathbb{P}}$ we simply take the expressions in (10.82) and (10.88) where the process X now has drift $\nu := \frac{(r-q-\frac{1}{2}\sigma^2)}{\sigma}$ and the time variable is τ, i.e., replace $\mu \to \nu$, $t \to \tau$, and $m \to w$ in (10.82) and (10.88) to give

$$\widetilde{F}_{M^X(\tau)}(w) := \widetilde{\mathbb{P}}(M^X(\tau) \leqslant w) = \mathcal{N}\left(\frac{w-\nu\tau}{\sqrt{\tau}}\right) - \mathrm{e}^{2\nu w}\,\mathcal{N}\left(\frac{-w-\nu\tau}{\sqrt{\tau}}\right), \quad w > 0, \tag{12.141}$$

$\widetilde{F}_{M^X(\tau)}(w) \equiv 0$ for $w \leqslant 0$, and

$$\widetilde{F}_{m^X(\tau)}(w) := \widetilde{\mathbb{P}}(m^X(\tau) \leqslant w) = \mathcal{N}\left(\frac{w-\nu\tau}{\sqrt{\tau}}\right) + \mathrm{e}^{2\nu w}\,\mathcal{N}\left(\frac{w+\nu\tau}{\sqrt{\tau}}\right), \quad w < 0, \tag{12.142}$$

$\widetilde{F}_{m^X(\tau)}(w) \equiv 1$ for $w \geqslant 0$. Differentiating these CDFs gives the densities, i.e., $\mathrm{d}\widetilde{F}_{M^X(\tau)}(w) = \widetilde{f}_{M^X(\tau)}(w)\,\mathrm{d}w$ and $\mathrm{d}\widetilde{F}_{m^X(\tau)}(w) = \widetilde{f}_{m^X(\tau)}(w)\,\mathrm{d}w$, where

$$\widetilde{f}_{M^X(\tau)}(w) = \frac{1}{\sqrt{\tau}}n\left(\frac{w-\nu\tau}{\sqrt{\tau}}\right) + \frac{\mathrm{e}^{2\nu w}}{\sqrt{\tau}}\,n\left(\frac{w+\nu\tau}{\sqrt{\tau}}\right) - 2\nu\mathrm{e}^{2\nu w}\,\mathcal{N}\left(\frac{-w-\nu\tau}{\sqrt{\tau}}\right), \quad w > 0, \tag{12.143}$$

$$\widetilde{f}_{m^X(\tau)}(w) = \frac{1}{\sqrt{\tau}}n\left(\frac{w-\nu\tau}{\sqrt{\tau}}\right) + \frac{\mathrm{e}^{2\nu w}}{\sqrt{\tau}}\,n\left(\frac{w+\nu\tau}{\sqrt{\tau}}\right) + 2\nu\mathrm{e}^{2\nu w}\,\mathcal{N}\left(\frac{w+\nu\tau}{\sqrt{\tau}}\right), \quad w < 0. \tag{12.144}$$

Alternatively, the reader can verify that these same expressions are obtained by successively integrating the respective joint PDFs in (12.108) and (12.109).

Based on (12.111) and (12.112), we can now derive the main pricing formulae for the above four types of lookback options. Consider case (a), where we denote the time-t pricing function for the LFS call by $C^{LFS}(t, S, m)$ for all $0 < m \leqslant S < \infty$. Substituting (12.137) into (12.112) gives this pricing function as a sum of three integrals involving the joint PDF $\widetilde{f}(w, x) \equiv \widetilde{f}_{m^X(\tau), X(\tau)}(w, x)$. The integrals are respectively reduced to single integrals involving the marginal PDF of $X(\tau)$ and of $m^X(\tau)$ as follows:

$$\begin{aligned} C^{LFS}(t, S, m) &= \mathrm{e}^{-r\tau} S \int_{\mathbb{R}} \left[\int_{\mathbb{R}} \widetilde{f}(w, x)\, \mathrm{d}w \right] \mathrm{e}^{\sigma x}\, \mathrm{d}x - \mathrm{e}^{-r\tau} S \int_{\mathbb{R}} \left[\int_{\mathbb{R}} \widetilde{f}(w, x)\, \mathrm{d}x \right] \mathrm{e}^{\sigma w} \mathbb{I}_{\{w < \widehat{m}\}}\, \mathrm{d}w \\ &\quad - \mathrm{e}^{-r\tau} m \int_{\mathbb{R}} \left[\int_{\mathbb{R}} \widetilde{f}(w, x)\, \mathrm{d}x \right] \mathbb{I}_{\{w > \widehat{m}\}}\, \mathrm{d}w \\ &= \mathrm{e}^{-r\tau} S \int_{\mathbb{R}} \widetilde{f}_{X(\tau)}(x)\, \mathrm{e}^{\sigma x}\, \mathrm{d}x - \mathrm{e}^{-r\tau} S \int_{-\infty}^{\widehat{m}} \widetilde{f}_{m^X(\tau)}(w)\, \mathrm{e}^{\sigma w}\, \mathrm{d}w \\ &\quad - \mathrm{e}^{-r\tau} m \int_{\widehat{m}}^{0} \widetilde{f}_{m^X(\tau)}(w)\, \mathrm{d}w \,. \end{aligned} \tag{12.145}$$

The integrals are recognized as expectations, where the third integral is $\widetilde{\mathbb{P}}(m^X(\tau) > \widehat{m}) = 1 - \widetilde{\mathbb{P}}(m^X(\tau) \leqslant \widehat{m}) \equiv 1 - \widetilde{F}_{m^X(\tau)}(\widehat{m})$:

$$\begin{aligned} C^{LFS}(t, S, m) &= \mathrm{e}^{-r\tau} S\, \widetilde{\mathrm{E}}[\mathrm{e}^{\sigma X(\tau)}] - \mathrm{e}^{-r\tau} S\, \widetilde{\mathrm{E}}[\mathrm{e}^{\sigma m^X(\tau)} \mathbb{I}_{\{m^X(\tau) < \widehat{m}\}}] \\ &\quad - \mathrm{e}^{-r\tau} m\, [1 - \widetilde{F}_{m^X(\tau)}(\widehat{m})] \,. \end{aligned} \tag{12.146}$$

The first expectation is computed simply as $\widetilde{\mathrm{E}}[\mathrm{e}^{\sigma X(\tau)}] = \mathrm{e}^{\sigma\nu\tau} \widetilde{\mathrm{E}}[\mathrm{e}^{\sigma \widetilde{W}(\tau)}] = \mathrm{e}^{\sigma\nu\tau} \cdot \mathrm{e}^{\frac{1}{2}\sigma^2\tau} = \mathrm{e}^{(r-q)\tau}$. This holds true even for more complex models as long as the stock price $S(t)$ discounted by $\mathrm{e}^{-(r-q)t}$ is a $\widetilde{\mathbb{P}}$-martingale. Hence, the pricing formula for the LFS call is given equivalently by:

$$\begin{aligned} C^{LFS}(t, S, m) &= \mathrm{e}^{-q\tau} S - \mathrm{e}^{-r\tau} m\, [1 - \widetilde{F}_{m^X(\tau)}(\widehat{m})] - \mathrm{e}^{-r\tau} S \int_{-\infty}^{\widehat{m}} \widetilde{f}_{m^X(\tau)}(w)\, \mathrm{e}^{\sigma w}\, \mathrm{d}w \quad (12.147) \\ &= \mathrm{e}^{-q\tau} S - \mathrm{e}^{-r\tau} m\, \widetilde{\mathbb{P}}(m^S(\tau) > m) - \mathrm{e}^{-r\tau} \widetilde{\mathrm{E}} \left[m^S(\tau)\, \mathbb{I}_{\{m^S(\tau) \leqslant m\}} \right] \\ &= \mathrm{e}^{-q\tau} S - \mathrm{e}^{-r\tau} m + \mathrm{e}^{-r\tau} \int_0^m \widetilde{\mathbb{P}}(m^S(\tau) \leqslant y)\, \mathrm{d}y \end{aligned}$$

where $\tau = T - t$. In the second equation line we have the respective quantities expressed in terms of $m^S(\tau)$, $\widetilde{\mathrm{E}}[S \mathrm{e}^{\sigma m^X(\tau)} \mathbb{I}_{\{m^X(\tau) \leqslant \widehat{m}\}}] = \widetilde{\mathrm{E}}[m^S(\tau)\, \mathbb{I}_{\{m^S(\tau) \leqslant m\}}]$ and $\widetilde{\mathbb{P}}(m^X(\tau) > \widehat{m}) = \widetilde{\mathbb{P}}(m^S(\tau) > m)$ since the sampled minimum of the stock price (started at spot value S) for a time interval τ is $m^S(\tau) = S \mathrm{e}^{\sigma m^X(\tau)}$. The third line is obtained by re-expressing the expectation in the second line upon using an integration by parts,

$$\begin{aligned} \widetilde{\mathrm{E}} \left[m^S(\tau)\, \mathbb{I}_{\{m^S(\tau) \leqslant m\}} \right] &= \int_0^m y\, \mathrm{d}\widetilde{F}_{m^S(\tau)}(y) = m \widetilde{F}_{m^S(\tau)}(m) - \int_0^m \widetilde{F}_{m^S(\tau)}(y)\, \mathrm{d}y \\ &= m \widetilde{\mathbb{P}}(m^S(\tau) \leqslant m) - \int_0^m \widetilde{\mathbb{P}}(m^S(\tau) \leqslant y)\, \mathrm{d}y \,. \end{aligned}$$

This identity is valid for any number $m \geqslant 0$.

A similar derivation follows for the floating strike lookback (LFS) put option defined by the payoff in case (b) above where we denote the time-t pricing function for the LFS put by $P^{LFS}(t, S, M)$, for all $0 < S \leqslant M < \infty$. We now substitute (12.138) into (12.111) and this leads to a sum of three integrals involving the joint PDF of $\widetilde{f}_{M^X(\tau),X(\tau)}(w,x)$. Using similar steps as above, the reader can verify that the resulting pricing formula takes the equivalent expressions:

$$\begin{aligned} P^{LFS}(t,S,M) &= M\mathrm{e}^{-r\tau}\,\widetilde{F}_{M^X(\tau)}(\widehat{M}) - \mathrm{e}^{-q\tau}S + \mathrm{e}^{-r\tau}S\int_{\widehat{M}}^{\infty} \mathrm{e}^{\sigma w}\,\mathrm{d}\widetilde{F}_{M^X(\tau)}(w) \qquad (12.148)\\ &= M\mathrm{e}^{-r\tau}\widetilde{\mathbb{P}}(M^S(\tau)\leqslant M) - \mathrm{e}^{-q\tau}S + \mathrm{e}^{-r\tau}\widetilde{\mathrm{E}}\left[M^S(\tau)\,\mathbb{I}_{\{M^S(\tau)>M\}}\right]\\ &= \mathrm{e}^{-r\tau}M - \mathrm{e}^{-q\tau}S + \mathrm{e}^{-r\tau}\int_M^{\infty}\widetilde{\mathbb{P}}(M^S(\tau)>y)\,\mathrm{d}y\,, \end{aligned}$$

$\tau = T-t$. In the second equation line we have the respective quantities expressed in terms of $M^S(\tau)$: $\widetilde{\mathrm{E}}[S\mathrm{e}^{\sigma M^X(\tau)}\mathbb{I}_{\{M^X(\tau)>\widehat{M}\}}] = \widetilde{\mathrm{E}}[M^S(\tau)\,\mathbb{I}_{\{M^S(\tau)>M\}}]$ and $\widetilde{\mathbb{P}}(M^X(\tau)>\widehat{M}) = \widetilde{\mathbb{P}}(M^S(\tau)>M)$ since the sampled maximum of the stock price is $M^S(\tau) = S\mathrm{e}^{\sigma M^X(\tau)}$. The third line is obtained by noting that $M^S(\tau)\,\mathbb{I}_{\{M^S(\tau)>M\}} = M^S(\tau) - M^S(\tau)\mathbb{I}_{\{M^S(\tau)\leqslant M\}}$, where $\mathbb{I}_{\{M^S(\tau)\leqslant M\}} = \mathbb{I}_{\{0\leqslant M^S(\tau)\leqslant M\}}$. The expected value of the positive random variable $M^S(\tau)$ can be represented as an integral over its right tail (risk-neutral) probability:

$$\widetilde{\mathrm{E}}[M^S(\tau)] = \int_0^{\infty}\widetilde{\mathbb{P}}(M^S(\tau)>y)\,\mathrm{d}y\,.$$

The expected value of $M^S(\tau)\mathbb{I}_{\{0\leqslant M^S(\tau)\leqslant M\}}$ can be expressed by applying an integration by parts procedure as above,

$$\begin{aligned} \widetilde{\mathrm{E}}\left[M^S(\tau)\mathbb{I}_{\{0\leqslant M^S(\tau)\leqslant M\}}\right] &= \int_0^M y\,\mathrm{d}\widetilde{F}_{M^S(\tau)}(y) = M\widetilde{F}_{M^S(\tau)}(M) - \int_0^M \widetilde{F}_{M^S(\tau)}(y)\,\mathrm{d}y\\ &= M\widetilde{\mathbb{P}}(M^S(\tau)\leqslant M) - \int_0^M \widetilde{\mathbb{P}}(M^S(\tau)\leqslant y)\,\mathrm{d}y\,. \end{aligned}$$

Since $\widetilde{\mathbb{P}}(M^S(\tau)\leqslant y) + \widetilde{\mathbb{P}}(M^S(\tau)>y) = 1$, for any $y\geqslant 0$, we can write the last integral as $\int_0^M \widetilde{\mathbb{P}}(M^S(\tau)\leqslant y)\,\mathrm{d}y = M - \int_0^M \widetilde{\mathbb{P}}(M^S(\tau)>y)\,\mathrm{d}y$, and then combine the above two expectations to establish the identity

$$\widetilde{\mathrm{E}}\left[M^S(\tau)\,\mathbb{I}_{\{M^S(\tau)>M\}}\right] = M\widetilde{\mathbb{P}}(M^S(\tau)>M) + \int_M^{\infty}\widetilde{\mathbb{P}}(M^S(\tau)>y)\,\mathrm{d}y$$

for any number $M\geqslant 0$. Substituting this into the second line of (12.148) gives the expression in the third line of (12.148). For the payoff in case (c) above we denote the time-t pricing function for the floating price lookback (LFP) call (on the maximum) with strike $K>0$ by $C^{LFP}(t,S,M;K)$, for all $0<S\leqslant M<\infty$. By using similar steps as in case (b) above, the reader can verify that the pricing formula takes on the equivalent expressions:

$$\begin{aligned} C^{LFP}(t,S,M;K) &= M\mathrm{e}^{-r\tau}\,\widetilde{F}_{M^X(\tau)}(\widehat{M}) - \mathrm{e}^{-r\tau}K + \mathrm{e}^{-r\tau}S\int_{\widehat{M}}^{\infty}\mathrm{e}^{\sigma w}\,\mathrm{d}\widetilde{F}_{M^X(\tau)}(w) \qquad (12.149)\\ &= M\mathrm{e}^{-r\tau}\widetilde{\mathbb{P}}(M^S(\tau)\leqslant M) - \mathrm{e}^{-r\tau}K + \mathrm{e}^{-r\tau}\widetilde{\mathrm{E}}\left[M^S(\tau)\,\mathbb{I}_{\{M^S(\tau)>M\}}\right]\\ &= \mathrm{e}^{-r\tau}\left[M - K + \int_M^{\infty}\widetilde{\mathbb{P}}(M^S(\tau)>y)\,\mathrm{d}y\right] \end{aligned}$$

for $M\geqslant K$, and

$$\begin{aligned} C^{LFP}(t,S,M;K) &= -\mathrm{e}^{-r\tau}K[1-\widetilde{F}_{M^X(\tau)}(\kappa)] + \mathrm{e}^{-r\tau}S\int_{\kappa}^{\infty}\mathrm{e}^{\sigma w}\,\mathrm{d}\widetilde{F}_{M^X(\tau)}(w) \qquad (12.150)\\ &= -\mathrm{e}^{-r\tau}K\,\widetilde{\mathbb{P}}(M^S(\tau)>K) + \mathrm{e}^{-r\tau}\widetilde{\mathrm{E}}\left[M^S(\tau)\,\mathbb{I}_{\{M^S(\tau)>K\}}\right]\\ &= \mathrm{e}^{-r\tau}\int_K^{\infty}\widetilde{\mathbb{P}}(M^S(\tau)>y)\,\mathrm{d}y \end{aligned}$$

for $M < K$, where $\tau = T - t$. Note that for $M < K$ the pricing function C^{LFP}, given by (12.150), is independent of the realized maximum M of the stock price at current time t.

In the last case (d) we denote the time-t pricing function for the floating price lookback (LFP) put (on the minimum) with strike $K > 0$ by $P^{LFP}(t, S, m; K)$, for all $0 < m \leqslant S < \infty$. By using similar steps as in case (a) above, the reader can verify that the pricing formula takes the equivalent forms:

$$\begin{aligned} P^{LFP}(t, S, m; K) &= K\mathrm{e}^{-r\tau}\, \widetilde{F}_{m^X(\tau)}(\kappa) - \mathrm{e}^{-r\tau} S \int_{-\infty}^{\kappa} \mathrm{e}^{\sigma w}\, \mathrm{d}\widetilde{F}_{m^X(\tau)}(w) \qquad (12.151)\\ &= K\mathrm{e}^{-r\tau}\, \widetilde{\mathbb{P}}(m^S(\tau) \leqslant K) - \mathrm{e}^{-r\tau}\widetilde{\mathrm{E}}\left[m^S(\tau)\, \mathbb{I}_{\{m^S(\tau)<K\}}\right] \\ &= \mathrm{e}^{-r\tau} \int_0^K \widetilde{\mathbb{P}}(m^S(\tau) \leqslant y)\, \mathrm{d}y \end{aligned}$$

for $m \geqslant K$, and

$$\begin{aligned} P^{LFP}(t, S, m; K) &= K\mathrm{e}^{-r\tau} - m\,\mathrm{e}^{-r\tau}[1 - \widetilde{F}_{m^X(\tau)}(\widehat{m})] - \mathrm{e}^{-r\tau} S \int_{-\infty}^{\widehat{m}} \mathrm{e}^{\sigma w}\, \mathrm{d}\widetilde{F}_{m^X(\tau)}(w) \\ & \qquad (12.152)\\ &= K\mathrm{e}^{-r\tau} - m\,\mathrm{e}^{-r\tau}\widetilde{\mathbb{P}}(m^S(\tau) > m) - \mathrm{e}^{-r\tau}\widetilde{\mathrm{E}}\left[m^S(\tau)\, \mathbb{I}_{\{m^S(\tau)\leqslant m\}}\right] \\ &= \mathrm{e}^{-r\tau}\left[K - m + \int_0^m \widetilde{\mathbb{P}}(m^S(\tau) \leqslant y)\, \mathrm{d}y\right] \end{aligned}$$

for $m < K$, where $\tau = T - t$. Note that for $m \geqslant K$ the pricing function P^{LFP} in (12.151) is independent of the realized minimum m of the stock price at current time t.

The relations in (12.147)–(12.152) can therefore be used to price all four main types of lookback options. These relations are valid for quite general (time-homogeneous Markov) models for the stock price with discounted process $\{\mathrm{e}^{-(r-q)t} S(t)\}_{t\geqslant 0}$ assumed to be a $\widetilde{\mathbb{P}}$-martingale. Of course, within the GBM model we have simple exact explicit formulae for all the necessary PDFs and CDFs that can now be used to derive analytically exact risk-neutral pricing formulae for all four types of lookback options. For instance, Example 12.5 below gives a derivation of $P^{LFS}(t, S, M)$ by implementing (12.148) within the GBM model for the stock price. Before presenting this example, we note that the pricing relations in (12.147), (12.148), (12.149), and (12.152) also further simplify in the case where the sampling of the maximum and minimum is started at the current time t: $S(t) = M^S(t) = m^S(t)$. This is seen by setting $M = S$ and $m = S$ and noting that $m^S(\tau) < S$ and $M^S(\tau) > S$ (a.s.), i.e., the probability $\widetilde{\mathbb{P}}(m^S(\tau) > m)$ becomes $\widetilde{\mathbb{P}}(m^S(\tau) > S) = 0$ and $\widetilde{\mathbb{P}}(M^S(\tau) > M)$ becomes $\widetilde{\mathbb{P}}(M^S(\tau) > S) = 1$. Moreover, the indicator functions simplify where $\mathbb{I}_{\{m^S(\tau)\leqslant m\}}$ becomes $\mathbb{I}_{\{m^S(\tau)\leqslant S\}} = 1$ and $\mathbb{I}_{\{M^S(\tau)>M\}}$ becomes $\mathbb{I}_{\{M^S(\tau)>S\}} = 1$. This is consistent with the fact that $\widehat{m} = \frac{1}{\sigma}\ln\frac{m}{S} = 0$ and $\widehat{M} = \frac{1}{\sigma}\ln\frac{M}{S} = 0$ when $M = m = S$. All the pricing formulae are then only functions of spot S and time t (or S and τ).

Example 12.5. (Floating Strike (LFS) Put Price) Derive the no-arbitrage pricing formula $P^{LFS}(t, S, M)$ for the lookback option with payoff (b) in (12.99). Assume the stock price is a GBM with a constant interest rate and a continuous dividend yield q.

Solution. It is convenient to obtain the pricing function $P^{LFS}(t, S, M)$ by using the first equation line in (12.148). The first term is evaluated explicitly by evaluating the CDF in

(12.141) at $w = \widehat{M} \equiv \frac{1}{\sigma}\ln\frac{M}{S}$ and using the drift parameter $\nu = (r - q - \frac{1}{2}\sigma^2)/\sigma$,

$$\begin{aligned}\widetilde{F}_{M^X(\tau)}(\widehat{M}) &= \mathcal{N}\left(\frac{\ln\frac{M}{S} - (r-q-\frac{1}{2}\sigma^2)\tau}{\sigma\sqrt{\tau}}\right) - \mathrm{e}^{\frac{2\nu}{\sigma}\ln\frac{M}{S}}\,\mathcal{N}\left(-\frac{\ln\frac{M}{S} + (r-q-\frac{1}{2}\sigma^2)\tau}{\sigma\sqrt{\tau}}\right) \\ &= \mathcal{N}\left(-d_-(\frac{S}{M},\tau)\right) - \frac{S}{M}\left(\frac{M}{S}\right)^{\frac{2(r-q)}{\sigma^2}}\mathcal{N}\left(-d_-(\frac{M}{S},\tau)\right), \end{aligned} \tag{12.153}$$

where we define $d_\pm(x,\tau) := \frac{\ln x + (r-q\pm\frac{1}{2}\sigma^2)\tau}{\sigma\sqrt{\tau}}$; $x > 0, \tau = T - t > 0$. Note that $\frac{2\nu}{\sigma} = \frac{2(r-q-\frac{1}{2}\sigma^2)}{\sigma^2} = \frac{2(r-q)}{\sigma^2} - 1$.

We now need to compute the (expectation) integral in (12.148). By substituting the density in (12.143), the integral is a sum of three integrals:

$$\begin{aligned}\int_{\widehat{M}}^{\infty} \mathrm{e}^{\sigma w}\widetilde{f}_{M^X(\tau)}(w)\,\mathrm{d}w &= \int_{\widehat{M}}^{\infty} \mathrm{e}^{\sigma w}\frac{1}{\sqrt{\tau}}n\left(\frac{w-\nu\tau}{\sqrt{\tau}}\right)\mathrm{d}w + \int_{\widehat{M}}^{\infty} \mathrm{e}^{(\sigma+2\nu)w}\frac{1}{\sqrt{\tau}}n\left(\frac{w+\nu\tau}{\sqrt{\tau}}\right)\mathrm{d}w \\ &\quad - 2\nu\int_{\widehat{M}}^{\infty} \mathrm{e}^{(\sigma+2\nu)w}\mathcal{N}\left(\frac{-w-\nu\tau}{\sqrt{\tau}}\right)\mathrm{d}w\,. \end{aligned} \tag{12.154}$$

The first two Gaussian integrals are readily evaluated by completing the square in the exponents, or simply by direct use of the integral identity (A.1) of the Appendix. It turns out that both integrals are given by

$$\int_{\widehat{M}}^{\infty} \mathrm{e}^{\sigma w}\frac{1}{\sqrt{\tau}}n\left(\frac{w-\nu\tau}{\sqrt{\tau}}\right)\mathrm{d}w = \int_{\widehat{M}}^{\infty} \mathrm{e}^{(\sigma+2\nu)w}\frac{1}{\sqrt{\tau}}n\left(\frac{w+\nu\tau}{\sqrt{\tau}}\right)\mathrm{d}w = \mathrm{e}^{(r-q)\tau}\mathcal{N}\left(d_+(\frac{S}{M},\tau)\right). \tag{12.155}$$

Note that $\sigma + 2\nu = \frac{2(r-q)}{\sigma}$. So the third integral can be evaluated in two separate cases: (i) $r - q = 0$ and (ii) $r - q \neq 0$. We will treat the latter case since the pricing formula for case (i) can be obtained by taking the limit $(r-q) \to 0$ in the pricing formula for case (ii).

We now evaluate the third integral using a change of variables, $x = (\sigma + 2\nu)(w - \widehat{M})$, and write $\mathrm{e}^{(\sigma+2\nu)\widehat{M}} = \exp(\frac{2(r-q)}{\sigma^2}\ln\frac{M}{S}) = (\frac{M}{S})^{\frac{2(r-q)}{\sigma^2}}$, giving

$$\begin{aligned}2\nu\int_{\widehat{M}}^{\infty} \mathrm{e}^{(\sigma+2\nu)w}\mathcal{N}\left(\frac{-w-\nu\tau}{\sqrt{\tau}}\right)\mathrm{d}w &= \frac{2\nu}{\sigma+2\nu}\mathrm{e}^{(\sigma+2\nu)\widehat{M}}\int_0^{\infty}\mathrm{e}^x\mathcal{N}(Ax+B)\,\mathrm{d}x \\ &= \left[1 - \frac{\sigma^2}{2(r-q)}\right]\left(\frac{M}{S}\right)^{\frac{2(r-q)}{\sigma^2}}\int_0^{\infty}\mathrm{e}^x\mathcal{N}(Ax+B)\,\mathrm{d}x\,. \end{aligned} \tag{12.156}$$

Note that $\frac{2\nu}{\sigma+2\nu} = 1 - \frac{\sigma^2}{2(r-q)}$. Here we define the constants $A \equiv -\frac{1}{(\sigma+2\nu)\sqrt{\tau}} = -\frac{\sigma}{2(r-q)\sqrt{\tau}}$ and $B \equiv -\frac{\widehat{M}+\nu\tau}{\sqrt{\tau}}$. We can assume that $r - q > 0$, i.e., $A < 0$, so that the integral identity in (A.5) of the Appendix can be directly applied. [We leave it to the reader to apply a change of variable and verify that the same result obtains by making use of an appropriate integral identity in the Appendix for the case that $r - q < 0$.] Applying (A.5) and simplifying the terms gives

$$\int_0^\infty e^x \mathcal{N}(Ax + B)\, dx = -\mathcal{N}(B) + e^{(1-2AB)/2A^2}\mathcal{N}\left(\frac{1-AB}{|A|}\right)$$
$$= -\mathcal{N}\left(-\frac{\widehat{M}+\nu\tau}{\sqrt{\tau}}\right) + \exp\left[\left(1 - 2\frac{(\widehat{M}+\nu\tau)/\sqrt{\tau}}{(\sigma+2\nu)\sqrt{\tau}}\right)\frac{(\sigma+2\nu)^2\tau}{2}\right]$$
$$\cdot \mathcal{N}\left(\left(1 - \frac{(\widehat{M}+\nu\tau)/\sqrt{\tau}}{(\sigma+2\nu)\sqrt{\tau}}\right)(\sigma+2\nu)\sqrt{\tau}\right)$$
$$= -\mathcal{N}\left(-\frac{\widehat{M}+\nu\tau}{\sqrt{\tau}}\right) + e^{\frac{\sigma}{2}(\sigma+2\nu)\tau - (\sigma+2\nu)\widehat{M}}\mathcal{N}\left(\frac{-\widehat{M}+(\sigma+\nu)\tau}{\sqrt{\tau}}\right)$$
$$= -\mathcal{N}\left(-d_-\left(\frac{M}{S},\tau\right)\right) + e^{(r-q)\tau}\left(\frac{S}{M}\right)^{\frac{2(r-q)}{\sigma^2}}\mathcal{N}\left(d_+\left(\frac{S}{M},\tau\right)\right).$$

Substituting this expression into (12.156) and summing the resulting expression with the two equal expressions in (12.155) gives the left-hand side integral in (12.154):

$$\int_{\widehat{M}}^\infty e^{\sigma w}\widetilde{f}_{M^X(\tau)}(w)\, dw = 2e^{(r-q)\tau}\mathcal{N}\left(d_+(\frac{S}{M},\tau)\right) - \left[1 - \frac{\sigma^2}{2(r-q)}\right]\left[e^{(r-q)\tau}\mathcal{N}\left(d_+\left(\frac{S}{M},\tau\right)\right)\right.$$
$$\left. - \left(\frac{M}{S}\right)^{\frac{2(r-q)}{\sigma^2}}\mathcal{N}\left(-d_-\left(\frac{M}{S},\tau\right)\right)\right]. \tag{12.157}$$

Finally, by inserting this expression and the CDF in (12.153) into (12.148) and cancelling out two terms gives the pricing function for $r - q \neq 0$:

$$P^{LFS}(t,S,M) = Me^{-r\tau}\widetilde{F}_{M^X(\tau)}(\widehat{M}) - e^{-q\tau}S + e^{-r\tau}S\int_{\widehat{M}}^\infty e^{\sigma w}\, d\widetilde{F}_{M^X(\tau)}(w)$$
$$= e^{-q\tau}S\left[1 + \frac{\sigma^2}{2(r-q)}\right]\mathcal{N}\left(d_+(\frac{S}{M},\tau)\right) + Me^{-r\tau}\mathcal{N}\left(-d_-(\frac{S}{M},\tau)\right)$$
$$- \frac{\sigma^2}{2(r-q)}e^{-r\tau}S\left(\frac{M}{S}\right)^{\frac{2(r-q)}{\sigma^2}}\mathcal{N}\left(-d_-(\frac{M}{S},\tau)\right) - e^{-q\tau}S. \tag{12.158}$$

The pricing formula for $r - q = 0$, i.e., when $r = q$, follows by taking the limit $r - q \to 0$. We leave it as a simple exercise in calculus (using L'Hôpital's Rule) to show that the sum of the two terms in $\frac{\sigma^2}{2(r-q)}$ cancel out when $r - q \to 0$. Then, using $r = q$, the final expression for the pricing function in case $r - q = 0$ simplifies to

$$P^{LFS}(t,S,M) = e^{-r\tau}\left[M\mathcal{N}\left(\frac{\ln\frac{M}{S} + \frac{1}{2}\sigma^2\tau}{\sigma\sqrt{\tau}}\right) - S\mathcal{N}\left(\frac{\ln\frac{M}{S} - \frac{1}{2}\sigma^2\tau}{\sigma\sqrt{\tau}}\right)\right]$$
$$+ \sigma\sqrt{\tau}e^{-r\tau}S\left[d_+(\frac{S}{M},\tau)\mathcal{N}\left(d_+(\frac{S}{M},\tau)\right) + n\left(d_+(\frac{S}{M},\tau)\right)\right], \tag{12.159}$$

where $d_\pm(x,\tau) := \frac{\ln x \pm \frac{1}{2}\sigma^2\tau}{\sigma\sqrt{\tau}}$. □

In the above example, if we assume that the sampling of the maximum starts at current time t, i.e., $S = M$, then the pricing formulae in (12.158) and (12.159) simplify to

$$P^{LFS}(t,S) = \mathrm{e}^{-q\tau} S \left[1 + \frac{\sigma^2}{2(r-q)}\right] \mathcal{N}\left(\frac{(r-q+\frac{1}{2}\sigma^2)}{\sigma}\sqrt{\tau}\right) - \mathrm{e}^{-q\tau} S + \mathrm{e}^{-r\tau} S \left[1 - \frac{\sigma^2}{2(r-q)}\right] \mathcal{N}\left(-\frac{(r-q-\frac{1}{2}\sigma^2)}{\sigma}\sqrt{\tau}\right) \quad (12.160)$$

for $r \neq q$ and

$$P^{LFS}(t,S) = \mathrm{e}^{-r\tau} S \left[(2 + \sigma^2\tau/2)\, \mathcal{N}\left(\sigma\sqrt{\tau}/2\right) - 1 + \sigma\sqrt{\tau} n(\sigma\sqrt{\tau}/2)\right] \quad (12.161)$$

for $r = q$, where we simply write $P^{LFS}(t,S) \equiv P^{LFS}(t,S,M = S)$. The pricing functions in (12.160) and (12.161) are simply linear functions in S.

The derivations of explicit pricing functions for the other lookback options are left as exercises at the end of this chapter (see Exercise 12.24).

12.4 Exercises

Exercise 12.1. Let the stock price process $\{S(t)\}_{t\geqslant 0}$ be a geometric Brownian motion (GBM) with constant volatility σ. Assume a constant continuous dividend yield q on the stock and a bank account with constant continuously compounded interest rate r. Fix $S(0) = S > 0$ and time $T > 0$. Let the process $\{\mathrm{e}^{(q-r)t} S(t)\}_{t\geqslant 0}$ be a $\widetilde{\mathbb{P}}$-martingale. Derive explicit analytical expressions for the risk-neutral probability of the following events:

(a) $\{S(T) < K\}$, with constant $K > 0$;

(b) $\{K_1 < S(T) < K_2\}$, with constants $K_2 > K_1 > 0$;

(c) $\{1/S^2(T) > K\}$, with constant $K > 0$;

(d) $\{S^\alpha(T_2) > S^\beta(T_1)\}$, with times $T_2 > T_1 > 0$ and constants $\alpha, \beta \neq 0$;

(e) $\{S(T_2) > S(T_1) > S(t)\}$, with times $T_2 > T_1 > t > 0$;

(f) $\{S(T_1) < K_1, S(T_2) > K_2\}$ for any $K_1, K_2 > 0$.

Exercise 12.2. Assume the standard Black–Scholes model in an economy with constant continuously compounded interest rate r and with stock price process $\{S(t)\}_{t\geqslant 0}$ as a GBM with constant volatility σ and constant continuous dividend yield q. Derive the no-arbitrage pricing formula for the European-style option with the corresponding payoff functions (a)–(c), where $K > 0$ is a fixed strike and $a > 0$ is a constant. Express your answer in terms of the spot S and the time to maturity.

(a) $\Lambda(S) = a(S - K)^2$;

(b) $\Lambda(S) = (aS^2 - K)^+$;

(c) $\Lambda(S) = a|S^\alpha - K|$ with nonzero real constant α.

Exercise 12.3. Assume the standard Black–Scholes model and stock price process as in Exercise 12.2. Consider a European option with payoff at expiry T:

$$\Lambda(S(T)) = \begin{cases} (S(T) - K_1)^+ & 0 \leqslant S(T) \leqslant X_1, \\ X_1 - K_1 & X_1 \leqslant S(T) \leqslant X_2, \\ (K_2 - S(T))^+ & S(T) \geqslant X_2. \end{cases}$$

where $K_1 < X_1 < X_2 < K_2$ and $X_2 = K_1 + K_2 - X_1$.

(a) Give a sketch of this payoff function and determine a replicating portfolio for $\Lambda(S(T))$ that consists of only calls or puts.

(b) Let $S(t) = S$ be the spot price of the stock at current time $t < T$. Derive the risk-neutral pricing formula for the time-t value of a European-style option with the above payoff function. Give an explicit answer in terms of all parameters in the model.

(c) Obtain a formula for the delta position in the stock at time $t < T$ that is required in a self-financing replicating strategy for the option.

Exercise 12.4. Assume the standard Black–Scholes model and stock price process as in Exercise 12.2. Let $S(t) = S > 0$ be the spot at time $t < T$, where T is the expiry date. Derive the corresponding arbitrage-free time-t pricing formula, $V(t, S)$, for a European option with the respective payoffs in (a) and (b) below.

(a) $\Lambda(S(T)) = \sum_{n=0}^{N} a_n S^n(T)$, $N \geqslant 1$, where a_n are real constant coefficients of the polynomial function.

(b) $\Lambda(S(T)) = \big(S^\alpha(T) - K\big)\mathbb{I}_{\{S(T)>K\}}$ where α is any nonzero real constant.

Exercise 12.5. Assume the standard Black–Scholes model and stock price process as in Exercise 12.2. A European *call spread* has payoff $\Lambda(S(T))$ equal to zero for $S(T) \leqslant K$, $S(T) - K$ for $K < S(T) < K + \epsilon$, and ϵ for $S(T) \geqslant K + \epsilon$, where K, ϵ are any positive values.

(a) Give a sketch of the payoff function.

(b) Derive a formula for the option's present value $V(t, S)$ and $\Delta(t, S) = \frac{\partial V}{\partial S}$. Express your answers in terms of spot S, time to maturity $T - t$, and parameters K,ϵ,r,q,σ.

(c) Find $V(t, S)$ in both limits $\epsilon \searrow 0$ and $\epsilon \to \infty$ and explain your results.

Exercise 12.6. Let $0 < K_1 < K_2 < K_3 < K_4 < K_5 < K_6$ and consider the payoff function:

$$\Lambda(S) = \begin{cases} (S - K_1)^+ & 0 \leqslant S \leqslant K_2, \\ K_2 - K_1 & K_2 \leqslant S \leqslant K_3, \\ K_2 - K_1 - (S - K_3) & K_3 \leqslant S \leqslant K_4, \\ K_2 - K_1 - (K_4 - K_3) & K_4 \leqslant S \leqslant K_5, \\ -(K_6 - S)^+ & S \geqslant K_5. \end{cases}$$

where we assume $K_4 - K_3 > K_2 - K_1$ and $K_2 - K_1 - (K_4 - K_3) = K_5 - K_6$.

(a) Give a sketch of this payoff function.

(b) Determine a replicating portfolio for $\Lambda(S)$ consisting of only calls (or puts, cash, and stock positions).

(c) Assuming a Black–Scholes economy as in Exercise 12.2, derive the no-arbitrage pricing formula for the European-style option with the above payoff.

Exercise 12.7. A so-called *range forward* European contract is specified as follows: at maturity T the holder must buy the underlying stock at price K_1 if $S(T) < K_1$, at price $S(T)$ if $K_1 \leqslant S(T) \leqslant K_2$, and at price K_2 for $S(T) > K_2$ where $K_1 < K_2$ are fixed strikes.

(a) Derive the explicit formula for the present value $V(t, S)$ of this contract with spot $S(t) = S$. Assume a Black–Scholes economy as in Exercise 12.2.

(b) Find the relationship between K_1 and K_2 such that the present value $V(t, S) = 0$.

Exercise 12.8. Consider a so-called *strangle* payoff function $\Lambda(S)$ defined by two strikes $0 < K_1 < K_2$:

$$\Lambda(S) = \begin{cases} K_2 - S & 0 \leqslant S \leqslant K_1, \\ K_2 - K_1 & K_1 \leqslant S \leqslant K_2, \\ S - K_1 & S \geqslant K_2. \end{cases}$$

(a) Give a sketch of $\Lambda(S)$.

(b) Determine a replicating portfolio for $\Lambda(S)$ in terms of positions in standard calls, stock, and cash.

(c) Assume the stock price $\{S(t)\}_{t\geqslant 0}$ is a GBM with constant volatility σ, constant continuous dividend q in an economy with constant interest rate r. Derive the arbitrage-free pricing formula for the present time $t < T$ value of a European-style derivative with the above payoff function.

(d) Provide an expression for the value, at calendar time $t < T$, of the position in the stock in the self-financed portfolio required to dynamically replicate the option value.

Exercise 12.9. A so-called *pay-later* European option costs the holder nothing (i.e., zero premium) to set up at present time $t = 0$. The payoff to the holder at maturity $T > 0$ is $(S(T) - K)^+$. Moreover, the holder must *pay out* X dollars to the writer in the case that $S(T) \geqslant K$. Derive an expression for the fair value of X. Determine the fair value for X in the limit of infinite volatility, $\sigma \to \infty$. Assume a Black–Scholes economy as in Exercise 12.2.

Exercise 12.10. Let $C(S, \tau)$ be the Black–Scholes pricing formula of a standard call option with spot S, strike K, fixed interest rate r, zero stock dividend, constant volatility σ, and time to maturity $\tau > 0$.

(a) Show that the respective limiting values of the call price for vanishing and infinite volatility are given by

$$\lim_{\sigma \searrow 0} C(S, \tau) = \left(S - \mathrm{e}^{-r\tau} K\right)^+ \quad \text{and} \quad \lim_{\sigma \to \infty} C(S, \tau) = S.$$

(b) Give a financial interpretation of both limits. Note that the second limit is independent of the strike value K; give a financial intuition of this fact.

Exercise 12.11. Consider the value of a European call option written by an issuer who only has a fraction $0 \leqslant \alpha < 1$ of the underlying asset. That is, at expiration time T the payoff of this type of call is given by

$$V_T = (S(T) - K)^+ \mathbb{I}_{\{\alpha S(T) \geqslant S(T) - K\}} + \alpha S(T) \mathbb{I}_{\{\alpha S(T) < S(T) - K\}} \,.$$

Let $C_L(S, \tau; K, \alpha)$ denote the value of such a European call, where $\tau = T - t$ is the time to expiry, $K > 0$ is the strike, $S(t) = S$ is the spot of the underlying. Show that

$$C_L(S, \tau; K, \alpha) = C(S, K, \tau) - (1 - \alpha)C\big(S, \frac{K}{1 - \alpha}, \tau\big)$$

where $C\big(S, \frac{K}{1-\alpha}, \tau\big)$ is the price of a standard European call with strike $\frac{K}{1-\alpha}$, spot S, and time to expiry τ. NOTE: You should not assume any model for the stock price process and therefore you need not provide any explicit formulas for any of the call price functions.

Exercise 12.12. Suppose that the cost of carry on a commodity is b and assume a bank account with constant interest rate r. Let the price of the commodity follow a GBM model with constant volatility σ. Let $V = V(t, S)$ be the value of a European option on this commodity, where $S > 0$ is the spot value of the commodity at calendar time t.

(a) Show that V satisfies the Black–Scholes PDE:

$$\frac{\partial V}{\partial t} + \frac{\sigma^2}{2} S^2 \frac{\partial^2 V}{\partial S^2} + bS \frac{\partial V}{\partial S} - rV = 0\,.$$

(b) Based on (a), find the put-call parity relation between the put price $P(t, S)$ and call price $C(t, S)$, with common strike K and time to maturity $\tau = T - t$.

Exercise 12.13. Consider a portfolio Π with fixed positions θ_i in N securities each with price f_i, $i = 1, \ldots, N$, respectively. Assume the ith security has price f_i as a function of the same spot S at current time t and that each $f_i = f_i(S, T_i - t)$ satisfies the time-homogeneous Black–Scholes PDE with constant interest rate and volatility. The contract maturity dates T_i are allowed to be distinct. Find the algebraic relationship among the portfolio Greeks: $\Theta \equiv \frac{\partial \Pi}{\partial t}$, $\Delta \equiv \frac{\partial \Pi}{\partial S}$, and $\Gamma \equiv \frac{\partial^2 \Pi}{\partial S^2}$.

Exercise 12.14. Apply integration by parts twice to show that

$$\Lambda_a(S) = \frac{1}{2a} \int_{K-a}^{K+a} (S - k)^+ \, \mathrm{d}k$$

is an integral representation of the payoff in (12.43). Hence, this shows that the soft-strike call option with given central strike K and strike width $a > 0$ is replicated as a uniform superposition of standard call options of all strikes in the interval $(K - a, K + a)$.

Exercise 12.15. Recall the pricing of the soft-strike call in Example 12.2.

(a) Provide the corresponding definition of the payoff (as in (12.43)) of the soft-strike put option with center strike K and width a. Provide a plot of the payoff and describe its main features as done in Example 12.2.

(b) Derive a corresponding put-call parity relation for the soft-strike call and soft-strike put options having common center strike K and strike width a. Assuming the standard GBM model as in Example 12.2, derive the formula for $P(t, S; K, a)$, the no-arbitrage time-t price of the soft-strike put option.

Exercise 12.16. Assume the risk-neutral pricing formula in (12.11) holds. By using a general formula for the expectation of a random variable conditional on an event, in this case the event $\{S(T) > K\}$, derive the following general representations for the respective prices of a standard call and put option with strike K:

$$C(t, S) = \mathrm{e}^{-r(T-t)} \widetilde{\mathbb{P}}_{t,S}\big(S(T) > K\big) \cdot \left(\widetilde{\mathrm{E}}_{t,S}\big[S(T) | S(T) > K\big] - K\right),$$

$$P(t, S) = e^{-r(T-t)}\widetilde{\mathbb{P}}_{t,S}(S(T) < K) \cdot \left(K - \widetilde{\mathrm{E}}_{t,S}\left[S(T)|S(T) < K\right]\right).$$

Give a probabilistic interpretation of these formulae.

Exercise 12.17. Assume a nondividend paying stock with price process $\{S(t)\}_{t\geqslant 0}$ as a GBM with constant volatility $\sigma > 0$ in the (B, S) economy with constant interest rate r. Let $C_t := C(t, S(t), K)$ be the price process at calendar time t, $0 \leqslant t < T$, with $C(t, S, K)$ as the pricing function of a standard European call option on the stock with given strike $K > 0$ and fixed maturity date T. It follows that C_t satisfies the SDE

$$\mathrm{d}C_t = \mu_c\, C_t\, \mathrm{d}t + \sigma_c\, C_t\, \mathrm{d}\widetilde{W}(t),$$

where $\{\widetilde{W}(t)\}_{t\geqslant 0}$ is a standard BM under the risk-neutral measure $\widetilde{\mathbb{P}}$.

(a) Find explicit expressions for μ_c and σ_c, i.e., the (log)-drift and (log)-volatility coefficient functions of the call price process.
NOTE: The (log)-volatility of the call price, σ_c, is a function of $S(t)$ (spot value) and parameters $K, \sigma, r, \tau = T - t$. Your expressions should be simplified as much as possible. Define all terms in your answer.

(b) Find the limiting expression for σ_c as $K \searrow 0$ (holding spot and other parameters fixed). Give a financial explanation of the resulting limit.

Exercise 12.18. Consider a stock price process as a GBM and assume *deterministic (non-random) time-dependent* volatility $\sigma(t)$, stock dividend yield $q(t)$, and interest rate $r(t)$. Assume that these are integrable functions of time $t \in [0, T]$.

(a) Derive the time-0 no-arbitrage pricing function $C(0, S_0, K)$ for the standard call option on this stock with $S(0) = S_0 > 0$ as spot, $K > 0$ as strike, and T as maturity.

(b) Derive the put-call parity relation between the time-0 values of the standard call and put prices $C(0, S_0, K)$ and $P(0, S_0, K)$.

(c) Now let the present time be any time $t \in [0, T)$ with spot $S(t) = S > 0$. Derive the time-t no-arbitrage pricing function for the standard call $C(t, S, K)$ and provide the corresponding put-call parity relation where $P(t, S, K)$ denotes the time-t pricing function for the standard put.

Exercise 12.19. A variant of the forward starting call option that we already considered in Section 12.2 is structured as follows. The holder receives at date $T_1 > t$ a call with strike $K_{T_1} = \alpha S(T_1)$ and maturity $T > T_1$. Here, α is a positive constant and $S(T_1)$ is the stock price realized at time T_1. Let the stock price be a GBM with constant interest rate r and dividend yield q.

(a) Let $S(t) = S$. Derive the time-t pricing formula $C(t, S; T_1, T)$ for this forward starting call and give a hedging strategy that applies up to time T_1. Is the strategy static or not?

(b) Show that the price of the forward starting call simplifies to that of a standard call struck at $K = \alpha S$ with time to maturity $T - t$ in the limiting case that $T_1 \to t$ (with t, T held fixed). On the other hand, show that in the limit $T \to T_1$ (with t, T_1 held fixed) the contract price is simply given by $S(1 - \alpha)^+$.

Exercise 12.20. Assume the stock price $\{S(t)\}_{t\geqslant 0}$ is a GBM with constant volatility σ and zero dividend in an economy with constant interest rate r. Let $t < T_0 < T$, i.e., T_0 is an arbitrary intermediate time before expiry time T, and consider a European-style option with payoff at time T:

$$V_T = \min\{S(T_0), S(T)\}.$$

(a) Show that the value V at any time $t \leqslant T_0$ of this option is given by

$$V = S[\mathcal{N}(-d_+) + \mathrm{e}^{-r(T-T_0)}\mathcal{N}(d_-)]$$

where $S(t) = S$ is the spot and $d_\pm := \dfrac{r(T-T_0) \pm \frac{1}{2}\sigma^2(T-T_0)}{\sigma\sqrt{T-T_0}} = \dfrac{(r \pm \frac{1}{2}\sigma^2)}{\sigma}\sqrt{T-T_0}$.

[Note: The option value is dependent on $T - T_0$ (where T and T_0 are fixed) and does not depend on t.]

(b) What is the position held in the stock at time $t \leqslant T_0$ in a self-financing replicating strategy? Is this position static over time? Justify whether or not a bank account is needed in the dynamic replication.

Exercise 12.21. Recall the derivation of the pricing function $V^{cc}(t, S)$ for the European call-on-a-call where $S(t) = S$ is the spot at calendar time $t < T_1 < T_2$. Assume the stock price is a GBM with constant interest rate r and constant dividend yield q.

(a) Using similar steps as in Section 12.2, derive the pricing formula for $V^{cp}(t, S)$, the arbitrage-free value of a *call-on-a-put*, with time-T_1 value $\left(P_{T_1} - K_1\right)^+$, where $P_{T_1} \equiv P_{T_1}(S(T_1), K_2, T_2)$ is the price of the (embedded) standard put with time to maturity $T_2 - T_1$ and strike K_2.

(b) Derive an expression for the delta position $\Delta_t = \Delta(t, S)$ in the stock at time $t < T_1$ needed to dynamically hedge the call-on-a-put option in (a).

Exercise 12.22. Consider the stock price process $\{S_{(B)}(t)\}_{t\geqslant 0}$ as a GBM that is killed at the first-hitting time $\mathcal{T}_B^S = \inf\{t \geqslant 0 : S(t) = B\}$ to level $B > 0$:

$$S_{(B)}(t) := \begin{cases} S(t) & \text{for } t < \mathcal{T}_B^S, \\ \partial^\dagger & \text{for } t \geqslant \mathcal{T}_B^S, \end{cases} \tag{12.162}$$

where $S(t)$ is given by (12.100) and $X(t)$ is given by (12.101). Note that the state space for the process is restricted to either interval $(0, B)$ or (B, ∞) corresponding to the two cases with $S(0) \in (0, B)$ or $S(0) \in (B, \infty)$, respectively.

(a) Show that the transition CDF for the process $S_{(B)}$ is given by

$$\widetilde{\mathbb{P}}(S_{(B)}(T) \leqslant y | S_{(B)}(t) = S) = \widetilde{\mathbb{P}}\left(X_{(b)}(\tau) \leqslant \frac{1}{\sigma}\ln\frac{y}{S}\right)$$

where $\tau = T - t > 0$ and $X_{(b)}$ is the BM process given by (12.101) with killing at the first-hitting time to level $b \equiv \frac{1}{\sigma}\ln\frac{B}{S}$ (see the definition in (10.75) of Chapter 10). By differentiating, obtain the (time-homogeneous) transition PDF:

$$\widetilde{p}^{S_{(B)}}(t, T; S, y) \equiv \widetilde{p}^{S_{(B)}}(\tau; S, y) = \frac{1}{\sigma y}\widetilde{p}^{X_{(b)}}(\tau; 0, x)$$

where $x = \frac{1}{\sigma}\ln\frac{y}{S}$ and $\widetilde{p}^{X_{(b)}}$ is the risk-neutral transition PDF for the killed BM process $X_{(b)}$. Then, using the density in (12.120), obtain the expression in (12.133).

(b) For a fixed y value, show that $v(\tau, S, y) := \mathrm{e}^{-r\tau}\widetilde{p}^{S_{(B)}}(\tau; S, y)$ solves the time-homogeneous BSPDE in (12.51) subject to the initial condition $v(0+, S, y) = \delta(S - y)$ and has zero boundary conditions $v(\tau, S = B, y) = 0$ and $v(\tau, S = 0+, y) = 0$ and $v(\tau, S = \infty, y) = 0$ for all $\tau > 0$ and all y.

Exercise 12.23. Using similar steps as in the derivation of the up-and-out call (and up-and-out put) pricing formula, derive explicit pricing functions and delta hedging positions for

(a) the down-and-out put and down-and-in put option;

(b) the down-and-out call and down-and-in call option.

Assume a barrier level $B > 0$, strike $K > 0$, time to maturity $\tau = T - t > 0$ and where the stock price is a GBM with constant interest rate r and stock dividend yield q.

Exercise 12.24. Use similar steps as in Example 12.5 and make appropriate use of integral identities in the Appendix to derive explicit pricing functions for

$$\text{(i) } C^{LFS}(t, S, m), \ \text{(ii) } C^{LFP}(t, S, M; K), \ \text{and (iii) } P^{LFP}(t, S, m; K).$$

Assume the stock price is a GBM with constant interest rate r and dividend yield q.

Exercise 12.25. Consider the discrete geometric averaging of a stock price process at evenly distributed discrete times $t_j = t_0 + j\,\delta t$, $j = 1, 2, \ldots, n$, with a time step $\delta t = (T - t_0)/n$; $t_n = T$ is the time of expiration. Define the discretely monitored geometric averaging by

$$G_k = \left[\prod_{j=1}^{k} S(t_j)\right]^{1/k}, \quad k = 1, 2, \ldots, n\,.$$

(a) Assuming that the stock price follows a GBM process, show that G_n is a log-normal random variable. Find the mean and variance of G_n.

(b) Derive the risk-neutral time-t_0 prices of the fixed strike Asian call and put options with respective payoff functions $(G_n - K)^+$ and $(K - G_n)^+$. Here, $K > 0$ is a strike price.

Exercise 12.26. Let the stock price process follow the GBM model in (12.46). Define the continuously monitored geometric average of $S(t)$ over a time period $[0, t]$ by

$$G(t) = \exp\left(\frac{1}{t}\int_0^t \ln S(u)\,\mathrm{d}u\right).$$

(a) Show that the process $\{\ln G(t)\}_{t\geqslant 0}$ is Gaussian.

(b) Find the mean and variance of $G(T)$ conditional on $G(t)$ and $S(t)$ for $0 \leqslant t \leqslant T$.

(c) Show that $G(T)$ can be written as

$$G(T) = G(t)^{t/T} S(t)^{(T-t)/T} \exp(\bar{\mu} + \bar{\sigma}\widetilde{Z})$$

for some $\bar{\mu}, \bar{\sigma} \in \mathbb{R}$, and where $\widetilde{Z} \sim Norm(0, 1)$ under measure $\widetilde{\mathbb{P}}$. Find the values of $\bar{\mu}$ and $\bar{\sigma}$.

(d) Derive the risk-neutral time-t pricing functions for the fixed strike Asian call and put options with respective payoff functions $(G(T) - K)^+$ and $(K - G(T))^+$ for $0 \leqslant t \leqslant T$. Express the pricing functions in terms of the spot values $G(t) = G > 0$, $S(t) = S > 0$, and times t and T.

(e) Establish the put-call parity relation for the fixed strike Asian call and put options.

Chapter 13

Risk-Neutral Pricing in a Multi-Asset Economy

In the previous chapter we considered continuous-time risk-neutral derivative pricing in the simplest so-called (B, S) economy consisting of only one risky asset (stock) S and a risk-free bond or money market account B. We now extend the (B, S) economy to a continuous-time model of an economy consisting of an arbitrary number $n \geqslant 1$ of tradable risky base assets as well as a money market account. The simplest way to extend the classical (B, S) model to include multiple risky base assets is to assume that they are all independent of one another. For example, we can let each base asset price process be an Itô process, such as a geometric Brownian motion (GBM), driven by an independent Brownian motion (BM). Of course, this is a trivial and not very interesting extension. A more realistic extension is a model where the base assets are correlated stochastic processes. In particular, in this chapter, we shall remain within the framework of multidimensional correlated Itô processes. The interdependence among the base asset price processes arises by forming correlations among BMs that drive each price process. As was learned in Chapter 11, such correlations can be constructed by taking linear combinations of several, say d, independent BMs. The multidimensional GBM process is one such model for describing n stock price processes driven by d independent BMs. In all derivations of explicit pricing formulae for multi-asset derivatives considered in this chapter, we assume the "classical" multidimensional GBM model. This model offers fairly simple analytical tractability for many standard options written on multiple stocks. However, we shall first develop the pricing and hedging theory in a more general multidimensional continuous-time framework where all base asset price processes are quite general correlated Itô processes. We then simplify the framework to the classical multidimensional GBM model for the risky assets and also assume nonrandom interest rates and stock dividends. This allows us to price several standard (as well as some path dependent) European-style derivatives whose payoffs are dependent on the prices of multiply correlated assets. As in the case of a single asset, for non-path-dependent European-style derivatives the Markov property reduces the pricing problem to a conditional expectation where the multidimensional (discounted) Feynman–Kac formula leads to the Black–Scholes–Merton PDE for multi-asset contracts. Section 11.10 of Chapter 11 gives us this connection between the SDEs of Itô processes and the corresponding PDE problem.

As in the case of the (B, S) model covered in Chapter 12, we use a self-financing replicating portfolio strategy involving all base assets (risky stocks and bank account) and thereby obtain a general risk-neutral pricing formula expressing prices of European-style derivatives with attainable payoffs as a conditional expectation of the discounted payoff under the risk-neutral measure $\widetilde{\mathbb{P}}$ with the bank account as the numéraire asset. For the multidimensional GBM model, we have already shown (see the end of Section 11.10.3 of Chapter 11 where we applied Girsanov's Theorem) that the so-called *price of risk equations* must have a solution in order to guarantee the existence of a risk-neutral measure $\widetilde{\mathbb{P}}$. Essentially the same linear system of equations follows for our more general multi-asset model and hence similar conditions must be imposed on the model for the existence of a risk-neutral measure $\widetilde{\mathbb{P}}$. As shown in the discrete-time models, we will also see that the existence of a risk-neutral measure $\widetilde{\mathbb{P}}$ implies that the model is arbitrage-free. This is the statement of the so-called *First Fundamental Theorem of Asset Pricing.* Then, by an application of the multidimensional Brownian Martingale Representation Theorem we shall answer the question of completeness of the multi-asset market model, i.e., the question of whether derivative claims can be hedged (replicated) by a self-financing portfolio strategy. This will lead us to another important result, called the *Second Fundamental Theorem of Asset Pricing*, stating that the market model is complete if the risk-neutral measure $\widetilde{\mathbb{P}}$ exists and is unique. It then follows that the multidimensional GBM model with $n = d$, i.e., with same number of assets as independent BMs, having a nonsingular log-volatility matrix is a complete and arbitrage-free multi-asset market model. This is the model that we shall adopt in our examples of pricing derivatives.

The last topic covered in this chapter discusses the risk-neutral pricing framework in the more general context of *equivalent martingale measures (EMMs)* that correspond to different choices of numéraire assets for pricing. As was shown in the discrete-time models, there is a general numéraire invariant risk-neutral derivative pricing formula. A change of numéraire for pricing is just a special kind of change of measure that takes us from one EMM to another. A numéraire asset is any positive asset, such as a base asset or even a derivative asset, that is an Itô process driven by the same multidimensional BM. Hence, Girsanov's Theorem is again the tool that allows us to price derivatives under different choices of numéraires. We will present some examples of how changing numéraires facilitates the pricing of some multi-asset derivatives.

13.1 General Multi-Asset Market Model: Replication and Risk-Neutral Pricing

For a given time $T > 0$, let $\{\mathbf{W}(t) = (W_1(t), \ldots, W_d(t))\}_{0 \leqslant t \leqslant T}$ be a standard d-dimensional Brownian motion under the physical (real-world) measure $\mathbb{P}$ and fix a filtered probability space $(\Omega, \mathcal{F}, \mathbb{P}, \mathbb{F})$ with $\mathbb{F} = \mathbb{F}^{\mathbf{W}}$ as the natural filtration generated by the BM. One of our *base assets* is the bank account with value process $\{B(t)\}_{t \geqslant 0}$ where

$$B(t) = \exp\left(\int_0^t r(u)\,\mathrm{d}u\right) \implies \mathrm{d}B(t) = r(t)B(t)\,\mathrm{d}t\,, \quad B(0) = 1. \tag{13.1}$$

Hence, $B(t)$ is the value of one dollar invested in the bank account at time 0 and accruing interest at the continuously compounded rate $r(u)$ for all times $u \in [0, t]$. We also conveniently define the discount factor $D(t) := 1/B(t) = \mathrm{e}^{-\int_0^t r(u)\,\mathrm{d}u}$. The instantaneous interest rate process $\{r(t)\}_{t \geqslant 0}$ is assumed to be $\mathbb{F}$-adapted. The other base assets in our market model have prices assumed to be Itô processes $\{\mathbf{S}(t) = (S_1(t), \ldots, S_n(t))\}_{t \geqslant 0}$ solving the system of SDEs written equivalently as

$$\begin{aligned} \mathrm{d}S_i(t) &= S_i(t)\left[\mu_i(t)\,\mathrm{d}t + \sum_{j=1}^{d} \sigma_{ij}(t)\,\mathrm{d}W_j(t)\right] \\ &\equiv S_i(t)\left[\mu_i(t)\,\mathrm{d}t + \boldsymbol{\sigma}_i(t)\cdot \mathrm{d}\mathbf{W}(t)\right], \quad i = 1, \ldots, n. \end{aligned} \tag{13.2}$$

At the moment we are assuming a zero dividend yield on all the risky base assets. Recalling our description of the multidimensional GBM process in Chapter 11, we note that (13.2) is of the same form as (11.191). However, the coefficients are now considered to be generally adapted processes. The multidimensional GBM process is recovered in the special case where we let all coefficients be constants.

The log-drift coefficients $\{\mu_i(t)\}_{t\geqslant 0}$ and the log-volatility coefficients $\{\sigma_{ij}\}_{t\geqslant 0}$ are assumed to be $\mathbb{F}$-adapted real-valued processes. The drift $\mu_i(t)$ gives the instantaneous physical rate of return (actual growth rate) of the ith asset. We denote the $n \times d$ matrix of log-volatilities by $\boldsymbol{\sigma}(t) := [\sigma_{ij}(t)]_{i=1,\ldots,n;\, j=1,\ldots,d}$. The ith row of $\boldsymbol{\sigma}(t)$ gives the log-volatility vector $\boldsymbol{\sigma}_i(t) = [\sigma_{i1}(t), \sigma_{i2}(t), \ldots, \sigma_{id}(t)]$ for the i-th asset price process. The magnitude of each ith vector is denoted (without boldface) by

$$\sigma_i(t) \equiv \|\boldsymbol{\sigma}_i(t)\| = \left(\sigma_{i1}^2(t) + \ldots + \sigma_{id}^2(t)\right)^{1/2} \tag{13.3}$$

where we assume $\sigma_i(t) > 0$. Each component $\sigma_{ik}(t)$ is the log-volatility of the ith base asset (e.g., ith stock) w.r.t. the kth BM (kth source of randomness). We define the adapted log-diffusion $n \times n$ matrix by $\mathbf{C}(t) := \boldsymbol{\sigma}(t)[\boldsymbol{\sigma}(t)]^\top$ and the corresponding instantaneous log-correlation $n \times n$ matrix by $\boldsymbol{\rho}(t) = [\rho_{ij}(t)]_{i,j=1,\ldots,n}$, where

$$C_{ij}(t) = \boldsymbol{\sigma}_i(t) \cdot \boldsymbol{\sigma}_j(t) = \rho_{ij}(t)\sigma_i(t)\sigma_j(t) \tag{13.4}$$

and

$$\rho_{ij}(t) := \frac{\boldsymbol{\sigma}_i(t) \cdot \boldsymbol{\sigma}_j(t)}{\sigma_i(t)\sigma_j(t)} = \sum_{k=1}^{d} \frac{\sigma_{ik}(t)\sigma_{jk}(t)}{\sigma_i(t)\sigma_j(t)}\,. \tag{13.5}$$

Note that this is (11.160), but now these coefficients are *instantaneous log-correlations.* That is, multiplying the stochastic differentials of any pair of prices (relative to their prices at time t) using (13.2) gives

$$\frac{\mathrm{d}S_i(t)}{S_i(t)}\frac{\mathrm{d}S_j(t)}{S_j(t)} = \boldsymbol{\sigma}_i(t)\cdot\boldsymbol{\sigma}_j(t)\,\mathrm{d}t = C_{ij}(t)\,\mathrm{d}t = \rho_{ij}(t)\sigma_i(t)\sigma_j(t)\,\mathrm{d}t\,, \tag{13.6}$$

or in terms of the product of the price differentials,

$$\mathrm{d}S_i(t)\,\mathrm{d}S_j(t) = \rho_{ij}(t)\sigma_i(t)\sigma_j(t)S_i(t)S_j(t)\,\mathrm{d}t\,. \tag{13.7}$$

Recall from Chapter 11 that each SDE in (13.2) is of the form in (11.148). By the multi-factor Itô formula, each process has the representation in (11.149). Each ith stock price solving the SDE in (13.2) has a representation as its initial value, times a drift term and a martingale in the ith volatility vector:

$$\begin{aligned} S_i(t) &= S_i(0)\exp\left(\int_0^t \left(\mu_i(s) - \frac{1}{2}\|\boldsymbol{\sigma}_i(s)\|^2\right)\mathrm{d}s + \int_0^t \boldsymbol{\sigma}_i(s)\cdot \mathrm{d}\mathbf{W}(s)\right) \\ &= S_i(0)\,\mathrm{e}^{\int_0^t \mu_i(s)\,\mathrm{d}s}\,\mathcal{E}_t(\boldsymbol{\sigma}_i\cdot\mathbf{W})\,, \end{aligned} \tag{13.8}$$

where

$$\mathcal{E}_t(\boldsymbol{\sigma}_i\cdot\mathbf{W}) := \exp\left(-\frac{1}{2}\int_0^t \|\boldsymbol{\sigma}_i(s)\|^2\,\mathrm{d}s + \int_0^t \boldsymbol{\sigma}_i(s)\cdot\mathrm{d}\mathbf{W}(s)\right). \tag{13.9}$$

Throughout we are assuming that all Itô integrals are martingales so the square-integrability condition is assumed to hold. This is assured by assuming the Novikov condition holds for each $i = 1, \ldots, n$,

$$\mathrm{E}\left[\exp\left(\frac{1}{2}\int_0^T \sigma_i^2(t)\,\mathrm{d}t\right)\right] < \infty\,, \tag{13.10}$$

where $\sigma_i^2(t) \equiv \|\boldsymbol{\sigma}_i(t)\|^2 = \sum_{j=1}^d \sigma_{ij}^2(t)$.

Let us take a small step back and briefly recall our application of Girsanov's Theorem in Section 11.10.3 of Chapter 11. There, we arrived at the conditions for the existence and uniqueness of a risk-neutral measure $\widetilde{\mathbb{P}}$ within the multidimensional GBM model. We showed that such a measure exists if and only if there is a solution to the linear system in (11.217) and then $\{e^{-rt}S_i(t)\}_{t\geqslant 0}$ are $\widetilde{\mathbb{P}}$-martingales, or equivalently all stock prices have the constant drift rate r as in (11.216). In the GBM model the drift and volatility coefficients are all real constant parameters and we assumed a constant interest rate r.

Let us now again apply Girsanov's Theorem by using the same steps as in Section 11.10.3. The coefficients are now generally adapted processes so we need to use Theorem 11.21 for a generally adapted vector process $\{\boldsymbol{\gamma}(t) = (\gamma_1(t), \ldots, \gamma_d(t))\}_{0\leqslant t\leqslant T}$ satisfying the condition in (11.210). The measure change is defined more generally in terms of the Radon–Nikodym derivative having the form in (11.211). Repeating the same step by using the stochastic differential $d\widetilde{\mathbf{W}}(t) = d\mathbf{W}(t) - \boldsymbol{\gamma}(t)\,dt$, we again have (11.215), but now the coefficients are adapted (denoted by the time t argument):

$$dS_i(t) = S_i(t)\left[(\mu_i(t) + \boldsymbol{\sigma}_i(t)\cdot\boldsymbol{\gamma}(t))\,dt + \boldsymbol{\sigma}_i(t)\cdot d\widetilde{\mathbf{W}}(t)\right]. \tag{13.11}$$

The measure $\widetilde{\mathbb{P}}$ is risk-neutral if $\boldsymbol{\gamma}(t)$ is such that

$$dS_i(t) = S_i(t)\left[r(t)\,dt + \boldsymbol{\sigma}_i(t)\cdot d\widetilde{\mathbf{W}}(t)\right] \tag{13.12}$$

for all $i = 1, \ldots, n$. Note that this SDE is equivalent to stating that *all discounted stock price processes* $\{\overline{S}_i(t) := D(t)S_i(t)\}_{0\leqslant t\leqslant T}$ *are* $\widetilde{\mathbb{P}}$*-martingales.* This is easily seen by applying the Itô product rule to compute the differential

$$d[D(t)S_i(t)] \equiv d\overline{S}_i(t) = \overline{S}_i(t)\boldsymbol{\sigma}_i(t)\cdot d\widetilde{\mathbf{W}}(t), \tag{13.13}$$

which has zero drift and is hence a $\widetilde{\mathbb{P}}$-martingale. Alternatively, the $\widetilde{\mathbb{P}}$-martingale property of $\overline{S}_i(t)$ is shown by using the stochastic exponential representation of (13.12) and then multiplying by $D(t)$ and using $B(t) = e^{\int_0^t r(s)\,ds} = 1/D(t)$,

$$\overline{S}_i(t) \equiv D(t)S_i(t) = D(t)S_i(0)\,e^{\int_0^t r(s)\,ds}\,\mathcal{E}_t(\boldsymbol{\sigma}_i\cdot\widetilde{\mathbf{W}}) = S_i(0)\mathcal{E}_t(\boldsymbol{\sigma}_i\cdot\widetilde{\mathbf{W}}) \tag{13.14}$$

with exponential $\widetilde{\mathbb{P}}$-martingale

$$\mathcal{E}_t(\boldsymbol{\sigma}_i\cdot\widetilde{\mathbf{W}}) = \exp\left(-\frac{1}{2}\int_0^t \|\boldsymbol{\sigma}_i(s)\|^2\,ds + \int_0^t \boldsymbol{\sigma}_i(s)\cdot d\widetilde{\mathbf{W}}(s)\right). \tag{13.15}$$

The $\widetilde{\mathbb{P}}$-martingale property of the discounted prices hence follows simply by (13.14):

$$\widetilde{\mathrm{E}}\left[\overline{S}_i(t) \mid \mathcal{F}_s\right] = \widetilde{\mathrm{E}}\left[D(t)S_i(t) \mid \mathcal{F}_s\right] = S_i(0)\widetilde{\mathrm{E}}\left[\mathcal{E}_t(\boldsymbol{\sigma}_i\cdot\widetilde{\mathbf{W}}) \,\middle|\, \mathcal{F}_s\right] = S_i(0)\mathcal{E}_s(\boldsymbol{\sigma}_i\cdot\widetilde{\mathbf{W}}) = \overline{S}_i(s)$$

for all $0 \leqslant s \leqslant t \leqslant T$.

Hence, a risk-neutral measure exists if the log-drift coefficient in (13.11) equals $r(t)$, i.e., if and only if there exists a vector process $\boldsymbol{\gamma}(t)$ solving

$$\mu_i(t) + \boldsymbol{\sigma}_i(t)\cdot\boldsymbol{\gamma}(t) = r(t)\,, \quad i = 1, \ldots, n,$$

or in matrix form

$$\boldsymbol{\sigma}(t)\,\boldsymbol{\gamma}^\top(t) = \mathbf{b}(t), \tag{13.16}$$

with $n \times 1$ vectors $\mathbf{b}(t) := r(t)\mathbf{1} - \boldsymbol{\mu}(t)$, $\boldsymbol{\mu}(t) = [\mu_1(t), \ldots, \mu_n(t)]^\top$, $\mathbf{1} = [1, \ldots, 1]^\top$, and $d \times 1$ vector $\boldsymbol{\gamma}^\top(t) = [\gamma(t)]^\top$. This is the same linear system of n equations in d unknowns as in (11.217). This system is commonly referred to as the "price of risk equations" since a solution guarantees the existence of a risk-neutral measure $\widetilde{\mathbb{P}}$ and (as seen later) the existence of such a $\widetilde{\mathbb{P}}$ implies no-arbitrage in the above market model. Otherwise, if (13.16) has no solution (i.e., a $\widetilde{\mathbb{P}}$ does not exist) then it can be shown that an arbitrage portfolio strategy exists.

The discussion that follows (11.217) in Section 11.10.3 also pertains to the system (13.16) where the conditions must be satisfied for all values of $t \in [0, T]$ and (almost) all realizations of the multidimensional BM since the quantities in (13.16) are now generally adapted processes. For almost all outcomes ω and time $t \in [0, T]$ we must have a solution to the linear system $\boldsymbol{\sigma}(t, \omega)\boldsymbol{\gamma}^\top(t, \omega) = r(t, \omega)\mathbf{1} - \boldsymbol{\mu}(t, \omega)$. Hence for almost all paths (t, ω) we require $\text{rank}(\boldsymbol{\sigma}(t, \omega)) = n$. We simply state this as the condition that $\text{rank}(\boldsymbol{\sigma}(t)) = n$. In the important case that $\text{rank}(\boldsymbol{\sigma}(t)) = n = d$, i.e., when the *number of stocks equals the number of independent BMs and the* $n \times n$ *matrix* $\boldsymbol{\sigma}(t)$ *has an inverse* $\boldsymbol{\sigma}^{-1}(t)$, then the risk-neutral measure $\widetilde{\mathbb{P}}$ exists and is uniquely given by the adapted vector $\boldsymbol{\gamma}^\top(t) = \boldsymbol{\sigma}^{-1}(t)(r(t)\mathbf{1} - \boldsymbol{\mu}(t))$. We shall adopt this case when pricing derivatives later within the multidimensional GBM model.

We now assume that the above conditions are satisfied, i.e., that $\text{rank}(\boldsymbol{\sigma}(t)) = n$, so a risk-neutral measure $\widetilde{\mathbb{P}}$ exists. As in Chapter 12, we consider the problem of no-arbitrage pricing of a derivative claim having value process $\{V_t\}_{0 \leqslant t \leqslant T}$. At maturity T, the derivative price equals the payoff V_T, which is an $\mathcal{F}_T$-measurable random variable. In our model, V_T is hence generally a functional of the multidimensional BM that is driving all of the $n+1$ base assets up to time T. In our multi-asset economy a portfolio (trading) strategy is a continuous sequence of portfolios in the $n+1$ base assets:

$$(\beta_t, \boldsymbol{\Delta}_t) \equiv (\beta_t, \Delta_t^1, \ldots, \Delta_t^n)$$

where $\{\beta_t\}_{0 \leqslant t \leqslant T}$ and $\{\Delta_t^i\}_{0 \leqslant t \leqslant T}$, $i = 1, \ldots, n$, are adapted. As in the single stock economy, β_t represents the time-t position in the bank account. Each hedge position Δ_t^i denotes the number of shares held in the ith base asset (e.g., ith stock) at time t where $\Delta_t^i < 0$ corresponds to shorting the ith stock and $\Delta_t^i > 0$ is a long position. In continuous time, trading (i.e., portfolio re-balancing) is allowed at any moment in time. An investor begins with a given initial wealth Π_0 financing the initial portfolio with positions $(\beta_0, \Delta_0^1, \ldots, \Delta_0^n)$. A trading strategy over time $t \in [0, T]$ consists of holding Δ_t^i shares in each ith base asset and β_t units (as a loan or investment) in the bank account at time t, i.e., the portfolio value at time $t \in [0, T]$ is

$$\Pi_t = \boldsymbol{\Delta}_t \cdot \mathbf{S}(t) + \beta_t B(t). \tag{13.17}$$

As in the single risky asset model of Chapter 12, the only admissible portfolio replicating strategies are self-financing. That is, the differential change in portfolio value is only due to differential changes in the prices of all base assets while holding all the positions fixed. Formally, a self-financing portfolio strategy is a strategy $(\beta_t, \boldsymbol{\Delta}_t)$, $0 \leqslant t \leqslant T$, with cumulative gain in portfolio value given by (a.s.)

$$\begin{aligned}
\Pi_t &= \Pi_0 + \int_0^t \beta_u \, \mathrm{d}B(u) + \int_0^t \boldsymbol{\Delta}_u \cdot \mathrm{d}\mathbf{S}(u) \\
&= \Pi_0 + \int_0^t \beta_u r(u) B(u) \, \mathrm{d}u + \sum_{i=1}^n \int_0^t \Delta_u^i \, \mathrm{d}S_i(u), \quad 0 \leqslant t \leqslant T.
\end{aligned} \tag{13.18}$$

In differential form,

$$\mathrm{d}\Pi_t = r(t)\beta_t B(t)\,\mathrm{d}t + \boldsymbol{\Delta}_t \cdot \mathrm{d}\mathbf{S}(t) \equiv r(t)\beta_t B(t)\,\mathrm{d}t + \sum_{i=1}^{n} \Delta_t^i\,\mathrm{d}S_i(t)\,.$$

Using (13.17) we express the bank account portion as $\beta_t B(t) = \Pi_t - \boldsymbol{\Delta}_t \cdot \mathbf{S}(t)$ and substituting this into the above differential gives

$$\mathrm{d}\Pi_t = r(t)(\Pi_t - \boldsymbol{\Delta}_t \cdot \mathbf{S}(t))\,\mathrm{d}t + \boldsymbol{\Delta}_t \cdot \mathrm{d}\mathbf{S}(t)\,. \tag{13.19}$$

This is the multidimensional extension of (12.6) in Chapter 12, giving the differential change in value of a self-financing portfolio in the $n+1$ base assets.

As in the single asset case, it is easy to show that the discounted self-financing portfolio value process, $\{\overline{\Pi}_t := D(t)\Pi_t\}_{0\leqslant t\leqslant T}$, is a $\widetilde{\mathbb{P}}$-martingale by computing its stochastic differential upon using (13.19) and simplifying (here we use compact vector notation):

$$\begin{aligned}
\mathrm{d}\overline{\Pi}_t \equiv \mathrm{d}(D(t)\Pi_t) &= -r(t)D(t)\Pi_t\,\mathrm{d}t + D(t)\,\mathrm{d}\Pi_t \\
&= -r(t)D(t)\Pi_t\,\mathrm{d}t + D(t)\left[r(t)\Pi_t\,\mathrm{d}t - r(t)\boldsymbol{\Delta}_t \cdot \mathbf{S}(t)\,\mathrm{d}t + \boldsymbol{\Delta}_t \cdot \mathrm{d}\mathbf{S}(t)\right] \\
&= \boldsymbol{\Delta}_t \cdot \left[-r(t)D(t)\mathbf{S}(t)\,\mathrm{d}t + D(t)\,\mathrm{d}\mathbf{S}(t)\right] \\
&= \boldsymbol{\Delta}_t \cdot \mathrm{d}\left(D(t)\mathbf{S}(t)\right) \\
&\equiv \sum_{i=1}^{n} \Delta_t^i\,\mathrm{d}\overline{S}_i(t) = \sum_{i=1}^{n} \Delta_t^i \overline{S}_i(t)\,\boldsymbol{\sigma}_i(t) \cdot \mathrm{d}\widetilde{\mathbf{W}}(t)\,.
\end{aligned} \tag{13.20}$$

In the last line we have two equivalent ways of expressing the $\widetilde{\mathbb{P}}$-martingale property, either as a linear combination of differentials in the discounted stock prices (which are all $\widetilde{\mathbb{P}}$-martingales) or, after substituting (13.13), as a linear combination of differentials in $\widetilde{\mathbf{W}}(t)$. Changes in the discounted (self-financing) portfolio are due only to changes in the discounted stock prices. The integral form of (13.20) expresses the cumulative change in the discounted portfolio value as

$$\overline{\Pi}_t = \overline{\Pi}_0 + \sum_{i=1}^{n} \int_0^t \Delta_u^i\,\mathrm{d}\overline{S}_i(u) = \Pi_0 + \sum_{i=1}^{n} \int_0^t \Delta_u^i \overline{S}_i(u)\,\boldsymbol{\sigma}_i(u) \cdot \mathrm{d}\widetilde{\mathbf{W}}(u)\,, \quad 0 \leqslant t \leqslant T. \tag{13.21}$$

We are generally assuming that the Itô integral is square integrable and so the process in (13.21) is a $\widetilde{\mathbb{P}}$-martingale.

As in the (B,S) model of Chapter 12, no-arbitrage pricing of derivative claims relies on replicating the claim's payoff V_T by using a self-financing replicating strategy. In our continuous-time multi-asset model we have the following definition of such a strategy and market completeness.

Definition 13.1. A self-financing strategy $(\beta_t, \boldsymbol{\Delta}_t), 0 \leqslant t \leqslant T$, is said to *replicate* the $\mathcal{F}_T$-measurable payoff V_T at maturity T if $\Pi_T = V_T$ (a.s.), i.e., $\mathbb{P}(\Pi_T = V_T) = 1$, and we say that the payoff V_T is *attainable*. A market model is *complete* if every $\mathcal{F}_T$-measurable payoff can be replicated, i.e., every derivative security can be hedged.

We now suppose that a self-financing replicating portfolio strategy for a given payoff V_T exists, i.e., that the payoff is attainable. Below we discuss the conditions for which this holds (i.e., market completeness) and the connection to the existence of risk-neutral measures. Repeating the same argument as in the (B,S) model of Chapter 12, the wealth Π_t at time $t \leqslant T$ needed to set up a self-financing replicating strategy up to time T must equal the time-t price of the derivative V_t. This is the fair price for an investor to hedge a short position in the derivative security with attainable payoff V_T at future time T, and hence avoid an obvious arbitrage. Therefore, $V_t = \Pi_t$, $0 \leqslant t \leqslant T$, for any attainable payoff V_T. Combining this with the fact that the value process of the discounted replicating portfolio is a $\widetilde{\mathbb{P}}$-martingale implies that the price process of the discounted derivative is a $\widetilde{\mathbb{P}}$-martingale. That is, $D(t)V_t = \widetilde{\mathrm{E}}[D(T)V_T \mid \mathcal{F}_t]$:

$$V_t = B(t)\,\widetilde{\mathrm{E}}\left[\frac{V_T}{B(T)}\,\middle|\,\mathcal{F}_t\right] = \widetilde{\mathrm{E}}\left[\mathrm{e}^{-\int_t^T r(u)\,\mathrm{d}u} V_T \mid \mathcal{F}_t\right], \quad 0 \leqslant t \leqslant T. \tag{13.22}$$

This is the risk-neutral pricing formula for the above multi-asset model with n base asset prices modelled as Itô processes obeying (13.11) and a bank account with a generally adapted (stochastic) interest rate process. Note that in the case of constant interest rate $r(t) \equiv r$, this recovers (12.9).

Before moving on to the actual implementation of (13.22), we discuss the two main theorems of asset pricing in the context of the above multi-asset model. The first theorem tells us when the model can be used for risk-neutral pricing without arbitrage in the market and the second relates the completeness of the model and arbitrage. We begin by giving a simple definition of arbitrage in the above multi-asset continuous-time model. The definition mirrors that given in Section 8.3.1 of Chapter 8. The basic meaning is that the strategy requires no cost to execute and has a potential to create a profit at no risk. A market model that admits such a strategy is said to have an arbitrage.

Definition 13.2. A self-financing portfolio strategy $(\beta_t, \boldsymbol{\Delta}_t)$, $t \geqslant 0$, is said to be an *arbitrage or arbitrage strategy* if it has zero initial value, $\Pi_0 = 0$, and satisfies the condition that $\Pi_T \geqslant 0$ (a.s), i.e., $\mathbb{P}(\Pi_T \geqslant 0) = 1$, and $\mathbb{P}(\Pi_T > 0) > 0$ for some $T > 0$.

Note that the probabilities in the definition are in the actual (physical) measure. The above definition implies that if arbitrage exists then an investor can use a trading strategy to beat the bank account at no risk. To see that this is implied by the above definition of arbitrage, assume that an investor begins with some positive capital, say $\Pi'_0 > 0$. The investor can then employ a strategy with value Π'_t, $t \geqslant 0$, in which the initial capital Π'_0 is invested in the bank account and executes a zero initial cost arbitrage strategy $(\beta_t, \boldsymbol{\Delta}_t)$ with value Π_t. At time $T > 0$ the investment in the bank account has value $B(T)\Pi'_0$ and the no-cost arbitrage strategy has value Π_T. The time-T value of the combined strategy is $\Pi'_T = \Pi_T + B(T)\Pi'_0$ and by the conditions in the above definition,

$$\mathbb{P}(\Pi_T \geqslant 0) = 1 \implies \mathbb{P}(\Pi'_T \geqslant B(T)\Pi'_0) = 1$$

and

$$\mathbb{P}(\Pi_T > 0) > 0 \implies \mathbb{P}(\Pi'_T > B(T)\Pi'_0) > 0\,.$$

The first condition states that (with certainty) the strategy has a return greater than or equal to that of the bank account investment. The second condition states that there is a positive probability that the strategy has a greater return than the bank account. We leave it as an exercise for the reader to prove the converse, i.e., that the existence of a strategy that beats the bank account at no risk implies the existence of an arbitrage portfolio strategy with zero initial cost.

Based on the above definition of arbitrage we can now state the so-called first fundamental theorem of asset (or derivative) pricing for the above continuous-time multi-asset market model as follows. The result is a simple consequence of the martingale property of any discounted self-financing portfolio strategy.

Theorem 13.1 (First Fundamental Theorem of Asset Pricing). *If there exists a risk-neutral measure $\widetilde{\mathbb{P}}$ then there are no arbitrage strategies in the market.*

Proof. The discounted value process of any (admissible) self-financing portfolio is a $\widetilde{\mathbb{P}}$-martingale. Let $\{\Pi_t\}_{t\geqslant 0}$ be the value process of an assumed arbitrage portfolio strategy with $\Pi_0 = 0$. The $\widetilde{\mathbb{P}}$-martingale property gives

$$\widetilde{\mathrm{E}}[\overline{\Pi}_T] = \widetilde{\mathrm{E}}[D(T)\Pi_T] = \overline{\Pi}_0 = \Pi_0 = 0 \quad \text{for any } T \geqslant 0.$$

By the definition of an arbitrage portfolio, $\Pi_T \geqslant 0$ and hence $D(T)\Pi_T \geqslant 0$ (a.s.). So $D(T)\Pi_T$ is nonnegative (a.s.) and has zero expectation under measure $\widetilde{\mathbb{P}}$. This implies that $\widetilde{\mathbb{P}}(D(T)\Pi_T > 0) = 0$, which implies $\widetilde{\mathbb{P}}(\Pi_T > 0) = 0$ since $D(T) > 0$. Since the measure $\widetilde{\mathbb{P}}$ is equivalent to $\mathbb{P}$, i.e., zero probability events w.r.t. $\widetilde{\mathbb{P}}$ are zero probability events w.r.t. $\mathbb{P}$, then $\mathbb{P}(\Pi_T > 0) = 0$. We conclude that there are no zero-cost and zero-risk self-financing portfolio strategies such that $\mathbb{P}(\Pi_T > 0) > 0$, implying no arbitrage strategies are possible. □

This theorem provides a way of verifying if the market model can be used to calculate no-arbitrage prices of attainable derivative securities. That is, the above multi-asset model can be used for no-arbitrage pricing if the price of risk equations have a solution, i.e., if $\operatorname{rank}(\boldsymbol{\sigma}(t)) = n$. For example, there is no arbitrage in the model when the number of BMs is the same as the number of stocks, $d = n$, and the log-volatility matrix $\boldsymbol{\sigma}(t)$ is invertible. The multidimensional GBM model for n stocks driven by n BMs is an important example of a market model having no arbitrage. The (B, S) model of Chapter 12, where $d = n = 1$, is the simplest case of a GBM model with no arbitrage.

We now consider the question of completeness of the multi-asset model. As in Chapter 12, we consider all payoffs V_T that are square integrable. As well, all payoffs V_T are assumed to be $\mathcal{F}_T^{\mathbf{W}}$-measurable. The process $\{D(t)V_t\}_{0\leqslant t\leqslant T}$ is then a square-integrable $(\widetilde{\mathbb{P}}, \mathbb{F}^{\mathbf{W}})$-martingale. We can hence make use of the multidimensional Brownian Martingale Representation Theorem 11.22 where we identify $\widehat{P} \equiv \widetilde{\mathbb{P}}$, $\widehat{\mathbf{W}} \equiv \widetilde{\mathbf{W}}$, and $\widehat{M}(t) \equiv D(t)V_t$. In particular, there exists an adapted d-dimensional process, $\{\widetilde{\boldsymbol{\theta}}(t) = (\widetilde{\theta}_1(t), \ldots, \widetilde{\theta}_d(t))\}_{0\leqslant t\leqslant T}$, such that (a.s.)

$$D(t)V_t = V_0 + \int_0^t \widetilde{\boldsymbol{\theta}}(u) \cdot \mathrm{d}\widetilde{\mathbf{W}}(u)\,, \quad 0 \leqslant t \leqslant T. \tag{13.23}$$

In order to replicate the claim, i.e., to have $\Pi_t = V_t$ for all $t \in [0, T]$, we must have the equivalence of this expression with that in (13.21) (note $\overline{\Pi}_t = D(t)\Pi_t$). In other words, replication is possible if and only if both Itô integrands are equal,

$$\sum_{i=1}^{n} \Delta_t^i \overline{S}_i(t)\, \boldsymbol{\sigma}_i(t) = \widetilde{\boldsymbol{\theta}}(t)\,, \quad 0 \leqslant t \leqslant T. \tag{13.24}$$

Equating each vector component gives a linear system of d equations in n unknowns $\Delta_t^1, \ldots, \Delta_t^n$,

$$\sum_{i=1}^{n} \sigma_{ij}(t)\overline{S}_i(t)\Delta_t^i = \widetilde{\theta}_j(t)\,, \quad j = 1, \ldots, d.$$

In matrix form, this system is written compactly as

$$[\boldsymbol{\sigma}(t)]^\top \mathbf{v}(t) = \widetilde{\boldsymbol{\theta}}(t) \tag{13.25}$$

where $[\boldsymbol{\sigma}(t)]^\top$ is the $d \times n$ matrix transpose of $\boldsymbol{\sigma}(t)$; $\mathbf{v}(t) = \left[\overline{S}_1(t)\Delta_t^1, \ldots, \overline{S}_n(t)\Delta_t^n\right]^\top$ is an $n \times 1$ vector of the delta hedge position in each stock, scaled by each (strictly positive) discounted stock price; $\widetilde{\boldsymbol{\theta}}(t) = \left[\widetilde{\theta}_1(t), \ldots, \widetilde{\theta}_d(t)\right]^\top$ is a $d \times 1$ vector. The market model is complete (and can hedge any derivative) if there exists a solution to (13.25) for *any vector* $\widetilde{\boldsymbol{\theta}}(t) \in \mathbb{R}^d$. Hence, the market is complete if and only if $\text{rank}(\boldsymbol{\sigma}(t)) = d$. Note that the column and row rank of a matrix are the same, $\text{rank}(\boldsymbol{\sigma}(t)) = \text{rank}([\boldsymbol{\sigma}(t)]^\top)$. Given a solution $\mathbf{v}(t) = [v_1(t), \ldots, v_n(t)]^\top$ to the system in (13.25), the hedging positions in the stocks are given by $\Delta_t^1 = v_1(t)/\overline{S}_1(t), \ldots, \Delta_t^n = v_n(t)/\overline{S}_n(t)$. We note that this is not generally a practical method for obtaining the hedging positions. We are only focusing here on the existence of the hedging positions for replicating any payoff.

Let us assume that at least one risk-neutral measure $\widetilde{\mathbb{P}}$ exists, i.e., $\text{rank}(\boldsymbol{\sigma}(t)) = n$. If the market is complete, $\text{rank}(\boldsymbol{\sigma}(t)) = d = n$ and hence the solution to (13.16) must be unique and therefore $\widetilde{\mathbb{P}}$ is unique. Conversely, if $\widetilde{\mathbb{P}}$ is unique then $\text{rank}(\boldsymbol{\sigma}(t)) = n = d$ and therefore the market model is complete. We have therefore proven the following result, which is commonly stated as the "Second Fundamental Theorem of Asset Pricing." In a model having a risk-neutral measure (no arbitrage), this theorem ties together the uniqueness of a risk-neutral measure and market completeness.

Theorem 13.2 (Second Fundamental Theorem of Asset Pricing). *Assume the above multi-asset continuous-time market model has a risk-neutral probability measure* $\widetilde{\mathbb{P}}$. *Then,* $\widetilde{\mathbb{P}}$ *is unique if and only if the market model is complete.*

13.2 Black–Scholes PDE and Delta Hedging for Standard Multi-Asset Derivatives within a General Diffusion Model

As remarked for the case of the simplest (B, S) model of Chapter 12, the practical implementation of (13.22) depends on the complexity of the market model as well as the type of payoff. The general model considered in Section 13.1, where all the coefficients are adapted processes, is useful for a general theoretical discussion of the main concepts of pricing and replication. However, a general model is nearly impossible to calibrate to the market since the coefficients are too general, with possibly infinite numbers of parameters. In practice, we make additional assumptions on the model that simplify its use in both calibration and derivative pricing. Here we simply formulate the problem of no-arbitrage pricing of *standard (non-path-dependent) options* within a market model where n risky base asset prices (e.g., n stock prices) are modelled as correlated positive diffusion processes and the interest rate $r(t)$ is assumed to be a nonrandom (ordinary) positive function of time. We also allow for a continuous dividend yield $q_i(t)$ on each stock i, where these are assumed to be ordinary functions of time. The general Black–Scholes PDE satisfied by the pricing function of such options then follows. The formula for computing the delta hedging positions in the multi-asset replicating strategy for a standard option is then derived using a simple application of the multidimensional Itô formula.

Let us consider $n \geqslant 1$ asset (stock) price processes $\{\mathbf{S}(t) = (S_1(t), \ldots, S_n(t))\}_{t \geqslant 0}$ solving

$$\mathrm{d}S_i(t) = S_i(t)\left[(\mu_i(t, \mathbf{S}(t)) - q_i(t))\,\mathrm{d}t + \boldsymbol{\sigma}_i(t, \mathbf{S}(t)) \cdot \mathrm{d}\mathbf{W}(t)\right], \quad i = 1, \ldots, n, \tag{13.26}$$

where $\mathbf{W}(t) = (W_1(t), \ldots, W_n(t))$, i.e., the number of BMs is the same as the number of base assets. Note that this system of SDEs is a special case of (13.2) where the adapted log-drift coefficients and log-volatility vectors are specified functions of the asset prices $\mathbf{S}(t)$ and time t: $\mu_i(t) \equiv \mu_i(t, \mathbf{S}(t))$ and $\boldsymbol{\sigma}_i(t) \equiv \boldsymbol{\sigma}_i(t, \mathbf{S}(t))$. We assume that the $n \times n$ log-diffusion matrix of elements $C_{ij}(t, \mathbf{S}(t)) := \boldsymbol{\sigma}_i(t, \mathbf{S}(t)) \cdot \boldsymbol{\sigma}_j(t, \mathbf{S}(t))$ is invertible (nonsingular). We also assume that all Itô integrals are square-integrable martingales and that the coefficient functions are such that a unique strong solution to (13.26) exists. Based on the analysis of the previous section we hence have a complete market model with a unique risk-neutral probability measure $\widetilde{\mathbb{P}}$ where

$$\mathrm{d}S_i(t) = S_i(t)\big[(r(t) - q_i(t))\,\mathrm{d}t + \boldsymbol{\sigma}_i(t, \mathbf{S}(t))\cdot \mathrm{d}\widetilde{\mathbf{W}}(t)\big]\,,\ \ i = 1, \ldots, n. \tag{13.27}$$

Our goal is to price a European option with a payoff being a function of the terminal random values of the asset prices

$$V_T = \Lambda(\mathbf{S}(T)) \equiv \Lambda(S_1(T), \ldots, S_n(T))\,. \tag{13.28}$$

This is referred to as a standard "basket option" on n assets or stocks. Later we derive explicit analytical pricing formulae for specific examples of such options within the multidimensional GBM model. By the (joint) Markov property of the time-T solution $\mathbf{S}(T)$ to (13.26), the conditioning on $\mathcal{F}_t$ in (13.22) is replaced with a conditioning on the time-t (random) prices $\mathbf{S}(t)$. The time-t derivative value is a function of the random vector $\mathbf{S}(t)$ and time t, $V_t = V(t, \mathbf{S}(t))$, where (13.22) now takes the form (note that $r(t)$ is nonrandom by assumption so it can be pulled out of the expectation):

$$V(t, \mathbf{S}(t)) = \mathrm{e}^{-\int_t^T r(u)\,\mathrm{d}u}\,\widetilde{\mathrm{E}}\big[V_T \mid \mathcal{F}_t\big] = \mathrm{e}^{-\int_t^T r(u)\,\mathrm{d}u}\,\widetilde{\mathrm{E}}\big[\Lambda(\mathbf{S}(T)) \mid \mathbf{S}(t)\big]\,. \tag{13.29}$$

The pricing function $V(t, \mathbf{S}) = V(t, S_1, \ldots, S_n)$ depends on calendar time t and the n spot values[1] (ordinary variables) $S_1 > 0, S_2 > 0, \ldots, S_n > 0$. This function is given by the discounted risk-neutral expectation of the payoff conditional on $\mathbf{S}(t) = \mathbf{S} \equiv (S_1, \ldots, S_n)$, i.e., with joint condition $S_1(t) = S_1, \ldots, S_n(t) = S_n$:

$$V(t, \mathbf{S}) = \mathrm{e}^{-\int_t^T r(u)\,\mathrm{d}u}\,\widetilde{\mathrm{E}}[\,\Lambda(\mathbf{S}(T)) \mid \mathbf{S}(t) = \mathbf{S}\,] \equiv \mathrm{e}^{-\int_t^T r(u)\,\mathrm{d}u}\,\widetilde{\mathrm{E}}_{t,\mathbf{S}}[\,\Lambda(\mathbf{S}(T))]\,. \tag{13.30}$$

By applying the discounted Feynman–Kac Theorem 11.20, we therefore have that the pricing function $V = V(t, S_1, \ldots, S_n)$ solves the *multi-asset Black–Scholes PDE (BSPDE)*

$$\frac{\partial V}{\partial t} + \frac{1}{2}\sum_{i=1}^{n}\sum_{j=1}^{n} C_{ij}(t, \mathbf{S}) S_i S_j \frac{\partial^2 V}{\partial S_i \partial S_j} + \sum_{i=1}^{n}(r(t) - q_i(t)) S_i \frac{\partial V}{\partial S_i} - r(t) V = 0 \tag{13.31}$$

subject to the terminal condition $V(T, \mathbf{S}) = \Lambda(\mathbf{S})$, where $\Lambda(\mathbf{S}) \equiv \Lambda(S_1, \ldots, S_n)$ is the payoff function. For continuous $\Lambda(\mathbf{S})$ we also have the $t \nearrow T$ limit $V(T-, \mathbf{S}) = V(T, \mathbf{S})$.

The BSPDE in (13.31) is also written compactly as $\frac{\partial V}{\partial t} + \mathcal{G}_{t,\mathbf{S}} V - r(t) V = 0$, where

$$\mathcal{G}_{t,\mathbf{S}} V := \frac{1}{2}\sum_{i=1}^{n}\sum_{j=1}^{n} C_{ij}(t, \mathbf{S}) S_i S_j \frac{\partial^2 V}{\partial S_i \partial S_j} + \sum_{i=1}^{n}(r(t) - q_i(t)) S_i \frac{\partial V}{\partial S_i} \tag{13.32}$$

is the generator for the system of SDEs in (13.27). The fundamental solution to (13.31) is the $\widetilde{\mathbb{P}}$-measure transition PDF, denoted by $\widetilde{p}(t, T; \mathbf{S}, \mathbf{y})$, multiplied by the discount factor $\mathrm{e}^{-\int_t^T r(u)\,\mathrm{d}u}$. In terms of this risk-neutral transition PDF, the price in (13.30) is given by an n-dimensional integral

$$V(t, \mathbf{S}) = \mathrm{e}^{-\int_t^T r(u)\,\mathrm{d}u} \int_{\mathbb{R}_+^n} \Lambda(\mathbf{y}) \widetilde{p}(t, T; \mathbf{S}, \mathbf{y})\,\mathrm{d}\mathbf{y}\,. \tag{13.33}$$

[1]As in Chapter 12, in pricing formulae we prefer to choose more appropriate letters for some of the ordinary variables. Here we denote the spot values by S_i, instead of using some other letter like x_i. The ordinary variables S_i should not be confused with the random variables such as $S_i(t)$ and $S_i(T)$.

We remark that if the log-volatility vectors are time independent, $\boldsymbol{\sigma}_i(t, \mathbf{S}(t)) \equiv \boldsymbol{\sigma}_i(\mathbf{S}(t))$, then $C_{ij}(t, \mathbf{S}) \equiv C_{ij}(\mathbf{S})$ is time independent. Moreover, if the interest rate $r(t) = r$ is constant and all dividend yields are constants, $q_i(t) = q_i$, then (13.27) is time homogeneous. The BSPDE is then time homogeneous and the pricing function can also be expressed as a function $v(\tau, \mathbf{S}) = V(t, \mathbf{S})$, where $\tau = T - t$ is time to maturity. The BSPDE in (13.31) is then equivalent to

$$\frac{\partial v}{\partial \tau} = \frac{1}{2}\sum_{i=1}^{n}\sum_{j=1}^{n} C_{ij}(\mathbf{S}) S_i S_j \frac{\partial^2 v}{\partial S_i \partial S_j} + \sum_{i=1}^{n}(r - q_i) S_i \frac{\partial v}{\partial S_i} - rv \tag{13.34}$$

subject to the initial condition $v(0, \mathbf{S}) = \Lambda(\mathbf{S})$. The risk-neutral transition PDF is then time homogeneous, i.e. $\widetilde{p}(t, T; \mathbf{S}, \mathbf{y}) = \widetilde{p}(\tau; \mathbf{S}, \mathbf{y})$, and the pricing function is given by

$$v(\tau, \mathbf{S}) = \mathrm{e}^{-r\tau}\, \widetilde{\mathrm{E}}[\, \Lambda(\mathbf{S}(T)) \mid \mathbf{S}(t) = \mathbf{S}\,] = \mathrm{e}^{-r\tau} \int_{\mathbb{R}_+^n} \Lambda(\mathbf{y}) \widetilde{p}(\tau; \mathbf{S}, \mathbf{y})\, \mathrm{d}\mathbf{y}\,. \tag{13.35}$$

Let us now turn to the derivation of the hedging positions required in a self-financing strategy that replicates the above standard European multi-asset option. What we will achieve shortly below is a more explicit form for the martingale representation in (13.23) since the option price process is a smooth (pricing) function of the stock price process. The first step is to apply the Itô formula and compute the stochastic differential $\mathrm{d}V_t = \mathrm{d}V(t, \mathbf{S}(t))$ of the option value process:

$$\begin{aligned}
\mathrm{d}V_t &= \left[\frac{\partial V}{\partial t}(t, \mathbf{S}(t)) + \frac{1}{2}\sum_{i=1}^{n}\sum_{j=1}^{n} C_{ij}(t, \mathbf{S}) S_i(t) S_j(t) \frac{\partial^2 V}{\partial S_i \partial S_j}(t, \mathbf{S}(t))\right] \mathrm{d}t + \sum_{i=1}^{n} \frac{\partial V}{\partial S_i}(t, \mathbf{S}(t))\, \mathrm{d}S_i(t) \\
&= \left(\frac{\partial}{\partial t} + \mathcal{G}_{t,\mathbf{S}}\right) V(t, \mathbf{S}(t))\, \mathrm{d}t + \sum_{i=1}^{n} S_i(t) \frac{\partial V}{\partial S_i}(t, \mathbf{S}(t))\, \boldsymbol{\sigma}_i(t, \mathbf{S}(t)) \cdot \mathrm{d}\widetilde{\mathbf{W}}(t) \\
&= r(t) V(t, \mathbf{S}(t))\, \mathrm{d}t + \sum_{i=1}^{n} S_i(t) \frac{\partial V}{\partial S_i}(t, \mathbf{S}(t))\, \boldsymbol{\sigma}_i(t, \mathbf{S}(t)) \cdot \mathrm{d}\widetilde{\mathbf{W}}(t)\,.
\end{aligned} \tag{13.36}$$

In the last line we used the fact that the pricing function $V(t, \mathbf{S})$ solves the BSPDE, i.e., $\frac{\partial V}{\partial t} + \mathcal{G}_{t,\mathbf{S}} V(t, \mathbf{S}) = r(t) V(t, \mathbf{S})$ for all $(t, \mathbf{S})$. The second step is to use the Itô product rule and obtain the stochastic differential of the discounted option value, which simplifies to a driftless expression upon using (13.36):

$$\begin{aligned}
\mathrm{d}[D(t) V_t] &= D(t)\, [\, \mathrm{d}V(t, \mathbf{S}(t)) - r(t) V(t, \mathbf{S}(t))\, \mathrm{d}t] \\
&= \sum_{i=1}^{n} \overline{S}_i(t) \frac{\partial V}{\partial S_i}(t, \mathbf{S}(t))\, \boldsymbol{\sigma}_i(t, \mathbf{S}(t)) \cdot \mathrm{d}\widetilde{\mathbf{W}}(t)\,.
\end{aligned} \tag{13.37}$$

In integral form,

$$D(t) V_t \equiv D(t) V(t, \mathbf{S}(t)) = V(0, \mathbf{S}(0)) + \int_0^t \underbrace{\left(\sum_{i=1}^{n} \overline{S}_i(u) \frac{\partial V}{\partial S_i}(u, \mathbf{S}(u))\, \boldsymbol{\sigma}_i(u, \mathbf{S}(u))\right)}_{= \widetilde{\boldsymbol{\theta}}(u)} \cdot \mathrm{d}\widetilde{\mathbf{W}}(u)\,. \tag{13.38}$$

This is the representation in (13.23) where we identify the adapted vector process $\widetilde{\boldsymbol{\theta}}(t)$, for all $0 \leqslant t < T$. Finally, by equating this vector with that in (13.24), we obtain the *delta hedging position in each ith asset as the partial derivative of the option pricing formula w.r.t. the ith spot variable*:

$$\Delta_t^i = \frac{\partial V}{\partial S_i}(t, \mathbf{S}(t)) \equiv \frac{\partial V}{\partial S_i}(t, \mathbf{S})\bigg|_{\mathbf{S}=\mathbf{S}(t)}, \quad 0 \leqslant t < T. \tag{13.39}$$

This clearly generalizes the delta hedging formula in (12.20) in Chapter 12 for a standard European option on a single asset (stock). As in the single asset case, we define the delta hedging functions $\Delta^i(t, \mathbf{S}) \equiv \Delta^i(t, S_1, \ldots, S_n)$,

$$\Delta^i(t, \mathbf{S}) := \frac{\partial V}{\partial S_i}(t, \mathbf{S}), \quad i = 1, \ldots, n,$$

where $\Delta^i(t, \mathbf{S})$ gives the time-t position in the ith base asset given the spot values $\mathbf{S} = (S_1, \ldots, S_n)$. Given the pricing function $V(t, \mathbf{S})$, the partial derivatives $\Delta^i(t, \mathbf{S})$ can be computed and the self-financing replicating portfolio in (13.17) is specified by taking a (long or short) position $\Delta_t^1 = \Delta^1(t, \mathbf{S}(t))$ in the first asset with share price $S_1(t)$, $\Delta_t^2 = \Delta^2(t, \mathbf{S}(t))$ in the second asset with share price $S_2(t)$, ..., $\Delta_t^n = \Delta^n(t, \mathbf{S}(t))$ in the nth asset with share price $S_n(t)$ at time t. The time-t position in the bank account is then given by $\beta_t = D(t)[V(t, \mathbf{S}(t)) - \mathbf{\Delta}_t \cdot \mathbf{S}(t)]$, i.e., the value of the bank account portion of the portfolio at time t is $V(t, \mathbf{S}(t)) - \mathbf{\Delta}_t \cdot \mathbf{S}(t)$, where $\mathbf{S}(t) \cdot \mathbf{\Delta}_t = \sum_{i=1}^n \Delta_t^i S_i(t)$.

13.2.1 Standard European Option Pricing for Multi-Stock GBM

In order to be able to derive explicit pricing and hedging formulae for several standard options that occur in practice, we now specialize Section 13.2 to the classical multidimensional GBM market model where all log-volatility vectors, and hence all matrix elements of the covariance matrix of log-returns, are assumed to be *constants*. We shall assume a nonsingular $n \times n$ covariance matrix of log-returns and set $d = n$, i.e., the number of independent BMs and the number of risky assets to be equal. Hence, our multidimensional GBM market model has a unique risk-neutral measure $\widetilde{\mathbb{P}}$. By the fundamental theorems of asset pricing, the market model is therefore complete with no arbitrage opportunities.

We refer the reader to Section 11.10 of Chapter 11 where the multidimensional GBM process was introduced. Here we also include a constant dividend yield q_i on each ith stock where the stock prices satisfy the system of SDEs

$$\mathrm{d}S_i(t) = S_i(t)\left[(r - q_i)\,\mathrm{d}t + \boldsymbol{\sigma}_i \cdot \mathrm{d}\widetilde{\mathbf{W}}(t)\right], \ i = 1, \ldots, n. \tag{13.40}$$

We recall our previous notation where the volatility vectors $\boldsymbol{\sigma}_i = [\sigma_{i1}, \ldots, \sigma_{in}]$ have square magnitude denoted by $\sigma_i^2 \equiv \|\boldsymbol{\sigma}_i\|^2$ and their inner product $\boldsymbol{\sigma}_i \cdot \boldsymbol{\sigma}_j = \rho_{ij}\sigma_i\sigma_j$. The correlation matrix $\boldsymbol{\rho}$ of constant elements $\rho_{ij} \in (-1, 1)$ gives the correlations among all pairwise log-returns, as shown in (11.201). The unique strong solution subject to initial prices $S_1(0), \ldots, S_n(0)$ is given by (see (11.194))

$$S_i(t) = S_i(0)\,\mathrm{e}^{(r - q_i - \frac{1}{2}\sigma_i^2)t + \boldsymbol{\sigma}_i \cdot \widetilde{\mathbf{W}}(t)} = S_i(0)\,\mathrm{e}^{(r - q_i)t}\mathcal{E}_t(\boldsymbol{\sigma}_i \cdot \widetilde{\mathbf{W}}) \tag{13.41}$$

where

$$\mathcal{E}_t(\boldsymbol{\sigma}_i \cdot \widetilde{\mathbf{W}}) = \exp\left[-\frac{1}{2}\sigma_i^2 t + \boldsymbol{\sigma}_i \cdot \widetilde{\mathbf{W}}(t)\right] = \exp\left[-\frac{1}{2}\sigma_i^2 t + \sum_{j=1}^{d} \sigma_{ij}\widetilde{W}_j(t)\right] \tag{13.42}$$

is an exponential $\widetilde{\mathbb{P}}$-martingale for each $i = 1, \ldots, n$. From (13.41) we have the stock price vector at time T expressed in terms of its value at time $t \leqslant T$:

$$\begin{aligned} S_i(T) &= S_i(t)\, \mathrm{e}^{(r-q_i-\frac{1}{2}\sigma_i^2)(T-t)+\boldsymbol{\sigma}_i\cdot(\widetilde{\mathbf{W}}(T)-\widetilde{\mathbf{W}}(t))} \\ &= S_i(t)\, \mathrm{e}^{(r-q_i-\frac{1}{2}\sigma_i^2)\tau+\sigma_i\sqrt{\tau}\widetilde{Z}_i} \end{aligned} \tag{13.43}$$

where $\tau = T - t$ is time to maturity and we define

$$\widetilde{Z}_i := \frac{\boldsymbol{\sigma}_i \cdot (\widetilde{\mathbf{W}}(T) - \widetilde{\mathbf{W}}(t))}{\sigma_i\sqrt{\tau}} = \frac{1}{\sigma_i\sqrt{\tau}} \sum_{j=1}^{d} \sigma_{ij}(\widetilde{W}_j(T) - \widetilde{W}_j(t)), \quad i = 1, \ldots, n.$$

Under measure $\widetilde{\mathbb{P}}$, these random variables are i.i.d. *Norm*$(0,1)$ and their joint distribution is the multivariate standard normal with correlation matrix $\boldsymbol{\rho}$,

$$\widetilde{\mathbf{Z}} = [\widetilde{Z}_1, \ldots, \widetilde{Z}_n]^\top \sim Norm_n\,(\mathbf{0}, \boldsymbol{\rho})\,. \tag{13.44}$$

The pricing function $V(t, \mathbf{S}) = V(t, S_1, \ldots, S_n)$ is given as in (13.30) but now the interest rate is constant,

$$V(t, \mathbf{S}) = \mathrm{e}^{-r(T-t)}\, \widetilde{\mathrm{E}}[\, \Lambda(\mathbf{S}(T)) \mid \mathbf{S}(t) = \mathbf{S}\,] \equiv \mathrm{e}^{-r(T-t)}\, \widetilde{\mathrm{E}}_{t,\mathbf{S}}[\, \Lambda(\mathbf{S}(T))]\,. \tag{13.45}$$

By substituting (13.43) into (13.45), using the independence of $\mathbf{S}(t)$ and all $\widetilde{\mathbf{Z}}$ and conditioning on the spot values $S_1(t) = S_1, \ldots, S_n(t) = S_n$, we obtain the pricing function as an n-dimensional integral in the standard Gaussian density:

$$\begin{aligned} V(t, \mathbf{S}) &= \mathrm{e}^{-r\tau}\, \widetilde{\mathrm{E}}[\, \Lambda(S_1 \mathrm{e}^{(r-q_1-\frac{1}{2}\sigma_1^2)\tau+\sigma_1\sqrt{\tau}\widetilde{Z}_1}, \ldots, S_n \mathrm{e}^{(r-q_n-\frac{1}{2}\sigma_n^2)\tau+\sigma_n\sqrt{\tau}\widetilde{Z}_n})\,] \\ &= \mathrm{e}^{-r\tau} \int \!\!\ldots\!\! \int_{\mathbb{R}^n} \Lambda(S_1 \mathrm{e}^{(r-q_1-\frac{1}{2}\sigma_1^2)\tau+\sigma_1\sqrt{\tau}z_1}, \ldots, S_n \mathrm{e}^{(r-q_n-\frac{1}{2}\sigma_n^2)\tau+\sigma_n\sqrt{\tau}z_n})\, n_n(\mathbf{z}; \boldsymbol{\rho})\, \mathrm{d}\mathbf{z} \end{aligned} \tag{13.46}$$

where $n_n(\mathbf{z}; \boldsymbol{\rho}) = n(z_1, \ldots, z_n; \boldsymbol{\rho})$ is the multivariate standard normal PDF in (11.203). This generalizes the pricing formula in (12.14) to the case of n stocks. By a similar change of integration variables, i.e., setting $y_i = S_i\, \mathrm{e}^{(r-q_i-\frac{1}{2}\sigma_i^2)\tau+\sigma_i\sqrt{\tau}z_i}$, we obtain the generalization of the pricing formula in (12.15),

$$V(t, \mathbf{S}) = \mathrm{e}^{-r\tau} \int_0^\infty \ldots \int_0^\infty \Lambda(\mathbf{y})\, \widetilde{p}(t, T; \mathbf{S}, \mathbf{y})\, \mathrm{d}\mathbf{y}\,. \tag{13.47}$$

The transition PDF in the risk-neutral measure, $\widetilde{p}(t, T; \mathbf{S}, \mathbf{y})$, is time homogeneous, i.e., $\widetilde{p}(t, T; \mathbf{S}, \mathbf{y}) \equiv \widetilde{p}(\tau; \mathbf{S}, \mathbf{y})$ is given by the multivariate log-normal expression in (11.204) for all positive vectors $\mathbf{S}, \mathbf{y}$, and where

$$a_i = \frac{\ln \frac{y_i}{S_i} - (r - q_i - \sigma_i^2/2)\tau}{\sigma_i\sqrt{\tau}}, \quad i = 1, \ldots, n.$$

The BSPDE in (13.31) simplifies to a parabolic PDE in the n spot variables:

$$\frac{\partial V}{\partial t} + \frac{1}{2} \sum_{i=1}^{n} \sum_{j=1}^{n} \rho_{ij}\sigma_i\sigma_j S_i S_j \frac{\partial^2 V}{\partial S_i \partial S_j} + \sum_{i=1}^{n} (r - q_i) S_i \frac{\partial V}{\partial S_i} - rV = 0 \tag{13.48}$$

subject to the terminal condition $V(T, \mathbf{S}) = \Lambda(\mathbf{S})$. In compact form this reads $\frac{\partial V}{\partial t} + \mathcal{G}_{\mathbf{S}} V - rV = 0$ where

$$\mathcal{G}_{\mathbf{S}} V := \frac{1}{2} \sum_{i=1}^{n} \sum_{j=1}^{n} \rho_{ij}\sigma_i\sigma_j S_i S_j \frac{\partial^2 V}{\partial S_i \partial S_j} + \sum_{i=1}^{n} (r - q_i) S_i \frac{\partial V}{\partial S_i} \tag{13.49}$$

is the (time-homogeneous) differential generator for the GBM price process solving (13.40). By time homogeneity, the pricing function is expressible as $v(\tau, \mathbf{S}) = V(t, \mathbf{S})$. The BSPDE in (13.48) is equivalent to

$$\frac{\partial v}{\partial \tau} = \frac{1}{2}\sum_{i=1}^{n}\sum_{j=1}^{n} \rho_{ij}\sigma_i\sigma_j S_i S_j \frac{\partial^2 v}{\partial S_i \partial S_j} + \sum_{i=1}^{n}(r - q_i)S_i \frac{\partial v}{\partial S_i} - rv \tag{13.50}$$

subject to the initial condition $v(0, \mathbf{S}) = \Lambda(\mathbf{S})$.

A common case is when $n = 2$ stocks. In this case there are two log-volatility vectors $\boldsymbol{\sigma}_1 = [\sigma_1, 0]$ and $\boldsymbol{\sigma}_2 = [\sigma_2\rho, \sigma_2\sqrt{1-\rho^2}]$, with volatility parameters $\sigma_1, \sigma_2 > 0$ and only one correlation coefficient $\rho_{12} = \rho_{21} = \rho$ where $\boldsymbol{\sigma}_1 \cdot \boldsymbol{\sigma}_2 = \rho\sigma_1\sigma_2$. The pricing function $V = V(t, S_1, S_2)$ solves the BSPDE

$$\begin{aligned} \frac{\partial V}{\partial t} + \frac{1}{2}\sigma_1^2 S_1^2 \frac{\partial^2 V}{\partial S_1^2} + \frac{1}{2}\sigma_2^2 S_2^2 \frac{\partial^2 V}{\partial S_2^2} + \rho\sigma_1\sigma_2 S_1 S_2 \frac{\partial^2 V}{\partial S_1 \partial S_2} \\ +(r - q_1)S_1 \frac{\partial V}{\partial S_1} + (r - q_2)S_2 \frac{\partial V}{\partial S_2} - rV = 0\,, \end{aligned} \tag{13.51}$$

subject to the terminal (payoff) condition $V(T, S_1, S_2) = \Lambda(S_1, S_2)$. The analogous BSPDE for $v = v(\tau, S_1, S_2)$ is given by (13.50) for $n = 2$.

We close this section by noting a useful relationship between the pricing function for zero stock dividends and that for dividends. In particular, we have the simple symmetry that extends (12.55) to the case of $n \geqslant 1$ stocks. From the pricing formula in (13.46) we observe that, for each i, the inclusion of a dividend q_i corresponds to changing the spot value $S_i \to S_i e^{q_i \tau}$ in the pricing formula with $q_i = 0$. In particular, let $V(t, \mathbf{S}; \mathbf{q}) \equiv V(t, S_1, \ldots, S_n; q_1, \ldots, q_n)$ be the pricing function for the case that the respective dividend yields on stocks $1, \ldots, n$ are $q_1, \ldots, q_n$ (which can have any real value including zero) and let $V(t, \mathbf{S}; \mathbf{0}) \equiv V(t, \mathbf{S})$ be the pricing function in the case that all stock dividends are zero. Then,

$$V(t, \mathbf{S}; \mathbf{q}) = V(t, e^{-q_1(T-t)}S_1, \ldots, e^{-q_n(T-t)}S_n) \tag{13.52}$$

or equivalently for the pricing functions expressed in terms of time to maturity τ, that is, $v(\tau, \mathbf{S}; \mathbf{q}) = v(\tau, e^{-q_1\tau}S_1, \ldots, e^{-q_n\tau}S_n)$. Hence, when deriving the pricing function for a standard (non-path-dependent) multi-stock European option for the case of constant dividends we can derive the pricing function for the case of all dividends being set to zero and then apply the above relation.

13.2.2 Explicit Pricing Formulae for the GBM Model

There are basically three analytical methods or approaches that we can combine or use separately in deriving pricing functions for multi-asset options. One main approach is to implement the risk-neutral pricing formula in (13.22) by computing the expectation by using nested conditioning. Another is the PDE approach, where we solve the BSPDE, which can involve some symmetry reduction technique whereby the original PDE is reduced to a lower dimensional PDE. The other is to use change of probability measures. The change of measure technique can be quite general and amounts to some judicious application of Girsanov's Theorem in order to simplify the evaluation of the expectation in the risk-neutral pricing formula. In Section 13.3 we will cover such an approach where measure changes correspond to a change of numéraires for risk-neutral pricing. In this section we consider some examples of the first two methods for pricing some standard multi-stock options where the stock prices are correlated GBMs as specified in the previous section. In all our examples we assume an economy with constant interest rate r and also allow for a constant continuous dividend yield q_i on each ith stock. The extension to the case of nonrandom time-dependent volatility vectors $\boldsymbol{\sigma}_i(t)$, interest rate $r(t)$, and dividend yields $q_i(t)$ is fairly straightforward as the price processes are still GBMs. We leave examples of such extensions as exercises for the reader.

13.2.2.1 Exchange and Other Related Options

A common type of standard European two-stock option is a so-called *exchange option* where the payoff is the positive part of the difference of the two asset prices at maturity. We now consider pricing this type of option with payoff

$$V_T = (\,S_2(T) - S_1(T)\,)^+ \tag{13.53}$$

where each random variable $S_i(T)$ is the terminal share price of stock $i = 1, 2$. In this example, we have $n = 2$ stocks where the payoff is nonzero only in the event that the terminal share price of the second stock is greater than that of the first stock, in which case the payoff is given by the difference of the two share prices. Let $\widetilde{\mathbb{P}}$ be the risk-neutral measure and let $\widetilde{\mathbf{W}}(t) = (\widetilde{W}_1(t), \widetilde{W}_2(t))$ be a standard two-dimensional vector Brownian motion under $\widetilde{\mathbb{P}}$, i.e., $\widetilde{W}_1(t)$ and $\widetilde{W}_2(t)$ are i.i.d. standard $\widetilde{\mathbb{P}}$-Brownian motions. The respective continuously compounded dividend yields on stocks 1 and 2 are denoted by q_1 and q_2. Defining $\tau := T - t > 0$, then by (13.43) we have

$$\begin{aligned} S_1(T) &= S_1(t)\,\mathrm{e}^{(r-q_1-\frac{1}{2}\sigma_1^2)\tau+\boldsymbol{\sigma}_1\cdot(\widetilde{\mathbf{W}}(T)-\widetilde{\mathbf{W}}(t))} \stackrel{d}{=} S_1(t)\,\mathrm{e}^{(r-q_1-\frac{1}{2}\sigma_1^2)\tau+\sigma_1\sqrt{\tau}\widetilde{Z}_1} \\ S_2(T) &= S_2(t)\,\mathrm{e}^{(r-q_2-\frac{1}{2}\sigma_2^2)\tau+\boldsymbol{\sigma}_2\cdot(\widetilde{\mathbf{W}}(T)-\widetilde{\mathbf{W}}(t))} \stackrel{d}{=} S_2(t)\,\mathrm{e}^{(r-q_2-\frac{1}{2}\sigma_2^2)\tau+\sigma_2\sqrt{\tau}(\rho\widetilde{Z}_1+\sqrt{1-\rho^2}\widetilde{Z}_2)} \end{aligned} \tag{13.54}$$

where

$$\widetilde{Z}_1 := \frac{\widetilde{W}_1(T) - \widetilde{W}_1(t)}{\sqrt{\tau}} \quad \text{and} \quad \widetilde{Z}_2 := \frac{\widetilde{W}_2(T) - \widetilde{W}_2(t)}{\sqrt{\tau}}$$

are independent $Norm(0,1)$ random variables in the measure $\widetilde{\mathbb{P}}$. Recall that $\boldsymbol{\sigma}_1 = [\sigma_1, 0]$ and $\boldsymbol{\sigma}_2 = [\sigma_2\rho, \sigma_2\sqrt{1-\rho^2}]$ are the respective volatility vectors of stocks 1 and 2 where ρ is the correlation coefficient of the log-returns of the two stocks, i.e., $\boldsymbol{\sigma}_1 \cdot \boldsymbol{\sigma}_2 = \rho\sigma_1\sigma_2$.

We denote the respective time-t spot prices of the two stocks by $S_1 > 0$ and $S_2 > 0$. The option pricing function $v(\tau, S_1, S_2)$ is a function of time to maturity τ and the spot variables S_1, S_2. From the risk-neutral pricing formula in (13.45),

$$\begin{aligned} v(\tau, S_1, S_2) &\equiv \mathrm{e}^{-r\tau}\widetilde{\mathrm{E}}\big[\,(S_2(T) - S_1(T))^+ \mid S_1(t) = S_1, S_2(t) = S_2\big] \\ &= \mathrm{e}^{-r\tau}\widetilde{\mathrm{E}}\Big[\big(S_2\mathrm{e}^{(r-q_2-\frac{1}{2}\sigma_2^2)\tau+\sigma_2\sqrt{\tau}(\rho\widetilde{Z}_1+\sqrt{1-\rho^2}\widetilde{Z}_2)} - S_1\mathrm{e}^{(r-q_1-\frac{1}{2}\sigma_1^2)\tau+\sigma_1\sqrt{\tau}\widetilde{Z}_1}\big)^+\Big]. \end{aligned} \tag{13.55}$$

The second equation line is obtained by substituting the expressions in (13.54), conditioning on $S_1(t) = S_1, S_2(t) = S_2$, and using the independence between $\widetilde{Z}_1, \widetilde{Z}_2$ and the joint random variables $S_1(t), S_2(t)$. This expectation is now readily computed by applying nested conditioning. We first condition on $\widetilde{Z}_1$ and thereby isolate the exponential in $\widetilde{Z}_2$, giving

$$\begin{aligned} v(\tau, S_1, S_2) &= \mathrm{e}^{-r\tau}\widetilde{\mathrm{E}}\big[\,\widetilde{\mathrm{E}}\big[\big(S_2\mathrm{e}^{(r-q_2-\frac{1}{2}\sigma_2^2)\tau+\sigma_2\sqrt{\tau}(\rho\widetilde{Z}_1+\sqrt{1-\rho^2}\widetilde{Z}_2)} \\ &\qquad - S_1\mathrm{e}^{(r-q_1-\frac{1}{2}\sigma_1^2)\tau+\sigma_1\sqrt{\tau}\widetilde{Z}_1}\big)^+ \mid \widetilde{Z}_1\big]\big] \\ &= \mathrm{e}^{-r\tau}S_2\mathrm{e}^{(r-q_2-\frac{1}{2}\sigma_2^2)\tau}\widetilde{\mathrm{E}}\big[\mathrm{e}^{\rho\sigma_2\sqrt{\tau}\widetilde{Z}_1}\,\widetilde{\mathrm{E}}\big[\big(\mathrm{e}^{\sigma_2\sqrt{1-\rho^2}\sqrt{\tau}\widetilde{Z}_2} - X_1\big)^+ \mid \widetilde{Z}_1\big]\big] \end{aligned} \tag{13.56}$$

where $X_1 \equiv \frac{S_1}{S_2} e^{(q_2 - q_1 + \frac{1}{2}(\sigma_2^2 - \sigma_1^2))\tau + (\sigma_1 - \rho\sigma_2)\sqrt{\tau}\widetilde{Z}_1}$ is a function of only random variable $\widetilde{Z}_1$ (not $\widetilde{Z}_2$) and hence has a fixed value within the inner expectation which is conditioned on $\widetilde{Z}_1$. Since $\widetilde{Z}_1$ and $\widetilde{Z}_2$ are independent, the inner conditional expectation is readily evaluated as an unconditional expectation, $\widetilde{\mathrm{E}}\big[\big(e^{\sigma_2\sqrt{1-\rho^2}\sqrt{\tau}\widetilde{Z}_2} - X_1\big)^+ \mid \widetilde{Z}_1\big] = g(X_1)$ where

$$\begin{aligned} g(x_1) &:= \widetilde{\mathrm{E}}\Big[\big(e^{\sigma_2\sqrt{1-\rho^2}\sqrt{\tau}\widetilde{Z}_2} - x_1\big)^+\Big] \\ &= \widetilde{\mathrm{E}}\left[e^{\sigma_2\sqrt{1-\rho^2}\sqrt{\tau}\widetilde{Z}_2}\mathbb{I}_{\{\widetilde{Z}_2 > \frac{\ln x_1}{\sigma_2\sqrt{1-\rho^2}\sqrt{\tau}}\}}\right] - x_1\widetilde{\mathrm{E}}\left[\mathbb{I}_{\{\widetilde{Z}_2 > \frac{\ln x_1}{\sigma_2\sqrt{1-\rho^2}\sqrt{\tau}}\}}\right] \\ &= e^{\frac{1}{2}\sigma_2^2(1-\rho^2)\tau}\mathcal{N}\left(\frac{-\ln x_1 + \sigma_2^2(1-\rho^2)\tau}{\sigma_2\sqrt{1-\rho^2}\sqrt{\tau}}\right) - x_1\mathcal{N}\left(\frac{-\ln x_1}{\sigma_2\sqrt{1-\rho^2}\sqrt{\tau}}\right). \end{aligned}$$

Note that in evaluating the above expectations we have applied the identity (A.1) in the Appendix, i.e. $\widetilde{\mathrm{E}}[e^{b\widetilde{Z}_2}\mathbb{I}_{\{\widetilde{Z}_2 > a\}}] = e^{\frac{1}{2}b^2}\mathcal{N}(b-a)$, for any real constants a, b, where $\widetilde{Z}_2 \sim N(0,1)$ under $\widetilde{\mathbb{P}}$. Therefore, setting $x_1 = X_1$ into $g(x_1)$:

$$\begin{aligned} &\widetilde{\mathrm{E}}\big[\big(e^{\sigma_2\sqrt{1-\rho^2}\sqrt{\tau}\widetilde{Z}_2} - X_1\big)^+ \mid \widetilde{Z}_1\big] \\ &= e^{\frac{1}{2}\sigma_2^2(1-\rho^2)\tau}\mathcal{N}\left(\frac{-\ln X_1 + \sigma_2^2(1-\rho^2)\tau}{\sigma_2\sqrt{1-\rho^2}\sqrt{\tau}}\right) - X_1\mathcal{N}\left(\frac{-\ln X_1}{\sigma_2\sqrt{1-\rho^2}\sqrt{\tau}}\right) \\ &= e^{\frac{1}{2}\sigma_2^2(1-\rho^2)\tau}\mathcal{N}(A\widetilde{Z}_1 + D) - X_1\mathcal{N}(A\widetilde{Z}_1 + C). \end{aligned} \tag{13.57}$$

Here we conveniently define the constants

$$A := \frac{\rho\sigma_2 - \sigma_1}{\sigma_2\sqrt{1-\rho^2}}, \quad C := \frac{\ln(S_2/S_1) + [q_1 - q_2 + \frac{1}{2}(\sigma_1^2 - \sigma_2^2)]\tau}{\sigma_2\sqrt{1-\rho^2}\sqrt{\tau}}, \quad D := C + \sigma_2\sqrt{1-\rho^2}\sqrt{\tau}.$$

Substituting the random variable expression on the right-hand side of (13.57) into the outer expectation in (13.56), and writing X_1 in terms of $\widetilde{Z}_1$ and simplifying the exponents, gives

$$\begin{aligned} v(\tau, S_1, S_2) &= S_2 e^{-(q_2 + \frac{1}{2}\sigma_2^2)\tau}\Big(e^{\frac{1}{2}\sigma_2^2(1-\rho^2)\tau}\widetilde{\mathrm{E}}\big[e^{\rho\sigma_2\sqrt{\tau}\widetilde{Z}_1}\mathcal{N}(A\widetilde{Z}_1 + D)\big] \\ &\qquad - \widetilde{\mathrm{E}}\big[e^{\rho\sigma_2\sqrt{\tau}\widetilde{Z}_1}X_1\mathcal{N}(A\widetilde{Z}_1 + C)\big]\Big) \\ &= S_2 e^{-(q_2 + \frac{1}{2}\rho^2\sigma_2^2)\tau}\widetilde{\mathrm{E}}\big[e^{\rho\sigma_2\sqrt{\tau}\widetilde{Z}_1}\mathcal{N}(A\widetilde{Z}_1 + D)\big] \\ &\quad - S_1 e^{-(q_1 + \frac{1}{2}\sigma_1^2)\tau}\widetilde{\mathrm{E}}\big[e^{\sigma_1\sqrt{\tau}\widetilde{Z}_1}\mathcal{N}(A\widetilde{Z}_1 + C)\big]. \end{aligned} \tag{13.58}$$

Note that both expectations in (13.58) can be exactly evaluated by using the integral identity (A.4) in the Appendix, i.e., $\widetilde{\mathrm{E}}\big[e^{b\widetilde{Z}_1}\mathcal{N}(a\widetilde{Z}_1 + c)\big] = e^{b^2/2}\mathcal{N}(\frac{ab+c}{\sqrt{1+a^2}})$, for any real constants a, b, c, and where $\widetilde{Z}_1 \sim N(0,1)$ under $\widetilde{\mathbb{P}}$. Using this identity twice, once for each expectation in (13.58), now gives

$$v(\tau, S_1, S_2) = S_2 e^{-q_2\tau}\mathcal{N}\left(\frac{\rho\sigma_2\sqrt{\tau}A + D}{\sqrt{1+A^2}}\right) - S_1 e^{-q_1\tau}\mathcal{N}\left(\frac{\sigma_1\sqrt{\tau}A + C}{\sqrt{1+A^2}}\right). \tag{13.59}$$

Using the definitions of A, B, C, D above, we now identity the two arguments in the standard normal CDF function in terms of the original model parameters. In particular,

$$\sqrt{1+A^2} = \sqrt{1 + \frac{(\rho\sigma_2 - \sigma_1)^2}{\sigma_2^2(1-\rho^2)}} = \sqrt{\frac{\sigma_2^2(1-\rho^2) + (\rho\sigma_2 - \sigma_1)^2}{\sigma_2^2(1-\rho^2)}} = \frac{\nu}{\sigma_2\sqrt{1-\rho^2}}$$

where $\nu := \sqrt{\sigma_1^2+\sigma_2^2-2\rho\sigma_1\sigma_2} = \|\boldsymbol{\sigma}_1-\boldsymbol{\sigma}_2\|$ is the magnitude of the difference of the volatility vectors of the two stocks. Then, the first argument is

$$\begin{aligned}\frac{\rho\sigma_2\sqrt{\tau}A+D}{\sqrt{1+A^2}} &= \frac{\sigma_2\sqrt{1-\rho^2}}{\nu}(\rho\sigma_2\sqrt{\tau}A+D)\\ &= \frac{\rho\sigma_2(\rho\sigma_2-\sigma_1)\tau+\ln(S_2/S_1)+[q_1-q_2+\frac{1}{2}(\sigma_1^2+\sigma_2^2-2\rho\sigma_2^2)]\tau}{\nu\sqrt{\tau}}\\ &= \frac{\ln(S_2/S_1)+[q_1-q_2+\frac{1}{2}(\sigma_1^2+\sigma_2^2-2\rho\sigma_1\sigma_2)]\tau}{\nu\sqrt{\tau}}\\ &= \frac{\ln(S_2/S_1)+(q_1-q_2+\frac{1}{2}\nu^2)\tau}{\nu\sqrt{\tau}}\end{aligned}$$

and the second argument is

$$\begin{aligned}\frac{\sigma_1\sqrt{\tau}A+C}{\sqrt{1+A^2}} &= \frac{\sigma_2\sqrt{1-\rho^2}}{\nu}\left(\frac{\sigma_1(\rho\sigma_2-\sigma_1)\tau+\ln(S_2/S_1)+[q_1-q_2+\frac{1}{2}(\sigma_1^2-\sigma_2^2)]\tau}{\sigma_2\sqrt{1-\rho^2}\sqrt{\tau}}\right)\\ &= \frac{\ln(S_2/S_1)+(q_1-q_2-\frac{1}{2}\nu^2)\tau}{\nu\sqrt{\tau}}.\end{aligned}$$

Inserting these expressions into (13.59) finally gives the option pricing formula

$$v(\tau,S_1,S_2) = S_2\mathrm{e}^{-q_2\tau}\mathcal{N}\left(d_+(\frac{S_2}{S_1},\tau)\right) - S_1\mathrm{e}^{-q_1\tau}\mathcal{N}\left(d_-(\frac{S_2}{S_1},\tau)\right) \tag{13.60}$$

where

$$d_\pm(x,\tau) := \frac{\ln x+(q_1-q_2\pm\frac{1}{2}\nu^2)\tau}{\nu\sqrt{\tau}},\quad x,\tau>0. \tag{13.61}$$

The delta (hedging) positions in the two stocks can be computed by directly differentiating (13.60) w.r.t. S_1 and S_2 while adapting the algebraic identity in (12.35). We now give an alternate derivation of the hedging positions by exploiting the symmetry of the pricing function. In particular, note that the pricing function in (13.60) can be expressed as a product of S_1 and a function of $x := S_2/S_1$:

$$v(\tau,S_1,S_2) = S_1\left[\mathrm{e}^{-q_2\tau}x\mathcal{N}(d_+(x,\tau)) - \mathrm{e}^{-q_1\tau}\mathcal{N}(d_-(x,\tau))\right] \equiv S_1C(x,1,\tau;q_1,q_2,\nu) \tag{13.62}$$

where we identify $C(x,1,\tau;q_1,q_2,\nu)$ as the Black–Scholes pricing function for a standard European call option on a single underlying with effective "spot" x, "strike" 1, time to maturity τ, "interest rate" q_1, and "dividend" q_2. Hence, at calendar time $t=T-\tau$, the hedging position $\Delta^1(t,S_1,S_2)$ in the first stock is readily derived by taking the partial derivative w.r.t. S_1 while using the known relation for $\Delta_c = \frac{\partial C}{\partial x} = \mathrm{e}^{-q_2\tau}\mathcal{N}(d_+(x,\tau))$ and $S_1\frac{\partial x}{\partial S_1} = -x$:

$$\begin{aligned}\Delta^1(t,S_1,S_2) = \frac{\partial v}{\partial S_1} = \frac{\partial}{\partial S_1}\left[S_1C(x,1,\tau;q_1,q_2,\nu)\right] &= C(x,1,\tau;q_1,q_2,\nu)+S_1\frac{\partial x}{\partial S_1}\Delta_c\\ &= C(x,1,\tau;q_1,q_2,\nu)-x\Delta_c\\ &= -\mathrm{e}^{-q_1\tau}\mathcal{N}(d_-(x,\tau))\,.\end{aligned} \tag{13.63}$$

The hedging position in the second stock is computed by taking the partial derivative w.r.t. S_2 of (13.62) while using $S_1 \frac{\partial x}{\partial S_2} = 1$,

$$\Delta^2(t, S_1, S_2) = \frac{\partial v}{\partial S_2} = \frac{\partial}{\partial S_2}\left[S_1 C(x, 1, \tau; q_1, q_2, \nu)\right] = S_1 \frac{\partial x}{\partial S_2} \Delta_c = \mathrm{e}^{-q_2 \tau} \mathcal{N}\big(d_+(x, \tau)\big) . \tag{13.64}$$

We note that for $t = T$ ($\tau = 0$) the respective positions are given by the limit $t \nearrow T$ ($\tau = 0+$) of the above expressions. Hence, from the positions (13.63) and (13.64) we see that the *stock portion of the self-financing replicating strategy completely replicates the exchange option.* That is, no bank account is needed (i.e., the position in the bank account $\beta(t) \equiv 0$) in order to replicate the option. The pricing function equals the value of the self-financing stock portfolio,

$$v(\tau, S_1, S_2) = S_1 \Delta^1(t, S_1, S_2) + S_2 \Delta^2(t, S_1, S_2). \tag{13.65}$$

As a stochastic price process we have $V_t = v(\tau, S_1(t), S_2(t)) = \Delta_t^1 S_1(t) + \Delta_t^2 S_2(t)$, where $\Delta_t^i = \Delta^i(t, S_1(t), S_2(t))$, $i = 1, 2$.

A European-style basket option whose payoff at some expiry date T is given by the maximum or minimum share price between two or more stocks can also be priced by the above methods. For example, in the case of the minimum price of two stocks at expiry T the payoff takes on the equivalent forms

$$V_T = \min\left(S_1(T), S_2(T)\right) = S_1(T) - \left(S_1(T) - S_2(T)\right)^+ = S_2(T) - \left(S_2(T) - S_1(T)\right)^+ . \tag{13.66}$$

Similarly, for the case of the maximum price of two stocks, the payoff is

$$V_T = \max\left(S_1(T), S_2(T)\right) = \left(S_2(T) - S_1(T)\right)^+ + S_1(T) = \left(S_1(T) - S_2(T)\right)^+ + S_2(T). \tag{13.67}$$

The expressions in (13.66) and (13.67) follow from $\max(x, y) = (y - x)^+ + x = (x - y)^+ + y$ and $\min(x, y) = x - (x - y)^+ = y - (y - x)^+$. We also make note of a related useful identity: $\min(x, y) + \max(x, y) = x + y$. As shown just below, the main point of (13.66) and (13.67) is that the problem of pricing a two-stock option with the payoff specified as either the minimum or maximum is immediately solved once we have priced the two-stock option with payoff in (13.53). The converse is also true. The pricing formula for the latter option was derived in the previous section with the explicit expression given in (13.60).

As in the previous section, let S_1, S_2 be the spot values, $\tau = T - t$ be the time to maturity, and let $v_{min}(\tau, S_1, S_2)$ denote the time-t price of the two-stock option on the minimum with payoff in (13.66). Then, by risk-neutral pricing:

$$\begin{aligned} v_{min}(\tau, S_1, S_2) &= \mathrm{e}^{-r\tau} \widetilde{\mathrm{E}}_{t,S_1,S_2}\left[\min\left(S_1(T), S_2(T)\right)\right] \\ &= \mathrm{e}^{-r\tau} \widetilde{\mathrm{E}}_{t,S_1,S_2}\left[S_2(T)\right] - \mathrm{e}^{-r\tau} \widetilde{\mathrm{E}}_{t,S_1,S_2}\left[\left(S_2(T) - S_1(T)\right)^+\right]. \end{aligned} \tag{13.68}$$

The second term in this last equation is simply the price $v(\tau, S_1, S_2)$ (see (13.55)) given by (13.60). The first term is readily evaluated by substituting the representation in (13.54) and using the independence of the Brownian increment $(\widetilde{\mathbf{W}}(T) - \widetilde{\mathbf{W}}(t))$ and the pair $(S_1(t), S_2(t))$, which are only functions of $\widetilde{\mathbf{W}}(t)$, i.e., $S_i(t) = S_i(0)\,\mathrm{e}^{(r - q_1 - \frac{1}{2}\sigma_i^2)t + \boldsymbol{\sigma}_i \cdot \widetilde{\mathbf{W}}(t)}$, $i = 1, 2$. Hence,

$$\begin{aligned} \widetilde{\mathrm{E}}_{t,S_1,S_2}\left[S_2(T)\right] &\equiv \widetilde{\mathrm{E}}\left[S_2(T) \mid S_1(t) = S_1, S_2(t) = S_2\right] \\ &= S_2\,\mathrm{e}^{(r - q_2 - \frac{1}{2}\sigma_2^2)\tau}\, \widetilde{\mathrm{E}}\left[\mathrm{e}^{\boldsymbol{\sigma}_2 \cdot (\widetilde{\mathbf{W}}(T) - \widetilde{\mathbf{W}}(t))} \mid S_1(t) = S_1, S_2(t) = S_2\right] \\ &= S_2\,\mathrm{e}^{(r - q_2 - \frac{1}{2}\sigma_2^2)\tau}\, \widetilde{\mathrm{E}}\left[\mathrm{e}^{\boldsymbol{\sigma}_2 \cdot (\widetilde{\mathbf{W}}(T) - \widetilde{\mathbf{W}}(t))}\right] \text{ (by independence)} \\ &= S_2\,\mathrm{e}^{(r - q_2 - \frac{1}{2}\sigma_2^2)\tau}\,\mathrm{e}^{\frac{1}{2}\sigma_2^2 \tau} = S_2\,\mathrm{e}^{(r - q_2)\tau} . \end{aligned}$$

The last line follows by the m.g.f. of $\boldsymbol{\sigma}_2 \cdot (\widetilde{\mathbf{W}}(T) - \widetilde{\mathbf{W}}(t)) \sim Norm(0, \sigma_2^2 \tau)$. We remark here that this conditional expectation also follows directly from the fact that the process $\{e^{(q_2-r)t} S_2(t)\}_{t\geqslant 0}$ is a $\widetilde{\mathbb{P}}$-martingale, and combining this with the Markov property of the joint process $\{(S_1(t), S_2(t))\}_{t\geqslant 0}$ gives

$$e^{(q_2-r)t} S_2(t) = \widetilde{E}[e^{(q_2-r)T} S_2(T) \mid \mathcal{F}_t] = \widetilde{E}[e^{(q_2-r)T} S_2(T) \mid S_1(t), S_2(t)] .$$

Setting $(S_1(t), S_2(t)) = (S_1, S_2)$ and grouping the exponential terms gives the above result: $\widetilde{E}\big[S_2(T) \mid S_1(t) = S_1, S_2(t) = S_2\big] = S_2 e^{(r-q_2)(T-t)}$. Hence, for each stock we have $\widetilde{E}_{t,S_1,S_2}\big[S_i(T)\big] = S_i e^{(r-q_i)\tau}$, $i = 1, 2$.

Substituting this into (13.68) gives us the explicit pricing formula

$$\begin{aligned} v_{min}(\tau, S_1, S_2) &= e^{-q_1\tau} S_1 \mathcal{N}\Big(d_-(\frac{S_2}{S_1}, \tau)\Big) + e^{-q_2\tau} S_2 \Big[1 - \mathcal{N}\Big(d_+(\frac{S_2}{S_1}, \tau)\Big)\Big] \\ &= e^{-q_1\tau} S_1 \mathcal{N}\Big(d_-(\frac{S_2}{S_1}, \tau)\Big) + e^{-q_2\tau} S_2 \mathcal{N}\Big(-d_+(\frac{S_2}{S_1}, \tau)\Big) \end{aligned} \tag{13.69}$$

with $d_\pm(x, \tau)$ defined in (13.61). In the last equation line we used the symmetry relation $\mathcal{N}(x) + \mathcal{N}(-x) = 1$.

Since $\max(S_1(T), S_2(T)) = S_1(T) + S_2(T) - \min(S_1(T), S_2(T))$, the value of the option with payoff in (13.67) follows simply by (13.69):

$$\begin{aligned} v_{max}(\tau, S_1, S_2) &= e^{-r\tau} \widetilde{E}_{t,S_1,S_2}\big[\max(S_1(T), S_2(T))\big] \\ &= e^{-r\tau} \widetilde{E}_{t,S_1,S_2}\big[S_1(T)\big] + e^{-r\tau} \widetilde{E}_{t,S_1,S_2}\big[S_2(T)\big] \\ &\quad - e^{-r\tau} \widetilde{E}_{S_1,S_2,t}\big[\min(S_1(T), S_2(T))\big] \\ &= e^{-q_1\tau} S_1 + e^{-q_2\tau} S_2 - v_{min}(\tau, S_1, S_2) \\ &= e^{-q_1\tau} S_1 \mathcal{N}\Big(-d_-(\frac{S_2}{S_1}, \tau)\Big) + e^{-q_2\tau} S_2 \mathcal{N}\Big(d_+(\frac{S_2}{S_1}, \tau)\Big) . \end{aligned} \tag{13.70}$$

Notice that the expressions in (13.69) and (13.70) are invariant with respect to interchanging all subscripts $1 \leftrightarrow 2$ on the dividends, volatilities, and spot values. This must be the case since the payoffs in (13.66) and (13.67) are invariant to the interchange $S_1(T) \leftrightarrow S_2(T)$. We leave it as an exercise for the reader (see Exercise 13.2) to derive expressions for the delta positions for the above two options on the maximum and minimum and to show that the relation in (13.65) holds, i.e., only the stocks are needed in the self-financing replicating strategies.

We now present a useful identity that can be used to more readily derive pricing functions for options whose payoff has a certain type of dependence on the terminal value of two correlated GBM processes. In particular, the payoffs in either (13.53), (13.66) or (13.67) can all be represented as a linear combination of *elemental payoffs*: $S_1(T)$, $S_2(T)$, $S_2(T)\mathbb{I}_{\{S_2(T)\geqslant S_1(T)\}}$, and $S_1(T)\mathbb{I}_{\{S_1(T)>S_2(T)\}}$ where $S_1(T)$ and $S_2(T)$ are correlated log-normal random variables. Other examples of payoffs that involve similar indicator functions in the terminal values of GBM processes occur when pricing foreign exchange options, as seen in Section 13.2.3.

Consider two log-normal random variables X and Y, represented by

$$X = x e^{(\mu_X - \frac{1}{2}\sigma_X^2)\tau + \sigma_X \sqrt{\tau} Z_1}, \quad Y = y e^{(\mu_Y - \frac{1}{2}\sigma_Y^2)\tau + \sigma_Y \sqrt{\tau} Z_2}$$

with positive parameters $x, y, \sigma_X, \sigma_Y, \tau$ and where Z_1, Z_2 are jointly bivariate standard normals (in a given measure $\mathbb{P}$) with $\mathrm{Corr}(Z_1, Z_2) = \rho$. Then, we have the following expectation formula (under measure $\mathbb{P}$):

$$\mathrm{E}\left[X\,\mathbb{I}_{\{X>Y\}}\right] = x\,\mathrm{e}^{\mu_X\tau}\,\mathcal{N}\left(\frac{\ln\frac{x}{y} + (\mu_X - \mu_Y + \frac{1}{2}\nu^2)\tau}{\nu\sqrt{\tau}}\right) \tag{13.71}$$

where $\nu^2 = \sigma_X^2 + \sigma_Y^2 - 2\rho\sigma_X\sigma_Y$. This identity is equivalent to considering any two correlated GBM processes

$$X(t) = X(0)\,\mathrm{e}^{(\mu_X - \frac{1}{2}\sigma_X^2)t + \boldsymbol{\sigma}^{(X)}\cdot\mathbf{W}(t)}, \quad Y(t) = Y(0)\,\mathrm{e}^{(\mu_Y - \frac{1}{2}\sigma_Y^2)t + \boldsymbol{\sigma}^{(Y)}\cdot\mathbf{W}(t)}, \quad t \geqslant 0,$$

where $\mathbf{W}(t)$ is a standard d-dimensional $\mathbb{P}$-BM, with $d \geqslant 2$, $\sigma_X = \|\boldsymbol{\sigma}^{(X)}\|$, $\sigma_Y = \|\boldsymbol{\sigma}^{(Y)}\|$, $\boldsymbol{\sigma}^{(X)} \cdot \boldsymbol{\sigma}^{(Y)} = \rho\sigma_X\sigma_Y$, and $\nu^2 = \|\boldsymbol{\sigma}^{(X)} - \boldsymbol{\sigma}^{(Y)}\|^2 = \sigma_X^2 + \sigma_Y^2 - 2\rho\sigma_X\sigma_Y$ with correlation coefficient ρ. Then, (13.71) is equivalent to the following conditional expectation formula:

$$\begin{aligned}\mathrm{E}_{t,x,y}\left[X(T)\,\mathbb{I}_{\{X(T)>Y(T)\}}\right] &\equiv \mathrm{E}\left[X(T)\,\mathbb{I}_{\{X(T)>Y(T)\}} \mid X(t) = x, Y(t) = y\right] \\ &= x\,\mathrm{e}^{\mu_X\tau}\,\mathcal{N}\left(\frac{\ln\frac{x}{y} + (\mu_X - \mu_Y + \frac{1}{2}\nu^2)\tau}{\nu\sqrt{\tau}}\right)\end{aligned} \tag{13.72}$$

where $\tau = T - t$.

Example 13.1. Use the identity in (13.72) to derive the pricing function $v_{max}(\tau, S_1, S_2)$ in (13.70).

Solution. The payoff has the form

$$V_T = \max\left(S_1(T), S_2(T)\right) = S_1(T)\mathbb{I}_{\{S_1(T)>S_2(T)\}} + S_2(T)\mathbb{I}_{\{S_2(T)\geqslant S_1(T)\}}\,.$$

We need only derive the price for the payoff $V_T^{(1)} \equiv S_1(T)\mathbb{I}_{\{S_1(T)>S_2(T)\}}$ since, by symmetry, the price for the second portion of the payoff obtains by interchanging the roles of the two stocks. That is, after deriving the expression for the pricing function $v^{(1)}(\tau, S_1, S_2)$ for payoff $V_T^{(1)}$ we interchange $S_1 \leftrightarrow S_2, \sigma_1 \leftrightarrow \sigma_2$, $q_1 \leftrightarrow q_2$ to obtain the pricing function $v^{(2)}(\tau, S_1, S_2)$ for payoff $V_T^{(2)} \equiv S_2(T)\mathbb{I}_{\{S_2(T)\geqslant S_1(T)\}}$. Finally, by linearity, we add the two pricing functions to obtain $v_{max}(\tau, S_1, S_2)$.

For payoff $V_T^{(1)}$ we have, by risk-neutral pricing,

$$v^{(1)}(\tau, S_1, S_2) = \mathrm{e}^{-r\tau}\widetilde{\mathrm{E}}_{t,S_1,S_2}\left[S_1(T)\mathbb{I}_{\{S_1(T)>S_2(T)\}}\right]$$

where the stock prices are two correlated GBM processes as in (13.41), i.e. as in (13.54). Hence, this conditional expectation is exactly of the form in (13.72), where now we are in the $\widetilde{\mathbb{P}}$ measure. We can therefore directly apply (13.72) once we identify the processes and the corresponding parameters. In this case, we identify the processes $X(t) = S_1(t), Y(t) = S_2(t)$, the spot variables $x = S_1, y = S_2$, the log-drift parameters $\mu_X = r - q_1, \mu_Y = r - q_2$, and log-volatility vectors $\boldsymbol{\sigma}^{(X)} = \boldsymbol{\sigma}_1 = [\sigma_1, 0], \boldsymbol{\sigma}^{(Y)} = \boldsymbol{\sigma}_2 = [\sigma_2\rho, \sigma_2\sqrt{1-\rho^2}]$, $\sigma_X = \sigma_1, \sigma_Y = \sigma_2$ where $\nu^2 = \|\boldsymbol{\sigma}_1 - \boldsymbol{\sigma}_2\|^2 = \sigma_1^2 + \sigma_2^2 - 2\rho\sigma_1\sigma_2$. Hence, (13.72) gives

$$\begin{aligned}v^{(1)}(\tau, S_1, S_2) &= \mathrm{e}^{-r\tau}S_1\,\mathrm{e}^{(r-q_1)\tau}\,\mathcal{N}\left(\frac{\ln\frac{S_1}{S_2} + ((r-q_1) - (r-q_2) + \frac{1}{2}\nu^2)\tau}{\nu\sqrt{\tau}}\right) \\ &= S_1\,\mathrm{e}^{-q_1\tau}\,\mathcal{N}\left(-d_-(\frac{S_2}{S_1}, \tau)\right)\end{aligned} \tag{13.73}$$

with $d_\pm(x,\tau)$ defined in (13.61). For the payoff $V_T^{(2)}$, the pricing function is given by applying the same identity or simply interchanging subscripts $1 \leftrightarrow 2$, giving (note that ν^2 remains the same)

$$v^{(2)}(\tau, S_1, S_2) = S_2\,\mathrm{e}^{-q_2\tau}\,\mathcal{N}\left(\frac{\ln\frac{S_2}{S_1} + (q_1 - q_2 + \frac{1}{2}\nu^2)\tau}{\nu\sqrt{\tau}}\right) = S_2\,\mathrm{e}^{-q_2\tau}\,\mathcal{N}\left(d_+(\frac{S_2}{S_1},\tau)\right). \tag{13.74}$$

Adding the pricing functions for the two elemental payoffs gives the previously derived price in (13.70),

$$v_{max}(\tau, S_1, S_2) = v^{(1)}(\tau, S_1, S_2) + v^{(2)}(\tau, S_1, S_2).$$

□

The above exercise shows that the identity in (13.72) immediately gives us the pricing functions for the elemental payoffs. The payoff in (13.53) for the exchange option is also a linear combination of elemental payoffs, e.g.,

$$\begin{aligned}(\,S_2(T) - S_1(T)\,)^+ &= \max(S_1(T), S_2(T)) - S_1(T)\\ &= S_1(T)\mathbb{I}_{\{S_1(T)>S_2(T)\}} + S_2(T)\mathbb{I}_{\{S_2(T)\geqslant S_1(T)\}} - S_1(T).\end{aligned}$$

Hence, the pricing function in (13.60) for this payoff is obtained immediately by adding the pricing functions in (13.73) and (13.74) and subtracting $S_1\mathrm{e}^{-q_1\tau}$.

In the next example we reconsider pricing the exchange option by solving the two-stock BSPDE problem. However, we do not solve the BSPDE by using the two-dimensional Feynman–Kac formula since this brings us right back to our previous methods. The key step in the method is to write the solution (i.e. the pricing function) in a form that reduces the original BSPDE in (13.51) to a *lower dimensional* PDE problem that is more readily solved. The functional form for the pricing function is dictated by the symmetry of the payoff function with respect to the spot variables. The lower dimensional PDE is then essentially like solving a simple pricing problem for a single asset with an effective payoff. This method is also useful in higher dimensions involving three or more stocks, assuming that the payoff has some simplifying (factoring) symmetry. The methodology is best demonstrated by example, as follows.

Example 13.2. (Symmetry Reduction in BSPDE) Derive the pricing formula for the exchange option with payoff function $\Lambda(S_1, S_2) = (S_2 - S_1)^+$ by solving the BSPDE in (13.51) while employing a symmetry reduction.

Solution. We note the symmetry of the payoff function, which can be written as a product of $S_2 > 0$ and a function of the ratio $x := S_1/S_2 > 0$, or as a product of S_1 and a function of the ratio S_2/S_1,

$$\Lambda(S_1, S_2) = (S_2 - S_1)^+ = S_2(1 - S_1/S_2)^+ = S_2(1 - x)^+. \tag{13.75}$$

[Note: we can also write $(S_2 - S_1)^+ = S_1(S_2/S_1 - 1)^+$.] Hence, the pricing function for the terminal value of time $t = T$ is given by $V(T, S_1, S_2) = S_2(1 - x)^+$, i.e. for $t = T$ it is in fact a product of the spot variable S_2 and a function of variable $x = S_1/S_2$. We therefore make an "Ansatz" and *seek a solution for all* $t \leqslant T$ in the form of a product of the variable S_2 and some function $f(t, x)$:

$$V(t, S_1, S_2) = S_2 f(t, x) = S_2 f(t, S_1/S_2). \tag{13.76}$$

By construction, for $t = T$ this relation holds by combining (13.76) and (13.75):

$$V(T, S_1, S_2) = S_2 f(T, x) = S_2(1 - x)^+ \implies f(T, x) = (1 - x)^+ \equiv \phi(x). \tag{13.77}$$

The condition on the right is viewed as a terminal (effective payoff) condition $f(T,x) = \phi(x)$.

The main step is now to substitute the form in (13.76) into (13.51) and compute all partial derivative terms. The form for the solution is justified once the BSPDE is *simplified into a PDE in only the variables t, x and the function $f(t,x)$*. The S_2 variable should factor out completely; otherwise, either the form for the solution is not correct or some errors were made, such as in the calculation of the derivative terms. The partial derivatives in (13.51) are calculated by simply using the chain rule of ordinary calculus on the pricing function V in (13.76) and using $\frac{\partial x}{\partial S_2} = -\frac{S_1}{S_2^2} = -\frac{x}{S_2}$, $\frac{\partial x}{\partial S_1} = \frac{1}{S_2}$. For the first partial derivatives:

$$\frac{\partial V}{\partial S_2} = \frac{\partial}{\partial S_2}(S_2 f) = f + S_2 \frac{\partial x}{\partial S_2}\frac{\partial f}{\partial x} = f - x\frac{\partial f}{\partial x}, \quad \frac{\partial V}{\partial S_1} = S_2 \frac{\partial x}{\partial S_1}\frac{\partial f}{\partial x} = \frac{\partial f}{\partial x},$$

and $\frac{\partial V}{\partial t} = S_2 \frac{\partial f}{\partial t}$. For the second partial derivatives:

$$\begin{aligned}
\frac{\partial^2 V}{\partial S_1 \partial S_2} &= \frac{\partial}{\partial S_1}\left(\frac{\partial V}{\partial S_2}\right) = \frac{\partial x}{\partial S_1}\frac{\partial}{\partial x}\left(f - x\frac{\partial f}{\partial x}\right) = -\frac{x}{S_2}\frac{\partial^2 f}{\partial x^2} \\
\frac{\partial^2 V}{\partial S_1^2} &= \frac{\partial}{\partial S_1}\left(\frac{\partial V}{\partial S_1}\right) = \frac{\partial x}{\partial S_1}\frac{\partial^2 f}{\partial x^2} = \frac{1}{S_2}\frac{\partial^2 f}{\partial x^2} \\
\frac{\partial^2 V}{\partial S_2^2} &= \frac{\partial}{\partial S_2}\left(\frac{\partial V}{\partial S_2}\right) = \frac{\partial x}{\partial S_2}\frac{\partial}{\partial x}\left(f - x\frac{\partial f}{\partial x}\right) = \frac{x^2}{S_2}\frac{\partial^2 f}{\partial x^2}
\end{aligned}$$

Substituting all respective terms into the BSPDE in (13.51) leads to

$$\begin{aligned}
S_2\frac{\partial f}{\partial t} &+ \frac{1}{2}\sigma_1^2 S_1^2 \frac{1}{S_2}\frac{\partial^2 f}{\partial x^2} + \frac{1}{2}\sigma_2^2 S_2^2 \frac{x^2}{S_2}\frac{\partial^2 f}{\partial x^2} - \rho\sigma_1\sigma_2 S_1 S_2 \frac{x}{S_2}\frac{\partial^2 f}{\partial x^2} \\
&+ (r - q_1)S_1\frac{\partial f}{\partial x} + (r - q_2)S_2\left(f - x\frac{\partial f}{\partial x}\right) - rS_2 f = 0\,.
\end{aligned}$$

We can finally simplify this equation by factoring out S_2 in all terms and substituting x for $\frac{S_1}{S_2}$. Some terms cancel out and we can group together all terms in $x^2\frac{\partial^2 f}{\partial x^2}$ and $x\frac{\partial f}{\partial x}$. What we obtain is in fact a PDE for f in the variables t, x:

$$\frac{\partial f}{\partial t} + \frac{1}{2}\nu^2 x^2 \frac{\partial^2 f}{\partial x^2} + (q_2 - q_1)x\frac{\partial f}{\partial x} - q_2 f = 0\,, \tag{13.78}$$

where $\nu^2 := \sigma_1^2 + \sigma_2^2 - 2\rho\sigma_1\sigma_2$, subject to the terminal condition in (13.77): $f(T,x) = (1-x)^+$.

The PDE in (13.78) is in only one spatial dimension instead of two. It can be solved by inspection at this point! This is because the PDE is a one-dimensional BSPDE for a GBM process corresponding to a single "asset or stock" with effective "spot variable" x, effective volatility parameter ν, effective "interest rate" q_2, and effective "stock dividend" q_1. The effective payoff function $\phi(x) = (1-x)^+$ is that of a put with strike $K = 1$. Hence, the function is given by the Black–Scholes pricing function for a put option on a dividend paying stock in (12.56) where we set $S = x$, $K = 1$, $\tau = T - t$, $r = q_2$, $\sigma = \nu$, and $q = q_1$:

$$f(t,x) = \mathrm{e}^{-q_2\tau}\mathcal{N}\left(-\frac{\ln x + (q_2 - q_1 - \frac{1}{2}\nu^2)\tau}{\nu\sqrt{\tau}}\right) - \mathrm{e}^{-q_1\tau}x\,\mathcal{N}\left(-\frac{\ln x + (q_2 - q_1 + \frac{1}{2}\nu^2)\tau}{\nu\sqrt{\tau}}\right). \tag{13.79}$$

[Alternatively this solution is found by using the one-dimensional discounted Feynman–Kac formula. We leave it to the reader to show this by using $f(t,x) = \mathrm{e}^{-q_2(T-t)}\mathrm{E}_{t,x}[\phi(X(T))]$ where $X(t)$ is the GBM process corresponding to the PDE in (13.78).]

The pricing function follows by using (13.79), with $x = S_1/S_2$, into (13.76):

$$V(t, S_1, S_2) = S_2 f(t, S_1/S_2) = S_2 \mathrm{e}^{-q_2\tau}\mathcal{N}\Big(d_+\big(\frac{S_2}{S_1}, \tau\big)\Big) - S_1 \mathrm{e}^{-q_1\tau}\mathcal{N}\Big(d_-\big(\frac{S_2}{S_1}, \tau\big)\Big) \quad (13.80)$$

Of course, this is the same expression as in (13.60) with $d_\pm$ defined in (13.61). □

13.2.2.2 Other Basket Options

The methods of the previous section can also be used to derive pricing functions for other standard European options where the payoff is a function of the terminal values of two or more stock prices. Within the multidimensional GBM model we know that this generally involves a multivariate integral as given by (13.46). For the case of exchange-type options on two stocks and for other similar payoffs, we saw that the calculations are simplified to single dimensional integrals and the resulting pricing functions involve the standard normal CDF. This simplification is not possible for other types of payoffs on two stocks involving some fixed strike level. One example is the so-called *chooser max call* option on two stocks with payoff

$$V_T = \big(\max\{S_1(T), S_2(T)\} - K\big)^+. \quad (13.81)$$

A related option is a *chooser min put* option on two stocks with payoff

$$V_T = \big(K - \min\{S_1(T), S_2(T)\}\big)^+. \quad (13.82)$$

Both options are worth more than the corresponding call or put on either single underlying with strike $K > 0$. The valuation of these options leads to explicit expressions involving the bivariate standard normal CDF. We leave these as Exercises 13.13 and 13.14 at the end of this chapter. Another type of two-stock option is a so-called *spread option* where the payoff can be call-like or put-like on the difference of the terminal values of two assets. For a call spread we have

$$V_T = \big(S_2(T) - S_1(T) - K\big)^+ \quad (13.83)$$

for a given strike $K > 0$. The pricing function for this option does not reduce to a combination of any known functions such as standard normal CDFs. However, the pricing function can still be written as an integral. In the limit of small strike $K \searrow 0$ this option becomes the simpler exchange option with payoff in (13.53).

Certain types of options whose payoff depends on the terminal values of three or more stocks can also be priced analytically. The derivation depends on some simplifying symmetry of the payoff function. In some cases the pricing function can be derived as an explicit expression involving the bivariate standard normal CDF. We refer the reader to some exercises at the end of this chapter.

13.2.3 Cross-Currency Option Valuation

We now consider the pricing of equity options whose payoff in domestic currency involves the prices of one or more assets denominated in a foreign currency as well as (possibly) the prices of domestic assets. These options are therefore subject to currency risk as well as foreign equity risk. For example, a *quanto option* refers generally to an option on some asset denominated in foreign currency and whose payoff is in domestic currency. There are several types of such options and their payoffs can be quite complex and path dependent. Four simple examples of call-like foreign exchange options (having obvious put-like analogues) include the following.

1. Foreign Equity Call struck in foreign currency: This is a call on a foreign asset (stock) S^f with a strike price K_f, both denominated in foreign currency, which is converted to domestic currency at the terminal value of the exchange rate $X(T)$ with payoff

$$C_T = X(T)\,(S^f(T) - K_f)^+ . \tag{13.84}$$

2. Foreign Equity Call struck in domestic currency: This is a call on a domestically converted foreign asset XS^f with a domestic strike price K with payoff

$$C_T = (X(T)S^f(T) - K)^+ . \tag{13.85}$$

3. Fixed Foreign Equity Rate Call: This is like the first call above, except that the foreign exchange rate is fixed to some preassigned value, say $\bar{X}$. The payoff is

$$C_T = \bar{X}\,(S^f(T) - K_f)^+ . \tag{13.86}$$

4. (Elf-X) Equity Linked Foreign Exchange Call: The holder has the right to purchase a foreign asset S^f by placing a lower value bound κ on the exchange rate for converting the asset value to domestic currency. The payoff is

$$C_T = S^f(T)(X(T) - \kappa)^+ . \tag{13.87}$$

Consider two markets or economies – a domestic one with assets denominated in domestic currency and a foreign one with assets denominated in a foreign currency. Examples of currencies are USD, CAD, EUR, GBP, JPY, etc. We therefore have two bank accounts, where one domestic dollar invested in the domestic account is worth $B(t) = e^{\int_0^t r(u)\,du}$ in domestic currency and one foreign dollar invested in the foreign account has value $B^f(t) = e^{\int_0^t r^f(u)\,du}$ in foreign currency. Note that we are still denoting the domestic interest rate process by $\{r(t)\}_{t\geqslant 0}$ while the foreign interest rate process is denoted by $\{r^f(t)\}_{t\geqslant 0}$. We first assume these processes are adapted and later simplify the model to make them constants or nonrandom functions of time so that we can derive simple pricing functions since we will work within the GBM model for all assets. Let $\{S^f(t)\}_{t\geqslant 0}$ be the price process for any foreign asset or stock and let $\{X^{f\to d}(t) \equiv X(t)\}_{t\geqslant 0}$ be the exchange rate process where $X(t)$ is the time-t exchange rate for converting an asset denominated in *foreign currency into domestic currency.* Hence, X has units of (domestic currency)/(foreign currency), e.g., CAD/USD, USD/CAD, USD/GBP, etc. In the above four examples we only have one foreign asset. However, we can and do allow for option payoffs involving multiple assets in both market currencies.

We assume the existence of a risk-neutral measure $\widetilde{\mathbb{P}} \equiv \widetilde{\mathbb{P}}^{(B)}$ for the domestic market with the domestic bank account as the numéraire asset. The exchange rate process is assumed to satisfy the SDE

$$dX(t) = X(t)\left[\widetilde{\mu}_X(t)\,dt + \boldsymbol{\sigma}^{(X)}(t)\cdot d\widetilde{\mathbf{W}}(t)\right], \tag{13.88}$$

where $\boldsymbol{\sigma}^{(X)}(t)$ is an adapted log-volatility vector and where $\widetilde{\mu}_X(t)$ is an adapted log-drift coefficient in the $\widetilde{\mathbb{P}}$-measure. Similarly, the foreign asset is assumed to satisfy the SDE

$$dS^f(t) = S^f(t)\left[\widetilde{\mu}_S(t)\,dt + \boldsymbol{\sigma}^{(S)}(t)\cdot d\widetilde{\mathbf{W}}(t)\right], \tag{13.89}$$

where $\boldsymbol{\sigma}^{(S)}(t)$ is an adapted log-volatility vector and where $\widetilde{\mu}_S(t)$ is an adapted log-drift in the measure $\widetilde{\mathbb{P}}$. As in previous sections, $\widetilde{\mathbf{W}}(t)$ is a vector standard Brownian motion w.r.t.

measure $\widetilde{\mathbb{P}}$. We determine the drifts $\widetilde{\mu}_X(t)$ and $\widetilde{\mu}_S(t)$ just below. First, let us recall that in the $\widetilde{\mathbb{P}}$-measure, all domestic non-dividend paying assets must have log-drift equal to the domestic interest rate $r(t)$ or otherwise an arbitrage exists. If the asset pays a dividend, then the log-drift in the $\widetilde{\mathbb{P}}$-measure must equal the interest rate minus the dividend yield. Note that $X(t)S^f(t)$ is the time-t price of a foreign asset converted to *domestic currency* and tradable in the domestic market. Hence, the process $\{X(t)S^f(t)\}_{t\geqslant 0}$ evolves as a domestic asset,

$$\mathrm{d}(X(t)S^f(t)) = X(t)S^f(t)\left[(r(t) - q_S(t))\,\mathrm{d}t + \boldsymbol{\sigma}^{(XS)}(t)\cdot \mathrm{d}\widetilde{\mathbf{W}}(t)\right], \tag{13.90}$$

where $\boldsymbol{\sigma}^{(XS)}(t) = \boldsymbol{\sigma}^{(X)}(t) + \boldsymbol{\sigma}^{(S)}(t)$ is the log-volatility vector of the product process XS. We recall from the Itô product rule that the log-volatility vector of a product of two processes is the sum of the log-volatility vectors of the two processes. We have also included a dividend yield $q_S(t)$ on the stock. Any additional domestic asset, say A, with adapted dividend yield $q_A(t)$ and log-volatility vector $\boldsymbol{\sigma}^{(A)}(t)$ has price process $\{A(t)\}_{t\geqslant 0}$ in domestic currency satisfying a similar SDE,

$$\mathrm{d}A(t) = A(t)\left[(r(t) - q_A(t))\,\mathrm{d}t + \boldsymbol{\sigma}^{(A)}(t)\cdot \mathrm{d}\widetilde{\mathbf{W}}(t)\right]. \tag{13.91}$$

We now determine $\widetilde{\mu}_X(t)$ in (13.88) by using the fact that the foreign bank account investment converted to domestic currency, $X(t)B^f(t)$, must have log-drift $r(t)$. Applying the Itô product rule and the fact that $\mathrm{d}B^f(t) = r^f(t)B^f(t)\,\mathrm{d}t$ and $\mathrm{d}X(t)\,\mathrm{d}B^f(t) = 0$,

$$\begin{aligned}\frac{\mathrm{d}(X(t)B^f(t))}{X(t)B^f(t)} &= \frac{\mathrm{d}B^f(t)}{B^f(t)} + \frac{\mathrm{d}X(t)}{X(t)} \\ &= \left(r^f(t) + \widetilde{\mu}_X(t)\right)\,\mathrm{d}t + \boldsymbol{\sigma}^{(X)}(t)\cdot \mathrm{d}\widetilde{\mathbf{W}}(t).\end{aligned} \tag{13.92}$$

Hence, $r^f(t) + \widetilde{\mu}_X(t) = r(t) \implies \widetilde{\mu}_X(t) = r(t) - r^f(t)$. We have hence determined the drift coefficient in (13.88), giving

$$\mathrm{d}X(t) = X(t)\left[(r(t) - r^f(t))\,\mathrm{d}t + \boldsymbol{\sigma}^{(X)}(t)\cdot \mathrm{d}\widetilde{\mathbf{W}}(t)\right]. \tag{13.93}$$

Next, we determine $\widetilde{\mu}_S(t)$ in (13.89) by first using the Itô product rule and combining (13.93) and (13.89) with the relation $\frac{\mathrm{d}X(t)}{X(t)}\frac{\mathrm{d}S^f(t)}{S^f(t)} = \boldsymbol{\sigma}^{(X)}(t)\cdot\boldsymbol{\sigma}^{(S)}(t)\,\mathrm{d}t$:

$$\begin{aligned}\frac{\mathrm{d}(X(t)S^f(t))}{X(t)S^f(t)} &= \frac{\mathrm{d}X(t)}{X(t)} + \frac{\mathrm{d}S^f(t)}{S^f(t)} + \frac{\mathrm{d}X(t)}{X(t)}\frac{\mathrm{d}S^f(t)}{S^f(t)} \\ = (\widetilde{\mu}_S(t) + r(t) - r^f(t) &+ \boldsymbol{\sigma}^{(X)}(t)\cdot\boldsymbol{\sigma}^{(S)}(t)\,)\,\mathrm{d}t + (\boldsymbol{\sigma}^{(X)}(t) + \boldsymbol{\sigma}^{(S)}(t))\cdot \mathrm{d}\widetilde{\mathbf{W}}(t).\end{aligned} \tag{13.94}$$

Equating this SDE with that in (13.90) gives $\widetilde{\mu}_S(t) = r^f(t) - q_S(t) - \boldsymbol{\sigma}^{(X)}(t)\cdot\boldsymbol{\sigma}^{(S)}(t)$, i.e., this determines the drift coefficient in (13.89):

$$\mathrm{d}S^f(t) = S^f(t)\left[\left(r^f(t) - q_S(t) - \boldsymbol{\sigma}^{(X)}(t)\cdot\boldsymbol{\sigma}^{(S)}(t)\right)\mathrm{d}t + \boldsymbol{\sigma}^{(S)}(t)\cdot \mathrm{d}\widetilde{\mathbf{W}}(t)\right]. \tag{13.95}$$

The solution to the system of SDEs in (13.93), (13.95), and (13.91) can be used for risk-neutral pricing of foreign exchange options with a payoff that is generally a function of the terminal values of the two assets and the foreign exchange rate, $V_T = \Lambda(S^f(T), A(T), X(T))$. Since the discounted domestic value process of the foreign exchange option (discounted by the domestic bank account) is a $\widetilde{\mathbb{P}}$-martingale, the no-arbitrage price of such an option is given by (13.22) where:

$$V_t = \widetilde{\mathrm{E}}\left[\mathrm{e}^{-\int_t^T r(u)\,\mathrm{d}u}\,\Lambda(S^f(T), A(T), X(T)) \mid \mathcal{F}_t\right], \quad 0 \leqslant t \leqslant T. \tag{13.96}$$

For quanto options with payoffs of the form $V_T = \Lambda(X(T), S^f(T))$, such as in (13.84)–(13.87), only (13.93) and (13.95) are needed. If the payoff is a function of the product $X(T)S^f(T)$, then (13.90) can be used directly.

For analytical tractability we now adopt the GBM model where asset prices $\{A(t)\}_{t\geqslant 0}$ and $\{S^f(t)\}_{t\geqslant 0}$ and the exchange rate $\{X(t)\}_{t\geqslant 0}$ are GBM processes. We can therefore consider all coefficients to be nonrandom functions of time. Here we simply assume constant interest rates $r(t) = r, r^f(t) = r^f$, constant dividend yields $q_S(t) = q_S, q_A(t) = q_A$, and constant log-volatility vectors $\boldsymbol{\sigma}^{(X)}(t) = \boldsymbol{\sigma}^{(X)}, \boldsymbol{\sigma}^{(S)}(t) = \boldsymbol{\sigma}^{(S)}, \boldsymbol{\sigma}^{(A)}(t) = \boldsymbol{\sigma}^{(A)}$. As in the case of basket options where the underlying assets are all domestic stocks modelled as GBM processes, the pricing formulation and pricing functions that follow also extend in the same fairly straightforward manner if we make these nonrandom functions of time rather than constants. The dot products of the log-volatility vectors are

$$\boldsymbol{\sigma}^{(X)}\cdot\boldsymbol{\sigma}^{(S)} = \rho_{XS}\sigma_X\sigma_S\,,\quad \boldsymbol{\sigma}^{(X)}\cdot\boldsymbol{\sigma}^{(A)} = \rho_{XA}\sigma_X\sigma_A\,,\quad \boldsymbol{\sigma}^{(A)}\cdot\boldsymbol{\sigma}^{(S)} = \rho_{AS}\sigma_A\sigma_S\,, \tag{13.97}$$

with constant correlation coefficients $\rho_{XS}, \rho_{XA}, \rho_{AS}$ and vector magnitudes $\sigma_X \equiv \|\boldsymbol{\sigma}^{(X)}\|$, $\sigma_S \equiv \|\boldsymbol{\sigma}^{(S)}\|$, $\sigma_A \equiv \|\boldsymbol{\sigma}^{(A)}\|$. To compact notation it is also useful to define the vector magnitudes

$$\sigma_{XS} \equiv \|\boldsymbol{\sigma}^{(X)} + \boldsymbol{\sigma}^{(S)}\| = \sqrt{\sigma_X^2 + \sigma_S^2 + 2\rho_{XS}\sigma_X\sigma_S}\,,$$

$$\sigma_{XA} \equiv \|\boldsymbol{\sigma}^{(X)} + \boldsymbol{\sigma}^{(A)}\| = \sqrt{\sigma_X^2 + \sigma_A^2 + 2\rho_{XA}\sigma_X\sigma_A}\,,$$

and $\sigma_{AS} \equiv \|\boldsymbol{\sigma}^{(A)} + \boldsymbol{\sigma}^{(S)}\| = \sqrt{\sigma_A^2 + \sigma_S^2 + 2\rho_{AS}\sigma_A\sigma_S}$.

As GBM processes, the above SDEs take the form

$$\frac{\mathrm{d}X(t)}{X(t)} = (r - r^f)\,\mathrm{d}t + \boldsymbol{\sigma}^{(X)}\cdot \mathrm{d}\widetilde{\mathbf{W}}(t)\,, \tag{13.98}$$

$$\frac{\mathrm{d}S^f(t)}{S^f(t)} = (r^f - q_S - \rho_{XS}\sigma_X\sigma_S)\,\mathrm{d}t + \boldsymbol{\sigma}^{(S)}\cdot \mathrm{d}\widetilde{\mathbf{W}}(t)\,, \tag{13.99}$$

$$\frac{\mathrm{d}\big[X(t)S^f(t)\big]}{X(t)S^f(t)} = (r - q_S)\,\mathrm{d}t + (\boldsymbol{\sigma}^{(X)} + \boldsymbol{\sigma}^{(S)})\cdot \mathrm{d}\widetilde{\mathbf{W}}(t)\,, \tag{13.100}$$

$$\frac{\mathrm{d}A(t)}{A(t)} = (r - q_A)\,\mathrm{d}t + \boldsymbol{\sigma}^{(A)}\cdot \mathrm{d}\widetilde{\mathbf{W}}(t)\,. \tag{13.101}$$

From the unique solutions to the above linear SDEs we have

$$X(T) = X(t)\,\mathrm{e}^{(r-r^f-\frac{1}{2}\sigma_X^2)(T-t)+\boldsymbol{\sigma}^{(X)}\cdot(\widetilde{\mathbf{W}}(T)-\widetilde{\mathbf{W}}(t))}\,, \tag{13.102}$$

$$S^f(T) = S^f(t)\,\mathrm{e}^{(r^f-q_S-\rho_{XS}\sigma_X\sigma_S-\frac{1}{2}\sigma_S^2)(T-t)+\boldsymbol{\sigma}^{(S)}\cdot(\widetilde{\mathbf{W}}(T)-\widetilde{\mathbf{W}}(t))}\,, \tag{13.103}$$

$$X(T)S^f(T) = X(t)S^f(t)\,\mathrm{e}^{(r-q_S-\frac{1}{2}\sigma_{XS}^2)(T-t)+(\boldsymbol{\sigma}^{(X)}+\boldsymbol{\sigma}^{(S)})\cdot(\widetilde{\mathbf{W}}(T)-\widetilde{\mathbf{W}}(t))}\,, \tag{13.104}$$

$$A(T) = A(t)\,\mathrm{e}^{(r-q_A-\frac{1}{2}\sigma_A^2)(T-t)+\boldsymbol{\sigma}^{(A)}\cdot(\widetilde{\mathbf{W}}(T)-\widetilde{\mathbf{W}}(t))}\,. \tag{13.105}$$

By the joint Markov property of the processes, and nonrandom interest rates, the conditioning on the filtration $\mathcal{F}_t$ in (13.96) is simply replaced by conditioning on the triplet $S^f(t), A(t), X(t)$, i.e. $V_t = V(t, S^f(t), A(t), X(t))$. Let $S^f(t) = S, A(t) = A, X(t) = x$ be the time-t *spot values of the foreign and domestic stocks and the exchange rate.* Then, (13.96) gives the pricing function $V(t, S, A, x)$ as the conditional expectation

$$V(t, S, A, x) = \mathrm{e}^{-r(T-t)}\,\widetilde{\mathrm{E}}\big[\Lambda(S^f(T), A(T), X(T)) \mid S^f(t) = S, A(t) = A, X(t) = x\big]\,. \tag{13.106}$$

Given a payoff function $\Lambda(S, A, x)$, we can therefore derive pricing functions by computing this expectation by whatever means. For a quanto option with payoff function $\Lambda(S, x)$, i.e., $V_T = \Lambda(S^f(T), X(T))$, the pricing function $V(t, S, x)$ is given by

$$V(t, S, x) = \mathrm{e}^{-r(T-t)}\, \widetilde{\mathrm{E}}\big[\Lambda(S^f(T), X(T)) \mid S^f(t) = S, X(t) = x\big]\,. \tag{13.107}$$

As in the case of multi-asset pricing in a domestic economy where all assets are denominated in a single domestic market, we can now also apply the discounted Feynman–Kac Theorem 11.20 to obtain a corresponding BSPDE for the above option pricing functions. Consider the vector process $\mathbf{X}(t) \equiv (X_1(t), X_2(t)) := (S^f(t), X(t))\}_{t\geqslant 0}$ satisfying the system of two SDEs (13.98) and (13.99). We identify the constant log-drifts and log-volatility vectors of these two processes by $\mu_1 \equiv r^f - q_S - \rho_{XS}\sigma_X\sigma_S$, $\mu_2 \equiv r - r^f$, $\boldsymbol{\sigma}_1 \equiv \boldsymbol{\sigma}^{(S)}$ and $\boldsymbol{\sigma}_2 \equiv \boldsymbol{\sigma}^{(X)}$. The corresponding time-homogeneous generator for this pair of GBM processes in the spot variables $x_1 \equiv S, x_2 = x$ is then

$$\mathcal{G}_{(x_1,x_2)} := \frac{1}{2}\sum_{i=1}^{2}\sum_{j=1}^{2} \boldsymbol{\sigma}_i \cdot \boldsymbol{\sigma}_j\, x_i x_j \frac{\partial^2}{\partial x_i \partial x_j} + \sum_{i=1}^{2} \mu_i \frac{\partial}{\partial x_i}\,. \tag{13.108}$$

Writing out all the partial derivative terms explicitly and using the spot variable names S, x in the place of x_1, x_2, then, according to the Feynman–Kac Theorem 11.20, the pricing function $V = V(t, S, x)$ in (13.107) satisfies the PDE

$$\begin{aligned} &\frac{\partial V}{\partial t} + \frac{1}{2}\sigma_S^2 S^2 \frac{\partial^2 V}{\partial S^2} + \frac{1}{2}\sigma_X^2 x^2 \frac{\partial^2 V}{\partial x^2} + \rho_{XS}\sigma_S\sigma_X\, Sx \frac{\partial^2 V}{\partial S \partial x} \\ &+ (r^f - q_S - \rho_{XS}\sigma_X\sigma_S)\, S\frac{\partial V}{\partial S} + (r - r^f)\, x \frac{\partial V}{\partial x} - rV = 0\,, \end{aligned} \tag{13.109}$$

subject to the terminal (payoff) condition $V(T, S, x) = \Lambda(S, x)$. This is the BSPDE for a quanto option where the payoff is a function of the share price of the foreign stock and the foreign exchange rate.

We leave it to the reader to show that by a similar application of the Feynman–Kac Theorem 11.20 the pricing function $V = V(t, S, A, x)$ in (13.106) satisfies the BSPDE

$$\begin{aligned} &\frac{\partial V}{\partial t} + \frac{1}{2}\sigma_S^2 S^2 \frac{\partial^2 V}{\partial S^2} + \frac{1}{2}\sigma_A^2 A^2 \frac{\partial^2 V}{\partial A^2} + \frac{1}{2}\sigma_X^2 x^2 \frac{\partial^2 V}{\partial x^2} \\ &+ \rho_{XS}\sigma_S\sigma_X\, Sx \frac{\partial^2 V}{\partial S \partial x} + \rho_{AS}\sigma_A\sigma_S\, AS \frac{\partial^2 V}{\partial A \partial S} + \rho_{XA}\sigma_X\sigma_A\, Ax \frac{\partial^2 V}{\partial A \partial x} \\ &+ (r^f - q_S - \rho_{XS}\sigma_X\sigma_S)\, S\frac{\partial V}{\partial S} + (r - q_A)\, A\frac{\partial V}{\partial A} + (r - r^f)\, x \frac{\partial V}{\partial x} - rV = 0\,, \end{aligned} \tag{13.110}$$

subject to the terminal (payoff) condition $V(T, S, A, x) = \Lambda(S, A, x)$. The PDEs in (13.109) and (13.110) can be readily solved in cases where the payoff functions allow for a symmetry reduction to be employed. As an example, consider a quanto option having a payoff in the form of a product $V(T, S, x) = \Lambda(S, x) = xg(S)$. Then, the pricing function, solving (13.109) for all $t \leqslant T$, can be written in the form of a product, $V(t, S, x) = xf(t, S)$. By substituting this form into (13.109), computing all partial derivatives and factoring out the exchange spot variable x, the reader can verify that this leads to a reduced PDE for $f(t, S)$ in the variables (t, S) subject to the terminal condition $f(T, S) = g(S)$. The fundamental solution of the reduced PDE is readily obtained. Explicit examples of pricing quanto options by symmetry reduction of the above BSPDE in (13.109) are left as exercises at the end of this chapter.

13.3 Equivalent Martingale Measures: Derivative Pricing with General Numéraire Assets

Consider an arbitrage-free multi-asset market model within a given domestic economy, as described in Section 13.1. In particular, we assume that there exists a solution to the price of risk equations so that a risk-neutral measure $\widetilde{\mathbb{P}}$ exists. If $n = d$, where d is the number of BMs driving the base asset prices, then $\widetilde{\mathbb{P}}$ is unique and the market is complete. We shall keep the formulation more general with $n \leqslant d$ and such that a $\widetilde{\mathbb{P}}$ exists. Assuming no dividends on the assets, we recall that the measure $\widetilde{\mathbb{P}}$ has the defining property that all discounted base asset price processes $\{\frac{S_0}{B(t)}, \frac{S_1(t)}{B(t)}, \ldots, \frac{S_n(t)}{B(t)}\}_{t\geqslant 0}$ are $\widetilde{\mathbb{P}}$-martingales. Here, $S_0(t) := B(t)$ and we note that $S_0(t)/B(t) \equiv 1$ is a trivial martingale. We recall that (13.12)–(13.15) hold. That is, for each base asset $i = 1, \ldots, n$,

$$\mathrm{d}\left(\frac{S_i(t)}{B(t)}\right) = \frac{S_i(t)}{B(t)}\boldsymbol{\sigma}_i(t)\cdot \mathrm{d}\widetilde{\mathbf{W}}(t)\,. \tag{13.111}$$

Equivalently, each base asset price relative to the bank account value $B(t)$ is a $\widetilde{\mathbb{P}}$-martingale,

$$\frac{S_i(t)}{B(t)} = \frac{S_i(0)}{B(0)}\,\mathcal{E}_t(\boldsymbol{\sigma}_i \cdot \widetilde{\mathbf{W}})\,. \tag{13.112}$$

Dividing the asset price ratio at time t with the ratio at time 0 gives the exponential $\widetilde{\mathbb{P}}$-martingale:

$$\frac{S_i(t)/B(t)}{S_i(0)/B(0)} = \mathcal{E}_t(\boldsymbol{\sigma}_i \cdot \widetilde{\mathbf{W}}) := \exp\left(-\frac{1}{2}\int_0^t \|\boldsymbol{\sigma}_i(s)\|^2\,\mathrm{d}s + \int_0^t \boldsymbol{\sigma}_i(s)\cdot \mathrm{d}\widetilde{\mathbf{W}}(s)\right)\,. \tag{13.113}$$

Hence, all base asset prices relative to $B(t)$, including $B(t)$, are $\widetilde{\mathbb{P}}$-martingales. When this property holds we say that $\widetilde{\mathbb{P}} \equiv \widetilde{\mathbb{P}}^{(B)}$ is an equivalent martingale measure (EMM) with bank account B as the numéraire (or numéraire asset). As in the discrete-time market models of Chapters 7 and 8, we can use different numéraire assets for discounting. We shall denote a generic *positive asset price process by* $\{g(t)\}_{t\geqslant 0}$. This can be any one of the base assets, a portfolio in the base assets, or a derivative asset in the market model. Such an asset can be chosen as a *numéraire asset* and its price process is called the numéraire asset price process. We now show that this leads to an EMM for any given choice of numéraire. Just as in the discrete-time market models, we shall see that we are not restricted to using the bank account as the choice of numéraire for no-arbitrage derivative pricing.

Let $\{g(t)\}_{t\geqslant 0}$ be a (non-dividend-paying) numéraire asset price process with given adapted log-volatility vector $\boldsymbol{\sigma}^{(g)}(t) = [\sigma_1^{(g)}(t), \ldots, \sigma_d^{(g)}(t)]$. Since any numéraire is itself an asset, $\{\frac{g(t)}{B(t)}\}_{t\geqslant 0}$ is a $\widetilde{\mathbb{P}}$-martingale, i.e., we equivalently have

$$\mathrm{d}g(t) = g(t)\big[r(t)\,\mathrm{d}t + \boldsymbol{\sigma}^{(g)}(t)\cdot \mathrm{d}\widetilde{\mathbf{W}}(t)\big]\,, \tag{13.114}$$

$$\mathrm{d}\left(\frac{g(t)}{B(t)}\right) = \frac{g(t)}{B(t)}\boldsymbol{\sigma}^{(g)}(t)\cdot \mathrm{d}\widetilde{\mathbf{W}}(t)\,, \tag{13.115}$$

and

$$\frac{g(t)}{B(t)} = \frac{g(0)}{B(0)}\,\mathcal{E}_t(\boldsymbol{\sigma}^{(g)} \cdot \widetilde{\mathbf{W}}) \implies \frac{g(t)/B(t)}{g(0)/B(0)} = \mathcal{E}_t(\boldsymbol{\sigma}^{(g)} \cdot \widetilde{\mathbf{W}})\,. \tag{13.116}$$

By dividing (13.116) into (13.113) we obtain a representation for the ratio process $\{\frac{S_i(t)}{g(t)}\}_{t\geqslant 0}$ for each $i = 0, 1, \ldots, n$:

$$\frac{S_i(t)/g(t)}{S_i(0)/g(0)} = \frac{S_i(t)/B(t)}{S_i(0)/B(0)}\left[\frac{g(t)/B(t)}{g(0)/B(0)}\right]^{-1} = \frac{\mathcal{E}_t(\boldsymbol{\sigma}_i \cdot \widetilde{\mathbf{W}})}{\mathcal{E}_t(\boldsymbol{\sigma}^{(g)} \cdot \widetilde{\mathbf{W}})}\,. \tag{13.117}$$

Note that the right-hand side ratio of exponential $\widetilde{\mathbb{P}}$-martingales is not a $\widetilde{\mathbb{P}}$-martingale. In particular, the ratio is *not* equal to $\mathcal{E}_t\big((\boldsymbol{\sigma}_i - \boldsymbol{\sigma}^{(g)}) \cdot \widetilde{\mathbf{W}}\big)$.

We now apply Girsanov's Theorem 11.21 to change measures such that the right-hand side in (13.117) is a martingale under a new measure. This is accomplished by defining a new measure $\widetilde{\mathbb{P}}^{(g)}$ via the Radon–Nikodym derivative process

$$\varrho_t^{B\to g} \equiv \left(\frac{\mathrm{d}\widetilde{\mathbb{P}}^{(g)}}{\mathrm{d}\widetilde{\mathbb{P}}^{(B)}}\right)_t := \exp\left(-\frac{1}{2}\int_0^t \|\boldsymbol{\sigma}^{(g)}(s)\|^2\,\mathrm{d}s + \int_0^t \boldsymbol{\sigma}^{(g)}(s)\cdot \mathrm{d}\widetilde{\mathbf{W}}(s)\right) \tag{13.118}$$

for all $t \in [0, T]$. Note that $\widetilde{\mathbb{P}}^{(B)} \equiv \widetilde{\mathbb{P}}$, so $\widetilde{\mathbf{W}}(t) \equiv \widetilde{\mathbf{W}}^{(B)}(t)$ denotes the d-dimensional standard $\widetilde{\mathbb{P}}$-BM. Hence,

$$\widetilde{\mathbf{W}}^{(g)}(t) := \widetilde{\mathbf{W}}(t) - \int_0^t \boldsymbol{\sigma}^{(g)}(s)\,\mathrm{d}s \tag{13.119}$$

is a d-dimensional standard $\widetilde{\mathbb{P}}^{(g)}$-BM. In differential form, we have[2]

$$\mathrm{d}\widetilde{\mathbf{W}}^{(g)}(t) = \mathrm{d}\widetilde{\mathbf{W}}(t) - \boldsymbol{\sigma}^{(g)}(t)\,\mathrm{d}t \quad \text{or} \quad \mathrm{d}\widetilde{\mathbf{W}}(t) = \mathrm{d}\widetilde{\mathbf{W}}^{(g)}(t) + \boldsymbol{\sigma}^{(g)}(t)\,\mathrm{d}t\,. \tag{13.120}$$

Expressing the ratio in (13.117) using (13.120), i.e. $\mathrm{d}\widetilde{\mathbf{W}}(s) = \mathrm{d}\widetilde{\mathbf{W}}^{(g)}(s) + \boldsymbol{\sigma}^{(g)}(s)\,\mathrm{d}s$,

$$\begin{aligned}
\frac{\mathcal{E}_t(\boldsymbol{\sigma}_i \cdot \widetilde{\mathbf{W}})}{\mathcal{E}_t(\boldsymbol{\sigma}^{(g)} \cdot \widetilde{\mathbf{W}})} &= \exp\left[-\frac{1}{2}\int_0^t \big(\|\boldsymbol{\sigma}_i(s)\|^2 - \|\boldsymbol{\sigma}^{(g)}(s)\|^2\big)\,\mathrm{d}s + \int_0^t \big(\boldsymbol{\sigma}_i(s) - \boldsymbol{\sigma}^{(g)}(s)\big)\cdot \mathrm{d}\widetilde{\mathbf{W}}(s)\right] \\
&= \exp\left[\int_0^t \left(-\frac{1}{2}\big(\|\boldsymbol{\sigma}_i(s)\|^2 - \|\boldsymbol{\sigma}^{(g)}(s)\|^2\big) + \big(\boldsymbol{\sigma}_i(s) - \boldsymbol{\sigma}^{(g)}(s)\big)\cdot\boldsymbol{\sigma}^{(g)}(s)\right)\mathrm{d}s \right.\\
&\qquad \left. + \int_0^t \big(\boldsymbol{\sigma}_i(s) - \boldsymbol{\sigma}^{(g)}(s)\big)\cdot \mathrm{d}\widetilde{\mathbf{W}}^{(g)}(s)\right] \\
&= \exp\left[-\frac{1}{2}\int_0^t \|\boldsymbol{\sigma}_i(s) - \boldsymbol{\sigma}^{(g)}(s)\|^2\,\mathrm{d}s + \int_0^t \big(\boldsymbol{\sigma}_i(s) - \boldsymbol{\sigma}^{(g)}(s)\big)\cdot \mathrm{d}\widetilde{\mathbf{W}}^{(g)}(s)\right] \\
&\equiv \mathcal{E}_t\big((\boldsymbol{\sigma}_i - \boldsymbol{\sigma}^{(g)}) \cdot \widetilde{\mathbf{W}}^{(g)}\big)\,.
\end{aligned} \tag{13.121}$$

In the third line we simplified the integrand by the vector identity $-\frac{1}{2}\big(\|\boldsymbol{\sigma}_i\|^2 - \|\boldsymbol{\sigma}^{(g)}\|^2\big) + \big(\boldsymbol{\sigma}_i - \boldsymbol{\sigma}^{(g)}\big)\cdot\boldsymbol{\sigma}^{(g)} = -\frac{1}{2}\|\boldsymbol{\sigma}_i - \boldsymbol{\sigma}^{(g)}\|^2$.

We have therefore shown that the *probability measure* $\widetilde{\mathbb{P}}^{(g)}$ *defined by (13.118) is an EMM where all base asset prices relative to the numéraire price* $g(t)$ *are* $\widetilde{\mathbb{P}}^{(g)}$*-martingales*:

$$\frac{S_i(t)}{g(t)} = \frac{S_i(0)}{g(0)}\,\mathcal{E}_t\big((\boldsymbol{\sigma}_i - \boldsymbol{\sigma}^{(g)}) \cdot \widetilde{\mathbf{W}}^{(g)}\big) \tag{13.122}$$

or as a differential expression

$$\mathrm{d}\left(\frac{S_i(t)}{g(t)}\right) = \frac{S_i(t)}{g(t)}(\boldsymbol{\sigma}_i - \boldsymbol{\sigma}^{(g)})\cdot \mathrm{d}\widetilde{\mathbf{W}}^{(g)}(t) \tag{13.123}$$

---

[2] We remark that in the trivial case where $g(t) = B(t)$, $\boldsymbol{\sigma}^{(g)}(t) \equiv \boldsymbol{\sigma}^{(B)}(t) \equiv \mathbf{0}$ and $\varrho_t^{B\to g} \equiv 1$ and $\widetilde{\mathbf{W}}^{(g)}(t) \equiv \widetilde{\mathbf{W}}(t)$.

for all $0 \leqslant t \leqslant T$.

Consider any *domestic non-dividend-paying asset* with a price process $\{A(t)\}_{t\geqslant 0}$ obeying the SDE

$$\mathrm{d}A(t) = A(t)\big[r(t)\,\mathrm{d}t + \boldsymbol{\sigma}^{(A)}(t)\cdot \mathrm{d}\widetilde{\mathbf{W}}(t)\big] \tag{13.124}$$

where $\boldsymbol{\sigma}^{(A)}(t)$ is an adapted log-volatility vector for asset A. For example, the asset price $A(t)$ can be any base asset price (or stock price) $S_i(t)$, any portfolio in the base assets, or any domestic derivative asset. In terms of the differential $\mathrm{d}\widetilde{\mathbf{W}}^{(g)}(t)$, the SDE in (13.124) gives

$$\mathrm{d}A(t) = A(t)\big[\,\big(r(t) + \boldsymbol{\sigma}^{(A)}(t)\cdot\boldsymbol{\sigma}^{(g)}(t)\big)\,\mathrm{d}t + \boldsymbol{\sigma}^{(A)}(t)\cdot \mathrm{d}\widetilde{\mathbf{W}}^{(g)}(t)\big]\,. \tag{13.125}$$

In particular, for the case where A is the ith domestic (non-dividend-paying) base asset (or stock), $A(t) = S_i(t)$:

$$\mathrm{d}S_i(t) = S_i(t)\big[\,\big(r(t) + \boldsymbol{\sigma}_i(t)\cdot\boldsymbol{\sigma}^{(g)}(t)\big)\,\mathrm{d}t + \boldsymbol{\sigma}_i(t)\cdot \mathrm{d}\widetilde{\mathbf{W}}^{(g)}(t)\big]\,. \tag{13.126}$$

Equation (13.125) is a useful SDE that gives the log-drift coefficient of an arbitrary (non-dividend-paying) domestic asset under the EMM $\widetilde{\mathbb{P}}^{(g)}$. We see that under the new measure $\widetilde{\mathbb{P}}^{(g)}$, the original ($\widetilde{\mathbb{P}}$-measure) risk-neutral drift $r(t)$ changes by an additional term given by the dot product of the log-volatility vector of asset A and that of the numéraire asset g. In particular, using (13.125) for $A(t) = g(t)$ gives the log-drift coefficient of the numéraire asset g under measure $\widetilde{\mathbb{P}}^{(g)}$,

$$\mathrm{d}g(t) = g(t)\big[\,\big(r(t) + \|\boldsymbol{\sigma}^{(g)}(t)\|^2\big)\,\mathrm{d}t + \boldsymbol{\sigma}^{(g)}(t)\cdot \mathrm{d}\widetilde{\mathbf{W}}^{(g)}(t)\big]\,. \tag{13.127}$$

Applying the Itô quotient rule and canceling terms in the drift gives the driftless SDE

$$\mathrm{d}\left(\frac{A(t)}{g(t)}\right) = \frac{A(t)}{g(t)}\big(\boldsymbol{\sigma}^{(A)}(t) - \boldsymbol{\sigma}^{(g)}(t)\big)\cdot \mathrm{d}\widetilde{\mathbf{W}}^{(g)}(t)\,. \tag{13.128}$$

This is consistent with the fact that the ratio process $\{\frac{A(t)}{g(t)}\}_{t\geqslant 0}$ must be a $\widetilde{\mathbb{P}}^{(g)}$-martingale.

From (13.116) and the definition in (13.118) we see that the Radon–Nikodym derivative process at any time $0 \leqslant t \leqslant T$ is given by the ratio of the numéraire asset price relative to the bank account value at times t and initial time 0, which has the equivalent expressions:

$$\varrho_t^{B\to g} := \mathcal{E}_t(\boldsymbol{\sigma}^{(g)}\cdot\widetilde{\mathbf{W}}) = \frac{g(t)/B(t)}{g(0)/B(0)} = \frac{g(t)}{g(0)}\frac{B(0)}{B(t)}\,. \tag{13.129}$$

Using this expression for any time $t \leqslant T$ and for time T gives us the ratio of the Radon–Nikodym derivative process at the two times,

$$\frac{\varrho_T^{B\to g}}{\varrho_t^{B\to g}} = \frac{g(T)/B(T)}{g(0)/B(0)}\frac{g(0)/B(0)}{g(t)/B(t)} = \frac{g(T)/B(T)}{g(t)/B(t)} = \frac{g(T)}{g(t)}\frac{B(t)}{B(T)}\,. \tag{13.130}$$

We also note that the change of measure in the opposing direction, $\widetilde{\mathbb{P}}^{(g)} \to \widetilde{\mathbb{P}}^{(B)}$, is specified by the Radon–Nikodym derivative process

$$\varrho_t^{g\to B} \equiv \left(\frac{\mathrm{d}\widetilde{\mathbb{P}}^{(B)}}{\mathrm{d}\widetilde{\mathbb{P}}^{(g)}}\right)_t = \left[\left(\frac{\mathrm{d}\widetilde{\mathbb{P}}^{(g)}}{\mathrm{d}\widetilde{\mathbb{P}}^{(B)}}\right)_t\right]^{-1} = \frac{1}{\varrho_t^{B\to g}} \tag{13.131}$$

and hence

$$\frac{\varrho_T^{g\to B}}{\varrho_t^{g\to B}} = \left(\frac{\varrho_T^{B\to g}}{\varrho_t^{B\to g}}\right)^{-1} = \frac{g(t)}{g(T)}\frac{B(T)}{B(t)}. \tag{13.132}$$

By applying the above measure change, $\widetilde{\mathbb{P}} \equiv \widetilde{\mathbb{P}}^{(B)} \to \widetilde{\mathbb{P}}^{(g)}$, to (13.22) while using the property in (11.85) with $\varrho_t \equiv \varrho_t^{g\to B}$, and noting that $\widetilde{\mathrm{E}}^{(B)}[\,] \equiv \widetilde{\mathrm{E}}[\,]$, we obtain

$$V_t = \widetilde{\mathrm{E}}^{(B)}\left[\frac{B(t)}{B(T)}V_T \,\middle|\, \mathcal{F}_t\right] = \widetilde{\mathrm{E}}^{(g)}\left[\frac{\varrho_T^{g\to B}}{\varrho_t^{g\to B}}\frac{B(t)}{B(T)}V_T \,\middle|\, \mathcal{F}_t\right] = \widetilde{\mathrm{E}}^{(g)}\left[\frac{g(t)}{g(T)}V_T \,\middle|\, \mathcal{F}_t\right]. \tag{13.133}$$

This is the *numéraire invariant form of the risk-neutral pricing formula for any attainable payoff V_T, given any choice of non-dividend-paying positive asset price process g as numéraire.* Since $g(t)$ is $\mathcal{F}_t$-measurable we can pull it out of the expectation, giving

$$V_t = g(t)\,\widetilde{\mathrm{E}}^{(g)}\left[\frac{V_T}{g(T)} \,\middle|\, \mathcal{F}_t\right]. \tag{13.134}$$

This general form of the asset pricing formula is also a statement of the fact that the derivative value relative to the numéraire asset price process, $\{\frac{V_t}{g(t)}\}_{0\leqslant t\leqslant T}$, is a $\widetilde{\mathbb{P}}^{(g)}$-martingale. The reader can also verify that the value of a self-financing replicating portfolio relative to the numéraire price, $\{\frac{\Pi_t}{g(t)}\}_{0\leqslant t\leqslant T}$, is a $\widetilde{\mathbb{P}}^{(g)}$-martingale. Then, since an attainable payoff can (by definition) be replicated, we have $\Pi_t = V_t$, which gives (13.134).

Note that the derivative price in (13.134) is expressed as an $\mathcal{F}_t$-measurable random variable. For most payoffs, as in standard and some path-dependent European derivatives, we can employ a (joint) Markov property, which then allows us to express V_t as a function of underlying random variables at time t. For example, in standard basket options with payoff $V_T = \Lambda(T, \mathbf{S}(T))$ we can use the Markov property of the vector stock price process $\{\mathbf{S}(t)\}_{t\geqslant 0}$. In particular, if the numéraire is also a function of the underlying stocks, i.e., $g(t) = g(t, \mathbf{S}(t))$, then the pricing formula gives $V_t = V(t, \mathbf{S}(t))$:

$$V(t, \mathbf{S}(t)) = g(t, \mathbf{S}(t))\,\widetilde{\mathrm{E}}^{(g)}\left[\frac{\Lambda(T, \mathbf{S}(T))}{g(T, \mathbf{S}(T))} \,\middle|\, \mathbf{S}(t)\right]. \tag{13.135}$$

Conditioning on $\mathcal{F}_t$ has been reduced to conditioning on the time-t stock price vector $\mathbf{S}(t)$. The pricing function, which is a function of time t and the spot variables $\mathbf{S} = (S_1, \ldots, S_n)$, is then given by setting $\mathbf{S}(t) = \mathbf{S}$,

$$V(t, \mathbf{S}) = g(t, \mathbf{S})\,\widetilde{\mathrm{E}}_{t,\mathbf{S}}^{(g)}\left[\frac{\Lambda(T, \mathbf{S}(T))}{g(T, \mathbf{S}(T))}\right]. \tag{13.136}$$

A similar relation holds for the foreign exchange derivatives, where the joint process $\{S^f(t), A(t), X(t)\}_{t\geqslant 0}$ is Markov and conditioning on the spot variables $S^f(t) = S, X(t) = X, A(t) = A$ gives

$$V(t, S, A, X) = g(t, S, A, X)\,\widetilde{\mathrm{E}}_{t,(S,A,X)}^{(g)}\left[\frac{\Lambda(T, S^f(T), A(T), X(T))}{g(t, S^f(T), A(T), X(T))}\right] \tag{13.137}$$

where we assume the numéraire is some function of the time and the joint process values, $g(t) = g(t, S^f(t), A(t), X(t))$.

We now extend the above formulation to the case where the assets can pay dividends which are adapted processes. Any *domestic dividend-paying asset* has a price process denoted by $\{A(t)\}_{t\geqslant 0}$ and obeys an SDE of the form

$$\mathrm{d}A(t) = A(t)\big[(r(t) - q_A(t))\,\mathrm{d}t + \boldsymbol{\sigma}^{(A)}(t)\cdot \mathrm{d}\widetilde{\mathbf{W}}(t)\big] \tag{13.138}$$

where $q_A(t)$ is an adapted dividend yield for asset A at time t. Note that this is equivalent to (13.124) when the dividend yield is zero. The asset price $A(t)$ can be any domestic base asset price (or stock price) $S_i(t)$, any portfolio in the base assets, or any domestic derivative asset. For a domestic stock price process $\{S_i(t)\}_{t\geqslant 0}$ (or a foreign stock price process converted to domestic currency) with a dividend yield process $\{q_i(t)\}_{t\geqslant 0}$ we have

$$\mathrm{d}S_i(t) = S_i(t)\big[\,\big(r(t) - q_i(t) + \boldsymbol{\sigma}_i(t)\cdot\boldsymbol{\sigma}^{(g)}(t)\big)\,\mathrm{d}t + \boldsymbol{\sigma}_i(t)\cdot \mathrm{d}\widetilde{\mathbf{W}}^{(g)}(t)\big]\,, \tag{13.139}$$

and, in particular,

$$\mathrm{d}S_i(t) = S_i(t)\big[\,\big(r(t) - q_i(t)\big)\,\mathrm{d}t + \boldsymbol{\sigma}_i(t)\cdot \mathrm{d}\widetilde{\mathbf{W}}(t)\big]\,. \tag{13.140}$$

For a foreign stock we have (upon using $\mathrm{d}\widetilde{\mathbf{W}}(t) = \mathrm{d}\widetilde{\mathbf{W}}^{(g)}(t) + \boldsymbol{\sigma}^{(g)}(t)\,\mathrm{d}t$ in (13.95))

$$\mathrm{d}S^f(t) = S^f(t)\big[(r^f(t) - q_S(t) - \boldsymbol{\sigma}^{(S)}(t)\cdot(\boldsymbol{\sigma}^{(X)}(t) - \boldsymbol{\sigma}^{(g)}(t)))\,\mathrm{d}t + \boldsymbol{\sigma}^{(S)}(t)\cdot \mathrm{d}\widetilde{\mathbf{W}}^{(g)}(t)\big]\,, \tag{13.141}$$

and the foreign exchange rate process in (13.93) satisfies

$$\mathrm{d}X(t) = X(t)\big[(r(t) - r^f(t) + \boldsymbol{\sigma}^{(X)}(t)\cdot\boldsymbol{\sigma}^{(g)}(t))\,\mathrm{d}t + \boldsymbol{\sigma}^{(X)}(t)\cdot \mathrm{d}\widetilde{\mathbf{W}}^{(g)}(t)\big]\,. \tag{13.142}$$

It is important to note that we are still defining the same measure $\widetilde{\mathbb{P}}^{(g)}$ given by the Radon–Nikodym derivative in (13.118). However, we now also allow the numéraire (domestic) asset g to have an adapted dividend yield $q_g(t)$ where

$$\mathrm{d}g(t) = g(t)\big[(r(t) - q_g(t))\,\mathrm{d}t + \boldsymbol{\sigma}^{(g)}(t)\cdot \mathrm{d}\widetilde{\mathbf{W}}(t)\big]\,. \tag{13.143}$$

The dividend drift portions can be "eliminated" by defining the price processes

$$\widehat{S}_i(t) := \mathrm{e}^{\int_0^t q_i(s)\,\mathrm{d}s} S_i(t),\ \ \widehat{A}(t) := \mathrm{e}^{\int_0^t q_A(s)\,\mathrm{d}s} A(t)\ ,\ \text{and}\ \ \widehat{g}(t) := \mathrm{e}^{\int_0^t q_g(s)\,\mathrm{d}s} g(t)\,.$$

Note that $\widehat{A}(0) = A(0), \widehat{g}(0) = g(0), \widehat{S}_i(0) = S_i(0)$. It follows that the ratio processes $\{\frac{\widehat{A}(t)}{B(t)}\}_{t\geqslant 0}$, $\{\frac{\widehat{S}_i(t)}{B(t)}\}_{t\geqslant 0}$, for all i, and $\{\frac{\widehat{g}(t)}{B(t)}\}_{t\geqslant 0}$ are all $\widetilde{\mathbb{P}}$-martingales:

$$\frac{\widehat{A}(t)/B(t)}{A(0)/B(0)} = \mathcal{E}_t(\boldsymbol{\sigma}^{(A)}\cdot\widetilde{\mathbf{W}})\,,\quad \frac{\widehat{S}_i(t)/B(t)}{S_i(0)/B(0)} = \mathcal{E}_t(\boldsymbol{\sigma}_i\cdot\widetilde{\mathbf{W}})\,,\quad \frac{\widehat{g}(t)/B(t)}{g(0)/B(0)} = \mathcal{E}_t(\boldsymbol{\sigma}^{(g)}\cdot\widetilde{\mathbf{W}})\,.$$

The ratio process $\{\frac{\widehat{A}(t)}{\widehat{g}(t)}\}_{t\geqslant 0}$ and $\{\frac{\widehat{S}_i(t)}{\widehat{g}(t)}\}_{t\geqslant 0}$, for all i, are $\widetilde{\mathbb{P}}^{(g)}$-martingales, i.e.,

$$\frac{\widehat{A}(t)}{\widehat{g}(t)} = \frac{A(0)}{g(0)}\,\mathcal{E}_t((\boldsymbol{\sigma}^{(A)} - \boldsymbol{\sigma}^{(g)})\cdot\widetilde{\mathbf{W}}^{(g)}) \tag{13.144}$$

and

$$\frac{\widehat{S}_i(t)}{\widehat{g}(t)} = \frac{S_i(0)}{g(0)}\,\mathcal{E}_t((\boldsymbol{\sigma}_i - \boldsymbol{\sigma}^{(g)})\cdot\widetilde{\mathbf{W}}^{(g)})\,. \tag{13.145}$$

These reproduce our earlier formulae in case the dividend yields are zero.

The expressions in (13.129)–(13.132) are now given in terms of the process $\widehat{g}$:

$$\varrho_t^{B\to g} := \mathcal{E}_t(\boldsymbol{\sigma}^{(g)} \cdot \widetilde{\mathbf{W}}) = \frac{\widehat{g}(t)/B(t)}{g(0)/B(0)} = \frac{\widehat{g}(t)}{g(0)}\frac{B(0)}{B(t)}, \tag{13.146}$$

$$\frac{\varrho_T^{B\to g}}{\varrho_t^{B\to g}} = \frac{\widehat{g}(T)/B(T)}{\widehat{g}(t)/B(t)} = \frac{\widehat{g}(T)}{\widehat{g}(t)}\frac{B(t)}{B(T)} \tag{13.147}$$

and

$$\frac{\varrho_T^{g\to B}}{\varrho_t^{g\to B}} = \left(\frac{\varrho_T^{B\to g}}{\varrho_t^{B\to g}}\right)^{-1} = \frac{\widehat{g}(t)}{\widehat{g}(T)}\frac{B(T)}{B(t)}. \tag{13.148}$$

Finally, the *numéraire invariant form of the risk-neutral pricing formula for any attainable payoff V_T, given any choice of dividend-paying positive asset price process g as numéraire,* takes the equivalent form:

$$V_t = \widehat{g}(t)\,\widetilde{\mathrm{E}}^{(g)}\left[\frac{V_T}{\widehat{g}(T)}\,\middle|\,\mathcal{F}_t\right] = g(t)\,\widetilde{\mathrm{E}}^{(g)}\left[\mathrm{e}^{-\int_t^T q_g(s)\,\mathrm{d}s}\frac{V_T}{g(T)}\,\middle|\,\mathcal{F}_t\right]. \tag{13.149}$$

This general form of the asset pricing formula is also a statement of the fact that the derivative value relative to $\widehat{g}(t)$, $\{\frac{V_t}{\widehat{g}(t)}\}_{0\leqslant t\leqslant T}$, is a $\widetilde{\mathbb{P}}^{(g)}$-martingale. We remark that this formula recovers (13.134) in case the numéraire has no dividends. When $g = B$, $\widetilde{\mathbb{P}}^{(g)} = \widetilde{\mathbb{P}}$, the domestic bank account is the numéraire and therefore the dividend $q_g = q_B \equiv 0$, $g(t)/g(T) = B(t)/B(T)$ and (13.22) is recovered.

We finally consider some examples of how appropriate choices of numéraire can simplify derivative pricing problems. We leave several other applications of the change of numéraire technique for pricing multi-asset and foreign exchange options as exercises at the end of this chapter. The key idea is to choose a numéraire that simplifies the effective payoff $\frac{V_T}{g(T)}$ in the pricing problem.

Example 13.3. (Pricing an Exchange Option with a Stock as Numéraire) Reconsider the option pricing problem solved in Section 13.2.2.1 with the payoff in (13.53). Assume the stock prices are GBM processes driven by two independent BMs in a domestic economy with constant interest rate r, but now set the stock dividends to zero. Derive the European option pricing formula using the asset pricing formula with the *numéraire asset chosen as one of the stocks.*

Solution. As numéraire, let us choose $g(t) = S_1(t)$, $t \geqslant 0$. Then,

$$\frac{V_T}{g(T)} = \frac{(S_2(T) - S_1(T))^+}{S_1(T)} = \left(\frac{S_2(T)}{S_1(T)} - 1\right)^+ = (Y(T) - 1)^+$$

where we define the process $Y(t) := \frac{S_2(t)}{S_1(t)}, t \geqslant 0$. We can substitute this into either of the pricing formulas in (13.134)–(13.136). In particular, using (13.134) gives $V_t = V(t, S_1(t), S_2(t))$ as

$$V_t = S_1(t)\,\widetilde{\mathrm{E}}^{(S_1)}\left[(Y(T) - 1)^+ \,|\, \mathcal{F}_t\right].$$

To simplify notation, we shall simply denote $\widetilde{\mathbb{P}}^{(g)} = \widetilde{\mathbb{P}}^{(S_1)} \equiv \widehat{\mathbb{P}}$, $\widetilde{\mathrm{E}}^{(S_1)} \equiv \widehat{\mathrm{E}}$ and $\widetilde{\mathbf{W}}^{(g)}(t) = \widetilde{\mathbf{W}}^{(S_1)}(t) \equiv \widehat{\mathbf{W}}(t)$. Note that the conditioning on $\mathcal{F}_t$ can also be replaced by $Y(t)$.

The GBM process $\{Y(t)\}_{t\geqslant 0}$ is a $\widehat{\mathbb{P}}$-martingale with representation in (13.122) [where $g(t) = S_1(t)$, $\boldsymbol{\sigma}^{(g)} = \boldsymbol{\sigma}_1$, $S_i(t) = S_2(t)$]:

$$\frac{S_2(t)}{S_1(t)} \equiv Y(t) = Y(0)\,\mathcal{E}_t\big((\boldsymbol{\sigma}_2 - \boldsymbol{\sigma}_1)\cdot\widehat{\mathbf{W}}\big), \quad Y(0) = \frac{S_2(0)}{S_1(0)}.$$

[We remark that this solution can also be arrived at by computing the stochastic differential $\mathrm{d}\left(\frac{S_2(t)}{S_1(t)}\right)$ by the Itô quotient rule and using (13.126), giving $\mathrm{d}Y(t) = Y(t)(\boldsymbol{\sigma}_2-\boldsymbol{\sigma}_1)\cdot\mathrm{d}\widehat{\mathbf{W}}(t)$, which has the solution given by the above expression.] By the above solution we have the random variable $X := \ln\frac{Y(T)}{Y(t)} = \ln\mathcal{E}_{T-t}\big((\boldsymbol{\sigma}_2 - \boldsymbol{\sigma}_1)\cdot\widehat{\mathbf{W}}\big)$, i.e.,

$$\begin{aligned}
X &= -\frac{1}{2}\|\boldsymbol{\sigma}_2 - \boldsymbol{\sigma}_1\|^2(T-t) + (\boldsymbol{\sigma}_2 - \boldsymbol{\sigma}_1)\cdot(\widehat{\mathbf{W}}(T) - \widehat{\mathbf{W}}(t)) \\
&= -\frac{1}{2}\nu^2(T-t) + (\boldsymbol{\sigma}_2 - \boldsymbol{\sigma}_1)\cdot(\widehat{\mathbf{W}}(T) - \widehat{\mathbf{W}}(t)) \\
&= -\frac{1}{2}\nu^2(T-t) + \nu\sqrt{T-t}\widehat{Z}
\end{aligned}$$

where $\nu^2 := \|\boldsymbol{\sigma}_2 - \boldsymbol{\sigma}_1\|^2 = \sigma_1^2 + \sigma_2^2 - 2\rho\sigma_1\sigma_2$ and $\widehat{Z} := \frac{(\boldsymbol{\sigma}_2-\boldsymbol{\sigma}_1)\cdot(\widehat{\mathbf{W}}(T)-\widehat{\mathbf{W}}(t))}{\nu\sqrt{T-t}} \sim Norm(0,1)$ under measure $\widehat{\mathbb{P}}$. The Brownian increments are independent of $\mathcal{F}_t$ and therefore $\widehat{Z}$ (and X) is independent of $\mathcal{F}_t$. Since $Y(T) = Y(t)\mathrm{e}^X$, the above expectation is simplified to an unconditional one by independence, where $Y(t)$ is $\mathcal{F}_t$-measurable. The expectation below is very easily computed by applying the usual identities. However, observe that the expectation is just like that of a standard call option on a single stock "Y" with zero "interest rate," "zero dividend," "volatility" ν, "spot" $Y(t)$, "strike" of 1, and time to maturity $T-t$:

$$\begin{aligned}
V_t &= S_1(t)\,\widehat{\mathrm{E}}\left[(Y(t)\mathrm{e}^X - 1)^+ \,|\, \mathcal{F}_t\right] \\
&= S_1(t)\,\left[Y(t)\mathcal{N}(d_+(Y(t), T-t) - 1\cdot\mathcal{N}(d_-(Y(t), T-t)\right] \\
&= S_1(t)\left[\frac{S_2(t)}{S_1(t)}\mathcal{N}\left(d_+(\frac{S_2(t)}{S_1(t)}, T-t)\right) - \mathcal{N}\left(d_-(\frac{S_2(t)}{S_1(t)}, T-t)\right)\right] \\
&= S_2(t)\mathcal{N}\left(d_+(\frac{S_2(t)}{S_1(t)}, T-t)\right) - S_1(t)\mathcal{N}\left(d_-(\frac{S_2(t)}{S_1(t)}, T-t)\right),
\end{aligned}$$

where $d_\pm(x,\tau)$ are defined as in (13.61) with $q_1 = q_2 = 0$. This expresses the option value as a random variable $V_t = V(t, S_1(t), S_2(t))$. Setting $S_1(t) = S_1, S_2(t) = S_2$, gives the pricing formula as a function of calendar time $t < T$ and the spot values $S_1, S_2 > 0$:

$$V(t, S_1, S_2) = S_2\mathcal{N}\left(d_+(\frac{S_2}{S_1}, T-t)\right) - S_1\mathcal{N}\left(d_-(\frac{S_2}{S_1}, T-t)\right).$$

Note: this formula is exactly that in (13.60) for $q_1 = q_2 = 0$, $\tau = T-t$. □

Example 13.4. (Pricing an Equity-Linked Foreign Exchange Call) By using a different choice of numéraire than the domestic bank account, derive the pricing function for the equity-linked foreign exchange call with domestic payoff in (13.87). Assume the exchange process and the foreign stock price process are GBMs as in (13.98) and (13.99) and let the stock dividend $q_S = 0$.

Solution. The payoff $V_T \equiv C_T$ is given by (13.87), which we can write as

$$C_T = (X(T)S^f(T) - \kappa S^f(T))^+ .$$

Note that the domestic asset price process $\{X(t)S^f(t)\}_{t\geqslant 0}$ qualifies as a non-dividend-paying numéraire asset, as seen by the SDE in (13.100) where $q_S = 0$.

By choosing $g(t) = X(t)S^f(t)$, the payoff relative to this numéraire simplifies to

$$\frac{C_T}{g(T)} = \frac{(X(T)S^f(T) - \kappa S^f(T))^+}{X(T)S^f(T)} = (1 - \kappa X^{-1}(T))^+ \equiv (1 - \kappa Y(T))^+ = \kappa(\kappa^{-1} - Y(T))^+$$

where we define the reciprocal of the exchange process $Y(t) := X^{-1}(t)$ for all $t \geqslant 0$. The numéraire asset price is given by (13.104), where $q_S = 0$. Hence, the pricing formula in (13.134) gives the time-t price $V_t = C_t$,

$$C_t = g(t)\,\widetilde{\mathrm{E}}^{(g)}\left[\frac{C_T}{g(T)}\,\Big|\,\mathcal{F}_t\right] = \kappa\,X(t)S^f(t)\,\widehat{\mathrm{E}}\left[(\kappa^{-1} - Y(T))^+\,\big|\,\mathcal{F}_t\right] \tag{13.150}$$

where we abbreviate $\widetilde{\mathrm{E}}^{(g)} \equiv \widetilde{\mathrm{E}}^{(XS^f)} \equiv \widehat{\mathrm{E}}$ and $\widetilde{\mathbf{W}}^{(g)}(t) \equiv \widehat{\mathbf{W}}(t)$. Computing this expectation is essentially like pricing a standard put with effective strike κ^{-1}.

[Remark: C_t above is kept as an $\mathcal{F}_t$-measurable random variable. We can also perform all the calculations by computing the (ordinary) pricing function $C(t, S, x)$ for arbitrary spot variables $S^f(t) = S, X(t) = x$, $Y(t) \equiv X^{-1}(t) = x^{-1}$, giving

$$C(t, S, x) = \kappa\,x S\,\widehat{\mathrm{E}}\left[(\kappa^{-1} - Y(T))^+\,\big|\,Y(t) = x^{-1}\right]. \tag{13.151}$$

This pricing function then also gives us $C_t = C(t, S^f(t), X(t))$. Alternatively, we could compute the expectation in (13.150), giving $C_t = C(t, S^f(t), X(t))$ and the pricing function then follows by setting $S^f(t) = S, X(t) = x$. The main point is that both calculations give us the price of the option, expressible as an ordinary function of the spot variables or as a random variable.]

We need to represent $Y(T)$ as a GBM random variable involving the $\widehat{\mathbb{P}}$-BM and then either of the above expectations is trivially computed. One way to do this[3] is to write the BM increment $\widetilde{\mathbf{W}}(T) - \widetilde{\mathbf{W}}(t)$, using (13.119) where $\boldsymbol{\sigma}^{(g)}$ is a constant vector, as

$$\widetilde{\mathbf{W}}^{(g)}(T) - \widetilde{\mathbf{W}}^{(g)}(t) + (T-t)\,\boldsymbol{\sigma}^{(g)} \equiv \widehat{\mathbf{W}}(T) - \widehat{\mathbf{W}}(t) + (T-t)\,\boldsymbol{\sigma}^{(XS^f)},$$

where $\boldsymbol{\sigma}^{(XS^f)} = \boldsymbol{\sigma}^{(X)} + \boldsymbol{\sigma}^{(S)}$. Substituting this BM increment into the representation (13.102) gives

$$X(T) = X(t)\,\mathrm{e}^{(\widehat{\mu}_X + \frac{1}{2}\sigma_X^2)(T-t) + \boldsymbol{\sigma}^{(X)}\cdot(\widehat{\mathbf{W}}(T) - \widehat{\mathbf{W}}(t))}$$

where $\widehat{\mu}_X = r - r^f + \boldsymbol{\sigma}^{(X)}\cdot\boldsymbol{\sigma}^{(S)} = r - r^f + \rho\sigma_X\sigma_S$, $\rho = \rho_{XS}$. Taking the reciprocal gives

$$Y(T) = Y(t)\,\mathrm{e}^{(\widehat{\mu}_Y - \frac{1}{2}\sigma_Y^2)(T-t) + \boldsymbol{\sigma}^{(Y)}\cdot(\widehat{\mathbf{W}}(T) - \widehat{\mathbf{W}}(t))}$$

where $\widehat{\mu}_Y = -\widehat{\mu}_X, \boldsymbol{\sigma}^{(Y)} = -\boldsymbol{\sigma}^{(X)}$, $\sigma_Y = \|\boldsymbol{\sigma}^{(Y)}\| = \|\boldsymbol{\sigma}^{(X)}\| = \sigma_X$. The random variable in the exponent

$$\boldsymbol{\sigma}^{(Y)}\cdot(\widehat{\mathbf{W}}(T) - \widehat{\mathbf{W}}(t)) \stackrel{d}{=} \sigma_Y\sqrt{T-t}\,\widehat{Z}\,, \quad \text{where } \widehat{Z} \sim Norm(0,1) \text{ under } \widehat{\mathbb{P}}$$

and $\widehat{Z}$ is $\mathcal{F}_t$-independent (independent of $Y(t)$) and $Y(t)$ is $\mathcal{F}_t$-measurable. Expressing $Y(T)$

[3] Alternatively, the SDE in (13.142) may be directly used to compute the log-drift and log-volatility vector of the reciprocal process $Y = 1/X$ upon applying the Itô quotient rule, with $\boldsymbol{\sigma}^{(g)} = \boldsymbol{\sigma}^{(XS^f)}$, giving $\mathrm{d}Y(t) = Y(t)[\widehat{\mu}_Y\,\mathrm{d}t + \boldsymbol{\sigma}^{(Y)}\cdot\mathrm{d}\widehat{\mathbf{W}}(t)]$, where $\widehat{\mu}_Y = r^f - r - \boldsymbol{\sigma}^{(X)}\cdot\boldsymbol{\sigma}^{(S)}$ and $\boldsymbol{\sigma}^{(Y)} = -\boldsymbol{\sigma}^{(X)}$.

in terms of $\widehat{Z}$, then by independence the expectation in either (13.150) or (13.151) reduces to an unconditional expectation. In particular, (13.151) gives

$$C(t,S,x) = \kappa x S \,\widehat{\mathrm{E}}\left[(\kappa^{-1} - x^{-1}\,\mathrm{e}^{(\widehat{\mu}_Y - \frac{1}{2}\sigma_Y^2)(T-t)+\sigma_Y\sqrt{T-t}\widehat{Z}})^+\right]. \tag{13.152}$$

This expectation is easily computed by applying the identity (A.2) of the Appendix or we can directly use the pricing function for a standard put. Let $P_{BS}(S,K,\tau;r,\sigma)$ be the Black–Scholes pricing function in (12.30) for a standard put with spot S, strike K, $\tau = T-t$, interest rate r, volatility σ, and zero dividend. Then, the above expectation equals $\mathrm{e}^{\widehat{\mu}_Y\tau}P_{BS}(x^{-1},\kappa^{-1},\tau;\widehat{\mu}_Y,\sigma_Y)$ and using the above relations for $\widehat{\mu}_Y$ and σ_Y in terms of the original parameters gives the pricing function

$$\begin{aligned}
C(t,S,x) &= \kappa x S \mathrm{e}^{\widehat{\mu}_Y\tau}P_{BS}(x^{-1},\kappa^{-1},T-t;\widehat{\mu}_Y,\sigma_Y)\\
&= \kappa x S\left[\kappa^{-1}\mathcal{N}\left(-\frac{\ln(x^{-1}/\kappa^{-1}) + (\widehat{\mu}_Y - \frac{1}{2}\sigma_Y^2)(T-t)}{\sigma_Y\sqrt{T-t}}\right)\right.\\
&\quad\left. - \mathrm{e}^{\widehat{\mu}_Y(T-t)}x^{-1}\mathcal{N}\left(-\frac{\ln(x^{-1}/\kappa^{-1}) + (\widehat{\mu}_Y + \frac{1}{2}\sigma_Y^2)(T-t)}{\sigma_Y\sqrt{T-t}}\right)\right]\\
&= S\left[x\mathcal{N}\left(d_+(\frac{x}{\kappa},T-t)\right) - \mathrm{e}^{(r^f - r - \rho\sigma_X\sigma_S)(T-t)}\kappa\mathcal{N}\left(d_-(\frac{x}{\kappa},T-t)\right)\right]
\end{aligned} \tag{13.153}$$

where

$$d_\pm(\frac{x}{\kappa},T-t) := \frac{\ln(x/\kappa) + (r - r^f + \rho\sigma_X\sigma_S \pm \frac{1}{2}\sigma_X^2)(T-t)}{\sigma_X\sqrt{T-t}}.$$

If the stock pays a constant dividend yield q_S, then the pricing function is given by (13.153) where the spot S is replaced by $S\mathrm{e}^{-q_S(T-t)}$. □

13.4 Exercises

Exercise 13.1. Consider three domestic stocks with prices satisfying correlated geometric Brownian motions:

$$\mathrm{d}S_i(t) = S_i(t)\left[(r - q_i)dt + \boldsymbol{\sigma}_i \cdot \mathrm{d}\widetilde{\mathbf{W}}(t)\right], \quad i = 1,2,3.$$

$\widetilde{\mathbf{W}}(t)$ is a standard three-dimensional standard BM under the risk-neutral measure $\widetilde{\mathbb{P}}$ with the bank account as the numéraire. The parameters $\|\boldsymbol{\sigma}_i\| = \sigma_i > 0$, and the correlation coefficients ρ_{ij}, where $\boldsymbol{\sigma}_i \cdot \boldsymbol{\sigma}_j = \rho_{ij}\sigma_i\sigma_j$, are all assumed constants. The interest rate r is constant and q_i are the respective constant dividend yields. Let $S_i(0) = S_i > 0$ be the initial prices for each respective stock $i = 1,2,3$.

(a) For $t \geqslant 0$, derive explicit expressions for:

(i) $\widetilde{\mathbb{P}}(S_1(t) < S_2(t) < S_3(t))$,

(ii) $\widetilde{\mathbb{P}}(S_1(t) < K < S_2(t))$, for any constant $K > 0$,

(iii) $\widetilde{\mathbb{P}}\left(S_3(t) > K\frac{S_2(t)}{S_1(t)}\right)$, for any constant $K > 0$.

(b) Derive the time-t ($t < T$) no-arbitrage pricing formula for the value of the European basket option with payoff at maturity T given by

$$V_T = \max\{S_3(T), S_1(T)\} - \min\{S_2(T), S_1(T)\}.$$

Exercise 13.2. Derive expressions for the hedge positions in the two stocks 1 and 2 for the options on the maximum and the minimum with pricing functions in (13.69) and (13.70). Show that the relation in (13.65) holds for both options.

Exercise 13.3. Derive the explicit time-t pricing formula $V(t, S_1, S_2)$ for the exchange option with payoff in (13.66) assuming the stock prices are GBM processes as in (13.40), but now let the interest rate $r(t)$, the dividend yields on the stocks $q_i(t)$, and the log-volatility vectors $\boldsymbol{\sigma}_i(t)$, $i = 1, 2$, be nonrandom integrable functions of time.

Exercise 13.4. A plain currency call option on a foreign exchange rate has payoff

$$C_T = (X(T) - \kappa)^+ .$$

(a) Derive the pricing function $C(t, x)$, $t < T$, for this call by evaluating the risk-neutral expectation formula in (13.107).

(b) Give the BSPDE for the pricing function $C(t, x)$.

Exercise 13.5. Assume that a foreign stock price process $\{S^f(t)\}_{t\geqslant 0}$ and the (foreign to domestic) exchange rate $\{X(t)\}_{t\geqslant 0}$ are correlated geometric Brownian motions with respective log-volatility vectors $\boldsymbol{\sigma}^{(S)} = [\sigma_S, 0]$ and $\boldsymbol{\sigma}^{(X)} = [\rho\sigma_X, \sqrt{1-\rho^2}\sigma_X]$. Assume the domestic and foreign interest rates r and r^f are constants and that the foreign stock pays no dividend.

(a) Derive a formula for the current time $t < T$ price C_t of a call option on foreign stock denominated in domestic currency with domestic payoff

$$C_T = \left(X(T)S^f(T) - K\right)^+.$$

(b) Similarly, derive a formula for the current price P_t of a put option with domestic payoff

$$P_T = \left(K - X(T)S^f(T)\right)^+.$$

(c) Derive a put-call parity formula relating the call and put prices C_t and P_t.

Exercise 13.6. Consider the fixed foreign equity rate call with payoff in (13.86). Assume the foreign stock price is a GBM with constant log-volatility vector $\boldsymbol{\sigma}^{(S)}$ and having a dividend yield q_S. Derive its pricing function $C(t, S, X)$ by

(a) evaluating the risk-neutral expectation formula in (13.107);

(b) solving the BSPDE in (13.109) by symmetry reduction.

Exercise 13.7. Consider the foreign equity call struck in foreign currency with payoff in (13.84). Assume the foreign stock price is a GBM with constant log-volatility vector $\boldsymbol{\sigma}^{(S)}$ and having a dividend yield q_S and the exchange rate is a GBM with constant log-volatility vector $\boldsymbol{\sigma}^{(X)}$. Derive its pricing function $C(t, S, X)$ by

(a) evaluating the risk-neutral expectation formula in (13.107);

(b) solving the BSPDE in (13.109) by symmetry reduction.

Exercise 13.8. Consider a self-financing portfolio process defined by

$$\Pi_t = \beta_t B(t) + \beta_t^f X(t) B^f(t) + \Delta_t^f X(t) S^f(t) + \Delta_t A(t) \tag{13.154}$$

where an agent in a domestic economy takes a time-t position β_t in the domestic bank account, β_t^f in the foreign bank account converted to domestic currency, a position Δ_t^f in the foreign asset (e.g., stock) converted to domestic currency, and a position Δ_t in the domestic asset A. Assume the domestic and foreign interest rates r and r^f are constant and that the processes are GBM satisfying (13.98)–(13.101). Let $V(t, S, A, x)$ be a smooth pricing function and $V_t = V(t, S^f(t), A(t), X(t)), t \geqslant 0$ be the price process of a standard (non-path-dependent) European foreign exchange derivative.

By applying a multidimensional Itô formula and assuming that the self-financing portfolio replicates the foreign exchange derivative, i.e. $\Pi_t = V_t$ and therefore $\mathrm{d}\Pi_t = \mathrm{d}V_t$, show that the dynamic *delta hedging positions* in (13.154) are given by

$$\Delta_t = \frac{\partial V}{\partial A}, \quad \Delta_t^f = \frac{1}{x}\frac{\partial V}{\partial S}, \tag{13.155}$$

and the *domestic value of the investment in the foreign bank account* is

$$X(t)\beta_t^f B^f(t) = x\frac{\partial V}{\partial x} - S\frac{\partial V}{\partial S}, \tag{13.156}$$

where S, A, x are the respective spot values of $S^f(t), A(t), X(t)$.

Exercise 13.9. Assume that a foreign stock price $\{S^f(t)\}_{t\geqslant 0}$, a foreign exchange rate process $\{X(t)\}_{t\geqslant 0}$, and a *domestic* asset price process $\{A(t)\}_{t\geqslant 0}$ (e.g., a stock denominated in domestic currency) are all geometric Brownian motions with respective constant log-volatility vectors $\boldsymbol{\sigma}^{(S)}$, $\boldsymbol{\sigma}^{(X)}$, and $\boldsymbol{\sigma}^{(A)}$:

$$\frac{dA(t)}{A(t)} = \mu_A dt + \boldsymbol{\sigma}^{(A)} \cdot d\mathbf{W}(t), \quad \frac{dS^f(t)}{S^f(t)} = \mu_S dt + \boldsymbol{\sigma}^{(S)} \cdot d\mathbf{W}(t), \quad \frac{dX(t)}{X(t)} = \mu_X dt + \boldsymbol{\sigma}^{(X)} \cdot d\mathbf{W}(t).$$

$\mathbf{W}(t)$ is a three-dimensional standard $\mathbb{P}$-BM in the physical measure $\mathbb{P}$ and the assets are correlated, where $\|\boldsymbol{\sigma}^{(S)}\| = \sigma_S$, $\|\boldsymbol{\sigma}^{(X)}\| = \sigma_X$, $\|\boldsymbol{\sigma}^{(A)}\| = \sigma_A$, $\boldsymbol{\sigma}^{(S)} \cdot \boldsymbol{\sigma}^A = \rho_{SA}\sigma_S\sigma_A$, $\boldsymbol{\sigma}^{(X)} \cdot \boldsymbol{\sigma}^{(A)} = \rho_{XA}\sigma_X\sigma_A$, $\boldsymbol{\sigma}^{(X)} \cdot \boldsymbol{\sigma}^{(S)} = \rho_{SX}\sigma_S\sigma_X$. Assume a domestic and a foreign economy with respective interest rates r and r^f as constants, zero dividends on all assets, and let $A(t) = A, S^f(t) = S, X(t) = X$ be the spot values.

(a) Derive the time $t < T$ pricing function for a domestic European option with payoff

$$V_T = \max\{X(T)S^f(T), A(T)\}.$$

(b) Based on part (a), provide formulas for the delta position in the domestic stock and the delta position in the foreign stock that are required in a dynamic hedging strategy for all $t < T$ (see Exercise 13.8).

(c) Derive the time $t < T$ pricing function for a domestic European-style option with payoff

$$V_T = X(T)S^f(T)\mathbb{I}_{\{X(T)\geqslant X_0\}} + A_T\mathbb{I}_{\{X(T)<X_0\}}$$

where $X(0) = X_0$ is a fixed positive initial exchange rate.

(d) Derive the time $t < T$ pricing function for a domestic European-style option with payoff

$$V_T = (aX(T)S^f(T) - bA(T))^+, \quad \text{with positive constants } a, b.$$

Exercise 13.10. Consider a market with domestic and foreign bank accounts where $r(t)$ and $r^f(t)$ are deterministic time-dependent interest rates in the respective domestic and foreign currencies. Assume a domestic asset with price process $\{A(t)\}_{t\geqslant 0}$, a foreign asset with price process $\{A^f(t)\}_{t\geqslant 0}$, and a foreign exchange rate process $\{X(t)\}_{t\geqslant 0}$. Assume zero dividend yield on the assets and all processes are GBM with respective constant log-volatility vectors $\boldsymbol{\sigma}^{(A)}$, $\boldsymbol{\sigma}^{(A^f)}$, and $\boldsymbol{\sigma}^{(X)}$, where $\boldsymbol{\sigma}^{(X)}\cdot\boldsymbol{\sigma}^{(A)} = \rho_{XA}\sigma_X\sigma_A$, $\boldsymbol{\sigma}^{(X)}\cdot\boldsymbol{\sigma}^{(A^f)} = \rho_{XA^f}\sigma_X\sigma_{A^f}$, $\boldsymbol{\sigma}^{(A)}\cdot\boldsymbol{\sigma}^{(A^f)} = \rho_{AA^f}\sigma_A\sigma_{A^f}$.

(a) Find the log-drift of each of the following separate processes: (i) $X(t)$, (ii) $X(t)B^f(t)$, (iii) $A(t)$, and (iv) $A^f(t)$ under the measure $\widetilde{\mathbb{P}}^{(g)}$ with $g(t) = A(t)$ as the numéraire asset price. Express your answers explicitly in terms of all the relevant parameters.

(b) Derive the pricing function for a European contract at current time $t = 0$ whose payoff at maturity T is

$$V_T = \max\{A(T), X_0 A^f(T)\}$$

where $X_0 = \frac{A(0)}{A^f(0)}$ is the current (known) exchange rate.

Hint: You can use the results of part (a).

Exercise 13.11. Assume $\{S^f(t)\}_{t\geqslant 0}$ and $\{(X(t)\}_{t\geqslant 0}$ are correlated GBM as in Exercise 13.5.

(a) By using any approach, derive the pricing function $P(t, S, X)$, $t < T$, for a put option on foreign stock denominated in domestic currency with domestic payoff:

$$P_T = X(T)\big(K_f - S^f(T)\big)^+, \quad K_f > 0.$$

(b) Determine the self-financed portfolio *delta position in the foreign stock* and the *domestic value of the investment in the foreign bank account* (as functions of the spot values and time) that an investor must hold at time $t < T$ in order to *dynamically replicate* the value P_t (see Exercise 13.8).

(c) Derive a put-call parity relation between the above put option value $P(t, S, X)$ and the corresponding call option price $C(t, S, K)$ with payoff $C_T = X(T)\big(S^f(T) - K_f\big)^+$.

Exercise 13.12. Assume a foreign stock S^f and exchange rate process X modelled as in Exercise 13.5 with time-0 (spot) values $S^f(0) = S$, $X(0) = X$. Derive the time-0 pricing formula, as a function of S, X, T, for a domestic European option having payoff at maturity $T > 0$ given by

$$V_T = \mathbb{I}_{\{M(T)<K\}}\, X(T)S^f(T)\,,$$

where $M(T)$ is the maximum realized value of the exchange rate up to time T:

$$M(T) := \max_{0\leqslant t\leqslant T} X(t)\,.$$

Hint: Use an appropriate choice of numéraire asset that simplifies the derivative pricing. You will need to make use of the CDF of $M(T)$.

Exercise 13.13. Consider a domestic economy with constant interest rate r and two correlated GBM stock price processes given by (13.54). Let $S_1(t) = S_1, S_2(t) = S_2$, $0 \leqslant t \leqslant T$, be the stock spot values. Derive the pricing function $V(t, S_1, S_2)$ for a European chooser max call with payoff in (13.81).
Hint: You may rewrite the payoff using the identity $(\max\{a,b\} - c)^+ = \mathbb{I}_{\{a\geqslant b\}}(a-c)^+ + \mathbb{I}_{\{b>a\}}(b-c)^+$.

Exercise 13.14. Consider a domestic economy with two stocks as in Exercise 13.13. Derive the pricing function for a European chooser min put with payoff in (13.82).
Hint: You may use similar identities as in Exercise 13.13 to rewrite the payoff.

Exercise 13.15. Consider three stocks with GBM price dynamics as in Exercise 13.1 with constant interest rate r and constant dividend yield q_i on each stock $i = 1, 2, 3$.

(a) Derive the pricing function $V(t, S_1, S_2, S_3)$, $t < T$, for a European option with payoff

$$V_T = S_3(T)\, \mathbb{I}_{\{S_3(T)>S_1(T), S_3(T)>S_2(T)\}}.$$

(b) Derive the pricing function for a European option with payoff

$$V_T = \max\{S_1(T), S_2(T), S_3(T)\}.$$

Hint: You can use the result of part (a).

(c) Derive the pricing function for a European option with payoff

$$V_T = K\, \mathbb{I}_{\{S_1(T)<S_3(T),\, S_2(T)<S_3(T)\}}$$

for constant $K > 0$.

Exercise 13.16. Consider an exchange option on two stocks having a payoff

$$V_T = (aS_2(T) - bS_1(T)\,)^+$$

with positive constants a, b. Assume the stocks are GBM processes as in (13.54). Derive the time-t price V_t for this option by explicitly using one of the stocks as the numéraire asset and by implementing the risk-neutral pricing formula in (13.149). Note: the derivation is similar to that in Example 13.3.

Exercise 13.17. Consider a domestic economy with constant interest rate r and two domestic stock price processes as in Exercise 13.1. Derive the time-0 pricing function $V_0 = V(T, S_1, S_2)$, in the spot variables $S_1(0) = S_1$, $S_2(0) = S_2$, for a European path-dependent option with payoff at maturity T given by

$$V_T = S_1(T) \min_{0\leqslant t\leqslant T} \frac{S_2(t)}{S_1(t)}.$$

Hint: Use an appropriate numéraire asset that simplifies the evaluation of the risk-neutral expectation. The CDF of the *minimum* of a GBM process can then be employed.

Chapter 14

American Options

In this chapter we briefly present the theory for pricing early-exercise (American) options in continuous time. Recall that American options were first introduced in Chapter 4, and the discrete-time case was dealt with in Chapter 7. Let us recall some of the properties of early-exercise options. The key difference between European-style and American-style options is that a holder of an American option can exercise her rights any time before the expiration date. This additional early exercise privilege should not be worthless. Thus, an American option is expected to be worth more than its European analogue. The extra premium paid on top of the price of the European option is called the *early-exercise premium.* We mainly focus our discussion on standard American call and put options with respective payoffs $(S-K)^+$ and $(K-S)^+$, although the theory can also be applied to other types of options. Throughout this chapter, we assume that we deal with only one underlying whose asset price process, $\{S(t)\}_{t\geqslant 0}$, follows the standard geometric Brownian motion model, the risk-neutral dynamics of which is given by

$$S(t) = S(0)\mathrm{e}^{(r-q-\sigma^2/2)t+\sigma\widetilde{W}(t)}, \quad t \geqslant 0. \tag{14.1}$$

Here, $r \geqslant 0$ is the risk-free interest rate, $q \geqslant 0$ is the continuous dividend yield, $\sigma > 0$ is the asset price volatility, and $\{\widetilde{W}(t)\}_{t\geqslant 0}$ is a Brownian motion considered under the risk-neutral probability measure $\widetilde{\mathbb{P}}$ with the bank account as the numéraire asset.

14.1 Basic Properties of Early-Exercise Options

Let $t_0 \geqslant 0$ be the contract inception time. American call and put options struck at K with expiration time $T > t_0$ are claims to payoffs $(S(t)-K)^+$ and $(K-S(t))^+$, respectively, that the holder can exercise at any intermediate time $t \in [t_0, T]$. As in previous chapters, the time-t value of an option with expiration date T is denoted by $V(t, S)$, where S is the current asset (spot) price and $t \in [t_0, T]$ is the calendar time. It is also convenient to express the option value as a function of the time to expiration $\tau = T - t$ when the asset price model is assumed to be a time-homogeneous stochastic process; the option value therefore depends on τ and, as in previous chapters, we also denote the pricing function as $v(\tau, S)$, where $v(T-t, S) = V(t, S)$.

An American option cannot be worth less than its corresponding intrinsic value, which is the payoff associated with immediate exercise. In contrast to the case of European calls and puts, whose no-arbitrage values satisfy a put-call parity, there exists a put-call estimate for American options, which gives lower and upper bounds on the difference of call and put values (see Equation (4.8) and Exercise 4.8):

$$S\mathrm{e}^{-q\tau} - K \leqslant C(\tau, S) - P(\tau, S) \leqslant S - K\mathrm{e}^{-r\tau},$$

where C and P denote the prices of the American call and put options, respectively, with spot S and time to maturity τ. In the absence of dividends on the underlying asset, there is no advantage in exercising an American call prior to the expiry date. Hence, the American call being exercised at the expiration date is equivalent to its European counterpart (with the same strike and expiration). As a result, the American and European calls on a stock without dividends have the same value. If the stock pays dividends, the arguments above are not valid. The optimal exercise date depends on the dividend process. If dividends occur occasionally at certain dates, it may be optimal to exercise an American call option right before a dividend payment that is large enough. Therefore, in the presence of dividends, the American call can be worth more than the European call. Recall that this situation was studied for a binomial tree model in Chapter 7. Let us generalize our findings about relationships between American and European call/put options in the following propositions.

Proposition 14.1. *Let V^E and V, respectively, denote the values of European and American options having the same payoff $\Lambda(S)$ and expiration date T. The following two conditions are equivalent:*

(i) $V^E(t,S) \geqslant \Lambda(S)$ *for all* $S > 0$ *and all* $t \in [t_0, T]$;

(ii) $V(t,S) = V^E(t,S)$ *for all* $S > 0$ *and all* $t \in [t_0, T]$.

That is, if the corresponding European price is always above the intrinsic value during the contract lifetime, then it is never optimal to exercise the American option at any time earlier than expiry, i.e., there is no early-exercise premium and $V \equiv V^E$.

Proof. For any time t, the value $V^E(t,S)$ is the arbitrage-free price of a contract exercised at the expiry time T. Condition (i) implies that it is not optimal to exercise earlier for a lower value. For every time t, it is more beneficial to wait until expiration. The optimal exercise (stopping) time is therefore at expiry T. Hence, (i) implies (ii). To prove the converse, observe that the American option value is always above the intrinsic value, i.e., $V(t,S) \geqslant \Lambda(S)$ for all (t,S). Hence, condition (ii) implies (i). This result is essentially a statement of the fact that an early-exercise opportunity (and premium) arises if and only if the corresponding European option value falls below the intrinsic (payoff function) value. □

As a corollary of Proposition 14.1, we have the rather well-known result:

Proposition 14.2.

(1) An American call has a nonzero early-exercise premium if and only if $q > 0$.

(2) An American put has a nonzero early-exercise premium if and only if $r > 0$.

This result follows from an arbitrage-free argument. However, a simple and instructive proof goes as follows.

Proof. Recall the put-call parity relation for European call and put options with common expiry T and strike price K:

$$C^E(t,S) - P^E(t,S) = \mathrm{e}^{-q(T-t)}S - \mathrm{e}^{-r(T-t)}K.$$

Using the fact that $P^E(t,S) \geqslant 0$ gives

$$C^E(t,S) = P^E(t,S) + \mathrm{e}^{-q(T-t)}S - \mathrm{e}^{-r(T-t)}K \geqslant \mathrm{e}^{-q(T-t)}S - K. \tag{14.2}$$

Then, for $q = 0$, (14.2) gives $C^E(t,S) \geqslant S - K$. Hence the European call value is always above its payoff function. From Proposition 14.1, we conclude that the European call value, $C^E(t,S)$, is equal to the American call value, $C(t,S)$, so that the early-exercise premium is zero. For the case $q > 0$, we use (14.2) and note that since the European put is a decreasing function of S, there exist large enough values of S such that $P^E(t,S) + \mathrm{e}^{-q(T-t)}S - \mathrm{e}^{-r(T-t)}K < 0$, i.e., $C^E(t,S) < S - K$ for some $S > K$. From the previous result we therefore have $C(t,S) > C^E(t,S)$ and hence conclude that the early-exercise premium is nonzero for $q > 0$. This proves (i), while statement (ii) is proved in a similar fashion by reversing the roles of S and q with K and r, respectively, and is left as an exercise. □

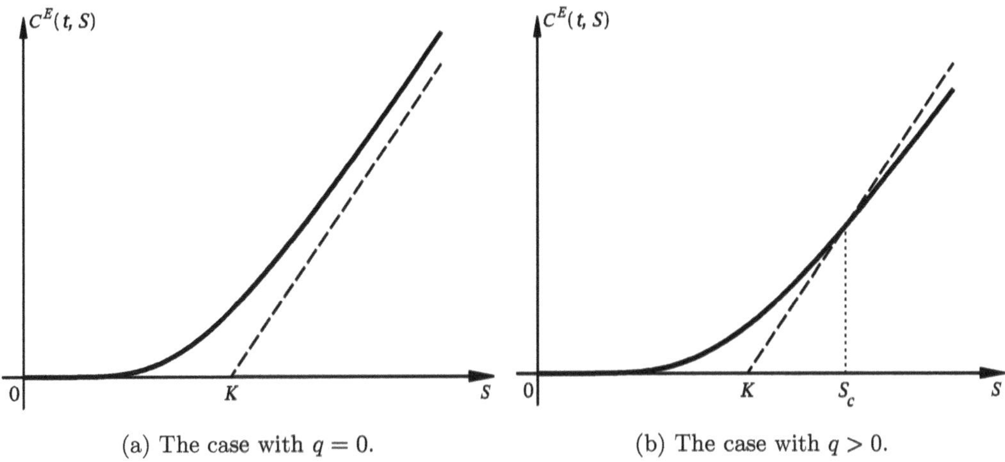

(a) The case with $q = 0$.

(b) The case with $q > 0$.

FIGURE 14.1: The value of a European call option.

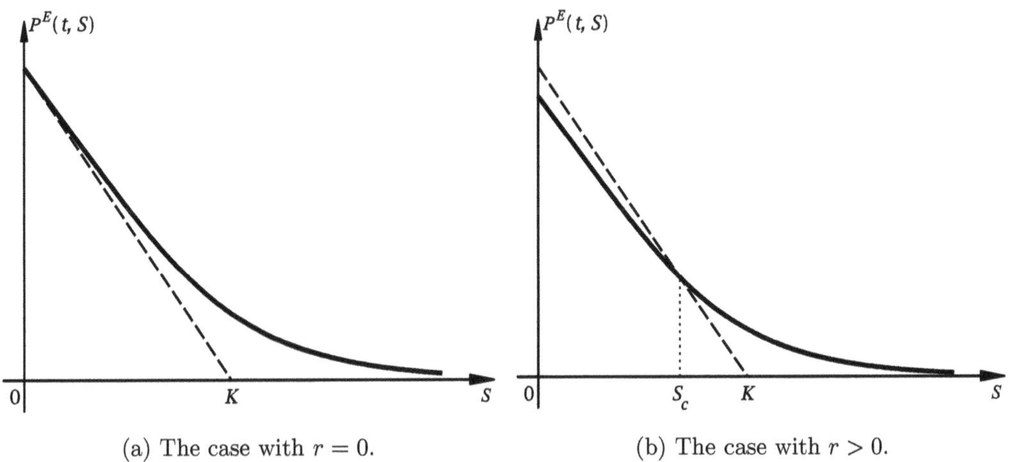

(a) The case with $r = 0$.

(b) The case with $r > 0$.

FIGURE 14.2: The value of a European put option.

Statements (i) and (ii) are illustrated in Figures 14.1 and 14.2, respectively. If $q > 0$ ($r > 0$), then the graph of the call (put) option value function intersects the payoff diagram at some positive point S_c. For values of S greater (less) than S_c, the call (put) option price is strictly less than the payoff value. Note that $P(t,0) = \mathrm{e}^{-r(T-t)}K$, which equals K if $r = 0$ and is less that K if $r > 0$, for all $T - t > 0$.

When we considered the no-arbitrage valuation of options under the binomial tree model, we analyzed the pricing of American options from both holder's and writer's points of view. The writer sells an option in exchange for some initial capital which can be used to hedge the short position in the derivative. This initial capital should provide the writer with sufficient funds to fulfil her obligations when the option is exercised. It is possible to find the best time for the holder (which is the worst time for the writer) to exercise the option. This time is called the *optimal exercise time.* The fair initial price of an American option is then defined as the smallest initial capital required to be hedged against exercise at the optimal time. Although the optimal exercise is not known in advance, the holder wishes to exercise the option as optimally as possible. Because the writer has no control over what exercise policy will be used by the buyer, the initial fair value of an American derivative with intrinsic value $\Lambda(S)$ is defined as the maximum of $\widetilde{\mathrm{E}}_0[\mathrm{e}^{-r\mathcal{T}}\Lambda(S(\mathcal{T}))]$ over all possible exercise polices $\mathcal{T} \in [t_0, T]$. It can be shown that both approaches give the same value. In the next section, we carry out this analysis in continuous time.

14.2 Arbitrage-Free Pricing of American Options

14.2.1 Optimal Stopping Formulation and Early-Exercise Boundary

Consider an American option issued at time $t_0 \geqslant 0$ with payoff (intrinsic) value function Λ and expiration time $T > t_0$. Suppose that the option has not been exercised before time $t \in [t_0, T]$. Additionally, suppose that the holder of the option follows some exercise policy $\mathcal{T}$, which is a stopping time taking values in the interval $[t, T]$ or taking the value ∞. Recall that $\mathcal{S}_{t,T}$ denotes the collection of all such stopping times. The no-arbitrage time-t value of the option, with underlying spot S, under a given exercise policy $\mathcal{T}$ is given by

$$\widetilde{\mathrm{E}}_{t,S}\big[\mathrm{e}^{-r(\mathcal{T}-t)}\,\Lambda(S(\mathcal{T}))\big].$$

Note that we are using our previous shorthand notation for the conditional expectation $\widetilde{\mathrm{E}}_{t,S}[\,\cdot\,] \equiv \widetilde{\mathrm{E}}[\cdot \mid S(t) = S]$. If $\mathcal{T}$ is infinite, we set $\mathrm{e}^{-r\infty}\,\Lambda(S(\infty))$ to be zero. Under uncertainty of what exercise rule will be used by the holder, the time-t value of an American option is defined to be

$$V(t,S) = \sup_{\mathcal{T}\in\mathcal{S}_{t,T}} \widetilde{\mathrm{E}}_{t,S}\big[\mathrm{e}^{-r(\mathcal{T}-t)}\,\Lambda(S(\mathcal{T}))\big]. \tag{14.3}$$

The optimal stopping time $\mathcal{T}^*$ maximizes the expectation on the right-hand side of (14.3). In other words, $\mathcal{T}^*$ is given implicitly by

$$V(t,S) = \widetilde{\mathrm{E}}_{t,S}\big[\mathrm{e}^{-r(\mathcal{T}^*-t)}\,\Lambda(S(\mathcal{T}^*))\big]. \tag{14.4}$$

By analogy with the case of the binomial tree model, the optimal exercise time $\mathcal{T}^*$ is the random variable corresponding to the first time when the option value equals the intrinsic value,

$$\mathcal{T}^* = \min\{u \;:\; t \leqslant u \leqslant T,\; V(u, S(u)) = \Lambda(S(u))\}. \tag{14.5}$$

That is, for maximal gain, along a stock price path $(u, S(u,\omega))_{t\leqslant u\leqslant T}$, the option should be exercised at the first time, say $u = t^* = \mathcal{T}^*(\omega)$, when $V(t^*, S(t^*)) = \Lambda(S(t^*))$. Thus, letting $S(t) = S$, the (t, S)-plane consisting of the time and spot value points is separated into two sub-domains: a stopping domain where the option is exercised early,

$$\mathcal{D} = \{(t,S) \;:\; t_0 \leqslant t \leqslant T,\; V(t,S) = \Lambda(S)\}, \tag{14.6}$$

and a continuation domain for which the option is not exercised,

$$\mathcal{D}^c = \{(t, S) \ : \ t_0 \leqslant t \leqslant T, \ V(t, S) > \Lambda(S)\}. \tag{14.7}$$

The continuation domain is the complement of the stopping domain within the rectangle $[t_0, T] \times [0, \infty)$. As is seen from (14.6), the continuation domain is the set of all points (t, S) such that the option value $V(t, S)$ exceeds the value of the payoff function $\Lambda(S)$.

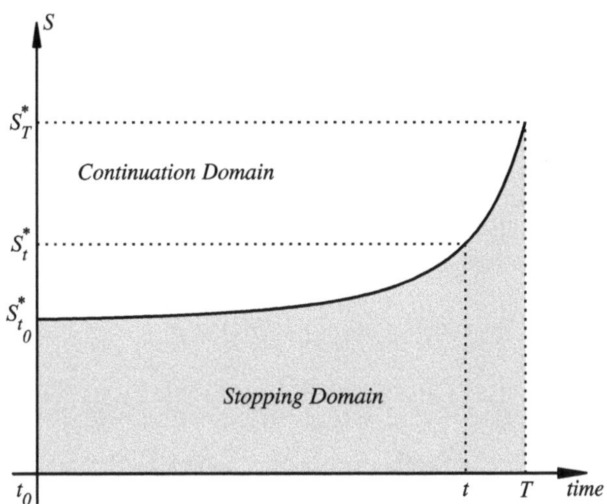

FIGURE 14.3: A typical stopping domain for an American put option. The option is exercised in the shaded area.

The geometry of the stopping domain may be quite complicated. It depends on the payoff function Λ and whether the underlying asset pays discrete dividends. However, for the case with a standard call/put payoff and continuous dividends, the stopping domain turns out to be simply connected. A typical shape of the stopping domain for a standard American put option is demonstrated in Figure 14.3. The *early-exercise boundary*, $\partial\mathcal{D}$, which separates the continuation and stopping domains, is a smooth curve on the (t, S)-plane and is of the form $\partial\mathcal{D} = \{(t, S) \ : \ 0 \leqslant t \leqslant T, \ S = S^*_t\}$, where the function S^*_t is given by

$$S^*_t = \min\{S > 0 \ : \ V(t, S) = (S - K)^+\} \tag{14.8}$$

for a call and

$$S^*_t = \max\{S > 0 \ : \ V(t, S) = (K - S)^+\} \tag{14.9}$$

for a put. Here, V represents the value of an American call, C, or put, P, respectively. Since the American option value is always nonnegative, the superscript $+$ signs in (14.8) and (14.9) are actually redundant. Because the option value is positive, the critical value S^*_t is larger than the strike K for the American call and it is less than K for the American put.

In general, the early-exercise curve S^*_t varies with time t. For a given calendar time $t \in [t_0, T]$, we define the continuation and stopping intervals in $[0, \infty)$ for spot price S, denoted by $\mathcal{D}^c_t$ and $\mathcal{D}_t$, respectively, so that $S \in \mathcal{D}_t$ iff $(t, S) \in \mathcal{D}$ and $\mathcal{D}^c_t := [0, \infty) \setminus \mathcal{D}_t$. For a given time t, the American option is exercised if $S \in \mathcal{D}_t$, and its value is equal to the payoff function in $\mathcal{D}_t$. For the American call, we have $\mathcal{D}^c_t = [0, S^*_t)$ and $\mathcal{D}_t = [S^*_t, \infty)$; while for the American put, we have $\mathcal{D}^c_t = (S^*_t, \infty)$ and $\mathcal{D}_t = [0, S^*_t]$. When S is in the stopping domain $\mathcal{D}_t$, the values of the American call and put are $C(t, S) = S - K$ and

$P(t, S) = K - S$, respectively. In what follows, it will be convenient to express quantities in terms of time to maturity $\tau = T - t$. Consequently, we adopt the notations $S^*(\tau) \equiv S_t^*$, $\mathcal{D}(\tau) \equiv \mathcal{D}_t$, and $\mathcal{D}^c(\tau) \equiv \mathcal{D}_t^c$.

An obvious consequence of Proposition 14.2 is that (a) for an American call on a non-dividend-paying stock the exercise boundary is trivial (i.e., it is never optimal to exercise early where $\mathcal{D}^c(\tau) = [0, \infty)$) and (b) for an American put on a non-dividend-paying stock the exercise boundary is nontrivial (i.e., there is an optimal early-exercise time) if the interest rate is positive.

Note that for a general payoff function, the stopping domain may consist of several disconnected regions; therefore, the early-exercise boundary may be a union of separate curves. For example, in the case of an American strangle whose payoff is a sum of standard call and put payoffs, the stopping domain consists of upper and lower regions (assuming that both r and q are strictly positive).

As follows from the definition of the stopping domain, the optimal exercise time $\mathcal{T}^*$ defined in (14.5) is also the first passage time of the stopping domain $\mathcal{D}$. That is, the time $\mathcal{T}^*$ is the first time when the stock price process reaches the early-exercise boundary S^* (see Figure 14.4):

$$\mathcal{T}^* = \min\{t \in [t_0, T] \; : \; S(t) = S_t^*\}. \tag{14.10}$$

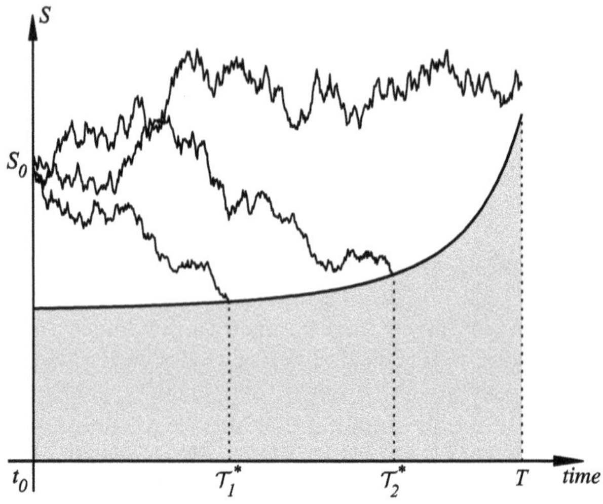

FIGURE 14.4: For two sample asset price paths, an American put is exercised early at times $\mathcal{T}_1^*$ and $\mathcal{T}_2^*$, respectively. The option is not exercised early when a sample path does not cross the early-exercise boundary S^*.

14.2.2 The Smooth Pasting Condition

The American option should be exercised whenever the asset price S is in the stopping domain $\mathcal{D}$. This implies that the American option value $V(t, S)$ can be written as a piecewise continuous function of S which is equal to the payoff for all spot values $S \in \mathcal{D}_t$. For standard American call and put options, we have that $C(t, S) = (S - K)^+ = S - K$ if $S \geqslant S_t^*$ and $P(t, S) = (K - S)^+ = K - S$ if $S \leqslant S_t^*$, respectively. Clearly, the option value functions have to be continuous at the exercise boundary point $S = S_t^*$ or otherwise there exists an arbitrage opportunity. Moreover, as is demonstrated below, the optimal exercise condition requires the curve of $V(t, S)$ to be tangent to the payoff at the point S_t^*. In other words, the pricing function for an American option has a continuous derivative at the exercise boundary; that is, the delta of the option price is continuous at the early-exercise boundary. This property is known as the *smooth pasting condition.*

We need to prove that the derivative of the option pricing function is equal to the derivative of the payoff function at the point $S = S_t^*$ for any $t \in [t_0, T]$. Let us first consider the case with a put option. Note that the argument presented below also readily follows for any continuous and monotonic payoff function such as the standard call and put payoffs. Suppose that the American put has not been exercised before time t. Consider the collection $\mathcal{B}$ of all possible early-exercise boundaries defined by continuous functions $b\colon [t,T] \to \mathbb{R}_+$. For each $b \in \mathcal{B}$, there is an exercise policy $\mathcal{T}_b$, which is the first passage time for the boundary b. Let

$$P(t, S; b) = \widetilde{\mathrm{E}}_{t,S}\left[\mathrm{e}^{-r(\mathcal{T}_b - t)}\,(K - S(\mathcal{T}_b))^+\right]$$

be the put value under the exercise policy $\mathcal{T}_b$. Since every policy $\mathcal{T}_b$ is a stopping time in the set $\mathcal{S}_{[t,T]}$ and the function S_t^* of time t, defining the optimal exercise boundary, belongs to the collection $\mathcal{B}$, the value of the American put in (14.3) is given by $P(t,S) = \sup_{b\in\mathcal{B}} P(t,S;b)$. The optimal exercise boundary $b = S^*$ maximizes the function $b \mapsto P(t,S;b)$. So, this is a problem in calculus of variations. In general, if u is a function and $F(u)$ is some functional of u, then the calculus of variations is a technique that tries to find such a u that optimizes F. If u^* is an optimal solution, the functional derivative $\frac{\mathrm{d}F(u)}{\mathrm{d}u}$ is zero at $u = u^*$. Since the optimal choice for the boundary b is the optimal early-exercise boundary S^*, the derivative of the put value P with respect to b is zero at $b = S^*$:

$$\left.\frac{\partial P(t,S;b)}{\partial b}\right|_{b=S^*} = 0.$$

Let us find the total derivative of the function P with respect to b along the boundary:

$$\frac{\mathrm{d}P}{\mathrm{d}b} = \left.\frac{\partial P(t,S;b)}{\partial S}\right|_{S=b} + \left.\frac{\partial P(t,S;b)}{\partial b}\right|_{S=b},$$

where we use the fact that $\frac{\partial S}{\partial b} = 1$ along the curve $S = b(t)$. On the one hand, when $b = S^*$, we have $\frac{\mathrm{d}P}{\mathrm{d}b} = \frac{\partial P(t,S;b)}{\partial S}\big|_{S=S^*}$. On the other hand, the option value is equal to the payoff function when $S = b$. Therefore, $P(t,b;b) = (K - b)$ and then

$$\frac{\mathrm{d}P}{\mathrm{d}b}(t,b;b) = \frac{\mathrm{d}}{\mathrm{d}b}(K - b) = -1.$$

Putting the results together gives

$$\left.\frac{\partial P(t,S)}{\partial S}\right|_{S=S^*} = \left.\frac{\partial P(t,S;S^*)}{\partial S}\right|_{S=S^*} = -1 = \left.\frac{\mathrm{d}\Lambda(S)}{\mathrm{d}S}\right|_{S=S^*}. \tag{14.11}$$

where $\Lambda(S) = (K - S)^+$ is the put payoff and equals $K - S$ for all $S < K$. Note that $S^* < K$ for a put.

An alternative proof of the smooth pasting condition (14.11) is based on the no-arbitrage argument. If $S < S_t^*$, then $P(t,S) = K - S$. Therefore, the limit from the left is $\lim_{S\nearrow S_t^*} \frac{\partial P(t,S)}{\partial S} = -1$. Now consider the limit from the right. Since the American put value is always above the payoff values, we have $\lim_{S\searrow S_t^*} \frac{\partial P(t,S)}{\partial S} \geqslant -1$. Our objective is to show that this inequality is actually a strict equality. We prove this by showing that there exists an arbitrage if the limiting value is strictly greater than -1. Suppose that the asset price at calendar time t is at the boundary, i.e., $S \equiv S(t) = S_t^*$. Consider a portfolio of one long put option and one share of stock. The portfolio value is

$$\Pi(t) = S + P(t, S) = S + K - S = K.$$

After a sufficiently small time lapse δt, the asset price can move downward into the exercise domain or upward into the continuation domain. We denote the change in the asset price by $\delta S = S(t+\delta t) - S(t)$, so $S(t+\delta t) = S+\delta S$. If $\delta S < 0$, then $P(t+\delta t, S+\delta S) = K-(S+\delta S)$ and the change in the portfolio value is $\delta\Pi = \delta P + \delta S = 0$. If $\delta S > 0$, the small move δS creates a profit opportunity since $\delta S > 0$ and $\delta P < 0$ but in absolute value δP is smaller than δS. We thus have $\delta\Pi > 0$. Let us find the order of magnitude of $\delta\Pi$. Using the stochastic differential equation for the asset price process gives $\delta S = \mu S\delta t + \sigma S\delta W$, where $\delta W \sim Norm(0, \delta t)$. Therefore, we have

$$\begin{aligned}\delta\Pi = \delta S + \delta P = \delta S + \left.\frac{\partial P}{\partial S}\right|_{S\searrow S_t^*} \delta S + \frac{1}{2}\left.\frac{\partial^2 P}{\partial S^2}\right|_{S\searrow S_t^*} (\delta S)^2 + o(\delta t) \\ = \mathcal{O}(\delta t) + \left(\sigma S + \left.\frac{\partial P}{\partial S}\right|_{S\searrow S_t^*} \sigma S\right)\delta W = \mathcal{O}(\delta t) + \left(1 + \left.\frac{\partial P}{\partial S}\right|_{S\searrow S_t^*}\right)\sigma S\sqrt{\delta t}|Z|,\end{aligned}$$

where $Z \sim Norm(0,1)$, since $\delta W > 0$ for an upward movement of the stock price. Thus,

$$\mathrm{E}_t[\delta\Pi \mid \delta S > 0] = \mathcal{O}(\delta t) + \left(1 + \left.\frac{\partial P}{\partial S}\right|_{S\searrow S_t^*}\right)\sigma S\sqrt{\delta t}.$$

This represents the change in portfolio value conditional on the stock price having a positive change within an arbitrarily small time interval δt. Since the term in $\sqrt{\delta t}$ is arbitrarily larger than all other terms of order $\mathcal{O}(\delta t)$, the change in portfolio value is positive if $\frac{\partial P}{\partial S}|_{S\searrow S_t^*} > -1$, i.e., the upward return on the portfolio is positive and of order $\sqrt{\delta t}$. This is an arbitrage opportunity since the risk-free return should be a smaller quantity. Therefore, the condition (14.11) holds. In fact, the smooth pasting condition is applicable to all types of American options. The general result is presented below without a proof.

Proposition 14.3. *Consider an American option with a differentiable payoff function Λ at any point (t, S_t^*) on the early-exercise boundary. Then, the American option pricing function V satisfies the smooth pasting condition:*

$$\left.\frac{\partial V(t,S)}{\partial S}\right|_{S=S_t^*} = \Lambda'(S_t^*), \tag{14.12}$$

that is, the (spacial) derivative of V is continuous at the boundary of the stopping and continuation domains (as is illustrated in Figure 14.5). Additionally, the option value V satisfies the zero time-decay condition on the early-exercise domain,

$$\frac{\partial V(t,S)}{\partial t} = 0, \quad \text{for all } S \in \mathcal{D}_t. \tag{14.13}$$

Proof. The proof is left as an exercise for the reader. □

Remarks:

1. The condition in (14.12) is also obviously valid for all $S \in \mathcal{D}_t$ since $V(t,S) = \Lambda(S)$ on that region.

2. For a call (or put) the property (14.12) simply gives

$$\left.\frac{\partial V(t,S)}{\partial S}\right|_{S=S_t^*} = 1 \text{ (or } -1).$$

This is illustrated in Figure 14.5.

3. The properties (14.12) and (14.13) are also valid under a general diffusion model.

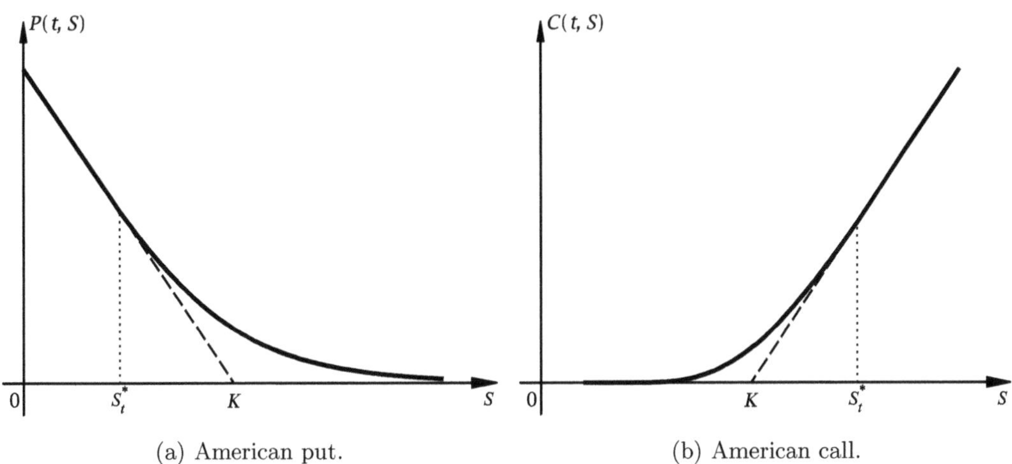

(a) American put. (b) American call.

FIGURE 14.5: The value functions for American put and call options satisfy the smooth pasting condition with slope equal to -1 and 1, respectively, at the optimal exercise boundary point S_t^*. The solid price curve touches the dashed payoff line tangentially at the point $(S_t^*, \Lambda(S_t^*))$, where $\Lambda(S)$ is the payoff function equal to $(S-K)^+$ and $(K-S)^+$ for the call and put options, respectively.

14.2.3 Put-Call Symmetry Relation

An American option can be considered as providing the right to exchange one asset for another, namely, one share of the underlying stock worth S dollars (asset one) and K dollars of cash (asset two). Such an exchange can take place any time before expiration. So, a call option gives the right to exchange cash for one unit of stock (i.e., exchange asset two for asset one) while a put option gives the right to exchange one unit of stock for cash (i.e., exchange asset one for asset two). Assets one and two have dividend yields q and r, respectively. Given this similarity we might expect call and put prices to be equal when we interchange the role of the underlying stock and cash. Let $P(\tau, S; K, r, q)$ and $C(\tau, S; K, r, q)$ respectively denote the price of the American put and call options with asset price S, strike price K, time to expiration τ, dividend yield q, and risk-free interest rate r. After interchanging the role of the underlying asset and cash, the price of the modified American put is $P(\tau, K; S, q, r)$. Since the modified put is equivalent to the American call, we have

$$P(\tau, K; S, q, r) = C(\tau, S; K, r, q). \tag{14.14}$$

This symmetry between the value function of an American call and put is called the *put-call symmetry relation*. The result is true for both European and American styles of options. It is also true for both perpetual and finite-expiration contracts. The symmetry relation

(14.14) implies that the prices of at-the-money (i.e., $S = K$) American call and put options are equal if $r = q$. A rigorous proof of this relation is left as an exercise for the reader (see Exercises 14.2–14.3).

14.2.4 Dynamic Programming Approach for Bermudan Options

Bermudans are contracts that essentially lie in between European and American derivatives. Exercise of a Bermudan option can only occur at a fixed set of dates. Let

$$\mathbf{T} := \{t_j \ : \ j = 1, 2, \ldots, M\}$$

be the set of allowable exercise dates, $t_0 \leqslant t_1 < t_2 < \cdots < t_M = T$. On the one hand, the value of a Bermudan option is defined as the maximum arbitrage-free price of the contract under all possible exercise policies. Every exercise policy is a stopping time taking its values in the set $\mathbf{T}$ or taking the value ∞. Let $\mathcal{S}_{\mathbf{T}}$ denote the collection of all stopping times taking values in $\mathbf{T} \cup \{\infty\}$. Assuming that the option has not been exercised before time t, its arbitrage-free value at time t is

$$V(t, S) = \sup_{\mathcal{T} \in \mathcal{S}_{\mathbf{T} \cap [t,T]}} \widetilde{\mathrm{E}}\left[\mathrm{e}^{-r(\mathcal{T}-t)}\, \Lambda(S(\mathcal{T})) \mid S(t) = S\right], \tag{14.15}$$

where the underlying asset price process $\{S(t)\}_{t \geqslant 0}$ is assumed to be Markovian. We expect (14.15) to be a good approximation to (14.3) for small durations $\delta t_j = t_j - t_{j-1}$ between the many exercise dates. As $\max \delta t_j \to 0$ (and hence $M \to \infty$), the Bermudan option value in (14.15) converges to the (continuous-time exercise) American option value (14.3).

Since the continuous-time process $\{S(t)\}_{t \geqslant 0}$ is Markovian, $S(t_1), S(t_2), \ldots, S(t_M)$ form a Markov chain. It can be shown that the Bermudan option values at the discrete exercise dates satisfy the recurrence relation

$$\begin{aligned} V(t_j, S) &= \max\left\{\Lambda(S), \widetilde{\mathrm{E}}_{t_j}\left[\mathrm{e}^{-r(t_{j+1}-t_j)} V(t_{j+1}, S(t_{j+1})) \mid S(t_j) = S\right]\right\} \\ &:= \max\left\{\Lambda(S), V^{cont}(t_j, S)\right\} \quad \text{for } j = M-1, M-2, \ldots, 1, \end{aligned} \tag{14.16}$$

where $V(T, S) = \Lambda(S)$ at the expiration time T. The conditional expectation in (14.16) denoted by $V^{cont}(t, S)$ represents the *continuation value* of the option at time t_i. It is the time-t value of the option that has not been exercised yet at time t.

Let the asset price process be a continuous-time stochastic process with assumed risk-neutral transition PDF $\tilde{p}(t, t'; S, S') := \frac{\widetilde{\mathbb{P}}(S(t') \in \mathrm{d}S' | S(t) = S)}{\mathrm{d}S'}$ with $S, S' > 0$ and $0 \leqslant t < t'$. Then, the continuation value in (14.16) is computed as

$$V^{cont}(t_j, S) = \mathrm{e}^{-r(t_{j+1}-t_j)} \int_0^\infty \tilde{p}(t_j, t_{j+1}; S, S') V(t_{j+1}, S') \,\mathrm{d}S'.$$

Recall that in the case of a time-homogeneous process, the transition PDF only depends on the duration $t' - t$ rather than on the individual time moments t and t', that is,

$$\tilde{p}(t, t'; S, S') = \tilde{p}(t' - t; S, S').$$

In particular, recall that for the geometric Brownian motion model in (14.1), $\tilde{p}$ is the log-normal density given by (12.16) with r replaced by $r - q$.

The dynamic-programming formulation (14.16) of the Bermudan option pricing problem does not require computation of the early-exercise boundary. However, the optimal exercise rule and early-exercise boundary can be obtained simultaneously while computing option values. In particular, the optimal exercise rule is

$$\mathcal{T}^* = \min\{t_j,\ j = 1, 2, \ldots, M \ :\ \Lambda(S(t_j)) = V(t_j, S(t_j))\}.$$

Essentially, the rule is defined in the same way as that for the binomial tree model in Chapter 6. Since the Bermudan option can only be exercised at times from the set $\mathbf{T}$, the stopping domain $\mathcal{D} \subset [0, T] \times \mathbb{R}_+$ is the union of lines

$$\mathcal{D} = \bigcup_{1 \leqslant j \leqslant M} \{(t_j, S) \ :\ \Lambda(S) \geqslant V^{cont}(t, S)\}.$$

The dynamic programming formulation of the American option pricing problem provides a basis for implementing a number of numerical methods for computing option prices using Monte Carlo simulations, finite-difference schemes, lattice methods, or a combination of such approaches. For a detailed exposition on the numerical methods for pricing American options, we refer the reader to the last chapter of this book.

14.3 Perpetual American Options

In this section, we consider a *perpetual option* with infinite time to expiration. That is, a perpetual option has no expiration date and can be exercised at any future time. Such options are not among traded securities. However, they are an instructive mathematical concept because they admit simple analytic solutions. In this section we derive closed-form pricing formulae for perpetual American calls and puts. Additionally, perpetual options represent an accurate approximation of finite-expiration American options when the time horizon of the expiry is very long.

Consider a perpetual derivative with intrinsic value $\Lambda(S)$. We are interested in its value as a function of the underlying asset price. The holder of a perpetual American option can exercise it at any time. Since the time to expiry is always infinite, the option value must be independent of time. That is, the value of a perpetual derivative, V, depends only on the asset price, i.e., we simply write $V(t, S) = V(S)$ for all $t \geqslant 0$. It is reasonable to expect that the early-exercise boundary that separates the continuation and stopping domains does not depend on the time variable t as well. The function S_t^* that describes the early-exercise boundary is constant, i.e., there is a fixed exercise level $S_t^* = S^*$ for all $t \geqslant 0$. The stopping domain (on the time and stock price plane) is $\mathcal{D} = \mathbb{R}_+ \times (0, S^*]$ for a perpetual put option and $\mathcal{D} = \mathbb{R}_+ \times [S^*, \infty)$ for a perpetual call option.

14.3.1 Pricing a Perpetual Put Option

Let us consider the case of a perpetual American put option with intrinsic value $(K-S)^+$. The holder will exercise the option when it is deep enough in the money. That is, there exists some *constant* level $S^* \leqslant K$ so that it is optimal to exercise as soon as the price $S(t)$ reaches the level S^*. From the definition of the optimal stopping time $\mathcal{T}^*$ as given in (14.4) and (14.10), we conclude that the value of a perpetual put, $P(S) \equiv P(S; K, r, q)$, is

$$P(S) = \widetilde{\mathrm{E}}_{0,S}\left[\mathrm{e}^{-r\mathcal{T}^*}\,(K - S(\mathcal{T}^*))\right] \tag{14.17}$$

$$= (K - S^*)\widetilde{\mathrm{E}}_S\left[\mathrm{e}^{-r\mathcal{T}^*}\right] \ \text{ for } S > S^* \tag{14.18}$$

where $\widetilde{\mathrm{E}}_{0,S} \equiv \widetilde{\mathrm{E}}_S$ is the risk-neutral expectation conditional on $S(0) = S$, and $\mathcal{T}^*$ is the first passage time to the level S^*, i.e.,

$$\mathcal{T}^* = \inf\{t \geqslant 0 \; : \; S(t) = S^*\}.$$

Under the assumption that the asset price follows the geometric Brownian motion model, the time $\mathcal{T}^*$ is the first passage time of the drifted Brownian motion $X(t) = \mu t + \widetilde{W}(t)$, with drift $\mu = \frac{r-q-\sigma^2/2}{\sigma}$, to the level $L^* = \frac{1}{\sigma}\ln\frac{S^*}{S}$, i.e.,

$$\mathcal{T}^* = \inf\{t \geqslant 0 \; : \; X(t) \leqslant L^*\}.$$

We recall from Section 10.4.2 of Chapter 10 that this corresponds to the first hitting time, $\tau^X_{L^*}$, of the drifted Brownian motion down to the level $L^* < 0$ in case $S > S^*$. The mathematical expectation in (14.18) is actually the Laplace transform of the PDF of $\mathcal{T}^*$, evaluated at r. This Laplace transform can be readily obtained analytically. However, it can also be computed via the Feynman–Kac formula for processes stopped at a first passage time. In this case the stopping time $\mathcal{T}^*$ is the first hitting time to the level S^*. In particular, the function $v(S) = \widetilde{\mathrm{E}}_S\left[\mathrm{e}^{-r\mathcal{T}^*}\right]$ satisfies the ordinary differential equation (ODE) $\mathcal{G}v - rv = 0$ where $\mathcal{G}$ is the Generator for the stock price process under the risk-neutral measure, i.e.,

$$\frac{1}{2}\sigma^2 S^2 \frac{\mathrm{d}^2 v}{\mathrm{d}S^2} + (r-q)S\frac{\mathrm{d}v}{\mathrm{d}S} - rv = 0, \quad S > S^*, \tag{14.19}$$

subject to the boundary conditions

$$\text{(i) } v(S^*) = 1, \quad \text{(ii) } \lim_{S\to\infty} v(S) = 0, \tag{14.20}$$

where S^* is yet unknown but uniquely determined once $v(S)$ is obtained in terms of S^*, as described just below. The first boundary condition corresponds to $\mathcal{T}^*$ being zero ($\mathrm{e}^{-r\mathcal{T}^*} = 1$) if the process starts at $S = S^*$. The second condition arises since $\mathcal{T}^*$ goes to infinity ($\mathrm{e}^{-r\mathcal{T}^*} \to 0$) as the process is started at an arbitrarily large value S (i.e., the point at infinity is a natural boundary for the stock price process).

The value of the perpetual put given by $P(S) = (K - S^*)v(S)$ satisfies the ODE (14.19) subject to

$$\text{(i) } P(S^*) = (K - S^*), \quad \text{(ii) } \lim_{S\to\infty} P(S) = 0. \tag{14.21}$$

Note that the ODE (14.19) is also obtained from the Black–Scholes partial differential equation (PDE) by setting the time derivative to zero. As is demonstrated below, the pricing function of a finite-expiration American option does indeed satisfy the Black–Scholes PDE (14.31) in the stopping domain. Since the perpetual option value V is time independent, the time derivative $\frac{\partial V}{\partial t}$ is identically zero. Hence, the Black–Scholes PDE (14.31) reduces to the ODE (14.19).

Equation (14.19) is of the Cauchy–Euler type, $ax^2y''(x) + bxy'(x) + cy(x) = 0$, with constants a, b, and c. The general solution has the form $y(x) = A_1x^{\lambda_1} + A_2x^{\lambda_2}$, where A_1 and A_2 are arbitrary constants. Putting the function $v(S) = S^\lambda$ into (14.19) gives the following auxiliary equation on the exponent λ:

$$\frac{\sigma^2}{2}\lambda(\lambda-1) + (r-q)\lambda - r = 0.$$

Solving it gives

$$\lambda = \lambda_\pm := \frac{-(r-q-\sigma^2/2) \pm \sqrt{(r-q-\sigma^2/2)^2 + 2\sigma^2 r}}{\sigma^2}. \tag{14.22}$$

Assuming positive interest rate r, it is easy to show that $\lambda_- < 0$ and $\lambda_+ > 0$. So the general solution to (14.19) is $v(S) = a_+ S^{\lambda_+} + a_- S^{\lambda_-}$. Now, we determine the coefficients $a_\pm$ in terms of S^*. To satisfy condition (14.20(ii)),

$$\lim_{S\to\infty} \left(a_+ S^{\lambda_+} + a_- S^{\lambda_-}\right) = 0,$$

we must have $a_+ = 0$. Then, a_- is determined from condition (14.20(i)):

$$v(S^*) = a_-(S^*)^{\lambda_-} = 1$$

for an arbitrary yet undetermined parameter S^*. Thus, the solution to the problem (14.19)–(14.20) is

$$v(S) = \left(\frac{S}{S^*}\right)^{\lambda_-}, \quad S > S^*.$$

In fact, the function $v(S)$ is the price of a so-called digital option that pays \$1 when the asset price breaches a lower level S^*. The perpetual put value is then

$$P(S) \equiv P(S; S^*) = \begin{cases} (K - S^*)\left(\frac{S}{S^*}\right)^{\lambda_-}, & S^* < S, \\ K - S, & 0 < S \leqslant S^*. \end{cases} \tag{14.23}$$

The last step is to determine S^*. Consider two methods: first, using the fact that S^* must be the optimal value that maximizes the option value $P(S)$ among all possible choices of $S^* > 0$; second, using the smooth pasting condition. Let us find the maximum of the function $S^* \mapsto P(S; S^*)$. Differentiating the option value with respect to S^* for $S^* < S$ gives

$$\frac{\partial P}{\partial S^*} = \frac{\partial}{\partial S^*}\left\{(K - S^*)\left(\frac{S}{S^*}\right)^{\lambda_-}\right\} = -\left(\frac{S}{S^*}\right)^{\lambda_-}\left(1 + \frac{K - S^*}{S^*}\lambda_-\right).$$

Set $\frac{\partial P}{\partial S^*} = 0$ to obtain the extremum

$$S^* = \frac{K\lambda_-}{\lambda_- - 1}. \tag{14.24}$$

Note that $\frac{\partial^2 P}{\partial (S^*)^2} = \frac{K\lambda_-}{(S^*)^2}\left(\frac{S}{S^*}\right)^{\lambda_-} < 0$ since $\lambda_- < 0$. Thus, S^* given by (14.24) is indeed a maximum. Finally, substituting (14.24) in (14.23) gives the perpetual put price:

$$P(S; K, r, q) = \begin{cases} \frac{K}{1-\lambda_-}\left(\frac{\lambda_- - 1}{\lambda_-}\right)^{\lambda_-}\left(\frac{S}{K}\right)^{\lambda_-} = -\frac{S^*}{\lambda_-}\left(\frac{S}{S^*}\right)^{\lambda_-}, & S^* < S, \\ K - S, & 0 < S \leqslant S^*. \end{cases} \tag{14.25}$$

The solution in (14.25) is also readily obtained by applying the smooth pasting condition to the function in (14.23). Namely, we differentiate the function for $S > S^*$ and set the derivative to -1 at $S = S^*$:

$$\left.\frac{\partial P}{\partial S}\right|_{S=S^*} = -\lambda_-\left(\frac{K - S^*}{S^*}\right) = -1.$$

Solving the above equation for S^* gives the optimal value in (14.24).

Let us examine what happens to the exercise boundary in the limiting case when the interest rate r is zero. From (14.22), we obtain $\lambda_- = 0$. Thus, from (14.24) we see that $S^* = 0$. Hence, for a zero interest rate, the perpetual put is never exercised early. This observation is consistent with the property of any finite-expiration American put.

14.3.2 Pricing a Perpetual Call Option

Now we consider the perpetual American call struck at K. Let $C(S) \equiv C(S; K, r, q)$ denote its price when the stock price is equal to S. As in the case of the perpetual put option, the call pricing function $C(S)$ satisfies the ODE (14.19) but now for $S \in (0, S^*)$. In the stopping domain $[S^*, \infty)$, the call value is equal to the payoff value, i.e.,

$$C(S) = S - K \quad \text{for} \quad S \geqslant S^*.$$

Hence, the boundary conditions are

$$\text{(i)} \lim_{S \to 0+} C(S) = 0, \quad \text{(ii)}\ C(S^*) = (S^* - K). \tag{14.26}$$

The first condition states that the call is worthless if the stock price is zero since it will remain at zero indefinitely, i.e., the payoff is always zero. The general solution to (14.19) is $C(S) = a_+ S^{\lambda_+} + a_- S^{\lambda_-}$, where the exponents $\lambda_\pm$ are given by (14.22). Since $C(0+) = 0$ and $\lambda_- < 0$, we must set $a_- = 0$. Imposing the condition (14.26(ii)) gives $a_+ = (S^* - K)/(S^*)^{\lambda_+}$. Following the same procedure as above, we obtain the optimal value of S^*:

$$S^* = \frac{K\lambda_+}{\lambda_+ - 1}. \tag{14.27}$$

The perpetual call price is then

$$C(S; K, r, q) = \begin{cases} \frac{K}{\lambda_+ - 1}\left(\frac{\lambda_+ - 1}{\lambda_+}\right)^{\lambda_+}\left(\frac{S}{K}\right)^{\lambda_+} = \frac{S^*}{\lambda_+}\left(\frac{S}{S^*}\right)^{\lambda_+}, & 0 < S < S^*, \\ S - K, & S^* \leqslant S. \end{cases} \tag{14.28}$$

A simple check by differentiation shows that this function satisfies the required smooth pasting condition for a call,

$$\left.\frac{\partial C}{\partial S}\right|_{S=S^*} = 1.$$

Let us examine what happens to the exercise boundary in the limiting case when the dividend yield q is zero. From (14.22) and (14.27), we see that $\lambda_+ \to 1$ and $S^* \to \infty$ as $q \to 0+$. Hence, for a zero dividend yield, the perpetual call is never exercised early. Again, this is consistent with the property of any finite-expiration American call.

14.4 Finite-Expiration American Options

14.4.1 The PDE Formulation

Pricing an American option can be formulated as a boundary initial value problem for a partial differential equation with a time-dependent free boundary. The solution domain is divisible into a union of a stopping region, $\mathcal{D}$, where the American option is exercised, and a continuation domain, $\mathcal{D}^c$, where the option is not exercised. Inside the continuation domain, the option value function satisfies the Black–Scholes PDE. On the early-exercise boundary, the option value is equal to the payoff function $\Lambda(S)$. The early-exercise boundary is an unknown function of time which must also be determined as part of the solution. Here we assume the payoff is time independent, although the formulation also extends to the case of a time-dependent payoff.

Delta hedging and continuous-time replication arguments apply to American options in the same way as they apply to European options. Therefore, within the continuation domain the option pricing function must satisfy the Black–Scholes PDE. The connection between the optimal stopping time formulation and the PDE approach can be shown using the following heuristic. Consider the recurrence relation (14.16) for a calendar time $t \in [t_0, T]$ and a small time step $\delta t > 0$:

$$V(t,S) = \max\left\{\Lambda(S),\, \mathrm{e}^{-r\delta t}\widetilde{\mathrm{E}}\big[V(t+\delta t, S(t+\delta t)) \mid S(t) = S\big]\right\}.$$

Assuming $V(t,S)$ is a sufficiently smooth function with continuous derivatives, we can expand $V(t+\delta t, S(t+\delta t))$ in a Taylor series while keeping terms up to $\mathcal{O}(\delta t)$. Assuming the underlying asset follows the GBM model in (14.1), we have:

$$\begin{aligned} V(t,S) &= \max\left\{\Lambda(S),\, (1 - r\delta t)\widetilde{\mathrm{E}}_{t,S}\left[V(t,S(t)) + \left(\frac{\partial V(t,S(t))}{\partial t} + (r-q)S(t)\frac{\partial V(t,S(t))}{\partial S}\right.\right.\right. \\ &\quad \left.\left.\left. + \frac{1}{2}\sigma^2 S^2(t)\frac{\partial^2 V(t,S(t))}{\partial S^2}\right)\delta t + \sigma S(t)\frac{\partial V(t,S(t))}{\partial S}\delta\widetilde{W}_t\right]\right\} + \mathcal{O}\left(\delta t^2\right) \\ &= \max\left\{\Lambda(S),\, V(t,S) + \left[\frac{\partial V(t,S)}{\partial t} + \mathcal{L}V(t,S)\right]\delta t\right\} + \mathcal{O}(\delta t^2). \end{aligned} \tag{14.29}$$

The second equation is obtained by evaluating the expectation conditional on $S(t) = S$ and then collecting terms up to $\mathcal{O}(\delta t)$. Note that the coefficient terms multiplying δt and $\delta\widetilde{W}_t$ are both $\mathcal{F}_t$-measurable with given spot value $S(t) = S$. Moreover, the conditional expectation term in $\delta\widetilde{W}_t$ vanishes by the independence of the Brownian increment $\delta\widetilde{W}_t = \widetilde{W}_{t+\delta t} - \widetilde{W}_t$ and $S(t)$, i.e., $\widetilde{\mathrm{E}}_{t,S}[S(t)\frac{\partial V(t,S(t))}{\partial S}\delta\widetilde{W}_t] = S\frac{\partial V(t,S)}{\partial S}\widetilde{\mathrm{E}}_{t,S}[\delta\widetilde{W}_t] = 0$ since $\widetilde{\mathrm{E}}_{t,S}[\delta\widetilde{W}_t] = \widetilde{\mathrm{E}}[\delta\widetilde{W}_t] = 0$. The expression in (14.29) has been written more compactly using the Black–Scholes differential operator $\mathcal{L} := \mathcal{G} - r$,

$$\mathcal{L}f := \frac{1}{2}\sigma^2 S^2\frac{\partial^2 f}{\partial S^2} + (r-q)S\frac{\partial f}{\partial S} - rf. \tag{14.30}$$

For values of S in the continuation domain $\mathcal{D}_t^c$ the inequality $V(t,S) > \Lambda(S)$ is satisfied. In (14.29), we require the right-hand side to equal the left to small order $o(\delta t)$. Hence, the coefficient in δt must be zero, which is the Black–Scholes PDE:

$$\frac{\partial V(t,S)}{\partial t} + \mathcal{L}V(t,S) = 0, \text{ for all } S \in \mathcal{D}_t^c \text{ and all } t \in [t_0, T].$$

Thanks to the time-homogeneous property of the solution, $V(t,S) = v(\tau, S)$, we have a PDE in terms of the time to maturity variable $\tau = T - t$ and spot S:

$$\frac{\partial v(\tau,S)}{\partial \tau} = \mathcal{L}v(\tau,S), \text{ for all } S \in \mathcal{D}^c(\tau) \text{ and all } \tau > 0. \tag{14.31}$$

The Black–Scholes PDE does not hold on the stopping domain, where the American option is given by the payoff function $V(S,\tau) = \Lambda(S)$. Since the payoff is time-independent, the derivative $\frac{\partial\Lambda(S)}{\partial\tau}$ is zero. Thus, the solution $v(\tau,S)$ on $\mathcal{D}(\tau)$ satisfies $\frac{\partial v}{\partial \tau} = 0$. Combining regions and assuming the payoff is a twice differentiable function gives a *nonhomogeneous* Black–Scholes PDE valid in the whole region:

$$\frac{\partial v(\tau,S)}{\partial \tau} = \mathcal{L}v(\tau,S) + f(\tau,S), \tag{14.32}$$

with function

$$f(\tau, S) = \begin{cases} 0, & S \in \mathcal{D}^c(\tau), \\ -\mathcal{L}\Lambda(S), & S \in \mathcal{D}(\tau). \end{cases} \tag{14.33}$$

Given the function $f(\tau, S)$ whose time dependence is determined in terms of the early-exercise boundary, the solution to (14.32), subject to the initial condition

$$v(0, S) = \Lambda(S) \tag{14.34}$$

and the boundary conditions

$$\lim_{S\to 0+} v(\tau, S) = \lim_{S\to 0+} \Lambda(S), \quad \lim_{S\to\infty} v(\tau, S) = \lim_{S\to\infty} \Lambda(S), \tag{14.35}$$

can be obtained in terms of the solution to the corresponding (homogeneous) Black–Scholes PDE, $\frac{\partial v(\tau,S)}{\partial \tau} = \mathcal{L}v(\tau, S)$.

For the American call and put, we recall the early-exercise curves $S = S^*(\tau)$ expressed as a function of time to maturity on the (S, τ)-plane. For the call, the set of spot values $S \in \mathcal{D}(\tau)$ is then equivalent to the set of values $S \geqslant S^*(\tau)$. For the put, the values $S \in \mathcal{D}(\tau)$ are the values $S \leqslant S^*(\tau)$. Hence, expressing the call and put option values as functions of τ and S, (14.32) and (14.33) specialize to

$$\frac{\partial C}{\partial \tau} - \frac{\sigma^2 S^2}{2}\frac{\partial^2 C}{\partial S^2} - (r-q)S\frac{\partial C}{\partial S} + rC = \begin{cases} 0 & S < S^*(\tau), \\ qS - rK, & S \geqslant S^*(\tau), \end{cases} \tag{14.36}$$

and

$$\frac{\partial P}{\partial \tau} - \frac{\sigma^2 S^2}{2}\frac{\partial^2 P}{\partial S^2} - (r-q)S\frac{\partial P}{\partial S} + rP = \begin{cases} rK - qS & S \leqslant S^*(\tau), \\ 0 & S > S^*(\tau), \end{cases} \tag{14.37}$$

respectively. For the call, we used the fact that $-\mathcal{L}\Lambda(S) = -\mathcal{L}(S-K) = -(rK - qS) = qS - rK$. For the put, $-\mathcal{L}\Lambda(S) = -\mathcal{L}(K - S) = rK - qS$. Here, we used $S^*(\tau)$ to denote the early-exercise boundary for the respective call and put with given strike K. The right-hand sides of these nonhomogeneous PDEs are nonzero only within the respective stopping regions.

We close this section by working out some other basic properties of the early-exercise boundary for a call and a put option. Let's consider the limit of infinitesimally small $\tau \searrow 0$. In particular, consider an American call struck at K with continuous dividend yield q. The pricing function $C(\tau, S; K)$, is an increasing function of τ, i.e., $C(\tau_2, S; K) \geqslant C(\tau_1, S; K)$ for $\tau_2 > \tau_1$. Also, the smooth pasting condition guarantees that the pricing functions join the intrinsic line at levels $S^*(\tau_1) - K$ and $S^*(\tau_2) - K$, respectively, giving $S^*(\tau_2) > S^*(\tau_1)$. We hence conclude that $S^*(\tau)$ is a continuously increasing function of $\tau > 0$ and attains the value of the early-exercise boundary of the corresponding perpetual call option, given by (14.27), in the limit $\tau \to \infty$. That is, an American call with greater time to maturity should be exercised deeper in the money to account for the loss of time value on the strike K. Since one would never prematurely exercise at a spot value that is below the strike level, the early-exercise boundary for an American call must satisfy the property $S^*(\tau) > K$ for all $\tau > 0$. To determine the boundary in the limit $\tau \searrow 0$, we note that the option value approaches the intrinsic value, i.e., at expiry the option value is given by the payoff $C(S, K, \tau = 0) = S - K$ for values on the exercise boundary. According to (14.36) we have

$$\frac{\partial C(S, K, 0+)}{\partial \tau} = rK - qS \tag{14.38}$$

for $S > K$. Since the condition $\partial C(S, K, 0+)/\partial \tau > 0$ ensures that the option is not yet exercised, the spot value S at which $\partial C(S, K, 0+)/\partial \tau$ becomes negative and hence for which the call is exercised at an instant just before expiry is given by $S = \frac{r}{q}K$. This is the case, however, if the value $\frac{r}{q}K$ is in the interval $S > K$; that is if $r > q > 0$. In this case just prior to expiry the call is not yet exercised if the spot is in the region $K < S < \frac{r}{q}K$, but would be exercised if $S \geqslant \frac{r}{q}K$. Hence, in the limiting case we have $S^*(0+) = \frac{r}{q}K$ for $r > q > 0$. For the other case we have $r \leqslant q$, so $\frac{r}{q}K \leqslant K$. Yet $S > K$, so that $S^*(0+) = K$ for $r > q$. Note that the condition $S^*(0+) > K$ is not possible in this case as this leads to a sub-optimal early exercise since the loss in dividends would have greater value than the interest earned over the infinitesimal time interval until expiry. Combining the above arguments we arrive at the general limiting condition for the exercise boundary of an American call just prior to expiry:

$$\lim_{\tau \searrow 0} S^*(\tau) = \max\left(K, \frac{r}{q}K\right) = K \max\left(1, \frac{r}{q}\right). \tag{14.39}$$

From this property we see that $S^*(0+) \to \infty$ as $q \to 0$. So, for zero dividend yield the American call is never exercised early which is consistent with the fact that the American call has exactly the same worth as the European call. By using similar arguments as above we readily work out $S^*(0+)$ for the Amercian put struck at K with continuous dividend yield q. For the put, $S^*(\tau)$ is a continuously decreasing function of $\tau > 0$ and attains the value in (14.24) in the limit $\tau \to \infty$. We leave it as an exercise for the reader to show that

$$\lim_{\tau \searrow 0} S^*(\tau) = K \min\left(1, \frac{r}{q}\right). \tag{14.40}$$

14.4.2 The Integral Equation Formulation

Recall that the time-homogeneous risk-neutral transition PDF, $\tilde{p} = \tilde{p}(\tau; S, S')$, solves the forward Kolmogorov PDE in the S' variable and the backward Kolmogorov PDE in the spot variable S with zero boundary conditions at $S = 0$ and $S = \infty$ for all $\tau > 0$. As mentioned above, for the process (14.1) the PDF $\tilde{p}$ is just the log-normal density. We also know that the function $\mathrm{e}^{-r\tau}\tilde{p}$ solves the (homogeneous) Black–Scholes PDE. Combining these facts and applying Laplace transforms, one arrives at the well-known Duhamel solution to (14.32) in the form

$$\begin{aligned} v(\tau, S) &= \mathrm{e}^{-r\tau} \int_0^\infty \tilde{p}(\tau; S, S')\Lambda(S')\,\mathrm{d}S' + \int_0^\tau \mathrm{e}^{-r\tau'} \left[\int_0^\infty \tilde{p}(\tau'; S, S') f(\tau - \tau', S')\,\mathrm{d}S'\right] \mathrm{d}\tau' \\ &\equiv v^E(\tau, S) + v^e(\tau, S) \end{aligned} \tag{14.41}$$

where $f(\tau, S) = -\mathcal{L}\Lambda(S) \cdot \mathbb{I}_{\{S \in \mathcal{D}(\tau)\}}$ is defined in (14.33). An important aspect of this result is the fact that the American option value is expressible as a sum of two components. The first term is simply the standard European option value v^E, as given by the discounted risk-neutral expectation of the payoff. Hence the second term, denoted by v^e, must represent the early-exercise premium which the holder must pay to have the additional liberty of early exercise, i.e., $v^e(\tau, S) = v(\tau, S) - v^E(\tau, S)$.

For a call, $f(\tau, S) = (qS - rK)\mathbb{I}_{\{S \geqslant S^*(\tau)\}}$, i.e., $f(\tau - \tau', S') = (qS' - rK)\mathbb{I}_{\{S' \geqslant S^*(\tau - \tau')\}}$. For a given value of τ', this restricts the inner integral in (14.41) to the interval $[S^*(\tau - \tau'), \infty)$. Similarly, for a put, $f(\tau - \tau', S') = (rK - qS')\mathbb{I}_{\{0 < S' \leqslant S^*(\tau - \tau')\}}$, the inner integral in S' is restricted to the interval $(0, S^*(\tau - \tau')]$. Using (14.41), the solutions to (14.36) and (14.37) for the American call and put prices are given by:

$$C(\tau, S; K) = C^E(\tau, S; K) + C^e(\tau, S; K)$$

and

$$P(\tau, S; K) = P^E(\tau, S; K) + P^e(\tau, S; K)$$

respectively, where the corresponding early-exercise premiums take on the integral forms

$$C^e(\tau, S; K) = \int_0^\tau e^{-r\tau'} \left[\int_{S^*(\tau-\tau')}^{\infty} \tilde{p}(\tau'; S, S')(qS' - rK)\, dS' \right] d\tau' \tag{14.42}$$

and

$$P^e(\tau, S; K) = \int_0^\tau e^{-r\tau'} \left[\int_0^{S^*(\tau-\tau')} \tilde{p}(\tau'; S, S')(rK - qS')\, dS' \right] d\tau' . \tag{14.43}$$

The above premiums can also be recast as

$$C^e(\tau, S; K) = \int_0^\tau e^{-r\tau'} \widetilde{\mathrm{E}}_{0,S} \left[(qS(\tau') - rK)\, \mathbb{I}_{\{S(\tau') \geqslant S^*(\tau-\tau')\}} \right] d\tau' \tag{14.44}$$

and

$$P^e(\tau, S; K) = \int_0^\tau e^{-r\tau'} \widetilde{\mathrm{E}}_{0,S} \left[(rK - qS(\tau'))\, \mathbb{I}_{\{S(\tau') \leqslant S^*(\tau-\tau')\}} \right] d\tau' , \tag{14.45}$$

where $\widetilde{\mathrm{E}}_{0,S}$ denotes the time-0 risk-neutral expectation conditional on asset paths starting at $S(0) = S$. The time integral is over all intermediate times to maturity and the indicator functions ensure that all asset paths fall within the early-exercise region. The properties of the early-exercise boundaries established in the previous section guarantee that the early-exercise premiums are nonnegative. Using (14.39), for a dividend paying call we have $S(\tau') \geqslant S^*(\tau)$, i.e., $S(\tau') \geqslant K \max\left(1, \frac{r}{q}\right) \geqslant \frac{r}{q} K$. Hence, $qS(\tau') - rK \geqslant 0$ and $C^e \geqslant 0$. A similar analysis follows for the put premium where (14.40) gives $rK - qS(\tau') \geqslant 0$ and $P^e \geqslant 0$. The exercise premiums hence involve a continuous stream of discounted expected cash flows from contract inception until maturity.

For the geometric Brownian motion model (with constants r, q, and σ) the function $\tilde{p}$ is given by the log-normal density and the above double integrals readily simplify to single time integrals in terms of standard cumulative normal functions. Namely, the expectations in the integrands of (14.44) and (14.45) are easily computed by making use of the stock price representation

$$S(\tau') = S(0) e^{(r-q-\sigma^2/2)\tau' + \sigma \widetilde{W}(\tau')} \stackrel{d}{=} S(0) e^{(r-q-\sigma^2/2)\tau' + \sigma\sqrt{\tau'}\widetilde{Z}}$$

where $\widetilde{Z} \sim Norm(0,1)$ under measure $\widetilde{\mathbb{P}}$. Note that the quantity $S^*(\tau - \tau')$ is simply a constant, i.e., it is the value of the early-exercise boundary for the time to maturity $\tau - \tau'$. By employing the identity in (A.1) of the Appendix, and using the usual steps as in the derivation of the standard European call, the expectation in (14.44) is given by

$$\begin{aligned}
\widetilde{\mathrm{E}}_{0,S} \left[(qS(\tau') - rK)\, \mathbb{I}_{\{S(\tau') \geqslant S^*(\tau-\tau')\}} \right] &= qS e^{(r-q-\sigma^2/2)\tau'}\, \widetilde{\mathrm{E}} \left[e^{\sigma\sqrt{\tau'}\widetilde{Z}} \mathbb{I}_{\{\widetilde{Z} \geqslant -d^*_-(\tau')\}} \right] \\
&\quad - rK \widetilde{\mathbb{P}}(\widetilde{Z} \geqslant -d^*_-(\tau')) \\
&= qS e^{(r-q)\tau'} \mathcal{N}(d^*_+(\tau')) - rK \mathcal{N}(d^*_-(\tau'))
\end{aligned}$$

where we define $d^*_\pm(\tau') := \dfrac{\ln \frac{S}{S^*(\tau-\tau')} + \left(r - q \pm \frac{1}{2}\sigma^2\right)\tau'}{\sigma\sqrt{\tau'}}$. By a similar calculation, the expectation in (14.45) is given by

$$\widetilde{\mathrm{E}}_{0,S}\left[(rK - qS(\tau'))\,\mathbb{I}_{\{S(\tau')\leqslant S^*(\tau-\tau')\}}\right] = rK\mathcal{N}(-d^*_-(\tau')) - qS\mathrm{e}^{(r-q)\tau'}\mathcal{N}(-d^*_+(\tau')).$$

By inserting these expressions into (14.44) and (14.45), and using the pricing functions of the standard European call and put options, we have the explicit integral representations for the price of the American call and put:

$$\begin{aligned} C(\tau, S; K) &= S\mathrm{e}^{-q\tau}\mathcal{N}(d_+) - K\mathrm{e}^{-r\tau}\mathcal{N}(d_-) \\ &\quad + \int_0^\tau \left[qS\mathrm{e}^{-q\tau'}\mathcal{N}(d^*_+(\tau')) - rK\mathrm{e}^{-r\tau'}\mathcal{N}(d^*_-(\tau'))\right] \mathrm{d}\tau', \end{aligned} \tag{14.46}$$

$$\begin{aligned} P(\tau, S; K) &= K\mathrm{e}^{-r\tau}\mathcal{N}(-d_-) - S\mathrm{e}^{-q\tau}\mathcal{N}(-d_+) \\ &\quad + \int_0^\tau \left[rK\mathrm{e}^{-r\tau'}\mathcal{N}(-d^*_-(\tau')) - qS\mathrm{e}^{-q\tau'}\mathcal{N}(-d^*_+(\tau'))\right] \mathrm{d}\tau', \end{aligned} \tag{14.47}$$

where $d_\pm = \frac{\ln \frac{S}{K} + \left(r-q\pm\frac{1}{2}\sigma^2\right)\tau}{\sigma\sqrt{\tau}}$.

These integral representations are valid for $S \in (0,\infty)$, $\tau \geqslant 0$. By setting the spot equal to the boundary value, $S = S^*(\tau)$, and applying the respective boundary conditions, $C(\tau, S^*(\tau); K) = S^*(\tau) - K$ for the call and $P(\tau, S^*(\tau); K) = K - S^*(\tau)$ for the put, (14.46) and (14.47) give rise to integral equations for the early-exercise boundary. For the call:

$$\begin{aligned} S^*(\tau) - K &= S^*(\tau)\mathrm{e}^{-q\tau}\mathcal{N}(\hat{d}_+) - K\mathrm{e}^{-r\tau}\mathcal{N}(\hat{d}_-) \\ &\quad + \int_0^\tau \left[qS^*(\tau)\mathrm{e}^{-q\tau'}\mathcal{N}(\hat{d}^*_+(\tau')) - rK\mathrm{e}^{-r\tau'}\mathcal{N}(\hat{d}^*_-(\tau'))\right] \mathrm{d}\tau', \end{aligned} \tag{14.48}$$

and separately for the put:

$$\begin{aligned} K - S^*(\tau) &= K\mathrm{e}^{-r\tau}\mathcal{N}(-\hat{d}_-) - S^*(\tau)\mathrm{e}^{-q\tau}\mathcal{N}(-\hat{d}_+) \\ &\quad + \int_0^\tau \left[rK\mathrm{e}^{-r\tau'}\mathcal{N}(-\hat{d}^*_-(\tau')) - qS^*(\tau)\mathrm{e}^{-q\tau'}\mathcal{N}(-\hat{d}^*_+(\tau'))\right] \mathrm{d}\tau', \end{aligned} \tag{14.49}$$

where

$$\hat{d}_\pm = \frac{\ln \frac{S^*(\tau)}{K} + \left(r - q \pm \frac{1}{2}\sigma^2\right)\tau}{\sigma\sqrt{\tau}} \quad \text{and} \quad \hat{d}^*_\pm(\tau') = \frac{\ln \frac{S^*(\tau)}{S^*(\tau-\tau')} + \left(r - q \pm \frac{1}{2}\sigma^2\right)\tau'}{\sigma\sqrt{\tau'}}.$$

Note that (14.48) and (14.49) involve a variable upper integration limit and the integrands are nonlinear functions of $S^*(\tau)$, $S^*(\tau')$, τ, and τ'. From the theory of integral equations, (14.48) and (14.49) are known as *nonlinear Volterra integral equations*. Note that the solution $S^*(\tau)$, at time to maturity τ, is dependent on the solution $S^*(\tau-\tau')$ from zero time to maturity up to time τ. Although (14.48) and (14.49) are not analytically tractable, simple and efficient algorithms can be employed to solve for $S^*(\tau)$ numerically. A typical procedure involves the division of the solution domain into a regular mesh: $\tau_0 = 0$, $\tau_i = ih$, $i = 1, \ldots, n$, with n steps spaced as $h = \tau/n$. By approximating the time integral via a quadrature rule (e.g., the trapezoidal rule) one can obtain a system of algebraic equations in the values $S^*(\tau_i)$ which can be iteratively solved starting from the known value $S^*(\tau_0) = S^*(\tau = 0+)$ at zero time to maturity. Once the early-exercise boundary is determined, the integral in (14.46) or

(14.47) for the respective call or put can be computed. In particular, a quadrature rule that makes use of the computed points $S^*(\tau_i)$ can be implemented. Accurate approximations to the early-exercise boundary are obtained by choosing the number n of points to be sufficiently large.

14.5 Exercises

Exercise 14.1. Prove Proposition 14.3 for an arbitrary American option with a differentiable payoff function Λ. In particular show the following.

(a) At any point (t, S_t^*) of the early-exercise boundary, the American option pricing function V satisfies the smooth pasting condition:

$$\left.\frac{\partial V(t,S)}{\partial S}\right|_{S=S_t^*} = \Lambda'(S_t^*).$$

(b) The option value V satisfies the zero time-decay condition on the early-exercise domain,

$$\frac{\partial V(t,S)}{\partial t} = 0, \quad \text{for all} \quad S \in \mathcal{D}_t.$$

Exercise 14.2. Let $P(S; K, r, q)$ and $C(S; K, r, q)$, respectively, denote the price functions of the perpetual American put and call options struck at K. The underlying asset price process follows geometric Brownian motion (14.1). Using the closed-form pricing formula (14.25) and (14.28), show that the option prices satisfy the put-call symmetry relation

$$P(K; S, q, r) = C(S; K, r, q).$$

Exercise 14.3. Let $P(S, \tau; K, r, q)$ and $C(S, \tau; K, r, q)$, respectively, denote the price function of the American put and call options with strike price K and time to expiration τ. The underlying asset price process follows geometric Brownian motion (14.1). Show that $P(K, \tau; S, q, r)$ satisfies the Black–Scholes PDE along with the auxiliary conditions

$$P(K,0;S,q,r) = (S-K)^+, \quad P(K,\tau;S,q,r) \geqslant (S-K)^+ \;\text{ for }\; \tau > 0.$$

Since the auxiliary conditions are identical to those of the American call option price, the put and call price functions satisfy the put-call symmetry relation

$$P(K, \tau; S, q, r) = C(S, \tau; K, r, q).$$

Exercise 14.4. Consider a call-like quadratic payoff $\Lambda(S) = a(S-K)^2 \mathbf{1}_{S \geqslant K}$, with fixed strike K and where a is some positive constant factor. Derive an analytical expression for the early-exercise boundary value S^* as well the perpetual American pricing function $V(S)$ for this payoff. Assume the geometric Brownian motion asset price model in (14.1). Express your answer in terms of the spot S and the parameters a, K, r, q, σ.

Exercise 14.5. Consider a butterfly spread option with payoff centred at $K > 0$ and some positive width w where $K - w > 0$.

(a) Let spot S be the vertical axis and τ (time to maturity) the horizontal axis. Provide a sketch that depicts the early-exercise boundary curve(s), and clearly include labels for shaded regions corresponding to the continuation and early-exercise domains. Determine the asymptotes and asymptotic values for all the boundary values corresponding to $S^*(\tau = 0^+)$ and $S^*(\tau = \infty)$.

(b) Provide a sketch of the American option value for this butterfly spread option as a function of spot S for a typical time to maturity $\tau > 0$. Include the payoff in your graph.

(c) Obtain an analytical pricing formula for this perpetual American butterfly option for all spot $S > 0$.

Exercise 14.6. Consider a strangle option with the payoff $\Lambda(S) = (K_1 - S)^+ + (S - K_2)^+$ where K_1 and K_2 are fixed strikes so that $0 < K_1 < K_2$. Derive an analytical expression for the early-exercise boundary value S^* as well the perpetual American pricing function $V(S)$ for this payoff. Assume the usual geometric Brownian motion asset price model in (14.1).

Exercise 14.7. Show that the pricing formulae for the American call and put in (14.46) and (14.47) satisfy the required boundary conditions at $S = 0$ and $S = \infty$.

Exercise 14.8. Using (14.46) and (14.47), derive integral representations for the delta, gamma, and vega sensitivities of the American call and put.

Exercise 14.9. Consider a Bermudan put option with strike K at maturity T with only a single intermediate early-exercise date $T_1 \in [0, T]$. Assume the underlying stock price process is a geometric Brownian motion within the risk-neutral measure, and let $P(S, T-t)$ denote the option value at calendar time t with spot S. Find an analytically closed-form expression for the present time $t = 0$ price $P(S_0, T)$. Hint: this problem is very closely related to the valuation of a compound option. In particular, proceed as follows. From backward recurrence show that

$$P(S_0, T) = \mathrm{e}^{-rT_1} \widetilde{\mathrm{E}}_0 \big[P(S(T_1), T - T_1) \big], \tag{14.50}$$

with

$$P(S(T_1), T - T_1) = \begin{cases} P^E(S(T_1), T - T_1), & S(T_1) > S^*_{T_1}, \\ K - S(T_1), & S(T_1) \leqslant S^*_{T_1}, \end{cases} \tag{14.51}$$

where P^E is the price of the European put option and the critical value $S^*_{T_1}$ for the early-exercise boundary at calendar time T_1 solves

$$P_E(S^*_{T_1}, T - T_1) = K - S^*_{T_1}.$$

Compute the above expectation as a sum of two integrals: one over the domain $S_{T_1} > S^*_{T_1}$ and the other over $0 < S(T_1) \leqslant S^*_{T_1}$, to finally arrive at the expression for $P(S_0, T)$ in terms of univariate and bivariate cumulative normal functions. Show whether $S^*_{T_1}$ is a strictly increasing or decreasing function of the volatility σ and explain your answer. What is this functional dependency for the case of a Bermudan call? Explain.

Chapter 15

Interest-Rate Modelling and Derivative Pricing

15.1 Basic Fixed Income Instruments

15.1.1 Bonds

The term *fixed income* refers to any type of investment that provides payments of a fixed amount on a fixed schedule. A most common type of fixed income investment is a bond. In Chapter 1, we introduced a zero-coupon bond (ZCB) that only returns the investor a redemption amount on the maturity date. Another example of a fixed income security is a coupon bond that provides regular payments (coupons) on a fixed schedule and a redemption value on the maturity date. Recall that $Z(t,T)$ denotes the time-t purchase price of a unit zero-coupon bond maturing at time T with $0 \leqslant t \leqslant T$, whose face value is equal to \$1 . The following notations were also introduced in Chapter 1.

y is the continuously compounded yield rate (also called the spot rate); it is generally a function of calendar time t and maturity T for a given investment time period $[t,T]$, i.e., $y = y(t,T)$. When regarded as a function of the time to maturity, $\tau = T - t$, i.e., $y(\tau)$ is the yield rate earned by money invested for a period of τ years.

r is the short rate; it is a function of calendar time t, i.e., $r = r(t)$. The short rate is the rate on instantaneous borrowing or lending and is defined as $r(t) = y(t,t)$.

The continuously compounded yield rate y and zero-coupon bond price Z are simply related by

$$Z(t,T) = e^{-y(t,T)(T-t)}\,. \tag{15.1}$$

Any cash flow stream of multiple coupon payments can be replicated by means of a portfolio of zero-coupon bonds. In particular, the time-t value $V(t;\mathbf{c},\mathbf{T})$ of a cash flow stream $(\mathbf{c},\mathbf{T})$ with $\mathbf{c} = [c_1, c_2, \ldots, c_n]$ and $\mathbf{T} = [t_1, t_2, \ldots, t_n]$, where c_i is the payment at time t_i and $t \leqslant t_1 < t_2 < \cdots < t_n$ holds, is equal to the sum of discounted cash flows,

$$V(t;\mathbf{c},\mathbf{T}) = \sum_{i=1}^{n} c_i\, Z(t,t_i) = \sum_{i=1}^{n} c_i\, \mathrm{e}^{-y(t,t_i)(t_i-t)}. \tag{15.2}$$

The above formula allows for pricing coupon-paying bonds. On the other hand, if we are given the price of a coupon-paying bond for each maturity $t_1, t_2, \ldots, t_n$, then using (15.2) we can solve recursively for the prices $Z(t,t_1), Z(t,t_2), \ldots, Z(t,t_n)$ and therefore obtain the yield rates $y(t,t_1), y(t,t_2), \ldots, y(t,t_n)$. Indeed, let the coupon bond value $V_i(t) = c_1 Z(t,t_1) + c_2 Z(t,t_2) + \cdots + c_i Z(t,t_i)$ be known for all $i = 1,2,\ldots,n$. Then the ZCB values are

$$Z(t,t_1) = \frac{V_1(t)}{c_1} \text{ and } Z(t,t_i) = \frac{V_i(t) - V_{i-1}(t)}{c_i} \text{ for } i = 2,3,\ldots,n.$$

This method of deducing zero-coupon bond values from coupon-paying bond prices is called *bootstrapping*.

15.1.2 Forward Rates

Suppose we wish to borrow an amount of A dollars for a period between times T and T' and we want to lock in the rate on the loan at time t with $t \leqslant T < T'$. To do this, at time t, we purchase A units of the zero-coupon (unit) bond maturing at time T and finance this purchase by selling short $A\frac{Z(t,T)}{Z(t,T')}$ units of a zero-coupon bond maturing at time T'. The cost of setting up this portfolio is zero. At time T, we receive \$$A$ from the long position in the T-maturity bonds. At time T', we are required to have $A\frac{Z(t,T)}{Z(t,T')}$ dollars to cover the short position in the T'-maturity bonds. Hence, to avoid arbitrage, A dollars invested during the time interval $[T, T']$ must yield an effective (continuously compounded) rate, denoted by $f(t; T, T')$, such that

$$A\mathrm{e}^{(T'-T)f(t;T,T')} = A\frac{Z(t,T)}{Z(t,T')} \implies f(t;T,T') = \frac{1}{T'-T}\ln\frac{Z(t,T)}{Z(t,T')}. \tag{15.3}$$

The rate $f(t; T, T')$ is called a *forward rate*. It is determined at time t for investing during the period $[T, T']$. Using (15.1), we express the forward rate in terms of yield rates as follows:

$$f(t;T,T') = \frac{1}{T'-T}\ln\left(\frac{\mathrm{e}^{-y(t,T)(T-t)}}{\mathrm{e}^{-y(t,T')(T'-t)}}\right) = y(t,T')\frac{T'-t}{T'-T} - y(t,T)\frac{T-t}{T'-T}. \tag{15.4}$$

Note that, when $t = T$ the forward rate is determined in the beginning of the investment interval and we have

$$f(T;T,T') = y(T,T')\frac{T'-T}{T'-T} - y(T,T)\frac{T-T}{T'-T} = y(T,T').$$

That is, yield rates are forward rates for immediate delivery.

The *instantaneous forward rate* of maturity T is defined as

$$f(t,T) := \lim_{T'\to T} f(t;T,T'). \tag{15.5}$$

By applying the definition of the derivative of $\ln Z(t,T)$ w.r.t. T, for fixed t, (15.3) gives the instantaneous forward rate at time T as

$$f(t,T) := \lim_{T'\to T} f(t;T,T') = -\lim_{T'\to T}\frac{\ln Z(t,T') - \ln Z(t,T)}{T'-T} = -\frac{\partial}{\partial T}\ln Z(t,T). \tag{15.6}$$

Hence, integrating this equation, where $f(t,u) = -\frac{\partial}{\partial u}\ln Z(t,u)$, for $u \in [t,T]$, while using $Z(t,t) = 1$, gives the zero-coupon bond price in terms of instantaneous forward rates,

$$Z(t,T) = \exp\left(-\int_t^T f(t,u)\,\mathrm{d}u\right). \tag{15.7}$$

Taking the logarithm of both sides of (15.7) gives

$$\int_t^T f(t,u)\,\mathrm{d}u = -\ln Z(t,T). \tag{15.8}$$

Using (15.8), we can find the integral of the instantaneous forward rate of maturity changing from T to T' with $t \leqslant T \leqslant T'$:

$$\int_T^{T'} f(t,u)\,\mathrm{d}u = \int_t^{T'} f(t,u)\,\mathrm{d}u - \int_t^{T} f(t,u)\,\mathrm{d}u = \ln Z(t,T) - \ln Z(t,T')$$
$$= f(t;T,T')(T'-T)\,. \tag{15.9}$$

The forward rate is also related to the forward price for a unit zero-coupon bond maturing at time T' with settlement at time T. Recall that a forward contract is a contract under which one party is obliged to sell a specified asset for an agreed price to the other party at a designated future date. A *fixed income forward contract* is an agreement between two parties to pay a specified delivery price for a fixed income security (a zero-coupon bond, for instance) at a given delivery date. The *forward price* of the underlying security is the value of the delivery price that makes the forward contract have no-arbitrage price zero at initiation.

15.1.3 Arbitrage-Free Pricing

The risk-free interest rate has been assumed to be a constant or a deterministic function for most of the option pricing models applied in previous chapters. In this chapter, we will deal with stochastic models of interest rates. Assume a filtered probability space $(\Omega, \mathcal{F}, \mathbb{P}, \mathbb{F})$, where $\mathbb{F} = \{\mathcal{F}_t\}_{0\leqslant t\leqslant T^*}$ is a filtration generated by the stochastic risk-free (short) rate process $\{r(t)\}_{0\leqslant t\leqslant T^*}$ for some $T^* > 0$. Our general assumptions are as follows:

1. the short rate process is Markovian;
2. the zero-coupon bond price process $\{Z(t,T)\}_{0\leqslant t\leqslant T}$ is adapted to $\mathbb{F}$ for every maturity T with $T \leqslant T^*$.

As in previous chapters, the bank account $\{B(t)\}_{t\geqslant 0}$ evolves according to

$$dB(t) = r(t)B(t)\,\mathrm{d}t, \quad B(0) = 1, \tag{15.10}$$

and is given by

$$B(t) = \exp\left(\int_0^t r(s)\,\mathrm{d}s\right).$$

Since $B(t)$ is a function of short rates $\{r(s)\}_{0\leqslant s\leqslant t}$, the bank account is adapted to the filtration $\mathbb{F}$. Additionally, let us define the (stochastic) discount factor $D(t,T)$ from time t to time T with $0 \leqslant t \leqslant T \leqslant T^*$ given by

$$D(t,T) = \frac{B(t)}{B(T)} = \exp\left(-\int_t^T r(s)\,\mathrm{d}s\right).$$

The discount factor has the property $D(t,T)D(T,T') = D(t,T')$ for all $0 \leqslant t \leqslant T \leqslant T'$. Also, we recall from previous chapters that $D(t) := D(0,t) = B^{-1}(t)$ for $t \geqslant 0$.

One of the main problems discussed in this chapter is the no-arbitrage pricing of bonds, options on interest rate, and other fixed income derivatives. Here, bonds can be regarded as derivative assets since any bond is derived from the knowledge of the short (risk-free) rate, $r(t)$, which takes on the role of the underlying. The Fundamental Theorem of Asset Pricing (FTAP) is a cornerstone of the no-arbitrage pricing. Let us state the FTAP for the model of stochastic interest rates.

Theorem 15.1. *The market is arbitrage free if there exists a probability measure $\widetilde{\mathbb{P}}$ equivalent to the real-world measure $\mathbb{P}$, under which the discounted zero-coupon bond price*

$$\overline{Z}(t,T) := Z(t,T)D(t) = Z(t,T)/B(t),\ 0 \leqslant t \leqslant T,$$

is a martingale for each $T > 0$. Assuming the absence of arbitrage, the market is complete iff the equivalent martingale measure $\widetilde{\mathbb{P}}$ is unique.

The definition of a zero-coupon bond and definition of a martingale imply that[1]

$$\overline{Z}(t,T) = \widetilde{\mathrm{E}}_t\left[\overline{Z}(T,T)\right] = \widetilde{\mathrm{E}}_t\left[Z(T,T)D(T)\right] = \widetilde{\mathrm{E}}_t\left[D(T)\right].$$

Thus, the time-t price of the zero-coupon bond is

$$Z(t,T) = \widetilde{\mathrm{E}}_t[B(t)D(T)] = \widetilde{\mathrm{E}}_t[D(t,T)] = \widetilde{\mathrm{E}}_t\left[\exp\left(-\int_t^T r(s)\,\mathrm{d}s\right)\right]. \tag{15.11}$$

By using the results of the FTAP, we can also price derivatives. The time-t no-arbitrage value $V(t)$ of a payoff $V(T)$ payable at time T with $0 \leqslant t \leqslant T$ (note that $V(T)$ is $\mathcal{F}_T$-measurable) is

$$V(t) = \widetilde{\mathrm{E}}_t\left[D(t,T)V(T)\right] = \widetilde{\mathrm{E}}_t\left[\exp\left(-\int_t^T r(s)\,\mathrm{d}s\right)V(T)\right]. \tag{15.12}$$

As an example, consider a forward contract under which \$$K$ will be paid at time T in return for a repayment of \$1 at time T' with $T < T'$. Equivalently, this contract is arranged as if a zero-coupon bond maturing at time T' is delivered at time T in return for K dollars paid at the same time T. The time-T payoff is $Z(T,T') - K$. According to (15.12), the price of the forward contract at time t (with $t \leqslant T$) is

$$\begin{aligned} V(t) &= \widetilde{\mathrm{E}}_t\left[D(t,T)(Z(T,T') - K)\right] = \widetilde{\mathrm{E}}_t\left[D(t,T)\left(\widetilde{\mathrm{E}}_T[D(T,T')] - K\right)\right] \\ &= \widetilde{\mathrm{E}}_t\left[D(t,T')\right] - K\,\widetilde{\mathrm{E}}_t\left[D(t,T)\right] = Z(t,T') - K\,Z(t,T). \end{aligned}$$

Here, we combined the tower property with $Z(T,T') = \widetilde{\mathrm{E}}_T[D(T,T')]$ and the identity $D(t,T)D(T,T') = D(t,T')$. Choosing $K = Z(t,T')/Z(t,T)$ ensures that $V(t) = 0$. This value is called the *T-forward price* of the zero-coupon bond with maturity $T' > T$. The forward price is expressed in terms of the forward rate $f(t;T,T')$ as given in (15.3).

Consider now an asset X with price process $\{X(t)\}_{0 \leqslant t \leqslant T}$ and a forward contract that delivers one unit of X at time T in return for K. According to (15.12), the time-t price of this contract is

$$V(t) = \widetilde{\mathrm{E}}_t\left[D(t,T)(X(T) - K)\right] = \frac{1}{D(t)}\widetilde{\mathrm{E}}_t\left[D(T)X(T)\right] - K\widetilde{\mathrm{E}}_t\left[D(t,T)\right].$$

Since the discounted process $D(t)X(t)$ is a $\widetilde{\mathbb{P}}$-martingale and $\widetilde{\mathrm{E}}_t\left[D(t,T)\right] = Z(t,T)$, the above equation reduces to

$$V(t) = X(t) - KZ(t,T).$$

The present (time-t) value is zero iff $K = \frac{X(t)}{Z(t,T)}$. We call $\frac{X(t)}{Z(t,T)}$ the *T-forward price* at time $t \in [0,T]$ of the asset X. Note that $X(t) = Z(t,T')$ is a special case where the chosen asset is the time-T' maturity zero-coupon bond.

---

[1]We note the shorthand notation used throughout this chapter where $\widetilde{\mathrm{E}}_t[\,\cdot\,] \equiv \widetilde{\mathrm{E}}[\,\cdot \mid \mathcal{F}_t]$ is the $\widetilde{\mathbb{P}}$-measure expectation conditional on $\mathcal{F}_t$.

15.1.4 Fixed Income Derivatives

15.1.4.1 Options on Bonds

A European-style option on a zero-coupon bond is defined in the same way as the option on any other underlying security. A call option with maturity T and strike K on a bond maturing at time $T' > T$ gives its holder the right but not the obligation to purchase the bond at time T for a fixed price K. The time-T payoff to the call option holder is $(Z(T,T') - K)^+$. Similarly, we define a put option with the time-T payoff $(K - Z(T,T'))^+$. If the joint (conditional) distribution of the discount factor $D(t,T)$ and bond price $Z(T,T')$ is known, then the no-arbitrage price at time t of a European option with payoff $V(T) = \Lambda(Z(T,T'))$ can be calculated using the risk-neutral pricing formula (15.12):

$$V(t) = \widetilde{\mathrm{E}}_t \left[D(t,T)\, \Lambda(Z(T,T')) \right], \quad 0 \leqslant t \leqslant T.$$

Note that many other common options on interest rates can be expressed as options on bonds. For example, a call option on the LIBOR rate considered in the last section of this chapter is equivalent to a put option on a zero-coupon bond.

We can also consider a European call option written on a coupon-bearing bond. The payoff of the option struck at exercise K, of maturity date T, can be written as $V(T) = (P(T) - K)^+$, where $P(T)$ is the value of the bond at maturity T:

$$P(T) = \sum_{j=1}^{n} c_j Z(T, T_j),$$

with cash flows $c_1, c_2, \ldots, c_n$ at times $T_1, T_2, \ldots, T_n$, respectively, with $T < T_1 < T_2 < \ldots < T_n$. Note that the sum involves all cash flows at future times past the maturity of the option, discounted back to time T.

15.1.4.2 Cap and Caplets

A *caplet* is a contract that gives its holder the right to pay the smaller of two simple interest rates: the floating rate f and fixed rate κ. The floating rate is typically the three- or six-month LIBOR. For the holder of a caplet over the interval $[T, T+\tau]$ with *tenor* τ, the rate of payment is capped at κ from T to $T+\tau$. So the simple interest paid on each dollar of the principal at time $T+\tau$ is the smaller of τf and $\tau\kappa$. Since without the caplet the interest payment would be τf, the caplet's worth to the holder is $\tau f - \min\{\tau f, \tau\kappa\} = (f - \kappa)^+\tau$. That is, the caplet pays $(f - \kappa)^+\tau$ to its holder at time $T + \tau$. In fact, we deal with a call payoff on the floating rate f with strike κ and maturity $T + \tau$. Under the equivalent martingale measure (EMM) $\widetilde{\mathbb{P}}$, the time-t value of the caplet is

$$\mathrm{Caplet}_{T+\tau}(t) = \tau \widetilde{\mathrm{E}}_t \left[D(t, T+\tau)(f - \kappa)^+ \right], \quad 0 \leqslant t \leqslant T. \tag{15.13}$$

As will be demonstrated below, the LIBOR rate is expressed in terms of zero-coupon bonds, and hence the application of the risk-neutral pricing formula in (15.13) is legitimate.

A *cap* is defined as a collection of caplets. Suppose that the payments are done at the times $T_1, T_2, \ldots, T_n$ with $T_{i+1} = T_i + \tau$. Let f_i denote the floating rate over $[T_{i-1}, T_i]$ for $i = 1, 2, \ldots, n$. A cap is defined as a stream of cash flows that pays to its holder $(f_i - \kappa)^+\tau$ at time T_i for all $i = 1, 2, \ldots, n$. The risk-neutral value of the cap with the tenor structure $\mathbf{T} := [T_1, T_2, \ldots, T_n]$ at time $t \leqslant T_0$ is

$$\mathrm{Cap}(t; \mathbf{T}) = \sum_{i=1}^{n} \mathrm{Caplet}_{T_i}(t) = \tau \widetilde{\mathrm{E}}_t \left[\sum_{i=1}^{n} D(t, T_i)(f_i - \kappa)^+ \right]. \tag{15.14}$$

15.1.4.3 Swap and Swaptions

Another basic fixed income instrument is an *interest rate swap*. It is an agreement between two parties in which one party makes fixed interest payments on some notional amount at regularly spaced dates in return for floating interest payments on the same principal at the same dates. The *payer swap* is a contract in which the floating rate f_i is swapped in arrears against a fixed rate κ at n intervals $[T_{i-1}, T_i]$ of length $\tau = T_i - T_{i-1}$ for all $i = 1, 2, \ldots, n$. The holder of the payer swap receives the cash flows $(f_1 - \kappa)\tau, \ldots, (f_n - \kappa)\tau$ at dates $T_1, \ldots, T_n$, respectively. The other party in the swap contract enters a *receiver swap*, in which a fixed rate is swapped against the floating rate. It suffices to only consider a payer swap which we simply call a swap. The time-t value of the swap is

$$\mathrm{Swap}(t; \mathbf{T}) = \tau \sum_{i=1}^{n} \widetilde{\mathrm{E}}_t \left[D(t, T_i)(f_i - \kappa) \right]. \tag{15.15}$$

The fixed rate κ can be defined so that the swap agreement costs zero at initiation. The *forward swap rate* is a fixed rate of interest that makes the swap contract worthless at current time $t \leqslant T_0$. In other words, the forward swap rate at time t is the rate κ that solves $\mathrm{Swap}(t; \mathbf{T}) = 0$.

Another contract called a *swaption* gives you the right but not the obligation to enter into a swap agreement with another party at time T_0. So a swaption delivers at time T_0 a swap when the swap value $\mathrm{Swap}(T_0; \mathbf{T})$ is positive. Thus, the time-t value of a swaption at time $t \leqslant T_0$ is

$$\mathrm{Swaption}(t; \mathbf{T}) = \widetilde{\mathrm{E}}_t \left[D(t, T_0) \left(\mathrm{Swap}(T_0; \mathbf{T}) \right)^+ \right]. \tag{15.16}$$

15.2 Single-Factor Models

A *single-factor model* is one that has a single, one-dimensional source of randomness affecting bond prices. Let a standard Brownian motion $\{W(t)\}_{t \geqslant 0}$ be such a source of randomness. Let the Brownian filtration $\mathbb{F}^W := \{\mathcal{F}_t^W\}_{0 \leqslant t \leqslant T^*}$ coincide with the filtration $\mathbb{F}$ generated by the short rate process $\{r(t)\}_{0 \leqslant t \leqslant T^*}$. This is the case when $r(t)$ is an Itô process governed by a nontrivial stochastic differential equation (SDE) of the form $\mathrm{d}r(t) = a(t)\,\mathrm{d}t + b(t)\,\mathrm{d}W(t)$ with $\mathbb{F}^W$-adapted processes a and b.

Let us derive an SDE for the zero-coupon bond $Z(t, T)$ under the EMM $\widetilde{\mathbb{P}}$. The discounted process $\overline{Z}(t, T) := D(t) Z(t, T)$ for $0 \leqslant t \leqslant T$, and a given fixed $T > 0$, is a $\widetilde{\mathbb{P}}$-martingale. By the Brownian Martingale Representation Theorem 11.14, there exists an $\mathbb{F}$-adapted process $\theta(t)$ such that $\overline{Z}(t, T) = \overline{Z}(0, T) + \int_0^t \theta(s)\,\mathrm{d}\widetilde{W}(s)$ (a.s.) for $0 \leqslant t \leqslant T$. Since $\overline{Z}(t, T)$ is strictly positive, we can define $\sigma_Z(t, T) = \theta(t)/\overline{Z}(t, T)$. Thus, we have

$$\mathrm{d}\overline{Z}(t, T) = \theta(t)\,\mathrm{d}\widetilde{W}(t) = \overline{Z}(t, T)\sigma_Z(t, T)\,\mathrm{d}\widetilde{W}(t).$$

By using the Itô product rule, we obtain the following expression for the stochastic differential of the discounted bond price:

$$\mathrm{d}\overline{Z}(t, T) = -\overline{Z}(t, T) r(t)\,\mathrm{d}t + D(t)\,\mathrm{d}Z(t, T).$$

As a result, we have the following SDE for $Z(t, T)$ under the EMM $\widetilde{\mathbb{P}}$:

$$\frac{\mathrm{d}Z(t, T)}{Z(t, T)} = r(t)\,\mathrm{d}t + \sigma_Z(t, T)\,\mathrm{d}\widetilde{W}(t). \tag{15.17}$$

This SDE has the equivalent integral form

$$Z(t,T) = Z(0,T)\mathrm{e}^{\int_0^t r(s)\,\mathrm{d}s - \frac{1}{2}\int_0^t \sigma_Z^2(s,T)\,\mathrm{d}s + \int_0^t \sigma_Z(s,T)\,\mathrm{d}\widetilde{W}(s)}. \tag{15.18}$$

By comparing SDEs (15.10) and (15.17), we notice that the bond Z is riskier than the bank account B since the SDE in (15.17) contains an extra Brownian term.

To find the SDE for $Z(t,T)$ under the original (physical) $\mathbb{P}$-measure, we use the equivalence of the measures $\mathbb{P}$ and $\widetilde{\mathbb{P}}$. By Girsanov's Theorem 11.13, there exists a change of measure generated by an $\mathbb{F}$-adapted process $\gamma(t)$ such that

$$\widetilde{W}(t) = W(t) + \int_0^t \gamma(s)\,\mathrm{d}s \tag{15.19}$$

is a standard $\widetilde{\mathbb{P}}$-BM, given $W(t)$ is a standard $\mathbb{P}$-BM, for all $t \in [0,T^*]$. Substituting $\mathrm{d}W(t) + \gamma(t)\,\mathrm{d}t$ in place of $\mathrm{d}\widetilde{W}(t)$ in (15.17) gives the following SDE under $\mathbb{P}$:

$$\frac{\mathrm{d}Z(t,T)}{Z(t,T)} = (r(t) + \gamma(t)\sigma_Z(t,T))\,\mathrm{d}t + \sigma_Z(t,T)\,\mathrm{d}W(t), \tag{15.20}$$

for all $t \in [0,T]$, $T \leqslant T^*$. The quantity $\gamma(t)$ is the excess rate of return over the risk-free rate of return $r(t)$ per one unit of volatility; it is known as the *market price of risk* or *risk premium*. This term represents the extra reward we receive for investing in the risky bond rather than in the risk-free bank account. Since $\gamma(t)$ is adapted to the σ-algebra $\mathcal{F}_t$ generated by the short rate process $\{r(t)\}_{t\geqslant 0}$ and since $\{r(t)\}_{t\geqslant 0}$ is a Markov process, the risk premium is a function of t and $r(t)$, i.e., $\gamma = \gamma(t, r(t))$. The market price of risk can be estimated from observable values of rate and bond prices. However, it is a common practice to assume that γ is constant.

15.2.1 Diffusion Models for the Short Rate Process

Suppose the short rate process $\{r(t)\}_{t\geqslant 0}$ is a diffusion described by an SDE

$$\mathrm{d}r(t) = a(t,r(t))\,\mathrm{d}t + b(t,r(t))\,\mathrm{d}W(t). \tag{15.21}$$

The coefficients a and b are smooth functions that meet standard conditions required to ensure the existence of solutions to (15.21). Here is a list of popular models of interest rates falling in the class of diffusions as in (15.21):

$$\mathrm{d}r(t) = \alpha\,\mathrm{d}t + \sigma\,\mathrm{d}W(t) \quad \text{(the } \textit{Merton model}\text{)} \tag{15.22}$$

$$\mathrm{d}r(t) = \beta r(t)\,\mathrm{d}t + \sigma r(t)\,\mathrm{d}W(t) \quad \text{(the } \textit{Dothan model}\text{)} \tag{15.23}$$

$$\mathrm{d}r(t) = (\alpha - \beta r(t))\,\mathrm{d}t + \sigma\,\mathrm{d}W(t) \quad \text{(the } \textit{Vasiček model}\text{)} \tag{15.24}$$

$$\mathrm{d}r(t) = (\alpha - \beta r(t))\,\mathrm{d}t + \sigma r(t)\,\mathrm{d}W(t) \quad \text{(the } \textit{Brennan–Schwartz model}\text{)} \tag{15.25}$$

$$\mathrm{d}r(t) = (\alpha - \beta r(t))\,\mathrm{d}t + \sigma\sqrt{r(t)}\,\mathrm{d}W(t) \quad \text{(the } \textit{Cox–Ingersoll–Ross model}\text{)} \tag{15.26}$$

$$\mathrm{d}r(t) = \alpha(t)\,\mathrm{d}t + \sigma\,\mathrm{d}W(t) \quad \text{(the } \textit{Ho–Lee model}\text{)} \tag{15.27}$$

$$\mathrm{d}r(t) = \alpha(t)r(t)\,\mathrm{d}t + \sigma(t)\,\mathrm{d}W(t) \quad \text{(the } \textit{Black–Derman–Toy model}\text{)} \tag{15.28}$$

$$\mathrm{d}r(t) = (\alpha(t) - \beta(t)r(t))\,\mathrm{d}t + \sigma(t)\,\mathrm{d}W(t) \quad \text{(the } \textit{Hull–White model}\text{)} \tag{15.29}$$

$$\mathrm{d}r(t) = r(t)(\alpha(t) - \beta(t)\ln r(t))\,\mathrm{d}t + \sigma(t)r(t)\,\mathrm{d}W(t) \quad \text{(the } \textit{Black–Karasinski model}\text{)} \tag{15.30}$$

Here, α, β, and σ are constants; $\alpha(t)$, $\beta(t)$, and $\sigma(t)$ are nonrandom continuous functions of time. Assuming that the coefficients a and b in (15.21) are independent of t, i.e., $a = a(r(t))$ and $b = b(r(t))$, we obtain a time-homogeneous model for short rates. The models (15.22)–(15.26) are time homogeneous, whereas the models (15.27)–(15.30) are time-inhomogeneous.

Each model has some motivation behind it. There are three main characteristics to be taken into account when a model for interest rates is designed.

- Positiveness of interest rates. For instance, the Merton model and the Vasiček model do not guarantee that $r(t)$ remains positive. However, these models can still be used for short intervals of time.
- The rate process $r(t)$ is mean reverting, meaning that $r(t)$ fluctuates near a fixed long-term mean level given by $\lim_{t\to\infty} \mathrm{E}[r(t)]$. So the process cannot wander off toward $+\infty$ (or to zero) since a negative drift (or a positive drift) will eventually pull the path back to the long-term level. For instance, the Vasiček model has the long-term mean level equal to α/β. The drift rate of the diffusion in (15.24) is positive for $r(t) < \alpha/\beta$ and is negative for $r(t) > \alpha/\beta$.
- The model has to be tractable. Ideally, formulae for bond prices and for prices of some derivatives can be derived in closed form.

Note that any model is only an approximation of reality, and a financial model is worthy of consideration if it gives a good approximation of what is observed on the market. The coefficient functions a and b involve parameters that need to be estimated. For instance, they can be chosen so that model values of bonds and other derivatives are as close to the respective market values as possible.

15.2.2 PDE for the Zero-Coupon Bond Value

A model for the term structure of interest rates and bond pricing can be developed from a model for the short rate process. Let us derive the governing differential equation for the zero-coupon bond price. Our first approach employs the (discounted) Feynman–Kac formula. The second method presented here is based on the no-arbitrage argument.

The modelling equation (15.21) for $r(t)$ is specified under the real-world measure. The price $Z(t,T)$ of a zero-coupon bond satisfies (15.11), where the expectation is taken under the EMM $\widetilde{\mathbb{P}}$. So we need to know the risk-neutral dynamics of the short rate process. Using (15.19), we change measure in (15.21) to obtain the SDE

$$\mathrm{d}r(t) = \big(a(t,r(t)) - \gamma(t,r(t))b(t,r(t))\big)\,\mathrm{d}t + b(t,r(t))\,\mathrm{d}\widetilde{W}(t)\,. \tag{15.31}$$

The short rate process $\{r(t)\}_{t\geqslant 0}$ is hence Markovian. The expectation conditional on $\mathcal{F}_t$ in the right-hand side of (15.11) is then simply a conditional expectation given the value of $r(t)$. That is, as a random variable the zero-coupon bond price is given as a function of the random variable $r(t)$, $Z(t,T) \equiv Z(t,T,r(t))$, where

$$Z(t,T,r(t)) = \widetilde{\mathrm{E}}[\mathrm{e}^{-\int_t^T r(s)\,\mathrm{d}s} \mid \mathcal{F}_t] = \widetilde{\mathrm{E}}[\mathrm{e}^{-\int_t^T r(s)\,\mathrm{d}s} \mid r(t)]\,.$$

Conditioning on the known (spot) value of the short rate, $r(t) = r$, gives the pricing function $Z(t,T,r)$ for a zero-coupon bond. Thus, the time-t value of a zero-coupon bond is an ordinary function (i.e., $f(t,r) = Z(t,T,r)$) of the current short rate r and time t (for fixed T) where

$$Z(t,T,r) = \widetilde{\mathrm{E}}_{t,r}[\mathrm{e}^{-\int_t^T r(s)\,\mathrm{d}s}] \equiv \widetilde{\mathrm{E}}[\mathrm{e}^{-\int_t^T r(s)\,\mathrm{d}s} \mid r(t) = r]\,. \tag{15.32}$$

By associating this expectation with that in (11.62), the (discounted) Feynman–Kac Theorem 11.9 shows that the pricing function $Z = Z(t,T,r)$ satisfies the following PDE[2] (see (11.63)):

$$\frac{\partial Z}{\partial t} + \frac{b^2(t,r)}{2}\frac{\partial^2 Z}{\partial r^2} + \big(a(t,r) - \gamma(t,r)b(t,r)\big)\frac{\partial Z}{\partial r} - rZ = 0 \tag{15.33}$$

for all $t < T$ and all r, subject to the terminal condition

$$Z(T,T,r) = 1 \text{ for all } r. \tag{15.34}$$

Once the short rate model and the market price of risk γ are specified, the bond price can be determined by solving (15.33) subject to (15.34).

An alternative derivation of the PDE (15.33) is based on hedging and no-arbitrage arguments. Applying the Itô formula to the bond price process $Z(t,T) \equiv Z(t,T,r(t))$, which is a function of the stochastic short rate $r(t)$ and time t, gives the following SDE for the bond price process:

$$\mathrm{d}Z(t,T) = \left(\frac{\partial Z}{\partial t} + a\frac{\partial Z}{\partial r} + \frac{b^2}{2}\frac{\partial^2 Z}{\partial r^2}\right)\mathrm{d}t + b\frac{\partial Z}{\partial r}\,\mathrm{d}W(t)$$

or, equivalently, in log-normal form

$$\frac{\mathrm{d}Z(t,T)}{Z(t,T)} = \mu\,\mathrm{d}t + \sigma\,\mathrm{d}W(t)\,,$$

where the log-drift rate μ and log-diffusion coefficient σ are, respectively,

$$\mu(t,T,r) = \frac{1}{Z(t,T,r)}\left[\frac{\partial Z(t,T,r)}{\partial t} + a(t,r)\frac{\partial Z(t,T,r)}{\partial r} + \frac{b^2(t,r)}{2}\frac{\partial^2 Z(t,T,r)}{\partial r^2}\right], \tag{15.35}$$

$$\sigma(t,T,r) = \frac{b(t,r)}{Z(t,T,r)}\frac{\partial Z(t,T,r)}{\partial r}\,. \tag{15.36}$$

The underlying short rate process is not a traded security and it cannot therefore be used for hedging bonds. Instead, we try to hedge one bond with another one of a different maturity. Consider two zero-coupon bonds maturing at times T_1 and T_2 with $T_1 < T_2$, respectively. At time t, we buy T_1-bonds of value $V_1(t)$ (a long position) and sell T_2-bonds of value $V_2(t)$ (a short position). The total value of this portfolio at time t is $\Pi(t) = V_1(t) - V_2(t)$. The change in portfolio value from t to $t + dt$ is

$$\begin{aligned}
\mathrm{d}\Pi(t) &= \frac{V_1(t)}{Z(t,T_1)}\,\mathrm{d}Z(t,T_1) - \frac{V_2(t)}{Z(t,T_2)}\,\mathrm{d}Z(t,T_2) \\
&= V_1(t)(\mu_1\,\mathrm{d}t + \sigma_1\,\mathrm{d}W(t)) - V_2(t)(\mu_2\,\mathrm{d}t + \sigma_2\,\mathrm{d}W(t)) \\
&= (V_1(t)\mu_1 - V_2(t)\mu_2)\,\mathrm{d}t + (V_1(t)\sigma_1 - V_2(t)\sigma_2)\,\mathrm{d}W(t)\,,
\end{aligned}$$

where, to compact notations, we denote

$$\mu_i = \mu(t,T_i,r(t)) \text{ and } \sigma_i = \sigma(t,T_i,r(t)) \text{ for } i = 1,2.$$

[2]We remark that the role of the process $X(t)$ with SDE in (11.24) is now played by $r(t)$ with SDE in (15.31) and the dummy variable x in (11.63) is now called r. The variable r is not to be confused with the function denoted by $r(t,x)$ in Theorem 11.9, which is now given by $r(t,x) \equiv x$, as seen by identifying (11.62) with (15.32) and where $\phi(x) \equiv 1$.

Suppose that V_1 and V_2 are chosen such that

$$\frac{V_1(t)}{V_2(t)} = \frac{\sigma_2}{\sigma_1} \equiv \frac{\sigma(t,T_2,r(t))}{\sigma(t,T_1,r(t))}$$

holds for all t. Then $V_1\sigma_1 - V_2\sigma_2 \equiv 0$ and hence the $\mathrm{d}W(t)$ term in the stochastic differential $\mathrm{d}\Pi$ vanishes. As a result, the equation for $\mathrm{d}\Pi$ becomes

$$\frac{\mathrm{d}\Pi(t)}{\Pi(t)} = \frac{\mu_1\sigma_2 - \mu_2\sigma_1}{\sigma_2 - \sigma_1} \equiv \frac{\mu(t,T_1,r(t))\sigma(t,T_2,r(t)) - \mu(t,T_2,r(t))\sigma(t,T_1,r(t))}{\sigma(t,T_2,r(t)) - \sigma(t,T_1,r(t))}\,\mathrm{d}t$$

Thus, we have a risk-free, self-financing portfolio strategy. To avoid arbitrage, the rate of return has to be equal to the risk-free rate $r(t)$; that is,

$$\frac{\mu_1\sigma_2 - \mu_2\sigma_1}{\sigma_2 - \sigma_1} = r \iff \frac{\mu_1 - r}{\sigma_1} = \frac{\mu_2 - r}{\sigma_2}.$$

The above relation is valid for arbitrary maturity dates T_1 and T_2, so the ratio $\frac{\mu(t,T,r)-r}{\sigma(t,T,r)}$ is independent of T for all $T > t$. We can hence define

$$\gamma(t,r) = \frac{\mu(t,T,r) - r}{\sigma(t,T,r)} \tag{15.37}$$

so that the drift rate of the bond price process is

$$\mu(t,T,r) = r + \gamma(t,r)\sigma(t,T,r)\,. \tag{15.38}$$

Comparing the above equation with (15.20), we conclude that γ is nothing but the market price of risk for the short rate process. Equating the formulae (15.35) and (15.38) gives the governing PDE (15.33) for the price of a zero-coupon bond.

15.2.3 Affine Term Structure Models

A short-rate model that produces the bond pricing function of the form

$$Z(t,T,r) = \mathrm{e}^{A(t,T)-C(t,T)r}, \tag{15.39}$$

where r is the short rate at time t, and the functions $A(t,T)$ and $C(t,T)$ are independent of r, is called an *affine term structure* model. Let the short rate process follow the SDE of the form in (15.31):

$$\mathrm{d}r(t) = \tilde{a}(t,r(t))\,\mathrm{d}t + b(t,r(t))\,\mathrm{d}\widetilde{W}(t)\,, \tag{15.40}$$

where $\tilde{a}(t,r) := a(t,r) - \gamma(t,r)b(t,r)$ and $\widetilde{W}(t)$ is a standard BM under the EMM $\widetilde{\mathbb{P}}$.

Certain conditions must be set on the short rate process $r(t)$ in order that the zero-coupon bond price admits the form (15.39). Applying the Itô formula to the bond price $Z(t,T,r(t)) = \mathrm{e}^{A(t,T)-C(t,T)r(t)}$, where $r(t)$ follows (15.40), gives

$$\frac{\mathrm{d}Z(t,T,r(t))}{Z(t,T,r(t))} = \left[\frac{\partial A(t,T)}{\partial t} - \frac{\partial C(t,T)}{\partial t}r(t) - C(t,T)\tilde{a}(t,r(t)) + \frac{1}{2}C^2(t,T)b^2(t,r(t))\right]\mathrm{d}t$$
$$- C(t,T)b(t,r(t))\,\mathrm{d}\widetilde{W}(t)\,.$$

We also know that the risk-neutral dynamics of the bond price is given by (15.17). That is, the drift rate in the above SDE has to be equal to the risk-neutral rate $r(t)$. It follows that the ordinary (nonrandom) function $g(t,r)$ defined by

$$g(t,r) = \frac{\partial A(t,T)}{\partial t} - \frac{\partial C(t,T)}{\partial t}r - C(t,T)\tilde{a}(t,r) + \frac{1}{2}C^2(t,T)b^2(t,r) - r$$

is identically zero for all t and r. Since A and C are independent of r, then $g(t,r) \equiv 0$ holds only if $\tilde{a}(t,r)$ and $b^2(t,r)$ are both linear functions of r. That is, for the bond pricing formula to be of the affine form (15.39) it is necessary that the risk-neutral drift and the square of the diffusion coefficient both be affine (i.e., linear) functions of r:

$$\tilde{a}(t,r) = a_0(t) + a_1(t)r \quad \text{and} \quad b^2(t,r) = b_0(t) + b_1(t)r \tag{15.41}$$

where $a_0(t)$, $a_1(t)$, $b_0(t)$, and $b_1(t)$ are only functions of time.

The zero-coupon bond pricing function $Z(t,T,r)$ solves the PDE (15.33) with terminal conditions $Z(T,T,r) = 1$. Substituting the solution in (15.39) into (15.33) yields

$$\frac{\partial A(t,T)}{\partial t} - \left(1 + \frac{\partial C(t,T)}{\partial t}\right) r + \frac{b^2(t,r)}{2} C^2(t,T) - \tilde{a}(t,r)C(t,T) = 0 \text{ for } t < T, \tag{15.42}$$

with terminal conditions $A(T,T) = 0$ and $C(T,T) = 0$.

Substituting (15.41) into (15.42) gives

$$\begin{aligned}&\frac{\partial A(t,T)}{\partial t} - a_0(t)C(t,T) + \frac{b_0(t)}{2} C^2(t,T) \\ &\quad - \left(\frac{\partial C(t,T)}{\partial t} + a_1(t)C(t,T) - \frac{b_1(t)}{2} C^2(t,T) + 1\right) r = 0\,.\end{aligned}$$

Since the left-hand side of the above equation is identically zero for *all* values of the rate r, the functions A and C must solve the following pair of differential equations:

$$\frac{\partial C(t,T)}{\partial t} + a_1(t)C(t,T) - \frac{b_1(t)}{2} C^2(t,T) + 1 = 0, \tag{15.43}$$

$$\frac{\partial A(t,T)}{\partial t} - a_0(t)C(t,T) + \frac{b_0(t)}{2} C^2(t,T) = 0, \tag{15.44}$$

for $t < T$, subject to the respective boundary conditions at $t = T$: $A(T,T) = 0$ and $C(T,T) = 0$. The equation in (15.43) is a first order nonlinear ODE and is known as the *Ricatti equation*. For some special cases of a_1 and b_1, it is possible to solve equation (15.43) in closed form. Once an analytic solution for C is available, the solution for A is obtained by integrating (15.44) with respect to t. In the next three subsections, we consider three short-rate models that admit the bond pricing formula as in (15.39) where A and C are given in analytically closed form.

15.2.4 The Ho–Lee Model

The short rate in the Ho–Lee model satisfies (15.27). We note that the Merton model is a special case of the Ho–Lee model with constant drift rate α. For pricing bonds, we need the short rate dynamics under the EMM $\widetilde{\mathbb{P}}$. Assume that the market price of risk γ is independent of r, then by (15.31) the SDE for $r(t)$ under $\widetilde{\mathbb{P}}$ is

$$\mathrm{d}r(t) = \tilde{\alpha}(t)\,\mathrm{d}t + \sigma\,\mathrm{d}\widetilde{W}(t)\,,$$

where $\tilde{\alpha}(t) = \alpha(t) - \gamma(t)\sigma$. The strong solution to this linear SDE is a drifted and scaled Brownian motion:

$$r(s) = r(t) + \int_t^s \tilde{\alpha}(u)\,\mathrm{d}u + \sigma(\widetilde{W}(s) - \widetilde{W}(t)) \tag{15.45}$$

for $0 \leqslant t \leqslant s$. The rate $r(s)$ conditional on $r(t) = r$ is normally distributed with mean $r + \int_t^s \tilde{\alpha}(u)\,\mathrm{d}u$ and variance $\sigma^2(s-t)$. Since $\widehat{W}(s-t) = \widetilde{W}(s) - \widetilde{W}(t)$ is independent of $\mathcal{F}_t$

for every $s > t$, the random variable $\int_t^T \widehat{W}(s-t)\,\mathrm{d}s \sim Norm(0, (T-t)^3/3)$ is independent of $\mathcal{F}_t$, i.e., it is independent of $r(t)$. Hence, the conditional expectation simplifies to an unconditional expectation and we obtain the no-arbitrage pricing function:

$$\begin{aligned}
Z(t,T,r) &= \widetilde{\mathrm{E}}_{t,r}\left[\mathrm{e}^{-\int_t^T r(s)\,\mathrm{d}s}\right] = \widetilde{\mathrm{E}}\left[\mathrm{e}^{-\int_t^T \left(r+\int_t^s \tilde{\alpha}(u)\,\mathrm{d}u+\sigma\widehat{W}(s-t)\right)\mathrm{d}s}\right] \\
&= \mathrm{e}^{-r(T-t)-\hat{\alpha}(t,T)(T-t)^2/2}\widetilde{\mathrm{E}}\left[\mathrm{e}^{-\sigma\int_t^T \widehat{W}(s-t)\,\mathrm{d}s}\right] \\
&= \mathrm{e}^{-r(T-t)-\hat{\alpha}(t,T)(T-t)^2/2+\sigma^2(T-t)^3/6}, \qquad (15.46)
\end{aligned}$$

where $\hat{\alpha}(t,T) := \frac{2}{(T-t)^2}\int_t^T\int_t^s \tilde{\alpha}(u)\,\mathrm{d}u\,\mathrm{d}s$. In the case of the Merton model with constant $\tilde{\alpha}$, we have $\hat{\alpha}(t,T) = \tilde{\alpha}$ and

$$Z(t,T,r) = \mathrm{e}^{-r(T-t)-\tilde{\alpha}(T-t)^2/2+\sigma^2(T-t)^3/6}. \qquad (15.47)$$

The yield rate is an affine function of r,

$$y(t,T,r) = -\frac{\ln Z(t,T,r)}{T-t} = r + \hat{\alpha}(t,T)\frac{(T-t)}{2} - \sigma^2\frac{(T-t)^2}{6}.$$

Since the distribution of the short rate $r(t)$ is normal, it then follows that $y(t,T,r(t))$ is also normally distributed and the distribution of $Z(t,T,r(t))$ is log-normal.

The Ho–Lee model has several serious shortcomings:

1. the short rate can become negative (with nonzero probability);
2. $Z(t,T,r) \to \infty$ and $y(t,T,r) \to -\infty$ as $T \to \infty$;
3. the yield rates $y(t,T_1,r(t))$ and $y(t,T_2,r(t))$ for different maturities T_1 and T_2 are both a linear function of the short rate $r(t)$, and they are hence perfectly correlated.

The Ho–Lee model is an affine model with $a_0(t) = \tilde{\alpha}(t)$, $a_1(t) \equiv 0$, $b_0(t) = \sigma^2$, and $b_1(t) \equiv 0$ in (15.41). Therefore, the bond pricing function in (15.47) can also be found by solving (15.43)–(15.44) for $A(t,T)$ and $C(t,T)$, which take the following simpler form:

$$\begin{aligned}
&\frac{\partial C(t,T)}{\partial t} + 1 = 0, \\
&\frac{\partial A(t,T)}{\partial t} - \tilde{\alpha}(t)C(t,T) + \frac{\sigma^2}{2}C^2(t,T) = 0,
\end{aligned}$$

subject to $A(T,T) = C(T,T) = 0$. The first equation is trivially integrated to give $C(t,T) = T-t$. Substituting this into the second equation, integrating and applying the condition $A(T,T) = 0$, gives

$$A(t,T) = -\int_t^T \tilde{\alpha}(u)(T-u)\,\mathrm{d}u + \frac{\sigma^2}{2}\int_t^T (T-u)^2\,\mathrm{d}u = -\hat{\alpha}(t,T)\frac{(T-t)^2}{2} + \sigma^2\frac{(T-t)^3}{6}.$$

This recovers the above same formula for the yield and zero-coupon bond price in (15.47). For the case of constant $\tilde{\alpha}(t) \equiv \tilde{\alpha}$, the formulae simplify where $\hat{\alpha}(t,T) = \tilde{\alpha}$.

15.2.5 The Vasiček Model

The short rate process in the Vasiček model satisfies SDE (15.24). Recall that the solution to (15.24) is also known as the Ornstein–Uhlenbeck process. Assuming the market price of risk γ is constant, we obtain the following risk-neutral dynamics of $r(t)$:

$$\mathrm{d}r(t) = (\tilde{\alpha} - \beta r(t))\,\mathrm{d}t + \sigma\,\mathrm{d}\widetilde{W}(t), \qquad (15.48)$$

where $\tilde{\alpha} = \alpha - \gamma\sigma$, β, and σ are positive parameters. The diffusion solving the above SDE is called a mean-reverting process since the instantaneous drift $\tilde{\alpha} - \beta r(t) = \beta(\tilde{\alpha}/\beta - r(t))$ pulls the process toward the constant mean level $\tilde{\alpha}/\beta$ with magnitude proportional to the deviation of the process from the mean.

Integrating the above SDE (see Examples 11.8 and 11.13) gives

$$r(T) = \mathrm{e}^{-\beta(T-t)} r(t) + \frac{\tilde{\alpha}}{\beta}\left(1 - \mathrm{e}^{-\beta(T-t)}\right) + \sigma \int_t^T \mathrm{e}^{-\beta(T-s)}\, \mathrm{d}\widetilde{W}(s)\,. \tag{15.49}$$

The probability distribution of $r(T)$ conditional on $r(t)$ is normal with mean

$$\widetilde{\mathrm{E}}[r(T) \mid r(t)] = \mathrm{e}^{-\beta(T-t)} r(t) + \frac{\tilde{\alpha}}{\beta}(1 - \mathrm{e}^{-\beta(T-t)})$$

and variance

$$\widetilde{\mathrm{Var}}(r(T) \mid r(t)) = \sigma^2 \frac{1 - \mathrm{e}^{-2\beta(T-t)}}{2\beta}\,.$$

Note that the long-term mean and variance (as $T - t \to \infty$) are given by the constants

$$\lim_{T\to\infty} \widetilde{\mathrm{E}}[r(T) \mid r(t)] = \frac{\tilde{\alpha}}{\beta} \quad \text{and} \quad \lim_{T\to\infty} \widetilde{\mathrm{Var}}(r(T) \mid r(t)) = \frac{\sigma^2}{2\beta}\,.$$

We now proceed to the calculation of bond prices. The drift and diffusion coefficients in (15.48) correspond to $a_0 = \tilde{\alpha}$, $a_1 = -\beta$, $b_0 = \sigma^2$, and $b_1 = 0$ in (15.41). We have the following pair of ODEs for $A(t,T)$ and $C(t,T)$:

$$\frac{\partial C(t,T)}{\partial t} - \beta C(t,T) + 1 = 0, \tag{15.50}$$

$$\frac{\partial A(t,T)}{\partial t} - \tilde{\alpha} C(t,T) + \frac{\sigma^2}{2} C^2(t,T) = 0, \tag{15.51}$$

for $t < T$, subject to $A(T,T) = C(T,T) = 0$. Solving the above system, we obtain

$$C(t,T) = \frac{1}{\beta}\left(1 - \mathrm{e}^{-\beta(T-t)}\right) \text{ and } A(t,T) = (C(t,T) - (T-t))y_\infty - \frac{\sigma^2}{4\beta} C^2(t,T) \tag{15.52}$$

where $y_\infty := \frac{\tilde{\alpha}}{\beta} - \frac{\sigma^2}{2\beta^2}$. Thus, according to (15.39), the bond pricing formula for the Vasiček model is

$$Z(t,T,r) = \exp\left[\left(\frac{1 - \mathrm{e}^{-\beta(T-t)}}{\beta}\right)(y_\infty - r) - y_\infty(T-t) - \frac{\sigma^2}{4\beta}\left(\frac{1 - \mathrm{e}^{-\beta(T-t)}}{\beta}\right)^2\right]. \tag{15.53}$$

The yield rate is found to be

$$y(t,T,r) = y_\infty - \frac{1}{\beta}(1 - \mathrm{e}^{-\beta(T-t)})\frac{y_\infty - r}{T-t} + \frac{\sigma^2}{4\beta(T-t)}\left(\frac{1 - \mathrm{e}^{-\beta(T-t)}}{\beta}\right)^2.$$

By taking $T \to \infty$, the last two terms of the above equation vanish so that the long-term yield rate is in fact equal to y_∞.

15.2.6 The Cox–Ingersoll–Ross Model

One common drawback of the Merton model and the Vasiček model is that the short rate can be negative due to its normal distribution. The first tractable model for the short rate process $r(t)$ that keeps rates positive was proposed by Cox, Ingersoll, and Ross (the CIR model). The short-rate model follows the square root diffusion described by the SDE in (15.26). Assuming the market price of risk is equal to $\gamma\sqrt{r}$ with constant γ, we obtain the following risk-neutral dynamics:

$$\mathrm{d}r(t) = (\alpha - \tilde{\beta}r(t))\,\mathrm{d}t + \sigma\sqrt{r(t)}\,\mathrm{d}\widetilde{W}(t)\,.$$

where $\tilde{\beta} = \beta + \gamma\sigma$. With a nonnegative initial interest rate, $r(t)$ stays nonnegative. Moreover, the CIR process is mean-reverting with the long-run mean level $\alpha/\tilde{\beta}$.

The CIR process $r(t)$ is reduced to the squared Bessel (SQB) process $X(t)$, which solves the SDE (16.14) by means of a scale and time transformation,

$$r(t) = e^{-\tilde{\beta}t}\frac{\sigma^2}{4}X(\tau_t)$$

where the (strictly increasing) time transformation τ_t is defined to be

$$\tau_t := \begin{cases} t & \text{if } \tilde{\beta} = 0, \\ \frac{\mathrm{e}^{\tilde{\beta}t}-1}{\tilde{\beta}} & \text{if } \tilde{\beta} \neq 0. \end{cases} \tag{15.54}$$

The index of the SQB process in (16.14) is $\mu = \frac{2\alpha}{\sigma^2} - 1$. In what follows, we assume that $\mu \geqslant 0$ (or, equivalently, $2\alpha \geqslant \sigma^2$) and hence the left-hand endpoint 0 is an entrance boundary for the short-rate CIR process.

The transition PDF for the CIR process relates to that of the SQB process. Under the risk-neutral measure $\widetilde{\mathbb{P}}$ it is given by

$$\tilde{p}(t; r_0, r) = c_t\mathrm{e}^{\tilde{\beta}t}\left(\frac{r\mathrm{e}^{\tilde{\beta}t}}{r_0}\right)^{\frac{\mu}{2}} \exp\left(-c_t(r\mathrm{e}^{\tilde{\beta}t} + r_0)\right) I_\mu\left(2c_t\sqrt{rr_0\mathrm{e}^{\tilde{\beta}t}}\right), \tag{15.55}$$

where $c_t := \frac{2}{\sigma^2\tau_t}$.

By expressing the above SDE in integral form, with $r(0) = r_0$, and taking expectations (under measure $\widetilde{\mathbb{P}}$) on both sides while denoting the mean by $\widetilde{\mathrm{E}}[r(t)] \equiv m(t)$, we have

$$m(t) = r_0 + \int_0^t (\alpha - \tilde{\beta}m(u))\,\mathrm{d}u\,.$$

Here we used the fact that the Itô integral $\int_0^t \sqrt{r(u)}\,\mathrm{d}\widetilde{W}(u)$ has zero expectation (i.e., it is a $\widetilde{\mathbb{P}}$-martingale). Differentiating gives a linear first order ODE

$$\frac{\mathrm{d}}{\mathrm{d}t}m(t) = \alpha - \tilde{\beta}m(t)\,, \quad t \geqslant 0\,,$$

subject to $m(0) = r_0$. Solving gives

$$m(t) \equiv \widetilde{\mathrm{E}}[r(t)] = \frac{\alpha}{\tilde{\beta}}(1 - \mathrm{e}^{-\tilde{\beta}t}) + r_0\mathrm{e}^{-\tilde{\beta}t}\,.$$

Here, we assume that $\tilde{\beta} \neq 0$. Note that the mean of the process converges to the long-term level $\alpha/\tilde{\beta}$, as $t \to \infty$.

The CIR model is within the class of affine term structure models with a bond pricing formula of the form in (15.39). The respective ODEs in (15.43) and (15.44) for $A(t,T)$ and $C(t,T)$ are

$$\frac{\partial C(t,T)}{\partial t} - \tilde{\beta} C(t,T) - \frac{\sigma^2}{2} C^2(t,T) + 1 = 0, \tag{15.56}$$

$$\frac{\partial A(t,T)}{\partial t} - \alpha C(t,T) = 0, \tag{15.57}$$

subject to $A(T,T) = C(T,T) = 0$. The first equation is a first order ODE with a quadratic nonlinear term. The trick in solving this equation is to turn it into a linear second order ODE by invoking the transformation

$$\psi(t) := \exp\left(\frac{\sigma^2}{2}\int_t^T C(s,T)\,\mathrm{d}s\right).$$

Taking derivatives gives (denoting $C'(t,T) \equiv \partial C(t,T)/\partial t$):

$$\frac{\psi'(t)}{\psi(t)} \equiv \frac{\mathrm{d}}{\mathrm{d}t}\ln\psi(t) = -\frac{\sigma^2}{2}C(t,T) \implies C(t,T) = -\frac{2}{\sigma^2}\frac{\psi'(t)}{\psi(t)},$$

$$\frac{\psi''(t)}{\psi(t)} = -\frac{\sigma^2}{2}C'(t,T) - \frac{\sigma^2}{2}C(t,T)\frac{\psi'(t)}{\psi(t)} \implies C'(t,T) = -\frac{2}{\sigma^2}\frac{\psi''(t)}{\psi(t)} + \frac{\sigma^2}{2}C^2(t,T).$$

Substituting the above two expressions for $C(t,T)$ and $C'(t,T)$ into (15.56), and simplifying, gives

$$\psi''(t) - \tilde{\beta}\psi'(t) - \frac{\sigma^2}{2}\psi(t) = 0. \tag{15.58}$$

This is a second order linear ODE with constant coefficients. Its solution is found by standard methods. In particular, this ODE has the general solution

$$\psi(t) = c_1 \mathrm{e}^{\frac{1}{2}(\tilde{\beta}+\vartheta)t} + c_2 \mathrm{e}^{\frac{1}{2}(\tilde{\beta}-\vartheta)t}$$

with derivative

$$\psi'(t) = \frac{c_1}{2}(\tilde{\beta}+\vartheta)\mathrm{e}^{\frac{1}{2}(\tilde{\beta}+\vartheta)t} + \frac{c_2}{2}(\tilde{\beta}-\vartheta)\mathrm{e}^{\frac{1}{2}(\tilde{\beta}-\vartheta)t},$$

where $\vartheta := \sqrt{\tilde{\beta}^2 + 2\sigma^2}$. The constants $c_{1,2}$ are determined by applying the boundary conditions, $\psi(T) = \mathrm{e}^0 = 1$ and $\psi'(T) = -\frac{\sigma^2}{2}C(T,T)\psi(T) = 0$, giving a 2×2 linear system in c_1 and c_2:

$$\mathrm{e}^{\frac{1}{2}(\tilde{\beta}+\vartheta)T}c_1 + \mathrm{e}^{\frac{1}{2}(\tilde{\beta}-\vartheta)T}c_2 = 1$$

$$\frac{1}{2}(\tilde{\beta}+\vartheta)\mathrm{e}^{\frac{1}{2}(\tilde{\beta}+\vartheta)T}c_1 + \frac{1}{2}(\tilde{\beta}-\vartheta)\mathrm{e}^{\frac{1}{2}(\tilde{\beta}-\vartheta)T}c_2 = 0.$$

Solving gives

$$c_1 = \left(\frac{1}{2} - \frac{\tilde{\beta}}{2\vartheta}\right)\mathrm{e}^{-\frac{1}{2}(\tilde{\beta}+\vartheta)T} \quad \text{and} \quad c_2 = \left(\frac{1}{2} + \frac{\tilde{\beta}}{2\vartheta}\right)\mathrm{e}^{-\frac{1}{2}(\tilde{\beta}-\vartheta)T}.$$

Using these coefficients gives the unique explicit expression for $\psi(t)$, which can be equivalently written as

$$\begin{aligned}\psi(t) &= \left(\frac{1}{2} - \frac{\tilde{\beta}}{2\vartheta}\right)\mathrm{e}^{-\frac{1}{2}(\tilde{\beta}+\vartheta)(T-t)} + \left(\frac{1}{2} + \frac{\tilde{\beta}}{2\vartheta}\right)\mathrm{e}^{-\frac{1}{2}(\tilde{\beta}-\vartheta)(T-t)} \\ &= \mathrm{e}^{-\frac{1}{2}\tilde{\beta}(T-t)}\left[\cosh\frac{\vartheta(T-t)}{2} + \frac{\tilde{\beta}}{\vartheta}\sinh\frac{\vartheta(T-t)}{2}\right].\end{aligned}$$

Differentiating this expression, and using the fact that $C(t,T) = -\frac{2}{\sigma^2}\frac{\psi'(t)}{\psi(t)}$, finally gives

$$C(t,T) = \frac{2\sinh\frac{\vartheta(T-t)}{2}}{\vartheta\cosh\frac{\vartheta(T-t)}{2} + \tilde{\beta}\sinh\frac{\vartheta(T-t)}{2}} = \frac{2\,\mathrm{e}^{\vartheta(T-t)} - 2}{(\tilde{\beta}+\vartheta)\left(\mathrm{e}^{\vartheta(T-t)} - 1\right) + 2\,\vartheta}. \tag{15.59}$$

Having solved for $C(t,T)$, the function $A(t,T)$ is obtained by integrating (15.57), with $\int_t^T \frac{\partial A(u,T)}{\partial u}\,\mathrm{d}u = A(T,T) - A(t,T) = -A(t,T)$, giving

$$A(t,T) = -\alpha\int_t^T C(u,T)\,\mathrm{d}u\,.$$

We can compute this integral using the fact that $-C(t,T) = \frac{2}{\sigma^2}\frac{\mathrm{d}}{\mathrm{d}t}\ln\psi(t)$, i.e.,

$$A(t,T) = \frac{2\alpha}{\sigma^2}\int_t^T \frac{\mathrm{d}}{\mathrm{d}u}\ln\psi(u)\,\mathrm{d}u = \frac{2\alpha}{\sigma^2}\ln\frac{\psi(T)}{\psi(t)} = \frac{2\alpha}{\sigma^2}\ln\frac{1}{\psi(t)}\,.$$

Note that $\psi(T) = 1$. Using the above expression for ψ gives the explicit form for $A(t,T)$, which we can write equivalently as

$$A(t,T) = \frac{2\alpha}{\sigma^2}\ln\left(\frac{\vartheta\,\mathrm{e}^{\frac{1}{2}\tilde{\beta}(T-t)}}{\vartheta\cosh\frac{\vartheta(T-t)}{2} + \tilde{\beta}\sinh\frac{\vartheta(T-t)}{2}}\right) = \frac{2\alpha}{\sigma^2}\ln\left(\frac{2\vartheta\,\mathrm{e}^{\frac{1}{2}(\tilde{\beta}+\vartheta)(T-t)}}{(\tilde{\beta}+\vartheta)(\mathrm{e}^{\vartheta(T-t)} - 1) + 2\vartheta}\right). \tag{15.60}$$

Inserting the coefficients in (15.60) and (15.59) into (15.39) gives us the closed form analytical expression for the price of a zero-coupon bond under the CIR model.

15.3 Heath–Jarrow–Morton Formulation

When we use a short-rate model such as one of those considered in previous subsections, the bond prices, yield rates, and forward rates are the output of the model. We usually find that the model bond prices $Z(t,T)$ do not perfectly match the market (observed) bond prices $Z_{\mathrm{obs}}(t,T)$. The usual approach is to calibrate the short-rate model parameters to achieve the best possible agreement between the model prices and the market prices. For example, we can use the least squares method and find the optimal model parameters by minimizing $\sum_i [Z(t,T_i) - Z_{\mathrm{obs}}(t,T_i)]^2$, the sum of squares of the differences between the model and observed bond prices across a given set of maturities $T_1, T_2, \ldots$. However, since the number of bonds with different maturities exceeds the number of model parameters, we will find that the observed prices are closer to the prices produced by the calibrated model but still differ to some extent. This issue can be dealt with by using one of the following two approaches. One way is to construct a time-inhomogeneous model with multiple random factors. Such a model will be more flexible than a single-factor model with constant parameters and can be calibrated to historical data with a greater degree of accuracy. Multifactor models are briefly discussed further in Section 15.4. Several time-inhomogeneous short-rate models, including the Ho–Lee model, the Black–Derman-Toy model, and the Hull–White model, are presented in the beginning of the previous section.

The other approach is to construct a no-arbitrage term-structure model where observed bond prices, yield rates, or forward rates are taken as input variables. The ideal result would be if the model prices $Z(t,T)$ precisely match the market prices $Z_{\mathrm{obs}}(t,T)$ at the time of calibration t. The Heath–Jarrow–Morton (HJM) framework attempts to construct a model for the family of forward rate curves $\{f(t,T)\}_{0\leqslant t\leqslant T}$ with $0 \leqslant T \leqslant T^*$. Under the HJM model, the forward rate $f(t,T)$ (for *arbitrarily fixed* $T \in [0,T^*]$) follows the SDE

$$\mathrm{d}f(t,T) = \alpha_F(t,T)\,\mathrm{d}t + \sigma_F(t,T)\,\mathrm{d}W(t)\,, \quad 0 \leqslant t \leqslant T\,, \tag{15.61}$$

where $\{W(t)\}_{t\geqslant 0}$ is a standard Brownian motion under the physical measure $\mathbb{P}$, and the processes $\alpha_F(t,T)$ and $\sigma_F(t,T)$ are adapted to its natural filtration $\mathbb{F}^W$. We note that the differential is taken w.r.t. the calendar time variable t and T acts as a fixed parameter. To simplify the analysis in what follows, we shall assume a single Brownian motion where the forward rate process starts with the initial rate $f(0,T)$. The framework readily extends to the case where the forward rates are driven by a multidimensional Brownian motion.

Let us assume that there exists an EMM (risk-neutral measure) $\widetilde{\mathbb{P}}$ such that the discounted bond price process $\{\overline{Z}(t,T)\}_{0\leqslant t\leqslant T}$ is a $\widetilde{\mathbb{P}}$-martingale. As is demonstrated below, the EMM assumption implies that the drift coefficient α_F is determined by the diffusion coefficient σ_F when the SDE (15.61) for forward rates is considered under $\widetilde{\mathbb{P}}$.

15.3.1 HJM under Risk-Neutral Measure

Let us work out the risk-neutral dynamics of the bond prices. The zero-coupon bond price can be expressed in terms of forward rates, as given in (15.7). The discounted bond price is

$$\overline{Z}(t,T) \equiv D(t)Z(t,T) = \exp\left(-\int_0^t r(u)\,\mathrm{d}u - \int_t^T f(t,u)\,\mathrm{d}u\right),$$

for $0 \leqslant t \leqslant T \leqslant T^*$. By the Itô formula, we have

$$\mathrm{d}\overline{Z}(t,T) = \overline{Z}(t,T)\left(\mathrm{d}X(t) + \frac{1}{2}\,\mathrm{d}[X,X](t) - r(t)\,\mathrm{d}t\right), \tag{15.62}$$

where $X(t)$ denotes the log-price of the bond and is given by

$$X(t) := \ln Z(t,T) = -\int_t^T f(t,u)\,\mathrm{d}u\,. \tag{15.63}$$

In (15.62) we used the Itô formula $\mathrm{d}e^{Y(t)} = e^{Y(t)}\left(\mathrm{d}Y(t) + \frac{1}{2}\,\mathrm{d}[Y,Y](t)\right)$, where $Y(t) = X(t) - \int_0^t r(u)\,\mathrm{d}u$, $\mathrm{d}Y(t) = \mathrm{d}X(t) - r(t)\,\mathrm{d}t$, and $\mathrm{d}[Y,Y](t) = \mathrm{d}[X,X](t) = \mathrm{d}X(t)\,\mathrm{d}X(t)$.

The next step is to derive an SDE for the discounted bond price in terms of the drift and diffusion functions that are driving the forward rate in (15.61). We need to compute the stochastic differential of $X(t)$ defined by (15.63), which is a Riemann integral of the process $f(t,u)$ w.r.t. the parameter $u \in [t,T]$. It can be shown that[3]

$$\mathrm{d}X(t) \equiv \mathrm{d}\left(-\int_t^T f(t,u)\,\mathrm{d}u\right) = f(t,t)\,\mathrm{d}t - \int_t^T \mathrm{d}f(t,u)\,\mathrm{d}u\,. \tag{15.64}$$

[3] If $f : \mathbb{R}^2 \to \mathbb{R}$ is a nonrandom (ordinary) function, then ordinary calculus gives the ordinary derivative w.r.t. t as $\frac{\mathrm{d}}{\mathrm{d}t}\left(-\int_t^T f(t,u)\,\mathrm{d}u\right) = f(t,t) - \int_t^T \frac{\partial f}{\partial t}(t,u)\,\mathrm{d}u$, i.e., multiplying both sides by $\mathrm{d}t$, with $\frac{\partial f}{\partial t}(t,u)\,\mathrm{d}t = \mathrm{d}f(t,u)$ for fixed parameter u, gives the ordinary differential $\mathrm{d}\left(-\int_t^T f(t,u)\,\mathrm{d}u\right) = f(t,t)\,\mathrm{d}t - \int_t^T \mathrm{d}f(t,u)\,\mathrm{d}u$. If $f(t,u) : \Omega \to \mathbb{R}$ is an $\mathcal{F}_t$-measurable Itô process (for all parameter values u), then this result applies where $\mathrm{d}f(t,u)$ is a stochastic differential w.r.t. t.

The first term is $r(t)\,\mathrm{d}t$ since the forward rate corresponds to the instantaneous short rate at time t when $T = t$, i.e., $f(t,t) = r(t)$ is the observed rate on any risk-free investment at current time t. The second term can now be written as a stochastic differential of an Itô process upon substituting the form in (15.61), using the parameter u in the place of T, within the integral and interchanging the order of the differentials (as follows by applying Fubini's Theorem twice):

$$\begin{aligned}\int_t^T \mathrm{d}f(t,u)\,\mathrm{d}u &= \int_t^T (\alpha_F(t,u)\,\mathrm{d}t + \sigma_F(t,u)\,\mathrm{d}W(t))\,\mathrm{d}u \\ &= \left(\int_t^T \alpha_F(t,u)\,\mathrm{d}u\right)\mathrm{d}t + \left(\int_t^T \sigma_F(t,u)\,\mathrm{d}u\right)\mathrm{d}W(t)\,.\end{aligned}$$

Defining the new drift and volatility functions (which are proportional the instantaneous drift and volatility functions of the forward price process averaged over all maturity times between t and T),

$$A_F(t,T) := \int_t^T \alpha_F(t,u)\,\mathrm{d}u \quad\text{and}\quad \Sigma_F(t,T) := \int_t^T \sigma_F(t,u)\,\mathrm{d}u\,, \tag{15.65}$$

gives, according to (15.64),

$$\mathrm{d}X(t) = [r(t) - A_F(t,T)]\,\mathrm{d}t - \Sigma_F(t,T)\,\mathrm{d}W(t)\,. \tag{15.66}$$

We note that, by differentiating the integrals in (15.65) w.r.t. the maturity variable T, we have

$$\frac{\partial}{\partial T}A_F(t,T) = \alpha_F(t,T) \quad\text{and}\quad \frac{\partial}{\partial T}\Sigma_F(t,T) = \sigma_F(t,T)\,. \tag{15.67}$$

Therefore, using (15.66) gives $\mathrm{d}[X,X](t) = \Sigma_F^2(t,T)\,\mathrm{d}t$ and the SDE in (15.62) for the discounted bond price takes the form

$$\frac{\mathrm{d}\overline{Z}(t,T)}{\overline{Z}(t,T)} = \left(\frac{1}{2}\Sigma_F^2(t,T) - A_F(t,T)\right)\mathrm{d}t - \Sigma_F(t,T)\,\mathrm{d}W(t)\,. \tag{15.68}$$

Note that this SDE is in terms of the Brownian increment in the physical measure $\mathbb{P}$.

The HJM model includes a zero-coupon bond for each maturity $T \in [0,T^*]$. Hence, to avoid arbitrage when trading in any of these bonds we make recourse to the First Fundamental Theorem of Chapter 13. Namely, if there exists of a risk-neutral measure $\widetilde{\mathbb{P}}$ under which all discounted bond price processes are $\widetilde{\mathbb{P}}$-martingales, then there are no arbitrage strategies in the model. The risk-neutral measure $\widetilde{\mathbb{P}}$ is the measure under which $\{\overline{Z}(t,T)\}_{0\leqslant t\leqslant T}$ are martingales for all choices of $T \in [0,T^*]$. By Girsanov's Theorem 11.13, we can change measures from $\mathbb{P}$ to $\widetilde{\mathbb{P}}$ where

$$\widetilde{W}(t) := W(t) + \int_0^t \theta(s)\,\mathrm{d}s\ ,\ t \geqslant 0,$$

is a standard $\widetilde{\mathbb{P}}$-BM with $\{\theta(t)\}_{0\leqslant t\leqslant T}$ as an adapted process. Using the differential form $\mathrm{d}W(t) = \mathrm{d}\widetilde{W}(t) - \theta(t)\,\mathrm{d}t$ into (15.68) gives

$$\frac{\mathrm{d}\overline{Z}(t,T)}{\overline{Z}(t,T)} = \left(\frac{1}{2}\Sigma_F^2(t,T) - A_F(t,T) + \Sigma_F(t,T)\,\theta(t)\right)\mathrm{d}t - \Sigma_F(t,T)\,\mathrm{d}\widetilde{W}(t)\,. \tag{15.69}$$

Hence, the measure $\widetilde{\mathbb{P}}$ defines the risk-neutral measure if the drift term in this expression is *identically zero*, i.e., if $\theta(t)$ satisfies the equation

$$\frac{1}{2}\Sigma_F^2(t,T) - A_F(t,T) + \Sigma_F(t,T)\,\theta(t) = 0, \tag{15.70}$$

for all $0 \leqslant t \leqslant T$, and for each $T \in [0, T^*]$. As we saw in Chapter 11, this corresponds to a so-called market price of risk equation where the unknown $\theta(t)$ is the market price of risk. Here, we note that $\theta(t)$ must solve the above equation for each maturity value $T \in [0, T^*]$. By taking partial derivatives w.r.t. the maturity T on both sides of (15.70), and using the derivatives defined in (15.67), gives:

$$\alpha_F(t,T) = \sigma_F(t,T)\left[\Sigma_F(t,T) + \theta(t)\right]. \tag{15.71}$$

In order to have no arbitrage in the HJM model, the drift and volatility functions and the averaged volatility $\Sigma_F(t,T)$ for the forward price process must necessarily satisfy a relation of this form for every maturity value T. In fact, if the relation in (15.71) holds, then the HJM model has no arbitrage, and assuming a nonzero volatility function $\sigma_F(t,T)$, the risk-neutral measure is given uniquely by solving for $\theta(t)$ in (15.71):

$$\theta(t) = \frac{\alpha_F(t,T)}{\sigma_F(t,T)} - \Sigma_F(t,T) \tag{15.72}$$

for all $t \in [0, T]$. This is the statement of the so-called Heath-Jarrow-Morton Theorem, which states that the above HJM model driven by a single Brownian motion admits no arbitrage if there exists an adapted process $\theta(t)$ solving (15.71) for all values of time t and maturities T.

The Heath-Jarrow-Morton Theorem is now readily proven by showing that (15.71), for all $0 \leqslant t \leqslant T \leqslant T^*$, implies (15.70), i.e., that $\widetilde{\mathbb{P}}$ exists. Moreover, if $\sigma_F(t,T) \neq 0$, then the risk-neutral measure $\widetilde{\mathbb{P}}$ is uniquely specified by (15.72). Indeed, by assumption, (15.71) holds if we replace the maturity T by the (dummy) variable s, $0 \leqslant s \leqslant T \leqslant T^*$, i.e.,

$$\alpha_F(t,s) = \sigma_F(t,s)\left[\Sigma_F(t,s) + \theta(t)\right].$$

Integrating both sides of this equation w.r.t. s, from $s = t$ to $s = T > t$, while fixing t:

$$\begin{aligned}
\int_t^T \alpha_F(t,s)\,\mathrm{d}s &= \int_t^T \sigma_F(t,s)\Sigma_F(t,s)\,\mathrm{d}s + \theta(t)\int_t^T \sigma_F(t,s)\,\mathrm{d}s \\
&= \int_t^T \frac{1}{2}\frac{\partial}{\partial s}(\Sigma_F^2(t,s))\,\mathrm{d}s + \theta(t)\int_t^T \frac{\partial}{\partial s}\Sigma_F(t,s)\,\mathrm{d}s \\
&= \frac{1}{2}\left[\Sigma_F^2(t,T) - \Sigma_F^2(t,t)\right] + \theta(t)\left[\Sigma_F(t,T) - \Sigma_F(t,t)\right] \\
&= \frac{1}{2}\Sigma_F^2(t,T) + \theta(t)\Sigma_F(t,T)
\end{aligned}$$

where $\Sigma_F(t,t) \equiv 0$. The integral on the left-hand side is, by definition, $A_F(t,T)$, and hence we recover the relation in (15.70).

Hence, assuming no arbitrage in the HJM model, i.e., assuming (15.71) holds, the SDE in (15.69) is driftless, i.e.,

$$\frac{\mathrm{d}\overline{Z}(t,T)}{\overline{Z}(t,T)} = -\Sigma_F(t,T)\,\mathrm{d}\widetilde{W}(t), \tag{15.73}$$

where discounted zero-coupon bond price processes of all maturities T are $\widetilde{\mathbb{P}}$-martingales. In particular, the ratio $\frac{\overline{Z}(t,T)}{\overline{Z}(0,T)}$ is the stochastic exponential of the process $\{-\Sigma_F(t,T)\}_{0 \leqslant t \leqslant T}$ w.r.t. the Brownian motion $\widetilde{W}$ on the time interval $[0, t]$:

$$\begin{aligned}
\overline{Z}(t,T) &= \overline{Z}(0,T)\mathcal{E}_t(-\Sigma_F \cdot \widetilde{W}) \\
&= Z(0,T)\exp\left[-\frac{1}{2}\int_0^t \Sigma_F^2(s,T)\,\mathrm{d}s - \int_0^t \Sigma_F(s,T)\,\mathrm{d}\widetilde{W}(s)\right].
\end{aligned} \tag{15.74}$$

Note that $\overline{Z}(0,T) = Z(0,T)$ since $B(0) = 1$. We recall that the unit bank account has value $B(t) = \exp\left(\int_0^t r(s)\,\mathrm{d}s\right)$. Hence, the risk-neutral value process of a zero-coupon bond takes the form

$$Z(t,T) = Z(0,T)\exp\left[-\frac{1}{2}\int_0^t \Sigma_F^2(s,T)\,\mathrm{d}s - \int_0^t \Sigma_F(s,T)\,\mathrm{d}\widetilde{W}(s) + \int_0^t r(s)\,\mathrm{d}s\right]. \quad (15.75)$$

The SDE for $Z(t,T)$ under $\widetilde{\mathbb{P}}$ has the form

$$\frac{\mathrm{d}Z(t,T)}{Z(t,T)} = r(t)\,\mathrm{d}t - \Sigma_F(t,T)\,\mathrm{d}\widetilde{W}(t). \quad (15.76)$$

We recall from Chapter 13 that a (domestic) nondividend paying asset has (log-)drift equal to $r(t)$ under the risk-neutral measure. Clearly, a zero-coupon bond is an example of a nondividend paying asset.

Finally, upon using (15.71), the SDE in (15.61) for the forward price process takes the form (w.r.t. the $\widetilde{\mathbb{P}}$-BM):

$$\begin{aligned}\mathrm{d}f(t,T) &= [\alpha_F(t,T) - \theta(t)\sigma_F(t,T)]\,\mathrm{d}t + \sigma_F(t,T)\,\mathrm{d}\widetilde{W}(t)\\ &= \sigma_F(t,T)\Sigma_F(t,T)\,\mathrm{d}t + \sigma_F(t,T)\,\mathrm{d}\widetilde{W}(t).\end{aligned} \quad (15.77)$$

This SDE shows us that the instantaneous (time-t) risk-neutral drift of the forward price process (for given maturity T) is determined by its instantaneous volatility and its (integrated) volatility across all times up to the maturity T. That is, the forward price has SDE of the form

$$\mathrm{d}f(t,T) = \widetilde{\alpha}_F(t,T)\,\mathrm{d}t + \sigma_F(t,T)\,\mathrm{d}\widetilde{W}(t), \quad (15.78)$$

with risk-neutral drift $\widetilde{\alpha}_F(t,T) = \sigma_F(t,T)\int_t^T \sigma_F(t,u)\,\mathrm{d}u$. This is necessarily the form for the drift of the forward price process under the risk-neutral measure $\widetilde{\mathbb{P}}$.

15.3.2 Relationship between HJM and Affine Yield Models

We can formulate any one-factor short-rate model within the HJM framework. Consider an affine term structure model. The short rate process follows the SDE (15.40) under $\widetilde{\mathbb{P}}$ with coefficients given by (15.41):

$$\mathrm{d}r(t) = \big(a_0(t) + a_1(t)r(t)\big)\,\mathrm{d}t + \sqrt{b_0(t) + b_1(t)r(t)}\,\mathrm{d}\widetilde{W}(t).$$

Note that $b(t,r) \equiv \sqrt{b_0(t) + b_1(t)r}$ defines the diffusion function of the short rate process, where $b^2(t,r(t)) = b_0(t) + b_1(t)r(t)$. The bond price is in the affine form (15.39):

$$Z(t,T) = \mathrm{e}^{A(t,T) - r(t)C(t,T)},$$

where the functions C and A solve the ODEs (15.43) and (15.44), respectively. According to (15.6), the forward rates are

$$f(t,T) = -\frac{\partial}{\partial T}\ln Z(t,T) = -\frac{\partial A(t,T)}{\partial T} + r(t)\frac{\partial C(t,T)}{\partial T}.$$

Applying the Itô formula, where $A(t,T)$ and $C(t,T)$ are nonrandom functions of time t, we obtain the stochastic differential of the forward rate in the form

$$\begin{aligned} \mathrm{d}f(t,T) &= \frac{\partial C(t,T)}{\partial T}\,\mathrm{d}r(t) + r(t)\frac{\partial^2 C(t,T)}{\partial t\partial T}\,\mathrm{d}t - \frac{\partial^2 A(t,T)}{\partial t\partial T}\,\mathrm{d}t \\ &= \left[\frac{\partial C(t,T)}{\partial T}\big(a_0(t)+a_1(t)r(t)\big) + r(t)\frac{\partial^2 C(t,T)}{\partial t\partial T} - \frac{\partial^2 A(t,T)}{\partial t\partial T}\right]\mathrm{d}t \\ &\quad + \frac{\partial C(t,T)}{\partial T} b(t,r(t))\,\mathrm{d}\widetilde{W}(t)\,. \end{aligned}$$

Equating the diffusion term in the above SDE with (15.77) gives

$$\sigma_F(t,T) = \frac{\partial C(t,T)}{\partial T} b(t,r(t)) = \frac{\partial C(t,T)}{\partial T}\sqrt{b_0(t)+b_1(t)r(t)}\,. \tag{15.79}$$

Equating the drift term with $\widetilde{\alpha}_F(t,T)$ in (15.77) gives:

$$\begin{aligned} &\frac{\partial C(t,T)}{\partial T}\big(a_0(t)+a_1(t)r(t)\big) + r(t)\frac{\partial^2 C(t,T)}{\partial t\partial T} - \frac{\partial^2 A(t,T)}{\partial t\partial T} \\ &= \frac{\partial C(t,T)}{\partial T} b^2(t,r(t)) \int_t^T \frac{\partial C(t,u)}{\partial u}\,\mathrm{d}u \\ &= b^2(t,r(t))\frac{\partial C(t,T)}{\partial T}\big(C(t,T)-C(t,t)\big) \\ &= \big(b_0(t)+b_1(t)r(t)\big)\frac{\partial C(t,T)}{\partial T}C(t,T). \end{aligned} \tag{15.80}$$

This is essentially the no-arbitrage condition that must be satisfied by any affine yield model for all times $t \leqslant T$.

15.3.2.1 The Ho–Lee Model in the HJM Framework

Suppose that the diffusion coefficient $\sigma_F(t,T)$ in the SDE (15.61) for forward rates is constant and equal to σ. The function Σ_F is then given by

$$\Sigma_F(t,T) = \int_t^T \sigma\,\mathrm{d}u = \sigma(T-t)\,.$$

The SDE (15.77) takes the form

$$\mathrm{d}f(t,T) = \sigma^2(T-t)\,\mathrm{d}t + \sigma\,\mathrm{d}\widetilde{W}(t)\,.$$

Integrating the above equation w.r.t. time t gives

$$f(t,T) = f(0,T) + \sigma^2 t(T-t/2) + \sigma\widetilde{W}(t)\,. \tag{15.81}$$

Setting $T=t$ gives the short rate (since $r(t)=f(t,t)$)

$$r(t) = f(0,t) + \sigma^2 t^2/2 + \sigma\widetilde{W}(t)\,.$$

In the above equation, we recognize the Ho–Lee model (15.45). Isolating the Brownian term in (15.81) in terms of the forward rates and substitution into the last equation gives

$$f(t,T) = r(t) + f(0,T) - f(0,t) + \sigma^2 t(T-t)\,. \tag{15.82}$$

As is seen from the above equation, the spot rate $r(t)$ and forward rate $f(t,T)$ are linearly dependent and hence are perfectly correlated.

Using (15.7), and (15.81) for $T=u$, we find the bond price:

$$\begin{aligned} Z(t,T) &= \exp\left[-\int_t^T \left(f(0,u)+\sigma^2 t(u-t/2)+\sigma\widetilde{W}(t)\right)\mathrm{d}u\right] \\ &= \exp\left[-\int_t^T f(0,u)\,\mathrm{d}u - \frac{1}{2}\sigma^2 tT(T-t) - \sigma(T-t)\widetilde{W}(t)\right]. \end{aligned}$$

From (15.9), we obtain $-\int_t^T f(0,u)\,\mathrm{d}u = \ln\frac{Z(0,T)}{Z(0,t)}$ and then

$$Z(t,T) = \frac{Z(0,T)}{Z(0,t)}\exp\left[-\frac{1}{2}\sigma^2 tT(T-t) - \sigma(T-t)\widetilde{W}(t)\right]. \tag{15.83}$$

Since $\frac{Z(0,T)}{Z(0,t)} = \mathrm{e}^{-f(0;t,T)(T-t)}$, the yield rate is

$$y(t,T) = -\frac{\ln Z(t,T)}{T-t} = f(0;t,T) + \frac{\sigma^2 tT}{2} + \sigma\widetilde{W}(t).$$

By eliminating the Brownian term using the equation just below (15.81), we express the yield rate in terms of forward rates (where $r(t)=f(t,t)$):

$$y(t,T) = f(t,t) - f(0,t) + f(0;t,T) + \frac{1}{2}\sigma^2 t(T-t). \tag{15.84}$$

15.3.2.2 The Vasiček Model in the HJM Framework

Let us derive the forward rate SDE and verify the no-arbitrage condition in (15.80) for the Vasiček model. The short-rate process (under $\widetilde{\mathbb{P}}$) is driven by the SDE (15.48) which we repeat here:

$$\mathrm{d}r(t) = (\tilde{\alpha} - \beta r(t))\,\mathrm{d}t + \sigma\,\mathrm{d}\widetilde{W}(t).$$

The functions A and C for the bond price are given in (15.52). Using (15.79) and the expression for $C(t,T)$ in (15.52), we obtain the diffusion coefficient of the forward rate:

$$\sigma_F(t,T) = \sigma\frac{\partial C(t,T)}{\partial T} = \sigma\frac{\partial}{\partial T}\left(\frac{1}{\beta}\left(1-\mathrm{e}^{-\beta(T-t)}\right)\right) = \sigma\mathrm{e}^{-\beta(T-t)}. \tag{15.85}$$

Let us verify that (15.80) holds. Using the expression for $C(t,T)$ and (15.51), we have

$$\begin{aligned} \frac{\partial^2 C(t,T)}{\partial t\partial T} &= \beta\mathrm{e}^{-\beta(T-t)}, \\ \frac{\partial^2 A(t,T)}{\partial t\partial T} &= \left(\tilde{\alpha} - \frac{\sigma^2}{\beta}\right)\mathrm{e}^{-\beta(T-t)} + \frac{\sigma^2}{\beta}\mathrm{e}^{-2\beta(T-t)}. \end{aligned}$$

For any given $r(t)=r$, the left-hand side of (15.80) is hence given by

$$\begin{aligned} \mathrm{e}^{-\beta(T-t)}(\tilde{\alpha}-\beta r) + r\beta\mathrm{e}^{-\beta(T-t)} - \tilde{\alpha}\mathrm{e}^{-\beta(T-t)} + \frac{\sigma^2}{\beta}\mathrm{e}^{-\beta(T-t)} - \frac{\sigma^2}{\beta}\mathrm{e}^{-2\beta(T-t)} \\ = \frac{\sigma^2}{\beta}\left(\mathrm{e}^{-\beta(T-t)} - \mathrm{e}^{-2\beta(T-t)}\right), \end{aligned} \tag{15.86}$$

and the right-hand side of (15.80) is

$$\frac{\sigma^2}{\beta}\mathrm{e}^{-\beta(T-t)}\left(1-\mathrm{e}^{-\beta(T-t)}\right). \tag{15.87}$$

Hence the expressions in (15.86) and (15.87) are equal, i.e., the Vasiček short-rate model is of the HJM type for which the no-arbitrage condition holds. Using (15.85), the risk-neutral drift of the forward rate is given by

$$\widetilde{\alpha}_F(t,T) \equiv \sigma_F(t,T)\Sigma_F(t,T) = \sigma\mathrm{e}^{-\beta(T-t)}\int_t^T \sigma\mathrm{e}^{-\beta(u-t)}\,\mathrm{d}u = \frac{\sigma^2}{\beta}\mathrm{e}^{-\beta(T-t)}\left(1-\mathrm{e}^{-\beta(T-t)}\right).$$

Thus, for the Vasiček model, the forward rates follow the SDE

$$\mathrm{d}f(t,T) = \frac{\sigma^2}{\beta}\left(\mathrm{e}^{-\beta(T-t)} - \mathrm{e}^{-2\beta(T-t)}\right)\mathrm{d}t + \sigma\mathrm{e}^{-\beta(T-t)}\,\mathrm{d}\widetilde{W}(t) \tag{15.88}$$

under the risk-neutral measure $\widetilde{\mathbb{P}}$.

15.4 Multifactor Affine Term Structure Models

In the previous two sections, we discussed one-factor models where Brownian motion is the only source of randomness. One-factor short-rate models offer good analytical tractability. For many models, a closed-form solution for the bond price can be found. However, one-factor models of interest rates have many drawbacks, including the fact that yield rates for different maturities are perfectly correlated. So the term structure of interest rates for a one-factor model is oversimplified. One of the possible solutions is to consider multifactor interest rate models that involve the short rate along with other random parameters. Consider the following example. Let the short rate follow

$$\mathrm{d}r(t) = \alpha\big(\bar{r} - \beta r(t)\big)\,\mathrm{d}t + \sigma\,\mathrm{d}W_1(t).$$

The *stochastic volatility model* assumes that the volatility σ of the short rate process is stochastic. For example, its square (or variance) is governed by a square-root model,

$$\mathrm{d}\sigma^2(t) = \big(\gamma - \delta\sigma^2(t)\big)\,\mathrm{d}t + \xi\sigma\,\mathrm{d}W_2(t),$$

where W_1 and W_2 are correlated Brownian motions, and α, γ, δ, ξ are constant parameters. So the interest rate volatility $\sigma(t)$ is included as the second random variable. This approach can be extended by including another additional factor: the stochastic mean level of the short rate $\bar{r}$. As a result, one can construct a three-factor model with three state variables: the short rate $r(t)$, the volatility $\sigma(t)$, and the mean level $\bar{r}(t)$.

The multifactor approach provides a greater flexibility in modelling the stochastic term structure of interest rates. However, increasing the number of random factors, reduces the analytical tractability of a model. Valuation of fixed income derivatives often relies on efficient numerical methods. Calibration of a multifactor model can also be a challenge. In this section we discuss a general multifactor affine term structure model and provide several examples of two- and three-factor models. These models are popular due to their good analytical tractability.

Consider a stochastic interest rate model with n state variables $X_1(t), X_2(t), \ldots, X_n(t)$

(or, as an n-by-1 vector, $\mathbf{X}(t) = [X_1(t), X_2(t), \ldots, X_n(t)]^\top$). The model is said to be *affine* if the zero-coupon bond prices admit the following exponential form:

$$Z(t,T) = \exp\left[A(t,T) + \sum_{j=1}^{n} c_j(t,T)X_j(t)\right] = \exp\left[A(t,T) + \mathbf{C}(t,T)^\top \mathbf{X}(t)\right], \tag{15.89}$$

where $\mathbf{C}(t,T)^\top := [c_1(t,T), c_2(t,T), \ldots, c_n(t,T)]$. The model is time homogeneous if the state variables $X_j(t)$ are all time-homogeneous processes and A and $\mathbf{C}$ are functions of the time to maturity $\tau = T - t$ only. In this case the bond price function takes the form

$$Z(t, t+\tau) = \exp\left[A(\tau) + \mathbf{C}(\tau)^\top \mathbf{X}(t)\right]. \tag{15.90}$$

Hence, the yield rate is

$$y(t, t+\tau) = -\frac{1}{\tau}\left(A(\tau) + \mathbf{C}(\tau)^\top \mathbf{X}(t)\right).$$

By taking the limit $\tau \searrow 0$, we obtain the short-rate process

$$r(t) = y(t,t) = -\frac{\mathrm{d}A}{\mathrm{d}\tau}(0) - \frac{\mathrm{d}\mathbf{C}^\top}{\mathrm{d}\tau}(0)\mathbf{X}(t).$$

As was demonstrated in Section 15.2, a single-factor time-homogeneous model admits a bond pricing function in the affine form, if the short-rate process follows an SDE with a linear drift and a linear squared diffusion coefficient,

$$\mathrm{d}r(t) = \left(\alpha + \beta r(t)\right)\mathrm{d}t + \sqrt{\lambda + \mu r(t)}\right)\mathrm{d}\widetilde{W}(t).$$

Clearly, certain conditions have to be set on the dynamics of the state vector $\mathbf{X}(t)$ in order that $Z(t, t+\tau)$ has the form (15.90). Duffie and Kan proved that the SDE for $\mathbf{X}(t)$ under $\widetilde{\mathbb{P}}$ has to be of the form

$$\mathrm{d}\mathbf{X}(t) = (\boldsymbol{\alpha} + \boldsymbol{B}\mathbf{X}(t))\,\mathrm{d}t + \boldsymbol{\Sigma}\mathbf{D}(\mathbf{X}(t))\,\mathrm{d}\widetilde{\mathbf{W}}(t), \quad t \geqslant 0, \tag{15.91}$$

where $\mathbf{D}$ is a diagonal matrix with

$$\sqrt{\lambda_1 + \boldsymbol{\mu}_1^\top \mathbf{X}(t)}, \sqrt{\lambda_2 + \boldsymbol{\mu}_2^\top \mathbf{X}(t)}, \ldots, \sqrt{\lambda_n + \boldsymbol{\mu}_n^\top \mathbf{X}(t)}$$

on the main diagonal, $\boldsymbol{\alpha}$ and $\boldsymbol{\mu}_i$, $i = 1, 2, \ldots, n$, are constant n-dimensional vectors, $\boldsymbol{B} = [\beta_{ij}]_{i,j=1}^n$ and $\boldsymbol{\Sigma} = [\sigma_{ij}]_{i,j=1}^n$ are constant n-by-n matrices, and $\{\widetilde{\mathbf{W}}(t)\}_{t\geqslant 0}$ is an n-dimensional Brownian motion under $\widetilde{\mathbb{P}}$. Certain conditions on the model parameters are required to ensure that each of the variance processes $\lambda_i + \boldsymbol{\mu}_i^\top \mathbf{X}(t)$, $i = 1, 2, \ldots, n$, remains positive.

15.4.1 Gaussian Multifactor Models

Let us set the coefficient vectors $\boldsymbol{\mu}_1, \boldsymbol{\mu}_2, \ldots, \boldsymbol{\mu}_n$ to be zero. The SDE (15.91) reduces to the Gaussian form

$$\mathrm{d}\mathbf{X}(t) = (\boldsymbol{\alpha} + \boldsymbol{B}\mathbf{X}(t))\,\mathrm{d}t + \boldsymbol{\Sigma}\,\mathrm{d}\widetilde{\mathbf{W}}(t), \tag{15.92}$$

where all parameters and matrices are constant. It can be shown that the distribution of $\mathbf{X}(t)$ is multivariate normal.

15.4.2 Equivalent Classes of Affine Models

The number of possible affine models as defined by (15.91) can be quite large. However, by a transformation of variables, a model can be represented in different ways. Dai and Singleton claimed that the models are considered equivalent if they generate identical prices for all contingent claims. Affine models can be classified according to the number of factors and the number of state variables appearing in the volatility matrix $\mathbf{D}$. There are only two equivalent classes of one-factor models: the Vasiček model and the CIR model. When the number of factors is two, we have three equivalent classes, listed below in their canonical forms.

1. The two-factor Vasiček model

$$\begin{aligned} \mathrm{d}X_1(t) &= -\beta_{11}X_1(t)\,\mathrm{d}t + \mathrm{d}\widetilde{W}_1(t),\\ \mathrm{d}X_2(t) &= \big(-\beta_{21}X_1(t) - \beta_{22}X_2(t)\big)\,\mathrm{d}t + \mathrm{d}\widetilde{W}_2(t). \end{aligned} \tag{15.93}$$

2. The two-factor CIR model (for example, the Longstaff–Schwartz model)

$$\begin{aligned} \mathrm{d}X_1(t) &= \big(\mu_1 - \beta_{11}X_1(t) - \beta_{12}X_2(t)\big)\,\mathrm{d}t + \sqrt{X_1(t)}\,\mathrm{d}\widetilde{W}_1(t),\\ \mathrm{d}X_2(t) &= \big(\mu_2 - \beta_{21}X_1(t) - \beta_{22}X_2(t)\big)\,\mathrm{d}t + \sqrt{X_2(t)}\,\mathrm{d}\widetilde{W}_2(t). \end{aligned} \tag{15.94}$$

3. The two-factor stochastic volatility model (for example, the Fong–Vasiček model)

$$\begin{aligned} \mathrm{d}X_1(t) &= (\mu_1 - \beta_{11}X_1(t))\,\mathrm{d}t + \sqrt{X_1(t)}\,\mathrm{d}\widetilde{W}_1(t),\\ \mathrm{d}X_2(t) &= \big(\mu_2 - \beta_{21}X_1(t) - \beta_{22}X_2(t)\big)\,\mathrm{d}t + \big(1 + \delta_{21}\sqrt{X_1(t)}\big)\,\mathrm{d}\widetilde{W}_2(t). \end{aligned} \tag{15.95}$$

In each of the above models, $(\widetilde{W}_1(t), \widetilde{W}_2(t))$ is a standard two-dimensional Brownian motion, under the risk-neutral measure $\widetilde{\mathbb{P}}$ with bank account as numéraire. The short rate is assumed to be an affine (linear) function of the two factors:

$$r(t) = a_0 + a_1X_1(t) + a_2X_2(t)\,. \tag{15.96}$$

The coefficients a_0, a_1, a_2 are typically assumed to be constants, although they can also be chosen as nonrandom functions of time t. These parameters, along with those arising from the SDE for each model, can be used to calibrate the affine model to the spot yield curve.

In the two-factor Vasiček model, the parameters in (15.93) and (15.96) are assumed to take on real values where $\beta_{11} > 0, \beta_{22} > 0$. It is readily shown that the factors $X_1(t), X_2(t)$ are jointly normal random variables and hence the short rate is a normal random variable as long as a_1, a_2 are not both zero. Hence, there is a positive probability that $r(t) < 0$ for any positive time $t > 0$. In the two-factor CIR model the parameters in (15.94) are chosen such that $\mu_1 \geqslant 0, \mu_2 \geqslant 0, \beta_{11} > 0, \beta_{22} > 0, \beta_{12} \leqslant 0, \beta_{21} \leqslant 0$. Under these conditions it can be shown that the factors are nonnegative processes. That is, if the factors start with nonnegative values $X_1(0) \geqslant 0, X_2(0) \geqslant 0$, then $X_1(t) \geqslant 0, X_2(t) \geqslant 0$ for all time $t \geqslant 0$ (almost surely). Assuming a nonnegative initial short rate $r(0) \geqslant 0$, together with the conditions that $a_0 \geqslant 0, a_1 > 0, a_2 > 0$ in (15.96), guarantees that $r(t) \geqslant 0$ for all time $t \geqslant 0$.

In the interest of space, we do not present the details for pricing bonds under these two-factor affine models. Rather, we summarize the basic steps that are similar to those given for the single-factor affine models and leave the rest of the details as exercises at the end of this chapter. For any of the above two-factor affine models we have prices of zero-coupon bonds that are driven by a two-dimensional system of SDEs for the vector $\mathbf{X}(t) = [X_1(t), X_2(t)]^\top$. Hence, the corresponding bond pricing function can be expressed as a function of the calendar time t and the time-t value of the two factors, i.e., the zero-coupon bond price process $Z(t,T) = V(t, X_1(t), X_2(t))$, where $V = V(t, x_1, x_2)$ is a smooth differentiable function of time t and twice differentiable function of the spot values $X_1(t) = x_1, X_2(t) = x_2$. Recalling our analysis in Chapter 11, we see that the pricing function can be determined by applying the (two-dimensional) discounted Feynman-Kac Theorem 11.20. We can simply use Theorem 11.20 to express $V(t, x_1, x_2)$ as a solution to a PDE for $t < T$, subject to the terminal condition $V(T, x_1, x_2) = Z(T,T) = 1$. For an affine model, the solution necessarily has the form given by (15.89) for $n = 2$, $X_1(t) = x_1, X_2(t) = x_2$:

$$V(t, x_1, x_2) = \mathrm{e}^{A(t,T)+c_1(t,T)x_1+c_2(t,T)x_2}\,. \tag{15.97}$$

Note that the exponent is linear in the factor variables x_1 and x_2. This is an extension of the form of the solution in (15.39) to include two factors rather than just the single short rate variable r. For a time-homogeneous model we have $V(t, x_1, x_2) = v(\tau, x_1, x_2)$, where the coefficients are functions of $\tau = T - t$: $A(t,T) = A(\tau)$, $c_1(t,T) = c_1(\tau)$, $c_2(t,T) = c_2(\tau)$. Substitution of the above exponential form for V into the corresponding bond-pricing PDE (for the particular model) leads to a system of first order ordinary differential equations in the functions $A(\tau)$, $c_1(\tau)$, and $c_2(\tau)$. This system can then be solved subject to the initial conditions: $A(0) = c_1(0) = c_2(0) = 0$. Given the solutions for these coefficient functions, the bond pricing function is then given by (15.97) (see Exercise 15.10).

15.5 Pricing Derivatives under Forward Measures

15.5.1 Forward Measures

Taking the zero-coupon bond Z rather than the bank account B as a numèraire asset allows us to simplify the derivative pricing formula (15.12). Let $\widehat{\mathbb{P}}_T^{(Z)}$ denote the EMM relative to the numèraire $g(t) = Z(t,T)$. The probability measure $\widehat{\mathbb{P}}_T^{(Z)}$ is called the T-*forward measure*. It is defined so that the process $\{B(t)/Z(t,T)\}_{0\leqslant t\leqslant T}$ is a $\widehat{\mathbb{P}}_T^{(Z)}$-martingale. The Radon–Nykodim derivative of $\widetilde{\mathbb{P}}^{(B)} \equiv \widetilde{\mathbb{P}}$ w.r.t. $\widetilde{\mathbb{P}}^{(g)} \equiv \widehat{\mathbb{P}}_T^{(Z)} \equiv \widehat{\mathbb{P}}$ is

$$\varrho \equiv \frac{d\widetilde{\mathbb{P}}}{d\widehat{\mathbb{P}}} = \frac{B(T)/B(0)}{Z(T,T)/Z(0,T)} = Z(0,T)B(T). \tag{15.98}$$

Note that $B(0) = Z(T,T) = 1$. The respective Radon–Nykodim derivative process is

$$\varrho_t \equiv \left(\frac{d\widetilde{\mathbb{P}}}{d\widehat{\mathbb{P}}}\right)_t = \frac{B(t)/B(0)}{Z(t,T)/Z(0,T)} = \frac{Z(0,T)B(t)}{Z(t,T)}, \quad 0 \leqslant t \leqslant T, \tag{15.99}$$

where $\varrho_T = \varrho$. We recall, from (13.129) in Chapter 13, that $\varrho_t = \varrho_t^{g\to B}$, where $\varrho_t^{B\to g} = 1/\varrho_t^{g\to B} = 1/\varrho_t$ corresponds to changing numèraires from the bank account to the zero-coupon bond in this case. Let payoff $V(T)$ with maturity time T be $\mathcal{F}_T$-measurable and integrable w.r.t. $\widetilde{\mathbb{P}}$. Applying the change of numèraire theorem, i.e., as follows immediately by the risk-neutral pricing formula (13.134) with $g(t) = Z(t,T)$ as numèraire asset price, the time-t price of the claim is given by the conditional expectation under the T-forward measure $\widehat{\mathbb{P}}$:

$$V(t) = g(t)\,\widehat{\mathrm{E}}_t\left[\frac{V(T)}{g(T)}\right] = Z(t,T)\widehat{\mathrm{E}}_t\left[V(T)\right]. \tag{15.100}$$

The expected value $\widehat{\mathrm{E}}_t[V(T)] = V(t)/Z(t,T)$ is called the forward price at time t of the payoff $V(T)$ with maturity time T. Recall that the forward price makes the time-t value of forward delivery of $V(T)$ at time T zero.

Example 15.1. Compute the expectation of the future short rate $r(T)$, conditional on $\mathcal{F}_t$, under the forward measure $\widehat{\mathbb{P}} \equiv \widehat{\mathbb{P}}_T^{(Z)}$.

Solution. We note that the Radon–Nykodim derivative process of $\widehat{\mathbb{P}} \equiv \widehat{\mathbb{P}}_T^{(Z)}$ w.r.t. $\widetilde{\mathbb{P}}$ is $\left(\frac{\mathrm{d}\widehat{\mathbb{P}}}{\mathrm{d}\widetilde{\mathbb{P}}}\right)_t = \frac{1}{\varrho_t}$. Hence, by (15.99),

$$\frac{\left(\frac{\mathrm{d}\widehat{\mathbb{P}}}{\mathrm{d}\widetilde{\mathbb{P}}}\right)_T}{\left(\frac{\mathrm{d}\widehat{\mathbb{P}}}{\mathrm{d}\widetilde{\mathbb{P}}}\right)_t} = \frac{1/\varrho_T}{1/\varrho_t} = \frac{\varrho_t}{\varrho_T} = \frac{1}{Z(t,T)}\frac{B(t)}{B(T)}.$$

Thus, we have (by the property(11.83) where $r(T)$ is $\mathcal{F}_T$-measurable),

$$\begin{aligned}\widehat{\mathrm{E}}_t[r(T)] &= \widetilde{\mathrm{E}}_t\left[\frac{1}{Z(t,T)}\frac{B(t)}{B(T)}r(T)\right] = \frac{1}{Z(t,T)}\widetilde{\mathrm{E}}_t\left[\exp\left(-\int_t^T r(u)\,\mathrm{d}u\right)r(T)\right]\\ &= \frac{1}{Z(t,T)}\widetilde{\mathrm{E}}_t\left[-\frac{\partial}{\partial T}\exp\left(-\int_t^T r(u)\,\mathrm{d}u\right)\right]\end{aligned}$$

(assume that we can interchange the operations of differentiation and integration)

$$= -\frac{1}{Z(t,T)}\frac{\partial}{\partial T}\left\{\widetilde{\mathrm{E}}_t\left[\exp\left(-\int_t^T r(u)\,\mathrm{d}u\right)\right]\right\} = -\frac{1}{Z(t,T)}\frac{\partial Z(t,T)}{\partial T},$$

where the bond price formula (15.11) is used. Using (15.6), we have

$$-\frac{1}{Z(t,T)}\frac{\partial Z(t,T)}{\partial T} = -\frac{\partial \ln Z(t,T)}{\partial T} = f(t,T).$$

Hence, the instantaneous forward rate $f(t,T)$ is equal to the mathematical expectation of the short rate $r(T) \equiv f(T,T)$ conditional on $\mathcal{F}_t$ under the T-forward measure:

$$\widehat{\mathrm{E}}_t[r(T)] = \widehat{\mathrm{E}}_t[f(T,T)] = f(t,T)$$

Thus, we conclude that the forward rate process $\{f(t,T)\}_{0\leqslant t\leqslant T}$ is a $\widehat{\mathbb{P}}_T^{(Z)}$-martingale. □

Let the dynamics of the T-maturity bond price be governed by the SDE (15.17) which we repeat here:

$$\frac{\mathrm{d}Z(t,T)}{Z(t,T)} = r(t)\,\mathrm{d}t + \sigma_Z(t,T)\,\mathrm{d}\widetilde{W}(t),$$

where $\widetilde{W}$ is a $\widetilde{\mathbb{P}}$-Brownian motion. Under the HJM framework with forward rates following the SDE (15.61), we have the SDE (15.76). Hence, equating the volatility terms gives

$$\sigma_Z(t,T) = -\Sigma_F(t,T) = -\int_t^T \sigma_F(t,u)\,\mathrm{d}u\,. \tag{15.101}$$

The bond price is given by (15.18). We rewrite it as follows:

$$Z(t,T) = Z(0,T)B(t)\exp\left(-\frac{1}{2}\int_0^t \sigma_Z^2(s,T)\,\mathrm{d}s + \int_0^t \sigma_Z(s,T)\,\mathrm{d}\widetilde{W}(s)\right). \tag{15.102}$$

The measure $\widehat{\mathbb{P}} \equiv \widehat{\mathbb{P}}_T^{(Z)}$ is defined by the Radon–Nykodim derivative $\frac{d\widehat{\mathbb{P}}}{d\widetilde{\mathbb{P}}} = \widehat{\varrho}_T$, where

$$\widehat{\varrho}_t := \frac{1}{\varrho_t} = \frac{Z(t,T)}{B(t)Z(0,T)}, \quad 0 \leqslant t \leqslant T$$

with ϱ_t given in (15.99). Using the bond price solution (15.102), we have

$$\widehat{\varrho}_t \equiv \left(\frac{d\widehat{\mathbb{P}}}{d\widetilde{\mathbb{P}}}\right)_t = \exp\left(-\frac{1}{2}\int_0^t \sigma_Z^2(s,T)\, ds + \int_0^t \sigma_Z(s,T)\, d\widetilde{W}(s)\right).$$

Note that this is precisely the definition of the change of measure as given by the exponential $\widetilde{\mathbb{P}}$-martingale in (13.118) since $\widehat{\varrho}_t = \varrho_t^{B\to g}$ with $g(t) = Z(t,T)$. Since we are assuming that asset prices are driven by a single Brownian component, the volatility vector of the numèraire asset is now simply a scalar, $\sigma^{(g)}(t) \equiv \sigma_Z(t,T)$ for fixed maturity T. By Girsanov's Theorem 11.13 the process

$$\widehat{W}(t) := \widetilde{W}(t) - \int_0^t \sigma_Z(s,T)\, ds, \quad 0 \leqslant t \leqslant T$$

is a standard $\widehat{\mathbb{P}}$-Brownian motion. Using

$$d\widetilde{W}(t) = d\widehat{W}(t) + \sigma_Z(t,T)\, dt\,, \tag{15.103}$$

we have the following SDE for $Z(t,T)$ under the T-forward measure $\widehat{\mathbb{P}}$:

$$\frac{dZ(t,T)}{Z(t,T)} = \left(r(t) + \sigma_Z^2(t,T)\right) dt + \sigma_Z(t,T)\, d\widehat{W}(t)\,.$$

Within the HJM framework, we hence obtain

$$\begin{aligned} df(t,T) &= \sigma_F(t,T)\Sigma_F(t,T)\, dt + \sigma_F(t,T)\, d\widetilde{W}(t) \\ &= \sigma_F(t,T)\Sigma_F(t,T)\, dt + \sigma_F(t,T)\left(d\widehat{W}(t) - \Sigma_F(t,T)\, dt\right) \\ &= \sigma_F(t,T)\, d\widehat{W}(t)\,. \end{aligned} \tag{15.104}$$

Thus, the forward rate satisfies the driftless SDE $df(t,T) = \sigma_F(t,T)\, d\widehat{W}(t)$ and is hence a $\widehat{\mathbb{P}}$-martingale. We arrived at the same conclusion in Example 15.1.

15.5.2 Pricing Stock Options under Stochastic Interest Rates

In this section we present a generalized Black–Scholes formula for option prices under an asset price model with stochastic interest rates. Consider a risky asset such as a stock without dividends. Let the price dynamics of a nondividend paying stock $S(t)$ and the bond price $Z(t,T)$ at time t be governed by

$$\frac{dS(t)}{S(t)} = r(t)\, dt + \sigma_S(t)\, d\widetilde{W}(t), \tag{15.105}$$

$$\frac{dZ(t,T)}{Z(t,T)} = r(t)\, dt + \sigma_Z(t,T)\, d\widetilde{W}(t), \tag{15.106}$$

with respective log-volatilities $\sigma_S(t)$ and $\sigma_Z(t,T)$ and where $\widetilde{W}$ is a standard (one-dimensional) $\widetilde{\mathbb{P}}$-Brownian motion. For a given maturity T, we define

$$F_S(t) = \frac{S(t)}{Z(t,T)}, \quad 0 \leqslant t \leqslant T,$$

which is the time-t price of the T-maturity forward of the stock. By the definition of the T-forward measure $\widehat{\mathbb{P}} \equiv \widehat{\mathbb{P}}_T^{(Z)}$, all (domestic) *nondividend paying asset prices divided by the bond price $Z(t,T)$ are $\widehat{\mathbb{P}}$-martingales.* That is, the forward price of the stock is a $\widehat{\mathbb{P}}$-martingale (see (13.128) where $A(t) = S(t), g(t) = Z(t,T)$). This is also easily shown by computing the stochastic differential of $F_S(t)$ expressed in terms of the Brownian increment $\mathrm{d}\widehat{W}(t)$ (see Exercise 15.8):

$$\frac{\mathrm{d}F_S(t)}{F_S(t)} = (\sigma_S(t) - \sigma_Z(t,T))\,\mathrm{d}\widehat{W}(t) = \sigma_{F_S}(t)\,\mathrm{d}\widehat{W}(t)\,, \tag{15.107}$$

where $\sigma_{F_S}(t) \equiv \sigma_S(t) - \sigma_Z(t,T)$ is the log-volatility of $F_S(t)$ and $\widehat{W}$ is a standard $\widehat{\mathbb{P}}$-Brownian motion. For the sake of analytical tractability, we now assume $\sigma_S(t)$ and $\sigma_Z(t,T)$ to be nonrandom (deterministic) continuous functions of time. Using the strong solution to the above SDE, i.e., $F_S(t) = F_S(0)\mathcal{E}_t(\sigma_{F_S} \cdot \widehat{W})$, we have

$$\begin{aligned} F_S(T) &= F_S(t)\exp\left[-\frac{1}{2}\int_t^T (\sigma_S(u) - \sigma_Z(u,T))^2\,\mathrm{d}u + \int_t^T (\sigma_S(u) - \sigma_Z(u,T))\,\mathrm{d}\widehat{W}(u)\right] \\ &\stackrel{d}{=} F_S(t)\exp\left[-\frac{1}{2}\bar{\sigma}_{t,T}^2(T-t) + \bar{\sigma}_{t,T}\sqrt{T-t}\widehat{Z}\right], \end{aligned} \tag{15.108}$$

where $\bar{\sigma}_{t,T}^2 := \frac{1}{T-t}\int_t^T \left(\sigma_S(u) - \sigma_Z(u,T)\right)^2 \mathrm{d}u$, and $\widehat{Z} \sim Norm(0,1)$ (under measure $\widehat{\mathbb{P}}$) is independent of $F_S(t)$.

According to (15.100), the time-t price of a standard European call on the stock with maturity T and strike K is

$$C(t) = Z(t,T)\widehat{\mathrm{E}}_t\left[(S(T) - K)^+\right].$$

Since $S(T) = S(T)/Z(T,T) = F_S(T)$, we have

$$\begin{aligned} C(t) &= Z(t,T)\,\widehat{\mathrm{E}}_t\left[(F_S(T) - K)^+\right] \\ &= Z(t,T)\,\widehat{\mathrm{E}}\left[\left(F_S(t)\mathrm{e}^{-\frac{1}{2}\bar{\sigma}_{t,T}^2(T-t)+\bar{\sigma}_{t,T}\sqrt{T-t}\widehat{Z}} - K\right)^+ \,\middle|\, F_S(t)\right] \end{aligned} \tag{15.109}$$

where the expectation is conditional on the time-t forward price $F_S(t) = S(t)/Z(t,T)$. Note that, since $F_S(t)$ is independent of $\widehat{Z}$, this expectation is computed by making use of the independence proposition. In particular, the expectation is the same as that of a standard call with strike K, zero "effective interest rate and dividend on the underlying", spot value $F_S(t)$, volatility $\bar{\sigma}_{t,T}$, and time to maturity $T-t$. Hence, we obtain the well-known *Black formula* for the value of a call option on a stock:

$$\begin{aligned} C(t) &= Z(t,T)\left[F_S(t)\mathcal{N}(d_+(t)) - K\,\mathcal{N}(d_-(t))\right] \\ &= S(t)\mathcal{N}(d_+(t)) - K Z(t,T)\,\mathcal{N}(d_-(t)) \end{aligned} \tag{15.110}$$

where

$$d_\pm(t) = \frac{1}{\bar{\sigma}_{t,T}\sqrt{T-t}}\left[\ln\left(\frac{S(t)}{K\,Z(t,T)}\right) \pm \frac{1}{2}\bar{\sigma}_{t,T}^2(T-t)\right].$$

Note that here we have expressed the time-t call price in terms of the time-t stock price and zero-coupon bond price $S(t)$ and $Z(t,T)$. Plugging in the time-t spot values for the stock and the bond gives the pricing function for the call. The above formulation is readily extended to the case with multiple stocks driven by a vector Brownian motion where the stocks are correlated with each other as well as being correlated with the bond price process (see Exercise 15.9).

15.5.3 Pricing Options on Zero-Coupon Bonds

The risk-neutral pricing formulae (15.12) and (15.100) allow for computing no-arbitrage prices of any attainable claim. We assume that an EMM (risk-neutral measure) exists, implying the absence of arbitrage. Consider a European-style claim with maturity T on a zero-coupon bond maturing at time $T' > T$. The payoff function $V(T)$ of such a claim has the form $\Lambda(Z(T,T'))$. The no-arbitrage value $V(t)$, $t \leqslant T$, of the claim is given by

$$V(t) = \widetilde{\mathrm{E}}_t\left[D(t,T)\Lambda(Z(T,T'))\right] = \widetilde{\mathrm{E}}_t\left[\exp\left(-\int_t^T r(u)\,\mathrm{d}u\right)\Lambda(Z(T,T'))\right] \tag{15.111}$$

under the risk-neutral measure $\widetilde{\mathbb{P}} \equiv \widetilde{\mathbb{P}}^{(B)}$, or

$$V(t) = Z(t,T)\widehat{\mathrm{E}}_t\left[\Lambda(Z(T,T'))\right] \tag{15.112}$$

when the equivalent T-forward measure $\widehat{\mathbb{P}} \equiv \widehat{\mathbb{P}}_T^{(Z)}$ is used. The main drawback of the pricing formula (15.111) is that it is necessary to find the joint distribution of $\int_t^T r(u)\,\mathrm{d}u$ and $Z(T,T')$ under $\widetilde{\mathbb{P}}$ to find the expectation. When using (15.112), we only need to find the dynamics of the bond price $Z(t,T')$ under the T-forward measure.

Let $F_Z(t) \equiv F_Z(t;T,T')$ denote the T-forward price of the T'-maturity bond at time t:

$$F_Z(t) = \frac{Z(t,T')}{Z(t,T)}, \quad 0 \leqslant t \leqslant T \leqslant T'.$$

Note that the bond price $Z(t,T')$ is a domestic (nondividend paying) asset. Hence, under the risk-neutral measure $\widetilde{\mathbb{P}}$, its price follows the SDE in (15.106) (now with maturity T' replacing T),

$$\mathrm{d}Z(t,T') = Z(t,T')\left[r(t)\,\mathrm{d}t + \sigma_Z(t,T')\,\mathrm{d}\widetilde{W}(t)\right].$$

In the T-forward measure, with $g(t) = Z(t,T)$ as numèraire asset price, the forward price $F_Z(t)$ is a $\widehat{\mathbb{P}}$-martingale satisfying a driftless SDE,

$$\frac{\mathrm{d}F_Z(t)}{F_Z(t)} = (\sigma_Z(t,T') - \sigma_Z(t,T))\,\mathrm{d}\widehat{W}(t). \tag{15.113}$$

The reader will recognize this as an application of (13.128) in Chapter 13, where $A(t) = Z(t,T'), g(t) = Z(t,T)$. Since the forward price $F_Z(t)$ is a $\widehat{\mathbb{P}}_T^{(Z)}$-martingale, we have $\widehat{\mathrm{E}}_t\left[F_Z(T)\right] = F_Z(t)$. By the above definition we also have $F_Z(T) = \frac{Z(T,T')}{Z(T,T)} = Z(T,T')$, which gives

$$\widehat{\mathrm{E}}_t\left[Z(T,T')\right] = F_Z(t). \tag{15.114}$$

In other words, $F_Z(t)$ is the time-t forward price for delivery of $Z(T,T')$ at time T.

Alternatively, to find the dynamics of $F_Z(t)$, we use the solution (15.18) for deducing a relation between bond prices with maturities T and T':

$$F_Z(t) = \frac{Z(t,T')}{Z(t,T)} = \frac{Z(0,T')}{Z(0,T)}\mathrm{e}^{-\frac{1}{2}\int_0^t\left(\sigma_Z^2(s,T')-\sigma_Z^2(s,T)\right)\mathrm{d}s+\int_0^t\left(\sigma_Z(s,T')-\sigma_Z(s,T)\right)\mathrm{d}\widetilde{W}(s)} \tag{15.115}$$

(now we use (15.103) to change the probability measure)

$$\begin{aligned}
&= \frac{Z(0,T')}{Z(0,T)}\mathrm{e}^{-\frac{1}{2}\int_0^t\left(\sigma_Z^2(s,T')-2\sigma_Z(s,T')\sigma_Z(s,T)+\sigma_Z^2(s,T)\right)\mathrm{d}s+\int_0^t\left(\sigma_Z(s,T')-\sigma_Z(s,T)\right)\mathrm{d}\widehat{W}(s)} \\
&= \frac{Z(0,T')}{Z(0,T)}\mathrm{e}^{-\frac{1}{2}\int_0^t\left(\sigma_Z(s,T')-\sigma_Z(s,T)\right)^2\mathrm{d}s+\int_0^t\left(\sigma_Z(s,T')-\sigma_Z(s,T)\right)\mathrm{d}\widehat{W}(s)}.
\end{aligned} \tag{15.116}$$

On the right hand side of (15.116) we recognize a stochastic exponential. Hence, (15.113) follows. In particular, we can relate the forward price of the T'-maturity bond at time t to that at time T by simply dividing (15.116) into the same expression for $t = T$:

$$\frac{F_Z(T)}{F_Z(t)} = \frac{Z(T,T')}{F_Z(t)} = \mathrm{e}^{-\frac{1}{2}\int_t^T (\sigma_Z(s,T')-\sigma_Z(s,T))^2\,\mathrm{d}s + \int_t^T (\sigma_Z(s,T')-\sigma_Z(s,T))\,\mathrm{d}\widehat{W}(s)}. \tag{15.117}$$

The reader will note that this also follows by directly integrating (15.113).

Let us consider pricing standard European call and put options on the bond. For example, the call option with maturity T and strike K on the bond $Z(T,T')$ gives the right to buy the bond at time T for K. For some term structure models, such as the Gaussian HJM model, one is able to derive the pricing formulae for standard European options in closed form.

Suppose that the bond volatility function $\sigma_Z(t,T)$ is a deterministic (nonrandom) function. Under the HJM framework we have the relation (15.101), which implies that the diffusion coefficient of the forward rate in the SDE (15.61) is also a deterministic function of time t. It follows that the drift and diffusion coefficients in both (15.77) and (15.104) are nonrandom functions of t. Hence, the forward rate is a Gaussian process in either measure $\widetilde{\mathbb{P}}$ or $\widehat{\mathbb{P}}$. We see from (15.117) that $\frac{F_Z(T)}{F_Z(t)}$ is a log-normal random variable. In particular, taking logarithms on both sides of (15.117) gives

$$\begin{aligned}\ln Z(T,T') &= \ln F_Z(t) - \frac{1}{2}\int_t^T \left(\sigma_Z(s,T') - \sigma_Z(s,T)\right)^2 \mathrm{d}s + \int_t^T \left(\sigma_Z(s,T') - \sigma_Z(s,T)\right) \mathrm{d}\widehat{W}(s) \\ &\stackrel{d}{=} \ln F_Z(t) - \frac{1}{2}\bar{\sigma}^2(T-t) + \bar{\sigma}\sqrt{T-t}\,\widehat{Z} \end{aligned} \tag{15.118}$$

where $\widehat{Z} \sim Norm(0,1)$ (under $\widehat{\mathbb{P}}$) and we conveniently define the constant (for given t, T, T')

$$\bar{\sigma}^2 \equiv \bar{\sigma}^2_{t,T,T'} := \frac{1}{T-t}\int_t^T \left(\sigma_Z(s,T') - \sigma_Z(s,T)\right)^2 \mathrm{d}s\,.$$

This corresponds to the time-averaged square difference of the volatilities of the bond price with respective maturities T' and T. Note that the last line in (15.118) is an equality in distribution where we used the fact that the Itô integral is a normal random variable with zero mean and variance $\bar{\sigma}^2(T-t)$. The reader can readily verify that this follows by Itô isometry, since the integrand is a nonrandom square-integrable function of s (for fixed T, T'). Moreover, the above Itô integral is independent of $\mathcal{F}_t$, i.e., $\widehat{Z}$ is independent of $\mathcal{F}_t$ (and, of course, also independent of $F_Z(t)$).

Based on the above properties, we can now readily price a call option with maturity T and strike K, which is written on the underlying bond with value $Z(T,T')$. The price of the call at time t is given by computing the conditional expectation in (15.112). The steps are exactly as those for computing the price of a standard European call. We condition on $\mathcal{F}_t$, where $F_Z(t)$ is $\mathcal{F}_t$-measurable and independent of $\widehat{Z}$. According to (15.118), we substitute $Z(T,T') = F_Z(t)\mathrm{e}^{-\frac{1}{2}\bar{\sigma}^2(T-t)+\bar{\sigma}\sqrt{T-t}\widehat{Z}}$ into the conditional expectation, and we use the independence proposition to reduce the calculation to an unconditional expectation. The latter is then computed by the usual expectation identities in the Appendix (or by simply recognizing the call pricing function at hand in the second line below):

$$\begin{aligned}
C(t) &= Z(t,T)\,\widehat{\mathrm{E}}[(Z(T,T')-K)^+ \mid \mathcal{F}_t] \\
&= Z(t,T)\,\widehat{\mathrm{E}}\left[\left(F_Z(t)\mathrm{e}^{-\frac{1}{2}\bar{\sigma}^2(T-t)+\bar{\sigma}\sqrt{T-t}\widehat{Z}} - K\right)^+ \middle| \mathcal{F}_t\right] \\
&= Z(t,T)\,\widehat{\mathrm{E}}\left[\left(x\mathrm{e}^{-\frac{1}{2}\bar{\sigma}^2(T-t)+\bar{\sigma}\sqrt{T-t}\widehat{Z}} - K\right)^+\right]\Bigg|_{x=F_Z(t)} \\
&= Z(t,T)\left[F_Z(t)\mathcal{N}(d_+(t)) - K\,\mathcal{N}(d_-(t))\right] \\
&= Z(t,T')\mathcal{N}(d_+(t)) - K\,Z(t,T)\mathcal{N}(d_-(t)),
\end{aligned} \tag{15.119}$$

where

$$d_\pm(t) := \frac{1}{\bar{\sigma}\sqrt{T-t}}\left[\ln\left(\frac{Z(t,T')}{K\,Z(t,T)}\right) \pm \frac{1}{2}\bar{\sigma}^2(T-t)\right].$$

Note: $Z(t,T)F_Z(t) = Z(t,T')$ and $\frac{F_Z(t)}{K} = \frac{Z(t,T')}{KZ(t,T)}$.

15.6 LIBOR Model

15.6.1 LIBOR Rates

LIBOR, which stands for London InterBank Offer Rate, refers to the market interest rate. Let $L(t;T,T+\tau)$ denote an annual simple rate of interest that is locked at time t for borrowing from time T to time $T+\tau$, where $0 \leqslant t \leqslant T \leqslant T+\tau$. That is, \$1 invested at time T will grow to $1+\tau L(t;T,T+\tau)$ dollars at time $T+\tau$. We call $L(t;T,T+\tau)$ the *forward LIBOR*.

In Section 15.1.2 we discussed how the rate on a loan for a period between times T and $T' = T+\tau$ can be locked in at time t. We purchase one zero-coupon bond maturing at time T and finance this purchase by selling short $\frac{Z(t,T)}{Z(t,T+\tau)}$ units of the bond maturing at time $T+\tau$. The cost of setting up this portfolio is zero. The forward LIBOR rate, which is the effective simple rate of interest applied over the interval $[T, T+\tau]$, is calculated as follows:

$$1+\tau L(t;T,T+\tau) = \frac{Z(t,T)}{Z(t,T+\tau)} \implies L(t;T,T+\tau) = \frac{1}{\tau}\left[\frac{Z(t,T)}{Z(t,T+\tau)} - 1\right]. \tag{15.120}$$

The interest period τ is often referred as the *tenor*, and it is typically equal to 0.25 for a three-month LIBOR or 0.5 for a six-month LIBOR. If $t = T$, then we call $L(T;T,T+\tau)$ the *spot LIBOR*. It is given by

$$L(T;T,T+\tau) = \frac{1}{\tau}\left[\frac{1}{Z(T,T+\tau)} - 1\right].$$

15.6.2 Brace–Gatarek–Musiela Model of LIBOR Rates

Pricing interest rate derivatives such as caps and swaps requires a tractable model of floating rates. To adapt the Black–Scholes formula for stock options to the case with fixed-income products, it is desirable to assume that the risk-neutral dynamics of floating rates is log-normal. Suppose that caps and swaps are written on forward rates $f(t,T)$. It can be shown that if the diffusion coefficient in the SDE (15.61) is proportional to the forward rate, i.e., if $\sigma_F(t,T) = \sigma(t,T)f(t,T)$ with $\sigma(t,T)$ assumed to be a nonrandom function, then the forward rates governed by the risk-neutral SDE (15.77) can explode in finite time. Such behaviour of forward rates is caused by the risk-neutral drift term $\widetilde{\alpha}(t,T)$ in (15.77), which takes the form

$$\widetilde{\alpha}(t,T) = \sigma(t,T)f(t,T)\int_t^T \sigma(t,u)f(t,u)\,du\,.$$

To overcome this problem, Brace, Gatarek, and Musiela (BGM) have suggested using the LIBOR forward rates $L(t;T,T+\tau)$, which are simple rates of interest, instead of continuously compounding forward rates.

Let us consider the case with a single maturity $T \leqslant T^*$ and tenor $\tau > 0$. For notational simplicity we write $L(t) \equiv L(t;T,T+\tau)$. Here we present the theory for LIBOR forward rates driven by a single Brownian factor, although the extension to multiple Brownian factors follows readily. Hence, suppose that the bond price process is governed by the SDE in the form of (15.17). As follows from (15.113), the $(T+\tau)$-forward price of the bond, $\frac{Z(t,T)}{Z(t,T+\tau)}$, is a $\widehat{\mathbb{P}}$-martingale, for all $0 \leqslant t \leqslant T$, where $\widehat{\mathbb{P}} \equiv \widehat{\mathbb{P}}^{(Z)}_{T+\tau}$ is the $(T+\tau)$-forward measure. From (15.120), we have that the LIBOR rate $L(t)$ is a strictly positive $\widehat{\mathbb{P}}$-martingale as well. According to the Brownian martingale representation theorem 11.14, there exists an $\mathbb{F}$-adapted process $v(t,T)$ such that

$$\mathrm{d}L(t) = v(t,T)L(t)\,\mathrm{d}\widehat{W}(t), \quad 0 \leqslant t \leqslant T, \tag{15.121}$$

where $\widehat{W}$ is a $\widehat{\mathbb{P}}$-Brownian motion. The process $v(t,T)$ relates to the zero-coupon volatilities as follows:

$$v(t,T) = \left(\frac{1+\tau L(t)}{\tau L(t)}\right)\big[\sigma_Z(t,T) - \sigma_Z(t,T+\tau)\big]. \tag{15.122}$$

The proof of (15.122) is left as an exercise at the end of this chapter. Notice that $v(t,T) \equiv \sigma_L^{(\tau)}(t,T)$ is the log-volatility of the process $L(t;T,T+\tau)$ for $0 \leqslant t \leqslant T$.

The Brace–Gatarek–Musiela (BGM) model for forward LIBOR rates is constructed such that $v(t,T)$ is a deterministic (nonrandom) function of time for $0 \leqslant t \leqslant T$. As a result, the forward LIBOR rate $L(T)$ conditional on $L(t)$ has a log-normal distribution under measure $\widehat{\mathbb{P}}$ with conditional mean and variance:

$$\widehat{\mathrm{E}}_t[\ln L(T)] = \ln L(t) - \frac{1}{2}(T-t)\bar{v}^2_{t,T}, \tag{15.123}$$

$$\widehat{\mathrm{Var}}_t(\ln L(T)) = \bar{v}^2_{t,T}(T-t) \tag{15.124}$$

with time-averaged variance $\bar{v}^2_{t,T} := \frac{1}{T-t}\int_t^T v^2(s,T)\,\mathrm{d}s$. These expressions follow readily by simply using the solution to the SDE (15.121) as an exponential $\widehat{\mathbb{P}}$-martingale for $L(t)$ at time t and $L(T)$ at time T,

$$L(T) = L(t)\mathrm{e}^{-\frac{1}{2}\int_t^T v^2(s,T)\,\mathrm{d}s + \int_t^T v(s,T)\,\mathrm{d}\widehat{W}(s)} \stackrel{d}{=} L(t)\mathrm{e}^{-\frac{1}{2}(T-t)\bar{v}^2_{t,T} + \bar{v}_{t,T}\sqrt{T-t}\,\widehat{Z}}. \tag{15.125}$$

The second equality is in distribution where $\int_t^T v(s,T)\,\mathrm{d}\widehat{W}(s) \sim Norm(0, \bar{v}_{t,T}\sqrt{T-t})$ under $\widehat{\mathbb{P}}$ and $\widehat{Z}$ denotes a standard normal random variable under $\widehat{\mathbb{P}}$. The log-normality of the forward LIBOR rate, i.e., the representation in (15.125), allows for the construction of pricing formulae for caps and swaps in closed form. One of the main results is the Black caplet formula presented in the next subsection.

15.6.3 Pricing Caplets, Caps, and Swaps

A cap is defined as a portfolio of caplets. Hence, to find the no-arbitrage price of a cap, it suffices to find the price of a single caplet, Caplet(t), with $0 \leqslant t \leqslant T$. Under the BGM model, a caplet is a European call on the spot LIBOR rate. Its payoff at maturity $T+\tau$ is $(L(T)-\kappa)^+\tau$, where $L(T) \equiv L(T;T,T+\tau)$ and $\kappa > 0$ is a strike rate. By (15.13), the time-t price of the caplet is

$$\begin{aligned}\text{Caplet}(t) &= \tau\widetilde{\mathrm{E}}_t\left[D(t,T+\tau)(L(T)-\kappa)^+\right]\\ &= \widetilde{\mathrm{E}}_t\left[\frac{B(t)}{B(T+\tau)}\left(\frac{1}{Z(T,T+\tau)}-1-\kappa\tau\right)^+\right].\end{aligned}$$

From the bond price formula (15.11), we have

$$Z(T,T+\tau) = \widetilde{\mathrm{E}}_T\left[\frac{B(T)}{B(T+\tau)}\right].$$

Using the tower property and $\mathcal{F}_T$-measurability of $Z(T,T+\tau)$ gives

$$\begin{aligned}\text{Caplet}(t) &= \widetilde{\mathrm{E}}_t\left[\widetilde{\mathrm{E}}_T\left[\frac{B(t)}{B(T+\tau)}\left(\frac{1}{Z(T,T+\tau)}-1-\kappa\tau\right)^+\right]\right]\\ &= \widetilde{\mathrm{E}}_t\left[\frac{B(t)}{B(T)}\left(\frac{1}{Z(T,T+\tau)}-1-\kappa\tau\right)^+\widetilde{\mathrm{E}}_T\left[\frac{B(T)}{B(T+\tau)}\right]\right]\\ &= \widetilde{\mathrm{E}}_t\left[\frac{B(t)}{B(T)}\left(\frac{1}{Z(T,T+\tau)}-1-\kappa\tau\right)^+ Z(T,T+\tau)\right]\\ &= (1+\kappa\tau)\,\widetilde{\mathrm{E}}_t\left[\frac{B(t)}{B(T)}\left(\frac{1}{1+\kappa\tau}-Z(T,T+\tau)\right)^+\right].\end{aligned}$$

That is, the caplet is equivalent to a put option on the zero-coupon bond $Z(T,T+\tau)$ with strike $\frac{1}{1+\kappa\tau}$ and maturity T.

The price of a caplet is easier to evaluate by using a forward measure. As was demonstrated in the previous subsection, the forward LIBOR rate $L(T)$ conditional on $L(t)$ has a log-normal distribution under the forward measure $\widehat{\mathbb{P}}^{(Z)}_{T+\tau}$. The price of the caplet at time t is

$$\text{Caplet}(t) = \tau Z(t,T+\tau)\widehat{\mathrm{E}}_t\left[(L(T)-\kappa)^+\right], \tag{15.126}$$

where $\widehat{\mathrm{E}}[\,\cdot\,]$ denotes the expectation under $\widehat{\mathbb{P}}^{(Z)}_{T+\tau}$. We leave it to the reader to show (by using similar steps as pricing a standard call with the use of (15.125)) that Black's caplet formula obtains:

$$\text{Caplet}(t) = \tau Z(t,T+\tau)\big[L(t)\mathcal{N}(d_+(t)) - \kappa\mathcal{N}(d_-(t))\big], \tag{15.127}$$

where

$$d_\pm(t) = \frac{1}{\bar{v}_{t,T}\sqrt{T-t}}\left[\ln\frac{L(t)}{\kappa} \pm \frac{1}{2}\bar{v}^2_{t,T}(T-t)\right].$$

A cap is a series of caplets that pays $\tau\big(L(T_{i-1};T_{i-1},T_i)-\kappa\big)^+$ at time $T_i = T_0 + i\tau$ for all $i = 1,2,\ldots,n$. The total value of the cap at time $t \leqslant T_0$ is equal to the sum of all caplet values. Each i-th caplet is valued by considering the expectation $\widehat{\mathrm{E}}_t^{(T_i)}[\,\cdot\,]$ conditional on $\mathcal{F}_t$ under the T_i-forward measure $\widehat{\mathbb{P}}_{T_i}^{(Z)}$. Summing each caplet value gives the price of the cap:

$$\begin{aligned}\mathrm{Cap}(t) &= \sum_{i=1}^{n} \tau Z(t,T_i)\,\widehat{\mathrm{E}}_t^{(T_i)}\left[(L(T_{i-1};T_{i-1},T_i)-\kappa)^+\right]\\ &= \tau\sum_{i=1}^{n} Z(t,T_i)\big[L(t;T_{i-1},T_i)\mathcal{N}(d_+^{(i-1)}(t)) - \kappa\mathcal{N}(d_-^{(i-1)}(t))\big] \qquad (15.128)\end{aligned}$$

where

$$d_{\pm}^{(i-1)}(t) = \frac{1}{\bar{\upsilon}_{t,T_{i-1}}\sqrt{T_{i-1}-t}}\left[\ln\frac{L(t;T_{i-1},T_i)}{\kappa} \pm \frac{1}{2}\bar{\upsilon}_{t,T_{i-1}}^2(T_{i-1}-t)\right]$$

and $\upsilon_{t,T_{i-1}}^2 = \frac{1}{T_{i-1}-t}\int_t^{T_{i-1}} \upsilon^2(s,T_{i-1})\,\mathrm{d}s$.

The price of a payer swap at time-t is also easy to evaluate by using forward measures. The holder of the swap receives $\tau\big(L(T_{i-1};T_{i-1},T_i)-\kappa\big)$ at time $T_i = T_0 + i\tau$ for all $i = 1,2,\ldots,n$. The no-arbitrage price at time t is now readily computed by choosing the T_i-forward measure $\widehat{\mathbb{P}}_{T_i}^{(Z)}$ for each i-th payoff term:

$$\mathrm{Swap}(t) = \sum_{i=1}^{n} \tau Z(t,T_i)\,\widehat{\mathrm{E}}_t^{(T_i)}\left[L(T_{i-1};T_{i-1},T_i)-\kappa\right]$$

(using the fact that the LIBOR rate $L(t;T_{i-1},T_i)$ is a $\widehat{\mathbb{P}}_{T_i}^{(Z)}$-martingale)

$$\begin{aligned}&= \tau\sum_{i=1}^{n} Z(t,T_i)\big[L(t;T_{i-1},T_i)-\kappa\big] = \sum_{i=1}^{n}\tau Z(t,T_i)\left[\frac{Z(t,T_{i-1})-Z(t,T_i)}{\tau Z(t,T_i)} - \kappa\right]\\ &= \sum_{i=1}^{n}\big[Z(t,T_{i-1})-Z(t,T_i)\big] - \tau\kappa\sum_{i=1}^{n} Z(t,T_i) \qquad (15.129)\\ &= \big[Z(t,T_0)-Z(t,T_n)\big] - \tau\kappa\sum_{i=1}^{n} Z(t,T_i)\,. \qquad (15.130)\end{aligned}$$

In obtaining (15.129) we used the definition in (15.120), i.e., $L(t;T_{i-1},T_i) = \frac{Z(t,T_{i-1})-Z(t,T_i)}{\tau Z(t,T_i)}$.

The forward swap rate that makes the swap contract have a no-arbitrage price of zero at time t is hence

$$\kappa(t;T_0,T_n) = \frac{Z(t,T_0)-Z(t,T_n)}{\tau\sum_{i=1}^{n} Z(t,T_i)}\,. \qquad (15.131)$$

Substituting the expression $Z(t,T_0) - Z(t,T_n)$, in terms of this forward swap rate, into (15.130) gives

$$\mathrm{Swap}(t) = \tau\big[\kappa(t;T_0,T_n)-\kappa\big]\sum_{i=1}^{n} Z(t,T_i)\,. \qquad (15.132)$$

15.7 Exercises

Exercise 15.1. Suppose that the (continuously compounded) spot rates for the next three years are

T	1	2	3
$y(0,T)$	3%	3.25%	3.5%

Find the forward rates $f(0;1,2)$, $f(0;1,3)$, and $f(0;2,3)$.

Exercise 15.2. Consider a derivative which has the payoff $(y(T,T+1)-y(T,T+5))^+$. Why might it be inappropriate to use a one-factor interest rate model to value this derivative?

Exercise 15.3. Consider the Hull–White model (15.29). Show that the short rate $r(t)$ is a Gaussian process. Find the mean and variance of $r(T)$ conditional on $r(t)$ for $0 \leqslant t \leqslant T$.

Exercise 15.4. The short-rate prices under the Pearson–Sun (PS) model follows the SDE

$$\mathrm{d}r(t) = \big(\alpha - \beta r(t)\big)\,\mathrm{d}t + \sigma\sqrt{r(t)-\lambda}\,\mathrm{d}W(t).$$

Make use of the CIR bond pricing formula to derive bond prices for the PS model.

Exercise 15.5. Find the forward rates $f(t,T)$ in the Vasiček model.

Exercise 15.6. Consider an extended CIR model where the short rate $r(t)$ follows the SDE

$$\mathrm{d}r(t) = \big(\alpha(t) - \beta(t)r(t)\big)\,\mathrm{d}t + \sigma(t)\sqrt{r(t)}\,\mathrm{d}\widetilde{W}(t)$$

under the risk-neutral measure $\widetilde{\mathbb{P}}$. Here, $\alpha(t)$, $\beta(t)$, and $\sigma(t) > 0$ are continuous ordinary (nonrandom) functions. Show that the price of a zero-coupon bond is

$$Z(t,T) = \exp\left(-\int_t^T C(s,T)\alpha(s)\,\mathrm{d}s - C(t,T)r(t)\right), \quad t \leqslant T,$$

where $C(t,T)$ solves the first order differential equation

$$C'(t,T) = \beta(t)C(t,T) + \frac{\sigma^2(t)}{2}C^2(t,T) - 1$$

for $t < T$, subject to $C(T,T) = 0$.

Exercise 15.7. A two-factor HJM model is given by the SDE

$$\mathrm{d}f(t,T) = \alpha(t,T)\,\mathrm{d}t + \sigma_1(t,T)\,\mathrm{d}W_1(t) + \sigma_2(t,T)\,\mathrm{d}W_2(t),$$

where W_1 and W_2 are independent Brownian motions.

(a) Find the SDE for the discounted zero-coupon bond price process $\overline{Z}(t,T) = D(t)Z(t,T)$ and show that

$$\overline{Z}(t,T) = \overline{Z}(0,T)\exp\left(-\int_0^t A(s,T)\,\mathrm{d}s - \int_0^t \Sigma_1(s,T)\,\mathrm{d}W_1(s) - \int_0^t \Sigma_2(s,T)\,\mathrm{d}W_2(s)\right),$$

where $\Sigma_i(t,T) = \int_t^T \sigma_i(t,u)\,\mathrm{d}u$ for $i = 1,2$ and $A(t,T) = \int_t^T \alpha(t,u)\,\mathrm{d}u$.

(b) Show that the no-arbitrage condition is given by

$$\widetilde{\alpha}(t,T) = \sigma_1(t,T)\int_t^T \sigma_1(t,s)\,\mathrm{d}s + \sigma_2(t,T)\int_t^T \sigma_2(t,s)\,\mathrm{d}s\,.$$

(c) Provide the multidimensional extension to the formulas in part (a) and (b) for the more general case of a d-factor HJM model for any $d \geqslant 1$.

Exercise 15.8. Let the dynamics of the stock price $S(t)$ and the bond price $Z(t,T)$ be respectively governed by (15.105) and (15.106). Show that the dynamics of the forward price $F(t) = \frac{S(t)}{Z(t,T)}$ under the T-forward measure $\widehat{\mathbb{P}} \equiv \widehat{\mathbb{P}}_T^{(Z)}$ is governed by (15.107).

Exercise 15.9. Consider the multidimensional extension of (15.105)-(15.106) where each stock price process $\{S_i(t)\}_{0\leqslant t\leqslant T}$, $i = 1,\ldots,n$, and the zero-coupon bond price $Z(t,T)$ are all driven by a vector Brownian motion $\mathbf{W}(t) = (W_1(t),\ldots,W_d(t))$. Assume the existence of a risk-neutral measure $\widetilde{\mathbb{P}} \equiv \widetilde{\mathbb{P}}^{(B)}$, i.e.,

$$\frac{\mathrm{d}S_i(t)}{S_i(t)} = r(t)\,\mathrm{d}t + \boldsymbol{\sigma}_i(t)\cdot \mathrm{d}\widetilde{\mathbf{W}}(t)\,,$$
$$\frac{\mathrm{d}Z(t,T)}{Z(t,T)} = r(t)\,\mathrm{d}t + \boldsymbol{\sigma}_Z(t)\cdot \mathrm{d}\widetilde{\mathbf{W}}(t)\,,$$

where $\boldsymbol{\sigma}_i(t)$ and $\boldsymbol{\sigma}_Z(t)$ are the respective log-volatility vectors of each stock and bond price. The forward price of each stock is defined by $F_i(t) = \frac{S_i(t)}{Z(t,T)}$.

(a) Obtain the analogue to the SDE in (15.107) for each forward price $F_i(t)$, $i = 1,\ldots,n$.

(b) Assume a given stock $S_1(t) \equiv S(t)$ has constant log-volatility vector $\boldsymbol{\sigma}_S(t) = \boldsymbol{\sigma}_S$ with magnitude $||\boldsymbol{\sigma}_S|| = \sigma_S$ and assume a constant vector $\boldsymbol{\sigma}_Z(t) = \boldsymbol{\sigma}_Z$ with magnitude $||\boldsymbol{\sigma}_Z|| = \sigma_Z$, where $\boldsymbol{\sigma}_S\cdot\boldsymbol{\sigma}_Z = \rho\,\sigma_S\sigma_Z$, $\rho \in (-1,1)$. By employing the measure $\widehat{\mathbb{P}} \equiv \widehat{\mathbb{P}}_T^{(Z)}$, derive a pricing formula for a European call with payoff $(S(T)-K)^+$.

(c) By employing the measure $\widehat{\mathbb{P}}$, obtain a pricing formula for an exchange option with payoff $(S_2(T) - S_1(T))^+$, assuming constant log-volatility vectors with $||\boldsymbol{\sigma}_i|| = \sigma_i$, $\boldsymbol{\sigma}_1\cdot\boldsymbol{\sigma}_2 = \rho\sigma_1\sigma_2$, $||\boldsymbol{\sigma}_Z|| = \sigma_Z$, and $\boldsymbol{\sigma}_Z\cdot\boldsymbol{\sigma}_i = \rho_i\sigma_i\sigma_Z$, $\rho\in(-1,1)$, $\rho_i \in (-1,1)$, $i = 1,2$.

Exercise 15.10. Consider the two-factor Vasiček model defined by the two-dimensional system of SDEs in (15.93).

(a) Provide the time-homogeneous PDE for the pricing function $V(t,x_1,x_2)$ of a zero-coupon bond under this model by directly applying Theorem 11.20 to the conditional expectation

$$V(t,x_1,x_2) = \widetilde{\mathrm{E}}_{t,x_1,x_2}\left[\mathrm{e}^{-\int_t^T r(u,\mathbf{X}(u))\,\mathrm{d}u}\right].$$

Note that the short rate is given by (15.96). Namely, as a process we have $r(t) = r(t,\mathbf{X}(t)) = a_0 + a_1X_1(t) + a_2X_2(t)$ with spot value $r(t,x_1,x_2) = a_0 + a_1x_1 + a_2x_2$.

(b) Let $V(t,x_1,x_2) = v(\tau,x_1,x_2)$, $\tau = T-t$, be the solution to the PDE in part (a) where $V(t,x_1,x_2)$ has the affine form in (15.97), i.e., let

$$v(\tau,x_1,x_2) = \mathrm{e}^{A(\tau)+c_1(\tau)x_1+c_2(\tau)x_2}$$

for $0 \leqslant \tau \leqslant T$. By substituting this form into the PDE of part (a), show that the coefficient functions must satisfy the first order system of ordinary differential equations:

$$\begin{aligned} A'(\tau) &= \frac{1}{2}\left(c_1^2(\tau) + c_2^2(\tau)\right) - a_0 \,, \\ c_1'(\tau) &= -\beta_{11}c_1(\tau) - \beta_{21}c_2(\tau) - a_1 \,, \\ c_2'(\tau) &= -\beta_{22}c_2(\tau) - a_2 \,, \end{aligned}$$

subject to the initial conditions $A(0) = c_1(0) = c_2(0) = 0$.

(c) Solve the system of equations in part (b) and hence obtain the pricing function for the zero-coupon bond.

Exercise 15.11. Consider the two-factor CIR model defined by the two-dimensional system of SDEs in (15.94). Repeat parts (a) and (b) of Exercise 15.10.

Exercise 15.12. Prove that the forward LIBOR rate volatility $\upsilon(t,T)$ and the T- and $(T+\tau)$-maturity bond volatilities $\sigma_Z(t,T)$ and $\sigma_Z(t,T+\tau)$ are related by

$$\upsilon(t,T) = \left(\frac{1+\tau L(t;T,T+\tau)}{\tau L(t;T,T+\tau)}\right)\left[\sigma_Z(t,T) - \sigma_Z(t,T+\tau)\right].$$

Exercise 15.13. A *floorlet* is similar to a caplet except that the floating rate is bounded from below. The effective payoff of a floorlet at time T is $\tau\left(\kappa - L(T;T,T+\tau)\right)^+$. Derive the Black pricing formula for a floorlet. Is there a relationship between a floorlet and a caplet?

Exercise 15.14. *Caps* and *floors* are collections of caplets and floorlets, respectively, applied to periods $[T_j, T_j+\tau]$ with $j = 1, 2, \ldots, n$. Show that a model-independent relationship *cap* = *floor* + *swap* exists.

Chapter 16

Alternative Models of Asset Price Dynamics

In the study of stochastic processes and their applications to finance, geometric Brownian motion (GBM) is the simplest model used for continuous-time asset pricing. The volatility in the GBM model is constant, i.e., the diffusion coefficient is a linear function of the underlying asset price, as is the drift coefficient. For a long period of time the GBM model has stood as one of the few known continuous-time stochastic models which admits exact analytically tractable transition probability density functions and closed-form pricing formulae for various standard, barrier, and lookback European-style options. However, despite its simplicity, it is commonly recognized that the GBM model only partially captures the complexity of financial markets. The log-normal assumption of asset price returns disagrees with most empirical evidence. For example, let us consider a series of options and then plot the value of implied volatility against strike and time to expiry. A very notable defect of the GBM model is the fact that, by construction, the implied volatility surface is supposed to be completely flat while market observed implied volatility surfaces of stock index options exhibit various pronounced smile and skew patterns. This phenomenon is commonly called "the volatility smile." These and other important market observations have spawned the development of more realistic pricing models based on alternative stochastic processes. In this chapter we demonstrate how the standard Black–Scholes model can be extended so as to make it consistent with the volatility smile. A variety of more sophisticated models has been proposed for asset prices, interest rates, and other financial variables in the mathematical finance literature. Examples of alternative models include processes with jumps, stochastic volatility models, and local (i.e., state-dependent) volatility diffusions. In this chapter, we focus our attention on a selection of alternative models of asset price dynamics, namely, the local volatility model, the constant elasticity of variance (CEV) diffusion model, the Heston stochastic volatility model, diffusions with Poisson-type jumps, and the variance gamma pure jump model.

16.1 Stochastic Volatility Diffusion Models

16.1.1 Local Volatility Models

A local volatility model is one which considers the volatility σ as a deterministic function of the current calendar time and the current asset value, i.e., $\sigma = \sigma(t, S)$. The volatility function can be chosen so that the model option values will precisely match the market counterparts. However, we cannot generally find closed-form pricing formulae for European options except for some special cases of σ.

Let the asset price process, considered under the risk-neutral probability measure $\widetilde{\mathbb{P}}$, follow the stochastic differential equation (SDE)

$$\frac{\mathrm{d}S(t)}{S(t)} = (r-q)\,\mathrm{d}t + \sigma(t, S(t))\,\mathrm{d}\widetilde{W}(t)\,,\ t \geqslant 0\,, \quad S(0) = S_0 > 0 \tag{16.1}$$

where $r \geqslant 0$ is a constant interest rate, $q \geqslant 0$ is a constant dividend yield and $\{\widetilde{W}(t)\}_{t\geqslant 0}$ is Brownian motion under the measure $\widetilde{\mathbb{P}}$. The *time- and state-dependent volatility* $\sigma(t, S)$ is sometimes called the *local volatility function.* In general, σ is a nonnegative continuous function. We are also assuming that the process $\{\mathrm{e}^{-(r-q)t}S(t)\}_{t\geqslant 0}$ is a $\widetilde{\mathbb{P}}$-martingale. For some special cases of σ, the corresponding distribution of the process with SDE (16.1) can be obtained analytically. The most known example of a solvable state-dependent volatility model is the constant elasticity of variance (CEV) diffusion model, which is discussed in the next subsection. As is shown below, the volatility surface $\sigma(t, S)$ can be calibrated to empirical data so that the model successfully produces option values consistent with all market prices across different strikes and maturities.

Let $\widetilde{p} = \widetilde{p}(t, t'; S, S')$ with $t < t'$ and $S, S' \in \mathbb{R}_+$ denote a risk-neutral transition probability function associated with (16.1). Under quite general conditions on σ, we recall that the function $\widetilde{p}$ satisfies both the backward and forward Kolmogorov equations. The backward partial differential equation (PDE) reads

$$\frac{\partial \widetilde{p}}{\partial t} + \frac{1}{2}\sigma^2(t, S)S^2 \frac{\partial^2 \widetilde{p}}{\partial S^2} + (r-q)S\frac{\partial \widetilde{p}}{\partial S} = 0\,, \tag{16.2}$$

and the corresponding forward PDE is

$$\frac{\partial \widetilde{p}}{\partial t'} = \frac{1}{2}\frac{\partial^2}{\partial S'^2}\left(\sigma^2(t', S')S'^2\,\widetilde{p}\right) - (r-q)\frac{\partial}{\partial S'}\left(S'\,\widetilde{p}\right), \tag{16.3}$$

with the initial (or final) time condition

$$\widetilde{p}(t, t; S, S') = \widetilde{p}(t', t'; S, S') = \delta(S - S')\,.$$

Consider a European-style derivative written at time $t_0 \geqslant 0$ on the underlying asset S. The no-arbitrage derivative value $V(t, S)$, which is a function of the current calendar time $t \in [t_0, T]$ and the current asset price $S = S(t)$, is given by the risk-neutral pricing formula

$$V(t, S) = \mathrm{e}^{-r(T-t)}\widetilde{\mathrm{E}}_{t,S}[V(T, S(T))]\,. \tag{16.4}$$

By the (discounted) Feynman–Kac Theorem, the pricing function $V(t, S)$ satisfies the Black–Scholes partial differential equation (BSPDE)

$$\frac{\partial V(t, S)}{\partial t} + \frac{1}{2}\sigma^2(t, S)S^2\frac{\partial^2 V(t, S)}{\partial S^2} + (r-q)S\frac{\partial V(t, S)}{\partial S} - rV(t, S) = 0\,. \tag{16.5}$$

Suppose that the European claim of interest is a standard call option with strike price K and expiration time T. The pricing function of a European call option, $C(t, S; T, K)$, satisfies the BSPDE in (16.5). Surprisingly, the function $C(t, S; T, K)$ regarded explicitly as a function of the strike and maturity time arguments (T, K) (instead of functions of the arguments (t, S) which are held fixed), also satisfies a PDE known as Dupire's equation.

Theorem 16.1 (Dupire's Equation). *The pricing function, $c(T, K) := C(t, S; T, K)$, of a European call option satisfies the PDE*

$$\frac{\partial c(T, K)}{\partial T} = \frac{1}{2}\sigma^2(T, K)K^2\frac{\partial^2 c(T, K)}{\partial K^2} + (q-r)K\frac{\partial c(T, K)}{\partial K} - qc(T, K). \tag{16.6}$$

We note that this is a PDE in the so-called dual variables (T, K), where $K > 0$ and $T > t$, and it is also sometimes called the dual BSPDE. The validity of this theorem also rests upon certain technical assumptions which are stated in the proof.

Proof. The conditional expectation in (16.4) is an integral of the product of the payoff function and the risk-neutral transition PDF $\widetilde{p}(t, T; S, S')$, i.e., the call option value is given by

$$c(T, K) = \mathrm{e}^{-r(T-t)}\widetilde{\mathrm{E}}_{t,S}[(S(T) - K)^+] = \mathrm{e}^{-r(T-t)} \int_K^\infty \widetilde{p}(t, T; S, S')(S' - K)\,\mathrm{d}S'. \quad (16.7)$$

The first and second derivatives of (16.7) with respect to K give

$$\frac{\partial c(T, K)}{\partial K} = -\mathrm{e}^{-r(T-t)} \int_K^\infty \widetilde{p}(t, T; S, S')\,\mathrm{d}S', \quad (16.8)$$

and

$$\frac{\partial^2 c(T, K)}{\partial K^2} = \mathrm{e}^{-r(T-t)}\widetilde{p}(t, T; S, K). \quad (16.9)$$

The derivative of the option price with respect to expiration T is

$$\begin{aligned} \frac{\partial c(T, K)}{\partial T} &= -r\mathrm{e}^{-r(T-t)} \int_K^\infty \widetilde{p}(t, T; S, S')(S' - K)\,\mathrm{d}S' \\ &\quad + \mathrm{e}^{-r(T-t)} \int_K^\infty \frac{\partial \widetilde{p}(t, T; S, S')}{\partial T}(S' - K)\,\mathrm{d}S'. \end{aligned} \quad (16.10)$$

Using the forward Kolmogorov equation (16.3) for $\widetilde{p}$ with $t' = T$ gives

$$\begin{aligned} \frac{\partial c(T, K)}{\partial T} &= -rc(T, K) + \mathrm{e}^{-r(T-t)} \int_K^\infty \Big[- (r - q)\frac{\partial}{\partial S'}(S'\widetilde{p}(t, T; S, S')) \\ &\quad + \frac{1}{2}\frac{\partial^2}{\partial S'^2}\left(\sigma^2(T, S')S'^2\widetilde{p}(t, T; S, S')\right) \Big](S' - K)\,\mathrm{d}S'. \end{aligned} \quad (16.11)$$

The integral containing the first derivative with respect to S' can be evaluated by parts as follows:

$$\begin{aligned} \int_K^\infty (S' - K)\frac{\partial}{\partial S'}(S'\widetilde{p})\,\mathrm{d}S' &= \widetilde{p}S'(S' - K)\Big|_{S'=K}^{S'=\infty} - \int_K^\infty S'\widetilde{p}\,\mathrm{d}S' \\ &= -\int_K^\infty (S' - K)\widetilde{p}\,\mathrm{d}S' - K\int_K^\infty \widetilde{p}\,\mathrm{d}S' \\ &= \left[- c + K\frac{\partial c}{\partial K}\right]\mathrm{e}^{r(T-t)}, \end{aligned}$$

where $\widetilde{p}S'(S' - K)|_{S'=K} = 0$, and we simply wrote $S' = (S' - K) + K$ in the second equation line and used (16.7) and (16.8). Its important to note here that we are assuming $\widetilde{p}S'(S' - K) \to 0$, as $S' \to \infty$. That is, the process is assumed to have a transition PDF that decays faster than $(S')^{-2}$, as $S' \to \infty$, i.e. we are assuming $\lim\limits_{S'\to\infty}(S')^2\widetilde{p}(t, T; S, S') = 0$. These limits are implied by the existence of the integral in (16.11), where it is also assumed that $S'\frac{\partial^2}{\partial S'^2}(\sigma^2(T, S')S'^2\widetilde{p})$ decays faster than $1/S'$, as $S' \to \infty$. The latter also implies that

$\frac{\partial}{\partial S'}(\sigma^2(T,S')S'^2\widetilde{p}) \to 0$, as $S' \to \infty$. Hence, evaluating the second integral term in (16.11) by parts gives

$$\begin{aligned}\int_K^{\infty} (S'-K)\frac{\partial^2}{\partial S'^2}(\sigma^2(T,S')S'^2\widetilde{p})\,\mathrm{d}S' &= -\int_K^{\infty} \frac{\partial}{\partial S'}(\sigma^2(T,S')S'^2\widetilde{p})\,\mathrm{d}S' \\ &= \sigma^2(T,K)K^2\widetilde{p}(t,T;S,K) \\ &= \mathrm{e}^{r(T-t)}\sigma^2(T,K)K^2\frac{\partial^2 c}{\partial K^2}\,.\end{aligned}$$

Collecting all the intermediate results obtained above into (16.11) gives the following dual Black–Scholes PDE:

$$\begin{aligned}\frac{\partial c}{\partial T} &= -rc + (r-q)c - (r-q)K\frac{\partial c}{\partial K} + \frac{1}{2}\sigma^2(T,K)K^2\frac{\partial^2 c}{\partial K^2} \\ &= -qc + (q-r)K\frac{\partial c}{\partial K} + \frac{1}{2}\sigma^2(T,K)K^2\frac{\partial^2 c}{\partial K^2}\,.\end{aligned}$$ □

A consequence of the above result is the following formula for the local volatility, which may be used in practice to calibrate a local volatility surface $\sigma(t,S)$ using market European call option prices across a range of maturities and strikes.

Theorem 16.2 (The Derman–Kani–Dupire Formula). *Let $q = 0$. The local volatility function is expressed in analytically closed form as follows in terms of call option prices:*

$$\sigma^2(T,K) = \frac{2}{K^2}\frac{\frac{\partial C}{\partial T} + rK\frac{\partial C}{\partial K}}{\frac{\partial^2 C}{\partial K^2}}. \tag{16.12}$$

16.1.2 Constant Elasticity of Variance Model

In the realm of state-dependent volatility models, the *constant elasticity of variance* (CEV) model has provided an introduction to nonlinear diffusion models that exhibit an implied volatility (half) smile as a function of strike. The CEV diffusion model has a power law volatility function with two adjustable parameters. The model admits closed-form pricing formulae for standard European, barrier, and lookback options. Spectral expansions and Laplace transform techniques are very useful in deriving analytical pricing formulae and transition densities for the CEV process as well as for other time-homogeneous models with more complex nonlinear local volatility functions.

16.1.2.1 Definition and Basic Properties

The constant elasticity of variance diffusion model assumes that the asset price is a time-homogeneous diffusion process $\{S(t)\}_{t\geqslant 0}$ that obeys the stochastic differential equation (SDE)

$$\mathrm{d}S(t) = \nu S(t)\,\mathrm{d}t + \alpha(S(t))^{\beta+1}\,\mathrm{d}W(t)\,,\ t\geqslant 0,\quad S(0) = S_0 > 0\,, \tag{16.13}$$

where ν, β, α are real parameters (with $\alpha > 0$) and $\{W(t)\}_{t\geqslant 0}$ is a standard Brownian motion under a given probability measure $\mathbb{P}$. The SDE (16.13) has a linear drift coefficient and power-type nonlinear diffusion coefficient $\alpha S^{\beta+1}$. If $\beta = 0$, then the CEV diffusion reduces to a geometric Brownian motion governed by the SDE $\mathrm{d}S(t) = \nu S(t)\,\mathrm{d}t + \alpha S(t)\,\mathrm{d}W(t)$. Thus the CEV model can be viewed as a generalization of the standard Black–Scholes model for asset prices, where the log-volatility σ is a deterministic nonlinear function of the asset price S given by $\sigma(S) := \alpha S^{\beta}$. We will refer to $\sigma(S)$ as the *local volatility function.*

The parameter $\beta > 0$ can be interpreted as the *elasticity*[1] *of the local volatility function* with the property $\sigma'(S) = \beta\sigma(S)/S$, and α is a scale parameter fixing the instantaneous volatility to equal a constant σ_0 at an initial spot value S_0, i.e., $\sigma_0 = \sigma(S_0) = \alpha S_0^\beta$. Typical values of the CEV elasticity implicit in equity index option markets are negative and can be as low as $\beta = -4$. For this and other reasons discussed later, β is assumed to take on negative values in what follows.

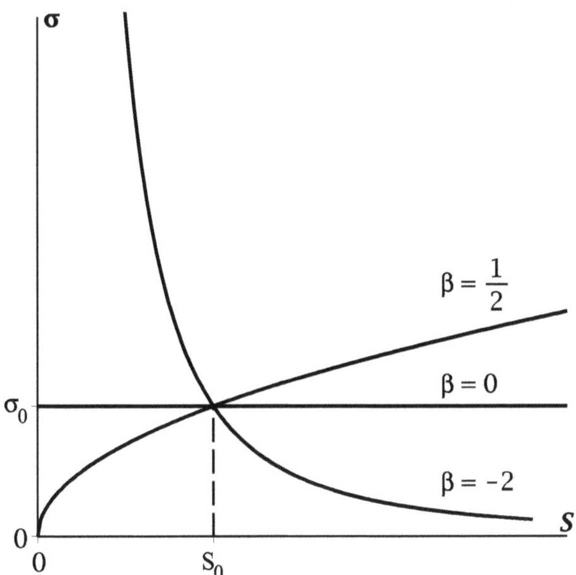

FIGURE 16.1: The local volatility function $\sigma(S) = \alpha S^\beta$ of the CEV model. The parameter α is chosen such that $\sigma(S_0) = \sigma_0$ is fixed.

The CEV process is a nonnegative process. To discuss its behaviour at the boundary points 0 and ∞, several definitions are required. For a diffusion process, $\{S(t)\}_{t\geqslant 0}$, defined on an interval (a, b), we say that $x = a$ and $x = b$ are boundary points of the *state space* of the process. We say that point $x \in \{a, b\}$ is

(a) an *entrance* boundary if the process S can enter the state space starting from x but cannot reach x in finite time starting from the interior of the state space;

(b) an *exit* boundary if the process S can reach x in finite time starting from the interior of the state space but cannot enter the state space starting from x;

(c) a *regular* boundary if it is both an exit and an entrance boundary;

(d) a *natural* boundary if it is neither an exit nor an entrance boundary.

We note that "entrance" really means entrance and not exit and "exit" means exit and not entrance. It is possible to impose a boundary condition for the diffusion S at a regular boundary. A regular boundary is called a *killing* (or *reflecting*) boundary when an instantaneously killing (or reflecting) boundary condition is imposed.

[1] In physics, elasticity is the ability of a deformed material body to return to its original shape and size when the forces causing the deformation are removed.

Given the CEV process satisfying (16.13), we can classify the boundary points 0 and ∞ according to the value of β. For $\beta < 0$, infinity is a natural boundary point. For $-\frac{1}{2} \leqslant \beta < 0$, the origin is an exit boundary point. For $\beta < -\frac{1}{2}$, the origin is a regular boundary point. For $\beta = 0$, i.e., for GBM, both boundary points 0 and ∞ are natural boundaries. For $\beta > 0$, the origin is a natural boundary and infinity is an entrance boundary point.

16.1.2.2 Transition Probability Law

A combination of a change of variables and a scale and time transformation allows us to represent the CEV diffusion process $\{S(t)\}_{t\geqslant 0}$ as a function of a squared Bessel (SQB) process $\{X(t)\}_{t\geqslant 0}$, which has generator $\mathcal{G}f(x) := (2\mu+2)f'(x)+2xf''(x)$ and hence satisfies the SDE

$$\mathrm{d}X(t) = (2\mu + 2)\,\mathrm{d}t + 2\sqrt{X(t)}\,\mathrm{d}W(t), \tag{16.14}$$

where μ is the so-called index of the process. The left-hand endpoint 0 is an entrance boundary if $\mu \geqslant 0$, a regular boundary if $-1 < \mu < 0$, or an exit boundary if $\mu \leqslant -1$. The right-hand endpoint ∞ is a natural boundary. The state space of the SQB process is the interval $[0, \infty)$ or $(0, \infty)$ depending on the value of μ and the boundary condition at zero. The SQB diffusion is a time-homogeneous Markov process. The transition probability density function (PDF) is given by

$$p_\mu(t; x_0, x) \equiv \frac{\mathbb{P}(X(t) \in \mathrm{d}x | X(0) = x_0)}{\mathrm{d}x} = \left(\frac{x}{x_0}\right)^{\frac{\mu}{2}} \frac{e^{-(x+x_0)/(2t)}}{2t} I_{\tilde{\mu}}\left(\frac{\sqrt{x x_0}}{t}\right), \tag{16.15}$$

where $\tilde{\mu} = \mu$ if $\mu \geqslant 0$ or if $\mu \in (-1, 0)$ and 0 is specified as a regular reflecting boundary, and $\tilde{\mu} = |\mu|$, $\mu < 0$, if 0 is an exit or a regular killing boundary. Here, $I_\mu(z)$ denotes the modified Bessel function of the first kind, of order μ and argument z.

Let us find the transition PDF of the CEV process by reducing it to the SQB process as follows. First, a CEV process with a nonzero drift parameter ν is obtained from a driftless CEV process denoted $\{F(t)\}_{t\geqslant 0}$ by means of a scale and time change:

$$S(t) = \mathrm{e}^{\nu t} F(\tau_t), \text{ where } \tau_t = \begin{cases} \frac{1}{2\nu\beta}\left(e^{2\nu\beta t} - 1\right) & \nu \neq 0, \\ t & \nu = 0. \end{cases} \tag{16.16}$$

Second, the change of variables via the monotonic mapping

$$\mathsf{X}\colon F \to F^{-2\beta}/(\alpha^2\beta^2) \tag{16.17}$$

gives us the SQB process $X(t) := \mathsf{X}(F(t)) = F(t)^{-2\beta}/(\alpha^2\beta^2)$ with the index $\mu = \frac{1}{2\beta}$. Note that this mapping is strictly increasing for $\beta < 0$. Thus, the CEV diffusion can be expressed in terms of the SQB process as follows:

$$S(t) = \mathrm{e}^{\nu t}\left(\alpha^2\beta^2 X(\tau_t)\right)^{-\frac{1}{2\beta}}, \tag{16.18}$$

where the process $\{X(t)\}_{t\geqslant 0}$ solves the SDE (16.14) with $\mu = \frac{1}{2\beta}$. Although the detailed mathematical proof of the representation (16.18) is left as an exercise for the reader, we briefly discuss the transformation. Two techniques are applied. First, the Itô formula is applied when transforming $F(t) \to X(t)$. Using the change of variables $x = \mathsf{X}(S)$ in the PDF (16.15) gives us the transition PDF of the driftless CEV process:

$$p^{(0)}(t; S_0, S) = p_\mu(t; \mathsf{X}(S_0), \mathsf{X}(S)) \cdot \mathsf{X}'(S). \tag{16.19}$$

Second, the removal of drift is done with the use of the Kolmogorov PDE for the transition PDF. Denote the transition PDF of the CEV process with the drift parameter ν by $p^{(\nu)}(t; S_0, S)$. The function $p^{(\nu)}$ solves the corresponding forward Kolmogorov PDE,

$$\frac{\partial p^{(\nu)}}{\partial t} = \frac{1}{2}\frac{\partial^2}{\partial S^2}\left(\sigma^2(S)\, p^{(\nu)}\right) - \frac{\partial}{\partial S}\left(\nu S\, p^{(\nu)}\right), \tag{16.20}$$

subject to the Dirac delta initial condition, $p^{(\nu)}(0; S_0, S) = \delta(S - S_0)$, and imposed homogeneous boundary conditions at the endpoints 0 and ∞. According to the transformation (16.16), the PDFs $p^{(0)}$ and $p^{(\nu)}$ (with $\nu \neq 0$) relate to each other as follows:

$$p^{(\nu)}(t; S_0, S) = \mathrm{e}^{-\nu t} p^{(0)}(\tau_t; S_0, \mathrm{e}^{-\nu t} S). \tag{16.21}$$

The reader may check that $p^{(\nu)}$ satisfies (16.20), assuming that $p^{(0)}$ solves Equation (16.20) with $\nu = 0$. This proves that the process $\{S(t)\}_{t\geqslant 0}$ in (16.16) satisfies the SDE (16.13) for the CEV process with drift. As a result, the PDF $p^{(\nu)}$ is expressed in terms of the transition PDF p_μ of the SQB process:

$$p^{(\nu)}(t; S_0, S) = \mathrm{e}^{-\nu t} p_\mu(\tau_t; \mathsf{X}(S_0), \mathsf{X}(\mathrm{e}^{-\nu t} S)) \cdot \mathsf{X}'(\mathrm{e}^{-\nu t} S). \tag{16.22}$$

Assume that the parameter β is negative. Here, we consider the case when the endpoint $S = 0$ is a regular killing boundary (if $\beta < -0.5$) or exit (if $-0.5 \leqslant \beta < 0$). The transition PDF $p^{(0)}(t; S_0, S)$ with $S_0, S > 0$ and $t > 0$ for the driftless CEV process $F(t)$ takes the form

$$p^{(0)}(t; S_0, S) = \frac{S^{-2\beta-\frac{3}{2}} S_0^{\frac{1}{2}}}{\alpha^2|\beta| t} \exp\left(-\frac{S^{-2\beta} + S_0^{-2\beta}}{2\alpha^2\beta^2 t}\right) I_{\frac{1}{2|\beta|}}\left(\frac{S^{-\beta} S_0^{-\beta}}{\alpha^2\beta^2 t}\right). \tag{16.23}$$

The density $p^{(0)}(t; S_0, S)$ does not integrate (with respect to S) to 1, since $S = 0$ is an absorbing point. However, the driftless CEV diffusion process satisfies the martingale property:

$$\mathrm{E}_t[F(t + u)] = F(t) \quad \text{for all } \; t, u \geqslant 0.$$

For $\beta < 0$, the probability of absorption at zero (bankruptcy) is

$$\mathbb{P}(S(t) = 0) = 1 - \mathbb{P}(S(t) > 0) = 1 - \int_0^\infty p^{(\nu)}(t; S_0, S)\,\mathrm{d}S = G\left(\frac{1}{2|\beta|}, \frac{\mathsf{X}(S_0)}{1 - \mathrm{e}^{2\nu\beta t}}\right), \tag{16.24}$$

where $G(\mu, a)$ is the complementary gamma function given by

$$G(\mu, a) = \frac{1}{\Gamma(\mu)} \int_a^\infty \mathrm{e}^{-t}\, t^{\mu-1}\,\mathrm{d}t.$$

Here, $\Gamma(x)$ is the gamma function defined by $\Gamma(x) = \int_0^\infty t^{x-1}\mathrm{e}^{-t}\,\mathrm{d}t$.

Note that if the reflecting boundary condition is imposed at 0, then there is no absorption at the endpoint. The corresponding transition density (for the case with $\beta < -0.5$) is given by (16.23) with the replacement $I_{\frac{1}{2|\beta|}} \to I_{\frac{1}{2\beta}}$. The driftless process $\{F(t)\}_{t\geqslant 0}$ becomes a strict submartingale.

16.1.2.3 Pricing European Options

We assume $\beta < 0$, so the driftless CEV process (with $\nu = 0$) obeys the martingale property. Generally, we have

$$\mathrm{E}_t[S(t + u)] = \mathrm{e}^{\nu(t+u)}\,\mathrm{E}_{\tau_t}[F(\tau_{t+u})] = \mathrm{e}^{\nu u}\mathrm{e}^{\nu t} F(\tau_t) = \mathrm{e}^{\nu u} S(t),$$

for all $t, u \geqslant 0$. Therefore, the forward price process $\{e^{-\nu t} S(t)\}_{t\geqslant 0}$ is a martingale. Under the risk-neutral probability measure $\widetilde{\mathbb{P}}$ we set $\nu = r-q$. The no-arbitrage price of a European call option (with expiry T and strike K) under the CEV process with the transition density in (16.19)–(16.23) is then given by

$$
\begin{aligned}
C(T, S_0; K) &= e^{-rT}\, \widetilde{E}_{0,S_0}\left[(S(T)-K)^+\right] = e^{-rT} \int_K^\infty (S-K)\, \widetilde{p}^{(\nu)}(T; S_0, S)\, dS \\
&= e^{-qT} S_0 \int_m^\infty \frac{1}{2}\left(\frac{y_0}{y}\right)^{\frac{1}{4\beta}} \exp\left(-\frac{y+y_0}{2}\right) I_{\frac{1}{2|\beta|}}(\sqrt{y y_0})\, dy \\
&\quad - e^{-rT} K \int_m^\infty \frac{1}{2}\left(\frac{y}{y_0}\right)^{\frac{1}{4\beta}} \exp\left(-\frac{y+y_0}{2}\right) I_{\frac{1}{2|\beta|}}(\sqrt{y y_0})\, dy \qquad (16.25)
\end{aligned}
$$

where

$$
m = \frac{2\nu K^{-2\beta}}{\alpha^2 \beta\,(1-e^{-2\nu\beta T})}, \quad y_0 = \frac{2\nu S_0^{-2\beta}}{\alpha^2\beta\,(e^{2\nu\beta T}-1)}
$$

for the case when $\nu = r-q \neq 0$. If $\nu = 0$, then $m = \frac{K^{-2\beta}}{\alpha^2\beta^2 T}$ and $y_0 = \frac{S_0^{-2\beta}}{\alpha^2\beta^2 T}$.

Proof. Using (16.23), we express the price of a European call under the CEV model as follows:

$$
\begin{aligned}
C \equiv C(T, S_0, K) &= e^{-rT} \int_K^\infty (S-K)\, \widetilde{p}^{(\nu)}(T; S_0, S)\, dS \\
&= e^{-rT} \int_K^\infty (S-K)\, e^{-\nu T} p^{(0)}(\tau_T; S_0, e^{-\nu T} S)\, dS\,.
\end{aligned}
$$

By changing variables defined by $S' = e^{-\nu T} S$, and then renaming the dummy variable S' as S,

$$
\begin{aligned}
C &= e^{-rT} \int_{Ke^{-\nu T}}^\infty \left(e^{\nu T} S - K\right)\, p^{(0)}(\tau; S_0, S)\, dS \\
&= e^{-qT} \int_{Ke^{-\nu T}}^\infty S\, p^{(0)}(\tau; S_0, S)\, dS - e^{-rT} K \int_{Ke^{-\nu T}}^\infty p^{(0)}(\tau; S_0, S)\, dS\,.
\end{aligned}
$$

Using (16.19) and applying the change of variables $x = \mathsf{X}(S) := \frac{S^{-2\beta}}{\alpha^2\beta^2}$ gives

$$
\begin{aligned}
C &= e^{-qT} \int_{Ke^{-\nu T}}^\infty S\, p_\mu(\tau; \mathsf{X}(S_0), \mathsf{X}(S)) \cdot \mathsf{X}'(S)\, dS \\
&\quad - e^{-rT} K \int_{Ke^{-\nu T}}^\infty p_\mu(\tau; \mathsf{X}(S_0), \mathsf{X}(S)) \cdot \mathsf{X}'(S)\, dS \\
&= e^{-qT} \int_k^\infty (\alpha^2\beta^2 x)^{-\frac{1}{2\beta}} p_\mu(\tau; x_0, x)\, dx - e^{-rT} K \int_k^\infty p_\mu(\tau; x_0, x)\, dx \\
&= e^{-qT} S_0 \int_k^\infty \left(\frac{x}{x_0}\right)^{-\frac{1}{2\beta}} p_\mu(\tau; x_0, x)\, dx - e^{-rT} K \int_k^\infty p_\mu(\tau; x_0, x)\, dx\,,
\end{aligned}
$$

where $\tau \equiv \tau_T$ is given by (16.16), $k = \mathsf{X}(e^{-\nu T} K) = \frac{(e^{-\nu T} K)^{-2\beta}}{\alpha^2\beta^2}$, $x_0 = \mathsf{X}(S_0) = \frac{S_0^{-2\beta}}{\alpha^2\beta^2}$, and $\mu = \frac{1}{2\beta}$. The formula (16.25) is deduced by applying (16.15) and changing variables in the above two integrals: $y = \frac{x}{\tau}$, $y_0 = \frac{x_0}{\tau}$, and $m = \frac{k}{\tau}$. □

The two integrals in (16.25) can be computed numerically with the use of some quadrature rule. Alternatively, the call option pricing formula can be expressed in terms of the complementary noncentral chi-square distribution function. The latter can be represented as a series of elementary functions. When the call option price is calculated, the put option value can be obtained by using a put-call parity.

The *noncentral chi-square distribution* denoted $\chi^2_v(\lambda)$ is a continuous probability distribution with parameters $v \in (0, \infty)$ and $\lambda \in [0, \infty)$ and with PDF

$$f(x; v, \lambda) = \frac{1}{2}\left(\frac{x}{\lambda}\right)^{(v-2)/4} I_{(v-2)/2}(\sqrt{\lambda x}) \exp\left(-\frac{\lambda + x}{2}\right), \quad x > 0. \tag{16.26}$$

For integer v, this distribution arises as a probability distribution of a sum of squared normal variables. Let $Z_1, Z_2, \ldots, Z_v$ be independent standard normal random variables, and $a_1, a_2, \ldots, a_v$ be some real constants, then $Y = \sum_{j=1}^{v}(Z_j + a_j)^2$ is said to have the noncentral chi-square distribution with v degrees of freedom and noncentrality parameter $\lambda = \sum_{j=1}^{v} a_j^2$. If all $a_j = 0$, then Y has the central chi-square distribution with v degrees of freedom, which is denoted as usual by χ^2_v.

Since the integrands in (16.25) are both the noncentral chi-square PDFs of the form (16.26), the integrals in (16.25) can be expressed in terms of the complementary distribution function for $\chi^2_v(\lambda)$:

$$Q(x; v, \lambda) = \int_x^\infty f(y; v, \lambda)\, dy. \tag{16.27}$$

Recall that the cumulative and complementary distribution functions, respectively denoted by F and Q, relate to each other as $F(x) + Q(x) = 1$.

The first integrand in (16.25) is the noncentral chi-square PDF $f\left(y; 2 + \frac{1}{|\beta|}, y_0\right)$ with $2 + \frac{1}{|\beta|}$ degrees of freedom and noncentrality parameter y_0. Integrating with respect to y from m to infinity gives the corresponding complementary distribution function $Q\left(m; 2 + \frac{1}{|\beta|}, y_0\right)$. The second integrand in (16.25) is equal to $f\left(y_0; 2 + \frac{1}{|\beta|}, y\right)$, where y is now the noncentrality parameter. The second integral is therefore equal to $1 - Q\left(y_0; \frac{1}{|\beta|}, m\right)$ (see Exercise 16.4). Thus, the call option value is

$$C(T, S_0; K) = \mathrm{e}^{-qT} S_0 Q(m; 2 + \frac{1}{|\beta|}, y_0) - \mathrm{e}^{-rT} K \left(1 - Q(y_0; \frac{1}{|\beta|}, m)\right). \tag{16.28}$$

The complementary noncentral chi-square distribution function $Q(x; v, \lambda)$ can be computed using the following formula:

$$Q(2x; 2v, 2\lambda) = 1 - \sum_{n=1}^{\infty} g\,(n + v, x) \sum_{j=1}^{n} g\,(j, \lambda), \tag{16.29}$$

where $g(m, x)$ is a gamma PDF given by

$$g(m, x) = \frac{\mathrm{e}^{-x}\, x^{m-1}}{\Gamma(m)}.$$

16.1.3 The Heston Model

While in local volatility models a more realistic behaviour of asset prices is obtained by introducing a nonlinear volatility function, in the stochastic volatility framework, volatility changes over time according to another random process correlated with the asset price process. The volatility process is usually assigned through a suitable stochastic differential equation. Therefore, we deal with a two-dimensional SDE for asset price modelling. Let us denote the asset price by $S(t)$ and the instantaneous variance (squared volatility) by $\mathrm{v}(t)$. In the *Heston model*, the asset price follows geometric Brownian motion and the variance is modelled as a square-root, mean-reverting diffusion with dynamics similar to the Cox–Ingersoll–Ross (CIR) model for interest rates. Thus, we consider the following two-dimensional model:

$$\mathrm{d}S(t) = \mu S(t)\,\mathrm{d}t + \sqrt{\mathrm{v}(t)}S(t)\,\mathrm{d}W_1(t), \tag{16.30}$$

$$\mathrm{dv}(t) = \kappa(\theta - \mathrm{v}(t))\,\mathrm{d}t + \xi\sqrt{\mathrm{v}(t)}\,\mathrm{d}W_2(t), \tag{16.31}$$

where the two Brownian motions are correlated, i.e., as $\mathrm{d}W_1(t)\,\mathrm{d}W_2(t) = \rho\,\mathrm{d}t$. In reality, the correlation coefficient ρ is typically negative and can be close to -1. If $\rho < 0$, then volatility will increase when the asset price decreases. The parameters in (16.30)–(16.31) represent the following:

μ is the rate of return on the asset.

θ is the long variance, or long run average price variance; as t approaches infinity, the expected value of the variance $\mathrm{v}(t)$ goes to θ.

κ is the rate at which $\mathrm{v}(t)$ reverts to θ.

ξ is the volatility of the volatility; as the name suggests, this determines the variance of $\mathrm{v}(t)$.

To guarantee the positiveness of the instantaneous variance $\mathrm{v}(t)$, the model parameters satisfy

$$2\kappa\theta \geqslant \xi^2.$$

It is actually not easy to price derivatives under the Heston model. First, the joint distribution of the process (16.30)–(16.31) is not available in closed form. Therefore, the risk-neutral expectation of a payoff function is hard to compute. If the price process and volatility process are uncorrelated, then the price of a standard European option can be written as an integral over Black–Scholes prices. To develop derivative prices in the general case, one can use Monte Carlo simulations, PDE solutions, or transform methods. Stochastic-volatility models are quite popular among practitioners. Such models are able to produce pronounced volatility smiles. Since the model has two sources of uncertainty, namely, the movement of stock price $S(t)$ and the movement of instantaneous variance $\mathrm{v}(t)$, the Heston model is incomplete. Further discussion of these techniques and the properties of the model is beyond the scope of this book.

16.2 Models with Jumps

16.2.1 The Poisson Process

Suppose that events of interest are occurring at random time moments. Let $N(t)$ be the number of occurrences from time 0 to time t. By varying $t \in \mathbb{R}_+$, we obtain a random process $\{N(t)\}_{t\geqslant 0}$. Initially, we set $N(0) = 0$. Assume that only an individual event may occur at a time. Therefore, each possible realization (sample path) of $N(t)$ is a nondecreasing step function that only changes by jumps of size 1. As a result, we obtain a fundamental pure jump process known as the *Poisson process*.

Definition 16.1. A Poisson process is an integer-valued stochastic process $\{N(t)\}_{t\geqslant 0}$ with $N(0) = 0$ that satisfies the following properties.

(a) The numbers of events that occur in disjoint time intervals are independent. That is, for all s_1, s_2, t_1, t_2 with $0 \leqslant s_1 < s_2 \leqslant t_1 < t_2$ the random variables $N(s_2) - N(s_1)$ and $N(t_2) - N(t_1)$ are independent.

(b) The distribution of the number of events that occur on a given time interval only depends on the length and does not depend on the location of the interval. That is, for all s and t with $0 \leqslant s < t$, the probability distribution of $N(t) - N(s)$ depends on $t - s$ and does not depend on t and s considered individually.

(c) There exists a constant $\lambda > 0$ such that for any small interval of length δt, the probability of having one event occurring in a given small time interval, $[t, t + \delta t]$, is approximately $\lambda \cdot \delta t$ plus an error of small order $o(\delta t)$. That is,

$$\lim_{\delta t \searrow 0} \frac{\mathbb{P}(N(t + \delta t) - N(t) = 1)}{\delta t} = \lambda,$$

whereas the probability to have two or more events occurring in $[t, t + \delta t]$ is $o(\delta t)$. That is,

$$\lim_{\delta t \searrow 0} \frac{\mathbb{P}(N(t + \delta t) - N(t) \geqslant 2)}{\delta t} = 0.$$

Properties (a) and (b) mean that $\{N(t)\}_{t\geqslant 0}$ is a process with independent and stationary increments. Additionally, as is shown just below, the above properties lead to the fact that all increments of the process $\{N(t)\}_{t\geqslant 0}$ are Poisson distributed.

Proposition 16.3. *For all s and t such that $0 \leqslant t < s$,*

$$N(s) - N(t) \sim \mathit{Pois}(\lambda(s - t)).$$

Proof. The proof is left as an exercise for the reader. □

Since $N(0) = 0$, Proposition 16.3 implies that the value $N(t)$ for any $t > 0$ has the Poisson probability distribution $\mathit{Pois}(\lambda t)$. Therefore, the mean and variance of $N(t)$ are equal,

$$\mathrm{E}[N(t)] = \mathrm{Var}(N(t)) = \lambda t,$$

respectively. In summary, the Poisson process is a process with independent, stationary, Poisson-distributed increments. The number λ is called the *rate* or *intensity*. For the Poisson process, the rate is equal to the average number of jumps per one unit of time, i.e.,

$$\lambda = \frac{\mathrm{E}[N(s) - N(t)]}{s - t} \quad \text{for} \quad 0 \leqslant t < s.$$

The Poisson process with rate λ is denoted by $N_\lambda(t)$. Note that although we assume here that λ is constant, in general, the intensity can be a function of time t defined by

$$\lambda(t) = \lim_{s \to t} \frac{\mathrm{E}[N(s) - N(t)]}{s - t}.$$

Let T_i be the occurrence time of the ith event or, equivalently, the moment when the Poisson process makes the ith jump. These random times $T_1, T_2, \ldots$ form an increasing sequence of positive reals. Let us find the probability distribution of T_1. The probability that $T_1 > t$ from some $t > 0$ is the same as the probability that no jumps occur in the interval $[0, t]$. So, we have

$$\mathbb{P}(T_1 > t) = \mathbb{P}(N(t) = 0) = \mathrm{e}^{-\lambda t}.$$

Therefore, the PDF of T_1 is

$$\frac{\partial \mathbb{P}(T_1 \leqslant t)}{\partial t} = \frac{\partial(1 - \mathrm{e}^{-\lambda t})}{\partial t} = \lambda \mathrm{e}^{-\lambda t},$$

for all $t > 0$, and zero otherwise, i.e., this is an exponential density where $T_1 \sim Exp(\lambda)$. Similarly, one can show that the increments $T_{i+1} - T_i$ are all $Exp(\lambda)$ distributed. Additionally, one can show that all increments $\tau_i = T_i - T_{i-1}$ with $i = 1, 2, \ldots$ are independent.

This observation leads us to the second definition of the Poisson process presented just below. It is more constructive than the first one since it gives us a simple algorithm for generating sample paths.

Definition 16.2. Consider a sequence of i.i.d. exponentially distributed random variables, $\{\tau_i\}_{i \geqslant 1}$, having common mean $\frac{1}{\lambda}$. The Poisson process $\{N(t)\}_{t \geqslant 0}$ is defined as

$$N(t) := \max\{k \ : \ T_k \leqslant t\} = \sum_{k \geqslant 1} \mathbb{I}_{\{T_k \leqslant t\}}, \tag{16.32}$$

where $T_k = \sum_{j=1}^{k} \tau_j$ for $k \geqslant 1$.

The process defined in (16.32) starts at zero; it makes a jump of size 1 at each time T_i, i.e., T_i is the moment of the ith jump. Each τ_i is a time lag between two successive jumps of the process. Note that the Poisson process $\{N(t)\}_{t \geqslant 0}$ is a process with independent and stationary increments. Let us find the distribution of $N(t)$ for $t > 0$. First, notice that for any $k \geqslant 1$ the occurrence time T_k has the gamma distribution $Gamma(k, \lambda)$ with PDF

$$f_{T_k}(x) = \frac{\lambda^k x^{k-1}}{(k-1)!} \mathrm{e}^{-\lambda x}, \quad x > 0.$$

The probability of having at least k jumps before time t is

$$\mathbb{P}(N(t) \geqslant k) = \mathbb{P}(T_k \leqslant t) = \int_0^t f_{T_k}(x)\, \mathrm{d}x = \int_0^t \frac{\lambda^k x^{k-1}}{(k-1)!} \mathrm{e}^{-\lambda x}\, \mathrm{d}x.$$

Therefore, the probability of having k jumps before time t is

$$\mathbb{P}(N(t) = k) = \mathbb{P}(N(t) \geqslant k) - \mathbb{P}(N(t) \geqslant k+1) = \int_0^t \frac{\lambda^k x^{k-1}}{(k-1)!} \mathrm{e}^{-\lambda x}\, \mathrm{d}x - \int_0^t \frac{\lambda^{k+1} x^k}{k!} \mathrm{e}^{-\lambda x}\, \mathrm{d}x.$$

Integration by parts gives

$$\int_0^t \frac{\lambda^{k+1} x^k}{k!} \mathrm{e}^{-\lambda x}\, \mathrm{d}x = \int_0^t \frac{\lambda^k x^{k-1}}{(k-1)!} \mathrm{e}^{-\lambda x}\, \mathrm{d}x - \frac{(\lambda t)^k}{k!} \mathrm{e}^{-\lambda t}.$$

Therefore, $\mathbb{P}(N(t) = k) = \frac{(\lambda t)^k}{k!} \mathrm{e}^{-\lambda t}$ for any $k = 0, 1, 2, \ldots$. That is, $N(t) \sim Pois(\lambda t)$. Similarly, using the memoryless property of the exponential distribution, one can show that

$N(t) - N(s) \sim \mathit{Pois}(\lambda(t-s))$ for all t and s with $0 \leqslant s < t$. The result also follows from time stationarity where $N(t) - N(s) \stackrel{d}{=} N(t-s)$.

The Poisson process only changes its value at the time moments $T_1, T_2, T_3, \ldots$ forming a strictly increasing sequence $0 < T_1 < T_2 < T_3 < \ldots$. The process is constant in between these jump times. According to the definition in (16.32), the sample paths are right-continuous, nondecreasing, piecewise-constant functions of time. A typical sample path is shown in Figure 16.2. The mean function of the Poisson process is a linear function of time: $\mathrm{E}[N_\lambda(t)] = \lambda t$. Finally, we define the *compensated Poisson process* $\{X(t) := N_\lambda(t) - \lambda t\}_{t \geqslant 0}$, which has zero mean function

$$\mathrm{E}[X(t)] = \mathrm{E}[N_\lambda(t)] - \lambda t = \lambda t - \lambda t = 0.$$

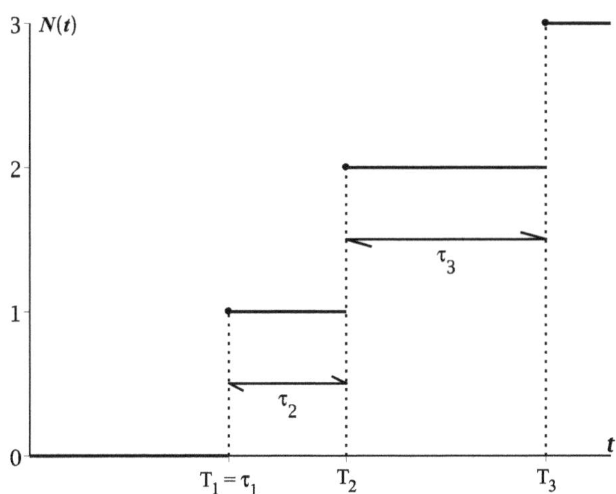

FIGURE 16.2: A sample path of a Poisson process.

16.2.2 Jump-Diffusion Models with a Compound Poisson Component

A standard Poisson process has deterministic jumps of size 1. A *compound Poisson process* is defined as a Poisson process with random jumps. Such a process is useful in insurance to model claims that arrive at time points generated by a Poisson process and the amounts claimed are positive random variables.

Let $\{N(t)\}_{t \geqslant 0}$ be the Poisson process with intensity λ; let $Y_1, Y_2, \ldots$ be a sequence of i.i.d. random variables with finite common mean $\mathrm{E}[Y]$. Suppose that they are all independent of the Poisson process. The compound Poisson process, denoted $\{Q(t)\}_{t \geqslant 0}$, is defined as

$$Q(t) = \sum_{k=1}^{N(t)} Y_k, \quad t \geqslant 0.$$

In the above example with claims made by policyholders to an insurance company, $\{Y_k\}_{k \geqslant 1}$ are the amounts claimed and $N(t)$ is the number of claims made by time t. Then, $Q(t)$ gives the total amount claimed by time t. The jumps in $\{Q(t)\}_{t \geqslant 0}$ occur at the same time as the jumps in $\{N(t)\}_{t \geqslant 0}$ but whereas the jumps in $\{N(t)\}_{t \geqslant 0}$ are always of size 1, the jumps in $\{Q(t)\}_{t \geqslant 0}$ are of random size. The kth jump that occurs at time T_k is of size Y_k.

Properties of the compound Poisson process are similar to those of the standard Poisson process. The mean function of the compound Poisson process is linear in time as well:

$$\mathrm{E}[Q(t)] = \mathrm{E}\left[\sum_{k=1}^{N(t)} Y_k\right] = \mathrm{E}[N(t)] \cdot \mathrm{E}[Y] = \lambda t\,\mathrm{E}[Y]\,.$$

Sample paths of the compound Poisson process are right-continuous, piecewise-constant functions of time. If the jumps Y_k are nonnegative (with probability one), then the sample paths are nondecreasing functions. Typical sample paths of a standard Poisson process and a compound Poisson process with jumps uniformly distributed in $(0,1)$ are given in Figures 16.3a and 16.3b, respectively.

The probability distribution of $Q(t)$ can be derived by conditioning on the number of occurrences. Set $X_m = \sum_{k=1}^{m} Y_k$ for $m \geqslant 1$. Then,

$$\mathbb{P}(Q(t) \leqslant x) = \sum_{m=0}^{\infty} \mathbb{P}(X_m \leqslant x) \cdot \mathrm{e}^{-\lambda t}\frac{(\lambda t)^m}{m!}, \quad \text{for } x \in \mathbb{R}.$$

For example, it is not difficult to show that if all $Y_k \sim Bin(1,p)$, $0 < p \leqslant 1$, then $\{Q(t)\}_{t\geqslant 0}$ is a Poisson process with intensity λp.

Let us consider the following jump-diffusion model for asset prices. We assume that the log-price of the underlying asset follows Brownian motion with superimposed jumps:

$$\ln S(t) = \ln S(0) + \mu t + \sigma W(t) + \sum_{k=1}^{N(t)} Y_k\,,\ t \geqslant 0\,, \quad S(0) = S_0 > 0\,, \tag{16.33}$$

where the standard Brownian motion $\{W(t)\}_{t\geqslant 0}$ is independent of the compound Poisson component. Taking the exponential of both parts of (16.33) gives the asset price formula:

$$S(t) = S_0 \exp\left(\mu t + \sigma W(t) + \sum_{k=1}^{N(t)} Y_k\right) = S_0 \mathrm{e}^{\mu t + \sigma W(t) + Q(t)}. \tag{16.34}$$

Since the exponential function takes strictly positive values, the price $S(t)$ is strictly positive at every time t. As seen from (16.34), the asset price process follows geometric Brownian motion (with continuous paths) between the moments of jumps. At the moment of a jump, the asset price is multiplied by the exponential of the jump size, $J \equiv \mathrm{e}^Y$. Thus, a sample path of $S(t)$ is a piecewise continuous functions.

The asset price process (16.34) can be viewed as a solution to the stochastic differential equation with a Poisson component. The log-price, $\ln S(t)$, satisfies the SDE

$$\mathrm{d}\ln S(t) = \mu\,\mathrm{d}t + \sigma\,\mathrm{d}W(t) + Y\,\mathrm{d}N(t)\,. \tag{16.35}$$

As is seen, if no jump occurs, then the log-price simply evolves as a log-normal continuous process; if a jump occurs then the log-jump, $Y \equiv \ln J$, is added to the log-price, i.e., the proportional change in the stock price is given by

$$\frac{\mathrm{d}S(t)}{S(t)} = (\mu + \sigma^2/2)\,\mathrm{d}t + \sigma\,\mathrm{d}W(t) + (J-1)\,\mathrm{d}N(t).$$

When a jump occurs, the asset price S changes by $S(J-1)$, which is equivalent to multiplying S by J.

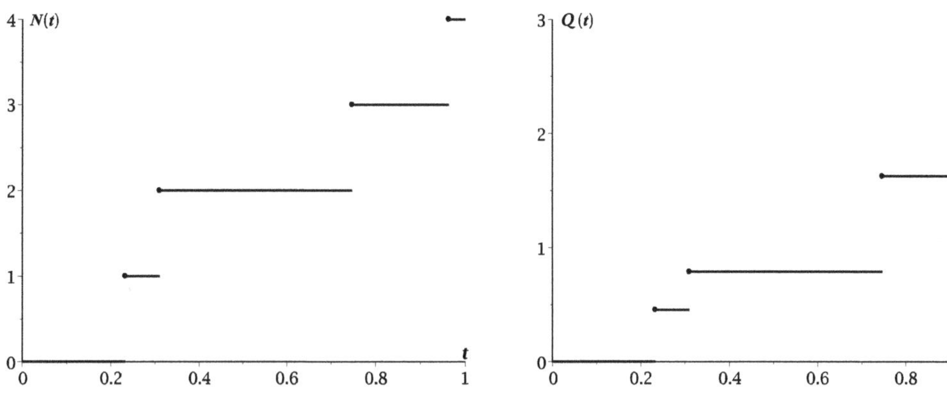

(a) A standard Poisson process with jumps of size 1.

(b) A compound Poisson process with uniformly distributed jumps.

FIGURE 16.3: Sample paths of standard and compound Poisson processes with $\lambda = 4$.

As we know, the asset price model is arbitrage free, if there exists an equivalent martingale measure (EMM). Under the EMM $\widetilde{\mathbb{P}}$, the coefficient μ in (16.34) has to be selected such that $\widetilde{\mathrm{E}}_{0,S_0}[S(t)] = \mathrm{e}^{(r-q)t}S_0$ holds for all $t \geqslant 0$. Since $W(t)$, $N(t)$, and $\{Y_i\}$ are mutually independent, we have

$$\begin{aligned}\mathrm{E}_{0,S_0}[S(t)] &= S_0\mathrm{e}^{\mu t}\mathrm{E}\left[\mathrm{e}^{\sigma W(t)}\right]\mathrm{E}\left[\mathrm{E}[\mathrm{e}^Y]^{N(t)}\right] \\ &= S_0\mathrm{e}^{\mu t}\mathrm{e}^{t\sigma^2/2}\mathrm{e}^{\lambda t(a-1)} = S_0\mathrm{e}^{(\mu+\sigma^2/2+\lambda(a-1))t}.\end{aligned}$$

where $a := \mathrm{E}[\mathrm{e}^Y]$. Therefore, under the EMM, the drift coefficient μ is given by

$$\mu = r - q - \sigma^2/2 + \lambda(1-a). \tag{16.36}$$

The jump size $J \equiv \mathrm{e}^Y$ can be taken to be a nonrandom number or a random variable. By assuming that J is log-normal (and hence the log-jump Y is normal), we obtain the *Merton model*. In the case of normal log-jumps, we can derive a closed-form pricing formula for standard (non-path-dependent) European options. Let all log-jumps Y_i be independent normally distributed random variables with common mean α and variance β^2. When the value of $N(t)$ in (16.33) is fixed, the log-price $\ln S(t)$ becomes a sum of normal random variables and is therefore a normal variable as well. The distribution of the log-price $\ln S(t)$ conditional on the value of the Poisson process $N(t)$ has mean

$$\mathrm{E}[\ln S(t) \mid N(t)] = \ln S_0 + \mu t + N(t)\mathrm{E}[Y] = \ln S_0 + \mu t + \alpha N(t)$$

and variance

$$\mathrm{Var}(\ln S(t) \mid N(t)) = \sigma^2 t + N(t)\,\mathrm{Var}(Y) = \sigma^2 t + \beta^2 N(t).$$

Thus, we have the conditional normal distribution

$$(\ln S(t) \mid N(t)) \stackrel{d}{=} \big(\ln S_0 + \mu t + \alpha N(t)\big) + \sqrt{\sigma^2 + \beta^2 N(t)/t}\,\sqrt{t}Z,$$

where $Z \sim \mathit{Norm}(0,1)$. Therefore, the value of $S(t)$ conditional on $N(t) = n$ can be rewritten as

$$S(t) = \hat{S}_n \exp\big((r - q - \hat{\sigma}_n^2/2)t + \hat{\sigma}_n W(t)\big) \tag{16.37}$$

where

$$\hat{S}_n = S_0 e^{(\mu - r + q + \hat{\sigma}_n^2/2)t + \alpha n} \text{ and } \hat{\sigma}_n^2 = \sigma^2 + n\beta^2/t. \tag{16.38}$$

Using the risk-neutral value of μ in (16.36) into (16.38) gives

$$\hat{S}_n = S_0 e^{(\lambda(1-a) + (\hat{\sigma}_n^2 - \sigma^2)/2)t + \alpha n} = S_0 e^{\lambda(1-a)t + (\alpha + \beta^2/2)n}$$

where $a \equiv \mathrm{E}[e^Y] = e^{\alpha + \beta^2/2}$ since $Y \sim Norm(\alpha, \beta^2)$. Hence, according to (16.37), the Merton asset price model (with fixed $N(t) = n$) can be reduced to the Black–Scholes model with an asset modelled as a GBM with an effective spot value $\hat{S}_n$ and volatility $\hat{\sigma}_n$.

The values of standard European options such as a call and put with strike K and expiry T are given by regular formulae. Let us denote the Black–Scholes (BS) option pricing function by $v_{BS}(T, \hat{S}_n; \hat{\sigma}_n)$. This is the pricing function for an underlying stock price following a GBM with spot value $\hat{S}_n$, volatility parameter $\hat{\sigma}_n$, interest rate r, and stock dividend yield q. Now the option pricing function under the Merton jump-diffusion model, denoted v_M, is obtained by taking the expectation of the BS option value with respect to $N(T)$:

$$v_M(T, S_0) = \mathrm{E}\big[v_{BS}(T, \hat{S}_{N(T)}; \hat{\sigma}_{N(T)})\big] = \sum_{n=0}^{\infty} e^{-\lambda T} \frac{(\lambda T)^n}{n!} v_{BS}(T, \hat{S}_n; \hat{\sigma}_n),$$

where $\hat{S}_n$ and $\hat{\sigma}_n$ are given by (16.38) with $t = T$. Note that here we used the probability mass function of the Poisson random variable, $\mathbb{P}(N(T) = n) = e^{-\lambda T} \frac{(\lambda T)^n}{n!}$.

16.2.3 The Variance Gamma Model

The *variance gamma* (VG) process is a three-parameter generalization of the Brownian motion model for the dynamics of the logarithm of the asset price. The VG process is obtained by evaluating the Brownian motion with drift at a random time given by a gamma process. The resulting process is a pure jump process.

Let $\{B(t; \theta, \sigma) \equiv W^{(\theta,\sigma)}(t) = \theta t + \sigma W(t)\}_{t \geqslant 0}$ denote a scaled Brownian motion with drift rate θ and scale parameter σ, as defined in (10.23) of Section 10.3.1. Hence, this is a normal random variable,

$$B(t; \theta, \sigma) \sim Norm(\theta t, \sigma^2 t).$$

The gamma process $\{G(t) \equiv G(t; \mu, \upsilon)\}_{t \geqslant 0}$ with mean rate μ and variance rate υ is a random process starting at zero and with independent gamma distributed increments over nonoverlapping intervals of time. The increment $\Delta G = G(t+h) - G(t)$ over time interval $[t, t+h]$ with $0 \leqslant t < t+h$ has the gamma distribution with mean μh and variance υh. Its probability density is

$$f_{\Delta G}(g; \mu, \upsilon, h) = \left(\frac{\mu}{\upsilon}\right)^{\mu^2 h/\upsilon} \frac{g^{\mu^2 h/\upsilon - 1} \exp\left(-\frac{\mu}{\upsilon} g\right)}{\Gamma\left(\frac{\mu^2 h}{\upsilon}\right)}, \quad g > 0. \tag{16.39}$$

Here, $\Gamma(x)$ is the gamma function defined below Equation (16.24).

The VG process $\{X(t) \equiv X(t; \sigma, \upsilon, \theta)\}_{t \geqslant 0}$ is obtained by evaluating the Brownian motion at a random time given by the gamma process with mean rate 1 and variance rate υ,

$$X(t; \sigma, \upsilon, \theta) := B(G(t; 1, \upsilon); \theta, \sigma) = \theta G(t; 1, \upsilon) + \sigma W(G(t; 1, \upsilon)).$$

The process starts at zero: $X(0) = 0$. The PDF of the VG process at time t can be expressed by conditioning on the realization of the gamma time change G as a normal density function:

$$f_{X(t)|G(t)}(x|g) = \frac{1}{\sigma\sqrt{2\pi g}} \exp\left(-\frac{(x - \theta g)^2}{2\sigma^2 g}\right).$$

The unconditional probability density $f_{X(t)}$ of $X(t)$ may then be obtained by employing the density (16.39), with $h = t$, for the time change g and integrating out g. This gives us the PDF in the following integral form:

$$f_{X(t)}(x) = \int_0^\infty f_{X(t)|G(t)}(x|g) f_{G(t)}(g)\,\mathrm{d}g$$
$$= \int_0^\infty \frac{1}{\sigma\sqrt{2\pi g}} \exp\left(-\frac{(x-\theta g)^2}{2\sigma^2 g}\right) \frac{g^{\frac{t}{v}-1}\exp(-\frac{g}{v})}{v^{\frac{t}{v}}\Gamma(\frac{t}{v})}\,\mathrm{d}g\,. \tag{16.40}$$

The new specification for the stock price dynamics is obtained by replacing the role of Brownian motion in the original Black–Scholes geometric Brownian motion model by the variance gamma process. The stock price process is given by the geometric VG law with parameters σ, v, and θ. The risk-neutral process for the asset price is given by

$$S(t) := S_0 \exp((r-q-\omega)t + X(t;\sigma,v,\theta)), \tag{16.41}$$

where r and q have the usual meaning, and the constant $\omega = \ln(1-\theta v - \sigma^2 v/2)/v$ is chosen so that the discounted asset price process $\{\mathrm{e}^{-(r-q)t}S(t)\}_{t\geqslant 0}$ is a martingale.

The price of a European-style derivative, $v(T,S)$ with spot $S_0 = S$, time to maturity T, and payoff Λ, is given by the usual risk-neutral pricing formula, $v(T,S) = \mathrm{e}^{-rT}\widetilde{\mathrm{E}}_{0,S}\left[\Lambda(S(T))\right]$, where the expectation is taken under the risk-neutral measure. The evaluation of the derivative price proceeds by conditioning on the random time change $G(T)$, which is independent of the Brownian motion. Conditional on the value of $G(T)$, the VG process $X(T)$ is normally distributed. Thus, the asset price $S(t)$ conditional on time $G(t) = g$ is log-normal:

$$S(t) = S_0\mathrm{e}^{(r-q-\omega)t+\theta g+\sigma W(g)} = \hat{S}_g\mathrm{e}^{(r-q-\sigma^2/2)g+\sigma W(g)},$$

where $\hat{S}_g(t) := S_0\mathrm{e}^{(r-q-\omega)t-(r-q-\sigma^2/2)g}$. Thus, the option value with the asset price S at the expiry time g is given by the Black–Scholes formula. Let v_{BS} denote the Black–Scholes derivative pricing function. The derivative price with expiry T under the VG risk-neutral dynamics, denoted v_{VG}, is then obtained by integrating this conditional Black–Scholes price with respect to the gamma time change:

$$v_{\mathrm{VG}}(T,S) = \mathrm{E}\big[v_{\mathrm{BS}}(G(T), \hat{S}_{G(T)}(T))\big] = \int_0^\infty v_{\mathrm{BS}}(g, \hat{S}_g(T)) f_{G(T)}(g)\,\mathrm{d}g.$$

The latter integral can be evaluated numerically.

16.3 Exercises

Exercise 16.1. Let the process $\{F(t)\}_{t\geqslant 0}$ solve the SDE

$$\mathrm{d}F(t) = \alpha(F(t))^{\beta+1}\,\mathrm{d}W(t).$$

Show that the process $\{X(t)\}_{t\geqslant 0}$ defined by $X(t) := (F(t))^{-2\beta}/(\alpha^2\beta^2)$ is an SQB process with index $\mu = \frac{1}{2\beta}$, i.e., it satisfies the SDE (16.14).

Exercise 16.2. Suppose that $p^{(0)}(t; S_0, S)$ solves the PDE

$$\frac{\partial p^{(0)}}{\partial t} = \frac{1}{2}\frac{\partial^2}{\partial S^2}\left(\sigma^2(S)\, p^{(0)}\right)$$

with $\sigma(S) = \alpha S^{\beta+1}$. Show that the function $p^{(\nu)}$ defined by

$$p^{(\nu)}(t; S_0, S) = \mathrm{e}^{-\nu t} p^{(0)}(\tau_t; S_0, \mathrm{e}^{-\nu t} S)$$

with $\tau_t = \frac{1}{2\nu\beta}\left(\mathrm{e}^{2\nu\beta t} - 1\right)$ and $\nu \neq 0$ satisfies the PDE

$$\frac{\partial p^{(\nu)}}{\partial t} = \frac{1}{2}\frac{\partial^2}{\partial S^2}\left(\sigma^2(S)\, p^{(\nu)}\right) - \frac{\partial}{\partial S}\left(\nu S\, p^{(\nu)}\right).$$

Exercise 16.3. Let $\{F(t)\}_{t\geqslant 0}$ be a forward price process solving the SDE

$$\mathrm{d}F(t) = \sigma(F(t))\, \mathrm{d}W(t)$$

with some nonlinear $\sigma(F)$. Consider a transformed process $S(t) = \mathrm{e}^{rt}F(\tau_t)$ with some strictly increasing smooth function τ_t such that

$$\tau_0 = 0 \ \text{ and } \ \tau_t'|_{t=0} = 1. \tag{16.42}$$

The function τ is a time change. The transition PDF $p^{(r)}$ for the process $\{S(t)\}_{t\geqslant 0}$ is given by

$$p^{(r)}(t; S_0, S) = \mathrm{e}^{-rt} p^{(0)}(\tau_t; S_0, \mathrm{e}^{-rt} S),$$

where $p^{(0)}$ is a transition PDF of the forward price F. Show the following.

(a) The PDF $p^{(r)}$ satisfies the forward Kolmogorov PDE

$$\frac{\partial p}{\partial t} = \frac{1}{2}\frac{\partial^2}{\partial S^2}\left(\tilde{\sigma}^2(t, S)\, p\right) - r\frac{\partial}{\partial S}\left(Sp\right),$$

where $\tilde{\sigma}(t, S) = \mathrm{e}^{rt}\sqrt{\tau'(t)}\sigma(\mathrm{e}^{-rt}S)$ is a time- and state-dependent function. The process $\{S(t)\}_{t\geqslant 0}$ hence solves the time-inhomogeneous SDE

$$\mathrm{d}S(t) = rS(t)\, \mathrm{d}t + \tilde{\sigma}(t, S(t))\, \mathrm{d}W(t).$$

(b) Show that $\frac{\partial \tilde{\sigma}}{\partial t} = 0$ holds (hence $\tilde{\sigma}$ is time homogeneous) only if τ_t solves the ordinary differential equation (ODE)

$$\tau''(t) + 2r\left(1 - \frac{\mathrm{e}^{-rt} S\sigma'(\mathrm{e}^{-rt}S)}{\sigma(\mathrm{e}^{-rt}S)}\right)\tau'(t) = 0 \tag{16.43}$$

subject to the initial conditions (16.42).

(c) The solution to (16.43) is independent of S only if the function $\sigma(F)$ satisfies

$$\left(\frac{F\sigma'(F)}{\sigma(F)}\right)' = 0.$$

The only solution to this ODE is a power function: $\sigma(F) = \alpha F^{\beta+1}$, where α and β are constants. The respective solution to (16.43) is $\tau_t = \frac{\mathrm{e}^{2r\beta t} - 1}{2r\beta}$.

Thus, the CEV process is the only model where the above scale and time change give a time-homogeneous diffusion with linear drift. Moreover, the transformation in this case is volatility preserving, i.e., $\tilde{\sigma}(S,t) = \sigma(S)$.

Exercise 16.4. Prove the following property of the noncentral chi-square PDF:

$$\int_a^\infty f(x; \upsilon, y)\,\mathrm{d}y = 1 - Q(x, \upsilon - 2, a).$$

Hint: Use the following identities:

$$(a)\ \ I_\nu(z) = \left(\frac{z}{2}\right)^\nu \sum_{j=0}^\infty \frac{(z^2/4)^j}{j!\Gamma(\nu + j + 1)}, \quad (b)\ \ \int_a^\infty g(n, y)\,\mathrm{d}y = \sum_{j=1}^n g(j, a)$$

where $g(m, x) = \frac{\mathrm{e}^{-x}\, x^{m-1}}{\Gamma(m)}$.

Exercise 16.5. Prove that the compensated Poisson process, $\{N_\lambda(t) - \lambda t\}_{t\geqslant 0}$, is a martingale with respect to its natural filtration.

Exercise 16.6. Consider the exponential process $\{\exp(at + N_\lambda(t))\}_{t\geqslant 0}$. For what value of a is this process a martingale with respect to its natural filtration?

Exercise 16.7. Apply the following integral representation of the modified Bessel function of the second kind (of order ν), denoted K_ν, to find a closed-form nonintegral expression for the transition PDF of the VG process:

$$K_\nu(z) = \frac{1}{2}\left(\frac{z}{2}\right)^\nu \int_0^\infty \exp\left(-t - \frac{z^2}{4t}\right) \frac{\mathrm{d}t}{t^{\nu+1}}.$$

Exercise 16.8. Consider the Heston model (16.30)–(16.31) with uncorrelated Brownian motions W_1 and W_2.

(a) Show that the instantaneous variance $\mathrm{v}(t)$ in (16.31) can be expressed in terms of the SQB process by means of a suitable scale and time change.

(b) Find the transition PDF of the variance process $\mathrm{v}(t)$.

(c) Write the price of a standard European option with payoff Λ as an integral over the Black–Scholes prices.

Part IV

Computational Techniques

Chapter 17

Introduction to Monte Carlo and Simulation Methods

17.1 Introduction

The Monte Carlo method is a numerical technique that allows scientists to analyze various natural phenomena and compute complicated quantities by means of repeated generation of random numbers. For example, if you make several thousand tosses of a fair (i.e., balanced) coin, then you may notice that the long-run ratio of a count of heads to the total number of tosses is approaching one half. If the limiting ratio is not close to one half, then you may conclude that the coin is not balanced. Similarly, researchers can compute other more complicated quantities, although in reality nobody tosses coins and throws dice. Instead, computer simulations are used. First, a computer experiment that produces some quantitative information about a random phenomenon of interest needs to be designed. After that, a computer is used to perform independent repeated random simulations and then to calculate averages of results. Such averages are used to approximate the quantity of interest. In particular, the two fundamental theorems, namely, the Law of Large Numbers (LLN) and the Central Limit Theorem (CLT), allow researchers to construct a valid approximation of the quantity of interest and estimate the approximation error.

The Monte Carlo method was coined in the 1940s by John von Neumann, Stanislaw Ulam, and Nicholas Metropolis, while they were working on nuclear weapons project at the Los Alamos National Laboratory. It was named after a famous casino in Monte Carlo, where Ulam's uncle often gambled away his money. The range of problems that can be analyzed by Monte Carlo methods is vast. Many quantities such as the probability of default of a financial company, distribution of galaxies in the universe, and characteristics of a nuclear reactor can be computed. Monte Carlo methods are especially useful for simulating complex systems with coupled degrees of freedom and with significant uncertainty in inputs, such as the calculation of risk in business. Monte Carlo methods are used to evaluate multidimensional integrals and to numerically solve large systems of equations. The success and popularity of Monte Carlo is partly explained by the enormous progress of computers that are so powerful and inexpensive these days.

The Monte Carlo method is a very popular computational tool in financial economics. Many typical problems such as the optimal allocation of financial assets, pricing of derivative contracts, and evaluation of business risk can be numerically solved by means of repeated generation of possible market scenarios. For example, we can calculate the no-arbitrage value of an option written on the asset by repeatedly generating sample paths of the underlying asset price process. The simplest possible model is the binomial price model. At each time step, the price may go up or down by a constant factor. Every possible scenario can be represented by a random walk on the binomial tree. An estimate of the option price is computed by averaging payoff values calculated for independent realizations of such a random walk.

17.1.1 The "Hit-or-Miss" Method

A typical application of the Monte Carlo method is the computation of areas and volumes of objects of complex shape and geometry. This type of Monte Carlo computation is based on two ideas: *geometric probability* and the *frequentist definition of probability.* The latter tells us that the likelihood of some event E can be calculated as a long-run ratio of the number of successful trials to the total number of trials. By a success we mean here a trial where the event E occurs. Count the number of times, $N_E(n)$, the event E occurs in the first n trials that are performed independently and under identical conditions. The probability $\mathbb{P}(E)$ is approximated as

$$\mathbb{P}(E) \approx \frac{N_E(n)}{n}.$$

Now let us link the notion of probability to the area of a figure (or the volume of a multidimensional domain in the general case). Consider a planar figure D contained completely within a unit square $S = [0,1]^2$. Select a point at random uniformly on the square. This can be done by sampling two independent identically distributed (i.i.d.) Cartesian coordinates uniformly in the interval $[0,1]$. According to the principle of *geometric probability*, the likelihood that such a randomly chosen point belongs to D is equal to the ratio of the area of D, denoted $|D|$, to the area of the unit square (which equals one). Choose at random n points independently and uniformly on the square. Let $N_D(n)$ points fall on the figure D. On the one hand, the ratio $N_D(n)/n$ is approximately equal to the probability that a random point chosen uniformly on the square lies on D. On the other hand, this probability equals the ratio of areas of D and S. Thus, we obtain the following approximation of the area of D:

$$|D| \approx \frac{N_D(n)}{n} \cdot |S|.$$

17.1.2 The Law of Large Numbers

The Monte Carlo method is based on two fundamental laws, namely, the Law of Large Numbers (LLN) and the Central Limit Theorem (CLT). There are several versions of the LLN. Here are some of them.

Theorem 17.1 (Borel's LLN). *Suppose that an experiment with an uncertain outcome is repeated a large number of times independently and under identical conditions. Then, for any event E we have*

$$\mathbb{P}(E) = \lim_{n\to\infty} \frac{N_E(n)}{n},$$

where $N_E(n)$ is the number of times the event E occurs in the first n trials.

Borel's LLN provides us with a theoretical foundation for the frequintist interpretation of probability. On the other hand, the probability of an event E can be expressed as a mathematical expectation of an indicator function of E: $\mathbb{P}(E) = \mathrm{E}[\mathbb{I}_E]$. Thus, Borel's LLN is a special case of a general LLN provided just below.

Let $\{X_k\}_{k\geqslant 1}$ be a sequence of i.i.d. random variables with common finite mean μ_X, which are all defined on some probability space $(\Omega, \mathcal{F}, \mathbb{P})$. Let $\overline{X}_n = \frac{1}{n}\sum_{k=1}^n X_k$ denote the arithmetic average of the first n variables.

Theorem 17.2 (Chebyshev's Weak LLN). *It is true that $\overline{X}_n \xrightarrow{p} \mu_X$, as $n \to \infty$. That is, $\forall \varepsilon > 0\ \mathbb{P}\left(|\overline{X}_n - \mu_X| > \varepsilon\right) \to 0$, as $n \to \infty$.*

Theorem 17.3 (Kolmogorov's Strong LLN). *It is true that $\overline{X}_n \xrightarrow{a.s.} \mu_X$, as $n \to \infty$. That is, $\mathbb{P}\big(\{\omega \mid \overline{X}_n(\omega) \to \mu_X \text{ as } n \to \infty\}\big) = 1$.*

The LLN provides us with a recipe for estimation of quantities of interest. For a given unknown Q, we first construct a random variable X so that $Q = \mu_X = \mathrm{E}[X]$, i.e., we find the *probabilistic representation* of the quantity Q. The random variable X is called an *estimator* of the quantity Q. It is an *unbiased* estimator of Q, meaning that $\mathrm{E}[X] = Q$. Let $X_1, X_2, \ldots, X_n$ be i.i.d. variates all having the same probability distribution as that of X. Following the LLN, construct a *sample estimator* of $Q, \overline{X}_n = \frac{1}{n}\sum_{k=1}^{n} X_k$, which converges almost surely to Q, as $n \to \infty$. That is, the quantity Q can be approximated by an average of n independent sample values:

$$Q \approx \bar{x}_n = \frac{1}{n}\sum_{k=1}^{n} x_k,$$

where $x_k = X_k(\omega)$, $1 \leqslant k \leqslant n$, for some outcome $\omega \in \Omega$. To obtain independent sample values $\{x_k\}_{k \geqslant 1}$, it suffices to construct a sequence of statistically independent numbers (or almost statistically independent pseudorandom numbers) sampled from the target distribution of X as follows. First, we sample from the uniform distribution *Unif*$(0,1)$. After that, the independent sample values (also called realizations or draws) x_k, $1 \leqslant k \leqslant n$, can be generated from a sequence of uniform random numbers by using a special transformation algorithm. One example of such algorithms is the inverse cumulative distribution function (CDF) method, which is based on the representation $X \stackrel{d}{=} F_X^{-1}(U)$, where F_X^{-1} is the (generalized) inverse of the CDF of X and $U \sim$ *Unif*$(0,1)$.

In what follows, we will distinguish the *sample mean estimator* $\overline{X}_n$, which is a random quantity equal to an average of n i.i.d. variates, and a *sample mean estimate* $\bar{x}_n$, which is a nonrandom quantity equal to an average of n statistically independent realizations of the estimator X. While the former is important for the theoretical analysis (e.g., to construct a confidence interval), the latter is used in practice to approximate Q.

17.1.3 Approximation Error and Confidence Interval

The next goal is to estimate the approximation error. The Central Limit Theorem (CLT) provides a solution to this problem.

Theorem 17.4 (The CLT). *Consider a sequence $\{X_k\}_{k \geqslant 1}$ of i.i.d. variates with finite common variance σ_X^2 and expected value μ_X. Then,*

$$\frac{\overline{X}_n - \mu_X}{\sigma_X/\sqrt{n}} \stackrel{d}{\to} Norm(0,1), \quad \text{as } n \to \infty.$$

That is, $\mathbb{P}\left(\frac{\overline{X}_n - \mu_X}{\sigma_X/\sqrt{n}} \leqslant z\right) \to \mathcal{N}(z)$, *as* $n \to \infty$, *for all* $z \in \mathbb{R}$.

With the help of the CLT, we can construct a *confidence interval* for the mathematical expectation μ_X. We have that

$$\mathbb{P}\left(\left|\frac{\overline{X}_n - \mu_X}{\sigma_X/\sqrt{n}}\right| \leqslant z\right) \to \mathbb{P}(|Z| \leqslant z) = 2\mathcal{N}(z) - 1 \text{ for all } z \in \mathbb{R}.$$

We wish to make this probability as close to one as possible. Let us fix a confidence level $1 - \alpha \in (0,1)$ with $\alpha \ll 1$. Solve the equation

$$2\mathcal{N}(z) - 1 = 1 - \alpha \iff \mathcal{N}(z) = 1 - \frac{\alpha}{2}$$

for z. Since $\mathcal{N}$ is a strictly monotone function of z, the solution is unique. It is a so-called

TABLE 17.1: Commonly used normal quantiles $z_{\alpha/2}$ that solve $1 - \mathcal{N}(z_{\alpha/2}) = \frac{\alpha}{2}$.

Confidence level (%)	α	$z_{\alpha/2}$
90	0.1	1.645
95	0.05	1.960
95.46	0.0454	2.000
99	0.01	2.576
99.74	0.0026	3.000
99.9	0.001	3.29

$(1 - \alpha/2)$-quantile of $Norm(0, 1)$ denoted by $z_{\alpha/2}$. Table 17.1 contains commonly used normal quantiles.

For large values of n, the distribution of $\overline{X}_n$ is approximately normal. By the CLT we obtain

$$|\overline{X}_n - \mu_X| \leqslant \frac{z_{\alpha/2}\,\sigma_X}{\sqrt{n}} \text{ with probability} \approx 1 - \alpha.$$

Hence, $\mathbb{P}\left((\overline{X}_n - \frac{z_{\alpha/2}\,\sigma_X}{\sqrt{n}}, \overline{X}_n + \frac{z_{\alpha/2}\,\sigma_X}{\sqrt{n}}) \ni \mu_X\right) \approx 1 - \alpha$. Replacing the average $\overline{X}_n$ by its sample value $\bar{x}_n$ gives us a confidence interval for μ_X:

$$\left(\bar{x}_n - \frac{z_{\alpha/2}\,\sigma_X}{\sqrt{n}}, \bar{x}_n + \frac{z_{\alpha/2}\,\sigma_X}{\sqrt{n}}\right) \ni \mu_X \text{ with the confidence level of } (1 - \alpha).$$

Typically, the variance σ_X^2 is unknown. However, it can be approximated by the *sample variance*:

$$\sigma_X^2 \approx s_n^2 := \frac{1}{n}\sum_{k=1}^{n}(x_k - \bar{x}_n)^2 = \frac{1}{n}\sum_{k=1}^{n} x_k^2 - \bar{x}_n^2.$$

As is seen from the formula of the confidence interval, the relative error $\frac{z_{\alpha/2}\,\sigma_X}{\sqrt{n}\mu_X}$ is of order $\mathcal{O}(n^{-0.5})$, as $n \to \infty$. This fact points out the main drawback of the Monte Carlo method—a slow rate of convergence. For example, to decrease the error by a factor of 10, the number of sample values needs to be increased to 100 times. The quantity $\frac{s_n/\sqrt{n}}{\bar{x}_n}$ is often referred to as an *accuracy measure* for the sample estimate $\bar{x}_n$.

17.1.4 Parallel Monte Carlo Methods

One of the main advantages of the Monte Carlo method is the ease of its parallelization. In the simplest case, independent CPUs can calculate partial expectations and then a head CPU computes the final average and the confidence interval. Consider a computational cluster, where CPUs are numbered from 1 through ℓ. Let CPU $\# i$ generate n_i independent draws, $\{x_1^i, \ldots, x_{n_i}^i\}$. The partial averages

$$\bar{x}^{(i)} = \frac{x_1^i + x_2^i + \cdots + x_{n_i}^i}{n_i} \text{ and } \bar{y}^{(i)} = \frac{(x_1^i)^2 + (x_2^i)^2 + \cdots + (x_{n_i}^i)^2}{n_i}$$

are calculated and then their values are sent to the head CPU. The total number of draws is $n = n_1 + n_2 + \cdots + n_k$. To construct a confidence interval for the mean value, we only need to compute the sample mean $\bar{x}_n$ and the sample variance s_n^2 as follows:

$$\bar{x}_n = \frac{1}{n}\sum_{i=1}^{\ell} n_i \bar{x}^{(i)} = \frac{1}{n}\sum_{i=1}^{\ell}\sum_{j=1}^{n_i} x_j^i$$

$$s_n^2 = \frac{1}{n}\sum_{i=1}^{\ell} n_i \bar{y}^{(i)} - \bar{x}_n^2 = \frac{1}{n}\sum_{i=1}^{\ell}\sum_{j=1}^{n_i} (x_j^i)^2 - \bar{x}_n^2.$$

Note that the numbers n_i, $i = 1, 2, \ldots, \ell$, can be chosen so that all CPUs will finish their jobs almost simultaneously. If the CPUs we deal with have a similar performance, then we can simply set $n_1 = n_2 = \cdots = n_\ell$.

17.1.5 One Monte Carlo Application: Numerical Integration

Suppose that we wish to evaluate a definite integral of an integrable function $g\colon \mathbb{R} \to \mathbb{R}$ on a finite interval $[a, b]$: $I = \int_a^b g(x)\,\mathrm{d}x$. Let us consider several Monte Carlo methods of approximating I.

The *"hit-or-miss" method* is based on the geometric interpretation of a definite integral. Suppose that the function g is nonnegative and bounded: $0 \leqslant g \leqslant M$. If a point with Cartesian coordinates (X, Y) is chosen uniformly on the rectangle $R = [a, b] \times [0, M]$, then the probability of the event $\{Y \leqslant g(X)\}$ (i.e., the event that (X, Y) belongs to the domain D bounded by the graph of g and the lines $x = a$, $x = b$, and $y = 0$) is equal to the ratio of two areas, $\frac{|D|}{|R|}$. Since $|D| = I$ and $|R| = (b - a)M$, we obtain the following approximation:

$$I \approx (b - a)M\frac{N_D(n)}{n},$$

where n statistically independent random points (x_k, y_k), $1 \leqslant k \leqslant n$, are sampled uniformly on the rectangle R, and $N_D(n)/n$ is the fraction of points belonging to D. This result can be viewed as an application of the LLN. Let us introduce an indicator function $\mathbb{I}_D(X, Y)$, which equals 1 if $(X, Y) \in D$ and equals 0 otherwise. Then, the expected value of this indicator is $\mathrm{E}\big[\mathbb{I}_D(X, Y)\big] = \mathbb{P}(Y \leqslant g(X)) = \frac{|D|}{|R|}$. Therefore, $\mathrm{E}\big[|R| \cdot \mathbb{I}_D(X, Y)\big] = |D| = I$ and applying the LLN gives

$$\frac{1}{n}\sum_{k=1}^{n} |R|\,\mathbb{I}_D(x_k, y_k) = (b - a)M\frac{N_D(n)}{n} \to I, \text{ as } n \to \infty.$$

For the *sample mean method*, the representation of the integral I in the form of a mathematical expectation w.r.t. a uniform probability distribution is used:

$$I = (b - a)\int_a^b g(x)\frac{1}{b - a}\,\mathrm{d}x = \mathrm{E}\big[(b - a)g(X)\big],$$

where $X \sim \mathit{Unif}(a, b)$. Therefore, by the LLN we have $I \approx \frac{1}{n}\sum\limits_{k=1}^{n}(b-a)g(x_k)$, where $\{x_k\}_{k\geqslant 1}$ are independent draws from $\mathit{Unif}(a, b)$. In comparison with the "hit-or-miss" method, g is only required to be integrable on the interval (a, b).

The *weighted method* generalizes the previous approach. Suppose that there exists a probability density function (PDF) f whose support is (a, b), where (a, b) can be a finite, semi-infinite, or infinite interval. Then, the integral I can be represented in the form of a mathematical expectation w.r.t. the PDF f:

$$I = \int_a^b \frac{g(x)}{f(x)}f(x)\,\mathrm{d}x = \mathrm{E}\left[\frac{g(X)}{f(X)}\right], \text{ where } X \sim f,$$

provided that the ratio $\frac{g(x)}{f(x)}$ is defined for all $x \in (a,b)$ and is an integrable function of x. The random variable $\frac{g(X)}{f(X)}$ is the estimator of I. Thus, we approximate $I \approx \frac{1}{n}\sum_{k=1}^{n}\frac{g(x_k)}{f(x_k)}$ where $\{x_k\}_{k\geqslant 1}$ are independent draws from the PDF f.

Clearly, there are many PDFs that can be used in the weighted method. The *importance sampling principle*, which is discussed in Section 17.5, explains how to chose f. The optimal PDF f that minimizes the variance of the random estimator $\frac{g(X)}{f(X)}$, where $X \sim f$, i.e., the PDF that solves the minimization problem

$$\mathrm{Var}\left(\frac{g(X)}{f(X)}\right) \to \min_f,$$

is a function proportional to $|g|$:

$$f(x) = \frac{|g(x)|}{\int_a^b |g(t)|\,\mathrm{d}t}, \qquad x \in (a,b).$$

17.2 Generation of Uniformly Distributed Random Numbers

A central part of every stochastic simulation algorithm is a generator of random numbers, which produces a sequence of statistically independent samples (or draws) from a given distribution. As we demonstrate in the following sections, the sampling from a nonuniform probability distribution can be reduced to the sampling from the uniform distribution on $(0,1)$, denoted *Unif*$(0,1)$, by applying certain procedures such as transformation methods and acceptance-rejection techniques. Our main priority is to have available a reliable method of sampling from *Unif*$(0,1)$. Therefore, we first concentrate on methods of generating statistically independent (pseudo-)random numbers uniformly distributed on $(0,1)$.

The process of obtaining truly random numbers can be quite complicated. There exist different types of physical generators of random numbers. The simplest ones are balanced coins and dice, and even a playing roulette. More advanced generators are based on the use of physical phenomena such as thermal noise. However, there are common drawbacks of such hardware generators such as their slow speed, lack of portability, possible non-uniformness of numbers generated, and impossibility of reproducing the same sequence of draws. Algorithmic (software) generators of random numbers solve most of these issues although they create new ones. Numbers generated by such algorithms only mimic truly random numbers, hence the generated draws are called *pseudo-random numbers* (PRNs). However, the statistical properties of software generators are put to scrutiny so we can trust the numbers obtained.

17.2.1 Uniform Probability Distributions

As is well known, the distribution of a continuous random variable X can be characterized by its PDF, denoted f_X, or a cumulative distribution function (CDF), denoted F_X. For a continuous random variable U uniformly distributed on $(0,1)$,

$$f_U(x) = \mathbb{I}_{(0,1)}(x) = \begin{cases} 1 & \text{if } 0 < x < 1, \\ 0 & \text{otherwise,} \end{cases} \quad F_U(x) = \int_{-\infty}^{x} = \begin{cases} 0 & x \leqslant 0, \\ x & 0 < x < 1, \\ 1 & 1 \leqslant x. \end{cases}$$

The mathematical expectation and variance of U are, respectively,

$$\mathrm{E}[U] = \int_{-\infty}^{\infty} x\, f_U(x)\, \mathrm{d}x = \int_0^1 x\, \mathrm{d}x = \frac{1}{2}, \quad \mathrm{Var}(U) = \mathrm{E}[U^2] - (\mathrm{E}[U])^2 = \frac{1}{3} - \frac{1}{4} = \frac{1}{12}.$$

The continuous uniform distribution on (a, b) with $a < b$, denoted *Unif*(a, b), reduces to the case with the interval $(0, 1)$ as follows. Let $X \sim$ *Unif*(a, b) and $U \sim$ *Unif*$(0, 1)$. Then, we have

$$X \stackrel{d}{=} a + (b-a)U, \quad F_X(x) = F_U\left(\frac{x-a}{b-a}\right), \quad f_X(x) = \frac{1}{b-a} f_U\left(\frac{x-a}{b-a}\right) = \frac{1}{b-a}\mathbb{I}_{(a,b)}(x).$$

Now consider the multidimensional case. Suppose that a point $\mathbf{X}$ is chosen at random in a domain $D \subset \mathbb{R}^m$ with a finite volume $|D|$ so that $\mathbf{X}$ is equally likely to lie anywhere in D. Then, the random vector $\mathbf{X} = [X_1, X_2, \ldots, X_m]^\top$ is said to have a multivariate uniform distribution in D; its multivariate PDF is

$$f_{\mathbf{X}}(\mathbf{x}) = \frac{1}{|D|}\,\mathbb{I}_D(\mathbf{x}) = \begin{cases} \frac{1}{|D|} & \mathbf{x} \in D, \\ 0 & \text{otherwise}, \end{cases} \qquad \mathbf{x} = \begin{bmatrix} x_1 & x_2 & \cdots & x_m \end{bmatrix}^\top \in D.$$

In the case of a hyperparallelepiped $D = (a_1, b_1) \times (a_2, b_2) \times \cdots \times (a_m, b_m)$ with $a_i < b_i$, $1 \leqslant i \leqslant m$, we have

$$f_{\mathbf{X}}(\mathbf{x}) = \prod_{i=1}^m f_{X_i}(x_i) = \prod_{i=1}^m \frac{1}{b_i - a_i}\,\mathbb{I}_{(a_i, b_i)}(x_i).$$

Since the multivariate PDF $f_{\mathbf{X}}$ is a product of m univariate PDFs $f_{X_i} = \frac{1}{b_i - a_i}\,\mathbb{I}_{(a_i,b_i)}$, the entries of the vector $\mathbf{X}$ are independent uniformly distributed random variables. Therefore, the simulation of a vector uniformly distributed in a hyperparallelepiped reduces to sampling from *Unif*$(0, 1)$.

17.2.2 Linear Congruential Generator

Many algorithms for generating pseudo-random numbers (PRNs for short) uniformly distributed in $[0, 1]$ have the form of an iterative rule, $u_{t+1} = F(u_t)$, $t = 0, 1, \ldots$, where both the range and the domain of the function F are $[0, 1]$ and the initial seed $u_0 \in [0, 1]$ is given. Suppose that a sequence of PRNs $\{u_t\}_{t\geqslant 0}$ is generated by such a rule. Combine the numbers in pairs to obtain points $(u_t, u_{t-1}) \in [0, 1]^2$, $t = 1, 3, 5, \ldots$. On the one hand, these points are situated on the curve $y = F(x)$. On the other hand, they have to be uniformly distributed in the unit square $[0, 1]^2$ so the points will fill out the square without leaving gaps. Therefore, the plot of F should provide a sufficiently dense filling of the square. One example of a function that has such a property is the mapping $y = \{Mx\}$, where $\{x\}$ denotes the fractional part of x, i.e., $\{x\} = x - \lfloor x \rfloor$, where $\lfloor \cdot \rfloor$ is the floor function, and M is a large positive number called a *multiplier*. The plot of $y = \{Mx\}$ consists of parallel segments and the distance between them goes to zero as $M \to \infty$ (see Figure 17.2).

Proposition 17.5 (Voitishek and Mikhaĭlov (2006)). *Consider the transformation $y = \{Mx + a\}$ where $\{x\}$ denotes the fractional part of x, and $M \in \mathbb{Z}$ and $a \in \mathbb{R}$ are positive constants.*

1. *If $U \sim$ Unif$(0, 1)$, then $\{MU + a\} \sim$ Unif$(0, 1)$.*
2. *Let $U_0 \sim$ Unif$(0, 1)$ and the sequence $\{U_k\}_{k\geqslant 1}$ be generated from the rule $U_{k+1} = \{MU_k\}$. Then, $U_k \sim$ Unif$(0, 1)$ and $\mathrm{Corr}(U_0, U_k) = \mathrm{Corr}(U_n, U_{n+k}) = M^{-k}$ for all $n \geqslant 0$ and $k \geqslant 0$.*

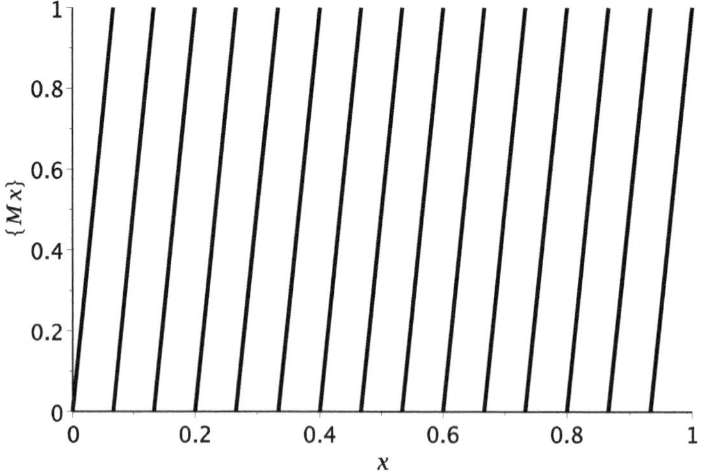

FIGURE 17.2: The plot of $y = \{Mx\}$ with $M = 15$.

Proof. First, we prove the fact that $X := \{MU\} \sim$ *Unif*$(0,1)$. By the definition of the fractional part function, we have $X \in [0,1)$. Moreover, $X = 0$ iff MU is an integer what happens with probability zero. Hence $\mathbb{P}(X \in (0,1)) = 1$. For $x \in (0,1)$, we have

$$\mathbb{P}(X \leqslant x) = \sum_{k=0}^{M-1} \mathbb{P}(k \leqslant MU \leqslant k + x) = \sum_{k=0}^{M-1} \mathbb{P}\left(\frac{k}{M} \leqslant U \leqslant \frac{k}{M} + \frac{x}{M}\right) = \sum_{k=0}^{M-1} \frac{x}{M} = x.$$

Here, we use the fact that $0 \leqslant MU \leqslant M$ and $\{z\} \in [0,x]$ iff $z \in \cup_{k\in\mathbb{Z}}[k, k+x]$. Thus, the CDF of X is $F_X(x) = x$ for $x \in (0,1)$. Therefore, $X \sim$ *Unif*$(0,1)$.

Second, let us prove that $\{MU + a\} \sim$ *Unif*$(0,1)$. Note that

$$\{MU + a\} = \{\{MU\} + \{a\}\} \stackrel{d}{=} \{U + b\},$$

where $b = \{a\} \in [0,1)$. Show that $\mathbb{P}(\{U+b\} \leqslant x) = x$ for $x \in (0,1)$. Consider two cases.

$(x \leqslant b)$: Since $\{U + b\} \leqslant x$ iff $1 \leqslant U + b \leqslant 1 + x$, we have

$$\mathbb{P}(\{U+b\} \leqslant x) = \mathbb{P}(\underbrace{1-b}_{\in(0,1)} \leqslant U \leqslant \underbrace{1+x-b}_{\in(0,1)}) = (1+x-b) - (1-b) = x.$$

$(b < x)$: Since $\{U + b\} \geqslant x$ iff $x \leqslant U + b \leqslant 1$, we have

$$\mathbb{P}(\{U+b\} \geqslant x) = \mathbb{P}(\underbrace{x-b}_{\in(0,1)} \leqslant U \leqslant \underbrace{1-b}_{\in(0,1)}) = (1-b) - (x-b) = 1 - x.$$

Therefore, $\mathbb{P}(\{U+b\} \leqslant x) = 1 - \mathbb{P}(\{U+b\} \geqslant x) = 1 - (1-x) = x$.

Finally, let us prove the last assertion. By induction, all $U_k \sim$ *Unif*$(0,1)$, $k = 0,1,2,\ldots$. Clearly, for any $k \geqslant 1$, the pairs (U_n, U_{n+k}), $n \geqslant 0$, all have the same distribution. Hence, we only need to prove that $\mathrm{Corr}(U_0, U_k) = M^{-k}$ for all $k \geqslant 0$. Denote $r_k = \mathrm{Corr}(U_0, U_k)$. Let us show that $r_k = \frac{1}{M} r_{k-1}$ for $k \geqslant 1$. Since $r_0 = \mathrm{Corr}(U_0, U_0) = 1$, the assertion will be proved by induction. A uniform random variable $MU \sim$ *Unif*$(0, M)$ can be expressed as a sum of its fractional and integer parts: $MU = \lfloor MU \rfloor + \{MU\}$. Clearly, $\lfloor MU \rfloor \sim \mathit{Unif}\{0, 1, \ldots, M-1\}$ (since $\mathbb{P}(\lfloor MU \rfloor = k) = \mathbb{P}(MU \in [k, k+1)) = \frac{1}{M}$ for $0 \leqslant k \leqslant M-1$) and $\{MU\} \sim \mathit{Unif}(0,1)$ (as proved above). For any $k = 0, 1, \ldots, M-1$ and any $x \in (0,1)$, we have

$$\mathbb{P}(\lfloor MU \rfloor = k; \{MU\} \leqslant x) = \mathbb{P}(MU \in [k, k+x]) = \frac{x}{M} = \mathbb{P}(\lfloor MU \rfloor = k)\, \mathbb{P}(\{MU\} \leqslant x).$$

Therefore, the fractional and integer parts are independent random variables. It is also true that $\mathrm{E}[\, \lfloor MU \rfloor \,] = {}^{(M-1)}/_{2}$ and $\mathrm{Var}\left(\lfloor MU \rfloor \right) = {}^{(M^2-1)}/_{\sqrt{12}}$. Moreover, we have

$$\begin{aligned}
\frac{U - \mathrm{E}[U]}{\sqrt{\mathrm{Var}(U)}} &= \frac{U - {}^{1}/_{2}}{{}^{1}/_{\sqrt{12}}} = \frac{MU - {}^{M}/_{2}}{{}^{M}/_{\sqrt{12}}} \\
&= \frac{\lfloor MU \rfloor + \{MU\} - {}^{(M-1)}/_{2} - {}^{1}/_{2}}{{}^{M}/_{\sqrt{12}}} \\
&= \sqrt{\frac{M^2-1}{M^2}} \left(\frac{\lfloor MU \rfloor - {}^{(M-1)}/_{2}}{\sqrt{{}^{(M^2-1)}/_{12}}} \right) + \frac{1}{M} \left(\frac{\{MU\} - {}^{1}/_{2}}{{}^{1}/_{\sqrt{12}}} \right).
\end{aligned}$$

Now we are ready to calculate r_k:

$$r_k = \mathrm{Corr}(U_0, U_k) = \sqrt{\frac{M^2-1}{M^2}}\, \mathrm{Corr}(\lfloor MU_0 \rfloor, U_k) + \frac{1}{M}\, \mathrm{Corr}(\{MU_0\}, U_k).$$

We proved that $\lfloor MU_0 \rfloor$ and $\{MU_0\} = U_1$ are independent random variables. Moreover, since $U_k = \{MU_{k-1}\} = \{M\{MU_{k-2}\}\} = \cdots = g(U_1)$ for some $g\colon \mathbb{R} \to \mathbb{R}$, the random variable U_k as a function of U_1 is independent of $\lfloor MU_0 \rfloor$ as well. Thus, $\mathrm{Corr}(\lfloor MU_0 \rfloor, U_k) = 0$ and

$$\begin{aligned}
r_k = \mathrm{Corr}(U_0, U_k) &= \frac{1}{M}\, \mathrm{Corr}(\{MU_0\}, U_k) = \frac{1}{M}\, \mathrm{Corr}(U_1, U_k) \\
&= \frac{1}{M}\, \mathrm{Corr}(U_0, U_{k-1}) = \frac{1}{M} r_{k-1}.
\end{aligned}$$ □

According to Proposition 17.5, a sequence $\{U_t,\ t = 0, 1, \ldots\}$ of random numbers uniformly distributed in $(0,1)$ can be generated from one random number $U_0 \sim \mathit{Unif}(0,1)$ by using the rule $U_t = \{MU_{t-1}\}$ for $t \geqslant 1$. We shall call this method a *multiplicative method.* Although the numbers obtained are dependent, the correlation between any two members of the sequence is negligible and rapidly goes to zero as the distance between the numbers in the sequence increases. Thus, the transformation $y = \{Mx + a\}$ with a suitable choice of M and a can be used to generate pseudo-random numbers having good statistical properties such as uniformity and independence of samples. A drawback of the multiplicative method is that multidimensional tuples formed from the sequence $\{U_t\}_{t \geqslant 0}$ lie on a family of multidimensional hyperplanes in the unit hypercube. For example, the points $(U_0, U_1), (U_2, U_3), \ldots$ lie on a family of M parallel lines.

The *linear congruential generator* (LCG) of PRNs is one of the oldest and simplest methods. It was proposed by Lehmer (1951). This algorithm is based on the mapping $y = \{Mx + a\}$ but to avoid round-off errors the generating rule is written in a different form. As an example, let us consider the function $y = \{Mx\}$ and suppose that x is a ratio of two integers: $x = \frac{s}{m}$. Then,

$$\left\{M \frac{s}{m}\right\} = \frac{Ms}{m} - \left\lfloor \frac{Ms}{m} \right\rfloor = \frac{Ms - \lfloor \frac{Ms}{m} \rfloor m}{m} = \frac{Ms \bmod m}{m}.$$

The operation ($s \bmod m$) returns the remainder of an integer s after division by m:

$$s \bmod m := s - \lfloor s/m \rfloor \cdot m$$

This operation is also called the *reduction of s by modulo m*; the result is called the *residue of s modulo m*.

The LCG works as follows. First, a sequence $\{s_t,\ t = 0, 1, \ldots\}$ of integers is generated. The initial number s_0 (called the initial seed) is chosen by the user and all subsequent numbers are calculated from the rule

$$s_t = (Ms_{t-1} + a) \bmod m, \quad t = 1, 2, \ldots \tag{17.1}$$

By construction, all integers s_t lie between 0 and $m-1$. Then, the pseudo-random numbers in $[0, 1)$ are obtained by dividing these integers by m:

$$u_t = \frac{s_t}{m}, \quad t = 0, 1, 2, \ldots \tag{17.2}$$

The LCG is equivalent to the recurrence $u_k = \{Mu_{t-1} + a\}$, where all u_t are of the form $\frac{s}{m}$ with integers s and m, but it avoids round-off errors. Here the *multiplier* M, the *increment* a, and the *modulus* m are integers so that $1 \leqslant M < m$, and $0 \leqslant a < m$. The initial seed s_0 is an integer from $\{0, 1, \ldots, m-1\}$. If $a = 0$, then the generator is called a *multiplicative congruential generator* (MCG). It operates on multiplicative group of integers modulo m. The generating rule of the MCG is

$$s_t = Ms_{t-1} \bmod m, \quad u_t = \frac{s_t}{m}, \quad t = 1, 2, \ldots \tag{17.3}$$

To guarantee that numbers produced from the rule (17.3) are all nonzero, the initial seed s_0 has to be nonzero. Otherwise, all $s_t = 0$, $t \geqslant 1$, if $s_0 = 0$.

Example 17.1. Construct PRNs generated by the LCG with $M = 11$, $m = 16$, $a = 5$, and $s_0 = 0$.

Solution. The LCG rule is $u_t = \frac{s_t}{16}$, where $s_t = (11s_{t-1} + 5) \bmod 16$. We first obtain the sequence of integers s_t, $t \geqslant 0$:

$$\begin{aligned}
s_1 &= (11 \cdot 0 + 5) \bmod 16 = 5 \bmod 16 = 5, & s_2 &= (11 \cdot 5 + 5) \bmod 16 = 60 \bmod 16 = 12,\\
s_3 &= (11 \cdot 12 + 5) \bmod 16 = 137 \bmod 16 = 9, & s_4 &= (11 \cdot 9 + 5) \bmod 16 = 104 \bmod 16 = 8,\\
s_5 &= (11 \cdot 8 + 5) \bmod 16 = 93 \bmod 16 = 13, & s_6 &= (11 \cdot 13 + 5) \bmod 16 = 148 \bmod 16 = 4,\\
s_7 &= (11 \cdot 4 + 5) \bmod 16 = 49 \bmod 16 = 1, & s_8 &= (11 \cdot 1 + 5) \bmod 16 = 16 \bmod 16 = 0,\\
s_9 &= (11 \cdot 0 + 5) \bmod 16 = 5 \bmod 16 = 5, \ldots
\end{aligned}$$

As is seen, the sequence obtained is periodic. The numbers repeat themselves after eight steps. The PRNs $u_t = \frac{s_t}{m}$, $t \geqslant 0$, are

$$0, \frac{5}{16}, \frac{12}{16}, \frac{9}{16}, \frac{8}{16}, \frac{13}{16}, \frac{4}{16}, \frac{1}{16}, 0, \frac{5}{16}, \ldots \qquad \square$$

The quality of the LCG depends on the choice of parameters M, a, and m. Often the modulus m is chosen as a prime number. Then, all calculations are done in the finite field $\mathbb{Z}_m$. Preferred moduli are Mersenne primes of the form $2^r - 1$, e.g., $2^{31} - 1 = 2{,}147{,}483{,}647$. Sometimes, m is a power of 2 since calculations can be done faster by exploiting the binary structure of computer arithmetic. Here are some choices of m and M for the MCG:

- $m = 2^{31} - 1$ and $M = 16807$ (Park and Miller (1988));
- $m = 2^{40}$ and $M = 5^{17}$ (Ermakov and Mikhaĭlov (1982));
- $m = 2^{128}$ and $M = 5^{100109} \bmod 2^{128}$ (Dyadkin and Kenneth (2000)).

Since the set $\{0, 1, \ldots, m-1\}$ is finite, the sequence generated by an LCG is periodic. That is, $s_{t+r} = s_t$ holds for all t and some integer $r > 0$ called a period of the sequence. Let ℓ be the least possible period, which is called the *length of period.* Because there are only m possible different values of s_t, the maximum possible period of an LCG is m (or $m-1$ if $a = 0$).

Proposition 17.6 (Ermakov (1975)). *The maximum possible length of period for the sequence $s_t = Ms_{t-1} \bmod 2^p$, $t \geqslant 1$, with $p \geqslant 3$ is $\ell = 2^{n-2}$. The maximum period length is achieved if $M \equiv 3 \bmod 8$ or $M \equiv 5 \bmod 8$ and the seed s_0 is odd.*

There are many ways of improving a classical linear congruential generator. Let us consider some of them (a more comprehensive review of modern PRNGs can be found in L'Ecuyer (2012)). To increase the maximum period, one can combine two (or more) LCG methods as follows. Let $\{u_t^{(1)}\}_{t\geqslant 1}$ and $\{u_t^{(2)}\}_{t\geqslant 1}$ be the outputs of two LCGs. A new sequence $\{u_t\}_{t\geqslant 1}$ is given by

$$u_t := \left\{u_t^{(1)} + u_t^{(2)}\right\}, \quad t \geqslant 0.$$

Another generalization of the classical LCG is a multiple recursive generator defined by:

$$s_t = (M_1 s_{t-1} + \cdots + M_k s_{t-k}) \bmod m, \quad u_t = \frac{s_t}{m}, \quad t \geqslant k,$$

where $M_1, \ldots, M_k$ are multipliers selected from $\mathcal{S} := \{0, 1, \ldots, m-1\}$; $k \geqslant 2$ is the order of recursion, and $(s_0, s_1, \ldots, s_{k-1})$ is the seed sequence with $s_i \in \mathcal{S}$ for $0 \leqslant i \leqslant k-1$. The maximum period length for this generator is $m^k - 1$.

One of important properties of the MCG is the ease of its parallelization. Suppose that a sequence $\{s_t\}_{t\geqslant 0}$ is generated by (17.3). To share this sequence among several independent processors, we split it into several disjoint subsequences. This can be achevied by using different seeds situated far apart along the original sequence. The seeds $s_0^{(j)}$, $j \geqslant 0$, that start respective subsequences of length K are generated by using the *leaping-frog generator*:

$$s_0^{(j)} = As_0^{(j-1)} \bmod m, \qquad \text{where } s_0^{(0)} \equiv s_0 \text{ and } A \equiv M^K \bmod m.$$

Once the seeds are generated, the original generator (17.3) is used to obtain disjoint subsequences starting from these seeds:

$$s_t^{(j)} = Ms_{t-1}^{(j)} \bmod m, \quad u_t^{(j)} = \frac{s_t^{(j)}}{m}, \quad t = 1, 2, \ldots, K, \quad j \geqslant 0.$$

The maximum number of processors that can be served by such a leaping-frog generator cannot exceed the ratio $\frac{\ell}{K}$ of the length of period to the length of an individual subsequence.

17.3 Generation of Nonuniformly Distributed Random Numbers

Suppose that we have a good PRN generator for the uniform distribution *Unif*$(0, 1)$. However, our ultimate goal is to be able to sample from any given probability distribution.

This can be achieved by transforming uniform PRNs into nonuniform random numbers. We are interested in general transformation methods that work for large classes of distributions. The transformation methods should be fast and efficient and should not use too much memory. In the sequel we are going to consider three main groups of sampling algorithms, namely, inversion methods, composition methods, and acceptance-rejection methods.

17.3.1 Transformations of Random Variables

Recall that a (univariate) *random variable* defined on a probability space $(\Omega, \mathcal{F}, P)$ is a measurable function from Ω to $\mathbb{R}$. The *support* of X is defined as the smallest closed set $\mathcal{S}$ whose compliment $\mathcal{S}^c$ has probability zero: $\mathbb{P}(X \in \mathcal{S}^c) = 0$. The *cumulative distribution function* (CDF) of a random variable X is a function F_X from $\mathbb{R}$ to $[0, 1]$ that is defined by $F_X(x) = \mathbb{P}(X \leqslant x)$ for $x \in \mathbb{R}$. A CDF F satisfies the following properties:

1. F is a nondecreasing, right-continuous (i.e., $F(x+) = F(x)$) function;
2. $F(-\infty) = 0$ and $F(+\infty) = 1$.

A random variable X (and its CDF F_X) is said to be

Discrete, if there exists a nonnegative function p_X called a *probability mass function* (PMF for short) with a countable support such that

$$F_X(x) = \sum_{y \leqslant x \,:\, p_X(y) \neq 0} p_X(y) \text{ for } x \in \mathbb{R}.$$

The CDF F_X is a piecewise constant function with jumps.

(Absolutely) continuous, if there exists a nonnegative integrable function f_X called a *probability density function* (PDF for short) such that

$$F_X(x) = \int_{-\infty}^{x} f_X(x)\, \mathrm{d}x \text{ for } x \in \mathbb{R}.$$

The CDF F_X is a continuous function (without jumps).

Since $F_X(\infty) = 1$, a PMF p satisfies $\sum_{x \in \mathcal{S}} p(x) = 1$, and a PDF $f \geqslant 0$ satisfies $\int_{-\infty}^{\infty} f(x)\, \mathrm{d}x = 1$. For a discrete random variable, the support is the smallest countable collection of points $\mathcal{S}$ so that $\mathbb{P}(X \in \mathcal{S}) = 1$. It is defined by $\mathcal{S} = \{x \in \mathbb{R} : p_X(x) \neq 0\}$. Typically, the support $\mathcal{S}$ of a univariate continuous probability distribution is an interval of the real line. Since for a continuous random variable X the mass probability $\mathbb{P}(X = x)$ is zero for any $x \in \mathbb{R}$, the support may be chosen to be an open interval.

Often, one random variable can be expressed as a function of another variate, say $Y = f(X)$. Then sample values of Y can be obtained by applying the mapping f to sample values of X: $y_i = f(x_i)$, $i \geqslant 1$. Let us consider several useful examples.

A linear (or affine) mapping $f(x) = \alpha + \beta x$. The density of $Y = \alpha + \beta X$ is given by $f_Y(x) = \frac{1}{\beta} f_X\left(\frac{x - \alpha}{\beta}\right)$. A normal random $X \sim Norm(\mu, \sigma^2)$ can be expressed in terms of a standard normal variate $Z \sim Norm(0, 1)$ as $X = \mu + \sigma Z$. Similarly, a uniform variable $X \sim Unif(a, b)$ with $a < b$ is a function of $U \sim Unif(0, 1)$ given by $X = a + (b - a)U$.

A power mapping $f(x) = x^c$ for $x > 0$. Let X be a nonnegative random variable. The density of $Y = X^c$ is then $f_Y(x) = \frac{1}{c} x^{\frac{1}{c} - 1} f(x^{\frac{1}{c}})$. For example, the density of U^c, where $U \sim Unif(0, 1)$, is $f(x) = \frac{1}{c} x^{\frac{1}{c} - 1} \mathbb{I}_{(0,1)}(x)$. A special case is a reciprocal of a nonzero variate: $Y = \frac{1}{X}$. The PDF is $f_Y(x) = \frac{1}{x^2} f\left(\frac{1}{x}\right)$.

An exponential mapping $f(x) = \exp(x)$. The PDF of $Y = e^X$ is $f_Y(x) = \frac{1}{x}f(\ln x)$ for $x > 0$. For example, the log-normal random variable is an exponential of a normal variate: $Y = e^{\mu+\sigma Z}$, where $Z \sim Norm(0,1)$.

17.3.2 Inversion Method

The inversion method of generating nonuniform random numbers is based on the analytical or numerical inversion of the CDF. This method is most preferable if we wish to use it with low-discrepancy (quasi-random) sequences of numbers. Also it is compatible with some control variate techniques such as the antithetic variate method. The inversion method can be applied to both discrete and continuous distributions; however, its efficiency (in comparison with other methods) depends on how fast the inverse CDF can be evaluated.

17.3.2.1 Inverse Distribution Function

Consider a random variable X that has a continuous and strictly increasing on its support CDF F. Thus, the inverse function F^{-1} is well-defined and is also a strictly increasing function. Let us find the distribution function of $Y := F(X)$. By using the definitions of a CDF and an inverse function, we obtain

$$F_Y(y) = \mathbb{P}(F(X) \leqslant y) = \mathbb{P}(F^{-1}(F(X)) \leqslant F^{-1}(y)) = \mathbb{P}(X \leqslant F^{-1}(y)) = F(F^{-1}(y)) = y$$

for all $y \in (0,1)$. As is seen, the function F_Y is the CDF of a continuous random variable uniformly distributed on $(0,1)$. That is, $F(X) \sim Unif(0,1)$. This result provides us with a simple algorithm for sampling from continuous strictly increasing CDFs. Algorithm 17.1 is very simple and transparent. However, it relies on the knowledge of the inverse CDF F^{-1} in closed form (or the ability to compute F^{-1} in efficient manner). Figure 17.3 illustrates the method.

Algorithm 17.1 The Inverse CDF Method.

(1) Obtain a draw u from the unform distribution $Unif(0,1)$.

(2) A draw x from a CDF F is given by $x = F^{-1}(u)$.

Example 17.2. Using the inverse CDF method, find generating formulae for

(a) the uniform distribution $Unif(a,b)$, $a < b$;

(b) the exponential distribution $Exp(\lambda)$, $\lambda > 0$;

(c) the power distribution with the PDF $f(x) = cx^{c-1}\mathbb{I}_{(0,1)}(x)$, $c > 0$;

(d) the Weibull distribution with the PDF $f(x) = \frac{\alpha}{\beta}x^{\alpha-1}e^{-x^\alpha/\beta}\mathbb{I}_{\mathbb{R}_+}(x)$, $\alpha, \beta > 0$.

Solution.

(a) The CDF of $Unif(a,b)$ is $F(x) = \frac{x-a}{b-a}$ for $x \in (a,b)$. Solve $\frac{x-a}{b-a} = u$ with $u \in (0,1)$ for x to find the inverse CDF: $F^{-1}(u) = a + (b-a)u$. So the generating formula is $X = a + (b-a)U$.

(b) The CDF of the exponential distribution with rate $\lambda > 0$ is $F(x) = 1 - e^{-\lambda x}$, $x > 0$. Solve $1 - e^{-\lambda x} = u$ for x to obtain $x = -\frac{1}{\lambda}\ln(1-u)$. If $U \sim Unif(0,1)$, then $1 - U \sim Unif(0,1)$. So the generating formula for $X \sim Exp(\lambda)$ simplifies: $X = -\frac{\ln U}{\lambda}$.

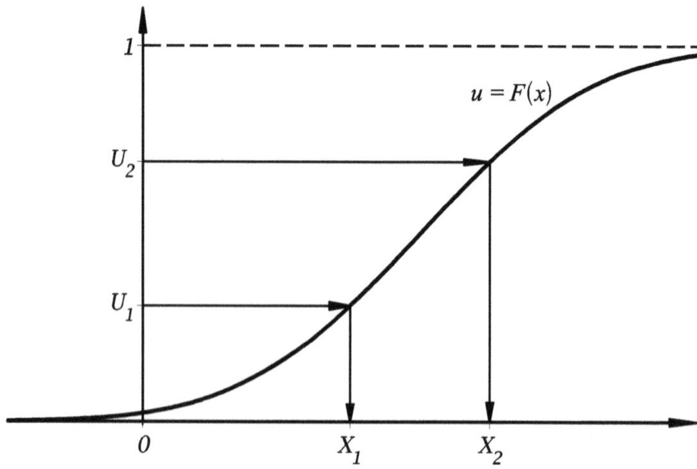

FIGURE 17.3: The inverse CDF method.

(c) Integrate $f(x) = cx^{c-1}\mathbb{I}_{(0,1)}(x)$ on $(0, x)$ for $0 < x < 1$, to obtain

$$F(x) = \int_0^x f(x)\,\mathrm{d}x = x^c.$$

Thus, $F^{-1}(u) = u^{1/c}$ and $X = U^{1/c}$.

(d) The CDF is

$$F(x) = \int_0^x \frac{\alpha}{\beta} x^{\alpha-1} \mathrm{e}^{-x^\alpha/\beta}\,\mathrm{d}x = 1 - \mathrm{e}^{-x^\alpha/\beta} \text{ for } x > 0.$$

Its inverse is $F^{-1}(u) = (-\beta \ln(1-u))^{1/\alpha}$. Thus, we obtain

$$X = F^{-1}(1-U) = (-\beta \ln U)^{1/\alpha}.$$

□

To generalize the inverse CDF sampling method to the case of any probability distribution, we define the generalized inverse CDF $F^{-1}\colon (0,1) \to \mathbb{R}$ by

$$F^{-1}(u) = \inf\{x \in \mathbb{R} : u \leqslant F(x)\}. \tag{17.4}$$

To justify this formula, let us consider two special cases. First, suppose that a CDF F of a random variable X has a jump discontinuity at x_0, i.e., $F(x_0-) < F(x_0)$. Therefore, $\mathbb{P}(X = x_0) = F(x_0) - F(x_0-) \neq 0$. The formula in (17.4) gives that

$$\begin{aligned}
&F^{-1}(u) = x_0 \text{ for } F(x_0-) \leqslant u \leqslant F(x_0),\\
&F^{-1}(u) < x_0 \text{ for } u < F(x_0-),\\
&F^{-1}(u) > x_0 \text{ for } u > F(x_0).
\end{aligned}$$

Thus, the random variable $F^{-1}(U)$ has a nonzero mass probability at x_0 and

$$\mathbb{P}(F^{-1}(U) = x_0) = F(x_0) - F(x_0-)$$

as expected. Now, suppose that F has a flat section on $[x_0, x_1]$ with $x_0 < x_1$. There exists $u_0 \in [0,1]$ such that $F(x) = u_0$ for $x \in (x_0, x_1)$, $F(x) \leqslant u_0$ for $x \leqslant x_0$, and $F(x) \geqslant u_0$ for $x \geqslant x_1$. In this case, we have $\mathbb{P}(x_0 < X < x_1) = F(x_1-) - F(x_0) = 0$. Then, the generalized inverse has a jump at u_0: $F^{-1}(u_0-) \leqslant x_0$ and $F^{-1}(u_0) \geqslant x_1$. Thus, with probability zero the random variable $F^{-1}(U)$ takes on a value in (x_0, x_1) as expected. Now, as we can see, the same sampling formula $X = F^{-1}(U)$, $U \sim \mathit{Unif}(0,1)$, with a generalized inverse CDF in (17.4), works for any CDF F, including those having a jump discontinuity or a flat section.

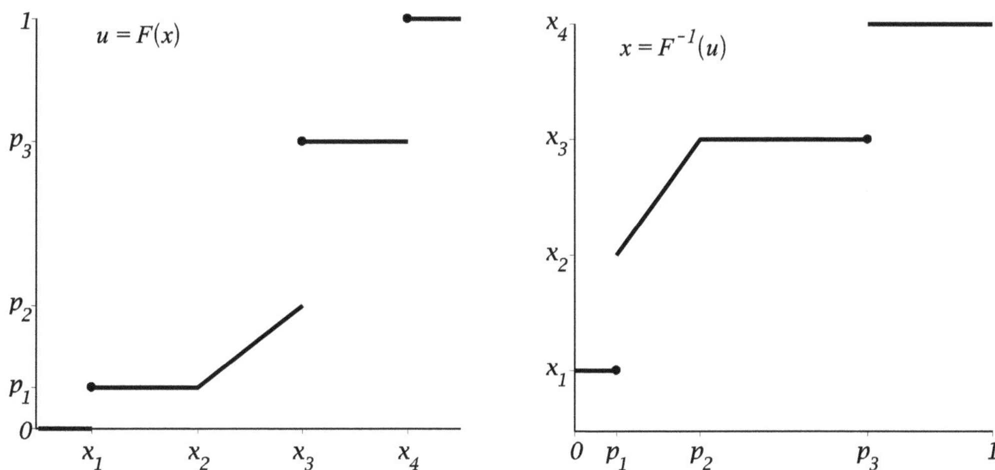

FIGURE 17.4: The plot of a CDF (the left plot) and its generalized inverse (the right plot). A mixture of a discrete distribution and a continuous distribution is considered. Note that a CDF is a right-continuous function and a generalized inverse CDF is a left-continuous function.

Example 17.3. Using the inverse CDF method, find a generating formula for

(a) a Bernoulli random variable $X = \begin{cases} 0 & \text{with probability } 1-p, \\ 1 & \text{with probability } p; \end{cases}$

(b) a random variable with the PMF

$$f(x) = p_1 \mathbb{I}_{\{x_1\}}(x) + p_2 \mathbb{I}_{\{x_2\}}(x) + p_3 \mathbb{I}_{\{x_3\}}(x),$$

where p_i, $i = 1, 2, 3$, are positive probabilities so that $p_1 + p_2 + p_3 = 1$.

Solution.

(a) Sample $U \sim \mathit{Unif}(0,1)$. If $U < p$, then set $X = 1$; otherwise set $X = 0$. Verify that X has the Bernoulli distribution: $\mathbb{P}(X = 1) = \mathbb{P}(U < p) = p$ and $\mathbb{P}(X = 0) = \mathbb{P}(U \geqslant p) = 1 - p$.

(b) Suppose that $x_1 < x_2 < x_3$. The CDF of $X \sim f$ and the generalized inverse CDF are, respectively,

$$F_X(x) = \begin{cases} 0 & \text{if } x < x_1, \\ p_1 & \text{if } x_1 \leqslant x < x_2, \\ p_1 + p_2 & \text{if } x_2 \leqslant x < x_3, \\ 1 & \text{if } x_3 \leqslant x, \end{cases} \qquad F_X^{-1}(u) = \begin{cases} x_1 & \text{if } u \leqslant p_1, \\ x_2 & \text{if } p_1 < u \leqslant p_1 + p_2, \\ x_3 & \text{if } p_1 + p_2 < u, \end{cases}$$

for $x \in \mathbb{R}$ and $u \in (0,1)$. Sample $U \sim \mathit{Unif}(0,1)$ and set

$$X = F_X^{-1}(U) = \begin{cases} x_1 & \text{if } U \leqslant p_1, \\ x_3 & \text{if } U > p_1 + p_2, \\ x_2 & \text{otherwise.} \end{cases}$$

□

Example 17.4. Justify the following method of sampling from the discrete uniform distribution on a set of N distinct numbers $x_1, x_2, \ldots, x_N$:

(i) generate $U \sim \mathit{Unif}(0,1)$;

(ii) set $X = x_K$, where $K = \lfloor N \cdot U + 1 \rfloor$.

Solution. We need to show that $\mathbb{P}(X = x_k) = \frac{1}{N}$ for any $k = 1, 2, \ldots, N$. Indeed,

$$\begin{aligned} \mathbb{P}(X = x_k) &= \mathbb{P}(\lfloor N \cdot U + 1 \rfloor = k) = \mathbb{P}(k \leqslant N \cdot U + 1 < k + 1) \\ &= \mathbb{P}\left(\frac{k-1}{N} \leqslant U < \frac{k}{N}\right) = \frac{k}{N} - \frac{k-1}{N} = \frac{1}{N}. \end{aligned}$$

□

17.3.2.2 The Chop-Down Search Method

Consider a general discrete random variable with a countable support $\mathcal{S} = \{x_j\}_{j \geqslant 1}$ and mass probabilities $\{p_j\}_{j \geqslant 1}$, where $p_j = \mathbb{P}(X = x_j) > 0$ and $\sum_{j \geqslant 1} p_j = 1$. The CDF F of such a discrete probability distribution is a piecewise-constant function. Let us assume that the mass points x_j are sorted in increasing order: $x_1 < x_2 < \ldots$. Then the CDF is given by $F(x) = \sum\limits_{j : x_j \leqslant x} p_j$. Hence the generalized inverse CDF is calculated as follows:

$$F^{-1}(u) = \inf \left\{ x_k \in S : u \leqslant \sum_{j=1}^{k} p_j \right\} = \left\{ x_k \in \mathcal{S} : \sum_{j=1}^{k-1} p_j < u \leqslant \sum_{j=1}^{k} p_j \right\} \tag{17.5}$$

The requirement that mass points x_j are sorted in increasing order is not necessary for the application of the inversion method. We can consider any arrangement for $\{x_j\}_{j \geqslant 1}$, since the sampling of X is equivalent to the sampling of a random index $K \in \mathbb{N}$ with probabilities $\mathbb{P}(K = j) = p_j$, $j \geqslant 1$. First, generate $U \sim \mathit{Unif}(0,1)$. Second, find the index $K \geqslant 1$ such that

$$\sum_{j=1}^{K-1} p_j < U \leqslant \sum_{j=1}^{K} p_j. \tag{17.6}$$

Finally, set $X = x_K$. Note that the probability of the event that U satisfies the above double inequality is exactly p_K. One of possible implementations of this approach is the chop-down search (CDS) algorithm (see Algorithm 17.2).

The number of cycles in the chop-down search method is equal to the expected value $\sum_{j \geqslant 1} j\, p_j$. Indeed, if $U \in (0, p_1]$, then the algorithm stops after one cycle, and this happens with probability $p_1 = \mathbb{P}(U \in (0, p_1])$. If $U \in (p_1, p_1 + p_2]$, then the algorithm stops after two cycles, and this happens with probability $p_2 = \mathbb{P}(U \in (p_1, p_1 + p_2])$, and so on. Let c_U be the computational cost of the generation of one draw from $\mathit{Unif}(0,1)$ and c_I the computational cost of one cycle of the method. Then, the total cost is

$$c_U + c_I \sum_{j \geqslant 1} j\, p_j.$$

Algorithm 17.2 The Chop-Down Search (CDS) Method.

input: the mass points $\{x_j\}_{j\geqslant 1}$ and probabilities $\{p_j\}_{j\geqslant 1}$
generate $U \leftarrow$ *Unif*$(0,1)$
set $K \leftarrow 0$
repeat
  set $K \leftarrow K+1$
  set $U \leftarrow U - p_K$
until $U \leqslant 0$
return $X = x_K$

Example 17.5. Find the computational cost of the CDS method for

(a) the geometric distribution *Geom*(p) with $\mathbb{P}(X=j) = (1-p)^{j-1}p$, $j \geqslant 1$, $0<p<1$;

(b) the Poisson distribution *Pois*(λ) with $\mathbb{P}(X=j) = \frac{\lambda^j}{j!}\mathrm{e}^{-\lambda}$, $j \geqslant 0$, $\lambda > 0$.

Solution. Compute the mathematical expectation $\mathcal{E} = \sum_{j=1}^{\infty} j\, p_j$ for both distributions. The total cost is then $c_U + c_I\,\mathcal{E}$.

$$\text{(a) } \mathcal{E} = \sum_{j=1}^{\infty} j\,(1-p)^{j-1}p = \frac{1}{p}; \quad \text{(b) } \mathcal{E} = \sum_{j=0}^{\infty} (j+1)\,\frac{\lambda^j}{j!}\mathrm{e}^{-\lambda} = \lambda + 1. \qquad \square$$

The computational cost can be reduced by rearranging elements of $\{(x_j, p_j)\}_{j\geqslant 1}$. Consider an arrangement $\{j_i\}_{i\geqslant 1}$ of the integers $1,2,3,\ldots$. Then, the sequence $\{(x_{j_i}, p_{j_i})\}_{i\geqslant 1}$ defines another discrete probability distribution equivalent to the original one.

Proposition 17.7. *The computational cost of the chop-down search method attains its minimal value iff the mass probabilities are arranged in the decreasing order*

$$p_1 \geqslant p_2 \geqslant p_3 \geqslant \cdots \tag{17.7}$$

Proof. Suppose that there exists another arrangement of the mass probabilities, $\{p'_j\}_{j\geqslant 1}$, for which (17.7) is violated and the expected value $\mathcal{E}' = \sum_{j\geqslant 1} j\, p'_j$ is *minimal.* There are two indices k and l, $k<l$, so that $p'_l < p'_k$. Let us construct another arrangement $\{p''_j\}_{j\geqslant 1}$, which is obtained from $\{p'_j\}_{j\geqslant 1}$ by swapping p'_l and p'_k, i.e., $p''_l = p'_k$, $p''_k = p'_l$, and $p''_j = p'_j$ for $j \notin \{k,l\}$. Let $\mathcal{E}'' = \sum_{j\geqslant 1} j\, p''_j$. We have

$$\mathcal{E}' - \mathcal{E}'' = l\,p'_l + k\,p'_k - l\,p''_l - k\,p''_k = (l-k)(p'_l - p'_k) > 0,$$

since $l<k$ and $p'_l < p'_k$. Hence $\mathcal{E}'' < \mathcal{E}'$. We arrive at a contradiction. $\square$

According to Proposition 17.7, the mass probabilities $\{p_j\}_{j\geqslant 1}$ should be arranged in decreasing order before applying the chop-down search method. Another way to speed up calculations is to use recurrence relations for mass probabilities. This allows us to reduce the parameter c_I—the cost of one cycle of the CDS method. For example, the mass probabilities of a geometric random variable $X \sim$ *Geom*(p) satisfy

$$\mathbb{P}(X=j+1) = \mathbb{P}(X=j)\cdot(1-p), \quad j=1,2,\ldots$$

For a Poisson random variable $X \sim$ *Pois*(λ), we have

$$\mathbb{P}(X=j) = \mathbb{P}(X=j-1)\cdot\frac{\lambda}{j}, \quad j=1,2,\ldots$$

Algorithm 17.3 The Binary Search Method.

input: the mass points $\{x_j\}_{1\leqslant j\leqslant N}$ and mass probabilities $\{p_j\}_{1\leqslant j\leqslant N}$
calculate $F_0 = 0, \quad F_k = \sum_{j=1}^{k} p_j$ for $1 \leqslant k \leqslant N-1, \quad F_N = 1.$
generate $U \leftarrow \mathit{Unif}(0,1)$
set $L \leftarrow 0$ and $R \leftarrow N$
repeat
  set $K \leftarrow \lfloor \frac{L+R}{2} \rfloor$
  if $F_K < U$ **then**
    set $L \leftarrow K$
  else
    set $R \leftarrow K$
  end if
until $R = L$
return $X = x_K$

17.3.2.3 The Binomial Search Method

The formula (17.5) of the generalized inverse CDF requires the cumulative probabilities $\sum_{j=1}^{k} p_j$, $k \geqslant 1$. Therefore, the inversion method can be speeded up if these cumulative probabilities are precalculated in advance and stored in memory. After that, to find K that satisfy (17.6), we can employ a fast search procedure such as the binary search method. Suppose that a random variable X takes on one of N distinct values $x_1 < x_2 < \cdots < x_N$ with the respective mass probabilities $p_1, p_2, \ldots, p_N$, i.e., $\mathbb{P}(X = x_j) = p_j$, $1 \leqslant j \leqslant N$. Note that a random variable with a countably infinite support can be reduced to the case with N mass points by suitable truncation of the support such that the total probability of removed mass points is very small. Suppose that $N = 2^r$, then $r = \log_2 N$ cycles are required to find K. For general N, the computation cost is proportional to $\lceil \log_2 N \rceil$, where $\lceil \cdot \rceil$ denotes the ceiling function.

17.3.3 Composition Methods

It is a well-known fact that a linear combination (a mixture) of CDFs, PDFs, or PMFs with positive weights summing up to one is again a CDF, PDF, or PMF, respectively. The composition method aims to represent the probability distribution of interest as a mixture of simpler-organized distributions. The sampling from a mixture distribution is a two-step procedure. First, one of the distributions which appear in the composition is selected at random; second, a sample is drawn from the distribution selected (e.g., by using an inversion algorithm). In comparison with the inverse CDF method that requires one draw from *Unif*$(0,1)$, a composition method needs at least two uniform random numbers. However, the composition method allows us to express a probability distribution in a simpler form that allows for simplifying the sampling algorithm.

17.3.3.1 Mixture of PDFs

Consider a continuous random variable X with PDF f. Suppose that f can be represented as a linear combination of m PDFs $f_1, \ldots, f_m$ with m positive weights $w_1, \ldots, w_m$ so that $w_1 + \cdots + w_m = 1$:

$$f(x) = \sum_{j=1}^{m} w_j f_j(x), \quad x \in \mathbb{R}.$$

The PDF f is called a *mixture PDF*. The support of f is a union of supports of f_j, $1 \leqslant j \leqslant m$. If all pairwise intersections of supports of f_j are empty sets, then such a mixture is called a *stratification*.

Algorithm 17.4 The Composition Sampling Method.

input: $\{w_j\}_{j\geqslant 1}$ and $\{f_j\}_{j\geqslant 1}$
generate K from the probabilities $\mathbb{P}(K = j) = w_j$, $j \geqslant 1$
generate X from the PDF f_K
return X

Proof of Algorithm 17.4. Let us find the distribution function of X generated by the algorithm. Applying the total probability law gives

$$\mathbb{P}(X \leqslant x) = \sum_{j=1}^{m} \mathbb{P}(X \leqslant x; K = j) = \sum_{j=1}^{m} \mathbb{P}(K = j)\,\mathbb{P}(X \leqslant x \mid K = j) = \sum_{j=1}^{m} w_j \int_{-\infty}^{x} f_j(x)\,\mathrm{d}x$$
$$= \int_{-\infty}^{x} \sum_{j=1}^{m} w_j f_j(x)\,\mathrm{d}x = \int_{-\infty}^{x} f(x)\,\mathrm{d}x = F_X(x)$$

for all $x \in \mathbb{R}$. □

Example 17.6. Develop a stratification method for sampling from the PDF

$$f(x) = \begin{cases} \frac{2}{3}x & \text{if } 0 < x \leqslant 1, \\ \frac{2}{3} & \text{if } 1 < x < 2, \\ 0 & \text{otherwise.} \end{cases}$$

Solution. As is seen, f is a piecewise linear function; it is constant on $[1, 2]$ and linear on $[0, 1]$. Introduce two new PDFs: $f_1(x) \propto x$ for $x \in (0, 1]$ and $f_2(x) \propto 1$ for $x \in (1, 2)$. The second function is a PDF for *Unif*$(1, 2)$. Hence, $f_2(x) = \mathbb{I}_{(1,2)}(x)$. Calculate a normalizing constant for f_1 to obtain

$$f_1(x) = \frac{x}{\int_0^1 x\,\mathrm{d}x}\,\mathbb{I}_{(0,1]}(x) = 2x\,\mathbb{I}_{(0,1]}(x).$$

Now, the PDF $f(x)$ can be decomposed as follows:

$$f(x) = \frac{2}{3}\,x\,\mathbb{I}_{(0,1]}(x) + \frac{2}{3}\,\mathbb{I}_{(1,2)}(x) = \frac{1}{3} \cdot \left(2x\,\mathbb{I}_{(0,1]}(x)\right) + \frac{2}{3} \cdot \mathbb{I}_{(1,2)}(x)$$
$$= \frac{1}{3}\,f_1(x) + \frac{2}{3}\,f_2(x) := w_1 f_1(x) + w_2 f_2(x).$$

Sampling from f_2 is easy (see Example 17.3): $X = 1 + U$ with $U \sim$ *Unif*$(0, 1)$. To sample from f_1, we apply the inverse CDF method. First, find the CDF:

$$F_1(x) = \int_0^x f_1(x)\,\mathrm{d}x = x^2, \quad x \in [0, 1].$$

The inverse CDF is $F_1^{-1}(u) = \sqrt{u}$, $u \in [0, 1]$. As a result, we obtain the following sampling algorithm:

(1) Generate two independent uniform random numbers $U_1, U_2 \leftarrow$ *Unif*$(0, 1)$.

(2) Sample $K \in \{1,2\}$ as follows. If $U_1 < \frac{1}{3}$, then $K = 1$, else $K = 2$.

(3) Generate X from f_K as follows:

   (i) if $K = 1$, then set $X \leftarrow 1 + U_2$;

   (ii) if $K = 2$, then set $X \leftarrow \sqrt{U_2}$. □

In general, the stratification method can be used with any partition of the support of a probability distribution. Suppose that A is the support of a PDF f. Let

$$A = \bigcup_{j=1}^{m} A_k, \text{ where } i \neq j \implies A_i \cap A_j = \emptyset.$$

Then, the PDF f admits the following representation:

$$f(x) = f(x)\,\mathbb{I}_A(x) = f(x)\sum_{j=1}^{m} \mathbb{I}_{A_j}(x) = \sum_{j=1}^{m} w_j\, f_j(x),$$

where $w_j = \int_{A_j} f(x)\,\mathrm{d}x$ and $f_j(x) = \frac{1}{w_j} f(x)\,\mathbb{I}_{A_j}(x)$, $1 \leqslant j \leqslant m$, $x \in \mathbb{R}$.

17.3.3.2 Randomized Gamma Distributions

Probability distributions whose PDFs contain special functions such as Bessel and other hypergeometric functions present a real challenge for sampling random numbers. By replacing a special function by its integral or series representation in terms of simpler functions, the PDF can be expressed as a mixture of more regular densities with known sampling algorithms.

For example, consider the noncentral chi-square distribution with $\kappa > 0$ degrees of freedom and noncentrality parameter $\lambda > 0$. Its PDF is

$$f(x;\kappa,\lambda) = \frac{1}{2}\mathrm{e}^{-(x+\lambda)/2}\left(\frac{x}{\lambda}\right)^{\frac{\kappa}{4}-\frac{1}{2}} I_{\frac{\kappa}{2}-1}(\sqrt{\lambda x}), \quad x > 0. \tag{17.8}$$

The modified Bessel function I_μ of the first kind (of order μ) admits the following series expansion:

$$I_\mu(x) = \left(\frac{x}{2}\right)^{\mu} \sum_{j=0}^{\infty} \frac{(x^2/4)^j}{j!\,\Gamma(\mu+j+1)}.$$

By using this expansion for $I_{\frac{\kappa}{2}-1}$ in (17.8), we can represent the noncentral chi-square PDF as a mixture of gamma densities with Poisson weights:

$$\begin{aligned} f(x;\kappa,\lambda) &= \frac{1}{2}\mathrm{e}^{-(x+\lambda)/2}\left(\frac{x}{\lambda}\right)^{\frac{\kappa}{4}-\frac{1}{2}} \left(\frac{\sqrt{\lambda x}}{2}\right)^{\frac{\kappa}{2}-1} \sum_{j=0}^{\infty} \frac{(\lambda x/4)^j}{j!\,\Gamma(\frac{\kappa}{2}+j)} \\ &= \sum_{j=0}^{\infty} \underbrace{\mathrm{e}^{-\lambda/2}\frac{(\lambda/2)^j}{j!}}_{=p_j \text{ (Poisson prob.)}} \underbrace{\frac{1}{2}\left(\frac{x}{2}\right)^{\kappa/2+j-1} \frac{\mathrm{e}^{-x/2}}{\Gamma(\frac{\kappa}{2}+j)}}_{=f_j(x) \text{ (gamma density)}}. \end{aligned} \tag{17.9}$$

Recall that a random variable X is said to be gamma-distributed with shape parameter α and scale parameter θ, denoted by *Gamma*(α,θ), if its PDF is

$$f_X(x) = \frac{\theta}{\Gamma(\alpha)}(\theta x)^{\alpha-1}\mathrm{e}^{-\theta x}, \quad x > 0. \tag{17.10}$$

In (17.9), the PDF $f(x;\kappa,\lambda)$ is expressed as a mixture $\sum_{j=0}^{\infty} p_j\, f_j(x)$, where $\{p_j\}_{j\geqslant 0}$ are mass probabilities of the Poisson distribution with intensity $\frac{\lambda}{2}$ and f_j is a gamma PDF of the form (17.10) with parameters $\alpha = \frac{\kappa}{2}+j$ and $\theta = \frac{1}{2}$ for all $j = 0, 1, 2, \ldots$. Such a mixture probability distribution is called a *randomized gamma distribution*, denoted $Gamma(Y_1 + \frac{\kappa}{2}, \frac{1}{2})$, where $Y_1 \sim Pois(\frac{\lambda}{2})$.

In general, we can consider a mixture gamma distribution $Gamma(Y + \nu + 1, \theta)$ with parameters $\nu > -1$ and $\theta > 0$, where the *randomizer* Y is a discrete random variable taking its values in the set of nonnegative integers with probabilities $\mathbb{P}(Y = j) = p_j$, $j = 0, 1, 2, \ldots$. The PDF f of such a randomized gamma distribution admits the form of a series expansion:

$$f(y) = \sum_{j=0}^{\infty} p_j \frac{\theta}{\Gamma(\nu+j+1)} (\theta y)^{\nu+j}\, \mathrm{e}^{-\theta y}.$$

Let us consider three choices for the randomizer Y. The resulting probability distributions are called the *randomized gamma distribution* of the first, second, and third types, respectively.

Let $Y_1 \sim Pois(\alpha)$ be a Poisson random variable with mean $\alpha > 0$. The randomized gamma distribution of the *first type* is $Gamma(Y_1 + \nu + 1, \theta)$ with the PDF

$$f_1(y) = \theta \left(\frac{\theta y}{\alpha}\right)^{\nu/2} \mathrm{e}^{-\alpha-\theta y} I_\nu(2\sqrt{\alpha\theta y}), \quad y > 0. \tag{17.11}$$

So the noncentral chi-square distribution with parameters $\kappa > 0$ and $\lambda > 0$ is the randomized gamma distribution of the first type with $\nu = \frac{\kappa}{2} - 1$, $\theta = \frac{1}{2}$, and $\alpha = \frac{\lambda}{2}$.

A discrete random variable Y_2 is said to have the *Bessel probability distribution*, denoted $Bes(\nu, b)$, with parameters $\nu > -1$ and $b > 0$ if

$$\mathbb{P}(Y_2 = j) = \frac{(b/2)^{2j+\nu}}{I_\nu(b)\, j!\, \Gamma(j+\nu+1)}, \quad j = 0, 1, 2, \ldots. \tag{17.12}$$

This distribution is related to many other distributions, where the modified Bessel function I is involved in the density, including the squared Bessel bridge distribution. The randomized gamma distribution of the *second type* is a mixture distribution $Gamma(Y_1+2Y_2+\nu+1, \theta)$, where $Y_1 \sim Pois(\frac{a+b}{4\theta})$ and $Y_2 \sim Bes(\nu, \frac{\sqrt{ab}}{2\theta})$ are independent Poisson and Bessel random variables, respectively. For any positive numbers θ, a, b, and $\nu > -1$, the PDF is

$$f_2(y) = \theta\, \mathrm{e}^{-\theta y-(a+b)/(4\theta)} \frac{I_\nu(\sqrt{ay}) I_\nu(\sqrt{by})}{I_\nu(\sqrt{ab}/(2\theta))}, \quad y > 0. \tag{17.13}$$

A discrete random variate Y_3 is said to follow an *incomplete gamma* probability distribution, which we simply denote by $I\Gamma(\nu, \lambda)$ with parameters $\lambda > 0$ and $\nu > 0$, if

$$\mathbb{P}(Y_3 = j) = \frac{\mathrm{e}^{-\lambda}\lambda^{j+\nu}}{\Gamma(j+\nu+1)} \frac{\Gamma(\nu)}{\gamma(\nu,\lambda)}, \quad j = 0, 1, 2, \ldots, \tag{17.14}$$

where $\gamma(a, x) := \int_0^x t^{a-1} \mathrm{e}{-t}\, \mathrm{d}t$ is the lower incomplete gamma function. Note that if ν is a nonnegative integer, then the distribution of Y_3 is simply a truncated and shifted Poisson distribution thanks to the property

$$\frac{\gamma(m, a)}{\Gamma(m)} = 1 - \left(1 + x + \cdots + \frac{x^{m-1}}{(m-1)!}\right) \mathrm{e}^{-x}, \quad m = 0, 1, 2, \ldots$$

We call a mixture probability distribution $Gamma(Y_3+1,\theta)$, $Y_3 \sim \mathrm{I}\Gamma(\nu,\lambda)$, the randomized gamma distribution of the *third type*. The PDF is

$$f_3(y) = \frac{\theta\,\Gamma(\nu)}{\gamma(\nu,\lambda)}\left(\frac{\theta y}{\lambda}\right)^{-\nu/2} \mathrm{e}^{-\lambda-\theta y} I_\nu(\sqrt{4\lambda\theta y}), \quad y > 0. \tag{17.15}$$

As we will see in the next chapter, randomized gamma distributions play a significant role in simulation of the so-called constant elasticity of variance diffusion model (see also Makarov and Glew (2010)).

17.3.3.3 The Alias Method by Walker

Let us study the case with discrete random variables. A mixture of PMFs is defined in the same manner as a mixture of PDFs. Consider m PMFs $p_j(x)$ and m weights $w_j > 0$, $1 \leqslant j \leqslant m$, such that $w_1 + w_2 + \cdots + w_m = 1$. The function p defined by $p(x) = \sum_{j=1}^m w_j p_j(x)$, $x \in \mathbb{R}$, is also a PMF called the mixture of the PMFs p_j, $1 \leqslant j \leqslant m$. To sample from p, we can apply Algorithm 17.4, where the sampling from the PDF f_K is replaced by sampling from the PMF p_K.

The alias method proposed by Walker (1977) allows us to represent any discrete probability distribution with m mass points $\mathcal{S} := \{x_1, x_2, \ldots, x_m\}$ as an equally weighted mixture of m two-point distributions. That is, there exist m two-point PMFs

$$p_j(x) = q_j \mathbb{I}_{\{x_j\}}(x) + (1-q_j)\mathbb{I}_{\{a_j\}}(x) \text{ with } a_j \in \mathcal{S} \text{ and } q_j \in [0,1],\ 1 \leqslant j \leqslant m,$$

such that

$$p(x) = \frac{1}{m}\sum_{j=1}^m p_j(x) \text{ for } x \in \mathbb{R}. \tag{17.16}$$

Since all weights in (17.16) are equal to $\frac{1}{m}$, Algorithm 17.4 is simplified and we obtain Algorithm 17.5.

Algorithm 17.5 The Alias Sampling Method.

---

input: m, $\{x_j\}_{1\leqslant j\leqslant m}$, $\{a_j\}_{1\leqslant j\leqslant m}$, and $\{q_j\}_{1\leqslant j\leqslant m}$
generate i.i.d. $U_1, U_2 \leftarrow Unif(0,1)$
set $K \leftarrow \lfloor m \cdot U_1 + 1 \rfloor$
if $U_2 \leqslant q_K$ **then**
  set $X = x_K$
else
  set $X = a_K$
end if
return X

To obtain such a decomposition of the PMF p, we need to construct two lists, namely, the list of probabilities $Q = (q_1, q_2, \ldots, q_m)$ and the list of *aliases* $A = (a_1, a_2, \ldots, a_m)$. This can be achieved by using the "leveling the histogram" procedure, which is described below. During this procedure the original histogram $\{w_j = p_j : 1 \leqslant j \leqslant m\}$ is transformed into an equally weighed histogram $\{w_j = \frac{1}{m} : 1 \leqslant j \leqslant m\}$; the lists Q and A are generated in the course of this process.

Step 1: Start with $w_j = p_j$, $q_j = 1$, and $a_j = x_j$ for all $j = 1, 2, \ldots, m$.

Step 2: Find two indices $\ell, u \in \{1, 2, \dots, m\}$ (ℓ and u stand for "lower" and "upper," respectively) such that

$$w_\ell = \min_{1 \leqslant j \leqslant m} \{w_j\} \text{ and } w_u = \max_{1 \leqslant j \leqslant m} \{w_j\}.$$

Step 3: If $w_\ell = w_u = \frac{1}{m}$, then the histogram is levelled and the algorithm is stopped; otherwise (i.e., $w_\ell < \frac{1}{m} < w_u$) we proceed with Step 4

Step 4: Set $q_\ell \leftarrow N\, w_\ell$ and $a_\ell \leftarrow x_u$. Change $w_u \leftarrow w_u - \left(\frac{1}{m} - w_\ell\right)$ and $w_\ell \leftarrow \frac{1}{m}$. Note that w_u can become less than $\frac{1}{m}$. Go back to Step 2.

As is seen, the alias method requires at most $m - 1$ iterations until all columns of the histogram (i.e., the weights w_j, $1 \leqslant j \leqslant m$) become the same height.

Example 17.7. Apply the alias method to the probability distribution with

$$p(x) = 0.1\, \mathbb{I}_{\{1\}}(x) + 0.2\, \mathbb{I}_{\{2\}}(x) + 0.3\, \mathbb{I}_{\{3\}}(x) + 0.4\, \mathbb{I}_{\{4\}}(x).$$

Solution. This is a four-point distribution with the list of mass points $(x_1, x_2, x_3, x_4) = (1, 2, 3, 4)$ and the list of mass probabilities $(p_1, p_2, p_3, p_4) = (0.1, 0.2, 0.3, 0.4)$. Initially, we set $(a_1, a_2, a_3, a_4) = (1, 2, 3, 4)$ and $(w_1, w_2, w_3, w_4) = (0.1, 0.2, 0.3, 0.4)$; all q_i are equal to 1.

(i) The smallest weight is $w_1 = 0.1$; the largest is $w_4 = 0.4$. Set

$$\begin{aligned} &w_4 \leftarrow w_4 - (0.25 - w_1) = 0.4 - (0.25 - 0.1) = 0.25, && q_1 \leftarrow 4w_1 = 0.4, \\ &w_1 \leftarrow 0.25, && a_1 \leftarrow x_4 = 4. \end{aligned}$$

(ii) Now, $(w_1, w_2, w_3, w_4) = (0.25, 0.2, 0.3, 0.25)$. The smallest weight is $w_2 = 0.2$; the largest is $w_3 = 0.3$. Set

$$\begin{aligned} &w_3 \leftarrow w_3 - (0.25 - w_1) = 0.3 - (0.25 - 0.2) = 0.25, && q_2 \leftarrow 4w_2 = 0.8, \\ &w_2 \leftarrow 0.25, && a_2 \leftarrow x_3 = 3. \end{aligned}$$

(iii) Now, $(w_1, w_2, w_3, w_4) = (0.25, 0.25, 0.25, 0.25)$. All weights are equal to 0.25. Stop the algorithm.

As a result, we obtained the following two lists:

$$(a_1, a_2, a_3, a_4) = (4, 3, 3, 4) \text{ and } (q_1, q_2, q_3, q_4) = (0.4, 0.8, 1, 1). \qquad \square$$

17.3.4 Acceptance-Rejection Methods

To generate realizations from some probability distribution, an acceptance-rejection method makes use of realizations of another random variable whose probability distribution is similar to the target one. The distribution from which the independent samples are generated is called a *proposal distribution.* Each sample can be accepted or rejected. The realizations being accepted have the target probability distribution. The computational cost of such a method is proportional to the average number of draws generated before one is accepted. Since the number of trials (draws) before the first success (acceptance of a draw) follows the geometric distribution, the average number of draws generated before an acceptance occurs is a reciprocal of the probability of the acceptance of a proposal draw.

Example 17.8. Propose an acceptance-rejection method of sampling a point uniformly distributed in the circle $C = \{(x, y) : x^2 + y^2 \leqslant 1\}$.

Solution. The circle C is contained in the square $S = [-1, 1]^2$. To obtain a draw from the uniform distribution *Unif*(C), proceed as follows.

(1) Sample a random point (X, Y) uniformly distributed in S as follows: $X = 2U_1 - 1$ and $Y = 2U_2 - 1$, where U_1 and U_2 are i.i.d. *Unif*$(0, 1)$-distributed random variables.

(2) Accept the point if $(X, Y) \in C$, i.e., $X^2 + Y^2 \leqslant 1$. Otherwise, the point is rejected and we return to (1). □

The formal justification of this example and of the general acceptance-rejection method is based on Propositions 17.8–17.10, which follow.

Proposition 17.8. *Let a random vector* $\mathbf{X}$ *be uniformly distributed in a domain* $D \subset \mathbb{R}^d$ *of a finite (d-dimensional) volume* $|D| < \infty$. *Let* Ω *be a subdomain of* D. *The distribution of* $\mathbf{X}$ *conditional on* $\mathbf{X} \in \Omega$ *is uniform in* Ω.

Proof. Let $\mathrm{d}\Omega$ be an arbitrary subdomain of Ω. Then,

$$\mathbb{P}(\mathbf{X} \in \mathrm{d}\Omega \mid \mathbf{X} \in \Omega) = \frac{\mathbb{P}(\mathbf{X} \in \mathrm{d}\Omega; \mathbf{X} \in \Omega)}{\mathbb{P}(\mathbf{X} \in \Omega)} = \frac{\mathbb{P}(\mathbf{X} \in \mathrm{d}\Omega)}{\mathbb{P}(\mathbf{X} \in \Omega)} = \frac{|\mathrm{d}\Omega|/|D|}{|\Omega|/|D|} = \frac{|\mathrm{d}\Omega|}{|\Omega|}.$$

Thus, the assertion is proved. □

In Example 17.8 we deal with a case covered by Proposition 17.8. Proposed points are sampled uniformly on the square S. A point (X, Y) is accepted if it lies in the circle C. According to Proposition 17.8, the probability distribution of (X, Y) conditional on $(X, Y) \in C$ is uniform in C. As a result, accepted points are uniformly distributed in C.

The simplest acceptance-rejection algorithm is the so-called *Neumann method.* Consider a bounded PDF $f \leqslant M$ with a support contained in a finite interval $[a, b]$. The plot of f on $[a, b]$ is contained in the rectangle $[a, b] \times [0, M]$. To sample from f we proceed as follows. First, sample (X, Y) uniformly in the rectangle $[a, b] \times [0, M]$. This point is accepted if $Y \leqslant f(X)$ and rejected otherwise. According to Proposition 17.8, the distribution of accepted points is uniform in the region bounded by the plot of $y = f(x)$ and the x-axis. Moreover, we can show that the distribution of the x-coordinate of an accepted point has the PDF f. That is, X conditional on $Y \leqslant f(X)$ has the distribution with the PDF f. This result will be proved in Proposition 17.9 in more general setting.

Definition 17.1. Consider a nonnegative integrable function g with support $D \subseteq \mathbb{R}^d$, that is, $g(\mathbf{x}) \geqslant 0$ for $\mathbf{x} \in D$, $g(\mathbf{x}) = 0$ for $\mathbf{x} \notin D$, and $I_D(g) := \int_{\mathbb{R}^n} g(\mathbf{x})\,\mathrm{d}\mathbf{x} = \int_D g(\mathbf{x})\,\mathrm{d}\mathbf{x} < \infty$ hold. The region

$$B_D(g) := \{(\mathbf{x}, y) \,:\, \mathbf{x} \in D,\, 0 < y < g(\mathbf{x})\} \subseteq \mathbb{R}^{d+1}$$

is called the *body* of the function g.

Proposition 17.9. *Suppose that the* $(d+1)$*-dimensional point* $(\mathbf{X}, Y)$, *with* $\mathbf{X} \in \mathbb{R}^d$ *and* $Y \in \mathbb{R}$, *is uniformly distributed in the body* $B_D(g)$ *of an integrable function* $g : D \to [0, \infty)$ *defined on* $D \subseteq \mathbb{R}^d$. *Then, the random vector* $\mathbf{X}$ *is distributed in* D *with the PDF proportional to* g:

$$p_{\mathbf{X}}(\mathbf{x}) = \frac{1}{I_D(g)}\, g(\mathbf{x}), \quad \mathbf{x} \in D.$$

Proof. The joint PDF of $(\mathbf{X}, Y)$ is

$$f_{\mathbf{X},Y}(\mathbf{x}, y) = \frac{1}{|B_D(g)|}\, \mathbb{I}_{B_D(g)}(\mathbf{x}, y) = \frac{1}{|B_D(g)|}\, \mathbb{I}_D(\mathbf{x})\, \mathbb{I}_{(0,g(\mathbf{x}))}(y).$$

The marginal PDF of $\mathbf{X}$ is

$$p_{\mathbf{X}}(\mathbf{x}) = \int_{-\infty}^{\infty} f_{\mathbf{X},Y}(\mathbf{x},y)\,\mathrm{d}y = \int_{0}^{g(\mathbf{x})} \frac{1}{|B_D(g)|}\,\mathbb{I}_D(\mathbf{x})\,\mathrm{d}y = \frac{1}{I_D(g)}\,g(\mathbf{x})\,\mathbb{I}_D(\mathbf{x}),$$

since $I_D(g) = |B_D(g)|$. □

In the Neumann method the proposed sample values are drawn from the uniform PDF $p(x) = \frac{1}{b-a}\mathbb{I}_{(a,b)}(x)$. This choice is explained by the fact that the PDF p majorizes the target PDF f up to a multiplicative constant: $f(x) \leqslant C\,p(x)$ with $C = M(b-a)$ (provided that $f(x) \leqslant M$ for all $x \in [a,b]$). A draw X is accepted if $Y \leqslant f(X)$ where $Y \sim \mathit{Unif}(0,M)$. Therefore, the Neumann method can be generalized to the case with an arbitrary PDF f as long as we can find a majorizing function for f.

In what follows, we will require the following proposition that explains how to sample a point uniformly distributed in a body of a nonnegative function.

Proposition 17.10. *Let p be a multivariate PDF with support D. Suppose that $\mathbf{X} \sim p$ and $(Y|\mathbf{X}=\mathbf{x}) \sim \mathit{Unif}(0, C\,p(\mathbf{x}))$ for some constant $C > 0$. Then, the point $(\mathbf{X}, Y)$ is uniformly distributed in the domain $B_D(C\,p)$.*

Proof. The joint PDF of $(\mathbf{X}, Y)$ is

$$\begin{aligned} f_{\mathbf{X},Y}(\mathbf{x},y) &= f_{\mathbf{X}}(\mathbf{x})\,f_{Y|\mathbf{X}}(y|\mathbf{x}) = p(\mathbf{x})\,\frac{1}{C\,p(\mathbf{x})}\,\mathbb{I}_{(0,C\,p(\mathbf{x}))}(y) \\ &= \frac{1}{C}\,\mathbb{I}_D(\mathbf{x})\,\mathbb{I}_{(0,C\,p(\mathbf{x}))}(y) = \frac{1}{|B_D(C\,p)|}\,\mathbb{I}_{B_D(C\,p)}(\mathbf{x}), \end{aligned}$$

since $|B_D(C\,p)| = \int_D C\,p(\mathbf{x})\,\mathrm{d}\mathbf{x} = C\int_D p(\mathbf{x})\,\mathrm{d}\mathbf{x} = C$. □

Consider an n-variate PDF f with support $D \subseteq \mathbb{R}^n$. Suppose that there exists another PDF p called a proposal PDF and a constant $C > 0$ such that $f(\mathbf{x}) \leqslant C\,p(\mathbf{x})$ for all $\mathbf{x} \in \mathbb{R}^n$. Often p is chosen to be a simple function like a piecewise-linear function so that sampling from p is feasible.

Algorithm 17.6 The Acceptance-Rejection Method (Version 1).

(1) Sample $\mathbf{X}$ from the proposal PDF p.

(2) Generate $U \sim \mathit{Unif}(0,1)$ independent of $\mathbf{X}$.

(3) Accept $\mathbf{X}$ if $U \leqslant \frac{f(\mathbf{X})}{C\,p(\mathbf{X})}$. Otherwise return to (1).

The acceptance-rejection method can be visualized as choosing a subsequence of draws from a sequence of i.i.d. realizations from the PDF p in such a way that the resulting subsequence consists of i.i.d. realizations from the target PDF f.

i.i.d. draws from p	$\widetilde{X}_1$	$\widetilde{X}_2$	$\widetilde{X}_3$	$\widetilde{X}_4$	$\widetilde{X}_5$	$\widetilde{X}_6$	...
Accept?	no	yes	no	no	yes	yes	...
i.i.d. draws from f		X_1			X_2	X_2	...

The proof of the acceptance-rejection method is based on Propositions 17.8–17.10. The steps of Algorithm 17.6 can be reformulated as follows. First, sample two random variables $\mathbf{X} \sim p$ and $Y \sim Unif(0, C\,p(\mathbf{X}))$. As is proved in Proposition 17.10, the point $(\mathbf{X}, Y)$ is uniformly distributed in $B_D(C\,p)$. If $(\mathbf{X}, Y) \in B_D(f)$, then this point is accepted. In accordance with Proposition 17.8, the accepted point is uniformly distributed in $B_D(f)$. Finally, as follows from Proposition 17.9, the $\mathbf{X}$-coordinate is distributed with the PDF f.

While justifying Algorithm 17.6, we did not use the fact that f is a *normalized* PDF. The acceptance-rejection method is often applied to complicated multivariate densities only known up to a multiplicative constant. Thus, the following generalization of Algorithm 17.6 is quite useful in dealing with such cases. Consider the sampling from a PDF *proportional* to some nonnegative integrable function f. Suppose that there exists another integrable function g so that it majorizes f, i.e., $f(\mathbf{x}) \leqslant g(\mathbf{x})$ for all $\mathbf{x}$. Let f and g have the same support D. The sampling algorithm is as follows.

Algorithm 17.7 The Acceptance-Rejection Method (Version 2).

(1) Sample $\mathbf{X}$ from the PDF $p \propto g$, that is, $p(\mathbf{x}) = \left(\int_{\mathbb{R}^n} g(\mathbf{x})\,d\mathbf{x}\right)^{-1} g(\mathbf{x})$.

(2) Generate $U \sim Unif(0, 1)$ independent of $\mathbf{X}$.

(3) If $U < \frac{f(\mathbf{X})}{g(\mathbf{X})}$, then accept $\mathbf{X}$, otherwise return to step 1.

A proposed draw $\mathbf{X}$ is accepted if the point $(\mathbf{X}, Y)$ being sampled uniformly in the body of g belongs to the body of f. Therefore, the probability of accepting $\mathbf{X}$ equals the ratio of the volumes of $B_D(f)$ and $B_D(g)$:

$$\mathbb{P}(\text{Accept}) = \frac{|B_D(f)|}{|B_D(g)|} = \frac{\int_D f(\mathbf{x})\,\mathrm{d}\mathbf{x}}{\int_D g(\mathbf{x})\,\mathrm{d}\mathbf{x}}.$$

To maximize this probability, we need to choose g as close to f as possible (see Figure 17.5). The average number of trials per one accepted draw (the computational cost of the acceptance-rejection method) is

$$\text{Cost} = \mathbb{P}(\text{Accept})^{-1} = \frac{\int_D g(\mathbf{x})\,\mathrm{d}\mathbf{x}}{\int_D f(\mathbf{x})\,\mathrm{d}\mathbf{x}}.$$

Example 17.9. Develop an acceptance-rejection method for the PDF

$$f(x) = \frac{2\,\arcsin x}{\pi - 2}, \quad 0 < x < 1.$$

Solution. By using the property that $\arcsin x \leqslant \frac{\pi}{2} x$ for $0 < x < 1$, we obtain that

$$f(x) \leqslant g(x) := \frac{\pi}{\pi - 2}\, x \text{ for } 0 < x < 1.$$

The ratio of f and g is $\frac{\arcsin x}{(\pi/2)x}$. The proposal PDF $p \propto g$ is given by $p(x) = 2x\,\mathbb{I}_{(0,1)}(x)$. The sampling algorithm is as follows.

(1) Generate two i.i.d. $U_1, U_2 \sim Unif(0, 1)$.

(2) Sample $X \sim p$ by using the inverse CDF method: $X = \sqrt{U_1}$.

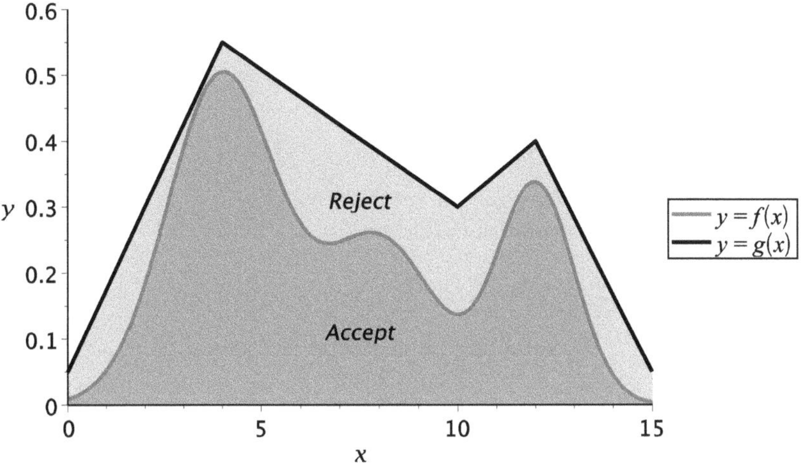

FIGURE 17.5: The acceptance-rejection method.

(3) Accept X if $U_2 \leqslant \frac{\arcsin X}{(\pi/2)X}$. Otherwise, return to (1). □

Example 17.10. Develop an acceptance-rejection method for the standard normal distribution using the double-exponential sampling distribution as a proposal one. Find the computational cost.

Solution. The standard normal PDF n is proportional to $f(x) = \mathrm{e}^{-x^2/2}$. We have the following upper bound for f:

$$\exp\left(-\frac{x^2}{2}\right) = \exp\left(-\frac{x^2 - 2|x| + 1}{2} - |x| + \frac{1}{2}\right) = \exp\left(-\frac{(|x|-1)^2}{2}\right)\sqrt{\mathrm{e}}\,\mathrm{e}^{-|x|} \leqslant \sqrt{\mathrm{e}}\,\mathrm{e}^{-|x|}.$$

So the majorizing function is $g(x) = \sqrt{\mathrm{e}}\,\mathrm{e}^{-|x|}$. Therefore, the proposal probability distribution is the double-exponential distribution with the PDF $p(x) = \frac{1}{2}\mathrm{e}^{-|x|}$, which can be expressed as a mixture:

$$p(x) = \frac{1}{2}\mathrm{e}^{-x}\,\mathbb{I}_{[0,\infty)}(x) + \frac{1}{2}\mathrm{e}^{-|x|}\,\mathbb{I}_{(-\infty,0)}(x).$$

To sample from p, we first obtain a draw from $Exp(1)$ and then assign a random sign to it:

$$X = \begin{cases} Y & \text{with probabilty } \frac{1}{2}, \\ -Y & \text{with probabilty } \frac{1}{2}, \end{cases} \quad \text{where } Y \sim Exp(1).$$

As a result, we obtain the following algorithm.

(1) Generate three i.i.d. $U_1, U_2, U_1 \sim Unif(0,1)$.

(2) Sample $X \sim p$ by using the composition method: $X = \operatorname{sgn}(U_1 - \frac{1}{2}) \ln U_2$.

(3) Accept X if $U_3 \leqslant \frac{f(X)}{g(X)} = \exp\left(-\frac{(|X|-1)^2}{2}\right)$. Otherwise, return to (1).

The probability of acceptance P_A is

$$P_A = \frac{\int_{-\infty}^{\infty} \mathrm{e}^{-x^2/2}\,\mathrm{d}x}{\sqrt{\mathrm{e}}\int_{-\infty}^{\infty} \mathrm{e}^{-|x|}\,\mathrm{d}x} = \frac{\sqrt{2\pi}}{2\sqrt{\mathrm{e}}} = \sqrt{\frac{\pi}{2\mathrm{e}}} \cong 0.7602.$$

Therefore, the computational cost is E[# trials per acceptance] $= \frac{1}{P_A} \cong 1.3155$. □

17.3.5 Multivariate Sampling

17.3.5.1 Sampling by Conditioning

A d-variate joint PDF $f_{\mathbf{X}}$ of the random vector $\mathbf{X} = [X_1, X_2, \ldots, X_d]^\top$ can be represented as a product of univariate conditional densities:

$$f_{\mathbf{X}}(\mathbf{x}) = f_{X_1}(x_1) f_{X_2|X_1}(x_2|x_1) \cdots f_{X_d|X_1,\ldots,X_{d-1}}(x_d|x_1,\ldots,x_{d-1}),$$

where $\mathbf{x} = [x_1, x_2, \ldots, x_d]^\top$. The sampling procedure is as follows:

Step 1. Generate $X_1 \sim f_{X_1}$.
Step 2. Generate X_2 conditional on X_1 from $f_{X_2|X_1}$.
⋮
Step d. Generate X_d conditional on $X_1, \ldots, X_{d-1}$ from $f_{X_d|X_1,\ldots,X_{d-1}}$.

This method is simplified if the components X_j, $1 \leqslant j \leqslant d$, are independent random variables. The joint PDF is then a product of marginal PDFs:

$$f_{\mathbf{X}}(\mathbf{x}) = \prod_{j=1}^{d} f_{X_j}(x_j).$$

Example 17.11. Construct a sampling algorithm for a random vector $\mathbf{X}$ uniformly distributed in a hyperparallelepiped $D = \prod_{j=1}^{d}(a_j, b_j)$, $a_j < b_j$, $1 \leqslant j \leqslant d$.

Solution. The joint PDF is a product of d marginal uniform densities:

$$\begin{aligned} f_{\mathbf{X}}(\mathbf{x}) &= \frac{1}{|D|}\mathbb{I}_D(\mathbf{x}) = \frac{1}{(b_1 - a_1)\cdots(b_d - a_d)}\mathbb{I}_{(a_1,b_1)}(x_1)\cdots\mathbb{I}_{(a_d,b_d)}(x_d) \\ &= \prod_{j=1}^{d}\frac{1}{b_j - a_j}\mathbb{I}_{(a_j,b_j)}(x_j) \equiv \prod_{j=1}^{d} f_{X_j}(x_j). \end{aligned}$$

Therefore, the vector $\mathbf{X}$ is formed of d i.i.d. uniformly distributed random variables:

$$X_j = a_j + (b_j - a_j)U_j, \quad U_j \sim \mathit{Unif}(0,1), \quad 1 \leqslant j \leqslant d.$$ □

17.3.5.2 The Box–Müller method

A pair of independent standard normal random variables Z_1 and Z_2 can be generated from two independent *Unif*$(0, 1)$-distributed random variables by using the following steps.

(1) Define the random variables R and Θ implicitly by

$$Z_1 = R\cos\Theta \text{ and } Z_2 = R\sin\Theta. \tag{17.17}$$

(2) One can show that R and Θ are independent random variables. Moreover, they can be simulated by the following formulae:

$$R = \sqrt{-2\ln U_1} \text{ and } \Theta = 2\pi U_2, \tag{17.18}$$

where U_1 and U_2 are independent *Unif*$(0, 1)$-distributed random variables.

(3) Therefore, Z_1 and Z_2 can be expressed in terms of U_1 and U_2 as follows:

$$Z_1 = \sqrt{-2\ln U_1}\cos(2\pi U_2) \text{ and } Z_2 = \sqrt{-2\ln U_1}\sin(2\pi U_2). \tag{17.19}$$

To justify the Box–Müller method we apply the following theorem.

Theorem 17.11 (Bivariate Transformation Theorem, e.g., See Gut (2009)). *Consider X and Y—jointly continuous random variables, and a one-to-one bivariate continuously differentiable transformation defined on the support of (X, Y) by $u = g(x, y)$ and $v = h(x, y)$. The joint PDF of $U := g(X, Y)$ and $V := h(X, Y)$ is $f_{U,V}(u, v) = \frac{1}{|J(x,y)|} f_{X,Y}(x, y)$, where (x, y) is a unique solution to $\begin{cases} g(x, y) = u \\ h(x, y) = v \end{cases}$ and $J(x, y)$ is the Jacobian determinant of the transformation defined by*

$$J(x, y) = \det \begin{bmatrix} \frac{\partial g}{\partial x}(x, y) & \frac{\partial g}{\partial y}(x, y) \\ \frac{\partial h}{\partial x}(x, y) & \frac{\partial h}{\partial y}(x, y) \end{bmatrix}.$$

Since Z_1 and Z_2 are independent standard normal random variables, their joint PDF is

$$f_{Z_1,Z_1}(z_1, z_2) = n(z_1)\, n(z_2) = \frac{1}{2\pi} e^{-\frac{1}{2}(z_1^2+z_2^2)}.$$

The bivariate transformation theorem allows us to obtain a joint PDF of the pair (R, Θ). The Jacobian determinant of the transformation in (17.17) is equal to r. Thus,

$$f_{Z_1,Z_2}(r\cos\theta, r\sin\theta) = \frac{1}{r} f_{R,\Theta}(r, \theta) \implies f_{R,\Theta}(r, \theta) = r e^{-r^2/2} \frac{1}{2\pi},$$

for $r > 0$ and $0 < \theta < 2\pi$. The joint PDF of R and Θ is a product of the marginal PDFs $f_R(r) = r e^{-r^2/2}\, \mathbb{I}_{[0,\infty)}(r)$ and $f_\Theta(\theta) = \frac{1}{2\pi} \mathbb{I}_{[0,2\pi)}(\theta)$. Therefore, R and Θ are independent random variable. Let us apply the inverse CDF method to generate R:

$$F_R(r) = 1 - e^{-r^2/2} \implies R = F_R^{-1}(1 - U_1) = \sqrt{-2\ln U_1},$$

where $U_1 \sim \mathit{Unif}(0, 1)$. Moreover, $\Theta \sim \mathit{Unif}(0, 2\pi)$, hence $\Theta = 2\pi U_2$ with $U_2 \sim \mathit{Unif}(0, 1)$. So, (17.18) is proved and (17.19) follows.

17.3.5.3 Simulation of Multivariate Normals

Consider a multivariate normal vector $\mathbf{X} = [X_1, X_2, \ldots, X_d]^\top \sim \mathit{Norm}_d(\boldsymbol{\mu}, \boldsymbol{\Sigma})$. If the covariance matrix $\boldsymbol{\Sigma}$ is a diagonal matrix $\text{diag}(\sigma_1^2, \ldots, \sigma_d^2)$, then X_j, $1 \leqslant j \leqslant d$, are all independent normals which can be expressed in terms of independent standard normal variables as follows:

$$X_j = \mu_j + \sigma_j Z_j, \quad Z_j \sim \mathit{Norm}(0, 1), \quad 1 \leqslant j \leqslant d.$$

Independent standard normals can be generated by the Box–Müller method, by the acceptance-rejection method, or by the inverse CDF method. In the latter case we set $Z = \mathcal{N}^{-1}(U)$ with $U \sim \mathit{Unif}(0, 1)$, where the inverse normal CDF $\mathcal{N}^{-1}(x)$ can be calculated numerically. One interesting application of standard normal random variables is the sampling of an isotropic vector in d dimensions.

(1) Sample i.i.d. $Z_1, \ldots, Z_d \sim \mathit{Norm}(0, 1)$

(2) Define the vector $\mathbf{X} = [X_1, X_2, \ldots, X_d]^\top$ by $X_j = Z_j/R_d$, $1 \leqslant j \leqslant d$, where $R_d^2 = Z_1^2+\cdots+Z_d^2$. [Note that R_d^2 is a chi-square random variable with d degrees of freedom.]

(3) As a result, $\mathbf{X}$ is uniformly distributed on a unit d-dimensional sphere.

Suppose that the covariance matrix $\mathbf{\Sigma}$ has nonzero off-diagonal elements. Consider the following two general methods of sampling $\mathbf{X}$:

(1) using the Cholesky factorization of the covariance matrix;

(2) using the conditional normal distribution.

Sampling by the Cholesky Factorization.
Let $\mathbf{L}$ be a lower-triangular matrix from the Cholesky factorization of $\mathbf{\Sigma}$, i.e., $\mathbf{\Sigma} = \mathbf{L}\,\mathbf{L}^\top$. Let $\mathbf{Z}$ be a d-dimensional vector formed by i.i.d. standard normals $Z_j \sim Norm(0,1)$, $j = 1,2,\ldots,d$. Then, we set

$$\mathbf{X} := \boldsymbol{\mu} + \mathbf{L}\,\mathbf{Z} \sim Norm_d(\boldsymbol{\mu}, \mathbf{\Sigma}).$$

Conditional Normal.
Let us split the vector $\mathbf{X}$ into two parts:

$$\mathbf{X} = \begin{bmatrix} \mathbf{X}_1 \\ \mathbf{X}_2 \end{bmatrix}, \text{ where } \mathbf{X}_1 \in \mathbb{R}^m \text{ and } \mathbf{X}_2 \in \mathbb{R}^{d-m}$$

for some $1 \leqslant m < d$. Split also the vector $\boldsymbol{\mu}$ and matrix $\mathbf{\Sigma}$ to represent them in block form:

$$\boldsymbol{\mu} = \begin{bmatrix} \boldsymbol{\mu}_1 \\ \boldsymbol{\mu}_2 \end{bmatrix} \text{ and } \mathbf{\Sigma} = \begin{bmatrix} \mathbf{\Sigma}_{11} & \mathbf{\Sigma}_{12} \\ \mathbf{\Sigma}_{21} & \mathbf{\Sigma}_{22} \end{bmatrix},$$

where $\boldsymbol{\mu}_1 \in \mathbb{R}^m$ and $\boldsymbol{\mu}_2 \in \mathbb{R}^{d-m}$ are vectors; $\mathbf{\Sigma}_{11} \in \mathbb{R}^{m\times m}$, $\mathbf{\Sigma}_{12} \in \mathbb{R}^{m\times(d-m)}$, $\mathbf{\Sigma}_{21} \in \mathbb{R}^{(d-m)\times m}$, and $\mathbf{\Sigma}_{22} \in \mathbb{R}^{(d-m)\times(d-m)}$ are matrices. Then, the conditional distribution of $\mathbf{X}_1$ given the value of $\mathbf{X}_2$ is normal:

$$\mathbf{X}_1|\{\mathbf{X}_2 = \mathbf{x}_2\} \sim Norm_m\left(\boldsymbol{\mu}_1 + \mathbf{\Sigma}_{12}\,\mathbf{\Sigma}_{22}^{-1}(\mathbf{x}_2 - \boldsymbol{\mu}_2), \mathbf{\Sigma}_{11} - \mathbf{\Sigma}_{12}\,\mathbf{\Sigma}_{22}^{-1}\mathbf{\Sigma}_{21}\right). \tag{17.20}$$

Example 17.12. Construct two methods of sampling from the trivariate normal distribution

$$Norm_3\left(\begin{bmatrix} 3 \\ 2 \\ 4 \end{bmatrix}, \begin{bmatrix} 9 & 0 & 0 \\ 0 & 4 & 2 \\ 0 & 2 & 3 \end{bmatrix}\right),$$

using the Cholesky factorization and the conditional sampling approach, respectively.

Solution.

1. Find the Cholesky factorization of the covariance matrix. Let us solve the matrix equation $\mathbf{\Sigma} = \mathbf{L}\,\mathbf{L}^\top$ to find $\mathbf{L}$:

$$\begin{bmatrix} 9 & 0 & 0 \\ 0 & 4 & 2 \\ 0 & 2 & 3 \end{bmatrix} = \begin{bmatrix} \ell_{11} & 0 & 0 \\ \ell_{21} & \ell_{22} & 0 \\ \ell_{31} & \ell_{32} & \ell_{33} \end{bmatrix} \begin{bmatrix} \ell_{11} & \ell_{21} & \ell_{31} \\ 0 & \ell_{22} & \ell_{32} \\ 0 & 0 & \ell_{33} \end{bmatrix} \implies \mathbf{L} = \begin{bmatrix} 3 & 0 & 0 \\ 0 & 2 & 0 \\ 0 & 1 & \sqrt{2} \end{bmatrix}.$$

Thus, we obtain the following sampling formulae:

$$\begin{bmatrix} X_1 \\ X_2 \\ X_3 \end{bmatrix} = \begin{bmatrix} 3 \\ 2 \\ 4 \end{bmatrix} + \begin{bmatrix} 3 & 0 & 0 \\ 0 & 2 & 0 \\ 0 & 1 & \sqrt{2} \end{bmatrix} \begin{bmatrix} Z_1 \\ Z_2 \\ Z_3 \end{bmatrix} \implies \begin{cases} X_1 = 3 + 3Z_1, \\ X_2 = 2 + 2Z_2, \\ X_3 = 4 + Z_2 + \sqrt{2}Z_3, \end{cases}$$

where Z_1, Z_2, Z_3 are i.i.d. standard normals.

2. First, sample $X_1 \sim Norm(3,9)$: $X_1 = 3+3Z_1$. Second, sample X_2 conditional on X_1. Recall that two normal variables are independent iff they are uncorrelated. Hence X_2 is independent of X_1 since $\text{Cov}(X_1, X_2) = 0$. Therefore, $(X_2|X_1) \stackrel{d}{=} X_2 \sim Norm(2,4)$ and $X_2 = 2 + 2Z_2$. Third, sample X_3 conditional on X_1 and X_2. Again, X_3 and X_1 are uncorrelated, hence $(X_3|X_1, X_2) \stackrel{d}{=} (X_3|X_2)$. We have
$$\begin{bmatrix} X_3 \\ X_2 \end{bmatrix} \sim Norm_2\left(\begin{bmatrix} 4 \\ 2 \end{bmatrix}, \begin{bmatrix} 3 & 2 \\ 2 & 4 \end{bmatrix}\right) \implies X_3|X_2 \sim Norm\left(3 + \frac{X_2}{2}, 2\right).$$
Thus, we have $X_3 = 3 + \frac{X_2}{2} + \sqrt{2}Z_3 = 4 + Z_2 + \sqrt{2}Z_3$. □

17.4 Simulation of Random Processes

A typical problem that requires simulation of sample paths of a stochastic process $\{X(t)\}_{t\geqslant 0}$ is the estimation of a mathematical expectation of the form
$$\mathrm{E}\big[g(\{X(t) : 0 \leqslant t \leqslant T\})\big] \tag{17.21}$$
with some function g of an X-path. There are several possible cases.

1. The function g depends on a discretely monitored skeleton of the process X:
$$g = g(X(t_1), X(t_2) \ldots, X(t_m)), \quad 0 \leqslant t_1 < t_2 \cdots < t_m \leqslant T.$$
One special case is when $g = g(X(T))$. For example, the estimation of $\mathrm{E}[g(X(T))]$ is required to price a European-style option.

2. The function g depends on path-dependent quantities such as the running maximum/minimum of the process and the first passage time. It may be possible to sample such path-dependent quantities directly from their distributions rather than calculate them from a sample path.

3. The function g depends on a full sample path of process X on $[0, T]$. Since it may be not feasible to generate a complete sample path of a continuous-time process (unless we deal with a Poisson process or a similar process with piecewise paths that changes at a finite number of time points), such a full path can only be obtained by applying an interpolation algorithm to a path skeleton.

So our goal is to sample a path skeleton
$$X(t_1), X(t_2), \ldots, X(t_m) \text{ for } 0 \leqslant t_1 < t_2 < \cdots < t_m \leqslant T.$$
The skeleton can be generated from its exact multivariate distribution. In this case, the problem (17.21) can be reduced to the estimation of a multivariate integral of the form
$$\int_{R^m} g(x_1, x_2, \ldots, x_m)\, f_{X(t_1),X(t_2),\ldots,X(t_m)}(x_1, x_2, \ldots, x_m)\, \mathrm{d}x_1\, \mathrm{d}x_2 \cdots \mathrm{d}x_m$$
where f is a joint PDF of $X(t_1), X(t_2), \ldots, X(t_m)$. Another approach is to sample an approximation path by applying some discretization scheme. Note that Brownian motion and other Gaussian processes as well as some jump processes can be sampled precisely form their path distributions. General diffusions can be simulated approximately by using, for example, the Euler approximation scheme.

17.4.1 Simulation of Brownian Processes

17.4.1.1 Sequential Sampling

The sequential sampling of Brownian motion (BM) and geometric Brownian motion is based on the property that Brownian increments on nonoverlapping intervals are independent. Consider a scaled Brownian motion with drift, $W_{x_0}^{(\mu,\sigma)}(t) := x_0+\mu t+\sigma W(t)$. The standard BM is recovered from the process $W_{x_0}^{(\mu,\sigma)}$ if we take $x_0 = 0$, $\mu = 0$, and $\sigma = 1$. Suppose that the process $W_{x_0}^{(\mu,\sigma)}$ is to be sampled at a set of time points $0 = t_0 < t_1 < t_2 < \cdots < t_m$. For all $j \geqslant 1$, we have

$$W_{x_0}^{(\mu,\sigma)}(t_j) = x_0 + \mu t_j + \sigma W(t_j) = W_{x_0}^{(\mu,\sigma)}(t_{j-1}) + \mu\,(t_j - t_{j-1}) + \sigma\,(W(t_j) - W(t_{j-1})).$$

Since the increment $W(t_j) - W(t_{j-1}) \sim \mathit{Norm}(0, t_j - t_{j-1})$ is independent of $W_{x_0}^{(\mu,\sigma)}(t_{j-1})$, we obtain the following simple algorithm.

Algorithm 17.8 Sequential Simulation of a Scaled BM with Drift.

input: x_0, μ, σ, $0 = t_0 < t_1 < t_2 < \cdots < t_m$
set $W_{x_0}^{(\mu,\sigma)}(0) = x_0$
for j **from** 1 **to** m **do**
  generate $Z_j \leftarrow \mathit{Norm}(0,1)$
  set $W_{x_0}^{(\mu,\sigma)}(t_j) \leftarrow W_{x_0}^{(\mu,\sigma)}(t_{j-1}) + \mu\,(t_j - t_{j-1}) + \sigma\,\sqrt{t_j - t_{j-1}}\,Z_j$
end for
return $\{W_{x_0}^{(\mu,\sigma)}(t_j)\}_{0\leqslant j\leqslant m}$

The sample path of a geometric Brownian motion $S(t) = S_0 \mathrm{e}^{\mu t+\sigma W(t)} = \mathrm{e}^{\ln S_0+\mu t+\sigma W(t)}$ can be obtained by taking the exponential function of a sample path of the scaled BM with drift $W_{x_0}^{(\mu,\sigma)}(t)$ that starts at $x_0 = \ln S_0$, i.e., $S(t) = \exp\left(W_{\ln S_0}^{(\mu,\sigma)}(t)\right)$.

Now we consider is a multidimensional BM $\mathbf{W}(t) = \left[W_1(t), W_2(t), \ldots, W_d(t)\right]^\top$. Each component of $\mathbf{W}(t)$ is a standard Brownian motion. Suppose that the processes W_j, $1 \leqslant j \leqslant d$, are correlated. For $1 \leqslant i,j \leqslant d$, the correlation coefficient between $W_i(t)$ and $W_j(t)$ is

$$\rho_{ij} = \mathrm{Corr}(W_i(t), W_j(t)) = \frac{\mathrm{E}[W_i(t)W_j(t)] - \mathrm{E}[W_i(t)]\mathrm{E}[W_j(t)]}{\sqrt{\mathrm{E}[W_i^2(t)]\,E[W_j^2(t)]}} = \frac{\mathrm{E}[W_i(t)W_j(t)]}{t}.$$

Let $\mathbf{R} = [\,\rho_{ij}\,]_{1\leqslant i,j\leqslant d}$ be the correlation matrix. $\mathbf{R}$ is a positive definite matrix with ones on the main diagonal. If we deal with independent Brownian motions then $\mathbf{R} = \mathbf{I}$. Apply the Cholesky factorization to find a lower triangular matrix $\mathbf{L}$ so that $\mathbf{R} = \mathbf{L}\,\mathbf{L}^\top$. For example, for the two-dimensional case we have

$$\mathbf{R} = \begin{bmatrix} 1 & \rho_{12} \\ \rho_{12} & 1 \end{bmatrix} = \mathbf{L}\,\mathbf{L}^\top \implies \mathbf{L} = \begin{bmatrix} 1 & 0 \\ \rho_{12} & \sqrt{1-\rho_{12}} \end{bmatrix}.$$

Algorithm 17.9 allows us to obtain a realization of $\mathbf{W}$ at time points $t_0, t_1, \ldots, t_m$ with $0 = t_0 < t_1 < \cdots < t_m$.

17.4.1.2 Bridge Sampling

Previously, we derived the probability distribution of Brownian motion pinned at the endpoints of a time interval. Recall that Brownian motion conditional on $W(0) = a$ and

Algorithm 17.9 Sequential Simulation of a Standard d-Dimensional BM.

input: $\mathbf{L}$ and $0 = t_0 < t_1 < t_2 < \cdots < t_m$
set $\mathbf{W}(0) = \mathbf{0}$
for j **from** 1 **to** m **do**
  generate d i.i.d. variates $Z_i^j \leftarrow Norm(0,1)$, $1 \leqslant i \leqslant d$
  set $\mathbf{W}(t_j) \leftarrow \mathbf{W}(t_{j-1}) + \sqrt{t_j - t_{j-1}}\,\mathbf{L}\,\mathbf{Z}^j$, where $\mathbf{Z}^j = [\,Z_1^j,\, Z_2^j, \ldots,\, Z_d^j\,]^\top$
end for
return $\{\mathbf{W}(t_j)\}_{0 \leqslant j \leqslant m}$

$W(T) = b$ is called a Brownian bridge from a to b on $[0, T]$. There exist several applications of the Brownian bridge. First, the bridge distribution can be used to refine a sample skeleton. Second, it can be used as an alternative to the sequential simulation method for sampling a Brownian trajectory.

Suppose that a standard BM is sampled at m time moments $0 = t_0 < t_1 < \cdots < t_m = T$ and we wish to sample $W(s)$ at some additional time moment $s \in (0, T)$ conditional on these values. Let $s \in (t_j, t_{j+1})$ for some $j \in \{0, 1, \ldots, m-1\}$. It follows from the Markov property of BM that

$$\big(W(s) \mid \{W(t_i) = x_i,\ 0 \leqslant i \leqslant m\}\big) \stackrel{d}{=} \big(W(s) \mid \{W(t_j) = x_j,\ W(t_{j+1}) = x_{j+1}\}\big).$$

Hence $W(s)$ can be sampled from the distribution of a Brownian bridge on $[t_i, t_{j+1}]$. By applying this procedure, a sample path can be refined without re-sampling its values at $t_1, t_2, \ldots, t_m$.

Consider the so-called *dyadic partition* of the time interval $[0, T]$ with $m = 2^k$ points $t_j = \frac{j}{m}T$, where $j = 0, 1, \ldots, m$ and $k \geqslant 1$. Let a realization of Brownian motion be sampled as follows:

Step 1: sample $W(t_m)$ conditional on $W(t_0) = 0$,
Step 2: sample $W(t_{m/2})$ conditional on $W(t_0), W(t_m)$,
Step 3: sample $W(t_{m/4})$ conditional on $W(t_0), W(t_{m/2})$,
Step 4: sample $W(t_{3m/4})$ conditional on $W(t_{m/2}), W(t_m)$,
⋮
Step m: sample $W(t_{m-1})$ conditional on $W(t_{m-2}), W(t_m)$.

In other words, first we sample $W(t_m)$ and after that for each t_j, $1 \leqslant j \leqslant m-1$, $W(t_j)$ is sampled conditionally on $W(t_\ell)$ and $W(t_k)$ previously generated, where the indices ℓ and k satisfy $0 \leqslant \ell < j < k \leqslant m$ and $j = \frac{\ell+k}{2}$. As a result, a trajectory of BM is sampled at the time points in the following order of generation:

$$\underbrace{t_m}\,,\underbrace{t_{m/2}},\underbrace{t_{m/4}, t_{3m/4}},\underbrace{t_{m/8}, t_{3m/8}, t_{5m/8}, t_{7m/8}}, \ldots \qquad \underbrace{t_2, t_6, t_{10}, \ldots, t_{m-2}},\underbrace{t_1, t_3, \ldots, t_{m-1}} \tag{17.22}$$

Bridge sampling with $m = 8$ time points is illustrated in Figure 17.6.

The bridge sampling algorithm is useful in pricing path-dependent financial instruments. Being applied with (randomized) low-discrepancy numbers, i.e., when the (randomized) quasi-Monte Carlo method is used, it allows us to reduce the variance of a path-dependent estimator. Another advantage is that the bridge sampling algorithm can be easily parallelized.

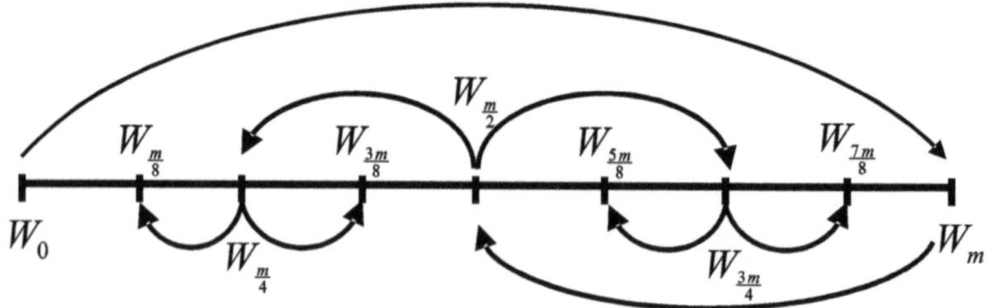

FIGURE 17.6: The bridge sampling. Here W_j denotes $W(t_j)$ for $0 \leqslant j \leqslant m$.

Algorithm 17.10 Brownian Bridge Sampling for a Dyadic Time Partition.

input: the time points $t_j = \frac{j}{m} T$, $j = 0, 1, \ldots, m$, where $m = 2^k$
generate $Z \leftarrow Norm(0, 1)$
set $W(t_0) \leftarrow 0$, $W(t_m) \leftarrow \sqrt{T}\, Z$, and $h \leftarrow T$
for ℓ **from** 1 **to** k **do**
  set $h \leftarrow h/2$
  for j **from** 1 **to** $2^{\ell-1}$ **do**
    generate $Z \leftarrow Norm(0, 1)$
    set $W(t_{(2j-1)2^{k-\ell}}) \leftarrow \frac{1}{2}\left(W(t_{(j-1)2^{k-\ell+1}}) + W(t_{j2^{k-\ell+1}})\right) + \sqrt{h}\, Z$
  end for
end for
return $\{W(t_j)\}_{0 \leqslant j \leqslant m}$

17.4.2 Simulation of Gaussian Processes

In accordance with the definition of a Gaussian process $\{X(t)\}_{t \geqslant 0}$, for any sequence of time points $0 < t_1 < t_2 < \cdots < t_m$, the vector $\mathbf{X} = [X(t_1), X(t_2), \ldots, X(t_m)]^\top$ has a multivariate normal distribution $Norm_m(\boldsymbol{\mu}_m, \boldsymbol{\Sigma}_m)$ with

$$\boldsymbol{\mu}_m = \begin{bmatrix} m_X(t_1) \\ m_X(t_2) \\ \vdots \\ m_X(t_m) \end{bmatrix} \quad \text{and} \quad \boldsymbol{\Sigma}_m = \begin{bmatrix} c_X(t_1,t_1) & c_X(t_1,t_2) & \cdots & c_X(t_1,t_m) \\ c_X(t_2,t_1) & c_X(t_2,t_2) & \cdots & c_X(t_2,t_m) \\ \vdots & \vdots & \ddots & \vdots \\ c_X(t_m,t_1) & c_X(t_m,t_2) & \cdots & c_X(t_m,t_m) \end{bmatrix},$$

where $m_X(t) = \mathrm{E}[X(t)]$ and $c_X(t,s) = \mathrm{Cov}(X(t), X(s))$ are, respectively, the mean and covariance functions. Therefore, the realization of $\mathbf{X}$ at time points $t_1, t_2, \ldots, t_m$ can be constructed by sampling from the m-variate normal distribution $Norm_m(\boldsymbol{\mu}_m, \boldsymbol{\Sigma}_m)$ as follows.

(1) Apply the Cholesky factorization to find a lower-triangular matrix $\mathbf{L}_m$ so that

$$\boldsymbol{\Sigma}_m = \mathbf{L}_m \mathbf{L}_m^\top.$$

(2) Sample m i.i.d. standard normals $Z_1, Z_2, \ldots, Z_m$.

(3) Set $\mathbf{X} = \boldsymbol{\mu}_m + \mathbf{L}_m \mathbf{Z}$ where $\mathbf{Z} = [Z_1, Z_2, \ldots, Z_m]^\top$.

This generic algorithm can be applied to any Gaussian process including those listed below.

Brownian Motion $W_{x_0}^{(\mu,\sigma)}(t) := x_0 + \mu t + \sigma W(t)$ with $m(t) = x_0 + \mu t$ and $c(t,s) = \sigma^2\,(t \wedge s)$ for $s, t \geqslant 0$.

Itô Processes $X(t) := \int_0^t \mu(s)\,\mathrm{d}s + \int_0^t \sigma(t)\,\mathrm{d}W(s)$ with $m(t) = \int_0^t \mu(u)\,du$ and $c(t,s) = \int_0^{t\wedge s} \sigma^2(u)\,du$ for $s,t \geqslant 0$.

Fractional Brownian Motion $W^{(H)}(t)$ with $c(t,s) = (t^{2H} + s^{2H} - |t-s|^{2H})/2$ and $m(t) = 0$. Here, $H \in [0,1]$ is the so-called Hurst parameter. Note that $W^{(1/2)}$ is a standard BM with $c(t,s) = (t+s-|t-s|)/2 = t\wedge s$ for $s,t \geqslant 0$.

Standard Brownian Bridge from a to b on $[0,T]$, denoted $\{B^{a,b}_{[0,T]}(t)\}_{0\leqslant t\leqslant T}$, with $m(t) = \frac{a\,(T-t)+b\,t}{T}$ and $c(t,s) = t\wedge s\,(T - t\vee s)$ for $0 \leqslant s,t \leqslant T$.

17.4.3 Diffusion Processes: Exact Simulation Methods

A diffusion process $\{X(t)\}_{t\geqslant 0}$ is a solution to an initial value problem for a stochastic differential equation (SDE):

$$\begin{cases} \mathrm{d}X(t) = \mu(t,X(t))\,\mathrm{d}t + \sigma(t,X(t))\,\mathrm{d}W(t),\ t\geqslant 0, \\ X(0) = X_0. \end{cases} \tag{17.23}$$

The process can also be written in the integral form

$$X(t) = X_0 + \int_0^t \mu(s,X(s))\,\mathrm{d}s + \int_0^t \sigma(s,X(s))\,\mathrm{d}W(s). \tag{17.24}$$

The functions μ and σ are called the drift coefficient and the diffusion coefficient, respectively.

The problem (17.23) (or (17.24)) can be solved analytically or numerically. One approach is to represent the process X as an explicit function of the underlying Brownian motion W:

$$X(t) = f(t, \{W(s)\ :\ 0\leqslant s\leqslant t\}). \tag{17.25}$$

Such an explicit representation is called a *strong solution* to (17.23). As a result of (17.25), a sample path of the X-process is obtained by transforming a Brownian trajectory. Another approach is to find the transition PDF for X by solving the Kolmogorov equation. The strong solution and/or the transition density can be used to generate a realization of the diffusion X from its exact finite-dimensional distribution. As usual, our goal is to generate a path skeleton for an arbitrary sequence of time points. Alternatively, being unable to analytically solve (17.23), one can apply a numerical scheme to find an approximate realization of the diffusion process. The Euler scheme, which is the simplest and most popular simulation method, is considered in Section 17.4.4.

17.4.3.1 The Stochastic Calculus Approach

Most of the SDEs for which we can find a strong solution are SDEs with linear (w.r.t. the space variable) drift and diffusion coefficients. Let us consider some examples of such SDEs and derive transition probability distributions of their solutions.

1. Geometric Brownian motion is a solution to

$$\mathrm{d}X(t) = \mu X(t)\,\mathrm{d}t + \sigma X(t)\,\mathrm{d}W(t), \quad X(0) = X_0.$$

Applying the Itô formula gives $X(t) = X_0 \mathrm{e}^{(\mu-\sigma^2/2)t+\sigma W(t)}$. To model the transition $X(s)\to X(t)$ for $0\leqslant s<t$, we use the following representation:

$$X(t) = X(s)\,\mathrm{e}^{(\mu-\sigma^2/2)(t-s)+\sigma\,(W(t)-W(s))} \stackrel{d}{=} X(s)\,\mathrm{e}^{(\mu-\sigma^2/2)(t-s)+\sigma\sqrt{t-s}Z},$$

where $Z \sim \mathit{Norm}(0,1)$ is independent of $X(s)$.

2. The solution to

$$\mathrm{d}X(t) = \mu(t)\,\mathrm{d}t + \sigma(t)\,\mathrm{d}W(t), \quad X(0) = X_0,$$

is a Gaussian process given by $X(t) = X_0 + \int_0^t \mu(u)\,\mathrm{d}u + \int_0^t \sigma(u)\,\mathrm{d}W(u)$. By representing $X(t)$ in terms of $X(s)$ for $0 \leqslant s < t$ and using the fact that the Itô integral $\int_s^t \sigma(u)\,\mathrm{d}W(u)$ is normally distributed with mean 0 and variance $\int_s^t \sigma^2(u)\,\mathrm{d}u$, we obtain that the increments of X are normal:

$$X(t) - X(s) \sim Norm\left(\int_s^t \mu(u)\,\mathrm{d}u, \int_s^t \sigma^2(u)\,\mathrm{d}u\right).$$

Since the increments of X are independent, we are able to model the transition $X(s) \to X(t)$ for all $0 \leqslant s < t$.

3. The Ornstein–Uhlenbeck process is a solution to the SDE with a constant diffusion coefficient and linear drift:

$$\mathrm{d}X(t) = \alpha(b - X(t))\,\mathrm{d}t + \sigma\,\mathrm{d}W(t), \quad X(0) = X_0.$$

The strong solution is

$$X(t) = \mathrm{e}^{-\alpha t} X_0 + \alpha b \int_0^t \mathrm{e}^{-\alpha(t-u)}\,\mathrm{d}u + \sigma \int_0^t \mathrm{e}^{-\alpha(t-u)}\,\mathrm{d}W(u).$$

The conditional distribution of $X(t)$ given $X(s)$ for $0 \leqslant s < t$ is normal:

$$(X(t)|X(s)) \sim Norm\left(\mathrm{e}^{-\alpha(t-s)}X(s) + \alpha b \int_s^t \mathrm{e}^{-\alpha(t-u)}\,\mathrm{d}u, \sigma^2 \int_s^t \mathrm{e}^{-2\alpha(t-u)}\,\mathrm{d}u\right).$$

17.4.3.2 The PDF Approach

Consider a Markov stochastic process $\{X(t)\}_{t \geqslant 0}$ starting at X_0 so that its transition PDF p given by $p(s,t;y,x)\,\mathrm{d}x = \mathbb{P}(X(t) \in \mathrm{d}x \mid X(s) = y)$, $0 \leqslant s < t$, $x, y \in \mathbb{R}$, is known in closed form. Moreover, suppose that it is feasible to sample from the PDF $p(s,t;y,\cdot)$. The general simulation algorithm is as follows.

Algorithm 17.11 Sequential Simulation of a Stochastic Process from Its Transition PDF.

input the PDF p, X_0, $0 = t_0 < t_1 < t_2 < \cdots < t_m$
set $X(0) = X_0$
for $j = 1 \to m$ **do**
  generate $X(t_j) \leftarrow p(t_{j-1}, t_j; X(t_{j-1}), x)$
end for
return $\{X(t_j)\}_{0 \leqslant j \leqslant m}$

For example, due to the time- and space-homogeneity property, the Brownian motion transition PDF $p(s,t;y,x)$ reduces to a two-variable function p_0 given by

$$p_0(t;x) = \frac{1}{\sqrt{2\pi t}} \exp\left(-\frac{x^2}{2t}\right)$$

as follows: $p(s,t;y,x) = p_0(t-s; x-y)$. The latter solves the PDE $\partial_t p_0 = \frac{1}{2}\partial_{xx} p_0$. To generate $(W(t_j)|\{W(t_{j-1}) = y\}) \sim p(t_{j-1}, t_j; y, x) = p_0(t_j - t_{j-1}; x - y)$, we first sample $Z \sim Norm(0,1)$ and then set $W(t_j) \leftarrow y + \sqrt{t_j - t_{j-1}}Z$.

To sample from a transition PDF, we can use a whole arsenal of sampling techniques such as the inversion method, composition approach, and acceptance-rejection technique. The main example considered in the current section is the exact simulation of a family of Bessel diffusions. We start with a squared Bessel (SQB) process and show that its transition PDF reduces to a randomized gamma distribution. Moreover, as demonstrated in Chapter 16, other processes such as the Cox–Ingersolll–Ross (CIR) process and the constant elasticity of variance (CEV) diffusion model can be obtained from the SQB process by means of a scale and time transformation and change of variable.

Let us consider a λ_0-dimensional squared Bessel (SQB) process $\{X(t) \in \mathbb{R}_+\}_{t\geqslant 0}$ obeying the stochastic differential equation (SDE)

$$dX(t) = \lambda_0 \, \mathrm{d}t + \nu\sqrt{X(t)}\, \mathrm{d}W(t), \tag{17.26}$$

with constant parameters λ_0 and $\nu > 0$. For simplicity of presentation, we assume here that $\nu = 2$. The process X is a time-homogeneous Markov process and its transition PDF is given by (16.15).

The transition PDF (16.15) looks very similar to the PDF (17.8) of the noncentral chi-square distribution. In fact, for the case when $\mu \geqslant 0$ or $\mu \in (-1, 0)$ and $x = 0$ is a reflecting boundary (so there is no absorption at the origin), the transition PDF $p(t; y, x)$ of the SQB process reduces to the PDF $f(x; \kappa, \lambda)$ in (17.8) as follows:

$$p(t; y, x) = \frac{1}{t} f\left(\frac{x}{t}; \kappa = 2\mu + 2, \lambda = \frac{y}{t}\right).$$

Following the composition method, the noncentral chi-square PDF (17.8) can be represented as a randomized gamma distribution (17.10). The value of $X(t)$ conditional on $X(s) = y$ has the randomized gamma distribution

$$\mathit{Gamma}\,(Y + \mu + 1, 2(t - s)), \text{ where } Y \sim \mathit{Pois}\left(\frac{y}{2(t-s)}\right), \tag{17.27}$$

for $0 \leqslant s < t$, $y > 0$. Therefore, we have the following sampling algorithm for the SQB process without absorption.

Algorithm 17.12 Sampling of an SQB Process without Absorption (Variant 1).

> **input** $X(0) = X_0 > 0$, $0 = t_0 < t_1 < \cdots < t_m$, $\mu > -1$
> **for** $j = 1 \to m$ **do**
>   **generate** $Y_j \leftarrow \mathit{Pois}\left(\dfrac{X(t_{j-1})}{2(t_j - t_{j-1})}\right)$
>   **generate** $X(t_j) \leftarrow \mathit{Gamma}\,(Y_j + \mu + 1, 2(t_j - t_{j-1}))$
> **end for**
> **return** $\{X(t_j)\}_{0\leqslant j\leqslant m}$

Now consider the case when $\mu < 0$ and $x = 0$ is an exit or a killing boundary. The stochastic process X admits absorption at the origin. It can be shown that the transition PDF p given by (16.15) with $\tilde{\mu} = |\mu|$, $\mu < 0$ does not integrate to one. Let us define the probability P_{sv} of surviving before time t and the probability P_{ab} of absorption before time t for the SQB process starting at $x_0 > 0$:

$$P_{sv}(x_0; t) = \int_0^\infty p(t; x_0, x)\, \mathrm{d}x > 0 \text{ and } P_{ab}(x_0; t) = 1 - P_{sv}(x_0; t) > 0.$$

The probabilities of surviving and absorption of the SQB process before time t are

$$P_{sv}(x;t) = \mathbb{P}\{\tau_0 > t\} = \frac{\gamma\left(|\mu|, \frac{x}{2t}\right)}{\Gamma(|\mu|)} \text{ and } P_{ab}(x;t) = \mathbb{P}\{\tau_0 \leqslant t\} = \frac{\Gamma\left(|\mu|, \frac{x}{2t}\right)}{\Gamma(|\mu|)},$$

respectively, where τ_0 is the first hitting time (FHT) at zero. Here,

$$\gamma(a,x) = \int_0^x t^{a-1}\mathrm{e}^{-t}\,\mathrm{d}t \text{ and } \Gamma(a,x) := \Gamma(a) - \gamma(a,x)$$

are, respectively, the lower and upper incomplete gamma functions.

Observe that the actual transition probability distribution is then a mixture of continuous and discrete probability distributions with the following generalized density:

$$f(X(s) \to X(t)) = P_{sv}(X(s);t-s) \cdot \left(\frac{p(t-s;X(s),X(t))}{P_{sv}(X(s);t-s)}\right) + P_{ab}(X(s);t-s)\cdot\delta(X(t)),$$

for $0 \leqslant s < t$. Here, δ denotes the Dirac delta function that can be viewed as a generalized density of the discrete distribution with an only mass point at zero. With the probability P_{ab}, the process is absorbed at zero. With the additional probability P_{sv}, the process survives. The normalized transition PDF of the SQB process conditioned on the survival of the process before time t is

$$\frac{p(t;x_0,x)}{P_{sv}(x_0;t)} = \frac{\Gamma\left(|\mu|\right)}{\gamma\left(|\mu|, \frac{x_0}{2t}\right)} \left(\frac{x}{x_0}\right)^{\frac{\mu}{2}} \frac{\mathrm{e}^{-(x+x_0)/(2t)}}{2t} I_{|\mu|}\left(\frac{\sqrt{x x_0}}{t}\right). \tag{17.28}$$

As is seen, the function on the right-hand side of (17.28) reduces to the form of (17.15) with $\nu = |\mu|$, $\lambda = x_0/(2t)$, and $\theta = 2t$. Thus, the above normalized transition PDF follows the randomized gamma distribution of the third kind, $Gamma(Y+1, 2t)$, where $Y \sim \mathrm{I}\Gamma(|\mu|, x_0/(2t))$. As a result, we obtain the following sampling algorithm that returns a sample path $[X(t_1), X(t_2), \ldots, X(t_m)]$ and an approximation $\tilde{\tau}_0 \in \{t_1, \ldots, t_m, \infty\}$ of the FHT τ_0.

17.4.4 Diffusion Processes: Approximation Schemes

Approximation schemes can be used whenever an exact method is not available or not feasible. In particular, approximation methods are efficient for the numerical solution of multidimensional stochastic differential equations. Computational schemes for SDEs are based on the same ideas as the numerical methods for solving deterministic differential equations. A discrete-time approximation solution is calculated on a time grid with small time steps. To illustrate the main idea, let us first consider a Cauchy problem for a system of ordinary differential equations written in a vector form:

$$\frac{\mathrm{d}\mathbf{x}(t)}{\mathrm{d}t} = \mathbf{a}(t,\mathbf{x}(t)), \quad t \in [0,T]; \quad \mathbf{x}(0) = \mathbf{x}_0.$$

The solution admits the following integral representation on a small time interval $[t, t+h]$:

$$\mathbf{x}(t+h) = \mathbf{x}(t) + \int_t^{t+h} \mathbf{a}(s,\mathbf{x}(s))\,\mathrm{d}s. \tag{17.29}$$

By applying a rectangle quadrature rule to the integral in (17.29), we derive the so-called Euler approximation scheme:

$$\mathbf{x}(t+h) \approx \mathbf{x}(t) + \mathbf{a}(t,\mathbf{x}(t))\,h.$$

Algorithm 17.13 Sampling of an SQB Process with Absorption (Variant 2).

```
input X_0 > 0, 0 = t_0 < t_1 < ··· < t_m, μ < 0
set X(0) ← X_0, τ̃_0 ← ∞
for j = 1 → m do
  if τ̃_0 = ∞ then
    set p_a ← Γ(|μ|, X(t_{j-1}) / (2(t_j − t_{j-1}))) / Γ(|μ|)
    generate U_j ← Unif(0, 1)
    if U_j < p_a then τ̃_0 ← t_j
  end if
  if t_j < τ̃_0 then
    generate Y_j ← IΓ(|μ|, X(t_{j-1}) / (2(t_j − t_{j-1})))
    generate X(t_j) ← Gamma(Y_j + 1, 2(t_j − t_{j-1}))
  else
    set X(t_j) ← 0
  end if
end for
return {X(t_j)}_{0⩽j⩽m} and τ̃_0
```

A discrete-time numerical solution $\{\mathbf{x}_k^h\}_{k=0,1,2,\ldots}$ on the time grid $\{t_k := kh\}_{k=0,1,2,\ldots}$ is given by

$$\mathbf{x}_0^h = \mathbf{x}_0, \quad \mathbf{x}_{k+1}^h = \mathbf{x}_k^h + \mathbf{a}(t_k, \mathbf{x}_k^h)\, h, \quad k = 0, 1, 2, \ldots$$

The numerical solution approximates the genuine solution: $\mathbf{x}_k^h \approx \mathbf{x}(t_k)$, $k \geqslant 1$; it converges to the exact one as the time step h goes to zero. Applying a similar approach to stochastic differential equations, we can construct a discrete-time approximate realization (a skeleton) of a sample path. Since we deal with stochastic processes, there are different types of convergence of the numerical solution to the exact one.

17.4.4.1 Types of Convergence

Consider a continuous-time d-dimensional stochastic process $\{\mathbf{X}(t)\}_{t\in[0,T]}$ and its discrete-time approximation $\{\mathbf{X}_k^h\}_{0\leqslant k\leqslant m}$ defined on a time grid $0 = t_0 < t_1 < \cdots < t_m = T$ with the maximum step size $h = \max\{t_k - t_{k-1} : k = 1, 2, \ldots, m\}$. The simplest time grid is an equally spaced grid with $t_k = kh$ with $h = \frac{T}{m}$ and $k = 0, 1, \ldots, m$. Let us analyze the convergence of the approximate solution $\mathbf{X}^h$ to the genuine solution $\mathbf{X}$, as $h \to \infty$, in the Euclidian vector norm $\|\mathbf{x}\|_2 = \sqrt{x_1^2 + \cdots + x_d^2}$, where $\mathbf{x} = [x_1, x_2, \ldots, x_d]^\top \in \mathbb{R}^d$. We say that the approximation $\mathbf{X}^h$ has:

- the *strong order of convergence* $\alpha > 0$ if there exists $C > 0$ so that
$$\mathrm{E}\left[\|\mathbf{X}_k^h - \mathbf{X}(t_k)\|\right] \leqslant C\, h^\alpha, \quad \forall k = 0, 1, 2, \ldots, m;$$
- the *mean-square order of convergence* $\beta > 0$ if there exists $C > 0$ so that
$$\left(\mathrm{E}\left[\|\mathbf{X}_k^h - \mathbf{X}(t_k)\|^2\right]\right)^{1/2} \leqslant C\, h^\beta, \quad \forall k = 0, 1, 2, \ldots, m;$$
- the *weak order of convergence* $\gamma > 0$ if for any real-valued function f selected from some large class of functions (usually, f is a sufficiently smooth function) $\exists\, C_f > 0$ so that
$$\left|\mathrm{E}[f(\mathbf{X}_m^h) - f(\mathbf{X}(T))]\right| \leqslant C_f\, h^\gamma.$$

17.4.4.2 The Euler Scheme

Consider a multivariate SDE

$$\mathrm{d}\mathbf{X}(t) = \mathbf{a}(t, \mathbf{X}(t))\,\mathrm{d}t + \mathbf{b}(t, \mathbf{X}(t))\,\mathrm{d}\mathbf{W}(t), \tag{17.30}$$

where $\mathbf{W}(t)$ is an d-dimensional standard Brownian motion with independent components, and $\mathbf{a}\colon \mathbb{R} \times \mathbb{R}^d \to \mathbb{R}^d$ and $\mathbf{b}\colon \mathbb{R} \times \mathbb{R}^d \to \mathbb{R}^{d\times d}$ are, respectively, the drift and diffusion coefficient functions. The SDE (17.30) can be written in the integral form on $[t, t+h]$, $t \geqslant 0$, $h > 0$:

$$\mathbf{X}(t+h) = \mathbf{X}(t) + \int_t^{t+h} \mathbf{a}(s, \mathbf{X}(s))\,\mathrm{d}s + \int_t^{t+h} \mathbf{b}(s, \mathbf{X}(s))\,\mathrm{d}\mathbf{W}(s). \tag{17.31}$$

Application of the rectangular approximation formula to each integral in (17.31) gives

$$\mathbf{X}(t+h) \approx \mathbf{X}(t) + \mathbf{a}(t, \mathbf{X}(t))h + \mathbf{b}(t, \mathbf{X}(t))(\mathbf{W}(t+h) - \mathbf{W}(t)). \tag{17.32}$$

So, the Euler discrete-time approximation on a time grid $0 = t_0 < t_1 < \cdots < t_m = T$ is given by

$$\mathbf{X}_0^h = \mathbf{X}_0, \quad \mathbf{X}_{k+1}^h = \mathbf{X}_k^h + \mathbf{a}(t_k, \mathbf{X}_k^h)\,h + \mathbf{b}(t_k, \mathbf{X}_k^h)\sqrt{t_k - t_{k-1}}\mathbf{Z}_k,\ 0 \leqslant k \leqslant m-1, \tag{17.33}$$

where $\{\mathbf{Z}_k\}_{k\geqslant 0}$ is a sequence of i.i.d. multivariate $Norm_d(\mathbf{0}, \mathbf{I})$-distributed vectors.

Theorem 17.12. *The Euler scheme has strong order* $\frac{1}{2}$ *and weak order* 1. *Moreover, the weak error admits the following expansion:*

$$\mathrm{E}[f(\mathbf{X}_m^h) - f(\mathbf{X}(T))] = C_f h + \mathcal{O}(h^2).$$

For a proof, see Kloeden and Platen (2011).

17.4.4.3 Extrapolation

The process of *extrapolation* uses two approximations computed from the same formula but with different step sizes to obtain higher-order approximation. The *Richardson extrapolation method* allows us to achieve second-order accuracy for a first-order scheme. For the Euler method with a constant time step h, we have

$$\mathrm{E}[f(\mathbf{X}_m^h)] = \mathrm{E}[f(\mathbf{X}(T))] + C_f h + \mathcal{O}(h^2)$$

for some constant C_f that depends on f. Suppose that the number of steps m is even. Apply the Euler method with steps $h = \frac{T}{m}$ and $2h = \frac{T}{m/2}$ to obtain approximations $\mathbf{X}_m^h$ and $\mathbf{X}_{m/2}^{2h}$ of the time-T realization $X(T)$, respectively. By combining

$$\mathrm{E}[f(\mathbf{X}_m^h)] = \mathrm{E}[f(\mathbf{X}(T))] + C_f h + \mathcal{O}(h^2) \quad\text{and}\quad \mathrm{E}[f(\mathbf{X}_{m/2}^{2h})] = \mathrm{E}[f(\mathbf{X}(T))] + C_f 2h + \mathcal{O}(h^2),$$

we can eliminate the leading term of the error expansion:

$$\mathrm{E}[2f(\mathbf{X}_m^h) - f(\mathbf{X}_{m/2}^{2h})] = \mathrm{E}[f(\mathbf{X}(T))] + \mathcal{O}(h^2).$$

The error of the combined estimate is of (weak) order 2. The variance of the combined estimator is

$$\mathrm{Var}\left(2f(\mathbf{X}_m^h) - f(\mathbf{X}_{m/2}^{2h})\right) = 4\,\mathrm{Var}\left(f(\mathbf{X}_m^h)\right) + \mathrm{Var}\left(f(\mathbf{X}_{m/2}^{2h})\right) - 4\,\mathrm{Cov}\left(f(\mathbf{X}_m^h), f(\mathbf{X}_{m/2}^{2h})\right).$$

By making $f(\mathbf{X}^h_m)$ and $f(\mathbf{X}^{2h}_{m/2})$ positively correlated, we can reduce the variance. This can be achieved by using consistent Brownian increments in simulating paths of $\mathbf{X}^h$ and $\mathbf{X}^{2h}$. Suppose that $\mathbf{X}^h_m$ is constructed from m i.i.d. normal vectors

$$\sqrt{h}\mathbf{Z}_0, \ldots, \sqrt{h}\mathbf{Z}_{m-1} \sim Norm_d(\mathbf{0}, \sqrt{h}\mathbf{I}).$$

The same normal increments can be used for sampling $\mathbf{X}^{2h}$. For example, to construct $\mathbf{X}^{2h}_{m/2}$ we use the normally distributed vectors

$$\sqrt{h}(\mathbf{Z}_0 + \mathbf{Z}_1), \ldots, \sqrt{h}(\mathbf{Z}_{m-2} + \mathbf{Z}_{m-1}) \sim Norm_d(\mathbf{0}, \sqrt{2h}\mathbf{I}).$$

Here, the property that a sum of two normal random variables is again normally distributed is applied.

17.4.4.4 Error Analysis

Suppose that we wish to approximate some quantity Q by using a biased estimator Y^h where h is a discretization parameter approaching zero. For example,

$$Q := \mathrm{E}[f(\mathbf{X}(T))] \approx \mathrm{E}[f(\mathbf{X}^h_m)],$$

where $\mathbf{X}^h$ is the Euler approximation of a diffusion $\mathbf{X}$ and $h = \frac{T}{m}$ is a time step size. Introduce the approximation error $\mathcal{E}(h) = Q - \mathrm{E}[Y^h]$. Suppose that

$$\mathcal{E}(h) \approx C_1\, h^\beta \tag{17.34}$$

holds for some positive constants C_1 and β. Taking the logarithm of both parts of (17.34) gives

$$\log \mathcal{E}(h) \approx \log C_1 + \beta \log h.$$

As is seen from the above equation, we can use the linear regression method to calculate C_1 and β. First, we compute the sample estimate of Q for a decreasing sequence of values of h:

$$Q \approx \bar{y}^{h_k}_n = \frac{1}{n}\sum_{j=1}^{n} y^{h_k}_j, \quad k = 1, 2, \ldots,$$

where $y^{h_k}_j$ are i.i.d. samples of Y^{h_k} and $h_1 > k_2 > h_3 > \ldots$. For example, we set $h_k = 2^{-k}$. Assuming that the number of draws, n, is sufficiently large so that the statistical error is negligible in comparison with the approximation error, we obtain

$$\bar{\mathcal{E}}(h_k) := Q - \bar{y}^{h_k}_n \approx Q - \mathrm{E}\left[Y^{h_k}\right] \approx C_1\, h^\beta \implies \log\bar{\mathcal{E}}(h_k) \approx \log C_1 + \beta \log h_k.$$

Now, we plot $\log\bar{\mathcal{E}}(h_k)$ versus $\log h_k$ for all h_k and then perform a linear regression. As a result, the slope of the regression line gives us the order of approximation β.

In fact, the error $\bar{\mathcal{E}}(h_k) = Q - \bar{y}^h_n$ includes two components: the approximation bias and the statistical (Monte Carlo) error. To optimize the method, it is helpful to separate these two errors:

$$Q - \bar{y}^h_n = \underbrace{Q - \mathrm{E}[Y^h]}_{\approx C_1\, h^\beta} + \mathrm{E}[Y^h] - \bar{y}^h_n.$$

Define the mean-square (statistical) error as

$$\mathrm{MSE}(h, n) = \mathrm{E}\left[\left(Q - \bar{Y}^h_n\right)^2\right],$$

where $\overline{Y}_n^h = \frac{1}{n}\sum_{j=1}^n Y_j^h$ is the sample estimator. The value of $\text{MSE}(h,n)$ is

$$\begin{aligned}
\text{E}\left[(Q-\overline{Y}_n^h)^2\right] &= \text{E}\left[((Q-\text{E}[Y^h]) + (\text{E}[Y^h]-\overline{Y}_n^h))^2\right] \\
&= \text{E}\left[(Q-\text{E}[Y^h])^2\right] + 2\text{E}\left[(Q-\text{E}[Y^h])(\text{E}[Y^h]-\overline{Y}_n^h)\right] + \text{E}\left[(\text{E}[Y^h]-\bar{Y}_n^h)^2\right] \\
&\approx (C_1\, h^\beta)^2 + 2(Q-\text{E}[Y^h])\text{E}\left[\text{E}[Y^h]-\overline{Y}_n^h\right] + \text{Var}\left(\overline{Y}_n^h\right) \\
&= (C_1\, h^\beta)^2 + \text{Var}\left(\overline{Y}_n^h\right) = C_1^2 h^{2\beta} + \frac{1}{n}\text{Var}(Y^h) \approx C_1^2 h^{2\beta} + \frac{1}{n}\text{Var}(Y^{0+}) \\
&= C_1^2 h^{2\beta} + \frac{C_2}{n}.
\end{aligned}$$

Here the constant $C_2 = \text{Var}(Y^{0+})$ is defined as the limiting value of the variance $\text{Var}(Y^h)$, as $h \searrow 0$. To find the optimal values of the sample volume n and the discretization parameter h, we will minimize the computational cost for a given level of error: $\text{MSE} = \epsilon^2$. Clearly, the computational cost (i.e., the number of operations) is directly proportional to m and n. The number of steps $m = \frac{T}{h}$ is inversely proportional to h. Hence, the computational cost (or the runtime) is given by $C_3\, n/h$, where C_3 is another positive constant. Let us minimize the computational cost under the constraint that the mean-square error is fixed and equal to ϵ. The optimization problem takes the following form:

$$\begin{cases} \dfrac{C_3\, n}{h} \to \min\limits_{n,h}, \\ \text{MSE}(h,n) = C_1^2 h^{2\beta} + \frac{C_2}{n} = \epsilon^2. \end{cases} \tag{17.35}$$

The problem (17.35) is easy to solve. As a result, we obtain that the computational cost is proportional to $\epsilon^{-2-1/\beta}$. Alternatively, we can solve the problem of minimizing MSE subject to a fixed computational cost s (see Exercise 17.23).

17.4.5 Simulation of Processes with Jumps

17.4.5.1 Poisson Processes

Recall that a Poisson process is a continuous-time stochastic process $\{N_\lambda(t)\}_{t\geqslant 0}$ with independent, stationary, Poisson distributed increments that starts at zero. In other words,

(a) $N_\lambda(0) = 0$;

(b) for all m and $0 \leqslant t_0 \leqslant t_1 \leqslant \cdots \leqslant t_m$, the increments $N_\lambda(t_k) - N_\lambda(t_{k-1})$, $1 \leqslant k \leqslant m$, are independent random variables;

(c) $N_\lambda(t) - N_\lambda(s) \sim \textit{Pois}(\lambda(t-s))$ for $0 \leqslant s < t$.

The parameter λ is called the intensity of the process.

Every realization of a Poisson process is a step function that starts at the origin. It stays at each level $k \geqslant 0$ for a random time period and then jumps to the next level $k+1$. The occurrence time T_k is the time when the process jumps from $k-1$ to k for $k \geqslant 1$. Set $\tau_1 = T_1$ and $\tau_k = T_k - T_{k-1}$ for $k \geqslant 2$. These variables $\{\tau_k\}_{k\geqslant 1}$ are called durations. We can express $\{T_k\}_{k\geqslant 1}$ in terms of $\{\tau_k\}_{k\geqslant 1}$ as

$$T_m = \sum_{k=1}^m \tau_k.$$

Thus, a Poisson process can be defined via durations as follows:

$$N_\lambda(t) = \sup\left\{ m : \sum_{k=1}^{m} \tau_k \leqslant t \right\}. \tag{17.36}$$

A realization of a Poisson process can be generated from the sample values of the durations. The probability distribution of the random variables $\{\tau_k\}_{k\geqslant 1}$ is given by the following result.

Proposition 17.13. *Consider a Poisson process with occurrence times T_k, $k \geqslant 1$, and durations $\tau_1 = T_1$, $\tau_k = T_k - T_{k-1}$, $k \geqslant 2$. Then,*

(a) τ_k, $k \geqslant 1$, are jointly independent, Exp(λ)-distributed random variables;

(b) $T_k \sim$ Gamma(k, λ), $k \geqslant 1$.

For a complete proof of Propositions 17.13 and 17.14, see Gut (2009).

By using Proposition 17.13 and Equation 17.36, we come up with the following sampling algorithm of the Poisson process on $[0, T]$.

Algorithm 17.14 Sampling a Poisson Process (Variant 1).

Input: $\lambda > 0$, $T > 0$.

(1) Simulate i.i.d. $\tau_1, \tau_2, \ldots, \tau_m \in$ Exp(λ) where $m = \sup\{k : \tau_1 + \cdots + \tau_k \leqslant T\}$.

(2) Set $T_k = \sum_{j=1}^{k} \tau_j$ for $k = 1, 2, \ldots, m$.

(3) Define $N_\lambda(t) = \sum_{k=1}^{m} \mathbb{I}_{\{T_k \leqslant t\}}$ for $0 \leqslant t \leqslant T$.

Another algorithm for sampling a Poisson process is based on conditioning on the number of occurrences in a time interval. As it turns out, the joint distribution of the occurrence times conditional on the number of occurrences is the same as that of the order statistics of a sample from a uniform distribution.

Proposition 17.14. *The joint density of $T_1, T_2, \ldots, T_m$ conditional on $N_\lambda(T) = m$ is*

$$f_{T_1,\ldots,T_m|N_\lambda(T)=m}(t_1, \ldots, t_m) = \begin{cases} \frac{m!}{T^m} & \text{for } 0 < t_1 < t_2 < \cdots < t_n < T, \\ 0 & \text{otherwise.} \end{cases}$$

In other words,

$$(T_1, T_2, \ldots, T_m | N_\lambda(T) = m) \stackrel{d}{=} (U_{(1)}, U_{(2)}, \ldots, U_{(m)}),$$

where $U_{(1)}, U_{(2)}, \ldots, U_{(m)}$ are the order statistics defined by sorting m i.i.d. Unif$(0, T)$-distributed random variables in increasing order.

Algorithms 17.14 and 17.15 allow us to sample a Poisson path on $[0, T]$. If we want to continue sampling on $(T, T + T']$, $T' > 0$, then the sample path on $[0, T]$ can be reused thanks to the following property.

Proposition 17.15. *If $\{N(t)\}_{t\geqslant 0}$ is a Poisson process, then so are*

1. *$\{N(t + s) - N(s)\}_{t\geqslant 0}$ for every fixed $s > 0$;*
2. *$\{N(t+T_k) - N(T_k)\}_{t\geqslant 0}$ for every fixed $k \geqslant 1$, where T_k is the time of the kth occurrence of the original Poisson process N.*

This result, along with the property of independence of Poisson increments, allows us to simulate a Poisson process individually on disjoint intervals. Suppose we have generated the process N_λ on $[0, T]$. To continue the sample path on $(T, T + T']$, we first generate another Poisson process $\{\widetilde{N}_\lambda(t)\}_{t\in[0,T']}$ independently of $\{N_\lambda(t)\}_{t\in[0,T]}$. Second, we set $N_\lambda(t) = N_\lambda(T) + \widetilde{N}_\lambda(t - T)$ for all $t \in (T, T + T']$.

The jumps of a Poisson process all have size one. A compound Poisson process is defined in such a way that the size of each of its jumps is random (see Section 16.2). Let $Y_1, Y_2, \ldots$ be i.i.d. random variables which are all independent of the Poisson process $N_\lambda(t)$. A compound Poisson process with jump sizes Y_k, $k \geqslant 1$, is

$$X(t) = \sum_{k=1}^{N_\lambda(t)} Y_k, \quad t \geqslant 0. \tag{17.37}$$

The simulation of a compound Poisson process on $[0, T]$ is straightforward. Applying Algorithm 17.14 or 17.15 gives us a sequence of occurrence times T_k, $1 \leqslant k \leqslant m$, where $m = N_\lambda(t) \sim Pois(\lambda T)$. After that we set

$$X(t) = \sum_{k=1}^{m} \mathbb{I}_{\{T_k \leqslant t\}} Y_k.$$

Algorithm 17.15 Sampling a Poisson Process (Variant 2).

Input: $\lambda > 0$, $T > 0$.

(1) Simulate $N_\lambda(T) \sim Pois(\lambda T)$. Set $m = N_\lambda(T)$.

(2) Simulate i.i.d. $U_1, U_2, \ldots, U_m \sim Unif(0, T)$. Sort them in increasing order:

$$0 \leqslant U_{(1)} \leqslant U_{(2)} \leqslant \cdots \leqslant U_{(m)}.$$

(3) Set $T_k = U_{(k)}$ for $k = 1, 2, \ldots, m$.

(4) Define $N_\lambda(t) = \sum_{k=1}^{m} \mathbb{I}_{\{T_k \leqslant t\}}$ for $0 \leqslant t \leqslant T$.

17.4.5.2 Subordinated Processes

The variance gamma (VG) process is a three-parameter generalization of the Brownian motion model for the dynamics of the logarithm of the stock price. It is obtained by evaluating a scaled Brownian motion with drift at a random time given by a gamma process (see Madan and Seneta (1990)). The gamma process $G(t; \mu, \upsilon)$ with mean rate $\mu > 0$ and variance rate $\upsilon > 0$ is a random process with independent gamma increments over nonoverlapping intervals of time.

The VG process $X(t; \sigma, \upsilon, \theta)$ is defined in terms of the scaled Brownian motion with drift, $B(t) \equiv W^{(\theta,\sigma)}(t) = \theta t + \sigma W(t)$, and the gamma process with unit mean rate, denoted $G(t) \equiv G(t; 1, \upsilon)$, as

$$X(t; \sigma, \upsilon, \theta) := B(G(t)), \quad t \geqslant 0.$$

The PDF of the VG process at time t can be expressed as a normal density function conditional on the realization of the gamma time change. The risk-neutral process for the asset price is given by

$$S(t) := S_0 \exp\big((r - q - \omega)t + X(t; \sigma, \upsilon, \theta)\big), \quad t \geqslant 0, \tag{17.38}$$

where r and q are, respectively, the risk-neutral interest rate and dividend yield. The parameter $\omega = \ln(1 - \theta\upsilon - \sigma^2\upsilon/2)/\upsilon$ is chosen so that the discounted asset price process $e^{-(r-q)t}S(t)$ is a true martingale.

Modelling of the variance gamma process relies on sampling from the normal and gamma probability distributions. One needs first to generate the gamma process and then to sample the Brownian motion conditional on the obtained values of the stochastic time process. Note that both the gamma process and Brownian motion are random processes with independent stationary increments. Thus, to sample a path of the variance gamma process at a discrete sequence $0 = t_0 < t_1 < t_2 < \cdots < t_N$, it is sufficient to generate the gamma increment $G(t_i) - G(t_{i-1})$ and then the Brownian increment $B(g_i) - B(g_{i-1})$ conditional on $G(t_i) = g_i$ and $G(t_{i-1}) = g_{i-1}$ for all $i = 1, 2, \ldots, N$. The sample values of $G(t_i)$ and $X(t_i) = B(G(t_i))$ can then be obtained by calculating cumulative sums of the respective increments. The increments of the gamma process G with mean rate one and variance rate θ and the Brownian process B with parameters θ and σ can be simulated as stated below:

$G(t_2) - G(t_1)$ has the $Gamma((t_2 - t_1)/\theta, \theta)$ distribution for any $0 < t_1 < t_2$;

$B(g_2) - B(g_1)$ has the $Norm(\mu(g_2 - g_1), \sigma^2(g_2 - g_1))$ distribution for any $0 < g_1 < g_2$.

By repeating the above procedure n times and taking the cumulative sums of the increments, we can obtain the values of the variance gamma process at a discrete sequence of time moments. Algorithm 17.16 is used for sampling paths of the variance gamma process.

Algorithm 17.16 Simulation of the Gamma and Variance Gamma Processes.

```
input X_0, 0 = t_0 < t_1 < t_2 < ··· < t_N, μ, σ, θ
G(0) ← 0
for k from 1 to N do
  ΔG_k ~ Gamma((t_k − t_{k−1})/θ, θ)
  ΔX_k ~ Norm(μ ΔG_k, σ^2 ΔG_k)
  G(t_k) ← G(t_{k−1}) + ΔG_k
  X(t_k) ← X(t_{k−1}) + ΔX_k
end for
return {G(t_k)}_{1⩽k⩽N} and {X(t_k)}_{1⩽k⩽N}
```

17.5 Variance Reduction Methods

Consider the evaluation of a mathematical expectation $\mathrm{E}[h(X)]$, where h is a real-valued function of a random variable X having a PDF f. We call $H \equiv h(X)$ a *direct Monte Carlo estimator* as opposed to an estimator with some variance reduction techniques embedded. The direct Monte Carlo *sample estimator* of $\mathrm{E}[h(X)]$ is

$$\overline{H}_n = \frac{1}{n}\sum_{i=1}^{n} h(X_i),$$

where $\{X_i\}_{i\geqslant 1}$ are i.i.d. random variables with the common PDF f. A direct *sample estimate* is the following average with statistically independent realizations $x_1, x_2, \ldots, x_n$

of X:

$$\bar{h}_n = \frac{1}{n} \sum_{i=1}^{n} h(x_i).$$

A large variance of the estimator $h(X)$ results in slow convergence of the sample estimate to $\mathrm{E}[h(X)]$. Any modification of the direct Monte Carlo method that results in a decrease of the variance is called a *variance reduction* method. One example of variance reduction techniques is the importance sampling method, which is discussed below. The goal of this section is to summarize various techniques used to improve the direct Monte Carlo method. Since a modification of the original estimator may result in an increase in computing time, we compare methods on the basis of their computational costs.

17.5.1 Numerical Integration by a Direct Monte Carlo Method

Consider a multidimensional integral $I(g) = \int_{\mathbb{R}^d} g(\mathbf{x})\, d\mathbf{x}$. If the number of dimensions d is small (say, $d \leqslant 3$) then deterministic quadrature rules can be successfully applied to evaluate $I(g)$. For larger dimensions, it is more beneficial to use stochastic methods. As is pointed out in the beginning of this chapter, to apply the Monte Carlo method (MCM), the quantity of interest, say Q, needs to be represented in the form of a mathematical expectation of a random variable called the estimator of Q. Select a d-variate PDF f such that $f(\mathbf{x}) \neq 0$ if $g(\mathbf{x}) \neq 0$ for all $\mathbf{x} \in \mathbb{R}^d$. If the integrand g has an integrable singularity at some point $\mathbf{x}_0$, then the PDF f should also be singular at $\mathbf{x}_0$ so that $\lim_{\mathbf{x}\to\mathbf{x}_0} \frac{g(\mathbf{x})}{f(\mathbf{x})}$ exists and is finite. Moreover, it is reasonable to select f having the same support as that of g. If the integral is taken on a manifold, e.g., a sphere, then the support of the PDF f has to be the same manifold. Rewrite the integral $I(g)$ as follows:

$$I(g) = \int_{\mathbb{R}^d} g(\mathbf{x})\, d\mathbf{x} = \int_{\mathbb{R}^d} \frac{g(\mathbf{x})}{f(\mathbf{x})} f(\mathbf{x})\, d\mathbf{x} = \mathrm{E}\left[h(\mathbf{X})\right], \tag{17.39}$$

where $\mathbf{X} \sim f$; $h(\mathbf{x}) := \frac{g(\mathbf{x})}{f(\mathbf{x})}$ if $g(\mathbf{x}) \neq 0$ and $g(\mathbf{x}) \neq \infty$, $h(\mathbf{x}) := 0$ if $g(\mathbf{x}) = 0$, and $h(\mathbf{x}_0) := \lim_{\mathbf{x}\to\mathbf{x}_0} \frac{g(\mathbf{x})}{f(\mathbf{x})}$ if $g(\mathbf{x}_0) = \infty$. The integral $I(g) = \mathrm{E}[h(\mathbf{X})]$ is estimated by the Monte Carlo method as follows.

(1) Generate n independent sample values $\{\mathbf{x}_i\}_{i=1}^n$ drawn from the PDF f.

(2) Construct a sample estimate of the integral $I(g)$:

$$I(g) \approx \bar{h}_n = \frac{1}{n} \sum_{i=1}^{n} h(\mathbf{x}_i).$$

(3) For $0 < \alpha < 1$, construct an asymptotically valid $100(1-\alpha)\%$ confidence interval for $I(g)$:

$$\left(\bar{h}_n - \frac{z_{\alpha/2} s_n}{\sqrt{n}}, \bar{h}_n + \frac{z_{\alpha/2} s_n}{\sqrt{n}}\right) \ni I(g),$$

where $s_n^2 = \frac{1}{n} \sum_{i=1}^{n} h(\mathbf{x}_i)^2 - \bar{h}_n^2$ is a sample variance for $h(\mathbf{X})$, and $z_{\alpha/2}$ is a $(1-\alpha/2)$-quantile of the standard normal distribution.

As is seen, the Monte Carlo method allows us to simultaneously construct an approximation of $I(g)$ and an upper bound for the error:

$$\left|I(g) - \bar{h}_n\right| \leqslant \frac{z_{\alpha/2}\sqrt{\mathrm{Var}(h(\mathbf{X}))}}{\sqrt{n}}, \tag{17.40}$$

which is valid with confidence $(1-\alpha)100\%$. The variance $\mathrm{Var}(h(\mathbf{X}))$ can be approximated by the sample variance s_n^2. As is seen from (17.40), the approximation error is of order $\mathcal{O}\left(n^{-1/2}\right)$, as $n \to \infty$. Since the number of sample values n is equal to the number of times the integrand g is evaluated, the Monte Carlo method can be compared with deterministic quadrature rules for which the approximation error is known in terms of n. For example, the midpoint quadrature rule provides the approximation error of order $\mathcal{O}(n^{-2/d})$, as $n \to \infty$. In contrast to the Monte Carlo method, the error of a quadrature rule depends the dimensionality d. The Monte Carlo method outperforms the midpoint rule for problems with $d > 4$. In general, the Monte Carlo method outperforms deterministic methods if $d \geqslant 10$. For lower dimensionality (say, $d \leqslant 3$), it is better to use a deterministic quadrature rule. For the case with $3 < d < 10$, a combination of stochastic and deterministic methods may be a more efficient approach to computing the integral. The Monte Carlo method is said to beat the curse of dimensionality since its performance is not considerably affected by the dimensionality of the problem. Another advantage of the Monte Carlo method is that it can be applied to integrals of nonsmooth functions, whereas deterministic quadrature rules require the smoothness of higher-order derivatives of integrands.

The computational time required to calculate the estimate $\bar{h}_n$ is a product of the number of samples, n, and the time t_H required to calculate one sample value of $H := h(\mathbf{X})$. Let δ be a given error level. Using (17.40), we obtain

$$\frac{z_{\alpha/2}\sqrt{\mathrm{Var}(h(\mathbf{X}))}}{\sqrt{n}} = \delta \iff n = \frac{z_{\alpha/2}^2\,\mathrm{Var}(h(\mathbf{X}))}{\delta^2}.$$

Therefore, the total computational time $t_H \cdot n$ is proportional to $\frac{t_H\ \mathrm{Var}(h(\mathbf{X}))}{\delta^2}$. The product $t_H \cdot \mathrm{Var}(h(\mathbf{X}))$ is called the *computational cost* of the stochastic estimator $h(\mathbf{X})$. Clearly, we can accelerate the Monte Carlo method by reducing the variance of the estimator. Another approach is to parallelize computations.

17.5.2 Importance Sampling Method

Clearly, there are many choices for the PDF f in (17.39) that fit for the estimation of $I(g)$ by the Monte Carlo method. One obvious requirement on the density f is that it should include all singularities of the integrand to guarantee that the variance of the random estimator is finite. For example, let us estimate the value of the integral

$$I = \int_0^1 \frac{g(x)}{\sqrt{x}}\,\mathrm{d}x, \text{ where } g \in C[0,1] \text{ and } g(0) \neq 0 \tag{17.41}$$

by the Monte Carlo method. Consider the following two choices for the PDF f.

Choice 1: Let $f(x) = \mathbb{I}_{(0,1)}(x)$, i.e., $X \sim$ *Unif*$(0,1)$. It is easy to verify that the variance of the random estimator $h(X) := \frac{g(X)}{\sqrt{X}}$ is infinite.

Choice 2: Let $f(x) \propto \frac{1}{\sqrt{x}}$ for $x \in (0,1)$. Since f integrates to one, we set $f(x) = \frac{1}{2\sqrt{x}}\mathbb{I}_{(0,1)}(x)$. Then, the random estimator

$$h(X) = \frac{g(X)}{\sqrt{X}f(X)} = \frac{g(X)}{2}$$

is bounded and hence has a finite variance.

Example 17.13. Construct a Monte Carlo estimator of the integral

$$I = \int_0^{\infty} \cdots \int_0^{\infty} \frac{e^{-x_1-x_2-\cdots-x_8}}{(1+x_1\cdots x_8)^2}\, dx_1 \cdots dx_8.$$

Solution. Notice that the integrand is a product of the function $e^{-x_1-x_2-\cdots-x_8}$, which is a product of exponential PDFs, and the positive, bounded function $(1+x_1\cdots x_8)^{-2}$. Let us select the PDF

$$f(x_1, x_2, \ldots, x_8) = e^{-x_1-x_2-\cdots-x_8} = e^{-x_1}e^{-x_2}\cdots e^{-x_8}, \quad x_1, x_2, \ldots, x_8 > 0.$$

That is, the entries of the random vector $\mathbf{X}$ are independent exponentially distributed variables $X_j \sim Exp(1)$, $1 \leqslant j \leqslant 8$. Set the random estimator to be

$$h(\mathbf{X}) = \frac{1}{(1+X_1X_2\cdots X_8)^2}.$$

To find an upper bound of $\mathrm{Var}(h(\mathbf{X}))$, use the following property (see Exercise 17.24). Suppose that Y is a bounded random variable such that $0 \leqslant m_1 \leqslant Y \leqslant m_2$ for some constants m_1 and m_2. Then,

$$\mathrm{Var}(Y) \leqslant \frac{(m_2-m_1)^2}{4}.$$

Applying this property to $Y = h(\mathbf{X}) \in [0,1]$, obtain that $\mathrm{Var}\,(h(\mathbf{X})) \leqslant \frac{1}{4}$. □

As is seen from the above examples, it is reasonable to select the PDF f as close to the integrand g as possible. This suggestion is confirmed by the *importance sampling principle*, which is presented just below. Since the computational cost is proportional to the variance of the estimator, the optimal PDF f solves the following optimization problem:

$$\mathrm{Var}\left(\frac{g(\mathbf{X})}{f(\mathbf{X})}\right) = \int_{\mathbb{R}^d} \frac{g^2(\mathbf{x})}{f(\mathbf{x})}\, d\mathbf{x} - I^2(g) \to \min_f. \tag{17.42}$$

Let us prove that the variance attains its minimum value if $f \propto |g|$. The proof is based on the Cauchy–Schwartz–Bunyakovsky inequality

$$\left(\int_{\mathbb{R}^d} |u(\mathbf{x})\, v(\mathbf{x})|\, d\mathbf{x}\right)^2 \leqslant \int_{\mathbb{R}^d} u^2(\mathbf{x})\, d\mathbf{x} \cdot \int_{\mathbb{R}^d} v^2(\mathbf{x})\, d\mathbf{x}, \tag{17.43}$$

for $u, v \in L^2(\mathbb{R}^d)$ (here L^2 is the set of square-integrable functions). Let us set $u(\mathbf{x}) = \frac{g(\mathbf{x})}{\sqrt{f(\mathbf{x})}}$ and $v(\mathbf{x}) = \sqrt{f(\mathbf{x})}$ to obtain

$$\left(\int_{\mathbb{R}^d} |g(\mathbf{x})|\, d\mathbf{x}\right)^2 = \left(\int_{\mathbb{R}^d} \left|\frac{g(\mathbf{x})}{\sqrt{f(\mathbf{x})}}\sqrt{f(\mathbf{x})}\right| d\mathbf{x}\right)^2$$
$$\leqslant \int_{\mathbb{R}^d} \frac{g^2(\mathbf{x})}{f(\mathbf{x})}\, d\mathbf{x} \cdot \underbrace{\int_{\mathbb{R}^d} f(\mathbf{x})\, d\mathbf{x}}_{=1} = \int_{\mathbb{R}^d} \frac{g^2(\mathbf{x})}{f(\mathbf{x})}\, d\mathbf{x}.$$

Therefore, we obtain a lower bound of the variance

$$\mathrm{Var}\left(\frac{g(\mathbf{X})}{f(\mathbf{X})}\right) \geqslant \left(\int_{\mathbb{R}^d} |g(\mathbf{x})|\, d\mathbf{x}\right)^2 - I^2(g).$$

Now, it remains to show that the variance attains the lower bound when the PDF is proportional to $|g|$, that is, when $f(\mathbf{x}) = \frac{1}{c}|g(\mathbf{x})|$ with $c = \int_{\mathbb{R}^d} |g(\mathbf{x})|\,d\mathbf{x}$. Indeed, for $f = \frac{1}{c}|g|$, we have

$$\int_{\mathbb{R}^d} \frac{g^2(\mathbf{x})}{f(\mathbf{x})}\,d\mathbf{x} = \int_{\mathbb{R}^d} \frac{c|g(\mathbf{x})|^2}{|g(\mathbf{x})|}\,d\mathbf{x} = c\int_{\mathbb{R}^d} |g(\mathbf{x})|\,d\mathbf{x} = \left(\int_{\mathbb{R}^d} |g(\mathbf{x})|\,d\mathbf{x}\right)^2.$$

In practice, it may be impossible to use the "best" PDF since the numerical evaluation of the normalizing constant $c = I(|g|)$ is equivalent to the original problem. Alternatively, we can use an acceptance-rejection method that does not require the normalizing constant to be calculated. If sampling from $f \propto |g|$ is not feasible or is computationally expensive, then one can use any approximation PDF that is close to the optimal PDF. For example, a piecewise approximation of $|g|$ can be used to construct the sampling density.

Example 17.14. Approximate the integral $I = \int_0^{\pi/2} \sin x\,dx = 1$ by the Monte Carlo method where the sampling PDF is

(a) $f_1(x) = \frac{2}{\pi}\,\mathbb{I}_{(0,\pi/2)}(x)$ (a constant approximation of g);

(b) $f_2(x) = \frac{8x}{\pi^2}\,\mathbb{I}_{(0,\pi/2)}(x)$ (a linear approximation of g).

Solution. Let us compare the variances $\sigma_i^2 = \mathrm{Var}(H_i)$ of the random estimators $H_i = \frac{g(X_i)}{f_i(X_i)}$ with $X_i \sim f_i$ for $i = 1, 2$. We have

$$\sigma_1^2 = \int_0^{\pi/2} \frac{\sin^2 x}{2/\pi}\,dx - 1 = \frac{\pi^2}{8} - 1 \cong 0.2337;$$
$$\sigma_2^2 = \int_0^{\pi/2} \frac{\sin^2 x}{8x/\pi}\,dx - 1 \cong 0.0168.$$

By using the PDF f_2 instead of f_1, we can reduce the computational cost by approximately $\frac{\sigma_1^2}{\sigma_2^2} \cong 14$ times. □

17.5.3 Change of Probability Measure

Let us consider another PDF $\widehat{f}$ such that $\widehat{f}(x) > 0$ for all $x \in \mathbb{R}$ with $f(x)h(x) \neq 0$. Apply the change of measure method to obtain:

$$Q = \mathrm{E}[h(X)] = \mathrm{E}\left[h(\widehat{X})\frac{f(\widehat{X})}{\widehat{f}(\widehat{X})}\right] \tag{17.44}$$

where the last mathematical expectation in (17.44) is relative to $\widehat{X} \sim \widehat{f}$. The weight function $f/\widehat{f}$ is called the *likelihood function*. According to the importance sampling method, the optimal PDF $\widehat{f}$ is proportional to the product $f \cdot |g|$. However, sampling from $\widehat{f} \propto f \cdot |g|$ may not be feasible; therefore, one may use one of the following simple methods to "improve" the sampling density.

(a) Shifting the PDF: $\widehat{f}_c(x) := f(x - c)$ for some $c \in \mathbb{R}$. The point c may be chosen in accordance with the maximum principle so that $\widehat{f}_c$ and $f \cdot |g|$ attain their maximum at the same point. For example, consider a normal density f which reaches its maximum value at the mean μ. In this case, set $c = \nu - \mu$, where $\nu = \arg\max\{f(x) \cdot |g(x)|\}$.

(b) Reshaping the PDF: $\widehat{f}_c(x) := \frac{1}{c} f(x/c)$ for some $c \in \mathbb{R}$.

17.5.4 Control Variate Method

The main idea of the control variate method is to represent the unknown quantity as a sum of two parts. One part can be calculated analytically and the other part is to be estimated by the Monte Carlo method. Such a splitting is expected to decrease the variance. For example, consider the evaluation of $I(g) = \int_{\mathbb{R}^d} g(\mathbf{x})\,d\mathbf{x}$. Suppose that there exists another function g_0, which is close to g and for which the integral $I(g_0) = \int_{\mathbb{R}^d} g_0(\mathbf{x})\,d\mathbf{x}$ can be calculated analytically. Write $I(g) = I(g_0) + I(g - g_0)$. To approximate the integral $I(g - g_0)$ we apply the Monte Carlo method with a PDF f:

$$I(g) = I(g_0) + \mathrm{E}\left[\frac{g(\mathbf{X}) - g_0(\mathbf{X})}{f(\mathbf{X})}\right], \quad \mathbf{X} \sim f. \tag{17.45}$$

If g_0 is close to g, then the difference $g - g_0$ is close to zero. So we expect that the Monte Carlo estimator of $I(g - g_0)$ has a smaller variance than that of the original integral $I(g)$. Rewrite (17.45) as follows:

$$I(g) = \mathrm{E}\left[\frac{g(\mathbf{X})}{f(\mathbf{X})} - \left(\frac{g_0(\mathbf{X})}{f(\mathbf{X})} - I(g_0)\right)\right] = \mathrm{E}[Y - (Z - I(g_0))], \tag{17.46}$$

where $Y := \frac{g(\mathbf{X})}{f(\mathbf{X})}$ and $Z := \frac{g_0(\mathbf{X})}{f(\mathbf{X})}$ are, respectively, unbiased estimators of $I(g)$ and $I(g_0)$.

Let us apply this method to the estimation of an arbitrary mathematical expectation $Q = \mathrm{E}[Y]$ of some random variable Y. Suppose that there exists another random variable Z with known mathematical expectation μ_Z such that Y and Z can be sampled simultaneously. Construct a new parametric family of random variables:

$$Y(b) := Y - b(Z - \mu_Z), \quad b \in \mathbb{R}.$$

Clearly, the mathematical expectation of $Y(b)$ is equal to Q for all $b \in \mathbb{R}$:

$$\mathrm{E}[Y(b)] = \mathrm{E}[Y] - b[Z - \mu_Z] = \mathrm{E}[Y] = Q.$$

Thus, for every $b \in \mathbb{R}$, $Y(b)$ is an unbiased estimator of Q. The random variable Z is called a *control variate*; $Y(b)$, $b \in \mathbb{R}$, is called a *controlled estimator*. The sample estimate $\bar{y}_n(b)$ is constructed as usual.

Algorithm 17.17 The Control Variate Method.

(1) Generate n independent realizations (y_j, z_j), $j = 1, 2, \ldots, n$.

(2) Set $\bar{y}_n(b) = \frac{1}{n}\sum_{j=1}^{n}(y_j - b \cdot (z_j - \mu_z))$.

Proposition 17.16. *The optimal value of b chosen so that* $\mathrm{Var}(Y(b))$ *reaches its minimum value is given by*

$$b_{opt} = \frac{\mathrm{Cov}(Y, Z)}{\mathrm{Var}(Z)}. \tag{17.47}$$

The minimum variance of the controlled estimator is

$$\mathrm{Var}(Y(b_{opt})) = \mathrm{Var}(Y) \cdot \left(1 - \mathrm{Corr}(Y, Z)^2\right).$$

Proof. The variance of the controlled estimator is a quadratic function of b:

$$\operatorname{Var}(Y(b)) = \operatorname{Var}(Z)\, b^2 - 2\operatorname{Cov}(Y,Z)\, b + \operatorname{Var}(Y).$$

Differentiate it with respect to b and equate the derivative obtained to zero:

$$\frac{\mathrm{d}\operatorname{Var}(Y(b))}{\mathrm{d}b} = 2\operatorname{Var}(Z)\, b - 2\operatorname{Cov}(Y,Z) = 0 \implies b = \frac{\operatorname{Cov}(Y,Z)}{\operatorname{Var}(Z)}.$$

Since the second derivative is positive, the variance attains its minimum value at the point b_{opt} given by (17.47). The variance of the controlled estimator $Y(b)$ for $b = b_{\text{opt}}$ is

$$\begin{aligned}\operatorname{Var}(Y(b_{\text{opt}})) &= \operatorname{Var}(Z)\,\frac{\operatorname{Cov}^2(Y,Z)}{\operatorname{Var}^2(Z)} - 2\operatorname{Cov}(Y,Z)\,\frac{\operatorname{Cov}(Y,Z)}{\operatorname{Var}(Z)} + \operatorname{Var}(Y)\\ &= \operatorname{Var}(Y) - \frac{\operatorname{Cov}^2(Y,Z)}{\operatorname{Var}(Z)} = \operatorname{Var}(Y)\cdot\left(1 - \frac{\operatorname{Cov}^2(Y,Z)}{\operatorname{Var}(Y)\operatorname{Var}(Z)}\right)\\ &= \operatorname{Var}(Y)\cdot\left(1 - \operatorname{Corr}(Y,Z)^2\right).\end{aligned}$$

Thus, $\operatorname{Var}(Y(b_{\text{opt}})) \leqslant \operatorname{Var}(Y)$ since $0 \leqslant \operatorname{Corr}(Y,Z)^2 \leqslant 1$. □

In practice, $\operatorname{Var}(Z)$ and/or $\operatorname{Cov}(Y,Z)$ are unknown in closed form. So, the optimal value of b can be approximated by the ratio of the sample covariance of Y and Z to the sample variance of Z. The variance reduction factor

$$\frac{\operatorname{Var}(Y)}{\operatorname{Var}(Y(b_{\text{opt}}))} = \frac{1}{1-\rho_{YZ}^2}, \tag{17.48}$$

increases very rapidly as $|\rho_{YZ}| \to 1$, where $\rho_{YZ} = \operatorname{Corr}(Y,Z)$. For example, $\frac{1}{1-\rho^2} = \frac{4}{3} \approx 1.3$ for $|\rho| = {}^1/_2$, but if $|\rho| = 0.99$ then $\frac{1}{1-\rho^2} \approx 50$. This observation implies that a very high degree of correlation between the original estimator and a control variate is required to yield a substantial variance reduction.

If a nonoptimal value of b is used, then the controlled estimator may have a larger variance than the original one, as is demonstrated in the following example. Let Z be a control variate for Y so that $\operatorname{Cov}(Y,Z) > \frac{1}{2}\operatorname{Var}(Z)$. The controlled estimator $Y(1) = Y - (Z - \mu_Z)$ has a smaller variance than Y:

$$\operatorname{Var}(Y(1)) = \operatorname{Var}(Y) - 2\operatorname{Cov}(Y,Z) + \operatorname{Var}(Z) < \operatorname{Var}(Y) - \operatorname{Var}(Z) + \operatorname{Var}(Z) = \operatorname{Var}(Y).$$

However, the variance of $Y(-1) = Y + (Z - \mu_Z)$ (i.e., replace Z by $-Z$) is larger than that of Y:

$$\operatorname{Var}(Y(-1)) = \operatorname{Var}(Y) + 2\operatorname{Cov}(Y,Z) + \operatorname{Var}(Z) > \operatorname{Var}(Y) + \operatorname{Var}(Z) + \operatorname{Var}(Z) > \operatorname{Var}(Y).$$

Example 17.15. Apply the control variate method to the integral

$$I = \int_{(0,1)^d} (\mathrm{e}^{x_1+x_2+\cdots+x_d} - 1)\,\mathrm{d}x_1\,\mathrm{d}x_2\cdots\mathrm{d}x_d = (\mathrm{e}-1)^d - 1.$$

Assume that a uniform sampling distribution is used in the direct Monte Carlo method. Calculate the variance reduction factor for the optimal control variate when $d = 1, 5, 10$.

Solution. The PDF of the uniform distribution in the unit hypercube $(0,1)^d$ is $f(\mathbf{x}) = \mathbb{I}_{(0,1)^d}(\mathbf{x})$. The direct estimator of the integral is $Y = \mathrm{e}^{U_1+U_2+\cdots+U_d} - 1$, where $U_1, U_2, \ldots, U_d$ are i.i.d. *Unif*$(0,1)$-distributed random variables. Applying Taylor's formula to the integrand gives us the following approximation: $e^{x_1+x_2+\cdots+x_d} - 1 \approx x_1 + x_2 + \cdots + x_d$. Let us use $Z = U_1 + U_2 + \cdots + U_d$ as a control variate. The expected value of Z is

$$\mathrm{E}[Z] = \int_{[0,1]^d} (x_1 + x_2 + \cdots + x_d)\, dx_1\, dx_2 \cdots dx_d = d \int_0^1 x\, dx = \frac{d}{2}.$$

So, the controlled estimator is

$$Y(b) = Y - b\,(Z - \mathrm{E}[Z]) = e^{U_1+U_2+\cdots+U_d} - 1 - b\,(U_1 + U_2 + \cdots + U_d - d/2)\,.$$

To find the optimal value of b and the respective variance reduction, we calculate $\mathrm{Cov}(Y, Z)$, $\mathrm{Var}(Y)$, and $\mathrm{Var}(Z)$:

$$\begin{aligned}
\mathrm{E}[YZ] &= \mathrm{E}\left[(e^{U_1+U_2+\cdots+U_d} - 1)\,(U_1 + U_2 + \cdots + U_d)\right] \\
&= d \int_0^1 x e^x\, dx \left(\int_0^1 e^y\, dy\right)^{d-1} - d \int_0^1 x\, dx = d(e-1)^{d-1} - d/2, \\
\mathrm{Cov}(Y, Z) &= \mathrm{E}[YZ] - \mathrm{E}[Y]\,\mathrm{E}[Z] = d(e-1)^{d-1} - \frac{d}{2} - ((e-1)^d - 1) \cdot \frac{d}{2} \\
&= d(e-1)^{d-1} \cdot \frac{3-e}{2}, \\
\mathrm{Var}(Y) &= \mathrm{E}[Y^2] - \mathrm{E}[Y]^2 = \left(\int_0^1 e^{2x}\, dx\right)^d - 2\left(\int_0^1 e^x\, dx\right)^d + 1 - ((e-1)^d - 1)^2 \\
&= \left(\frac{e^2-1}{2}\right)^d - 2(e-1)^d + 1 - (e-1)^{2d} + 2(e-1)^d - 1 \\
&= \left(\frac{e^2-1}{2}\right)^d - (e-1)^{2d}, \\
\mathrm{Var}(Z) &= \mathrm{Var}(U_1 + U_2 + \cdots + U_d) = d\,\mathrm{Var}(U_1) = \frac{d}{12}.
\end{aligned}$$

Therefore, the optimal value of b is $b_{\mathrm{opt}} = \frac{\mathrm{Cov}(Y,Z)}{\mathrm{Var}(Z)} = 6(e-1)^{d-1}(3-e)$. The results for $d = 1, 5, 10$ are summarized in Table 17.7. As is seen from the table, the efficiency of this control variate method is decreasing with the growth of d. □

TABLE 17.7: Efficiency of the control variate method.

d	1	5	10
b_{opt}	1.69	14.73	220.71
$\mathrm{Corr}(Y, Z)$	99.18%	91.38%	82.02%
$\mathrm{Var}(Y)/\mathrm{Var}(Y(b_{\mathrm{opt}}))$	61.43	6.06	3.06

The above example suggests a universal approach to the construction of a control variate. Consider the estimation of $Q = \mathrm{E}[Y]$ with $Y = h(X)$, $X \sim f$. Suppose that $h\colon \mathbb{R} \to \mathbb{R}$ has continuous derivatives and it can be approximated by the Taylor series expansion about x_0:

$$h(x) \approx \sum_{j=1}^{k} \frac{h^{(j)}(x_0)}{j!}(x - x_0)^j.$$

Suppose that we are able to exactly calculate the moments $\mathrm{E}\left[(X - x_0)^j\right]$, $1 \leqslant j \leqslant k$. Then, the random variable $Z := \sum_{j=1}^{k} \frac{h^{(j)}(x_0)}{j!}(X - x_0)^j$ can be used as a control variate. The point x_0 should be selected in a way such that $|\,\mathrm{Corr}(Y, Z)|$ is maximized.

17.5.5 Antithetic Variate

The *antithetic variate method* attempts to reduce the variance by exploiting a symmetry of a probability distribution. Consider the estimation of $Q = \mathrm{E}[h(U)]$, where $U \sim \mathit{Unif}(0,1)$. For example, $Q = \int_a^b g(x)\,dx = \mathrm{E}[h(U)]$ with $h(U) = (b-a)g(a+(b-a)U)$. Since U and $1-U$ have the same distribution, the estimators $h(U)$ and $h(1-U)$ are equally distributed as well. The *antithetic estimator* is defined as $Y_{\text{anti}} = \frac{1}{2}(h(U)+h(1-U))$. Compare the variances of the direct estimator $Y = h(U)$ and the antithetic estimator Y_{anti}:

$$\begin{aligned}\operatorname{Var}(Y_{\text{anti}}) &= \frac{1}{4}\left(\operatorname{Var}(h(U)) + 2\operatorname{Cov}(h(U),h(1-U)) + \operatorname{Var}(h(1-U))\right)\\ &= \frac{1}{2}\operatorname{Var}(h(U)) + \frac{1}{2}\operatorname{Cov}(h(U),h(1-U)).\end{aligned}$$

If $h(U)$ and $h(1-U)$ are negatively correlated, then $\operatorname{Var}(Y_{\text{anti}}) \leqslant \frac{1}{2}\operatorname{Var}(Y)$. Since every realization of Y_{anti} requires two evaluations of h, the computational time is doubled when we apply the antithetic variate method. However, we can still expect a reduction in the computational cost if $\operatorname{Var}(Y_{\text{anti}})$ is less than $\operatorname{Var}(Y)$ by half. This is the case if h is a monotone function, as is proved in the following proposition.

Proposition 17.17. *Let h be a monotone function defined on $[0,1]$; let $\operatorname{Cov}(h(U),h(1-U))$, where $U \sim \mathit{Unif}(0,1)$, be finite. Then, $\operatorname{Cov}(h(U),h(1-U)) \leqslant 0$.*

Proof. Without loss of generality, assume that the function h is nondecreasing. We need to show that

$$\mathrm{E}[h(U)h(1-U)] \leqslant \mathrm{E}[h(U)]^2.$$

Since h is nondecreasing on $[0,1]$, the value Q lies between the values of h at the endpoints on the unit interval, i.e., $h(0) \leqslant Q \leqslant h(1)$. Consider the function $f(y) := \int_0^y h(1-x)\,dx - Qy$ defined on the unit interval, $[0,1]$. The function f is zero at the endpoints, $f(0) = f(1) = 0$. Its derivative, $f'(y) = h(1-y) - Q$, is a nonincreasing function. Since $f'(0) = h(1) - Q \geqslant 0$ and $f'(1) = h(0) - Q \leqslant 0$, the function f is nonnegative everywhere on $[0,1]$. Therefore, $\int_0^1 f(y)h'(y)\,dy \geqslant 0$. Integration by parts gives

$$\int_0^1 f(y)h'(y)\,dy = f(y)h(y)\big|_0^1 - \int_0^1 f'(y)h(y)\,dy = -\int_0^1 f'(y)h(y)\,dy.$$

Therefore, we have

$$\int_0^1 f'(y)h(y)\,dy = \int_0^1 (h(y)h(1-y) - h(y)Q)\,dy = \int_0^1 h(y)h(1-y)\,dy - Q^2 \leqslant 0.$$

Hence, $\int_0^1 h(y)h(1-y)\,dy \leqslant Q^2$. □

In summary, the antithetic variate method attempts to reduce the variance by introducing negative correlation between pairs of realizations. Here are some examples of antithetic pairs.

- U and $1-U$ are both $\mathit{Unif}(0,1)$-distributed. $\operatorname{Corr}(U, 1-U) = -1$.
- Let F be a CDF and F^{-1} its generalized inverse. The random variables $F^{-1}(U)$ and $F^{-1}(1-U)$ have the same CDF F, and they are antithetic to each other (F and F^{-1} are both monotone functions).
- Z and $-Z$ are both standard normal variables, and $\operatorname{Corr}(Z,-Z) = -1$.
- Z and $2\mu - Z$, where $Z \sim \mathit{Norm}(\mu,\sigma^2)$, form an antithetic pair.

17.5.6 Conditional Sampling

Consider the estimation of $\mathrm{E}[h(X)]$ where h is a function of a random variable X. Suppose that there exists another random variable V correlated with X so that the conditional expectation $\hat{h}(V) = \mathrm{E}[h(X) \mid V]$ can be calculated exactly. Applying the double expectation formula gives us

$$\mathrm{E}[h(X)] = \mathrm{E}\big[\mathrm{E}[h(X) \mid V]\big] = \mathrm{E}[\hat{h}(V)].$$

As is well known from a standard course on probability theory, for any pair of square-integrable random variables (Y, V), we have $\mathrm{Var}(Y) = \mathrm{E}[\mathrm{Var}(Y \mid V)] + \mathrm{Var}(\mathrm{E}[Y \mid V])$. Setting $Y = h(X)$ gives that

$$\mathrm{Var}(h(X)) = \mathrm{E}[\mathrm{Var}(h(X) \mid V)] + \mathrm{Var}(\mathrm{E}[h(X) \mid V]) \geqslant \mathrm{Var}(\mathrm{E}[h(X) \mid V]).$$

Therefore, using the estimator $\hat{h}(V) := \mathrm{E}[h(X) \mid V]$ instead of $h(X)$ always leads to a variance reduction.

Algorithm 17.18 The Conditional Sampling Method.

---

(1) Generate n independent samples $v_1, v_2, \ldots, v_n$.

(2) Calculate $\hat{h}(v_i) = \mathrm{E}[h(X) \mid V = v_i]$, $i = 1, 2, \ldots, n$.

(3) Estimate $Q = \mathrm{E}[h(X)] \approx \frac{1}{n} \sum_{i=1}^{n} \hat{h}(v_i)$.

For this algorithm to be of practical use, the following conditions must be met.

(a) It is easy to generate V.

(b) $\mathrm{E}[h(X) \mid V = v]$ is readily computable for all v.

(c) $\mathrm{E}[\mathrm{Var}(h(X) \mid V)]$ is large relative to $\mathrm{Var}(\mathrm{E}[h(X) \mid V])$.

Example 17.16. Let $\{X_i\}_{i \geqslant 1}$ be i.i.d. random variables with a common CDF F; let R be a positive-integer-valued random variable independent of all X_i. Apply the conditional sampling method to estimate

$$q(x) = \mathbb{P}(S_R \leqslant x) = \mathrm{E}\left[\mathbb{I}_{\{S_R \leqslant x\}}\right], \text{ where } S_R = \sum_{i=1}^{R} X_i.$$

Solution. We can try to improve the direct Monte Carlo estimator by isolating X_1:

$$\begin{aligned}\mathbb{P}(S_R \leqslant x \mid R = r) &= \mathbb{P}\left(X_1 \leqslant x - \sum_{i=2}^{r} X_i\right) \\ &= \mathrm{E}\left[\mathbb{P}\left(X_1 \leqslant x - \sum_{i=2}^{r} X_i \mid X_2, X_3, \ldots\right)\right] = \mathrm{E}\left[F\left(x - \sum_{i=2}^{r} X_i\right)\right],\end{aligned}$$

where the mathematical expectation is relative to $X_2, X_3, \ldots$. Therefore,

$$q(x) = \mathrm{E}\left[\mathbb{P}(S_R \leqslant x \mid R)\right] = \mathrm{E}\left[\mathrm{E}\left[F\left(x - \sum_{i=2}^{R} X_i\right)\right]\right],$$

where the external expectation is relative to R and the internal expectation is taken with respect to $X_2, X_3, \ldots$. This method is efficient if $\mathbb{P}(R = 1)$ is large. □

17.5.7 Stratified Sampling

Stratified sampling is a special case of conditional sampling. Consider again the estimation of $Q = \mathrm{E}[h(X)]$. We construct V such that $Q = \mathrm{E}\big[\mathrm{E}[h(X) \mid V]\big]$ by stratifying the support of X as follows. Introduce mutually disjoint subsets $A_1, A_2, \ldots, A_k$ so that $\mathbb{P}(X \in \cup_{i\geqslant 1} A_i) = 1$. Define a discrete random variable $V \in \{1, 2, \ldots, k\}$ so that $V = i$ if $X \in A_i$, $1 \leqslant i \leqslant k$. Set $p_i = \mathbb{P}(V = i) = \mathbb{P}(X \in A_i)$, $1 \leqslant i \leqslant k$. The quantity of interest Q is given as a double expectation:

$$Q = \mathrm{E}[h(X)] = \sum_{i=1}^{k} p_i \,\mathrm{E}[h(X) \mid X \in A_i] = \mathrm{E}\big[\mathrm{E}[h(X) \mid V]\big].$$

The stratified sample estimator of Q is

$$\overline{H}_n^{\text{strat}} = \sum_{i=1}^{k} \frac{p_i}{n_i} \sum_{j=1}^{n_i} h(X_{i,j}),$$

where $X_{i,1}, \ldots, X_{i,n_i}$ are i.i.d. samples of X conditional on $V = i$, for $1 \leqslant i \leqslant k$. The total sample size is $n = n_1 + n_2 + \cdots + n_k$. A crucial assumption is that sampling of X conditional on V is possible. The variance of the stratified estimator is

$$\mathrm{Var}(\overline{H}_n^{\text{strat}}) = \sum_{i=1}^{k} \frac{p_i^2}{n_i^2} \sum_{j=1}^{n_i} \sigma_i^2 = \sum_{i=1}^{k} \frac{p_i^2 \sigma_i^2}{n_i}, \tag{17.49}$$

where $\sigma_i^2 = \mathrm{Var}(h(X) \mid V = i)$.

The most important question is how to select optimal allocation. The following two results provide a simple method and an optimal method for choosing the sample sizes n_i, $1 \leqslant i \leqslant k$, respectively.

Proposition 17.18. *Let the sample sizes n_i be proportional to the probabilities p_i, i.e., $n_i = n \cdot p_i$, for $i = 1, 2, \ldots, k$. Then, the variance of the stratified sample estimator $\overline{H}_n^{strat}$ does not exceed that of the direct sample estimator $\overline{H}_n^{direct}$:*

$$\mathrm{Var}\left(\sum_{i=1}^{k} \frac{p_i}{n_i} \sum_{j=1}^{n_i} h(X_{i,j})\right) \leqslant \mathrm{Var}\left(\frac{1}{n} \sum_{i=1}^{k} \sum_{j=1}^{n_i} h(X_{i,j})\right),$$

where $n = n_1 + n_2 + \cdots + n_k$.

Proof.

$$\begin{aligned} n \cdot \mathrm{Var}\left(\overline{H}_n^{\text{strat}}\right) &= n \cdot \sum_{i=1}^{k} \frac{p_i^2 \sigma_i^2}{n_i} = \sum_{i=1}^{k} p_i \sigma_i^2 \quad (np_i = n_i \implies np_i^2/n_i = p_i) \\ &= \mathrm{E}[\mathrm{Var}(h(X) \mid V)] \leqslant \mathrm{Var}(h(X)) = n \cdot \mathrm{Var}\left(\overline{H}_n^{\text{direct}}\right). \end{aligned}$$ □

Proposition 17.19. *The optimal allocations are $n_i^* = n \frac{p_i \sigma_i}{\sum_{j=1}^{k} p_j \sigma_j}$, which give*

$$\mathrm{Var}\left(\overline{H}_n^{strat}\right) = \frac{1}{n} \left(\sum_{i=1}^{k} p_i \sigma_i\right)^2.$$

Proof. The proof is left as an exercise for the reader (see Exercise 17.35). □

The standard deviations σ_i, $1 \leqslant i \leqslant k$, are usually unknown. In practice, one may estimate them from "pilot" runs and then estimate the optimal sample sizes $n_1^*, n_2^*, \ldots, n_k^*$.

A typical problem where the stratified sampling method is efficient is numerical integration. Consider an integral of some function g on $D \subseteq \mathbb{R}^d$. Introduce a partition of D into k disjoint subdomains:

$$D = \bigcup_{i=1}^{k} D_i, \quad D_i \cap D_j = \emptyset \text{ for } i \neq j.$$

Apply the stratified sampling method to the integral to obtain

$$\begin{aligned}\int_D g(\mathbf{x})\,\mathrm{d}\mathbf{x} &= \sum_{i=1}^{k} \int_{D_i} g(\mathbf{x})\,\mathrm{d}\mathbf{x} = \sum_{i=1}^{k} \int_{D_i} h(\mathbf{x}) f_{\mathbf{X}}(\mathbf{x})\,\mathrm{d}\mathbf{x} \quad (\text{where } h(\mathbf{x}) := \frac{g(\mathbf{x})}{f_{\mathbf{X}}(\mathbf{x})}) \\ &= \sum_{i=1}^{k} p_i \int_{D_i} h(\mathbf{x}) f_i(\mathbf{x})\,\mathrm{d}\mathbf{x} = \mathrm{E}\big[\mathrm{E}[h(\mathbf{X}) \mid V]\big],\end{aligned}$$

where $V = i$ with probability $p_i = \int_{D_i} f(\mathbf{x})\,\mathrm{d}\mathbf{x}$ for $1 \leqslant i \leqslant k$, and $f_i(\mathbf{x}) := \frac{f_{\mathbf{X}}(\mathbf{x})}{p_i} \mathbb{I}_{D_i}(\mathbf{x})$ is the conditional density of $\mathbf{X}$ given $V = i$ for $1 \leqslant i \leqslant k$.

Example 17.17. Construct a direct sample estimator and a stratified sample estimator with two equal subintervals for the integral $I = \int_0^1 \mathrm{e}^x\,dx$. Use 10 sample points for both estimators. Compare the variances of the estimators obtained.

Solution. Let us construct an estimator using a uniform sampling distribution, i.e., $I = \mathrm{E}[\mathrm{e}^U]$ with $U \sim \mathit{Unif}(0,1)$. The direct Monte Carlo estimator with $n = 10$ sample points is

$$\overline{H}_{10}^{\text{direct}} = \frac{1}{10}\left(\mathrm{e}^{U_1} + \cdots + \mathrm{e}^{U_{10}}\right),$$

where $U_1, \ldots, U_{10}$ are i.i.d. $\mathit{Unif}(0,1)$-distributed random variables. The variance of the direct MC estimator is

$$\mathrm{Var}\left(\overline{H}_{10}^{\text{direct}}\right) = \frac{1}{10}\mathrm{Var}\left(\mathrm{e}^{U_1}\right) = \frac{1}{10}\left(\int_0^1 \mathrm{e}^{2x}\,dx - I^2\right) = \frac{4\mathrm{e} - \mathrm{e}^2 - 3}{20} \cong 0.02421.$$

Divide the integration interval into two subintervals ($k = 2$): $(0,1) = (0, {}^1\!/_2) \cup [{}^1\!/_2, 1)$. Now for each interval we need to find the conditional distribution of U and the probability, as is discussed above. The distribution of $U \sim \mathit{Unif}(0,1)$ conditional on $\{U \in (0, {}^1\!/_2)\}$ (or $\{U \in [{}^1\!/_2, 1)\}$) is uniform:

$$(U | \{U \in (0, {}^1\!/_2)\}) \stackrel{d}{=} \frac{U}{2} \sim \mathit{Unif}(0, {}^1\!/_2), \quad (U | \{U \in [{}^1\!/_2, 1)\}) \stackrel{d}{=} \frac{U+1}{2} \sim \mathit{Unif}({}^1\!/_2, 1).$$

The probabilities are $p_1 = \mathbb{P}(U \in (0, {}^1\!/_2)) = {}^1\!/_2$ and $p_2 = \mathbb{P}(U \in [{}^1\!/_2, 1)) = {}^1\!/_2$. Let n_1 and n_2 be sample sizes so that $n_1 + n_2 = 10$. The stratified sample estimator is

$$\overline{H}_{10}^{\text{strat}} = \frac{1}{2n_1}\left(\mathrm{e}^{U_1/2} + \cdots + \mathrm{e}^{U_{n_1}/2}\right) + \frac{1}{2n_2}\left(\mathrm{e}^{(V_1+1)/2} + \cdots + \mathrm{e}^{(V_{n_2}+1)/2}\right),$$

where $U_1, \ldots, U_{n_1}$ and $V_1, \ldots, V_{n_2}$ are i.i.d. $U(0,1)$-distributed random variables. The variance of the stratified estimator is

$$\begin{aligned}\operatorname{Var}\left(\overline{H}_{10}^{\text{strat}}\right) &= \frac{1}{4n_1}\operatorname{Var}\left(\mathrm{e}^{U_1/2}\right) + \frac{\mathrm{e}}{4n_2}\operatorname{Var}\left(\mathrm{e}^{V_1/2}\right)\\ &= \left(\frac{1}{4n_1}+\frac{\mathrm{e}}{4n_2}\right)\operatorname{Var}(\mathrm{e}^{U_1/2}) = \left(\frac{1}{4n_1}+\frac{\mathrm{e}}{4n_2}\right)\left(\int_0^1 \mathrm{e}^x\,\mathrm{d}x - \left(\int_0^1 \mathrm{e}^{x/2}\,\mathrm{d}x\right)^2\right)\\ &= \left(\frac{1}{4n_1}+\frac{\mathrm{e}}{4n_2}\right)(8\sqrt{\mathrm{e}}-3\mathrm{e}-5) \cong \frac{0.008731}{n_1} + \frac{0.02373}{n_2}.\end{aligned}$$

If $n_1 = n_2 = 5$, then $\operatorname{Var}\left(\overline{H}_{10}^{\text{strat}}\right) \cong 0.006493 < 0.02421 \cong \operatorname{Var}\left(\overline{H}_{10}^{\text{direct}}\right)$. □

17.6 Exercises

Exercise 17.1.

In the "hit-or-miss" method, the value of π is estimated as follows. A point is sampled uniformly in the square $[-1,1]\times[-1,1]$, i.e., sample two independent Cartesian coordinates $X, Y \sim \mathit{Unif}(-1,1)$. A point (X,Y) is accepted if it lies inside the circle inscribed in the square, i.e., $X^2+Y^2<1$. The experiment is repeated N times, and the number of accepted points, N_H, is recorded. The ratio $\frac{N_H}{N}$ converges to a limiting value whose expression involves π.

(a) Construct a random estimator of π based on the experiment described above.

(b) Construct a sample estimate of π with N trials. Express it in terms of the ratio $\frac{N_H}{N}$.

(c) Construct a 99% confidence interval for π. Find the number of trials, N, required to obtain the confidence interval of length less than 10^{-3}.

Exercise 17.2. Let $U \sim \mathsf{Unif}(0,1)$ admit the following binary representation:

$$U = (0.B_1\,B_2\ldots B_k\ldots)_2 = \sum_{k=1}^{\infty} B_k \cdot 2^{-k}.$$

Show that the digits B_k, $k = 1,2,\ldots$, are independent Bernoulli random variables uniformly distributed in $\{0,1\}$.

Exercise 17.3. Propose an algorithm for simulating the occurrence of two dependent events A and B that uses only one uniform random variable $U \sim \mathit{Unif}(0,1)$. Assume that the probabilities $\mathbb{P}(A) = p_A$, $\mathbb{P}(B) = p_B$, and $\mathbb{P}(A\cup B) = q$ are given.

Exercise 17.4. Using the inverse CDF method, find generating formulae for the following PDFs:

(a) $f(x) = \dfrac{3\,e^{-x}}{(e^{-x}+3)\sqrt{9-e^{-2x}}}$, $-\ln 3 < x < +\infty$;

(b) $f(x) = \frac{2\cos x}{\sin^3 x}$, $\frac{\pi}{4} \leqslant x \leqslant \frac{\pi}{2}$;

(c) $f(x) = \frac{8x}{(x+1)^3}$, $0 \leqslant x \leqslant 1$;

(d) $f(x) = x \sin x^2$, $0 \leqslant x \leqslant \sqrt{\pi}$;

(e) $f(x) = \frac{3}{2\,(2x+1)^{3/2}}$, $0 \leqslant x \leqslant 4$;

(f) $f(x) = \frac{1}{\cos^2 x}$, $0 \leqslant x \leqslant \frac{\pi}{4}$.

Exercise 17.5. Using the inverse CDF method, find generating formulae for the Rayleigh distribution with the CDF $F(x) = 1 - \mathrm{e}^{-2x(x-b)}$, $x > b$.

Exercise 17.6. Let X be an absolutely continuous random variable such that the inverse function of its distribution function F is well-defined. Consider the problem of sampling X conditional on $a < X < b$, with $F(a) < F(b)$.

(a) Find the CDF for such a conditional distribution.

(b) Show that the generating formula $X = F^{-1}(F(a) + (F(b) - F(a))U)$ with $U \sim \mathsf{Unif}(0,1)$ produces the desired conditional distribution.

Exercise 17.7. Justify the following sampling formula for the geometric random variable X with parameter $p \in (0,1)$ (i.e., $\mathbb{P}(X = k) = (1-p)^{k-1}p$ for $k = 1, 2, \ldots$):

$$X = \left\lfloor \frac{\ln(1-U)}{\ln(1-p)} \right\rfloor + 1, \quad U \in \mathit{Unif}(0,1)\,.$$

[Hint: Evaluate the probability $\mathbb{P}(X = k)$, $k = 1, 2, \ldots$.]

Exercise 17.8. Consider a random variable with the piecewise-constant PDF

$$f(x) = \begin{cases} \frac{1}{2}, & 0 < x \leqslant 1, \\ \frac{1}{8}, & 1 < x \leqslant 3, \\ \frac{1}{12}, & 3 < x < 6. \end{cases}$$

Design a simulation algorithm using the inverse-transform method.

Exercise 17.9. Consider a random variable with a PDF f, which is proportional to the piecewise function

$$g(x) = \begin{cases} \frac{x}{3}, & 0 < x \leqslant 2, \\ \frac{1}{3}, & 2 < x \leqslant 4, \\ 1 - \frac{x}{6}, & 4 < x < 6. \end{cases}$$

Find the constant c such that $f(x) = cg(x)$ is a probability density. Design a simulation algorithm using the composition method.

Exercise 17.10. Present an acceptance-rejection method for a random variable on $(0,1)$ with the PDF of the form

$$f(x) = g(x)/x^a, \quad 0 < g(x) \leqslant M, \quad 0 < a < 1\,.$$

Exercise 17.11. Construct the Neumann method (and estimate its computational cost) for sampling a random variable X whose density is proportional to

(a) $f(u) = 3 - \sqrt[3]{2\,u}\,, \quad 0 \leqslant u \leqslant 1$;

(b) $f(u) = -u^2 + 2\,u + 3\,, \quad 0 < u < 3$;

(c) $f(u) = u^{5/3}\,(1-u)^{3/2}\,, \quad 0 \leqslant u \leqslant 1$;

(d) $f(u) = u^{5/3}\,\mathrm{e}^{-u}\,, \quad u > 0$.

For each case, compute the probability of acceptance.

Exercise 17.12. Consider the triangular probability distribution with the PDF

$$f(x) = \frac{x}{2}\,\mathbb{I}_{(0,1]}(x) + \frac{4-x}{6}\,\mathbb{I}_{(1,4)}(x).$$

(a) Obtain the CDF F and then its inverse F^{-1}. Describe the inverse CDF sampling method.

(b) Develop the decomposition method using the strata $\{(0,1],(1,4)\}$.

(c) Determine which of the following majorizing functions provides the largest acceptance probability in the acceptance-rejection method:

$$g_1(x) = \frac{x}{2}, \quad g_2(x) = \frac{4-x}{6}, \quad g_3(x) = \frac{1}{2}, \quad x \in (0,4).$$

Find the acceptance probability for each case.

Exercise 17.13. Let X be a positive random variable with the probability density function $f(x) = \sqrt{\frac{2}{\pi\sigma^2}}e^{-x^2/(2\sigma^2)}$, i.e., $X = |Z|$ where $Z \sim \mathsf{Norm}(0,\sigma^2)$. Develop an acceptance-rejection algorithm for generating a random variable from the PDF f using the exponential distribution $\mathsf{Exp}(\lambda)$ as the majorizing distribution. Which λ gives the largest acceptance probability (give the answer in terms of σ)?

Exercise 17.14. Consider a bivariate distribution with the joint density

$$f(x,y) = \begin{cases} cxy & \text{if } x>0,\ y<1,\ y-x>0, \\ 0 & \text{otherwise.} \end{cases}$$

(a) Find the constant c.

(b) Represent the joint density as a product of marginal and conditional univariate densities.

(c) Construct the exact simulation algorithm based on the inverse CDF method. Consider the following approach: first model X and then Y.

Exercise 17.15. The density of a random variable X is represented in the integral form:

$$f_X(u) = c\int_1^{\infty} v^{-c}\mathrm{e}^{-uv}\,dv, \quad u>0, \quad c>0\,.$$

(a) Show that f_X is a density function.

(b) For the joint density

$$f_{X,Y}(u,v) = cv^{-c}\mathrm{e}^{-uv}, \quad u>0, \quad v>1\,,$$

find a representation of the form $f_{X,Y}(u,v) = f_Y(v)f_{X|Y}(u|v)\,.$

(c) Based on the above representation of the marginal density

$$f_X(u) = \int_{-\infty}^{\infty} f_Y(v)f_{X|Y}(u|v)\,dv,$$

construct an algorithm for generating X, first by sampling Y from $f_Y(v)$, and then by sampling from the conditional density $f_{X|Y}(u|v)$.

Exercise 17.16. Demonstrate how three independent tosses of a balanced coin can be modelled by two rolls of a balanced die (with six faces). In other words, develop an exact algorithm for sampling a vector $[X_1, X_2, X_3]$ formed by three independent discrete random variables uniformly distributed in $\{0, 1\}$ by using two independent discrete random variables Y_1 and Y_2 uniformly distributed in $\{1, 2, 3, 4, 5, 6\}$.

Exercise 17.17. Develop an algorithm that allows you to "replace" a fair roulette wheel (with 37 pockets numbered from 0 to 36) by a balanced die (with six faces labelled from 1 to 6). In other words, develop an exact algorithm for sampling a discrete random variable X uniformly distributed in $\{0, 1, 2, \ldots, 36\}$ by using a sequence $(Y_k)_{k\geqslant 0}$ of i.i.d. discrete random variables uniformly distributed in $\{1, 2, 3, 4, 5, 6\}$.

Exercise 17.18. Consider acceptance-rejection sampling from a standard normal distribution.

(a) Find the range for $\lambda > 0$ and $c > 0$ so that the function

$$g(x) = \begin{cases} 1 & \text{if } |x| < c, \\ \mathrm{e}^{-\lambda(|x|-c)} & \text{if } |x| \geqslant c, \end{cases}$$

majorizes $\mathrm{e}^{-x^2/2}$ on $(-\infty, \infty)$.

(b) Calculate the acceptance probability (as a function of c and λ).

(c) Let us set $c = 1/\sqrt{2}$ and $\lambda = 2c = \sqrt{2}$. Show that this choice of parameters maximizes the acceptance probability equal to $\frac{\sqrt{\pi}}{2}$.

(d) Show that the proposal CDF G is given by

$$G(x) = \begin{cases} \frac{1}{4}\mathrm{e}^{\sqrt{2}x+1} & \text{if } x < -\frac{\sqrt{2}}{2}, \\ \frac{\sqrt{2}x+2}{4} & \text{if } |x| \leqslant \frac{\sqrt{2}}{2}, \\ 1 - \frac{1}{4}\mathrm{e}^{-\sqrt{2}x+1} & \text{if } x > \frac{\sqrt{2}}{2}. \end{cases}$$

(e) Develop computational algorithms

(i) for sampling from the inverse of G;

(ii) for sampling from the standard normal distribution by using the acceptance-rejection method with the proposal CDF G.

Exercise 17.19. Construct two methods of sampling from the trivariate normal distribution

$$Norm_3\left(\begin{bmatrix} 1 \\ 0 \\ -1 \end{bmatrix}, \begin{bmatrix} 1 & 1 & 1 \\ 1 & 4 & 1 \\ 1 & 1 & 9 \end{bmatrix}\right).$$

using the Cholesky factorization and the conditional sampling approach, respectively.

Exercise 17.20. By applying the conditioning formula for a multivariate normal distribution, find the conditional (bridge) distribution of a Brownian motion with drift

$$B(t) = \mu t + \sigma W(t), \quad t \geqslant 0.$$

In particular, show that for every $0 \leqslant s < u < t$, $B(u)$ conditional on values of $B(s)$ and $B(t)$ has a normal distribution. Find the mean and variance of this conditional distribution.

Exercise 17.21. Consider a time-homogeneous random process $\{X(t)\}_{t\geqslant 0}$ with the transition PDF $p(t;y,x)$ defined by

$$p(t;y,x)\,\mathrm{d}x = \mathbb{P}(X(t+s)\in[x,x+\mathrm{d}x)\mid X(s)=y) \text{ for } t,s>0$$

with infinitesimally small $\mathrm{d}x$. Using the notion of conditional probability, show that for every $0\leqslant t_1<t<t_2$ the bridge density $b(t;x|t_1,t_2;x_1,x_2)$ of $X(t)$ conditional on $X(t_1)=x_1$ and $X(t_2)=x_2$ is given by

$$b(t;x|t_1,t_2;x_1,x_2) = \frac{p(t-t_1;x_1,x)\,p(t_2-t;x,x_2)}{p(t_2-t_1;x_1,x_2)}.$$

Exercise 17.22. To simplify the Euler scheme, the Brownian increments $\Delta W = W(t+h)-W(t)$ can be replaced by other random variables $\widehat{\Delta W}$ with moments up to order 5 that are within $\mathcal{O}(h^3)$ of those of ΔW. Find the moments of the three point distribution

$$P(\widehat{\Delta W}=\pm\sqrt{3h})=\frac{1}{6}, \qquad P(\widehat{\Delta W}=0)=\frac{2}{3}$$

and compare them with those of ΔW.

Exercise 17.23. Solve the problem of minimizing the mean-square error subject to a computational budget s as follows:

$$\frac{C_2}{n}+C_1^2h^{2\beta}\to\min \quad \text{subject to} \quad \frac{nC_3}{h}=s,$$

where h is the discretization parameter, n is the sample size, and C_1, C_2, C_3 are positive constants. Find the optimal values of h and n. Find the order (w.r.t. s) of the optimal MSE.

Exercise 17.24. Let a random variable Y satisfy $0\leqslant m_1\leqslant Y\leqslant m_2<\infty$ a.s. Show that the variance of Y satisfies

$$\mathrm{Var}(Y)\leqslant (m_2-m_1)^2/4.$$

[Hint: Consider $\mathrm{E}\left[(Y-\frac{m_1+m_2}{2})^2\right]$.]

Exercise 17.25. Construct an importance sampling Monte Carlo method for evaluating the integral

$$I=\int_0^\pi\cdots\int_0^\pi \cos(1+\sin(u_1\ldots u_6))\sin u_1\cdots\sin u_6\,\mathrm{d}u_1\cdots\mathrm{d}u_6.$$

Choose a nonconstant density function f so that it is close enough to the integrand and, on the other hand, can be easily sampled from.

Exercise 17.26. Suppose that the integral $I=\int_0^1 \mathrm{e}^x\,\mathrm{d}x$ is evaluated by the Monte Carlo method using the density $f(x)\propto(1+cx)$, $c>0$. Find the optimal value of c so that the variance of the standard estimator $\mathrm{e}^X/f(X)$, $X\sim f$, attains its minimum value.

Exercise 17.27. Show the advantage of using the control variate method for evaluating the integral

$$I=\int_0^1\cdots\int_0^1\sqrt{1+u_1}\cdots\sqrt{1+u_8}\,\mathrm{d}u_1\cdots\mathrm{d}u_8.$$

Use the first two terms of a Taylor series of $\sqrt{1+u}$ to construct the control variate.

Exercise 17.28. Let the integral $I = \int_0^1 \sin\left(\frac{\pi x}{2}\right)\, dx$ be evaluated using the Monte Carlo method.

(a) Suppose that a combination of stratified random sampling and an antithetic variate method of the form

$$Y = {}^1\!/_2\big(f(U/2) + f(1 - U/2)\big), \quad U \sim \mathit{Unif}(0,1),$$

where $f(x) = \sin(\pi x/2)$, is applied to evaluate $I = \mathrm{E}[Y]$. What is the efficiency gain of this estimator as compared to the direct Monte Carlo estimator $X = f(U)$, $U \sim \mathit{Unif}(0,1)$.

(b) How large a sample size do you need if you use the direct Monte Carlo or the antithetic stratified method of (a), respectively, in order to estimate the above integral, correct to four decimal places (i.e., the error does not exceed $10^{-4}/2$) with a confidence level of 95%?

Exercise 17.29. Suppose the integral

$$I = \int_0^2 x^m\, dx \text{ for } m = 2, 3, \ldots$$

is approximated by the Monte Carlo method.

(a) Find the variance of the direct Monte Carlo estimator $X = 2f(U)$, $U \sim \mathit{Unif}(0,2)$, where $f(x) = x^m$.

(b) Suppose that a combination of stratified random sampling and an antithetic variate method of the form $Y = f(U) + f(2 - U)$, $U \sim \mathit{Unif}(0,2)$, is applied to evaluate the integral.

  (i) Show that $I = \mathrm{E}[Y]$.

  (ii) Find the variance $\mathrm{Var}(Y)$.

  (iii) What is the efficiency gain of this estimator as compared to the direct estimator?

  (iv) How does the efficiency gain behave with the increase of m?

(c) Suppose $Z = U$ is used as a control variate. Give the formula of the controlled estimator. Find the variance reduction factor for the optimal controlled estimator in comparison with the direct Monte Carlo estimator. How does the efficiency gain behave with the increase of m? [Note: You are not required to compute the optimal value of the control variate parameter.]

(d) How large a sample size do you need if you use the direct Monte Carlo or the control variate method of (c), respectively, to approximate the above integral, correct to two decimal places (i.e., the absolute error does not exceed $10^{-2}/2$) with a confidence level of 95%?

Exercise 17.30. Let $R \sim \mathit{Geom}(p)$, and $X_1, X_2, \ldots$ be i.i.d. $\mathit{Exp}(\lambda)$ random variables. Find the distribution of $S_R = \sum_{i=1}^R X_i$.

Exercise 17.31. Construct an importance sampling Monte Carlo method for evaluating the integral

$$I = \int_0^1 \cdots \int_0^1 \ln(3 + u_1 u_2 \cdots u_{11}) u_1^2 u_2^2 \cdots u_{11}^2\, du_1\, du_2 \cdots du_{11}.$$

Choose the density function f so that it is close enough to the integrand and, on the other hand, can be easily sampled from. A constant density is not acceptable!

Exercise 17.32. Show the advantage of using the control variate method for evaluating the integral

$$I = \int_0^{\pi/2} \int_0^{\pi/2} \cdots \int_0^{\pi/2} \sin u_1 \sin u_2 \cdots \sin u_5 \, \mathrm{d}u_1 \, \mathrm{d}u_2 \cdots \mathrm{d}u_5.$$

Use $u_1 \, u_2 \cdots u_5$ to construct a control variate.

Exercise 17.33. Prove that for any pair of random variables (U, V),

$$\mathrm{Var}(U) = \mathrm{E}[\mathrm{Var}(U|V)] + \mathrm{Var}(E[U|V])$$

[Hint: Use the fact that $\mathrm{E}[U^2] = \mathrm{E}\big[\mathrm{E}[U^2|V]\big]$ and $\mathrm{Var}(U) = \mathrm{E}[U^2] - (\mathrm{E}[U])^2$.]

Exercise 17.34. Let $X_1, X_2, \ldots, X_n$ be independent random variables with expected values $\mathrm{E}[X_i] = \mu_i$ and nonzero variances. Consider the following estimator of $\mathrm{E}[Y]$:

$$Z = Y + \sum_{i=1}^{n} c_i (X_i - \mu_i).$$

Show that the values of $c_1, c_2, \ldots, c_n$ that minimize $\mathrm{Var}(Z)$ are

$$c_i = -\frac{\mathrm{Cov}(Y, X_i)}{\mathrm{Var}(X_i)}, \quad i = 1, 2, \ldots, n.$$

Exercise 17.35. Show that the solution to the minimization problem

$$\min_{n_1, \ldots, n_k} \sum_{i=1}^{k} \frac{p_i^2 \sigma_i^2}{n_i}, \text{ such that } n_1 + n_2 + \cdots + n_k = N,$$

is given by

$$n_i = N \frac{p_i \sigma_i}{\sum_{j=1}^{k} p_j \sigma_j}, \quad i = 1, 2, \ldots, k.$$

[Hint: Use Lagrange multipliers.]

Exercise 17.36. Show that the solution of

$$\operatorname*{arg\,min}_{g} \mathrm{Var}_g \left(h(X) \frac{f(X)}{g(X)} \right), \quad X \sim g,$$

where g is a PDF, is given by

$$g(x) = \frac{|h(x)| f(x)}{\int_{-\infty}^{\infty} |h(x)| f(x) \, \mathrm{d}x}.$$

References

I. G. Dyadkin and G. H. Kenneth. A study of 128-bit multipliers for congruential pseudo-random number generators. *Computer Physics Communications*, 125(13):239–258, 2000.

S. M. Ermakov. *Metod Monte-Karlo i smezhnye voprosy [The Monte Carlo Method and Related Questions]*. Izdat. "Nauka," Moscow, 1975. Second edition, augmented, Teoriya Veroyatnostei i Matematicheskaya Statistika. [Monographs in Probability and Mathematical Statistics].

S. M. Ermakov and G. A. Mikhaĭlov. *Statisticheskoe modelirovanie [Statistical Modelling]*. "Nauka", Moscow, second edition, 1982.

A. Gut. *An Intermediate Course in Probability*. Springer Texts in Statistics. Springer, New York, second edition, 2009.

P. E. Kloeden and E. Platen. *Numerical Solution of Stochastic Differential Equations*. Applications of Mathematics. Springer, 2011.

P. L'Ecuyer. Random number generation. In *Handbook of Computational Statistics*, pages 35–71. Springer, 2012.

D. H. Lehmer. Mathematical methods in large-scale computing units. In *Proceedings of a Second Symposium on Large-Scale Digital Calculating Machinery, 1949*, pages 141–146, Cambridge, MA, 1951. Harvard University Press.

D. B. Madan and E. Seneta. The Variance Gamma model for share market returns. *Journal of Business*, 63(4):511–524, 1990.

R. N. Makarov and D. Glew. Exact simulation of Bessel diffusions. *Monte Carlo Methods and Applications*, 16(3):283–306, 2010.

S. K. Park and K. W. Miller. Random number generators: good ones are hard to find. *Commun. ACM*, 31(10):1192–1201, October 1988.

A. V. Voitishek and G. A. Mikhaĭlov. *Numerical Statistical Modeling: Monte Carlo Methods*. Akademiya, Moscow, 2006.

A. J. Walker. An efficient method for generating discrete random variables with general distributions. *ACM Transactions on Mathematical Software*, 3(3):253–256, 1977.

Chapter 18

Numerical Applications to Derivative Pricing

18.1 Overview of Deterministic Numerical Methods

18.1.1 Quadrature Formulae

One of the fundamental problems in numerical analysis is the evaluation of integrals. Consider a one-dimensional definite integral of a function $f\colon [a,b] \to \mathbb{R}$,

$$I(f) \equiv I(f;[a,b]) := \int_a^b f(x)\,\mathrm{d}x \text{ with } -\infty < a < b < \infty.$$

The function f is said to be integrable (and the integral of f is defined) if the integral $I(|f|)$ exists and is finite. Assuming that a closed-form expression for $I(f)$ is unavailable or intractable, we rely on a numerical evaluation. Any explicit formula that is suitable for providing an approximation of $I(f)$ is called a *quadrature formula*, or a *quadrature rule*, or a *numerical integration formula*. A typical quadrature takes the form

$$Q_n(f) \equiv Q_n(f;[a,b]) := \sum_{i=1}^{n} w_i f(x_i), \tag{18.1}$$

where the *nodes* $x_1, x_2, \ldots, x_n \in [a,b]$ and positive weights $w_1, w_2, \ldots, w_n$ are selected such that $Q(f)$ approximates $I(f)$ for any sufficiently smooth function f. In this section we focus on the approximation of one-dimensional integrals with bounded integration intervals and sufficiently smooth integrands. We consider two main approaches to the construction of a quadrature formula (18.1), namely, the Newton–Cotes quadrature formula and the Gaussian quadrature formula, which are, respectively, covered in the following two subsections.

18.1.1.1 Newton–Cotes Quadrature Formulae

In the *Newton–Cotes quadrature formulae*, the nodes $x_1, x_2, \ldots, x_n$ are fixed and equally spaced within $[a,b]$. A *closed* Newton–Cotes formula includes the endpoints of the integration interval among the nodes, whereas all nodes of an *open* Newton–Cotes formula are internal points of the integration interval. The weights $w_1, w_2, \ldots, w_n$, are selected in such a way that the quadrature rule (18.1) is precise for polynomials of the highest possible degree. Suppose that (18.1) is exact for every polynomial of degree no greater than d and is not exact for some polynomial of degree $d+1$. Then, the quadrature formula Q is said to have *degree of precision* d. Since the quadrature rule $Q(f)$ in (18.1) is linear, i.e., $Q(f+g) = Q(f) + Q(g)$ holds for any functions f and g, we need only test it on monomials x^k, $k = 0, 1, 2, \ldots$ to find d. Thus, the degree of precision of a quadrature formula Q is the largest integer d such that $I(x^k) = Q(x^k)$ for all $k = 0, 1, \ldots, d$ and $I(x^{d+1}) \neq Q(x^{d+1})$.

Example 18.1. Construct Newton–Cotes formulae with (a) $n = 1$ node and (b) $n = 2$ nodes.

Solution.
(a) There only exists an open Newton–Cotes rule with one node $x_1 = \frac{a+b}{2}$ located in the middle of the integration interval $[a, b]$. We find the weight w_1 so that the rule $Q_1(f) = w_1 f(x_1)$ is exact for monomials of the highest possible degree. Let us verify whether $\int_a^b x^k \, dx = w_1 x_1^k$ holds for $k = 0, 1, 2$.

If $k = 0$, then $I(1) = \int_a^b 1 \, dx = b - a = Q_1(1) = w_1$. Thus, $w_1 = b - a$.

If $k = 1$, then $I(x) = \int_a^b x \, dx = \frac{1}{2}(b^2 - a^2) = Q_1(x) = (b - a)\frac{a+b}{2}$.

If $k = 2$, then $I(x^2) = \int_a^b x^2 \, dx = \frac{1}{3}(b^3 - a^3) \neq Q_1(x^2) = (b - a)\left(\frac{a+b}{2}\right)^2$.

So the degree of precision of the open Newton–Cotes formula with one node is 1. This formula is called the *midpoint rule*. It is given by

$$I(f) \approx Q_1(f) = (b - a) f\left(\frac{a + b}{2}\right).$$

(b) There exist two Newton–Cotes formulae with two nodes, namely, an open formula and a closed formula. Let us consider the case of the closed quadrature formula with the nodes $x_1 = a$ and $x_2 = b$. The weights w_1 and w_2 are as follows:

$$\begin{cases} I(1) = Q_1(1) \\ I(x) = Q_1(x) \end{cases} \implies \begin{cases} w_1 + w_2 = b - a \\ w_1 a + w_2 b = \frac{b^2 - a^2}{2} \end{cases} \implies \begin{cases} w_1 = \frac{b-a}{2} \\ w_2 = \frac{b-a}{2} \end{cases}$$

The closed Newton–Cotes formula with two nodes denoted $Q_2(f)$ is called the *trapezoidal rule* and is given by

$$I(f) \approx Q_2(f) = \frac{b - a}{2} \left(f(a) + f(b)\right).$$

Let us apply it to $f(x) = x^2$:

$$I(x^2) = \frac{b^3 - a^3}{3} \neq Q_2(x^2) = \frac{b - a}{2} \left(a^2 + b^2\right).$$

Thus, the degree of precision of the trapezoidal rule is 1. □

Alternatively, quadrature formulae can be constructed with the use of the Lagrange interpolation method. Every sufficiently smooth function f can be approximated by the Lagrange polynomial

$$p(x) = \sum_{i=1}^{n} L_{n,i}(x) f(x_i), \text{ where } L_{n,i}(x) = \prod_{\substack{j=1 \\ j \neq i}}^{n} \frac{x - x_j}{x_i - x_j}.$$

It interpolates f at the nodes $x_1, x_2, \ldots, x_n$, i.e., $p(x_i) = f(x_i)$ for all $i = 1, 2, \ldots, n$. The integral of f can be approximated by the integral of the interpolating polynomial:

$$I(f) \approx I(p) = I\left(\sum_{i=1}^{n} L_{n,i}(x) f(x_i)\right) = \sum_{i=1}^{n} I(L_{n,i}(x)) f(x_i) = \sum_{i=1}^{n} w_i f(x_i).$$

As a result, a quadrature formula with weights $w_i := \int_a^b L_{n,i}(x) \, dx$ is defined.

For a general integrand f, a quadrature formula is not exact. The error term $I(f) - Q(f)$ can be found by expanding the function f in a Taylor series. The following theorem generalizes our findings related to the degree of precision of a quadrature formula from the above example and provides the formula for the error term.

Theorem 18.1. *Let $Q_n(f)$ denote a Newton–Cotes quadrature rule (open or closed) with n nodes.*

(a) If n is odd and f has $n+1$ continuous derivatives, then the degree of precision of Q_n is equal to n and the error is $I(f) - Q_n(f) = C(b-a)^{n+2} f^{(n+1)}(\xi)$ for some constant C independent of f and some point $\xi \in (a,b)$.

(b) If n is even and f has n continuous derivatives, then the degree of precision of Q_n is equal to $n-1$ and the error is $I(f) - Q_n(f) = C(b-a)^{n+1} f^{(n)}(\xi)$ for some constant C independent of f and some point $\xi \in (a,b)$.

The constant of the error term $I(f) - Q_n(f)$ can be found using the fact that the quadrature formula is not exact for $f(x) = x^{d+1}$:

$$I(x^{d+1}) - Q_n(x^{d+1}) = C(b-a)^{d+1} \left(x^{d+1}\right)^{(d+1)} \Big|_{x=\xi} = C(b-a)^{d+1}(d+1)!$$

$$\implies C = \frac{1}{(b-a)^{d+1}(d+1)!} \left(\frac{b^{d+2} - a^{d+2}}{d+2} - Q_n(x^{d+1}) \right).$$

Let us list some of the Newton–Cotes quadrature rules together with their error terms.

- The midpoint rule ($n = 1$):

$$I(f) = (b-a) f\left(\frac{a+b}{2}\right) + \frac{(b-a)^3}{24} f''(\xi). \tag{18.2}$$

- The trapezoidal rule ($n = 2$):

$$I(f) = \frac{(b-a)^2}{2} \left[f(a) + f(b)\right] - \frac{(b-a)^3}{12} f''(\xi). \tag{18.3}$$

- Simpson's rule ($n = 3$):

$$I(f) = \frac{(b-a)^2}{6} \left[f(a) + 4f\left(\frac{a+b}{2}\right) + f(b)\right] - \frac{(b-a)^5}{2880} f^{(4)}(\xi). \tag{18.4}$$

18.1.1.2 Gaussian Quadrature Formulae

For the *Gaussian quadrature formulae*, the nodes and weights in (18.1) are selected so as to achieve the highest possible degree of precision. Fixing n, we determine the nodes $x_1, \ldots, x_n$ and the weights $w_1, \ldots, w_n$ such that the rule $Q_n(f)$ provides the exact value of $I(f)$ for all $f(x) \in \{1, x, x^2, \ldots, x^{2n-1}\}$. That is, the degree of precision of the quadrature formula is $2n-1$. To determine the $2n$ unknowns, we form a system of $2n$ equations (nonlinear equations, in general). For example, if $n = 1$, then we obtain two simultaneous equations

$$w_1 = b - a \text{ and } (b-a)x_1 = \frac{(b-a)^2}{2}.$$

Solving them, we obtain the midpoint rule.

Example 18.2. Find the Gaussian quadrature rule for $n = 2$.

Solution. First, we apply the change of variables $x = \frac{b-a}{2} t + \frac{a+b}{2}$ to obtain an integral over $[-1, 1]$:

$$\int_a^b f(x)\,\mathrm{d}x = \int_{-1}^1 g(t)\,\mathrm{d}t, \text{ where } g(t) := \frac{b-a}{2} \cdot f\left(\frac{b-a}{2} t + \frac{a+b}{2}\right).$$

Second, find the nodes t_1, t_2 and weights w_1, w_2 such that $\int_{-1}^{1} g(t)\,\mathrm{d}t = w_1 g(t_1) + w_2 g(t_2)$ holds for all $g(t) \in \{1, t, t^2, t^3\}$:

$$\begin{aligned} g(t) &= 1 \quad \to 2 = w_1 + w_2, \\ g(t) &= t \quad \to 0 = w_1 t_1 + w_2 t_2, \\ g(t) &= t^2 \quad \to \frac{2}{3} = w_1 t_1^2 + w_2 t_2^2, \\ g(t) &= t^3 \quad \to 0 = w_1 t_1^3 + w_2 t_2^3. \end{aligned}$$

Solving the above simultaneous equations gives $w_1 = w_2 = 1$, $t_1 = -\frac{1}{\sqrt{3}}$, and $t_2 = \frac{1}{\sqrt{3}}$. Applying the inverse change of variables, we obtain the following quadrature rule for the integral of $f(x)$ over $[a, b]$:

$$\int_a^b f(x)\,\mathrm{d}x \approx \frac{b-a}{2}\left[f\left(\frac{a+b}{2} - \frac{b-a}{\sqrt{3}}\right) + f\left(\frac{a+b}{2} + \frac{b-a}{\sqrt{3}}\right)\right]. \qquad \square$$

18.1.1.3 Composite Quadrature Formulae

Newton–Cotes integration formulae and Gaussian integration rules can only be applied to integrals with small integration intervals since the error term grows as a power function of the length $b-a$. Large integration intervals can be first partitioned into smaller subintervals. After this, some simple rule is applied on each subinterval and the final result is given by a sum of approximations for the subintervals:

$$\int_a^b f(x)\,\mathrm{d}x = \sum_{j=1}^{m} \int_{c_{j-1}}^{c_j} f(x)\,\mathrm{d}x \approx \sum_{j=1}^{m} Q(f; [c_{j-1}, c_j]),$$

where $c_1, c_2, \ldots, c_m$ are points partitioning $[a, b]$ so that $a = c_0 < c_1 < \cdots < c_m = b$. The resulting formula is called a *composite quadrature rule*. For example, let us obtain the composite trapezoidal rule. By splitting the integration interval into m equal subintervals of step size $h = \frac{b-a}{m}$ and applying the trapezoidal rule on each subinterval $[c_{j-1}, c_j]$ with $c_j = a + jh$, $j = 0, 1, \ldots, m$, we have

$$\begin{aligned} \int_a^b f(x)\,\mathrm{d}x &= \sum_{j=1}^{m} \int_{c_{j-1}}^{c_j} f(x)\,\mathrm{d}x \\ &= \sum_{j=1}^{m} \frac{c_j - c_{j-1}}{2}\left[f(c_{j-1}) + f(c_j)\right] - \sum_{j=1}^{m} \frac{(c_j - c_{j-1})^3}{12} f''(\xi_j) \\ &= \underbrace{\frac{h}{2}\left[f(c_0) + 2\sum_{j=1}^{m-1} f(c_j) + f(c_m)\right]}_{=T_h(f) \text{ (the composite trapezoidal rule)}} \underbrace{- \frac{h^3}{12}\sum_{j=1}^{m} f''(\xi_j)}_{\text{the error term}}, \end{aligned}$$

where $\xi_j \in (c_{j-1}, c_j)$ for each $j = 1, 2, \ldots, m$. By applying the intermediate value theorem, we obtain a more compact formula for the error term: $-\frac{(b-a)h^2}{12} f''(\xi) = \mathcal{O}(h^2)$, as $h \to 0$, where $\xi \in (a, b)$.

18.1.1.4 Extrapolation and Romberg Integration

Suppose that an integral $I(f)$ is approximated by some quadrature formula $Q_h(f)$ which is a function of a small parameter h (e.g., the distance between two neighbouring nodes in a

Newton–Cotes integration formula with uniformly spaced nodes). The Richardson extrapolation method is a general procedure that takes two approximations computed with different values of parameter h and combines them in such a way that the resulting approximation is of higher order than the original one. To apply this method, the exponent of h in the leading term of the error of the original approximations must be known. For example, the error term of the composite trapezoidal rule admits the following asymptotic expansion:

$$I(f) = \int_a^b f(x)\,\mathrm{d}x = T_h(f) + C_2h^2 + C_4h^4 + C_6h^6 + \cdots$$

provided that f is a sufficiently smooth function. Here $C_2, C_4, C_6, \ldots$ are some constants independent of h, and T_h denotes the composite trapezoidal rule with step size h. Let us take two trapezoidal approximations with steps h and $2h$, respectively,

$$I(f) = T_h(f) + C_2h^2 + C_4h^4 + C_6h^6 + \cdots$$
$$I(f) = T_{2h}(f) + 4C_2h^2 + 16C_4h^4 + 64C_6h^6 + \cdots$$

Multiplying the former by 4 and subtracting the latter from the product, we have

$$3I(f) = [4T_h(f) - T_{2h}(f)] + (4C_2 - 4C_2)h^2 + (4C_4 - 16C_4)h^4 + (4C_6 - 64C_6)h^6 + \cdots$$

Dividing both sides by 3 gives

$$I(f) = \underbrace{\left[\frac{4T_h(f) - T_{2h}(f)}{3}\right]}_{=S_h(f)} - 4C_4h^4 - 20C_6h^6 + \cdots$$

As a result, we have derived a new approximation denoted $S_h(f)$, whose approximation error is of order $\mathcal{O}(h^4)$, instead of $\mathcal{O}(h^2)$, as that of the composite trapezoidal rule. In fact, the combination of two trapezoidal rules with respective step sizes h and $2h$ gives us *Simpson's rule.* Since the exponent of h appearing in the leading term of the approximation error is known, we can apply the extrapolation procedure again to obtain another approximation with an error of order $\mathcal{O}(h^6)$, as $h \to 0$. By applying the Richardson extrapolation method repeatedly on the composite trapezoidal quadrature rule, a sequence of improved approximations of the integral $I(f)$ is generated. This method is known as *Romberg integration.*

Define the approximation $R(k,0)$ to be the composite trapezoidal rule with $n = 2^k + 1$ quadrature nodes, i.e., $R(k,0) = T_{h_k}(f)$ with step size $h_k = \frac{b-a}{2^k}$. We begin with the approximation $R(L,0)$ for some positive integer L. By successively halving the step size, we obtain a series of trapezoidal approximations $R(L,0), R(L+1,0), R(L+2,0), \ldots$. Each successive approximation $R(k,0)$ is obtained from its predecessor $R(k-1,0)$ by adding contributions from the midpoints of every subinterval:

$$R(k,0) = \frac{1}{2}R(k-1,0) + h_k \sum_{i=1}^{2^{k-1}} f(a + (2i-1)h_k).$$

The approximation error of the trapezoidal approximations is $I(f) - R(k,0) = \mathcal{O}(h_k^2)$. As for the next step, we generate a series of improvements. First, we calculate the extrapolations of the composite trapezoidal approximations:

$$R(k,1) = \frac{4^1R(k,0) - R(k-1,0)}{4^1 - 1} = \frac{4R(k,0) - R(k-1,0)}{3}, \quad k = L+1, L+2, \ldots$$

The rule $R(k,1)$, is equivalent to the composite Simpson rule with 2^k+1 points. The approximation error of the first improvement is $I(f) - R(k,1) = \mathcal{O}(h_k^4)$. Now generate the second improvement:

$$R(k,2) = \frac{4^2 R(k,1) - R(k-1,1)}{4^2-1} = \frac{16R(k,1) - R(k-1,1)}{15}, \quad k = L+2, L+3, \ldots$$

The second extrapolation, $R(k,2)$, is equivalent to Boole's rule with 2^k+1 points. The approximation error is $I(f) - R(k,2) = \mathcal{O}(h_k^6)$. The third improvement is

$$R(k,3) = \frac{4^3 R(k,2) - R(k-1,2)}{4^3-1} = \frac{48R(k,2) - R(k-1,2)}{47}, \quad k = L+3, L+4, \ldots$$

The approximation error is $I(f) - R(k,3) = \mathcal{O}(h_k^8)$. In a similar manner, we can generate further improvements provided that the integrand f has continuous high-order derivatives. The jth improvement is given by

$$R(k,j) = \frac{4^j R(k,j-1) - R(k-1,j-1)}{4^j-1}, \quad k = L+j, L+j+1, \ldots$$

It is convenient to organize the Romberg method in the form of a lower-triangular table, where the first column is filled with the composite trapezoidal approximations $R(L+j,0)$, $j \geqslant 0$, and the next column is defined as the extrapolations of the previous one:

$$\begin{array}{l|llll}
n = 2^L + 1 & R(L,0) & & & \\
n = 2^{L+1} + 1 & R(L+1,0) & R(L+1,1) & & \\
n = 2^{L+2} + 1 & R(L+2,0) & R(L+2,1) & R(L+2,2) & \\
n = 2^{L+3} + 1 & R(L+3,0) & R(L+3,1) & R(L+3,2) & R(L+3,3) \\
\vdots & \vdots & & & \ddots
\end{array}$$

In the above table, n denotes the number of nodes. This table can be filled row by row. It is a common practice to compute new rows until two successive diagonal entries, say $R(L+j,j)$ and $R(L+j-1,j-1)$, differ by less than a given error tolerance.

18.1.2 Finite-Difference Methods

18.1.2.1 Finite-Difference Approximations for ODEs

Finite differences are used to numerically solve (initial-)boundary value problems for differential equations by reducing them to systems of linear or nonlinear algebraic equations. As an illustrative example, let us consider the following boundary value problem (BVP) for a second-order ordinary differential equation (ODE):

$$y''(x) = f(x, y(x), y'(x)), \quad l < x < r, \tag{18.5}$$

$$y(l) = \tilde{y}_l, \quad y(r) = \tilde{y}_r, \tag{18.6}$$

where f is a continuous function of its arguments, $\tilde{y}_l$ and $\tilde{y}_r$ are constant boundary values, and $y(x)$ is an unknown function to be computed on the interval $[l, r]$. Let us discretize the BVP (18.5)–(18.6) as follows.

1. Introduce a mesh $x_j = l + j\,\delta x$, $j = 0, 1, \ldots, n+1$, with the step size $\delta x = \frac{r-l}{n+1}$. The exact solution $y(\cdot)$ is approximated by a mesh solution $y.$ at the points x_j; that is, $y(x_j) \approx y_j$ for $j = 0, 1, \ldots, n+1$.

TABLE 18.1: Commonly used finite-difference approximations of first and second derivatives.

Derivative	Finite Difference + Order	Type
$f'(x)$	$\frac{f(x+\delta x)-f(x)}{\delta x} + \mathcal{O}(\delta x)$	Forward
$f'(x)$	$\frac{f(x)-f(x-\delta x)}{\delta x} + \mathcal{O}(\delta x)$	Backward
$f'(x)$	$\frac{f(x+\delta x)-f(x-\delta x)}{2h} + \mathcal{O}(\delta x^2)$	Central
$f''(x)$	$\frac{f(x+\delta x)-2f(x)+f(x-\delta x)}{\delta x^2} + \mathcal{O}(\delta x^2)$	Central

2. Discretize the ODE (18.5) using finite-difference (FD) approximations of derivatives (see Table 18.1):

$$y''(x) = f(x, y(x), y'(x))$$

$$\Big\downarrow \text{(replace all derivatives by their FD approximations)}$$

$$\frac{y(x_j + \delta x) - 2y(x_j) + y(x_j - \delta x)}{\delta x^2} \approx f\left(x_j, y(x_j), \frac{y(x_j + \delta x) - y(x_j - \delta x)}{2\delta x}\right)$$

$$\Big\downarrow$$

$$\frac{y(x_{j+1}) - 2y(x_j) + y(x_{j-1})}{\delta x^2} \approx f\left(x_j, y(x_j), \frac{y(x_{j+1}) - y(x_{j-1})}{2\delta x}\right)$$

$$\Big\downarrow \text{(obtain a difference equation for the mesh solution)}$$

$$\frac{y_{j+1} - 2y_j + y_{j-1}}{\delta x^2} = f\left(x_j, y_j, \frac{y_{j+1} - y_{j-1}}{2\delta x}\right).$$

3. The boundary conditions (18.6) give the values of the mesh solution y_0 and y_{n+1}:

$$y_0 = \tilde{y}_l, \quad y_{n+1} = \tilde{y}_r.$$

As a result, we obtain a system of n difference equations with n unknowns $y_1, y_2, \dots, y_n$. In general, the resulting difference equations are nonlinear, and the system can be solved by the multivariate Newton method. If the function f in (18.5) is a linear function of y and y' (i.e., in the case of a linear second order ODE), the resulting system is linear and it can be solved by the Gaussian elimination approach or by an iterative method such the Gauss–Seidel and Jacobi methods. For example, consider the following linear ODE: $y''(x) = a(x) + b(x)y(x) + c(x)y'(x)$. The respective system of difference equations for the mesh solution has a triangular matrix of coefficients and is given by

$$\begin{cases} -(2 + b_1\delta x^2)y_1 + (1 - c_1\delta x/2)y_2 = \delta x^2 a_1 - (1 + c_1\delta x/2)\tilde{y}_l, \\ (1 + c_i\delta x/2)y_{i-1} - (2 + b_i\delta x^2)y_i + (1 - c_i\delta x/2)y_{i+1} = \delta x^2 a_i, \quad i = 2, 3, \dots, n-1, \\ (1 + c_n\delta x/2)y_{n-1} - (2 + b_n\delta x^2)y_n = \delta x^2 a_n - (1 - c_n\delta x/2)\tilde{y}_r, \end{cases}$$

where $a_i := a(x_i)$, $b_i := b(x_i)$, $c_i := c(x_i)$ for $i = 1, 2, \dots, n$. The system of linear equations (for the mesh solution $y_\cdot$) can be solved by the Gaussian elimination method. Since the matrix of coefficients is tridiagonal, the application of the Gaussian elimination method requires $\mathcal{O}(n)$ arithmetic operations.

18.1.2.2 Second-Order Linear PDEs

The finite-difference approach can be applied to solve numerically a partial differential equation (PDE) that involves more than one independent variable. By replacing all partial derivatives with their respective finite-difference approximations, we can reduce a differential equation to a system of difference equations. Consider a PDE of the form

$$A\frac{\partial^2 u}{\partial x^2} + 2B\frac{\partial^2 u}{\partial x \partial y} + C\frac{\partial^2 u}{\partial y^2} + D\frac{\partial u}{\partial x} + E\frac{\partial u}{\partial y} + Fu = G(x, y), \tag{18.7}$$

where $A, B, C, \ldots, F$ are constants, $G(x, y)$ is a known function, and $u = u(x, y)$ is the solution to be determined. The PDE (18.7) is called

- an *elliptic*-type PDE, if $AC - B^2 > 0$ (e.g., the Laplace equation $\frac{\partial^2 u}{\partial x^2} + \frac{\partial^2 u}{\partial y^2} = 0$);
- a *parabolic*-type PDE, if $AC - B^2 = 0$ (e.g., the heat equation $\frac{\partial u}{\partial t} = c^2 \frac{\partial^2 u}{\partial x^2}$);
- a *hyperbolic*-type PDE, if $AC - B^2 < 0$ (e.g., the wave equation $\frac{\partial^2 u}{\partial t^2} = c^2 \frac{\partial^2 u}{\partial x^2}$).

Let $B = 0$ and $C = 0$, then Equation (18.7) is of a parabolic type. Relabel $y \to t$ to obtain the following equation:

$$\frac{\partial u}{\partial t} = \widetilde{A}\frac{\partial^2 u}{\partial x^2} + \widetilde{D}\frac{\partial u}{\partial x} + \widetilde{F}u + \widetilde{G}(x, y).$$

The simplest parabolic-type PDE is the heat equation

$$\frac{\partial u}{\partial t} = c^2 \frac{\partial^2 u}{\partial x^2}. \tag{18.8}$$

This equation models the temperature in a thin insulated rod. Here x and t are called spacial and temporal variables, respectively. As is well known, the Black–Scholes equation for the GBM model can be reduced to the heat equation by applying a change of variables and a scale transformation. The heat equation can be solved numerically or analytically using finite-differences or spectral expansions, respectively.

A differential equation may have multiple solutions, and hence a differential problem becomes well-formulated only when a PDE is combined with initial and boundary conditions. We consider a typical initial boundary value (IBV) problem for the heat equation:

$$\begin{cases} \frac{\partial u(t,x)}{\partial t} = c^2 \frac{\partial^2 u(t,x)}{\partial x^2} & \text{for } 0 \leqslant x \leqslant \ell,\ 0 < t \leqslant T, \\ u(0, x) = f(x) & \text{for } 0 \leqslant x \leqslant \ell, \\ u(t, 0) = g_0(t) & \text{for } 0 \leqslant t \leqslant T, \\ u(t, \ell) = g_1(t) & \text{for } 0 \leqslant t \leqslant T, \end{cases} \tag{18.9}$$

where f, g_0, and g_1 are continuous functions, and c is a constant. The unknown solution is a function $u(t, x)$ defined on the rectangle $D := [0, T] \times [0, \ell]$. In the following section, we describe three finite-difference schemes for solving (18.9) numerically. The computational solution is a function defined on a two-dimensional grid (or mesh) of points in the domain D. Being given a mesh solution, the approximation solution can be calculated at every point of the domain D using interpolation methods.

18.1.2.3 Finite-Difference Approximations for the Heat Equation

Our goal is to approximate the solution u to the IBV problem (18.9) on a spacial-temporal mesh of points. First, we define a spacial mesh at which the solution is to be approximated. Divide $[0, \ell]$ into $n+1$ equal subintervals of length $\delta x = \frac{\ell}{n+1}$. Introduce the spacial mesh points $x_j = j\,\delta x$, $j = 0, 1, \ldots, n+1$. Second, the derivatives with respect to the spacial variable x are approximated by their central finite-difference approximations (with the order of approximation $\mathcal{O}(\delta x^2)$). Inserting the finite-difference formula for u_{xx} in the heat equation (18.8) gives

$$\frac{\partial u(t,x)}{\partial t} = c^2 \frac{u(t, x-\delta x) - 2u(t,x) + u(t, x+\delta x)}{\delta x^2} + \mathcal{O}(\delta x^2). \tag{18.10}$$

Let $u_j(t)$ denote the semi-discrete approximation to $u(t, x_j)$ for all $j = 0, 1, \ldots, n+1$. Each u_j is a function of time, not a function of the spacial variable x. Discarding the error term and writing (18.10) at every internal spacial node x_j, we have a system of ordinary differential equations

$$\frac{\mathrm{d}u_j(t)}{\mathrm{d}t} = c^2 \frac{u_{j-1}(t) - 2u_j(t) + u_{j+1}(t)}{\delta x^2}, \quad j = 1, 2, \ldots, n, \tag{18.11}$$

supplemented by the boundary conditions $u_0(t) = g_0(t)$ and $u_{n+1}(t) = g_1(t)$. Employing the initial condition on $u(t,x)$ gives

$$u_j(0) = f(x_j), \quad j = 0, 1, \ldots, n+1.$$

As a result, we obtain a Cauchy problem for a system of ODEs which can be written in matrix-vector form as follows:

$$\frac{\mathrm{d}\mathbf{u}(t)}{\mathrm{d}t} = -\frac{c^2}{\delta x^2}\mathbf{A}\mathbf{u}(t) + \frac{c^2}{\delta x^2}\mathbf{g}(t) \text{ subject to } \mathbf{u}(0) = \mathbf{f}, \text{ where} \tag{18.12}$$

$$\mathbf{u}(t) := \begin{bmatrix} u_1(t) \\ u_2(t) \\ \vdots \\ u_n(t) \end{bmatrix} \in \mathbb{R}^n, \quad \mathbf{f} := \begin{bmatrix} f(x_1) \\ f(x_2) \\ \vdots \\ f(x_n) \end{bmatrix} \in \mathbb{R}^n, \quad \mathbf{g}(t) := \begin{bmatrix} g_0(t) \\ 0 \\ \vdots \\ 0 \\ g_1(t) \end{bmatrix} \in \mathbb{R}^n, \tag{18.13}$$

$$\mathbf{A} = \begin{bmatrix} 2 & -1 & 0 & \cdots & 0 & 0 \\ -1 & 2 & -1 & \cdots & 0 & 0 \\ 0 & -1 & 2 & \cdots & 0 & 0 \\ \vdots & \vdots & \vdots & \ddots & \vdots & \vdots \\ 0 & 0 & 0 & \cdots & 2 & -1 \\ 0 & 0 & 0 & \cdots & -1 & 2 \end{bmatrix} \in \mathbb{R}^{n\times n}. \tag{18.14}$$

The coefficient matrix $\mathbf{A}$ is called the *discretization matrix*. The key feature of $\mathbf{A}$ is that it is a sparse matrix. All its nonzero elements are located on its main diagonal and upper and lower adjacent diagonals. Such a matrix is called a tridiagonal matrix or a band matrix with bandwidth 3.

For the numerical solution of the system of ODEs (18.12), we can use a time-discretization scheme by introducing a temporal mesh $t_i = i\,\delta t$, $i = 0, 1, \ldots, m$ with a mesh size $\delta t := \frac{T}{m}$. Using different time discretizations, we derive three computational schemes, namely, the *explicit Euler method*, the *implicit Euler method*, and the *Crank–Nicolson method*. Applying one of these three schemes, we obtain a fully discretized approximation solution

$$u_{i,j} \approx u_j(t_i) \approx u(t_i, x_j), \quad i = 0, 1, \ldots, m, \; j = 0, 1, \ldots, n+1,$$

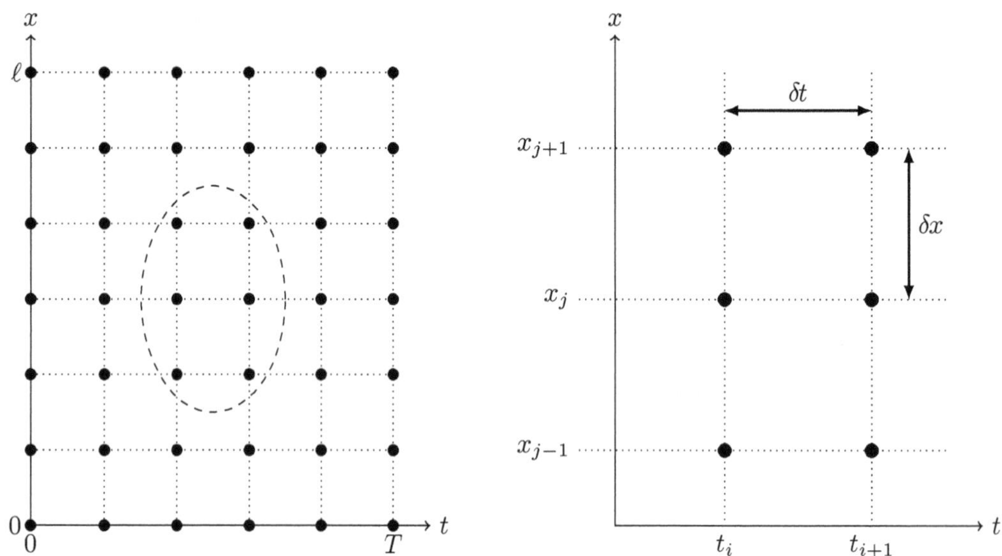

FIGURE 18.2: The temporal-spatial mesh and the grid points.

defined on the two-dimensional temporal-spatial mesh (see Figure 18.2). This is called a *mesh solution* to the IBV problem (18.9).

The Explicit Method

The explicit Euler method is obtained by replacing the time derivative by a forward finite-difference approximation. For a continuously differentiable function f, we have

$$\frac{\mathrm{d}f(t)}{\mathrm{d}t} = \frac{f(t+\delta t) - f(t)}{\delta t} + \mathcal{O}(\delta t).$$

Applying this formula for $\frac{\partial u}{\partial t}$ in (18.10) gives

$$\frac{u(t_{i+1}, x_j) - u(t_i, x_j)}{\delta t} = c^2 \frac{u(t_i, x_{j-1}) - 2u(t_i, x_j) + u(t_i, x_{j+1})}{\delta x^2} + \mathcal{O}(\delta t + \delta x^2) \quad (18.15)$$

for each $i = 0, 1, \ldots, m-1$ and $j = 1, 2, \ldots, n$. Discarding the error term $\mathcal{O}(\delta t + \delta x^2)$ and replacing every $u(t_i, x_j)$ by its approximation $u_{i,j}$, we obtain the following finite-difference approximation to the heat equation (18.8):

$$u_{i+1,j} = \alpha u_{i,j-1} + (1 - 2\alpha) u_{i,j} + \alpha u_{i,j+1}, \quad i = 1, 2, \ldots, n,\ j = 0, 1, \ldots, m-1, \quad (18.16)$$

where $\alpha := c^2 \frac{\delta t}{\delta x^2}$. The explicit method starts with the values $u_{0,j}$ (for $i = 0$), which are given by the initial time condition:

$$u_{0,j} = f(x_j), \quad j = 0, 1, \ldots, n+1. \quad (18.17)$$

Additionally, the values $u_{i,0}$ and $u_{i,n+1}$ for all $i = 1, 2, \ldots, m$ are known thanks to the boundary conditions:

$$u_{i,0} = g_0(t_i), \quad u_{i,n+1} = g_1(t_i), \quad i = 1, 2, \ldots, m. \quad (18.18)$$

The formula (18.16) is an explicit expression for every approximation $u_{i+1,j}$ at time t_{i+1} in terms of the solution at the previous time t_i. Therefore, this method is called the explicit Euler method. For a fixed i, the values $u_{i+1,j}$ for all $j = 0, 1, \ldots, n+1$ are computed from the values $\{u_{i,j}\}_{j=0,1,\ldots,n+1}$ of the ith time level. So the explicit rule (18.16) allows us to advance from the ith time level to the $(i+1)$st time level. The accuracy of the explicit method is $\mathcal{O}(\delta t + \delta x^2)$.

The explicit method (18.16) can also be written in matrix-vector form. Let us denote $\mathbf{u}^{(i)} := [u_{i,1}, u_{i,2}, \ldots, u_{i,n}]^\top$. For each $i = 0, 1, 2, \ldots, m$, the vector $\mathbf{u}^{(i)}$ is an approximation to $\mathbf{u}(t_i)$. Inserting the forward finite-difference approximation of the time derivative $\frac{\mathrm{d}\mathbf{u}(t)}{\mathrm{d}t}$,

$$\left.\frac{\mathrm{d}\mathbf{u}(t)}{\mathrm{d}t}\right|_{t=t_i} = \frac{\mathbf{u}(t_{i+1}) - \mathbf{u}(t_i)}{\delta t} + \mathcal{O}(\delta t) \approx \frac{\mathbf{u}^{(i+1)} - \mathbf{u}^{(i)}}{\delta t},$$

into Equation (18.12) written at $t = t_i$ gives

$$\frac{\mathbf{u}^{(i+1)} - \mathbf{u}^{(i)}}{\delta t} = -\frac{c^2}{\delta x^2}\mathbf{A}\mathbf{u}^{(i)} + \frac{c^2}{\delta x^2}\mathbf{g}(t_i),$$

which leads to the matrix-vector equations

$$\mathbf{u}^{(i+1)} = (\mathbf{I}_n - \alpha\mathbf{A})\,\mathbf{u}^{(i)} + \alpha\mathbf{g}^{(i)}, \quad i \geqslant 0, \tag{18.19}$$

where $\mathbf{I}_n$ denotes the n-by-n identity matrix and $\mathbf{g}^{(i)} := \mathbf{g}(t_i)$. The formula (18.19) is an iterative rule which starts at

$$\mathbf{u}^{(0)} = \mathbf{f}. \tag{18.20}$$

The problem (18.19)-(18.20) admits a closed-form solution

$$\mathbf{u}^{(k)} = (\mathbf{I}_n - \alpha\mathbf{A})^k\,\mathbf{f} + \alpha\sum_{i=1}^{k-1}(\mathbf{I}_n - \alpha\mathbf{A})^{k-i-1}\,\mathbf{g}^{(i)}, \quad 0 \leqslant k \leqslant m.$$

The Implicit Method

To derive the implicit method, we replace the time derivative of the solution by a *backward* finite-difference approximation:

$$\frac{\mathrm{d}f(t)}{\mathrm{d}t} = \frac{f(t) - f(t-\delta t)}{\delta t} + \mathcal{O}(\delta t).$$

Applying this formula for $\frac{\partial f(t)}{\partial t}$ in (18.10) gives

$$\frac{u(t_{i+1}, x_j) - u(t_i, x_j)}{\delta t} = c^2\frac{u(t_{i+1}, x_{j-1}) - 2u(t_{i+1}, x_j) + u(t_{i+1}, x_{j+1})}{\delta x^2} + \mathcal{O}(\delta t + \delta x^2) \tag{18.21}$$

for each $i = 0, 1, \ldots, m-1$ and $j = 1, 2, \ldots, n$. Discarding the error term $\mathcal{O}(\delta t + \delta x^2)$ and replacing every $u(t_i, x_j)$ by its approximation $u_{i,j}$, we have

$$-\alpha u_{i+1,j-1} + (2\alpha+1)u_{i+1,j} - \alpha u_{i+1,j+1} = u_{i,j}, \quad i = 1, 2, \ldots, n, \; j = 0, 1, \ldots, m-1, \tag{18.22}$$

subject to the initial condition (18.17) and boundary condition (18.18). As is seen from (18.22), the approximations $u_{i+1,j}$ of the $(i+1)$st time level are coupled and cannot be calculated explicitly from approximations $u_{i,j}$ of the ith time level. Hence, this scheme is called the *implicit method.*

The implicit method (18.22) can be written in matrix-vector form. Inserting the backward finite-difference approximation of the time derivative $\frac{\mathrm{d}\mathbf{u}(t)}{\mathrm{d}t}$,

$$\left.\frac{\mathrm{d}\mathbf{u}(t)}{\mathrm{d}t}\right|_{t=t_{i+1}} = \frac{\mathbf{u}(t_{i+1}) - \mathbf{u}(t_i)}{\delta t} + \mathcal{O}(\delta t) \approx \frac{\mathbf{u}^{(i+1)} - \mathbf{u}^{(i)}}{\delta t},$$

into Equation (18.12) written at $t = t_{i+1}$ gives

$$\frac{\mathbf{u}^{(i+1)} - \mathbf{u}^{(i)}}{\delta t} = -\frac{c^2}{\delta x^2}\mathbf{A}\mathbf{u}^{(i+1)} + \frac{c^2}{\delta x^2}\mathbf{g}^{(i+1)}.$$

As a result, we obtain the following computational formula:

$$(\mathbf{I}_n + \alpha\mathbf{A})\,\mathbf{u}^{(i+1)} = \mathbf{u}^{(i)} + \alpha\mathbf{g}^{(i+1)}, \quad i \geqslant 0, \tag{18.23}$$

subject to the initial condition (18.20). The order of approximation is $\mathcal{O}(\delta t + \delta x^2)$ which is the same as that of the explicit method.

To advance to the $(i+1)$st time level and find the mesh solution $\mathbf{u}^{(i+1)}$, we need to solve a system of linear equations with a tridiagonal coefficient matrix $\mathbf{C} = \mathbf{I}_n + \alpha\mathbf{A}$. In general, the numerical solution of a system with n equations and a band coefficient matrix requires $\mathcal{O}(n)$ operations. So the linear system $\mathbf{C}\mathbf{x} = \mathbf{b}$ for $\mathbf{x} = \mathbf{u}^{(i+1)}$ with the tridiagonal matrix $\mathbf{C}$ and vector $\mathbf{b} = \mathbf{u}^{(i)} + \alpha\mathbf{g}^{(i+1)}$ can be solved by the Gaussian elimination method with $n-1$ eliminations (during the forward step) and $n-1$ substitutions (during the backward step).

The Crank–Nicolson Method

The Crank–Nicolson method is one of the most well-known implicit finite-difference methods thanks to its stability properties and high order of approximation. To derive the Crank–Nicolson method, we first approximate the time derivative of $\mathbf{u}(t)$ at point $t = t_i + \delta t/2$ with the central finite-difference formula that has the second order of approximation:

$$\frac{\mathbf{u}(t_{i+1}) - \mathbf{u}(t_i)}{\delta t} + \mathcal{O}\left(\delta t^2\right) = \left.\frac{\mathrm{d}\mathbf{u}(t)}{\mathrm{d}t}\right|_{t=t_i+\delta t/2} = -\frac{c^2}{\delta x^2}\mathbf{A}\mathbf{u}(t_i + \delta t/2) + \frac{c^2}{\delta x^2}\mathbf{g}(t_i + \delta t/2).$$

Second, we use the average of the spatial approximations at times t_{i+1} and t_i in the right-hand side of the above equation:

$$\frac{\mathbf{u}(t_{i+1}) - \mathbf{u}(t_i)}{\delta t} = -\frac{c^2}{2\delta x^2}\mathbf{A}\left(\mathbf{u}(t_i) + \mathbf{u}(t_{i+1})\right) + \frac{c^2}{2\delta x^2}\left(\mathbf{g}(t_i) + \mathbf{g}(t_{i+1})\right) + \mathcal{O}\left(\delta t^2 + \delta x^2\right).$$

As a result, we obtain the following computational scheme, called the Crank–Nicolson method:

$$\left(\mathbf{I}_n + \frac{\alpha}{2}\mathbf{A}\right)\mathbf{u}^{(i+1)} = \left(\mathbf{I}_n - \frac{\alpha}{2}\mathbf{A}\right)\mathbf{u}^{(i)} + \frac{\alpha}{2}\left(\mathbf{g}^{(i)} + \mathbf{g}^{(i+1)}\right), \quad i \geqslant 0. \tag{18.24}$$

The method starts with the initial condition (18.20). As is seen from (18.24), the Crank–Nicolson method is an implicit computational scheme since we need to solve a system of linear equations to advance the mesh solution. However, the order of approximation of the Crank–Nicolson method is $\mathcal{O}\left(\delta t^2 + \delta x^2\right)$, which is higher than the order of approximation of both explicit and implicit Euler methods. Additionally, the Crank–Nicolson method is unconditionally stable, meaning that it is not sensitive to roundoff errors regardless of the values of the mesh sizes δt and δx.

18.1.2.4 Stability Analysis

The stability of a computational method is closely associated with numerical errors. A finite-difference scheme is *stable* if the roundoff error made at one iteration does not amplify as the computations are continued. That is, if the errors decay or remain bounded, the numerical method is stable. If, on the contrary, the errors grow with time, the numerical method is said to be *unstable*. To investigate the stability of the explicit and implicit Euler methods and the Crank–Nicolson scheme, we represent them in the form of an iteration rule:

$$\mathbf{u}^{(i+1)} = \mathbf{Q}\mathbf{u}^{(i)} + \mathbf{c}^{(i)}, \quad i \geqslant 0, \tag{18.25}$$

where

$$\mathbf{Q} = \begin{cases} \mathbf{I}_n - \alpha\mathbf{A} & \text{(the explicit method)} \\ (\mathbf{I}_n + \alpha\mathbf{A})^{-1} & \text{(the implicit method)} \\ \left(\mathbf{I}_n + \frac{\alpha}{2}\mathbf{A}\right)^{-1}\left(\mathbf{I}_n - \frac{\alpha}{2}\mathbf{A}\right) & \text{(the Crank–Nicolson method)} \end{cases}$$

$$\mathbf{c}^{(i)} = \begin{cases} \alpha\mathbf{g}^{(i)} & \text{(the explicit method)} \\ \alpha\left(\mathbf{I}_n + \alpha\mathbf{A}\right)^{-1}\mathbf{g}^{(i+1)} & \text{(the implicit method)} \\ \frac{\alpha}{2}\left(\mathbf{I}_n + \alpha\mathbf{A}\right)^{-1}\left(\mathbf{g}^{(i)} + \mathbf{g}^{(i+1)}\right) & \text{(the Crank–Nicolson method)} \end{cases}$$

Let $\mathbf{e}^{(0)}$ be the error introduced to the method initially. That is, in actual computations we start with the vector $\tilde{\mathbf{u}}^{(0)} = \mathbf{u}^{(0)} + \mathbf{e}^{(0)}$, which includes the introduced error. Of interest is the propagation of this error during computations as we advance the solution in time. Let $\tilde{\mathbf{u}}^{(i)}$ denote the mesh solution (of the ith time step) that includes the error. This perturbed mesh solution satisfies the rule (18.25):

$$\tilde{\mathbf{u}}^{(i+1)} = \mathbf{Q}\tilde{\mathbf{u}}^{(i)} + \mathbf{c}^{(i)}.$$

Subtracting (18.25) from this equation gives us the following recurrence formula for the accumulated error $\mathbf{e}^{(i)} := \tilde{\mathbf{u}}^{(i)} - \mathbf{u}^{(i)}$:

$$\mathbf{e}^{(i+1)} = \mathbf{Q}\tilde{\mathbf{u}}^{(i)} - \mathbf{Q}\mathbf{u}^{(i)} = \mathbf{Q}\mathbf{e}^{(i)}.$$

Hence, the error at the kth time step is

$$\mathbf{e}^{(k)} = \mathbf{Q}^k\mathbf{e}^{(0)}, \quad k \geqslant 0. \tag{18.26}$$

Since the approximation error is a vector, we need matrix and vector norms to measure the magnitude of the error. Examples of matrix norms include

- $\|\mathbf{Q}\|_\infty := \max_i \sum_j |Q_{ij}| = \|\mathbf{Q}^\top\|_1$ (the biggest row-sum of magnitudes);
- $\|\mathbf{Q}\|_1 := \max_j \sum_i |Q_{ij}| = \|\mathbf{Q}^\top\|_\infty$ (the biggest column-sum of magnitudes);
- $\|\mathbf{Q}\|_2 := \sqrt{\lambda_{\max}(\mathbf{Q}^\top\mathbf{Q})}$ (the spectral norm given by the square root of the largest eigenvalue of the positive-semidefinite matrix $\mathbf{Q}^\top\mathbf{Q}$).

Here $\mathbf{Q}$ is an n-by-n matrix of real elements Q_{ij}. Recall that a matrix norm $\|\cdot\|_*$ on $\mathbb{R}^{n\times n}$ is said to be *compatible* with a vector norm $\|\cdot\|$ on $\mathbb{R}^n$ if

$$\|\mathbf{Q}\,\mathbf{x}\| \leqslant \|\mathbf{Q}\|_* \cdot \|\mathbf{x}\|$$

holds for all $\mathbf{Q} \in \mathbb{R}^{n\times n}$ and all $\mathbf{x} \in \mathbb{R}^n$.

According to the definition given in the beginning of this subsection, a finite-difference method with the coefficient matrix $\mathbf{Q}$ is stable if the errors decay during the computations. For the error $\mathbf{e}^{(k)}$ to decay and eventually vanish as $k \to \infty$, that is, $\lim_{k\to\infty} \|\mathbf{e}^{(k)}\| = 0$, it is necessary and sufficient that

$$\|\mathbf{Q}^K\|_* < 1 \tag{18.27}$$

for some matrix norm $\|\cdot\|_*$ and some positive integer K. Indeed, using properties of matrix and vector norms and the representation (18.26), we obtain

$$0 \leqslant \|\mathbf{e}^{(k)}\| \leqslant \|\mathbf{Q}\|_*^b \cdot \|\mathbf{Q}^K\|_*^a \cdot \|\mathbf{e}^{(0)}\|, \tag{18.28}$$

where the nonnegative integers a and $b < K$ are defined so that $k = a \cdot K + b$. Provided that (18.27) holds, the right-hand side of (18.28) converges to 0 as $k \to \infty$. Note that if $\|\mathbf{Q}^K\|_* = 1$, then the errors remain bounded as $k \to \infty$, i.e., $\limsup_{k\to\infty} \|\mathbf{e}^{(k)}\| < \infty$.

The stability condition can be formulated with the use of the spectral radius. The spectral radius $\rho(\mathbf{Q})$ of an n-by-n matrix $\mathbf{Q}$ is defined by the maximum magnitude of the eigenvalues $\lambda_1, \lambda_2, \ldots, \lambda_n$ of $\mathbf{Q}$. That is,

$$\rho(\mathbf{Q}) := \max\{|\lambda_1|, |\lambda_2|, \ldots, |\lambda_n|\}.$$

Note that for a symmetric matrix, the 2-norm is equal to the spectral radius of the matrix, i.e., $\|\mathbf{Q}\|_2 = \rho(\mathbf{Q})$ if $\mathbf{Q} = \mathbf{Q}^\top$.

Proposition 18.2. *A finite-difference method is stable if the spectral radius of the matrix* $\mathbf{Q}$ *be less than one, i.e.,*

$$\rho(\mathbf{Q}) < 1. \tag{18.29}$$

Proof. Let us prove that the conditions (18.27) and (18.29) are equivalent. First, use the fact that the spectral radius does not exceed $\|\mathbf{Q}^k\|_*^{1/k}$ for any matrix norm $\|\cdot\|_*$ and any positive integer k. That is, $\rho(\mathbf{Q}) \leqslant \|\mathbf{Q}^k\|_*^{1/k}$. Therefore, $\|\mathbf{Q}^k\|_* < 1$ implies that $\rho(\mathbf{Q}) < 1$. On the other hand, according to Gelfand's formula, for any matrix norm $\|\cdot\|_*$ we have $\lim_{k\to\infty} \|\mathbf{Q}^k\|_*^{1/k} = \rho(\mathbf{Q})$. Let $\rho(\mathbf{Q}) = \alpha_0 < 1$. Then, there exists k so that we have

$$\left|\|\mathbf{Q}^k\|_*^{1/k} - \rho(\mathbf{Q})\right| < \frac{1-\alpha_0}{2} \implies \|\mathbf{Q}^k\|_*^{1/k} < \rho(\mathbf{Q}) + \frac{1-\alpha_0}{2} = \frac{1+\alpha_0}{2} < 1. \qquad \square$$

Stability Analysis of the Explicit Method

Let us calculate the spectral radius of $\mathbf{Q} = \mathbf{I}_n - \alpha\mathbf{A}$, where $\alpha = c^2 \frac{\delta t}{\delta x^2} > 0$ and $\mathbf{A}$ is given by (18.14). Let λ be an eigenvalue of $\mathbf{A}$ and $\mathbf{x}$ be a respective eigenvector, i.e., $\mathbf{A}\,\mathbf{x} = \lambda\mathbf{x}$ holds.[1] We have

$$\mathbf{Qx} = (\mathbf{I}_n - \alpha\mathbf{A})\,\mathbf{x} = \mathbf{I}_n\mathbf{x} - \alpha\mathbf{A}\mathbf{x} = \mathbf{x} - \alpha\lambda\mathbf{x} = (1 - \alpha\lambda)\,\mathbf{x}.$$

That is, all eigenvalues of the matrix $\mathbf{Q}$ are of the form $\mu = 1 - \alpha\lambda$. Therefore, to calculate $\rho(\mathbf{Q})$ we only need to know the smallest and largest eigenvalues of the matrix $\mathbf{A}$.

Proposition 18.3. *The eigenvalues of the* n*-by-*n *matrix* $\mathbf{A}$ *defined by (18.14) are*

$$\lambda_j = 2 - 2\cos\theta_j,$$

which are associated with the respective n*-by-1 eigenvectors*

$$\mathbf{x}_j = [\sin(\theta_j),\, \sin(2\theta_j),\, \ldots,\, \sin(n\theta_j)]^\top, \tag{18.30}$$

where $\theta_j = \frac{j\pi}{n+1}$ *for all* $j = 1, 2, \ldots, n$.

[1] Recall that the eigenvalues are zeros of the polynomial $\det(\mathbf{A} - \lambda\mathbf{I}_n)$.

Proof. It is left as an exercise for the reader to verify this. □

Due to Proposition 18.3, the eigenvalues of $\mathbf{Q}$ are

$$\mu_j = 1 - \alpha\lambda_j = 1 - \alpha(2 - 2\cos\theta_j) = 1 - 4\alpha\sin^2\left(\frac{j\pi}{2(n+1)}\right), \quad j = 1, 2, \ldots, n.$$

Hence, the stability requirement, $\rho(\mathbf{Q}) = \max_j |\mu_j| < 1$, can be written as follows:

$$-1 < 1 - 4\alpha\sin^2\left(\frac{j\pi}{2(n+1)}\right) < 1, \quad j = 1, 2, \ldots, n.$$

Since $\alpha > 0$, the right inequality is automatically satisfied. Rearranging the left inequality gives

$$\frac{1}{2} > \alpha\sin^2\left(\frac{j\pi}{2(n+1)}\right), \quad j = 1, 2, \ldots, n.$$

The sequence $\left\{\sin^2\left(\frac{j\pi}{2(n+1)}\right); j = 1, 2, \ldots, n\right\}$ attains its maximum value at $j = n$. The maximum approaches 1 from below as $n \to \infty$. Thus, the above restrictions on α are satisfied for any $n \geqslant 1$ iff $\alpha \leqslant \frac{1}{2}$. To insure the stability of the explicit method, we require that

$$c^2 \frac{\delta t}{\delta x^2} \leqslant \frac{1}{2} \tag{18.31}$$

holds. The explicit method is said to be conditionally stable. The condition (18.31) can also be interpreted as follows: every time we double the number of mesh points along the x-axis (and hence halve the mesh size δx), we need to quadruple the number of mesh points along the t-axis.

The stability of an explicit finite-difference method can also be investigated using the matrix norm. The explicit Euler method is stable if $\|\mathbf{Q}\|_* \leqslant 1$ for some norm $\|\cdot\|_*$. Calculating the biggest row-sum of magnitudes of Q_{ij} we have

$$\|\mathbf{Q}\|_\infty = |1 - 2\alpha| + 2|\alpha|.$$

If $\alpha > \frac{1}{2}$, then $\|\mathbf{Q}\|_\infty = 4\alpha - 1 > 1$ and the scheme is unstable. If $\alpha \leqslant \frac{1}{2}$, then $\|\mathbf{Q}\|_\infty = 1$ and the scheme is stable.

Stability Analysis of the Implicit Method

The eigenvalues ν_j of the matrix $\mathbf{I}_n + \alpha\mathbf{A}$ can be calculated from those of $\mathbf{A}$ as follows:

$$\nu_j = 1 + \alpha\lambda_j = 1 + 4\alpha\sin^2\left(\frac{j\pi}{2(n+1)}\right), \quad j = 1, 2, \ldots, n.$$

All of them are strictly greater than one. In particular this fact ensures that the inverse $\mathbf{Q} = (\mathbf{I}_n + \alpha\mathbf{A})^{-1}$ exists. Since $\mathbf{Q}$ is a nonsingular matrix, its eigenvalues μ_j are reciprocals of the eigenvalues ν_j of $\mathbf{I}_n + \alpha\mathbf{A}$. Therefore, the implicit method is stable since

$$0 < \frac{1}{1 + 4\alpha\sin^2\left(\frac{j\pi}{2(n+1)}\right)} < 1$$

for all $j = 1, 2, \ldots, n$ and for all $\alpha > 0$. The implicit method is said to be unconditionally stable since there is no restriction on the choice of the mesh sizes δx and δt.

Stability Analysis of the Crank–Nicolson Method
An important observation is that the matrices $\mathbf{B} \equiv \mathbf{I}_n + \frac{\alpha}{2}\mathbf{A}$ and $\mathbf{C} \equiv \mathbf{I}_n - \frac{\alpha}{2}\mathbf{A}$ have the same eigenvectors as those of the matrix $\mathbf{A}$. The eigenvalues ν_j^B of $\mathbf{B}$ and ν_j^C of $\mathbf{C}$ associated with the eigenvector $\mathbf{x}_j$ defined by (18.30) are, respectively, given by

$$\nu_j^B = 1 + \frac{\alpha}{2}\lambda_j \text{ and } \nu_j^C = 1 - \frac{\alpha}{2}\lambda_j.$$

Moreover, since $\alpha > 0$, all eigenvalues of $\mathbf{B}$ are strictly positive; in fact, they all lie in the interval $(1, 1+\alpha)$. Hence the matrix $\mathbf{B}$ is nonsingular. The eigenvalues of the inverse $\mathbf{B}^{-1}$ are of the form $\frac{1}{\nu_j^B}$ since

$$\mathbf{B}\mathbf{x}_j = \nu_j^B \mathbf{x} \iff \frac{1}{\nu_j^B}\mathbf{x}_j = \mathbf{B}^{-1}\mathbf{x}_j.$$

Therefore, the eigenvectors of $\mathbf{Q} = \mathbf{B}^{-1}\mathbf{C}$ are the same as those of $\mathbf{A}$. The eigenvalues μ_j of $\mathbf{Q}$ are

$$\mu_j = \frac{1 - \frac{\alpha}{2}\lambda_j}{1 + \frac{\alpha}{2}\lambda_j} = \frac{1 - 2\alpha \sin^2\left(\frac{j\pi}{2(n+1)}\right)}{1 + 2\alpha \sin^2\left(\frac{j\pi}{2(n+1)}\right)}$$

for $j = 1, 2, \ldots, n$. Indeed, we have

$$\mathbf{Q}\mathbf{x}_j = \mathbf{B}^{-1}(\mathbf{C}\mathbf{x}_j) = \nu_j^C \mathbf{B}^{-1}\mathbf{x}_j = \frac{\nu_j^C}{\nu_j^B}\mathbf{x}_j.$$

The magnitude of every eigenvalue μ_j is strictly less than one. Therefore, the Crank–Nicolson method is unconditionally stable.

18.2 Pricing European Options

Valuation of a European derivative is probably the most elementary problem of computational finance. In the case of a single asset price model, this problem reduces to the numerical evaluation of a one-dimensional integral. Hence, the result can be achieved with the use of quadrature rules. On the other hand, according to the risk-neutral pricing formula, the no-arbitrage derivative value is the expected value of the discounted payoff function. The mathematical expectation can be evaluated by the Monte Carlo method, which is especially efficient for pricing basket (i.e., multi-asset) options and path-dependent derivatives, the numerical evaluation of which is discussed in later sections. The PDE formulation of the derivative price problem allows for the application of finite-difference schemes. Finally, we may use multinomial tree methods (also known as lattice methods) that can be derived as a result of a discrete-time approximation of a continuous-time asset price model. In fact, tree methods are closely related to Monte Carlo methods. Furthermore, they can also be considered as a special case of explicit finite-difference methods.

18.2.1 Pricing European Options by Quadrature Rules

Let us consider a standard European-style option maturing at time T with payoff function $\Lambda(S(T))$, where $S(T)$ is the time-T price of the underlying asset. Here, we assume that the underlying price process $\{S(t)\}_{t\geqslant 0}$ is time homogeneous and Markovian and that the (risk-neutral) probability law of the price $S(T)$ conditional on $S(t) = S$ for $0 \leqslant t < T$ is known. We can therefore express the option value $v(\tau, S)$ as function of the time to maturity $\tau = T - t$ and spot $S = S(t)$. As we have seen in previous chapters, the no-arbitrage option value is given by the risk-neutral pricing formula and can be evaluated as a one-dimensional integral (using the risk-free bank account $B(t) = \mathrm{e}^{rt}$ as numéraire):

$$v(\tau, S) = \mathrm{e}^{-r\tau}\widetilde{\mathrm{E}}_{t,S}[\Lambda(S(T))] = \mathrm{e}^{-r\tau}\int_0^{\infty} \Lambda(S')\tilde{p}(\tau; S, S')\,\mathrm{d}S'. \tag{18.32}$$

Here, $\tilde{p}(\tau; S, S')$ denotes the risk-neutral transition PDF of the asset price process. Some examples of processes for which the PDF $\tilde{p}$ is known in closed form include the log-normal (GBM) model and the constant elasticity of variance (CEV) diffusion model. There are also other interesting families of more generally state-dependent nonlinear volatility diffusion models that admit analytically closed-form expressions for the transition PDF. Some jump-diffusion processes such as the variance gamma (VG) model and the normal inverse Gaussian (NIG) model, which are constructed from Brownian motion with the use of a stochastic time change, also admit transition PDFs in closed form.

We can also compute the Greeks of the contract such as the delta $\Delta(\tau, S) = \frac{\partial v}{\partial S}(\tau, S)$. Provided that we can change the order of integration and differentiation and that the integrals obtained exist, we have

$$\Delta(\tau, S) = \mathrm{e}^{-r\tau}\frac{\partial}{\partial S}\int_0^{\infty} \Lambda(S')\tilde{p}(\tau; S, S')\,\mathrm{d}S' = \mathrm{e}^{-r\tau}\int_0^{\infty} \Lambda(S')\frac{\partial \tilde{p}}{\partial S}(\tau; S, S')\,\mathrm{d}S'. \tag{18.33}$$

In some cases, the integrals in (18.32)–(18.33) can be computed in closed form. Explicit formulas for the Greeks in the case of the log-normal (GBM) asset price model were derived in previous chapters. In general, we rely on numerical methods. The integral in (18.32) can be computed with the use of quadrature formulae like those of Subsection 18.1.1:

$$v(\tau, S) \approx \mathrm{e}^{-r\tau}\sum_{i=1}^{n} w_i\Lambda(s_i)\tilde{p}(\tau; S, s_i),$$

where the summation is taken over a set of nodes $s_1, s_2, \ldots, s_n \in \mathbb{R}_+$. The integral in (18.32) is an improper one. To use a quadrature formula, we can first apply a change of variables to transform the integration interval into a finite one. Alternatively, we can truncate the integration region and integrate the function $\Lambda(S')\tilde{p}(\tau; S, S')$ w.r.t. S' over a bounded interval $(0, S_{\max})$. One case is where the payoff Λ is nonzero on a finite interval such as the put option payoff $\Lambda(S') = (K - S')^+ = (K - S)\mathbb{I}_{\{S' \leqslant K\}}$. If $\Lambda(S') = 0$ for all $S' \geqslant K$, then we take $S_{\max} = K$. In many other cases we may observe that the integrand goes to zero exponentially fast as $S' \to \infty$. If this is the case, then for any given $\varepsilon > 0$ one can find $S_{\max} > 0$ such that $\int_{S_{\max}}^{\infty} \Lambda(S')\tilde{p}(\tau; S, S')\,\mathrm{d}S' < \varepsilon$.

Let us calculate the initial (set $t = 0$) value of a European call option under two asset price models that are more involved than the GBM model, namely, the constant elasticity of variance (CEV) diffusion model and the geometric variance gamma (GVG) model. First, we need to identify the transition probability density of each asset price process. Second, we apply a quadrature rule to evaluate the integral (18.32), which can be written for the case with the European call option with time to maturity T, strike price K, and spot S_0 as

$$v(T, S_0) = \mathrm{e}^{-rT}\int_0^{\infty} \Lambda(S)\tilde{p}(T; S_0, S)\,\mathrm{d}S \tag{18.34}$$

$$= \mathrm{e}^{-rT}\int_K^{\infty} (S - K)\tilde{p}(T; S_0, S)\,\mathrm{d}S. \tag{18.35}$$

In the following calculations, we assume that the risk-free interest rate r is 5%, the dividend yield q is 0, the initial asset price S_0 is 100, and the time to maturity T is 0.5 (years).

Example 18.3 (Option Pricing in the CEV model). Consider the CEV diffusion model $\{S(t) \in \mathbb{R}_+\}_{t \geqslant 0}$, which follows

$$\mathrm{d}S(t) = \nu S(t)\mathrm{d}t + \alpha S(t)^{\beta+1}\mathrm{d}\widetilde{W}(t) \tag{18.36}$$

with real constants $\alpha > 0$, β, $\nu = r - q$ and where $\{\widetilde{W}(t)\}_{t\geqslant 0}$ is a standard Brownian motion in a risk-neutral measure $\widetilde{\mathbb{P}}$. The local volatility is a nonlinear (power) function: $\sigma(S)/S = \alpha S^{\beta}$. We consider the case with $\beta < 0$. Hence, the origin is attainable in finite time, i.e., the stock can attain a zero price in finite time. In particular, the origin is an exit boundary for $\beta \in [-1/2, 0)$ and it is a regular boundary (which we specify as killing) for $\beta < -1/2$. Note that the discounted process $\{\mathrm{e}^{-\nu t}S(t)\}_{t\geqslant 0}$ is a $\widetilde{\mathbb{P}}$-martingale. For numerical illustration, we take the following parameter values: $\beta = -2$ and $\alpha = 2500$. The instantaneous volatility at $S_0 = 100$ is $\alpha S_0^{\beta} = \frac{2500}{100^2} = 25\%$. The transition density has the known explicit representation

$$\tilde{p}(t; S_0, S) = \mathrm{e}^{-\nu t}\frac{(\mathrm{e}^{-\nu t}S)^{-2\beta-\frac{3}{2}}S_0^{\frac{1}{2}}}{\alpha^2|\beta|\tau_t}\mathrm{e}^{-\frac{(\mathrm{e}^{-\nu t}S)^{-2\beta}+S_0^{-2\beta}}{2\alpha^2\beta^2\tau_t}} I_{\frac{1}{2|\beta|}}\left(\frac{(\mathrm{e}^{-\nu t}S)^{-\beta}S_0^{-\beta}}{\alpha^2\beta^2\tau_t}\right), \tag{18.37}$$

for $S_0, S, t > 0$, $\tau_t = \frac{1}{2\nu\beta}\left(\mathrm{e}^{2\nu\beta t} - 1\right)$ if $\nu \neq 0$ and $\tau_t = t$ if $\nu = 0$. This PDF involves $I_\alpha(z)$ which is the modified Bessel function of the first kind of order α and argument z.

To simplify our calculations, we recall how the above CEV process is related to the simpler known *squared Bessel process* (SQB) $\{X(t) \in \mathbb{R}_+\}_{t\geqslant 0}$ with index $\mu \equiv 1/(2\beta)$ and having the transition PDF

$$p_\mu(t; x_0, x) = \left(\frac{x}{x_0}\right)^{\frac{\mu}{2}} \frac{\mathrm{e}^{-(x+x_0)/2t}}{2t} I_{|\mu|}\left(\sqrt{x x_0}/t\right). \tag{18.38}$$

By defining the invertible smooth mapping $\mathsf{X}\colon \mathbb{R}_+ \to \mathbb{R}_+$

$$\mathsf{X}(S) := S^{-2\beta}/(\alpha^2\beta^2) = S^{2|\beta|}/(\alpha^2\beta^2),$$

with inverse $\mathsf{X}^{-1}(x) \equiv \mathsf{F}(x) := (\alpha^2\beta^2 x)^{1/2|\beta|}$, we obtain the driftless, i.e., $\nu = 0$, CEV process $\{S^{(0)}(t) := \mathsf{F}(X(t)),\ t \geqslant 0\}$. The CEV processes with and without drift are simply related by a scale and time change:

$$S(t) = \mathrm{e}^{\nu t}S^{(0)}(\tau_t) = \mathrm{e}^{\nu t}\mathsf{F}(X(\tau_t)). \tag{18.39}$$

It can be verified that the risk-neutral transition PDF of the CEV process, $\tilde{p}$, is related to that of the SQB process as

$$\tilde{p}(t; S_0, S) = \mathrm{e}^{-\nu t}|\mathsf{X}'(\mathrm{e}^{-\nu t}S)|\, p_\mu(\tau_t; \mathsf{X}(S_0), \mathsf{X}(\mathrm{e}^{-\nu t}S)). \tag{18.40}$$

By substituting the expression in (18.38) for p_μ, we see that (18.40) is exactly Equation (18.37).

Using (18.40), for $t = T$ and $\nu = r$, and changing integration variables with $S = \mathrm{e}^{rT}\mathsf{F}(x)$, $x = \mathsf{X}(\mathrm{e}^{-rT}S)$, and $\mathrm{d}S = \frac{\mathrm{e}^{rT}}{\mathsf{X}'(\mathsf{F}(x))}\,\mathrm{d}x$, the integral in (18.35) takes the form

$$v(T, S_0) = \int_{\mathsf{X}(\mathrm{e}^{-rT}K)}^{\infty} (\mathsf{F}(x) - \mathrm{e}^{-rT}K)\, p_\mu(\tau_T; \mathsf{X}(S_0), x)\,\mathrm{d}x, \tag{18.41}$$

where $\mu = \frac{1}{2\beta}$. This is an improper integral over a half-line which can be transformed into an integral over a finite interval by a change of variables. For example,

$$\int_a^\infty f(x)\,\mathrm{d}x = \int_0^1 f\left(a + \frac{y}{1-y}\right)\frac{\mathrm{d}y}{(1-y)^2}. \tag{18.42}$$

To compute the call price, the integral in (18.41) is transformed as in (18.42) where

$$a = \mathsf{X}(\mathrm{e}^{-rt}K) \quad \text{and} \quad f(x) = \big(\mathsf{F}(x) - \mathrm{e}^{-rT}K\big)\, p_\mu(\tau_T; \mathsf{X}(S_0), x).$$

Alternatively, we can first evaluate the corresponding European put option (with same strike and maturity) and then obtain the call option value with the use of the put-call parity. The evaluation of a put option reduces to computing the integral (18.34) with $\Lambda(S) = (K-S)^+$ over a finite interval $(0, K)$.

With the above chosen model parameters, we employ Romberg integration with 101 initial nodes and four improvements. First, we evaluate the call option with strike $K = 100$. The output of the Romberg method is in shown Table 18.3, where the best approximation for the definite integral is the bottom rightmost boxed entry. The approximation error is given by the magnitude of the difference between the diagonal elements in the two bottom rows of the table. Second, we evaluate the call option for strikes $K \in \{90, 100, 110\}$ and compare the results obtained with the call option values calculated under the GBM model with volatility $\sigma = 0.25$ (see Table 18.4). For comparison, we also calculated the Black–Scholes implied volatility for the CEV call option values. As is seen, the implied volatility is a decreasing function of strike K.

TABLE 18.3: Application of the Romberg numerical integration method to pricing a European call option with $T = 0.5$ and $K = 100$ under the CEV model. The approximation error is estimated at $1.057 \cdot 10^{-11}$.

8.2978647600469				
8.2978711263121	8.2978732484004			
8.2978727104861	8.2978732385441	8.2978732378870		
8.2978731065344	8.2978732385506	8.2978732385511	8.2978732385616	
8.2978732055469	8.2978732385510	8.2978732385511	8.2978732385511	8.2978732385510

TABLE 18.4: Comparison of the call values computed under the Black–Scholes (GBM) model and the CEV model for different strikes. The implied Black–Scholes volatility (Impl. BS Vol.) is computed for the CEV option values.

Strike K	Call Value under GBM	Call Value under CEV	Impl. BS Vol.
90	14.437116236461	15.033304012884	27.89%
100	8.260015199343	8.297873238551	25.14%
110	4.225782392960	3.642151895619	22.81%

Example 18.4 (Pricing in the GVG model)**.** The variance gamma (VG) process $\{X(t)\}_{t\geqslant 0}$ is obtained by evaluating a scaled Brownian motion with drift $W^{(\theta,\sigma)}(t)$ at a random time given by a gamma process $G(t)$:

$$X(t) := W^{(\theta,\sigma)}(G(t)) = \theta\, G(t) + \sigma W(G(t)), \quad t \geqslant 0.$$

Recall that the gamma process is characterized by two positive parameters, μ and υ. The parameter μ controls the rate of jump arrivals and the scaling parameter υ inversely controls the jump size. It is assumed that the process starts from 0 at time 0. The marginal distribution of a gamma process at time t is a gamma distribution with shape parameter $\frac{\mu t}{\upsilon}$ and rate parameter $\frac{1}{\upsilon}$. The mean and variance of $G(t)$ are, respectively, $\frac{\mu t}{\upsilon}$ and $\frac{\mu t}{\upsilon^2}$. In the VG model, the parameter μ is set to 1 thus $G(t) \sim Gamma(\frac{t}{\upsilon}, \frac{1}{\upsilon})$.

The risk-neutral asset price dynamics is given by the geometric variance gamma (GVG) process as

$$S(t) = S_0 \mathrm{e}^{(r-q+\omega)t+X(t)},$$

where $\omega = \frac{1}{\upsilon}\ln(1-\theta\upsilon-\sigma^2\upsilon/2)$ is a correction constant ensuring that the mean rate of return on the asset is $r-q$. The risk-neutral transition probability distribution is a mixture of the normal and gamma probability distributions. The PDF of $Z(t) := \ln(S(t)/S(0))$ is given by

$$p_{Z(t)}(z) = \frac{2\mathrm{e}^{\theta x/\sigma^2}}{\upsilon^{t/\upsilon}\sqrt{2\pi}\sigma\Gamma(t/\upsilon)} \left(\frac{x^2}{2\sigma^2/\upsilon+\theta^2}\right)^{\frac{t}{2\upsilon}-\frac{1}{4}} K_{\frac{t}{\upsilon}-\frac{1}{2}}\left(\frac{|x|}{\sigma^2}\sqrt{(2\sigma^2/\upsilon+\theta^2)}\right), \quad (18.43)$$

for all $t > 0, z \in \mathbb{R}$, where Γ is the gamma function, K_α is the modified Bessel function of the second kind of order α, and $x = z-(r-q)t-\omega t$. Since the asset price is given by an exponential of the log price ratio $Z(t)$, i.e., $S(t) = S_0\mathrm{e}^{Z(t)}$, the value of a European option maturing at time T with payoff $\Lambda(S)$ is given by

$$v(T, S_0) = \mathrm{e}^{-rT}\int_{-\infty}^{\infty} \Lambda(S_0\mathrm{e}^z) p_{Z(T)}(z)\,\mathrm{d}z. \quad (18.44)$$

For a standard European call with strike K we have

$$v(T, S_0) = \mathrm{e}^{-rT}\int_{\ln(K/S_0)}^{\infty} (S_0\mathrm{e}^z - K) p_{Z(T)}(z)\,\mathrm{d}z.$$

By using the transformation in (18.42), we can change the semi-infinite interval into the finite interval $(0, 1)$ and then apply Romberg integration to evaluate the European call with $K = 100$. For numerical illustration, we take parameter values from Madan et al. (1998): $\theta = 0.1436$, $\sigma = 0.12136$, and $\upsilon = 0.3$. The other parameters are the same as those in the previous example. The output of Romberg method is given in Table 18.5, where the best approximation for the definite integral is the bottom rightmost boxed value.

TABLE 18.5: Application of the Romberg numerical integration method to pricing a European call option with $T = 0.5$ and $K = 100$ under the GVG model. The approximation error is estimated at $8.663 \cdot 10^{-15}$.

5.0807555357638				
5.0835997925475	5.0845478781420			
5.0843105384813	5.0845474537926	5.0845474255026		
5.0844882050319	5.0845474272154	5.0845474254436	5.0845474254426	
5.0845326204230	5.0845474255534	5.0845474254426	5.0845474254426	5.0845474254426

Algorithm 18.1 The Monte Carlo Method for Estimation of the No-Arbitrage Value of a European-Style Contract.

1. Sample n i.i.d. asset prices $\{S_k\}_{k=1,2,\ldots,n}$ from the probability distribution of $S(T)$ conditional on $S(0) = S_0$. Calculate the sample values of the discounted payoff function: $\Lambda_k := \mathrm{e}^{-rT}\Lambda(S_k)$ for $k = 1, 2, \ldots, n$.

2. Compute the sample estimate for the price $V_0 = v(T, S_0) = \widetilde{\mathrm{E}}[\mathrm{e}^{-rT}\Lambda(S(T))]$:
$$V_0 \approx \overline{\Lambda}_n := \frac{1}{n}\sum_{k=1}^{n} \Lambda_k.$$

3. Construct an asymptotically valid $100(1-\alpha)\%$ confidence interval for V_0:
$$(\overline{\Lambda}_n - z_{\alpha/2}\sigma_n, \overline{\Lambda}_n + z_{\alpha/2}\sigma_n) \ni V_0,$$
where $z_{\alpha/2}$ is the upper $(1-\alpha/2)$-quantile of the standard normal distribution, $\sigma_n = s_n/\sqrt{n}$ is the stochastic error, and $s_n^2 = \frac{1}{n}\sum_{k=1}^{n}\Lambda_k^2 - \left(\overline{\Lambda}_n\right)^2$ is the sample variance.

Algorithm 18.2 Simulation of the CEV Diffusion Model with Absorption at the Origin.

input $S_0 > 0$, $T > 0$, $\beta < 0$, $\alpha > 0$, $\nu = r - q$

set $\mu \leftarrow \frac{1}{2\beta}$, $X_0 \leftarrow \frac{S_0^{-2\beta}}{\alpha^2\beta^2}$, $\tau \leftarrow \begin{cases} \frac{1}{2\nu\beta}\left(\mathrm{e}^{2\nu\beta T} - 1\right), & \nu \neq 0, \\ T, & \nu = 0, \end{cases}$ $p_a \leftarrow \Gamma\left(|\mu|, \frac{X_0}{2\tau}\right)/\Gamma(|\mu|)$

generate $U \leftarrow \mathit{Unif}(0,1)$

if $U > p_a$ **then**

  generate $Y \leftarrow \mathrm{I}\Gamma\left(|\mu|, \frac{X_0}{2\tau}\right)$, $X \leftarrow \mathit{Gamma}\left(Y, \frac{1}{2\tau}\right)$

else

  set $X \leftarrow 0$

end if

return $S(T) \leftarrow \mathrm{e}^{\nu T}\left(\alpha^2\beta^2 X\right)^{-0.5/\beta}$

18.2.2 Pricing European Options by the Monte Carlo Method

Let $\{S(t)\}_{t\geqslant 0}$ be a stochastic asset price process starting at S_0. Suppose that it can be simulated exactly from its transition distribution. The no-arbitrage price $v(T, S_0)$ of a European-style contract with the time to maturing $T > 0$ and payoff $\Lambda(S(T))$ can be estimated by the Monte Carlo method as described in Algorithm 18.1.

To illustrate the use of the Monte Carlo algorithm, we apply Algorithm 18.1 to the CEV diffusion asset price model, the geometric variance gamma pure jump model, and a multidimensional geometric Brownian motion model.

Example 18.5 (Pricing Options under the CEV Diffusion Model). The Monte Carlo simulation of the constant elasticity of variance (CEV) diffusion process that follows the SDE (18.36) is based on the reduction of its transition probability distribution to one of the randomized gamma distributions studied in Chapter 17. Assume the risk-neutral probability measure so that $\nu = r - q$ in (18.36). Recall from (18.39) that the CEV diffusion

TABLE 18.6: Estimation of the standard European call option value. The MCM estimate, respective standard stochastic error $\sigma_n = s_n/\sqrt{n}$, and 95% confidence interval are reported for several values of the sample size n. Computations are performed for $K = 100$ and $T = \frac{1}{2}$ under the CEV model with parameters as specified in Example 18.3.

Size n	MCM Estimate	Stoch. Error σ_n	95% Confidence Interval
20,000	8.35181	0.14084	(8.21097, 8.49265)
40,000	8.25013	0.09896	(8.15117, 8.34908)
60,000	8.20935	0.08049	(8.12886, 8.28984)
80,000	8.24294	0.06985	(8.17309, 8.31280)
100,000	8.24361	0.06241	(8.18120, 8.30602)

$\{S(t)\}_{t\geqslant 0}$ can be represented as a function of a time-changed SQB process $\{X(t)\}_{t\geqslant 0}$:

$$S(t) = \mathrm{e}^{\nu t}\left(\alpha^2\beta^2 X(\tau_t)\right)^{-1/2\beta}, \text{ where } \tau_t = \begin{cases} \frac{1}{2\nu\beta}\left(\mathrm{e}^{2\nu\beta t} - 1\right), & \nu \neq 0, \\ t, & \nu = 0. \end{cases} \tag{18.45}$$

Again, we consider the case with $\beta < 0$ so that the origin is an exit for $\beta \in [-1/2, 0)$ and specified as a killing regular boundary in the case $\beta < -1/2$. The CEV model reduces to the SQB process as given in (18.45). First, Algorithm 17.13 is used to sample the value $X(\tau_t)$ of the SQB process. Second, the terminal stock value $S(T)$ of the CEV process is obtained by applying the mapping (18.45) with $t = T$. As a result, we obtain Algorithm 18.2, which returns the sample value of $S(T)$ (for $T > 0$) conditional on $S(0) = S_0 > 0$.

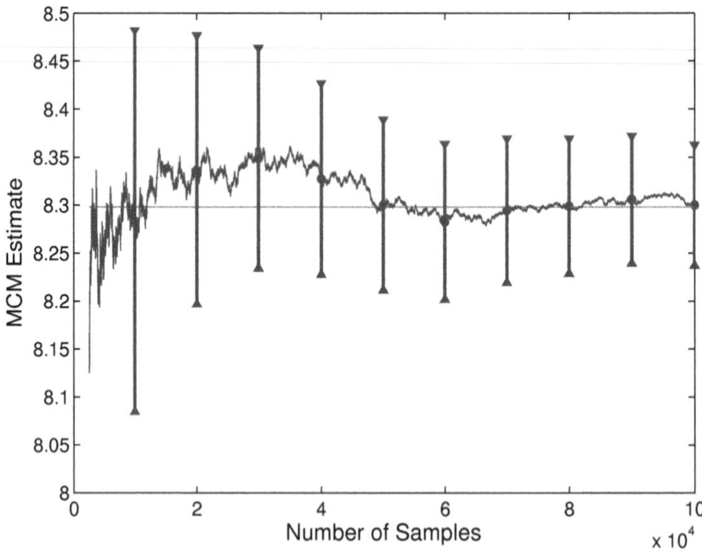

FIGURE 18.7: Convergence of the unbiased Monte Carlo estimate to the exact price of the European call option (represented by a horizontal line) as the sample size n increases. For each value of n equal to a multiple of 10^4, the 95% confidence interval is constructed. Computations are performed for $K = 100$ and $T = \frac{1}{2}$ under the CEV model with parameters as specified in Example 18.3.

In the Monte Carlo experiment, we estimate the initial value of a European call option with maturity $T = \frac{1}{2}$ and strike price $K = 100$. The CEV model parameters used here are the same as those in Example 18.3. In Table 18.6 we report the Monte Carlo estimate, standard error, and 95% confidence interval computed for several values of the sample size n. Figure 18.7 demonstrates the convergence of the sample estimate to the exact value of the call option as n increases.

Example 18.6 (Pricing Options under the Variance Gamma Model)**.** The simulation of the VG process relies on sampling normal and gamma random variables. To simulate $X(t) = B(G(t))$ at some time $t > 0$, we first sample the gamma time $G(t) \sim Gamma(t/\nu, 1/\nu)$ and then sample a Brownian increment $B(g) \sim Norm(\theta\, g, \sigma^2\, g)$ given $g = G(t)$. Thus, we can express the GVG asset price $S(T)$ at maturity T (under the risk-neutral measure) as a function of independent gamma and normal random variables:

$$S(T) \stackrel{d}{=} S_0 e^{(r-q+w)T+\theta\, G/\nu+\sigma\sqrt{G/\nu}\, Z}, \text{ where } Z \sim Norm(0,1) \text{ and } G \sim Gamma(T/\nu, 1).$$

In the Monte Carlo experiment, we estimate the value of a European call option with maturity $T = \frac{1}{2}$ and strike price $K = 100$. The GVG model parameters used here are the same as those in Example 18.4. In Table 18.8 we report the Monte Carlo estimate, standard error, and 95% confidence interval computed for several values of the sample size n. Figure 18.9 demonstrates the convergence of the sample estimate to the exact value of the call option as n increases.

TABLE 18.8: Estimation of the standard European call option value. The MCM estimate, respective standard stochastic error $\sigma_n = s_n/\sqrt{n}$, and 95% confidence interval are reported for several values of the sample size n. Computations are performed for $K = 100$ and $T = \frac{1}{2}$ under the GVG model with parameters as specified in Example 18.4.

Size n	MCM Estimate	Stoch. Error σ_n	95% Confidence Interval
20,000	5.06430	0.12338	$(4.94091, 5.18768)$
40,000	5.03148	0.08679	$(4.94469, 5.11827)$
60,000	5.06728	0.07091	$(4.99637, 5.13819)$
80,000	5.07949	0.06160	$(5.01789, 5.14109)$
100,000	5.10104	0.05521	$(5.04583, 5.15624)$

Example 18.7 (Pricing Basket Options under the Multi-Asset GBM Model)**.** Consider the pricing of derivative contracts under an N-asset GBM model with M Brownian factors. Let us recall the risk-free dynamics of the ith asset price described by

$$S_i(t) = S_i(0) \exp\left((r - q_i - \sigma_i^2/2)t + \sigma_i \sum_{j=1}^{M} \ell_{ij} \widetilde{W}_j(t) \right), \quad t \geqslant 0,$$

where σ_i is the volatility of the ith asset, q_i is the dividend yield on the ith asset, $\{\widetilde{W}_j(t)\}_{t\geqslant 0}$ are i.i.d. Brownian motions (within the assumed risk-neutral measure $\widetilde{\mathbb{P}}$) labelled by $j = 1, 2, \ldots, M$, and the coefficients ℓ_{ij} are chosen so that $\sum_{j=1}^{M} \ell_{ij}^2 = 1$ for all $i = 1, 2, \ldots, N$. For each $i = 1, 2, \ldots, N$, the log-return $X_i(t) := \ln\left(\frac{S_i(t)}{S_i(0)}\right)$ is a normal random variable with mean $(r - q_i - \sigma_i^2/2)t$ and variance $\sigma_i^2 t$. The joint distribution of the log-returns

$X_1(t), X_2(t) \ldots, X_N(t)$ is multivariate normal with N-by-N correlation matrix $\boldsymbol{\rho} = \mathbf{L}\mathbf{L}^\top$, where $\mathbf{L}$ is an N-by-M matrix with entries ℓ_{ij}. The covariance among the log-returns is given by the N-by-N matrix with entries

$$\operatorname{Cov}(X_i(t), X_j(t)) = t\,\mathbf{\Sigma}_{ij} \text{ where } \mathbf{\Sigma} = \mathbf{D}\boldsymbol{\rho}\mathbf{D} \text{ and } \mathbf{D} = \operatorname{diag}(\sigma_1, \sigma_2, \ldots, \sigma_N).$$

Consider a multi-asset (basket) European derivative contract with time to maturity T and payoff $\Lambda(S_1(T), S_2(T), \ldots, S_N(T))$. We have priced some examples of such options, including options on the maximum with payoff $\max_i\{w_i S_i(T)\}$, or the minimum with payoff $\min_i\{w_i S_i(T)\}$, etc., where $w_1, w_2, \ldots, w_N$ are positive coefficients. Given time-0 spot values $S_i(0) = s_i > 0$, $i = 1, 2, \ldots, N$, the initial no-arbitrage value of this derivative is

$$V(s_1, s_2, \ldots, s_N; T) = \mathrm{e}^{-rT}\widetilde{\mathrm{E}}_0\left[\Lambda(S_1(T), S_2(T), \ldots, S_N(T)) \mid S_i(0) = s_i,\ 1 \leqslant i \leqslant N\right].$$

For large values of N and M, we can only rely on Monte Carlo methods for pricing. To apply Algorithm 18.1, we need to generate n i.i.d. sample values of the payoff function. The completion of part 1 in Algorithm 18.1 involves the following three steps:

1.i. Sample $n \times M$ i.i.d. random variables $Z_j^{(k)}$, $1 \leqslant k \leqslant n$, $1 \leqslant j \leqslant M$, from the standard normal distribution.

1.ii. For each $i = 1, 2, \ldots, N$, obtain n sample values of the terminal stock value $S_i(T)$,

$$S_i^{(k)} = s_i \exp\left((r - q_i - \sigma_i^2/2)\,T + \sigma_i\sqrt{T}\sum_{j=1}^{M} \ell_{ij} Z_j^{(k)} \right), \quad k = 1, 2, \ldots, n.$$

1.iii. Calculate the sample values of the discounted payoff function

$$\Lambda_k = \mathrm{e}^{-rT}\Lambda\left(S_1^{(k)}, S_2^{(k)}, \ldots, S_N^{(k)}\right), \quad k = 1, 2, \ldots, n.$$

18.2.3 Pricing European Options by Tree Methods

Assume a log-normal model for stock prices in an economy with fixed interest rate r and zero stock dividend. Recall that under the risk-neutral measure $\widetilde{\mathbb{P}}$ with $\widetilde{W}$ as standard Brownian motion, the stock price has the representation

$$S(t) = S_0\mathrm{e}^{(r-\sigma^2/2)t + \sigma\widetilde{W}(t)}.$$

As was discussed in Chapter 2, the log-normal continuous-time model can be derived as a limiting case of discrete-time binomial tree models. Therefore, we can use a binomial tree to approximate the GBM dynamics. To guarantee a good order of approximation, the length of time periods has to be small enough. Multinomial tree models allow for improving both the accuracy and speed of the binomial method. In an n-nomial tree model, the stock price can jump to one of n possible branches over each time period. As a result, we can attain a higher order of approximation without having to increase the computational work by much.

18.2.3.1 Binomial Model

Consider a recombining binomial tree model with $N + 1$ trading dates $t_n = n\,\delta t$, $n = 0, 1, \ldots, N$ uniformly spread throughout the interval $[0, T]$ with step size $\delta t = \frac{T}{N}$ (i.e., the number of periods per year equals $\frac{1}{\delta t}$). The upward and downward factors, u and d, of the

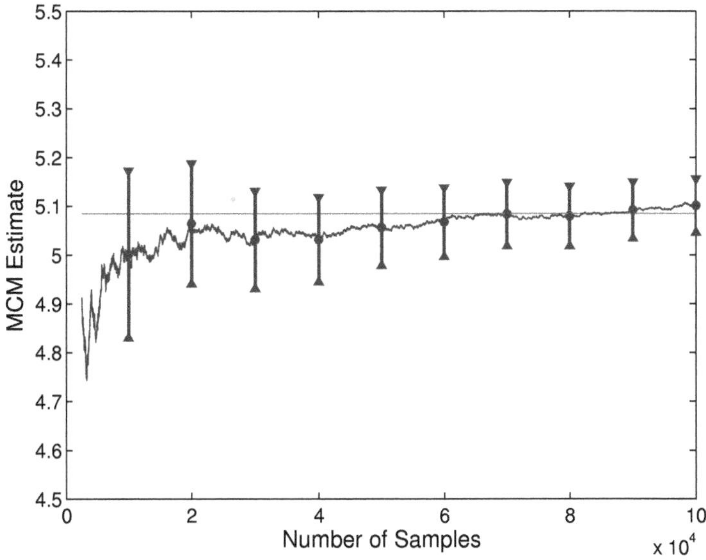

FIGURE 18.9: Convergence of the unbiased Monte Carlo estimate to the exact price of the European call option (represented by a horizontal line) as the sample size n increases. For each value of n equal to a multiple of 10^4, a 95% confidence interval is constructed. Computations are performed for $K = 100$ and $T = \frac{1}{2}$ under the GVG model with parameters as specified in Example 18.4.

standard binomial tree model can be obtained by matching the first two moments of the stock price returns. The following solution is the most commonly used in binomial models:

$$u = e^{\sigma\sqrt{\delta t}} \quad \text{and} \quad d = \frac{1}{u} = e^{-\sigma\sqrt{\delta t}}. \tag{18.46}$$

The risk-neutral probability is given by the usual formula:

$$\tilde{p} = \frac{\mathrm{e}^{r\,\delta t} - d}{u - d}. \tag{18.47}$$

For this choice of parameters, the variances of a single period return in the continuous- and discrete-time models agree up to $\mathcal{O}\left(\delta t^2\right)$. Note that other solutions for u, d, $\tilde{p}$ are possible; however, their analytical expressions are more cumbersome.

Consider the application of a binomial tree to pricing a European option with maturity T and payoff $\Lambda(S)$. The continuous-time log-normal stock price process is approximated by a discrete-time binomial price process on the N-period lattice with the nodes $S_{n,k} = S_0 u^n d^{n-k}$, $k = 0, 1, \ldots, n$ and $n = 0, 1, \ldots, N$. We calculate the option values $V_{n,k} := V(n, S_{n,k})$ using the backward-in-time recursion as follows.

1. At maturity, set $V_{N,k} = \Lambda(S_{N,k})$ for each $k = 0, 1, \ldots, N$.
2. For each $n = N-1, N-2, \ldots, 0$, compute the option values $V_{n,k}$ as follows:

$$V_{n,k} = \mathrm{e}^{-r\,\delta t}\left(\tilde{p}V_{n+1,k+1} + (1-\tilde{p})V_{n+1,k}\right), \quad k = 0, 1, \ldots, n.$$

Example 18.8. In this example, we study whether the binomial option price formula

TABLE 18.10: Errors of binomial approximations for the butterfly option value and delta.

N	δt	Price Error	Delta Error
50	0.001	0.08652	0.00013
100	0.0005	0.03047	0.00048
200	0.0025	0.01606	0.00022
400	0.00125	0.01018	0.00002

approximates the respective continuous-time solution with accuracy $\mathcal{O}(\delta t)$. Consider a butterfly option centred at 100 with the spread $[80, 120]$. The payoff is

$$\Lambda(S) = \begin{cases} S - 80 & \text{if } 80 < S \leqslant 100, \\ 120 - S & \text{if } 100 < S < 120, \\ 0 & \text{if } S \leqslant 80 \text{ or } S \geqslant 120. \end{cases} \tag{18.48}$$

Assume that the annual volatility σ is 0.25 and the annual risk-free rate of interest r is 0.05. Let us compute and compare the no-arbitrage prices and deltas of the butterfly option with $T = \frac{1}{2}$ for spot value $S_0 = 100$ under the following asset price models:

(a) the log-normal (GBM) model,

(b) the binomial model with $N = 50, 100, 200, 400$ where u and d are given in (18.46).

The payoff of a butterfly option is equivalent to that of a portfolio in three standard European calls. In particular, the payoff in (18.48) is equivalent to a portfolio in one long call with strike 80, two short calls with strike 100, and one long call with strike 120. Therefore, the initial value and delta of a butterfly option with the payoff in (18.48) calculated under the log-normal model are, respectively,

$$V = C_{BS}(K = 80) - 2C_{BS}(K = 100) + C_{BS}(K = 120),$$
$$\Delta = \Delta_{BS}(K = 80) - 2\Delta_{BS}(K = 100) + \Delta_{BS}(K = 120),$$

where $C_{BS}(K)$ and $\Delta_{BS}(K)$ are, respectively, the Black–Scholes call value and call delta for strike K and the same maturity and spot value as those used to price the butterfly option. For $T = 0.5$ and $S_0 = 100$, we have the following value and delta of the butterfly option under the GBM model:

$$V = 7.97318602436266 \quad \text{and} \quad \Delta = -0.0381920926996022.$$

The errors of binomial approximations are reported in Table 18.10. As is seen from the table, the binomial model approximation error is proportional to δt.

18.2.3.2 Multinomial Models

In the binomial model, the asset price can jump to one of two possible states in each time step. To improve the accuracy of option valuation, we allow a jump to multiple states for the asset price. Consider a European-style option with a payoff function $\Lambda(S)$ and maturity date $T > 0$. The no-arbitrage option value $V(t, S)$ as a function of (actual) calendar time t and spot S satisfies

$$V(t, S) = \mathrm{e}^{-r\,\delta t}\widetilde{\mathrm{E}}\left[V(t + \delta, S(t + \delta t)) \mid S(t) = S\right], \quad 0 \leqslant t < t + \delta t \leqslant T. \tag{18.49}$$

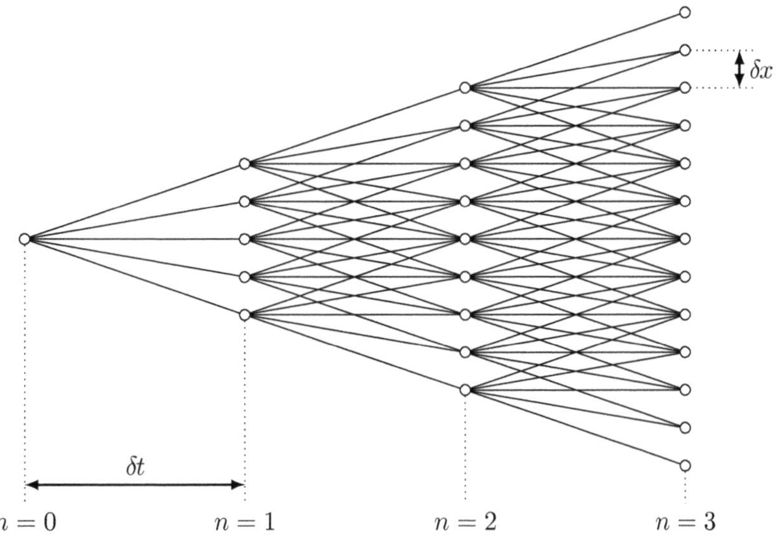

FIGURE 18.11: A pentanomial tree ($b = 2$) with $4n + 1$ nodes at time $n\,\delta t$.

This equation can be used to recursively compute option prices

$$V(T - \delta t, \cdot), V(T - 2\delta t, \cdot), \ldots, V(0, \cdot)$$

at every time $t = n\,\delta t$ starting at maturity T, where $V(T, S) = \Lambda(S)$, and then advancing backward in time until time 0 is reached. To make this approach feasible, we need to introduce a discrete grid in the spatial variable S and then approximate the mathematical expectation in (18.49) by a finite sum evaluated on the grid.

Let us fix the time step δt and introduce the time grid $t_n = n\,\delta t$ for $n = 0, 1, \ldots, N$, where $N = \frac{T}{\delta t}$. For every period $[t_n, t_{n+1}]$, the ratio of asset prices $S(t_{n+1})$ and $S(t_n)$ is a function of the Brownian increment $\delta\widetilde{W}_{t_{n+1}} := \widetilde{W}(t_{n+1}) - \widetilde{W}(t_n) \sim Norm(0, \delta t)$:

$$\frac{S(t_{n+1})}{S(t_n)} = \mathrm{e}^{(r-\sigma^2/2)\delta t + \sigma\, \delta t \widetilde{W}_{t_{n+1}}}.$$

Each Brownian increment $\delta\widetilde{W}_{t_n}$ is approximated by a discrete random variable $\delta\widetilde{W}_n$ that takes values $-b\,\delta x, -(b-1)\,\delta x, \ldots, b\,\delta x$ with respective probabilities $p_{-b}, p_{-b+1}, \ldots, p_b$, summing to unity, where $b \in \mathbb{N}$ is a branching parameter and $\delta x > 0$ is a mesh size. As a result, the Brownian motion $\{\widetilde{W}(t)\}_{t\geqslant 0}$ is approximated by a Markov chain $\{\widetilde{W}_{t_n}\}_{n\geqslant 0}$, which is defined at discrete times $t_n = n\,\delta t$, $n \geqslant 0$, and takes discrete values $k\,\delta x$, $k \in \mathbb{Z}$. For each time t_n, the variable $\widetilde{W}_{t_n}$ is given by a sum of n i.i.d. increments,

$$\widetilde{W}_{t_n} = \delta\widetilde{W}_1 + \delta\widetilde{W}_2 + \cdots + \delta\widetilde{W}_n;$$

it takes values in the set $\{k\,\delta x \mid k = -nb, -nb+1, \ldots, nb\}$. The value of $\widetilde{W}_{t_{n+1}}$ conditional on $\widetilde{W}_{t_n} = k\,\delta x$ is in the set $\{(k-b)\,\delta x, (k-b+1)\,\delta x, \ldots, (k+b)\,\delta x\}$. The process $\{\widetilde{W}_{t_n}\}_{n\geqslant 0}$ is a random walk whose realizations are paths in a recombining multinomial tree. The tree is organized such that there are $2bn+1$ nodes at time t_n and each node has $2b+1$ branches. By setting b equal to 1 or 2, we can obtain a trinomial or pentanomial tree, respectively. The latter is shown in Figure 18.11.

The approximation of the process $\{\widetilde{W}(t)\}_{t\geqslant 0}$ by $\{\widetilde{W}_{t_n}\}_{n\geqslant 0}$ leads to an approximation of the continuous-time asset price process $\{S(t)\}_{t\geqslant 0}$ by a discrete-time Markov chain $\{S_{t_n}\}_{n\geqslant 0}$ defined by

$$S_{t_n} = \exp\left((r-\sigma^2/2)t_n + \sigma\widetilde{W}_{t_n}\right).$$

Denote the value of S_{t_n} at the node (n,k) of the multinomial tree by

$$S_{n,k} := \exp\left((r-\sigma^2/2)\, n\, \delta t + \sigma\, k\, \delta x\right).$$

The conditional expectation in (18.49) can now be approximated as follows:

$$\begin{aligned} e^{r\,\delta t}V(t_n, S_{n,k}) &= \widetilde{\mathrm{E}}\left[V(t_{n+1}, S(t_{n+1})) \mid S(t_n) = S_{n,k}\right] \\ &= \widetilde{\mathrm{E}}\left[V\left(t_{n+1}, S_{n,k}e^{(r-\sigma^2/2)\delta t + \sigma\delta\widetilde{W}_{t_{n+1}}}\right)\right] \\ &\approx \widetilde{\mathrm{E}}\left[V\left(t_{n+1}, S_{n,k}e^{(r-\sigma^2/2)\delta t + \sigma\delta\widetilde{W}_{n+1}}\right)\right] \\ &\approx \sum_{j=-b}^{b} p_j V(t_{n+1}, S_{n+1,k+j}). \end{aligned} \tag{18.50}$$

This gives us the following computational scheme:

$$V_{n,k} = e^{-r\,\delta t}\sum_{j=-b}^{b} p_j V_{n+1,k+j}, \quad n = N-1, N-2, \ldots, 0, \quad k = -nb, -nb+1, \ldots, nb$$

where $V_{n,k}$ denotes the approximation of the option price $V(t_n, S_{n,k})$ at the node (n,k). As usual, the option values at maturity $T = N\,\delta t$ are given by the payoff function:

$$V_{N,k} = \Lambda(S_{N,k}), \quad k = -Nb, -Nb+1, \ldots, Nb.$$

To attain a high order of convergence in multinomial methods, the probabilities $\{p_j;\ j = -b, -b+1, \ldots, b\}$ should be chosen such that $\widetilde{W}_{t_n}$ is a good approximation of $\widetilde{W}(t_n)$. This can be achieved by using the moment matching method. According to Heston and Zhou (2000), if the first k moments of the Brownian increment δW_n match those of $\delta\widetilde{W}_{t_n} = \widetilde{W}_{t_n} - \widetilde{W}_{t_{n-1}}$, that is,

$$\sum_{j=-b}^{b} (j\,\delta x)^l p_j = 0, \qquad \text{for all odd } l \leqslant k \tag{18.51}$$

and

$$\sum_{j=-b}^{b} (j\,\delta x)^l p_j = \frac{\delta t^{l/2} l!}{2^{l/2}(l/2)!}, \qquad \text{for all even } l \leqslant k, \tag{18.52}$$

and if the terminal payoff function is $2k$ times continuously differentiable, then the multinomial approximation (18.50) has a local error of $\mathcal{O}\left(\delta t^{(k+1)/2}\right)$, and the associated discrete-time solution converges to the continuous-time solution at a rate of $\mathcal{O}\left(\delta t^{(k-1)/2}\right)$, that is,

$$V(t_n, S_{n,k}) = V_{n,k} + \mathcal{O}\left(\delta t^{(k-1)/2}\right).$$

We can easily match all the odd moments by putting $p_{-j} = p_j$ for all j. To match the first b even moments, we must solve the simultaneous equations

$$\sum_{j=0}^{b} 2(j\,\delta x)^{2k} p_j = \frac{\delta t^k (2k)!}{2^k k!}, \qquad k = 1, 2, \ldots, b. \tag{18.53}$$

Additionally, we require the probabilities p_j to sum up to one,

$$p_0 + 2\sum_{j=1}^{b} p_j = 1. \tag{18.54}$$

The left- and right-hand sides of (18.53) are, respectively, proportional to δx^{2k} and δt^k. Thus, if assume that $\delta x = \sqrt{a\,\delta t}$ holds for some constant $a > 0$, then we can cancel out δx and δt in (18.53) to obtain

$$\sum_{j=0}^{b} 2(j\,a)^{2k} p_j = \frac{(2k)!}{2^k k!}, \quad k = 1, 2, \ldots, b. \tag{18.55}$$

Solving these equations, we find the probabilities p_k that do not depend on δt or δx. The parameter a can be chosen to match an additional moment constraint

$$\sum_{j=1}^{b} 2(j\,a)^{2b+2} p_j = \frac{(2(b+1))!}{2^{b+1}(b+1)!} \cdot \delta t^{b+1}. \tag{18.56}$$

Combining (18.54), (18.55), and (18.56), we obtain the following matrix equation on the probabilities $p_0, p_1, \ldots, p_b$ and parameter $\ell = \frac{1}{a}$:

$$\begin{bmatrix} 1 & 2 & 2 & \cdots & 2 \\ 0 & 2 & 2\cdot 2^2 & \cdots & 2\cdot b^2 \\ 0 & 2 & 2\cdot 2^4 & \cdots & 2\cdot b^4 \\ \vdots & \vdots & \vdots & \ddots & \vdots \\ 0 & 2 & 2\cdot 2^{2b} & \cdots & 2\cdot b^{2b} \\ 0 & 2 & 2\cdot 2^{2(b+1)} & \cdots & 2\cdot b^{2(b+1)} \end{bmatrix} \begin{bmatrix} p_0 \\ p_1 \\ p_2 \\ \vdots \\ p_b \end{bmatrix} = \begin{bmatrix} 1 \\ \delta t \\ 3\cdot \ell^2 \\ \vdots \\ \frac{(2b)!}{2^b b!}\cdot \ell^b \\ \frac{(2b+2)!}{2^{b+1}(b+1)!}\cdot \ell^{b+1} \end{bmatrix}. \tag{18.57}$$

For small values of b, the system (18.57) can be solved explicitly. For example, for $b = 1$, we obtain a trinomial tree with Brownian increments

$$\delta \widetilde{W}_n \stackrel{d}{=} \begin{cases} \sqrt{a\,\delta t} & \text{with probability } p_1 = \frac{1}{2a}, \\ 0 & \text{with probability } p_0 = \frac{a-1}{a}, \\ -\sqrt{a\,\delta t} & \text{with probability } p_{-1} = \frac{1}{2a}, \end{cases}$$

where $a = 3$.

In Alford and Webber (2001), this equation was solved numerically for odd b. No feasible solutions appear to exist for even b. It was confirmed that multinomial methods for $b = 3, 7, 11, 15, 19$ have higher rates of convergence for European options than those of the binomial method. Note that as b increases the moment matching probabilities take values very close to those determined by the transition density function for Brownian increments:

$$p_j \approx \frac{\delta x}{\sqrt{\delta t}}\, n\left(\frac{j\,\delta x}{\sqrt{\delta t}}\right) = \sqrt{a}\, n(j\sqrt{a}),$$

where a can be found by solving (18.56). This approximate solution satisfies (18.55) with a great degree of accuracy when b is large and δx is small.

Example 18.9. In this example, we compute discrete-time approximations for the Black–Scholes price of a standard European call option with $T = 0.5$ (years), $K = 100$, $S_0 = 100$,

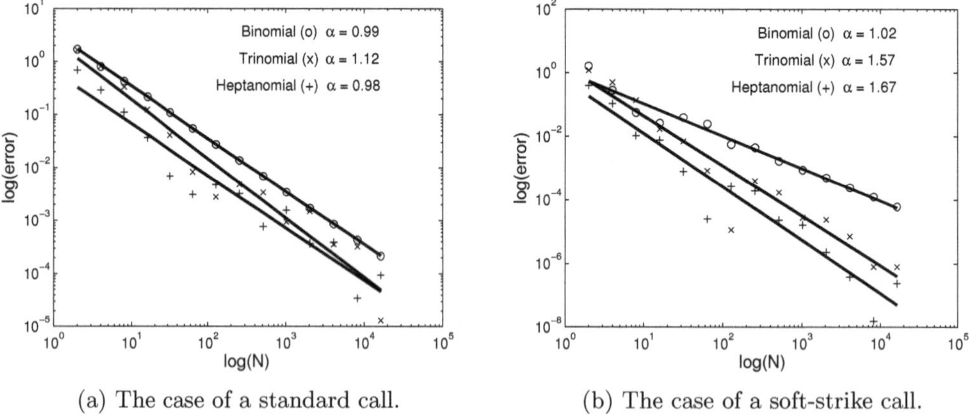

(a) The case of a standard call. (b) The case of a soft-strike call.

FIGURE 18.12: Convergence of multinomial approximations to the Black–Scholes prices of European call options.

$r = 0.05$, $q = 0$, and $\sigma = 0.25$. The exact option value is 8.260014598601677. We apply three models, namely, the binomial, trinomial, and heptanomial tree models. In theory, the highest rate of convergence of discrete-time approximations to the continuous-time solution as the number of periods N goes to ∞ is provided by the heptanomial tree model. However, a higher rate of convergence is achieved for options with smooth payoffs. The derivative of a standard European payoff has a discontinuity at $S = K$. Thus, we should not expect that the trinomial and heptanomial tree models will outperform the binomial one. Indeed, as is demonstrated in Table 18.13a and Figure 18.12a, the rate of convergence of each of these three models is just linear. To study the rate of convergence of a tree model, we compute approximations of the option price for an increasing sequence of values of the number of periods N. Assuming that the approximation error is proportional to $\delta t^\alpha = N^{-\alpha}$, we perform a linear regression to estimate the exponent α. The results for the three models are reported in Table 18.13a and Figure 18.12a.

Example 18.10 (Soft-Strike Call Option under GBM)**.** Power options differ from vanilla European options in that the payoff function is not linear but raised to some power function. In a previous chapter we learned how to analytically price such options under the GBM model. Typically, the terminal payoff of a power option is a quadratic function of the underlying. The widest possible application of power options is for addressing the nonlinear risk of option sellers. There was proposed a class of soft-strike options which do not have a single fixed strike price but a range of strike spread over an interval. Such options allow for addressing limitations of a standard delta hedging when the underlying asset is close to the strike at expiration of the option and any re-balancing of the hedging portfolio causes the underlying asset to become "pinned" to the strike price. Recall that this is related to the fact that the delta position for a standard call approaches a unit step (Heaviside) function as time to maturity goes to zero.

Consider the soft strike European call option with payoff

$$\Lambda_a(S) = \begin{cases} 0 & \text{if } S < K - a, \\ \frac{1}{4a}(S - K + a)^2 & \text{if } K - a \leqslant S < K + a, \\ S - K & \text{if } S > K + a. \end{cases} \tag{18.58}$$

We have obtained an exact analytical (Black–Scholes) pricing formula for this option in a

TABLE 18.13: Errors of multinomial approximations of Black–Scholes prices.

N	Binomial	Trinomial	Heptanomial
1024	0.003428	0.000935	0.001577
2048	0.001714	0.001477	0.000368
4096	0.000857	0.000354	0.000388
8192	0.000428	0.000321	0.000034
16384	0.000214	0.000013	0.000094

(a) The case of a standard call.

N	Binomial	Trinomial	Heptanomial
1024	$8.85 \cdot 10^{-4}$	$2.77 \cdot 10^{-5}$	$1.66 \cdot 10^{-5}$
2048	$5.02 \cdot 10^{-4}$	$2.39 \cdot 10^{-5}$	$2.30 \cdot 10^{-6}$
4096	$2.43 \cdot 10^{-4}$	$7.13 \cdot 10^{-6}$	$3.81 \cdot 10^{-7}$
8192	$1.26 \cdot 10^{-4}$	$7.79 \cdot 10^{-7}$	$1.57 \cdot 10^{-8}$
16384	$6.15 \cdot 10^{-5}$	$7.92 \cdot 10^{-7}$	$2.40 \cdot 10^{-7}$

(b) The case of a soft-strike call.

previous chapter, i.e., see equation (12.45). Note that in the limit as $a \to 0$, the soft-strike payoff converges to the call payoff $(S - K)^+$.

Since the payoff of a soft strike option is a continuously differentiable function of spot, the trinomial and heptanomial model will demonstrate a higher rate of convergence than that of the binomial tree model. Let us compute discrete-time approximations of the exact analytical price of the soft strike call option with $K = 100$ and $a = 5$. The other parameters are the same as those of the previous example. By evaluating the standard normal CDF expressions of the analytical pricing formula in (12.45), we obtain the exact option value (within 15 decimals) as 8.351245148287005. Notice that it is slightly larger than the value of a standard European call with the same maturity T and strike K from the previous example. This is expected since the payoff $\Lambda_a(S)$ is strictly larger than the call payoff $(S - K)^+$ for $K - a < S < K + a$. The errors for the multinomial approximations are reported in Table 18.13b and Figure 18.12b. Note that the rate of convergence of the binomial model is linear. The trinomial and heptanomial models demonstrate a superlinear rate of convergence.

18.2.4 Pricing European Options by PDEs

Let us assume the stock price process is a time-homogeneous diffusion with linear drift and local volatility function $\sigma(S)$, i.e., $S(t)$ satisfies the SDE

$$\frac{\mathrm{d}S(t)}{S(t)} = (r - q)\mathrm{d}t + \sigma(S(t))\mathrm{d}\widetilde{W}(t)\,.$$

A standard European-style derivative has price $V = V(t, S)$ as a function of calendar time t and spot S, where V satisfies the corresponding Black–Scholes PDE

$$\frac{\partial V}{\partial t} + \frac{1}{2}\sigma^2(S)S^2\frac{\partial^2 V}{\partial S^2} + (r - q)S\frac{\partial V}{\partial S} - rV = 0, \quad 0 \leqslant t \leqslant T, \quad 0 < S < \infty, \tag{18.59}$$

subject to the terminal (payoff) condition at maturity as $t \nearrow T$:

$$V(T-, S) \equiv V(T, S) = \Lambda(S), \quad 0 < S < \infty, \tag{18.60}$$

Here, Λ is the payoff function, r and q are, respectively, the risk-free interest rate and dividend rate, the volatility term is a state-dependent function expressed as a product of the spot and local volatility: $\sigma(S)S$. Note that this generalizes the case $\sigma(S) = \sigma$, where $\sigma > 0$ is the usual constant volatility parameter in the GBM model. Below, we focus on the case of pricing standard (vanilla) European options. The values of European call and put options, denoted below by V_C and V_P, respectively, satisfy the terminal conditions:

$$V_C(T, S) = (S - K)^+ \equiv \max\{S - K, 0\}, \tag{18.61}$$

$$V_P(T, S) = (K - S)^+ \equiv \max\{K - S, 0\}. \tag{18.62}$$

The Black–Scholes PDE is defined on the domain

$$D = \{(t, S) \mid 0 \leqslant t \leqslant T,\ 0 < S < \infty,\} \tag{18.63}$$

which is unbounded as $S \to \infty$. To complete the formulation of the differential problem (18.59)–(18.60), we need to impose a boundary condition on the solution V for $S = 0$ and an asymptotic condition on V as $S \to \infty$. These conditions depend on the option. For standard European options, we can use the put-call parity and the fact that an option becomes worthless when it is out of the money. The unboundedness of D makes the application of finite-difference numerical schemes unfeasible. One possible solution is to map the semi-infinite interval $S \in (0, \infty)$ onto a finite one by applying a suitable change of variables. Another approach is to truncate D to obtain a bounded rectangular domain. Let us introduce a lower bound S_{min} that represents small asset prices and an upper bound S_{max} that represents large asset prices. The asset price S is allowed to change in the interval $[S_{\text{min}}, S_{\text{max}}]$. Consequently, the Black–Scholes PDE is considered in the truncated domain

$$\widehat{D} = \{(t, S) \mid 0 \leqslant t \leqslant T,\ S_{\text{min}} \leqslant S \leqslant S_{\text{max}}\}. \tag{18.64}$$

The selection of S_{min} and S_{max} is also dictated by the type of option. For knock-out barrier options we may set $S_{\text{min}} = B_L$ and/or $S_{\text{max}} = B_U$, where B_L and B_U are, respectively, the lower and upper barriers.

We now need to specify the boundary conditions on V at $S = S_{\text{min}}$ and $S = S_{\text{max}}$. Consider European options for which we set S_{min} close to zero and S_{max} very large. A European call option is out of the money (and hence worthless) when $S = S_{\text{min}}$ (as long as $S_{\text{min}} \approx 0$). Therefore, for the call price V_C we adopt the following boundary condition at the left boundary:

$$V_C(t, S_{\text{min}}) = 0 \text{ for all } 0 \leqslant t \leqslant T. \tag{18.65}$$

Similarly, a European put option is out of the money (and hence worthless) when $S = S_{\text{max}}$ (as long as $S_{\text{max}} \gg K$). Hence, we set

$$V_P(t, S_{\text{max}}) = 0 \text{ for all } 0 \leqslant t \leqslant T. \tag{18.66}$$

The boundary condition on the other side of $\widehat{D}$ can be obtained by considering the put-call parity

$$V_C(t, S) - V_P(t, S) = S\mathrm{e}^{-q(T-t)} - K\mathrm{e}^{-r(T-t)} \text{ for all } 0 \leqslant t \leqslant T \text{ and } S > 0. \tag{18.67}$$

Combining (18.65) and (18.67) at $S = S_{\text{min}}$ gives us the left boundary condition for the put value:

$$V_P(t, S_{\text{min}}) = K\mathrm{e}^{-r(T-t)} - S_{\text{min}}\mathrm{e}^{-q(T-t)} \text{ for all } 0 \leqslant t \leqslant T. \tag{18.68}$$

Combining (18.66) and (18.67) at $S = S_{\max}$ gives us the right boundary condition for the call value:

$$V_C(t, S_{\max}) = S_{\max}\mathrm{e}^{-q(T-t)} - K\mathrm{e}^{-r(T-t)} \text{ for all } 0 \leqslant t \leqslant T. \tag{18.69}$$

Since the payoff functions of a call and put option are linear in the underlying, we can also use the boundary conditions

$$\frac{\partial^2}{\partial S^2} V_C(t, S_{\max}) = 0 \quad \text{and} \quad \frac{\partial^2}{\partial S^2} V_P(t, S_{\min}) = 0. \tag{18.70}$$

18.2.4.1 Pricing by the Heat Equation

As we already saw, the Black–Scholes PDE (18.59) with constant volatility σ reduces to the classical heat equation (18.8) in the spatial variable x and time to maturity τ,

$$\frac{\partial u(\tau, x)}{\partial \tau} = \frac{\partial^2 u(\tau, x)}{\partial x^2},$$

upon applying the following transformation to Equation (18.59):

$$\begin{aligned} S &= K\,\mathrm{e}^{x}, \quad t = T - \tfrac{\tau}{\sigma^2/2}, \\ V(t, S) &= K\,\mathrm{e}^{-\gamma x - (\beta^2 + \ell)\tau} u(\tau, x), \end{aligned} \tag{18.71}$$

where the constants γ, β, and ℓ are defined as follows:

$$\begin{aligned} \gamma &= \tfrac{1}{2}(\kappa - 1), & \beta &= \tfrac{1}{2}(\kappa + 1), \\ \ell &= \tfrac{q}{\sigma^2/2}, & \kappa &= \tfrac{r-q}{\sigma^2/2}. \end{aligned} \tag{18.72}$$

The change of variable in (18.71) maps the original domain D, defined in (18.63), and the truncated domain $\widehat{D}$, defined in (18.64), to

$$D_u = \{(\tau, x) \mid 0 \leqslant \tau \leqslant \frac{\sigma^2}{2} T,\ -\infty < x < \infty\}, \tag{18.73}$$

$$\widehat{D}_u = \left\{(\tau, x) \mid 0 \leqslant \tau \leqslant \frac{\sigma^2}{2} T,\ x_{\min} \leqslant x \leqslant x_{\max}\right\}, \tag{18.74}$$

respectively, where $x_{\min} := \ln\left(\frac{S_{\min}}{K}\right)$ and $x_{\max} := \ln\left(\frac{S_{\max}}{K}\right)$. The terminal conditions (18.61) and (18.62) are transformed into the initial conditions

$$u_C(0, x) = \max\{\mathrm{e}^{\beta x} - \mathrm{e}^{\gamma x}, 0\} \qquad \text{(for the call option)}, \tag{18.75}$$

$$u_P(0, x) = \max\{\mathrm{e}^{\gamma x} - \mathrm{e}^{\beta x}, 0\} \qquad \text{(for the put option)}, \tag{18.76}$$

respectively, for all $x \in [x_{\min}, x_{\max}]$. Finally, the boundary conditions for the heat equation on the truncated domain $\widehat{D}_u$ are as follows:

$$\begin{cases} u_C(\tau, x_{\min}) = 0, \\ u_C(\tau, x_{\max}) = \mathrm{e}^{\beta x_{\max} + \beta^2 \tau} - \mathrm{e}^{\gamma x_{\max} + \gamma^2 \tau} \end{cases} \qquad \text{(for the call option)}, \tag{18.77}$$

$$\begin{cases} u_P(\tau, x_{\min}) = \mathrm{e}^{\gamma x_{\min} + \gamma^2 \tau} - \mathrm{e}^{\beta x_{\min} + \beta^2 \tau}, \\ u_P(\tau, x_{\max}) = 0 \end{cases} \qquad \text{(for the put option)}, \tag{18.78}$$

for all $\tau \in [0, \frac{\sigma^2}{2} T]$. To find the numerical solution, we apply one of the finite-difference schemes to the initial boundary value problem obtained.

18.2.4.2 Pricing by the Black–Scholes PDE

Before we introduce a mesh in the domain $\widehat{D}$ and discretize the PDE (18.59), let us rewrite the terminal value problem (18.59)–(18.60) as an initial value problem. We replace the actual time t by the time to maturity $\tau = T - t$ in Equations (18.59), (18.60), (18.68), and (18.69) to obtain the following differential equation:

$$\frac{\partial v}{\partial \tau} = \frac{1}{2}\sigma^2(S)S^2\frac{\partial^2 v}{\partial S^2} + (r-q)S\frac{\partial v}{\partial S} - rv, \quad \tau > 0, \quad S_{\min} < S < S_{\max} \tag{18.79}$$

subject to the initial condition

$$v(0, S) = \Lambda(S), \tag{18.80}$$

and the boundary conditions

$$v_P(\tau, S_{\min}) = K\mathrm{e}^{-r\tau} - S_{\min}\mathrm{e}^{-q\tau} \text{ and } v_P(\tau, S_{\max}) = 0 \tag{18.81}$$

and

$$v_C(\tau, S_{\max}) = S_{\max}\mathrm{e}^{-q\tau} - K\mathrm{e}^{-r\tau} \text{ and } v_C(\tau, S_{\min}) = 0, \tag{18.82}$$

for the European put and call options, respectively. Here, the option value $v = v(\tau, S)$ is a function of the spot price S and the time to maturity τ.

Consider a uniform spacial mesh on the interval $[S_{\min}, S_{\max}]$:

$$s_j = S_{\min} + j\delta s,\ j = 0, 1, \ldots, n+1, \text{ where } \delta s = \frac{S_{\max} - S_{\min}}{n+1}.$$

Replacing all derivatives with respect to S by their central finite-difference approximations, we obtain the following approximation to the Black–Scholes PDE (18.79):

$$\begin{aligned}\frac{\partial v(\tau, S)}{\partial \tau} &= \frac{1}{2}\sigma^2(S)S^2\frac{v(\tau, S+\delta s) - 2v(\tau, S) + v(\tau, S-\delta s)}{\delta s^2}\\ &\quad + (r-q)S\frac{v(\tau, S+\delta s) - v(\tau, S-\delta s)}{2\delta s} - rv(\tau, S) + \mathcal{O}(\delta s^2).\end{aligned} \tag{18.83}$$

Let $V_j(\tau)$ denote the semi-discrete approximation to $v(\tau, s_j)$. Applying (18.83) at each internal node s_j, we obtain the following system of first-order ordinary differential equations:

$$\begin{aligned}\frac{\mathrm{d}V_j(\tau)}{\mathrm{d}\tau} &= \underbrace{\frac{1}{2}\left(\left(\frac{\sigma(s_j)s_j}{\delta s}\right)^2 - \frac{(r-q)s_j}{\delta s}\right)}_{=:L_{j,j-1}} V_{j-1}(\tau) + \underbrace{\left(-\left(\frac{\sigma(s_j)s_j}{\delta s}\right)^2 - r\right)}_{=:L_{j,j}} V_j(\tau)\\ &\quad + \underbrace{\frac{1}{2}\left(\left(\frac{\sigma(s_j)s_j}{\delta s}\right)^2 + \frac{(r-q)s_j}{\delta s}\right)}_{=:L_{j,j+1}} V_{j+1}(\tau), \quad j = 1, 2, \ldots, n.\end{aligned} \tag{18.84}$$

System (18.84) has n equations in $n+2$ unknown functions $V_0(\tau), V_1(\tau), \ldots, V_n(\tau), V_{n+1}(\tau)$. Using the boundary conditions we have the functions $V_0(\tau)$ and $V_{n+1}(\tau)$, which, respectively, approximate the solution at the boundary nodes $s_0 = S_{\min}$ and $s_{n+1} = S_{\max}$. As a result, the system of differential equations (18.84) can be written as the following matrix-vector differential equation with an n-by-n tridiagonal coefficient matrix $\mathbf{L}$ whose entries are defined in (18.84):

$$\frac{\mathrm{d}\mathbf{V}(\tau)}{\mathrm{d}\tau} = \mathbf{L}\mathbf{V}(\tau) + \mathbf{G}(\tau), \tag{18.85}$$

subject to the initial condition

$$\mathbf{V}(0) = \mathbf{\Lambda} := [\Lambda(s_1), \Lambda(s_2), \ldots, \Lambda(s_n)]^\top. \tag{18.86}$$

Here we use the notation:

$$\mathbf{L} = \begin{bmatrix} L_{11} & L_{12} & 0 & \cdots & 0 & 0 \\ L_{21} & L_{22} & L_{23} & \cdots & 0 & 0 \\ 0 & L_{32} & L_{33} & \cdots & 0 & 0 \\ \vdots & \vdots & \vdots & \ddots & \vdots & \vdots \\ 0 & 0 & 0 & \cdots & L_{n-1,n-1} & L_{n-1,n} \\ 0 & 0 & 0 & \cdots & L_{n,n-1} & L_{n,n} \end{bmatrix}, \quad \mathbf{V}(\tau) = \begin{bmatrix} V_1(\tau) \\ V_2(\tau) \\ \vdots \\ V_{n-1} \\ V_n(\tau) \end{bmatrix}.$$

The vector $\mathbf{G}(\tau) \in \mathbb{R}^n$ is given by

$$\left[\left(\frac{\sigma^2(s_0)s_0^2}{2\delta s^2} - \frac{(r-q)s_0}{2\delta s}\right) V_0(\tau), 0, \ldots, 0, \left(\frac{\sigma^2(s_{n+1})s_{n+1}^2}{2\delta s^2} + \frac{(r-q)s_{n+1}}{2\delta s}\right) V_{n+1}(\tau)\right]^\top.$$

$\mathbf{G}(\tau)$ contains boundary values of the mesh solution. For vanilla European options we can use the boundary conditions (18.81) and (18.82) to obtain the following:

$$V_0(\tau) = K\mathrm{e}^{-r\tau} - S_{\min}\mathrm{e}^{-q\tau} \text{ and } V_{n+1}(\tau) = 0 \text{ for European put options} \tag{18.87}$$

and

$$V_{n+1}(\tau) = S_{\max}\mathrm{e}^{-q\tau} - K\mathrm{e}^{-r\tau} \text{ and } V_0(\tau) = 0 \text{ for European call options.} \tag{18.88}$$

Applying finite differences in time, we derive the explicit method, the implicit method, and the Crank–Nicolson method as follows, by introducing the fully discretized approximate solution (the mesh solution) to the Black–Scholes equation:

$$V_{i,j} \approx V_j(\tau_i) \approx V(x_j, \tau_i), \quad i = 0, 1, \ldots, m, \ j = 0, 1, \ldots, n+1.$$

The approximation solution at the ith time level will be denoted by

$$\mathbf{V}^{(i)} := [V_{i,1}, V_{i,2}, \ldots, V_{i,n}]^\top \approx \mathbf{V}(t_i)$$

for each $i = 0, 1, 2, \ldots, m$.

The Explicit Method

Using a forward finite difference in (18.85) gives

$$\frac{\mathbf{V}^{(i+1)} - \mathbf{V}^{(i)}}{\delta t} = \mathbf{L}\mathbf{V}^{(i)} + \mathbf{G}(t_i),$$

which leads to the iterative rule

$$\mathbf{V}^{(i+1)} = (\mathbf{I}_n + \delta t\mathbf{L})\,\mathbf{V}^{(i)} + \delta t\mathbf{G}(t_i), \quad i \geqslant 0. \tag{18.89}$$

The Implicit Method

Approximating the time derivative in (18.85) with a backward finite difference gives

$$\frac{\mathbf{V}^{(i+1)} - \mathbf{V}^{(i)}}{\delta t} = \mathbf{L}\mathbf{V}^{(i+1)} + \mathbf{G}(t_{i+1}),$$

which leads to the iterative rule

$$(\mathbf{I}_n - \delta t \mathbf{L})\, \mathbf{V}^{(i+1)} = \mathbf{V}^{(i)} + \delta t \mathbf{G}(t_{i+1}), \quad i \geqslant 0. \tag{18.90}$$

The Crank–Nicolson Method

The Crank–Nicolson method is derived by taking the arithmetic average of the explicit (18.90) and the implicit (18.90) methods to give the following:

$$\left(\mathbf{I}_n - \frac{\delta t}{2}\mathbf{L}\right) \mathbf{V}^{(i+1)} = \left(\mathbf{I}_n + \frac{\delta t}{2}\mathbf{L}\right) \mathbf{V}^{(i)} + \frac{\delta t}{2}\left(\mathbf{G}(t_i) + \mathbf{G}(t_{i+1})\right), \quad i \geqslant 0. \tag{18.91}$$

All the above schemes are subject to the initial condition

$$\mathbf{V}^{(0)} = \mathbf{\Lambda}$$

with $\mathbf{\Lambda}$ defined in (18.86).

Example 18.11. In this example we apply the implicit and Crank–Nicolson schemes to price a standard call option under the Black–Scholes (log-normal) model. The parameters used are the same as those of Example 18.9. The truncated domain $\widehat{D}$ has the lower bound $S_{\min} = 0$ and the upper bound

$$S_{\max} = S_0 \exp\left((r - q - \sigma^2/2)T + 5\sigma\sqrt{T}\right) \cong 244.3077.$$

To discretize the PDE problem, we split the space and time intervals in 5001 and 1001 subintervals, respectively. Thus, the time and space discretization steps are

$$\delta t = \frac{0.5}{1001} \cong 0.0005 \text{ and } \delta s \cong \frac{244.3077}{5001} \cong 0.05.$$

The results for the two schemes are reported in Figure 18.14a. As is seen from the graphs in Figure18.14b, the rate of convergence of the Crank–Nicolson method is higher than that of the implicit method.

18.2.5 Calibration of Asset Price Models to Empirical Data

The term calibration refers to the computational process of fitting an asset price model to historical financial data. First, we select an asset price price model that seems to explain well empirical observations. There are two typical calibration procedures. The least squares method (LSM) allows for fitting model derivative values to the respective market prices. To apply this method, we assume that a derivative pricing formula is available in closed form or that the prices can be computed numerically in a fast and efficient manner. The maximum likelihood estimation (MLE) method allows for fitting the probability distribution to historical asset prices (or their returns). To apply it we assume that a transition probability density is known in closed form. Suppose the model can be characterized by a vector of parameters, say $\boldsymbol{\xi} = (\xi_1, \xi_2, \ldots, \xi_k)$. The objective is find the best-fitted vector $\hat{\boldsymbol{\xi}}$ that solves the respective optimization problem. Typically, there are two levels of calibration: first, an initial full calibration of all parameters of the model and, second, a faster re-calibration conditional on the vector $\hat{\boldsymbol{\xi}}$ computed previously that can be used as soon as new data have arrived. The second calibration scheme may be used throughout the day or even for longer periods, while the full calibration only needs to be executed at the outset or if markets move considerably.

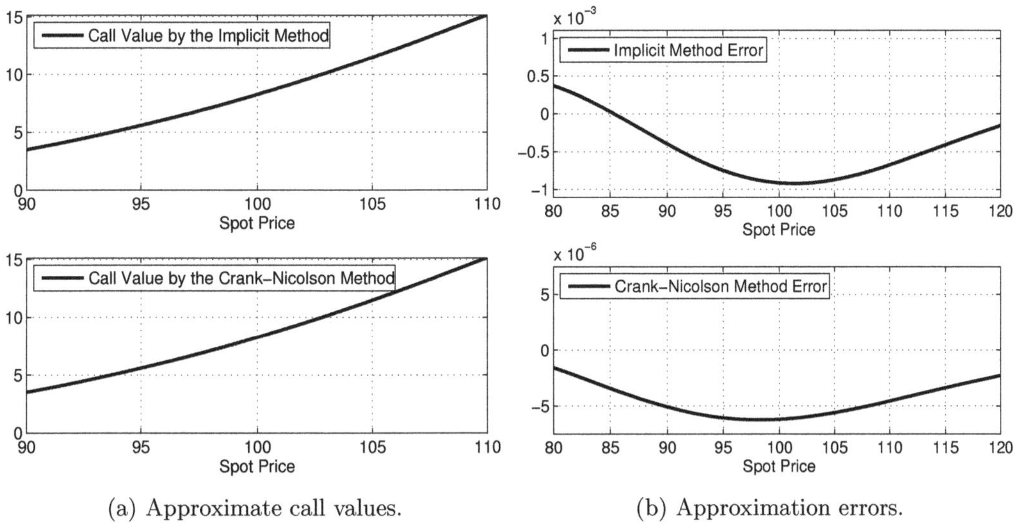

(a) Approximate call values.

(b) Approximation errors.

FIGURE 18.14: Computing Black–Scholes call option values with the use of the implicit and Crank–Nicolson methods.

18.2.5.1 Least Squares Method

Suppose we wish to calibrate an asset price model to historical European call and put option prices. Let the option with strike K_i and maturity T_i have an *observed price* O_i. Thus, we have two arrays of observations, i.e., a pair of variables (K_i, T_i) specifying the option and market prices O_i, where $i = 1, 2, \ldots, m$. Let the model produce a price $C_i = C(K_i, T_i; \boldsymbol{\xi})$ for the same ith option. The goal of the calibration process is to minimize the least squares error between the model and market prices for the m option values:

$$F(\boldsymbol{\xi}) := \sum_{i=1}^{m} w_i \left(C(K_i, T_i; \boldsymbol{\xi}) - O_i\right)^2 \to \min_{\boldsymbol{\xi}}, \tag{18.92}$$

where w_i is a weight that reflects the relative significance of reproducing the ith option price precisely. The suitable choice of the weights w_i, $i = 1, 2, \ldots, m$, is important for good calibration results. The simplest choice is to set $w_i = \frac{1}{m}$. The confidence in individual data points is determined by the liquidity of the option. Thus, the weights can also be evaluated from the bid-ask spreads:

$$w_i = \frac{1}{|O_i^{\text{ask}} - O_i^{\text{bid}}|}.$$

Alternatively, as suggested by Cont and Tankov (2003), one may use the Black–Scholes (BS) "Vegas" evaluated at the implied volatilities of the market option prices to compute the weights:

$$w_i = \left(\partial C^{\text{BS}}(\sigma_i^{\text{BS}})/\partial\sigma\right)^{-2},$$

where $\partial C^{\text{BS}}/\partial\sigma$ denotes the derivative of the BS option pricing formula with respect to the volatility σ, evaluated at $\sigma = \sigma_i^{\text{BS}}$, and $\sigma_i^{\text{BS}} = \sigma^{\text{BS}}(O_i, K_i, T_i)$ is the BS implied volatility for the observed market price O_i. For the observed market prices O_i we may also use the midpoint of the bid and ask prices, i.e., $O_i = (O_i^{\text{ask}} + O_i^{\text{bid}})/2$.

18.2.5.2 Maximum Likelihood Estimation

Suppose the asset price process $\{S(t)\}_{t\geqslant 0}$ is a Markov process with a transition PDF $p_{\boldsymbol{\xi}}(u,t;S,S')$. The PDF depends on the vector of parameters $\boldsymbol{\xi}$. Typically, empirical data are available to us in the form of a truncated series of historical asset prices. Consider the array of observations $(t_i, \hat{s}_i)$, $i = 0, 1, 2, \ldots, m$, where $\hat{s}_i$ is a market asset (stock) price at calendar time t_i. In general, the MLE method selects values of the model parameters that produce a probability distribution that gives the observed data the greatest probability. To use the method of maximum likelihood, we first need to specify the joint density function for all observations. Let the times t_i form an increasing sequence. The joint probability density of the model asset prices at $S(t_0) = s_0, S(t_1) = s_1, \ldots, S(t_m) = s_m$ is then simply a product of transition PDFs

$$\prod_{i=1}^{m} p_{\boldsymbol{\xi}}(t_{i-1}, t_i; s_{i-1}, s_i).$$

Now, by substituting the time series of observed asset prices $s_0 = \hat{s}_0, s_1 = \hat{s}_1, \ldots, s_m = \hat{s}_m$, we obtain the *likelihood function*

$$L(\boldsymbol{\xi}) := \prod_{i=1}^{m} p_{\boldsymbol{\xi}}(t_{i-1}, t_i; \hat{s}_{i-1}, \hat{s}_i).$$

The method of maximum likelihood estimates the best-fitted $\hat{\boldsymbol{\xi}}$ by finding a value of $\boldsymbol{\xi}$ that maximizes $L(\boldsymbol{\xi})$:

$$\hat{\boldsymbol{\xi}} = \operatorname{argmax} L(\boldsymbol{\xi}). \tag{18.93}$$

In practice it is often more convenient to work with the (natural) logarithm of the likelihood function, called the log-likelihood. The result is the same regardless of whether we maximize the likelihood or the log-likelihood function, since the natural logarithm is a monotonically increasing function.

In solving the problems (18.92) and (18.93), we apply a numerical optimization routine such as the Nelder–Mead method (a nongradient based algorithm) or gradient search methods. In practice, the parameters $\xi_1, \xi_2, \ldots, \xi_k$ have lower and/or upper constraints. However, by changing variables, a constrained optimization problem can be transformed into one without constrains.

18.3 Pricing Early-Exercise and Path-Dependent Options

Pricing exotic derivatives such as American and Asian options is a much less trivial problem in comparison to pricing standard European options. There are no closed-form formulae for the American and the arithmetic Asian options within the Black–Scholes model. In this section we focus on pricing options with an underlying asset price following the log-normal model or another Markov process whose transition distribution is known in closed form.

18.3.1 Pricing American and Bermudan Options

18.3.1.1 Pricing American Options by Tree Methods

Consider the application of the binomial tree method to pricing an American option with time to maturity $T > 0$ and having the intrinsic payoff value $\Lambda(S)$. The continuous-time asset price process, which follows the log-normal model, is approximated by a discrete-time binomial model with $N+1$ trading dates $\{n\,\delta t \,:\, n = 0, 1, \ldots, N\}$, where δt is the length of one period and $N\,\delta t = T$. Let the factors u and d be specified as in (18.46). The one-period interest rate is $e^{r\,\delta t} - 1$. The risk-neutral probability $\tilde{p}$ of the upward move is given in (18.47).

Let $V_{n,k}$ denote the American option value at time $n\delta t$ for the asset price $S_{n,k} = S_0 u^n d^{n-k}$ with $k = 0, 1, \ldots, n$ and $n = 0, 1, \ldots, N$. From Chapter 7, we know that a backward-in-time recursion relation can be used to compute the option value at each node of the binomial tree as follows.

1. At maturity, set $V_{N,k} = \Lambda(S_{N,k})$ for each $k = 0, 1, \ldots, N$.
2. For each $n = N-1, N-2, \ldots, 0$ and $k = 0, 1, \ldots, n$, the option value $V_{n,k}$ is given by

$$V_{n,k} = \max\left\{\Lambda(S_{n,k}), e^{-r\,\delta t}\left(\tilde{p}V_{n+1,k+1} + (1-\tilde{p})V_{n+1,k}\right)\right\}.$$

Let us see how the above algorithm can be used to compute prices of a Bermudan derivative that can only be exercised at discrete times $0 < t_1 < t_2 < \cdots < t_M = T$ with $M \ll N$. Assume that the exercise dates are all of the form $n\delta t$ for some integer n. At each exercise date, the option value is a maximum of the continuation and the intrinsic values. In between the exercise dates, we treat the derivative as a European option. The steps are as follows.

1. At maturity, set $V_{N,k} = \Lambda(S_{N,k})$ for each $k = 0, 1, \ldots, N$.
2. For each $n = N-1, N-2, \ldots, 0$, compute the option values $V_{n,k}$, $k = 0, 1, \ldots, n$, as follows:

   (i) Compute the continuation values

$$V_{n,k}^{\text{cont}} = e^{-r\,\delta t}\left(\tilde{p}V_{n+1,k+1} + (1-\tilde{p})V_{n+1,k}\right).$$

   (ii) If $n\delta t$ is an exercise date, i.e., $n\delta t = t_m$ for some $m = 1, 2, \ldots, M$, then

$$V_{n,k} = \max\left\{\Lambda(S_{n,k}), V_{n,k}^{\text{cont}}\right\};$$

   otherwise, we set

$$V_{n,k} = V_{n,k}^{\text{cont}}.$$

Example 18.12. In this example, we compute discrete-time approximations for the Black–Scholes price of American and Bermudan put options with $T = 1$, $K = 95$, $S_0 = 100$, $r = 0.05$, $q = 0$, and $\sigma = 0.25$. As for the Bermudan options, we assume that the option can only be exercised at one of the M times $t_m = m\frac{T}{M}$, $m = 1, 2, \ldots, M$. The results of our calculations are reported in Tables 18.15a and 18.15b.

18.3.1.2 Pricing Bermudan Options by the Monte Carlo Method

Instead of a continuously exercisable American option, we consider a Bermudan option that can only be exercised at a fixed set of times $0 < t_1 < t_2 < \cdots < t_M = T$. The option expires at time T and the payoff to the holder is $\Lambda(S)$. The American option can be viewed as a limiting case of Bermudan options as $M \to \infty$ and $\max_{1\leqslant k\leqslant M}|t_k - t_{k-1}| \to 0$. Suppose that the number M of exercise times is not too large. Even in this case the early-exercise option price problem poses a significant challenge to the Monte Carlo method.

TABLE 18.15: Pricing options in the binomial tree model.

N	δt	European	American
100	0.01	5.3957684	5.7494428
200	0.005	5.4240877	5.7388323
500	0.002	5.4165327	5.7584385
1000	0.001	5.4147939	5.7517360
2000	0.0005	5.4148298	5.7502178
5000	0.0002	5.4140541	5.7501685

(a) Pricing European and American puts with N periods.

M	Bermudan
2	5.5609302
5	5.6629469
10	5.7042806
20	5.7262399
50	5.7400010
5000	5.7501685

(b) Pricing Bermudan puts with $N = 5000$ periods and M exercise dates.

The initial price of a Bermudan option can be calculated using a backward-in-time recursion. Let $V(t, S)$ denote the no-arbitrage price at time t given that $S(t) = S$ and the option has not been exercised before time t. It is more convenient to write the equation on the option values using discounted quantities. Consider the discounted payoff function $\Lambda_k(S) = \mathrm{e}^{-rt_k}\Lambda(S)$ at the exercise time t_k. Let $V_k(S)$ denote the present value of the time-t_k option price. That is, $V_k(S) = \mathrm{e}^{-rt_k}V(t_k, S)$ is the no-arbitrage option value calculated at time t_k and discounted to time 0. We also denote $S_k = S(t_k)$ for all $k = 0, 1, \ldots, M$. At the maturity time $T = t_M$ we have $V_M(S) = \Lambda_M(S)$. At any exercise time $t_k < t_M$ the holder of the option can exercise the option and immediately receive the payoff (the intrinsic value), or continue to hold the option. In the latter case, the average proceeds to the holder are given by the expected time-t_{k+1} option value (the continuation value). Therefore, the discounted option values V_k satisfy the following dynamic programming problem:

$$V_M(S_M) = \Lambda_M(S_M) \tag{18.94}$$

$$V_k(S_k) = \max\left\{\Lambda_k(S_k), \widetilde{\mathrm{E}}[V_{k+1}(S_{k+1}) \mid S_k]\right\}, \quad 0 \leqslant k \leqslant M-1. \tag{18.95}$$

The mathematical expectations considered here are computed under the risk-neutral probability measure $\widetilde{\mathbb{P}}$:

$$\widetilde{\mathrm{E}}[V_{k+1}(S_{k+1}) \mid S_k] = \int_0^\infty V_{k+1}(S_{k+1})\, \tilde{p}(t_{k+1} - t_k; S_k, S_{k+1})\, \mathrm{d}S_{k+1}.$$

The main difficulty of the problem is the computation of the continuation value. Let us consider three Monte Carlo approaches to dealing with this issue. Note that the asset price process $\{S(t)\}_{t\geqslant 0}$ does not necessarily follow geometric Brownian motion. The only assumption regarding the underlying asset process we will require is that the Markov chain $S_0 \to S_1 \to S_2 \to \cdots \to S_M$ can be simulated exactly from the transition probability distribution. Although we only discuss the single-asset case, all the methods presented below can also be used for the evaluation of multi-asset Bermudan options.

Stochastic Tree Method. For each t_k, the continuation value $\widetilde{\mathrm{E}}[V_{k+1}(S_{k+1}) \mid S_k]$ can be approximated by the Monte Carlo method. First, sample b successors $S^1_{k+1}, S^2_{k+1}, \ldots, S^b_{k+1}$ from the exact probability distribution of S_{k+1} conditional on S_k. Second, approximate the

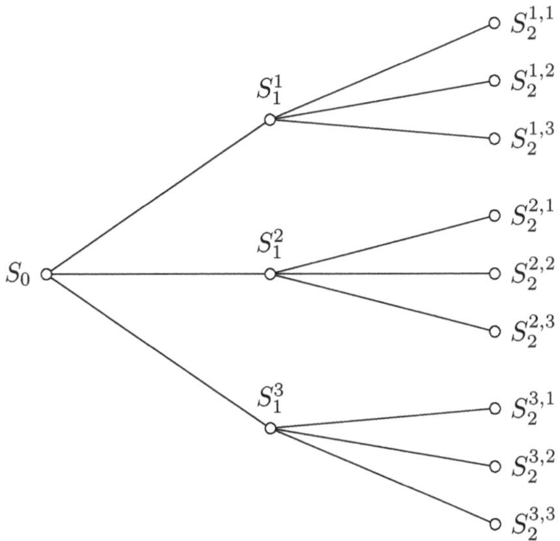

FIGURE 18.16: A schematic illustration of a two-period stochastic tree.

mathematical expectation by the sample mean:

$$\widetilde{\mathrm{E}}[V_{k+1}(S_{k+1}) \mid S_k] \approx \frac{1}{b}\sum_{i=1}^{b} V_{k+1}(S_{k+1}^i).$$

The main idea of the stochastic tree method is to construct a nonrecombining stochastic tree with M levels (so that each level corresponds to one of M exercise times) and b branches for each node. As a result we obtain a tree with b^M branches at the last level (see Figure 18.16). Since the number of branches grows exponentially fast as M increases, this method is feasible only if both b and M are not very large. Algorithm 18.3 describes the construction of a stochastic tree in detail.

Stochastic Mesh Method. The main shortcoming of the stochastic tree method is that the number of branches of a stochastic tree grows exponentially-fast as the number M of exercise times increases. The stochastic mesh method proposed by Broadie and Glasserman (2004) also operates with a random tree, but there are two key distinctions. First, there is no branching at intermediate times and the total number of sample paths does not explicitly depend on the number of exercise times. We start with sampling n paths of the Markov chain $S_0 \to S_1 \to S_2 \to \cdots \to S_M$, where $S_k \equiv S(t_k)$ for all k. These paths form a stochastic tree with n branches (and $n \cdot M$ nodes) all originated at time 0. Second, in computing the continuation value at some time t_k we use time-t_{k+1} option values from all branches regardless of whether or not the values are computed for the successor of the current node. Algorithm 18.4 describes the stochastic mesh method step by step. In comparison with the stochastic tree method, we additionally assume that the risk-neutral transition PDF $\tilde{p}(t; S, S')$ (under the risk-neutral measure) is known in closed form. Note that, for simplicity, we are assuming a time-homogeneous process.

To explain the formula for the weights, let us consider the computation of the continuation value in more detail. In computing the Monte Carlo approximation of the continuation value $\widetilde{\mathrm{E}}\left[V_{k+1}(S_{k+1}) \mid S_k = S_k^i\right]$, the asset values $S = S_{k+1}^j$, $j \neq i$ at time t_{k+1} are not sampled from the single time-step transition PDF $\tilde{p}(t_{k+1} - t_k; S_k^i, S)$. Since all paths begin with

Algorithm 18.3 The Stochastic Tree Method for Estimation of the No-Arbitrage Value of a Bermudan Option.

(1) Starting from the initial asset price value S_0 we simulate b independent successors $S_1^1, S_1^2, \ldots, S_1^b$ all having the probability law of $S_1 \equiv S(t_1)$.

(2) For each $S_1^{i_1}$, $i_1 = 1, 2, \ldots, b$, simulate b independent successors

$$S_2^{i_1,1}, S_2^{i_1,2}, \ldots, S_2^{i_1,b}$$

all having the law of $S_2 \equiv S(t_2)$ conditional on $S_1 = S_1^{i_1}$.

(3) For each $S_2^{i_1,i_2}$, $i_1, i_2 = 1, 2, \ldots, b$, simulate b independent successors

$$S_3^{i_1,i_2,1}, S_3^{i_1,i_2,2}, \ldots, S_3^{i_1,i_2,b}$$

all having the law of $S_3 \equiv S(t_3)$ conditional on $S_2 = S_2^{i_1,i_2}$.

⋮

(M) For each $S_{M-1}^{i_1,i_2,\ldots,i_{M-1}}$, $i_1, i_2, \ldots, i_{M-1} = 1, 2, \ldots, b$, simulate b independent successors

$$S_M^{i_1,i_2,\ldots,i_{M-1},1}, S_M^{i_1,i_2,\ldots,i_{M-1},2}, \ldots, S_M^{i_1,i_2,\ldots,i_{M-1},b}$$

all having the law of $S_M \equiv S(t_M)$ conditional on $S_{M-1} = S_{M-1}^{i_1,i_2,\ldots,i_{M-1}}$.

Let $\widehat{V}_k^{i_1,\ldots,i_k}$ denote the approximation of the option value at time t_k for the asset price $S(t_k) = S_k^{i_1,\ldots,i_k}$. At the terminal nodes we set the option value equal to the payoff value:

$$\widehat{V}_M^{i_1,\ldots,i_M} = \Lambda_M(S_M^{i_1,\ldots,i_M}).$$

Working backward in time throughout the tree, we can compute option values for each node of the tree. Starting with $k = M-1$, for each $k = M-1, M-2, \ldots, 1$, we set

$$\widehat{V}_k^{i_1,\ldots,i_k} = \max\left\{\Lambda_k\left(S_k^{i_1,\ldots,i_k}\right), \frac{1}{b}\sum_{j=1}^{b} \widehat{V}_{k+1}^{i_1,\ldots,i_k,j}\right\}.$$

Finally, we calculate the approximation of the option price at current time $t = 0$ by

$$\widehat{V}_0 = \frac{1}{b}\sum_{i=1}^{b} \widehat{V}_1^i.$$

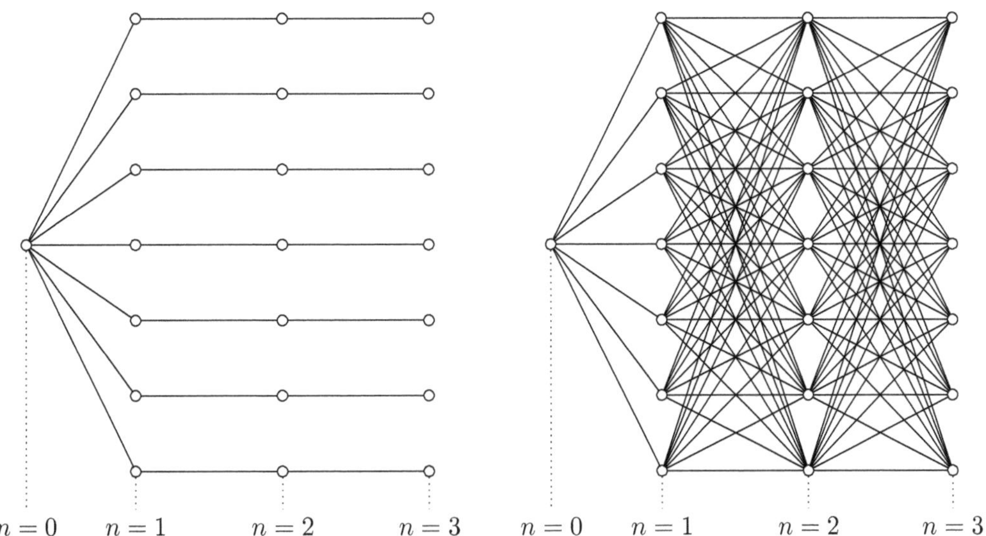

FIGURE 18.17: Construction of stochastic mesh from independent paths. The figure on the left shows the nodes and the figure on the right contains all the paths up to three time steps.

S_0, the values $S = S^j_{k+1}$ at time t_{k+1} are sampled from the marginal PDF $\tilde{p}(t_{k+1}; S_0, S)$. At the node (t_k, S^i_k) we have

$$\widetilde{\mathrm{E}}\left[V_{k+1}(S_{k+1}) \mid S_k = S^i_k\right] = \int_0^\infty V_{k+1}(S)\, \tilde{p}(t_{k+1} - t_k; S^i_k, S)\, \mathrm{d}S$$

(multiply and divide the integrand by $p(t_{k+1}; S_0, S)$)

$$\begin{aligned} &= \int_0^\infty V_{k+1}(S) \frac{\tilde{p}(t_{k+1} - t_k; S^i_k, S)}{\tilde{p}(t_{k+1}; S_0, S)} \tilde{p}(t_{k+1}; S_0, S)\, \mathrm{d}S \\ &= \widetilde{\mathrm{E}}\left[V_{k+1}(S_{k+1}) \frac{\tilde{p}(t_{k+1} - t_k; S^i_k, S_{k+1})}{\tilde{p}(t_{k+1}; S_0, S_{k+1})}\right], \end{aligned}$$

where the time-t_{k+1} asset price S_{k+1} has probability density $\tilde{p}(t_{k+1}; S_0, S_{k+1})$ in the last mathematical expectation. As we can see, in computing the option value at the node (t_k, S^i_k) we use the option values from all nodes at the next time layer t_{k+1}, as is illustrated in Figure 18.17, which gives a complete picture of all relations between the nodes of the stochastic tree. The tree looks more like a net or mesh. This structure explains the name of the stochastic mesh method.

Theorem 18.4 (Broadie and Glasserman (2004)). *The mesh estimator $\hat{V}_0$ is biased high, i.e.,* $\mathrm{E}[\hat{V}_0] \geqslant V(0, S_0)$ *for all* $n \geqslant 1$.

Proof. Let us prove the assertion by using Jensen's inequality and induction. At the terminal time $t_M = T$ we have $\hat{V}^i_M = \Lambda_M(S^i_M) = V(T, S^i_M)$ for all $i = 1, 2, \ldots, n$. Assume $E[\hat{V}^i_{k+1}] \geqslant V(t_{k+1}, S^i_{k+1})$ for some $k = 0, 1, \ldots, M-1$ and all $i = 1, 2, \ldots, n$. At time t_k we have

$$\begin{aligned}
\mathrm{E}[\hat{V}_k^i] &= \mathrm{E}\left[\max\left\{\Lambda_k(S_k^i), \frac{1}{n}\sum_{j=1}^{n}\frac{\tilde{p}(t_{k+1}-t_k; S_k^i, S_{k+1}^j)}{\tilde{p}(t_{k+1}; S_0, S_{k+1}^j)}\widehat{V}_{k+1}^j\right\}\right] \\
&\geqslant \max\left\{\Lambda_k(S_k^i), \mathrm{E}\left[\frac{1}{n}\sum_{j=1}^{n}\frac{\tilde{p}(t_{k+1}-t_k; S_k^i, S_{k+1}^j)}{\tilde{p}(t_{k+1}; S_0, S_{k+1}^j)}\widehat{V}_{k+1}^j\right]\right\} \\
&\geqslant \max\left\{\Lambda_k(S_k^i), \mathrm{E}\left[\frac{\tilde{p}(t_{k+1}-t_k; S_k^i, S_{k+1})}{\tilde{p}(t_{k+1}; S_0, S_{k+1})}V(t_{k+1}, S_{k+1})\right]\right\} \\
&= \max\left\{\Lambda_k(S_k^i), V(t_k, S_k^i)\right\} \\
&= V(t_k, S_k^i).
\end{aligned}$$

□

Algorithm 18.4 The Stochastic Mesh Method for Estimation of the No-Arbitrage Price of a Bermudan Option.

1. Generate n independent paths $S_0^j, S_1^j, S_2^j \ldots, S_M^j$ using the transition PDF p. That is, S_k^j is sampled from the density $p(t_k - t_{k-1}; S_{k-1}^j, S_k^j)$ for every $k = 1, 2, \ldots, M$ and every $j = 1, 2, \ldots, n$. Here all S_0^j equal S_0. The sample paths form a stochastic tree starting at S_0 with $n \cdot M$ nodes. The node (k, j) corresponds to the exercise time t_k and asset price S_k^j.

2. At the terminal nodes (corresponding to the time $t_M \equiv T$), we set

$$\widehat{V}_M^j = \Lambda_M(S_M^j) \text{ for } j = 1, 2, \ldots, n.$$

3. Working recursively backward in time, we set

$$\widehat{V}_k^i = \max\left\{\Lambda_k(S_k^i), \frac{1}{n}\sum_{j=1}^{n} W_k^{ij}\widehat{V}_{k+1}^j\right\},$$

for $k = M-1, M-2, \ldots, 1$ and $i = 1, 2, \ldots, n$. Here W_k^{ij} are weights given by

$$W_k^{ij} = \frac{\tilde{p}(t_{k+1}-t_k; S_k^i, S_{k+1}^j)}{\tilde{p}(t_{k+1}; S_0, S_{k+1}^j)}.$$

4. Calculate the approximation of the initial option value, $\widehat{V}_0 = \frac{1}{n}\sum_{i=1}^{n}\widehat{V}_1^i$.

In conclusion, we mention another popular Monte Carlo technique for pricing Bermudan options based on the least squares (regression) method by Longstaff and Schwartz (2001). Regression methods rely on the representation of the continuation value as a linear combination of some basis functions $\psi_\ell\colon \mathbb{R} \to \mathbb{R}$, $\ell = 1, 2, \ldots, L$. That is, the single time-step conditional expectation is written as a linear combination of a chosen set of so-called basis functions:

$$\widetilde{\mathrm{E}}[V_{k+1}(S_{k+1}) \mid S_k = S] = \sum_{\ell=1}^{L}\beta_{k\ell}\psi_\ell(S).$$

18.3.2 Pricing Asian Options

Asian options are averaging options, where the payoff functions depend on some form of average of the underlying asset prices over the life of the option. By considering different types of averaging, such as arithmetic or geometric, continuous or discrete, one can obtain different types of options. The difficulty with the pricing of Asian options depends both on the underlying asset price model and on the type of the option. If we assume the asset price follows the GBM process, then one can derive closed-form pricing formulae for Asian options with geometric averaging. Pricing arithmetic Asian options is a computational challenge even under the GBM model.

18.3.2.1 Pricing Discrete-Time Asian Options by the Monte Carlo Method

Here we analyze only one type of averaging, namely, discrete arithmetic averaging at M time observation points equally spaced throughout $[0, T]$, with random variables

$$A_M := \frac{1}{M}\sum_{k=1}^{M} S_k, \quad S_k := S\left(k \cdot \frac{T}{M}\right). \tag{18.96}$$

By using methods of extrapolation, one can also approximate values of Asian options for continuous-time arithmetic averaging based on option values computed using discrete averaging. Most Asian options are of European style (i.e., not early exercise). There are two main examples of Asian options: the arithmetic average price (AAP) or floating price option and the arithmetic average strike (AAS) or floating strike option. The payoff functions of Asian calls (C) and puts (P) with time to maturity T are as follows:

$$\begin{aligned} C_M^{AAP} &:= \max\{A_M - K, 0\}, \quad & P_M^{AAP} &:= \max\{K - A_M, 0\}, \\ C_M^{AAS} &:= \max\{S_M - A_M, 0\}, \quad & P_M^{AAS} &:= \max\{A_M - S_M, 0\}. \end{aligned} \tag{18.97}$$

The plain Monte Carlo evaluation of the no-arbitrage price of an Asian option with a payoff function of the form $\Lambda(A_M, S_M)$ is straightforward.

1. Generate n independent sample paths $(S_1^j, S_2^j, \ldots, S_M^j)$, $j = 1, 2, \ldots, n$, all starting at the initial spot S_0.

2. For each path, calculate the arithmetic average

$$A_M^j = \frac{1}{M}\sum_{i=1}^{M} S_i^j, \quad j = 1, 2, \ldots, n.$$

3. The no-arbitrage price $V_0 = \mathrm{e}^{-rT}\widetilde{\mathrm{E}}\left[\Lambda(A_M, S_M)\right]$ of the option is approximated by the discounted sample mean of the payoff values

$$V_0 \approx \overline{V}_{0,n} = \mathrm{e}^{-rT}\frac{1}{n}\sum_{j=1}^{n} \Lambda(A_M^j, S_M^j).$$

4. Additionally, we can estimate the variance

$$\mathrm{Var}\left(\mathrm{e}^{-rT}\Lambda(A_M, S_M)\right) \approx s_n^2 = \mathrm{e}^{-2rT}\frac{1}{n}\sum_{j=1}^{n} \Lambda^2(A_M^j, S_M^j) - \overline{V}_{0,n}^2$$

to construct an asymptotically valid $100(1-\alpha)\%$ confidence interval for V_0:

$$(\overline{V}_{0,n} - z_{\alpha/2}s_n/\sqrt{n}, \overline{V}_{0,n} + z_{\alpha/2}s_n/\sqrt{n}).$$

To implement the above algorithm, we only assume that sample paths can be generated exactly or approximately with high accuracy. Though the plain Monte Carlo approach is easy to implement, it is not very efficient. Several variance reduction methods have been proposed for valuing derivatives. In Boyle et al. (1997), the performances of the antithetic variate method, control variate (CV) method, and moment matching method were compared to that of the direct Monte Carlo method. They showed by simulation that the control variate method is the most efficient for valuing an Asian option. The CV method, considered in Boyle et al. (1997), was used to improve upon a naive Monte Carlo estimator for an Asian call option by choosing as analytical control variate the corresponding Asian call option with geometric averaging. We focus on this method of variance reduction.

We can achieve a larger variance reduction by considering multiple control variates. Suppose that there are two random variables $Y^{(1)}$ and $Y^{(2)}$ with respective means ν_1 and ν_2 known analytically. We construct the controlled Monte Carlo estimator of the mean of a random variable X, $\mathrm{E}[X] = \mu$, by using the random variable defined by

$$Z = X + c_1(\nu_1 - Y^{(1)}) + c_2(\nu_2 - Y^{(2)}). \tag{18.98}$$

For all values of the parameters c_1 and c_2, the mean of Z coincides with the mean of X. The variance of the controlled estimator Z is

$$\begin{aligned}\mathrm{Var}(Z) &= \mathrm{Var}(X) - 2c_1 \mathrm{Cov}(X, Y^{(1)}) - 2c_2 \mathrm{Cov}(X, Y^{(2)}) \\ &\quad + c_1^2 \mathrm{Var}(Y^{(1)}) + 2c_1c_2 \mathrm{Cov}(Y^{(1)}, Y^{(2)}) + c_2^2 \mathrm{Var}(Y^{(2)}) \\ &= \mathrm{Var}(X) - 2\mathbf{c}^\top \mathbf{\Sigma}_{X,\mathbf{Y}} + \mathbf{c}^\top \mathbf{\Sigma}_{\mathbf{Y},\mathbf{Y}}\mathbf{c},\end{aligned} \tag{18.99}$$

where

$$\mathbf{c} := \begin{bmatrix} c_1 \\ c_2 \end{bmatrix}, \quad \mathbf{\Sigma}_{X,\mathbf{Y}} := \begin{bmatrix} \mathrm{Cov}(X, Y^{(1)}) \\ \mathrm{Cov}(X, Y^{(2)}) \end{bmatrix}, \quad \mathbf{\Sigma}_{\mathbf{Y},\mathbf{Y}} := \begin{bmatrix} \mathrm{Var}(Y^{(1)}) & \mathrm{Cov}(Y^{(1)}, Y^{(2)}) \\ \mathrm{Cov}(Y^{(1)}, Y^{(2)}) & \mathrm{Var}(Y^{(2)}) \end{bmatrix}.$$

The optimal values of c_1 and c_2 that minimize $\mathrm{Var}(Z)$ can be found by solving simultaneous linear equations $\frac{\partial \mathrm{Var}(Z)}{\partial c_i} = 0$ for $i = 1, 2$:

$$\begin{cases} c_1 \mathrm{Var}(Y^{(1)}) + c_2 \mathrm{Cov}(Y^{(1)}, Y^{(2)}) = \mathrm{Cov}(X, Y^{(1)}) \\ c_1 \mathrm{Cov}(Y^{(1)}, Y^{(2)}) + c_2 \mathrm{Var}(Y^{(2)}) = \mathrm{Cov}(X, Y^{(2)}) \end{cases} \iff \mathbf{\Sigma}_{\mathbf{Y},\mathbf{Y}}\,\mathbf{c} = \mathbf{\Sigma}_{X,\mathbf{Y}}. \tag{18.100}$$

The solution to (18.100) is the optimal vector

$$\mathbf{c}_{\mathrm{opt}} = \mathbf{\Sigma}_{\mathbf{Y},\mathbf{Y}}^{-1} \mathbf{\Sigma}_{X,\mathbf{Y}}. \tag{18.101}$$

Substituting it back into (18.99) gives

$$\begin{aligned}\mathrm{Var}(Z) &= \mathrm{Var}(X) - 2\mathbf{\Sigma}_{X,\mathbf{Y}}^\top \mathbf{\Sigma}_{\mathbf{Y},\mathbf{Y}}^{-1} \mathbf{\Sigma}_{X,\mathbf{Y}} + \mathbf{\Sigma}_{X,\mathbf{Y}}^\top \mathbf{\Sigma}_{\mathbf{Y},\mathbf{Y}}^{-1} \mathbf{\Sigma}_{\mathbf{Y},\mathbf{Y}} \mathbf{\Sigma}_{\mathbf{Y},\mathbf{Y}}^{-1} \mathbf{\Sigma}_{X,\mathbf{Y}} \\ &= \mathrm{Var}(X) - \mathbf{\Sigma}_{X,\mathbf{Y}}^\top \mathbf{\Sigma}_{\mathbf{Y},\mathbf{Y}}^{-1} \mathbf{\Sigma}_{X,\mathbf{Y}} \\ &= \mathrm{Var}(X) \left(1 - \frac{\mathbf{\Sigma}_{X,\mathbf{Y}}^\top \mathbf{\Sigma}_{\mathbf{Y},\mathbf{Y}}^{-1} \mathbf{\Sigma}_{X,\mathbf{Y}}}{\mathrm{Var}(X)} \right).\end{aligned}$$

Here the ratio $\frac{\mathbf{\Sigma}_{X,\mathbf{Y}}^\top \mathbf{\Sigma}_{\mathbf{Y},\mathbf{Y}}^{-1} \mathbf{\Sigma}_{X,\mathbf{Y}}}{\mathrm{Var}(X)}$ measures the strength of the linear relation between X and $\mathbf{Y}$. It generalizes $\mathrm{Corr}^2(X, Y)$ for X and scalar Y. As in the scalar case, the optimal vector of coefficients $\mathbf{c}_{\mathrm{opt}}$ can be estimated by using sample estimates:

$$\mathbf{c}_{\mathrm{opt}} \approx \mathbf{S}_{\mathbf{Y},\mathbf{Y}}^{-1} \mathbf{S}_{X,\mathbf{Y}},$$

where the (i,k)th entry of the square matrix $\mathbf{S}_{\mathbf{Y},\mathbf{Y}}$ is

$$\frac{1}{n}\sum_{j=1}^{n} Y_j^{(i)} Y_j^{(k)} - \overline{Y}^{(i)}\overline{Y}^{(k)},$$

and the ith entry of the column vector $\mathbf{S}_{X,\mathbf{Y}}$ is

$$\frac{1}{n}\sum_{j=1}^{n} X_j Y_j^{(i)} - \overline{X}\,\overline{Y}^{(i)},$$

where $(X_j, Y_j^{(1)}, Y_j^{(2)})$, $j = 1, 2, \ldots, n$ are i.i.d. samples, and $\overline{X}$ and $\overline{Y}^{(i)}$ are sample means given by

$$\overline{X} = \frac{1}{n}\sum_{j=1}^{n} X_j \text{ and } \overline{Y}^{(i)} = \frac{1}{n}\sum_{j=1}^{n} Y_j^{(i)},\ i = 1, 2,$$

respectively.

Let us consider the arithmetic average price call option with the payoff C_M^{AAP} given by (18.97) within the GBM model. The variance of the naive Monte Carlo estimator can be significantly reduced with the use of multiple control variates. One control variate is the geometric Asian option, which can be evaluated analytically. The other control variate is an option whose payoff is an average of European payoffs with different maturity times. Let us find out how the payoffs of these control variates relate to the payoff of the average price call. According to the inequality of arithmetic and geometric means, the arithmetic mean of a set of nonnegative real numbers is greater than or equal to the geometric mean of the same set. That is,

$$G_M \leqslant A_M \text{ where } G_M := \left(\prod_{k=1}^{M} S_k\right)^{\frac{1}{M}}.$$

Using this and the fact that $\max\{x, 0\}$ is increasing in x, we have a lower bound for the payoff function:

$$\max\{G_M - K, 0\} \leqslant \max\{A_M - K, 0\}. \tag{18.102}$$

If the asset price $\{S(t)\}_{t\geqslant 0}$ follows GBM, then G_M is log-normally distributed and hence the geometric option with payoff $\max\{G_M - K, 0\}$ can be priced analytically. Let C_0^{GAP} denote the no-arbitrage time-0 value of the geometric average price call with the payoff $C_M^{GAP} := \max\{G_M - K, 0\}$.

Since $(x)^+ \equiv \max\{x, 0\}$ is a convex function of x, we have the following upper bound for C_M^{AAP}:

$$\max\{A_M - K, 0\} \leqslant \frac{1}{M}\sum_{k=1}^{M} \max\{S_k - K, 0\}, \tag{18.103}$$

which leads us to consider a new option with the payoff

$$C_M^{AAC} := \frac{1}{M}\sum_{k=1}^{M} \max\{S_k - K, 0\}$$

as a control variate for the average price call. As is readily observed, the payoff C_M^{AAE} is just an average of European call payoff functions at exercise times t_k, $k = 1, 2, \ldots, M$. The price $C_0^{AAC} = \mathrm{e}^{-rT}\widetilde{\mathrm{E}}[C_M^{AAC}]$ can be computed numerically using only univariate integrals:

$$C_0^{AAC} = \mathrm{e}^{-rT}\frac{1}{M}\sum_{k=1}^{M}\int_{-\infty}^{\infty}\left(\mathrm{e}^{(r-q-\sigma^2/2)t_k+\sigma\sqrt{t_k}\,z} - K\right)^{+}\frac{\mathrm{e}^{-z^2/2}}{\sqrt{2\pi}}\,\mathrm{d}z. \tag{18.104}$$

As a result, we can improve the plain Monte Carlo algorithm, as follows.

1. Generate n independent sample paths $(S_1^j, S_2^j, \ldots, S_M^j)$, $j = 1, 2, \ldots, n$, all starting at spot S_0.

2. For each path, labelled $j = 1, 2, \ldots, n$, calculate
   (a) the arithmetic average $A_M^j = \frac{1}{M}\sum_{k=1}^{M} S_k^j$,
   (b) the geometric average $G_M^j = \left(\prod_{k=1}^{M} S_k^j\right)^{\frac{1}{M}}$,
   (c) the arithmetic average price payoff $X_j = \max\left\{A_M^j - K, 0\right\}$,
   (d) the geometric average price payoff $Y_j^{(1)} = \max\left\{G_M^j - K, 0\right\}$,
   (e) the arithmetic average of European call payoffs $Y_j^{(2)} = \frac{1}{M}\sum_{k=1}^{M} \max\{S_k^j - K, 0\}$.

3. Estimate the optimal coefficients c_1 and c_2: $\mathbf{c} = \mathbf{S}_{\mathbf{Y},\mathbf{Y}}^{-1}\mathbf{S}_{X,\mathbf{Y}}$, with the respective 2-by-2 matrix and 2-by-1 vector:

$$\mathbf{S}_{\mathbf{Y},\mathbf{Y}} = \begin{bmatrix} \frac{1}{n}\sum_{j=1}^{n} (Y_j^{(1)})^2 - (\overline{Y}^{(1)})^2 & \frac{1}{n}\sum_{j=1}^{n} Y_j^{(1)} Y_j^{(2)} - \overline{Y}^{(1)}\overline{Y}^{(2)} \\ \frac{1}{n}\sum_{j=1}^{n} Y_j^{(1)} Y_j^{(2)} - \overline{Y}^{(1)}\overline{Y}^{(2)} & \frac{1}{n}\sum_{j=1}^{n} (Y_j^{(2)})^2 - (\overline{Y}^{(2)})^2 \end{bmatrix}$$

$$\mathbf{S}_{X,\mathbf{Y}} = \begin{bmatrix} \frac{1}{n}\sum_{j=1}^{n} X_j Y_j^{(1)} - \overline{X}\overline{Y}^{(1)} \\ \frac{1}{n}\sum_{j=1}^{n} X_j Y_j^{(2)} - \overline{X}\overline{Y}^{(2)} \end{bmatrix}.$$

4. Calculate the sample values of the controlled estimator

$$Z_j = X_j + c_1(\nu_1 - Y_j^{(1)}) + c_2(\nu_2 - Y_j^{(2)}), \quad j = 1, 2, \ldots, n,$$

   where $\nu_1 = C_0^{GAP}$ and $\nu_2 = C_0^{AAE}$.

5. The approximate no-arbitrage price C_0^{AAP} of the AAP call option is then:

$$C_0^{AAP} \approx \overline{C}_{0,n}^{AAP} = \mathrm{e}^{-rT}\frac{1}{n}\sum_{j=1}^{n} Z_j.$$

6. Estimate the variance

$$\mathrm{Var}\left(\mathrm{e}^{-rT} C_M^{AAP}\right) \approx s_n^2 = \mathrm{e}^{-2rT}\frac{1}{n}\sum_{j=1}^{n} Z_j^2 - \left(\overline{C}_{0,n}^{AAP}\right)^2$$

   and construct the confidence interval for C_0^{AAP}:

$$\left(\overline{C}_{0,n}^{AAP} - z_{\alpha/2}s_n/\sqrt{n}, \overline{C}_{0,n}^{AAP} + z_{\alpha/2}s_n/\sqrt{n}\right)$$

   with an asymptotically valid $100(1-\alpha)\%$ confidence level.

TABLE 18.18: Pricing an arithmetic average price call option using control variate Monte Carlo methods.

Control	Estimate $\overline{C}_{0,n}^{AAP}$	Error $s_n/\sqrt{n}$	Speedup
None	6.050039	0.037420	
AAC	6.052996	0.003871	93.45
GAP	6.051365	0.001051	1266.68
$GAP{+}AAC$	6.051600	0.000845	1961.39

Example 18.13. Let us study the efficiency of the single and multiple control variate methods for Monte Carlo pricing of an AAP Asian call option. The underlying asset price follows the geometric Brownian motion model with $\sigma = 0.25$, $r = 0.05$, $q = 0$, $T = 1$. The spot value and strike price are, respectively, $S_0 = 100$ and $K = 100$. The number of observation time moments is $M = 128$. Using the Black–Scholes pricing formula, we first compute the expected values of the control variates:

$$C_0^{AAC} = 6.755961403884071 \text{ and } C_0^{GAP} = 5.820772970849206.$$

After that, the Asian call option price is estimated by the Monte Carlo method using $n = 10{,}000$ sample paths. In Table 18.18 we report the simulation results for the crude estimate and three control variate estimates using, respectively, the geometric average price call (GAP), arithmetic average of European calls (AAC), and both averages as controls. The control variate method is very efficient. This observation is also supported by high correlation between the crude estimate and the GAP/AAC control estimates. The respective correlations coefficients are

$$\rho_{GAP} \approx 99.96\%, \quad \rho_{AAC} \approx 99.46\%.$$

References

J. Alford and N. Webber. Very high order lattice methods for one factor models. *Available at SSRN 259478*, 2001.

P. Boyle, M. Broadie, and P. Glasserman. Monte Carlo methods for security pricing. *Journal of Economic Dynamics and Control*, 21(8):1267–1321, 1997.

M. Broadie and P. Glasserman. A stochastic mesh method for pricing high-dimensional American options. *Journal of Computational Finance*, 7:35–72, 2004.

R. Cont and P. Tankov. *Financial Modelling with Jump Processes.* Chapman & Hall/CRC, 2003.

S. Heston and G. Zhou. On the rate of convergence of discrete-time contingent claims. *Mathematical Finance*, 10(1):53–75, 2000.

F. A. Longstaff and E. S. Schwartz. Valuing American options by simulation: A simple least-squares approach. *Review of Financial Studies*, 14(1):113–147, 2001.

D. B. Madan, P. Carr, and E. C. Chang. The Variance Gamma process and option pricing. *European Finance Review*, 2(1):79–105, 1998.

Appendix: Some Useful Integral Identities and Symmetry Properties of Normal Random Variables

Throughout all the formulas below, a, b, c, A, B, C are any real constants and X is a normal random variable with mean $\mu \in \mathbb{R}$ and standard deviation $\sigma > 0$, i.e., $X \sim Norm(\mu, \sigma^2)$ with PDF $\varphi_{\mu,\sigma}(x) \equiv \frac{e^{-(x-\mu)^2/2\sigma^2}}{\sigma\sqrt{2\pi}}$, $-\infty < x < \infty$. The functions $\mathcal{N}(x)$ and $\mathcal{N}_2(x, y; \rho)$ are the standard normal univariate and bivariate cumulative distribution functions, respectively.

$$\mathrm{E}\left[e^{BX}\mathbb{I}_{\{X>A\}}\right] \equiv \int_A^\infty e^{Bx}\varphi_{\mu,\sigma}(x)\,\mathrm{d}x = e^{\mu B+\frac{1}{2}\sigma^2B^2}\mathcal{N}\left(\sigma B + \frac{\mu - A}{\sigma}\right) \tag{A.1}$$

$$\mathrm{E}\left[e^{BX}\mathbb{I}_{\{X<A\}}\right] \equiv \int_{-\infty}^A e^{Bx}\varphi_{\mu,\sigma}(x)\,\mathrm{d}x = e^{\mu B+\frac{1}{2}\sigma^2B^2}\mathcal{N}\left(-\sigma B + \frac{A - \mu}{\sigma}\right) \tag{A.2}$$

$$\mathrm{E}\left[\mathcal{N}(AX + C)\right] \equiv \int_{-\infty}^\infty \mathcal{N}(Ax + C)\varphi_{\mu,\sigma}(x)\,\mathrm{d}x = \mathcal{N}\left(\frac{\mu A + C}{\sqrt{1+\sigma^2A^2}}\right) \tag{A.3}$$

$$\begin{aligned}\mathrm{E}\left[e^{BX}\mathcal{N}(AX + C)\right] &\equiv \int_{-\infty}^\infty e^{Bx}\mathcal{N}(Ax + C)\varphi_{\mu,\sigma}(x)\,\mathrm{d}x \\ &= e^{\mu B+\frac{1}{2}\sigma^2B^2}\mathcal{N}\left(\frac{\mu A + C + \sigma^2AB}{\sqrt{1+\sigma^2A^2}}\right)\end{aligned} \tag{A.4}$$

$$\int_0^\infty \mathcal{N}(Ax + B)e^x\,\mathrm{d}x = -\mathcal{N}(B) + e^{(1-2AB)/2A^2}\mathcal{N}\left(\frac{1 - AB}{|A|}\right) \quad (\text{for } A < 0) \tag{A.5}$$

$$\int_0^\infty \mathcal{N}(Ax + B)e^{-x}\,\mathrm{d}x = \mathcal{N}(B) + \mathrm{sgn}(A)\,e^{(1+2AB)/2A^2}\mathcal{N}\left(-\frac{1 + AB}{|A|}\right) \tag{A.6}$$

$$\begin{aligned}\int_{-\infty}^\infty \mathcal{N}(Ax + B)e^x\,\mathrm{d}x &= e^{(1-2AB)/2A^2}\mathcal{N}\left(\frac{1 - AB}{|A|}\right) \\ &\quad - e^{(1+2AB)/2A^2}\mathcal{N}\left(-\frac{1 + AB}{|A|}\right) \quad (\text{for } A < 0)\end{aligned} \tag{A.7}$$

$$\int_0^\infty \mathcal{N}(Ax + B)\,\mathrm{d}x = \frac{1}{|A|}\left[B\mathcal{N}(B) + \frac{e^{-B^2/2}}{\sqrt{2\pi}}\right] \quad (\text{for } A < 0) \tag{A.8}$$

$$\mathrm{E}\left[\mathcal{N}(AX+C)\mathbb{I}_{\{X<B\}}\right] \equiv \int_{-\infty}^{B} \mathcal{N}(Ax+C)\varphi_{\mu,\sigma}(x)\,\mathrm{d}x$$
$$= \mathcal{N}_2\left(\frac{B-\mu}{\sigma}, \frac{\mu A + C}{\sqrt{1+\sigma^2 A^2}}; \frac{-\sigma A}{\sqrt{1+\sigma^2 A^2}}\right) \tag{A.9}$$

$$\mathrm{E}\left[\mathcal{N}(AX+C)\mathbb{I}_{\{X>B\}}\right] \equiv \int_{B}^{\infty} \mathcal{N}(Ax+C)\varphi_{\mu,\sigma}(x)\,\mathrm{d}x$$
$$= \mathcal{N}_2\left(\frac{\mu-B}{\sigma}, \frac{\mu A + C}{\sqrt{1+\sigma^2 A^2}}; \frac{\sigma A}{\sqrt{1+\sigma^2 A^2}}\right) \tag{A.10}$$

Note that formulas (A.1)–(A.10) simplify in the special case when $X \stackrel{d}{=} Z \sim Norm(0,1)$ by setting $\mu = 0, \sigma = 1$.

The bivariate normal CDF also has useful symmetry relations such as

$$\mathcal{N}_2(a,b;\rho) = \mathcal{N}_2(b,a;\rho)\,, \tag{A.11}$$

$$\mathcal{N}_2(a,b;\rho) + \mathcal{N}_2(a,-b;-\rho) = \mathcal{N}(a)\,. \tag{A.12}$$

In the expectation formulas below, Z_1, Z_2 are i.i.d. standard normal random variables with covariance $\mathrm{Cov}(Z_1,Z_2) = \rho$, $|\rho| < 1$, i.e., with joint PDF $n_2(x,y;\rho) \equiv \frac{\partial^2}{\partial x \partial y}\mathcal{N}_2(x,y;\rho) = \frac{1}{2\pi\sqrt{1-\rho^2}}\mathrm{e}^{-(x^2+y^2-2\rho xy)/2(1-\rho^2)}$.

$$\mathrm{E}\left[\mathrm{e}^{-BZ_2}\,\mathbb{I}_{\{Z_1<a,Z_2<b\}}\right] \equiv \iint_{\mathbb{R}^2} \mathrm{e}^{-By}\, n_2(x,y;\rho)\mathbb{I}_{\{x<a,y<b\}}\,\mathrm{d}x\,\mathrm{d}y$$
$$= \mathrm{e}^{\frac{1}{2}B^2}\mathcal{N}_2\left(a+\rho B, b+B;\rho\right) \tag{A.13}$$

$$\mathrm{E}\left[\mathrm{e}^{-BZ_2}\,\mathbb{I}_{\{Z_1>a,Z_2<b\}}\right] = \mathrm{e}^{\frac{1}{2}B^2}\mathcal{N}_2\left(-(a+\rho B), b+B;-\rho\right) \tag{A.14}$$

$$\mathrm{E}\left[\mathrm{e}^{-BZ_2}\,\mathbb{I}_{\{Z_1<a,Z_2>b\}}\right] = \mathrm{e}^{\frac{1}{2}B^2}\mathcal{N}_2\left(a+\rho B, -(b+B);-\rho\right) \tag{A.15}$$

$$\mathrm{E}\left[\mathrm{e}^{-BZ_2}\,\mathbb{I}_{\{Z_1>a,Z_2>b\}}\right] = \mathrm{e}^{\frac{1}{2}B^2}\mathcal{N}_2\left(-(a+\rho B), -(b+B);\rho\right) \tag{A.16}$$

Note: interchanging $a \leftrightarrow b$ in (A.13)–(A.16) gives respectively equivalent formulas for the expectations $\mathrm{E}\left[\mathrm{e}^{-BZ_1}\,\mathbb{I}_{\{Z_1<a,Z_2<b\}}\right]$, $\mathrm{E}\left[\mathrm{e}^{-BZ_1}\,\mathbb{I}_{\{Z_1<a,Z_2>b\}}\right]$, $\mathrm{E}\left[\mathrm{e}^{-BZ_1}\,\mathbb{I}_{\{Z_1>a,Z_2<b\}}\right]$, and $\mathrm{E}\left[\mathrm{e}^{-BZ_1}\,\mathbb{I}_{\{Z_1>a,Z_2>b\}}\right]$.

Glossary of Symbols and Abbreviations

$a_{\overline{n}\rceil j}$	discount factor for n payments and rate j
$B(t)$ or B_t	value of a risk-free security (e.g. a bond or a bank account) at time t
$\mathscr{b}(x; n, p)$	probability mass function for the binomial law $Bin(n, p)$
$\mathcal{B}(x; n, p)$	(cumulative) probability distribution function for the binomial law $Bin(n, p)$
$Bin(n, p)$	binomial probability distribution with number of trials n and success probability p
C^A	American call
C^E	European call
CDF	cumulative (probability) distribution function
$\text{Corr}(X, Y)$	correlation coefficient of X and Y
$\text{Cov}(X, Y)$	covariance of X and Y
CRR	Cox–Ross–Rubinstein
D	downward move in a binomial tree
div	dividend
$D(t, T)$	discount factor from time t to time T
$D(t) \equiv D(0, t)$	discount factor from time 0 to time t
$\frac{\mathrm{d}\mathbb{P}}{\mathrm{d}\mathbb{Q}}$	Radon–Nikodym derivative of $\mathbb{P}$ w.r.t. $\mathbb{Q}$
$\left(\frac{\mathrm{d}\mathbb{P}}{\mathrm{d}\mathbb{Q}}\right)_t$	Radon–Nikodym derivative process at time t
EMM	equivalent martingale measure
$Exp(\lambda)$	exponential probability distribution with rate λ
$\mathrm{E}[X]$	mathematical expectation of X
$\widetilde{\mathrm{E}}[X]$	risk-neutral mathematical expectation of X
$\mathrm{E}[X \mid \mathcal{F}]$	mathematical expectation of X conditional on a σ-algebra $\mathcal{F}$
$\mathrm{E}_t[X]$	mathematical expectation of X conditional on $\mathcal{F}_t$
$\widetilde{\mathrm{E}}^{(g)}[X]$	mathematical expectation of X w.r.t. the probability measure $\mathbb{P}^{(g)}$
$\widetilde{\mathrm{E}}_t^{(g)}[X]$	mathematical expectation of X conditional on $\mathcal{F}_t$ w.r.t. the probability measure $\mathbb{P}^{(g)}$

$\mathrm{E}_{t,x}[X]$	mathematical expectation of X conditional on an underlying process having value x at time t
$\mathrm{E}_{t,\mathbf{x}}[X]$	mathematical expectation of X conditional on a underlying vector process having value $\mathbf{x}$ at time t
$\mathcal{E}_t(\gamma \cdot W)$	exponential martingale process of an adapted process γ w.r.t. Brownian motion W
$\mathcal{E}_t(\gamma \cdot \mathbf{W})$	exponential martingale process of an adapted vector process γ w.r.t. vector Brownian motion W
$\mathcal{F}_t$	σ-algebra generated by information available at time t
$\mathbb{F}$	filtration
$f(t;T,T')$	forward rate at time t for interval $[T,T']$
$f(t,T)$	instantaneous forward rate at time t for maturity T
$F(t,T)$	forward price at time t for maturity T
f_X, f_D	probability density function (of random variable X or probability distribution D)
F_X, F_D	(cumulative) distribution function (of random variable X or probability distribution D)
$Gamma(\kappa,\lambda)$	gamma probability distribution with shape parameter κ and rate parameter λ
GBM	geometric Brownian motion
$i^{(m)}$	nominal interest rate compounded at frequency m
$\mathbb{I}_A$	indicator of event (or set) A
iff	if and only if
i.i.d.	independent and identically distributed
K	strike price
$\Lambda(\cdot)$	payoff function
m_t^X	minimum over $[0,t]$ of the process X
M_t^X	maximum over $[0,t]$ of the process X
$n(x)$	probability density function for a standard normal law
$n_2(x,y;\rho)$	joint probability density function for two standard normal random variables with correlation coefficient ρ
$n_n(x_1,\ldots,x_n;\boldsymbol{\rho})$	joint probability density function for n standard normal random variables with correlation matrix $\boldsymbol{\rho}$
$\mathcal{N}(x)$	(cumulative) probability distribution function for a standard normal law

$\mathcal{N}_2(x, y; \rho)$	joint probability distribution function for two standard normal random variables with correlation coefficient ρ
$\mathcal{N}_n(x_1, \ldots, x_n; \boldsymbol{\rho})$	joint probability distribution function for n standard normal random variables with correlation matrix $\boldsymbol{\rho}$
$Norm(\mu, \sigma^2)$	normal probability distribution with mean μ and variance σ^2
$Norm_n(\mathbf{m}, \boldsymbol{\Sigma})$	n-variate normal probability distribution with mean vector $\mathbf{m}$ and covariance matrix $\boldsymbol{\Sigma}$
NPV	net present value
ω	scenario (element of a state space)
Ω	state space
P	present value, or principal, or purchase price
$\mathbb{P}$	probability measure
$\widetilde{\mathbb{P}} \equiv \widetilde{\mathbb{P}}^{(B)}$	risk-neutral probability measure with bank account as numéraire
$\widetilde{\mathbb{P}}^{(g)}$	risk-neutral probability measure (or EMM) with asset g as numéraire
$\mathbb{P}(A)$	probability of event A
$\mathbb{P}(A \mid B)$	probability of event A conditional on event B
$\mathcal{P}_t$	partition of Ω generated by information available at time t
$\mathcal{P}(X)$	partition of Ω generated by random variable X
$\mathcal{P}(X_1, \ldots, X_n)$	partition of Ω generated by random variables $X_1, \ldots, X_n$
P_A	present (discounted) value of an annuity
P^A	American put
P^E	European put
$\Pi(t)$ or Π_t	portfolio value at time t
$\overline{\Pi}(t)$ or $\overline{\Pi}_t$	discounted portfolio value at time t
$p(s, t; x, y)$	transition PDF for a one-dimensional diffusion
$p(t; x, y)$	time-homogeneous transition PDF for a one-dimensional diffusion
$p(s, t; \mathbf{x}, \mathbf{y})$	transition PDF for a multidimensional diffusion
$p(t; \mathbf{x}, \mathbf{y})$	time-homogeneous transition PDF for a multidimensional diffusion
PDF	probability density function
$Pois(\lambda)$	Poisson probability distribution with rate λ
ϱ	Radon–Nikodym derivative
ϱ_t	Radon–Nikodym derivative process at time t

ρ	correlation coefficient
r	interest rate
$r(t)$	instantaneous interest rate at time t
$r_{[t_1,t_2]}$	rate of return from time t_1 to time t_2
$R_{[t_1,t_2]}$	total return from time t_1 to time t_2
$\mathbb{R}$	set of real numbers
$\mathbb{R}_+$	set of nonnegative real numbers
$\sigma(X)$	σ-algebra generated by random variable X
$\sigma(\{X_\lambda\})$	σ-algebra generated by a collection $\{X_\lambda\}$
$S(t)$ or S_t	price of a risky asset (e.g. a stock) at time t
$S_i(t)$ or S_t^i	price of the ith risky asset at time t
SDE	stochastic differential equation
SQB	squared Bessel process
T	maturity time; expiry time; exercise time
$\mathcal{T}_b^X$	first hitting time of X at level b
$\mathcal{T}_{(a,b)}^X$	first exit time of X from the interval (a,b)
U	upward move in a binomial tree
$Unif(a,b)$	uniform probability distribution on an interval (a,b)
$V(t)$	(accumulated) value function at time t
$v(\tau,S)$	derivative pricing function of time to maturity τ and spot S
$V(t,S)$ or $V_t(S)$	derivative pricing function of calendar time t and spot S
$\overline{V}(t,S)$ or $\overline{V}_t(S)$	discounted derivative pricing function
V_A	future (accumulated) value of an annuity
$\mathrm{Var}(X)$	variance of X
VaR	Value at Risk
w.r.t.	with respect to
W	Brownian motion
$W^{(\mu,\sigma)}$	scaled Brownian motion with a linear drift
$y(\tau)$	yield rate for time to maturity τ
$y(t,T)$	yield rate at time t for maturity T
$Z(t,T)$	zero-coupon bond price at time t for maturity T
ZCB	zero-coupon bond

References

Theory of Probability and Stochastic Processes

D. Applebaum. *Lévy Processes and Stochastic Calculus.* Cambridge University Press, 2009.

K.B. Athreya and S.N. Lahiri. *Measure Theory and Probability Theory.* Springer Texts in Statistics. Springer-Verlag, 2006.

Z. Brzeźniak and T. Zastawniak. *Basic Stochastic Processes: A Course Through Exercices.* Springer-Verlag, 1999.

M. Capinski and P.E. Kopp. *Measure, Integral and Probability.* Springer Undergraduate Mathematics Series. Springer-Verlag, 2004.

R. Cont and P. Tankov. *Financial Modelling with Jump Processes.* Chapman & Hall/CRC Financial Mathematics Series. Taylor & Francis, 2004.

W. Feller. *An Introduction to Probability Theory and Its Applications*, volume 1. John Wiley & Sons, 1971a.

W. Feller. *An Introduction to Probability Theory and Its Applications*, volume 2. John Wiley & Sons, 1971b.

A. Gut. *Probability: a Graduate Course.* Springer-Verlag, 2005.

A. Gut. *An Intermediate Course in Probability.* Springer-Verlag, 2009.

R.V. Hogg and E.A. Tanis. *Probability and Statistical Inference.* Pearson/Prentice Hall, 8th edition, 2010.

M. Jeanblanc, M. Yor, and M. Chesney. *Mathematical Methods for Financial Markets.* Springer Finance. Springer-Verlag, 2009.

I. Karatzas and S.E. Shreve. *Brownian Motion and Stochastic Calculus.* Graduate Texts in Mathematics. Springer New York, 1991.

S. Karlin and H.M. Taylor. *A First Course in Stochastic Processes.* Academic Press, 1975.

S. Karlin and H.M. Taylor. *A Second Course in Stochastic Processes.* Academic Press, 1981.

F.C. Klebaner. *Introduction to Stochastic Calculus with applications.* Imperial College Press, 2005.

H.H. Kuo. *Introduction to Stochastic Integration.* Springer-Verlag, 2006.

B.K. Øksendal. *Stochastic Differential Equations: An Introduction with Applications.* Springer-Verlag, 2010.

M.M. Rao and R.J. Swift. *Probability Theory with Applications.* Mathematics and Its Applications. Springer-Verlag, 2006.

A.N. Shiryaev. *Probability.* Graduate Texts in Mathematics. Springer-Verlag, 1996.

Introduction to Mathematics of Finance

R. Brown, S. Kopp, and P. Zima. *Mathematics of Finance.* McGraw-Hill Ryerson Limited, 7th edition, 2011.

J.R. Buchanan. *An Undergraduate Introduction to Financial Mathematics.* World Scientific, 2008.

M. Capiński and T. Zastawniak. *Mathematics for Finance: An Introduction to Financial Engineering.* Springer Undergraduate Mathematics Series. Springer-Verlag, 2003.

M. Davis, L. Bachelier, A. Etheridge, and P.A. Samuelson. *Louis Bachelier's Theory of Speculation: The Origins of Modern Finance.* Princeton University Press, 2011.

D. Lovelock, M. Mendel, and A.L. Wright. *An Introduction to the Mathematics of Money: Saving and Investing.* Springer-Verlag, 2007.

T. Mikosch. *Elementary Stochastic Calculus: with Finance in View.* World Scientific, 1998.

S. Roman. *Introduction to the Mathematics of Finance: From Risk Management to Options Pricing.* Undergraduate Texts in Mathematics. Springer-Verlag, 2004.

S.M. Ross. *An Elementary Introduction to Mathematical Finance.* Cambridge University Press, 2011.

P. Wilmott, S. Howison, and J. Dewynne. *The Mathematics of Financial Derivatives: A Student Introduction.* Cambridge University Press, 1995.

Mathematics of Finance (Discrete-Time)

M. Capiński and E. Kopp. *Discrete Models of Financial Markets.* Mastering Mathematical Finance. Cambridge University Press, 2012.

H. Föllmer and A. Schied. *Stochastic Finance: An Introduction in Discrete Time.* De Gruyter Textbook Series. De Gruyter, 2011.

P.K. Medina and S. Merino. *Mathematical Finance and Probability. A Discrete Introduction.* Birkhauser, 2004.

S.R. Pliska. *Introduction to Mathematical Finance: Discrete Time Models.* John Wiley & Sons, 1997.

S.E. Shreve. *Stochastic Calculus for Finance I: The Binomial Asset Pricing Model.* Springer Finance. Springer-Verlag, 2012.

Mathematics of Finance (Continuous-Time)

C. Albanese and G. Campolieti. *Advanced Derivatives Pricing and Risk Management: Theory, Tools and Hands-on Programming Application.* Academic Press advanced finance series. Elsevier Academic Press, 2006.

M. Avellaneda and P. Laurence. *Quantitative Modeling of Derivative Securities: From Theory To Practice.* Chapman & Hall/CRC, 1999.

K. Back. *A Course in Derivative Securities: Introduction to Theory and Computation.* Springer Finance. Springer-Verlag, 2005.

M. Baxter and A. Rennie. *Financial Calculus: An Introduction to Derivative Pricing.* Cambridge University Press, 1996.

D. Brigo and F. Mercurio. *Interest Rate Models - Theory and Practice: With Smile, Inflation and Credit.* Springer Finance. Springer-Verlag, 2007.

A.J.G. Cairns. *Interest Rate Models: An Introduction.* Princeton University Press, 2004.

R.-A. Dana and M. Jeanblanc. *Financial Markets in Continuous Time.* Springer Finance. Springer-Verlag, 2003.

R.J. Elliott and P.E. Kopp. *Mathematics of Financial Markets.* Springer-Verlag, 2005.

A. Etheridge. *A Course in Financial Calculus.* Cambridge University Press, 2002.

J.-P. Fouque, G. Papanicolaou, and K.R. Sircar. *Derivatives in Financial Markets with Stochastic Volatility.* Cambridge University Press, 2000.

Y.K. Kwok. *Mathematical Models of Financial Derivatives.* Springer Finance. Springer-Verlag, 2008.

A.L. Lewis. *Option valuation under stochastic volatility: with Mathematica code.* Finance Press, 2000.

A.N. Shiryaev. *Essentials of Stochastic Finance: Facts, Models, Theory.* Advanced series on statistical science & applied probability. World Scientific, 1999.

S.E. Shreve. *Stochastic Calculus for Finance II: Continuous-Time Models.* Springer Finance. Springer-Verlag, 2010.

P. Wilmott. *Derivatives: the Theory and Practice of Financial Engineering.* Wiley Frontiers in Finance Series. John Wiley & Sons, 1998.

Computational Methods

G. Fusai and A. Roncoroni. *Implementing Models in Quantitative Finance: Methods and Cases: Methods and Cases.* Springer Finance. Springer-Verlag, 2007.

P. Glasserman. *Monte Carlo Methods in Financial Engineering.* Applications of mathematics: stochastic modelling and applied probability. Springer-Verlag, 2004.

A. Hirsa. *Computational Methods in Finance.* Chapman & Hall/CRC Financial Mathematics Series. Taylor & Francis, 2012.

P.E. Kloeden and E. Platen. *Numerical Solution of Stochastic Differential Equations.* Applications of mathematics: stochastic modelling and applied probability. Springer-Verlag, 1992.

R. Korn, E. Korn, and G. Kroisandt. *Monte Carlo Methods and Models in Finance and Insurance.* Chapman & Hall/CRC Financial Mathematics Series. Taylor & Francis, 2010.

D.L. McLeish. *Monte Carlo Simulation and Finance.* Wiley Finance. John Wiley & Sons, 2011.

E. Platen and N. Bruti-Liberati. *Numerical Solution of Stochastic Differential Equations with Jumps in Finance.* Stochastic modelling and applied probability. Springer-Verlag, 2010.

E. Platen and D. Heath. *A Benchmark Approach to Quantitative Finance.* Springer Finance. Springer-Verlag, 2006.

D. Tavella. *Quantitative Methods in Derivatives Pricing: An Introduction to Computational Finance.* Wiley Finance. John Wiley & Sons, 2003.

Financial Economics

S.L. Allen. *Financial Risk Management: A Practitioner's Guide to Managing Market and Credit Risk.* Wiley Finance. John Wiley & Sons, 2012.

J.P. Danthine and J.B. Donaldson. *Intermediate Financial Theory.* Academic Press advanced finance series. Elsevier Academic Press, 2005.

J. Gatheral. *The Volatility Surface: A Practitioner's Guide.* John Wiley & Sons, 2011.

J.C. Hull. *Options, Futures, and Other Derivatives.* Pearson Education, 8th edition, 2011.

D.G. Luenberger. *Investment Science.* Oxford University Press, 1997.

R.L. McDonald. *Derivatives markets.* Addison-Wesley series in finance. Addison Wesley, 2003.

H.H. Panjer and P.P. Boyle. *Financial economics: with applications to investments, insurance, and pensions.* Actuarial Foundation, 1998.

Index

Lightning Source UK Ltd.
Milton Keynes UK
UKHW050818200219
337584UK00016B/250/P